FISKE
GUIDE TO COLLEGES

费思克选校指南

Edward B. Fiske

爱德华·费思克/著

世界图书出版公司

北京·广州·上海·西安

图书在版编目(CIP)数据

费思克选校指南＝Fiske Guide to Colleges 2008：英文/（美）费思克（Fiske，E. B.）著. —北京：世界图书出版公司北京公司，2007.8

ISBN 978-7-5062-8339-7

Ⅰ. 费… Ⅱ. 费… Ⅲ. 高等学校－概况－美国－英文 Ⅳ. G649.712.8

中国版本图书馆 CIP 数据核字（2007）第 125164 号

本书的出版通过姚氏顾问社与作者以及原出版社 Sourcebooks 公司签订了相关的授权和约。

费思克选校指南
Fiske Guide to Colleges 2008

作　者：Edward B. Fiske	
责任编辑：张颖颖	
出　版：世界图书出版公司北京公司　　http://www.wpcbj.com.cn	
发　行：世界图书出版公司北京公司	
（地址：北京朝内大街 137 号　邮编：100010　电话：64077922）	
销　售：各地新华书店及外文书店	
印　刷：三河国英印务有限公司	
开　本：880×1230　　1/16	
印　张：50	
字　数：1993 千	
版　次：2007 年 8 月第 1 版　　2007 年 8 月第 1 次印刷	
版权登记：京权图字 01-2007-0740	

ISBN 978-7-5062-8339-7/H·908　　　　　　　　　　　　定价：96.00 元

FISKE GUIDE TO COLLEGES

2008

<u>Also by Edward B. Fiske</u>

Fiske Guide to Getting into the Right College with Bruce G. Hammond

Nailing the New SAT with Bruce G. Hammond

Fiske What to do When for College with Bruce G. Hammond

Fiske Real College Essays That Work

Fiske Word Power with Jane Mallison and Margery Mandell

Smart Schools, Smart Kids: Why Do Some Schools Work?

Using Both Hands: Women and Education in Cambodia

Decentralization of Education: Politics and Consensus

When Schools Compete: A Cautionary Tale with Helen F. Ladd

Elusive Equity: Education Reform in Post-Apartheid South Africa with Helen F. Ladd

Handbook of Research on Education Finance and Policy, ed. with Helen F. Ladd

To Sunny

Contents

Contents

Index by State and Country

The colleges in this guide are listed alphabetically and cross-referenced for your convenience. Below is a list of the selected colleges grouped by state. Following this listing, you will find a second listing in which the colleges are categorized by the yearly cost of attending each school.

Index by Price

	PUBLIC	PRIVATE
$$$$	More than $8,000	More than $34,000
$$$	$6,000–$8,000	$32,000–$34,000
$$	$5,000–$6,000	$27,000–$32,000
$	Less than $5,000	Less than $27,000

Price categories are based on current tuition and fees and do not include room, board, transportation, and other expenses.

PUBLIC COLLEGES AND UNIVERSITIES

Inexpensive—$
Aberdeen, University of
Arizona State University
Arizona, University of
British Columbia, University of
Evergreen State College
Florida State University
Florida, University of
Georgia Institute of Technology
Georgia, University of
Hawaii, University of–Manoa
Louisiana State University
McGill University
New College of Florida
New Mexico Institute of Mining and Technology
New Mexico, University of
North Carolina State
North Carolina, University of–Asheville
North Carolina, University of–Greensboro
Queen's University
St. Andrews, University of
Toronto, University of
Utah, University of
West Virginia University

Moderate—$$
Alabama, University of
Arkansas, University of
Auburn University
Colorado, University of–Boulder
Iowa State University
Iowa, University of
Montana Tech of the University of Montana
Nebraska, University of–Lincoln
North Carolina, University of–Chapel

Hill
Oklahoma, University of
Oregon State University
Oregon, University of
SUNY–University at Albany
SUNY–Binghamton University
SUNY–Geneseo
SUNY–Purchase College
SUNY–Stony Brook
Tennessee, University of
Washington, University of

Expensive—$$$
College of Charleston
Delaware, University of
George Mason University
Indiana University
James Madison University
Kansas, University of
Kentucky, University of
Maine, University of–Orono
Mary Washington, University of
Maryland, University of–College Park
Missouri, University of–Columbia
Purdue University
Rhode Island, University of
South Carolina, University of
SUNY–Buffalo
Texas A&M University
Texas Tech University
Texas, University of–Austin
Truman State University
California, University of–Berkeley
California, University of–Davis
California, University of–Irvine
California, University of–Los Angeles
California, University of–Riverside
California, University of–San Diego
California, University of–Santa

Barbara
California, University of–Santa Cruz
Virginia Polytechnic Institute
Virginia, University of
Wisconsin, University of–Madison

Very Expensive—$$$$
Cincinnati, University of
Clemson University
College of New Jersey
Colorado School of Mines
Connecticut, University of
Illinois, University of–Urbana-Champaign
Maryland, University of–Baltimore County
Massachusetts, University of–Amherst
Miami University (OH)
Michigan State University
Michigan, University of
Minnesota, University of–Morris
Minnesota, University of–Twin Cities
New Hampshire, University of
New Jersey Institute of Technology
Ohio State University
Ohio University
Pennsylvania State University
Pittsburgh, University of
Rutgers University
St. Mary's College of Maryland
Vermont, University of
William and Mary, College of

PRIVATE COLLEGES AND UNIVERSITIES

Inexpensive—$

Adelphi University
Agnes Scott College
Albertson College
Albion College
Alfred University
Alma College
Alverno College
Austin College
Baylor University
Birmingham-Southern College
Brigham Young University
Calvin College
Cooper Union
Cornell College
Dallas, University of
Dayton, University of
Deep Springs
DePaul University
Drexel University
Elon University
Emerson College
Gordon College
Guilford College
Gustavus Adolphus
Hampden-Sydney College
Hendrix College
Hiram College
Hofstra University
Hollins University
Hood College
Hope College
Houghton College
Howard University
Illinois Institute of Technology
Ithaca College
Loyola University–New Orleans
Manhattanville College
Marquette University
Millsaps College
Morehouse College
Oglethorpe University
Olin College of Engineering
Presbyterian College
Prescott College
Principia College
Randolph College
Rice University
Ripon College
Rochester Institute of Technology
Saint Louis University

Southwestern University
Spelman College
St. Benedict, College of, and St. John's University
Sweet Briar College
Texas Christian University
Trinity University
Tulsa, University of
Wabash College
Warren Wilson College
Wells College
Wheaton College (IL)
Wofford College
Xavier University of Lousiana

Moderate—$$

Allegheny College
American University
Antioch College
Atlantic, College of the
Beloit College
California Institute of Technology
Case Western Reserve University
Catholic University of America, The
Chapman University
Clark University
Clarkson University
Davidson College
Denison University
Denver, University of
DePauw University
Earlham College
Eckerd College
Eugene Lang College, The New School for Liberal Arts
Fairfield University
Florida Institute of Technology
Fordham University
Furman University
Goucher College
Grinnell College
Hartwick College
Illinois Wesleyan University
Kalamazoo College
Knox College
Lafayette College
Lake Forest College
Lawrence University
Lewis & Clark College
Loyola Marymount University
Macalester College

Marlboro College
Miami, University of (FL)
Mills College
Muhlenberg College
Northeastern University
Ohio Wesleyan University
Pacific, University of the
Pomona College
Puget Sound, University of
Redlands, University of
Rhode Island School of Design
Rhodes College
Rollins College
Rose-Hulman Institute of Technology
San Francisco, University of
Santa Clara University
South, University of the (Sewanee)
Southern Methodist University
St. Olaf College
Stetson University
Susquehanna University
Syracuse University
Washington and Jefferson College
Washington and Lee University
Whitman College
Whittier College
Willamette University
Wittenberg University
Wooster, The College of

Expensive—$$$

Babson College
Barnard College
Boston College
Bryn Mawr College
Centre College
Claremont McKenna
Cornell University
Dartmouth College
Dickinson College
Drew University
Duke University
Emory University
Georgetown University
Harvard University
Harvey Mudd
Haverford College
Holy Cross, College of the
Lehigh University
Massachusetts Institute of Technology
New York University

PRIVATE COLLEGES AND UNIVERSITIES

Northwestern University
Notre Dame, University of
Occidental College
Pepperdine University
Princeton University
Rensselaer Polytechnic Institute
Rochester, University of
Scripps College
Smith College
Southern California, University of
St. Lawrence University
Stanford University
Stevens Institute of Technology
Swarthmore College
Ursinus College
Vanderbilt University
Villanova University
Wake Forest University
Washington University in St. Louis
Wellesley College
Williams College
Worcester Polytechnic Institute

Yale University

Very Expensive—$$$$
Amherst College
Bard College
Bates College
Bennington College
Bowdoin College
Brandeis University
Brown University
Bucknell University
Carleton College
Carnegie Mellon University
Chicago, University of
Colby College
Colgate University
Colorado College
Columbia College
Connecticut College
Franklin and Marshall College
George Washington University
Gettysburg College

Hamilton College
Hampshire College
Hobart and William Smith
Johns Hopkins University
Kenyon College
Middlebury College
Mount Holyoke College
Oberlin College
Pennsylvania, University of
Pitzer College
Reed College
Richmond, University of
Sarah Lawrence College
Skidmore College
St. John's College
Trinity College
Tufts University
Tulane University
Union College
Vassar College
Wesleyan University
Wheaton College (MA)

The Best Buys of 2008

Following is a list of 46 colleges and universities
that qualify as Best Buys based on the quality of their academic offerings
in relation to the cost of attendance.
(See page xviii for an explanation of how Best Buys were identified.)

Public

University of Aberdeen (Great Britain)
University of Arizona
University of British Columbia (Canada)
University of Colorado
Evergreen State
University of Florida
Georgia Institute of Technology
University of Georgia
University of Iowa
Iowa State University
McGill University (Canada)
New College of Florida
University of North Carolina at Asheville
University of North Carolina at Chapel Hill
University of St. Andrews (Great Britain)
SUNY–Binghamton
SUNY–Geneseo
SUNY–Stony Brook
University of Toronto (Canada)
University of Washington

Private

Adelphi University
Baylor University
Birmingham–Southern University
Brigham Young University
Case Western Reserve
Cooper Union
Deep Springs College
Earlham College
Elon University
Grinnell College
Hendrix College
Howard University
Illinois Institute of Technology
Macalester College
Morehouse College
Olin College of Engineering
Randolph College
Rice University
University of the South (Sewanee)
Southwestern University
Spelman College
Trinity University (TX)
Wabash College
Warren Wilson College
Washington and Lee University
Wheaton College (IL)

Introduction

FISKE GUIDE TO COLLEGES—AND HOW TO USE IT

The 2008 edition of *Fiske Guide to Colleges* is a revised and updated version of a book that has been a bestseller since it first appeared two decades ago and is universally regarded as the definitive college guide of its type. Features of the new edition include:

- Updated write-ups on more than 300 of the country's best and most interesting colleges and universities
- A list of schools that no longer require the SAT or ACT of all applicants
- A section titled "Sizing Yourself Up," with a questionnaire that will help you figure out what kind of school is best for you
- "A Guide for Preprofessionals," which lists colleges and universities strong in nine preprofessional areas
- A list of schools with strong programs for students with learning disabilities
- Designation of the 46 schools that constitute this year's Best Buys
- Statistical summaries that give you the numbers you need, but spare you those that you do not
- Authoritative rankings of each institution by academics, social life, and quality of life
- The unique "If You Apply…" feature, which summarizes the vital information you need about each college's admission policies—including deadlines and essay topics
- A section on the top Canadian and British universities to help the growing number of students and families in the United States seeking the educational bargains lurking just across the border to the north. These universities offer first-rate academics—easily the equivalent of the flagship public institutions in the U.S.—but at a fraction of the cost.

Picking the right college—one that will coincide with your particular needs, goals, interests, talents, and personality—is one of the most important decisions any young person will ever make. It is also a major investment. Tuition and fees alone now run at least $5,000 at a typical public university and $27,000 at a typical private college, and the overall tab at the most selective and expensive schools tops $50,000. Obviously, a major investment like that should be approached with as much information as possible.

That's where *Fiske Guide to Colleges* fits in. It is a tool to help you make the most intelligent educational investment you can.

WHAT IS THE *FISKE GUIDE TO COLLEGES*?

Fiske Guide to Colleges mirrors a process familiar to any college-bound student and his or her family. If you are wondering whether to consider a particular college, it is logical to seek out friends or acquaintances who go there and ask them to tell you about their experiences. We have done exactly that—but on a far broader and more systematic basis than any individual or family could do alone.

In using the *Fiske Guide*, some special features should be kept in mind:

- The guide is **selective**. We have not tried to cover all four-year colleges and universities. Rather, we have taken more than 300 of the best and most interesting institutions in the nation—the ones that students most want to know about—and written descriptive essays of 1,000 to 2,500 words about each of them.
- Since choosing a college is a matter of making a calculated and informed judgment, this guide is also **subjective**. It makes judgments about the strengths and weaknesses of each institution, and it contains a unique set of ratings of each college or university on the basis of academic strength, social life, and overall quality of life. No institution is a good fit for every student. The underlying assumption of the *Fiske Guide* is that each of the colleges chosen for inclusion is the right place for some students but not a good bet for others. Like finding the right husband or wife, college admissions is a matching process. You know your own interests and needs; the *Fiske Guide* will tell you something about those needs that each college seems to serve best.
- Finally, the *Fiske Guide* is systematic. Each write-up is carefully constructed to cover specific topics—from the academic climate and the makeup of the student body to the social scene—in a systematic order. This

means that you can easily take a specific topic, such as the level of academic pressure or the role of fraternities and sororities on campus, and trace it through all of the colleges that interest you.

HOW THE COLLEGES WERE SELECTED

How do you single out "the best and most interesting" of the more than 2,200 four-year colleges in the United States? Obviously, many fine institutions are not included. Space limitations simply require that some hard decisions be made.

The selection was done with several broad principles in mind, beginning with academic quality. Depending on how you define the term, there are about 175 "selective" colleges and universities in the nation, and by and large these constitute the best institutions academically. All of these are included in the *Fiske Guide*. In addition, an effort was made to achieve geographic diversity and a balance of public and private schools. Special efforts were made to include a good selection of three types of institutions that seem to be enjoying special popularity at present: engineering and technical schools, those with a religious emphasis, and those located along the Sunbelt, where the cost of education is considerably less than at its Northern counterparts.

Finally, in a few cases we exercised the journalist's prerogative of writing about schools that are simply interesting. The tiny College of the Atlantic, for example, would hardly qualify on the basis of superior academic program or national significance, but it offers an unusual and fascinating brand of liberal arts within the context of environmental studies. Likewise, Deep Springs College, the only two-year school in the *Fiske Guide*, is a unique institution of intrinsic interest.

HOW THE *FISKE GUIDE* WAS COMPILED

Each college or university selected for inclusion in the *Fiske Guide to Colleges* was sent a packet of questionnaires. The first was directed to the administration and covered topics ranging from their perception of the institution's mission to the demographics of the student body. Administrators were also asked to distribute a set of questionnaires to a cross section of students.

The questions for students, all open-ended and requiring short essays as responses, covered a series of topics ranging from the accessibility of professors and the quality of housing and dining facilities to the type of nightlife and weekend entertainment available in the area. By and large, students responded enthusiastically to the challenge we offered them. The quality of the information in the write-ups is a tribute to their diligence and openness. American college students, we learned, are a candid lot. They are proud of their institutions, but also critical—in the positive sense of the word.

Other sources of information were also employed. Administrators were invited to attach to their questionnaires any catalogs, in-house research, or other documents that would contribute to an understanding of the institution and to comment on their write-up in the last edition. Also, staff members have visited many of the colleges, and in some cases, additional information was solicited through published materials, telephone interviews, and other contacts with students and administrators.

The information from these various questionnaires was then incorporated into write-ups by staff members under the editorial direction of Edward B. Fiske, former education editor of the *New York Times*.

THE FORMAT

Each essay covers certain broad subjects in roughly the same order. They are as follows:

Academics	**Housing**
Campus setting	**Food**
Student body	**Social life**
Financial aid	**Extracurricular activities**

Certain subtopics are covered in all of the essays. The sections on academics, for example, always discuss the departments (or, in the case of large universities, schools) that are particularly strong or weak, while the sections on housing contain information on whether the dorms are co-ed or single-sex and how students get the rooms they want. Other topics, however, such as class size, the need for a car, or the number of volumes in the library, are mentioned only if they constitute a particular strength or weakness at that institution.

We paid particular attention to the effect of the 21-year-old drinking age on campus life. Also, we noted efforts some schools' administrations have been making to change or improve the social and residential life on campuses through such measures as creating learning communities, restricting fraternities, and constructing new recreational facilities.

BEST BUYS

One of the lesser-known facts of life about higher education in the U.S. is that price and quality do not always go hand-in-hand. The college or university with the jumbo price tag may or may not offer a better education than the institution across town with much lower tuition. The relationship between the cost paid by the consumer and the quality of the education is affected by factors ranging from the size of an institution's endowment to calculations by college officials about what the market will bear.

In the face of today's skyrocketing tuition rates, students and families in all economic circumstances are looking for ways to get the best value for their education dollar. Fortunately, there are some bargains to be found in higher education; it just takes a bit of shopping around with a little guidance along the way. The *Fiske Guide* has an Index by Price that groups public and private institutions into four price categories, from inexpensive to very expensive. We also go one step further and suggest a number of schools that offer outstanding academics with relatively modest prices. This year we have designated 46 such institutions —20 public and 26 private—as Best Buys. Look for the Best Buy graphic next to the college name. (A list of all 2008 Best Buys appears on page xv.)

All of our Best Buys fall into the inexpensive or moderate price category, and most have four- or five-star academic ratings. But there are bargains to be found among all levels and types of institutions. For example, some of the best values in American higher education are public colleges and universities that have remained relatively small and offer the smaller classes and personalized approach to academics that are typically found only in expensive private liberal arts colleges. Several of these are included as Best Buys.

STATISTICS

At the beginning of each write-up are basic statistics about the college or university—the ones that are relevant to applicants. These include the address, type of location (urban, small town, rural, etc.), enrollment, male/female ratio, SAT or ACT score ranges of the middle 50 percent of the students, percentage of students receiving need-based financial aid, relative cost, whether or not the institution has a chapter of Phi Beta Kappa, the number of students who apply and the percentage of those who are accepted, the percentage of accepted students who enroll, the number of freshmen who graduate within six years, and the number of freshmen who return for their sophomore year. For convenience, we include the telephone number of the admissions office and the school's website and email and mailing addresses.

Unlike some guides, we have intentionally not published figures on the student/faculty ratio because colleges use different—and often self-serving—methods to calculate the ratio, thus making this particular statistic virtually meaningless.

Within the statistics, you will sometimes encounter the letters "N/A." In most cases, this means that the statistic was not available. In other cases, however, such as schools that do not require standardized tests, it means "not applicable." The write-up should make it clear which meaning is the relevant one.

We have included information on whether the school has a chapter of Phi Beta Kappa because this academic honorary society is a sign of broad intellectual distinction. Keep in mind, though, that even the very best engineering schools, because of their relatively narrow focus, do not usually qualify under the society's standards.

Tuition and fees are constantly increasing at American colleges, but for the most part, the cost of various institutions in relation to one another does not change. Rather than put in specific cost figures that would immediately become out of date, we have classified colleges into four groups ranging from inexpensive ($) to very expensive ($$$$) based on estimated costs of tuition and fees for the 2007–2008 academic year. The results for each college can be found in the Index by Price pages xii–xiv. Separate scales were used for public and private institutions, and the ratings for the public institutions are based on cost for residents of the state; out-of-staters should expect to pay more. If a public institution has a particularly low or high surcharge for out-of-staters, this is noted in the essay. The categories are defined as follows:

	PUBLIC	PRIVATE
$$$$	More than $8,000	More than $34,000
$$$	$6,000–$8,000	$32,000–$34,000
$$	$5,000–$6,000	$27,000–$32,000
$	Less than $5,000	Less than $27,000

We also include an index that groups colleges by their relative cost (see Index by Price pages xii–xiv).

SAT and ACT SCORES

A special word needs to be said about standardized test scores. Some publications follow the practice of giving the median or average score registered by entering freshmen. Such figures, however, are easily misinterpreted as thresholds rather than averages. Many applicants forget that if a school reports average SAT–Verbal scores of 500, this means that, by definition, about half of the students scored below this number and half scored above. An applicant with a 480 would still have lots of company.

To avoid such confusion, we report the range of scores of the middle half of freshmen—or, to put it another way, the scores achieved by those in the 25th and 75th percentiles. For example, that college where the SAT–Verbal average was 500 might have a range of 440 to 560. So if you scored within this range, you would have joined the middle 50 percent of last year's freshmen. If your score was above 560, you would have been in the top quarter and could probably look forward to a relatively easy time; if it was below 440, you would have been struggling along with the bottom quarter of students.

The reporting of ranges rather than a single average is an increasingly common practice, but some colleges do not calculate ranges. These are indicated by "N/A." Keep in mind, as well, that score ranges (and averages, for that matter) are misleading at the growing number of colleges that no longer require test scores from all applicants (see the section on SAT- and ACT-Optional Schools below). The ranges given for these colleges typically represents the range of scores of students who choose to submit their test scores, although they are not required to do so.

Unfortunately, another problem that arises with SAT and ACT scores is that, in their zeal to make themselves look good in a competitive market, some colleges and universities have been known to be less than honest in the numbers they release. They inflate their scores by not counting certain categories of students at the low end of the scale, such as athletes, certain types of transfer students, or students admitted under affirmative action programs. Some colleges have gone to such extremes as reporting the relatively high math scores of foreign students, but not their relatively low verbal scores. Aside from the sheer dishonesty of such practices, they can also be misleading. A student whose own scores are below the 25th percentile of a particular institution needs to know whether his profile matches that of the lower quarter of the student body as a whole, or whether there is an unreported pool of students with lower scores.

Even when dealing with a range rather than a single score, keep in mind that standardized tests are an imprecise measure of academic ability, and comparisons of scores that differ by less than 50 or 60 points on a scale of 200 to 800 have little meaning. According to the laws of statistics, there is one chance in three that the 550 that arrived in the little envelope from ETS should really be at least 580 or no more than 520. On the other hand, median scores offer some indication of your chances to get into a particular institution and the intellectual level of the company you will be keeping—or, if you prefer, competing against. Remember, too, that the most competitive schools have the largest and most sophisticated admissions staffs and are well aware of the limitations of standardized tests. A strong high school average or achievement in a field such as music will usually counteract the negative effects of modest SAT or ACT scores.

SCHOLARSHIP INFORMATION

Since the first edition of *Fiske Guide to Colleges* appeared, the problems of financing college have become increasingly critical, mainly because of the rising cost of education and a shift from grants to loans as the basis for financial aid packages.

In response to these developments, many colleges and universities have begun to devise their own plans to help students pay for college. These range from subsidized loan programs to merit scholarships that are awarded without reference to financial need. Most of these programs are aimed at retaining the middle class.

Some colleges advertise that they are "need blind" in their admissions, meaning that they accept or reject applicants without reference to their financial situation and then guarantee to meet the "demonstrated need" of all students whom they accept. Others say they are need blind in their admissions decisions, but do not guarantee to provide the financial aid required of all those who are accepted. Still others agree to meet the demonstrated need of all students, but they package their offers so that students they really want receive a higher percentage of their aid in the form of outright grants than in repayable loans.

"Demonstrated need" is itself a slippery term. In theory, the figure is determined when students and families fill out a needs-analysis form, which leads to an estimate of how much the family can afford to pay. Demonstrated need is then calculated by subtracting that figure from the cost at a particular institution. In practice, however, various colleges make their own adjustments to the standard figure.

Students and parents should not assume that their family's six-figure annual income automatically disqualifies them from some kind of subsidized financial aid. In cases of doubt, they should fill out a needs-analysis form to determine their eligibility. Whether they qualify or not, they are also eligible for a variety of awards made without regard to financial need.

Inasmuch as need-based awards are universal at the colleges in this guide, the awards generally singled out for special mention in the write-ups in *Fiske Guide to Colleges* are the merit scholarships. We have not mentioned awards of a purely local nature—restricted to residents of a particular county, for example—but all college applicants should search out these awards through their guidance offices and the bulletins of the colleges that are of interest to them. Similarly, we have not duplicated the information on federally guaranteed loan programs that is readily available through both high school and college counseling offices, but we cite novel and often less-expensive variants of the federal loan programs that are offered by individual colleges.

For more information on the ever-changing financial aid scene, we suggest that you consult the companion book to this guide, *Fiske Guide to Getting into the Right College*.

RATINGS

Much of the fierce controversy that greeted the first edition of *Fiske Guide to Colleges* a quarter century ago revolved around its unique system of rating colleges in three areas: academics, social life, and quality of life. In each case, the ratings are done on a system of one to five, with three considered normal for colleges included in the *Fiske Guide*. If a college receives a rating higher or lower than three in any category, the reasons should be apparent from the narrative description of that college.

Students and parents should keep in mind that these ratings are obviously general in nature and inherently subjective. No complex institution can be described in terms of a single number or other symbol, and different people will have different views of how various institutions should be rated in the three categories. They should not be viewed as either precise or infallible judgments about any given college. On the other hand, the ratings are a helpful tool in using this book. The core of the *Fiske Guide* is the essays on each of the colleges, and the ratings represent a summary—an index, if you will—of these write-ups. Our hope is that each student, having decided on the kind of configuration that suits his or her needs, will then thumb through the book looking for other institutions with a similar set of ratings. The three categories, defined as follows, are academics, social life, and quality of life.

Academics 🖎

This is a judgment about the overall academic climate of the institution, including its reputation in the academic world, the quality of the faculty, the level of teaching and research, the academic ability of students, the quality of libraries and other facilities, and the level of academic seriousness among students and faculty members.

Although the same basic criteria have been applied to all institutions, it should be evident that an outstanding small liberal arts college will by definition differ significantly from an outstanding major public university. No one would expect the former to have massive library facilities, but one would look for a high-quality faculty that combines research with a good deal of attention to the individual needs of students. Likewise, public universities, because of their implicit commitment to serving a broad cross-section of society, might have a broader range of curriculum offerings but somewhat lower average SAT scores than a large private counterpart. Readers may find the ratings most useful when comparing colleges and universities of the same type.

In general, an academics rating of three pens suggests that the institution is a solid one that easily meets the criteria for inclusion in a guide devoted to the top 10 percent of colleges and universities in the nation.

An academics rating of four pens suggests that the institution is above average even by these standards and that it has some particularly distinguishing academic feature, such as especially rich course offerings or an especially serious academic atmosphere.

A rating of five pens for academics indicates that the college or university is among the handful of top institutions of its type in the nation on a broad variety of criteria. Those in the private sector will normally attract students with combined SAT scores of at least 1300, and those in the public sector are invariably magnets for the top students in their states. All can be assumed to have outstanding faculties and other academic resources.

In response to the suggestion that the range of colleges within a single category has been too broad, we have introduced some half-steps into the ratings.

Social Life ☎

This is primarily a judgment about the amount of social life that is readily available. A rating of three telephones suggests a typical college social life, while four telephones means that students devote an above-average amount of time to socializing. It can be assumed that a college with a rating of five is something of a party school, which may or may not detract from the academic quality. Colleges with a rating below three have some impediment to a strong social life, such as geographic isolation, a high percentage of commuting students, or a disproportionate number of nerds who never leave the library. Once again, the reason should be evident from the write-up.

Quality of Life ★

This category grew out of the fact that schools with good academic credentials and plenty of social life may not, for one reason or another, be particularly wholesome places to spend four years. The term "quality of life" is one that has gained currency in social science circles, and, in most cases, the rating for a particular college will be similar to the academic and/or social ratings. The reader, though, should be alert to exceptions to this pattern. A liberal arts college, for example, might attract bright students who study hard during the week and party hard on weekends, and thus earn high ratings for academics and social life. If the academic pressure is cutthroat rather than constructive, though, and the social system is manipulative of women, this college might get an apparently anomalous two stars for quality of life. By contrast, a small college with modest academic programs and relatively few organized social opportunities might have developed a strong sense of supportive community, have a beautiful campus, and be located near a wonderful city—and thus be rated four stars for quality of life. As in the other categories, the reason can be found in the essay to which the ratings point.

OVERLAPS

Most colleges and universities operate within fairly defined "niche markets." That is, they compete for students against other institutions with whom they share important characteristics, such as academic quality, size, geographic location, and the overall tone and style of campus life. Not surprisingly, students who apply to College X also tend to apply to the other institutions in its particular niche. For example, "alternative" colleges such as Bard, Bennington, Hampshire, Marlboro, Oberlin, Reed, and Sarah Lawrence share many common applications, as do those with an evangelical flavor, such as Calvin, Hope, and Wheaton (IL).

As a service to readers, we ask each school to give us the names of the five colleges with which they share the most common applications, and these are listed in the Overlaps section at the end of each write-up. We encourage students who know they are interested in a particular institution to check out the schools with which it competes—and perhaps then check out the "overlaps of the overlaps." This method of systematic browsing should yield a list of 15 or 20 schools that, based on the behavior of thousands of past applicants, would constitute a good starting point for the college search.

IF YOU APPLY...

An extremely helpful feature is the "If You Apply" section at the end of each write-up. This is designed for students who become seriously interested in a particular college and want to know more specifics about what it takes to get in.

This section begins with the deadlines for early admissions or early decision (if the college has such a program), regular admissions, and financial aid. If the college operates on a rolling-admissions basis—making decisions as the applications are received—this is indicated.

"If You Apply" gives a snapshot of the institution's financial aid policies. It indicates whether the college or university guarantees to meet the demonstrated need of applicants and, if so, the percentage of students whose financial need is actually met. The phrase "guarantees to meet demonstrated need" means that the institution for all practical purposes makes every effort possible to come up with the aid for which all of its students qualify.

Colleges have widely varying policies regarding interviews, both on campus and with alumni, so we indicate whether each of these is required, recommended, or optional. We also indicate whether reports from the person doing the interview are used in evaluating students or whether, as in many cases, the interview is seen only as a means of conveying information about the institution and answering applicants' questions. You are within your rights to ask the admissions office to explain how they view interviews.

This section also describes what standardized tests—SAT, ACT, or achievement—are required, and whether applicants are asked to write one or more essays. In the latter case, the topics are given.

The admissions policies of most colleges are fairly similar, at least among competing clusters of institutions. In some cases, however, a school will have its own special priorities. Some don't care that much about test scores. Others are looking for students with special talent in math or science, while others pay special attention to personal characteristics such as leadership in extracurricular activities. We asked each institution to tell us if its admission policies are in any ways "unique or unusual," and their answers are reported.

CONSORTIA

Many colleges expand the range of their offerings by banding together with other institutions to offer unusual programs that they could not support on their own. These options range from foreign study programs around the world to semesters at sea, and keeping such arrangements in mind is a way of expanding the list of institutions that might meet your particular interests and needs. The final section of the *Fiske Guide* describes 16 of these consortia and lists the member institutions.

MOVING FORWARD

Students will find the *Fiske Guide* useful at various points in the college selection process—from deciding whether to visit a particular campus to selecting among institutions that have accepted them. To make it easy to find a particular college, the write-ups are arranged in alphabetical order in the index. An Index by State and Country and the Index by Price can be found on pages ix and xii, respectively.

While most people are not likely to start reading at Adelphi and keep going until they reach Yale (though some tell us they do), we encourage you to browse. This country has an enormously rich and varied network of colleges and universities, and there are dozens of institutions out there that can meet the needs of any particular student. Too many students approach the college selection process wearing blinders, limiting their sights to local institutions, the pet schools of their parents or guidance counselors, or to ones they know only by possibly outdated reputations.

But applicants need not be bound by such limitations. Once you have decided on the type of school you think you want—a small liberal arts college, an engineering school, or whatever—we hope you will thumb through the book looking for similar institutions that might not have occurred to you. As already noted above, one way to do this is to look at the overlaps of schools you like and then check out those schools' overlaps. Many students have found this worthwhile, and quite frankly, we view the widening of students' horizons about American higher education as one of the most important purposes of the book. Perhaps the most gratifying remark we hear comes when a student tells us, as many have, that she is attending a school that she first heard about while browsing through *Fiske Guide to Colleges*.

Picking a college is a tricky business. But given the current buyer's market, there is no reason why you should not be able to find the right one. That's what *Fiske Guide to Colleges* is designed to help you do. Happy college hunting.

Sizing Yourself Up

The college search is a game of matchmaking. You have interests and needs; the colleges have programs to meet those needs. If all goes according to plan, you'll find the right one and live happily ever after——or at least for four years. It ought to be simple, but today's admissions process resembles a high-stakes obstacle course.

Many colleges are more interested in making a sale than they are in making a match. Under intense competitive pressure, many won't hesitate to sell you a bill of goods if they can get their hands on your tuition dollars. Guidance counselors generally mean well, but they are often under duress from principals and trustees to steer students toward prestigious schools regardless of whether the fit is right. Your friends won't be shy with advice on where to go, but their knowledge is generally limited to a small group of hot colleges that everyone is talking about. National publications rake in millions by playing on the public's fascination with rankings, but a close look at their criteria reveals distinctions without a difference.

Before you find yourself spinning headlong on this merry-go-round, take a step back. This is your life and your college career. What are you looking for in a college? Think hard and don't answer right away. Before you throw yourself and your life history on the mercy of college admissions officers, you need to take some time to objectively and honestly evaluate your needs, likes and dislikes, strengths and weaknesses. What do you have to offer a college? What can a college do for you? Unlike the high school selection process, which is usually predetermined by your parents' property lines, income level, or religious affiliation, picking a college isn't a procedure you can brush off on dear ol' Mom and Dad. You have to take some initiative. You're the best judge of how well each school fits your personal needs and academic goals.

We encourage you to view the college selection process as the first semester in your higher education. Life's transitions often call for extra energy and focus. The college search is no exception. For the first time, you'll be contemplating a life away from home that can unfold in any direction you choose. Visions of majors and careers will dance in your head as you sample various institutions of higher learning, each with hundreds of millions of dollars in academic resources; it is hard to imagine a better hands-on seminar in research and matchmaking than the college search. The main impact, however, will be measured by what you learn about yourself. Piqued by new worlds of learning and tested by the competition of the admissions process, you'll be pushed as never before to show your accomplishments, clarify your interests, and chart a course for the future. More than one parent has watched in amazement as an erstwhile teenager suddenly emerged as an adult during the course of a college tour. Be ready when your time comes.

DEVELOP YOUR CRITERIA

One strategy is to begin the search with a personal inventory of your own strengths and weaknesses and your "wish list" for a college. This method tends to work well for compulsive list-makers and other highly organized people. What sorts of things are you especially good at? Do you have a list of skills or interests that you would like to explore further? What sort of personality are you looking for in a college? Mainstream? Conservative? Offbeat? What about extracurriculars? If you are really into riding horses, you might include a strong equestrian program in your criteria. The main problem won't be thinking of qualities to look for—you could probably name dozens—but rather figuring out what criteria should play a defining role in your search. Serious students should think carefully about the intellectual climate they are seeking. At some schools, students routinely stay up until 3:00 a.m. talking about topics like the value of deconstructing literary texts or the pros and cons of free trade. These same students would be viewed as geeks or weirdos on less cosmopolitan campuses. Athletes should take a hard look at whether they really want to play college ball and, if so, whether they want to go for an athletic scholarship or play at the less-pressured Division III level. Either way, intercollegiate sports require a huge time commitment.

Young women have an opportunity all to themselves—the chance to study at a women's college. The *Fiske Guide* profiles 14 such campuses, a vastly underappreciated resource on today's higher education scene. With small classes and strong encouragement from faculty, students at women's colleges move on to graduate study in significantly higher numbers than their counterparts at co-ed schools, especially in the natural sciences. Males seeking an all-male experience will find two options in the *Fiske Guide*, Hampden-Sydney College and Wabash College.

Students with a firm career goal will want to look for a course of study that matches their needs. If you want to major in aerospace engineering, your search will be limited to schools that have the program. Outside of specialized areas like this, many applicants overestimate the importance of their anticipated major in choosing a college. If you're interested in a liberal arts field, your expected major should probably have little to do with your college selection. A big purpose of college is to develop interests and set goals. Most students change their intentions regarding a major at least two or three times before graduation, and once out in the working world, they often end up in jobs bearing no relation to their academic specialty. Even those with a firm career goal may not need as much specialization as they think at the undergraduate level. If you want to be a lawyer, don't worry yourself looking for something labeled "prelaw." Follow your interests, get the best liberal arts education available, and then apply to law school.

Naturally, it is never a bad idea to check out the department(s) of any likely major, and occasionally your choice of major will suggest a direction for your search. If you're really into national politics, it may make sense to look at some schools in or near Washington, D.C. If you think you're interested in a relatively specialized field, say, oceanography, then be sure to look for some colleges that are a good match for you and also have programs in oceanography. But for the most part, rumors about top-ranked departments in this or that should be no more than a tie-breaker between schools you like for more important reasons. There are good professors (and bad ones) in any department. You'll have plenty of time to figure out who is who once you've enrolled. Being undecided about your career path as a senior in high school is often a sign of intelligence. Don't feel bad if you have absolutely no idea what you're going to do when you "grow up." One of the reasons you'll be paying megabucks to the college of your choice is the prospect that it will open some new doors for you and expand your horizons. Instead of worrying about particular departments, try to keep the focus on big-picture items, such as: What's the academic climate? How big are the freshman classes? Do I like it here? and Are these my kind of people?

KEEP AN OPEN MIND

The biggest mistake of beginning applicants is hyper-choosiness. At the extreme is the "perfect-school syndrome," which comes in two basic forms.

In one category are the applicants who refuse to consider any school that doesn't have every little thing they want in a college. If you're one who begins the process with a detailed picture of Perfect U. in mind, you may want to remember the oft-quoted advice, "Two out of three ain't bad." If a college seems to have most of the qualities you seek, give it a chance. You may come to realize that some things you thought were absolutely essential are really not that crucial after all.

The other strain of perfect-school syndrome is the applicant who gets stuck on a "dream" school at the beginning and then won't look anywhere else. With those 2,200 four-year colleges out there (not counting those in Canada and Great Britain), it is just a bit silly to insist that only one will meet your needs. Having a first choice is okay, but the whole purpose of the search is to consider new options and uncover new possibilities. A student who has only one dream school—especially if it is a highly selective one—could be headed for disappointment.

As you begin the college search, don't expect any quick revelations. The answers will unfold in due time. Our advice? Be patient. Set priorities. Keep an open mind. Reexamine priorities. Again, be patient.

To get the ball rolling, move on to the Sizing-Yourself-Up Survey.

FISKE'S SIZING-YOURSELF-UP SURVEY

With apologies to Socrates, knowing thyself is easier said than done. Most high school students can analyze a differential equation or a Shakespearean play with the greatest of ease, but when it comes to cataloging their own strengths, weaknesses, likes, and dislikes, many draw a blank. But self-knowledge is crucial to the matching process at the heart of a successful college search. The 30-item survey below offers a simple way to get a handle on some crucial issues in college selection—and what sort of college may fit your preferences.

In the space beside each statement, rate your feelings on a scale of 1 to 10, with 10 = Strongly Agree, 1 = Strongly Disagree, and 5 = Not Sure/Don't Have Strong Feelings. (For instance, a rating of 7 would mean that you agree with the statement, but that the issue is a lower priority than those you rated 8, 9, or 10.) After you're done, read on to "Grading Yourself" to find out what it all means.

Fiske's Sizing-Yourself-Up Survey

Size

____ 1. I enjoy participating in many activities.

____ 2. I would like to have a prominent place in my community.

____ 3. Individual attention from teachers is important to me.

____ 4. I learn best when I can speak out in class and ask questions.

____ 5. I am undecided about what I will study.

____ 6. I want to earn a PhD in my chosen field of study.

____ 7. I learn best by listening and writing down what I hear.

____ 8. I would like to be in a place where I can be anonymous if I choose.

____ 9. I prefer devoting my time to one or two activities rather than many.

____ 10. I want to attend a college that most people have heard of.

____ 11. I am interested in a career-oriented major.

____ 12. I like to be on my own.

Location

____ 13. I prefer a college in a warm or hot climate.

____ 14. I prefer a college in a cool or cold climate.

____ 15. I want to be near the mountains.

____ 16. I want to be near a lake or ocean.

____ 17. I prefer to attend a college in a particular state or region.

____ 18. I prefer to attend a college near my family.

____ 19. I want city life within walking distance of my campus.

____ 20. I want city life within driving distance of my campus.

____ 21. I want my campus to be surrounded by natural beauty.

Academics and Extracurriculars

____ 22. I like to be surrounded by people who are free-thinkers and nonconformists.

____ 23. I like the idea of joining a fraternity or sorority.

____ 24. I like rubbing shoulders with people who are bright and talented.

____ 25. I like being one of the smartest people in my class.

____ 26. I want to go to a prestigious college.

____ 27. I want to go to a college where I can get an excellent education.

____ 28. I want to try for an academic scholarship.

____ 29. I want a diverse college.

____ 30. I want a college where the students are serious about ideas.

Grading Yourself

Picking a college is not an exact science. People who are total opposites can be equally happy at the same college. Nevertheless, particular types tend to do better at some colleges than others. Each item in the survey is designed to test your feelings on an important issue related to college selection. Sizing Up the Survey (below) offers commentary on each item.

Taken together, your responses may help you construct a tentative blueprint for your college search. Statements 1–12 deal with the issue of size. Would you be happier at a large university or a small college? Here's the trick: Add the sum of your responses to questions 1–6. Then make a second tally of your responses to 7–12. If the sum of 1–6 is larger, you may want to consider a small college. If 7–12 is greater, then perhaps a big school would be more to your liking. If the totals are roughly equal, you should probably consider colleges of various sizes.

Statements 13–21 deal with location. The key in this section is the intensity of your feeling. If you replied to No. 13 with a 10, does that mean you are going to look only at schools in warm climates? Think hard. If you consider only schools within a certain region or state, you'll be eliminating hundreds of possibilities. By examining your most intense responses—the 1s, 2s, 9s, and 10s—you'll be able to create a geographic profile of likely options.

Statements 22–30 deal with big-picture issues related to the character and personality of the college that may be in your future. As before, pay attention to your most intense responses. Read on for a look at the significance of each question.

Sizing Up the Survey

1. **I enjoy participating in many activities.** Students at small colleges tend to have more opportunities to be involved in many activities. Fewer students means less competition for spots.

2. **I would like to have a prominent place in my community.** Student-council presidents and other would-be leaders take note: It is easier to be a big fish if you're swimming in a small pond.

3. **Individual attention from teachers is important to me.** Small colleges generally offer more one-on-one with faculty both in the classroom and the laboratory.

4. **I learn best when I can speak out in class and ask questions.** Students who learn from interaction and participation would be well-advised to consider a small college.

5. **I am undecided about what I will study.** Small colleges generally offer more guidance and support to students who are undecided. The exception: students who are considering a preprofessional or highly specialized major.

6. **I want to earn a PhD in my chosen field of study.** A higher percentage of students at selective small colleges earn a PhD than those who attend large institutions of similar quality.

7. **I learn best by listening and writing what I hear.** Students who prefer lecture courses will find more of them at large institutions.

8. **I would like to be in a place where I can be anonymous if I choose to be.** At a large university, the supply of new faces is never-ending. Students who have the initiative can always reinvent themselves.

9. **I prefer devoting my time to one or two activities rather than many.** Students who are passionate about one activity—say, writing for the college newspaper—will often find higher quality at a bigger school.

10. **I want to attend a college that most people have heard of.** Big schools have more name recognition because they're bigger and have Division I athletic programs. Even the finest small colleges are relatively anonymous among the general public

11. **I am interested in a career-oriented major.** More large institutions offer business, engineering, nursing, etc., though some excellent small institutions do so as well (depending on the field).

12. **I like to be on my own.** A higher percentage of students live off campus at large schools, which are more likely to be in urban areas than their smaller counterparts.

13. **I prefer a college in a warm or hot climate.** Keep in mind that the Southeast and the Southwest have far different personalities (not to mention humidity levels).

14. **I prefer a college in a cool or cold climate.** Consider the Midwest, where there are many fine schools that are notably less selective than those in the Northeast.

15. **I want to be near the mountains.** You're probably thinking Colorado or Vermont, but don't zero in too quickly. States from Maine to Georgia and Arkansas to Arizona have easy access to mountains.

16. **I want to be near a lake or ocean.** Oceans are only on the coasts, but keep in mind the Great Lakes, the Finger Lakes, etc. Think about whether you want to be on the water or, say, within a two-hour drive.

17. **I prefer to attend a college in a particular state or region.** Geographical blinders limit options. Even if you think you want a certain area of the country, consider at least one college located elsewhere just to be sure.

18. **I prefer to attend a college close to home.** Unless you're planning to live with Mom and Dad, it may not matter whether your college is a two-hour drive or a two-hour plane ride.

19. **I want city life within walking distance of my campus.** Check out the neighborhood(s) surrounding your campus. Urban campuses—even in the same city—can be wildly different.

20. **I want city life within driving distance of my campus.** Unless you're a hardcore urban-dweller, a suburban perch near a city may beat living in the thick of one. Does public transportation or a campus shuttle help students get around?

21. **I want my campus to be surrounded by natural beauty.** A college viewbook will take you only so far. To really know if you'll fall in love with the campus, visiting is a must.

22. **I like to be surrounded by free-thinkers and nonconformists.** Plenty of schools cater specifically to students who buck the mainstream. Talk to your counselor or browse the *Fiske Guide to Colleges* to find some.

23. **I like the idea of joining a fraternity or sorority.** Greek life is strongest at mainstream and conservative-leaning schools. Find out if there is a split between Greeks and non-Greeks.

24. **I like rubbing shoulders with people who are bright and talented.** This is perhaps the best reason to aim for a highly selective institution, especially if you're the type who rises to the level of the competition.

25. **I like being one of the smartest people in my class.** If so, maybe you should skip the highly selective rat race. Star students get the best a college has to offer.

26. **I want to go to a prestigious college.** There is nothing wrong with wanting prestige. Think honestly about how badly you want a big-name school and act accordingly.

27. **I want to go to a college where I can get an excellent education.** Throw out the *U.S. News* rankings and think about which colleges will best meet your needs as a student.

28. **I want to try for an academic scholarship.** Students in this category should consider less-selective alternatives. Scholarships are more likely if you rank high in the applicant pool.

29. **I want a diverse college.** All colleges pay lip service to diversity. To get the truth, see the campus for yourself and take a hard look at the student-body statistics in the *Fiske Guide*'s write-ups.

30. **I want a college where students are serious about ideas.** Don't assume that a college necessarily attracts true intellectuals merely because it is highly selective. Some top schools are known for their intellectual climate—and others for their lack of it.

A Guide for Preprofessionals

The lists that follow include colleges and universities with unusual strength in each of nine preprofessional areas: engineering, architecture, business, art/design, drama, dance, music, communications/journalism, and film/television. We also offer lists covering two of today's hottest interdisciplinary majors: environmental studies and international studies. In compiling the lists, we drew on data from the thousands of surveys used to compile the *Fiske Guide*. We examined the strongest majors at each college as reported in student and administrative questionnaires, and then weighed these against the selectivity and overall academic quality of each institution. After compiling tentative lists in each subject, we queried our counselor advisory group, listed on page 759, for additional suggestions and feedback. To make the lists as useful as possible, we have included some schools that do not receive full-length write-ups in the *Fiske Guide*. Moreover, while the lists are suggestive, they are by no means all-inclusive, and there are other institutions in the *Fiske Guide* that offer fine programs in these areas. Nevertheless, we hope the lists will be a starting place for students interested in these fields.

If you are planning a career in one of the subjects below, your college search may focus largely on finding the best programs for you in that particular area. But we also recommend that you shop for a school that will give you an adequate dose of liberal arts. For that matter, you might consider a double major (or minor) in a liberal arts field to complement your area of technical expertise. If you allow yourself to get too specialized too soon, you may end up as tomorrow's equivalent of the typewriter repairman. In a rapidly changing job market, nothing is so practical as the ability to read, write, and think.

ARCHITECTURE

Private Universities Strong in Architecture

Carnegie Mellon University
Catholic University of America
Columbia University
Cooper Union
Cornell University (NY)
Drexel University
Hobart and William Smith Colleges
Howard University
Lehigh University
Massachusetts Institute of Technology
University of Miami (FL)
New Jersey Institute of Technology
Northeastern University
University of Notre Dame
Princeton University
Rensselaer Polytechnic Institute
Rice University
Temple University
Tuskegee University
Tulane University
Washington University in St. Louis

Public Universities Strong in Architecture

University of Arizona
University of California–Berkeley
University of Cincinnati
Clemson University
University of Florida
Georgia Institute of Technology
University of Illinois–Urbana-Champaign
University of Kansas
Kansas State University
University Maryland
Miami University (OH)
University of Michigan
University of Nebraska
State University of New York–Buffalo
University of Oregon
Pennsylvania State University
Texas A&M University
University of Texas–Austin
Virginia Polytechnic Institute
University of Washington

A Few Arts-Oriented Architecture Programs

Barnard College
Bennington College
Pratt Institute
Rhode Island School of Design
Savannah School of Art and Design
Wellesley College
Yale University

ART/DESIGN

Top Schools of Art and Design

Art Center College of Design
California College of the Arts
California Institute of the Arts
Cooper Union
Kansas City Art Institute
Maryland Institute, College of Art
Massachusetts College of Art
Moore College of Art and Design
North Carolina School of the Arts
Otis Institute of Art and Design
Parsons School of Design
Pratt Institute
Rhode Island School of Design
Ringling School of Art and Design
San Francisco Art Institute
Savannah College of Art and Design
School of the Art Institute of Chicago
School of the Museum of Fine Arts (MA)
School of Visual Arts (NY)

Major Universities Strong in Art or Design

American University
Boston College
Boston University

Carnegie Mellon University
University of Cincinnati
Cornell University
Drexel University
Harvard University
University of Michigan
New York University
University of North Carolina/
  Greensboro
University of Pennsylvania
University of the Arts (PA)
Syracuse University
Washington University in St. Louis
University of Washington
Yale University

Small Colleges and Universities Strong in Art or Design

Alfred University
Bard College
Brown University
Centre College
Cornell College
Dartmouth College
Furman University
Hollins University
Kenyon College
Lake Forest College
Lewis and Clark College
Manhattanville College
Mills College
State University of New
  York–Purchase
Randolph College
University of North
  Carolina–Greensboro
Sarah Lawrence College
Scripps College
Skidmore College
Smith College
Southwestern University
Vassar College
Wheaton College (MA)
Willamette University
Williams College

BUSINESS

Major Private Universities Strong in Business

American University
Baylor University
Boston College
Boston University
Carnegie Mellon University

Case Western Reserve University
University of Dayton
Emory University
Fordham University
Georgetown University
Howard University
Ithaca College
Lehigh University
Massachusetts Institute of
  Technology
New York University
University of Notre Dame
University of Pennsylvania
Pepperdine University
Rensselaer Polytechnic Institute
University of San Francisco
Santa Clara University
University of Southern California
Southern Methodist University
Syracuse University
Texas Christian University
Tulane University
Villanova University
Wake Forest University
Washington University in St. Louis

Public Universities Strong in Business

University of Arizona
University of California–Berkeley
University of Cincinnati
University of Connecticut
University of Florida
University of Georgia
University of Illinois–Urbana-
  Champaign
Indiana University
James Madison University
University of Kansas
University of Maryland
University of
  Massachusetts–Amherst
Miami University (OH)
University of Michigan
University of Missouri
University of North
  Carolina–Chapel Hill
Ohio State University
University of Ohio
University of Oregon
Pennsylvania State University
University of Pittsburgh
Rutgers, The State University of
  New Jersey

University of South Carolina
SUNY–University at Albany
SUNY–Binghamton University
SUNY–Buffalo
SUNY–Geneseo
University of Tennessee
Texas A&M University
University of Texas–Austin
University of Vermont
University of Virginia
University of Washington
University of Wisconsin
College of William and Mary

Small Colleges and Universities Strong in Business

Agnes Scott College
Babson College
Bucknell University
Calvin College
Claremont McKenna College
Clarkson University
Eckerd College
Fairfield University
Franklin and Marshall College
Furman University
Gettysburg College
Guilford College
Hendrix College
Lafayette College
Lake Forest College
Lehigh University
Lewis and Clark College
Millsaps College
Morehouse College
Muhlenberg College
Oglethorpe College
Ohio Wesleyan University
Presbyterian College
Rhodes College
University of Richmond
Ripon College
Skidmore College
Southwestern University
Stetson College
Susquehanna University
Trinity University (TX)
Washington and Jefferson College
Washington and Lee University
Whittier College
Wofford College
Worcester Polytechnic Institute
Xavier University of Louisiana

COMMUNICATIONS/ JOURNALISM

Major Universities Strong in Communications/Journalism

American University
Arizona State University
Boston University
University of California–Los Angeles
University of California–San Diego
University of Florida
University of Georgia
University of Illinois–Urbana-Champaign
Indiana University
Ithaca College
University of Kansas
University of Maryland
University of Michigan
University of Missouri–Columbia
University of Nebraska
University of North Carolina–Chapel Hill
Northwestern University
Ohio University
University of Oregon
Pepperdine University
St. Lawrence University
University of San Francisco
University of Southern California
Stanford University
Syracuse University
Texas Christian University
University of Utah
University of Wisconsin–Madison

ENGINEERING

Top Technical Institutes

California Institute of Technology
California Polytechnic Institute–San Luis Obispo
Colorado School of Mines
Cooper Union
Florida Institute of Technology
Georgia Institute of Technology
Harvey Mudd College
Illinois Institute of Technology
Massachusetts Institute of Technology
Michigan Technological University
Montana Tech of the University of Montana
New Jersey Institute of Technology
New Mexico Institute of Mining and Technology
Rensselaer Polytechnic Institute
Rochester Institute of Technology
Rose-Hulman Institute of Technology
Stevens Institute of Technology
Worcester Polytechnic Institute

Private Universities Strong in Engineering

Boston University
Bradley University
Brigham Young University
Brown University
Carnegie Mellon University
Case Western Reserve University
Catholic University of America
Columbia University
Cornell University
Drexel University
Duke University
George Washington University
Johns Hopkins University
Northeastern University
Northwestern University
University of Notre Dame
Olin College of Engineering
University of Pennsylvania
Princeton University
University of Rochester
Rochester Institute of Technology
Santa Clara University
University of Southern California
Southern Methodist University
Stanford University
Syracuse University
Tufts University
Tulane University
University of Tulsa
Vanderbilt University
Villanova University
Washington University in St. Louis

Public Universities Strong in Engineering

University of Arizona
University of California–Berkeley
University of California–Davis
University of California–Los Angeles
University of California–San Diego
University of Cincinnati
Clemson University
University of Connecticut
University of Delaware
University of Florida
University of Illinois–Urbana-Champaign
Iowa State University
University of Kansas
McGill University
University of Maryland
University of Massachusetts–Amherst
University of Michigan
Michigan State University
University of Missouri–Rolla
University of New Hampshire
College of New Jersey
North Carolina State University
Ohio State University
Oregon State University
Pennsylvania State University
Purdue University
Queen's University (CA)
University of Rhode Island
Rutgers, The State University of New Jersey
SUNY–Binghamton University
SUNY–Buffalo
Texas A&M University
Texas Tech University
University of Texas–Austin
University of Toronto
Virginia Polytechnic Institute
University of Virginia
University of Washington
University of Wisconsin

Small Colleges and Universities Strong in Engineering

Alfred University
Bucknell University
Butler University
Calvin College
Clarkson University
Dartmouth College
Lafayette College
Lehigh University
Loyola University (MD)
University of the Pacific
Rice University
Smith College
Spelman College
Swarthmore College

Trinity College (CT)
Trinity University (TX)
University of Tulsa
Tuskegee University
Union College

FILM/TELEVISION

Major Universities Strong in Film/Television

Arizona State University
Boston University
University of California–Los Angeles
University of Cincinnati
Drexel University
Emerson College
University of Florida
Ithaca College
University of Kansas
Memphis State University
University of Michigan
New York University
Northwestern University
Quinnipiac University
Pennsylvania State University
University of Southern California
Syracuse University
University of Texas–Austin
Wayne State University

Small Colleges and Universities Strong in Film/Television

Bard College
Beloit College
Brown University
California Institute of the Arts
Columbia College (CA)
Columbia College (IL)
The Evergreen State College
Hampshire College
Hofstra University
Hollins University
Occidental College
Pitzer College
Pomona College
Sarah Lawrence College
School of Visual Arts
SUNY–Purchase
Wesleyan University

PERFORMING ARTS—MUSIC

Top Music Conservatories

Berklee College of Music

Boston Conservatory
California Institute of the Arts
Cleveland Institute of Music
Curtis Institute of Music
Eastman School of Music
Juilliard School
Manhattan School of Music
New England Conservatory of Music
North Carolina School of the Arts
Peabody Conservatory of Music
San Francisco Conservatory of Music

Major Universities Strong in Music

Baylor University
Boston College
Boston University
University of California–Los Angeles
Carnegie Mellon University
Case Western Reserve University
University of Cincinnati
University of Colorado–Boulder
University of Denver
Depaul University
Florida State University
Harvard University
Indiana University
Ithaca College
Miami University (OH)
University of Miami (FL)
University of Michigan
University of Nebraska–Lincoln
New York University
Northwestern University
University of Oklahoma
University of Southern California
Southern Methodist University
Vanderbilt University
Yale University

Small Colleges and Universities Strong in Music

Bard College
Bennington College
Bucknell University
Butler University
DePauw University
Furman University
Gordon College
Illinois Wesleyan University
Knox College

Lawrence University*
Loyola University–New Orleans
Manhattanville College
Mills College
Oberlin College*
University of the Pacific
Rice University
St. Mary's College of Maryland
St. Olaf College
Sarah Lawrence College
Skidmore College
Smith College
University of Southern California
Stetson University
SUNY–Geneseo
SUNY–Purchase
Wesleyan University
Wheaton College (IL)

* These two schools are unusual because they combine a world-class conservatory with a top-notch liberal arts college.

PERFORMING ARTS—DRAMA

Major Universities Strong in Drama

Boston College
Boston University
University of California–Los Angeles
Carnegie Mellon University
The Catholic University of America
DePaul University
Emerson College
Florida State University
Fordham University
Indiana University
University of Iowa
Ithaca College
University of Minnesota
New York University
Northwestern University
University of North Carolina–Chapel Hill
University of Southern California
Southern Methodist University
Syracuse University
Texas Christian University
University of Washington
Yale University

Small Colleges and Universities Strong in Drama

Beloit College
Bennington College
Centre College
Colorado College
Connecticut College
Drew University
Ithaca College
Juilliard School
Kenyon College
Lawrence University
Macalester College
Middlebury College
Muhlenberg College
Occidental College
Otterbein College
Princeton University
Rollins College
Sarah Lawrence College
Skidmore College
SUNY–Purchase
Vassar College
Whitman College
Wittenberg University

PERFORMING ARTS— DANCE

Major Universities Strong in Dance

Arizona State University
University of California–Irvine
University of California–Los Angeles
University of California–Riverside
Case Western Reserve University
Florida State University
George Washington University
Howard University
Indiana University
University of Iowa
University Minnesota
New York University
Ohio University
Southern Methodist University
Texas Christian University
University of Texas–Austin
University of Utah
Washington University in St. Louis

Small Colleges and Universities Strong in Dance

Amherst College
Barnard College
Bennington College
Butler University
Connecticut College
Dartmouth College
Goucher College
Hollins University
Juilliard School
Kenyon College
Middlebury College
Mills College
Muhlenberg College
North Carolina School of the Arts
Princeton University
Sarah Lawrence College
Smith College
SUNY–Purchase

ENVIRONMENTAL STUDIES

Allegheny College
College of the Atlantic
Bowdoin College
University of California–Davis
University of California–Santa Barbara
Clark University
Colby College
University of Colorado–Boulder
Dartmouth College
Deep Springs College
Eckerd College
The Evergreen State College
Hampshire College
Hiram College
Hobart and William Smith Colleges
McGill University
Middlebury College
University of New Hampshire
University of New Mexico
University of North Carolina–Asheville
University of North Carolina–Greensboro
Oberlin College
Prescott College
St. Lawrence University
Tulane University
University of Vermont

University of Washington
Williams College
University of Wisconsin–Madison

INTERNATIONAL STUDIES

American University
Austin College
Brandeis University
University of British Columbia
Brown University
Bucknell University
Claremont McKenna College
Clark University
Colby College
Connecticut College
Dartmouth College
Davidson College
Denison University
University of Denver
Dickinson College
Earlham College
Eckerd College
Georgetown University
George Washington University
Goucher College
Hiram College
The Johns Hopkins University
Kalamazoo College
Lewis and Clark College
Mary Washington College
University of Massachusetts–Amherst
Middlebury College
Mount Holyoke College
Occidental College
University of the Pacific
University of Pittsburgh
Pomona College
Princeton University
University of Puget Sound
Randolph College
Reed College
Rhodes College
University Richmond
St. Olaf College
Scripps College
University of South Carolina
Sweet Briar College
Tufts University
Wesleyan University
College of William and Mary

Learning Disabilities

Accommodation for students with learning disabilities is one of the fastest-growing academic areas in higher education. Colleges and universities recognize that a significant segment of the population may suffer problems that qualify as learning disabilities, and the range of support services offered to such students is increasing. Assistance ranges from counseling services to accommodations such as tapes of lectures or extended time on exams.

Following are two lists—the first of major universities, the second of smaller colleges—that offer particularly strong services for LD students. If you qualify for such support, you should be diligent in checking out the services at each college on your list. If possible, pay a visit to the LD support office or have a phone conversation with one of the administrators. Since many such programs depend on the expertise of one or two people, the quality of the services can change abruptly with changes in staff.

Keep in mind also that many colleges are becoming increasingly skeptical of requests for LD services, especially when the initial diagnosis is made on the eve of the college search.

Strong Support for Students with Learning Disabilities

Major Universities	Small Colleges
American University	Bard College
University of Arizona	Curry College
University of California–Berkeley	Landmark College
Clark University	Loras College
University of Colorado–Boulder	Lynn University
University of Denver	Marist College
DePaul University	Mercyhurst College
University of Georgia	Mitchell College
Hofstra University	Muskingum College
Purdue University	University of New England
Rochester Institute of Technology	St. Thomas Acquinas College (NY)
Syracuse University	Southern Vermont College
University of Vermont	Westminster College (MO)
University of Virginia	West Virginia Wesleyan College

SAT AND ACT OPTIONAL SCHOOLS

Two decades ago a small number of U.S. colleges and universities, including Bates and Bowdoin, decided that they would no longer require all applicants to submit SAT or ACT scores. They reasoned that there is a significant pool of bright students who can do quality academic work but who for one reason or another do not test well. A "test optional" policy would allow them to tap into this market.

Over the years the number of "test optional" schools has grown dramatically. The National Center for Fair and Open Testing (FairTest), a Cambridge, Massachusetts-based advocacy organization that is critical of standardized testing in general, has tracked this growth, and at press time its website (www.fairtest.org) listed 739 such colleges and universities. Reasons for this growing aversion to college admissions tests are many. The early test optional schools have been happy with the way the policy has worked out. The SAT has been a focus of repeated controversy, especially around incidents of scoring error. And perhaps most importantly, the whole field of "test prep" has spiraled out of control. Students and parents alike are tired of the anxiety surrounding prep courses—not to mention the financial cost of helping bolster the coffers of Kaplan or Princeton Review.

Until recently there was not much that students could do—especially if they hoped to be able to choose among a range of quality colleges. Over the last two or three years, however, a critical mass has emerged of quality liberal arts colleges and major state universities that are "test optional." There are now 50 such institutions covered in the *Fiske Guide*. For the first time, students who wish to avoid getting involved in the admissions test rat race can do so but while still enjoying a range of colleges and universities from which to choose.

Accordingly, we have decided to begin publishing a list of those colleges and universities in the Guide that are "test optional." We are not recommending that any particular student eschew college admissions tests and apply only to these schools. As a resource designed to help students and parents, we are simply pointing out that applicants now have that option.

In looking over the list below of "test optional" colleges and universities described in the *Fiske Guide*, please keep a couple of things in mind. First, most of them are large state universities or small liberal arts colleges. You won't find many other types, including the Ivies or flagship publics. Second, keep in mind that there are different ways of being "test optional." Some schools, for example, only exempt students who meet certain GPA or class rank criteria. Qualifications are noted in the footnotes. Finally, the test optional field is changing daily, so go to www.fairtest.org for updated information and, above all, confirm current policy with any school to which you are thinking of applying.

Arizona State University, Tempe (3)
Arizona, University of, Tucson, (2,3)
Bard College
Bates College
Bennington College
Bowdoin College
College of the Atlantic
College of the Holy Cross
Connecticut College (5)
Denison University
Dickinson College
Drew University
Franklin and Marshall College
George Mason University (3)
Gettysburg College
Guilford College
Gustavus Adolphus College
Hamilton College (5)
Hampshire College

Hartwick College
Hobart and William Smith Colleges
Iowa State University (1,3)
University of Iowa (3)
University of Kansas (2,3,4)
Knox College
Lake Forest College
Lawrence University
Lewis and Clark College
Louisiana State University (1,3,4)
Middlebury College(5)
University of Minnesota—Morris (1,3)
University of Minnesota—Twin
  Cities (1,3)
Montana Tech of the University of
  Montana (1,3)
Mount Holyoke College
Muhlenberg College
University of Nebraska—Lincoln (3)

University of Oklahoma (3)
Oregon State University (3)
University of Oregon (3)
Pitzer College
Prescott College
Rollins College
Saint John's College, MD
Saint John's College, NM
Saint Lawrence University
Sarah Lawrence College
Susquehanna University
Texas A&M University (3)
Texas Tech University (3)
University of Texas—Austin (3)
Union College
Ursinus College (3)
Wheaton College (MA)
Worcester Polytechnic Institute

Key:

1 = SAT/ACT used only for placement and/or academic advising

2 = SAT/ACT required only from out-of-state applicants

3 = SAT/ACT used only when minimum GPA and/or class rank is not met

4 = SAT/ACT required for some programs

5 = SAT/ACT not required if applicant submits SAT Subject Test, Advancement Placement, International Baccalaureate, or other exams

Adelphi University

One South Avenue, Garden City, NY 11530

Situated in a comfortable Long Island suburb within shouting distance of Manhattan, Adelphi lets you taste urban life without being overwhelmed. Strong on pre-professional programs with a grounding in the liberal arts.

Everyone loves a great comeback story. In the mid-1990s, this small Long Island university was strapped for both cash and students because of mismanagement by a free-spending president and a negligent board of trustees. Now, thanks to a revamped mission, Adelphi University is experiencing a renaissance of sorts. Enrollment has increased, new facilities are sprouting up all around campus, and students speak of an almost palpable sense of energy among students and faculty.

Located just 20 minutes from the urgency of Manhattan, Adelphi's campus occupies 75 acres in an attractive residential suburb replete with Gothic cathedrals and stately homes. A new three-story dormitory with 74 rooms opened in 2003, offering students private bathrooms, spacious student lounges, high-speed Internet access, and a communal kitchen.

Adelphi's most popular majors have a decidedly preprofessional bent: business management, nursing, psychology, physical education, and biology. The honors college offers a rigorous liberal arts program for exceptional students, who must complete all the requirements for their major within that department or school and supplement their learning with intensive honors courses. Recent honors seminar topics included Genetic Disease and Genetic Engineering, The Decline and Fall of Certainty, Censorship and Morality, The Fate of the Earth, and Equality and Inequality. Joint degree programs have been established in a number of disciplines, including physics, dentistry, law, and physical therapy. The combined degree program in engineering with Columbia allows students to earn a BS in physics from Adelphi and a BS in engineering from Columbia in five years, or an undergraduate and graduate degree in six years.

> **"There are many competitive classes that challenge students, but the students do not seem to be competing with each other."**

The general education curriculum was reorganized in 1999. Students must complete a 24-credit distribution requirement in several areas (arts, humanities and languages, natural sciences and mathematics, and social sciences) as well courses in composition and English, a foreign language, statistics, computer programming, critical thinking, or public speaking. Freshmen take part in an orientation course and a three-credit freshman seminar that introduces students to life at Adelphi.

The academic climate is challenging but not cutthroat. "Adelphi is a pretty laid-back university," says a senior. "There are many competitive classes that challenge students, but the students do not seem to be competing with each other." Nearly two-thirds of all classes have 25 or fewer students, and professors are commended for their accessibility and knowledge. "Professors at Adelphi are excellent," says one student. "They know their material and are intelligent. They are always available during office hours."

Website: www.adelphi.edu
Location: Suburban
Total Enrollment: 4,161
Undergraduates: 4,161
Male/Female: 29/71
SAT Ranges: V 480–590
   M 500–600
ACT Range: 21–26
Financial Aid: 68%
Expense: Pr $
Phi Beta Kappa: Yes
Applicants: 4,379
Accepted: 71%
Enrolled: 26%
Grad in 6 Years: 53%
Returning Freshmen: 79%
Academics: 🖊 🖊 🖊
Social: ☎ ☎
Q of L: ★ ★ ★
Admissions: (516) 877-3050
Email Address:
   admissions@adelphi.edu

Strongest Programs:
Nursing
Elementary Education
Biology
Accounting
Performing Arts
Psychology
Physical Education

Signs of rebirth and renewal are everywhere, from the campus facilities to the burgeoning enrollment.

"No two students at Adelphi are alike," says a sociology major. "There are students of all backgrounds, cultures, and places around the world." Ninety-one percent of the university's student body hails from New York, and just over half graduated in the top quarter of their high school class. Minority enrollment is consistent with the university's proximity to the Big Apple. Sixteen percent of students are African American, 13 percent Hispanic, and 5 percent Asian American. Political issues don't dominate campus conversations, but students say recent debates include issues of homosexuality, the importance of voting, and the war in Iraq. Adelphi offers nearly 1,000 merit scholarships worth an average of $6,971 to qualified students. Sixty-five athletic scholarships are available in a variety of sports.

Despite "comfortable and well maintained" residence halls, Adelphi remains primarily a suitcase school. Seventy-five percent of students live off campus, "even if they live close to school," says one student. There are six residence halls, and freshmen and sophomores receive first priority in arranging housing. Though some grumble about the price of food on campus, most admit that there is plenty to choose from. "Adelphi makes attempts

The combined degree program in engineering with Columbia allows students to earn a BS in physics from Adelphi and a BS in engineering from Columbia in five years, or an undergraduate and graduate degree in six years.

"There are days in which sushi, Caribbean food, and other foods are served. Vegetarians have a good selection, too."

to have things for everyone," says a senior. "There are days in which sushi, Caribbean food, and other foods are served. Vegetarians have a good selection, too." Campus security maintains a presence and the university recently installed additional cameras and emergency boxes throughout campus.

Since many students trek home on weekends, they budget their party time for Thursday and Friday nights. "Social life at Adelphi is pretty good," says one student. "There are usually events (parties, lectures, casino nights, bingo nights, poetry contests, etc.) every week." The school sponsors 70 clubs and organizations, including seven sororities and fraternities that attract 4 percent of the women and 2 percent of the men. Adelphi is a dry campus and students say that the policy is strictly enforced. Still, "As a resident assistant, I know that the 'no alcohol' policies don't work," says a senior. "Students will be tricksters!" The glitz of New York City is just 40 minutes away by train and the Long Island beaches draw crowds when it's warm.

Adelphi fields several competitive teams, including men's and women's basketball, men's and women's soccer, and softball. The women's lacrosse team captured two of the last three national championships and the men's basketball team is a perennial tournament competitor. In addition, the university has won the coveted New York Collegiate Athletic Conference Commissioner's Cup, awarded to the most outstanding program, for several years running. Intramurals are offered in a dozen sports.

"There are students of all backgrounds, cultures, and places around the world."

"Adelphi has changed tremendously," says one happy student. "It has become more lively and fun." Indeed, this newfound optimism is infectious and seems to permeate the campus. Signs of rebirth and renewal are everywhere, from the campus facilities to the burgeoning enrollment. Although the university's commuter heritage can leave some wanting for more social opportunities, many find that this small Northeastern school fits the bill.

Overlaps

Hofstra, St. John's, Long Island University, SUNY–Stony Brook, SUNY–Queens, Pace

If You Apply To ➤

Adelphi: Early action: Dec. 1. Financial aid: Jan. 1 (recommended). Does not guarantee to meet demonstrated need. Campus interviews: recommended, informational. No alumni interviews. SATs or ACTs: required. SAT IIs: optional. Accepts the Common Application and electronic applications. Essay question.

Agnes Scott College

141 East College Avenue, Decatur, GA 30030

Combines the tree-lined seclusion of Decatur with the bustle of Atlanta. More money in the bank than most selective schools, and enrollment is up 50 percent since 1990. Small classes, sisterhood, and a more exciting location than the Sweet Briars of the world.

Agnes Scott College, founded in 1889, offers a small town campus atmosphere and provides women with an intellectually challenging institution—absent the distractions of men. The college is known for its science and math programs, but it also produces skilled writers and artists and continues to be one of the South's leading women's schools. ASC's climate as a small, single-sex institution leads to close relationships with the faculty and very involved students—both academically and socially. "Consider the benefits of a single-sex institution," advises a first-year. "A woman will learn more here than anywhere else."

The Agnes Scott campus sits on 100 acres in the historic district of Decatur, just outside of Atlanta. The well-maintained Gothic and Victorian buildings are surrounded by gardens filled with rare shrubs, bushes, and trees—all evidence of strong alumnae support. A $125 million capital campaign has paid for a new campus center, library expansion and renovation, new tennis courts, and three renovated theme houses in the past few years. The $36.5 million Science Center at Agnes Scott includes an x-ray spectrometer, nuclear magnetic resonance imaging equipment, and a scanning tunneling microscope. The school's Delafield Planetarium has a computer-controlled Zeiss projector, one of only 10 in the United States.

Aside from outstanding instruction in the sciences, Agnes Scott provides students with solid grounding in the liberal arts. First-year students may choose seminars on special topics such as U.S. foreign policy, religion and human rights in Atlanta, or life in Roman Pompeii. To graduate, students must complete one semester of English composition and literature, plus courses in math, historical studies or classical civilization, fine arts, science, social and cultural analysis, and physical education. Students must also attain intermediate-level proficiency in a foreign language and, because ASC is affiliated with the Presbyterian Church, take one course in religion or philosophical thought.

"A woman will learn more here than anywhere else."

Academically, Agnes Scott excels in biology and math, producing numerous Goldwater scholars in the past four years. The school also offers a top-notch German program. The college also recently added a dual-degree program in nursing and a major in economics and organizational management. Those participating in research may attend or present results at an annual conference held in the spring. Entrepreneurs and students interested in nonprofit work may take advantage of the Kauffman and Hubert internship programs. The overall academic climate at ASC is competitive. "The courses are extremely challenging," says a sophomore, "but the teachers make it very easy to comprehend." Eighty percent of the classes taken by first-years and nearly 90 percent of classes overall have 25 or fewer students, which encourages close student-faculty interactions in the classroom. "The instructors are friendly and helpful," says one student. "They flavor their courses with personality and creativity. It's really difficult not to like them."

When Agnes Scott's idyllic campus feels too small, students may enroll in the Atlanta Semester, focusing on women, leadership, and social change, or the Global Awareness and Global Connections programs, which offer a semester of crosscultural study before sending students out around the globe. If it is sun and fun you

Website: www.agnesscott.edu
Location: Urban
Total Enrollment: 858
Undergraduates: 848
Male/Female: 0/100
SAT Ranges: V 570–685
  M 540–650
ACT Range: 24–29
Financial Aid: 63%
Expense: Pr $
Phi Beta Kappa: Yes
Applicants: 1,526
Accepted: 53%
Enrolled: 28%
Grad in 6 Years: 66%
Returning Freshmen: 84%
Academics: ✍ ✍ ✍
Social: ☎ ☎
Q of L: ★ ★ ★ ★
Admissions: (404) 471-6285
Email Address:
  admission@agnesscott.edu

Strongest Programs:
Astrophysics
Biology
Mathematics
German
Psychology
Economics
English

want, take a term at all-female Mills College in Oakland, California. Engineers and architects may complete their degrees through a 3–2 program with Georgia Tech or a 3–4 program with Washington University in St. Louis. As a member of the Atlanta Regional Consortium for Higher Education*, Agnes Scott shares facilities and resources with 18 other schools in the area through a crossregistration program.

The ASC student body hails mainly from the Southeast; more than half are Georgia natives. One student describes the student body as "inquisitive, bright, involved, activism-oriented, and focused. Sometimes they are characterized as neurotic or too intense. "The campus is decidedly liberal, and the politically conservative students are not well accepted," says a sophomore. Scott students are hard working, but also volunteer in the community.

"Agnes Scott is distinctive in that it is a bubble of academia in the middle of Atlanta."

Despite the school's small size, its campus is quite diverse, with 22 percent of the student body African American, 3 percent Hispanic, and 5 percent Asian American. An honor system, enforced by a student judiciary, allows for self-scheduled and unproctored exams. Agnes Scott awards merit scholarships averaging more than $12,000 annually, based on academic performance, leadership, or musical ability.

Ninety-two percent of Agnes Scott students live in the dorms, which are linked by tree-lined brick walks. "Students aren't allowed to move off campus, which creates a close campus community, but can be a barrier financially because room and board is so expensive," says a first-year student. Dorms are described as being spacious and well maintained compared to other colleges. Juniors and seniors can live in Avery Glen, the college-owned apartment complex, while first-years are assigned to the two dorms (out of six total). Agnes Scott has no sororities, but the college itself is a close-knit sisterhood. The selection of food at Agnes Scott dining facilities takes every type of diet into consideration, including vegetarian, low-carb, low-fat, and dairy-free. "The food is exquisite," says one well-fed sophomore.

ASC has many campus clubs and events, but when the weekend comes, "off campus is where the party's at," one student says. "Our social life revolves around metro Atlanta," adds a biology major. Convenient public transportation serves cultural landmarks like the High Art Museum and provides access to the social scene in nearby Atlanta. Decatur itself is not really a "college town," but there are some attractions for Agnes Scott students. "There are still nice little venues and coffee shops; very hip and fun," a student says. A lot of ASC students get involved with community service both on and off campus, with Habitat for Humanity, the DeKalb Rape Crisis Center, Girl Scouts, Hands Across Atlanta, and Best Buddies.

"Students aren't allowed to move off campus, which creates a close campus community."

Back on campus, underage students may not imbibe, in accordance with Georgia state law; enforcement falls under the honor code. One student says the restrictions work well, and 21-year-olds can enjoy alcohol in their dorms and at certain functions. But the bulk of the socializing at ASC is off campus, one student says. "Either you have a clique of friends or hardly any at all and inviting your boyfriend to campus is often difficult because the policies make many men uncomfortable." When traveling, popular road trips include Stone Mountain and Six Flags, or New Orleans for Mardi Gras. Every October, students invite dates to a formal dance known as Black Cat. It follows a week of class competitions and marks the end of new-student orientation. Other quaint old traditions survive, too, such as throwing recently engaged classmates into the alumnae pond. And seniors who get into grad school or find jobs go to the top of the college bell tower, pealing out chimes to share the good news.

Agnes Scott is a member of the Great South Athletic Conference, and both of the Agnes Scott tennis and cross-country teams brought home a conference trophy in recent years. Also strong are soccer, swimming, basketball, and softball. About 10 percent of students participate in intramural activities, which include the standard roundup of sports plus a few more exotic options, such as Frisbee.

Small but mighty, ASC stands out for the little touches that make students feel they're part of a close-knit community. "Agnes Scott is distinctive in that it is a bubble of academia in the middle of Atlanta," says a sophomore. Academically, the school offers challenging courses in science and math and supplements its instruction in many other fields with a variety of programs that allow students to explore new opportunities both on campus, in conjunction with other institutions, and abroad.

Overlaps

University of Georgia, Emory, Rhodes, Tulane, University of Virginia, Furman

If You Apply To ➤

Agnes Scott: Early decision: Nov. 15. Regular admissions: Mar. 1 (Jan. 15 for scholarship applicants). Financial aid: Feb. 15. Meets demonstrated need of 70%. Campus interviews: recommended, evaluative. Alumni interviews: optional, informational. SATs or ACTs: required. SAT IIs: optional. Accepts the Common Application and prefers electronic applications. Essay question: impact of significant experience, risk or ethical dilemma; issue of personal, local, national or international concern; influential person, fictional character, historical figure, or creative work; topic of your choice.

University of Alabama

Box 870132, Tuscaloosa, AL 35487-0166

"Roll, Tide, Roll" says a lot. Not exactly a hotbed of intellectual energy, 'Bama has been left in the dust by University of Georgia and its Hope Scholarship. Look for pockets of excellence in the professional programs, the Blount Undergraduate Initiative, and honors programs.

Tuscaloosa, Alabama, remains home to many Old South staples, including well-preserved antebellum homes, down-home Dreamland barbecue, and world-renowned blues music. But Tuscaloosa has also evolved to include a bevy of international companies that provide internship opportunities for students. The state's first university is evolving, too, by updating its facilities and making the education it offers more global and high-tech.

'Bama's thousand-acre campus combines classical, revival-style buildings (several of which survived the Civil War) with modern structures. One of the most stunning in the South, the campus wraps around a shaded quadrangle, the home of the main library and "Denny Chimes," a campanile carillon that rings the Westminster Chimes on the quarter hour. Newer additions to campus include Shelby Hall, a classroom and research building; a student recreation center, pool and other athletic facilities; a child development center; and additional student housing.

The University of Alabama is organized into eight undergraduate colleges and schools, which together offer 82 undergraduate degree programs. The Culverhouse College of Commerce and Business offers strong programs in marketing and management information systems, with finance and accounting as two of the most popular majors on campus. Nursing, public relations, and advertising are also popular. The College of Communication and Information Sciences has one of the country's top journalism schools, while respected programs in the College of Human Environmental Sciences include athletic training and family financial planning. The School of Music is a regional standout, attracting guest artists such as Jean-Pierre Rampal and Midori.

> **"Any Alabama football game is a festival."**

Website: www.ua.edu
Location: Small city
Total Enrollment: 18,399
Undergraduates: 15,735
Male/Female: 47/53
SAT Ranges: V 500–630
  M 500–630
ACT Range: 21–27
Financial Aid: 38%
Expense: Pub $ $
Phi Beta Kappa: Yes
Applicants: 10,707
Accepted: 72%
Enrolled: 35%
Grad in 6 Years: 63%
Returning Freshmen: 86%
Academics: ✍ ✍ ✍
Social: ☎ ☎ ☎
Q of L: ★ ★ ★
Admissions: (205) 348-5666
Email Address:
  admissions@ua.edu

The Culverhouse College of Commerce and Business offers strong programs in marketing and management information systems, with finance and accounting as two of the most popular majors on campus.

The Honors College houses the three university-wide honors programs, which feature smaller classes, early registration privileges, and the opportunity to write a senior thesis. Students of any major—mostly engineers—may participate in the computer-based honors program, which pays them to develop software programs in their field of study. The Blount Undergraduate Initiative is a living/learning program within the College of Arts and Sciences, where freshmen are housed with a faculty director and fellows. Other innovative offerings include 'Bama's Weekend College, the continuing education division, which has attracted a large undergraduate following, and the interim term in May, when students focus on one course in depth.

About 20 percent of Alabama freshmen take part in the Arts and Sciences Mentoring Program, which pairs them with faculty mentors who ease the adjustment to college through informal counseling and enrichment activities such as concerts, movies, or lectures. All students may also enroll in the two-credit Academic Potential Seminar, which covers self-assessment, motivation, personal responsibility, time management, memory, textbook reading, note-taking, and test preparation. The only course 'Bama requires students to take during the first year on campus is a two-term English composition sequence. Before graduation, students must also take courses in writing, natural sciences, math, humanities and social sciences, and either two semesters of a foreign language or one of computer science.

While UA's core curriculum has been streamlined and the math requirement has been reduced, don't expect to party all the time and still pass, students say. Courses are "quite challenging," says a student, who reports that there is little competition among classmates. Professors teach lectures and many seminar courses, including some classes for freshmen. "The teachers I have had have been wonderful," a sophomore says. "My experience has been filled with quality teachers," notes a junior.

Seventy-nine percent of 'Bama's students are homegrown and 1 percent are international students representing more than 79 countries. "Most of our students are middle/upper class with Southern values," says one senior. A classmate adds, "This is not the place for really artsy or eclectic people." Alabama is a leader in diversity among Southern flagship universities with 16 percent minority enrollment: 12 percent African American, 1 percent Asian American, 1 percent Native American, and 2 percent Hispanic.

> **"Most of our students are middle/upper class with Southern values."**

"The biggest strife is between Greek and non-Greek members," a junior says, noting the "majority of UA is united." 'Bama awards 343 athletic scholarships as well as merit scholarships worth an average of $5,094.

Most Alabama students live in off-campus apartments in the Tuscaloosa area; only 26 percent remain in campus residence halls, where options range from private rooms and suites to apartment-style living. "Housing is great and always improving," a senior says. New dorms opened recently, but a sophomore still advises, "If a student wants a room for the fall semester, they should apply early in the spring." Another freshman gives high marks to the living/learning communities, some of which are for first-year students and others that bring together students from a particular school or college. Twenty-one percent of men pledge fraternities and 27 percent of women join sororities, and they may live in chapter houses. 'Bama's campus is safe, thanks to visible school police and a free school escort service.

Most Alabama students live in off-campus apartments in the Tuscaloosa area; only 26 percent remain in campus residence halls, where options range from private rooms and suites to apartment-style living.

Much of 'Bama's social life—which students describe as "very active" and "second to none"—revolves around the Greek system and athletic events. Participation has been steady in recent years, despite administrators' efforts to weaken it by prohibiting fraternities and sororities from having parties on campus. Those under 21 can't have alcohol in the dorms—or elsewhere, for that matter, per state law—but a sophomore says, "UA is very lenient compared to other schools." "I think the rules are enforced, but there are plenty of underage people with alcohol," a senior says. Those who don't go Greek, or who don't wish to drink, will find everything from the Society for Creative Anachronism (medievalists) to Bible study groups. City Fest, featuring music, food, beer, and a German theme, is a popular annual event, as is Capstone, a Saturday dedicated to community service. A modern trolley service connects the 'Bama campus to the city's thriving downtown. Tuscaloosa is described as "a grand college town," that is "big enough to always find something to do," but "small enough to not get too lost." Road trips to New Orleans (for Mardi Gras and Greek weekend formals), Atlanta, Nashville, Birmingham, and the Gulf Coast and Florida beaches are popular, too, but "many people never leave UA!" says a sophomore.

Much of 'Bama's social life—which students describe as "very active" and "second to none"—revolves around the Greek system and athletic events.

Despite several inconsistent seasons of late, 'Bama football remains the cornerstone of the university's competitive athletic programs. The annual Auburn–Alabama game—the Iron Bowl, one of the most intense rivalries in college sports—is the highlight of the school year. "Any Alabama football game is a festival," a sophomore says. Alabama competes in Division I, and the women's gymnastics team and men's basketball team brought home sectional championships in 2003. Football, softball and men's baseball, tennis, swimming, and outdoor track and field are also competitive.

"My experience has been filled with quality teachers."

'Bama's administrators are no longer satisfied with being a purely athletic and social school. (In fact, they brag, the school's debate team has won 14 national championships—two more than the football team!) The emphasis is now on technology, honors programs, global perspectives, and undergraduate research. "We have made great attempts to become a more academically known school," a senior says. "The number of students enrolled has been steadily growing with every freshman class."

Overlaps

Auburn, University of Alabama–Birmingham, University of Georgia, University of Mississippi, University of Tennessee, Mississippi State

If You Apply To ➢

'Bama: Rolling admissions. Meets demonstrated need of 30%. Campus and alumni interviews: optional, informational. SATs or ACTs: required. SAT IIs: optional. Accepts the Common Application and prefers electronic applications. No essay question.

Albertson College

Caldwell, Idaho 83605

Got a map? You'll need a sharp eye to spot Albertson, the *Fiske Guide*'s only liberal arts school other than Colorado College in the interior of the Mountain West. Innovative programs include Leadership Studies and the Center for Experiential Learning.

With an emphasis on technology, education, and experiential learning, Albertson College of Idaho, the state's oldest four-year university, offers students an opportunity to earn a solid liberal arts education through small classes in a small town. "The professors know you by heart, and every face is familiar," a sophomore psychology major says. Outside class, the school's scenic environment allows sports and nature enthusiasts to romp freely before heading back into the classrooms.

The college is in the small town of Caldwell, where the atmosphere is calm and serene. For those looking for a little excitement, the state capitol of Boise is a short drive from campus. Also nearby are some of Idaho's most scenic locations such as beautiful mountains, deserts, and whitewater rivers. The school, originally a Presby-

Website: www.albertson.edu
Location: Small town
Total Enrollment: 777
Undergraduates: 761
Male/Female: 42/58
SAT Ranges: V 520–643
  M 530–630
ACT Range: 23–28
Financial Aid: 61%

(Continued)

Expense: Pr $
Phi Beta Kappa: No
Applicants: 923
Accepted: 84%
Enrolled: 25%
Grad in 6 Years: 55%
Returning Freshmen: 77%
Academics: ✍ ✍ ✍
Social: ☎ ☎ ☎
Q of L: ★ ★ ★
Admissions: (208) 459-5305
Email Address:
  admissions@albertson.edu

Strongest Programs:
Biology
Business
Psychology
English
Political Science

Biology, English, and history are among the majors recommended by students, and preprofessional majors, such as premed, prevet, and prelaw are also popular and strong.

terian college, first planted roots in downtown Caldwell in 1891 and then moved to its present site in 1910, where its 21 buildings now inhabit 43 acres. For more than 80 years, it was called the College of Idaho, but officials changed the college's name to honor the Albertson family (of supermarket fame), who gave $13.5 million for new facilities.

The school's academic schedule is composed of 12-week semesters, spring and fall, separated by a six-week winter session, during which students can assist professors with research, take an internship, volunteer, or travel abroad. The general education requirements include natural sciences, writing, mathematics, civilization, cultural diversity, physical activity, social science, fine arts, and humanities.

"We really are an academic community."

Freshmen go through a first-year program that includes reading a common book, a junior or senior mentor, a team of advisers, and a weeklong orientation that includes an off-campus overnight stay. First-year students demonstrating leadership potential are invited to a series of seminars to draw them into the leadership studies program, a business minor. The two libraries have 178,000 volumes and are accessible online 24 hours a day.

Biology, English, and history are among the majors recommended by students, and preprofessional majors, such as premed, prevet, and prelaw are also popular and strong. Weaker departments are the smaller ones, such as foreign languages, music, and education. Students can choose such specializations as sports and fitness-center management. Undergraduate research opportunities are available in biology, chemistry, and psychology, and students present their findings at state and regional conferences. Albertson has also overhauled its education program and is adding a five-year master's degree. The college cooperates with Columbia University, the University of Idaho, Boise State University, and Washington University in St. Louis to offer a five-year course of study in engineering. The Center for Experiential Learning coordinates out-of-classroom experiences, such as international education and service learning. For those who want to venture abroad (physically or mentally), the International Education Program offers several options, including attending a foreign university, traveling overseas during the summer and winter breaks, and taking international studies on campus. Travel has really taken off, with 25 percent of students accepting opportunities to study in such places as Cuba, Greece, and Germany.

"Our campus is concerned about environmental issues, and we're evenly divided between conservatives and liberals."

Most classes at Albertson have 25 or fewer students, and all are taught by full professors. "Our academic program is pretty competitive," says a sophomore, "but professors are always willing to help and we get a lot of one-on-one attention." The school's small size also allows students to skip the registration hassles that plague larger institutions. Albertson's price tag is lower than that of most private colleges. It offers merit scholarships to nearly all students, as well as 177 athletic scholarships. Not surprisingly, some of those scholarship dollars are set aside for skiers.

"Students at ACI are so close," chirps a psychology major. "We see each other almost every day, so everyone knows everyone." Seventy-one percent of the students are from Idaho; 73 percent are white. Five percent are Hispanic and Asian Americans add another 3 percent. Only 2 percent hail from foreign nations. "PC is an issue," says a junior. "Our campus is concerned about environmental issues, and we're evenly divided between conservatives and liberals." Environmental issues also get students riled up.

Fifty-two percent of students live on campus. Each room in the five residence halls has individual heating and cooling and hookups for Internet access, and each hall has a computer lab. Dorms are "decent-sized, not any better or worse than a typical college," a junior comments. For those looking to get their grub on, Albertson provides "a grill deli, salad bar, pizza, and vegetarian options" that are "to die for," according to one student. Students feel safe on campus; the college recently implemented a round-the-clock escort system, and there are many emergency stations around campus.

Twelve percent of men and 13 percent of women participate in the Greek system, which dominates campus social life. Annual social highlights include Winterfest, Spring Fling and homecoming week. Games against rival Northwest Nazarene also attract attention. Caldwell, with 24,000 people, is not a great spot for college students, but students get involved by helping out the local school district. "Finals breakfasts" offer something for bleary-eyed students to look forward to during finals week. At midnight on Tuesday, faculty and staff cook breakfast for students. Nearby Boise is a popular destination for shopping, dining, cultural events, and volunteering. Many students take advantage of hiking, kayaking, and skiing in the surrounding area, and many hit the road during a weeklong break taken every six weeks.

Men's basketball is a crowd-pleaser, and the team has made Albertson students proud, chalking up victories in the NAIA Division II competition. The women's team was a runner-up in 2001 and advanced to the Cascade Conference Tournament championship game in 2005. The skiing teams have claimed 28 individual or team championships, and men's baseball and soccer are also competitive. For those who enjoy the game but might not make the team, there is an active intramurals program, an outdoors program that offers instruction in areas such as rock climbing and fly fishing, and the large J. A. Albertson Activities Center. "Intramurals are huge at our school," says a sophomore.

Albertson College has much to offer its "Yotes" (translation: "We are the coyotes"). They enjoy a solid liberal arts education and personal academic attention on a campus striving to keep its offerings on the cutting edge. "The close, intimate nature of ACI is what makes it unique," a sophomore says. "We really are an academic community."

> **"The close, intimate nature of ACI is what makes it unique."**

For those who want to venture abroad (physically or mentally), the International Education Program offers several options, including attending a foreign university, traveling overseas during the summer and winter breaks, and taking international studies on campus.

Overlaps

Boise State, University of Idaho, Linfield, Brigham Young, Idaho State, Reed

If You Apply To ➤

Albertson: Rolling admissions. Early action: Nov. 15. Meets demonstrated need of 17%. Campus interviews: recommended, informational. Alumni interviews: optional, informational. SATs or ACTs: required, writing test required with ACT. SAT IIs: optional. Accepts the Common Application and electronic applications. Essay question: matter you have learned to question; how you would succeed in winter term; more about you and your values.

Albion College

Albion, MI 49224

Next to evangelical Hope and Calvin, and out-there Kalamazoo, Albion is Michigan's middle-of-the road liberal arts college. Think Gerald Ford, the moderate Republican president who is the namesake of Albion's signature Institute for Public Service. Future doctors, lawyers, and businesspeople will be well served.

Albion is a small, private college in Michigan whose motto is "Liberal arts at work." The school's motto emphasizes the importance Albion places on combining learning with hands-on experience. Students at Albion often participate in leadership and service-learning seminars. And when the work is through, students here enjoy a close-knit social life. "Albion is where I have built lifetime friendships," says a

Website: www.albion.edu
Location: Small town
Total Enrollment: 1,548
Undergraduates: 1,548

senior. Another student notes, "Since Albion is a small school, it gives each student personal attention in every aspect of their education."

Founded in 1835 by the Methodist Church, Albion is located near the banks of the Kalamazoo River. In addition to its newer Georgian-style architecture, the college has retained and restored several of its 19th-century buildings. The campus is spacious with statuesque oaks and a beautiful nature center. Robinson Hall, the campus centerpiece, houses myriad departments, including the Ford Institute for Public Service, the Gerstacker Liberal Arts Program in Professional Management, and the Anna Howard Shaw Women's Center. The campus continues to expand with the addition of the Ferguson Student Services Building.

> **"Since Albion is a small school, it gives each student personal attention in every aspect of their education."**

Academically, Albion is as sound as its buildings. It was the first private college in Michigan to have a Phi Beta Kappa chapter (1940) and has produced three Rhodes Scholars. On their journey to such success, students are required to take core courses distributed among humanities, natural sciences, social sciences, fine arts, and math. They must also satisfy requirements in environmental science and gender and ethnicity studies. Freshmen must take first-year seminars designed to provide a "stimulating learning environment" in a small-class setting, while seniors participate in a capstone experience.

Albion's most distinguishing feature is the emphasis placed on citizenship and service. The Gerald R. Ford Institute for Public Service takes a unique approach for future civic leaders. Students participate in a simulation of city government in which they play the roles of community leaders. Visiting speakers include senators and congressmen, governors and state legislators, and interest group representatives. The premedical and prelaw programs draw dedicated undergrads, and the English and economics departments are well respected. Another option is the Summer Research Program, which allows students to remain on campus during the summer to work with faculty members on different projects.

The academic climate at Albion is described as competitive but not cutthroat. One student says, "The average student who attends is the high achiever who is involved in school." Top-notch academic and career counseling and low student/faculty ratios keep students on track and motivated. Class size varies, but the average class is under 25 students. Professors are interested in students' academic performance and their emotional well-being. While "the courses are carefully prepared and the professors expect diligence from the students, the professors really care and take an interest in the students" and it is not "uncommon to have class at a professor's house with pizza and holiday cookies." The professors seem to know the secret to motivating college students—feed them and they will work. Teaching assistants are used for tutoring, not teaching. Albion's libraries feature computer facilities, an interlibrary loan service,

> **"It's not uncommon to have class at a professor's house with pizza and holiday cookies."**

a listening lab for language or music study, and a helpful staff. If you can't find what you need at Albion's libraries, weekly bus trips to the University of Michigan libraries in Ann Arbor provide access to even more resources.

Albion continues to attract an ambitious, involved group of students. Michigan residents make up 90 percent of the student population. Eighty-five percent are Caucasian, 2 percent African American, and 1 percent Hispanic. The remainder come from abroad. Up to now, there has been little deviation from the white, upper-middle-class norm. In an effort to change this, a host-family program matches minority students with families from within the community. There are a number of merit scholarships available, based on academic records, extracurricular involvement, and demonstrated leadership abilities. There are no athletic scholarships.

The premedical and prelaw programs draw dedicated undergrads, and the English and economics departments are well respected.

Eighty-nine percent of Albion students call the residence halls home, which are described as "sufficient" and "not exactly modern." The majority of the freshman class inhabits Wesley Hall. During their sophomore year, many students move to Seaton or Whitehouse halls; seniors enjoy apartment-style housing called The Mae. Dorms are co-ed by hall or floor, and the information each student provides in their housing request form is used to assign rooms and roommates. One student claims, "The dorms are extremely well-maintained and have maintenance boards located on every floor so that the students can voice their concerns. Most of the concerns are taken care of by the next day." Other housing options include apartment annexes and fraternity houses. Sororities do not have houses; they hold their meetings in lodges. Two large dining rooms feed campus residents on an "eat all day" meal plan.

Forty percent of Albion men and women belong to one of the school's six national fraternities and seven sororities. Greek parties draw large crowds, composed of Greeks and non-Greeks, making them a primary part of many students' social lives. Controlling the alcoholic intake of students has become a priority of the administration. Students say "underage drinking does happen" but the college "has strict consequences if caught." Those who insist on imbibing can do it at Gina's or Cascarelli's, popular bars in town.

Road trips are a big part of weekends for many students. Ann Arbor, East Lansing, and Canada are frequent destinations. A well-run union board organizes all sorts of activities—films, lectures, plays, comics, and concerts—to keep students occupied in their spare time. Several students report that the town movie theatre shows "free movies if you show a valid student ID!"

Still, students complain that there are not many social outlets available at Albion. Students focus some of their energy by working for groups supported by the Student Volunteer Bureau; in fact, half of the students volunteer on a regular basis. They are very involved in the community, including "city clean-up day, Habitat for Humanity, and volunteering at nursing homes and schools." Some traditional events that offer a nice break from academics are the Briton Bash, a fair that familiarizes students with clubs and organizations, and the Day of Woden, which is a picnic held in the spring on the last day of class.

"Albion is where I have built lifetime friendships."

The varsity football team has won nine conference championships in the past decade. Recently, the women's soccer team brought home the Conference MIAA Championship. Men's track, baseball, and golf, along with women's swimming, also receive a lot of attention on campus. Hope College is a hated rival, as is Alma College.

At Albion, professors are accessible and interested, and academics are challenging without being overwhelming. Students agree that the "small campus with friendly students, caring faculty, and kind staff members" make this college an appealing place.

Some traditional events that offer a nice break from academics are the Briton Bash, a fair that familiarizes students with clubs and organizations, and the Day of Woden, which is a picnic held in the spring on the last day of class.

Overlaps

University of Michigan, Michigan State, Hope, Kalamazoo, Western Michigan

If You Apply To ➤

Albion: Early decision: Nov. 15. Financial aid: Feb. 15. Does not guarantee to meet demonstrated need. Campus interviews: recommended, evaluative. Alumni interviews: optional, informational. SATs or ACTs: required. SAT IIs: optional. Accepts the Common Application and electronic applications. Essay question.

Alfred University

Alumni Hall, Saxon Drive, Alfred, NY 14802-1205

Talk about an unusual combination—Alfred combines a nationally renowned college of ceramics, a school of art and design, an engineering program, and a business school wrapped up in a university of just more than 2,200 students. It takes elbow grease to pry coastal types to the hinterlands of western New York.

Website: www.alfred.edu
Location: Rural
Total Enrollment: 2,235
Undergraduates: 1,853
Male/Female: 51/49
SAT Ranges: V 500–620
    M 520–610
ACT Range: 22–27
Financial Aid: 80%
Expense: Pr $
Phi Beta Kappa: Yes
Applicants: 2,134
Accepted: 77%
Enrolled: 26%
Grad in 6 Years: 65%
Returning Freshmen: 79%
Academics: ✏️ ✏️ ✏️
Social: ☎ ☎ ☎
Q of L: ★ ★ ★
Admissions: (607) 871-2115
Email Address:
    admissions@alfred.edu

Strongest Programs:
Engineering
Art & Design

Alfred University boasts highly respected programs in art and design, as well as ceramic engineering. Innovation not only shapes the curriculum, but also has a profound effect on campus life. Small classes and friendly competition support this diversity while encouraging individuals to succeed. With 2,200 students, Alfred isn't a bustling academic factory; it's a quiet, cloistered, self-described "educational village" in a tiny town wholly dedicated to the "industry of learning." Along with being able to handle the academic rigors of the college, students also have to weather brutal winters that dump snow by the foot on the region.

Alfred's campus consists of a charming, close-knit group of modern and Georgian brick buildings, along with a stone castle. The Kanakadea Creek runs right through campus, and the town of Alfred consists of two colleges (the other is the Alfred State College) and a main street with one stoplight. There are a few shops and restaurants, but certainly no malls, parking lots, or tall buildings. A new 52-stall equestrian center opened in 2005, and the university has added the Fasano Welcome Center, which houses office space and serves as a central location where visitors and alumni can congregate.

The university and its students share a no-nonsense approach to education. Although prospective students apply directly to one of four colleges and declare a tentative major, half of all requirements for a bachelor's degree are earned in the liberal arts college. Requirements are quite different in each school. However, the mix usually includes coursework in oral and written communication, foreign language and culture, social sciences, history, literature, philosophy, and religion.

Alfred, though private, is actually the "host" school for the New York State College of Ceramics, which is a unit of the state university system and comes with a public university price tag. Ceramic engineering (the development and refinement of ceramic materials) is the academic cornerstone and the program that brings Alfred international recognition. All engineering programs are found within the School of Engineering, and students may choose from six engineering majors. The art department, with its programs in ceramics, glass, printmaking, sculpture, video, and teacher certification, is also highly regarded. The School of Art and Design offers a graphic design major in which students use electronic and computer equipment. The business administration school also gets good reviews from students and provides undergraduates with work experience through a small business institute

"Most classes I've taken have engaged in extremely intellectual and stimulating discussions."

where students have real clients. The computer science major has been dropped, while majors in marketing and finance have been added.

"The atmosphere is relaxed but the courses are still rigorous," says a student. Another adds, "Most classes I've taken have engaged in extremely intellectual and stimulating discussions." The major in foreign language and culture sponsors trips abroad, and exchange programs are available in England, Germany, Italy, France, Japan, China, and the Czech Republic. The Track II program enables students to design their own interdisciplinary majors with personal guidance from top faculty members.

Whatever their major, all students enjoy small classes (average size is 18 students), and the quality of teaching is described as very high. "They will challenge you and they will make you think," says a freshman. Most classes are taught by full professors, with graduate students and teaching assistants helping out only in lab sessions. The university stresses its commitment to helping undergrads plan their future, and the academic-advising and career-planning services are strong enough for Alfred to deliver on its promise. Students say faculty members really want to see them succeed, both in class and in the real world. "Most teachers have high expectations, resulting in greater student performance," says one junior.

> **"The dining facilities are very nice and well-equipped and the food is both diverse and edible."**

Alfred students are "extremely down to earth, individualistic, accepting, and friendly," says a fine arts major. Sixty-five percent of the students at Alfred are from New York State, and 19 percent graduated from high school in the top tenth of their class. Students are mostly white and from public schools: 5 percent are African American, 3 percent are Hispanic, and 2 percent are Asian American. Financial aid is available and outstanding students can apply for many merit scholarships worth an average of $6,700, and National Merit finalists receive Alfred's Award of Merit. There are no athletic scholarships.

No one seems to mind the two-year on-campus residency requirement because the rooms are large and comfortable, and the dorms are equipped with lounges, kitchens, and laundry facilities. Upperclassmen have a choice of dorms that are co-ed by floor with single rooms, suites, or apartments. Freshmen enjoy their own housing divided into doubles. Sixty-seven percent of all students choose to live on campus, but some juniors and about half of the seniors opt to live off campus. The school has two dining halls and students say the offerings include plenty of selections for vegetarians and vegans. "The dining facilities are very nice and well-equipped and the food is both diverse and edible," says a junior. Campus security is good, according to most students. "There are blue lights all over campus and AU security will provide rides or walking escorts if you feel unsafe walking alone," a biology major says.

The Track II program enables students to design their own interdisciplinary majors with personal guidance from top faculty members.

Alfred's location in the Finger Lakes region, almost two hours from Buffalo and an hour and a half from Rochester, is isolated. The other chief complaint is the chilly, snowy weather. Social life is difficult due to the rural atmosphere and lack of Greek organizations, but the Student Activities Board brings many events to campus, including musicians, comedians, lecturers, and movies. Favorite road trips are to Letchworth and Stony Brook state parks, and to Ithaca, Rochester, Buffalo, and Toronto. Many students are skiing, hunting, camping, and rock-climbing enthusiasts. Friendly games of hackeysack, Frisbee, football, softball, and other sports can often be found on campus. Alfred is a dry campus. "The campus banned Greek life to curb drinking," says a student. "They seriously prohibit underage drinking." The alcohol policy is enforced in the dorms and students say it is largely respected.

> **"It is a small town, very close community, but nice."**

Because Alfred shares the town with Alfred State University, the dominant student population makes Alfred a good college town. "It is a small town, very close community, but nice," observes a biology major. The downtown scene provides students with an adequate number of movie theaters and eateries. Every spring brings the annual Hot Dog Weekend, a big carnival-like event that fills Main Street with game booths, bands, and lots of hot dog stands. Alfred's Division III Saxons are ominous opponents on the football, soccer, and lacrosse fields, and alpine skiing, equestrian, and men's swimming have all won spots in the nation's top 10.

Although small and somewhat secluded, Alfred University is a good choice for those students who want to concentrate on the ABCs of arts, business, and ceramic engineering—just be sure to bundle up for the long, snowy winters.

Overlaps
Rochester Institute of Technology, SUNY–Brockport, Ithaca, SUNY–Geneseo, Clarkson, Hartwick

Allegheny College

520 North Main Street, Meadville, PA 16335

An unpretentious cousin to more well-heeled places like Dickinson and Bucknell. Draws heavily from the Buffalo–Cleveland–Pittsburgh area. The college's powerhouse athletic teams clean up on Division III competition. If you've ever wondered what lake-effect snow is, you'll find out here.

Website: www.allegheny.edu
Location: Small city
Total Enrollment: 2,007
Undergraduates: 2,007
Male/Female: 47/53
SAT Ranges: V 570–660
    M 570–660
ACT Range: 23–28
Financial Aid: 68%
Expense: Pr $ $
Phi Beta Kappa: Yes
Applicants: 3,540
Accepted: 62%
Enrolled: 25%
Grad in 6 Years: 73%
Returning Freshmen: 90%
Academics: ✑ ✑ ✑
Social: ☎ ☎ ☎
Q of L: ★ ★ ★
Admissions: (800) 521-5293
Email Address:
    admissions@allegheny.edu

Strongest Programs:
Physical and Biological
    Sciences
Political Science
English
International Studies

Allegheny College is a down-to-earth Eastern liberal arts school boasting a rich history of academic excellence in an intimate setting. Administrators here understand the importance of providing students with real-world experience to complement their classroom work. The school's innovative May term offers time for internships or other off-campus work and study, and a commitment to civic responsibility has spurred several new programs. Allegheny's small size means students don't suffer from lack of attention, and despite the heavy workload, anyone struggling academically will get help before the situation becomes dire. "Professors really care about the students and will do anything they can to help them succeed at college and in life," says a sophomore.

Allegheny's 72-acre campus is tucked away in Meadville, Pennsylvania, 90 miles north of Pittsburgh. Founded in 1815, nestled in the Norman Rockwell–esque rolling hills of northwestern Pennsylvania, the campus is home to traditional, ivy-covered buildings and redbrick streets, as well as new apartment-style housing for upperclassmen. A nationally acclaimed science complex supports already-strong programs, and students who wish to pump iron can visit the new fitness center. The college also owns a 182-acre outdoor recreational complex, a 283-acre research reserve, and an 80-acre protected forest. North Village—a community of townhouse-style apartments—opened recently and houses 110 students.

Students at Allegheny work hard and do well, as evidenced by the fact that 52 percent of a recent senior class immediately went on to graduate school. "The courses are hard work but you learn a lot during the semester," says a junior. The school's strongest programs are in environmental studies, the sciences, and international studies, and students give high marks to English, too. Though the college eliminated its degree program in education some years ago, it enjoys

"The students do a lot of volunteer work and help immensely throughout the community."

articulation programs with the University of Pittsburgh School of Education and the Columbia University Teachers College, helping graduates get that much sought-after graduate degree. There is a major in biochemistry, a chemistry curriculum, and a track for students interested in entrepreneurial and managerial economics. Co-op programs include a 3–1, 3–2, 3–4 option leading to degrees in engineering, nursing, and allied health. The Allegheny College Center for Experiential Learning (ACCEL) is a clearinghouse for internship opportunities, service learning, and overseas study.

Allegheny has two 15-week semesters each year, and its general education requirements keep students busy. Juniors take a seminar in their major field, and seniors complete an intensive capstone project in their major. All students must take courses in each of three major divisions (humanities and the natural and social sciences) and must finish a divisional minor in a subject area outside their major. Despite Allegheny's small size—67 percent of courses have fewer than 25 students—most of those enrolled graduate in four years.

Students praise Allegheny's faculty for their passion, knowledge, and accessibility. "Most professors will meet you over lunch if there is no other time available," says one education major. You won't find a TA at the lectern in any Allegheny classroom, and the college's honor code allows students to take unproctored exams. Off campus, Allegheny offers study in several U.S. cities and abroad, an on-campus independent study option, and semester internships or "externships" (a chance to observe a professional at work during the winter vacation). There's also a three- or four-week Experiential Learning term following spring semester for study abroad and internships not available during the year. The library holds more than 1,009,000 volumes and the campus offers more than 325 computer workstations.

Sixty-four percent of Allegheny's students hail from Pennsylvania, and sizable contingents come from nearby Ohio, New York, and New England. While Allegheny isn't a terribly diverse campus—minority students make up only 6 percent of the total—the school is committed to increasing awareness of and appreciation for diversity. "No one is afraid to express their opinion," claims one senior. The college's new Center for Political Participation engages students by fostering an appreciation for the vital link between an engaged, active citizenry and a healthy democracy. Merit scholarships are available, with an average award of $10,835.

> "Most professors will meet you over lunch if there is no other time available."

Residence halls and houses (including TV and study rooms) accommodate undergraduates in relative style and comfort with a variety of living situations: all-freshmen dorms; co-ed and single-sex halls; small houses; and single, double, and triple rooms and suites. Dorms are "clean and of reasonable size," says a senior, and "there is no trouble getting a room," says a sophomore. Housing is guaranteed for four years, and 77 percent of students stay on campus. The most popular dorm is the townhouse-style College Court complex, which holds 80 students in suites with four single bedrooms each. Students are happy with their on-campus food choices. Officers from campus security are available 24 hours a day, but they aren't much needed.

Greek organizations draw 24 percent of the men and 26 percent of the women, and provide a great deal of nightlife. Five fraternities have their own houses, while the four sororities are relegated to special dorm suites. There are also two campus theater series, two-dollar movie nights, and comedians, ventriloquists, and live bands provided by the center for student activities. Large-scale philanthropic events like Make a Difference Day and the Month of Service, in March, are also popular. "Campus parties exist but don't dominate social life [neither does Greek life]. There is always something going on and most students belong to some club or organization that hosts special events," says one student. College policy states that students must be 21 to have or consume alcohol on campus, but one student says those who want alcohol can get it, and "as long as you're reasonably responsible, you probably won't get caught." Homecoming, Greek Sing, Wingfest in the fall (featuring free wings), Springfest (a day full of bands, activities, and food), and Winter Carnival break up the monotony of studying, and midnight breakfasts served by faculty help ease end-of-semester stress.

Downtown Meadville, lovingly referred to as Mudville, is a 10-minute walk from campus and worlds away from a "college town." It has a four-screen movie theater and several community playhouses, as well as schools, hospitals, children's

Students at Allegheny work hard and do well, as evidenced by the fact that 52 percent of a recent senior class immediately went on to graduate school.

Off campus, Allegheny offers study in several U.S. cities and abroad, an on-campus independent study option, and semester internships or "externships" (a chance to observe a professional at work during the winter vacation).

homes, animal shelters, and other organizations that benefit from the more than 25,000 hours of service students contribute each year. "The students do a lot of volunteer work and help immensely throughout the community," says an economics major. When more time in Meadville is too much to bear, students hit the road, venturing to factory outlets in nearby Grove City, Pennsylvania, or heading toward the bright lights of Pittsburgh, Buffalo, or Cleveland. "What's nice is that even though Meadville is small, larger cities are close by so you can get away if need be," a senior says. Nearby state parks at Conneaut Lake and Lake Erie offer water-skiing and boating in warm weather and cross-country skiing in the winter.

As for Allegheny traditions, there's the somewhat suspect "13th Plank" ritual, which states that all freshman women must be kissed on the 13th plank of the campus bridge by an upperclassman to be considered a "true Allegheny co-ed." Of course, a group of freshmen men steal the plank every year at the beginning of the first semester to prevent that from happening.

> **"What's nice is that even though Meadville is small, larger cities are close by so you can get away if need be."**

Athletics play a big role in Allegheny life, although there are no athletic scholarships, and the sports and fitness complex gives students reason to cheer. Twenty-five percent of Allegheny students participate in varsity athletics, while 70 percent participate in nine intramural sports, with basketball, soccer and volleyball drawing the most interest. The softball, men's and women's soccer, and men's and women's track and field squads (both indoor and outdoor) captured North Coast Athletic Conference championships in 2004–05.

Allegheny College boasts a rich history of academic excellence in an intimate setting, augmented by a new emphasis on extracurricular experiences designed to produce well-rounded alumni. The campus's natural beauty and the genuine affection students feel for it and for each other remain unchanged. As one student says: "There is a sense that Allegheny enables you to do anything you put your mind to—no matter how unusual."

If You Apply To ➤

Allegheny: Early decision: Nov. 15. Regular admissions: Feb 15. Meets demonstrated need of 45%. Campus interview: recommended, informational. Alumni interviews: optional, informational. SATs or ACTs: required. SAT IIs: optional. Accepts the Common Application and electronic applications. Essay question: significant experience, important issue, or person of significant influence.

Alma College

Alma, MI 48801-1599

The college that put the "Alma" back in "alma mater." As friendly a campus as you'll find, Alma combines the liberal arts with distinctive offerings in health and pre-professional fields. Diversity is an issue and few out-of-staters enroll. If central Michigan drives you stir-crazy, join the hordes who go abroad.

The mission of Alma College, a tiny gem on Michigan's lower peninsula, is fourfold: "To prepare graduates who think critically, serve generously, lead purposefully and live responsibly as stewards of the world they bequeath to future generations." Alma has a wide array of choices for their undergraduates, including distinctive offerings in health and preprofessional fields, as well as the opportunity to learn abroad.

Alma students have taken part in everything from the U.S. mission to the United Nations to the reclamation of a Jewish cemetery in Poland. And the annual Highland Festival, with kilt-wearing bagpipers and competitions, has led some to dub the school "Scotland, USA."

Alma's campus features 25 Prairie-style buildings of redbrick and limestone surrounding a scenic central mall. Alma was founded over a hundred years ago, but most of the buildings have been built or renovated in recent years. There are lots of trees and open places to sit, at least in the warmer months. Administrators are

> "Few classes at Alma are considered easy."

looking at how best to expand toward the Pine River, because the current campus is bound by residential neighborhoods, making it difficult to find room for new academic, athletic, and residential facilities. An environmentally friendly residence hall opened in 2005.

The Alma experience begins with the seven-day Preterm Orientation, built around a one-credit academic seminar that includes readings, discussions, and research and introduces students to computer resources, on-campus life, and extracurricular activities. Trustee Honors Scholarship recipients are invited to attend a two-credit Freshman Honors Seminar during their first year. To graduate, students must demonstrate proficiency in communication, computation, and foreign language, and must complete 16 credits in each of three areas: arts and humanities, social sciences, and natural sciences. In addition, students must complete a comprehensive series of English courses. "The academic climate is very rigorous," says one junior. "Few classes at Alma are considered easy," adds a senior. Administrators candidly admit that weaker programs include the preengineering 3–2 program and classical languages.

Alma's top majors, per student enrollment, are business and education. Biology and psychology prepare students for work with wellness intervention programs and public health agencies, or graduate study in medicine, nursing, or physical therapy. Athletic training was changed from an area of concentration in the Exercise and Health Sciences program to a major. Alma's Service Learning Program gives students academic credit for work with nonprofit economic development organizations or educational, environmental, and social service agencies. Alma's Model United Nations teams have received the top award for 11 of the last 12 years at the world's largest and most prestigious collegiate Model UN confer-

> "There is not much to do, but there is a large college town only 15 minutes away so it is never an issue."

ence. Students interested in the Scottish arts find a range of opportunities to work with nationally known instructors of bagpipe and highland dance, or join in seminars offered by award-winning Scottish authors. A new peer-mentoring program places successful upperclass students in contact with new students to help them adapt to the opportunities and expectations of an engaged college community.

Despite Alma's small size and the fact that 94 percent of students are Michigan natives, the terms "provincial" and "insular" just don't apply here. In fact, one of the college's selling points is its wide variety of study abroad opportunities. During the one-month spring term, students enroll in a single intensive course that often includes off-campus study. Past programs have taken students to Australia to study culture and trade in the Pacific Rim, to Paris to learn language and culture, and to Israel for archaeological fieldwork. Alma also offers foreign study programs in Bolivia, Ecuador, England, France, Germany, Ireland, New Zealand, Peru, Scotland, South Korea, and Spain.

Back on campus, Alma's professors win praise. "The professors are dedicated, knowledgeable, and enthusiastic, as well as excellent communicators," says a junior. None of the classes taken by freshmen have more than 100 students; 55 percent have

(Continued)
Male/Female: 41/59
SAT Ranges: V 520-670
  M 510-650
ACT Range: 21-27
Financial Aid: 76%
Expense: Pr $
Phi Beta Kappa: Yes
Applicants: 1,471
Accepted: 81%
Enrolled: 29%
Grad in 6 Years: 70%
Returning Freshmen: 80%
Academics: ✍ ✍ ✍
Social: ☎ ☎ ☎
Q of L: ★ ★ ★
Admissions: (800) 321-2562
Email Address:
  admissions@alma.edu

Strongest Programs:
Business Administration
Education
Biology
History
Psychology
Exercise and Health Science
Music
Sociology and Anthropology

A new peer-mentoring program places successful upperclass students in contact with new students to help them adapt to the opportunities and expectations of an engaged college community.

The Alma softball team took first place in the Michigan Intercollegiate Athletic Association nine consecutive times (1997 through 2005), and the volleyball and football teams have also brought home titles in recent years.

25 or fewer. Students say graduating in four years is seldom a problem, except for education majors, who may need a ninth semester to complete their student teaching.

Alma students are quite involved around campus and in the community. "We have a lot of service organizations, service learning courses, and a program called Discovering Vocation that funds service-based programs," says one business major. The campus is 93 percent white, with African Americans and Hispanics comprising 2 percent each. Asians Americans, Native Americans and foreign students make up 1 percent each. "Alma students tend to be high achievers and just a little obsessive-compulsive," a student says. Brainy types can vie for an unlimited number of merit scholarships worth an average of $16,112.

Eighty-three percent of Alma's students reside on campus. "Dorms are dorms; no more and no less than what you would expect from campus housing," admits one junior. Freshmen are assigned rooms in single-sex and co-ed halls (some co-ed by room, others by floor), while upperclassmen play the lottery and usually get suites. Other options include an international house, a Model UN house, and a Women's Resource Center. There are fraternities and sororities for the 17 percent of the men and 26 percent of women who go Greek. Everyone buys 14 or 19 meals a week, and chows down at the all-you-can-eat Commons or at Joe's Place, a snack bar. Students say they feel safe on campus thanks to the presence of patrolling security guards.

Alma students may live in a small town, but "it is very friendly and proud of Alma," according to one student. "There is not much to do, but there is a large college town only 15 minutes away so it is never an issue." Some students go to clubs or the movie theatre but most students claim there is a lot to do on campus, too. Officially, the campus alcohol policy follows Michigan law: No one under 21 can drink. Most students tend to agree that drinking is not a big problem on campus. Underage students caught imbibing are written up and fined.

Town/gown relations at Alma are strong, with students volunteering, taking part-time jobs, and otherwise getting involved in the community. The annual Highland Festival features bagpipers and Scottish dancing; for members of the Alma marching band, who strut in kilts stitched from the college's own registered Alma College tartan, every performance might as well be a festival. Students with wheels will find diversions within easy reach, as Mt. Pleasant, Saginaw, and the East Lansing campus of Michigan State are less than an hour away, and ski slopes are just a bit farther. In the warmer months, the beaches of two Great Lakes, Huron and Michigan, are also two hours away. The Alma softball team took first place in the Michigan Intercollegiate Athletic Association nine consecutive times (1997 through 2005), and the volleyball and football teams have also brought home titles in recent years. For nonvarsity types, there is an active intramural program.

"We have a lot of service organizations, service learning courses, and a program called Discovering Vocation that funds service-based programs."

Helpful advisors and caring faculty members help Alma students chart a course on campus and in the broader world, allowing them to leave well-prepared for today's competitive job market. Michigan natives and newcomers alike are sure to find a warm welcome and friendly faces at Alma. Confirms one senior, "Everyone here is ridiculously friendly. Nowhere else on earth have I seen so many smiling faces!"

Overlaps

Central Michigan, Grand Valley State, Michigan State, Hope, Albion, Western Michigan

If You Apply To ➤

Alma: Rolling admissions. Meets demonstrated need of 86%. Campus and alumni interviews: optional, informational. SATs or ACTs: required. SAT IIs: optional. Accepts the Common Application and prefers electronic applications. Optional essay question.

Alverno College

3401 South 39th Street, P.O. Box 343922, Milwaukee, WI 53234-3922

At last, a college that evaluates students on what they can do rather than how well they can memorize. Forget oval-blackening; students here show mastery in their chosen fields. Practical and hands-on, Alverno is at its best in preprofessional programs. Only women need apply.

If you're the type of student who obsesses over your GPA, take heed: At Alverno College, you can forget about earning an A. That's because this small Roman Catholic women's college emphasizes ability-based learning instead of letter grades. Students are required to show "mastery" in a range of liberal arts courses, as well as demonstrated ability in eight broad areas: communications, analysis, problem solving, values in decision making, social interaction, global perspectives, effective citizenship, and aesthetic responsiveness. Students move through interdisciplinary progressive levels toward a degree by being "validated" in these areas. For example, a course in sociology might contribute to validation in communication and social interaction, as well as in making independent value judgments. The learning environment isn't competitive, though the ability-based method can "create a lot of work requiring much thought," says a professional communications major. As one senior puts it: "We learn in a way that gives us a cutting edge."

Alverno is located in a quiet, well-kept residential area. The parklike 46-acre campus is just 15 minutes from downtown Milwaukee and a 10-minute walk from shops and restaurants. Its three main academic buildings, of 1950s and '60s vintage, feature brick and stone exteriors and stained-glass windows. The Teaching, Learning, and Technology Center houses 73,000 square feet of science labs,

> "We learn in a way that gives us a cutting edge."

multimedia production areas, and computer facilities. Parking issues have been addressed with the recent opening of a 500-car parking structure, and new soccer and softball fields are nearly complete.

Alverno is really two colleges rolled into one. There is a regular weekday program that attracts mostly traditional college-age women, the majority of whom are from Milwaukee. The Weekend College allows women, most of them older with full-time jobs, to earn a degree in four years by attending classes every other weekend. Students in the weekday college may earn credit by spending four to eight hours a week in internships related to their field of study. The college ditched grades long ago and administrators find innovative, "real-life" ways to assess students' mastery of subject matter and specific abilities.

First-year students take an orientation seminar and introductory courses in the arts and humanities, science, psychology and sociology, communications, and math. The student body is diverse—in age, ethnic background, and religion. "It would be very hard to make generalizations," one junior says. Religious studies aren't required, but for those who seek it, a Catholic liturgy is available on a daily basis. Alverno's business and management programs are well established. "The level of professionalism that Alverno students have compared to those at other colleges or universities is amazing," one junior says. Students praise the professional communications and teacher education programs and the strong nursing program. Newer majors include marketing, political science, and sociology. One art education major says the art department "has produced some of the best art instructors, local artists, and art therapists that Milwaukee has to offer." The library holds 350,000 titles, and for major research projects students can use the online library/media cen-

Website: www.alverno.edu
Location: Suburban
Total Enrollment: 1,523
Undergraduates: 1,508
Male/Female: 0/100
ACT Range: 17-22
Financial Aid: 98%
Expense: Pr $
Phi Beta Kappa: No
Applicants: 886
Accepted: 55%
Enrolled: 58%
Grad in 6 Years: 47%
Returning Freshmen: 76%
Academics: ✍ ✍ ✍
Social: ☎
Q of L: ★★★★
Admissions: (414) 382-6100
Email Address:
   admissions@alverno.edu

Strongest Programs:
Nursing
Elementary Education
Business and Management
Psychology
Biology
Professional Communication
Interdisciplinary Arts and
   Humanities

ter catalog made available through a consortial arrangement with five Milwaukee colleges. Bali, Paris, Tokyo, and London are just a few of the places where students have taken advantage of the school's study abroad program.

Many professors at Alverno teach all levels of classes, so "there is no distinct difference" in the quality of teaching for freshmen or seniors, one international business major says. The instructors don't generally focus on publishing research. "The faculty and staff really care whether you are successful," says a social science major. "They want to see you achieve and are willing to go over and beyond to make sure you do." Academic counseling and individual attention run throughout students' academic careers to keep them on track. "We couldn't have more help," one junior says.

Ninety-seven percent of Alverno students hail from Wisconsin and many are older than the traditional college age; quite a few have children. "Students are extraordinarily driven and they know what it takes to succeed in the real world," says a junior. There are dozens of student groups and cultural groups, like Women of Color, active on campus. Students and faculty often engage in roundtable discussions to look at political or social issues,

"The level of professionalism that Alverno students have compared to those at other colleges or universities is amazing."

according to a sophomore. African Americans comprise 19 percent of the student body, Asian Americans 5 percent, and Hispanics 11 percent. "This is a very liberal campus," says a professional communications major, "with lots of open-minded people." Ninety percent of the students receive financial aid, and merit scholarships are awarded based on a personal evaluation of each incoming student.

A three-day orientation program serves freshmen, transfer, resident, and commuter students. The majority of students are commuters, though a limited number of dorm rooms house 13 percent of students, who say the residence halls offer clean, spacious rooms with fully equipped lounges, laundry, and cooking facilities available on each floor. "Dorms are comfortable and well maintained every day, including weekends," says one junior. Male visitors are allowed, but they must sign in and be out by midnight on weekdays and 2:00 a.m. on weekends.

Most of the social life takes place off campus at local clubs, bars, coffee bars, and nearby colleges, but the student union, called the Pipeline, frequently offers on-campus activities. The campus also has an on-site daycare center, a fitness center, and a jogging track, and sponsors dance and theater groups. "Milwaukee is a thriving city of the arts—visual, theatrical, and performance—not to mention the festivals that go on every year," says one art education major. There are also myriad parks and shopping centers, a Performing Arts Center, professional sports teams, ethnic festivals, and free outdoor concerts. Students look forward to the annual Rotunda Ball and homecoming

"This is a very liberal campus with lots of open-minded people."

festivities. When it comes to alcohol on campus, the combination of strict policies and a generally low amount of drinking means Alverno is rarely the home of the Animal House booze fest.

Alverno competes in Division III and fields varsity basketball, volleyball, softball, soccer, and cross-country teams. Intramurals are popular as well. Nonsporting annual events include Student Seminar Day, which allows students and faculty to change places so that students can "share their experiences" with the Alverno community.

Attending a school like Alverno promises an experience far afield in some ways from the traditional college world. The lack of grades and emphasis on real-world applications builds confidence in one's actual ability to perform, rather than their ability to score an A. Students and faculty are often on a first-name basis from the start and build relationships that help students find their "own unique style of learning," one senior says. It's a method that obviously works.

If You Apply To ➤	**Alverno:** Rolling admissions. Does not guarantee to meet demonstrated need. Campus interviews: recommended, evaluative. No alumni interviews. ACTs: required. Accepts the Common Application and prefers electronic applications. Essay question: recent activities and/or work history, academic goals and abilities, and most important reason for applying to Alverno.

American University

4400 Massachusetts Avenue NW, Washington, DC 20016-8001

If the odds are against you at Georgetown and you can't see yourself on GW's highly urban campus, welcome to American University. The allure of AU is simple: Washington, D.C. American has a nice campus in a nice neighborhood with easy access to the Metro and the city. It is about a third smaller than GW.

Located just a few miles from where our country's leaders make decisions of national and global impact, American University is a breeding ground for the next generation of reporters, diplomats, lobbyists, and political leaders who will shape domestic and international policy. Alongside these eager buzz hounds is a host of students taking advantage of AU's strong programs in the arts and sciences and business. "American University is a diverse, pulsing, and dynamic school driven by some of the best faculty, staff, scholars, and students in the world," a senior says.

AU's 84-acre residential campus is located in the safe northwest corner of Washington, D.C., in a neighborhood called Tenleytown just minutes from downtown; free shuttle buses transport students to the nearby Metro (subway) station. There's a mix of classical and modern architecture and flower gardens alongside the parking lots. The quads have numerous sitting areas for reflection and study and the campus has gone totally wireless. Newer additions include the Katzen Arts Center, which provides studio and performing arts spaces as well as galleries.

All AU undergraduates must demonstrate competency in writing and English, either through two courses or an exam; for math or statistics, it's one semester of class or placing out through a test. The general education program requires 30 credit hours from five areas: the creative arts, traditions that shape the Western world, global and multicultural perspectives, social institutions and behavior, and the natural sciences. The requirements are typically completed during the first two years so that upperclassmen can study abroad or participate in an internship or co-op—of which there are many, thanks to the school's relationships with more than 900 private, nonprofit, or government institutions. The school also uses these connections in its Washington Semester* program, which attracts a wide range of majors.

> "Everyone has an opinion on something."

In the classroom, AU has outstanding programs in political science and government, international studies, business, and communications. An honors program offers the top 10 percent of entering students small seminars, special sections of many courses, and designated floors in the residence halls, plus specialized work in their major and a senior capstone experience. In all, students may choose from more than 70 programs and have the option to design their own interdisciplinary major.

Classes are "challenging but not overwhelming," a senior political science major says. "Courses are extremely demanding, but professors are always willing to work with students if they are struggling," a junior international studies major adds.

Website: www.admissions.american.edu
Location: Urban
Total Enrollment: 11,279
Undergraduates: 5,921
Male/Female: 38/62
SAT Ranges: V 580–690 M 570–660
ACT Range: 26–30
Financial Aid: 38%
Expense: Pr $ $
Phi Beta Kappa: Yes
Applicants: 15,004
Accepted: 53%
Enrolled: 18%
Grad in 6 Years: 70%
Returning Freshmen: 88%
Academics: ✑ ✑ ✑ ½
Social: ☎ ☎ ☎
Q of L: ★ ★ ★
Admissions: (202) 885-6000
Email Address: admissions@american.edu

Strongest Programs:
International Studies
Political Science/Government
Justice, Law, and Society
Studio and Performing Arts
Premed
Journalism
Business administration

The average class size is 23 and 96 percent of professors have the highest degree in their fields.

Students who attend AU are "preppy" yet "quirky," and "at parties kids talk about politics rather than your average party conversation," a sophomore says. AU prides itself on drawing students from every state and more than 140 foreign countries; just 5 percent hail from the District of Columbia. Six percent of the student body is African American, 5 percent is Hispanic, and 5 percent is Asian American.

> "We also sometimes have Secret Service agents on campus since we're so close to the vice president's house."

Unlike many college campuses where apathy reigns, AU is very active politically—after all, this is Washington, D.C. "Everyone has an opinion on something," notes a junior who says AU is a "pretty liberal campus, but the conservatives make themselves known." AU's hefty price tag draws complaints, but the school offers more than 700 merit scholarships averaging $14,153 and 152 athletic scholarships in six women's and four men's sports.

Sixty-eight percent of AU students, mostly freshmen and sophomores, live on campus. "The dorms are roomy, well-maintained, and clean," says a political science major. "South Side dorms are known to be louder and more social than North Side dorms," advises a junior. There is off-campus housing for upperclassmen in newly renovated luxury apartments, and a shuttle bus connects them to campus. "Housing draw is difficult, since it not only goes by current dorm status, but also by credits, making it more difficult for those with fewer credits," says a junior. Students say they generally feel safe on campus, noting public safety officers are visible and "we also sometimes have Secret Service agents on campus since we're so close to the vice president's house."

A good deal of the social life at AU revolves around campus-related functions, such as room and frat parties; 17 percent of men and 18 percent of women go Greek, though the women lament that the bottom-heavy male/female ratio is "a little ridiculous." The immediate area around AU has restaurants and shops, but you need to get a bit farther away for true nightlife in Dupont Circle and Georgetown. While greater D.C. certainly has its share of clubs and bars, they're largely off-limits to students under 21. The AU campus is officially dry, and most students take that seriously. Happily, there is so much other stuff to do in D.C., and much of it is free—the new art house movie theaters, gallery openings, pro soccer games, museums and monuments, and funky live music. "You just jump on the Metro to get anywhere in the city," says

> "We are a small campus, which gives the feeling of being out of the city, but yet the city is at our fingertips."

a communication major. Each year, Family Weekend brings games, rides, and popular bands to campus, along with a carnival on the quad. Homecoming and Founder's Week are also campus favorites. Popular road trips include Baltimore, Annapolis, Williamsburg, Richmond, the Ocean City shore, and nearby amusement parks and outlets.

American competes in Division I, but sports are an afterthought for most students. There's no football team, but students are enthusiastic about Eagles basketball, where games against Bucknell, Holy Cross, and the Naval Academy top the schedule. Strong teams include women's volleyball, basketball, field hockey and lacrosse, men's swimming and diving, men's and women's soccer, and men's and women's cross-country. AU athletics also earned the distinction of the highest combined GPA in AU history, proving that brains and brawn are not mutually exclusive. There's a slew of intramural and club sports, which are divided into different levels of competitiveness.

AU is heaven on earth for C-SPAN junkies. (AU made headlines of its own last year for the inept way that its board dealt with the university's free-spending former

General education requirements are typically completed during the first two years so that upperclassmen can study abroad or participate in an internship or co-op—of which there are many, thanks to the school's relationships with more than 900 private, nonprofit, or government institutions.

Overlaps

George Washington, Georgetown, Boston University, NYU, Boston College, University of Maryland–College Park president.) But even if you are not addicted to following current events, AU and Washington, D.C. are still a top combo for a rich college life. The opportunities for real-world experience—in fields ranging from business to international studies to political science—are outstanding. But AU is small enough to keep students from feeling lost in the fast-paced world inside the beltway. "We are a small campus, which gives the feeling of being out of the city, but yet the city is at our fingertips," a junior says.

Amherst College

Amherst, MA 01002-5000

Original home to the well-rounded, superachieving, gentle-person jock. Compare to Williams, Middlebury, and Colby. Not Swarthmore, not Wesleyan. Amherst has always been the king in its category—mainly because there are four other major institutions in easy reach to add diversity and depth.

Amherst offers a dynamic curriculum in the traditional academic disciplines and in numerous interdisciplinary fields. There are no core curriculum or distribution requirements, so students choose their program based on their own individual interests and plans for the future. Indeed, the focus isn't on racking up high grade point averages. Instead, students focus on becoming people who base their thinking on a strong foundation in the liberal arts. Emphasizing "freedom to explore," the spotlight here is on learning, not competing for grades. "Students encourage others through teamwork and discussion," confirms a senior. "Rarely do you ever hear students boasting about their grades."

Amherst's 1,000 acres overlook the picturesque town of Amherst and the Pioneer Valley and offer a panoramic view of the Holyoke Range and the Pelham Hills. On campus, a plot of open land housing a wildlife sanctuary and a forest shares space with academic and residential buildings, athletic fields, and facilities. While Amherst's predominant architectural style remains 19th-century academia—redbrick is key—everything from a "pale yellow octagonal structure to a garish, modern new dorm" can be found here. Amherst looks like a college is supposed to look, with trees and paths winding through the buildings to offer long, contemplative walks. Of course, the flip side is snowy winters that don't end soon enough when spring break has come and gone.

The most popular majors are English, economics, psychology, law, jurisprudence and social thought, and political science. Students may mix and match among these subjects to form dual-degree programs. About one-third of students pursue double majors, and a few overachievers even triple major. Students may create their own courses of study from Special Topics classes if the subject of their interest is not available. Amherst's unique Law, Jurisprudence, and Social Thought is not a prelaw major; instead, it's an interdisciplinary study of the law, drawing on fields as diverse as psychology, history, philosophy, and literature, with a strong theoretical focus. The dance program is also strong, although it requires courses at each school in the Five

Website: www.amherst.edu
Location: Small town
Total Enrollment: 1,638
Undergraduates: 1,638
Male/Female: 52/48
SAT Ranges: V 670–780
  M 670–770
ACT Range: 30-33
Financial Aid: 44%
Expense: Pr $ $ $ $
Phi Beta Kappa: Yes
Applicants: 6,284
Accepted: 19%
Enrolled: 37%
Grad in 6 Years: 96%
Returning Freshmen: 98%
Academics: 🎓 🎓 🎓 🎓 🎓
Social: ☎ ☎ ☎
Q of L: ★ ★ ★ ★
Admissions: (413) 542-2328
Email Address:
  admission@amherst.edu

Strongest Programs:
English
Economics
Psychology

College* Consortium. To house all of these programs, Amherst has spent millions of dollars in recent years renovating facilities, upgrading technological capabilities, and improving spaces for studying, exhibits, performances, and sports.

Amherst is home to a rich intellectual environment that centers on a wealth of acclaimed instructors. "The professors are articulate and kind both in and out of class," says a freshman. A classmate adds, "Professors are almost always willing to respond to lengthy emails and will arrange meeting times outside their designated office hours." On such a small campus without graduate students, interaction with professors is encouraged. "The quality of instruction at Amherst is phenomenal and stellar. Professors are not only wise and accomplished but they pursue teaching at Amherst because they are interested in teaching," explains a sophomore.

To graduate, students must take a first-year seminar, declare a major at the end of sophomore year, fulfill departmental program requirements, pass the requisite number of electives, and perform satisfactorily on comprehensive exams in their major field. First-year seminars, taught by two or more professors, help foster interdisciplinary approaches across topics and are offered in several subject areas. "Most courses have a rigorous workload," a psychology major reports, but "the climate is not competitive."

"Amherst students are motivated, open, and approachable."

In addition to being one of the Five Colleges*, Amherst also belongs to the Maritime Studies Program* and the Twelve College Exchange*. All-female Smith and Mount Holyoke also add to the social life, and numerous cultural and artistic events at the other schools are open to Amherst students. Each year, about one-third of the junior class spends a semester or year abroad; recent students have chosen from among 72 programs in 35 countries, ranging from a math program in Budapest to analyzing architecture in Rome. Amherst also has a program in Kyoto, Japan, where one of the college's Colonial-style buildings has been duplicated.

"Amherst students are motivated, open, and approachable," says a freshman. Eighty-one percent of Amherst students hail from outside Massachusetts. Eighty-one percent were in the top tenth of their high school class, and, as one junior explains, "Amherst College is populated by nearly every sort of person. Whether you are a geek, jock, hippie, or any other hard-to-define type of human being, you will find kindred spirits." The student body is unusually diverse; 9 percent are African American, 13 percent Asian American, 7 percent Hispanic, and 6 percent foreign. Amherst actively works to educate its community about issues that affect student life. Residential Life, through its staff of Resident Counselors, sponsors programs that both educate and encourage open discussion about many issues on campus, including race and gender relations, issues of sexual respect, and alcohol/drug abuse. The Diversity Educators group offers workshops during orientation and in all first-year student dorms, raising student awareness about issues pertaining to race. Discussions on political issues and hot topics are common. "Amherst prides itself on its diversity [of] not only color, but nationality, faith, socioeconomic background, interests, ideas, and political thought," says a sophomore.

Sports are taken seriously, both varsity and intramurals, and "impact" athletes get favored treatment from the admissions office.

"The social life is diverse—from parties to board games, you can find something that suits you."

Housing at Amherst is guaranteed for four years, and 98 percent of students live on campus. "The rooms are huge with lots of storage space," raves a student. Everyone who lives on campus, and anyone else who wants to, eats in Valentine Hall, which includes a central serving station and five dining rooms. The selection is diverse but the food overall gets mixed reviews. "We have vegetarian, vegan, and kosher options," notes a junior.

Although frats are nothing more than a distant memory, the legacy of beer-drenched partying at Amherst lingers. Social activities are conducted almost entirely on campus. They range from quiet gatherings of friends to dorm study breaks to campus-wide parties. "The social life is diverse," says a student, "from parties to board games, you can find something that suits you." The biggest party of the year, thrown in February, is Casino Night, which includes gambling with real money. The weekend-long Bavaria festival in the spring offers a pig roast and big-wheel joust, while a lip-sync contest that offers winners their first pick during room draw usually attracts hilarious entries. The Campus Center includes outdoor terraces, a formal living room, a game room, a snack bar, a small theater, and a student-run co-op coffeehouse, open three nights a week with live entertainment.

Amherst is "the quintessential college town, full of academics, old hippies, small shops, and cheap restaurants," says an interdisciplinary studies major. Students take part in community service projects, including "Big Brothers Big Sisters, and Habitat for Humanity, just to name a few," says a freshman. For the many outdoorsy types, good skiing in Vermont is not far, and Boston (an hour and a half) and New York (a little over three hours) are close enough to be convenient road trip destinations.

Sports are taken seriously, both varsity and intramurals, and "impact" athletes get favored treatment from the admissions office. Amherst's intramural program currently consists of six sports (flag football and soccer in the fall; basketball and volleyball in the winter; and softball and indoor soccer in the spring). Each sport is co-ed and open to all students, staff, and faculty. Amherst competes in Division III, but the strong baseball team takes on Division I opponents as

> "I see our community as made up of people who are comfortable with themselves and consequently are some of the kindest and most laid-back people I can imagine."

well. Men's soccer is strong, and recent NCAA champions include women's tennis, men's basketball, men's swimming, baseball, track, and the women's lacrosse team. Any showdown with archrival Williams is inevitably the biggest game of the season, drawing fans from all corners of campus.

Combine a lack of restrictive requirements with a cadre of professors who are focused on teaching, and it becomes clear why students here so love their institution. Says a proud student: "I see our community as made up of people who are comfortable with themselves and consequently are some of the kindest and most laid-back people I can imagine. Yet this does not deter from the fact that Amherst students are always doing extraordinary things: academically, athletically, artistically, and for the benefit of the world."

Amherst's unique Law, Jurisprudence, and Social Thought is not a prelaw major; instead, it's an interdisciplinary study of the law, drawing on fields as diverse as psychology, history, philosophy, and literature, with a strong theoretical focus.

Overlaps
Yale, Harvard, Princeton, Dartmouth, Brown, Williams

If You Apply To ➤ **Amherst:** Early decision: Nov. 15. Regular admissions: Jan.1. Financial aid: Feb. 15. Guarantees to meet demonstrated need. No campus or alumni interviews. SATs or ACTs: required. SAT II: any two subject tests. Accepts the Common Application and electronic applications. Essay question: one Common Application question and respond to one of five quotations.

Antioch College

795 Livermore Street, Yellow Springs, OH 45387

Part Goth, part granola, and part anarchist—with plenty of none-of-the-above mixed in—Antioch is a haven for square pegs. Yet Antioch offers something very practical: the chance for 16-week co-op experiences interspersed with academic study. March and protest, then get a job. Cool.

Website: www.antioch-college.edu

Location: Rural

Total Enrollment: 470

Undergraduates: 464

Male/Female: 43/57

SAT Ranges: V 600-710
M 520-630

ACT Range: 19-29

Financial Aid: 87%

Expense: Pr $ $

Phi Beta Kappa: No

Applicants: 368

Accepted: 51%

Enrolled: 34%

Grad in 6 Years: 48%

Returning Freshmen: 63%

Academics: ✍ ✍ ✍

Social: ☎ ☎

Q of L: ★ ★ ★ ★

Admissions: (800) 543-9436

Email Address:
admissions@antioch-college.edu

Strongest Program:
Cultural and Interdisciplinary Studies

Come to Yellow Springs, Ohio, and you may think you've stepped out of a time machine. But no, you haven't spun back to 1969—you're simply at Antioch College, where the humanistic messages of that era are still taken to heart and put into action. The students, many of them products of "alternative" high schools, discuss feminism, gay rights, and nuclear proliferation over vegetarian meals, and they are more likely to take road trips to Washington for an environmental rally than to show up at a neighboring school's fraternity party.

Founded in 1852 by abolitionist and social reformer Horace Mann as its first president, Antioch remains a haven for outspoken and independent students who thrive under the rigors of a refreshingly nontraditional education. Antioch pioneered the idea that students should alternate time in the classroom with jobs in the "real world," and this idea has remained the foundation for Antioch's unique approach to training students. Under the college's famed co-op program, students spend nearly half of their college years out in the real world, be it selling fresh-squeezed orange juice on a street corner in California, studying Buddhism in India, or working in a Fortune 500 company in New York City. Antioch does not use standardized test scores for admission. In classes, written faculty evaluations take the place of grades, and students are required to submit self-evaluations of their class performance.

In completing Antioch's 32-credit general education program, students spend their first year pursuing a core of courses that blends the traditional liberal arts with examination of the "social, historical, philosophical, and economic" nature of work. In addition, there are distribution requirements in the humanities, social and behavioral sciences, natural sciences, and cultural studies. Physical education is also required. "The academic climate at Antioch is what you decide to make it...There are competitive classes, as well as laid-back classes; it all depends on what you are interested in," says one freshman.

Because classes are usually no larger than 20, students must always be prepared to participate. Close relationships often develop between students and faculty; representatives of both groups sit on several of the influential governing committees, including the administrative council, the housing board, and the community council. "Professors at Antioch are some of the best in the country. The teachers are very

> **"The academic climate at Antioch is what you decide to make it...There are competitive classes, as well as laid-back classes; it all depends on what you are interested in."**

dedicated to their students and their level of understanding," explains a freshman. Each student is assigned a co-op advisor to help with the nearly continuous job hunt, and a network of alumni offering jobs is one major resource that students can depend on in their search. You can't miss with the career counselors. "It's not just counseling; its cooperative education that teaches you more about yourself than you could imagine. It's the best reason to come to Antioch," says a student.

Antioch's trimester system lasts 15 weeks, with co-op terms lasting 16 weeks. The college helps place students in co-op programs and credit is earned after the student completes a paper or project demonstrating what he or she learned during the co-op experience. Antioch's mission lies in its "commitment to undergraduate experiential learning and to preparing students to face the challenges and opportunities of the 21st century." In order to receive a crosscultural experience, Antioch has all students spend three to 12 months in a significantly different cultural environment. In addition to those offered by the school, study abroad options are available through the Great Lakes College Association.* There is a downside to this: students blame the high attrition rate on the rigors of the co-op program. Friendships and involvement in extracurricular activities at the Yellow Springs campus often suffer because of the on-again, off-again attendance schedules. In an effort to address this, the college has placed an increased emphasis on learning communities, which cover all aspects of campus life. "We have just started a first-year core program," explains a student. "All incoming freshmen attend these core classes that have a centralized theme. They are meant to create a support system."

Antioch's traditional academic programs are somewhat uneven. Although there are only eight official majors, each allows a student to concentrate on a more specialized area. The major in physical sciences is weak based on current student interests. Concentrations in political science, psychology, and many of the arts offerings are established strongholds. Environmental and

"Students at Antioch are encouraged to speak their mind. The issues on campus are influenced by national issues."

life sciences are excellent, in part because of the proximity of a 1,000-acre forest preserve, Glen Helen, and a nature museum. The communications program benefits from a major public radio station operated by Antioch, which gives students experience in the broadcasting field. Being different may be the only thing Antioch students have in common, and diversity is a given on this campus. Thirty percent of the students come from Ohio; the rest hail from points throughout the nation. Three percent of students are African American, 1 percent Asian American, 2 percent Hispanic, and 1 percent Native American. A variety of merit scholarships are renewable for four years.

Ninety-five percent of the students are housed on campus in apartment-style dorms. Co-ed and single-sex dorms are available, and all come equipped with kitchens. Everyone is guaranteed on-campus housing if they want it, and students can choose from a number of special options that include a quiet hall, a moderate-noise hall, and even a substance-free hall, which bars smoking and drinking. "Housing is an issue. Some of the dorms are in disrepair, but most of the students live on campus despite this. A strong sense of community draws students here and this is more important than the living conditions," says one student. The Spalt International Center houses 60 students in foreign language living/learning halls. Meal plans are available at the Caf, which features vegetarian entrees, a salad bar, and a popcorn machine. One student complains, "The food all seems to be deep fried, extremely cheesy, or lots of potato products. They have normal vegan and vegetarian offerings."

Without Greek organizations, social life tends to be rather spontaneous and limited to on-campus activities—some parties are held at the dance space, a student says. Drinking is said to be less of a problem among students than among the high school students who make the scene at campus parties. There is a student coffeehouse, which "is the best hangout space," while stargazers frequently congregate on the roof of the science building. Each fall, a three-day blues festival hits town. "Community Day" is actually two days a year where

"Be prepared to go through a lot of self-transformation."

everyone takes the day off to mellow out and relax. Many student organizations, including the Anarchist Study Group and Third World Alliance, draw widespread student interest.

The town of Yellow Springs is "quaint, but not for someone craving nightlife. You have to mooch a car and go to Dayton or somewhere for that," says a student. The town hosts a variety of health food stores, a pizza joint that makes its pies with whole wheat crust, and an assortment of bars and restaurants. Yellow Springs may become limiting for some students, but that problem is usually solved by the next co-op trimester. And because they are world-minded, many Antioch students volunteer in the community and care about social issues on campus. "Students at Antioch are encouraged to speak their mind. The issues on campus are influenced by national issues," explains a political science major.

Antioch does not use standardized test scores for admission. In classes, written faculty evaluations take the place of grades, and students are required to submit self-evaluations of their class performance.

Being different may be the only thing Antioch students have in common, and diversity is a given on this campus.

Traditionally considered taboo, varsity sports have enjoyed a rousing resurgence, thanks to the women's rugby team. Gym classes are offered in kayaking, rafting, and horseback riding. The 1,000-acre nature preserve ("the Glen") across the street is also near John Bryan State Park. A nearby reservoir is a popular place for swimming and windsurfing, and Clifton Gorge offers rock climbing.

For those who think cooperative education would be a "far out" experience, Antioch offers a unique opportunity. Even with an enrollment smaller than many high schools, the college still manages to offer an outside-the-box approach to academics. A junior offers would-be students this piece of advice: "Be prepared to go through a lot of self-transformation." Horace Mann would be proud.

Overlaps

Oberlin, Hampshire, Earlham, Evergreen State, Bard

If You Apply To ➤

Antioch: Rolling admissions. Does not guarantee to meet demonstrated need. Campus interviews: recommended, informational. No alumni interviews. SATs: optional. SAT IIs: optional. Accepts the Common Application and prefers electronic applications. Essay question.

University of Arizona

Robert L. Nugent Building, Tucson, AZ 85704

Tucson is an increasingly popular destination, and it isn't just because of the UA basketball team. A well-devoted honors program attracts top students, as do excellent programs in the sciences and engineering. Bring plenty of shorts and sunscreen.

Website: www.arizona.edu
Location: Urban
Total Enrollment: 36,805
Undergraduates: 28,010
Male/Female: 47/53
SAT Ranges: V 490–600
   M 500–630
ACT Range: 20–26
Financial Aid: 42%
Expense: Pub $
Phi Beta Kappa: Yes
Applicants: 17,904
Accepted: 80%
Enrolled: 47%
Grad in 6 Years: 58%
Returning Freshmen: 79%
Academics: ✐ ✐ ✐ ½
Social: 🐻 🐻 🐻 🐻
Q of L: ★ ★ ★ ★
Admissions: (520) 621-3237
Email Address:
   appinfo@arizona.edu

With a campus that's encircled by mountain ranges and the beautiful Sonoran Desert, lined with palm trees and cacti, and set against a backdrop of stunning Tucson sunsets, it's no surprise that students at the University of Arizona love to hang out at the mall. Not the shopping center, mind you—but a huge grassy area in the middle of campus where the nearly 35,000 Wildcats gather between classes. Judging by numbers alone, that's enough people to fill a medium-sized town. But students are quick to point out that UA has a strong sense of community and offers a genuinely friendly campus. "Nobody else has a huge central meeting place like we do," says a senior marketing major. "I always see familiar and friendly faces around the mall area." With all the natural beauty that surrounds them, many Wildcats simply purr through four satisfying years.

Architecturally, the UA campus distinguishes itself from the city's regiment of adobe buildings with a design that seems a study in the versatility of redbrick. Old Main, the university's first building, is into its second century, but others verge on high-tech science facilities. Hard hats and heavy machinery have become commonplace on campus; recent construction includes a new student union, facilities for first-year students and learning-disabled students, the Athletic Hall of Fame and weight training facility, a residence hall, and a 1,300-space parking structure.

> **"Nobody else has a huge central meeting place like we do."**

Sciences are unquestionably the school's forte—the astronomy department is among the nation's best, helped by those clear night skies. Students have access not only to leading astronomers, but also to the most up-to-date equipment, including a huge 176-inch telescope operated jointly by the university and the Smithsonian. A $28 million aerospace and mechanical engineering building has a state-of-the-art subsonic wind tunnel and rocket-combustion test facility. The history and English departments are standouts, as are several of the social science programs. Eager shutterbugs can pore through photographer Ansel Adams's personal collection, and the Center for Creative Photography offers one of the leading photographic collections in the world. Students in the popular business and public administration school can pick racetrack management as their area of expertise, while interested anthropology students can delve into garbage research. Two newer programs are optics and public health. Areas getting low marks from students are the language programs and the journalism department.

Under the core curriculum, students take 10 general education courses in common. They fall under the broad categories of arts, humanities, traditions and cultures, natural sciences, and individuals and societies. In addition, almost everyone gets a healthy dose of freshman composition, math, and foreign language. Academic competition, according to most students, is left up to both the individual and the specific concentration. "The academic program that I have chosen to take part in is a very competitive and challenging program," explains a physiology major. The University Honors Center offers one of the nation's largest and most selective honors programs (students must maintain a grade point average of 3.5 to remain in the program). In addition to offering 200 honors courses per year, the center features smaller classes, personalized advising, special library privileges, and great research opportunities. The Undergraduate Biology Research Program also has a national reputation. Teaching is well regarded, with some freshman courses taught by graduate students. "The quality of teaching here is pretty amazing," a senior says. "Professors are able to combine necessary information while also making the course laid-back."

> "In the past, housing was a problem, but three new dorms have opened up in the past year."

One student describes his peers as "go-getters" who "bleed red and blue." Despite tougher admission standards, the administration cites a sharp increase in freshman applications over the past few years, especially from out-of-staters, who constitute 33 percent of the student body. In addition to various merit scholarships, all the athletic scholarships allowed by the NCAA are available. Hispanics account for 16 percent of the student body, African Americans 3 percent, and Asian Americans 6 percent. A diversity action council, a newly developed student minority advisory committee, and cultural resource centers help students deal with race-relation issues. An active and popular student government runs a free legal service and a tenants' complaint center, and the university has instituted many programs to help those with learning disabilities.

Dorm rooms tend to be small but well maintained. "In the past, housing was a problem," confides one student, "but three new dorms have opened up in the past year." Only 20 percent of undergraduates live in the dorms; 20 percent of freshmen and most upperclassmen flock to the abundant and inexpensive apartments near the school. The best way to enjoy the excellent food service at the student

> "Professors are able to combine necessary information while also making the course laid-back."

union's seven restaurants is to use the university-issued All Aboard credit card, which helps students take advantage of the wealth of different gustatory options and frees them from carrying cash.

Despite the high percentage of off-campus residents, students stream back onto campus on weekends for parties, sports, and cultural events. Ten percent of the men belong to fraternities and 11 percent of the women belong to sororities. The campus is technically alcohol-free, though some question whether the frats have realized that yet. Still, most social life takes place off campus. There are a lot of different dance clubs around town, and some do have after-hours for underage people. Those

(Continued)
Strongest Programs:
Management Information
  Systems
Nursing
Astronomy
Pharmacy
Creative Writing
Aerospace Engineering

Division I football and baseball enjoy national prominence, generate lots of money for other men's and women's sports teams, and provide great weekend entertainment, especially when the opposing team is big-time rival Arizona State.

The Undergraduate Biology Research Program also has a national reputation.

who feel they must go elsewhere need only head to the Mexican town of Nogales (one hour away), where there is no drinking age. Many students are content remaining in Tucson because it offers "the most incredible sunrises and sunsets, and delightful temperatures year-round."

One of the UA's most time-honored traditions is Spring Fling, said to be the largest student-run carnival in the country. Athletics is also somewhat of a tradition here. The basketball Wildcats have been among the nation's leaders in recent years. Division I football and baseball enjoy national prominence, generate lots of money for other men's and women's sports teams, and provide great weekend entertainment, especially when the opposing team is big-time rival Arizona State. UA's battle cry, "Bear Down!"—frequently heard at sporting events—dates back to the 1930s, when a campus football hero, fatally injured in a car crash, whispered his last message to his teammates: "Tell them, tell them to bear down." More than 70 years later, the enigmatic slogan still appears on T-shirts and in a gym on the central campus.

The University of Arizona offers a wide variety of academic opportunities along with spectacular weather. Prospective students are warned to honestly evaluate how that will affect their ability to concentrate. UA is the place to go to engage in the pursuit of truth, knowledge, and a good tan.

Overlaps

Arizona State, Northern Arizona, UCLA, University of Colorado, UC–Santa Barbara

If You Apply To ➤

Arizona: Regular admissions. Campus interviews: optional, informational. Alumni interviews: optional. SATs or ACTs: recommended. SAT IIs: optional. Accepts electronic applications. No essay question.

Arizona State University

Box 870112, Tempe, AZ 85287-0112

Want to get lost in the crowd? ASU is the biggest university in the Southwest—apologies to UCLA. No matter how appealing the thought of 47,000 new faces, you'd better find the right program to get a good education. Try the professional schools and the honors college.

Website: www.asu.edu
Location: Urban
Total Enrollment: 46,679
Undergraduates: 39,690
Male/Female: 47/53
SAT Ranges: V 490–600
  M 500–620
ACT Range: 20–26
Financial Aid: 44%
Expense: Pub $
Phi Beta Kappa: Yes
Applicants: 24,727
Accepted: 80%
Enrolled: 43%
Grad in 6 Years: 55%

Whether climbing "A" Mountain with a lantern, enjoying water sports on Town Lake, or puzzling over the term "dry heat," there's no escaping the fact that Arizona State University sits smack in the middle of an oasis in the desert. A very large oasis. With ample educational opportunity and the promise of fun in the sun, it just might be that ASU is the perfect personal challenge. Students can find their niche in a variety of stellar programs, including ASU's anthropology department, which is ranked within the top five nationwide. Its Barnett Honors College allows exceptional students to live, work, and study among their peers, and to write a senior thesis, creating a small-school atmosphere on this mammoth campus.

ASU's north Tempe campus offers a beautiful blend of palm-lined walkways, desert landscapes, and public art displays. Architectural styles range from turn-of-the-century historic to more modern, as the school renovates and expands Memorial Union to add four restaurants and more seating and program space. Lattie F. Coor Hall, has state-of-the-art classroom space, and the 17,000-square-foot Arizona Biodesign Institute provides lab and office space for research in neural rehabilitation, genomics, molecular biophysics, and neutraceuticals. Other new construction includes Adelphi Commons II, fraternity housing that will replace older houses. The campus is officially listed as an arboretum, and ASU groundskeepers tend to more than 115 species of trees that thrive in Arizona's arid climate.

ASU has eight undergraduate schools—business, liberal arts and sciences, engineering and applied sciences, architecture and environmental design, education, fine arts, nursing, and public programs (justice studies, leisure studies, communication, social work, and public affairs)—though students apply to the institution as a whole. The most popular majors are business, psychology, and communication, followed by education and nursing. The college of business ranks second in placing graduates with the Big Four accounting firms, while the fine arts college fea-

> "It offers the intellectual stimulation and individual attention of a small liberal arts college at a large research university with innumerable opportunities."

tures an innovative child drama program, and nationally recognized majors in art, music, and dance. Engineering programs, especially microelectronics, robotics, and computer-assisted manufacturing, are sure bets; the facility for high-resolution microscopy allows students to get a uniquely close-up view of atomic structures. Students in the robust Creative Writing Program will benefit from a $10 million gift from The Virginia G. Piper Charitable Trust. The gift is the largest ever to an ASU humanities program and the funds will establish a new center for creative writing.

The sciences (including solar energy, physical science, geology, and biology) and social sciences also boast first-class facilities, notably the largest university-owned meteorite collection in the world. Planetary science is out of this world; a team of ASU students, faculty, and staff, led by geology professor Philip Christensen, designed the Thermal Emission Imaging System and are conducting research with the system as it orbits Mars. ASU is also a founding member of the NASA Astrobiology Institute, which will focus on studying the origin of life on Earth and elsewhere. Anthropology benefits from its association with the Institute of Human Origins' Donald C. Johannson, who discovered the 3.2-million-year-old fossil skeleton named Lucy. The Walter Cronkite School of Journalism and Telecommunication finished first in the annual Hearst writing competition twice. Each year, the legendary CBS newsman visits campus to lecture in individual classes.

"The best academic program would have to be the honors college," offers one student. "It's amazing! It offers the intellectual stimulation and individual attention of a small liberal arts college at a large research university with innumerable opportunities." Regardless of major, students must fulfill requirements in

> "ASU is competitive enough to drive excellence but laid-back enough to take a breather and relax."

literacy and critical inquiry (including composition), mathematical studies (including college-level algebra or higher), humanities and fine arts, social and behavioral sciences, and natural sciences. Students must also complete courses in three awareness areas: global, historical, and U.S. cultural diversity. An automated phone system helps speed registration. "ASU is competitive enough to drive excellence but laid-back enough to take a breather and relax," says a freshman. Professors get high marks for their knowledge and accessibility. "I think that majority of professors out here are great," says a junior.

One-quarter of the students at Arizona State come from elsewhere. "Tons of rich kids from California come to the Southwest," grumbles one sophomore. "Blonde and fashion-conscious is a pretty accurate stereotype." They must meet tougher standards for automatic admission: While in-state students need to be in the top quarter of their class or have a 3.0 GPA, as well as a total SAT score of 1040 or an ACT score of 22, out-of-staters must meet the same class-rank or GPA requirements and have a cumulative SAT score of 1110 or ACT score of 24. Because ASU draws so

(Continued)
Returning Freshmen: 79%
Academics: ✍ ✍ ✍
Social: 🐯 🐯 🐯 🐯 🐯
Q of L: ★ ★ ★ ★ ★
Admissions: (480) 965-7788
Email Address:
  askasu@asu.edu

Strongest Programs:
Business
History
Psychology
Geology
Anthropology
Music
Landscape Architecture
Accountancy

The college of business ranks second in placing graduates with the Big Four accounting firms, while the fine arts college features an innovative child drama program, and nationally recognized majors in art, music, and dance.

heavily from within Arizona, 13 percent of the student body is Hispanic; African Americans contribute 4 percent, Asian Americans 5 percent, and Native Americans 3 percent. The Intergroup Relations Center works to overcome racial, religious, gender, and other differences, including the gulf between Greeks and independents, and athletes and academics. ASU offers merit scholarships to qualified students and awards 340 athletic scholarships annually to male and female athletes. "ASU Advantage" offers families earning less than $18,850 per year a combination of financial resources that includes tuition, fees, books, and room and board.

Only 17 percent of ASU students live in the co-ed dorms, which fill up quickly and are generally available only to freshmen. "Everything is clean and comfortable, if you are quick enough to get a dorm," advises a freshman. After freshman year, students live off campus in nearby apartments and houses. Those lucky enough to get a bed on campus jockey for one of the three residence halls with their own swimming pools and volleyball courts. No matter where they live, students don't have to buy a meal plan and one student says the "food is great."

"Blonde and fashion-conscious is a pretty accurate stereotype."

ASU's Greek system attracts 6 percent of the students and "small kick-backs in dorms are just as common as huge house parties," says a sophomore. The campus is officially dry and those under the legal drinking age are also warned against drinking in student housing: "There is a strict enforcement of a 'three-strikes-you're-out' policy." Perhaps that's why students head off campus on weekends—often way off campus. Many have cars, giving them access to the mountains of Colorado, the beaches of San Diego, the natural beauty of the Grand Canyon, or the bright lights of Las Vegas.

Arizona State's Division I athletics department is consistently ranked among the nation's best. In 2004, ASU launched the Pat Tillman Scholarship Fund to honor the former football star who died in combat. The baseball team has made 8 consecutive trips to the NCAA tournament and the wrestling squad has won the past two PAC-10 titles. Women's track and field is also strong.

Arizona State may seem like an overwhelmingly big school to some, but many students come here looking for the enormity of options the campus offers, both academically and socially. "You meet people from all over the world, which definitely adds to your college experience," says a sophomore.

Overlaps

University of Arizona, Northern Arizona, San Diego State, UCLA, University of Southern California, UC–San Diego

---

If You Apply To ➤

ASU: Rolling admissions. Financial aid: Mar. 1. Does not guarantee to meet demonstrated need. Campus interviews: optional, informational. No alumni interviews. SATs or ACTs: required. SAT II: optional. Accepts electronic applications. No essay question.

University of Arkansas

200 Hunt Hall, Fayetteville, AR 72701

University of Arkansas rates in the second tier of Southern public universities alongside Alabama, LSU, and Ole' Miss. With traditional strength in agriculture, U of A has also developed programs in business, engineering, and other professional fields. U of A's highest-ranked program takes the field on Saturday afternoons in the fall.

The state of Arkansas is working to transform its flagship public university into a nationally competitive, student-centered research institution in an effort to help stop the flight of the state's young and talented, and to jump-start the Arkansas economy. A $300 million cash gift from the family of Wal-Mart founder Sam Walton, the largest ever made to an American public university, put his name on the U of A's business college, and also helped endow the undergraduate Honors College, which will eventually enroll 2,000 of the brightest young scholars on campus. The Walton grant also strengthens and improves the U of A's graduate school. Aspiring politicos might consider Arkansas, as it's the alma mater of Sen. J. William Fulbright and the first employer of one-time law professors Bill and Hillary Clinton. No matter what your aspirations, you're likely to find something to study among Arkansas' 208 academic programs.

The Arkansas campus is nestled among the mountains, lakes, and streams of the Ozarks, in the extreme northwest corner of the state. The community is friendly and safe, and the moderate climate means recreational opportunities abound. Architectural styles range from modern concrete to buildings that date from the Depression. The center of campus is the stately brick Old Main, which once housed the entire university. There are two new greenhouses for plant science majors and an Innovation Center for engineers, while the Pat Walker Health Facility has been renovated. The $46 million Northwest Quad project includes housing for 600 students, as well as dining areas, classrooms, computer labs, and other amenities.

> "I enjoy having graduate students as teachers for some classes because often they relate to students better."

Established as a land grant institution in 1871, with agricultural and mechanical roots, the U of A includes six colleges and professional schools. U of A's core requirements include six credits each in English and Fine Arts, three credits each in U.S. History and Math, eight in Science and nine in Social Sciences. Students in the Fulbright College of Arts and Sciences must also achieve foreign language proficiency. The Walton College of Business offers two of the most popular majors on campus: marketing and finance. Other popular programs are elementary education and journalism; architecture and engineering are likewise strong. The Dale Bumpers College of Agricultural, Food, and Life Sciences, another part of the institution named for a former senator, includes the Poultry Health Center, a national leader in research on poultry epidemics. The master's degree in physical education has been phased out, but undergraduate degrees have been added in landscape architecture, public service, and biomedical engineering.

Students are quick to point out that although the academic climate is generally stress-free, it is competitive at times. "The academic atmosphere is competitive in that students are always trying to beat their friends on the next test. They strive for that sense of pride," explains a junior. Professors are lauded for their teaching skills, but students are also likely to

> "We have students from many nationalities, states, and backgrounds."

have graduate students leading classes. "I enjoy having graduate students as teachers for some classes," says a freshman, "because often they relate to students better."

A management major claims, "We have students from many nationalities, states, and backgrounds," but the numbers tell a different story: 85 percent of U of A students are homegrown, and minorities comprise 12 percent of the student body. African Americans make up 5 percent of the student body, Asian Americans 3 percent, and Hispanics 2 percent. Still, there are commonalities. One student describes classmates as having "a strong sense of school spirit" and being "intelligent, but not snobs." The university continues to emphasize race relations, and the chancellor has personally chaired a campus task force to help boost success rates of students from

Website: www.uark.edu
Location: Small city
Total Enrollment: 17,821
Undergraduates: 11,559
Male/Female: 51/49
SAT Ranges: V 510–640
   M 520–640
ACT Range: 22–28
Financial Aid: 39%
Expense: Pub $ $
Phi Beta Kappa: Yes
Applicants: 6,041
Accepted: 87%
Enrolled: 52%
Grad in 6 Years: 56%
Returning Freshmen: 81%
Academics: ✎ ✎ ✎
Social: ☎ ☎ ☎ ☎
Q of L: ★ ★ ★
Admissions: (800) 377-8632
Email Address:
   uofa@uark.edu

Strongest Programs:
Marketing
Management
Finance
Elementary Education

The Walton College of Business offers two of the most popular majors on campus: marketing and finance.

Aside from the revelry that accompanies Razorback football and basketball, students say Greek parties are pretty much the only game in town on weekends.

underrepresented minority groups. In addition, increasing diversity is one of the school's top five priorities. Arkansas hands out thousands of merit scholarships each year, worth an average of $5,796. There are also 411 athletic scholarships, representing all U of A sports teams. Additionally, the Good Neighbor program lets students from nearby states with GPAs of 3.0 or higher and ACT scores of at least 24 enroll at in-state rates.

Twenty-eight percent of the undergrads at Arkansas live in the dorms, where "it is always too hot or too cold," according to a picky sophomore, who adds, "Sharing a bathroom with that many people is unsanitary." All halls are single-sex, except for one that's co-ed by floor. Students recommend Gregson and Holcombe for freshmen, and note that as part of the First-Year Experience, Arkansas strives to put freshmen in residence halls where peers surround them. A junior recommends "Rock Camp," an optional orientation weekend in Oklahoma. "It is a great way to meet new students, get on email lists, and start making connections." When it comes to chow, you can get everything from salad and burgers to sushi.

Arkansas' 25 Greek chapters attract 19 percent of the women and 11 percent of the men. Aside from the revelry that accompanies Razorback football and basketball, students say Greek parties are pretty much the only game in town on weekends.

"The student section is always packed, and you must arrive one to two hours early to get a seat."

Dixon Street, the main drag in the town of Fayetteville (population 55,000), is full of bars and restaurants; the town also offers drive-in movies, live music at local clubs, and touring Broadway shows at the Walton Arts Center. Those with cars will find Dallas, Tulsa, Oklahoma City, Memphis, and St. Louis all within six hours' drive.

And who could forget Razorback sports? Cries of "Woooooo! Pig sooie!" ring out during football and basketball weekends, and red Razorback logos are all over town—on T-shirts, napkins, book covers, license plates, and on game day, the cheeks of ecstatic fans. "The student section is always packed, and you must arrive one to two hours early to get a seat," says a junior. Men's indoor and outdoor track and field brought home SEC Championships in 2005, and the women's volleyball team has won SEC Western Division titles for the past two years. The football team reached the conference championship during the 2006-07 season. Eighty-two percent of Arkansas students participate in intramurals, where sports include everything from flag football and soccer to dominoes, putt-putt golf, and trivia. (Who knew using your brain was a competitive sport?)

The University of Arkansas boasts "fun-loving and free-spirited" students who are "genuinely friendly," says a senior. "If they encounter someone who needs help, they help." This kind of Southern hospitality means poultry science students aren't the only ones flocking to Arkansas for a solid education at a bargain price. Northerners may feel out of their element, and those who dislike football should keep their feelings quiet. Others may look forward to graduation day, when their names will join forever those of 120,000 other alumni, etched into the five-mile network of sidewalks on campus.

Overlaps

Oklahoma State, University of Mississippi, University of Missouri, University of Oklahoma, University of Texas

If You Apply To ➤

Arkansas: Rolling admissions. Early action: Nov. 15. Does not guarantee to meet demonstrated need. No campus or alumni interviews. SATs or ACTs: required. No SAT II. Accepts the Common Application and prefers electronic applications. No essay question.

Atlanta University Center

Atlanta is viewed as the preeminent city in the country for bright, talented, and successful African Americans. It became the capital of the civil rights movement in the 1960s—a town described by its leaders as "too busy to hate."

At the heart of this extraordinary culture is the Atlanta University Center, the largest African American educational complex in the world, replete with its own central library and computing center. The seven component institutions have educated generations of African American leaders. The Reverend Martin Luther King Jr. went to Morehouse College; his grandmother, mother, sister, and daughter went to Spelman College. Graduates spread across the country in a pattern that developed when these were among the best of the few colleges to which talented African Americans could aspire. Even now, when the options are almost limitless, alumni continue to send their children back for more.

The center consists of three undergraduate colleges (Morris Brown, Morehouse, and Spelman) and three graduate institutions (Clark Atlanta University, the Interdenominational Theological Seminary, and the Morehouse College of Medicine) on adjoining campuses in the center of Atlanta, three miles from downtown. Students at these affiliated schools can enjoy the quiet pace of their beautiful magnolia-studded campuses or plunge into all the culture and excitement of this most dynamic of Deep South cities. The six original schools—all but the medical school—became affiliated in 1929 using the model of California's Claremont Colleges, but they remain fiercely independent. Each has its own administration, board of trustees, and academic specialties, and each maintains its own dorms, cafeterias, and other facilities. There is crossregistration among the institutions (Morehouse students, for example, go to Spelman for drama and art courses) and with Georgia State and Emory University as well. The governing body of the consortium, the Atlanta University Center, Inc., administers a centerwide dual-degree program in engineering in conjunction with Georgia Tech—and it runs campus security, a student crisis center, and a joint institute of science research. There is also a centerwide service of career planning and placement, where recruiters may come and interview students from all six of the institutions.

Dating and social life at the coeducational institutions tend to take place within the individual schools, though Morehouse, a men's college, and Spelman, a women's college, maintain a close academic and social relationship. The Morehouse–Spelman Glee Club takes its abundance of talent around the nation, and its annual Christmas concert on the Spelman campus is a standing-room-only event.

Morehouse and Spelman (see full write-ups) constitute the Ivy League of historically African American colleges. The following are sketches of the other two institutions offering undergraduate degrees.

Clark Atlanta University (www.cau.edu)
Formed by the consolidation of Clark College, a four-year liberal arts institution, and Atlanta University, which offered only graduate degrees, CAU is a comprehensive coeducational institution that offers undergraduate, graduate, and professional degrees. The university draws on the former strengths of both schools, offering quality programs in the health professions, public policy, and mass communications (including print journalism, radio and television production, and filmmaking). Graduate and professional programs include education, business, library information studies, social work, and arts and sciences. Undergraduate enrollment: 4,000.

Morris Brown College (www.morrisbrown.edu)
An open-admission, four-year undergraduate institution that is related to the African Methodist Episcopal Church, MBC lost most of its students in the spring of 2003 after the college lost its accreditation. A new president and a restructured board of trustees, including the Reverend Jesse Jackson, is working hard to restore its financial and academic viability. Its most popular programs are education and business administration. Morris Brown also offers evening courses for employed adults, as well as a program of co-op work-study education.

Morehouse College

830 Westview Drive, Atlanta, GA 30314

Along with sister school Spelman, Morehouse is the most selective of the historically black schools. Alumni list reads like a Who's Who of African American leaders. Best known for business and popular 3–2 engineering program with Georgia Tech. Built on a Civil War battlefield, Morehouse is a symbol of the new South.

Website:
www.morehouse.edu
Location: Urban
Total Enrollment: 2,970
Undergraduates: 2,970
Male/Female: 100/0
SAT Ranges: V 440–680
 M 470–680
ACT Range: 19–32
Financial Aid: N/A
Expense: Pr $
Phi Beta Kappa: Yes
Applicants: 2,079
Accepted: 75%
Enrolled: 36%
Grad in 6 Years: 63%
Returning Freshmen: 83%
Academics: ✐ ✐ ✐
Social: ☎ ☎ ☎ ☎
Q of L: ★ ★ ★ ★
Admissions: (404) 215-2632
 or (800) 851-1254
Email Address: admissions
 @morehouse.edu

Strongest Programs:
Economics
Business
Biology
Political Science
Psychology

Founded in 1867, Morehouse College has the distinction of being the nation's only historically African American, four-year liberal arts college for men. If its sister school, Spelman, was once the "Vassar of African American society," Morehouse was the Harvard or Yale, attracting male students from the upper echelons of society around the country. Top students come to Morehouse because they want an institution with a strong academic program and a supportive atmosphere in which to cultivate their success-orientation and leadership skills without facing the additional barriers they might encounter at a predominantly white institution. "Morehouse is a college of young, assertive, ambitious black men," says a psychology major.

> **"Morehouse is a college of young, assertive, ambitious black men."**

Notable alumni include the Reverend Martin Luther King Jr., Samuel L. Jackson, Spike Lee, and Dr. Louis Sullivan, current president of the Morehouse School of Medicine and former U.S. Secretary of Health and Human Services.

Located near downtown Atlanta, the 61-acre Morehouse campus is home to 35 buildings, including the Martin Luther King Jr. International Chapel. In a little more than a decade, the college has enriched its academic program, conducted a successful multimillion-dollar national fund-raising campaign, increased student scholarships and faculty salaries, doubled its endowment, improved its physical plant, and acquired additional acres of land.

The general education program includes not only 68 semester hours in four major disciplines (humanities, natural sciences, math, and social sciences), but also the study of "the unique African and African American heritage on which so much of our modern American culture is built." In fact, appreciation of this culture is one of the college's main drawing cards. "Many students are here to get a greater understanding of their heritage and to promote it," attests one student. The academic climate at the House can get intense, with students learning and challenging themselves for the sake of learning and not just to bust a curve. "Morehouse offers an academic structure that is both competitive and rigorous," states a freshman. Counseling, including career counseling, is considered quite strong.

Undergraduate programs include the traditional liberal arts majors in the humanities and social and natural sciences. While the sciences have been traditionally strong at Morehouse, business courses have risen in prominence. The college has obtained accreditation of the undergraduate business department by the American Assembly of Collegiate Schools of Business,

> **"Many students are here to get a greater understanding of their heritage and to promote it."**

and current students are linked to graduates who serve as mentors in the ways of the business world. The most popular major is business administration. Engineering, which trails shortly behind in popularity, is actually a 3–2 program in conjunction with Georgia Tech and other larger universities. The school also runs a program with NASA that allows students to engage in independent research. Programs that receive less favorable reviews from students are English, art, and drama, and the administration admits that physical education and some of the humanities offerings could use some strengthening. Study abroad options include programs offered through the Associated Colleges of the South consortium.* The school also offers courses and additional resources as a member of the Atlanta Regional Consortium for Higher Education.* Newer options include a major in applied physics and minors in public health sciences and telecommunications.

Sixty-seven percent of Morehouse students come from outside the state, with a sizable number from New York and California. Sixty-seven percent graduated in the top quarter of their high school class. More than 600 merit scholarships are available, many providing full tuition. There are 121 scholarships for athletes in football, basketball, track, soccer, and tennis.

There's limited housing, leaving half of the student body to find their own off-campus accommodations. For freshmen, students recommend Graves Hall, the college's oldest building, built in 1889. Those who do get campus housing sometimes wish they hadn't. Complaints range from "too small" to "not well-maintained." Most upperclassmen live off campus. The meal plan at Morehouse is mandatory for students living on campus and draws its share of complaints.

Morehouse's homecoming is a joint effort between Morehouse and Spelman. The queen elected by Morehouse men has traditionally been a Spelman woman, as are the cheerleaders and majorettes. The four fraternities, which sign up a very small percentage of the students, hold popular parties; "drinking is not a big deal here," most students concur. Going out on the town in Atlanta is a popular evening activity, and on-campus football games, concerts, movies, and religious programs all draw crowds. In its early years, Morehouse left much to be desired in the area of varsity sports, but it now competes well in NCAA Division II. Track, cross-country, tennis, basketball, football, and soccer are all strong, but it is the strong intramural program

> **"Morehouse offers an academic structure that is both competitive and rigorous."**

that allows students a chance to become the superstars they know are lurking within them. During football season, Morehouse men road trip to follow the games at Howard, Hampton, and Tuskegee universities.

Morehouse is well equipped to serve the modern heirs of a distinguished tradition. Morehouse students don't just attend Morehouse. They become part of what amounts to a network of "Morehouse Men" who share the bonds of having had the Morehouse experience, and graduates find that alumni stand ready and willing to help them with jobs and other needs.

Morehouse's homecoming is a joint effort between Morehouse and Spelman. The queen elected by Morehouse men has traditionally been a Spelman woman, as are the cheerleaders and majorettes.

While the sciences have been traditionally strong at Morehouse, business courses have risen in prominence.

Overlaps
Clark Atlanta, Howard, Georgia Tech, Hampton

If You Apply To ➤ **Morehouse:** Early action: Nov. 1. Regular admissions: Feb. 15. Financial aid: Apr. 1. Does not guarantee to meet demonstrated need. Campus and alumni interviews: recommended, informational. SATs or ACTs: required. SAT IIs: optional. Essay question: greatest influence on your life; why Morehouse?

Spelman College

350 Spelman Lane, Atlanta, GA 30314

The Wellesley of the black college world. Reputation draws students from all corners of the country. Unusually strong in the sciences with particular emphasis on undergraduate research. Wooded 42-acre Atlanta campus offers easy access to urban attractions.

Website: www.spelman.edu
Location: Urban
Total Enrollment: 2,134
Undergraduates: 2,134
Male/Female: 0/100
SAT Ranges: V 510–600
    M 500–580
ACT Range: 21–25
Financial Aid: 55%
Expense: Pr $
Phi Beta Kappa: No
Applicants: 4,534
Accepted: 39%
Enrolled: 30%
Grad in 6 Years: 74%
Returning Freshmen: 90%
Academics: ✍ ✍ ✍
Social: ☎ ☎ ☎ ☎
Q of L: ★ ★ ★ ★ ★
Admissions: (800) 982-2411
Email Address:
    mrodgers@spelman.edu

Strongest Programs:
Biology
Engineering
Natural Sciences
Premed
Prelaw

Today's emphasis is on getting Spelman grads into the courtrooms, boardrooms, and engineering labs.

As one of only two surviving African American women's colleges in the United States, Spelman College holds a special appeal for African American women seeking to become leaders in fields ranging from science to the arts. Students flock here for that something special that the predominantly Caucasian schools lack: an environment with first-rate academics where African American women can develop self-confidence and leadership skills before venturing into a world where they will once again be in the minority.

Founded in 1881 by two white women from New England (it was named after John D. Rockefeller's in-laws, Mr. and Mrs. Harvey Buel Spelman), the school was traditionally the starting point for teachers, nurses, and other African American female leaders. Today's emphasis is on getting Spelman grads into the courtrooms, boardrooms, and engineering labs. Honing women for leadership is the main mission, and that nurturing takes place on a classic collegiate-green campus with a $140 million endowment.

These are heady times for Spelman. Although the college finds itself competing head-on with the Seven Sisters and other prestigious and predominantly Caucasian institutions that are eager to recruit talented African American women, the college is holding its own. The college offers a well-rounded liberal arts curriculum that emphasizes the importance of critical and analytical thinking and problem solving. Usually by the end of sophomore year, students are expected to complete 34 credit hours of core requirements, including English composition, foreign language, health and physical education, mathematics, African diaspora and the world, African American women's studies, and computer literacy. In addition, freshmen are required to take First Year Orientation, and sophomores must take Sophomore Assembly. Spelman's liberal arts program introduces students to the principal branches of learning, specifically languages, literature, English, the natural sciences, humanities, social sciences, and fine arts.

> **"The school is made up of the top students from around the country and the courses are designed to be a challenge for the best of the best."**

Spelman's established strengths lie in the natural sciences (especially biology) and the humanities, both of which have outstanding faculties. Over the last decade, the college has greatly strengthened its offerings in math and the natural sciences; extensive undergraduate research programs provide students with publishing opportunities, and many end up attending grad school to become researchers. Students have moved beyond the popular majors of the early '70s—education and the fine arts—in favor of premed and prelaw programs, and these programs remain strong. The dual-degree program in engineering (in cooperation with Georgia Tech) is also a standout. The Women's Research and Resource Center specializes in women's studies and community outreach to women.

"The academic climate is very competitive," says an English major. "The school is made up of the top students from around the country and the courses are designed to be a challenge for the best of the best." Individual attention is the hallmark of a Spelman education. About 70 percent of the faculty have doctorates, and many are African American and/or female—and thus, excellent role models, ones the students find very accessible. Except for some of the required courses, classes are small; most have fewer than 25 students. Students who want to spread their wings can venture abroad through a variety of programs, or try one of the domestic exchange arrangements with Wellesley, Mount Holyoke, Vassar, or Mills. The school also offers courses and additional resources as a member of the Atlanta Regional Consortium for Higher Education.*

> **"No alcohol on campus—period."**

Spelman's reputation continues to attract African American women from all over the country, including a high proportion of alumnae children. Three-quarters of the students come from outside Georgia. Students represented here include high achievers looking for a supportive environment and those women with high potential who performed relatively poorly in high school. Only 5 percent of the student body is not African American. Spelman does not guarantee to meet the financial need of all those admitted, but it does offer merit scholarships averaging $6,700. There are no athletic scholarships.

Fifty-seven percent of students live on campus, and housing is "dated," reports a biology major. "We have little to no air conditioning." Still, the older dorms certainly can add to the school's historical charm, and students report having little trouble in getting a room. There are 11 dorms, and students recommend that freshmen check out the Howard Harreld dorm. The meal plan is mandatory for campus-dwellers and food is described as "edible" if not diverse.

Largely because of the Atlanta University Center, students also have plenty of chances for social interaction with other nearby colleges. "Students mingle in the student centers of all four schools all the time, especially on Fridays," a veteran explains. "Atlanta is a great college town!" gushes one junior. "If there is any place

> **"Students mingle in the student centers of all four schools all the time, especially on Fridays."**

that a student can be academically enriched, it is here." Spelmanites do take advantage of the big-city nightlife; they attend plays, symphonies, and the hot Atlanta nightclubs such as Ethiopian Vibrations and Lenox Mall. Sororities are present but only in small numbers—3 percent of the students go Greek. The attitude on drinking leans toward the conservative. Says one student, "No alcohol on campus—period." The most anticipated annual events include sisterhood initiation ceremonies and the Founders Day celebration. Although varsity sports are not the highlight here, the school boasts fine volleyball, basketball, and tennis teams. Athletic facilities are poor, but there are several organized intramurals, including flag football and bowling.

Spelman College has spent almost 125 years furthering the education and opportunities of African American women. It has adapted its curriculum to meet the career aspirations of today's youth, built up its bankroll, and successfully met the challenge posed by affirmative action in other universities. Still an elite institution in African American society, Spelman is staking its future on its ability to provide a unique kind of education that allows its graduates to compete with anyone.

Overlaps

Hampton, Howard, Clark Atlanta, Georgia State, Florida A&M

If You Apply To ➤ **Spelman:** Early decision: Nov. 1. Early action: Nov. 15. Regular admissions: Feb. 1. Housing: May 10. Does not guarantee to meet demonstrated need. Campus interviews: optional, informational. No alumni interviews. SATs or ACTs: required. SAT IIs: optional. Accepts the Common Application and electronic applications. Essay question: personal statement. Seeks women who are active in school, church, or community.

College of the Atlantic

105 Eden Street, Bar Harbor, ME 04609

In today's conservative world, COA is as out-there as it gets. A haven for communal, earthy, vegetarian types who would rather save the world than make a buck. Cozy is an understatement; with fewer than 300 students, it is the second smallest institution in the _Fiske Guide_.

The College of the Atlantic attracts rugged individualists troubled by the same issues that so worried the founders of this "mission-oriented" school, notably pollution, environmental damage, and troubled inner cities. The college's curriculum is focused on human ecology—the study of the relationship between humans and their natural and social environments—which is the only major offered. "The interdisciplinary approach that this creates is the most amazing opportunity you can have in an undergraduate program," a sophomore says.

The 31-acre campus, covered in lush flowers, vegetable gardens, and lawns, sits on the island of Mount Desert, along the shoreline of Frenchman Bay and adjacent to the magnificent Acadia National Park. In addition, the college has acquired two offshore island research centers and an 86-acre organic farm, and recently opened new classrooms and an arts building.

Most courses focus on a single aspect of humans' relationships with the world. Instead of traditional academic departments, the school has three broad resource areas: environmental science, arts and design, and applied human studies. Many students choose to concentrate on more narrowly defined topics within human ecology, such as marine studies, biological and environmental sciences, public policy, visual and performing arts, environmental design, or education.

> "The interdisciplinary approach that this creates is the most amazing opportunity you can have in an undergraduate program."

With advisors and resource specialists, each student designs an individual course of study. "I really appreciate the school's encouragement in designing an academic program that suits me," says a junior human ecology/teacher certification student. "You have so much freedom and support here," adds another junior. The natural sciences are stellar, with excellent instruction in ecology, zoology, and marine biology. The arts are catching up, with a more formal video and performance art program created in recent years. COA has added new classes in areas such as marketing, business, and grant writing "to help idealists realize their ideal." Allied Whale, the school's marine study arm, offers hands-on research opportunities, and specializes in training and research in marine mammalogy. Founded in 1972, the non-profit program conducts research into effective methods. COA also offers exchange programs with the Olin College of Engineering and the SALT Institute.

Student life at COA is intense and semi-communal, beginning with a rugged five-day wilderness orientation preceding the first trimester. Before graduating, students must also complete a 10-week off-campus internship and 10-week final project. Other requirements are few: freshmen must take a human ecology core course, and two courses are required in environmental sciences, human studies, and arts and design. Sophomores must submit a writing portfolio for evaluation. All students incorporate research of the community into their studies, whether it is a development impact study for the local government or a study on the aggression of fire ants for Acadia National Park.

> "We are moving toward a 'green' campus and are involved in many organizations that work on these issues professionally."

Some departments only have a professor or two. Since the student body is small, scholars can become close to faculty members. "The teachers are some of the best in their fields," a sophomore says. "The reason professors are here is to teach, that's it. They love teaching and it shows in the quality of the classes." Ninety-nine percent of classes have 25 or fewer students. In lieu of grades, students receive in-depth written evaluations of their work, although they may request grades as well. They must reciprocate with an evaluation of the course and their performance in it. The unusual advising system, a three-person student/faculty team chosen by the advisee, further promotes close contact between students and professors.

Student life at COA is intense and semi-communal, beginning with a rugged five-day wilderness orientation preceding the first trimester.

Students attracted to this quirky academic gem and its unique curriculum tend to be bright and idealistic; many worked with Americorps or traveled the world before beginning school. "We've got a hippie side," says a freshman, "but also a serious academic and politically active mindset." The student body is 26 percent white, 1 percent African American, 1 percent Asian American, and 2 percent Hispanic. Twenty-one percent of the students are international students. "Most students get along very well and there is a huge diversity of cultures, religions, countries, beliefs, financial status, and thought," says a freshman. The college's governance system gives students and administrators almost equal voices in how it's run; anyone may voice concerns or vote on policy-change proposals or the hiring of new faculty at the All College Meeting. Students aren't shy about also speaking out on more worldly issues, "from 'students for a free Tibet,' to antiwar protests, to environmental justice, to the global AIDS campaign," says a sophomore. "The whole school is concerned with ongoing environmental and social issues," a sophomore says. "We are moving toward a 'green' campus and are involved in many organizations that work on these issues professionally."

> **"We've got a hippie side, but also a serious academic and politically active mindset."**

Forty percent of students—freshmen, international students, and upper-class resident advisors—live on campus, while the balance find cozy, inexpensive apartments or houses in the nearby town of Bar Harbor. "The dorms are super comfortable, from the new, spacious Blair/Tyson, to the eccentric, ocean-view, bay-windowed Sea Fox. Peach House houses only eight people—it's a little cabin in the woods," a junior says. The dining options draw praise, too. "Food is fresh, delicious, and healthy," says a sophomore. "I look forward to every meal!"

Though the tiny tourist town of Bar Harbor is packed with visitors during the warmer months, it largely shuts down in the winter. One junior admits, "It's nice to have the town to ourselves." A sophomore adds, "Bar Harbor is nice, but the whole island is one of the best places to go to school in the world." Students get to know the townspeople through the required 30 hours of community service. "People do lots of stuff for local organizations, including work on farms, parks, the Downeast AIDS Network, and the YMCA," says a student. On weekends, few students leave campus, since Portland, the nearest urban center,

> **"Social life centers around people, not substances."**

is three hours away. Campus social functions revolve around nature and the seasons, including biking, hiking, boating, cross-country skiing, skating, and rock climbing, often in Acadia National Park. There are no fraternities or sororities, and students kick back at off-campus house parties, which are generally alcohol-free. There's no drinking on campus, and it's tough for underage students to get served in town. "Social life centers around people, not substances," quips a student. There are no varsity sports, but about half of the students sign up for the intramural program, which offers sports such as cricket, ice hockey, kayaking, and soccer.

"When I visited COA I found myself one minute in an intense discussion about the philosophy of language and the next in the middle of a snowball fight," says one student. Indeed, College of the Atlantic is a place where Earth Day really is cause for celebration, where students ride nude through the cafeteria and on nearby streets during Bike Week, and where everyone from students to trustees jumps into frigid Frenchman Bay on the first Friday of the fall term to try to swim from the school's docks to Bar Island. "I visited Hampshire, Marlboro, and Colby before visiting COA," says a sophomore. "I swear it was like Goldilocks and the Three Bears—COA just felt right."

On weekends, few students leave campus, since Portland, the nearest urban center, is three hours away.

The college's governance system gives students and administrators almost equal voices in how it's run; anyone may voice concerns or vote on policy-change proposals or the hiring of new faculty at the All College Meeting.

Overlaps
Hampshire, Colby, Marlboro, Bard, Bowdoin, University of Maine, Middlebury

If You
Apply
To ➤

COA: Early decision: Dec. 1, Jan. 10. Regular admissions, financial aid and housing: Feb. 15. Does not guarantee to meet demonstrated need. Campus interviews: recommended, evaluative. Alumni interviews: optional, evaluative. SAT I or ACT: optional. SAT IIs: optional. Accepts the Common Application and electronic applications. Essay question: essay on any topic (five suggested topics are provided). Looks for students committed to improving the quality of life on Earth.

Auburn University

202 Mary Martin Hall, Auburn, AL 36849-5111

Sweet Home Alabama, where the skies are so blue and the spirit of football lasts year-round. Auburn was once called Alabama Polytechnic, and today AU's programs in engineering, agriculture, and the health fields are still among its best.

Website: www.auburn.edu
Location: Small town
Total Enrollment: 23,333
Undergraduates: 19,254
Male/Female: 52/48
SAT Ranges: V 500-600
 M 520–620
ACT Range: 21–27
Financial Aid: 45%
Expense: Pub $ $
Phi Beta Kappa: Yes
Applicants: 14,249
Accepted: 82%
Enrolled: 36%
Grad in 6 Years: 62%
Returning Freshmen: 85%
Academics: ✍ ✍
Social: ☎ ☎ ☎
Q of L: ★ ★ ★
Admissions: (334) 844-4080
Email Address:
 admissions@auburn.edu

Strongest Programs:
Information Technology
Biological Sciences
Software Engineering
Computer Science
Agriculture

Auburn University may be home to more than 20,000 football-crazy students, but students quickly learn they're here for more than games. Once known as Alabama Polytechnic, Auburn is a public land grant university that still excels in professional and technical fields such as architecture, engineering, and agriculture. But the school also welcomes students with warm and cozy hospitality and charm. "The students at Auburn are extremely friendly. It's one of the things we're known for," says one junior. "In general, students also really care about the university and love being part of the Auburn family."

The town of Auburn, which grew up amid miles of forest and farmland largely to serve the university, is depicted in an Oliver Goldsmith poem as the "loveliest village of the plain." The campus stretches for nearly 2,000 acres, graced by mossy trees, lush lawns, and majestic colonnades. Most buildings are redbrick and Georgian in style with some more modern facilities grouped in a compact central location. Recent additions include the Jule Collins Smith Museum of Fine Art and the 71,500-square-foot Vaughan Large Animal Teaching Hospital. A new $50 million Student Activities Center broke ground last year and is scheduled for completion in 2008.

Auburn's core curriculum requires six semester hours of composition; six semester hours each of history, literature, and social sciences; three semester hours each of philosophy and fine arts; three semester hours of math; and eight semester hours of a lab science. To ease the transition into college life, freshmen undergo the three-day "Camp War Eagle" session and transfer students spend a day in orientation. The academic climate varies by department, and students say it runs the gamut from easy to difficult. "In the college of engineering students have to work hard through the semester to receive a passing or better grade," says a senior. Regardless of the rigor, students say professors are always willing to go the extra mile for students. "They are helpful and readily available for any problems a student might have," a computer science major says.

> **"The students at Auburn are extremely friendly. It's one of the things we're known for."**

Students agree that the engineering, architecture, agriculture, and pharmacy programs are stellar; the humanities sometimes suffer because they are not perceived as useful in the work-a-day world. Many Auburn students are eager to get started on their careers, so the co-op program, which provides pay and credit in several professional fields, is increasingly popular. Auburn has also established a first-of-its-kind program in wireless engineering for students who want to design network hardware or software for cell phones and other mobile devices. Seven areas designated as Peaks of Excellence compete for millions of dollars in special funding; these include cell and molecular biosciences, food safety, fisheries and allied aquacultures, and forest sustainability. A new study abroad program called Human Odyssey in Italy allows students to complete core courses in world history while studying abroad as early as the summer of freshman year. In addition, the College of Liberal Arts has created a new music major with a performance option.

Auburn students are "very friendly, enthusiastic, and open-minded," says a junior. Sixty-six percent of Auburn students are Alabama natives, and many are legacies—the second or third generation in their families to attend the school. African Americans account for 8 percent of the student body, Hispanics and Asian Americans each add 2 percent, and nearly 2 percent are Native American or foreign.

"Auburn women are probably the best-looking in the South!"

The conservative tone of this Bible Belt campus makes it hospitable for many Christian groups, and Auburn is home to one of the largest chapters in the United States of the Campus Crusade for Christ. Students are also among the friendliest you'll find anywhere, and a public relations major—already using his hard-won knowledge—claims, "Auburn women are probably the best-looking in the South!" Each year, the university awards hundreds of merit scholarships and 388 athletic scholarships.

Auburn's 25 dorms are single-sex, and visiting hours are restricted to weekends. Fourteen of the halls have been renovated, but only 16 percent of the students live in them. "Get on a waiting list ASAP," advises a junior. First-year students compete on a first-come, first-served basis with returning students. Twenty percent of Auburn men join fraternities, and 32 percent of the women join sororities, perhaps because chapters get space in the best dorms. There are food courts at each end of campus and dining facilities in the Foy Student Union as well. Students generally give low marks to the diversity of the menu and accessibility of the facilities.

Aside from sporting events and fraternity parties, Auburn sponsors concerts, free movies, and plenty of intramural sports. "Social life is great whether you are Greek or not," says a freshman. "On campus, there are events available for students to participate in," says one communication major, including movies, concerts, and other organized activities. The campus is officially dry, except on game days, and students say the alcohol policy is enforced. Long-standing traditions include the Burn the (Georgia) Bulldogs Parade and Hey Day, when everyone wears a nametag and walks around saying, "Hey!"

Auburn is a football powerhouse, and on fall Saturdays, 86,000 screaming fans turn the place into Alabama's fourth-largest city. The rallying cry "Warrrrr Eagle!" rocks the place each time an Auburn back runs to daylight. Unfortunately, the football program was also rocked recently by yet another scandal—high profile players were allegedly allowed to pad their grades and take part in classes with dubious attendance requirements.

"Auburn has become more focused on the future."

On a brighter note, the men's and women's swimming teams brought home SEC championships in 2004, 2005, and 2006. The McWhorter Center for Women's Athletics is one of the finest gymnastics training facilities in the country, and intramural programs are numerous and popular.

Auburn is working hard to increase the caliber of its students and academic programs, and especially to achieve a top 20 national ranking for its college of engineering. "Auburn has become more focused on the future," one senior says. But students say certain key characteristics have stayed the same—and that's a good thing. Says one student, "We just keep getting cooler."

Overlaps

University of Georgia, University of Alabama, Georgia Tech, Clemson, University of Florida, University of Tennessee

Austin College

900 N. Grand Avenue, Sherman, TX 75090

The second-most famous institution in Texas with Austin in its name. Half the size of Trinity (Texas), runs neck-in-neck with Southwestern to be the leading Texas college with fewer than 2,000 students. Combines the liberal arts with strong programs in business, education, and the health fields.

Website:
www.austincollege.edu
Location: Small town
Total Enrollment: 1,315
Undergraduates: 1,286
Male/Female: 45/55
SAT Ranges: V 580–680
M 580–670
ACT Range: 23–28
Financial Aid: 54%
Expense: Pr $
Phi Beta Kappa: Yes
Applicants: 1,530
Accepted: 67%
Enrolled: 34%
Grad in 6 Years: 75%
Returning Freshmen: 88%
Academics: ✍ ✍ ✍
Social: ☎ ☎ ☎
Q of L: ★ ★ ★
Admissions: (903) 813-3000
Email Address: admission
@austincollege.edu

Strongest Programs:
Biology
Chemistry/Biochemistry
Political Science/International
Studies
Psychology
Business Administration
English
History

Only an hour away from the 10-gallon hats and gleaming skyscrapers of Dallas, Austin College is a small but warm institution where students know their professors personally and have a broad array of majors to choose from. AC's preprofessional programs, most notably premed, are among the strongest in the state. Professors here even serve students breakfast at 10 p.m. the night before finals. It's just another example of the personal style that is typical of this charming Southern institution.

Austin's 65-acre campus is in a residential area in the city of Sherman. The campus is designed in the traditional quadrangle style and comprises beige sandstone buildings, tree-lined plazas, decorative fountains, and an impressive 70-ton sculptured solstice calendar. Dorms are conveniently located approximately 200 yards from most classrooms, which eases the pain of first-period classes. The newest additions to campus are the Roo Suites residence halls, which house 152 students in four-person suites.

The core curriculum begins with a freshman seminar called Communication/Inquiry. Each professor who teaches the course becomes the mentor for the 20 freshmen in his or her class. Next is a three-course sequence on the Heritage of Western Culture. Then students select from courses in three categories in humanities, social, and natural sciences. Seventy-eight percent of all classes have 25 students or fewer and no class has more than 50. "The courses are fairly rigorous," says a student, "but not to the point that they become unable to handle." A freshman adds, "Exams are very difficult."

> **"The courses are fairly rigorous, but not to the point that they become unable to handle."**

Preprofessional areas are Austin College's specialties. When it comes time to apply to grad school, premed and predental students at this little college have one of the highest acceptance rates of any Texas school, and aspiring lawyers also do well. AC's five-year teaching program grants students both a bachelor's and a master's degree. Science and education receive high marks from students, as do international studies, political science and foreign languages, but psychology is the most popular major. The Jordan Language House boards 48 students studying French, German, Japanese, and Spanish, along with a native speaker of each language, and students have to speak the language in all common areas. Students can combine three of the school's 26 majors into an interdisciplinary degree. Students must complete one major and a minor or a double major to graduate. A cooperative engineering program links the college with other schools. Majors in environmental studies and an international relations track of political science have been added while international studies and policy studies majors have been dropped. A minor in Southwestern/Mexican studies has been added.

The college also offers its students independent study, directed research, junior year abroad, and departmental honors programs. The Leadership Institute is open to just 15 students of each entering class, and five more can get in after their first year. Participating students enjoy a suite of privileges. Austin also provides three research areas in Grayson County. Students can focus on just one course during the January term, and many use that time to study abroad. Back on campus, students heap praise on the professors. "Professors are available and have a way of making students understand concepts."

Ninety percent of Austin students hail from the Lone Star State. Hispanics and African Americans comprise 9 and 4 percent of the student body, respectively, and Asian Americans make up 11 percent. "For the most part, those who choose to attend AC are academically driven," a freshman says. Students also say there is a wide range of political views on campus. Austin has been tied to the Presbyterian Church in the United States since 1849; this affiliation manifests itself in the emphasis on values in the core courses, participation in service activities, and limited residence-hall visitation hours—a chief complaint of students. AC offers merit scholarships worth an average of $9,199.

"Dorms are very well kept and you are guaranteed housing for three years."

As for dorm life, there are six residence halls, and 70 percent of undergraduates live in traditional dorm housing. "Dorms are very well kept and you are guaranteed housing for three years, after that it's 'find your own place' or hope someone will not show up," a junior says. One dorm is co-ed, one houses language studies students, one is men-only, and two are women-only. Dean (the only co-ed residence hall) seems to be a popular choice for freshmen, despite (or perhaps because of) its reputation as being loud and social. Others say Clyce is the best bet for freshman women. The new Roo Suites house juniors and seniors, with priority given to juniors. Residence hall access is computerized, and students say safety is not an issue. Nearly all students take advantage of the three-meal-a-day plan, though not all take advantage of the all-you-can-eat option. "The dining facilities are good, though sometimes the food becomes monotonous," a sophomore says. The Pouch Club, an on-campus joint, serves pizza, burgers, beer and wine for those students 21 and over.

Most of the social life is either on or near campus, with the Greeks taking the lion's share of credit. Twenty-nine percent of the men and 28 percent of the women belong to fraternities and sororities, but the Greeks are not school-funded and are not allowed to advertise off-campus parties without the college's permission. Not everyone depends on the Greek system for a good time. Students get an eyeful during the Baker Bun Run, in which the men of Baker Hall

"AC is awesome because it's small and comfortable."

strip to their boxers and cavort around the campus on the Monday night before finals. There are spring and fall festivals, and the Final Blowout party before finals. Students can have alcohol in dorm rooms only if they are 21 or older, and school policy prohibits booze at campus events. Popular weekend excursions are a drive to Dallas or to the college's 28-acre recreational spot on Lake Texoma (a half-hour north). Sherman is "quaint" and "historic," and is becoming a better college town, students say. "There are lots of restaurants," a junior says, "but for the most part there is nothing to do late at night."

Even without athletic scholarships, varsity sports are generating increasing support. The school's teams compete in Division III, and the women's volleyball team won two recent division championships. The Robert T. Mason Athletic/Recreation Complex provides facilities for student athletes and the fitness-conscious. There's also an intramural program, with basketball, softball, flag football, and lacrosse proving popular; 31 percent of undergrads participate.

"AC is awesome because it's small and comfortable," a senior says. "The professors are incredibly approachable and helpful." The preprofessional programs are among the academic strengths at this college with roots in the Presbyterian Church. And while Sherman may be a sleepy little place, Austin College certainly isn't.

If You Apply To ➤ **Austin:** Early decision: Dec. 1. Early action: Jan. 15. Regular admissions: Mar. 1. Financial aid: Apr. 1. Housing: May 1. Campus and alumni interviews: recommended, evaluative. SATs or ACTs: required. SAT IIs: optional. Accepts the Common Application and electronic applications. Essay question: significant experience, achievement, risk, or ethical dilemma; important issue; influential person; influential character, historical figure, or creative work: experience that shows what you would bring to college's diversity or showed you the importance of diversity; topic of your choice.

Babson College

Babson Park, MA 02457-0310

The only college in the *Fiske Guide* devoted entirely to business. Only 11 miles from College Student Mecca, a.k.a. Boston, and tougher to get into now than at any time in its history. The one college in Massachusetts where it is possible to be a Republican with head held high.

The budding corporate leaders who choose Babson College can be summed up in a single word: focused. They're already committed to business careers, and some arrive on campus with their own ventures up and running. Once school starts, they work hard, put their social lives second, and "are determined to be successful," says a sophomore. "Many students are perfectionists, and not easily satisfied." Hands-on experience is the norm; students get school funding to start businesses during their first years, and may hone their stock-picking skills by managing part of the college's endowment. Sound like a four-year stint on Donald Trump's reality show, *The Apprentice*? For the budding tycoons and entrepreneurs from around the globe who choose Babson, that's the reason they're here.

Founded in 1919, the college sits on 370 acres near the sedate Boston suburb of Wellesley. The tract features open green spaces, gently rolling hills, and heavily wooded areas. Buildings are gently shaded and parking lots (filled with expensive foreign cars) are relatively hidden. Architecturally, the campus is mainly neo-Georgian and modern. Several dorms and the admissions office have recently been renovated, and two newer dining facilities—Jazzman's Café and Pandini's Italian bistro—provide students with tasty meals.

Although Babson is a business school, about half of students' classes are in the liberal arts, and in 2002, Babson won the Hesburgh Award for curricular innovation. General education requirements emphasize rhetoric (public speaking); ethics and social responsibility; international and multicultural perspectives; and leadership, teamwork, and creativity. In the Foundation Management Experience, students are split into groups of 30 to develop business plans; each group gets up to $3,000 in seed money from the college to get

"Many students are perfectionists, and not easily satisfied."

their concept up and running. At year-end, the business is liquidated, and profits go to charity. Recently, FME groups have developed Babsonopoly (a Babson-themed version of Monopoly), opened a Krispy Kreme Doughnuts franchise, and sold customized fleece blankets. "There are 14 businesses all around campus, all trying to sell to the same 1,700 students," says a sophomore. "That's competitive, but it gives us real-world experience."

All Babson students major in business and then select a concentration, such as management, finance, or marketing—or even gender studies or literary and visual arts. (The Sorenson Visual Arts Center has painting, ceramics, and sculpture studios, labs for photography and digital art, a student art gallery, and workspace for artists-in-residence.) The entrepreneurship program is one of Babson's strongest, bringing in venture capitalists and executives (from companies such as Dunkin' Donuts and Jiffy Lube) for how-to lectures. Courses are rigorous and most are small. "Not only textbook: teamwork," says a junior. For projects, students can use more than 250 workstations dotting five computer labs or their brand-new IBM laptops, which are included with tuition. Since the computers are leased, upgrades are guaranteed after two years.

In the classroom, Babson relies on the case-study approach more typically employed by MBA programs. (At 60/40, the school's lopsided male/female ratio is also more similar to those of MBA programs than those at undergraduate business schools.) In the case-study method, students break into groups or act as officers of pseudo corporations to address specific business situations and solve marketplace problems. Professors teach all courses, and a junior says most have 10 to 20 years of experience in their fields. Accounting students may take graduate classes at Babson in the summer and fall after finishing their bachelor's degrees, letting them sit for the CPA exam about one year earlier than most other programs. Babson offers 30 study abroad programs in 19 countries, as well as the Semester at Sea.* Students can go away for an entire semester or the two-week winter session, and not all programs are business-focused; the London Theatre Program, for example, focuses on arts appreciation.

"Not only textbook: teamwork."

Babson has partnered with the Posse Foundation to help increase diversity on campus, welcoming 10 Posse scholars from urban public high schools in the class of 2009. Still, the campus remains largely white and wealthy, with African Americans making up 3 percent of the student body, Hispanics 7 percent, and Asian Americans 10 percent. Foreign students comprise about 18 percent. Twenty-one percent of Babson students are Massachusetts natives. "I call Babson the mini–United Nations because we have representation from so many countries," says a junior. "People are conservative, yet trendy in dress, and very accepting." No one seems to care about politics, but students do want to end the self-segregation of various ethnic and racial groups on campus. Merit scholarships are available but there are no athletic scholarships.

Babson guarantees housing for four years, and 83 percent of undergraduates live on campus, resulting in high demand for singles and suites. Dorms are air-conditioned and carpeted, and most upper-class rooms have their own bathrooms. Most halls are co-ed, but one dorm is reserved for men, and floors and wings of other buildings are reserved for women. After the first year, the large rooms are assigned by lottery, and standing is based on credits earned. "You can easily switch rooms if you are not happy," says a junior. At the main dining hall, you'll find sushi and make your own stir fry stations; every Wednesday is gourmet night, and the menu may include fresh lobster, Italian specialties, or turkey with all the trimmings.

Students say social life at Babson is on campus during the first two years; after that, most students are 21 and have cars, so they head to the clubs and bars of Boston proper, about 20 minutes away. (It helps that many upperclassmen do not

Babson guarantees housing for four years, and 83 percent of undergraduates live on campus, resulting in high demand for singles and suites.

While making money may be the most popular "sport" at Babson, students recognize the importance of keeping their bodies in competitive condition, too.

have class on Fridays.) For those who aren't of age, or who lack wheels, the Campus Activities Board brings in comedians and organizes parties, as do Greek organizations, which attract 10 percent of the women and men. The school also sponsors trips to Celtics and Red Sox games. If you're caught with booze while underage, or sent to the hospital because of overindulgence, you get one strike; rack up two more, and you're out of the dorms. Popular road trips include the beaches of Cape Cod and Martha's Vineyard, the ski slopes of Vermont and New Hampshire, and the bright lights of New York City and Montreal.

The "very affluent" town of Wellesley has shops and restaurants and there is a subway stop. Students can take the T's Green Line into the city to explore Quincy Market, Fanueil Hall, or the campuses of Harvard, Northeastern, Emerson, and Boston Universities. Wellesley is also home to Wellesley College, and it's not uncommon for Babson students to socialize with Wellesley women; Babson also offers crossregistration at Wellesley, Brandeis, and the Olin College of Engineering.

"I call Babson the mini–United Nations because we have representation from so many countries."

Favorite campus festivals include homecoming (great networking opportunities), Oktoberfest, and Winter and Spring Weekends, when bands come to play and parties are thrown. April 8 is Founders Day, and classes are cancelled so everyone can celebrate entrepreneurship.

While making money may be the most popular "sport" at Babson, students recognize the importance of keeping their bodies in competitive condition, too. Popular intramural sports include volleyball, rugby and ice hockey, and on the varsity level, the Beavers play in NCAA Division III. Any match against archrival Bentley and soccer games against Brandeis and Colby draw crowds, and the baseball and men's basketball teams brought home conference titles recently.

If Babson College students have one complaint, it's the workload. "However, the benefits far exceed the load and so do the three-day weekends," rationalizes a junior. But perhaps that's the point. After all, learning how to balance work with everything else that's important in life is a prerequisite to climbing the corporate ladder. And thanks to small classes, a laser-like focus on all things financial, and plenty of hands-on experience, students leave Babson well equipped to begin scampering up those rungs—without once being subjected to The Donald's hair.

Bard College

Annandale-on-Hudson, NY 12504

Welcome to Nonconformity-Central-on-Hudson. Like Reed on the West Coast, Bard combines unabashed individuality with rigorous traditional academics. More selective than Hampshire and with a better male/female ratio than Sarah Lawrence, Bard is an up-and-comer among the nontraditional liberal arts colleges.

Bard College has come a long way since its 1860 founding, by 12 men studying to enter the seminaries of the Episcopal Church. Those pioneers would no doubt be surprised at the eclectic mix of students now running around Annandale-on-Hudson. "Bard students were the 'weird kids' in high school, and many struggle their first year, when they realize everyone is just as unique as they are, and no one cares what kind of radical statement they are trying to make, because it's not revolutionary when everyone is trying to be different," says a senior. The idea that Bard is strictly a school for artists and social studies majors is slowly disappearing, and the result is a school with lots of intellectual depth—and a higher national profile.

> **"Bard students were the 'weird kids' in high school."**

Bard's campus occupies 600 well-landscaped acres in upstate New York's Washington Irving country. There's no prevailing architectural theme, so each ivy-covered brick building stands out—especially the dorms, which range from cottages in the woods to Russian Colonial in style. Renowned architect Frank Gehry designed the $62 million Fisher Center for the Performing Arts. The 110,00-square-foot facility includes two theaters, and rehearsal and teaching space for Bard's theater and dance programs. The chemistry department has suffered from a lack of up-to-date facilities, but that should change with a new building on the drawing board from another renowned architect, Rafael Vinoly.

Despite Bard's reputation for nonconformism, the list of requirements is extensive. Freshmen show up three weeks before classes start for the Workshop in Language and Thinking, where they read extensively in several genres, and meet in small groups to discuss reading and writing. (A literature major calls L&T "the best three weeks of my life.") Students then take the two-semester First-Year Seminar, which introduces the intellectual, artistic, and cultural ideas at the core of a liberal arts education. At the end of the second year, students write their educational autobiography and declare a major; the autobiography is more like a proposal, presented for discussion to a board of professors in the relevant area. During the junior year, students take a tutorial to prepare for their senior project—while some students run and report on a scientific experiment, or complete a 100- to 300-page critical review of literary works, others write a play or novel or compose a piece of music, and still others organize a show of their own art, or choreograph a dance performance.

> **"Students tend to know who the good professors are and rush to take classes with them."**

Bard's academic climate is "intellectual and consistently challenging," says a senior. "Bard is great for motivated students who take their education into their own hands." Ninety percent of classes have 25 or fewer students, and if students want more individual attention, they can devise a syllabus for their own course and find a professor to sponsor it. Bard also considers visual and performing arts as equal to other academic disciplines; as a result, photography is one of the toughest majors to get into. There are no teaching assistants here, and "students tend to know who the good professors are and rush to take classes with them." With authors such as John Ashbery, Ann Lauterbach, Mark Danner, and Elizabeth Frank teaching creative writing at Bard, literature is among the school's best programs. Film and political science also draw praise. Each fall, science majors may take a class on human disease at New York City's Rockefeller University, which also reserves spots for Bard students as Summer Undergraduate Research Fellows. The distinguished-scientist scholars program offers full tuition to top applicants who plan to major in science or math.

Bard has established what administrators believe is the first collegiate program in human rights, which should improve the sociology major, while the new Vinoly building will strengthen the chemistry major. A five-year, dual-degree conservatory program for music students began in 2005; other new programs include Jewish studies and Middle East studies. Though Bard is far from preprofessional, it does

Website: www.bard.edu
Location: Rural
Total Enrollment: 1,745
Undergraduates: 1,521
Male/Female: 43/57
SAT Ranges: V 640–720
  M 580–690
Financial Aid: 59%
Expense: Pr $ $ $ $
Phi Beta Kappa: No
Applicants: 4,142
Accepted: 32%
Enrolled: 39%
Grad in 6 Years: 72%
Returning Freshmen: 88%
Academics: ✍ ✍ ✍ ✍
Social: ☎ ☎ ☎
Q of L: ★ ★ ★ ★
Admissions: (845) 758-7472
Email Address:
  admission@bard.edu

Strongest Programs:
Political Science
Economics
Human Rights
Fine Arts
Social Sciences
Literature and Languages

At the end of the second year, students write their educational autobiography and declare a major.

offer combined programs with other schools in engineering, architecture, city planning, social work, public health, business and public administration, forestry, and environmental science. Study abroad is available in Russia, China, Greece, India, Senegal, and South Africa; there are also language programs in France, Germany, Italy, and Mexico. Those interested in globalization and international affairs may participate in a residential Bard program in New York City, headquartered near Lincoln Center. The Trustee Leader Scholar program provides grants and support for student-run community service projects.

Bardians take pride in diversity, whether racial, geographical (only 29 percent are New Yorkers) or ideological, though they admit the latter can be lacking. "Bard students are highly motivated, creative, independent, and intellectual," says a senior. Another student adds, "If you're a Republican or conservative, please come and add some dimension to our conversation. I'm sick of agreeing with everyone." African Americans make up 2 percent of the student body, while Hispanics and Asian Americans add 4 percent each. Bard offers academic scholarships but no athletic awards. Under the Excellence and Equal Cost program, qualified high school students may apply to attend Bard for the price of a public-school education in their home state. About 200 students vie for the program's 25 available slots.

Seventy-seven percent of Bard students live on campus, and freshmen are required to do so. Residence halls "vary from small, quiet dorms to large, community-oriented buildings," says a theater major. A classmate adds, "Some are old Victorian mansions, some are new modern buildings that are eco-friendly, one looks like a castle, and others are big cement monsters from the 1950s." Bard still has

"If you're a Republican or conservative, please come and add some dimension to our conversation."

dorms where smoking is allowed, and the cafeteria, Kline, caters to vegetarians and vegans. Room draw can be chaotic, and since juniors and seniors are not guaranteed beds on campus, many upperclassmen move off campus. To help ease their commute, Bard runs a shuttle to the nearby towns of Red Hook and Tivoli. "Tivoli is home to a variety of restaurants (sushi, Mexican, and Cajun), a few bars (the most popular is the Black Swan), a bookstore, and a laundromat," says one student. "Red Hook is home to the Golden Wok (take-out Chinese, a student favorite), the Curry House, grocery stores, Mexican food, and more."

Social life at Bard is almost an afterthought, since New York City is just two to three hours by train. The school does offer concerts and movies, with indie films and alternative rock and hip-hop particularly popular. The Student Activities Board plans Urban Cowboy Night, the Valentine's Day Swing Dance, and Spring Fling, though there are no fraternities or sororities. To the dismay of many, the annual Drag Race party—a celebration of sexuality—has been cancelled, and en masse crossdressing on Parents Weekend is no longer the norm. When it comes to alcohol, policies are focused on safety and respect. "Bard is very liberal," says a student, "but I believe that the administration is trying to stop underage drinking." Bard's hometown of Annandale-on-Hudson is 20 miles from the crafts and antiques meccas of Woodstock and Rhinebeck, and not much farther from the ski slopes of the Catskills and the Berkshires. Having a car helps to prevent occasional attacks of claustrophobia.

Teams compete in the Division III North East Atlantic Conference. Bard is virtually devoid of dedicated jocks, but one student notes, "There are plenty of pseudo jocks and intellectuals in good shape." A third of the students get involved in intramurals, such as floor hockey, bowling, and table tennis, which emphasize participation and fun. Close to campus, five miles of trails stretch through the woods along the Hudson, perfect for everything from raspberry picking to jogging and hiking. "If you like the woods, it's amazing," sighs an anthropology major. "If you like the city, you'll go stir-crazy."

Bard College isn't for everyone, but thanks to the iconoclastic vision of President Leon Botstein, also erstwhile conductor of the American and Jerusalem symphonies and known by everyone simply as "Leon," it now offers strong programs in more than just the arts. "If you show that you have interests, and that you pursue them actively and can explain in an articulate manner what is important to you, Bard will accept you," one student explains. "You don't need perfect grades. You just need an adventurous spirit, an ambitious attitude toward self-improvement, and an ability to evaluate your experiences and capabilities."

> **Overlaps**
>
> NYU, Vassar, Oberlin, Reed, Wesleyan, Sarah Lawrence

If You Apply To ➤ **Bard:** Regular admissions: Jan. 15. Financial aid: Feb. 15. Meets demonstrated need of 31%. Campus interviews: optional, informational. No alumni interviews. SATs and ACTs: optional. SAT IIs: optional. Accepts the Common Application. Essay question.

Barnard College

New York, NY 10027

With applications running double what they were 10 years ago, Barnard has eclipsed Wellesley as the nation's most popular women's college. Barnard women are a little more artsy and a bit more city-ish than their female counterparts at Columbia College. Step outside and you're on Broadway.

Barnard students get the best of both worlds—the small, close-knit atmosphere of a liberal arts school, along with the limitless opportunities of Columbia College, the Ivy League research institution just across the street. Whether they are passionate about art and music or urban studies and politics, women seeking a high-energy environment with top-notch academics are likely to find a niche here. "With Barnard women, I can discuss Van Gogh's importance in the scheme of 19th-century art and then move on to debate whether or not Lars Von Trier is misogynistic," says one first-year at this New York City women's college.

Barnard's campus is on the Upper West Side of Manhattan, in the Morningside Heights neighborhood. It's just blocks from Riverside Drive, which has a lovely path parallel to the Hudson River for running, biking, or rollerblading. Trees and other greenery shade grand prewar apartment buildings, and grassy medians break up the wide expanse of Broadway itself. Barnard's architecturally diverse buildings are more modern than Columbia's, and in recent years, the college has invested to upgrade labs, classrooms, animal research facilities, and the residence halls with new bathrooms, heat and air-conditioning systems, elevators,

> "With Barnard women, I can discuss Van Gogh's importance in the scheme of 19th-century art and then move on to debate whether or not Lars Von Trier is misogynistic."

and windows. Several classrooms have also been upgraded with multimedia capabilities, and a new state-of-the-art organic chemistry lab was built. A campus master plan has been commissioned to focus on future space and building needs.

Barnard's requirements are designed to reflect the changing nature of our technological society, and the fact that graduates are increasingly pursuing law, business, and other professions rather than academic careers. To that end, students take two first-year foundation courses and fulfill nine "area requirements" in reason and

Website: www.barnard.edu
Location: Urban
Total Enrollment: 2,287
Undergraduates: 2,287
Male/Female: 0/100
SAT Ranges: V 650–730
  M 620–700
ACT Range: 28–31
Financial Aid: 56%
Expense: Pr $ $ $
Phi Beta Kappa: Yes
Applicants: 4,380
Accepted: 27%
Enrolled: 46%
Grad in 6 Years: 89%
Returning Freshmen: 96%
Academics: 🎓🎓🎓🎓🎓
Social: 🐿🐿🐿
Q of L: ★★★
Admissions: (212) 854-2014
Email Address:
  admissions@barnard.edu

Strongest Programs:
Psychology

value, social analysis, historical studies, cultures in comparison, laboratory science, quantitative and deductive reasoning, language, literature, and the visual and performing arts. Students must also take two physical education courses, though most get plenty of exercise running for the bus or dashing up and down subway stairs.

While curricular requirements guarantee Barnard graduates have intellectual breadth, the mandatory senior thesis project or comprehensive examination ensures academic depth. Barnard students may crossregister at Columbia if they find more courses of interest there, or enroll in graduate courses in any of Columbia's schools. Dual-degree and joint-degree programs are also available with Columbia and the Jewish Theological Seminary, and music students may also take classes at Juilliard and the Manhattan School of Music. Another innovative program offers women the chance to concentrate on dance, music, theater, visual arts, or writing—while also completing a degree in liberal arts. The school also offers study abroad opportunities at Oxford, Cambridge, the University of London, and other institutions in England, France, Germany, Italy, Japan, and other countries around the world.

Barnard's most popular majors are psychology, English, economics, political science and art history, all of which happen to be among the school's best departments. Also well-subscribed are biology (there's a healthy contingent of premeds) and history. "I think all departments are strong, and there is a department for everybody," says one enthusiastic student. "While renowned for our English and political science departments, even lesser-known areas, such as American studies, have phenomenal professors who truly care about their students," boasts a peer. Women's studies and education also draw praise, though students in these programs must also choose another major. The administration cites statistics and linguistics as weaker programs.

Barnard shares two co-ed dorms with Columbia.

Many students come to Barnard because of its low student/faculty ratio. Seventy percent of classes taken by first-years have no more than 25 students. Another plus: Barnard has no graduate teaching assistants. In fact, Barnard professors enjoy Columbia's proximity almost as much as undergraduates, and each year one-third of the full-time faculty teaches in graduate departments throughout the university. Still, faculty members focus on their teaching responsibilities first. "These are people who, in typical Barnard fashion, are not just amazing educators, but all-around wildly successful people in their individual fields, and are more like Renaissance men (and women) than teachers," says one student. Undergraduate research is also a priority at Barnard. The Centennial Scholars program allows students to conduct extensive research under the guidance of a faculty member for up to three semesters. The innovative Mentor Center program matches students with alumnae who have agreed to help with internships and career advisement.

> **"Barnard students are known to be creative thinkers, who think 'out of the box.'"**

Barnard's requirements are designed to reflect the changing nature of our technological society, and the fact that graduates are increasingly pursuing law, business, and other professions rather than academic careers.

Thirty-three percent of Barnard students are New York natives, including a sizable contingent from the East Side of Manhattan, and 48 percent attended private high school. A whopping 93 percent ranked in the top quarter of their high school class. Asian Americans make up 17 percent of the student body, while African Americans add 5 percent and Hispanics make up another 7 percent. Barnard now competes head-to-head with Columbia in admissions, an interesting dilemma because Barnard is just another division of Columbia University, similar to the engineering school, the medical school, the business school—or Columbia College. In general, women looking for a more traditional "rah-rah" experience may prefer Columbia. Those seeking flexibility might do better at Barnard, a hotbed of liberalism where students don't shy away from rallies and protests. "Barnard students are known to be creative thinkers, who think 'out of the box,'" a junior says. "They are socially aware, compassionate, passionate, highly intelligent women." Women's rights, the war in Iraq, and race relations are among the issues that have gotten students stirred up recently. Barnard students do share a first-year orientation program with Columbia, where they mix together in small groups and take tours of the campus and city. Students can also take part in a preorientation backpacking trip.

Ninety percent of Barnard students live in the dorms, which have come a long way since the college's beginning as a commuter school: There's an 18-story Barnard dormitory tower, plus one dorm complex and five off-campus apartment buildings; nonresidents must be signed in by a resident, and entries are always guarded, so students say they feel safe. In addition, Barnard shares two co-ed dorms with Columbia. With New York's notoriously high rents and broker fees, demand for dorm beds remains high. The Quad, which houses all first-years, is pretty well-kept, although individual floors, bathrooms, and common areas are "only as clean as the people who use them," according to one student. Seniors get the best rooms through a lottery system, though an undecided major says, "Res Life can be very unorganized and difficult to work with." Housing is guaranteed as long as you stay in the dorms; choose to move out for a year, and you may have difficulty getting back in. Dorm-dwellers must buy a meal plan, which may also be used at Columbia's John Jay cafeteria, though students say Barnard's food is better. "Though by no means gourmet, the variety and availability makes up for the mediocre quality," says a senior. "They make provisions for literally every special taste," another senior adds.

> **"RAs build community programming within the residence halls but usually students opt to explore New York."**

When it comes to social life, students tend to divide their time between on campus and off. "RAs build community programming within the residence halls but usually students opt to explore New York," a junior says. Traditions on campus include Midnight Breakfast the night before finals begin, when deans and administrators serve up eggs and waffles in the gym. Women in the arts are celebrated in the annual Winterfest, while the Books Etc. series brought Barnard alumnae authors like Jhumpa Lahiri '89 and Anna Quindlen '74 back to campus for lectures in the fall of 2003. Less academic pursuits are available, too. "If you still don't like what you see, you can head out to one of the bars in Morningside Heights," a student says. Fake IDs are easy to come by in Manhattan, and while they often work, Barnard students aren't focused on drinking themselves into oblivion. "Because there is so much else to do in NYC, alcohol is not a primary amusement," explains an English and creative writing double major. Many of the city's offerings are free to students with their school ID.

> **"I think all departments are strong, and there is a department for everybody."**

Road trips are infrequent—as not many students have cars—but when they happen, destinations range from Washington, D.C., to Boston, easily reached by train and plane, to skiing and snowboarding in Vermont, or spring break on the beaches of South Carolina.

Barnard athletes compete alongside their peers enrolled at Columbia, and the field hockey, soccer, lacrosse, archery, and crew teams have the largest number of participants. The fencing team is also strong. Columbia's marvelous gym and co-ed intramurals are also available to Barnard women, but many prefer to exercise their minds.

How do students see Barnard? As an "all women's college, located in a prime city, affiliated with a large research institution, with distinguished faculty, alums, and intelligent and driven students," one proud Barnard woman says. For some, the supportive community created by those students and faculty is what makes this a special place. "Barnard is different because although the students here are really intelligent, they are also really down to earth, and are supported by each other and the academic community," another student says.

Barnard athletes compete alongside their peers enrolled at Columbia, and the field hockey, soccer, lacrosse, archery, and crew teams have the largest number of participants.

Housing is guaranteed as long as you stay in the dorms; choose to move out for a year, and you may have difficulty getting back in.

Overlaps
NYU, Columbia, Brown, Yale, Vassar, Harvard

Bates College

23 Campus Avenue, Lewiston, ME 04240

Bowdoin got rid of its frats, Bates never had them, and therein hangs a tale. With its long-held tradition of egalitarianism, Bates is a kindred spirit to Quaker institutions such as Haverford and Swarthmore. A month-long spring term helps make Bates a leader in studying abroad.

Website: www.bates.edu
Location: Small city
Total Enrollment: 1,684
Undergraduates: 1,684
Male/Female: 49/51
SAT Ranges: V 640–710
  M 640–700
Financial Aid: 40%
Expense: Pr $ $ $ $
Phi Beta Kappa: Yes
Applicants: 4,356
Accepted: 29%
Enrolled: 39%
Grad in 6 Years: 89%
Returning Freshmen: 94%
Academics: 🖉 🖉 🖉 🖉 ½
Social: ☎ ☎ ☎
Q of L: ★ ★ ★
Admissions: (207) 786-6000
Email Address:
  admissions@bates.edu

Strongest Programs:
Economics
Biology
Psychology
History
Political Science

While Bates, Bowdoin, and Colby share a rugged Maine location, a mania for ice hockey, and popularity among prep-schoolers clad in Patagonia and L.L. Bean, Bates has more in common ideologically with Quaker colleges such as Haverford and Swarthmore. Founded by abolitionists in 1855, Bates takes pride in its heritage as a haven for seekers of guidance, freedom, and justice. Its 4–4–1 calendar offers ample opportunity for study abroad, even for just one month at year's end. The school's small size also means student/faculty interaction is plentiful, and close friendships are easily formed.

The Bates campus features a mix of Georgian and Federal buildings and Victorian homes spread out over the grassy lawns of Lewiston. Renovations to Merrill Gymnasium have added two dance studios, a cardiovascular workout room, a multipurpose room, and an ergometer room. A 150-bed residential hall is slated to open this year and is designed as a residential cluster, which is a hybrid of traditional dorm and suite layouts. In addition, a new dining commons will open shortly and will serve as the central dining facility for all students.

> **"One unique part of Bates is that just about all seniors write a thesis."**

Bates emphasizes a broad-based education in the liberal arts that encompasses the humanities, sciences and mathematics, social sciences, and the arts. Although there are no core course requirements, students are expected to select a major and two thematic concentrations, each consisting of four interrelated courses structured on the basis of a central organizing principle. These concentrations may fall within one department or program, or may focus on a particular topic or area of inquiry designed by faculty from different disciplines. "One unique part of Bates is that just about all seniors write a thesis," says a chemistry major. "Some are semester-long, while others are year-long, depending on the department and what you want to do." The Ladd Library is often crowded, as 85 percent of students write a thesis or produce an equivalent research, service, performance, or studio project. Ladd has almost 600,000 volumes, plus an all-night study room, computer labs, and an audio/visual room with everything from Bach to Bruce Springsteen.

While Bates doesn't require standardized tests for admission, don't expect to coast through. "Bates is a very rigorous college," says a junior. "The students tend to be hard workers and driven in the classroom." The most popular majors include economics, political science, psychology, and English, and these are also among

Bates' best. The music and art departments benefit from the Olin Arts Center, which houses a performance hall, gallery, recording studio, art studios, and practice rooms. Interdisciplinary programs at Bates include American cultural studies, neuroscience, and women and gender studies. Professors teach all courses, including lab and discussion sections. "The professors here really love the subject they are teaching and want the students to fall in love with it, too," says one student. "They make the material come alive and learning becomes fun."

Barbecues and clambakes are big when the weather is nice, and the annual Winter Carnival includes ice skating, snow sculpting, and a semiformal dance.

Bates offers study abroad opportunities in more than 70 foreign locations such as China, Japan, Croatia, Chile, and Austria, and more than two-thirds of students take advantage of them. The school also participates in the Washington Semester,* the 10-college Venture Program,* and the American Maritime Studies Program at Mystic Seaport*—all attractive options for students seeking real-world experience. Bates' 4-4-1 calendar allows for a five week short term at the end of the academic year and students may use this term to focus on a single subject of interest, frequently off campus. Recent examples include marine biological studies at stations on the coast of Maine; art, theater, and music studies in New York City and Europe; and field projects in economics, sociology, and psychology.

"Almost every big political headline creates a stir with Batesies."

Eighty-three percent of Bates students come from outside Maine, many from Massachusetts, Connecticut, and New York. "The students are friendly, outgoing, and just seem to love life in general," a student says. "Students do not feel that it is their right to be at the college, but rather a privilege." While the administration is trying to make Bates more diverse, minorities remain a tiny fragment of the student population, with African Americans adding 3 percent, Hispanics 2 percent and Asian Americans 4 percent. "Almost every big political headline creates a stir with Batesies," says a sophomore, noting that most students are "very liberal." There are no merit or athletic scholarships available, although the college does guarantee to meet the demonstrated need of all students.

Bates offers study abroad opportunities in more than 70 foreign locations such as China, Japan, Croatia, Chile, and Austria, and more than two-thirds of students take advantage of them.

All but 7 percent of Bates students live on campus. "Most of the dorms and houses are very large and comfortable," a student says, "although some rooms can be a little cramped." Housing is guaranteed for four years, but awarded by lottery after the first year. Food in the Commons, where everyone eats, is "great, but it gets monotonous," says a biology major. Students say campus security is visible and more than adequate. "Given the nature of a small campus, students know a majority of the officer's names and do not feel intimidated to approach them about a problem," explains a history major.

Since there's not much to do in Lewiston, parties, concerts, and other weekend diversions mostly occur on campus. "Whether it is parties, comedians, movies, or bands, there is always something to do for everyone on campus," a freshman says. Without a Greek system, college alcohol policies are fairly loose, student says, and a ban on hard liquor is often ignored. Barbecues and clambakes are big when the weather is nice, and the annual Winter Carnival includes ice skating, snow sculpting, and a semiformal dance. During the St. Patrick's Day Puddle Jump, students of Irish descent—and all those who want to

"Most of the dorms and houses are very large and comfortable."

be Irish for the day—cut a hole in the ice on Lake Andrews and plunge in. Students with cars can easily road trip to the outlet stores in Freeport and Kittery, Maine. Other popular destinations include Bar Harbor in Acadia National Park, or "Portland, for great food," says a senior. Montreal and Boston are not far, and neither are the ski slopes of Vermont and New Hampshire.

Bates' varsity teams compete in Division III, except for the ski team, which is Division I. Everyone gets excited for matches against Bowdoin and Colby, especially when they involve ice hockey. Basketball, football, and lacrosse are also popular among spectators. The intramural program, organized by the students and supervised by faculty members, is "strong and spirited," and attracts a large number of students. Intramural softball is popular, along with ultimate Frisbee, soccer, and basketball.

If you can stand the cold and the silent, starry nights, Bates may be a good choice. With caring professors, a small student body, and a focus on the liberal arts, students quickly become big fans. "Bates feels like home."

If You Apply To ➤

Bates: Early decision: Nov. 15, Jan. 1. Regular admissions: Jan. 1. Financial aid: Feb. 1. Guarantees to meet full demonstrated need. Campus and alumni interviews: recommended, evaluative. SATs and ACTs: optional. SAT IIs: optional. Accepts the Common Application and electronic applications. Essay question: Why Bates? Common Application essays.

Baylor University

Waco, TX 76798

BEST BUY

Come to Baylor and Mom can rest easy. Baylor is a Christian university in the Baptist tradition, which means less of the debauchery prevalent at many schools. Dad will like it, too: Baylor is one of the least expensive private universities in the country. Prayers come in handy on the football field, where the team is still praying to beat UT.

Website: www.baylor.edu
Location: Center city
Total Enrollment: 13,261
Undergraduates: 11,465
Male/Female: 42/58
SAT Ranges: V 540–650
  M 550–660
ACT Range: 22–27
Financial Aid: 51%
Expense: Pr $
Phi Beta Kappa: Yes
Applicants: 15,443
Accepted: 66%
Enrolled: 31%
Grad in 6 Years: 72%
Returning Freshmen: 83%
Academics: 🐽 🐽 🐽
Social: ☎ ☎ ☎
Q of L: ★ ★ ★
Admissions: (254) 710-3435
Email Address:
  Admissions@Baylor.edu

Strongest Programs:
Premed

Baylor University offers students a solid Christian-influenced education at a bargain price. The university's Baptist tradition fosters a strong sense of community among students and faculty, and the school's 2012 vision plan promises a slew of strategic changes such as lowering the student/teacher ratio and building new residence halls. It's a challenging mission, and one that students here are eager to accept. Without a doubt, success at Baylor requires a healthy dose of faith and fortitude.

The 432-acre Baylor campus, nicknamed Jerusalem on the Brazos, abuts the historic Brazos River near downtown Waco, Texas (population 110,000). The architectural style emphasizes the gracious tradition of the Old South, and the central part of campus, the quadrangle, was built when Baylor moved from Independence, Texas, in 1886. The campus has been witness to a number of renovations and new construction, including renovations to the Baylor Bookstore and the new Ed Crenshaw Student Foundation building.

Students pursue their major in arts and sciences or one of Baylor's five other schools: business, education, engineering and computer sciences, music, and nursing. "The Hankamer School of Business is always ranked among the top business schools in the nation," says a sophomore. "It offers students great insight into the world of business and ethics." Pre-med also draws praise. Core requirements include four English courses and four semesters of human performance. All students also take two religion courses and two semesters of Chapel Forum, a series of lectures and meetings on various issues or Christian testimonies. The Honors College oversees the honors program (which offers opportunities for course integration and independent research) and the University Scholars Program (which waives most distribution requirements). Students may major or minor in Great Texts, an interdisciplinary program exploring "the richness and diversity of the Western intellectual heritage."

> "Students at Baylor have a good sense of where they are in life."

Business is Baylor's most popular major, followed by biology, psychology, political science, and journalism. More unusual options include church-state studies and museum studies, and institutes focusing on environmental studies and childhood learning disorders. The archeology and geology departments benefit from fossil- and mineral-rich Texas prairies. An increasing number of Baylor students are traveling on study abroad programs, which send them to 15 countries, including England, Mexico, and the Netherlands. New undergraduate majors include exercise physiology, and new concentrations include church music, fabric design, and studio art. The BA in business administration has been discontinued.

One of Baylor's greatest strengths is the sense of campus community, fostered by the emphasis on Christianity and by the administration's efforts to focus faculty members on teaching, rather than on research and other activities, students say. Baylor also strives to keep classes small—students say most have 50 or fewer enrolled. "The majority of classes

> **"Baylor is a very conservative and Republican campus, especially being in such close proximity to Bush's Crawford ranch."**

demand time, effort, and hard work," says one student. "However, classes are enjoyable with the right professor." Full professors often teach freshmen courses, although "grad students are present to grade papers and hold study sessions," says a junior.

"Students at Baylor have a good sense of where they are in life," says an accounting major. "They are humble, eager, and waiting for an opportunity to improve themselves." They are also primarily middle- to upper-middle-class, white, and Christian. Eighty percent are Texans. Minorities total about one-fourth of the student body, with Hispanics the largest group at 10 percent, Asian Americans at 7 percent, and African Americans at 8 percent. "Baylor is a very conservative and Republican campus," says a sophomore, "especially being in such close proximity to Bush's Crawford ranch." Students vie for numerous merit scholarships and nearly 340 athletic scholarships.

As might be expected on a conservative and religious campus like Baylor's, dorms are single sex and have restrictive visitation privileges, which is a big complaint among the 34 percent of students who call them home. "The housing situation varies by hall," confides one junior. "Some are relatively new and well maintained. Others have air conditioning issues and room sizes that leave something to be desired." Upperclassmen look off campus for cheaper housing with private rooms and fewer rules, but there is a push for more students to stay on campus

> **"Social life is vibrant during the weekdays, but dramatically reduced on the weekends."**

with the construction of a newer residence hall with apartment-style rooms, notes one co-ed. Those who've moved out of the dorms appreciate the presence of 20 fully commissioned police officers patrolling the area on bikes and in patrol cars. "Safety on campus is generally good," says a student.

Thirteen percent of Baylor's men and 17 percent of the women belong to a fraternity or sorority, providing a party scene for those who want it, as well as community service outlets. With so many students residing off campus, the social life is decent, but not a party atmosphere. "Social life is vibrant during the weekdays, but dramatically reduced on the weekends," a sophomore says. Easy road trips include Dallas, Austin, San Antonio, Bryan/College Station, and beaches at Galveston, South Padre Island, and Corpus Christi. Most destinations are within a two-and-a-half hour drive, students say, making a set of wheels a big help, if not a necessity.

Alcohol isn't served on campus or at campus-sponsored events. "Alcohol is much less prevalent at Baylor than at most schools," says a senior. However, another student adds, "There are enough opportunities to get alcohol off campus." Highlights of Baylor's social calendar include the weekly Dr. Pepper Hour with free soda

(Continued)
Strongest Programs:
Communications Sciences and
  Disorders
Theater Arts
Baylor Interdisciplinary
  Program
Accounting
Entrepreneurship

One of Baylor's greatest strengths is the sense of campus community, fostered by the emphasis on Christianity and by the administration's efforts to focus faculty members on teaching, rather than on research and other activities.

As might be expected on a conservative and religious campus like Baylor's, dorms are single sex and have restrictive visitation privileges, which is a big complaint among the 34 percent of students who call them home.

floats and the Dia del Oso (Day of the Bear), when classes are cancelled for a day in April in favor of a campuswide carnival. The Fiesta on the River, organized by the Residence Hall Association, lets student-run organizations set up booths to raise funds for the causes they support. The school also has the largest collegiate homecoming parade in the nation.

When it comes to football, remember: You're in Texas. Freshmen wear team jerseys to games and take the field before the players, then sit together as a pack. "It's a very awesome part of the freshman experience," one student says. Baylor is in Division I, and both the men's and women's tennis teams brought home recent championships. Women's track and field also draws fans. For weekend warriors, the McLane Student Life Center offers the tallest rock-climbing wall in Texas. The university maintains a small marina for swimming and paddle boating, and several lakes with good beaches are nearby. Generally, though, religious groups are more popular than the intramural sports program, with chapters of Campus Crusade for Christ and the Fellowship of Christian Athletes very much alive and well.

While keeping true to its traditional Christian roots, Baylor recognizes it must remain open to new ways of thinking to achieve its goal of becoming a top-tier university by 2012. Baylor students may party less than their counterparts at other Texas schools. Instead, they focus on academics, spiritual nourishment, community involvement, and finding their vocational calling. "Get involved in a couple of organizations your freshman year and get out of your comfort zone," says a junior. "This is college!"

Overlaps

University of Texas–Austin, Texas A&M, Texas Tech, Texas Christian, University of North Texas, Texas State

If You Apply To ➤

Baylor: Rolling admissions. Guarantees to meet demonstrated need of 19%. SATs or ACTs: required. SAT IIs: optional. Campus interviews: optional, informational. No alumni interviews. Accepts the Common Application and prefers electronic applications. Essay question.

Beloit College

700 College Street, Beloit, WI 53511

Tiny Midwestern college known for freethinking students and international focus. Has steered back toward the mainstream after its heyday as an alternative school in the '60s and '70s. Wisconsin location makes Beloit easier to get into than comparable schools in sexier places.

Website: www.beloit.edu
Location: Small city
Total Enrollment: 1,320
Undergraduates: 1,320
Male/Female: 41/59
SAT Ranges: V 580–700
   M 560–660
ACT Range: 25–29
Financial Aid: 77%
Expense: Pr $ $
Phi Beta Kappa: Yes

Beloit College urges students to "Invent Yourself," encouraging intellectual curiosity and personal initiative by giving students freedom to explore. Known for attracting liberal freethinkers in the 1960s and '70s, the school is now steering back toward the mainstream. "The joke around campus is you can be anything here but a Republican," says a senior. What hasn't changed is its emphasis on tolerance, understanding, and the world beyond the United States. Beloit students are "taught to have a global perspective and encouraged to try new things," says a freshman. "Conformity is not encouraged."

Beloit's 40-acre campus is a Northeastern-style oasis an hour's drive from Madison and Milwaukee and 90 minutes from Chicago. Academic and administrative buildings sit on one side, with residence halls on the other. Two architectural themes dominate, says one student: "1850s Colonial and obtuse 1930s." The college's Turtle

Creek Bookstore is located three blocks away in downtown Beloit. A cozy coffee bar, selection of general books and magazines, and a patio for relaxing, reading, or studying augment the typical stacks of textbooks. As part of a new campus master plan, two new projects will be completed by late 2008, including a new science center and renovations to the World Affairs Center. Renovations are also being made to the Sports Center and Strong Stadium.

All Beloit freshmen complete a First Year Initiative (FYI) seminar led by a faculty member who serves as their advisor until they declare majors the next year. Courses under the recent FYI theme "Where have you been? Where are 'we' going?" included "'Hell No, We Won't Go!': Ritual Chant in the Modern World" and "Lions and Tigers and Bears! Democracy, Development, and Disenchantment in a Rapidly Changing World." The FYI program flows into the Sophomore Year Program, which offers a retreat, Exploration Week, and the charting of a Comprehensive Academic Plan for the final two years, which often includes internships, research, and off-campus study. If fact, 80 percent of Beloit students undertake some type of independent project, presenting their findings in the annual Student Symposium.

> **"The joke around campus is you can be anything here but a Republican."**

"Courses are challenging and very engaging," says one student, "but students are not competitive with one another." All students must also complete three writing courses, at least two units in each of three subject areas (natural science and math, social sciences, and arts and humanities), and a unit of interdisciplinary studies. Students are likewise required to complete at least two units involving a different culture or language, or focusing on relations between nations.

Teaching is the faculty's first priority, and 90 percent of classes have 25 or fewer students. "Professors here truly care," a student says. "Personal attention is the norm." Another adds that professors are "more than happy to jump in and work with students." English is the most popular major, followed by psychology, biology, sociology, economics, and political science. Among more unusual options are a museum studies minor, with hands-on restoration experience, and the rhetoric and discourse major, which asks students to reflect on current nonfiction writing while

> **"Personal attention is the norm."**

producing their own prose. Classics is Beloit's one weak spot, administrators say, and students cite education as less than stellar. Environmental studies has been added as a major and a minor, and the education studies major is now known as education and youth studies.

To satisfy Beloit's experiential learning and global diversity requirements, Venture Grants offer $500 to $1,500 for "entrepreneurial, self-testing activities" that benefit the community; recent awardees traveled to Sedona, Arizona to collect stories of extraterrestrial contact and studied peace and conflict resolution in Washington, D.C. through the Boabab Tree Project, Inc. An internship program provides up to $2,000 for projects that address a community need, and students may conduct biological and biomedical research at Northwestern and Rush Universities in Chicago. More than half of Beloit's students study or do research abroad through the International Education Program, which sends students to 35 countries. The Center for Language Studies complements Beloit's own foreign language programs with intensive summer study in Chinese, Czech, Japanese, Russian, Portuguese, Spanish, and English as a second language. Beloit is also a member of the Associated Colleges of the Midwest* consortium, increasing students' choices.

Students here are "generally down to earth, friendly, and a bit quirky," says a sophomore. Nineteen percent of Beloit's student body is homegrown; 81 percent hail from out-of-state, including 7 percent from abroad. Hispanics comprise 2 percent of the total, African Americans 3 percent, and Asian Americans represent 3 percent.

(Continued)
Applicants: 2,045
Accepted: 64%
Enrolled: 25%
Grad in 6 Years: 72%
Returning Freshmen: 89%
Academics: ✐ ✐ ✐
Social: ☎ ☎ ☎
Q of L: ★ ★ ★ ★
Admissions: (608) 363-2500
Email Address:
  admiss@beloit.edu

Strongest Programs:
Anthropology
Economics
English
Geology
Modern Languages and
  Literatures
Natural and Social Sciences
Theatre Arts

Sports at Beloit are played more for fun than glory.

Hot-button issues include "sexual assault awareness, volunteerism, and peace and justice rallies," says a student. Liberalism is the norm here, much to the chagrin of some. "We need to discuss conservative ideas before silencing them outright," says a junior. Merit scholarships worth an average of $13,410 are available.

Ninety-three percent of Beloit students live on campus, where they're required to remain for three years. The Haven, Wood, and 815 residence halls boast new carpeting, furniture, and central air-conditioning and heat; one student says other good choices include Chapin and Aldrich. Four fraternities attract 15 percent of the men, and two sororities draw 5 percent of the women; members may live in their chapter houses. Special-interest houses cater to those interested in foreign languages, music, anthropology, and other disciplines. A sophomore says food is "not stellar" and "very Midwestern—a lot of meat and potatoes," although vegetarian and vegan options are available.

"No one leaves on the weekend," says one student. "There is too much good stuff going on on campus." This "good stuff" includes on-campus movies, concerts, art shows, dance recitals, and parties at the frats. In addition, special-interest houses tie up many Friday and Saturday nights at Beloit. Two all-campus festivals liven up

> **"No one leaves on the weekend. There is too much good stuff going on on campus."**

the calendar: the Folk and Blues Fall Music Festival brings jazz, reggae, folk, and blues bands to campus, while on Spring Day, classes give way to concerts and everyone kicks back to enjoy the (finally!) warmer weather. The Beloit Science Fiction/Fantasy Association offers movie marathons, dramatic readings, roleplaying games, board and video games, and other activities for members and nonmembers alike. The school's alcohol policy is lax, students say. "We have an alcohol philosophy," says a sociology major. "Beloit would never go so far as to say 'policy.'"

Having wheels here will definitely raise your social standing, as they make it easier to take off from secluded Beloit for Chicago or the college town of Madison, also easily reached through a cheap local bus service. For the outdoors-minded, the nearby Dells offer camping and water parks. "Beloit is the perfect college town in that it encourages students to stay on campus and develop a strong community." That said, the town has basic necessities, such as a few bars (check out the excellent burgers at Hanson's Pub, says a senior), a bowling alley, a movie theater, and a Wal-Mart. "Groups such as Habitat for Humanity, Beloit Interaction Committee, and the Outreach Center work hard to integrate students into the community," says a sophomore.

Sports at Beloit are played more for fun than glory. Among the school's Division III squads, standouts include baseball, men's and women's basketball, men's and women's cross-country, and football, especially against rival Ripon College. The women's tennis team won Midwest Conference championships recently, and intramural ultimate Frisbee typically draws hundreds of players and spectators. Two-thirds of students participate in the intramural program, with basketball, volleyball, and soccer also popular.

Beloit is a bundle of contradictions: a small liberal arts college in the heart of Big Ten state university country, where the academic program has an East Coast rigor but the laid-back classroom vibe reflects the free-and-easy spirit of the Midwest. "Beloit College is not for the student who wants to be told what classes to take," warns a sophomore. "Beloit is for the student who wants to be asked, 'What do you want to do?'"

Overlaps

Lawrence, Knox, Oberlin, Grinnell, Macalester, Earlham, Carleton

If You Apply To ➤

Beloit: Early action: Dec. 1. Rolling admissions: Jan. 15. Financial aid: Mar. 1. Guarantees to meet full demonstrated need. Campus interviews: recommended, informational. Alumni interviews: optional, informational. SATs or ACTs: required. SAT IIs: optional. Accepts the Common Application and prefers electronic applications. Essay question: How you heard about Beloit; what factors led you to apply.

Bennington College

Bennington, VT 05201-6003

Known for top-notch performing arts and lavish attention to every student. Arts programs rely heavily on faculty who are practitioners in their field. Less competitive than Bard and Sarah Lawrence, comparable to Hampshire. Enrollment is robust after a dip in the '90s.

Bennington College is "a crazy bunch of absolutely brilliant geniuses," says one art and Spanish major. A school where architects are teachers, biologists sculpt, and a sociologist might work on Wall Street or in graphic design, it's no wonder they strive to abandon the theory of regimented knowledge. Bennington's focus is on learning by doing. The emphasis on self-direction, field work, and personal relationships with professors sets it apart even from other liberal arts colleges of similar (small) size. "We like to make our own paths and are in a constant state of evaluation and reflection," says one junior.

Bennington sits on 470 acres at the foot of Vermont's Green Mountains. The campus was once an active dairy farm, and a converted barn houses the main classroom and administrative spaces. But don't let the quaint New England setting fool you. The Dickinson Science Building offers high-tech equipment for aspiring chemists, biologists, environmental scientists, and geneticists. The building is also home to a media lab dedicated to the study of languages, including Chinese, French, Italian, Japanese, and Spanish. A new 7,500-square-foot student center offers students a snack bar, grill, convenience store, and multipurpose spaces.

> "A crazy bunch of absolutely brilliant geniuses."

Thanks to its focus on experiential learning, Bennington's academic structure differs from that of a typical college or university. Each student designs a major, although there are some academic requirements, including a seven-week internship each January and February in a field of interest and a location of the student's choice. "Our professors are so involved in their teaching that students can't help but be sucked in," an architecture major explains. Even without grades, students push themselves to learn, grow, and achieve. "While there is no sense of overwhelming competitiveness within the community, students push themselves creatively and academically," says a junior.

Without academic departments, the faculty works to provide students with a well-rounded academic foundation. The most popular area of study is visual and performing arts, followed by English, social sciences, math, and liberal arts. "Students fight tooth and nail for literature classes, but that is due to the fact that they come here for the literature department," says one sophomore. Social and biological sciences are likewise popular. For those undergrads seeking to deepen their understanding of the world, there's the Democracy Project, where students explore and experience democracy in its historical, philosophical, political, cultural,

> "Students fight tooth and nail for literature classes."

Website:
  www.bennington.edu
Location: Rural
Total Enrollment: 703
Undergraduates: 567
Male/Female: 34/66
SAT Ranges: V 610–700
  M 540–640
ACT Range: 25-28
Financial Aid: 69%
Expense: Pr $ $ $ $
Phi Beta Kappa: No
Applicants: 723
Accepted: 62%
Enrolled: 31%
Grad in 6 Years: 66%
Returning Freshmen: 87%
Academics: ✍ ✍ ✍
Social: ☎ ☎ ☎
Q of L: ★ ★ ★ ★
Admissions: (800) 833-6845
Email Address: admissions
  @bennington.edu

Strongest Programs:
Literature
Writing
Visual Arts
Drama
Music
Dance
Sciences

and social dimensions. Led by scholars and practitioners, the project draws on a curriculum that provides a broad range of perspectives and intellectual orientations.

"Everyone at Bennington is an artist, either in practice or philosophy," says one student. Another adds, "Bennington students are self-motivated, creative, innovative, deeply individual, and good humored." Four percent are from Vermont and 3

Each student designs a major, although there are some academic requirements, including a seven-week internship each January and February in a field of interest and a location of the student's choice.

> **"Everyone at Bennington is an artist, either in practice or philosophy."**

percent are foreign nationals. One literature major describes students as "vocal, engaged, opinionated, political, artistic, and very individualistic." Curiosity and excitement about exploration and experimentation will take you far here, and if you lean liberal in the voting booth, so much the better. While only moderately active, Bennington students are characterized by one freshman as "mostly anti-Bush, pro-choice, and environmentally conscious." Together, African Americans, Hispanics, and Asian Americans account for 8 percent of the student body. Merit scholarships worth an average of $7,498 are awarded annually; there are no athletic scholarships available.

As Bennington lacks traditional departments, requirements, and even faculty tenure, it's probably not surprising that the school also lacks dorms. Ninety-eight percent of students live in one of the college's 18 co-ed houses; 12 are white New England clapboard and three are more modern. Each holds 25 to 30 people with an elected chair to govern house affairs. "Our houses are houses, not dorms," says a junior. "They're spacious, well furnished, and comfortable." Freshmen and sophomores share rooms—those of the opposite sex may live together if both parties request it—and juniors and seniors are often placed in singles. The college food service provides plenty of options, from vegetarian and vegan choices to a salad bar, wok station, and pizza machine. "As college food goes, it's some of the best," says one student. "There are plenty of vegan and vegetarian options, large salad and sandwich bars, even woks to make your own stir-fry." Security is good and "the campus is extremely secure and virtually free of all crime," confides a literature major.

Although the vibe on Bennington's campus is liberal, sophisticated, and cosmopolitan, the neighboring town of the same name—four miles away—is far more conservative, typical of rural New England. "The town of Bennington is small and quaint with a few coffee shops, some specialty shops, an art store, and a few grocery

> **"Our houses are houses, not dorms."**

stores," says a freshman. Students are trying to mend the town/gown rift through volunteer work in local schools and homeless shelters, though such programs can be tough because of the mandatory midyear internship term, which takes many students away from campus.

Social life centers on rehearsals, performances, films, and lectures. "There is always something going on," assures one freshman. That said, the annual theme parties always draw raves—recent themes have included Gatsby's Funeral and Mods vs. Rockers. The alcohol policies on campus are fairly standard: no one under 21 is allowed to imbibe. "Underage drinkers are usually busted at parties," says a student. Road trips to Montreal and New York are fun but infrequent diversions.

Without academic departments, the faculty works to provide students with a well-rounded academic foundation.

Given Bennington's rugged location, hiking, rock climbing, caving, camping, and canoeing keep students moving. Ski slopes beckon in the colder months. Twice each year, the college turns part of its huge Visual and Performing Arts complex into an indoor roller rink for a Rollerama party. For 12 hours one day each May, the campus celebrates spring with Sunfest, which includes "amazing, crazy bands, a foam pit, sometimes Jello wrestling, and other games," says a sophomore. Transvestite Night livens things up during Parents' Weekend. And during finals week each term, the blaring of fire-truck sirens tells weary students to head to the dining hall, where professors, staff, and the college president serve up French toast and other breakfast favorites. Sports aren't a big focus, but Bennington does compete in an intramural co-ed soccer league that also includes other Northeastern colleges. The way one student sees it, "Bennington kids prefer going to a dance performance or poetry reading over playing sports."

As the first school in the nation to grant the arts equal status with other disciplines, Bennington offers a novel, participatory, and hands-on approach. Whether they're painters or writers, musicians or scientists, sculptors, dancers or some combination thereof, what Bennington students have in common is self-motivation and a real thirst for knowledge. Crossing disciplines is encouraged, and forget about taking the road less traveled; each student here charts his or her own course. And besides, there's a tree house—and milk and cookies served at every meal. Who says going to college means growing up?

If You Apply To ➤

Bennington: Early decision: Nov. 15. Regular admissions: Jan. 3. Financial aid: Mar. 1. Campus interviews: recommended, evaluative. Alumni interviews: optional, evaluative. SATs or ACTs: optional. SAT IIs: optional. Accepts the Common Application. Essay question.

Birmingham-Southern College

Box A18, Birmingham, AL 35254

One of the Deep South's best liberal arts colleges. The vast majority of students come from Alabama and the surrounding states. With fewer than 1,400 students, BSC is roughly the same size as Rhodes (Tennessee) and Millsaps (Missouri). Strong fraternity system and a throwback to the way college used to be.

Once an old-school conservative Southern institution, BSC is now striving to prepare students for all aspects the modern world, with high-tech facilities and a more global curriculum. Birmingham-Southern stresses service, effectively preserving the school's image as a strong liberal arts institution with its own brand of community involvement. More than half the student body participates in community service through Southern Volunteer Services, as well as claiming active membership in fraternities and sororities. Caring and attentive faculty add to a sense of commitment to both personal and community growth. "Our school really cares about the individual students on our campus," a sophomore says. "They try to cater to everyone's needs to make sure our college experience is the best it can be."

> "The courses are fairly rigorous because they require a lot of critical thinking and writing skills."

Known as the Hilltop for obvious reasons, BSC is the result of the 1918 merger of two smaller colleges: Birmingham College and Southern University. The campus, a green and shady oasis in an urban neighborhood, contains a pleasing hodgepodge of traditional and modern architecture, all surrounded by a security fence for added safety. Recent additions to campus include a new Humanities Building and $7.2 million six-house Fraternity Row.

The courses at BSC are described as rigorous by most. "I feel like the academic climate is more competitive than laid-back," says a Spanish major. "The courses are fairly rigorous because they require a lot of critical thinking and writing skills." Each student is assigned a faculty member who serves as his or her academic advisor from freshman convocation to graduation, an arrangement that students praise for its

Website: www.bsc.edu
Location: Urban
Total Enrollment: 1,381
Undergraduates: 1,335
Male/Female: 43/57
SAT Ranges: V 550–670
  M 540–640
ACT Range: 24–29
Financial Aid: 45%
Expense: Pr $
Phi Beta Kappa: Yes
Applicants: 1,157
Accepted: 83%
Enrolled: 32%
Grad in 6 Years: 75%
Returning Freshmen: 87%
Academics: ✍ ✍ ✍
Social: ☎ ☎ ☎
Q of L: ★ ★ ★
Admissions: (205) 226-4696
Email Address:
  admission@bsc.edu

(Continued)
Strongest Programs:
Biology
English
Business
Humanities
Psychology

effectiveness. Equal praise goes out to faculty in the classrooms, where 94 percent of classes have 25 or fewer students. A freshman says professors are "always willing to offer personal, one-on-one assistance."

Business, a division that includes programs ranging from accounting to international issues, is the most popular major, enrolling about a quarter of the students. Accounting and psychology majors are also popular, and the many premed students cite biology as a major drawing card. The Stephens Science Center gives this program, as well as the chemistry and physics departments, a further boost. English is also one of the school's strongest programs. The art, drama, dance, and music programs are all among the best in the South. Students stage several major productions each year, often including American and world premieres. An interdisciplinary major has been added in English/theatre arts.

The Foundations general education curriculum encompasses half of the credits needed to graduate. Students are required to take the first-year courses, five arts units, five science units, and one unit in each of the following: creative or performing arts, foreign language and culture, math and writing, an intercultural course, and a Senior Conference course, which can be a scholarly seminar, term project, or independent study. Students also must attend 40 approved intellectual and cultural events, 30 of which must be on campus. BSC, a member of the Associated Colleges of the South* consortium, also offers a wide variety of special programs. The January interim term allows students to explore new areas of study, from cooking lessons to travel in China. The international studies program offers students the chance to study abroad in several different countries, and the honors program allows 30 exceptional first-year students to take small seminars with one or more professors.

Forty-eight percent of the men and 52 percent of the women are members of Greek organizations, which means that much of the social activity at BSC revolves around the Greek system.

Seventy percent of the students are homegrown Alabamians, and practically all the rest hail from Deep South states, many with family ties to 'Southern. Though moderate by Alabama standards, the student body is quite conservative. Thirty-eight percent of the students belong to the Methodist Church. "Students are hard-working and driven to succeed," says a history major. Five percent of BSC students are African American, 1 percent are Hispanic, and 3 percent are Asian American. "We have an expanding international student base and multicultural affairs and the beginning of some African American sororities. The college is taking steps to diversify," a student says. BSC offers various merit scholarships to 83 percent of students, with an average grant of more than $11,000. National Merit Scholars who list 'Southern as their first choice receive an automatic scholarship of $500 to $2,000, and up to 10 get full-tuition awards.

"The college is taking steps to diversify."

Eighty percent of the students live on campus, including many of those whose families reside in Birmingham. Co-ed housing hasn't filtered down to 'Southern yet, so all seven dorms are single-sex with a variety of visitation policies depending on student preferences. Daniel Hall and New Mens are generally the most desired men's residences, while Bruno Hall is the preferred choice for women. Dorms are described as comfortable and convenient. "You can wake up 10 minutes before class, get ready, walk to class, and still have two or three minutes to spare," says a business administration major. More importantly, "The water pressure is great in the hall showers," cheers one student. Dining facilities get mixed reviews. The quality is decent and special tastes are accommodated, but the food can get repetitive. Campus security is quite visible and students praise its effectiveness in keeping the campus safe.

Known as the Hilltop for obvious reasons, BSC is the result of the 1918 merger of two smaller colleges: Birmingham College and Southern University.

Forty-eight percent of the men and 52 percent of the women are members of Greek organizations, which means that much of the social activity at BSC revolves around the Greek system, though plenty of opportunities are available for independents as well. "There are always fraternity parties every weekend, and they are open to everyone," a junior says. As for alcohol, it's not allowed on the quad, and elsewhere it must be in an opaque container, a policy most students find reasonable, described by one senior as a "don't see it, ignore it policy."

The biggest social event of the year is Soco, a two-day festival. Freshmen take part in a square dance during orientation. Other popular events include E-Fest and Halloween on the Hilltop, where students "dress up in costumes and the neighborhood kids go trick-or-treating." When social opportunities on campus dry up, many students take the shuttle to Birmingham for the

> "You can wake up 10 minutes before class, get ready, walk to class, and still have two or three minutes to spare."

city's nightlife. After learning that two of its students were arrested for a string of church arsons, university students, faculty, and staff rallied together to support the community and raised $350,000 for the Alabama Churches Rebuilding and Restoration Fund. Road trips to Auburn, Nashville, and Atlanta are popular, and beaches and mountains are less than five hours away.

BSC has made the move from Division I to Division III and will be adding football, women's lacrosse, and men's and women's track and field. The school already fields strong soccer, baseball, and basketball programs. Seventy percent of students take part in the intramural program. Basketball, flag football, and soccer are popular, but less traditional sports, such as dodgeball and table tennis, are also offered.

Students at BSC are focused on academics, but balance that with community service and an active social scene. Small classes, a caring faculty, and an expanding menu of academic offerings continue to draw attention to this close-knit liberal arts school.

Overlaps

Auburn, Millsaps, Rhodes, University of the South, University of Alabama

If You Apply To ➢

Birmingham-Southern: Rolling admissions. Financial aid: Mar. 1 (preferred). Does not guarantee to meet demonstrated need. Campus interviews: recommended, informational. Alumni interviews: optional, informational. SATs or ACTs: required. SAT IIs: optional. Accepts the Common Application and electronic applications. Essay question.

Boston College

140 Commonwealth Avenue, Devlin Hall, Room 208, Chestnut Hill, MA 02467

Many students clamoring for a spot at Boston College are surprised to learn that it is affiliated with the Roman Catholic church. Set on a quiet hilltop at the end of a "T" (subway) line, BC's location is solid gold. A close second in the pecking order among true-blue Catholics.

Boston College is a study in contrasts. Both the academics and athletic teams are well respected. The environment is safely suburban, but barely 20 minutes from Boston, the hub of the Eastern seaboard's college scene. The Jesuit influence on the college, one of the largest Roman Catholic schools in the country, provides a guiding spirit for campus life, but the social opportunities still seem endless.

Don't let the name fool you. Boston College is actually a university with nine schools and colleges. It has two campuses: the main campus at Chestnut Hill and the Newton campus a mile and a half away. The dominant architecture of the main campus (known as "the Heights") is Gothic Revival, with modern additions over the past few years, including a new science building. There's lots of grass and trees, not to mention a large, peaceful reservoir (perfect to jog around) right in the front yard.

Website: www.bc.edu
Location: Suburban
Total Enrollment: 13,755
Undergraduates: 9,019
Male/Female: 48/52
SAT Ranges: V 610–700
  M 640–720
ACT Range: 27–31
Financial Aid: 40%
Expense: Pr $ $ $

(Continued)

Phi Beta Kappa: Yes

Applicants: 23,823

Accepted: 31%

Enrolled: 30%

Grad in 6 Years: 89%

Returning Freshmen: 96%

Academics: ✏ ✏ ✏ ½

Social: ☎ ☎ ☎ ☎

Q of L: ★ ★ ★

Admissions: (617) 552-3100

Email Address:
ugadmis@bc.edu

Strongest Programs:
Chemistry
Economics
English
Finance
Political Science
Physics
History

When students are admitted, they are notified whether they will get on-campus housing for three or four years, and most juniors with three-year guarantees live off campus or study abroad in the fourth year.

Recent additions include renovations to the 66 Commonwealth Avenue Dorm and completion of the Yawkey Athletics Center.

Boston College was originally founded by the Jesuits to teach the sons of Irish immigrants. These days, the college's mission is to "educate skilled, knowledgeable, and responsible leaders within each new generation." To accomplish this goal, the Core Curriculum requires courses not only in literature, science, history, philosophy, social science, and theology, but also writing, mathematics, the arts, and cultural diversity, in addition to specific requirements set by each undergraduate school. "Core Curriculum forces you to take classes you might not want to take but end up enjoying," says a senior. Students in arts and sciences must also show proficiency in a modern foreign language or classical language before graduation. Freshmen are required to take a writing workshop in which each student develops a portfolio of personal and academic writing and reads a wide range of texts. Seniors participate in the University Capstone program, a series of seminars aiming to give a "big picture" perspective to the college experience.

> "Core Curriculum forces you to take classes you might not want to take but end up enjoying."

The academic climate is challenging and courses rigorous. "People at BC are extremely driven and ambitious," a junior says. "At the same time, they do not compete against each other." Professors are described as "phenomenal" and "unbelievable," and their classes are called "very interactive and discussion-based." The Jesuits on BC's faculty (about 60 out of 900) exert an influence out of proportion to their numbers. "The philosophy, theology, and ethics departments are the most important in setting the tone of the campus because they keep the students encouraged to be open-minded," says a freshman. Another student says, "Our teachers are dedicated to students and scholarly research, and they are easily accessible through regular office hours."

The schools of arts and sciences, management, nursing, and education award bachelor's degrees. Communications, finance, English, political science, and history are the most popular majors. Outside the traditional classroom, at the McMullen Museum of Art in Devlin Hall, students find exhibitions, lectures, and gallery tours. The Music Guild sponsors professional concerts throughout the year, and music students emphasizing performance can take advantage of facilities equipped with Steinways and Yamahas. Theater majors find a home in the 600-seat E. Paul Robsham Theater Arts Center, which produces eight student-directed productions each year.

Students searching for out-of-the-ordinary offerings will be happy at BC. The PULSE program provides participants with the opportunity to fulfill their philosophy and theology requirements while engaging in social-service fieldwork at any of about 35 Boston organizations, and sometimes leads students to major in those areas. Perspectives, a four-part freshman program, attempts to illustrate how great thinkers from the past have made us who we are. There's also a Freshman-Year Experience program, which offers seminars and services to help students adjust to college life and take advantage of the school and the city. An honors program allows students to work at a more intensive pace and requires a senior thesis.

> "THE college town."

About 27 percent of BC students come from the greater Boston area, and Catholics comprise about 70 percent of the student body. African Americans constitute 6 percent of the student body, while Asian Americans make up another 9 percent and Hispanics 8 percent. "There are many diversity initiatives but these are frivolous since BC is not diverse and minorities stick together," a sophomore says. "All the students here are identical. Most are wealthy and dress up for class daily." Still, the Jesuit appeal for tolerance means that students can find support and interaction even when approaching hot-button issues that Catholicism will not condone, such as homosexuality.

When students are admitted, they are notified whether they will get on-campus housing for three or four years, and most juniors with three-year guarantees live off campus or study abroad in the fourth year. The city of Boston has a fairly reliable bus and subway system to bring distant residents to campus; the few students that drive to school are required to show they need to park on campus. Another lottery system determines where on-campus residents hang their hats. Freshman dorms are described as "not great and not very modern," but accouterments in upper-class suites include private baths, dishwashers, and full kitchens. Students pay in advance for a certain number of dining hall meals, served a la carte.

BC students are serious about their work, but not excessively so. There is time and plenty of places to party. Yet BC's reputation as a hardcore party school is diminishing, now that no kegs or cases of beer are allowed on campus grounds. Those of legal age can carry in only enough beer for personal consumption. Bars and clubs in Boston ("THE college town," gushes a junior) are a big draw, along with Fenway Park. On weekends, especially in the winter, the mountains of Vermont and New Hampshire beckon outdoorsy types. The campus is replete with sporting events, movies, festivals, concerts, and plays. As at other Jesuit institutions, there is no Greek system at BC.

Athletic events become social events too, with tailgate and victory parties common. Football games are a big draw—the contest with Notre Dame is jokingly referred to as the "Holy War," and makes for a popular road trip. The football program has been recognized for achieving the highest graduation rate in the College Football Association, and the women's field hockey team made it to the quarterfinals of the NCAA tournament. The Silvio O. Conte Forum

> "The philosophy, theology, and ethics departments are the most important in setting the tone of the campus because they keep the students encouraged to be open-minded."

Sports Arena is well-attended, and BC meets fierce competition from Atlantic Coast Conference rivals Duke, Miami, Florida State, Virginia Tech, and others. Students even get the day off from classes to line the edge of campus and cheer Boston Marathon runners up "Heartbreak Hill." Intramural sports are huge here. About 4,300 undergrads play on 34 teams—from basketball and volleyball to skiing and golf. Students rave about Boston College's recreational complex and the Yawkey Athletics Center.

BC students spend four years fine-tuning the art of the delicate balance, finding ways to make old-fashioned morals relevant to life in the 21st century, and finding time for fun while still tending to their academic performance.

Boston College was originally founded by the Jesuits to teach the sons of Irish immigrants. These days, the college's mission is to "educate skilled, knowledgeable, and responsible leaders within each new generation."

Overlaps
Georgetown, Harvard, Yale, Notre Dame, Penn, Boston University

If You Apply To ➤ | **BC:** Early action: Nov. 1. Regular admissions: Jan. 2. Financial aid: Feb. 1. Housing: May 1. No campus or alumni interviews. SATs or ACTs: required, SAT preferred. SAT IIs: A minimum of two subject tests required. Apply to particular schools or programs. Guarantees to meet demonstrated need of all students. Accepts the Common Application and electronic applications.

Boston University

121 Bay State Road, Boston, MA 02215

One of the nation's biggest private universities, but easy to miss amid the bustle of the city. Boston's Back Bay neighborhood is the promised land for hordes of students nationwide seeking a funky, artsy, youth-oriented urban setting that is less in-your-face than New York City.

Website: www.bu.edu

Location: Center city

Total Enrollment: 29,596

Undergraduates: 17,740

Male/Female: 40/60

SAT Ranges: V 600–690
  M 621–700

ACT Range: 26–30

Financial Aid: 46%

Expense: Pr $ $ $ $

Phi Beta Kappa: Yes

Applicants: 28,240

Accepted: 55%

Enrolled: 28%

Grad in 6 Years: 75%

Returning Freshmen: 90%

Academics: ✐ ✐ ✐ ✐

Social: ☎ ☎ ☎ ☎

Q of L: ★ ★ ★

Admissions: (617) 353-2300

Email Address:
  admissions@bu.edu

International admissions:
  intadmis@bu.edu

Strongest Programs:

Communications

Management

Biomedical Engineering

Natural Sciences

Psychology

International Relations

Fine Arts

Boston University urges students to just "Be You" and most are happy to do so, but they warn that coming here is not for the faint of heart.

Like George Washington University and NYU, Boston University is an integral part of the city it calls home. The school's mammoth collection of nondescript high-rises straddles bustling, six-lane Commonwealth Avenue—and so do thousands upon thousands of students. Whether they're aspiring actors, musicians, journalists, and filmmakers, or wanna-be doctors, dentists, and hotel managers, BU seems to offers something for all of them. "Diversity is very present, but less focused around ethnicity than on personal backgrounds, political views, and social class," says a psychology major. "Generally, BU students are self-motivated and proactive, and they tend to be much more worldly and less insulated" than the students at rival Boston College. Even better, the quality of programs at BU is finally catching up to their quantity.

The BU campus is practically indistinguishable from the city that surrounds it, and while Boston may be America's town, BU is anything but bucolic. A measure of relief is available on the tree-lined side streets, which feature quaint Victorian brownstones. Recent construction includes a 35,000-square-foot Hillel House, a multi-level fitness center, a hockey arena that doubles as a concert hall, and a new life science and engineering building that allows the biology, chemistry, bioinformatics, and biomedical engineering faculty to operate under one roof. The John Hancock Student Village includes a track and tennis center, as well as an apartment-style dorm.

BU's 11 schools include the College of Arts and Sciences, the largest undergraduate division, which caters to the premeds and prelaw students. The College of Communication combines theory and hands-on training—some of it by adjunct professors with day jobs at major newspapers and TV networks. It also houses the

"Diversity is very present, but less focused around ethnicity than on personal backgrounds, political views, and social class."

nation's only center for the study of political disinformation. The School of Music benefits from its own concert hall and from faculty who also belong to the Boston Symphony Orchestra. The physical therapy program at the Sargent College of Health and Rehabilitation Sciences takes six years, culminating in BS and DPT degrees. Students in the School of Education can test their ideas for curricular reform in the public schools of nearby Chelsea, while those in School of Visual Arts may show their work in one of three campus galleries.

The School of Management, regarded as one of Boston University's top colleges, offers a four-year honors program and minors in law and hospitality administration; the College of Engineering boasts a robotics and biomedical engineering lab. Future employers of students in the School of Hospitality Administration offer paid internships in exotic locales such as Brussels and Britain. (Students in other fields who wish to work abroad may vie for jobs from Australia to Moscow.) Also intriguing is the University Professors Program, which begins with a two-year integrated core focusing on major authors and central themes of Western thought. Juniors and seniors in the program create individualized, interdisciplinary courses of study. BU also offers highly competitive seven- and eight-year programs admitting qualified students simultaneously to the undergraduate program and the university's medical or dental school.

Each of BU's schools and colleges sets its own general education requirements, but all students take the two-semester Freshman Writing Program. Students rave about FYSOP, the First-Year Student Outreach Project, which brings freshmen to campus a week early to do community service. For a break from brutal Boston winters, BU offers 44 study abroad programs, including internships, field work, research, language study, and liberal arts programs. There's also a marine science program at the Woods Hole Institute and the Semester at Sea,* a program welcoming students from many schools to spend a term living and studying on a cruise ship as it travels the world.

The academic climate at BU encourages both cooperation and competition. "The courses are extremely challenging and rigorous," reports one junior. "The science and math classes, in general, tend to be more competitive but the climate is a supportive and helpful one." Amazingly for such a large school, administrators say that 92 percent of the classes taken by freshmen have 25 or fewer students. "The quality of teaching here has been stellar," says a student. "Every professor has amazing experience that they bring to their subjects."

The physical therapy program at the Sargent College of Health and Rehabilitation Sciences takes six years, culminating in BS and DPT degrees.

"While the majority of students tend to lean towards 'preppy,' we do have a wide variety of students expressing different styles and interests," says a sophomore. More than three-quarters of BU undergrads are from outside Massachusetts and nearly 7 percent come from foreign countries. Asian Americans are the largest minority group on campus, at 13 percent of the total; African Americans add 2 percent and Hispanics 5 percent. "BU is a very liberal campus," a student reports. In all, the university offers more than 2,000 merit scholarships each year, with an average value of $13,403. There are also 210 athletic scholarships in seven men's and nine women's sports.

Three-quarters of BU students live in campus housing, which is guaranteed for four years. "Options are vast, from dorms to suites to apartments," says a journalism major. "Since we are on a lottery system it is possible to get a not-so-great room," explains a psychology major, "Freshman/sophomore

"Options are vast, from dorms to suites to apartments."

dorms are fairly typical, but housing gets even better as you get to be a senior." The luxury apartments, known as the Residences at 10 Buick Street, house 814 students on 18 floors, in apartments with four single bedrooms each. Each bedroom is wired for phone and fast Internet service, and each air-conditioned apartment also includes a kitchen, living and dining area, and two full bathrooms. Meal plans are flexible, and one of the six dining halls on campus is Kosher. There's also a food court with chains such as Burger King, Starbucks, and D'Angelo's, a local sub shop.

There are more than 450 student clubs and organizations at BU, and social life is "split about 50/50 between campus and city," says a psychology major. Three percent of the men and 5 percent of the women go Greek, and parties at neighboring schools are an option as well. Owing to Boston's heavily Irish heritage, St. Patrick's Day is also an occasion for revelry. Drinking is fairly common, though not in the dorms, because state laws are strictly enforced, and violators may find themselves without university housing. The Splash party in September, homecoming in October, and Culture Fest in March round out the social calendar. Possible road trips include Cape Cod, Cape Anne, and Providence, Rhode Island, but "Boston is where most people go on weekends," says a senior who calls it "the best college town in the world." The T's Green Line squiggles through the center of BU's

By far the most popular intramural sport is broomball, which is like ice hockey on sneakers, with a ball instead of a puck, and a broom instead of a stick.

"Fenway Park, downtown, Landsdowne Street, and Boston Common are all within walking distance."

campus, putting the entire city within easy reach. Even better, "Fenway Park, downtown, Landsdowne Street, and Boston Common are all within walking distance," says a marine biology major.

By far the most popular intramural sport is broomball, which is like ice hockey on sneakers, with a ball instead of a puck, and a broom instead of a stick. BU has won 26 of 53 titles in the annual Beanpot ice hockey tournament, which pits BU against Harvard, Northeastern, and archrival Boston College. Since BU doesn't field a football team, it's the athletic highlight of the school year. The women's tennis, soccer, and lacrosse teams are all conference standouts, along with men's basketball. All teams compete in NCAA Division I. The Head of the Charles regatta, which starts at BU's crew house each fall, draws college crew teams from across the country.

Boston University urges students to just "Be You" and most are happy to do so, but they warn that coming here is not for the faint of heart. The school is "a great

---

Overlaps
NYU, Boston College, Northeastern, George Washington, Tufts

place, with lots of academic and social opportunities, but it's not for the timid student," agrees a geophysics and planetary sciences major. "You have to be proactive about finding out what's going on around campus, so that you can find your niche."

If You Apply To ➤

BU: Early decision: Nov. 1. Regular admissions: Jan. 1 (Dec. 1 for the BA/MD and BA/DMD programs and applicants for some scholarships). Financial aid: Feb. 15. Housing: May 1. Meets demonstrated need of 51%. Campus interviews: required, evaluative for BA/MD and BA/DMD programs and some theatre arts programs; not available to other students. Alumni interviews: optional, informational. SATs or ACTs: required (including optional ACT Writing Test). SAT IIs: required (specific tests vary by program). Apply to particular school or program. Accepts the Common Application and electronic applications; prefers electronic to paper applications. Essay question: 500 words on a personally meaningful topic; 500 words on the experiences that have led you to select your professional field and objective; how you became interested in BU.

Bowdoin College

Brunswick, ME 04011

Rates with Amherst, Williams, and Wesleyan for liberal arts excellence and does not require the SAT. Bowdoin has strong science programs, and outdoor enthusiasts benefit from proximity to the Atlantic coast. Smaller than some of its competitors, with less overt competition among students.

Website: www.bowdoin.edu
Location: Midsize town
Total Enrollment: 1,660
Undergraduates: 1,660
Male/Female: 50/50
SAT Ranges: V 660–740
 M 660–730
Financial Aid: 45%
Expense: Pr $ $ $ $
Phi Beta Kappa: Yes
Applicants: 5,026
Accepted: 24%
Enrolled: 39%
Grad in 6 Years: 94%
Returning Freshmen: 97%
Academics: ✍ ✍ ✍ ✍ ✍
Social: ☎ ☎ ☎
Q of L: ★ ★ ★
Admissions: (207) 725-3100
Email Address:
 admissions@bowdoin.edu

Strongest Programs:
Natural Sciences
Classics
German
Anthropology

For more than two centuries, Bowdoin College sought to make nature, art, and friendship as integral to the student experience as the world of books. This is, after all, the alma mater of the great American poets Longfellow and Hawthorne. In fact, when they matriculate, new students sign their names in a book on Hawthorne's very desk. Though the New England weather can be brutal ("Be prepared for long, cold winters," warns a junior), students are quick to point out that good food and friendships that "transcend labels" help make campus a warm and friendly place.

Bowdoin's 205-acre campus sits in Brunswick, Maine, the state's largest town. Hidden amid the pine groves and athletic fields are 124 buildings, in styles from German Romanesque, Colonial, medieval, and neoclassical to neo-Georgian, modern, and postmodern. Former fraternity houses now house academic and administrative offices, since Greek groups were phased out. New campus additions include the Schwartz Outdoor Leadership Center, home to the Bowdoin Outing Club, and Kanbar Hall, which houses the psychology, education, and neuroscience departments. Two new dorms also recently opened, allowing the college to renovate first-year dorms and convert triple rooms to doubles.

To graduate, Bowdoin students must complete 32 courses, including one each in natural sciences and math, social and behavioral sciences, and fine arts and humanities, in addition to a required course in the visual and performing arts. New distribution requirements emphasize issues vital to a liberal education in the 21st century and include courses such as Exploring Social Differences; Mathematical, Computational, or Statistical Reasoning; and International Perspectives. Freshmen also have their choice of seminars, capped at 16 students each, which emphasize reading and writing; recent topics included Cultural Difference and the Crime Film, The Cuban Revolution, and The Economics of Art. Academic strengths include the sciences, specifically biology, chemistry, and environmental studies. Bowdoin also offers coursework in Arctic Studies (its mascot

"We always say that Bowdoin is a family."

is the polar bear), as well as opportunities for Arctic archeological research in Labrador or ecological research at the Kent Island Scientific Station in Canada. Premeds of all persuasions will find top-of-the-line lab equipment and outstanding faculty; the field of microscale organic chemistry was developed and advanced here.

(Continued)
Economics
English
Government
Environmental Studies

Students also praise the art history and English departments, and say the popularity of government and economics, the majors with the highest enrollment, is well deserved. "The less popular departments include physics and classics," says a senior, "but not because they aren't good—just because fewer students are interested in these fields." Newer majors include Latin American studies, Eurasian and East European studies, and English and theater. There's also an increasing emphasis on service learning; 57 percent of Bowdoin students apply their classroom work to real-world problems faced by local community groups. Undergraduate research is a priority, and it's common for juniors and seniors to conduct independent study with faculty members, then publish their results in professional journals.

Those same professors teach all Bowdoin classes—there are no graduate students here, and thus no TAs—and their skills in the classroom draw raves. "We always say that Bowdoin is a family," explains a junior, "and that comes across most clearly in the quality and type of professors we have." "Not only are they smart," adds a sophomore, "but they can actually teach a class and adapt material to the different types of learners." The recent shift to a plus-minus grading system has made the academic climate more competitive. "It is extraordinarily rigorous," says a senior. "Students work hard and are individually driven."

"Now there is much more socioeconomic, ethnic, racial, and regional diversity."

Before school begins, 75 percent of the entering class take preorientation hiking, canoeing, or sea kayaking trips, which teach them about the people and landscape of Maine. There's a community service experience in Brunswick for students less interested in the outdoors. The entering class also reads the same book before arriving, to start the year with a common academic experience.

Eighty-seven percent of students hail from outside Maine; most are generally hardworking, fun-loving, athletic types. "The student body has changed dramatically in the past four years," reports one senior. "It traditionally consisted of upper-middle class students from just outside Boston. Now there is much more socioeconomic, ethnic, racial, and regional diversity." African Americans make up 6 percent of the student body, Hispanics 6 percent, and Asian Americans add 12 percent. Bowdoin was the first U.S. institution to make SAT I scores an optional part of the admissions process, shifting the emphasis to a student's whole body of work. Merit scholarships worth an average of $1,000 are available to qualified students.

"We are spoiled in many ways," says a junior, "especially with regards to housing." Ninety-four percent of Bowdoin students live on campus, where "freshmen start off in a two-room triple (or if you're lucky, a two-room double), so you'll never be kept up by your roommate typing a paper at 3 a.m.," says a chemistry major. After that, students try their luck with the lottery, although members of the social houses, which have replaced sororities and fraternities, can escape by living with these groups. Upperclassmen may choose four-bedroom quads. Students give rave reviews to dining service workers and the food they serve. Students also love the lobster bake that kicks off each school year, and vegans and vegetarians are happy with the options available to them. "The food is unreal," raves one government major.

With the Greeks long gone, social life at Bowdoin centers around two groups: sports teams and social houses.

With the Greeks long gone, social life at Bowdoin centers around two groups: sports teams and social houses. "The college does bring in campuswide entertainment, such as hypnotists, bands, and comedians," says a senior. "In addition, there is usually a party open to the entire campus at one or more of the social houses every weekend." All parties and kegs must be registered and most students report that while underage drinking does occur, it doesn't dominate the social scene. "Students look forward to homecoming, the BearAIDS benefit concert, and Ivies Weekend, one last blast of fun before spring finals. The latter celebrates the fact that Bowdoin didn't join the Ivy League, with bands and games in the quad.

One student says Brunswick (population 21,000) is "a cute coastal town" with "a funky feel, nice restaurants, and a little shopping." A car comes in handy for the 15-minute drive to the outlets of Freeport (including L.L. Bean's 24/7 factory store) or a quick trip to Portland for a "real" night out. A school shuttle takes students to Boston, less than three hours' away, and ski bums will find several resorts even closer. Habitat for Humanity and various mentoring programs help build bridges between local residents and students. For those who get really stir-crazy, study abroad programs are available in more than 100 countries, including warm ones like Ecuador (a welcome treat when you consider the normally brutal New England winters). Sixty percent of students take advantage of such programs.

While "the long winters are certainly not a favorite," according to a senior, they do bring out school spirit, with any sporting event against Colby—especially hockey games—inspiring excitement. "That rivalry is almost out of control," says a biology major. "The chanting is brutal, and dead fish have been known to fly onto the ice." Bowdoin's Bears compete in the Division III New England Small College Athletic Conference, and students "get blacked out" to demonstrate support, wearing all black when they attend games. The women's basketball team fields perennially strong squads and has captured the NESCAC championship multiple times, and the women's field hockey squad captured the championship in 2006. Students have a big say in the start-up of intramural programs, and about 40 percent play an intramural sport in a given semester.

"We are spoiled in many ways, especially with regards to housing."

Outdoorsy types and those who can brave the cold will find warm and inviting academics at Bowdoin, where close friendships with peers and professors are easily forged. "Being able to make friends with my professors is something I'll treasure forever," says an art history major. For those considering Bowdoin, a senior offers this assessment: "The education and opportunities will be beyond anything you'd imagine."

If You Apply To ➤   **Bowdoin:** Early decision: Nov. 15, Jan. 1. Regular admissions: Jan. 1. Financial aid: Jan. 1 (international applicants), Feb. 15 (regular-admissions candidates). Guarantees to meet full demonstrated need. Campus and alumni interviews: recommended, evaluative. SATs or ACTs: optional (SATs preferred). SAT IIs: optional. Accepts the Common Application and electronic applications. Essay question.

Brandeis University

Waltham, MA 02454-9110

Founded in 1948 by Jews who wanted an elite institution to call their own. Now down to 55 percent Jewish and seeking top students of all faiths. Academic specialties include the natural sciences, the Middle East, and Jewish studies. Competes with Tufts in the Boston area.

Brandeis University, founded to provide educational opportunities to those facing discrimination, has always had a reputation for intense progressive thought. Now, it's being recognized as a rising star among research institutions. The only nonsectarian

Jewish-sponsored college in the nation, Brandeis continues its struggle to maintain its Jewish identity while attracting a well-rounded, eclectic group of students.

Set on a hilltop in a pleasant residential neighborhood nine miles west of Boston, Brandeis's attractively landscaped 270-acre campus boasts many distinctive buildings. The music building, for example, is shaped like a grand piano; the theater looks like a top hat. The 24-hour Carl and Ruth Shapiro Campus Center includes a student theater, electronic library and bookstore, and the Rose Art Museum recently received a 7,300-square-foot addition that doubled its exhibition space. The Abraham Shapiro Academic Building houses a state-of-the-art distance-learning classroom, conference rooms, the International Center for Ethics, Justice, and Public Life, the Center for Middle Eastern Studies, the Mandel Center for Jewish Education, and faculty offices. A new 34,000-square-foot addition to the Heller School for Social Policy and Management will include classrooms, offices, and a forum for lectures and events, as well as a cafe and grouped lounge seating on several levels.

> "Who needs Waltham for excitement when Boston is a short shuttle ride away?"

Biochemistry, chemistry, neuroscience, and physics are top-notch programs, while economics, biology, psychology, and politics enroll the most students. Dedicated premeds are catered to hand and foot, with special advisors, internships, and their own premedical center, with specialized laboratories designed to provide would-be MDs with research opportunities. Grad school acceptance rates are impressive; 80 percent.

With the largest faculty in the field outside of Israel, the university is virtually unrivaled in Near Eastern and Judaic studies; Hebrew is a Brandeis specialty. East Asian studies gives students a broad yet intimate knowledge of the history, politics, economics, art, and language of the major areas of East Asia. A growing number of popular interdisciplinary programs including business, journalism, and legal, environmental, Latin American, peace, and women's and gender studies add spice to the academic menu. Brandeis also maintains a commitment to the creative arts, with strong theater offerings and a theory-based music program founded by the late Leonard Bernstein. Over two-thirds of the classes have 25 or fewer students. Newer majors include history, politics, and Yiddish and Eastern European Jewish Culture.

The Brandeis core curriculum is rooted in a commitment to developing strong writing, foreign language, and quantitative-reasoning skills and an interdisciplinary and crosscultural perspective. Rising sophomores and juniors now have the opportunity to earn credit through summer internships related to their studies, and more than 250 study abroad programs are offered in 70 countries; 28 percent of each class takes advantage of these opportunities. The Schiff Undergraduate Fellows Program enables 10 Fellows each year to work with faculty mentors on research and teaching projects.

Twenty-five percent of the Brandeis student population are from Massachusetts and heavily bicoastal otherwise, with sizable numbers of New York, New Jersey, and California residents. The group is also very bright; 96 percent graduated in the top quarter of their high school class, and professors want them to continue working hard. Students say the academic climate here is intense. "Brandeis takes its academic integrity seriously," says a creative writing and English major. There is an out for those in need of respite; the Flex 3 option allows students to take three classes one semester if an especially rough course is required, and five the next, to stay on track for four-year graduation.

Though more than half the student body is Jewish, there are three chapels on campus—Catholic, Jewish, and Protestant—built so that the shadow of one never crosses the shadow of another. It's an architectural symbol that students say reflects the realities of the campus community. Muslim students, with an enrollment of over 200, have their own dedicated prayer space. African Americans make up 4 percent of

(Continued)

Total Enrollment: 4,606
Undergraduates: 3,212
Male/Female: 45/55
SAT Ranges: V 630–720
   M 640–720
ACT Range: 28-33
Financial Aid: 45%
Expense: Pr $ $ $ $
Phi Beta Kappa: Yes
Applicants: 7,343
Accepted: 38%
Enrolled: 26%
Grad in 6 Years: 88%
Returning Freshmen: 94%
Academics: ✑ ✑ ✑ ✑
Social: ☎ ☎ ☎
Q of L: ★ ★ ★
Admissions: (781) 736-3500
Email Address:
   admissions@brandeis.edu

Strongest Programs:
Neuroscience
Biology
Near Eastern and Judaic
   Studies
English and American
   Literature
Theater Arts
Economics
Psychology

the student body, Hispanics 4 percent, and Asian Americans 8 percent. "Brandeis's character offers a great opportunity to either gain a totally new insight and experience, or to connect in a small community," explains a sophomore. Gays and lesbians have an established presence, and throw some of the liveliest parties. The unofficial fraternities and sororities that have colonized at Brandeis are clamoring for recognition from the school. Other hot-button issues include political correctness, rape awareness, and environmental causes.

Even with one of the highest tuition rates in the country, Brandeis does not guarantee to meet each student's full demonstrated need, but help is generally available to those who apply on time. The level of support remains fairly constant over four years, students report. The university also offers merit scholarships, averaging $12,952. The eight-day freshman orientation program is one of the most extensive in the nation, and includes a broad spectrum of events, such as a Boston Harbor cruise and special programs for minority, international, commuter, and transfer students.

As befits its mold-breaking heritage, Brandeis is the only school in the nation where you can live in a replica of a Scottish castle with pie-shaped rooms and stairways leading to nowhere. More pedestrian housing options include traditional quadrangle dormitories, where freshmen and sophomores live in doubles, and juniors live in singles. The Foster Living Center, or the "Mods," are coed, university-owned town houses reserved for seniors. The newest option is The Village, which offers singles and doubles clustered around family-style kitchens, semi-private bathrooms, and lounges. Freshmen and sophomores are guaranteed housing, while upperclassmen play the lottery each spring. Eighty percent of students live on campus, and the rest find affordable off-campus housing nearby. Brandeis boasts the best college food in the Boston area, as well as the most appetizing set-ups, students say, thanks to a decision to outsource dining services. Campus meal tickets buy lunch or dinner in a fast-food joint, the pub, a country store, a kosher dining hall with vegetarian selections, or the Boulevard, a cafeteria where "the salad bars are huge."

> **"Brandeis students are more focused on grades and academics, and less on big parties or Greek life."**

Social life at Brandeis offers lots of options for those ready to relax. There are 257 campus clubs to keep students busy, but "Brandeis students are more focused on grades and academics, and less on big parties or Greek life," says a student. Weekends begin on Thursday, with live entertainment at the on-campus Stein pub. Students can party at will in the dorms providing they don't get too rambunctious, but suites are officially "dry" unless a majority of the residents are over 21. Major events on the campus calendar include a Tropics Night dance (where beachwear is required in February), the massive Bronstein Weekend festival just before spring finals, and the "Screw Your Roommate" dance, where dormies set up their roommates on blind dates. Also well attended are the homecoming soccer match and the annual lacrosse tilt against crosstown rival Bentley College. The possibilities for off-campus diversion are nearly infinite, thanks to the proximity of Boston and Cambridge, which are accessible by the free Brandeis shuttle bus or a nearby commuter train. (A car is more trouble than it's worth.) And what about Waltham, Brandeis's host town? Waltham receives lukewarm reviews from the students but one global studies major asks, "Who needs Waltham for excitement when Boston is a short shuttle ride away?"

> **"Brandeis takes its academic integrity seriously."**

Though the school does not field a football team, Brandeis has developed strong men's baseball and swimming and women's swimming, fencing, and cross-country squads, all of which have taken regional championships in recent years. In 2004, the women's basketball team won the ECAC conference title. The athletic program gets a boost from its membership in the NCAA Division III University Athletic Association, a neo–Ivy League for high-powered academic institutions such as the University of Chicago, Johns Hopkins, and Carnegie Mellon. Brandeis's sports facilities include the 70,000-square-foot Gosman Sports and Convocation Center, which has hosted the NCAA fencing championships three times in the past 10 years.

Few private universities have come as far as Brandeis so quickly, evolving from a bare 270-acre site with the leftovers of a failed veterinary school to a modern research university of more than 100 buildings, a $520 million endowment, and ever-evolving academic opportunities. Landscaping, dining services, health services, and the campus computer network have all been dramatically improved in the past few years, students say, adding to their feelings of pride in the school. One student sums it up this way: "Brandeis is not only an awesome place to get an education, it's also an open, accepting place where anyone can feel at home."

<table>
<tr><td>

Overlaps

Boston, Brown, NYU, Tufts, Cornell
</td></tr>
</table>

If You Apply To ➤   **Brandeis:** Early decision: Nov. 15. Regular admissions and financial aid: Jan. 15. Does not guarantee to meet demonstrated need. Campus and alumni interviews: recommended, informational. SATs or ACTs: required (SAT I preferred). SAT IIs: required (two different subject areas). Accepts the Common Application and electronic applications. Essay question.

Brigham Young University

Provo, UT 84602

From the time they are knee-high, Mormons in all corners of the country dream about coming to BYU. Most men and some women do a two-year stint as a missionary. The atmosphere is generally mild-mannered and conservative. Goes absolutely bonkers for its sports teams.

Brigham Young University's strong ties with the Church of Jesus Christ of Latter-Day Saints means "BYU has high morals and a wholesome environment, which makes students feel safe and comfortable," a senior says. A sense of spirituality pervades most everything at BYU, where faith and academia are intertwined and life is governed by a strict code of ethics, covering everything from dating to academic dishonesty. Indeed, the school's commitment to church values is the reason most students choose it. "The students who attend BYU are unique," says a communication major. "Everyone is clean-cut, shaven, modestly dressed, and proper in their etiquette."

The church's values of prosperity, chastity, and obedience are strongly evidenced on BYU's 557-acre campus, where the utilitarian buildings, like everything else, are "clean, modern, and orderly." The campus sits 4,600 feet above sea level, between the shores of Utah Lake and Mount Timpanogos, with breathtaking sunsets and easy access to magnificent skiing, camping, and hiking areas. Days begin early; church bells rouse students at 8 a.m. with the first four bars of the church hymn "Come, Come Ye Saints." (The same bells also peal every hour throughout the day.) The Joseph S. Smith building, new student housing, an indoor practice facility, and a student athlete building have recently been completed.

The church's influence continues when students set their schedules; students must take one religion course per term to graduate, and offerings include, of course, the *Book of Mormon*. BYU requires students to demonstrate proficiency in math, writing (first-year and advanced), and advanced languages, a catch-all category that can be satisfied with coursework in a foreign language or in statistics, advanced

<table>
<tr><td>

Website: www.byu.edu

Location: City

Total Enrollment: 29,400

Undergraduates: 27,460

Male/Female: 50/50

SAT Ranges: V 550–660
   M 570–670

ACT Range: 25–29

Financial Aid: 35%

Expense: Pr $

Phi Beta Kappa: No

Applicants: 8,696

Accepted: 78%

Enrolled: 79%

Grad in 6 Years: 70%

Returning Freshmen: 95%

Academics: ✍ ✍ ✍

Social: ☎ ☎ ☎

Q of L: ★ ★ ★ ★

Admissions: (801) 422-2507
</td></tr>
</table>

(Continued)

Email Address:
admissions@byu.edu

Strongest Programs:
Business
Law
Engineering
Languages
Nursing

math or advanced music. Students must also complete an extensive liberal arts core, which includes work in civilization, American heritage, biology, physical sciences, and wellness (three physical education or dance activity courses), and electives in the natural sciences, social and behavioral sciences, and arts and letters. Students agree that the academic climate is demanding and students can be competitive. "Some of the hardest and most competitive can be the religion courses," a music major says.

BYU's academic offerings run the gamut, from liberal arts and sciences to professional programs in engineering, nursing, business, and law. Students say the strongest offerings include the J. Rueben Clark Law School and most departments in the Marriott School of Management, especially accounting. Newer programs include degrees in public health education, school health education, and Ancient Near Eastern Studies. Brigham Young also has campuses in Idaho and Hawaii, a center in Jerusalem, and study abroad programs in Vienna, London, and elsewhere.

Freshmen are usually taught by full-time professors, who generally get good marks. "The professors are all very nice and compassionate," says a sophomore. General education courses can be quite large, but the honors program, open to highly motivated students, offers small seminars with more faculty interaction, and is "an excellent

"Everyone is clean-cut, shaven, modestly dressed, and proper in their etiquette."

way to get more out of your college experience," one participant says. The strength of the faculty is one reason BYU has more full-time students than any other church-sponsored university in the U.S. Still, with thousands of students to accommodate, registration can be a chore. Approximately 80 percent of the men and 12 percent of the women interrupt their studies—typically after the freshman year—to serve two years as a missionary.

Not surprisingly, the typical BYU student is conservative. "Religious dedication is the top priority of most students," confides a chemical engineering major. Forty-seven percent of BYU students are from Utah. Many others hail from California and Idaho, and 3 percent come from more than 100 other countries, testifying to the effective and far-flung Latter-day Saint missionary effort. Despite the heavy international presence, the student body remains largely white, with Hispanic students contributing 4 percent, and African Americans, Asian Americans and Native Americans less than 1 percent each. Their Honor Code requires students to eschew drinking, smoking, and drugs; a city ordinance in Provo takes care of the "vice" of dancing. Politically, "many of the students have the same views so there is not much activism on campus," says an English major. Tuition for church members is lower than for nonmembers, because Latter-day Saint families contribute to BYU through their tithes. Academic scholarships and 387 athletic scholarships, in several sports, are available.

Twenty percent of BYU students—primarily freshmen—live in the single-sex residence halls, where the "very valuable" Freshman Academy program allows them to take courses and eat meals with fellow dorm-dwellers and professors. "Although the dorms are older, they are well kept and well organized to provide an enjoyable social experi-

"Some of the hardest and most competitive can be the religion courses."

ence," says a student. Upperclassmen typically opt for cheaper off-campus apartments, which are also single-sex (remember the Honor Code?). When it comes to chow, the student dining outlets on campus are described as adequate. "The food is diverse and well balanced," says a public relations major.

Whether it's work with the homeless or disabled, dances, concerts, plays, or sporting events, most of BYU's social life is organized through or linked to the church. Community service is big, with students visiting patients at hospitals and

Physical fitness is big here, and the intramural facilities are some of the country's best, with indoor and outdoor jogging tracks; courts for tennis, racquetball, and handball; and a pool.

care centers, performing at local festivals, and building and refurbishing houses. Dating is common, within the church's bounds of propriety, and is given a light-hearted feeling with groups that encourage "creative dating and lots of dating, period," says a senior. There are no fraternities and sororities to provide housing or parties, which is just fine with most students, since alcoholic and caffeinated drinks are both banned. "There's a reason we're voted the No. 1 Stone-Cold Sober University every year," says a student. Don't be deterred by this as you will still find plenty to do in the social scene. The college does what it can to keep the students too busy to realize that they're still sober, with activities like dances, firesides, and "tons of clubs for every interest," according to a senior. Road trips include Vegas or southern Utah and, with the mountains being so close, you'll find plenty of skiing and camping. Provo itself has plenty of places to eat, shop, and play for those needing a quick getaway from the strict administration enforcing the even stricter honor code. "The town is very supportive of the university, especially the athletic programs," a senior says. "Many of the stores hang up the school flag in the window."

Physical fitness is big here, and the intramural facilities are some of the country's best, with indoor and outdoor jogging tracks; courts for tennis, racquetball, and handball; and a pool. Also important are varsity sports; the church philosophy of obedience

> **"There's a reason we're voted the No. 1 Stone-Cold Sober University every year."**

has worked wonders for Cougar teams and the football rivalry against the University of Utah provides some serious end-of-season intensity. One of the most popular courses offered at BYU is ballroom dancing, partly because many participants aspire to join BYU's award-winning dance team. The ESPN television network has dubbed the BYU–Utah rivalry the "Holy War."

To most Americans, BYU probably seems old-fashioned or like a step back in time. But for young members of The Church of Jesus Christ, that may be just what the elder ordered. "BYU's dedicated faculty, devout atmosphere, and beautiful, clean campus set it apart from all other universities," a satisfied senior says.

Approximately 80 percent of the men and 12 percent of the women interrupt their studies—typically after the freshman year—to serve two years as a missionary.

Overlaps

BYU–Idaho, Utah Valley State, University of Utah, Utah State, BYU–Hawaii

If You Apply To > | **BYU:** Rolling admissions: Feb. 15. Housing: First-come, first-served; students are encouraged to apply a year in advance, with housing contingent on acceptance at the university. Does not guarantee to meet demonstrated need. Campus and alumni interviews: optional, informational. SATs or ACTs: required (ACTs preferred). SAT IIs: optional. Prefers electronic applications. Essay question.

University of British Columbia: See page 330.

Brown University

45 Prospect Street, Providence, RI 02912

To today's stressed-out students, the thought of taking every course pass/fail is a dream come true. In reality, nobody does, but the pass/fail option, combined with Brown's lack of distribution requirements, gives it the freewheeling image that students love. Bashed by conservatives as a hotbed of political correctness.

Brown University is a perennial "hot college," with an overwhelming number of happy students and many more clamoring to join their ranks. Once here, students not only receive a prestigious and quality education, but a chance to explore their creative sides at a liberal arts college that does not emphasize grades and preprofessionalism and shuns required courses. Brown's environment and policies have drawn both praise and criticism over the years, but its students thrive on this discussion and lively debate. "The freedom of shaping one's own education is both frightening and exhilarating, since the possibilities for good and ill are almost endless," says one student.

Located atop College Hill on the east side of Providence, Brown's 140-acre campus affords an excellent view of downtown Providence that is especially pleasing at sunset. Campus architecture is a composite of old and new—plenty of grassy lawns surrounded by historic buildings that offer students refuge from the city streets beyond. One student describes it as a "melting pot of architecture's finest. We have a building that resembles a Greek temple [and] buildings in the Richardsonian tradition." The neighborhoods that surround the campus lie within a national historic district and boast beautiful tree-lined streets full of ethnic charm. Newer additions include the Sidney E. Frank Hall for Life Sciences and the English and creative writing department buildings.

Brown's faculty has successfully resisted the notion that somewhere in their collective wisdom and experience lies a core of knowledge that every educated person should possess. As a result, aside from completing courses in a major, the only universitywide requirements for graduation are to demonstrate writing competency and complete the 30-course minimum satisfactorily. (The assumption is that students will take four courses a

"No one is going to tell you what to take."

term for a total of 32 in four years.) Freshmen have no requirements. Those with interests in interdisciplinary fields will enjoy Brown's wide range of concentrations that cross departmental lines and cover everything from cognitive science to public policy. Indeed, there are bona fide departments in cognitive and linguistic sciences and media and modern culture. Students can also create their own concentration from the array of goodies offered. Brown also offers group independent-study projects, a popular alternative for students with the gumption to take a course they have to construct primarily by themselves. Particularly adventurous students can choose to spend time in one of Brown's 57 study abroad programs in 15 countries, including Brazil, Great Britain, France, Tanzania, Japan, Denmark, and Egypt. Closer to home, students can cross-register with Rhode Island School of Design, also on College Hill, or participate in the Venture Program.*

Students can take their classes one of two ways: for traditional marks of A, B, C, or No Credit; or for Satisfactory/No Credit. The NC is not recorded on the transcript, while the letter grade or Satisfactory can be supplemented by a written evaluation from the professor. A habit of NCs, however, lands students in academic hot water. Any fewer than seven courses passed in two consecutive semesters makes for an academic "warning" that does find its way onto the transcript, and means potential dismissal from the university.

The most popular majors are biology, history, international relations, English, and political science. History and geology are some of the university's best, and students also praise computer science, religious studies, and applied math. Other top-notch programs include comparative literature, classics, modern languages, and the writing program in the English department. Among the sciences, engineering and the premed curriculum are standouts. Future doctors can try for a competitive eight-year liberal medical education program where students can earn an MD without having to sacrifice their humanity. Fields related to scientific technology have very good facilities, including an instructional technology center, while minority issues are studied at the Center for Race and Ethnicity. Newer degrees include Commerce, Organizations & Entrepreneurship, Literary Arts, and Science & Society.

Brown prides itself on undergraduate teaching and considers skill in the classroom as much as the usual scholarly credentials when making tenure decisions. Younger professors can receive fellowships for outstanding teaching, and the administration's interest in interdisciplinary instruction and imaginative course design help cultivate high-quality instruction. The size of the faculty ranks is being increased by 20 percent, and investment in university libraries has also risen. The advising system reflects the administration's commitment to treating students as adults. The lack of predetermined requirements is supposed to challenge students, so "no one is going to tell you what to take." The advising system pairs each freshman with a professor and a peer advisor, and resident counselors in the dorms are also available to lend an ear. Sophomores utilize special advising resources, upperclassmen are assigned an advisor in their concentration, and a pool of interdisciplinary faculty counselors is on hand for general academic advising problems.

Brown offers more than 100 freshman courses via the Curricular Advising Program (CAP), and the professors in these courses officially serve as academic advisors for their students' first year. This program receives mixed reviews, but some professors are highly praised by students for their abilities and availability. "The quality of teaching is superb," one student says. "Each course is taught by professors who are supported by graduate and undergraduate teaching assistants." Upper-level classes are usually in the teens, CAP courses are limited to 20, and only 12 percent of introductory lectures have more than 50 students. Especially popular courses are usually jammed with students, and often there aren't enough teaching assistants to staff them effectively. Some popular smaller courses, especially writing courses in the English department and studio art courses, can be nearly impossible to get into, although the administration claims that perseverance makes perfect—in other words, show up the first day and beg shamelessly. Compared with the other Ivies, Brown's academic climate is relatively casual and students don't fret about competition. "We are very focused on personal success in our academic arenas and strive to challenge ourselves," says a senior.

"The students who attend Brown aren't all hippies," chides a junior. "There is no typical Brown student. The jock will do theater, the geek will play ultimate Frisbee, and the average Joe/Jane will do at least two extracurricular activities." With a mere 4 percent of students hailing from Rhode Island, geographical diversity is one of Brown's hallmarks. Brown is one of the few remaining hotspots of student activism in the nation; nary has a semester passed without at least one demonstration about the issue of the day. Students of color account for 29 percent of the population, and foreigners make up 9 percent. Minorities rarely miss an opportunity to speak out on issues of concern. The gay and lesbian community is also prominent. "They throw the best dances on campus," says one science major. Ninety-nine percent of students were in the top quarter of their high school class, and 40 percent hail from private or parochial schools.

> "The jock will do theater, the geek will play ultimate Frisbee, and the average Joe/Jane will do at least two extracurricular activities."

Brown admits all students regardless of their financial need, and although it doesn't offer athletic or academic merit scholarships, it does guarantee to meet the full demonstrated need of everyone admitted. Fifteen Starr National Service scholarships, ranging from $1,000 to $2,000, are awarded each year to students who devote a year or more to volunteer public service jobs. About 120 other "academically superlative" students, called University Scholars, will find their financial aid package sweetened with extra grant money. Brown's use of binding early decision, also used by Princeton, has made some waves.

Brown prides itself on undergraduate teaching and considers skill in the classroom as much as the usual scholarly credentials when making tenure decisions.

Freshmen arrive on campus a few days before everyone else for orientation, which includes a trip to Newport, and there is also a Third World Transition Program. About half the freshmen are assigned to one of the eight co-ed Keeney Quad dorms, in "loud and rambunctious" units of 30 to 40 with several sophomore or junior dorm counselors. The other half live in the quieter Pembroke campus dorms or in a few other scattered locations. After their freshman year, students seeking on-campus housing enter a lottery. The lottery is based on seniority, and sometimes the leftovers for sophomores can be a little skimpy, though there are some special program houses set aside to give them a chance to focus their interests in residential halls.

The dorms themselves are fairly nondescript. "There are no fireplaces or engraved wood trim a la Princeton," observes one student, but nevertheless there are many options from which to choose, including apartment-like suites with kitchens, three sororities, two social dorms, and three co-ed fraternities. Brown guarantees housing all four years, and a dorm with suites of singles ensures that there is room for all. A significant number of upperclassmen get "off-campus permission." Places nearby are becoming more plentiful and more expensive as the area gentrifies. Brown's food service, which gets high marks from students for tastiness and variety, offers meal plans ranging from seven to 20 meals a week. Everyone on a meal plan gets a credit card that allows the student to do what students at every other school only wish they could: use the meal ticket for nocturnal visits to snack bars should they miss a regular meal in one of Brown's two dining halls. Campus security is described as "very good." Says a junior, "I feel safer here than I do at home."

"I feel safer here than I do at home."

Providence is an old industrial city that recently underwent a renaissance. It is still the butt of student jokes—"Be prepared to wear your proletarian disguise," cautions one—but extensive renovations of the downtown area have had a positive impact. Providence is Rhode Island's capital, so many internship opportunities in state government are available, as are a few good music joints, lively bars, and a number of fine, inexpensive restaurants. For the couch-potato set, there are plenty of good things right in the neighborhood. "Downtown is a 10-minute walk, but why bother when you can buy anything from Cap'n Crunch to cowboy boots on Thayer Street, which runs through the east side of campus?" asks a philosophy major. For a change of scenery, many students head to Boston or the beaches of Newport, each an hour away.

The few residential Greek organizations are generally considered much too unmellow for Brown's taste (only 12 percent of the men and 2 percent of the women sign up), and hence freshmen and sophomores are their chief clientele. The nonresidential black fraternities and sororities serve a more comprehensive student life function. Tighter drinking rules have curtailed campus drinking somewhat. The university sponsors frequent campuswide parties and plays, concerts, and special events. Funk Nite every Thursday night at the Underground, a campus pub, draws a mixed bag of dancing fools. The biggest annual bash of the year is Spring Weekend, which includes plenty of parties and a big-name band. Strong theater and dance programs, daily and weekly newspapers, a skydiving club, political organizations, and "even a Scrabble club and a successful croquet team" represent just a few of the ways Brown students manage to keep themselves entertained. One other is the campus student center, which has been thoroughly renovated. For those interested in community outreach—and there are many at Brown who are—the university's nationally recognized public service center helps place students in a variety of volunteer positions. The Brown Community Outreach, in fact, is the largest student organization on campus.

Brown isn't an especially sports-minded school, but a number of teams nevertheless manage to excel. Of the 37 varsity teams, recent Ivy League champions include the volleyball, women's crew, men's tennis, and men's soccer teams. The women's cross-country squad finished third in the NCAA Regional Championships in 2002. Athletic facilities include an Olympic-size swimming pool and an indoor athletic complex with everything from tennis courts to weight rooms. There's also a basketball arena for those trying to perfect their slam dunks. The intramural program is solid, mixing fun with competitiveness.

Ever since the days of Roger Williams, Rhode Island has been known as a land of tolerance, and Brown certainly is a 21st-century embodiment of this tradition. The education offered at this university is decidedly different from that provided by the rest of the Ivy League, or for that matter, by most of the country's top universities. Brown is content to gather a talented bunch of students, offer a diverse and imaginative array of courses, and then let the undergraduates, with a little help, make sense of it all. It takes an enormous amount of initiative, maturity, and self-confidence to thrive at Brown, but most students feel they are up to the challenge. "You get four years of choice," says one student. "Deal with it."

Overlaps

Harvard, Yale, Stanford, Columbia, Penn, Dartmouth

If You Apply To ➤ **Brown:** Early decision: Nov. 1. Regular admissions: Jan. 1. Financial aid: Feb. 1 for regular admissions. Guarantees to meet demonstrated need. No campus interviews. Alumni interviews: optional, evaluative. SATs or ACTs: required. SAT IIs: required (any two). Accepts electronic applications. Essay question: personal statement.

Bryn Mawr College

101 North Merion Avenue, Bryn Mawr, PA 19010-2899

BMC has the most brainpower per capita of the elite women's colleges. Politics range from liberal to radical. Do Bryn Mawrtyrs take themselves a little too seriously? The college still benefits from ties to nearby Haverford, though the relationship is not as close as in the days when Haverford was all male.

Leafy suburban enclaves are a dime a dozen around Philadelphia, but only one is home to Bryn Mawr College, a top-notch liberal arts school. On this campus, students find a range of academic pursuits from archeology to film studies to physics, and a diverse yet community-oriented student body. Founded in 1885, Bryn Mawr has evolved into a place that prepares students for life and work in a global environment. Although students here abide by a strict academic honor code and participate in a host of loopy and long-standing campus traditions, they remain doggedly individualistic.

Bryn Mawr's lovely campus is a path-laced oasis set among trees (many carefully labeled with Latin and English names) and lush green hills, perfect for an afternoon walk, bike ride, or jog. Just a 20-minute train ride to downtown Philadelphia, Bryn Mawr provides a country setting with a vital and exciting city nearby. The predominant architecture is collegiate Gothic, a combination of the Gothic architecture of Oxford and Cambridge Universities and the local landscape, a style that Bryn Mawr introduced to the United States. Ten of Bryn Mawr's buildings are listed in the National Register of Historic Places. The M. Carey Thomas Library, which was named after the school's first dean and second president, a pioneer in women's education, is also a National Historic Landmark. Variations on the collegiate Gothic theme include a sprinkling of modern buildings, such as Louis Kahn's slate-and-concrete residence hall and the redbrick foreign language dormitory. Recent campus

Website: www.brynmawr.edu
Location: Suburban
Total Enrollment: 1,564
Undergraduates: 1,293
Male/Female: 2/98
SAT Ranges: V 620–720
  M 600–690
ACT Range: 25–30
Financial Aid: 54%
Expense: Pr $ $ $
Phi Beta Kappa: No
Applicants: 1,926
Accepted: 47%
Enrolled: 40%
Grad in 6 Years: 85%
Returning Freshmen: 91%
Academics: ✍ ✍ ✍ ✍ ✍
Social: ☎ ☎ ☎

(Continued)

Q of L: ★ ★ ★

Admissions: (610) 526-5152

Email Address:
admissions@brynmawr.edu

Strongest Programs:
Archaeology
Growth and Structure of Cities
Physics
Mathematics
Art History
Classics
Foreign Languages

additions include a 30,000-square-foot building for the departments of education and psychology and the renovation of four former faculty houses to create a student activity village called Cambrian Row.

Out of respect for their academic honor code, students refrain from discussing their grades, but they freely admit that they work hard. "The atmosphere is intense, and the classes are rigorous, but very interesting," says a political science major. A biology major adds, "Mawrtyrs have a commitment to engaging in intellectually stimulating reading, writing, and discussion." Most departments are strong, especially the sciences, classics, archaeology, art history, and the foreign languages, including Russian and Chinese. The Fine Arts Department, however, is relatively small. Doing serious work in music, art, photography, or astronomy requires a hike over to Haverford, Bryn Mawr's nearby partner in the "bi-college" system. Bryn Mawr handles the theater, dance, creative writing, geology, art history, Italian, and Russian programs for the two colleges, and the departments of German and French are joint efforts. Bryn Mawr also offers a rich variety of special programs. Approximately one-third of students study overseas during their junior year. Projects range from fieldwork in the Aleutian Islands with the Anthropology Department to studying Viennese architecture with the Growth and Structure of Cities Department.

> "The atmosphere is intense, and the classes are rigorous, but very interesting."

The general education requirements include two classes in each of the three divisions (social sciences, natural sciences, and the humanities), one semester of "quantitative" work, an intermediate level of competency in a foreign language, and the requirements of a major. Students are also required to take eight half-semesters of physical education and must also pass a swimming test. In addition, all freshmen are required to take two College Seminars to develop their critical thinking, writing, and discussion skills.

The quality of teaching at Bryn Mawr is unquestionably high. "Teachers are very accessible and endlessly interested in their material and in making sure that we learn it and appreciate it fundamentally," a senior says. "The teachers are brilliant and always available," a classmate adds. Freshmen and transfer students are initiated to the Bryn Mawr experience during Customs Week, which includes a variety of seminars and workshops as well as a tour of the campus and town. For those looking ahead to see what the steep tuition will buy in the long term, the campus has a career resource center that offers information on interviewing and building a résumé. It also brings recruiters to campus, offers mock interviews, and keeps students posted on internships.

> "We are willing to roll up our sleeves, whether it is to solve a complex math proof or fight poverty."

"Students at Bryn Mawr are extremely self-motivated, ambitious, intelligent, and confident," says a sophomore. African Americans make up 3 percent of the student body, Hispanics 3 percent, and Asian Americans 11 percent. To encourage diversity and harmony on campus, freshmen can take an intensive four-hour session during orientation on pluralism, which teaches students to examine assumptions about class, race, and sexual orientation. "Mawrtyrs are intellectually curious and stimulating," says a biology major, who adds, "We are willing to roll up our sleeves, whether it is to solve a complex math proof or fight poverty. Many of us are idealists, and we are not afraid to admit it or fight for what we believe in." Scholarships are not available, but the school does guarantee to meet the demonstrated financial need of everyone admitted.

Another guarantee is quality on-campus housing for all four years. "They usually do a stellar job of placing roommates together freshman year," says a senior. "After that, it is almost always possible to get a single if you want one." Dorm features

All freshmen are required to take two College Seminars to develop their critical thinking, writing, and discussion skills.

include hardwood floors, window seats, and fireplaces, says another senior. For good reason, the food service has received a national award from Restaurants and Institutions magazine. "It's not unusual to see the entire baseball team from neighboring Haverford College piling into Bryn Mawr's dining halls in the evenings. "The food is really that good!" says a student. A classmates adds, "I am a vegetarian, and I am in heaven every time I go into the dining hall."

Bryn Mawr is located on suburban Philly's wealthy Main Line (named after a railroad) and the campus is two blocks from the train station. "The town of Bryn Mawr is very much upper-class suburban," says an English major. "Coffee shops, bookstores, and restaurants provide student hangouts, while 'Main Line moms' frequent the gourmet grocery stores and overpriced gyms." A 20-minute train ride provides students with easy access to cultural attractions, as well as social and academic events at the nearby University of Pennsylvania. Students can also catch a bus to the mall or take a weekend trip to New York, the Jersey shore, or even Hershey Park. In addition to campus events, students can attend social events at nearby Haverford and Swarthmore colleges. "Much of the co-ed social life takes place on Haverford's campus," says a senior. Aside from being basically a women's college, "the social aspect of Bryn Mawr is

> "They usually do a stellar job of placing roommates together freshman year."

very hard at times considering how much work there is," says a political science major, "but that just means students need to take an active role in making a social life for themselves."

Tradition is a very important part of the campus social scene. The Elizabethan-style May Day festivities are held the Sunday after classes end in May. Everyone wears white, eats strawberries, and watches Greek plays. Students are known to skinny-dip in the fountains and drink champagne on the lawn. The presentation of lanterns and class colors to incoming freshmen on Lantern Night, and regal pageants, such as Parade Night, Hell Week, and Step-Sings, fill life with a Gothic sense of wonder and school spirit. Says a student, "They play a big role in uniting all four classes and give students a role in the greater history of the college." As for athletics, Bryn Mawr students are active in 12 intramural sports, including rugby, cross-country, volleyball, and field hockey. And, of course, there's always the champion badminton team. Club sports range from all-female teams, such as rugby, squash, and figure skating, to co-ed teams shared with Haverford College, such as ultimate Frisbee and fencing.

Bryn Mawr is a study in contrasts: the campus is in suburbia, but steps from a major city. Humanities programs are very strong, but science majors are also enormously popular. The students are independent but revel in college traditions. The result is overwhelmingly positive. Says a sophomore, "When I sit in a math or science course with 15 other women who are smart and eager to learn, I realize that I am surrounded by women who are going places."

Approximately one-third of students study overseas during their junior year. Projects range from fieldwork in the Aleutian Islands with the Anthropology Department to studying Viennese architecture with the Growth and Structure of Cities Department.

Overlaps

Smith, Wellesley, Swarthmore, Haverford, Brown, University of Chicago

If You Apply To ➢ **Bryn Mawr:** Early decision: Nov. 15. Regular admissions: Jan. 15. Housing: Jun. 1. Guarantees to meet demonstrated need. Campus and alumnae interviews: recommended, evaluative. SATs: required. SAT IIs: required (English and two others). Accepts the Common Application and electronic applications. Essay question: Common Application; what will you gain from the Bryn Mawr experience?

Bucknell University

Lewisburg, PA 17837

Bucknell, Colgate, Hamilton, Lafayette—all a little more conservative than the Ivy schools and nipping at their heels. Bucknell is the biggest of this bunch. (Perhaps Lehigh is a better comparison.) The central Pennsylvania campus is isolated but one of the most beautiful anywhere.

The students at Bucknell University strike a healthy balance between hitting the books and hitting the bars—or the frat houses of their pastoral central Pennsylvania campus. Yes, they tend to be preppy and outdoorsy—"Bucknell students are mostly upper middle class, relatively conservative, and materially conscious; however, they are also highly motivated and eager to succeed," says one junior. With small classes, caring faculty, and not a sub par dorm to be found, it's no wonder the only complaint is that "four years at Bucknell go by way too fast," says a junior.

In addition to being comfortable and friendly, Bucknell is physically beautiful. Located on a hill just south of quaint Lewisburg, the campus overlooks the scenic Susquehanna River valley. Playing fields, shaded by leafy trees, are sprinkled among the Greek Revival buildings. While some structures date from the 19th century, lending a fairy-tale quality, others are far more modern, including an $8 million engineering building, finished in 2004, with 38,600 square feet of office, lab, and classroom space. Bucknell is also nearing completion of a seven-year effort to renovate and refurbish 82 classrooms and 10 auditoriums. A $5 million engineering building opened in 2004 and provides additional office, laboratory, and classroom spaces.

Students in Bucknell's College of Arts and Sciences must complete four courses in the humanities, two in social science, three in natural science and math, and a first-year Foundation Seminar, designed to strengthen research, computing, and writing skills. Two of these courses must address "broadened perspectives" on human diversity and on the natural and fabricated world. In the College of Engineering, students have a common first semester, including a special course that introduces them to all five engineering disciplines. Along with major-related requirements, each student completes a capstone project during senior year, and must demonstrate competence in writing in order to graduate. Newer offerings include a new MS in environmental engineering as well as a major in biomedical engineering.

After fulfilling Bucknell's many requirements, students select from a variety of courses, including the popular Management 101, where students create and sell a product and donate their profits to charity. Approximately 40 percent of each graduating class studies abroad, and programs staffed by Bucknell professors take them to England, France, Spain, and Barbados. Relationships with other colleges and universities enable students—including engineers—to travel to more than 60 other nations, from Japan and Sweden to China, Argentina, and Australia. Independent study is common; in the past five years, 73 percent of chemistry majors have taken at least one research course for credit. The two-summer Institute of Leadership in Technology and Management allows engineering and business students to learn new ways to solve problems, while building their teamwork and communication skills. On-campus study the first summer is followed by an off-campus internship during the second. Honors programs attract scholars and the Presidential Scholars Program offers talented undergrads an opportunity to fulfill the work-study portion of their financial aid packages at higher wages than those for regular student employees.

> **"Four years at Bucknell go by way too fast."**

Back on campus, 95 percent of courses have fewer than 50 students, and the emphasis is on discussion and group work. "While students are competitive," explains a junior, "it is uncommon for students to try to undercut each other or not be supportive of classmates." Courses are challenging, but it helps that professors come to Bucknell because they want to teach: "It is clear that professors are dedicated to designing interesting courses that are also intellectually stimulating," says one student. While Bucknell is known for engineering, management, and the natural sciences, students say academics are strong across the curriculum and weaker programs are hard to find. "Even the smaller departments have dedicated professors and loyal students," says a junior. Administrators highlight programs in animal behavior, which benefits from an outdoor naturalistic primate facility for teaching and research, and in environmental relations, which includes not only science courses, but courses in the humanities, social policy, and civil engineering, too.

> **"It is uncommon for students to try to undercut each other or not be supportive of classmates."**

"The average Bucknell student is very involved," says a junior. "Everyone is brainy, but really has a good time as well." Twenty-eight percent are Pennsylvanians, and 72 percent went to public high school. Diversity has been slow in coming; Asian Americans are the largest minority group at 7 percent of the student body, while African Americans add 3 percent and Hispanics 2 percent. "A good number of students are active in political organizations on campus," reports one student. Another adds, "People debate fair trade and debt relief, and liberal versus conservative economic policies." Each year, Bucknell awards several merit scholarships, worth an average of $12,356 each, and six athletic scholarships—three each in men's and women's basketball.

Eighty-eight percent of Bucknellians live on campus; all first-years are required to do so, and since upperclassmen must obtain permission to leave, more than 300 live in five college-owned apartment buildings. "No one ever complains about the living situation," says a junior. "I have lived in three different dorms in my three years here and have liked them all for different reasons," a classmate adds. Beds have loft-style frames, helping students gain space, and though most freshmen live in doubles, students can usually get singles when they return as sophomores. About 25 percent of each entering class affiliates with one of the six intellectually focused "colleges": Arts, Humanities, Environmental, Global, Social Justice, and Society and Technology; a seminar on the theme of their college replaces the required Foundation Seminar that their classmates take. Students report that Bucknell dining is better than your average institutional fare and includes "local produce and lots of healthy and vegetarian options."

> **"Everyone is brainy, but really has a good time as well."**

Bucknell's Greek system draws 37 percent of the men and 39 percent of the women, though rush is delayed to the start of sophomore year. And while the Greeks are a driving force in campus social life, there are alternatives. "Most of our events happen on campus, but that's how we like it," says a French major. "We host concerts and speakers, show movies, and attend sporting events." Two student organizations arrange everything from carnivals to hypnotists and religious retreats, while the nonalcoholic, school-run Uptown nightclub offers dancing 'til dawn. When it comes to drinking, a point system hasn't stopped those younger than 21 from drinking, but it does provide for clarity and proportionality when offenses occur. "I think it is a decent deterrent for really dumb things like fighting or driving while drunk," offers one junior. For those who choose not to imbibe, support may be found with CALVIN and HOBBES—that's Creating a Lively, Valuable, Ingenious, and New Habit of Being at Bucknell and Enjoying Sobriety.

Market Street, in downtown Lewisburg, has an old-style movie theatre that serves up first-run flicks and food, and is also "welcoming and warm and perfect for Bucknell

Bucknell students get the best of several worlds: excellence in engineering and the liberal arts, abundant research opportunities, and a healthy social life.

In the College of Engineering, students have a common first semester, including a special course that introduces them to all five engineering disciplines.

students," says one junior. The nearby town of Bloomsburg offers a more ethnic feel, with Indian and Thai cuisine. Through the "I Serve 2" campaign, Bucknell is trying to get all students to complete at least two hours of community service each semester. Projects are coordinated through BISON, Bucknellians in Service to Our Neighbors. When students get claustrophobic, New York, Philadelphia, and Washington/Baltimore are less than three hours away; the main campus of Penn State, in State College, Pennsylvania, is even closer. Bucknell also sponsors road trips to these communities for students who lack wheels. Favorite traditions include Midnight Mania (the official start of the basketball season), House Party Weekend and the formal Chrysalis Ball in the spring, and First Night, "a ceremony congratulating first-year students on the completion of their first semester," says a junior. "They learn the alma mater and serenade the president and his wife." "A great tradition occurs during orientation when the entire first-year class walks through the Christy Mathewson Gates. Four years later, at graduation, you walk through the gates in the opposite direction," says a sophomore.

Bucknell has captured the Patriot League Presidents' Cup, for the league's all-sports champion, 11 times in 15 years—including seven of the past eight. Men's and women's cross-country and track and field are perennially strong, and men's lacrosse has won four of the past six league titles. Men's and women's swimming and diving are also notable, while the women's crew team won a gold medal at Philadelphia's Dad Vail Regatta in recent years. Bucknell's biggest rivalries are with Lafayette and Lehigh, though these aren't a tremendous focus. More than 50 intramural sports draw about a third of the students, while others praise the weights and cardio machines at the new Kenneth Langone Athletic and Recreation Center, named for the former Home Depot executive and New York Stock Exchange board member.

Bucknell students get the best of several worlds: excellence in engineering and the liberal arts, abundant research opportunities, and a healthy social life. The school's central Pennsylvania location is lovely but isolated and the preponderance of preppies may seem stifling, but this campus is slowly becoming both more liberal and more diverse. If you're seeking small classes and professors who really care, in a supportive environment with plenty of school spirit, Bucknell may be a good fit.

Overlaps

Lehigh, Colgate, Boston College, Cornell University (NY), Villanova, Tufts

If You Apply To ➤

Bucknell: Early decision: Nov. 15, Jan. 1. Regular admissions and financial aid: Jan. 1. Does not guarantee to meet full demonstrated need. Campus and alumni interviews: recommended, informational. SATs or ACTs: required. SAT IIs: optional (foreign language required for any student planning to major in that language). Accepts the Common Application and electronic applications. Essay question: significant experience, achievement, risk or ethical dilemma; issue of personal, local, national or international concern; influential person, fictional character, historical figure, or creative work; or topic of your choice.

California Colleges and Universities

California's three-tiered system of colleges and universities has long been viewed as a model of excellence by other public higher education institutions nationwide and even around the world. Many have attempted to emulate its revered status, which offers a wealth of educational riches, including world-class research universities, enough Nobel Prize winners to fill a seminar room, and colleges on the cutting edge of everything from film to viticulture. Underlying the creation of this remarkable system was a commitment to the notion that all qualified Californians, whatever their economic status, were entitled to the benefits of a college education. In pursuing this ideal, California led the nation in opening up access to higher education for African Americans, Hispanics, and other previously disenfranchised groups.

Unfortunately, in the early 1990s, this golden dream started fading due to the state's recession, population growth, and many other contributing factors. As a result, California's public universities and colleges received reduced tax support, student fees shot up, student/faculty ratios increased, fewer classes were offered, and, in some cases, entire academic programs were eliminated. Although it still remained relatively lower than most states, the total cost to attend a UC campus started to climb.

The system is composed of the ten combined research and teaching units of the University of California (UC) and 23 state universities (CSU), including the newest CSU campus at Channel Islands, that focus primarily on undergraduate teaching. It also includes 106 two-year community colleges that offer both terminal degrees and the possibility of transferring into four-year institutions.

Admissions requirements to the three tiers and the institutions within them vary widely. Community colleges are open to virtually all high school graduates. The top third of California high school graduates (as measured statewide by a combination of SAT scores and grade point average) may attend units of the CSU; all applicants must have taken a course in the fine or performing arts to be considered for admission. In the past, students in the top 12.5 percent of their class have been eligible to attend the University of California. In-state students graduating in the top 4 percent of their high school class will be guaranteed admission to the UC system, although not to a particular campus. The 4 percent proposal is part of a plan to broaden the representation of California applicants and to give more weight to GPA and SAT Subject Tests. Out-of-state students continue to face ferocious competition for a limited number of spots, and still pay more.

The University of California (UC) system boasts 209,000 students, 170,000 faculty and staff and 1.4 million living alumni. Although one university system, the nine undergraduate UC campuses (San Francisco is for upper division and graduate students) each offer a full range of academic programs, and each has its own distinctive character. The most recent addition, Merced, opened in 2005 as the first American research university to be founded in the 21st century. In recent years, UC has moved from relying primarily on statistical academic information to a "comprehensive review" that takes into consideration not only coursework and test scores, but also leadership, special talent, and the educational opportunities available to each student. Despite state laws that prohibit the university from considering race in admission, the system remains dedicated to achieving a diverse student body. The university offers a number of outreach programs designed to assist low-income or educationally disadvantaged students who have promising academic potential with admissions and support services.

To apply for admission to the University of California, complete the electronic application available at www.universityofcalifornia.edu/admissions. Prospective students may apply to as many as nine UC undergraduate campuses on the application. It should be understood that each campus to which a student applies reviews the application using its own methodology and criteria. Decisions are thus campus-unique. (One campus does not know what another campus is planning to do with any given applicant.) Each of the major undergraduate UC campuses receives a full-length summary in the following pages.

The California State University System is totally separate from the University of California; in fact, the two institutions have historically competed for funds as well as students. The largest system of senior higher education in the nation, CSU focuses on undergraduate education; while its members can offer master's degrees, they can award doctorates only in collaboration with a UC institution. Research in the state-university system is severely restricted, a blow to CSU's national prestige but a big plus for students. Unlike UC, where the mandate to publish or perish is alive and well, teachers in the state system are there to teach. CSU's biggest problem is the success of UC, and its frequent lament—"Anywhere else we'd be number one"—is not without justification.

The 23-campus system caters to more than 417,000 students a year. And while most of the campuses serve mainly commuters, Chico, Humboldt, Monterey Bay, San Luis Obispo, and Sonoma stand out as residential campuses. While a solid liberal arts education is offered, the stress is usually on career-oriented professional training. Size varies dramatically, from more than 32,000 students at San Diego State and Long Beach to fewer than 4,000 at several other branches, like Channel Islands and Monterey Bay. Each campus has its own specific strengths, although in most cases a student's choice of school is dictated by location rather than by academic specialties. For those with a wider choice, some of the more distinctive campuses are profiled below.

Chico (enrollment 16,000), situated in the beautiful Sacramento Valley, draws a large majority of its students from outside a 100-mile radius. The on-campus undergraduate life is strong and the social life is great. Bakersfield (7,800) and San Bernardino (16,500) boast residential villages along with more conventional dorms. The former is in a living/learning center with affiliated faculty members; the latter has its own swimming pool. California Polytechnic at San Luis Obispo (19,900) is the toughest state university to get into. It provides excellent training in the applied branches of such fields as agriculture, architecture, business, and engineering. Fresno (20,400), located in the Verdant Central Valley, has the only viticulture school in the state outside of UC–Davis, and undergraduates can work in the school winery. Yosemite, Kings Canyon, and Sequoia national parks are nearby.

San Diego State (32,700) is the balmiest of the campuses, and since it has a more residential, outdoorsy, and campus-oriented social scene, it appeals more to traditional-age undergraduates. "You could go for the weather alone—some do," says one former student. Contrasted with most other state schools, athletics are very important, and the academic offerings are almost as oriented to the liberal arts as at its UC neighbor at San Diego.

Humboldt State (7,500) is perched at the top of the state near the Oregon border in the heart of the redwoods. Humboldt's forestry and wildlife departments have national reputations, and the natural sciences are, in general, strong. Students have the run of excellent laboratory facilities and Redwood National Park. Most in-staters come here to get away from Los Angeles and enjoy the rugged coastline north of San Francisco.

California Maritime Academy, located 30 miles northeast of San Francisco with 800 students, specializes in marine transportation, engineering, and maritime technology, and requires summer cruises on the T.S. Gold Bear. Monterey Bay, one mile from the beach with 3,000 students, 65 percent of whom live on campus, offers an interdisciplinary focus with global perspective and opportunities for internships.

To apply to a California State University, complete the electronic application available at www.CSUMentor.edu. Students can apply to as many campuses as they wish, but each application will require a separate fee.

UC–Berkeley

110 Sproul Hall # 5800, Berkeley, CA 94720-5800

Like everything else, the academic side of Berkeley can be overwhelming. With more than 23,000 undergraduate overachievers crammed into such a small space, it is no wonder that the academic climate is about as intense as you can get at a public university.

Website: www.berkeley.edu
Location: Urban
Total Enrollment: 33,558
Undergraduates: 23,482
Male/Female: 44/56
SAT Ranges: V 590–710
  M 630–740
ACT Range: N/A
Financial Aid: 49%
Expense: Pub $ $ $
Phi Beta Kappa: Yes
Applicants: 36,989
Accepted: 27%
Enrolled: 41%
Grad in 6 Years: 87%
Returning Freshmen: 97%
Academics: 🖋🖋🖋🖋🖋
Social: 🐻🐻🐻🐻
Q of L: ★★★
Admissions: (510) 642-3175
Email Address:
  ouars@berkley.edu

Berkeley. Mention the name, and even down-to-earth students get stars in their eyes. Students who come here want the biggest and best of everything, though sometimes that ideal runs headlong into budget cuts, tuition increases, and housing shortages. Never mind. Berkeley is where the action is. If you want a quick indicator of Berkeley's academic prowess, look no farther than the parking lot. The campus is dotted with spots marked "NL"—spots reserved for resident Nobel laureates. The last time anyone counted, Berkeley boasted seven Nobel Prize winners, 391 Guggenheim fellows, and a bevy of Pulitzer Prize recipients, MacArthur fellows, and Fulbright scholars. Is it any wonder that this radical institution of the '60s still maintains the kind of reputation that makes the top private universities take note? The social climate at this mother of UC schools is not as explosive as it once seemed to be, but don't expect anything tame on today's campus. Flower children and granola chompers still abound, as do fledgling Marxists, young Republicans, and body-pierced activists.

Spread across 1,200 scenic acres on a hill overlooking San Francisco Bay, the Berkeley campus is a parklike oasis in a small city. The startlingly wide variety of architectural styles ranges from the stunning classical amphitheater to the modern University Art Museum draped in neon sculpture. Large expanses of grass dot the campus and are just "perfect for playing Frisbee or lying in the sun." The oaks along Strawberry Creek and the eucalyptus grove date back to Berkeley's beginnings nearly 140 years ago. Sproul Plaza, in the heart of the campus, is one of the great people-watching sites of the world.

But of course, as well as gorgeous, Berkeley is also academically intense. "Everyone was the top student in his or her high school class so they can't settle for anything less than number one," says one student. A classmate concedes that "it can be a stressful environment, especially during the first years." Another says tersely, "Expect very little sleep." Some introductory courses, particularly in the sciences, have as many as 800 students, and professors, who must publish or perish from the university's highly competitive teaching ranks, devote a great deal of time to research. After all, Berkeley has made a large part of its reputation on its research and graduate programs, many of which rank among the best in the nation.

And while the undergraduate education is excellent, students take a gamble with the trickle-down theory, which holds out the promise that the intellectual might of those in the ivory towers will drip down to them eventually. As a political science major explains, "This system has allowed me to hear outstanding lectures from amazing professors who write the books we read, while allowing far more personal attention by the graduate-student instructors." Another student opines, "It's better to stand 50 feet from brilliance than five feet from mediocrity." Evidence of such gravitation is

"Expect very little sleep."

seen in the promising curriculums designed specifically for freshmen and sophomores that include interdisciplinary courses in writing, public speaking, and the history of civilization, and an offering of small student seminars (enrollment is limited to 15) taught by regular faculty. Despite these attempts at catering to undergraduates, the sheer number of students at Berkeley makes it difficult to treat each student as an individual. As a result, such things as academic counseling suffer. "Advising? You mean to tell me they have advising here?" asks one student.

Each college or school has its own set of general education requirements, which are generally not extensive, and many can be fulfilled through advanced placement exams in high school. All students, however, must take English composition and literature, and one term each of American history and American institutions. Also, undergrads have an American Cultures requirement for graduation—an original approach (via courses offered in several departments) to comparative study of ethnic groups in the United States.

Most of the departments here are noteworthy, and some are about the best anywhere (like engineering and architecture). Business, sociology, mathematics, physics, chemistry, history, economics, and English are just a handful of the truly dazzling departments. Berkeley offers seven departments and seven interdisciplinary programs in engineering, and the biological sciences department integrates several undergraduate majors in biochemistry, biophysics, botany, zoology, and others into more interdisciplinary programs such as integrative biology and molecular and cell biology. Stanley Hall is being rebuilt to serve as the Berkeley headquarters for the California Institute for Quantitative Biomedical Research (QB3). The office and lab complex will support interdisciplinary teaching and research as part of the campus's Health Science Initiative. The College of Natural Resources has streamlined its eight departments into five, and participates in five interdisciplinary research centers.

Special programs abound at Berkeley, though it's up to the student to find out about them. "Our class enrollment system is much like playing a low-risk lottery," opines one undergrad. "Maybe you'll win, or maybe you won't. If anything, adding courses will

"It's better to stand 50 feet from brilliance than five feet from mediocrity."

definitely toughen up any person." Students may study abroad on fellowships at one of 50 centers around the world, or spend time in various internships around the country. If all you want to do is study, the library system, with more than eight million volumes, is one of the largest in the nation and maintains open stacks. The system consists of the main library (Doe-Moffitt) and more than 20 branch libraries, including the Hargrove Music Library.

Forty-one percent of the student population is Asian American, 4 percent African American, and 11 percent Hispanic. The Coalition for Excellence and Diversity in

(Continued)
Strongest Programs:
Engineering
Architecture
Business
Theoretical Physics
Molecular and Cell Biology
Political Science
English

The common denominator in the Berkeley community is academic motivation, along with the self-reliance that emerges from trying to make your mark among upward of 23,000 peers.

Mathematics, Science, and Engineering, which provides women and minorities with undergraduate mentors in these fields, is highly regarded. The university also provides a variety of other programs to promote diversity, including Project DARE (Diversity Awareness through Resources and Education), the Center for Racial Education, and a Sexual Harassment Peer Education Program. Despite Berkeley's liberal reputation, the recent trend is away from the legacy of the free speech movement. Business majors and fraternity members outnumber young Communists and peaceniks, though the school does produce a large number of Peace Corps volunteers. Recently, chain stores have drawn the ire of students and the cost of tuition remains an issue of concern for nearly everyone on campus. In the past few years, outrageous fee hikes and severe budget cuts had some students wondering if a first-rate, affordable education had gone the way of the dinosaurs.

Though dorms have room for only a quarter of the students, freshmen are guaranteed housing for their first year. After that, the Cal Rentals is a good resource for finding an apartment in town. Many students live a couple of miles off campus, where "apartments are cheaper," says one student. About two-thirds of the university's highly prized dorm rooms are reserved for freshmen, and the few singles go to resident assistants. A number of new student housing projects have opened in recent years, offering a variety of rooms in low-rise and high-rise settings. In the absence of a mandatory meal plan, everybody eats "wherever and whenever they wish," including in the dorms.

"Social life at UC–Berkeley is killer!"

Though the housing shortage can get you down, the beautiful California weather will probably take your mind off it in time. The BART subway system provides easy access to San Francisco, by far one of the most pleasant cities in the world and a cultural and countercultural mecca. The Bay Area boasts myriad professional sports teams, including the Oakland A's and the San Francisco 49ers. From opera to camping, San Francisco has a wide variety of activities to offer. Get yourself a car, and you can hike in Yosemite National Park, ski and gamble in Nevada, taste wine in the Napa Valley, or visit the aquarium at Monterey. But be advised that a car is only an asset when you want to go out of town—students warn that parking in Berkeley is difficult, to say the least.

"Social life at UC–Berkeley is killer!" exclaims one geography major. Weekends are generally spent in Berkeley, hanging out at the many bookstores, coffeehouses, and sidewalk cafes, heading to a fraternity or sorority party, or taking advantage of the many events right on campus. Berkeley is a quintessential college town ("kind of a crazy little town," opines one anthropology major), and of course, there's always the people-watching; where else can an individual meet people trying to convert pedestrians to strange New Age religions or revolutionary political causes on every street corner? Nearby Telegraph Avenue is famous (infamous?) for such antics every weekend. More than 350 student groups are registered on campus, which ensures that there is an outlet for just about any interest and that no one group will ever dominate campus life.

Despite all this activity, many students use the weekend to catch up on studying. Greeks have become more popular, with 10 percent of the men and 10 percent of the women in a fraternity or sorority. Varsity athletics have always been important, with strengths in the men's gymnastics and crew teams. A surge in popularity for the basketball team probably has to do with its great performance in the PAC 10. And just about everyone turns out for the "Big Game," (football) where the favorite activity on the home side of the bleachers is badmouthing the rival school to the south: Stanford. Intramurals are popular, and the personal fitness craze is fed by an extensive recreational facility and gorgeous weather year-round.

The common denominator in the Berkeley community is academic motivation, along with the self-reliance that emerges from trying to make your mark among upward of 23,000 peers. Beyond that, the diversity of town and campus makes an extraordinarily free and exciting college environment for almost anyone. "It makes one feel free to dress, say, think, or do anything and not be chastised for being unorthodox," explains a student. "At Berkeley, it is worse to be dull than odd."

<table>
<tr><td>If You Apply To ➢</td><td>Berkeley: Regular admissions: Nov. 30. Guarantees to meet demonstrated need of in-state students. No campus or alumni interviews. SATs or ACTS: required. SAT IIs: required (writing, math I or II, and one other). Essay question: personal statement. Apply to particular school or program.</td></tr>
</table>

UC–Davis

175 MRAK Hall, Davis, CA 95616

The agricultural and engineering branch of the UC system. Premed, prevet, food science—you name it. If the subject is living things, you can study it here. A small town alternative to the bright lights of UC–Berkeley and UCLA. As is often true of science-oriented schools, the work is hard.

At the University of California–Davis, environmental studies and most everything that has to do with agriculture or biological science is noteworthy. The Aggies' cup truly runneth over. Originally known as the University of California Farm, the campus maintains its sprawling, verdant beauty, replete with native and imported forestry, charming bike paths, and mooing cows. But lest you assume this environmentally oriented university is full of quaint country folk, think again. Davis has become an international leader in the agricultural, biological, biotechnical, and environmental sciences.

Located 15 miles west of Sacramento and 72 miles north of San Francisco, the 6,000-acre campus is in the middle of a stretch of flat farmland that even Dorothy and Toto could mistake for Kansas. It features nearly 1,000 buildings with a blend of architectural styles, from traditional dairy barn to modern concrete. The hub of the university is a central area known as the Quad, one of many grassy open spaces on campus. Newer facilities include the Center for Comparative Medicine and a variety of seismic renovations.

Though it has added programs in many disciplines over the past few years— including Chinese, Japanese, food engineering, and biological systems engineering— its biological and agricultural science departments are still the ones that shine. Animal science and engineering are strong departments, and the botany program is one of the best in the country. The school is "the No. 1 choice for any prevet," and it's great for premeds, too. The food sciences major is also stellar, and not for the faint of heart or those afraid of chemistry. It was Davis food scientists who gave us the square tomato (better for packing into boxes), as well as more useful things such as the method for creating orange juice concentrate. Studio art, boasting several internationally known artists, is also among the top in the nation, while history and English are generally good but not up to par with the sciences. Noteworthy special programs include the Inter-Disciplinary Electronics Arts (IDEA) Lab, which allows students to create electronically based productions by integrating photography, video, digital editing, and the Internet. Internships and co-op programs are well-established, which is why many students remain for more than four years.

Website: www.ucdavis.edu
Location: Small city
Total Enrollment: 26,094
Undergraduates: 20,388
Male/Female: 44/56
SAT Ranges: V 510–630
  M 550–660
ACT Range: 22–27
Financial Aid: 62%
Expense: Pub $ $ $
Phi Beta Kappa: Yes
Applicants: 22,224
Accepted: 63%
Enrolled: 27%
Grad in 6 Years: 73%
Returning Freshmen: 90%
Academics: ✍ ✍ ✍ ✍
Social: ☎ ☎ ☎
Q of L: ★ ★ ★ ★
Admissions: (530) 752-2971
Email Address:
  thinkucd@ucdavis.edu

Strongest Programs:
Environmental Studies
Botany
Animal Science
Viticulture
Agricultural Sciences

(Continued)
Studio Art
Biological Sciences
Engineering

Faculty members here are expected to do top-level research as well as teach, so Davis is charged with both education and research. These two are uniquely blended when undergraduate students contribute to first-class research groups as paid technicians or volunteer interns. Davis also offers the innovative Washington Program, which gives undergraduates academic credits for internships in Congress, at federal agencies, and the like. Many introductory courses are quite large, but Davis also offers 40 freshman seminars taught by the best instructors. The academic advising system gets generally high marks, but you must seek out their assistance. "They helped me plan a four-year college schedule and always kept me on track."

"They helped me plan a four-year college schedule and always kept me on track."

General education requirements stipulate that all students take courses in three broad areas: topical breadth, social–cultural diversity, and writing experience. These areas include courses in the arts and humanities, science and engineering, and the social sciences. Students may elect to take a general education theme option (sets of general education courses that share a common intellectual theme).

The academic demands are intense, and the students are high achievers. Many students describe the atmosphere as competitive if not cutthroat (especially in the biological sciences). "It is not rare to find many students in the library on Saturday night," testifies one student. Another student reports, "Professors expect students to learn vast amounts of information in a 10 week span." For students who still want more, the Davis Honors Challenge is designed for highly motivated, academically talented first- and second-year students who want to enhance their education through special courses. A famous campus saying claims that "Davis students take notes at graduation." Maybe they're taking notes for job interviews: 64 percent of Davis students get jobs after college and 38 percent prefer more class time in graduate school. African Americans account for 3 percent of the students, Asian Americans 35 percent, and Hispanics 10 percent. Students are slightly more conservative than in past years, and most are characterized as "friendly and open-minded." Campus hot topics include fair labor practices and political correctness. In its pledge to foster awareness of diversity issues, the university has established an Office of Campus Diversity and a Crosscultural Center. UC–Davis boasts the highest graduation rate in the UC system. Davis awards merit scholarships, but there are no athletic awards.

A famous campus saying claims that "Davis students take notes at graduation."

Virtually all freshmen inhabit campus housing, which is well-maintained and includes a number of theme houses. The vast majority of upperclassmen live off campus in nearby houses or apartments. Housing is guaranteed for freshmen and transfer students if applications are received by the deadline. Six different meal plans for the dining halls are available, and one student says, "The dorm food is very good (better than at most colleges)." A variety of nearby eating establishments serve the student clientele, but a car can come in handy if you are looking for a good meal in Sacramento (15 minutes) or a great one in San Francisco (a little more than an hour). Beaches are a two-hour drive from the campus, and the ski slopes and hiking trails of Lake Tahoe and the Sierra Nevada are a little closer. But if you feel, as most Davis students do, that studies are too important to be abandoned on weekends, the town has restaurants, activities, and entertainment enough to keep the stay-at-homes happy.

"It is not rare to find many students in the library on Saturday night."

In between quizzes and cram sessions, the outlying countryside offers a welcome change of pace. The town of Davis itself is small, about 50,000, and students make up half the population. If some call it a cowtown, others call it peaceful, with its tree-lined streets and quiet nights. The relationship between college and town is one of rare cooperation (partly because the students are a significant voting bloc in local elections). Health and energy consciousness runs high in town and on the vast, architecturally diverse campus, where bicycles are the main form of transportation on the incredible 46 miles of bike paths that crisscross the campus and environs. "Bicycles are the norm at Davis. Don't come without one," advises one psych major. The university has encouraged environmental awareness by sponsoring solar energy projects and promoting such novelties as contests between dorms for the lowest heating and electric bills.

The university's varsity athletic teams compete in Division II and attract relatively scant attention compared to those at most other state universities.

On-campus activities are varied, and many university-sponsored events fill the calendar. One rhetoric major points out that "social functions are hard to avoid at Davis." Active drama and music departments provide frequent entertainment, and there is plenty of room for homegrown talent in the coffeehouses, which offer mellow live entertainment and poetry readings on a regular basis. The 1,800-seat Mondavi Center for the Arts features international and local groups. Fraternities and sororities attract 7 percent of the men and 6 percent of the women. Alcohol is allowed in the dorms for those over 21 years old; those too young to imbibe have trouble finding booze, unless it's supplied

"Bicycles are the norm at Davis. Don't come without one."

by peers. Major annual social events include Picnic Day, in which alumni join current students in a massive outdoor shindig; African American Week; and the Whole Earth Festival, "an earthy, tie-dyed sort of event" in celebration of the '60s.

The university's varsity athletic teams compete in Division II and attract relatively scant attention compared to those at most other state universities. Nevertheless, Davis won the Sears Directors' Cup, a trophy symbolic of overall excellence in intercollegiate athletics, and cross-country, basketball, and track and field have brought home NCAA championships. The annual Causeway Classic against rival Sacramento State does create a measure of excitement. Intramurals, however, are much more popular than spectator sports, with 65 percent of students participating. On this outdoor campus, almost everyone does something athletic—jogging, softball, tennis, swimming, or Frisbee—if only to break up the monotony of studies with a different kind of competition.

Proud of its small town atmosphere, Davis is not for the lazy or faint of heart. As one man says, "There's no free ride. You are going to have to work for everything you get." And most students get a lot out of their four or more years at Davis. It's the ideal spot to combine high-powered work in science and agriculture with that famous easygoing California lifestyle.

> ## Overlaps
> **UC–Berkeley, UCLA, UC–San Diego, UC–Santa Barbara, UC–Irvine**

If You Apply To ➤ **Davis:** Regular admissions: Nov. 30. Financial aid: Mar. 3. Does not guarantee to meet demonstrated need. No campus or alumni interviews. SATs or ACTs: required. SAT IIs: required (writing, math, and one other). Accepts the Common Application and electronic applications. Essay question: personal statement. Apply to particular program.

UC–Irvine

260 ADM, Irvine, CA 92697

Irvine sits in the midst of one of the nation's biggest suburbs, combining funky modern architecture with perhaps the most conservative student body in the UC system. Premed is the featured attraction, along with various other health-related offerings. Not quite as close to the beach as Santa Barbara—but close enough.

Website: www.uci.edu
Location: Urban
Total Enrollment: 23,032
Undergraduates: 19,201
Male/Female: 50/50
SAT Ranges: V 519–620
  M 567–675
ACT Range: N/A
Financial Aid: 49%
Expense: Pub $ $ $
Phi Beta Kappa: Yes
Applicants: 34,417
Accepted: 54%
Enrolled: 22%
Grad in 6 Years: 79%
Returning Freshmen: 92%
Academics: ✑ ✑ ✑ ✑
Social: ☎ ☎
Q of L: ★ ★ ★
Admissions: (949) 824-6703
Email Address:
  admissions@uci.edu

Strongest Programs:
Biological Sciences
Economics
Information and Computer
  Science
Chemistry

There are 18 sororities and 18 fraternities, and each has something going on every weekend.

On the surface, UC–Irvine's clean, contemporary campus appears to be home to students who study diligently in the busy library, wear sensible shoes to biology lab, and resist that double shot of espresso at the local coffeehouse. But that image starts to dissipate as soon as you hear that bizarre noise: "Zot! Zot! Zot!" Then a UCI student explains that "it's the sound that an anteater supposedly makes when it swipes an ant with its tongue." Hey, any school that has an anteater as a mascot can't be completely straight-laced. The university is, however, straight on its reputation as a school with stellar programs in biology and creative writing. The current academic climate can be quite serious and challenging, but as one UCI student swears, the Anteaters are "also surprisingly cooperative."

"It is the perfect size."

Located in the heart of Orange County and founded in 1965, UCI is among the newest of the UC campuses. Although enrollment is up and the administration has dreams of further expansion, "it is the perfect size," says one English major. UCI is liberally supplied with trees and shrubs from all over the world. Futuristic buildings are arranged in a circle around a large park, "giving it the appearance of a relaxed art school," says one observer. Undergraduates have long quipped that UCI stood for "Under Construction Indefinitely," and current campus construction does little to challenge the moniker.

A "premed mentality" reigns at Irvine, since the School of Biological Sciences is the best and most competitive academic division. The School of Arts offers nationally ranked programs in dance, drama, music, studio art, and musical theatre, as well as a minor in digital arts. The Beall Center for Art and Technology in the Claire Trevor School of the Arts enables students to explore the relationship between digital technology and the arts and sciences. The popular interdisciplinary School of Social Ecology offers courses combining criminology, environmental and legal studies, and psychology and social behavior, and strongly emphasizes teacher/student relationships. Like most of the other UC campuses, UCI is on a 10-week quarter system, so the pace is fast and furious. Students should face registration with the same determination, too; it's a tough fight to get into the science classes of choice as a sophomore.

Languages are strong at UCI, as are the biggest nonbiology majors: economics; information and computer science; psychology; social behavior; criminology, law and society; and a fiction-writing program that is gaining national recognition. UCI has added a plethora of new programs, including majors in global cultures, literary journalism, biomedical engineering, and German studies.

"UCI is fairly competitive and the courses are moderately rigorous," says a junior. Students may be overwhelmed by the size of most classes. Even seniors find their classes packed with 100 undergrads. "Graduate students teach lower-division writing courses," says one student, adding that "most classes are overcrowded, leaving little room for personal attention."

"UCI is fairly competitive and the courses are moderately rigorous."

The Center for Health Sciences focuses on five areas of research, including: neuroscience, genetics, cancer, infectious diseases, and aging. The university also houses the Reeve-Irvine Research Center, which supports the study of spinal cord trauma and disease with emphasis on finding a cure. The "breadth requirement" means that students must take three courses each in writing, natural sciences, social and behavioral sciences, and humanities in order to graduate. There is also a foreign language requirement, and one in math, statistics, or computer science, as well as requirements in multicultural and international/global issues. Honors programs are available in humanities, economics, psychology, political science, physics, cognitive sciences, anthropology, and mathematics.

Ninety-seven percent of the student body are in-staters, the majority from Southern California and many of those from wealthy Orange County. The students are in general "much more conservative than at the other UC campuses," says one applied math major. Minorities account for well over half the student body, with Asian Americans comprising 49 percent, African Americans 2 percent, and Hispanics 12 percent. "Cultural groups seem to segregate from each other more than I really like," says a senior.

Condominium-style dorms, both single-sex and co-ed, are "exceptional compared to the high-rise dormitories of other institutions," says one senior. Others agree that the homey campus dwellings provide a good experience for freshmen, though finding a room can be a challenge. "If you really want on-campus housing," warns a student, "you need to make sure you meet the deadlines." Newly added housing, including those with academic themes and ones especially for fraternities and sororities, opens more rooms for students, but most opt to move off campus after their first year. Currently, 22 percent of freshmen live off campus—many on the beach—giving the campus a commuter-school atmosphere. One student laments, "You have to find the social life on this campus. It won't find you."

Still, the Greek scene is vigorous, attracting 8 percent of UCI men and women. There are 18 sororities and 18 fraternities, and each has something going on every weekend. As for booze, UCI is a dry campus and students say finding a drink on campus without proper ID is difficult. Irvine touts many festivals that seem to attest to a celebration of diversity: the Rainbow Festival (cultural heritage), Asian Heritage week, Black History month, Cinco de Mayo, and rush week. The one event that brings everybody out is the daylong Wayzgoose, when the campus is transformed into a medieval fair complete with mimes, jugglers, and performers dressed up in medieval costumes.

> "If you really want on-campus housing, you need to make sure you meet the deadlines."

But if life on campus is slow, beyond it is not. That's because the campus is located just 50 miles from L.A., five miles from the beach, and a little more than an hour from the ski slopes. Catalina Island, with beaches and hiking trails, is a quick boat trip off Newport Harbor; Mexico is two hours away. While some students treasure the quiet setting of Irvine, others lament its "lackluster, homogeneous communities." Notes one student, "UCI and the city of Irvine seem like completely different entities; the former is slightly liberal while the latter is ultraconservative."

Irvine fields 20 athletic teams and competes in Division I of the NCAA. Tennis and cross-country are perennial Big West powerhouses, and men's water polo has been ranked in the top five nationally for 23 of the last 31 years. There is no football team, but intramurals are extremely popular, as is the 5,000-seat multipurpose gym.

What lures students to UCI is its top-name professors, innovative academic programs, and the chance to be a part of its cutting-edge research. For the students who come here prepared to keep their heads buried in a book for a few years, the reward will be an exceptional education.

> *Currently, 22 percent of freshmen live off campus—many on the beach—giving the campus a commuter-school atmosphere.*

Overlaps

UCLA, UC–San Diego, UC–Santa Barbara, UC–Berkeley, University of Southern California

If You Apply To >

Irvine: Regular admissions: Nov. 30. Financial aid: Mar. 2. Housing: May 1. No campus or alumni interviews. SATs or ACTs: required. SAT IIs: required (English composition, math, and one other). Accepts electronic applications. Essay question: autobiographical statement.

UC–Los Angeles

1147 Murphy Hall, 405 Hillgard Avenue, Los Angeles, CA 90095

Tucked into exclusive Beverly Hills with the beach, the mountains, and chic Hollywood hangouts all within easy reach. Practically everything is offered here, but the programs in arts and media are some of the best in the world. More conservative than Berkeley and nearly as difficult to get into.

Website: www.ucla.edu
Location: Urban
Total Enrollment: 37,221
Undergraduates: 24,811
Male/Female: 44/56
SAT Ranges: V 570–690
  M 600–720
ACT Range: 24–30
Financial Aid: 53%
Expense: Pub $ $ $
Phi Beta Kappa: Yes
Applicants: 42,227
Accepted: 27%
Enrolled: 39%
Grad in 6 Years: 87%
Returning Freshmen: 97%
Academics: ✑ ✑ ✑ ✑ ✑
Social: ☎ ☎ ☎
Q of L: ★ ★ ★
Admissions: (310) 825-3101
Email Address:
  ugadm@saonet.ucla.edu

Strongest Programs:
Music
Engineering
Political Science
Dance
Economics
Psychology
Biology
Film/television

With stellar programs in music, film and television, journalism/communication, dance, and drama, you'd think UCLA was some kind of incubator for truly talented and gifted people. Or with alumni such as Kareem Abdul-Jabbar, Troy Aikman, and Arthur Ashe, maybe UCLA's some sort of farm that grows superstar athletes. Well, UCLA is all that and more. A superb faculty, a reputation for outstanding academics, and a powerful athletics program make this university the ultimate place to study.

UCLA's prime location—sandwiched between two glamorous neighborhoods (Beverly Hills and Bel Air) and a short drive away from Hollywood, the Sunset Strip, and downtown Los Angeles—makes it appealing for students who want more from their college experience than what classes offer. The beautifully landscaped 419-acre campus features a range of architectural styles, with Romanesque/Italian Renaissance as the dominant motif, providing only one of a number of reasons students enjoy staying on campus. A wealth of gardens—botanical, Japanese, and sculpture—adds a touch of quiet elegance to the campus. New facilities include the California Nanosystems Institute, an array of studios and theaters, and new residence halls.

Strong programs abound at UCLA, and many are considered among the best in the nation. The School of Engineering and Applied Science, especially electrical engineering, is generally regarded as the leading department. The School of Film, Theater, and Television is first-rate, and its students have the opportunity to study in Verona, Italy, with the Theater Overseas program. The popular music department offers a course in jazz studies, and the biological sciences are also highly regarded. Global studies is the newest undergraduate major. Research opportunities abound at UCLA, and the university ranks seventh in the nation in federal funding for research. "I think the best department here is psychology," says a student, "since it is so highly ranked and is the most popular among students."

Freshmen are encouraged to participate in a three-day summer orientation, which provides workshops, counseling, and a general introduction to the campus and community. Freshmen can also take a yearlong cluster of courses on topics such as The History of Modern Thought, or seminars with titles such as Asian American Youth: Culture, Identity and Ethnicity. During their first two years, most students take required core classes that are sometimes jammed with 300 to 4,000 people. But administrators are quick to point out that nearly two-thirds of all undergraduate classes have fewer than 50 students. Savvy students come to UCLA with advanced courses in their high school backgrounds and test out of the intro courses. First-year students are required to take a course involving quantitative reasoning unless they hit 600 or higher on their math SAT, and English composition requirements should also be met during the freshman year. Lab science and a language requirement are also required for a liberal arts degree. Simply getting into classes here can be a big challenge. Students register by phone in a sequence of two scheduled "passes" based on their class standing.

UCLA's academic environment is extremely intense. "All of the students here are very intelligent and it is hard to really establish yourself as an elite student," a

> "There is a group for every interest and different cultures and beliefs are accepted."

senior says. The climate can be competitive, due to the sheer number of students fighting to get in—UCLA gets more applications than any other college in the country. The faculty is also impressive. "I feel all of our professors are extremely knowledgeable about their subjects, which shows in their lectures," a political science major says. On the other hand, there is a widespread sense here that undergraduate teaching is often sacrificed on behalf of scholarly research. "It isn't until upper divisions that you really get to know professors," confides a biology major. The UCLA library ranks in the top 10 of all research libraries, public or private, and contains more than 7.2 million volumes. The campus newspaper, the *Daily Bruin*, is the third-largest daily in Los Angeles.

> "The trend right now is for students to live on campus two years then move to apartments close by."

Asian Americans account for 40 percent of UCLA's student population, Hispanics make up 16 percent, African Americans 3 percent, and Native Americans 1 percent. "Students here come from very diverse backgrounds," explains a senior. "There is a group for every interest and different cultures and beliefs are accepted." Minority representation, environmental issues and gay and lesbian rights are the largest political issues on campus. UCLA has several student-run newsmagazines, including the feminist *Together* and the Asian American newsmagazine *Pacific Ties*. UCLA is one of the few universities in the nation with a gay fraternity and a lesbian sorority.

Thirty-eight percent of the students live in university housing; freshmen and sophomores are guaranteed housing, but for everyone else it's strictly a waiting list. Overcrowding is a concern, though recent housing construction should provide enough space for all students to live on campus all four years. "The trend right now is for students to live on campus two years then move to apartments close by," a student says. The campus is philosophically divided into North and South. North attracts more liberal arts aficionados, while those in math and science tend to favor South. Fifteen dining halls, restaurants, and snack bars serve meals that students rave about. "There is an abundance of healthy and fresh choices, including vegetarian and vegan dishes," a senior says.

> "There is a lot of school spirit, and everyone is very friendly."

UCLA has won a nation-leading number of collegiate championships, including nearly 100 NCAA titles, and has produced more than 250 Olympians.

Owing to the gargantuan size of UCLA, there is no shortage of social options on campus. "Whether a fan of the big party scene or more of a Friday-night-movie kind of person, there are opportunities both on and off campus," a sophomore says. The hopping Westwood suburb, which borders the university, has at least 15 movie theaters and scores of restaurants, but the shops cater to the upper class. "There's nowhere to dance and only two bars, but a lot of coffee and cheap food," a junior says. UCLA's Ocean Discovery Center on the Santa Monica Pier is an innovative, hands-on ocean classroom for students and the public. The beach is five miles away, and the mountains are only a short drive. Although public transportation is cheap, it's also inconvenient, making a car almost a necessity for going outside of Westwood. Unfortunately, parking is expensive and difficult to obtain. The easiest solution is to live close to campus and bike it.

With all the attractions of the City of Angels at its doorstep, the campus tends to empty out on the weekends (except when the football Bruins have a home game). Fourteen percent of the men and 11 percent of the women join one of UCLA's 50 fraternities and sororities. The university's alcohol policy is similar to that of other UC schools—open consumption is a no-no. But according to one student, "It is extremely easy for undergrads to be served, especially at fraternities." Top-name entertainers, political figures, and speakers of all kinds come to the campus; film and theater presentations are frequent, and the air is thick with live music.

Freshmen are encouraged to participate in a three-day summer orientation, which provides workshops, counseling, and a general introduction to the campus and community.

UCLA has won a nation-leading number of collegiate championships, including nearly 100 NCAA titles, and has produced more than 250 Olympians. The men's football, basketball, baseball, and tennis teams are the undeniable superstars as are the women's gymnastic and water polo teams. Beating crosstown rival USC is the name of the game in any sport; UCLA fans regard their intracity rivals with passionate feelings. Beat SC Week, the week leading up to the football game between the two, is an event in itself, featuring a bonfire, concert, and blood drive.

A leading research center, 190 fields of study, distinguished faculty members, and outstanding athletics make UCLA one of the most prestigious universities in the nation. And despite the large size, students still feel they are part of a tight-knit community. "There is a lot of school spirit, and everyone is very friendly," a junior says.

<table>
<tr><td>

Overlaps

UC–Berkeley, UC–San Diego, UC–Irvine, UC–Santa Barbara, USC

</td></tr>
</table>

<table>
<tr>
<td>

If You Apply To ➤

</td>
<td>

UCLA: Regular admissions: Nov. 30. Meets demonstrated need of 40%. No campus or alumni interviews. SATs or ACTs: required, ACT with writing component. SAT IIs: required. Apply to particular school or program. Prefers electronic applications. Essay question.

</td>
</tr>
</table>

UC–Riverside

Riverside, CA 92521

Social life is relatively tame, since so many of the students commute. While some complain of a lack of nightlife in Riverside, they readily agree that activities on campus make up for it. Returning students are welcomed back every year with a campuswide block party, and Spring Splash brings in hot bands.

<table>
<tr><td>

Website: www.ucr.edu
Location: City outskirts
Total Enrollment: 16,622
Undergraduates: 14,555
Male/Female: 47/53
SAT Ranges: V 460–570
  M 490–630
ACT Range: 18–23
Financial Aid: 61%
Expense: Pub $ $ $
Phi Beta Kappa: Yes
Applicants: 19,060
Accepted: 75%
Enrolled: 20%
Grad in 6 Years: 65%
Returning Freshmen: 86%
Academics: ✍ ✍ ✍ ½
Social: ☎ ☎
Q of L: ★ ★ ★
Admissions: (951) 827-3411
Email Address:
  discover@ucr.edu

</td></tr>
</table>

Lacking the big-name reputation and booming athletic programs of the other UC schools, UC–Riverside has chosen to place its emphasis on something that not all institutions consider to be an important component of higher education: the student. Riverside offers one of the lowest student/faculty ratios in the UC system, strong programs with personalized attention, and a sense of academic community that seems to have been forgotten at other UC schools. "Students are well taken care of and get personal attention," says one satisfied senior. Though part of the UC system, UC–Riverside is a breed apart.

Located 60 miles east of Los Angeles, UCR is surrounded by mountains on the outskirts of the city of Riverside. The beautifully landscaped,1,200-acre campus consists of mainly modern architecture, with a 160-foot bell tower (with a 48-bell carillon) marking its center. Wide lawns and clusters of oaks create "a veritable botanical garden," where students and faculty enjoy relaxing between classes. Acres of citrus groves form a half-circle on the outer edges of campus and perfume the air. New facilities include residence halls, an international village, a large lecture hall, a plant genomics research center, an entomology building, and science laboratories. Current construction includes a four-level student commons area.

"Students are well taken care of and get personal attention."

Decades ago, researchers at the Citrus Experiment Station in Riverside perfected the growing methods for the imported navel orange, making discoveries to protect the fruit from disease and pests and saving California's citrus industry. Riverside continues to excel in plant sciences and entomology. But the campus has grown since its founding in 1954 to include excellent programs in engineering, natural sciences, social sciences, humanities, the arts, business, and education. The biomedical sciences program is UCR's most prestigious and demanding course of study, and its most successful students can earn a BS/MD in partnership with the medical school at UCLA. The engineering program also is quite selective, more so than the campus as a whole, which generally accepts students who are ranked in the top 12 percent of the state's high school graduates. One of the few undergraduate environmental engineering programs is at UCR, as is an undergraduate program in creative writing, the only one in the UC system.

Graduate programs in the arts are strong, with an MFA in writing for the performing arts and the nation's first doctoral program in dance history and theory. Academic weaknesses include journalism and geography. New undergrad majors include bioengineering, Asian Literatures and Cultures, and interdisciplinary studies. The University Honors Program offers exceptional students further academic challenges in addition to extracurricular activities and special seminars for freshmen. And talented student singers, dancers, and actors can earn stipends for performing in the community through an arts outreach program funded by the Maxwell H. Gluck Foundation.

> "UCR is one of the most diverse universities in the nation."

(Continued)
Strongest Programs:
Plant Sciences and
   Entomology
Engineering
Natural Sciences
Social Sciences
Biomedical Sciences
Humanities and Arts
Business
Education
Engineering

All students are required to meet extensive "breadth requirements" that include courses in English composition, natural sciences and math, humanities, and social sciences. Some majors include a foreign language requirement. Students do not encounter much difficulty in getting the courses they want. The campus libraries have an impressive two million volumes, an interlibrary loan system within the UC system, and vast electronic databases. A specialized research collection in science fiction is world-class. UCR's museum of photography, located in a downtown Riverside mall and available on the Web, has grown in stature.

One of the few undergraduate environmental engineering programs is at UCR, as is an undergraduate program in creative writing, the only one in the UC system.

Students say the academic climate is cooperative rather than competitive. "Instead of being super competitive," says a student, "I see more students working together to get the job done." Research is an institutional priority for faculty, but professors continue to dedicate much of their time and attention to their students. "Most of my professors were great at presenting and teaching to me in a relevant, understandable manner," says one junior. A senior adds, "The teachers are smart and passionate." Plus, UCR has a tradition of undergraduate and faculty interaction with a wide range of undergraduate research grants available during the academic year. This may be why one in six graduates goes on to get a PhD. State funding woes have not gone unnoticed on campus. "We've lost a lot of core classes and financial aid does not offer as much," grumbles one student.

> "There is always something going on, whether it be a concert, lecture, or sorority/fraternity party."

Ninety-nine percent of the UCR student body is from California, mainly L.A., Riverside, San Bernardino, and Orange County. Ninety-four percent of the students graduated in the top tenth of their public high school class. Asian Americans account for 42 percent of the students, and Hispanics and African Americans 24 percent and 7 percent, respectively. As part of the UC commitment to diversity, Riverside upholds policies prohibiting sexual harassment, hazing, and physical and verbal abuses. It supports centers for various ethnicities, for women, and for gay and lesbian students. "UCR is one of the most diverse universities in the nation," a political science major says. "Because of this, there is a wide range of students at UCR that make a blended environment of different cultures, nationalities, and social statuses." Numerous merit scholarships, averaging $5,600, are doled out every year, as well as Division I athletic scholarships in baseball, softball, basketball, tennis, volleyball, track, soccer, and golf. Scholarships also are available in specific academic departments.

While some complain of a lack of nightlife in Riverside, they readily agree that activities on campus make up for it.

Housing is reasonably priced and relatively easy to obtain, but the quality varies greatly. "While West Lothian looks like a prison, Pentland Hills is like a resort," says one student. Twenty-seven percent of the students live in the well-maintained dorms, where freshmen are guaranteed a spot. Campus dining is described as adequate. "I could eat their tater tots forever," gushes one student. Students feel safe on campus; security measures include an escort service and patrolling security officers.

Fraternities and sororities lure 4 percent of men and women on campus. The groups usually hold campuswide parties once a quarter. "There is always something going on, whether it be a concert, lecture, or sorority/fraternity party," one sophomore says. Campus hangouts, including "The Barn," have live bands and comedy nights. Every Wednesday the campus can enjoy a "nooner," where live bands play during lunch. University Village is a commercial center offering a movie theater, restaurants, and an arcade right on the edge of campus. The campus runs a cultural arts program that brings professional shows to campus, such as Laurie Anderson and Margaret Cho.

Riverside weather is temperate except during the summer months, when the heat and haze combine to make a trip to the coast look really inviting. The coast is only about 45 minutes by freeway and the desert is an hour east. Big Bear and numerous ski resorts are also within an hour's drive.

Athletics is generating more interest on campus since the switch to NCAA Division I competition. Successful teams include women's volleyball, men's and women's golf, and men's basketball. A recreational program in men's and women's karate has turned out national champions. A student recreation center offers a health-club atmosphere with sand volleyball, weight and workout machines, and intramural leagues.

All in all, Riverside is growing and improving. Although half the size of some sister UC campuses, it offers more personal attention to its students. UCR is fast becoming a nationally recognized research institution, from which students surely will benefit. "UCR has grown immensely over the past few years," one sophomore says. "The emphasis for the future is to establish a name for UCR, to let the nation know what a wonderful university this is."

Overlaps

UC–Irvine, UC–Santa Cruz, UC–Berkeley, UC–Davis, UCLA, UC–Santa Barbara

If You Apply To ➤

Riverside: Regular admissions: Nov. 30. Financial aid: Mar. 1. Housing: Jun. 1. Guarantees to meet full demonstrated need. No campus or alumni interviews. SATs or ACTs: required. SAT IIs: required (writing, math, and one other). Prefers electronic applications. Essay question: personal statement.

UC–San Diego

9500 Gilman Drive, Department 0021-A, La Jolla, CA 92093-0021

Applications have doubled in the past ten years at this seaside paradise. UCSD now rivals better-known Berkeley and UCLA as the Cal campus of choice for top students. Five undergraduate colleges break down UCSD to a more manageable size. Best known for science and engineering.

Website: www.ucsd.edu
Location: Suburban
Total Enrollment: 25,278
Undergraduates: 20,339

Some say that looking good is better than feeling good, but at UC–San Diego, they're doing a lot of both. Set against the serene beauty of La Jolla's beaches, students catch as much relaxation time as they do study time. But it's not all fun and games around this campus. The research star of the UC system, UCSD's faculty rates high nationally among public institutions in science productivity. And within each of the five undergraduate colleges, a system that offers undergraduates more intimate settings, students are honing their minds with the classics and the cutting edge in academics. Sure, San Diegans tend to be more mellow than the average Southern Californian, and UCSD students follow suit. But beneath the frown-free foreheads and bright smiles, UCSD's bubbling with intellectual energy and the healthy desire to be at the top of the UC system.

San Diego's tree-lined campus sits high on a bluff overlooking the Pacific in the seaside resort of La Jolla. Each of the six colleges has its own flavor, but the predominant architectural theme is contemporary, with a few out-of-the-ordinary structures, including a library that looks like an inverted pyramid. Another tinge of the postmodern is the nation's largest neon sculpture, which wraps around one of the high-rise academic buildings and consists of seven-foot-tall letters that spell out the seven virtues superimposed over the seven vices. Construction is taking place all around campus, including the first building for the Rady School of Management.

UCSD's programs in science and engineering are "not for the faint of heart," says one student. Engineering requires a B average in entry-level courses for acceptance into the major. The Scripps Institute of Oceanography is also excellent, due to the university's advantageous location. Computer science and chemistry also get strong recommendations, but you really can't go

"The residence halls are very nice, with all the amenities."

wrong in any of the hard sciences. Although the humanities and social sciences are not as solid in comparison, political science and psychology get strong backing from students. The math department, however, is less than adequate. To address its lack of a business program, the School of Management began accepting students in its MBA program in 2004. Imaginative interdisciplinary offerings include computer music, urban planning, ethnic studies, and a psychology/computer science program in artificial intelligence, as well as majors devised by students themselves. The School of Pharmacy and Pharmaceutical Sciences enrolled its first students in 2002. UCSD has also opened a 109,000-square-foot building for the Jacobs School of Engineering, called the Powell-Focht Bioengineering Hall, a state-of-the-art research and teaching facility.

San Diego operates on the quarter system, which makes for a semester's worth of work crammed into ten weeks. Science students find the load intense. "The courses here are challenging and intellectually stimulating, but there are also a lot of fun classes, too," says one senior. Students have a choice of six libraries, some good for research, others better for socializing. Despite the quality of research done by the faculty, half a dozen of whom are Nobel laureates, students find that the typical scenario of research over teaching seen at most large research universities is not as common at UC–San Diego. "Professors here are brilliant and conduct research throughout the year, but they also have a desire to share their knowledge with their students," says a communication major.

UC–San Diego's six undergraduate colleges have their own sets of general education requirements, their own personalities, and differing ideals on which they are based. Prospective freshmen apply to UC–San Diego—the admissions requirements are identical for each college—but students must indicate their college preference. Revelle College, the oldest, is the most rigorous and mandates that students become equally acquainted with a certain level of coursework in the humanities, sciences, and social sciences, as well as fulfill a language requirement. Muir allows more flexibility in the distribution requirements. Thurgood Marshall College was founded to emphasize and encourage social awareness; like Revelle, it places equal weight on sciences, social sciences, and humanities. However, it stresses a liberal arts education based on "an examination of the human condition in a multicultural society." Warren has developed a highly organized internship program that gives its undergraduates more

(Continued)
Male/Female: 48/52
SAT Ranges: V 540–660
  M 590–700
ACT Range: 23–29
Financial Aid: 89%
Expense: Pub $ $ $
Phi Beta Kappa: Yes
Applicants: 41,330
Accepted: 42%
Enrolled: 22%
Grad in 6 Years: 83%
Returning Freshmen: 94%
Academics: ✍ ✍ ✍ ✍ ✍
Social: ☎ ☎ ☎ ☎
Q of L: ★ ★ ★ ★
Admissions: (858) 534-4831
Email Address:
  admissionsinfo@ucsd.edu

Strongest Programs:
Biology
Engineering and
  Bioengineering
Cognitive Science
Economics
Political Science
Oceanography
Communication

Computer science and chemistry get strong recommendations, but you really can't go wrong in any of the hard sciences.

practical experience than the others do. Eleanor Roosevelt College ("Fifth") devotes its curriculum to international and crosscultural studies. The newest college—imaginatively dubbed Sixth College—focuses on art, culture, and technology. Its goal is to graduate multicultural students who can work collaboratively and enjoy working in their communities.

A theater major notes that UCSD's academic intensity "does not mean that all the students here are nerdy. They enjoy athletics and extracurricular activities, but academic excellence is their priority." A short walk to the beach, however, reveals the student body's wild and crazy half-surfers and their fans, who celebrate the "kick back." Students jumping curbs on skateboards are common on this campus. Yet these beach babies are no scholastic slouches. Most of them placed in the top 10 percent of their high school class. The average student pulls a 3.0 GPA while at UCSD. Many students here choose to take five years to graduate in order to gain a higher GPA, and many of the scientists continue their studies after graduation. UCSD also ranks high among public colleges and universities in the percentage of graduates who go on to earn a PhD, and in the percentage of students accepted to medical school. Only 2 percent of students are from out of state, and another 3 percent are foreign students. Minority representation is high, with 38 percent of the student body Asian American, 10 percent Hispanic, and 1 percent African American. Diversity education includes a Crosscultural Center for students, faculty, and staff that provides activities, brown-bag luncheons, and programs on race relations.

> "Most students hang out at the dance clubs, jazz bars, and great restaurants in the Gaslamp Quarter."

Each of the university's colleges has its own housing complex, with either dorms or apartments. Most freshmen live on campus and are guaranteed housing for their first two years. "The residence halls are very nice, with all the amenities, including Ethernet hookups in every room," says an animal physiology major. By junior year, students usually decide to take up residence in La Jolla proper or nearby Del Mar, often in beachside apartments; only 35 percent of all the students live on campus. But that can be costly: the price ends up being inversely proportional to proximity to the beach. If you are willing to relinquish the luxury of a five-minute walk to the beach, a short commute will bring you relatively affordable housing.

The immediate surroundings of UCSD, however, are definitely not affordable. "La Jolla is a rich, conservative, retired, white, snobbish community," one sophomore says. "Not a college town!" Cars are, of course, an inescapable part of Southern California life, and owning one—many people do—makes off-campus living even more pleasant. "No car equals no fun," one international studies major says. Unfortunately, trying to park on campus can be difficult, though at least one student says that "parking is not nearly as bad here as it is at other schools." Dorm residents are required to buy a meal card, which gets them into any of the four campus cafeterias as well as the campus deli and burger joints.

> "Professors here are brilliant and conduct research throughout the year."

The university is dry and most of the real socializing seems to take place off campus. "Most students hang out at the dance clubs, jazz bars, and great restaurants in the Gaslamp Quarter," says a senior. Annual festivals include the Open House, Renaissance Faire, UnOlympics, and the Reggae Festival. Another annual festival pays tribute to a hideously loud and colorful statue of the Sun God, which is the unofficial mascot for this sun-streaked student body. Ten percent of both the men and the women try to beat the blahs by joining a fraternity or sorority. Alcoholic parties are banned in the residence halls, though students say lax RAs and good fake IDs make for easy underage drinking. Although campus life is relatively tame, students rely heavily on the surrounding area—but not La Jolla—for their entertainment.

Students go to nearby Pacific Beach and downtown San Diego with the zoo, Sea World, and Balboa Park all only 12 miles away. Torrey Pines Natural Reserves are great for outdoor enthusiasts. Mexico—and the $5 lobster—is a half-hour drive (even nearer than the desert, where many students go hiking), and the two-hour trip to Los Angeles makes for a nice weekend jaunt.

Although San Diego is the farthest thing imaginable from a rah-rah school, it is rapidly becoming a Division II powerhouse, most notably in women's sports. The women's volleyball and tennis teams have won numerous national championships, and the men's water polo and volleyball teams have also done well. For weekend competitors, classes are available in windsurfing, sailing, scuba diving, and kayaking at the nearby Mission Bay Aquatic Center. Everyone participates in one intramural league or another, and if you're not on a team, "you're not a true UCSD student." The RIMAC, an impressive sports facility for students, gets even the couch potatoes off their Barcaloungers.

The students at UCSD are exceptionally serious and out for an excellent education. But the pace (study, party, relax, study more) and the props (sun, sand, Frisbees, and flip-flops) give the rigorous curriculum offered by UCSD's six colleges an inimitable flavor that undergraduates would not change. Indeed, many believe they have the best setup in higher education: "a beautiful beach-front environment that eases a life of academic rigor."

> ## Overlaps
> **UCLA, UC–Berkeley, Stanford, University of Southern California, Harvard, MIT**

If You Apply To ➤ **San Diego:** Regular admissions: Nov. 30. Financial aid: Mar. 2. Housing: May 4. Guarantees to meet full demonstrated need. No campus or alumni interviews. SATs or ACTs: required. SAT IIs: required (English, math level 1 or 2). Essay question: personal statement.

UC–Santa Barbara

Santa Barbara, CA 93106

Willpower is the word at UC–Santa Barbara. On a beautiful day with the sound of waves crashing in the distance, it takes willpower to hang in there with pen, paper, and book. Fairly or not, Santa Barbara is known as the party animal of the UC system. In the classroom, science is the best bet.

For students at UC–Santa Barbara, California's famed beaches serve as both classroom and playground. On weekends, sun-worshipping students grab surfboards and don bikinis and head to the water for some serious fun. During the week, those same students can likely be found studying technology rather than tan lines. UCSB provides a comfortable mixture of work and play that is unique to the UC system and draws praise from its students. Says a senior, "I love the fact that I am getting a highly rated UC education in such a relaxing location."

Located just a stone's throw from the beach, UC–Santa Barbara's 989-acre campus is bordered on two sides by the Pacific Ocean, with a clear view of the Channel Islands. On the landward side are a nature preserve and the predominantly student community of Isla Vista (IV), and five miles to the north lie the Santa Ynez Mountains. "We are definitely a college town," one senior says. "Isla Vista is almost all college students, and it is a really relaxed atmosphere." The campus itself features mainly 1950s Southern California architecture with a Southern California atmosphere to match. Recent

Website: www.admit.ucsb.edu
Location: City outskirts
Total Enrollment: 21,016
Undergraduates: 18,077
Male/Female: 45/55
SAT Ranges: V 530–650
   M 560–670
ACT Range: 22–28
Financial Aid: 65%
Expense: Pub $ $ $
Phi Beta Kappa: Yes
Applicants: 37,522
Accepted: 53%

construction includes a 56,000-square-foot recreation center and a 54,000-square-foot engineering building, featuring 19 research labs.

Not surprisingly, the marine biology department capitalizes on the school's aquatic resources and stands out among the university's best. Other favorites include physics, ecology, engineering materials, and chemistry. The accounting program is very strong, and the courses are geared toward taking and passing the CPA exam, so graduation is usually followed by a mass recruitment by California's big accounting firms. In addition, history, English, communications, and geological sciences are

> **"I love the fact that I am getting a highly rated UC education in such a relaxing location."**

solid, but students say political science and math are considerably weaker. The College of Creative Studies offers an unstructured curriculum to about 400 self-starters ready for advanced and independent work in the arts, math, or the sciences. An interdisciplinary program called the Global Peace and Security Program combines aspects of physics, anthropology, and military science. The National Science Foundation provides funding for the $5.5 million National Center for Geographic Information and Analysis program. The Bren School of Environmental Science and Management is open for business, and the science departments are world-renowned—the college boasts Nobel prize winners in chemistry and physics.

UCSB's general education program requires all students to fulfill four subject areas: writing, non-Western cultures, quantitative relationships, and ethnicity. Other required courses include English reading and composition, foreign languages, social sciences, and art. For those who crave time away, Santa Barbara is the headquarters of the UC system's Education Abroad program, which sends students to any of 100 host universities worldwide. In order to graduate, all students must take courses in English Composition and American History and Institutions, must fulfill a unit requirement, and must also meet the requirements of their individual majors. In addition, students must be registered at UCSB for a minimum of three regular quarters.

UCSB students are traditionally public-spirited; the fraternities and sororities, which attract 4 percent of men and 7 percent of women, are known for their philanthropy. The students, 94 percent of whom are California residents, are laid-back. "Everyone says 'hi' to other students, we ride bikes around our campus, and people are generally really friendly," one senior says. "Our students differ from students at other UC's because we are probably the most relaxed UC campus." Asian Americans comprise 18 percent of the student body; African Americans make up 3 percent, and Hispanics account for 20 percent. "Being of Latino background, I have never felt like

> **"We are definitely a college town."**

a minority on campus," one student says. "I associate with a lot of Latinos. We are a huge family, all know each other, and create great programs that help other Latinos get hyped up for college." The campus's beach locale inspires many students to be environmentally friendly. Merit scholarships (worth an average of $5,771) and various athletic scholarships are available for those who qualify.

University housing, which includes both dorms and privately run residence halls, is comfortable, well maintained, and much sought after. "Our on-campus housing is amazing, right in front of the beach," a junior says. "They come fully furnished, with high-speed Internet, cable, telephone lines, and a great atmosphere." Unfortunately, there is a waiting list to get into the dorms—even with the addition of the new Manzanita Village Student Housing. Only 29 percent of students, most of whom are freshmen, snag on-campus housing. The rest find a home in neighboring Isla Vista, which has welcomed its student population—after all, most of its population is UCSB students. As a result, students are very active in the community. "Community service and maintaining our little community of Isla Vista is very important to students," one film studies/Chicano studies major says. "Isla Vista is

Although the Greeks are strong and growing, there's an ample selection of other organizations from which to choose.

the best college town." Meals in the dorms are available to residents and nonresidents alike, and are, according to most students, more than simply edible. "When one thinks of cafeteria food they think of nasty food, but not at UCSB's dining commons," one student says. While all students say they feel extremely safe on campus, one frequently used motto is "four years, four bikes," because of the frequency of bicycle thefts.

Because Isla Vista is predominantly made up of students, it's become what some students consider Party Central. But don't call UCSB a party school—students bristle at what they say is a misnomer. "It's just because we have so many students living in such a small area," one senior says. Alcohol isn't allowed on campus, but many students say the rule is easy to get around. The local bars are off-limits to those under 21, but when the long-awaited birthday arrives, students celebrate with a quaint little ritual known as the State Street Crawl, imbibing at all the numerous establishments on the "main drag" of Santa Barbara. Movies and concerts are also available, and the mountains, Los Padres National Forest, and L.A. are all an easy drive away. The annual Extravaganza is an all-day, free concert, and students are known to go wild on Halloween and dress up for the entire weekend.

> **"Our on-campus housing is amazing, right in front of the beach."**

An interdisciplinary program called the Global Peace and Security Program combines aspects of physics, anthropology, and military science.

Although the Greeks are strong and growing, there's an ample selection of other organizations from which to choose. A never-ending rotation of intramurals is available on and off the beach. The most successful varsity teams include soccer, water polo, baseball, volleyball, swimming, and basketball. All of UCSB's varsity teams compete in the NCAA's Division I. Ultimate Frisbee is also quite popular, as well as nationally competitive.

UCSB students love to work and play. They rave about their professors and the academic challenges they face. But they also know a good thing when they see it: not everyone gets to spend four years on the beach and come away with a degree. "No matter what college you go to, you find people who you relate with," one student says. "At UCSB you find the social, happy, outgoing crowd. I love it here."

Overlaps
UC–San Diego, UCLA, UC–Irvine, UC–Berkeley, UC–Davis

If You Apply To >

Santa Barbara: Regular admissions: Nov. 30. Financial aid: Mar. 2. Does not guarantee to meet demonstrated need. Campus interviews: required for dramatic arts and dance, evaluative. No alumni interviews. SATs or ACTs: required (SAT preferred). SAT IIs: required (writing). Prefers electronic applications; College of Creative Studies requires additional application. Essay question: personal statement.

UC–Santa Cruz

1156 High Street, Santa Cruz, CA 95064

With its flower-child beginnings, UC–Santa Cruz has come back toward the mainstream. The yoga mats and surfboards still abound, but the students are a lot more conventional than in its earlier incarnation, and UCSC is not quite the intellectual powerhouse of yore. Santa Cruz's relatively small size and residential college system give it a homey feel.

UC–Santa Cruz, still a baby in the UC system, was born during the radical '60s when it reigned as the ultimate alternative school. The founding vision of an integrated learning environment remains to this day, and every undergraduate affiliates with

Website: www.ucsc.edu
Location: Suburban

(Continued)

Total Enrollment: 14,119
Undergraduates: 13,106
Male/Female: 46/54
SAT Ranges: V 520–630
   M 530–640
ACT Range: 21–27
Financial Aid: 44%
Expense: Pub $ $ $
Phi Beta Kappa: Yes
Applicants: 23,003
Accepted: 75%
Enrolled: 17%
Grad in 6 Years: 70%
Returning Freshmen: 89%
Academics: ✏ ✏ ✏ ✏
Social: ☎ ☎ ☎
Q of L: ★ ★ ★ ★ ★
Admissions: (831) 459-4008
Email Address:
   admissions@cats.ucsc.edu

Strongest Programs:
Marine Sciences
Biology
Psychology
Linguistics

Men's tennis is strong, winning its fifth team championship in 2005. The women's rugby squad is a national champ, too.

one of the residential colleges. Progressive thought continues to flourish, as does a strong academic program that strives to focus on undergraduate education. Students still come to UCSC to do their own thing.

The campus, among the most beautiful in the nation, is set on a 2,000-acre expanse of meadowland and redwood forest overlooking Monterey Bay. Bike paths and hiking trails wind throughout the redwood-tree-filled campus, and the beach is a quick drive away—or a spectacular bike ride or scenic hike. The buildings range from 1860 Cowell Ranch farm structures to the multi-award-winning modern colleges, whose styles range from Mediterranean to Japanese to sleek concrete block.

"I've been very impressed with how accessible professors are."

Thanks to a unique building code, nothing may be built taller than two-thirds the height of the nearest redwood tree. Newest additions include a student union, an award-winning second engineering building, an interdisciplinary sciences building, and an expanded bookstore.

The surroundings are deceptive. "Courses are very rigorous, in my experience," warns one undergrad. Santa Cruz's academic offerings range as widely as its architecture and feature both traditional and innovative programs. In an effort to become what one official calls a "near-perfect hybrid" between the large university and the small college, campus life revolves around the residential colleges. Whatever one's specialty, the curriculum can be demanding. Led by marine sciences and biology, the sciences are Santa Cruz's strongest suit, and frequently give students the opportunity to coauthor published research with their professors. Science facilities include state-of-the-art laboratories; the Institute of Marine Sciences, which boasts one of the largest groups of experts on marine mammals in the nation; and the nearby Lick Observatory for budding stargazers. UCSC also includes the Jack Baskin School of Engineering, which was developed to accommodate the growing needs of engineering students. UCSC has recently added majors in bioinformatics, health sciences, and applied physics.

While the majority of students now pursue traditional majors, the possibility is still there for eclectically minded students to pursue "history of consciousness" or just about anything else they can get a faculty member to OK. One of UCSC's most unusual features is that professors provide written evaluations for each student in their class and also provide letter grades. UCSC boasts more than the average number of interdisciplinary programs, including environmental, community, and feminist studies; bioinformatics; and creative writing.

"It's still a school with a social conscience."

Field study and internships are encouraged. Overall, the emphasis is on the liberal arts, and students will find few programs with a vocational emphasis. Newer programs include a combined Latin American studies/sociology degree and an applied physics major.

To meet general education requirements, students must complete courses in quantitative methods, U.S. ethnic minorities/non-Western society, arts, writing, humanities and arts, natural sciences, social sciences, and three topical courses. In addition, American History and Institutions and English Composition are required, as is a senior thesis or comprehensive exam. The main library, McHenry, houses more than a 1.5 million books and 25,000 periodicals, and students have access to books at other UC campuses through an online catalog system and interlibrary loans. The science library houses an additional 300,000 volumes.

Though the curriculum is demanding and the quarter system keeps the academic pace fast, the atmosphere is emphatically noncompetitive. Such competition as there is tends to be internalized. A majority of the students eventually go on to graduate study. All UC campuses insist on faculty research, but most Santa Cruz professors are there to teach. "I've been very impressed with how accessible professors are," says a sophomore. "Whether it's via email or regular office hours, I feel very comfortable approaching and talking to all of my professors."

Santa Cruz remains the most liberal of the UC campuses, and, according to one student, is "still a school with a social conscience." "Before I came here I was told that UCSC was a 'hippie-dippie' college," says one student, "but this is not true at all." Ninety-six percent of the students are Californians, though Santa Cruz always manages to lure a few Easterners. More than one-third of the students are members of minority groups, with Asian Americans accounting for 19 percent of the students, Hispanics 15 percent, and African Americans 3 percent. "Racial, ethnic, and cultural diversity is celebrated and strongly encouraged by the majority of the students here," reports a politics major. Santa Cruz offers more than 400 merit scholarships worth an average of $6,081, but there are no athletic scholarships.

Led by marine sciences and biology, the sciences are Santa Cruz's strongest suit, and frequently give students the opportunity to coauthor published research with their professors.

Forty-five percent of the undergraduate student population lives in university-sponsored housing. Some dorms have their own dining halls with reasonably good food; students may also opt to join a food co-op. Freshmen and transfer students are guaranteed on-campus housing for two years. Upperclassmen can take their chances in the lottery or move off campus.

There are a dozen fraternities and sororities (attracting 1 percent of the student population), as well as countless established student groups, to provide an active social life. The beach and resort town of Santa Cruz, with its board-walk and amusement park, are only

"Racial, ethnic, and cultural diversity is celebrated and strongly encouraged."

10 minutes away from campus by bike, but pedaling back up the hill takes much longer. Those looking for city lights can take the windy, mountainous highway to San Jose (35 miles away) or the slow, scenic coastal highway to San Francisco (75 miles), or ride a bus to either city. If you have a car, destinations such as Monterrey, Big Sur, the Napa Valley, and the Sierras are easily accessible.

Although Santa Cruz fields only a few varsity teams, students love their school mascot, Sammy the banana slug. Men's tennis is strong, winning its fifth team championship in 2005. The women's rugby squad is a national champ, too. Participation in intramurals ("Friendship through Competition" is the motto) is widespread, with rugby in particular growing in popularity. Sailing and scuba diving are among the many physical education classes offered, and the student recreation department sponsors everything from white-water rafting to cooking classes.

Santa Cruz is a progressive school with a gorgeous campus and innovative academic programs, where the main priority is the education of undergraduates. Many students are concerned that UCSC is growing too fast, and an ambitious proposal for future expansion has threatened its heretofore cozy relationship with local citizens. Still, as long as UCSC retains its belief in "to each his or her own," it will remain uniquely Santa Cruz.

Overlaps
UC–Santa Barbara, UC–San Diego, UC–Davis

If You Apply To ➤

Santa Cruz: Regular admissions: Nov. 30. Financial aid: Mar. 2. Housing: Nov. 30. Does not guarantee to meet demonstrated need. No campus or alumni interviews. SATs or ACTs: required (plus writing). SAT IIs: required. Prefers electronic applications. Essay question: personal statement.

California Institute of Technology

Mail Code 328-87, 1200 East California Boulevard, Pasadena, CA 91125

If you're armed with a perfect SAT score, a burning desire to study math, science, or engineering, and some independent research or published papers already under your belt, maybe you'll have a fighting chance of getting into and out of the California Institute of Technology.

Website: www.caltech.edu
Location: Suburban
Total Enrollment: 2,120
Undergraduates: 939
Male/Female: 67/33
SAT Ranges: V 710–780
   M 760–800
Financial Aid: 57%
Expense: Pr $ $
Phi Beta Kappa: No
Applicants: 2,615
Accepted: 21%
Enrolled: 45%
Grad in 6 Years: 85%
Returning Freshmen: 95%
Academics: ✍ ✍ ✍ ✍ ✍
Social: ☎
Q of L: ★ ★ ★
Admissions: (626) 395-6341
Email Address:
   ugadmissions@caltech.edu

Strongest Programs:
Engineering
Physics
Applied Sciences

The school counts 27 Nobel Prize winners among its faculty and alumni, and with administrators' permission—which is easy to obtain—students may tap into that brilliance by taking as many classes as they can cram in each semester. Expectations are high; "Techers" are fond of saying that "the admissions office doesn't make mistakes," and it's fairly common to take time off to deal with stress and avoid burnout. "The atmosphere promotes a love of science, learning, and discovery that is truly exhilarating," says a biology major. No doubt about it—if you prefer particle physics to parytying, Caltech may be the place for you.

Caltech's 124-acre campus is located in Pasadena, "A wealthy suburban town about 15 miles outside Los Angeles," says a senior. "It's not a college town at all." The distance from downtown means the school is relatively isolated from the glitz, glamour, and good times that many people associate with "La La Land." Outside the classroom, at least, tranquility prevails, with olive trees, lily ponds, and plenty of flowers breaking up clusters of older Spanish-mission style buildings. Leafy courtyards and arcades link these with the more modern, "block institutional" structures. The new Broad Center for the Biological Sciences offers 120,000 square feet of lab, classroom, and office space at the northwest corner of campus. It was designed by Pei Cobb Freed & Associates, the firm behind the U.S. Holocaust Memorial Museum in Washington, D.C.

Caltech's mission, one official says, is "to train the creative type of scientist or engineer urgently needed in our educational, governmental, and industrial development." After all, it was here that Albert Einstein abandoned his concept of a static cosmos and endorsed the expanding-universe model. This is also where physicist Carl Anderson discovered the positron. With these luminaries as their models, students plunge right into the demanding general requirements, which include five terms each of math and physics, three terms of chemistry with lab, one term of biology, two terms of science communication, and courses in the humanities and social sciences to round things out. Students complain about these, "and usually take no more than absolutely required," says a biology major. Still, they can

"The unique student body, how available professors are (I call almost all of them by their first names), and how much we learn make Caltech a special place,"

be tough to get into come registration time, says a computer science major, since enrollment is limited "to allow for discussion among a small group." The pass/fail grading system in the freshman year goes a long way toward easing the acclimation period for new arrivals. And the honor system, which mandates that "no one shall take unfair advantage of any other member of the Caltech community," helps discourage competition for grades. Professors give take-home exams, and if violations of the honor code are suspected, "students decide if a violation was indeed made," one student explains.

Caltech made its name in physics, and students say that program remains strong. A junior says, "I love mechanical engineering. The profs are great, the subject is fun, and you get to do fun contests." Regardless of major, Caltech students benefit from state-of-the-art facilities, including the Beckman Institute, a center for fundamental research in biology and chemistry, and the Keck telescope, the largest optical telescope in the world. The Moore Laboratory has 90,000 square feet of the latest equipment for engineering and communications majors studying fiber optics and the like. Summer Undergraduate Research Fellowships give 300 undergraduates the chance to get a head start on their own discoveries, with help from a faculty sponsor. Some 20 percent of these students publish results from their endeavors in scientific journals.

Despite Caltech's reputation for brilliance, students say the quality of teaching is hit or miss. "At times, you get lucky and get amazing professors," says a computer science major. "Other times, you get professors who either don't care about the class they teach, or are so advanced in their field that they are unable to convey 'simple' concepts." A senior adds, "The quality of teaching improves as you get into your major." Here, the student says, professors in the humanities and social sciences really shine, since they actually want to teach, rather than hole up in a lab with mass spectrometers and computer simulations of atomic fission. Another student describes the academic climate as "collaborative, intense, and busy." While teaching assistants do lead some recitation sections affiliated with large lectures, it's not uncommon for professors to lead them, too—even for freshmen, says a sophomore. "If you don't like your TA, switching sections is a breeze," the student says.

Twenty-nine percent of Techers come from California, and the same fraction are Asian American; other minorities are less well represented, with Hispanics making up 7 percent of the student body and African Americans 1 percent. Students are "intense, hardworking, and motivated,"

> "If you don't like your TA, switching sections is a breeze."

says an engineering major. Another student adds, "The girls aren't catty or ditzy, the guys aren't macho or aggressive. There's room for everyone." Foreign students account for 8 percent of the student body, but even in a year of political strife and war, "social and political issues are not a big deal on campus, period," says a junior. "People live in a Tech bubble, where they care about nothing more than 50 meters from campus," a sophomore agrees. The school awards merit scholarships of $11,000 to $23,000 a year, but "unless you're absolutely at the top of the incoming freshman class—which many think they are, yet most are not even close—you won't get any," a junior says.

Caltech guarantees on-campus or school-affiliated housing for all four years, and 87 percent of students live in the "comfortable and convenient" dorms. "The housing system is great," cheers one student, who describes residence life as "social support in an academically intense environment." While there are no fraternities or sororities, the seven co-ed on-campus houses inspire a loyalty worthy of the Greeks. The four older houses, which have been renovated, offer mostly single rooms, while the three newer dorms have doubles. Freshmen select their house during Rotation Week, after spending an evening of partying at each one, and indicating at week's end the four they like the most. Resident upperclassmen take it from there in a professional-sports-type draft, which places each new student in one of his or her top choices. Business-minded types, for example, may choose Avery House, which focuses on entrepreneurship. Each dorm has a dining hall, and those who live on campus must buy a meal plan, which a junior calls "quite expensive for the quality of food." A vegetarian calls the chow "awful," and says that "by the end of the week, I am often wondering if we're being served the same spinach for five days in a row."

The houses are the emotional center of Caltech life, and the scene of innumerable practical jokes. On Ditch Day, seniors barricade their dorm rooms using everything from steel bars to electronic codes, leave clues as to how to overcome the obstacles, and disappear from campus. Underclassmen spend the day figuring out how to break in, using "cleverness, brute force, and finesse," to claim a reward inside, which can range from the edible to...well, anything is possible. Perhaps the best student prank

The houses are the emotional center of Caltech life, and the scene of innumerable practical jokes. On Ditch Day, seniors barricade their dorm rooms using everything from steel bars to electronic codes, leave clues as to how to overcome the obstacles, and disappear from campus.

Caltech made its name in physics, and students say that program remains strong.

occurred during the 1984 Rose Bowl game, when crosstown rival UCLA played Illinois. A group of Caltech whiz kids spent months devising a radio-control device that would allow them to take control of the scoreboard in the second half, to gain national exposure for Caltech by flashing pictures of their school's mascot, the beaver, and a new version of the score that had Caltech leading MIT by a mile.

While drinking might seem a reasonable escape from the pressure of all that work, Caltech requires any organization hosting a party to hire a professional bartender—"and they card," says a senior. "Ask any local bartender for a Caltech Cocktail and you will get three ounces of straight water," quips a sophomore. Social life at Caltech "is horrible," agrees a junior. "There are occasional parties, but the administration does not allow students from other colleges to attend, unless accompanied by a Caltech student." So students head off campus—to Old Pasadena, nearby schools like USC, Occidental, and the Claremont colleges, or to downtown L.A., now easily reachable on the Metro's gold line. Disneyland and Hollywood are

"The girls aren't catty or ditzy, the guys aren't macho or aggressive."

always options, and road trips to the beach, mountains, or desert—or south of the border, to Tijuana—are options for those with cars. "From yoga studios to death metal concerts, L.A. has it all," one student says. But some Caltech students still prefer to make their own fun. The annual Pumpkin Drop (on Halloween, of course) involves immersing a gourd in liquid nitrogen, and then dropping it from the library roof, so that it shatters into a zillion frozen shards. During finals week, stereos blast "The Ride of the Valkyries" at seven o'clock each morning, just the thing to get you going after that all-nighter.

Caltech fields 18 Division III teams, and the most popular include men's soccer, men's and women's track and field, and men's cross-country. The school also offers more unusual sports such as water polo and fencing. Perhaps more popular than varsity competition, though, are the intramural matches between the houses, in nine sports every year. Also popular is the annual design competition that's the culmination of Mechanical Engineering 72; it helped inspire the TV shows Battle Bots and Robot Wars. The city of Pasadena is also home to the granddaddy of postseason college football competition—the storied Rose Bowl.

Caltech students must learn to thrive under intense pressure, thanks to the school's tremendous workload and lackluster social life. But students say they appreciate the freedom to think and explore—and the trust administrators place in them because of the honor code. "The unique student body, how available professors are (I call almost all of them by their first names), and how much we learn make Caltech a special place," says a sophomore.

If You Apply To ➤ **Caltech:** Early action: Nov. 1. Regular admissions: Jan 1. Guarantees to meet demonstrated need. No campus or alumni interviews. SATs: required. SAT IIs: required (writing, math, and either physics, biology, or chemistry). Accepts electronic applications. Looks for math/science aptitude as well as research orientation or unusual academic potential. Essay question: areas of interest; personal statement.

Calvin College

3201 Burton, Grand Rapids, MI 49546

An evangelical Christian institution that ranks high on the private-college bargain list. Nearly half the students are members of the Christian Reformed Church. Archrival of Michigan neighbor Hope and Illinois cousin Wheaton. Best known in the humanities.

Michigan's Calvin College prides itself on being "distinctively academic, strikingly Christian." Along with Wheaton College in Illinois, it is regarded as one of the country's top evangelical colleges. Though no one is required to attend the school's daily chapel services, classes stop when worship starts, and most students view Christian values as central to the academic experience. "We want to actively engage the world and discern it," says a freshman. "People at Calvin are not just numbers or simply students," adds a senior. "They are people looking to grow and develop in all areas of life."

Calvin was founded in 1876, as the educational wing of the Christian Reformed Church in North America. After outgrowing one of its first homes, the college bought a tract of land on the edge of Grand Rapids, and built its present campus. Calvin spreads out over 400 beautifully landscaped acres, including playing fields and three ponds. The campus also includes a 90-acre woodland and wetland ecosystem preserve used for classes, research, and recreation. Most facilities are less than 35 years old, and were designed by a student of famed architect Frank Lloyd Wright. The east campus, dedicated in 2002, includes the Prince Conference Center, DeVos Communication Center, Gainey Athletic Facility, and the award-winning Bunker Interpretive Center, powered primarily by student-designed solar energy technology.

Calvin's core curriculum has four components: core gateway, studies, competencies, and capstone courses. All first-year students must take the two linked gateway courses, Prelude and Developing a Christian Mind. Students then tackle the liberal arts core, entitled "An Engagement with God's World," which challenges them to develop knowledge,

"The best departments are nursing, education, and social work."

skills, and Christian character. Studies courses include The Physical World, Societal Structures in North America, and Biblical Foundations. Competencies courses cover foreign languages and Rhetoric in Culture. Newer programs include majors in international development studies, youth ministry leadership, and African and African Diaspora studies.

Preprofessional programs, such as teacher education, engineering, nursing, and business, tend to be Calvin's best bets, students say—perhaps that's why those programs, along with English, are the college's most popular majors. Biology, chemistry, philosophy, and religion are also regarded as strong, though faculty members in those departments are said to be tough. "The best departments are nursing, education, and social work," offers one senior. "They all have excellent, dedicated professors." Internships and small-business consulting opportunities are available, whether on campus or through Calvin's membership in the Council for Christian Colleges and Universities. Students who receive bachelor's degrees in accounting from Calvin pass the CPA exam at rates 15 to 20 percent higher than the national average.

Academically, Calvin's atmosphere is competitive. "The climate is very competitive and rigorous," one student explains. "Three weeks into any given semester you can hear every other student talking about the papers and tests piling up." The school believes that every subject—even the sciences, or mass media and popular culture—can be approached from a Christian perspective, and faculty members work hard to integrate faith and learning. "I really appreciate the mindset of the professors here," says a freshman, "because they maintain their professionalism yet still invest in the students on a personal level. They all seem very open and friendly."

Almost half of the freshman courses at Calvin have 25 or fewer students. Faculty members must be committed to Christian teachings, and there are no teaching assistants, so professors are expected to reserve about 10 hours per week for advising and assisting students outside of class. Students use the interim month of January to pursue a variety of creative, low-pressure enrichment experiences, such as art and

Website: www.calvin.edu
Location: Suburban
Total Enrollment: 4,042
Undergraduates: 3,969
Male/Female: 55/45
SAT Ranges: V 540–670
   M 540–670
ACT Range: 23–29
Financial Aid: 66%
Expense: Pr $
Phi Beta Kappa: No
Applicants: 1,721
Accepted: 98%
Enrolled: 54%
Grad in 6 Years: 76%
Returning Freshmen: 86%
Academics: ✍ ✍ ✍
Social: ☎ ☎ ☎
Q of L: ★ ★ ★ ★
Admissions: (800) 668-0122
Email Address:
   admissions@calvin.edu

Strongest Programs:
Natural Sciences
Education
Nursing
Engineering
Communication Arts and
   Sciences
English
Philosophy
Mathematics
History

Calvin has no Greek system, and—owing to its emphasis on Christianity and character—the campus is officially dry.

theater study in England, language study in Germany, Canada, or the Dominican Republic, or courses taught by Calvin profs in other cities and countries.

Though Calvin still has a strongly Dutch heritage, the fraction of students who are members of the Christian Reformed Church is now less than half of the student body. "The students here aim at walking in Jesus' footsteps," says a senior. Forty percent of Calvin students are legacies and most are white Michigan natives—African Americans and Hispanics each make up 1 percent of

"The students here aim at walking in Jesus' footsteps."

the total, and Asian Americans add 3 percent. A sophomore says, "Students here think for themselves, are open-minded, discern the world, and have purpose in life." Forty-five percent of Calvin students receive scholarships based on academic merit, ranging from $1,000 to $10,000 each. There are no athletic awards.

Fifty-six percent of Calvin students live on campus in the single-sex dorms. Freshmen and sophomores bunk in suites with two bedrooms, connected by a bathroom, while juniors and seniors move off campus or into the on-campus apartments. "Dorms are a key feature of Calvin," offers a nursing major. "They influence community and friendships." Each residence hall and two of the apartment buildings have computer rooms in the basement, along with free washers and dryers. The food gets mostly positive reviews, too—with options such as pizza, cereal, fresh fruit, made-to-order sandwiches, and ice cream and waffles available at every meal. "The food is not home cooking," an English major admits, "but it is better than what I've had at other colleges."

Calvin has no Greek system, and—owing to its emphasis on Christianity and character—the campus is officially dry. That's no great loss, students say, because there is so much to do on campus, including movies, speakers, concerts, and dances. Downtown Grand Rapids has ice-skating and coffee shops, and the college offers a substantial concert series; recent performers include Nickel Creek and the Victor Wooten Band. Road trips include the beaches of Lake Michigan (a one-hour drive) or Chicago (three hours distant), and even Florida or California for spring break. A popular annual event is Chaos Day, which brings the dorms together for a day of athletic contests. The Airband lip-sync competition each February is also a favorite, as are athletic contests versus Hope College.

ESPN2 ranked Calvin's rivalry with Hope as one of the top 10 college rivalries in the U.S.; the Calvin men's basketball team went to the Division III Final Four in 2005. "Any game against Hope always turns into a big deal," says a speech pathology major, especially in soccer, volley-

"Dorms are a key feature of Calvin."

ball, men's ice hockey, or women's basketball. Calvin's men's cross-country team is also strong, although the Knights don't field a football squad. The college's intramural program offers classes, leagues, and tournaments in sports from dodgeball to ultimate Frisbee and fantasy football. Thirty to 40 percent of students participate.

The students who come to Calvin College aren't seeking the traditional beer-soaked four years away from home. Instead, they're looking to build community with friends and faculty members who share their already-strong Christian faith. "Calvin has a distinctively Christian character and atmosphere," says a sophomore. "This college, its people, place, and mission, all revolve around a commitment to Christ and the furthering of His kingdom. Faith plays an integral part in the classrooms, offices, and dorm rooms."

Overlaps

Hope, Grand Valley State, Wheaton (IL), Taylor (IN), University of Michigan, Michigan State

Carleton College

Northfield, MN 55057

Less selective than Amherst, Williams, and Swarthmore because of its chilly Minnesota location. Yet Carleton retains its position as the premier liberal arts college in the upper Midwest. Predominately liberal, but not to the extremes of its more antiestablishment cousins.

Minnesota is many things: the land of 10,000 lakes, home to the massive Mall of America, birthplace of lore from Hiawatha to Paul Bunyan, and proud parent of the Mississippi River. Beyond all that history-book stuff, tucked into a small town in the southeastern corner of the state is Carleton College, arguably the best liberal arts school in the expansive Midwest. Add to this a midwinter carnival complete with human bowling, badminton competitions that raise money to fight cancer, and an expulsion of Coca Cola from campus for human rights violations, and you have one all-around unique institution.

Surrounded by rolling farmland, Carleton's 955-acre campus is in the small town of Northfield, whose one-time status as the center of the Holstein cattle industry brought it the motto "The City of Cows, Colleges, and Contentment." Lakes, woods, and streams abound, and you can traverse them on 12 miles of hiking and cross-country skiing trails. The city boasts of fragrant lilacs in spring, rich summer greens, red maples in the fall, and a glistening blanket of white in winter. There's even an 800-acre arboretum, put to good use by everyone from jogging jocks to bird-watching nature lovers. Carleton's architectural style is somewhat eclectic—everything from Victorian to contemporary, but mostly redbrick. When it's minus eight degrees, the indoor recreation center provides a rock-climbing wall, gym, putting green, sports courts, track, and dance studio.

Carleton's top-notch academic programs are no less varied: the sciences—biology, physics, astronomy, chemistry, geology, and computer science—are among the best anywhere, and scores of Carleton graduates go on to earn PhDs in these areas. Of all the liberal arts schools in the country, Carleton's undergrads were recently awarded the highest number of National Science Foundation fellowships for graduate studies. English, history, economics, and biology get high marks, too. Engineers can opt for a 3–2 program with Columbia University or Washington University in St. Louis, and for geologists seeking fieldwork—and maybe wanting to thaw out after a long Minnesota winter—Carleton sponsors a program in Death Valley. Closer to home at the "arb," as the arboretum is affectionately known, environmental studies majors have their own wilderness field station, which includes a prairie-restoration site. At the opposite end of the academic spectrum, the arts also flourish. Music and studio art majors routinely get into top graduate programs, and may take advantage of expanded offerings in dance and theatre.

Distribution requirements ensure that a Carleton education exposes students not only to rigor and depth in their chosen field, but also to "a wide range of subjects and

Website: www.carleton.edu
Location: Small city
Total Enrollment: 1,936
Undergraduates: 1,936
Male/Female: 48/52
SAT Ranges: V 660–760
  M 660–740
ACT Range: 27–32
Financial Aid: 60%
Expense: Pr $ $ $ $
Phi Beta Kappa: Yes
Applicants: 5,036
Accepted: 29%
Enrolled: 37%
Grad in 6 Years: 87%
Returning Freshmen: 97%
Academics: ✍ ✍ ✍ ✍ ✍
Social: ☎ ☎ ☎
Q of L: ★ ★ ★
Admissions: (507) 646-4190
Email Address: admissions
  @acs.carleton.edu

Strongest Programs:
Mathematics
Computer Science
Chemistry
Physics
English
History
Economics
Psychology

methods of studying them," administrators say. All students must show proficiency in English composition and a foreign language while fulfilling requirements in four broad areas: arts and literature; history, philosophy, and religion; social sciences; and

"There is a growing international community."

math and natural sciences. There's also a Recognition of Affirmation and Difference requirement, under which students must take at least one course

dealing with a non-Western culture, and a senior comprehensive project is required in every major field. Carleton offers interdisciplinary programs in Asian, Jewish, urban, African and African American, and women's studies. A concentration in crosscultural studies brings in foreign students to discuss global issues and dynamics with their American counterparts. Approximately 70 percent of students spend at least one term abroad, and many take advantage of programs available through numerous organizations, including Carleton and the Associated Colleges of the Midwest.* The school's 799,000-volume library is bright, airy, and—much to the delight of caffeine-stoked night owls—open until 1:00 a.m.

With highly motivated students and a heavy workload, Carleton isn't your typical mellow Midwestern liberal arts college. The trimester calendar means finals may be just three months apart and almost everyone feels the pressure. The six-week Christmas vacation is Carleton's way of dealing with the cold winters. "The courses are quite demanding and demand dedication and a great deal of time," says a student, "but they are manageable." Ninety-nine percent of all classes have fewer than 50 students, so Carls are expected to participate actively. Carleton's faculty members are very committed to teaching. "The teaching is stellar," says a junior.

"For the most part kids are fairly affluent and white," a sophomore says, "but there is a growing international community." Seventy percent of Carleton's students hail from outside Minnesota, half are from outside the Midwest, and most attended public schools. Both coasts are heavily represented, and international students account for 6 percent of the student body. African Americans and Hispanics account for 11 percent of the total student body, and Asian Americans for another

"Anyone who wants to live on campus will get a room on campus."

10 percent. But most Carls have a few things in common, such as being intellectually curious, yet laid-back;

individualistic, but concerned about building a community feeling on campus. Their earthy dress and attitude are a sharp contrast from that of their more traditional crosstown cousins at rival St. Olaf College. The Carleton campus is rather left of center, concerned with issues including the environment, multiculturalism, affirmative action, gay rights, and sexism. "Students are ambitions, aware, and ready to 'save the world'," says a sophomore. Qualified students receive Carleton-sponsored National Merit and National Achievement scholarships every year, and students call financial aid packages "definitely adequate."

Campus accommodations range from comfortable old townhouses to modern hotel-like residence halls. "Dorms are typically comfortable and well-maintained," says a junior. "Anyone who wants to live on campus will get a room on campus." Best of all are the 10 college-owned off-campus "theme" houses, which focus on special interests such as foreign languages, the outdoors, or nuclear-power issues. With the exception of the Farm House, an environmental studies house sitting on the edge of the arb, all the theme houses (including Women's Awareness House) are situated in an attractive residential section of town close to campus. Students who wish to live off campus must apply for a slot. Dorms are co-ed by room, but there are two halls with single-sex floors. Davis is the recommended dorm, although Burton enjoys a "fun" reputation. Everyone who stays on campus must submit to a meal plan. "The food isn't great," reports a student. "The dining staff is great, though, and they try hard."

Absent a Greek system, Carleton's social life tends to be relaxed and informal, and often centers on going out with friends. People go to parties on campus, or if they are of drinking age, bar hop around town. "In a 10-week term, there's not enough time to just go home—so much goes on every weekend between concerts, festivals, carnivals, parties, and cool speakers," says a student. "You really miss out by going away." There are activities for those who pass on imbibing; a group called Co-op sponsors dances and Wednesday socials every two weeks, free movies, and special events like Comedy Night. Students agree that Carleton makes little more than token efforts to enforce the drinking age. "The administration respects the Carleton students' responsibility to make his/her own decisions about drinking, and the alcohol policy is very relaxed," says a freshman.

"Enough coffee shops and restaurants for any college student."

Northfield itself is a quaint, history-filled town with a population of about 17,000. There are old-style shops and a beautiful old hotel. "Northfield is a beautiful, small, Midwestern town," a senior sighs. "Enough coffee shops and restaurants for any college student." Students often frequent the St. Olaf College campus and a nightspot known as the Reub'n'Stein. Minneapolis–St. Paul, 35 miles to the north, is a popular road trip destination. Since students aren't allowed to have cars on campus, Carleton charters buses on weekends.

About a third of the students play on varsity teams, and about two-thirds play intramurals. The track, swimming, tennis, basketball, and baseball teams are competitive, as are the championship cross-country ski teams. Popular events include the Winter Carnival, the Spring Concert, and Mai-F'te, a gala celebrated on an island in one of the two lakes on campus. Traditions include the weeklong freshman orientation program, where—during opening convocation—students bombard professors with bubbles as the faculty members process. There's also the annual spring softball game that begins at 5:30 a.m. and runs as many innings as there are years in Carleton's existence. The all-campus 10:00 p.m. scream on the eve of final exams keeps fatigued studiers awake. Notorious outlaw Jesse James failed to rob the Northfield Bank those many years ago, and Northfield still celebrates with a Wild West bank raid reenactment every year. (The robbery was thwarted by brave townsfolk, and the gang broke up immediately afterward.)

It can be cold in Minnesota, in a face-stinging, bone-chilling kind of way. And the classes are far from easy. But Carleton is a warm campus, and the academics are challenging without being impossible. Carls toe the line between individuality and community, which makes for personal growth and lifelong friendships. "Get ready to live in rural Minnesota with a bunch of crazy people that like to work hard and have fun," a junior says.

There's also the annual spring softball game that begins at 5:30 a.m. and runs as many innings as there are years in Carleton's existence.

Overlaps

Brown, Harvard, Macalester, Williams, Yale, Swarthmore

If You Apply To ➤ **Carleton:** Early decision: Nov. 15, Jan. 15. Regular admissions and financial aid: Jan. 15. Guarantees to meet full demonstrated need. Campus and alumni interviews: recommended, informational. SATs or ACTs: required. SAT IIs: recommended. Accepts the Common Application and perfers electronic applications. Essay question: significant experience; personal issue; person of significant influence; your background; or your own question.

5000 Forbes Avenue, Pittsburgh, PA 15213-3890

Carnegie Mellon is the only premier technical university that also happens to be equally strong in the arts. Applications nearly doubled in the past 10 years, so it must be doing something right. One of the few institutions that openly matches better financial aid awards from competitor schools.

Website: www.cmu.edu
Location: City outskirts
Total Enrollment: 10,120
Undergraduates: 5,580
Male/Female: 63/35
SAT Ranges: V 610–710
   M 690–780
ACT Range: 28–32
Financial Aid: 81%
Expense: Pr $ $ $ $
Phi Beta Kappa: Yes
Applicants: 18,864
Accepted: 34%
Enrolled: 22%
Grad in 6 Years: 86%
Returning Freshmen: 94%
Academics: 🖉 🖉 🖉 🖉
Social: 🐿 🐿 🐿
Q of L: ★ ★ ★
Admissions: (412) 268-2082
Email Address:
   undergraduate-admissions
   @andrew.cmu.edu

Strongest Programs:
Computer Science
Engineering
Drama
Music
Industrial Management
Business
Architecture

Students at Carnegie Mellon don't have to choose between soaking up the high drama of Shakespeare and plunging into the fast-paced dot-com world. The university is known for both its science offerings and strong drama and music programs. But scholars can't be focused on just their own course of study—Carnegie Mellon continues to strive to offer both its technical and liberal arts students a well-rounded education that requires a lot of hard work, but promises great results.

Carnegie Mellon was formed by the merger of Carnegie Institute of Technology and the Mellon Institute of Research in 1967, resulting in a self-contained 144-acre campus attractively situated in Pittsburgh's affluent Oakland section. Next door is the city's largest park and its major museum, named after—you guessed it—Andrew Carnegie. Henry Hornbostel won a competition in 1904 to design the Carnegie Technical Schools, now Carnegie Mellon University. Hornbostel, who attended the Ecole des Beaux-Arts in the 1890s, created a campus plan that is a modification of the Jefferson plan for the University of Virginia with the Beaux-Arts device of creating primary and secondary axes and grouping buildings around significant open spaces. Buildings are designed in a Renaissance style, with buff-colored brick arches and piers, tile roofs and terra cotta and granite details.

Carnegie Mellon is divided into seven colleges, six that offer undergraduate courses: the College of Fine Arts, the College of Humanities and Social Sciences, the Carnegie Institute of Technology, the Mellon College of Science, the School of Computer Science, and the Tepper School of Business. Each college has its own distinct character and admission requirements, which applicants may want to contact the admissions office to find out about. All the colleges, however, share the university's commitment to what it calls a "liberal-professional" education, which shows the relevance of the liberal arts while stressing courses that develop technical skills and good job prospects. Humanities and social science types can major in applied history, professional writing, or information systems, for example, instead of traditional disciplinary concentrations. The Science and Humanities Scholars program allows talented undergrads to develop a course of study based

"Expect to work hard if you come here."

on their interests in the humanities, natural sciences, math, or social sciences. In addition, the Fifth-Year Scholars program provides full tuition for outstanding students who want to remain at Carnegie Mellon for an additional year to pursue more studies that interest them.

Most departments at Carnegie Mellon are strong, but exceptional ones include chemical engineering and electrical and computer engineering. While some humanities courses are praised, most students agree Carnegie Mellon is definitely more of a science-oriented school. Each college requires core work from freshmen; in the College of Humanities and Social Science, for example, students are introduced to computers in a required first-year philosophy course, using the machines to work on problems of logic. Two majors—logic and computation in the philosophy department and cognitive science through the psychology department—combine computer science technology with such fields as artificial intelligence and linguistics. H&SS is home to the Department of Modern Languages, and a Bachelor of Science and Arts degree program was added recently along with a Science and Humanities Scholars Program and a Humanities Scholars Program, the latter of which is for students who desire an interdisciplinary humanities education.

As one student bluntly puts it, the courses at Carnegie Mellon are "extremely rigorous with many hours expected outside of the classroom. Expect to work hard if you come here." Students at Carnegie Mellon work hard, no doubt about it. "Carnegie Mellon is very intense, so lots of time is dedicated to class assignments and group projects," a student says. However, nearly all the classes are small, with fewer than 30 students. Most students agree that the Carnegie Institute of Technology is by far the most difficult college. Professors rate high with most students, who praise their availability and willingness to help. "Most professors are very eager to help and make sure that material is understood," says a sophomore psychology major.

Carnegie Mellon's professional focus shows through in its internship program. A number of five-year, dual-degree options exist, including a joint BS/MS or an industrial internship co-op program in materials science and engineering, which places students in the industrial environment. Beyond traditional student exchanges with major universities around the globe, Carnegie Mellon has established a campus in the Arabian Gulf nation of Qatar. The

"We are very diverse and therefore very culturally aware."

campus offers business and undergraduate computer science degrees, and is led by renowned roboticist Charles Thorpe. The university also has a campus in California and a rapidly expanding presence in Europe and Asia, with programs and educational partnerships in locations such as Portugal, Singapore, Japan, and Greece.

Carnegie Mellon remains one of the most fragmented campuses in the nation. Students divide themselves between actors, dancers, and other artsy types and engineers, scientists, and architects. In any case, students are united in their quest for a good job after graduation. Still, many complain about the fact that students will give up sleep to study, and that this kind of academic orientation can often hinder social life. "Sometimes there just isn't very much social life, but many students prefer it that way," says one student. Every day includes a designated "meeting-free" time for students, allowing them time to study or participate in student activities.

Once a largely regional institution, drawing mostly Pennsylvania residents, Carnegie Mellon now counts about 76 percent of its students from out of state. About one-third are from minority groups, including 24 percent Asian American, 5 percent African American, and 5 percent Hispanic. "We are very diverse and therefore very culturally aware," a biology major says. Some students say a big issue is the nearly two-to-one male/female ratio. The university says it remains committed to need-blind admissions, but it provides larger proportions of outright grants in financial aid packages to "academic superstars." Carnegie Mellon has stopped guaranteeing to meet the financial need of all accepted students, but now offers an early evaluation of financial aid eligibility for interested prospective students. The financial aid office encourages students who have received more generous packages from competing schools to let Carnegie Mellon know so they have an opportunity to match or better them.

Housing, guaranteed for all four years, offers old and newer buildings, the most popular being university-owned apartments. Upperclassmen get first pick, with freshman assignments coming from a lottery of the remainder. "The dorms are generally very comfortable and well-maintained, with ample living space," says one student. The best dorms for freshmen are New House, Donner, Resnik, and Morewood Gardens. Most halls are co-ed, but a few are men-only. Students can remain in campus housing as long as they wish, and about 66 percent do so each year. The Carnegie Mellon Café, which opened in 2006, was the first cafeteria in the nation to

Once a largely regional institution, drawing mostly Pennsylvania residents, Carnegie Mellon now counts about 76 percent of its students from out of state.

Carnegie Mellon is divided into seven colleges, six that offer undergraduate courses: the College of Fine Arts, the College of Humanities and Social Sciences, the Carnegie Institute of Technology, the Mellon College of Science, the School of Computer Science, and the Tepper School of Business.

earn the Leadership in Energy and Environmental Design (LEED) certification and offers students a state-of-the-art dining space.

With all the academic pressure at Carnegie Mellon, it's a good thing there are so many opportunities to unwind, especially with the entire city of Pittsburgh close at hand. The Greek system provides the most visible form of on-campus social life, and 8 percent of the students belong. For those who choose not to fraternize, coffee-houses, inexpensive films, dances, and concerts in nearby Oakland, plus downtown Pittsburgh itself (opera, ballet, symphony, concerts, and sporting events just 20 minutes away by bus) provide plenty of alternatives. The administration is desperately trying to curtail underage drinking. Some students say the penalties for being caught are harsh, but others maintain that the rules are only "vaguely known."

One event that brings everyone together is the Spring Carnival, when the school shuts down for a day. Students set up booths with electronic games, and student groups race in buggies made of lightweight alloys designed by engineering majors. Students put on original "Scotch and Soda" presentations, two of which—Pippin and Godspell—went on to become Broadway hits. The Jill Watson Festival Across the Arts—named for an alumna and former faculty member who died on TWA Flight 800—specifically targets artists who cross boundaries in their work.

"The dorms are generally very comfortable and well-maintained, with ample living space."

Carnegie Mellon competes in Division III. Both the women's soccer and volleyball teams recently won tournament championships, and the swim team has earned many AAU honors as well.

Carnegie Mellon appeals to those yearning for the bright lights of Broadway or the glowing computer screens of the scientific and business worlds. And with a broad range of liberal arts and technical courses not only available, but required, there's no doubt students leave Carnegie Mellon with a well-rounded education—and an impressive diploma. "It's driven, but there's no better place to be interacting with people of so many disciplines who are so focused," one satisfied senior says.

Overlaps

MIT, Cornell University, Penn, Princeton, Harvard

If You Apply To ➤   **Carnegie Mellon:** Early decision: Nov. 1, Dec. 1. Regular admissions: Jan. 1 (Dec. 1 for fine arts applicants). Financial aid: Feb. 15. Does not guarantee to meet demonstrated need. Campus interviews: recommended, evaluative. Alumni interviews: optional, evaluative. SATs or ACTs: required. SAT IIs: required (varies by college). Essay question: personal statement.

Case Western Reserve University

103 Tomlinson Hall, 10900 Euclid Avenue, Cleveland, OH 44106-7055

The Cleveland Browns always lose to the Pittsburgh Steelers and Case is still trying to catch up with Carnegie Mellon. Students may sing its praises, but Cleveland isn't exactly Boston, or even Pittsburgh. On the plus side, students can get an outstanding technical education at Case with solid offerings in other areas.

Website: www.case.edu
Location: Urban
Total Enrollment: 7,423
Undergraduates: 3,674

Cleveland's Case Western Reserve University has long been in the shadow of Pittsburgh's Carnegie Mellon, but the schools also have much in common. Both are the product of mergers between a technical college, known for excellence in engineering, and a more traditional university, focused on the arts and sciences. Both are located in aging rust-belt cities, which have struggled to reinvent themselves. And both tend to attract brainy students more concerned with studying than socializing. Case has increased its investment in the arts, humanities, and social sciences, with an aim toward helping students connect these disciplines with their technical studies. "We're not a big party school," says a psychology major. "But there are parties, concerts, and cultural and club events every weekend."

Case is located on the eastern edge of Cleveland, at University Circle. This 550-acre area of parks and gardens is home to more than 40 cultural, educational, medical, and research institutions, including the city's museums of art and natural history, its botanical gardens, and Severance Hall, home of the symphony orchestra. Campus buildings are an eclectic mix of architectural styles, and several are listed on the National Register of Historic Places. The Lewis Building, designed by Frank Gehry, is home for the Weatherhead School of Management. It features undulating walls similar to those of Gehry's Guggenheim Museum in Bilbao, Spain, along with offices, classrooms, and meeting rooms on every floor, to encourage informal student/faculty interaction.

The product of the 1967 marriage between Case Institute of Technology and Western Reserve University, Case has four undergraduate schools: the College of Arts and Sciences, the Case School of Engineering, the Bolton School of Nursing, and the aforementioned Weatherhead School; all also offer graduate programs. All Case students participate in a new general education program, SAGES, introduced in the fall of 2005. The acronym stands for the Seminar Approach to General Education and Scholarship, and the program emphasizes small seminars, critical thinking, and writing. A student's first seminar focuses on one of four themes: The Life of the Mind; Thinking about the Symbolic World; Thinking about the Natural/Technological World; and Thinking about the Social World. This is followed by more seminars in the sophomore year, then a departmental seminar in the student's major, and finally a senior project.

> **"We're not a big party school."**

Case's strongest programs include engineering, especially biomedical engineering, and the school's polymer science major is one of the few such undergraduate programs in the country. Biology, chemistry, physics, and mathematics are all strong. "Humanities have been increasing in reputation," says an anthropology major. Other strengths in the College of Arts and Sciences include music (a joint program with the nearby Cleveland Institute of Music), anthropology (especially medical anthropology), and psychology. Nursing, management, and accounting are also strong. One of the newest majors is Public Health Science, which students may pursue as part of a five-year program leading to a Master of Public Health degree. A major in cognitive sciences has been introduced as well.

Courses tend to be rigorous, but according to a management major, "Students are competitive with themselves, not with one another so much." While professors do focus on their research, "they take time to meet with students individually, and are easily accessible," says a chemistry major. Nearly 42 percent of freshman classes are taught by tenured professors. "It is not uncommon for students to overload to complete a second major," says an electrical engineering student. Even the non-engineers are tech-savvy and focused on life after college. "Students attend Case because they want an education that will empower them to have a fantastic and successful career," one student reasons. Combined bachelor's and master's programs are popular, as is the Preprofessional Scholars program, which gives top freshmen conditional acceptance to Case's law, social work, medical, or dental schools, assuming satisfactory progress through prerequisite courses.

About 55 percent of Case's students are Ohio natives, and 70 percent come from a public high school. African Americans make up 5 percent of the student body, Asian Americans 15 percent, and Hispanics 2 percent. In addition to need-based

(Continued)
Male/Female: 49/41
SAT Ranges: V 600–700
   M 640–740
ACT Range: 27–31
Financial Aid: 60%
Expense: Pr $ $
Phi Beta Kappa: Yes
Applicants: 7,181
Accepted: 68%
Enrolled: 24%
Grad in 6 Years: 77%
Returning Freshmen: 92%
Academics: ✍ ✍ ✍ ✍
Social: ☎ ☎
Q of L: ★ ★ ★
Admissions: (216) 368-4450
Email Address:
   admission@case.edu

Strongest Programs:
Engineering
Management
Biology
Chemistry
Music
Nursing
Psychology
Medical Anthropology

Given the lopsided male/female ratio, when a man at Case pledges his undying love and devotion, it can as easily be to his laptop as to a woman.

While students give the older dorms mixed reviews, they are awed by the apartment-based accommodations for upper classmen. The complex houses about 700 upperclassmen in apartment-style suites with two to nine students each.

financial aid, about half the students at Case receive scholarships based on academic merit, with merit-based awards averaging $18,590. Those attract "smart, self-motivated people," one student notes. "They are very studious and mostly middle class, and largely from the Midwest." Issues that get students riled up are those with a scientific bent, such as fuel cells and stem cell research. Mostly, though, kids at Case are too busy to care about the vagaries of politics, according to a psychology major: "The typical Case student has an opinion on controversial issues, but would never protest or write to the newspaper about it."

Seventy-five percent of students live on campus; freshmen and sophomores are required to do so. While students give the older dorms mixed reviews, they are awed by the apartment-based accommodations for upper classmen. The complex houses about 700 upperclassmen in apartment-style suites with two to nine students each. Students enjoy full kitchens, living rooms, single rooms, and double beds. "The new apartments are simply stunning, and the community on campus is unparalleled," says a sophomore. In addition to dishwashers and extra-large windows, the apartments include study rooms, laundry rooms, music practice rooms, a fitness center, and a cyber cafe. Another choice for upperclassmen is off-campus housing, where there are many desirable apartments within walking distance of campus. Among campus meal-plan options, students say the two dining halls are okay but best on weekdays. There are also plenty of quick-service options on campus, such as Starbucks, Subway, and Einstein's Bagels. The Silver Spartan, a classic diner named by students in an online poll, is open seven days, until midnight during the week and until 3 a.m. on weekends.

"The new apartments are simply stunning, and the community on campus is unparalleled."

Cleveland is no Boston—heck, it's not even Pittsburgh—but students seem to love it anyway, especially since Case is located in a quasi-suburban area five miles from downtown. "You have all the luxuries of a big city without the big cost," says a sophomore. "Right off campus, you can find Little Italy, home of the best Italian food in Cleveland, and a variety of bars and restaurants in surrounding areas, such as Coventry," says another sophomore. "Downtown, the Flats and the Warehouse District offer upscale dining and numerous clubs that are always packed with students and young professionals." There's also the Rock & Roll Hall of Fame, and—in the warmer months—Cleveland Indians games at Jacobs Field.

On campus, there are dances and fraternity parties (Greek groups draw 34 percent of the men and 23 percent of the women), but "Greek life at Case is atypical," says a senior. "The organizations are well-respected, active, often have high GPAs, and are civic-minded." Community service projects are also big in the Greek community. Having a car is helpful, especially for road trips to Cedar Point (an amusement park about an hour away, in Sandusky), or for longer jaunts to Chicago or Windsor, Ontario, where students under 21 are free to drink and gamble. In Cleveland, students get unlimited bus service for around $25 a semester.

Given the lopsided male/female ratio, when a man at Case pledges his undying love and devotion, it can as easily be to his laptop as to a woman. Rationalizes one student: "You may have trouble finding a date on most weekends, but if your computer crashes on a Friday night, it will be up and running by Saturday morning." Popular campus traditions are Greek Week, which includes nearly everyone on campus; the Spring Fest to celebrate the end of classes; and "Study Overs," where students gather during finals week for free food, massages, study groups, and more. The annual sci-fi movie marathon is a rite of passage, while Engineering Week features the Mousetrap Car Race (all cars must run on a one-mousetrap engine) and the Egg Drop (a foolproof protective package for the tossed egg is key).

"You have all the luxuries of a big city without the big cost."

Although sports are not a major focus on campus, the Tartans versus Spartans football game against Carnegie Mellon is big, and intramural competition draws 55 percent of undergrads. "Ultimate Frisbee is huge," says a junior. Also popular are intramural soccer, basketball, softball, volleyball, flag football, inner-tube water polo, floor hockey, and other sports. Swimming has been one of the most competitive varsity teams at Case. In recent years, individual swimmers, wrestlers, and track and cross-country athletes have won NCAA and UAA honors. The 26-mile Hudson Relay, held the last week of the spring semester to commemorate Case's relocation from Hudson to Cleveland, pits teams of runners from the four classes against one another, with each person running a half-mile. There's also a racquetball and squash complex, and a field house with an Olympic-size pool.

If you come to Case, students say, prepare to work hard. Studying takes priority, and one student says his favorite tradition is the Midnight Scream: "The night before you have a final, you go out on the balcony of your dorm, or lean out the window, and scream at midnight. It's great to hear other people screaming, to remind you that you aren't alone." For a student seeking a first-rate technical or premedical education in a friendly and vibrant environment, Case may be just the right fit.

If You Apply To ➤ **Case:** Early action: Nov. 1. Regular admissions: Jan. 15. Housing: May 1. Meets demonstrated need of 88%. Campus interviews: recommended, informational. Alumni interviews: optional, informational. SATs or ACTs: required (including optional ACT Writing Test). SAT subject tests not required. Accepts the Common Application and electronic applications. Essay question. Audition required for artists and musicians.

The Catholic University of America

Washington, DC 20064

There are other Roman Catholic–affiliated universities, but this is the Catholic University. Catholics make up 80 percent of the student body here (versus roughly half at nearby Georgetown). If you can't be in Rome, there is no better place than D.C. to work and play. CUA even has a Metro stop right on campus.

Founded in 1887 under a charter from Pope Leo XIII, The Catholic University of America was the brainchild of United States bishops who wanted to provide an American institution where the curriculum was guided by the tenets of Christian thought. Over time, the university has garnered a reputation as a research-oriented school that also provides a strong undergraduate, preprofessional education and an appreciation for the arts.

Catholic's campus comprises 145 tree-lined acres, an impressive layout for an urban university. Buildings range from ivy-covered brownstone and brick to ultra-modern, giving the place a true collegiate feel. Catholic is one of the few colleges in the country that began as a graduate institution (others are Clark University and The Johns Hopkins University), and grad students still outnumber their younger counterparts. Six of its 10 schools (arts and sciences, engineering, architecture, nursing, music, and philosophy) now admit undergrads, while two others (social service and religious studies) provide undergraduate programs through arts and sciences.

Students have excellent options in almost any department at CUA. Apart from politics (which all agree sets the tone on campus), the history, English, drama, psychology, and physics departments are very strong. Philosophy and religious studies

Website: www.cua.edu
Location: City outskirts
Total Enrollment: 6,130
Undergraduates: 3,053
Male/Female: 44/56
SAT Ranges: V 530–630
  M 510–620
ACT Range: 21-27
Financial Aid: 48%
Expense: Pr $ $
Phi Beta Kappa: Yes
Applicants: 3,152
Accepted: 81%
Enrolled: 31%
Grad in 6 Years: 73%
Returning Freshmen: 82%

are highly regarded and have outstanding faculty members as well. The School of Nursing is one of the best in the nation, and engineering and architecture are also highly regarded. Architecture and physics have outstanding facilities, the latter enjoying a modern vitreous-state lab, a boon for both research and hands-on undergraduate instruction. Also, a high-speed fiber-optic network connects the entire campus to the Internet. For students interested in the arts, CUA's School of Music offers excellent vocal and instrumental training, and there's also a program in musical theater. Administrators candidly cite sociology and art as weaker than others. CUA's library offers approximately 1.49 million volumes, modest for a school of this size.

Students at CUA need at least 40 courses in order to graduate. In the School of Arts and Sciences, approximately 25 of these must be from a core curriculum spread across the humanities, social and behavioral sciences, philosophy, environmental studies, religion, math and natural sciences, and languages and literature. English composition also is required. The brightest students can enroll in a 12-course interdisciplinary honors program that offers sequences in the humanities, philosophy, and social sciences. Standard off-campus opportunities are augmented by internships at the British

"The teachers push us to work hard but at the same time apply the subjects to everyday living."

and Irish parliaments, NASA, the National Institutes of Health, the Pentagon, and the Library of Congress. The School of Architecture and Planning hosts summer classes for seniors and grad students to study in Italy and a host of European and Mediterranean countries. The Rome study abroad program for design students incorporates design studio, field study, history, theory, and the Italian language. Finally, two students are chosen each year to spend a fall semester at the Fondazione Architetto Rancilio (FAAR) in Milan to study themes including architecture, urban studies, and technology. CUA also offers accelerated degree programs in which students can earn bachelor's and master's degrees in five years, or six years for a joint BA–JD. CUA is part of the 11-university Consortium of Universities of the Washington Metropolitan Area and the Oak Ridge Associated Universities consortium. The latter is comprised of 87 U.S. colleges and a contractor for the federal Energy Department. The program gives students access to federal research facilities.

Although Catholic University is research-oriented, most classes have fewer than 25 students. That means special attention from faculty members. It also means that there's no place to hide. "The teachers push us to work hard but at the same time apply the subjects to everyday living," a history major reports. The faculty is given high marks by students. "The quality of teaching is excellent. Professors are usually very willing to help students," praises a veteran. Clergy are at the helm of certain graduate schools, but the School of Arts and Sciences has a primarily lay faculty, with priests occupying less than 16 percent of the teaching posts. Its chancellor is the archbishop of Washington, and Catholic churches across the country donate a fraction of their annual collections to the university. One downside of being the only Catholic school with a papal charter is that officials in Rome, who do not always warm up to American traditions of academic freedom, keep a sharp eye on the theology department.

Catholicism is clearly the tie that binds the student body. Sunday masses are so well attended that extra services must be offered in the dorms. Says one administrator, "We like students to leave with Catholic values, but most of them come here with those values in the first place." Most are from the Northeast, and are primarily white. African Americans and Hispanics comprise 7 and 6 percent of the student body, respectively. Another 3 percent are Asian American, and foreign students account for 2 percent. Ninety-one percent are from out of state, and 54 percent of freshmen rank in the top quarter of their high school class. Politically, students are fairly conservative, and the big issues on campus include abortion and gay/lesbian

Thirty-one lucky students—one from each archdiocese in the nation—receive a full-tuition merit scholarship.

rights. "Everyone is basically the same at Catholic—white, upper-middle-class Catholics," says one student. The university took a dim view of efforts to establish a chapter of the National Association for the Advancement of Colored People on campus. Initially, the university rejected the group on the grounds that it would favor abortion rights, but changed its mind when the NAACP agreed not to advocate issues that oppose Catholic's values.

The university maintains a need-blind admissions policy. It does not guarantee to meet the full demonstrated need of all admitted, but 52 percent of aid recipients are offered full demonstrated need. Thirty-one lucky students—one from each archdiocese in the nation—receive a full-tuition merit scholarship. There are several hundred additional merit scholarships available worth an average of $8,962, along with various types of financial aid. There are no athletic scholarships.

Sixty-eight percent of the students live in the dorms, seven of which are co-ed by floor. The spacious and ultramodern Centennial Village (eight dorms and 600 beds laid out in suites) is available to all students, many of whom flee CUA's strict visitation (no guests past 2:00 a.m.) and alcohol policies and move into apartments of their own. The best dorms for freshmen are Spellman and Flather (co-ed) and Conaty (women). Dorm food is fairly tasty, and there's always the Rathskeller (or "Rat"), the campus bar and grill where students can get a late-night meal or brunch and dinner on weekends. Emergency phones, shuttle buses, and escort services are provided as part of campus security, and students agree that they always feel safe on campus as long as they are careful.

When students want to explore the city, they need only walk to the campus Metro stop and then enjoy the ride. Capitol Hill is 15 minutes away; the stylish Georgetown area, with its chic restaurants and nightspots, is only a half-hour away.

CUA students do indulge in some serious partying. Some of their favorite locales include the Irish Times, Colonel Brook Tavern, the Tune Inn, and Kitty's. It's no wonder most students agree that the social scene is "off campus at various bars, clubs, and coffeehouses in D.C." Only 1 percent of the men and women join the Greek system. No one under 21 can drink on campus, and most students agree that this policy is effective in curbing underage drinking. Also, alcohol "abuse" has been added as an offense in addition to use, possession, and distribution. For those eager to repent the weekend's excesses, there are student ministry retreats.

> "We like students to leave with Catholic values."

Annual festivals on the campus calendar include a weeklong homecoming celebration, Beaux Arts Ball, Christmas Holly Hop (held in New York City), and Spring Fling. A time-honored winter tradition is sledding down Flather Hill on cafeteria trays.

Sports on campus include varsity and intramural competition. CUA's athletic teams compete in Division III. Solid men's teams include basketball, baseball, swimming, and lacrosse. Volleyball, field hockey, softball, and swimming are the strongest women's teams. Intramural and varsity athletes alike enjoy the beautiful $10 million sports complex. Many students use their strength for community service, including the Christian-based Habitat for Humanity, in which they build houses for needy families.

When discussions first raised the idea of a Catholic university, the man who would become the university's first rector, Bishop John Joseph Keane, argued for an institution that would "exercise a dominant influence in the world's future" with a superior intellectual foundation. Now, more than 100 years later, CUA offers students a wealth of preprofessional courses spanning the arts and sciences. The founders' quest for "a higher synthesis of knowledge" is constantly being realized at CUA, a unique university and a capital destination.

When students want to explore the city, they need only walk to the campus Metro stop and then enjoy the ride.

The School of Nursing is one of the best in the nation, and engineering and architecture are also highly regarded.

Overlaps

American, Boston College, Georgetown, George Washington, Loyola (MD)

Catholic: Early action: Nov. 15. Regular admissions: Feb. 1. Financial aid: Feb. 1. Housing: May 15. Does not guarantee to meet demonstrated need. SATs or ACTs: required. SAT IIs: recommended (writing and foreign language). Accepts the Common Application and electronic applications. Essay question.

Centre College

600 West Walnut, Danville, KY 40422

Centre is college the way it used to be—gentleman scholars, football games, and fraternity pranks (preferably done in the nude). There is also the unparalleled closeness between students and faculty that comes with a student body of just over 1,000. Compare to Sewanee, DePauw, and Kenyon.

Website: www.centre.edu
Location: Small town
Total Enrollment: 1,127
Undergraduates: 1,127
Male/Female: 49/51
SAT Ranges: V 580–678
  M 600–670
ACT Range: 25–29
Financial Aid: 61%
Expense: Pr $
Phi Beta Kappa: Yes
Applicants: 1,989
Accepted: 63%
Enrolled: 25%
Grad in 6 Years: 79%
Returning Freshmen: 92%
Academics: ✍ ✍ ✍ ½
Social: ☎ ☎ ☎
Q of L: ★ ★ ★
Admissions: (800) 423-6236
Email Address:
  admission@centre.edu

Strongest Programs:
English
History
Biology
Biochemistry
Economics
Art

Centre College, the only independent school in Kentucky with a Phi Beta Kappa chapter, has produced two-thirds of the state's Rhodes Scholars over the last 40 years. But the school is not all work and no play. It's also a throwback to the way college used to be, with Friday night parties on fraternity row and Saturday afternoon football games. Centre's small size offers "the ability to get involved and have a direct hand in making improvements," says a biology major. And its liberal arts focus means that despite Centre's Southern location, students are progressive, intellectual, and perhaps more well-rounded than their peers at neighboring schools. "We have an amazing balance of 'northern academics' paired with 'southern hospitalities,'" says a sophomore.

Located in the heart of Kentucky Bluegrass country, Centre's campus is a mix of old Greek Revival and attractive modern buildings. More than 14 of them are listed on the National Registry of Historic Places, a fact that's less surprising when you know that Centre is the 48th oldest college in the United States. The College Centre, a multimillion-dollar renovation and expansion of library, classroom, faculty offices, and athletic facilities, opened in 2005, and a new dorm is due for completion in 2009.

General education requirements include basic skills in expository writing, math, and foreign language; and two courses in four contexts—aesthetic, social, scientific, and fundamental questions. Students also are required to take a computer seminar. The freshman seminar is offered during the three-week January "CentreTerm." These courses are required and are capped at 15 students each and offer a chance to explore topics such as cloning, baseball in American politics, and coffee and culture.

Centre's most popular majors are economics, biology, history, English, and government; not coincidentally, these are also among the school's best departments. Art is also strong, and glassblowing enthusiasts will find one of the few fully equipped undergraduate facilities for their pursuit in the nation. The climate is "very intense,"

> **"We have an amazing balance of 'northern academics' paired with 'southern hospitalities.'"**

says a senior, "but it is supportive and nurturing as well." "Centre is a very rigorous college," adds a sophomore, "but the rewards for hard work are immense." No course taken by a Centre freshman has more than 50 students, and more than three-quarters have 25 or fewer. Students report few problems with scheduling, and Centre guarantees graduation in four years. It also promises that interested students can study abroad or do internships. Professors are "experts in their fields and are also excellent at transmitting their knowledge to students," says a psychology and Spanish double major.

Eighty percent of any given class take advantage of those study abroad programs, which offer travel to London, France, Mexico, Japan, and Ireland during the semester, and to Australia, Russia, Turkey, Vietnam, and Nicaragua during the January term. Centre also belongs to the Associated Colleges of the South,* through which students may select programs in Central America. A 3–2 program sends aspiring engineers on to one of four major universities, including Columbia and Vanderbilt. About one-fifth of students perform collaborative research with faculty, and the John C. Young Scholars program allows select seniors to participate in a year of guided research.

"The people and just the feel of the campus were what really won me over."

"Centre students are hard working and excel inside the classroom and out in the community," a government major says. Two-thirds of Centre students hail from Kentucky, and 73 percent graduated from public high school. African American, Asian American, and Hispanic students together make up 6 percent of the largely homogeneous student body. "For the most part, Centre students seem open-minded and willing to respect differences whether or not they agree with others' opinions," a junior says. Centre does not offer athletic scholarships, though it does award hundreds of merit scholarships a year, averaging $10,381.

Ninety-four percent of students live in Centre's dorms, which are "comfortable and clean," according to a senior. There are three main clusters of halls on campus— North Side, Old Centre, and the Old Quad—and each has all of the necessities of college life. Freshmen live in single-sex halls, while upperclassmen may choose buildings that are co-ed by floor. Centre has also purchased and remodeled an apartment building to provide additional upperclass housing, though the 37 percent of men and 38 percent of women who go Greek may bunk in one of 10 fraternity and sorority houses. Everyone eats together in Cowan Dining Commons, and meal plan credits are also good at the Grille in the student union, and the House of Brews coffee shop. While the quality of food is a common complaint, students say it's no better or worse than other college fare. "The food is pretty good for cafeteria food," admits one student.

Social life at Centre is largely on campus, and revolves around Fraternity Row. "Campus organized events are relatively popular," says a sophomore, "but Greek parties draw the largest crowds." When it comes to alcohol, Centre follows federal and state law—no one under 21 can drink. "The policies generally work pretty well," a senior says. "However, alcohol is a large part of the weekend social life here for many students." That said, the town of Danville is located in a dry county, and

"Campus organized events are relatively popular, but Greek parties draw the largest crowds."

local restaurants only recently won permission to serve alcohol. Thankfully, Lexington and Louisville are within an hour's drive, and it's easy to get to the countryside for camping, fishing, and other outdoor pursuits. Students also get free admission to Centre's separately endowed Norton Center for the Arts, which brings touring musicals, plays, and other performances to campus. Eighty percent of the student body does community service, through the Greek system, Habitat for Humanity, and the Humane Society.

Centre's football team has been around for more than a century, and while it now competes against regional opponents in Division III, that wasn't always the case. In 1921, Centre beat then-powerhouse Harvard, 6–0, a triumph that has been called the greatest sports upset in the first half of the 20th century. Men's and women's cross-country and men's basketball are notable teams, and women's volleyball and softball are up-and-comers. Seventy percent of the student body regularly

Eighty percent of any given class take advantage of those study abroad programs, which offer travel to London, France, Mexico, Japan, and Ireland during the semester, and to Australia, Russia, Turkey, Vietnam, and Nicaragua during the January term.

Centre's most popular majors are economics, biology, history, English, and government; not coincidentally, these are also among the school's best departments.

take advantage of the intramural program, with flag football, softball, and soccer drawing a lot of players. Centre's archrival is nearby Transylvania University, but it's other traditions that really get students going. Those include a serenade for the president by senior women clad in bath towels, and faculty Christmas caroling for the freshmen. And don't forget "Running the Flame," which has students dashing from the fraternity houses, around a sculpture, and back—"naked, of course."

What Centre College lacks in size, it more than makes up for in quality. With a safe, bucolic campus, an emphasis on academic excellence, and faculty and students who care about forming lasting friendships with each other, this undiscovered gem may be worth a look. "The people and just the feel of the campus were what really won me over," one satisfied junior says.

Overlaps

University of Kentucky, Transylvania, University of Louisville, Vanderbilt, Rhodes, Sewanee

If You Apply To ➤

Centre: Early action: Dec.1. Regular admissions: Feb. 1. Financial aid: Mar. 1. Does not guarantee to meet full demonstrated need. Campus interviews: recommended, evaluative. Alumni interviews: optional, evaluative. SATs or ACTs: required. SAT IIs: recommended. Accepts the Common Application and electronic applications. Essay question: experience, achievement, or risk you have taken; issue of local, national, or international concern; influential person; or influential fictional character or historical figure.

Chapman University

One University Drive, Orange, CA 92866

The biggest thing to hit The O.C. since a certain racy TV show, Chapman sits at the hub of Orange County and a stone's throw away from L.A. Aspiring actors flock to marquee programs such as film, television, and the arts. Those without showbiz aspirations can opt for biology, journalism, and business.

Website: www.chapman.edu
Location: Suburban
Total Enrollment: 5,732
Undergraduates: 3,657
Male/Female: 42/58
SAT Ranges: V 543–656
  M 551–662
ACT Range: 23–29
Financial Aid: 60%
Expense: Pr $ $
Phi Beta Kappa: Yes
Applicants: 3,895
Accepted: 52%
Enrolled: 41%
Grad in 6 Years: 65%
Returning Freshmen: 88%
Academics: ✍ ✍ ✍
Social: ☎ ☎ ☎
Q of L: ★ ★ ★
Admissions: (714) 997-6711
Email Address:
  admit@chapman.edu

"Lights, camera, Chapman!" Although Southern California has more than its share of budding starlets and Brad Pitt wannabes, it's also home to Chapman University, where future filmmakers and other burgeoning artists flock to hone their crafts under the watchful eye of seasoned faculty. The university offers solid programs in film, television, and music, and sends students out into the world via countless internships. Even those who steer clear of show business find reason to cheer: Chapman has stellar programs in biology and journalism as well.

Founded in 1861, Chapman University is one of the oldest private universities in California. Originally called Hesperian College, the school later merged with California Christian College in Los Angeles. In 1934, the institution was renamed in honor of C.C. Chapman, an Orange County entrepreneur and benefactor of the school. In 1991, the college again changed its name to Chapman University, reflecting its evolution into a comprehensive institution of higher learning. The beautiful residential campus, situated on 75 tree-lined acres, features a mixture of landmark historic buildings and state-of-the-art facilities. It is located in the historic Old Towne district of Orange, near outstanding beaches, Disneyland, and the world-class cultural offerings of Orange County and Los Angeles. It's also home to the largest freestanding marble staircase west of the Mississippi River, as well as the largest piece of the Berlin Wall owned by an American university. New facilities include Oliphant Hall—featuring 14 teaching studios, a 60-seat lecture hall, and an orchestra hall—and Marion Knott Studios, which features a 500-seat theatre, digital arts center, and two full-sized sound stages. Current construction includes the Lastinger Athletics Complex and an aquatics center and stadium.

The most popular majors are film, business, theatre and dance, music, and education, and these are among Chapman's best. Budding filmmakers may enter the Dodge College of Film and Media Studies, a comprehensive, production-based program that includes majors in film production, television and broadcast journalism, screenwriting, public relations and advertising, and film studies, as well as internships and other active learning opportunities. Future entrepreneurs and business tycoons can take advantage of a well-stocked portfolio of business programs through the Argyos School of Business and Economics. For more artsy types, Chapman offers solid programs in theater and dance, each with frequent national and international per-

"There are over 60 clubs and organizations that take advantage of the facilities on campus."

formance components. The university's peace studies and legal studies programs are tied to internships and study abroad experiences. The movement and exercise science major emphasizes athletic training and physical education.

Regardless of major, all students must complete series of distribution requirements that includes credits in fine and performing arts, humanities, history, social sciences, and natural sciences. In addition, students are expected to fulfill requirements in quantitative reasoning, world cultures, human diversity, and foreign language, as well as pass a junior writing proficiency exam. "It is a competitive college that requires a student to work hard to succeed," says a sophomore. A senior adds, "I found that my lower division courses were more competitive and rigorous and once I got into my upper division work, everything became more laid-back." Freshmen are eased into college life by Chapman's comprehensive first-year program, which provides extensive contact with peers, student mentors, faculty, advisors, as well as an orientation. Freshmen are also assigned a "coach" with whom they meet weekly to discuss goals, plan and organize, and develop critical skills. Finally, new students may take part in the First-Year Experience (FYE), where they "have their own floor in one of the residence halls and participate in special programs aimed at helping them adjust to their first year of college," explains a sophomore.

Ninety percent of classes have 50 or fewer students and the majority have fewer than 25. Freshmen are taught by professors; there are no TAs. "Professors are very personable and always available for outside help," says a creative writing major. The 144,000-square-foot Leatherby Libraries feature state-of-the-art amenities, including nine discipline-specific libraries, a 24-hour study commons, and Internet-based learning environments.

Chapman attracts a friendly, largely affluent student body. More than two-thirds hail from California and 3 percent are international students. African Americans comprise 2 percent of the population, Asian Americans 8 percent, and Hispanics 6 percent. "Our student body is fairly active and most seem very driven," a senior says. Neither liberals nor conserva-

"Our student body is fairly active and most seem very driven."

tives dominate the political scene and hot topics center more on national events than campus politics. Twenty-one percent of undergraduates receive merit scholarships worth an average of $14,755 and music students with less than stellar standardized test scores needn't lose hope—the school places great emphasis on auditions. There are no athletic scholarships.

Fifty-five percent of Chapman students call the five residence halls home and "the dorms are comfortable, with each room getting its own bathroom or sharing with only one other room," according to a theatre major. Braden Hall has 70 rooms that are paired into suites, while Henley Hall features a substance-free community as well as a First Year Experience (FYE) floor for incoming students. Other university-owned options include off-campus houses (for families) and nearby apartments. Dining options are described as "decent, but nothing spectacular." As for safety, students

At Chapman, the spotlight is definitely on those looking to build careers in film and television, whether it's on a movie set, behind the camera, or on a stage.

say that campus security is a constant presence. "I've never felt unsafe on campus," says a junior.

Twenty-five percent of the men join fraternities and sororities attract 35 percent of the women, but the Greeks don't dominate the social scene. Students are apt to hang out with friend on campus or take part in a school-sponsored event. "There are over 60 clubs and organizations that take advantage of the facilities on campus," says a student. "There is always something going on," adds a biology major. Although alcohol is readily available, "The RAs and Public Safety do a good job of controlling alcohol consumption" by students, according to a theatre major. When students grow weary of campus life, they hit the local shops and bars or take trips to "Mexico, Disneyland, Knottsberry Farm, the mall, and L.A.," according to a film major. Back on campus, students flock to the homecoming celebration and the annual Greek Week festival.

> **"We are a fun school with excellent academic possibilities and opportunities."**

The city of Orange (pop. 135,000) is a college town "by the sole means of being a town in which a college is located," quips one student. It's also home to the usual litany of restaurants and shops, as well as a district known as "Old Towne Orange, the Antique Capital of California." Most of the town closes by midnight and students looking for fun generally head into Los Angeles (40 minutes away).

The Chapman Panthers compete in Division III and the most competitive sports include baseball (2003 national champions) and women's softball (2002 national champions). Intramurals are popular and students can usually be spotted playing ultimate Frisbee, volleyball, or soccer in the pristine Southern California weather.

At Chapman, the spotlight is definitely on those looking to build careers in film and television, whether it's on a movie set, behind the camera, or on a stage. Students are not only expected to hit the books, but to actively express their creativity through hands-on learning and forays into the real world. All the while, they're encouraged to build relationships with peers and faculty and enjoy themselves, too. "We have a lot of pride," a film major says. "We are a fun school with excellent academic possibilities and opportunities."

Overlaps

Loyola Marymount, University of San Diego, University of Southern California, UC–Irvine, California State, NYU

If You Apply To ➤

Chapman: Rolling admissions. Early action: Nov. 30. Financial aid: Mar. 1. Housing: May 1. Guarantees to meet full demonstrated need. Campus interviews: recommended, evaluative. No alumni interviews. SATs or ACTs: required. SAT IIs: optional. Accepts the Common Application and electronic applications. Essay question: personal statement.

College of Charleston

Charleston, SC 29424

A public school about half the size of the University of South Carolina that offers business, education, and the liberal arts. College of Charleston compares to William and Mary in both scale and historic surroundings but is far less rigorous academically. Is addressing its housing crunch to help reach the next level.

Website: www.cofc.edu
Location: Urban

Whether sampling the traditional Low Country cuisine or delving into the wide range of courses offered at this strong liberal arts institution, students at the College of Charleston know they are getting a solid education based on creative expression and intellectual freedom. Founded in 1770 as Colonial South Carolina's first college, C of C's original commitment to the liberal arts and to the citizens of the region has helped it become a well-respected institution throughout the Southeast. And the location only adds to the experience, providing opportunities for volunteering and interning and a second-to-none social scene. "This place is amazing, unique, cultural, historical, modern, and it feels like home, all at the same time," a senior says.

Located in Charleston's famous Historic District, the campus features many of the city's most historic and venerable buildings. More than 80 of its buildings are former private residences ranging from the typical Charleston "single" house to the Victorian. The wooded area in front of Randolph Hall, known as the Cistern, is a student gathering point and the site of graduation ceremonies. The campus has received countless awards for its design, and has been designated a national arboretum and a National Historic Landmark. Recently, the college has ramped up the expansion and improvement of campus facilities. A spate of new buildings and renovations has already been completed and current projects include a residence hall with 440 beds for freshmen and 208 apartments for upperclassmen, a cafeteria, and a new athletic center.

C of C has a core curriculum based strongly in the liberal arts and professional programs and focused on the development of writing, computing, language acquisition, and thinking skills. Each student is required to complete six hours in English, history, mathematics or logic, and social science; eight hours

"This place is amazing, unique, cultural, historical, modern, and it feels like home, all at the same time."

in natural sciences; and 12 hours in humanities. Students must also show proficiency in a foreign language. Biology and chemistry are two of the strongest programs; many of the graduates end up at the Medical University of South Carolina a few blocks down the street. Several programs have been awarded with Commendations of Excellence, including all those in the School of Sciences and Mathematics.

"If you don't like to study, read, or write papers," warns a student, "then don't come here." Most agree that the courses can be rigorous, but collaboration is much more common than competition. The most popular major is business administration, and communication, biology, psychology, and political science are also popular. Three new majors were recently added: Latin American and Caribbean studies, hospitality and tourism management, and discovery informatics, an interdisciplinary program that integrates statistics, social sciences, math, computer science, learning theory, logic, information theory, and artificial intelligence. It's the only degree of its type in the country. Many new performing arts majors take advantage of internship opportunities with Spoleto Festival USA, Charleston's annual arts festival. Study abroad options include the International Student Exchange Program and the Sea Semester.* Pro-

"If you don't like to study, read, or write papers, then don't come here."

fessors get high marks in and out of the classroom. "The quality of teaching is exceptionally good, with a major focus on each individual," a senior biology major says. All new students attend Convocation, where they are introduced to the college's academic traditions and freshmen minority students participate in several support programs designed to ensure their successful transition to college.

"Students here cover a spectrum," muses a senior, including "preppy pop-collars, hardcore hippies, emo punk rockers, jocks, semi-goths, and Dungeons & Dragoners." One-third of the students hail from out of state, and 25 percent graduated in the top tenth of their high school class. Asian Americans and Hispanics each make up 2 percent of the student body, African Americans make up 7 percent and 2 percent are foreign. Students are passionate and opinionated, says a biology major, and hot issues include the war in Iraq, campus smoking policies, abortion, and the environment.

(Continued)

Total Enrollment: 9,222
Undergraduates: 8,931
Male/Female: 36/64
SAT Ranges: V 570–650
  M 570–640
ACT Range: 22–25
Financial Aid: 37%
Expense: Pub $ $ $
Phi Beta Kappa: No
Applicants: 8,217
Accepted: 66%
Enrolled: 37%
Grad in 6 Years: 58%
Returning Freshmen: 83%
Academics: ✍ ✍ ✍
Social: ☎ ☎ ☎ ☎
Q of L: ★ ★ ★ ★
Admissions: (843) 953-5670
Email Address:
  admissions@cofc.edu

Strongest Programs:
Education
Biochemistry
Computer Science
Geology
Marine Biology
Environmental Science
Physics
Languages

Women far outnumber men, but females looking to beat the odds can always go to the Medical University of South Carolina or the Citadel Military College, both in Charleston.

The college offers hundreds of merit scholarships averaging $10,293 and athletic scholarships in nine sports.

Lack of dorm space has been one of C of C's biggest challenges as of late, but the forthcoming dorms will help alleviate the crunch. "Students don't live in shoebox rooms," says a sophomore. "They have suites with bathrooms, a kitchen area, and even a small family room." The percentage of students living on campus has slowly increased, and is now 30 percent. Parking is limited but food is plentiful; there are options for all types of eaters, including "homestyle, grill, deli, salad bar, Greek, dessert, and cereal bar" selections. Vegetarian fare is available as well.

No matter where they live, students enjoy Charleston, with its festivals, plays, and scenic plantations and gardens. "Charleston is filled with history, culture, the arts, and wonderful people," a senior says. "No one ever wants to leave!" Students party off campus in local clubs and apartments as well as on campus, where 13 percent of the men and 16 percent of the women belong to frats and sororities, respectively. Due to a well-enforced policy on drinking, students report that it is difficult to be served on campus if you are not 21, but off campus it is not a problem. "I think the policies keep drinking to a minimum in dorms," a senior says. Women far outnumber men, but females looking to beat the odds can always go to the Medical University of South Carolina or the Citadel Military College, both in Charleston.

"No one ever wants to leave!"

While Charleston is primarily a historic and tourist town, "The people of Charleston love the students and welcome them back every fall," says one student. On weekends, students can head to beaches such as Folly Beach, Sullivan's Island, and Isle of Palms, which are merely minutes away. For those who don't mind a drive, there's "the Grand Strand," Myrtle Beach, 90 miles north, Savannah and Hilton Head to the south, and Atlanta and Clemson University to the West.

The absence of a football team is a common gripe among students, but other athletics are relatively popular. C of C is a Division I school and several teams have claimed recent conference championships, including volleyball, softball, and men's baseball. The solid men's basketball program recently welcomed new head coach Bobby Cremins. There is a lengthy roster of intramural and club sports to choose from, including indoor soccer, tennis, and racquetball.

The College of Charleston has set its sights on becoming the finest public liberal arts and sciences institution in South Carolina, and it seems to be on its way, propelled by an honors college, opportunities to do research and study abroad, and new living and learning facilities. "Our small college feel with large college advantages bring a great atmosphere to C of C," a senior says.

Overlaps
Clemson, USC–Columbia, Georgia, North Carolina, Furman, James Madison

If You Apply To ➤

C of C: Early action: Nov. 1. Regular admissions: Apr. 1. Campus interviews: optional, informational (campus visit strongly encouraged). No alumni interviews. SATs or ACTs: required. SAT IIs: optional. Accepts the Common Application and prefers electronic applications. Essay question: something unique you will bring to the college; impact of a significant experience, achievement, or risk; how the college will help you reach your goals; other topic of choice.

University of Chicago

1116 East 59th Street, Chicago, IL 60637

Periodically, the news media reports that students at the University of Chicago are finally loosening up and having some fun. Don't believe it. This place is for true

intellectuals who don't mind working hard for their degrees. Less selective than the top Ivies, but just as good. Social climbers apply elsewhere.

The University of Chicago attracts students eager to move beyond the cliquishness of high school and the superficial trappings of Ivy League prestige—the kids more concerned about learning for learning's sake than about getting a job after graduation, though they're certainly capable of the latter. "We're all nerds at heart," says a senior. Still, administrators have realized that in the 21st century, even the best schools cannot survive on intellectual might alone. To make the U of C more attractive, they've made the core curriculum less restrictive, built a new dorm and sports center, resurrected varsity football, and even opened an office dedicated to fostering school spirit and improved alumni relations. "Everyone came to Chicago despite its reputation as the place where fun comes to die," agrees a freshman. "The fact that college here is a good time just makes us that much happier."

The university's 190-acre tree-lined campus is in Hyde Park, an eclectic community on Chicago's South Side, surrounded by low-income neighborhoods on three sides and Lake Michigan on the fourth. One of 77 city neighborhoods, Hyde Park "is pretty intellectual," says one student, noting that "two-thirds of our faculty live here." Streets are lined with brownstones, row houses, and townhouses, giving way to luxury high rises with beautiful views as you get closer to the lake; the city's Museum of Science and Industry is within spitting distance. The campus itself is self-contained and architecturally magnificent. The main quads are steel-gray Gothic—gargoyles and all—and newer buildings are by Eero Saarinen, Mies van der Rohe, and Frank Lloyd Wright, some of Chicago's favorite architectural sons. The $200 million Gordon Center for Integrative Sciences opened in 2006.

Historically, Chicago has drawn praise for its graduate programs. But administrators and faculty have realized that they must pay attention to undergraduates, too, if Chicago has any hope of staying competitive with schools like Stanford, Harvard, and Princeton. To that end, Chicago remains unequivocally committed to the view that a solid foundation in the liberal arts is the best preparation for future study or work and that, further, theory is better than practice. Thus, music students study musicology, but also learn calculus, along with everyone else. Regardless of major, 15 to 18 of a student's 42 courses fall under general education requirements called the Common Core. (That's down from 21 in years past; the precise number of courses in the core depends on how much foreign language instruction a student needs to reach proficiency.)

Other core requirements include courses in the sciences and math, humanities, social sciences, and European civilization, a recent substitution for a long-standing Western Civ sequence. Sound intense? Students say it is, especially because Chicago pioneered the quarter system. Class material is presented over 11 weeks, rather than the 13 or 14 weeks in a typical semester; the first term starts in late September and is over by Christ- **"We're all nerds at heart."** mas. Practically, this means virtually uninterrupted work through the year, punctuated by a long summer vacation and three exam weeks. "The courses are extremely rigorous because that's the way we like them," says a junior. A freshman adds, "People here have bookshelves overflowing with books they have actually read. Passionate discussion and intellectual curiosity are everywhere." Despite the academic rigor, collaboration is the norm. Seniors are also encouraged to do final-year projects. Helpings students learn those skills are brilliant and distinguished faculty members who've won Nobel Prizes, Guggenheim Fellowships (aka "genius grants"), and other prestigious awards. "I don't know how professors here could get better," says a senior. "They make themselves available to students and seem genuinely interested in our ideas."

Website: www.uchicago.edu
Location: Urban
Total Enrollment: 11,832
Undergraduates: 4,671
Male/Female: 50/50
SAT Ranges: V 680–770
   M 670–760
ACT Range: 29–33
Financial Aid: 45%
Expense: Pr $ $ $ $
Phi Beta Kappa: Yes
Applicants: 9,011
Accepted: 40%
Enrolled: 33%
Grad in 6 Years: 87%
Returning Freshmen: 97%
Academics: ✍ ✍ ✍ ✍ ✍
Social: ☎ ☎
Q of L: ★ ★ ★
Admissions: (773) 702-8650
Email Address:
   collegeadmissions@
   uchicago.edu

Strongest Programs:
Economics
English
Sociology
Anthropology
Political Science
Geography
History
Linguistics

The economics department, a bastion of neoliberal or New Right thinkers, is Chicago's main academic claim to fame. Popular majors include social sciences, biological and biomedical sciences, math, psychology, and history. The university also prides itself on interdisciplinary and area studies programs, such as those focusing on East Asia, South Asia, the Middle East, and the Slavic countries. Undergrads may also take courses in any of the university's graduate and professional schools—law, divinity, social service, public policy, humanities, social sciences, biological and physical sciences, and business. While you may have to fight through a thicket of PhD students to get a professor's attention, you'll be rewarded by an abundance of research assistantships—and opportunities for publication, even before you graduate. When Chicago gets too cold and snowy, students may take advantage of study abroad programs, which reach most corners of the globe.

"U of C students are intellectual and proud of it," says an anthropology major. "I've even discussed Max Weber at a frat party." Seventy-four percent of Chicago's students come from out of state, including many East Coasters with academic parents; another 7 percent are foreigners. Asian Americans represent 14 percent of the total; Hispanics account for 8 percent, and African Americans add 4 percent. Politically, "the conservative voice is allowed a presence on campus—the effect is to enliven debate and save us liberals from easy self-assurance," says a freshman. "Last year, students wanted to kick Taco Bell off campus because they underpay their tomato pickers in California," adds a junior. Students have fond memories of freshman orientation, known as O Week, an event administrators claim was invented at U of C in 1924. The school hands out merit scholarships each year, worth an average of $11,579. There are no athletic awards.

"Passionate discussion and intellectual curiosity are everywhere."

Though Chicago guarantees campus housing for four years, only 56 percent of undergrads live in the dorms. "Dorms are divided into units of 30 to 100 people called houses," says a junior. "These houses become the center of social life, at least for first-years." Each dorm is different—some house less than 100 people in traditional, shared double rooms without kitchens, while another has 700 beds organized into new and colorful suites. Two dorms used to be hotels, including the Shoreland, the largest and most social dorm, at the edge of Lake Michigan; another used to be a nursing home. All halls are co-ed by room or by floor. "Hyde Park has tons of really cute, cheap apartments," reports one student, so the more "independent-minded" students usually move off campus.

Despite its antisocial reputation, Chicago has nine fraternities and two sororities, which attract 10 percent of the men and 5 percent of the women. Aside from frat parties, Chicago, the school, and Chicago, the city, offer what a freshman calls "infinite options: bars downtown, a film festival at DOC, a White Sox game at Comiskey Park, a spoken-word performance at a Belmont coffeehouse, open-mic night at the student center." (Lest you get bored, the city also offers world-class symphony, opera, dance, and theater, along with museums galore, and great shopping and dining on Michigan Avenue.) Though everything is accessible by public transportation, cars are a nice luxury (if you can find a parking place). When it comes to drinking, campus policies are "very, um, accommodating," says one student. "The university treats us as responsible adults with common sense." Off campus, it's much harder for the underage to imbibe. Road trips are infrequent, but one popular destination is Ann Arbor, about five hours away, for concerts and more traditional collegiate fun at the University of Michigan.

Tradition is a hallmark at Chicago, and each fall, students look forward to Scavenger Hunt, "a pumped-up version of a regular scavenger hunt, with a list of 300 bizarre items," says a sociology major. "If you walked onto campus during those few

Class material is presented over 11 weeks, rather than the 13 or 14 weeks in a typical semester; the first term starts in late September and is over by Christmas.

days, you'd think most people had lost their minds." In the winter, the Midway, site of the 1893 World's Fair, is flooded for skating. Students also celebrate the festival of Kangeiko, which features their naked or semi-naked peers dashing across campus during the Polar Bear Run. The Festival of the Arts features concerts, a fashion show, special lectures, museum exhibits, and "funky

"Last year, students wanted to kick Taco Bell off campus because they underpay their tomato pickers in California."

installations on the quads." Come summer, students can be found "doing Jell-O wrestling and other carnival activities" as part of Summer Breeze, which also includes a concert. Past artists have included Maroon 5, Bela Fleck and the Flecktones, Talib Kweli, and Method Man.

Each dorm is different— some house less than 100 people in traditional, shared double rooms without kitchens, while another has 700 beds organized into new and colorful suites.

Chicago's Maroons compete in NCAA Division III, and the school belongs to the University Athletic Association, where rivals include Johns Hopkins and NYU. Aside from hitting the gridiron or the basketball court, "varsity athletes are Phi Beta Kappa (that is, very smart) and involved with university theater," a junior marvels. In fact, athletes here have a higher overall GPA than the student body as a whole. The wrestling team and women's soccer squad have boasted All-Americans in recent years, and both men's and women's soccer teams have been to the Division III Final Four. Women's softball earned a number one ranking in the Midwest regional in 2005. To everyone's surprise, the football team has also had a couple of winning seasons. When it comes to intramurals, 70 percent of undergraduates participate in sports ranging from the traditional (football, soccer, and broomball) to the offbeat (inner-tube water polo, badminton, and archery).

Robert Maynard Hutchins, who led Chicago from 1929 to 1951, once said, "Having fun is a form of intelligence." A half-century later, it seems his message is finally getting through. While fun has been slow in coming to the U of C, students are beginning to learn that letting their hair down on the weekends can complement, not compromise, the life of the mind. The quarter system is fast-paced, and not everyone will be happy with the high level of intellectualism that prevails here. But for those eager to be in a big city, and to focus on learning for its own sake, Chicago may be worth a look.

Overlaps

Columbia, Harvard, Northwestern, University of Pennsylvania, Yale

If You Apply To ➤ **Chicago:** Early action: Nov. 1. Regular admissions: Jan. 2. Financial aid: Feb. 1. Guarantees to meet full demonstrated need. Campus and alumni interviews: recommended, informational. SATs or ACTs: required. SAT IIs: optional. Accepts the Common Application and electronic applications. Essay question.

University of Cincinnati

P.O. Box 210091, Cincinnati, OH 45221-0091

In most states, UC would be the big enchilada. But with Ohio State two hours up the road and Miami U even closer, Cincinnati has to hustle to get its name out there. The inventor of co-op education, it offers quality programs in everything from engineering to art—and a competitive men's basketball team to boot.

Many first-time visitors to Cincinnati are surprised to find an attractive and very livable city. As they traverse the city's hilly roads, they are in for another surprise—its university. Not only is the University of Cincinnati renowned for its extensive

Website: www.uc.edu
Location: City outskirts

research programs, the school's co-op program is also one of the largest of any public college or university in the country.

The compact campus is a mile uphill from Cincinnati's downtown area. Ultra-modern buildings rise up next to traditional ivy-covered Georgian halls. A $233 million construction project to create a "Main Street" in the center of campus and consolidate all student activities was recently completed. A $113 million campus recreation center and the $100 million Richard E. Lindner Varsity Village, consisting of athletic facilities, both opened in 2006.

Research is a UC specialty. Campus scientists have given the world antiknock gasoline, the electronic organ, antihistamines, and the U.S. Weather Bureau. UC is also the place where, in 1906, cooperative education was born, allowing students to earn while they learn. Across the Cincinnati curriculum, there is an abundance of co-op opportunities available. More than 3,000 students take advantage of them. In all, more than 40 programs offer the popular five-year professional-practice option.

The colleges of engineering; business administration; and design, architecture, art, and planning (the schools with the most co-op students) are the best bets at UC. The university's music conservatory, one of the best state-run programs in the field,

> "UC students tend to be from larger cities or the suburbs of larger cities in the Midwest."

also offers broadcasting training. The schools of nursing and pharmacy are well known and benefit from UC's health center and graduate medical school. The most popular major is marketing, followed by psychology, communication, criminal justice, and mechanical engineering. In addition, education is a strong program. Education students earn two bachelor's degrees: one in education and one in a liberal arts subject. Newer initiatives include a culinary arts and science degree program offered jointly with Cincinnati State, and the state's first baccalaureate program in facilities and hospitality management.

The academic grind is determined largely by the major. Fields such as engineering, business, and nursing require a substantially larger academic commitment. "I would say that the most rigorous courses are those that are non-traditional," offers one student, citing "study abroad, capstones, projects with corporate partners, and advanced topics classes." Some courses end up being quite large (in popular design courses, two people to a desk is not unusual); 96 percent of classes have 50 or fewer students. One fine asset is the school's huge library, which has 3 million volumes and is completely computerized.

UC has taken steps to improve the quality of the undergraduate education by strengthening its general education requirements to focus on critical thinking and expression and expanding its honors program. Freshmen must take English and math as well as a contemporary issues class; other requirements vary by college. Additionally, UC has adopted a strategic plan known as UC/121 to place students at the center of campus life. A third of the faculty members hold outside jobs, bringing fresh practical experience to the classroom.

"UC students tend to be from larger cities or the suburbs of larger cities in the Midwest," says a student. Ten percent of the student body comes from out of state, which makes the student body fairly heterogeneous. African Americans, Asian Americans, and Hispanics comprise 14, 3, and 2 percent of the student body, respec-

> "We have new suite-style dorms that were opened three years ago and are very nice and spacious."

tively. Diversity, feminist issues, campus construction, and rising tuition are the hot topics on campus. The school offers merit scholarships averaging $4,696 and hundreds of athletic scholarships for men and women. While students say they have noticed the budget squeeze in terms of services being cut and the hiring of new personnel curtailed, the school is growing.

(Continued)
Total Enrollment: 20,964
Undergraduates: 15,960
Male/Female: 50/50
SAT Ranges: V 500–620
   M 500–640
ACT Range: 21–27
Financial Aid: 25%
Expense: Pub $ $ $ $
Phi Beta Kappa: Yes
Applicants: 11,813
Accepted: 76%
Enrolled: 35%
Grad in 6 Years: 50%
Returning Freshmen: 79%
Academics: ✐ ✐
Social: ☎ ☎ ☎
Q of L: ★ ★ ★
Admissions: (513) 556-1100
Email Address:
   admissions@uc.edu

Strongest Programs:
Architecture
Design
Electrical and computer
   engineering
Musical theater
Nursing

The schools of nursing and pharmacy are well known and benefit from UC's health center and graduate medical school.

Twenty percent of UC students live on campus. "We have new suite-style dorms that were opened three years ago and are very nice and spacious," reports a senior. "These are clearly difficult to get into." Many upperclassmen, especially the older and married students, consider off-campus living far better than dorm life, and inexpensive apartments can usually be found. A student says, "The Clifton area is also in the process of developing and building several new complexes that offer apartments for students." Dining facilities offer a wide variety of "edible and diverse" fare, according to a junior.

Merchants have turned the area surrounding UC, called Clifton, into a mini college town with plenty to do. Nine nearby bus lines take undergraduates into the heart of the "Queen City" of Cincinnati in minutes. There the students find museums, a ballet,

"I would say that the most rigorous courses are those that are non-traditional."

professional sports teams, parks, rivers, hills, and as many large and small shops as anyone could want. On-campus activities include 450 student clubs, with everything from mountaineering to clubs in various majors. Fraternities and sororities attract 8 percent of them men and 5 percent of the women, but are still the most active places to party on campus, usually opening their functions to everyone. The university sponsors some events, such as WorldFest and Greek Week. The most popular road trips are the city of Cleveland and white-water rafting in West Virginia.

In sports, men's and women's basketball, football, and volleyball all have seen recent postseason play. Track, soccer, baseball, and women's crew are also popular. Everyone mentions the football rivalry with Miami (of Ohio) as a game you won't want to miss, and the same holds true when the men's basketball squad takes on Xavier University. Weekend athletes also take advantage of UC's first-rate sports center.

Cooperative education is the name of the game at this Ohio school. Students get to take their degrees out for a test drive before graduation thanks to the work-study co-op programs. UC also offers students a lively social scene, both on campus and minutes away in downtown Cincinnati.

Overlaps

Ohio State, Miami University (OH), Ohio University, Bowling Green State, Wright State, Xavier University

If You Apply To ➤ **UC:** Rolling admissions: varies by college. Does not guarantee to meet full demonstrated need. Campus interviews: optional, informational. No alumni interviews. SATs or ACTs: required (ACTs preferred). SAT IIs: optional. Accepts the Common Application and electronic applications. Essay question: personal statement. Apply to particular program.

Claremont Colleges

In 1887, James A. Blaisdell had the vision to create a group of colleges patterned after Oxford and Cambridge in England. More than a century later, the five schools that comprise the Claremont Colleges thrive as a consortium of separate and distinct undergraduate colleges with two adjoining graduate institutions, a theological seminary, and botanical gardens. Like families, the colleges coexist, interact, and experience their share of both cooperation and tension. Ultimately, however, the Claremont College Consortium forms a mutually beneficial partnership that offers its students the vast resources and facilities one might only expect to find at a large university.

The colleges are located on 317 acres in the Los Angeles suburb of Claremont, a peaceful neighborhood replete with palm trees, Spanish architecture, and the nearby San Gabriel Mountains. The picture-perfect California weather is marred by smog, courtesy of the neighbors in nearby L.A., but the administration claims the smog level has declined dramatically in the past few years.

None of the five undergraduate colleges that make up the Claremont Colleges Consortium—Claremont McKenna, Harvey Mudd, Pitzer, Pomona, and Scripps—is larger than a medium-size dorm at a state school. Each school retains its own institutional identity, with its own faculty, administration, admissions, and curriculum, although the boundaries of both academic work and extracurricular activities are somewhat flexible. Each of the schools also tends to specialize in a particular area that complements the offerings of all the others. Claremont McKenna, which caters mainly to students planning careers in economics, business, law, or government, has eight research institutes located on its campus, while Harvey Mudd is the choice for future scientists. Pitzer, the most liberal of the five, excels mainly in the behavioral sciences, and at the all-women Scripps, the best offerings are in art and foreign languages. The oldest of the five colleges, Pomona ranks as one of the top liberal arts colleges anywhere, and is the one Claremont school that is strong across the board, and especially superb in the humanities.

Collectively, the colleges share many services and facilities, including art studios, a student newspaper, laboratories, an extensive biological field station, a health center, auditoriums, a 2,500-seat concert hall, a 350-seat theater, bookstores, a maintenance department, and a business office. The Claremont library system makes more than 1.9 million volumes available to all students, though each campus also has a library of its own. Faculties and administrations are free to arrange joint programs or classes between all or just some of the schools. Courses at any college are open to students from the others (approximately 1,200 courses in all), but each college sets limits on the number of classes that can be taken elsewhere. Perhaps the best example of academic cooperation is the team-taught interdisciplinary courses, which are organized by instructors from the different schools and appeal to a mix of different academic interests.

The Claremont Colleges draw large numbers of students from within California, although their national reputation is growing. These days, about half the students hail from other Western and non-Western states, with a sizable contingent from the East Coast. The tone at Claremont is decidedly intellectual—more so than at Stanford or any other place in the West—and graduate programs in the arts and sciences are more common goals than business or law school. Anyone who is bright and hardworking can find a niche at one of the five schools. Unfortunately, despite their excellence, the Claremonts are also among the most underrated colleges in the nation.

The local community of Claremont is geared more to senior citizens than college seniors. "Quiet town of rich white people—boring," yawns an English major. A sophomore says, "Most of the stores have strange granny knick-knacks or cosmic aura trinkets." Still, "the Village," a quaint cluster of specialty shops (including truly remarkable candy stores), is an easy skateboard ride from any campus, though the shades come down and the sidewalks roll up well before sunset. Students report that the endless list of social activities offered at the colleges make up for the ho-hum town of Claremont. For hot times, Hollywood's glamour and UCLA-dominated Westwood are within sniffing distance, and a convenient shuttle bus makes them even closer for Claremont students without cars. Nearby mountains and the fabled surfing beaches make this collegiate paradise's backyard complete. Mount Baldy ski lifts, for instance, are only 15 miles away, and you'll reach Laguna Beach before the end of your favorite CD. For spring break, Mexico is cheap and a great change of pace.

On campus, extracurricular life maintains a balance between cooperation and independence. Claremont McKenna, Harvey Mudd, and Scripps field joint athletic teams, and the men's teams especially are Division III powers, due to the exploits of CMC athletes. Pomona and Pitzer also compete together. Each of the five colleges has its own dorms, and since off-campus housing is limited in Claremont proper, the social life of students revolves around their dorms. "Scripps itself is quiet, but parties at Harvey Mudd and Claremont McKenna can get pretty wild," admits a Scripps student. There are no fraternities, except at Pomona, where joining one is far from *de rigueur*. All cafeterias are open to all students, and most big events—films, concerts, etcetera—are advertised throughout the campus. Large five-school parties are regular Thursday, Friday, and Saturday night fare. Social interaction among students at different schools, be it for meals or dates, is not what it might be. Pomona is seen as elitist, and its admissions office has been known to try to distance itself from the other colleges. Occasional political squabbles break out between liberal faculty and students at Pitzer and their conservative counterparts at Claremont McKenna. For the most part, students benefit not only from the nurturing and support within their own schools, each of which has its own academic or extracurricular emphasis, but also from the abundant resources the Claremont College Consortium offers as a whole.

Following are profiles of each undergraduate Claremont College.

Claremont McKenna College

890 Columbia Avenue, Claremont, CA 91711

Make way, Pomona—this up-and-comer now has the lowest acceptance rate in the Claremont Colleges and is no longer content with being a social sciences specialty school. CMC is half the size of a typical liberal arts college and 30 percent smaller than Pomona. Still developing a national reputation.

As a member of the Claremont Colleges consortium, Claremont McKenna College boasts top programs in government, economics, business, and international relations. In addition, CMC has eleven research institutes located on campus, which offer its undergraduates ample opportunities to study everything from political demographics to the environment. The arts and humanities are also available, but Claremont McKenna is better suited to those with high ambitions in business leadership and public affairs.

The fifty-acre campus, located thirty-five miles east of Los Angeles, is mostly "California modern" architecture with lots of Spanish tile roofs and picture windows that look out on the San Gabriel Mountains. Described by one student as "more functional than aesthetic," the physical layout fits right in with the school's pragmatic attitude. Roberts Hall is a state-of-the-art academic center that houses classrooms, seminar rooms, a computer laboratory, and faculty offices.

Claremont McKenna offers top programs in economics and government, but the international relations, psychology, and history programs are also considered strong. The biology, chemistry, and physics departments are greatly enhanced through the use of Keck Science Center, an outstanding facility providing students with hands-on access to a variety of equipment. In addition, the 85-acre Bernard Biological Field Station is located just north of the CMC campus and is available to students for field work.

CMC's extensive general education requirements include two semesters in the humanities; three in the social sciences; two in the natural sciences; a semester each in mathematics, English composition and literary analysis, and Questions of Civilization; and a Senior Thesis. The college offers popular 3–2 programs in management engineering and economics and engineering, and a 4–1 MBA program in conjunction with the Claremont Graduate University. Over 40 percent of Claremont McKenna students take advantage of study abroad programs in Australia, Brazil, Costa Rica, Japan, and other exotic locales around the globe. CMC also offers active campus exchange programs with Haverford, Colby, Spelman, and Morehouse. Another popular program is the Washington Semester program, in which students can intern with E-Span, the State Department, the White House, and lobbying groups. New majors include organismal biology and ecology, classical studies, and molecular biology.

The academic climate is fairly strenuous at Claremont McKenna, but not overwhelming. "We may be laid-back on Saturday nights," says a student, "but come Monday, it's all about achieving the highest GPA and perfecting the résumé." Professors are described as "outstanding" and praised for their accessibility. "We have no teaching assistants," says a junior. "Professors are brilliant and enormously accessible." Eighty-two percent of classes have 25 or fewer students ("one of the perks of a small college," says a neuroscience major).

A junior describes CMC students as "professional, career-oriented, smart jocks." The CMC student body is 44 percent Californian, with nearly 20 percent from the

> **"Professors are brilliant and enormously accessible."**

Website:
www.claremontmckenna.edu

Location: Suburban
Total Enrollment: 1,139
Undergraduates: 1,139
Male/Female: 54/46
SAT Ranges: V 650–750
M 660–740
ACT Range: 29–33
Financial Aid: 49%
Expense: Pr $ $ $
Phi Beta Kappa: Yes
Applicants: 3,734
Accepted: 21%
Enrolled: 35%
Grad in 6 Years: 88%
Returning Freshmen: 93%
Academics: ✍ ✍ ✍ ✍ ½
Social: ☎ ☎ ☎
Q of L: ★ ★ ★
Admissions: (909) 621-8088
Email Address: admission@claremontmckenna.edu

Strongest Programs:
Economics
Government
Psychology
International Relations
History
Sciences

east coast. Many attended public high school and 84 percent graduated in the top tenth of their class. Asian Americans comprise the largest minority at 15 percent. Hispanics comprise 12 percent, while African Americans make up 4 percent. A junior says, "I think our school definitely lives up to its reputation of being very politically active and aware. It was crazy during the last presidential election." All freshmen take part in a five-day orientation program that includes a beach trip and a reception with the president and department chairs. The school guarantees to meet the demonstrated need of accepted applicants and offers merit scholarships averaging $5,518 a year. There are no athletic scholarships.

Ninety-six percent of CMC students live on campus "because of the social life." The maid service probably doesn't hurt. "They dust and vacuum our rooms and clean our bathrooms! We do nothing (except study, of course)!" declares a happy resident.

> **"I think our school definitely lives up to its reputation of being very politically active and aware."**

All the residence halls are co-ed; freshmen are guaranteed a room. Stark Hall, a substance-free dorm, gives students more living options. A cluster of on-campus apartments equipped with kitchen facilities is a popular option for upperclassmen. "The dorms can be like palaces," asserts a junior. "Hot palaces in August and May. Dirty palaces on weekends." Dorm food is said to be quite good, and students can eat in dining halls at any of the other four colleges, though the best bet may be CMC's Collins Dining Hall. "They have a large spread with lots of different options," says one student, including vegetarian, vegan, and organic fare.

Most students agree that the social life at CMC is more than adequate, thanks to the five-college system. "CMC's campus is often the center of the social life for all of the Claremont colleges," says a junior. "There are always parties, club events, barbeques, movie screenings, and other events." In addition to the usual forms of revelry, a calendar full of annual bashes includes Monte Carlo Night, Disco Inferno, Oktoberfest, Chez Hub, and the Christmas Madrigal Feast. Ponding, another unusual CMC tradition, involves being thrown into one of the two campus fountains on one's birthday. The college sponsors an outstanding lecture series at the Marion

> **"CMC's campus is often the center of the social life for all of the Claremont colleges."**

Minor Cook Athenaeum on Monday through Thursday nights each week. Before each lecture, students and faculty can enjoy a formal gourmet dinner together and engage in intellectual debates. Road trips to Joshua Tree, San Francisco, Las Vegas, and Mount Baldy are highly recommended by the students.

Athletics are an important part of life at Claremont McKenna and the school has an overflowing trophy case to prove it. Men's and women's cross country, men's and women's track and field, women's basketball, and men's and women's tennis are especially competitive. A third of the students play varsity sports, and CMC students tend to dominate the teams jointly fielded with Harvey Mudd and Scripps. Top rivalries include Pomona, both in athletics and academics, one student claims. "Basketball games rock this campus," another student says.

CMC has embraced its mission to produce great leaders by providing students with ample opportunities for research and study abroad, as well as top-notch programs such as government and economics. "Leadership pervades almost everything that goes on here," says a junior. "Claremont McKenna builds character, fosters a sense of ambition among its students, and drives them to set their sights high."

| If You Apply To ➤ | **Claremont McKenna:** Early decision: Nov. 15. Regular admissions: Jan. 2. Financial aid: Feb 1. Guarantees to meet full demonstrated need. Campus interviews: recommended, evaluative. Alumni interviews: optional, informational. SATs or ACTs: required (SATs preferred). SAT IIs: optional. Accepts the Common Application and prefers electronic applications. Essay question: personal statement and discussion of leader. Places "strong emphasis on caliber of an applicant's extracurricular activities." |

Harvey Mudd College

301 East 12th Street, Kingston Hall, Claremont, CA 91711

The finest institution that few outside of the science and engineering world have ever heard of. Future PhDs graduate from here in droves. Rivals Caltech for sheer brainpower and access to outstanding faculty. Offers more exposure to the liberal arts than most science- and technology-oriented schools.

A top-ranked technical school, Harvey Mudd College strives to give its students a sense of academic balance. Although it's a leading provider of high-quality programs in science and engineering, it also emphasizes a well-rounded education with knowledge in the humanities. "It's challenging," says a junior. "The way I see it, everyone here—from faculty to staff to students—is cheering you to succeed."

HMC's mid-'50s vintage campus of cinder-block buildings even "looks like an engineering college; it's very symmetrical and there's no romance." In addition, the buildings have little splotches all over their surfaces that students have dubbed "warts"—not a very attractive picture. The latest campus additions include a new dorm and a new dining hall.

While most technology schools tend to have a narrow focus, HMC has come up with the novel idea that even scientists and engineers "need to know and appreciate poetry, philosophy, and non-Western thought," says an administrator. Students here take a third of their courses in the humanities, the most of any engineering college in the nation. Up to half of them can be taken by walking over to another Claremont school. To ensure breadth in the sciences, students take another third of their work in math, physics, chemistry, biology, engineering design, and computer science. The last third of a student's courses must be in one of six major areas: biology, computer science, chemistry, physics, engineering, or math. And finally, to cap off their HMC experience, all students must complete a research project in their major, as well as a senior thesis in the humanities

> "One semester, when I was struggling in a course, a full professor tutored me for about three to five hours per week."

and social sciences. "We wish we could focus more," complains a math major, but in the end it pays off. "All the hard work makes us the successful people that we are, and there are amazing people at this school," says an engineering student.

Of the six majors, engineering is considered the strongest and ranks as the most popular by students, followed by physics and computer science. HMC has one of the nation's top computer science programs and an award-winning math department. In recent years, the number of biology faculty has more than doubled. Students rave about the engineering clinic program, which plops real-life engineering tasks (sponsored by major corporations and government agencies to the tune of more than $30,000 per project) into the laps of students. There's also a Freshman Project that allows neophytes to tackle "some real-world engineering problems." And where else can you take a freshman seminar in integrated-circuit chip design?

Website: www.hmc.edu
Location: Suburban
Total Enrollment: 743
Undergraduates: 743
Male/Female: 70/30
SAT Ranges: V 670–760
　M 710–800
Financial Aid: 53%
Expense: Pr $ $ $
Phi Beta Kappa: No
Applicants: 1,898
Accepted: 36%
Enrolled: 29%
Grad in 6 Years: 88%
Returning Freshmen: 95%
Academics: ✎ ✎ ✎ ✎ ½
Social: ☎ ☎ ☎
Q of L: ★ ★ ★
Admissions: (909) 621-8011
Email Address:
　admission@hmc.edu

Strongest Programs:
Engineering
Computer Science
Math
Physics
Chemistry

The absence of graduate programs means that undergraduates get uncommon amounts of attention, even from top faculty. "Professors are always available," says a math major. "One semester, when I was struggling in a course, a full professor tutored me for about three to five hours per week."

These budding technology leaders are also top achievers: 91 percent graduated in the top 10 percent of their high school class. "Prospective students should know that it can be temporarily damaging to their egos to come to a school with so many bright students," says a junior. Forty-three percent of the students are homegrown Californians. African Americans represent only 2 percent of the student body; Hispanics account for 7 percent and Asian Americans comprise 18 percent. "Everyone here has interests beyond the sciences," says a chemistry major, "but we share a passion for learning and a talent for technical thinking." A handful of merit scholarships help with the hefty tuition bill, but as an NCAA Division III college, Mudd offers no athletic scholarships.

Ninety-eight percent of undergrads live on campus, and "the dorms are great and have the biggest rooms I've ever heard of," says a junior. Five older dorms and two newer, more modern ones are all co-ed and mix the classes. "They are definitely the engineering-school type—functional and efficient," notes a freshman. They range from Atwood ("study hard, party hard") to North ("way cool, so very"). The dorms are also ideal for computer whizzes: all are wired for online access to the HMC mainframe.

"Everyone here has interests beyond the sciences."

Despite their heavy workload, most HMC students find abundant social outlets, even if it's just joining the parade of unicycles that has overrun the campus. The college has no Greek life, and most social life takes place in and around the dorms, where there are parties every weekend. "People tend to know each other, so there is plenty of social life," says an engineering major. A student describes the town of Claremont as "a wonderful place if you're married or about to die." However, most students say there is always fun to be had on one of the five Claremont college campuses. Down-and-dirty types often frequent the Mudd Hole, a pizza/pinball/Ping-Pong hangout. Underage drinking is "compliments of a peer over 21," as one student puts it.

Mudd celebrated its 50th birthday in 2005, and for a school so young, it is rife with tradition. In the annual "pumpkin caroling" trip on Halloween, students serenade professors' homes with doctored-up Christmas carols. Another night of screwball fun is the Women's Pizza Party, in which men don dresses and crash a meeting of the Society of Women Engineers. There is also an annual Five Class competition among the four classes and the handful of fifth-year students, complete with amoebae soccer and relay races that include unicycles (backward), peanut butter and jelly, and slide-rule problem solving. Engineering pranks are popular but must be reversible within 24 hours.

"The dorms are great and have the biggest rooms I've ever heard of."

Mudd fields varsity sports teams together with Claremont McKenna and Scripps, and mainly because of all the CMC jocks, the teams do extremely well. Men's track and field has won 15 straight conference championships, while the women's tennis team has won 12 of the past 13. Men's tennis, women's cross country, and women's swimming and diving are also top-notch. Not long ago, some enterprising Mudders stole archrival Caltech's cannon, elevating the Mudd–Caltech rivalry to include a soccer game dubbed the Cannon Bowl. Intramurals, also in conjunction with Scripps and CMC, are even more popular. Traditional sporting events include the Black and Blue Bowl, an interdorm game of tackle football, and the Freshman–Sophomore Games, which climax in a massive tug-of-war across a pit of vile stuff.

A common student complaint is "too much work," and students would welcome more time to reflect on what they are learning, but the work tends to pay off in grad school and in the job world. HMC is right on the heels of Caltech as the best technical school in the West. Mudd doesn't promise you'll end up with a hot job in Silicon Valley, but some students certainly do, and the college offers a gem of a technical education perfectly blended with a dash of humanities and social sciences. HMC's intimate setting also offers something bigger schools can't: a sense of family.

> **If You Apply To ➤** **Harvey Mudd:** Early decision: Nov. 15. Regular admissions: Jan. 15. Guarantees to meet full demonstrated need. Campus interviews: recommended, informational. No alumni interviews. SATs: required (ACT not accepted). SAT IIs: required (math II, and one other). Accepts the Common Application and electronic applications. Essay question: fun, cool, or interesting things about yourself; recent moral, ethical, or personal decision; what troubles you about the world around you?

Pitzer College

1050 North Mills Avenue, Claremont, CA 91711

Offers a haven for the otherwise-minded without the hard edge of nonconformity at places like Evergreen and Bard. Traditional strengths lie in the social and behavioral sciences. Still nearly 60/40 female, but much more selective than it was 10 years ago.

As the most laid-back of the Claremont colleges, Pitzer College offers students a creative milieu, abundant opportunities for intellectual exploration, and a sense of fierce individualism. Founded in the '60s, this small school has changed with the times but continues its emphasis on progressive thought, social responsibility, and open social attitude.

Even the campus is, well, different. The classroom buildings are modernistic octagons, and the grass-covered "mounds" that distinguish the grounds "are perfect for sunbathing and Frisbee," says one student. The college is in the first phase of construction of new, environmentally friendly dorms.

In keeping with Pitzer's philosophy of student autonomy, each student has the maximum freedom to choose which classes he or she would like to take. A lively freshman seminar program sharpens students' learning skills, especially writing. Students select from 40 majors in sciences, humanities, arts, and social sciences. Almost anything in the social and behavioral sciences is a sure bet, especially psychology (the most popular major), anthropology, sociology, political science, and biology. Most courses in Pitzer's weaker areas can be picked up at one of the other Claremont schools. The Firestone Center for Restoration Ecology in Costa Rica is home to programs in science, language, and international studies and provides opportunities for research. There are also 35 exchange programs and study abroad opportunities in diverse locations such as Ecuador, Italy, and China.

> "The courses, for the most part, are rigorous and challenging."

The academic climate is demanding but cooperative. "The courses, for the most part, are rigorous and challenging," says a junior. Interdisciplinary inquiry is encouraged and original research is common. Class size is generally small, promoting close interaction between students and faculty. "People work together with their professors to create a classroom climate that is supportive and engaging," says a psychology major. "Professors take an active interest in you as a person and that shows in the classroom," a political studies major says.

Website: www.pitzer.edu
Location: Suburban
Total Enrollment: 958
Undergraduates: 958
Male/Female: 41/59
SAT Ranges: V 570–680
  M 560–660
ACT Range: N/A
Financial Aid: 50%
Expense: Pr $ $ $ $
Phi Beta Kappa: No
Applicants: 3,251
Accepted: 37%
Enrolled: 18%
Grad in 6 Years: 70%
Returning Freshmen: 88%
Academics: ✍ ✍ ✍
Social: ☎ ☎ ☎
Q of L: ★ ★ ★
Admissions: (909) 621-8129
Email Address:
  admission@pitzer.edu

Strongest Programs:
Psychology
Sociology
Political Studies

Individualism is a prized characteristic among Pitzer students but one junior says the oft-bandied "hippie" label is unfair: "Pitzer people are genuinely socially conscious and academically adventurous, but do go on to good jobs." Fifty-seven percent are from California and the college has a substantial minority community: African Americans comprise 5 percent of the student body, Hispanics 19 percent, and Asian Americans 8 percent. Pitzer offers merit scholarships worth an average of $8,125.

Seventy-six percent of students live on campus. "The dorms are starting to get old but are well maintained," says a student. Boarders can choose from a variety of meal plans in the dining hall (which never fails to have a vegetarian plate). Campus security is ever-present; a student says, "Pitzer seems to be more safe than other campuses because campus security is so effective."

> **"Pitzer people are genuinely socially conscious and academically adventurous, but do go on to good jobs."**

One interesting campus curiosity is Grove House, a California craftsman-style house students saved from the wrecking ball nearly two decades ago and moved to campus. It houses a dining room, study areas, and art exhibits.

And what about the social scene? "Most students stay on campus to attend events or go to parties," says a senior. Pitzer has no Greek organizations, nor does it want any, and social life tends to be fairly low-key. Kohoutek is the big party; activities include bands, food, and a "whole week of hoopla." The college enforces the 21-year-old drinking age, and all parties that serve alcohol must be registered. Dances, cocktail parties, and cultural events do much to occupy students' leisure time, but without a car things can get claustrophobic.

The Pomona–Pitzer football team—"The Sagehens"—has had winning seasons and the school fields a variety of competitive teams within the Southern California Intercollegiate Athletic Conference, including women's soccer and swimming. The women's water polo team has won four consecutive national championships. Strong men's teams include basketball and cross-country. Students play a large role in Pitzer's community government and sit on all policy committees, including those on curriculum and faculty promotion.

> **"Pitzer is an amalgamation of every color of the spectrum."**

Pitzer attracts open-minded students looking for the freedom to go their own way. Notes one student: "Pitzer is the only Claremont school that can claim to be genuinely different, in terms of race, religion, sexual orientation, and political belief. Pitzer is an amalgamation of every color of the spectrum."

Pitzer's football team— "The Sagehens"—has had winning seasons and the school fields a variety of competitive teams within the Southern California Intercollegiate Athletic Conference.

Overlaps

Occidental, Pomona, Claremont McKenna, University of Southern California, Scripps

If You Apply To ➤ **Pitzer:** Regular admissions: Jan. 1. Guarantees to meet full demonstrated need. Campus interviews: recommended, informational. No alumni interviews. SATs or ACTs: optional. SAT IIs: optional (English and two others). Accepts the Common Application and electronic applications. Essay question: Common Application.

Pomona College

333 North College Way, Claremont, CA 91711

The great Eastern-style liberal arts college of the West. Offers twice the resources of stand-alone competitors plus access to the other Claremonts. Location an hour east

of L.A. would be ideal except for the choking smog that hangs over the area during the warmer parts of the year.

Pomona College, located just 35 miles east of the glitz and glamour of Hollywood, is the undisputed star of the Claremont College Consortium and one of the top small liberal arts colleges anywhere. This small, elite institution is the best liberal arts college in the West, and its media studies program (film and television) gets top billing. But the school's prestigious reputation doesn't go to the heads of Pomona's friendly students. "Students here are very open about different types of people—[Pomona] prides itself on its diverse community," chirps one Sage Hen (the school's mascot).

The architecture is variously described as Spanish Mediterranean, pseudo-Italian, or, as a sophomore puts it, "a perfect mix of Northeastern Ivy and Southern California Modern." The administration building, Alexander Hall, is described as "postmodern with Mediterranean influences," and one notices more than one stucco building cloaked in ivy and topped with a red-tile roof on campus, as well as eucalyptus trees, canyon live oaks, and an occasional "secretive courtyard lined with flowers." By virtue of its location and beauty, Pomona's campus has served as the quintessential collegiate milieu in various Hollywood movies. New construction includes the Seaver Life Sciences Building and the Lincoln and Edmunds buildings.

"We do not have graduate students or TAs teaching class."

Economics, English, biology, politics, and math are the most popular majors at Pomona. Although no specific course or department is prescribed for graduation, students must take at least one course in each of five areas: creative expression; social institutions and human behavior; history, values, ethics and cultural studies; physical and biological sciences; and mathematical reasoning. The Critical Inquiry seminar emphasizes thoughtful reading, logical reasoning, and graceful writing through subjects such as "Living With Our Genes" and "Penguins, Polar Bears, People and Politics." Students must also complete foreign language and physical education requirements.

Educational opportunities abound at Pomona. Students can spend a semester at Colby or Swarthmore, pursue a 3–2 engineering plan with the California Institute of Technology, or spend a semester in Washington, D.C., working for a congressperson. Nearly one-half of the students take advantage of study abroad programs offered in 24 foreign countries, and many others participate in programs focusing on six cultures and languages at the Oldenborg Center. In addition, the Summer Undergraduate Research Program (SURP) provides students with the opportunity to conduct funded research with a faculty member in their area of study.

"At Pomona, living on campus is an integral part of the college experience."

Classes at Pomona are challenging. "Introductory classes are usually not very work-intensive," says one student. However, "Upper-level courses are very rigorous as the professors are attempting to teach a great deal of difficult material in a short period of time." Students often form study groups in an effort to help one another through the demanding curriculum. One undergrad estimates the average student spends 20 to 30 hours a week studying outside the classroom.

Classes are small at Pomona—the average is 14 students—and the faculty makes a point of being accessible. It's not uncommon for professors to hold study sessions at their houses. "I've had some really incredible professors who have taught me a lot," a neuroscience major says. An ever-popular take-a-professor-to-lunch program gives students free meals when they arrive with a faculty member in tow, and there is even a prof who leads aerobics classes open to all interested parties. Better still, "We do not have graduate students or TAs teaching class," says a senior. "Thus, students do not have to wait until they are upperclassmen to enjoy the benefits of working with and learning from brilliant professors."

Website: www.pomona.edu
Location: Suburban
Total Enrollment: 1,532
Undergraduates: 1,532
Male/Female: 50/50
SAT Ranges: V 690–770
   M 680–760
ACT Range: 30–34
Financial Aid: 53%
Expense: Pr $ $ $
Phi Beta Kappa: Yes
Applicants: 4,927
Accepted: 20%
Enrolled: 41%
Grad in 6 Years: 92%
Returning Freshmen: 99%
Academics: ✍ ✍ ✍ ✍
Social: ☎ ☎ ☎
Q of L: ★ ★ ★ ★
Admissions: (909) 621-8134
Email Address:
   admissions@pomona.edu

Strongest Programs:
English
International Relations
Economics
Neuroscience
Foreign Languages
Media Studies
Chemistry
Politics

Pomona is unique among the Claremont Colleges in that it has three non-national fraternities (two co-ed; there are no sororities), each with its own party rooms on campus.

Pomona students "tend to be high-achieving, confident, verbal students with a fairly liberal political ideology," says a senior. "You'll find people who are really into sports, people who are very artsy, computer geeks, talented musicians, people who like to party every night, or people who like to study on the weekends." Thirty-three percent of the students are Californians, and a growing percentage venture from the East Coast. Pomona is proud of its diverse student body: 7 percent are African American, 11 percent are Hispanic, and 14 percent are Asian American. There is a healthy mix of liberals and conservatives on campus, though the leftists, especially the feminist wing, are much more vocal. One interesting way students voice their issues is by painting the Walker Wall. Anyone is allowed to paint any message they want on the wall, and four-letter words and descriptions of alternative sexual practices show up on a regular basis. The student government is active, and the administration is credited with respecting students' opinions.

Pomona is need-blind in admissions and meets the full demonstrated need of all those who attend. Admissions officers are on the lookout for anyone with special talents and are more than willing to waive the usual grade-and-score standards for such finds. A five-day freshman orientation program divides the new arrivals into groups of six to 12 students headed by a sophomore. "We provide a great deal of support in acclimating students to a college environment," says a senior.

The majority of Pomona students (98 percent) live on campus all four years. The dorms are co-ed, student-governed, and divided into two distinct groups. Those on South campus are family-like, fairly quiet, and offer spacious rooms, and those on the North end have smaller rooms with a livelier social scene. "At Pomona, living on campus is an integral part of the college experience," says a student. The open courtyards and gardens are popular study spots. A handful of students isolate themselves in Claremont proper, where apartments are scarce and expensive. Boarders must buy at least partial meal plans. The food is good, with steak dinners on Saturday and ice cream for dessert every day. Students with common interests can occupy one of the large university houses; there is a vegetarian group and a kosher kitchen, both of which serve meals to other undergraduates. Pomona has a well-established language dorm with wings for speakers of French, German, Spanish, Russian, and Chinese, as well as language tables at lunch. Students generally feel safe on campus. "The worst that usually happens are bike thefts," says a junior. "We're in a pretty nice suburb, so there really aren't many problems with crime."

Students at Pomona often spend Friday afternoons relaxing with friends over a brew at the Greek Theater. Social life begins in the dorms, where barbecues, parties, and study breaks are organized. There are movies five nights a week, and students also enjoy just tossing a Frisbee on the lawn. One student wanted to be sure that incoming freshmen and transfers knew of the Coop's (student union) "best milkshakes west of the Mississippi" and its "game room, with pool, Ping-Pong, pinball, and assorted video games." "I appreciate the diversity and depth that the Five-College community brings to the social life," says a student. "You are guaranteed to meet new and interesting people whenever you step off campus." Five-college parties happen nearly every weekend. During midterms and finals, however, the campus is a "social ghost town." Of more concern is the poor air quality, described by a junior as "oppressive on hot days during the fall semester." Most students are willing to live with the smog, but at least one student complains, "I'd prefer to breathe clean air."

Pomona is unique among the Claremont Colleges in that it has three non-national fraternities (two co-ed; there are no sororities), each with its own party rooms on campus. There is "no peer pressure to join frats," and no fraternity rivalry. As for booze, "I haven't noticed any pressure to drink here," reports one student,

"I haven't noticed any pressure to drink here."

Admissions officers are on the lookout for anyone with special talents and are more than willing to waive the usual grade-and-score standards for such finds.

but "alcohol is definitely present in the social scene." Harwood dorm throws the five-college costume party every Halloween, and interdorm Jell-O fights keep things lively. Freshman orientation gets interesting, too. "First-years have to run through the gates of Pomona with blue and white carnations while upperclassmen throw water balloons and shoot water at them," says a student.

There was a time when Pomona was an athletic powerhouse; the football team even knocked off mighty USC on Thanksgiving Day back in 1899. Currently, women's basketball, tennis, soccer, and swimming, and men's football, baseball, soccer, track, and water polo are strong programs. Intense rivalry exists between the colleges in the Claremont consortium; basketball games between Pomona, Pitzer, and CMS are "particularly heated." Intramurals, including hotly contested inner-tube water polo matches, attract many participants, and Pomona's $14 million athletic complex makes its facilities the best of the Claremonts.

"Pomona offers a unique and desirable juxtaposition of rigorous academics and comfortable social atmosphere," says a student. Another student says, "Once you take advantage of the five-college system, you realize how cool it is." The strongest link in an extremely attractive chain, Pomona continues to symbolize the rising status of the Claremont Colleges—and the West in general—in the world of higher education. There are few regrets about coming to Pomona. Says a senior, "We're in California. The sun is always shining. What's the problem?"

If You Apply To ➤ **Pomona:** Early decision: Nov. 15, Dec. 28. Regular admissions: Jan. 2. Financial aid: Feb. 1. Guarantees to meet full demonstrated need. Campus and alumni interviews: recommended, evaluative. SATs or ACTs: required. SAT IIs: required. Accepts the Common Application and electronic applications. Essay question.

Scripps College

1030 Columbia Avenue, Claremont, CA 91711

Scripps is a tiny, close-knit women's college with co-ed institutions literally right next door. Only Barnard and Spelman offer the same combination of single-sex and co-ed. Innovative Core Curriculum takes an interdisciplinary approach to learning.

Scripps College offers the best of both worlds—a close-knit women's college, where traditions include weekly tea and fresh-baked cookies, and the size and scope of a major research institution, thanks to its membership in the Claremont Colleges.* Founded in 1926 by newspaper publisher Ellen Browning Scripps, the college continues to pursue her mission: "To educate women by developing their intellects and talents through active participation in a community of scholars." Students tend to be outgoing, articulate, and serious about their studies, though they still know how to have fun. "The atmosphere is helpful, not competitive and scary," says one student. "It is impossible to be depressed in this beautiful place."

> **"It is impossible to be depressed in this beautiful place."**

Indeed, Scripps's scenic 30-acre campus, listed on the National Register of Historic Places, offers a tranquil, safe, and comfortable environment. The architecture is Spanish and Mediterranean, with tiled roofs and elegant landscaping. A performing arts center opened in 2003, with permanent space for the Claremont Concert

Website:
www.scrippscollege.edu
Location: Suburban
Total Enrollment: 873
Undergraduates: 873
Male/Female: 0/100
SAT Ranges: V 650–740
M 620–710
ACT Range: 26–31
Financial Aid: 43%
Expense: Pr $ $ $
Phi Beta Kappa: Yes
Applicants: 1,836
Accepted: 46%

Athletic rivalries aren't the focus here, but Scripps does field joint teams with CMC and Harvey Mudd, and when those teams face off against Pomona and Pitzer, students pay attention.

The academic experience at Scripps emphasizes cooperation.

Orchestra and Concert Choir. In addition to a 700-seat theater, the center offers a music library, recital hall, practice rooms, faculty offices, and classrooms.

The academic experience at Scripps emphasizes cooperation. "Scripps is an extremely challenging school," a freshman says, "not only in terms of the course load, but also because of the way a Scripps education forces you to question your beliefs." Everyone takes the Core Curriculum in interdisciplinary humanities, focusing on ideas about the world and the methods used to generate them. Other requirements include courses in fine arts, letters, natural and social sciences, women's studies, race and ethnic studies, foreign language, and math. "The quality of teaching is wonderful," says one student. "Although there are always the more dry classes, for the most part the energy of the professors make the subjects interesting."

> **"Our students are so open-minded, politically very knowledgeable, and active."**

Popular and well-regarded majors at Scripps include psychology, English, studio art, politics and biology. The Millard Sheets Art Center offers a state-of-the-art studio and freestanding museum-quality gallery for aspiring painters and sculptors. Premeds benefit from the Keck Science Center, a joint facility for students at Scripps, Claremont McKenna (CMC), and Pitzer (the other Claremont Colleges are Pomona and Harvey Mudd). The Scripps Humanities Institute offers seminars and lectures open to the general public, along with fellowships for juniors; recently, the institute explored how and why empathy is a shared focal point in disciplines as broad as neuroscience, literary criticism, and legal studies. While Scripps doesn't offer business, students can take economics courses at the college and accounting across the street at Claremont McKenna.

Forty-one percent of Scripps women are from California. African Americans account for 3 percent of the student body, Hispanics 5 percent, and Asian Americans make up another 13 percent. "Our students are so open-minded, politically very knowledgeable, and active," says a student. SCORE, the Scripps Communities of Resources and Empowerment, provides support and funding to organizations that further promote social and political awareness, with respect to issues of class, ethnicity, gender, race, religion, sexuality, and sexual orientation. There are no athletic scholarships, though more grant money has been made available for students with financial need, so that each student has to take out no more than $4,000 in loans each year.

Ninety-six percent of Scripps students live in one of the eight "beautiful" dorms, where options range "from singles to seven-person suites, apartment-style living arrangements to charming Spanish Mediterranean residence halls built in the 1920s." Many dorms have reflecting pools and inner courtyards; some rooms have balconies and are furnished with antiques and beautiful rugs, and juniors are guaranteed single rooms. The cafeteria garners rave reviews as well: "This is not traditional college food," says a bioethics major. "The salad bar is gourmet, the bread comes from a local bakery, and the pizza is made in a wood-fired brick oven. Don't get me started about the hot cookies!"

> **"The quality of teaching is wonderful."**

Social life at Scripps centers on the residence halls, which take turns throwing parties. The school's alcohol policy complies with state law—those under 21 can't drink—but also specifies that if an underage student is drunk at a party, she will be helped and kept safe, rather than written up and punished. Students also receive daily emails detailing what's going on at all of the Claremont institutions, organized by "things to do" and "music, art, lectures, theater, film, and dance." The town of Claremont also offers a farmer's market on Sundays, with fresh fruit, flowers, and gifts. For students with cars, popular road trips include Pasadena, Mt. Baldy, San Diego, and even Las Vegas and Mexico; students without wheels can hop on the MetroLink commuter train and get to and from Los Angeles for less than $5.

Athletic rivalries aren't the focus here, but Scripps does field joint teams with CMC and Harvey Mudd, and when those teams face off against Pomona and Pitzer, students pay attention. All of the teams compete in Division III, and the Scripps water polo, tennis, cross country, track and field, and swimming and diving teams brought home conference championships in 2005–06. Intramural sports are also played jointly, and popular options include inner-tube water polo, soccer, flag football, volleyball, rugby, and lacrosse. Traditions are also important at Scripps, including the Matriculation ceremony at the start of each year, and the signing of the "graffiti wall" by each class before graduation.

Scripps offers a winning combination of outstanding academics and personal attention, with a cooperative, non-threatening feel. And should the women-only environment of Scripps begin to feel claustrophobic, the other Claremont Colleges beckon, with parties, intramural sports and crossregistration privileges. Scripps students want to achieve great things, but not if that requires stepping on their classmates' toes.

Overlaps

Pomona, UCLA, UC–San Diego, University of Southern California, UC–Berkeley, Claremont McKenna

If You Apply To ➤

Scripps: Early decision: Nov. 1, Jan. 1. Regular admissions: Jan. 1. Financial aid: Jan. 15. Guarantees to meet demonstrated need, but is "need-sensitive" for the bottom 10 percent of the class. Campus interviews: recommended, informational. Alumni interviews: optional, informational. SATs or ACTs: required. SAT IIs: recommended. Accepts the Common Application and prefers electronic applications. Essay question. Applicants must also submit a graded analytical paper from the junior or senior year of high school.

Clark University

950 Main Street, Worcester, MA 01610-1477

If Clark were located an hour to the east, it would have become the hottest thing since Harvard. Worcester is not Boston, but Clarkies bring a sense of mission to their relationship with this old industrial town. Clark is liberal, tolerant, and world-renowned in psychology and geography.

A classic Clark University poster distills the school's philosophy into a single photograph: A normal green peapod, filled with multicolored peas. "Categorizing people," the poster says, "isn't something you can do here." And indeed, it's not. "At Clark you have to be prepared to be open-minded and accepting since there are so many different types of people and ideas," says a junior. Clark started as an all-graduate school, excelling in disciplines including psychology and geography, and now welcomes undergraduates of all backgrounds and interests with small classes and no shortage of faculty attention.

Clark's compact, 50-acre campus has "enough ivy, tall maples, and collegiate brick buildings to make a traditionalist happy," even though it's located in the gritty Main South section of Worcester. Buildings range from remodeled Victorian-era residences—former homes of prosperous merchants—to the award-winning Robert Hutchings Goddard Library. Clark is always renovating something, and careful restoration has brought a renewed sense of history to the area. Clark is the only American university where famed psychoanalyst Sigmund Freud lectured, and his statue adorns the spot in the center of campus where he spoke.

While Clark now serves primarily undergraduates, its history of graduate education is evident in its classrooms. Most courses are seminars, and 64 percent have 25 or fewer students. First-year seminars are even smaller, limited to 16 students each.

Website: www.clarku.edu
Location: Center city
Total Enrollment: 3,433
Undergraduates: 1,964
Male/Female: 40/60
SAT Ranges: V 560–660
  M 540–650
ACT Range: 24–28
Financial Aid: 54%
Expense: Pr $ $
Phi Beta Kappa: Yes
Applicants: 4,463
Accepted: 58%
Enrolled: 19%
Grad in 6 Years: 71%
Returning Freshmen: 86%
Academics: ✑ ✑ ✑ ✑
Social: ☎ ☎ ☎

(Continued)

Q of L: ★ ★ ★

Admissions: (508) 793-7431

Email Address:

admissions@clarku.edu

Strongest Programs:

Psychology

Geography

Physics

Chemistry

Biology

Government and International
  Relations

Business Management

Communication and Culture

They permit students to explore issues in depth in their first or second semesters, and the faculty member teaching the course acts as an academic advisor until students declare a major. "It is a nice way to immediately meet people and continue having contact with them throughout the semester, and your advisor gets to know you better," a junior communication and culture major says. A newer program for incoming students, Clark Trek, is an optional orientation program held outdoors.

The foundation of a Clark education is the Program of Liberal Studies, which promotes the habits, skills, and perspectives essential to lifelong learning. Each student must complete eight courses: one in verbal expression, one in formal analysis, and six in perspectives—aesthetic, comparative, historical, language, scientific, and values. International Studies students take courses in those areas with an international focus. Interdisciplinary programs are popular, and students may design their own majors. Those who finish with a grade point average of 3.25 or better may take a fifth year for free to obtain a master's degree.

> "At Clark you have to be prepared to be open-minded and accepting."

Clark's historically strong psychology and geography departments continue to burnish their national reputations, the latter having churned out more PhDs in the field than any other school in the nation, plus four members of the National Academy of Science (the most of any geography program). Clark is the birthplace of the American Psychological Association, and the concept of adolescence as distinct from childhood. Also strong are the sciences and programs in management, government and international relations, and communication and culture. There are new majors in women and gender studies and environmental science.

Regardless of major, Clark encourages students to take internships, through the university itself or the 14-school Colleges of Worcester Consortium.* More than 20 percent of all Clark students spend at least one semester studying abroad at one of 15 programs in 10 countries. Even those who don't go abroad can get a taste of foreign culture by taking courses or attending international research conferences at the Clark University Center in Luxembourg. Back on campus, the academic climate is challenging but students don't compete with one another. "Courses are rigorous here," says a senior. "Academics are very important at Clark." Professors are "extremely knowledgeable and most are great resources outside of class as well," a junior reports. About 40 percent of students participate in undergraduate research.

> "There are a lot of hippies and people who love to speak their minds and protest."

Thirty-six percent of Clark students are from Massachusetts, and international students from about 90 countries make up another 8 percent. African Americans account for 2 percent of the student body; Hispanics and Asian Americans combine for another 6 percent. Clarkies are "laid-back, chill, friendly, liberal, and from the Northeast," a senior says. The university has a collaborative educational program with historically African American Howard University in Washington, D.C. Politically, "there are a lot of hippies and people who love to speak their minds and protest," a sophomore says. Many Clark students do community service, and the Community Engagement and Volunteering Center is a hub for the many social activism groups. Clark also offers merit scholarships averaging $11,591 annually, plus special "Making a Difference" scholarships for students interested in community service that are worth $44,000 each.

Freshmen and sophomores at Clark are required to live in the dorms, and all except one are co-ed by floor or wing. In all, 77 percent of the student body bunks on campus. "Dorms are fairly spacious and comfy to live in, although a few are due for new carpet and some paint," a junior says. The rest of the students—mostly juniors and seniors—find apartments and group houses nearby. After the first year, students

The foundation of a Clark education is the Program of Liberal Studies, which promotes the habits, skills, and perspectives essential to lifelong learning.

must play the lottery to get rooms. Campus dwellers must buy the meal plan, which always offers student favorites, along with vegan, vegetarian, and kosher options. "I enjoy the food," a student says, "but many dishes are often repeated."

Clark has no Greek life, but 80 student-run organizations offer "a ton of concerts, bands, comedians" and other programs, says a student. "Clark is not a big party school, but there is always something to do," one junior says. First-year dorms are dry, and mixed-class halls "are dry if you are under 21," says a Spanish major. A government major adds, "In general, alcohol isn't a huge problem."

Worcester and the vicinity host 14 colleges, but the area's history—a manufacturing and industrial hub that's fallen on hard times—means it's hardly a "college town," students say. That said, Worcester does have movie theaters, restaurants with every conceivable type of cuisine, and small clubs where bands can play, as well as the Centrum, a 13,000-seat arena. "Main South Worcester is not the prettiest, quietest locale for a college, but it's got flavor and spice, and you'll either love it or hate it," a junior biology major says. Students mix with the townspeople through volunteer programs such as Clark University Brothers and Sisters. "If you like volunteering someplace that needs it, or fighting for human rights, this is your school," a junior says. To get away, Clarkies head to the larger cities of Boston and Providence, or the rural wilds of Vermont, New Hampshire, and Maine—all easily reachable by car or public transit.

Coping with the frigid New England winters includes quaffing cups of hot chocolate and dreaming about Spree Day—which one student describes as "better than your birthday and Christmas combined." On Spree Day, classes are spontaneously cancelled one spring day, and everyone celebrates spring with a carnival on the green, including food, music, and games (with prizes!). Completely different is Academic Spree Day, which celebrates undergraduate research. Clark competes in Division III, and the men's and women's basketball teams have each won conference championships recently. Also competitive are women's field hockey, crew, soccer, and softball, and men's basketball and crew. About half of the students participate in intramural sports, which range from soccer and

"Academics are very important at Clark."

flag football to ultimate Frisbee and volleyball. The Dolan Field House provides facilities and locker rooms for spring and fall teams, plus lighted outdoor fields. The crew team has a new boathouse on Lake Quinsigamond, and the cross-country teams recently broke in a new course.

Clark started out differently, serving only graduate students. Though it now caters mainly to undergraduates, the school continues to challenge convention, pioneering new teaching methods, pursuing new fields of knowledge, and finding new ways to connect thinking and doing. "Its urban location gives it a distinct identity," a senior says. "It is also fairly progressive, making room for students to satisfy a wide variety of interests."

Clark's historically strong psychology and geography departments continue to burnish their national reputations, the latter having churned out more PhDs in the field than any other school in the nation.

Overlaps

Boston University, Northeastern, University of Massachusetts–Amherst, Boston College, American, University of Vermont

If You Apply To ➤ **Clark:** Early decision: Nov. 15. Regular admissions: Jan. 15. Meets demonstrated need of 68%. Campus and alumni interviews: recommended, evaluative. SATs or ACTs: required. SAT IIs: optional. Accepts the Common Application and electronic applications. Essay question: significant experience, achievement or risk and its impact on you; issue of personal, local, national, or intellectual concern; influential person, fictional character, historical figure, or creative work; what you would bring to the diversity of a college community; or a topic of your choice.

Clarkson University

Holcroft House, Box 5605, Potsdam, NY 13699

Website: www.clarkson.edu
Location: Small town
Total Enrollment: 2,993
Undergraduates: 2,633
Male/Female: 75/25
SAT Ranges: V 520–620
    M 580–670
ACT Range: 22–28
Financial Aid: 65%
Expense: Pr $ $
Phi Beta Kappa: No
Applicants: 2,405
Accepted: 87%
Enrolled: 31%
Grad in 6 Years: 75%
Returning Freshmen: 87%
Academics: 🐾 🐾 🐾
Social: ☎ ☎
Q of L: ★ ★ ★
Admissions: (315) 268-6479
Email Address:
    admission@clarkson.edu

Strongest Programs:
Engineering
Chemistry
Business

You know you're in the north country when the nearest major city is Montreal. Clarkson lies over the river and through the woods. With an informal and close-knit atmosphere, Clarkson is one of the few small, undergraduate-oriented technical universities in the nation. Compare to Lehigh, Bucknell, and Union.

At Clarkson University, engineering and ice hockey reign supreme. About half of the student body is enrolled in the engineering program, and the hockey team is a perennial contender for top honors. Students at this tiny school get a quality technical education in a small town environment that offers plenty to do, especially during the sled dog days of winter.

The village of Potsdam, New York, is cloistered away between the Adirondacks and the St. Lawrence River. The "hill campus," where most freshmen and sophomores live and take classes, relies mainly on modern architecture and lots of woods and wildlife. Recent additions to campus include a 10,000 square foot addition to the science building. The new wing features specialty labs to support the biomolecular science program.

Engineering isn't the only academic offering at Clarkson, but it certainly gets top billing; 49 percent of the students are in the program, and it includes the top five most popular majors. The combined programs in electrical/computer engineering and mechanical/aeronautical engineering earn the highest marks from students. Clarkson's School of Business has several majors to choose from, and all first-year business students actually start and run a business. Project Arete allows students to earn a double major in management and a liberal arts discipline. Physics and chemistry are the strongest offerings in the sciences, and would-be doctors have the benefit of a joint program combining biology and—you guessed it—engineering. Students say the liberal arts and humanities pale in comparison to the strong engineering programs.

> **"The courses are hard, and every grade received is well-earned."**

As part of the Clarkson Common Experience, all students are required to complete a set of courses and a professional experience. The program emphasizes four components that serve as common threads through multiple courses: learning to communicate effectively; developing an appreciation for diversity in both living and working environments; recognizing the importance of personal, societal, and professional ethics; and understanding how technology can be used to serve humanity. Study abroad opportunities are available at 20 different schools in 12 countries, and the honors program includes 120 students who participate in an intensive four-year curriculum.

Clarkson prides itself on intimacy and personalized instruction; 45 percent of classes taken by freshman have 25 or fewer students. Eighty percent of the Foundation courses are taught by full-fledged faculty members, and nearly 200 students conduct research with a faculty mentor each year. "I feel that all of my teachers have been excellent," says a senior. "The professors really care about the students." Some students say Clarkson isn't the academic pressure cooker that many technical institutes are, but others disagree. "The climate is very academic-oriented and very competitive," a senior says. "The courses are hard, and every grade received is well-earned." The bottom line of a Clarkson education is getting a job after graduation, and students uniformly praise the career counseling office and proudly note Clarkson's high placement rate.

At Clarkson, the students are friendly, serious-minded, and down-to-earth; radicals are notably absent.

At Clarkson, the students are friendly, serious-minded, and down-to-earth; radicals are notably absent. Sixty-nine percent of the students graduated in the top quarter of their class. Seventy-two percent are New Yorkers, and 3 percent hail from abroad. Clarkson has trouble luring minorities to its remote locale; African Americans, Hispanics, and Asian Americans account for 8 percent of the student population. "Our campus definitely needs more minority students," says a senior. Many applicants demonstrating financial need are offered some aid, but not necessarily enough to meet full need. Clarkson awards a handful of merit scholarships each year, averaging $7,628. Thirty-three athletic scholarships are offered, but only to ice hockey players.

Seventy-nine percent of students live in campus housing. Students are required to reside on campus all four years, unless exempted to live in a Greek house. Most dorms are centrally located and are cleaned every day. "The dorms are comfortable and very well maintained," says a student, but another notes, "There is a constant dorm shortage." Four of the dorms are all-men and the rest are co-ed by floor. All freshmen are housed with students in their major areas of study and in some instances in their department, giving them the chance to study and learn together. Many underclassmen are housed in conventional dorms, but university-owned townhouse apartments offer more gracious living. "The dorm situation is handled by the lottery, and housing improves based on your class year," a senior reports.

"The professors really care about the students."

In keeping with Clarkson's "come as you are" atmosphere, the social scene is low key. "There are activities, such as comedians, picnics, and other types of entertainment on campus," says one senior. "However, a good deal of socializing also takes place off campus." Fifteen percent of the men and 13 percent of the women join the Greek system. Fraternity beer blasts are the staple of weekend life, and those not into the Greek scene (and over 21) can head to the handful of bars in downtown Potsdam. Nearby SUNY–Potsdam is also a source of social life, especially for men frustrated by Clarkson's three-to-one male/female ratio. Drinking is prohibited on campus, but students report that it's still easy to imbibe and for underage students to get served off campus. For those who crave the bustle of city nightlife, Ottawa and Montreal are each about an hour and a half away by car.

When it comes to sports, hockey is first and foremost in the hearts of Clarkson students—the only one in which the university competes in Division I. The team has been ECAC champ in recent years, and contends for the national championship with other blue-chip teams like archrivals Cornell and St. Lawrence. A Division I women's ice hockey program has been initiated. In addition, Clarkson offers 19 Division III sports; women's volleyball and men's lacrosse, basketball, and baseball have been successful of late. About three-quarters of students take advantage of the intramural program, with soccer and volleyball proving popular. For weekend athletes, an abundance of skiing and other outdoor and winter sports is within easy driving distance.

"The dorms are comfortable and very well maintained."

Overlaps

Cornell, Northeastern, Rochester Institute of Technology, Rensselaer Polytechnic, Syracuse, Worcester Polytechnic

While other majors are offered, Clarkson's bread and butter are its technical programs, particularly its slew of engineering majors. Students here gain exposure to the ever-growing variety of specialties in the field. And the extended snowy winters in Potsdam are great for ice hockey fans or ski bunnies looking for fresh powder.

If You Apply To ➤ **Clarkson:** Rolling admissions. Early decision: Dec. 1, Jan 15. Does not guarantee to meet demonstrated need. Campus interviews: recommended, informational. Alumni interviews: optional, informational. SATs or ACTs: required. SAT II: optional. Accepts the Common Application and prefers electronic applications. Essay question: optional essay.

Clemson University

Clemson, SC 29634

Clemson is a technically oriented public university in the mold of Georgia Tech, Virginia Tech, and North Carolina State. Smaller than the latter two and more focused on undergraduates than Georgia Tech, Clemson serves up its education with ample helpings of school spirit and small town hospitality.

Website: www.clemson.edu
Location: Small town
Total Enrollment: 17,016
Undergraduates: 14,740
Male/Female: 55/45
SAT Ranges: V 550–630
    M 570–660
ACT Range: 24–29
Financial Aid: 38%
Expense: Pub $ $ $ $
Phi Beta Kappa: No
Applicants: 11,419
Accepted: 61%
Enrolled: 40%
Grad in 6 Years: 72%
Returning Freshmen: 89%
Academics: ✍ ✍ ✍
Social: ☎ ☎ ☎ ☎
Q of L: ★ ★ ★ ★
Admissions: (864) 656-2287
Email Address:
    cuadmissions@clemson.edu

Strongest Programs:
Engineering
Architecture
Landscape Architecture
Economics
Genetics

Nestled in the foothills of the Blue Ridge Mountains, Clemson University is a place where Southern spirit continues to flourish. The campus occupies terrain that once was walked by John C. Calhoun, former Southern senator and Civil War–era rabble-rouser of the first degree. Today Clemson features quality academics in technical areas such as engineering and biology, and big-time athletics. Tiger spirit is as strong as ever, as evidenced by the ubiquitous orange tiger paws that decorate the campus, and students here are happy to make tracks of their own.

CU's 1,400-acre campus is situated on what was once Fort Hill Plantation, the homestead of Thomas Green Clemson. The campus is surrounded by 17,000 acres of university farms and woodlands and offers a spectacular view of the nearby lake and mountains. Architectural styles are an eclectic mix of modern and 19th-century collegiate. Newer additions to the campus include an on-campus research facility for biotechnology and the McAdams Hall annex for computer science. Renovations have also been completed on Hardin Hall, with the addition of "smart classrooms" and the student recreation facility, featuring a fitness center focusing on health education. The Campbell Graduate Engineering Center and the research facility for advanced materials are currently under construction.

Electrical engineering is the university's largest department, and computer engineering is among the nation's best in research on large-scale integrated computer circuitry and robotics. The College of Architecture, one of the school's most selective programs, offers intensive semesters at the Overseas Center for Building Research and Urban Study in Genoa, Italy. A fantastic resource for science enthusiasts and history buffs is the library's collection of first editions of the scientific works of Galileo and Newton. Because of the prevailing technical emphasis, most students interested in the liberal arts head "down country" to the University of South Carolina.

"Students may get more competitive over co-op opportunities than over classes."

Undergraduate teaching has always been one of Clemson's strong points, and for students interested in pursuing a liberal arts curriculum, the school has degrees in fine arts, philosophy, and languages and enjoys a strong regional reputation for its history program. Most students agree that engineering and architecture are the school's strongest programs, but also give high marks to business and agriculture. Highly motivated students should consider Calhoun College, Clemson's honors program—the oldest in South Carolina—open to freshmen who scored 1200 or above on their SATs and ranked in the top 10 percent of their high school graduating class. Clemson also offers exchange programs in Mexico, Scotland, Ecuador, Spain, England, Australia, and Italy.

General education requirements have recently been reorganized. They include courses in advanced writing, oral communications, mathematical, scientific, and technological literacy, social sciences, arts and humanities, crosscultural awareness, and science and technology in society. An electronic portfolio pilot was instituted in 2004; freshmen build their portfolio, demonstrating changes in competencies throughout their experiences at Clemson. Academically, the level of difficulty varies.

"Students may get more competitive over co-op opportunities than over classes," one junior says. Professors run the gamut from average to stellar. "The quality of teaching varies," says a junior. "The best advice is to choose classes based on the professor's reputation, which is easy to find out." Students report some problems finishing a degree in four years, and class registration can be a hassle.

Clemson's student body has a decidedly Southern air, as 78 percent of the undergrads hail from South Carolina, with most of the rest from neighboring states. The average Clemson student is friendly and conservative, and though, as a public institution, the school isn't affiliated with any church, there is a strong Southern Baptist presence on campus. Forty-two percent were in the top tenth of their high school graduating class. African Americans make up 7 percent of the student body, Hispanics account for 1 percent, and Asian Americans 2 percent. The university offers over 4,100 merit scholarships and nearly 400 athletic scholarships. Parking is the overwhelming complaint among the students.

> "The west campus high-rise dorms are awesome."

Housing gets positive reviews, and 46 percent of the students live on campus, usually during their first two years. Most of the dorms are single-sex, though co-ed university-owned apartment complexes are also an option. "Apartment housing is competitive, but rooms on campus are usually available," says a student. Clemson House and Calhoun Courts, the co-ed halls, are considered the best places to be. "The west campus high-rise dorms are awesome," raves a junior. The dining facilities have been improving, according to students. "They have specials and even ask students to contribute recipes," according to one student. Upperclassmen can cook for themselves, and each dorm has kitchen facilities.

After class, many students hop on their bikes and head to nearby Lake Hartwell. The beautiful Blue Ridge mountain range is also close by for hiking and camping, and beaches and ski slopes are both within driving distance. Atlanta and Charlotte are only two hours away by car, and Charleston is four hours away on the coast. Aside from the sports teams, fraternities and sororities provide most of the social life. Sixteen percent of Clemson men and 30 percent of women go Greek. The town itself is pretty small, with a few bars and movie theaters, but some students love it. "Downtown is across the street from campus and has many bars," a student affairs major says.

Sports still help make the world go 'round at Clemson, and on weekends (when the Tiger teams are playing) there are pep rallies, cookouts, dances, and parties for the mobs of excited fans. The roads leading to campus are painted with large orange pawprints, an insignia that symbolizes great enthusiasm for Clemson sports. So, too, are half the fans at an athletic event, making the stands look like an orange grove. Football fever starts with the annual First Friday Parade, held before the first home game, and on every game day the campus dissolves into a sea of Tiger orange. Clemson has regained its former gridiron glory under the leadership of Coach Tommy Bowden. Hordes of Tiger fans

> "Downtown is across the street from campus and has many bars."

cram "Death Valley" for every game and are especially rowdy when the reviled University of South Carolina Gamecocks are in town. Other very competitive athletic teams include basketball, baseball, and men's track.

No haven for carpetbaggers or liberals, Clemson is best at serving those whose interests lie in technical fields. School spirit is contagious, fueled by a love of big-time college sports, and becomes lifelong for many Clemson students. Everyone can become part of the Clemson family, from Southern belle to Northern Yankee, as long as they're friendly, easygoing, and enthusiastic about life in general and the Tigers in particular.

Highly motivated students should consider Calhoun College, Clemson's honors program—the oldest in South Carolina—open to freshmen who scored 1200 or above on their SATs and ranked in the top 10 percent of their high school graduating class.

No haven for carpetbaggers or liberals, Clemson is best at serving those whose interests lie in technical fields.

Overlaps

South Carolina, UNC–Chapel Hill, Georgia, Georgia Tech, Duke, Charleston

Colby College

Waterville, ME 04901

The northernmost outpost of private higher education in New England. Colby's picturesque small town setting is a short hop from the sea coast or the Maine wilderness. No frats since the college abolished them 20 years ago. A well-toned, outdoorsy student body in the mold of Middlebury, Williams, and Dartmouth.

Website: www.colby.edu
Location: Small city
Total Enrollment: 1,871
Undergraduates: 1,871
Male/Female: 47/53
SAT Ranges: V 640–720
    M 640–710
ACT Range: 27–31
Financial Aid: 38%
Expense: Pr $ $ $ $
Phi Beta Kappa: Yes
Applicants: 3,874
Accepted: 38%
Enrolled: 35%
Grad in 6 Years: 88%
Returning Freshmen: 94%
Academics: ✍ ✍ ✍ ✍½
Social: ☎ ☎ ☎
Q of L: ★ ★ ★ ★
Admissions: (800) 723-3032
Email Address:
    admissions@colby.edu

Strongest Programs:
Art
Economics
Government
English
International Studies
Environmental Studies
Natural Sciences
Music

Colby College draws students who like to work hard and play harder, whether in the classroom or on the ski slopes. The nearby town of Waterville, Maine (population 20,000), offers few distractions, and close friendships with peers and professors help ward off the winter chill. Colby's top study abroad program offers students an opportunity to explore the world, and even those who don't spend a semester or year away can get a taste during the month of January, when Jan-Plan trips send Colby students far and wide. "Colby allows me to explore educational possibilities and experiences of all kinds, from taking a class with a top U.S. economist, to sea kayaking, to mentoring needy children in local schools," says a sophomore. "The setting is picturesque, and the faculty and students alike are friendly and warm."

Colby sits high on a hill, with beautiful views of the surrounding city and countryside. Its 714 acres include a wildlife preserve, miles of cross-country trails, and a pond used in winter as an ice-skating rink. Georgian architecture predominates, and the oldest buildings are redbrick with white trim, ivy, and brass nameplates above their hunter-green doors. The more contemporary buildings lend a touch of modernity. One of the most iconic Colby buildings is the library tower, which is topped with a blue light. A $44 million plan to renovate Colby's residence halls and three dining halls continues with the reopening of the Roberts dining hall.

As a small college with a history of innovation and educational excellence, Colby encourages students to learn for learning's sake, rather than for a good grade. "Colby isn't competitive, but you're expected to do well in your courses," says a senior. Students must complete distribution requirements in English composition, foreign language, "Areas" (one course each in arts, historical studies, literature, quantitative reasoning and social sciences, and two courses in natural sciences), "Diversity" (two courses focusing on how diversity has contributed to the human experience), and "Wellness" (five supper seminars over the first two semesters). Freshmen eager to fulfill that language requirement can ship off to Salamanca or Dijon to take care of it, delay-

"The setting is picturesque, and the faculty and students alike are friendly and warm."

ing on-campus enrollment until the second semester. Popular and well-regarded programs include economics and biology, followed closely by English, government, and history. A major in science, technology, and society has been added, while majors in geology have been restructured. In all, Colby offers 53 majors.

Colby's faculty is unusually devoted to undergraduate teaching. "Faculty members care deeply about the success of their students," says a sophomore. A philosophy major adds, "The profs will cook you dinner and get you into graduate school." Forty-eight percent of all classes have 25 or fewer students, allowing those highly lauded profs to spend more time with each student.

Colby was the first men's college in New England to admit women, and also the first to establish a special January term. Students must take three such terms for credit to graduate. Motivated students might use the month off to serve an internship, study abroad, or prepare an in-depth report. Less serious types head for the ski slopes or Southern beaches, and write a

"The profs will cook you dinner and get you into graduate school."

quick paper at the end of the month. The school also sponsors Jan-Plan trips to everywhere from Nicaragua to Vietnam, including Bermuda (for biology). Other programs include Connecticut's Mystic Seaport (for marine biology), Kyoto, Japan, and the great cities of Europe. For would-be engineers, there is a joint 3–2 program with Dartmouth, and others may take exchange programs with Clark Atlanta and Howard. Given all these options, it's no surprise that half of the school's majors have an international component and more than two-thirds of Colby students spend some time abroad taking advantage of more than 50 approved international programs. A high proportion of graduates enter the Peace Corps and the Foreign Service.

Colby students are "multi-taskers who were the presidents of their high school and captains of varsity sports," offers one student. "They work as hard as they play." Only 11 percent of Colby students are Mainers; the rest learn to act like natives during the COOT program (Colby Outdoor Orientation Trips). These four-day excursions by bicycle, canoe, or foot introduce them to the beauty of the Maine wilderness or to service or theater experiences. African Americans account for 2 percent of the student body, Hispanics 3 percent, and Asian Americans contribute 6 percent. There are no athletic or merit scholarships available.

Ninety-four percent of Colby students live on campus, where residence halls have live-in faculty members. "Dorms range from palatial apartments to closet-like doubles," says a junior. About 100 seniors live off campus each year in apartment-style buildings. Dining halls cater to the various tastes, lifestyles, religions, and even holidays throughout the year. "The three dining halls do a wonderful job of providing delicious food and unique atmospheres. Students are free to eat at whichever one they want," states a senior.

When the weekend comes, you'll find most Colby students staying close to campus. "As a college town, it isn't much," confides a senior, "but the restaurants are okay and the new Starbucks helps." Although fraternities are a thing of the past, students maintain an active party life. In response to student requests, administration has instituted a program allowing a glass of wine or beer at dinner on Friday

"As a college town, it isn't much, but the restaurants are okay and the new Starbucks helps."

nights in Dana dining hall for students of legal age. Still, there are options for those who choose not to imbibe, including "parties, shows, plays, talks, and concerts," according to a junior. Popular road trips include Augusta, Portland, and Freeport, Maine (home to the L.L. Bean factory and store). Also easy to reach are the bright lights of Boston and Montreal, and the slopes of Sugarloaf, Maine.

The Colby administration likes to share two "big secrets" about Maine winters: they're beautiful, and they're a lot harsher in the telling than in the living. Still, an enthusiasm for chilly weather and outdoor sports are the major nonacademic credentials needed to find contentment here. Everyone looks forward to football, lacrosse, and hockey games, as well as the annual winter carnival and snow-sculpture contest. Athletics have come a long way since the first intercollegiate croquet game, played at Colby in 1860. Nonvarsity athletes are eager participants in 10 club teams and six intramural sports.

Overlaps

Bowdoin, Middlebury, Bates, Dartmouth, Colgate, Hamilton

Colby's traditional New England liberal arts college vibe extends far beyond its small town setting and historic, ivy-covered buildings. It permeates the air, punctuated by the long-standing traditions, abundant school spirit, and caring faculty members who focus on developing their students' minds.

<table>
<tr><td>If You Apply To ➤</td><td>Colby: Early decision: Nov. 15, Jan. 1. Regular admissions: Jan. 1. Financial aid: Feb. 1. Campus and alumni interviews: recommended, informational. SATs or ACTs: required. SAT IIs: optional. Accepts the Common Application and electronic applications. Essay question.</td></tr>
</table>

Colgate University

13 Oak Drive, Hamilton, NY 13346

At less than 3,000 students, Colgate is smaller than Bucknell and Dartmouth but bigger than Hamilton and Williams. Like the other four, it offers small town living and close interaction between students and faculty. Greek organizations and jock mentality are still well entrenched despite administrative efforts to neutralize them.

Website: www.colgate.edu
Location: Rural
Total Enrollment: 2,779
Undergraduates: 2,740
Male/Female: 49/51
SAT Ranges: V 630–710
    M 650–720
ACT Range: 29–32
Financial Aid: 39%
Expense: Pr $ $ $ $
Phi Beta Kappa: Yes
Applicants: 8,008
Accepted: 27%
Enrolled: 34%
Grad in 6 Years: 91%
Returning Freshmen: 92%
Academics: ✍ ✍ ✍ ✍ ½
Social: ☎ ☎ ☎
Q of L: ★ ★ ★
Admissions: (315) 228-7401
Email Address:
    admission@mail.colgate.edu

Strongest Programs:
Biology
Economics
Foreign Language and
    Literature
History

Colgate University offers small town living, super skiing, and close student/faculty interaction. While you may see the same North Face or Patagonia fleece coming and going (and coming and going) as you stroll across campus, all students here aren't spun from the same recycled soda bottles. "There are the athletes, the bookworms, the partiers, the all-around kids, the preppy kids, the neat freaks, the drama kings and queens, the activists—the list goes on and on," says a computer science and physics major. "But there are two things which all the students share: an enjoyment of learning for the sake of learning, and acquiring knowledge because it is intriguing and interesting." From the herbarium to the Devonian fossils to the 16-inch reflecting telescope, it's clear that Colgate has more to offer than just its picture-postcard setting.

Colgate's 13 founders started the school with 13 prayers and 13 dollars. Their prayers were answered in 1880, when toothpaste mogul William Colgate gave $50,000 to the fledgling university, enough to get the name changed from Madison to his own. Today, the 515-acre campus sits on a hillside in rural New York, overlooking the village of Hamilton. Ivy-covered limestone buildings peek out from tree-lined drives; lush green spaces are perfect for rugby, Frisbee, or other outdoor diversions, at least in the warmer months. Rolling hills and farmland surround the campus, making for stunning vistas during the snowy season, which stretches from mid-October to mid-March. The university recently completed a $52 million renovation to the library and a new interdisciplinary science center is expected to open in late 2007.

Aside from blazing a trail to rural New York, Colgate has led its peers in emphasizing interdisciplinary study. The faculty first established an interdisciplinary core program in 1928, and it's been a foundation of the curriculum ever since. Even now, all freshmen take a first-year seminar that introduces liberal arts topics, skills, and ways of learning. The seminars are capped at 18 students each, and there are more than 40 topics, ranging from the history of rock and roll to the advent of the atomic bomb. The seminars focus on individual needs and strengths, learning from classmates, and learning from resources beyond the classroom; academic advising also comes from the seminar instructor, since students don't declare majors until the sophomore year. Students also complete two courses from each of Colgate's three academic divisions: natural sciences and math, social sciences, and humanities. Four courses in the liberal arts core—on Western traditions, the challenge of modernity, cultures, and scientific perspectives—round out the requirements and equip students to contemplate the issues they will face throughout their lives. Aside from a major (or two), Colgate also mandates foreign language proficiency, four physical education classes, and a swimming test.

Students give high marks to the natural and social sciences, including economics, political science, and history, all of which are also among the most popular majors. English is likewise strong, and sociology and anthropology round out the list of programs with the highest enrollments. Befitting Colgate's rugged location, there are four concentrations within environmental studies: environmental biology, geography, geology, and economics. Classrooms and labs devoted to foreign language study help students gain comfort with another tongue—a good thing, since more than half participate in Colgate's off-campus study programs. Aside from programs led by faculty members in foreign locales, from England, Japan, and Nigeria to Russia and Central America, there are three domestic programs, such as one at the National Institutes of Health in Bethesda, Maryland. Colgate also participates in the Maritime Studies Program* and the Semester at Sea.*

Classes at Colgate are small, and 97 percent of those taken by freshman have 50 or fewer students. "Considering the high level of intellectual talent and stimulation at Colgate, we have a very laid-back campus," says a senior. A sophomore agrees, "Colgate's courses are rigorous but by no means overwhelming." Students report that most people work together and help

> **"Considering the high level of intellectual talent and stimulation at Colgate, we have a very laid-back campus."**

one another, and that professors are friendly and accessible. "The professors challenge their students to think in new ways, voice their opinions, and approach the world with an inquisitive attitude," a student says. Undergraduate research also wins raves, and each summer, more than 100 students work under faculty members as research assistants. The 81 percent of students who go straight into the workforce credit Colgate's strong and loyal alumni network with helping them land their first job.

"Colgate students are passionate, dedicated, motivated, and concerned," says a sociology major. They're also overwhelmingly white graduates of public high schools, though only about a third are New Yorkers. African Americans and Hispanics each make up 4 percent of the student body, and Asian Americans add 6 percent—better than in the past, but still insufficient, students say. "The campus is fairly evenly split between liberal and conservative," says a student, "but the campus tends to be apathetic." The few issues that draw attention include gay and lesbian rights, genocide, and the university's increased involvement in Greek life. There are no merit scholarships, but Colgate offers a varying number of athletic awards.

Eighty-nine percent of Colgate students live in the dorms, which range from traditional buildings with fireplaces to newer facilities that seem more like hotels. "Even as a freshman, I lived in a spacious four-person room, with two doubles and a common room," says a sophomore. "Options range from one to five people per room the first year, to up to six as a sophomore (as well as a sophomore house on Broad Street), to theme housing, Greek housing, and apartment living as a junior or senior." About 250 upperclassmen are allowed to live off campus each year. Three of the four dining halls serve buffet-style, and the other is *a la carte*, with salad and sandwich bars, soup, cereal, and bagels always available; take-out food is provided at the campus center, known as the Coop. "The chefs love special requests," says a history major, "so if you have a favorite recipe from home, bring it along!"

(Continued)
Philosophy
Political Science

All freshmen take a first-year seminar that introduces liberal arts topics, skills, and ways of learning. The seminars are capped at 18 students each, and there are more than 40 topics, ranging from the history of rock and roll to the advent of the atomic bomb.

Classrooms and labs devoted to foreign language study help students gain comfort with another tongue—a good thing, since more than half participate in Colgate's off-campus study programs.

Hamilton is "the epitome of a college town—small, quaint, and very much a part of the university," says a junior. "We have three delis, two pizza places, two fancy places, and a coffee shop, as well as a number of small stores," a classmate adds. "When school is in session, the population more than doubles; the town really caters to the students, and vice versa." The town is within walking distance of campus, but there's also a free bus that cycles through every half-hour, especially nice in the dead of winter. Students enjoy free "Take Two" movies on Friday and Saturday nights, a cappella concerts featuring their friends, and open-mic nights at Donovan's Pub or the Barge Canal Coffee Company, which Colgate opened in a downtown storefront. The coffeehouse is open to all, including townspeople, and has become very popular. "The atmosphere is terrific, and there are games, puzzles, magazines, and books to occupy your hours," gushes a senior. The Palace draws crowds with music, dancing, a bar, and a Mexican restaurant; it's also located downtown.

> "Colgate students are passionate, dedicated, motivated, and concerned."

Back on campus, 33 percent of the men and 29 percent of the women join the Greek system, while debates continue over the proper balance between academics and social life. In an effort to cut down on alcohol consumption, hazing, and other problems that regularly get out of hand at Colgate, the administration forced fraternities and sororities to sell their off-campus houses to the university, resulting in an ongoing legal battle. For the outdoorsy, "Base Camp can direct you to campgrounds and other facilities to try out, and they will also rent you any gear you might need," says a computer science and physics major. In addition to the required four-day orientation program, freshmen may also participate in Wilderness Adventure, where groups of six to eight canoe and hike in the Adirondacks, or in Outreach, which involves three days of community service in the surrounding area. Colgate students remain involved in the community through work as tutors, mentors, and student teachers, on Habitat for Humanity projects, and with the elderly. For those with wheels, skiing is 45 minutes away in Toggenburg, and the malls and city lights of Syracuse and Utica are roughly the same distance. Everyone looks forward to Spring Party Weekend, a last blast before finals, which celebrates the thaw with a carnival, barbecues, fireworks, and multiple bands. Winterfest is a traditional field day... played in two feet of snow.

Colgate students remain involved in the community through work as tutors, mentors, and student teachers, on Habitat for Humanity projects, and with the elderly.

While Colgate students participate in intramurals ranging from bowling to ultimate Frisbee, their most fervent cheers are reserved for Division I-AA football against Bucknell, and Division I men's lacrosse against Cornell. The women's basketball and lacrosse teams and the football and men's water polo squads all competed in their national championship tournaments recently, and women's soccer and men's hockey won league titles. Even weekend warriors may take advantage of the Sanford Field House, the Lineberry natatorium, and the Seven Oaks golf course, which is ranked among the top five collegiate courses nationally. There's also a trap-shooting range, a quarry for rock-climbing, miles of trails for running and cycling, and sailing and rowing facilities at Lake Moraine, five minutes away.

> "The town really caters to the students, and vice versa."

Colgate has come a long way. The school has led the way in interdisciplinary work and continues to do so now, both literally—with torchlight processions for incoming freshmen and graduating seniors—and figuratively, through its first-year seminar program. What else has remained constant? A senior offers this assessment: "Whether it's academics, athletics, community service, the arts, or any of the other 1,001 things they are involved in, Colgate students are driven to succeed."

Overlaps

Dartmouth, Cornell University, Boston College, Middlebury, Williams, Bucknell

If You
Apply
To ➤

Colgate: Early decision: Nov. 15, Jan. 15. Regular admissions and financial aid: Jan. 15. Guarantees to meet full demonstrated need. Campus and alumni interviews: optional, informational. SATs or ACTs: required. SAT IIs: optional. Accepts the Common Application and electronic applications. Essay question: personal statement.

Colorado College

14 East Cache La Poudre Street, Colorado Springs, CO 80903

The Block Plan is CC's calling card. It is great for in-depth study and field trips but less suited to projects that take an extended period of time. The Rockies draw outdoor enthusiasts and East Coasters who want to ski. CC is the only top liberal arts college between Iowa and the Pacific.

CC is one of the few U.S. schools offering block scheduling, also known as the "One-Course-At-A-Time" method. For more than a century, CC's focus on creative approaches to academics and its breathtaking location in the heart of the Rocky Mountains have drawn liberal-leaning liberal arts enthusiasts who also like to go out and play.

Founded in 1874, the college campus lies at the foot of Pike's Peak, in the politically conservative town of Colorado Springs. The surrounding neighborhood is on the National Register of Historic Places, as are many CC buildings, including its first, Cutler Hall (1879), and Palmer Hall, named after town founder William J. Palmer, a major force behind the establishment of the college. The prevailing architectural styles are Romanesque and English Gothic, with some more modern structures thrown in. The Western Ridge dorm complex opened in fall 2002, offering apartment-style living for 290 students. A 54,000-square-foot building for the psychology, math, and geology departments, and the environmental sciences program, was finished in 2003. The Cornerstone Arts Building, with teaching and performance space, is being built.

> **"You decide how rigorous the course is by how much effort you give."**

CC requires students to take 32 courses, at least 18 outside their major department. Within those 32 courses, two must focus on the Western tradition; three on the non-Western tradition, minority culture, or gender studies; and two on the natural sciences, including lab or field study. Foreign language proficiency is also required, and students must complete either a six-course thematic minor, which examines an issue or theme, cultural group, geographic area or historical era, or six social sciences courses outside their major. What really defines the academic climate, though, is the block schedule (also see Cornell College in Iowa). Students take eight courses between early September and mid-May, but focus on each one, in turn, for three and a half weeks. Some courses, such as neuroscience, are two blocks long. Four-and-a-half-day breaks separate the terms. The plan helps students stay focused, eliminating the temptation to let one course slide so that they can catch up in another. But there are trade-offs. Students say it can be hard to integrate material from courses taken one at a time. There's also the danger of burnout, because so much material is crammed into such a short span. Still, the prevailing vibe is low-key. "You decide how rigorous the course is by how much effort you give," explains a freshman. "The only person you are ever really competing with is yourself." The First Year Experience program, with a student mentor and two advisors, helps students adjust.

Website:
www.coloradocollege.edu
Location: Urban
Total Enrollment: 1,998
Undergraduates: 1,970
Male/Female: 46/54
SAT Ranges: V 610–690
   M 620–700
ACT Range: 27–30
Financial Aid: 47%
Expense: Pr $ $ $
Phi Beta Kappa: Yes
Applicants: 4,386
Accepted: 34%
Enrolled: 33%
Grad in 6 Years: 84%
Returning Freshmen: 91%
Academics: ✑ ✑ ✑ ✑
Social: ☎ ☎ ☎ ☎
Q of L: ★ ★ ★ ★
Admissions: (800) 542-7214
Email Address: admissions
   @coloradocollege.edu

Strongest Programs:
Biology
Political Science
English
Psychology

Students at Colorado tend to be bright and independent; they say the school's best programs include the sciences and English. "The sciences are great," says a physics major, "especially geology." The block schedule permits some classes at unique times and in unique places—for instance, astronomy at midnight, or coral biology work in the Caribbean. The college's popular program in Southwest studies includes time at its Baca campus, 175 miles away in the historic San Luis Valley. Other interesting interdisciplinary programs include Asian studies, studies in war and peace, and American-ethnic studies. CC also offers study abroad in locations ranging from France and Germany to Japan, China, Tanzania, and Zimbabwe. Still

"The students are laid-back, nature-loving hippies."

more options are available through the Associated Colleges of the Midwest.* More than half of the class of 2005 spent time abroad. And for students who want to see more of the U.S., there are arts and urban studies programs in Chicago, a Washington Semester for budding politicos, and a science semester at Tennessee's Oak Ridge National Laboratory.

Students take eight courses between early September and mid-May, but focus on each one, in turn, for three and a half weeks. Some courses, such as neuroscience, are two blocks long.

Back on campus, the average class size is 13. Required courses aren't hard to get, since spots are secured with an auction system. At the beginning of each year, students get 80 points to "bid" on the classes they want. Those who bid the most for a particular class get a seat. And if you're going to take only one class at a time, it helps to like the teacher. Students say that's no problem here. "The professors are very accessible, and it is easy to have a good working relationship," says a history and political science major.

Just 29 percent of Colorado College students are in-staters; 3 percent are foreign, and the rest are from elsewhere in the U.S. "The students are laid-back, nature-loving hippies," says a student. Minorities account for less than 15 percent of the student body—2 percent are African American, 7 percent Hispanic, and 4 percent Asian American—and the school is trying to attract more. The Queer Straight Alliance, the Feminist Collective, the College Republicans, the Jewish Chaverim, and the Black Student Union also provide support to students of varied backgrounds and viewpoints. "Environmental issues are especially big on campus," says a sophomore. The admissions office places great weight on students' high school records, class rank, and essay. Only seniors are permitted to live off campus at Colorado College, so the other 73 percent of the student body bunks in the dorms. And while seniors don't have to move, a junior says it's easy to see why they do: "There's housing for about 80 percent of our study body, but a lot of it sucks." Architecturally, dorms range from

"I really enjoy the people and the unique learning environment."

large brick halls to small houses; some are for freshmen only, others are same-sex, and still others offer a language or cultural theme. Campus residents give the dining hall chow a universal thumbs-down: "The food is not a highlight," quips a student. "It is very bland and there are not many options." The dining options include an organic café, coffee bar, traditional dining hall, and a coffee house.

CC offers study abroad in locations ranging from France and Germany to Japan, China, Tanzania, and Zimbabwe.

When the weekend comes, students stay on campus for parties in friends' rooms or events sponsored by the "low-key" Greek system, which attracts 10 percent of the men and 16 percent of the women. Officially, no one under 21 is permitted to have alcohol, in the dorms or elsewhere, but students say enforcement is lax. "It's fairly easy to get alcohol from an upperclassman," says a biology major. For those who don't, won't, or can't imbibe, Herb 'n' Farm offers great smoothies. Each spring, the outdoors Llamapalooza festival features bands from on and off campus. For those seeking a bit of urban culture, Denver and Boulder are a short drive away. Most CC students love heading off campus to ski or hike, either at nearby resorts, or in Utah, New Mexico, or the Grand Canyon area. (Freshman Outdoor Orientation Trips help out-of-staters sort out the options, from backpacking and hiking to rafting, bicycling, and windsurfing. Students may even reserve a college-owned mountainside cabin.) Service trips are sponsored during block breaks, and 84 percent of students do some type of community service.

When it comes to sports, most Colorado College teams compete in Division III, except for women's soccer and men's ice hockey, which are Division I. Men's ice hockey reached the playoffs in 2005, and women's lacrosse and volleyball, men's soccer, and men's and women's track and field all have made national tournament appearances in recent years. There's a huge rivalry with the University of Denver.

The block plan made Colorado College what it is today, and the school continues to build on this reputation. While CC is intense, and the schedule is not for everyone, "I really enjoy the people and the unique learning environment," a freshman says. "It's small and conducive to the way I learn."

Overlaps

University of Colorado, Middlebury, University of Denver, Whitman, Colby

If You Apply To ➤

Colorado College: Early decision: Nov. 15, Jan. 1. Guarantees to meet full demonstrated need. Campus interviews: optional, informational. No alumni interviews. SATs or ACTs: required. No SAT IIs. Accepts the Common Application and electronic applications. Essay questions: why Colorado College, and design your own three-and-a-half-week intellectual adventure.

University of Colorado at Boulder

Office of Admissions, 552 UCB, Boulder, CO 80309-0552

Boulder is a legendary place that draws everyone from East Coast ski bums to California refugees. The scenery is breathtaking and the science programs are first-rate. The University of Arizona is the only public university of similar stature in the Mountain West. Check out the residential academic programs.

Wild buffalo may be all but extinct on America's Great Plains, but they're in boisterous residence, proudly wearing gold and black, at the University of Colorado at Boulder. A raft of scholars' programs, learning communities, and academic neighborhoods give the campus a community feel, and stress is no problem thanks to the active social scene, which emphasizes nature, fitness, sports, and outdoor pursuits. With more than 300 days of sunshine a year, is it any wonder students here are a happy lot?

Tree-shaded walkways, winding bike paths, open spaces, and an incredible view of the dramatic Flatirons rock formation makes CU's 600-acre Boulder campus a haven for students from both coasts and for Colorado residents eager to pursue knowledge in a snowy paradise. Campus buildings, in a rural Tuscany style, consist of red tile roofs made from native Colorado sandstone. The 45,000-square-foot Discovery Learning Center gives engineering students nine labs in which to tackle society's challenges with videoconferencing and other high-tech capabilities. The University Memorial Center has received a $27 million facelift and expansion, and Folsom Stadium has a 125,000-square-foot addition. The Williams Village residence hall complex has added more than 1,900 apartment-style beds and the most recent addition is the ATLAS Center which houses modern technology-enhanced teaching, learning, and research facilities.

Entering freshmen at CU choose from four colleges and one school: arts and sciences (where most students enter), architecture and planning, music, engineering and applied science (the hardest to enter, students say), and business. Transfers may

Website: www.colorado.edu
Location: Small city
Total Enrollment: 28,624
Undergraduates: 24,223
Male/Female: 53/47
SAT Ranges: V 530–630
  M 550–650
ACT Range: 23–28
Financial Aid: 32%
Expense: Pub $ $
Phi Beta Kappa: Yes
Applicants: 17,111
Accepted: 88%
Enrolled: 31%
Grad in 6 Years: 66%
Returning Freshmen: 83%
Academics: ✑ ✑ ✑ ✑
Social: ☎ ☎ ☎ ☎ ☎
Q of L: ★ ★ ★
Admissions: (303) 492-6301
Email Address: N/A

apply to two additional schools: journalism and education. Each has different entrance standards and requirements; music, for example, requires an audition. General education requirements for the 70 percent of students who enroll in arts and sciences are designed to provide a broad background in the liberal arts to complement their major specialization. The requirements cover four skills-acquisition areas—writing, quantitative reasoning and math, critical thinking, and foreign language—and seven content areas: historical context, culture and gender diversity, U.S. context, natural sciences, contemporary societies, literature and the arts, and ideals and values.

In a typical semester, CU–Boulder may offer 2,500 undergraduate courses in 100 fields. Among the best choices are molecular, cellular, and developmental biology, which take advantage of state-of-the-art electron microscopes, psychology, and integrative physiology, which are two of the most popular majors. CU–Boulder also receives the second-highest NASA funding of any university in the nation, leading to unparalleled opportunities for the design, construction, and flight of model spacecraft—and to 17 CU alumni having worked as astronauts. In 2005, CU physicist John Hall became the university's fourth faculty member to win a Nobel Prize and its third to win the Nobel Prize in physics. Business, engineering, sociology, and journalism are likewise strong. Students say the quality of teaching is very good. "The faculty at Boulder is extremely knowledgeable on their subject matter," says a junior. "Most importantly, they are always more than willing to share their knowledge with all students." A new department, aimed at preparing undergrads for careers in health care, has quickly become the second most popular on campus.

> "The Residential Academic Program for freshmen is essential for gaining a well-rounded experience at CU."

Boulder has tried to make its mammoth campus seem smaller through "academic neighborhoods" focusing on topics such as leadership, diversity, natural or social sciences, international studies, engineering, music, and the American West. Through these programs, students take one or two courses, each limited to 25 students, in their residence halls. "The Residential Academic Program for freshmen is essential for gaining a well-rounded experience at CU," advises a senior. The Presidents Leadership Class is a four-year scholarship program that exposes the most promising students to political, business, and community leaders through seminars, work and study trips, and site visits. The Undergraduate Academy offers special activities and advising for 150 to 200 of CU's most "intellectually committed" students, chosen for their excitement about learning and academic success.

Aside from skiing, exercise is the leading extracurricular activity at CU.

Seventy percent of CU's student body comes from Colorado, and by state regulation that fraction can be no lower than 55 percent, on average, over a three-year period. "I cannot characterize CU as a very diverse campus," laments one philosophy major. Hispanics and Asian Americans each comprise 6 percent of the total, and African Americans make up 2 percent. CU leans liberal, and important campus issues include environmental awareness, ethnic issues, and the war in Iraq. Thirty-six percent of undergrads receive merit scholarships worth an average of $2,667, and 241 athletes receive scholarships as well.

> "The dorms on campus are very comfortable, well maintained, and well supervised."

First-year students are required to live on campus, where rooms get good ratings. "The dorms on campus are very comfortable, well maintained, and well supervised," says one student. Most sophomores, juniors, and seniors, however, find off-campus digs in Boulder. Those who want to stay on campus are advised to make early reservations for Farrand, Sewall, or Kittredge halls. All rooms come with microwaves, refrigerators, cable TV, and high-speed Internet hook-ups. Generally, students say campus is safe, helped by 30 emergency telephones along walkways and paths, and 16 more in the parking structures. CU also offers walking and riding escorts at night via a service called CU NightRide. An alternative to the dining hall is the student-run Alferd Packer Memorial Grill in the student center, which provides fast food under innocent auspices. Boulder students and trivia buffs know, however, that Packer was a controversial 19th-century folk figure known as the "Colorado Cannibal." *Bon appetit.*

Seven percent of CU men and 10 percent of women go Greek, though fraternity and sorority parties have changed dramatically since CU's sorority chapters became the first in the nation to voluntarily make their houses dry. On campus, the ban on alcohol is taken seriously, and dorms are officially "substance-free." Get caught with booze two times while underage, and you'll be booted from school housing. Even if you don't drink, though, you'll surely find something to do. "Social life is great because Boulder is one of the best college towns in America," says a biochemistry major. "Only in Boulder can you go skiing, hiking, boating and to a play, all in the same day," boasts another student. For the culturally minded, the university and the city of Boulder offer films and plays, the renowned Colorado Shakespeare Festival, and concerts by top rock bands. Even better, Denver is only 30 miles southeast, reachable by a free bus service. Day trips to ski resorts like Breckenridge, Vail, and Aspen largely replace weekend getaways here, but for those who've got to get out of the cold, Las Vegas isn't so far, says one student. For a quick drive to the slopes, the Eldora ski area—with 12 lifts, 53 trails, runs up to three miles long, and a vertical drop of as much as 1,400 feet—is just a half hour from campus.

Aside from skiing, exercise is the leading extracurricular activity at CU. The CU–Boulder club sports program is consistently ranked among the top five in the nation for both the athletic and academic performance of its teams. Just $150 a year (part

> "Only in Boulder can you go skiing, hiking, boating and to a play, all in the same day."

of the mandatory student fee) gives students access to the Student Recreation Center, with swimming pools, basketball and tennis courts, three weight rooms, an indoor rock climbing wall, and an ice arena. The football program, which competes in the Big 12, has had its share of success on the field but is still reeling from a national scandal involving allegations of sexual assault and recruiting violations that raised embarrassing questions about the university's real values. Each year, fans flock to Denver for the game against Colorado State, and any showdown with Nebraska is sure to get students riled up. Ralphie, the live buffalo who acts as CU's mascot, doesn't miss a game—and neither do many students.

"CU–Boulder is energy. Every student, teacher and department always has something new and exciting going on. It would be very hard to get bored here," claims one senior. If you want to exercise your body as well as your mind, forget the ivy-covered bricks and gray city skies endemic to so many Eastern institutions, and consider going West instead.

In a typical semester, CU–Boulder may offer 2,500 undergraduate courses in 100 fields. Among the best choices are molecular, cellular, and developmental biology.

Overlaps

Colorado State, UCLA, University of Arizona, University of Illinois, University of Minnesota, University of Northern Colorado

If You Apply To ➤

CU–Boulder: Rolling admissions: Sept. 9. Regular admissions: Jan. 15. Financial aid: Apr. 1. Housing: May 1. Campus interviews: optional, informational. No alumni interviews. SATs or ACTs: required. SAT II: optional. Apply to one of four colleges and one school. Accepts electronic applications. Essay question: optional.

Colorado School of Mines

1811 Elm Street, Golden, CO 80401-1842

The preeminent technical institute in the Mountain West. Twice as big as New Mexico Tech, but one-tenth the size of Texas Tech. Best-known for mining-related fields but strong in many areas of engineering. Men outnumber women three to one, and Golden provides little other than a nice view of the mountains.

Website: www.mines.edu
Location: Small town
Total Enrollment: 3,921
Undergraduates: 3,098
Male/Female: 77/23
SAT Ranges: V 540–640
    M 600–690
ACT Range: 25–29
Financial Aid: 67%
Expense: Pub $ $ $ $
Phi Beta Kappa: Yes
Applicants: 3,686
Accepted: 71%
Enrolled: 29%
Grad in 6 Years: 67%
Returning Freshmen: 81%
Academics: 🖉 🖉 🖉 ½
Social: ☎ ☎ ☎
Q of L: ★ ★
Admissions: (303) 273-3220
Email Address:
    admit@mines.edu

Strongest Programs:
Geology/Geophysics
Mining Engineering
Petroleum Engineering
Metallurgy and Materials
    Engineering
Chemical Engineering
Engineering Physics

If you're a bit of a geek whose only dilemma is what type of engineer to become, and you want to spend your scarce free time hiking, biking, and skiing with friends, then Colorado School of Mines may be the place for you. The school's small size and rugged location endear it to the mostly male students who shoulder heavy workloads to earn their degrees—and truly enjoy starting salaries averaging $55,700 a year. "There are often fun and entertaining conversations that could only be possible with the types of students here," says a sophomore mechanical engineering major. Just down the road from Coors Brewing Co., which taps the Rockies for its legendary brews, students at Mines learn to tap the same mountains for coal, oil, and other natural resources.

CSM's 373-acre campus sits in the shadow of the spectacular Rocky Mountains in tiny Golden, Colorado. Architectural styles range from turn-of-the-century gold dome to present-day modern, and native trees and greenery punctuate lush lawns. New construction includes a $25 million Recreation Center and a $2.5 million expansion of a classroom building.

Academics at Mines are rigorous. All freshmen take the same first-year program, which includes chemistry, calculus, physical education, physics, design, earth and environmental systems, quantitative chemical measurement, nature and human values, and the Freshman Success Seminar, an advising and mentoring course designed to increase retention. Because of CSM's narrow focus, the undergraduate majors—or "options," as they're called—are quite good. There's plenty of variety, as long as you like engineering; programs range from geophysical, geological, and petroleum to civil, electrical, and mechanical. Courses in a student's option start in the second semester of sophomore year, after yet more calculus, physics, and differential equations. Mines offers the only BS degree in economics in Colorado. Physics has grown, now enrolling 8 percent of undergraduates, and the school has been investing more in humanities and social sciences.

> **"There are often fun and entertaining conversations that could only be possible with the types of students here."**

Pass/fail grading is unheard of at Mines, but failing grades are not. "The courses are very rigorous," a junior says, "but after the smoke clears, the push is worth it." Professors are well-qualified and helpful and adjunct professors, who work in the fields they teach, draw raves for bringing real-world application into the classroom. Ninety percent of freshman classes have 50 or fewer students, but 5 percent are packed with more than 100 students. The required two-semester EPICS program—the acronym stands for Engineering Practices Introductory Course Sequence—helps develop communications, teamwork, and problem-solving skills with weekly presentations and written reports. Students say teamwork helps soften the load a bit. "Because the courses are so difficult, the students tend to help each other rather than compete. In my experience, as long as everyone passes a course, we all won," one student says.

CSM supplements coursework with a required six-week summer field session, enabling students to gain hands-on experience. About 120 undergraduates participate in the McBride Honors Program in Public Affairs, which includes seminars and off-campus activities that encourage them to think differently about the implications of technology. CSM also offers the opportunity to live and study at more than 50 universities in Europe, Australia, Latin America, Asia, and the Middle East, though only 5 to 7 percent of students take part. Each year, 60 to 80 undergraduates participate in research with faculty members or on their own.

CSM is a state school, making it a good deal for homegrown students, who comprise 80 percent of the student body. "The students at Mines are very academically focused and determined," says a chemical engineering major. A classmate is more blunt, saying Mines attracts "a lot of nerds." Hispanics comprise 7 percent

> "Because the courses are so difficult, the students tend to help each other rather than compete."

of the student body, Asian Americans 5 percent, and African Americans and Asian Americans combine for 3 percent. Merit scholarships averaging $5,100 are available to qualified students, and student-athletes may vie for 230 athletic scholarships.

Forty-five percent of CSM students—mostly freshmen—live in the residence halls. Most buildings are co-ed, though the preponderance of men results in a few single-sex dorms. "All the residence halls have been refurbished and are looking better than ever," a sophomore says. Most upperclassmen move to fraternity or sorority housing, college-owned apartments, or off-campus condos and houses. There's only one cafeteria, and a junior says its "better to make your own food."

There is life outside of the library here. "There is a lot that goes on on campus," a junior says. "There is always a club putting together an event or just students throwing parties." CU–Boulder offers more partying 20 minutes away. Mines also has an active Greek system, with fraternities and sororities attracting 20 percent of the men and women. Still, rush is dry, and, owing to Mines' small size, those serving the alcohol almost always know the age of those trying to obtain it, making it tough for the underage to imbibe. Social life also includes comedy shows, homecoming, and Engineering Days—a three-day party with fireworks, a pig roast, tricycle races, taco-eating contests, and 25-cent beers. New student orientation includes the M-climb, in which freshmen hike up Mount Zion lugging a 10-pound rock, "and whitewash it, and each other," says one participant. The rock is added to an M formation atop the mountain, then "seniors return to take down a rock, completing the cycle."

CSM's location at the base of the Rockies means gorgeous Colorado weather (make sure to bring sunscreen) and easy access to skiing, hiking, mountain climbing, and biking. Denver is also nearby, and aside from its museums, concerts, and sports teams, the city is home to many government agencies and businesses involved in natural resources, computers, and technology, including the regional offices of the U.S. Geological Survey and Bureau of Mines. Golden hosts the National Earthquake Center, the National Renewable Energy Laboratory, and, of course, the Coors Brewery. (The 3,000-foot pipeline that runs from the Coors plant to campus is

> "All the residence halls have been refurbished and are looking better than ever."

there to convert excess steam from the brewery into heat for the school—not to supply the frats with the foamy stuff.) The biggest complaints are too much homework and not enough girls. Road trips to Las Vegas or Texas provide some respite.

Despite the abundance of wacky school traditions, CSM competes in Division II, and the program continues to thrive. In 2005, Mines ranked second in NCAA Directors' Cup standings. Women's cross-country placed fifth in the nation and the men's cross-country squad finished sixth. Women's volleyball and men's soccer qualified for the Division II Championships and the men's and women's basketball teams qualified for postseason tournaments. The intramural program has grown dramatically, with 70 percent of students now participating.

Pass/fail grading is unheard of at Mines, but failing grades are not.

CSM supplements coursework with a required six-week summer field session, enabling students to gain hands-on experience.

Overlaps

University of Colorado, Colorado State, Texas A&M, University of Texas, University of Denver, MIT

While time spent in the classroom at Mines may not be fun, for those who are focused on engineering, educational options don't get much better than those offered here. "Professors are extremely smart and more than willing to help," a junior math and computer science major says. But that doesn't mean academics are the only thing CSM has to offer. "Our school is academically challenging, but it has all the social events and an environment like any college."

If You Apply To ➤

CSM: Rolling admissions. Regular admissions: Jun. 1. Financial aid: Mar. 1. Housing: May 1. Guarantees to meet full demonstrated need. Campus and alumni interviews: optional, informational. SATs or ACTs: required. SAT IIs: optional. Prefers electronic applications. No essay question.

Columbia College

212 Hamilton Hall, New York, NY 10027

Columbia may soon leave Yale in the dust as the third most selective university in the Ivy League. Applications have doubled in the past 10 years for one simple reason: Manhattan trumps New Haven, Providence, Ithaca, and every other Ivy League city, with the possible exception of Boston. The heart of Columbia is still the core.

Website: www.columbia.edu
Location: Urban
Total Enrollment: 24,417
Undergraduates: 5,559
Male/Female: 54/46
SAT Ranges: V 670–760
  M 670–780
ACT Range: 28–33
Financial Aid: 49%
Expense: Pr $ $ $ $
Phi Beta Kappa: Yes
Applicants: 19,851
Accepted: 12%
Enrolled: 58%
Grad in 6 Years: 94%
Returning Freshmen: 98%
Academics: 🐾 🐾 🐾 🐾 🐾
Social: ☎ ☎ ☎
Q of L: ★ ★ ★ ★
Admissions: (212) 854-2522
Email Address: ugrad-admiss@columbia.edu

Strongest Programs:
English
History
Political Science

Though students entering Columbia will, of course, expect the rigorous academic program they'll encounter at this Ivy League school, there's no room here in the heart of Manhattan for the bookish nerd. Students must be streetwise, urbane, and together enough to handle one of the most cosmopolitan cities in the world. "It's an Ivy League school with a campus in the leading cultural center of the United States," says a sophomore. Columbia lets its students experience life in the Big Apple, but serves as a refuge when it becomes necessary to escape from New York; ideally, Columbians can easily be part of the "real world" while simultaneously immersing themselves in the best academia has to offer.

Although Columbia is among the smallest colleges in the Ivy League, its atmosphere is far from intimate. With a total universitywide enrollment of 24,000 students, says one, "It's easy to feel lost." Still, the college is the jewel in the university's crown and the focus is "unquestionably oriented toward undergraduate education," reports a classics major. Columbia's campus has a large central quadrangle in front of Butler Library and at the foot of the steps leading past the statue of Alma Mater to Low Library, which is now the administration building. The redbrick, copper-roofed neoclassical buildings are "stunning," and the layout, says an undergrad, "is well thought out and manages to provide a beautiful setting with an economy of space."

Columbia is an intellectual school, not a preprofessional one, and even though 60 percent of the students aspire to law or medical school (they enjoy a 90 percent acceptance rate), "we are mostly content to be liberal artists for as long as possible," says an English major. Even the 30 percent of undergraduates who enroll in the School of Engineering and Applied Sciences pursue "technical education" with a liberal arts base. Almost all departments that offer undergraduate majors are strong, notably English, history, political science, and psychology. Chemistry and biology are among the best of Columbia's high-quality science offerings. The Earth and Environmental Science department owns 200 acres in Rockland County, home to many rocks and much seismographic equipment. There are 35 offerings in foreign languages, ranging from Serbo-Croatian to Uzbek to Hausa. The fine arts are not fabulous, but are improving, thanks to departmental reorganization, new facilities, and joint offerings with schools such as the Juilliard School of Music. And while the administration admits that the economics and computer science departments are geared too much toward graduate students, at least the comp-sci undergrads benefit from an abundance of equipment. Columbia offers many challenging combined majors such as philosophy/economics and biology/psychology. The East Asian languages and cultures department is one of the best anywhere. There is also an African American studies major, and a women's studies major that delves into topics from the Asian woman's perspective to the lesbian experience in literature.

The kernel of the undergraduate experience is the college's renowned core curriculum. While these courses occupy most of the first two years and can become laborious, students generally praise them as worthwhile and enriching: "You learn how to read analytically, write sharply, and speak succinctly, and you are exposed to the greatest ideas in Western art, music, literature, and philosophy," exclaims an enlightened sociology major. The value of the Western emphasis of the core, however, is a subject of perennial debate. "Why should we study the Western tradition when it represents sexism, racism, imperialism, and exploitation?" asks one incensed student. "The canon is composed almost exclusively of dead European males." Yet, as it has since World War I, the college remains committed to the core while at the same time expanding the diversity of the canon and requiring core classes on non-Western cultures.

Two of the most demanding introductory courses in the Ivy League—Contemporary Civilization (CC) and Literature Humanities—form the basis of the core. Both are yearlong and taught in small sections, generally by full profs. "Nearly everything I'd grown up believing was questioned in one way or another. They forced me to examine my life and to ponder how I fit into the big picture," states an art history major. LitHum

"It's an Ivy League school with a campus in the leading cultural center of the United States."

(as it is affectionately called) covers about 26 masterpieces of literature from Homer to Dostoyevsky, usually with some Sappho, Jane Austen, and Virginia Woolf thrown in for alternative perspectives. CC examines political and moral philosophy from Plato to Camus, though professors have some leeway in choosing 20th-century selections. One semester each of art and humanities is required and, while not given the same reverence as their literary counterparts, are eye-opening all the same. Foreign language proficiency is required, as are two semesters of science; two semesters of "extended core" classes dealing in cultures not covered in the other core requirements; two semesters of phys ed; and logic and rhetoric, a one-semester, argumentative writing class that first-year students reportedly "either love or hate." Students at the School of Engineering complete approximately half of the core curriculum.

Columbia is tough, and students always have something to read or write. Student/faculty interaction is largely dependent on student initiative. Additional interaction stems from professorial involvement in campus politics and forums and from the faculty-in-residence program, which houses professors and their families in spruced-up apartments in several of the residence halls. First-year students are assigned an advisor and receive a faculty advisor when they declare a major at the end of sophomore year. Columbia students can take classes at Barnard, which maintains its own faculty, reported to be "more caring and involved than Columbia's." As at Barnard, students can also take graduate-level courses in several departments, notably political science, gaining access to the resources of the School of International and Public Affairs and its multitude of regional institutes. For students wishing to spend time away from New York, Columbia offers credit through more than 150 programs in 100 countries.

Columbia has the largest percentage of students of color in the Ivy League.

To call Columbia diverse would be "a gross understatement," says a sophomore. "We make Noah's Ark look homogeneous." In fact, Columbia has the largest percentage of students of color in the Ivy League; 8 percent are African American, 10 percent are Hispanic, and 22 percent are Asian American. Twenty-seven percent of the students come from New York. Socially, the campus is also diverse. In a city such as New York, "diversity is assumed," says one student. Students from low-income families may take advantage of a new grant program; the university's goal is for these students to graduate without the burdensome debt associated with a college degree.

Columbia remains one of the nation's most liberal campuses. "Columbia has always been known for its tradition of social and political activism," says a junior. "Students are not afraid to protest to get what they want." No one group dominates campus life. Although 15 percent of the men and 9 percent of the women go Greek, Columbia is hardly a Hellenocentric campus, namely because, as a junior argues, "the frats are chock-full of athletic recruits, the organizations—even the co-ed ones—are deemed elitist and politically incorrect, and there are too many better things to do in NYC on a Friday night than getting trashed in the basement of some random house." The advent of co-ed houses has raised interest in Greek life as has the arrival of sororities open to both Columbia and Barnard women.

"We are mostly content to be liberal artists for as long as possible."

With the New York housing market out of control, 99 percent of Columbia students live in university housing, which is guaranteed for four years. Security at the dorms is rated as excellent by students, and everyone entering has to flash an ID to the guard at the front door or be signed in by a resident of the building. One exciting aspect of Columbia housing is that many rooms are singles, and it is possible to go all four years without a roommate. Carman Hall is the exclusively first-year dorm and "the fact that you get to meet your classmates compensates for the noise and hideous cinder-block walls," says a music major. First-year students can also live in buildings with students of all years. "Living with upperclasspeople was great. They knew the ins and outs of the university and the neighborhood. It wasn't the blind leading the blind," offers a junior. First-year students are automatically placed on a 19-meal-a-week plan and take most of those meals at John Jay, an all-you-can-eat "binge-a-rama with salad bar, deli, grill, and huge dessert bar." Many soon-bloated students scale down their meal plans or convert to points, a buy-what-you-want arrangement with account information stored electronically on their ID cards. Several dorms have kitchens, allowing students to do their own cooking. Kosher dining is also available.

Social life on campus is best described as mellow. Rarely are there big all-inclusive bashes, the exceptions being fall's '60s throwback, Realityfest, and spring's Columbiafest. "The social scene here is well-balanced between school events, concerts, and dances, and the variety of activities the city offers," says a senior.

Columbia athletics don't inspire the rabid loyalty of, say, a Florida State, because "Columbia students are individualists," according to one sophomore. "This is not a school that rallies together at football games." Still, the fencing teams are superlative, and men's soccer and basketball are also strong. As an urban school, Columbia lacks team field facilities on campus; however, a mere 100 blocks to the north is the modern Baker Field, home of the football stadium, the soccer fields, an Olympic track, and the crew boathouse. On campus, the Dodge Gymnasium, an underground facility, houses four levels of basketball courts, swimming pools, weight rooms, and exercise equipment. The gym is often crowded and not all the stuff is wonderful. "It does the job, as well as providing for the best pickup basketball this side of Riverside Park," notes a sophomore. Intramural and club sports are popular, with men's and women's ultimate Frisbee both national competitors.

"We make Noah's Ark look homogeneous."

Overlaps

Harvard, Yale, MIT, Princeton, Stanford, University of Pennsylvania

Columbians are proud that they are going to college in New York City, and most would have it no other way. Explains an art history major: "Choosing to isolate oneself in the middle of nowhere for four years isn't what college is about. It's about taking one's place as an adult in an adult society. Columbia is the perfect place for that."

<table>
<tr><td>If You
Apply
To ➤</td><td>Columbia: Early decision: Nov. 1. Regular admissions: Jan. 1. Financial aid: Jan. 15. Guarantees to meet full demonstrated need. Campus interviews: not available. Alumni interviews: optional, evaluative. SATs or ACTs: required. SAT IIs: required (any two). Essay: personal statement.</td></tr>
</table>

University of Connecticut

Tasker Building, 2131 Hillside Road, Unit 3088, Storrs, CT 06269-3088

Squeezed in among the likes of Yale, Brown, Wesleyan, Trinity, and UMass—all within a two-hour drive—UConn could be forgiven for having an inferiority complex. But championship basketball teams, both men's and women's, have ignited Husky pride, and the university's mammoth rebuilding project is boosting its appeal.

The top public university in New England, UConn has seen billions of dollars poured into improving and expanding its facilities in the past decade. Couple the new buildings with the glow of two championship basketball teams, a wealth of research opportunities, and more than 250 clubs and organizations, and it's clear why students who in the past might have dismissed it as a "cow college" are choosing UConn, even when they have other options. "Students have great opportunities to explore their interests," says a senior. "It makes the campus seem smaller and more of a close-knit community."

UConn's 4,000-acre campus is about 23 miles northeast of Hartford. Building styles range from collegiate Gothic and neoclassical to half-century-old redbrick. Dense woods surround the campus, which also boasts two lakes, Swan and Mirror. Ongoing renovations are the norm, sparking jokes about the "University of Construction," but the results are impressive: an expanded and renovated student union, the newly expanded Neag School of Education, a new School of Pharmacy, a state-of-the-art biophysics building, a new five-story Information Technologies Engineering Building, a 40,000-seat football stadium, new student housing (with more on the way), and other projects.

Students say UConn's strongest offerings are preprofessional, including business, engineering, education, pharmacy, and allied health, including nursing and physical therapy. "I am a pharmacy major and I love it," says one senior. Also notable are basic sciences, history, linguistics, psychology, and, of course, agriculture. (UConn was founded more than a century ago as a farm school; it's where America learned to get more eggs per chicken by leaving the lights on in the coops.) In addition to new buildings, UConn continues to add new curricula, including an innovative program that combines biodiversity and conservation biology with public policy. There are also programs in biomolecular engineering, neurosciences, cognitive science, American studies, aquaculture, survey research, coastal studies, urban and community studies, and human rights. Engineering is demanding, and, as at many schools, it has a relatively high attrition rate, with many students switching to the less-rigorous major in management information systems. UConn is also the only

Website: www.uconn.edu
Location: Rural
Total Enrollment: 19,213
Undergraduates: 15,196
Male/Female: 48/52
SAT Ranges: V 540–630
  M 550–650
ACT Range: 23–27
Financial Aid: 47%
Expense: Pub $ $ $ $
Phi Beta Kappa: Yes
Applicants: 18,608
Accepted: 51%
Enrolled: 34%
Grad in 6 Years: 72%
Returning Freshmen: 92%
Academics: ✍ ✍ ✍ ✍
Social: ☎ ☎ ☎
Q of L: ★ ★ ★
Admissions: (860) 486-3137
Email Address:
  beahusky@uconn.edu

Strongest Programs:
Biosciences
Communication Sciences
Business
Education
Engineering

public university in New England to offer majors in environmental engineering, computer engineering, computer science, and metallurgy and materials engineering. A special program in medicine and dentistry allows students to earn bachelor's degrees in any of UConn's more than 100 disciplines, and guarantees admission to the School of Medicine or Dental Medicine if they meet all criteria.

"UConn has become increasingly competitive and the academic climate reflects this," says an accounting major. UConn's core requirements include courses in four basic areas: arts and humanities, social sciences, diversity/multiculturalism, and science and technology. Also required are

> "UConn has become increasingly competitive and the academic climate reflects this."

two foreign language courses, waived if a student has studied three years of a single language in high school, and competency in computer technology and information literacy. Seminar-style writing classes are available to all freshmen, and 75 percent also take one or more First Year Experience courses focusing on time management, how to use the library, and other useful skills. The Academic Center for Exploratory Students helps freshmen and sophomores who still need to decide on a major. Students applaud the enthusiasm of their professors—and the graduate teaching assistants who administer tests, collect assignments, and run labs and discussion groups. "The professors are very knowledgeable, enthusiastic, and genuinely interested in their students," one senior says. "Freshman are often taught by full professors, typically in a lecture hall setting."

UConn's engineering, business, pharmacy, and honors students are required to undertake research projects, and each year two teams of finance majors run the $1 million student-managed investment fund. Students who aspire to graduate school in academic fields, rather than professional certification, may win grants to work independently under faculty members through the undergraduate summer research program. The 7 percent of students who qualify for the honors program gain access to special floors and dorms; several programs for disadvantaged students are also available. In addition, about 300 students a year participate in the study abroad program,

> "The students who attend UConn are very enthusiastic and involved."

which offers programs in 65 countries. UConn's five campuses around Connecticut offer the first two years of the undergraduate program and some four-year degree programs. Students who satisfactorily complete work at these schools are automatically accepted at the Storrs campus for their last two years.

"The students who attend UConn are very enthusiastic and involved," says a senior. "There is a strong sense of school spirit." Three-quarters of UConn students are from Connecticut, and 16 percent are minorities—5 percent African American, 4 percent Hispanic, and 7 percent Asian American. There are cultural centers for African American, Asian American, Latin American, and Puerto Rican students, as well as the Rainbow Center, a resource for gay, lesbian, bisexual, and transgender students. "I like that students are really open-minded here," a junior says, "and that no issue is really one-sided or so heated that it divides the campus." Twenty percent of UConn students receive merit scholarships averaging $5,951, and 371 athletic scholarships are available, too, in seven men's sports and 11 women's sports.

Seventy-two percent of the students live in university housing, which is available to all undergraduates. "Halls are clean and spacious," says one student. "I love my dorm," says another. Though a few dorms are single-sex, most are co-ed by floor. In addition, "We have awesome suites and on-campus apartments," says a secondary education major. Nearly all campus housing has high-speed Internet, cable, and data networking service in the rooms, not just in student lounges. Ten dining halls offer plenty of choices, even for vegetarians and vegans, though many students would just as soon visit the snack bar for some ice cream, freshly made with

Favorite annual campus events include the mud volleyball tournament, carnival-style "UConn Late Nights," Midnight breakfasts during finals, homecoming, Winter Weekend, and Midnight Madness—the first official day of basketball practice.

help from the cows grazing nearby. "The food is very good and the type of food changes daily," one student says.

The freshman dorms are officially dry, and students in other dorms are allowed to possess no more than a six-pack of beer, one bottle of wine, or a small bottle of liquor. Students under 21 who are caught with booze may be evicted from campus housing. Late-night activities at the student union and other campus events provide a lot of alternatives to alcohol use. Fraternities attract 10 percent of the men, and sororities claim 7 percent of the women; members can live in chapter housing at the Husky Village. On weekends, there are busses to Hartford (only 30 minutes away), Boston, New Haven, New York, and Providence. Cape Cod and the Vermont ski slopes are within weekend driving distance. Still, "Most of the social life takes place on campus," explains a senior. "There are concerts, sporting events, and other activities."

The town of Storrs "is no college town, but it's rural and we're proud of it," says one student. The university provides transportation for students who volunteer in area schools and hospitals. Legend also holds that UConn also offers one diversion most other colleges can't: cow tipping—that is, sneaking up on unsuspecting cows, who sleep standing up, and tipping them over. The administration contends that this is a myth, though students always claim to "know someone who did it."

UConn's teams are known as the Huskies (Get it? *Yukon.*) and basketball is by far the most popular sport. In a state without professional sports teams, the UConn women's team routinely sells out the Hartford Civic Center. In 2004, both the men's and women's basketball teams won the ultimate prize: the NCAA national championship—the first time any university has pulled off this feat. The field hockey, football, men's and women's soccer, men's and women's track and field, women's lacrosse, and women's volleyball teams are also recent award-winners. Intramurals are offered at three levels, from recreational to competitive. Popular offerings range from underwater hockey and inner-tube water polo to basketball, volleyball, and flag football. Favorite annual campus events include the mud volleyball tournament, carnival-style "UConn Late Nights," midnight breakfasts during finals, homecoming, Winter Weekend, and Midnight Madness—the first official day of basketball practice. In addition to cheering for the Huskies, "it is good luck to rub the nose of the bronze statue of our mascot, Jonathan," says a sophomore.

"We have awesome suites and on-campus apartments."

Despite the school's agricultural roots, UConn students aren't "cowed" by the plethora of offerings. Those seeking greener pastures will be hard-pressed to find a more dynamic public institution. "We are a well-rounded campus with students from every background," a junior pharmacy student says. And with the campus undergoing a complete facelift, a student says it's an exciting time to be at UConn.

Overlaps

Northeastern, Boston University, University of Massachusetts, University of Delaware, Boston College

If You Apply To ➤ **UConn:** Early action: Dec. 1. Regular admissions: Feb. 1. Campus interviews: optional, informational. No alumni interviews. SATs or ACTs: required. No SAT IIs. Prefers electronic applications. Essay question: what makes you unique and how that will benefit the UConn community; an influential person or event; your choice.

Like Vassar and Skidmore, Connecticut College made a successful transition from women's college to co-ed. The college is strong in the humanities and renowned for its study abroad programs. It is also an SAT-optional school. New London does not offer much, but at least it is on the water.

Website: www
   .connecticutcollege.edu
Location: Suburban
Total Enrollment: 1,997
Undergraduates: 1,757
Male/Female: 40/60
SAT Ranges: V 620–720
   M 610–700
ACT Range: 25–29
Financial Aid: 40%
Expense: Pr $ $ $ $
Phi Beta Kappa: Yes
Applicants: 4,278
Accepted: 38%
Enrolled: 30%
Grad in 6 Years: 86%
Returning Freshmen: 91%
Academics: 🐫 🐫 🐫 🐫
Social: ☎ ☎ ☎
Q of L: ★ ★ ★ ★
Admissions: (860) 439-2200
Email Address:
   admission@conncoll.edu

Strongest Programs:
Fine and Performing Arts
Environmental Studies
Economics
Psychology
Anthropology
International Studies
Government

Students at Connecticut College follow the example of their mascot, the camel—they take pride in drinking up and storing knowledge. The student-run honor code means finals are not proctored; they're even self-scheduled, whenever students prefer, during a 10-day window. Thanks to the code, students also feel comfortable leaving doors and bikes unlocked. Utopian? Perhaps. Still, "there really isn't much to complain about," a sophomore says. "The location is right between Boston and New York, the campus is beautiful, the people are extremely friendly, and the teachers are great."

Placed majestically atop a hill, the Conn College campus sits within a 750-acre arboretum with a pond, wetlands, wooded areas, and hiking trails. It offers beautiful views of the Thames River (pronounced the way it looks, not like the "Temz" that Wordsworth so dearly loved) on one side and Long Island Sound on the other. The granite campus buildings are a mixture of modern and collegiate Gothic in style, with some neo-Gothic and neoclassical architecture thrown in for good measure.

Conn was founded in 1911 as a women's college, and since then, it's been dedicated to the liberal arts, broadly defined. The general education requirements are aimed at fostering intellectual breadth, critical thinking, and acquisition of the fundamental skills and habits of minds conducive to lifelong inquiry, engaged citizenship, and personal growth. To that end, students are required to complete a series of at least seven courses that introduce them to the natural and social sciences, humanities, and arts. Academics are definitely the focus

> **"If you want to be a number, Conn is not the place for you."**

here. "Conn's academic atmosphere is a very supportive one," reports one sophomore. "Practically all of the students work hard for their grades, yet grades are not what we're all about." Professors "are always available outside of class to help you," a student says. "They strive to make each class interesting and helpful for students," adds a senior.

Conn's dance and drama departments are superb, and it's not uncommon for dancers to take time off to study with professional companies. Aspiring actors, directors, and stagehands may work with the Eugene O'Neill Theater Institute, named for New London's best-known literary son. Chemistry majors may use high-tech gas chromatograms and mass spectrometers from their very first day, and students say Conn also offers excellent programs in biology and physics. The Ammerman Center for Arts and Technology allows students to examine the connections between artistic pursuits and the worlds of math and computer science. The most popular majors are English, government, economics, psychology, and international relations; not coincidentally, students give all of those departments high marks. The teacher certification program also wins raves. To escape Conn's small size and occasionally claustrophobic feel, the Study Away/Teach Away initiative allows groups of 15 to 30 Conn students and two faculty members to spend a semester living and working together at an overseas university, in locations as far-flung as Egypt, Ghana, Tanzania, and Vietnam. Conn also participates in the Twelve College Exchange* bringing the total number of foreign study programs to more than 40. A gift from Conn helps students secure extraordinary summer internships; everyone who participates in a set of workshops is guaranteed one $3,000 grant during his or her four years to help cover housing or other costs incurred while gaining real-world work experience.

"Conn attracts a certain kind of 'up-for-anything' attitude person," says a student, "so they're usually fairly open-minded about those with whom they're spending their four years." Only 18 percent of Conn College students come from Connecticut, and 55 percent graduated from public high schools. African Americans and Asian Americans each make up 4 percent of the student body, and Hispanic students add 6 percent. Freshmen must attend a session on issues of race, class, and gender, run by a panel of peers representing different cultures, socioeconomic backgrounds, sexual orientations, and physical disabilities. Efforts to improve diversity have been helped

"They strive to make each class interesting and helpful for students."

by the school's decision to emphasize high school transcripts, rather than the SAT, as a measure of achievement and potential in the admissions process. There are no merit or athletic scholarships; admissions decisions are need-aware.

Ninety-nine percent of Conn College students live on campus, where "there's something for everyone." Dorms house students of all ages, and are run by seniors who apply to be "house fellows." Roommates tend to be well matched because incoming students complete a three-page questionnaire about personal habits before coming to campus, says a sophomore. "All of the dorms are beautiful, comfortable, and well-maintained," says a freshman. "The amenities are plenty, and the facilities are spectacular." Among the seven specialty houses are Earth House (environmental awareness), the Abbey House co-op (where students cook their own meals), and houses dedicated to substance-free living, quiet lifestyles, and international languages. The campus dining hall offers traditional main courses, as well as "fast food, stir-fry, pizza, pasta, vegetarian options, deli sandwiches, salad bars, and an ice cream bar," says a psychology major.

Conn's dance and drama departments are superb, and it's not uncommon for dancers to take time off to study with professional companies.

Because Conn lacks a Greek system, most activities—including co-ed intramural sports—revolve around the dorms, which sponsor weekly keg and theme parties. "There is a strong sense of community that goes with a residential college," confirms a sophomore. Also keeping students busy are movie nights, comedy shows, student productions, and dances—sometimes with out-of-town bands and DJs. The alcohol policy falls under the honor code, so those under 21 can't imbibe at the campus bar, and students take that prohibition seriously. Conn is helping to revitalize New London, where defense-contractor General Dynamics and drug maker Pfizer both have operations. Students volunteer at

"Conn attracts a certain kind of 'up–for–anything' attitude person."

the local schools, aquarium, youth community center, and women's center; a college van makes it easy to get to and from work sites. When students get the urge to roam, the beaches of Mystic and other shore towns are 20 minutes from campus, and the Mohegan Sun casino is also very close. Trains go to Providence, Rhode Island, New York City, or Boston, while Vermont and upstate New York offer camping, hiking, and skiing. Conn's traditions include October's Camelympics, which pit dorms against each other in a 24-hour marathon of games from Scrabble to Capture the Flag; the winter Festivus ("the festival for the rest of us"); and Floralia, an all-day music festival the weekend before spring finals, recently headlined by the Dave Matthews Band.

Because Conn lacks a Greek system, most activities—including co-ed intramural sports—revolve around the dorms, which sponsor weekly keg and theme parties.

The Conn Camels compete in NCAA Division III, and men's ice hockey games against NESCAC rival Wesleyan draw crowds, though students say Conn doesn't really have any true athletic rivals. A T-shirt brags that Conn football has been undefeated since 1911; of course, Conn—as a former women's college—has never had a team. More than 700 students participate in intramural and club sports, the most popular of which are ultimate Frisbee and broomball. Between classes or at the end of the day, all students may use the natatorium's pool and fitness center, and the rowing tanks and climbing walls at the field house.

On its friendly campus, Conn College fosters strong student/faculty bonds and takes pride in its ability to challenge—and trust—students, both in and out of the classroom. But getting the most out of the Conn experience depends on being receptive—and on taking initiative, students say. "If you want to be a number, Conn is not the place for you."

If You Apply To ➤ **Conn College:** Early decision: Nov. 15, Jan.1. Regular admissions: Jan.1. Financial aid: Feb. 1. Guarantees to meet demonstrated need. Campus interviews: recommended, evaluative. Alumni interviews: optional, evaluative. SATs: optional. SAT IIs: required (any two or the ACT). Accepts the Common Application (and Connecticut College Supplement). Prefers electronic applications. Essay question: Common Application; why Connecticut College?

Cooper Union

30 Cooper Square, New York, NY 10003

As college costs skyrocket, so does the popularity of Cooper Union's free education in art, architecture, and engineering. Expect Ivy-level competition for a place in the class here. Instead of a conventional campus, Cooper Union has the East Village—which is quite a deal.

Some say the best things in life are free. In most cases, they're probably wrong. But not in the case of the Cooper Union for the Advancement of Science and Art. If you manage to get accepted into this technical institute, you get a full-tuition scholarship and some of the nation's finest academic offerings in architecture, engineering, and art. With cool and funky Greenwich Village in the background and rigorous studying in the forefront, college life at Cooper Union may seem to be faster than a New York minute. Whatever the pace, though, no one can deny that a CU education is one of the best bargains around—probably the best anywhere. The only problem is that its acceptance rate is comparable to the Ivies.

The school was founded in 1859 by entrepreneur Peter Cooper, who believed that education should be "as free as water and air." With hefty contributions from J.P. Morgan, Frederick Vanderbilt, Andrew Carnegie, and various other assorted robber barons, the school was able to stay afloat in order to recruit poor students of "strong moral character." Today, students must pay a few hundred dollars for nonacademic expenses, but tuition is still free.

In place of a traditional collegiate setting are three academic buildings and one dorm plunked down in one of New York's most eclectic and exciting neighborhoods. The stately brick art and architecture building is a beautiful historic landmark. Built of brick and topped by a classic water tower, the dorm blends right in with the neighborhood. The Great Hall was the site of Lincoln's "Right Makes Might" speech and the birthplace of the NAACP, the American Red Cross, and the national women's suffrage movement. Wedged between two busy avenues in the East Village, Cooper Union offers an environment for survivors.

The academic climate is intense: A junior says, "You don't know the meaning of stress until you've been through Cooper." A senior agrees, "Many students fail or drop out or take a year off." The curriculum is highly structured, and all students

Website: www.cooper.edu
Location: Urban
Total Enrollment: 982
Undergraduates: 928
Male/Female: 66/34
SAT Ranges: V 610–690
    M 600–760
Financial Aid: 44%
Expense: Pr $
Phi Beta Kappa: No
Applicants: 2,301
Accepted: 13%
Enrolled: 74%
Grad in 6 Years: 85%
Returning Freshmen: 97%
Academics: ✑ ✑ ✑ ½
Social: ☎
Q of L: ★ ★ ★
Admissions: (212) 353-4120
Email Address:
    admissions@cooper.edu

Strongest Programs:
Architecture
Fine Art
Engineering

must take a sequence of required courses in the humanities and social sciences. The first year is devoted to language and literature and the second to the making of the modern world. In some special circumstances, students are allowed to take courses at nearby New York University and the New School for Social Research. The nationally renowned engineering school, under the tutelage of the nation's first female engineering dean, offers

"You don't know the meaning of stress until you've been through Cooper."

both bachelor's and master's degrees in chemical, electrical, mechanical, and civil engineering as well as a bachelor of science in general engineering. "Architecture and engineering are the most acclaimed, but then, these occupations are more mainstream, and graduates get big money and success," reflects an art major. "It's harder to measure success in the art school." The art school offers a broad-based generalist curriculum that includes graphic design, painting, sculpture, photography, and video but is considered weak by some students. The architecture school, in the words of one pleased participant, is "phenomenal—even unparalleled." Requirements for getting into each of these schools vary widely—each looks for different strengths and talents—hence the differences in test-score ranges.

The Cooper Union library is small (90,000 volumes), but contains more than 100,000 graphic materials. Classes are small and, with a little persistence, are not too difficult to get into. Professors are engaging and accessible. "One of the best aspects of this college is that everyone is taught by full professors," reports a junior. A professional counseling and referral service is available, as is academic counseling, but the school's small size and its rigorously structured academic programs set the classes the students take and eliminate a lot of confusion or decision making. Reactions to career counseling vary between "horrible" from an art major to "excellent" from an engineering major. Students "tend to talk to other students, recommending or insulting various classes and profs around registration time," notes a senior.

Strong moral character is no longer a prerequisite for admission, but an outstanding high school academic average most certainly is. Prospective applicants should note, however, that art and architecture students are picked primarily on the basis of a faculty evaluation of their creative works. For engineering students, admission is based on a formula that gives roughly equal weight to the high school record, SAT scores, and the SAT IIs in mathematics and physics or chemistry.

"The students here tend to be incredibly driven people," says one student, noting that some classmates have spent "literally 24 hours at the drafting desk." Sixty percent of the students are from New York State, and more than half of those grew up in the city. Most are from public schools, and many are the first in their family to attend college. Thirty-five percent of the students are from minority groups, most of them Asian Americans (20 percent); 5 percent are African American, and 9 percent are Hispanic. One student attests that diversity is not an issue at CU: "We are a racially mixed student body that stays mixed. There's no overt hostility and rare self-segregation." The campus is home to ethnically based student clubs, but, according to one student, membership is not exclusive: "In

"The students here tend to be incredibly driven people."

other words, you can be white and be a member of Onyx—a student group promoting black awareness." According to one senior, CU is a very liberal place: "If you can't accept different kinds of people, you shouldn't come here." For students who demonstrate financial need, help with living expenses is available.

Students love the dorm, a 15-story residence hall that saves many students from commuting into the Village or cramming themselves into expensive apartments. It is noteworthy that housing here is guaranteed only to freshmen. The facility is composed of furnished apartments with kitchenettes and bathrooms complete with showers or tubs and is "in great condition and well maintained," states one resident.

The school was founded in 1859 by entrepreneur Peter Cooper, who believed that education should be "as free as water and air."

The curriculum is highly structured, and all students must take a sequence of required courses in the humanities and social sciences.

The heart of the Village, with its abundance of theaters, art galleries, and cafes, is just a few blocks to the west.

A less enraptured dweller notes, "Rooms are barely big enough to fit a bed, a table, and a clothes cabinet." Still, each apartment does have enough space for a stove, microwave, and refrigerator. So you can cook for yourself or eat at the unexciting but affordable school cafeteria or at one of the myriad nearby delis and coffee bars.

The combination of intense workload and CU's location means that campus social life is limited, though the administration hopes the dorm will promote more on-campus social activities. "Many students will say that Cooper social life is dead," notes a junior. "In many ways they are right." On the other hand, as one senior puts it, "The East Village is a great place to be young, with tons of bars and culture." About 10 percent of the men and 8 percent of the women belong to professional societies. Drinking on campus is allowed during school-sponsored parties for adult students—otherwise, no alcohol on campus. But as one student puts it, "This is New York; one can be served anywhere."

The intramural sports program is held in several different facilities in the city, and the games are popular. Students organize clubs and outings around interests such as soccer, basketball, skiing, fencing, Ping-Pong, classical music, religion, and drama. And of course, the colorful neighborhood is ideal for sketching and browsing. McSorley's bar is right around the corner, the Grassroots Tavern is just down the block, and nearby Chinatown and Little Italy are also popular destinations. The heart of the Village, with its abundance of theaters, art galleries, and cafes, is just a few blocks to the west. The Bowery and SoHo's galleries and restaurants are due south; all of midtown Manhattan spreads to the northern horizon.

"If you can't accept different kinds of people, you shouldn't come here."

Getting into Cooper Union is tough, and once admitted, students find that dealing with the onslaught of city and school is plenty tough as well. But most students like the challenge. "The workload, living alone in New York, and the administrative policies force you to act like an adult and take care of yourself," explains a senior. Surviving the school's academic rigors requires talent, self-sufficiency, and a clear sense of one's career objectives. Students who don't have it all can be sure that there are six or seven people in line ready to take their places. That's quite an incentive to succeed.

Overlaps

NYU, Columbia, Cornell University, RISD, MIT, Carnegie-Mellon

If You Apply To ➤

Cooper Union: Early decision: Dec. 1. Financial aid and housing: May 1. All students receive full-tuition scholarships. Campus interviews: optional, informational. No alumni interviews. (Portfolio Day strongly recommended for art applicants.) SATs: required. SAT IIs: required for engineering (math and physics or chemistry). Apply to particular program. Essay question: varies by school.

Cornell College

600 First Street West, Mount Vernon, IA 52314-1098

The one-course-at-a-time model is Cornell's calling card. Cornell's main challenge: trying to lure students to rural Iowa. With a student body of about 1,000, Cornell lavishes its students with personal attention. Though primarily a liberal arts institution, Cornell has small programs in business and education.

Website:
www.cornellcollege.edu

Cornell College attracts a unique type of student who seeks an intense yet flexible, self-designed program and a liberal, progressive atmosphere in which to solidify strict habits and routines. If you're not satisfied with easy answers, don't mind heading to the rural Midwest, and want loads of personal attention while you focus on one class, Cornell College may be worth a look. "This has been hands-down the best experience of my life," says a junior. "I plan on sending my children to Cornell someday."

Aside from its distinctive schedule, Cornell has one of only two U.S. college or university campuses listed in its entirety on the National Register of Historic Places. The majestic bell tower of King Chapel offers an unparalleled view of the Cedar River valley. A pedestrian mall runs through campus, and other recent additions include an eight-lane, all-weather track and an outdoor amphitheater. A state-of-the-art, 280-seat theatre has been added to the performing arts building.

"Courses are completed in 18 days, so each class is pretty intense."

Cornell awards the Bachelor of Arts (BA) in more than 40 academic majors, as well as an extensive group of preprofessional programs, resulting in a variety of degrees. The college also offers the Bachelor of Special Studies, pursued by 16 percent of students, which administrators describe as "an opportunity which permits students to combine courses in an individualized fashion and to broaden or deepen their studies beyond the traditional framework of the Bachelor of Arts. The BSS has no general education requirements and no restrictions as to either the number of courses that may be taken in any one department or the level of such courses, or even that a student complete traditional course work." The One-Course method can be problematic here because "it is difficult to learn a language in three and a half weeks," a junior says. Another sums it up. "OCAAT: You either love it or hate it." Also available is the Bachelor of Music degree, which attracts just 2 percent of students. An interdisciplinary major in ethnic studies has been added recently.

Block scheduling makes it easier for some students to graduate early; others use the flexibility to finish with a double major. If that sounds intimidating, it can be. But administrators say it also improves the quality of Cornell's liberal arts education by helping students acclimate to the business world, where "what needs to be done needs to be done quickly and done well." The One-Course method also helps in academic advising—with grades every four weeks, signs of trouble are quickly apparent. Block scheduling does have drawbacks, though. "Courses are completed in 18 days, so each class is pretty intense," says a junior biology and Spanish major.

The most popular programs are psychology, economics and business, English, biochemistry and molecular biology, and art, and other strong options include philosophy, politics, and theater. Weaker areas include German, Latin American studies, and medieval and early modern studies. Classes usually

"Students here are independent thinkers."

have fewer than 25 students, and most have around 15; you won't find graduate students at the lectern, since Cornell doesn't hire them. "The quality of teaching at Cornell is unmatched," raves a junior, though this praise is not extended to visiting profs. "All classes I have taken by visiting professors have been extremely tedious."

When Mount Vernon gets too small, Cornell students may choose from study programs around the world, such as Marine Science Research in the Bahamas, Advanced Spanish in Spain and Mexico, or Greek Archaeology in Greece. They can also spend a semester at sea or in one of 36 countries through the Associated Colleges of the Midwest* consortium. During the short breaks between courses, students can take advantage of symposia, Music Mondays, carnivals, and athletic events. The school's Cole Library is also the town of Mount Vernon's public library, one of only two such libraries in the country.

"Students here are independent thinkers, ambitious and passionate leaders, and motivated students," a math and biology double major says. "You have to be to get through a semester's worth of material in 18 days!" One of Cornell's biggest challenges is drawing students to its rural Iowa campus, but administrators are

On campus, the fraternities draw 30 percent of the men and 32 percent of the women, though they are not associated with national Greek systems.

(Continued)
Location: Rural
Total Enrollment: 1,160
Undergraduates: 1,160
Male/Female: 45/55
SAT Ranges: V 560–680
  M 550–680
ACT Range: 23–29
Financial Aid: 70%
Expense: Pr $
Phi Beta Kappa: Yes
Applicants: 1,653
Accepted: 66%
Enrolled: 29%
Grad in 6 Years: 65%
Returning Freshmen: 79%
Academics: ✍ ✍ ✍
Social: ☎ ☎ ☎
Q of L: ★ ★ ★
Admissions: (319) 895-4477
Email Address: admissions
  @cornellcollege.edu

Strongest Programs:
Art
Biology
Education
English
Philosophy
Psychology
Politics
Theater

doing well on that score: only a third of students are homegrown. African Americans represent 4 percent of the student population, Hispanics 3 percent, and Asian Americans 1 percent. Three percent hail from other nations. Politically, Cornell leans liberal. "The social and political issues on campus reflect the issues of society as a whole," another student explains. Merit scholarships are offered to qualified students; there are no athletic awards.

Ninety percent of students live on campus. "About half the dorms are comfortable and well maintained," says a student. "The other half are an embarrassment." Nearly 10 percent of students participate in living/learning communities. The largest co-ed residence hall, which houses Cornell's living/learning communities, has been renovated. Everyone eats together in the Commons. Sodexho Marriott runs the kitchen, and the food is "an ongoing struggle," a junior says. Vegetarian options are always available.

On campus, the fraternities draw 30 percent of the men and 32 percent of the women, though they are not associated with national Greek systems. Some dorms and floors are substance-free; on the others, only students over 21 may have alcohol. "They treat underage drinkers as children and aim to punish but this does not reduce underage drinking," says a senior. Mount Vernon itself is "small, but very welcoming," says a physical education major. "Many students knit themselves into the community, whether it be with jobs downtown or going to the local

> "It's a tradition that we throw rolls of toilet paper out on the floor after we score our first basket."

churches," a junior agrees. Students either love the town's idyllic pace—a few local bars, an acclaimed restaurant, some funky shops, and a lot of peace, quiet, and safety—or long for more excitement. The latter is available in Cedar Rapids (home of archrival Coe College) or Iowa City (home to the University of Iowa), each less than half an hour away. Chicago is less than four hours away.

On the field or on the court, Cornell's competition with Coe "is intense, and the entire student body is involved," says one student, especially when it comes to football or basketball. "Every time a rival plays us at home in men's basketball, it's a tradition that we throw rolls of toilet paper out on the floor after we score our first basket," adds another—perhaps to show how Cornell plans to "clean up" its opponent. The baseball field was upgraded recently, and the Meyer Strength Training Facility helps student athletes improve their speed and power. Cornell teams compete in the Iowa Intercollegiate Athletic Conference and have been to the NCAA men's basketball tournament five times. Women's tennis and soccer are also competitive. Forty-four percent of students, and some faculty and staff, participate in intramural sports.

Cornell offers a top-notch education and a supportive community—if you can take the bitter winters and relative isolation of rural Iowa. And while the curriculum requires students to focus on just one course at a time, "It is the most practical and interesting way to learn," declares a senior. Indeed, life here allows them to explore just about anything.

If You Apply To ➤

Cornell: Early action: Dec. 1. Regular admissions: Mar. 1. Meets demonstrated need of 50%. Campus interviews: recommended, informational. Alumni interviews: optional, informational. SATs or ACTs: required. SAT IIs: optional. Accepts the Common Application and electronic applications. Essay question: significant experience, achievement, risk, or ethical dilemma; important issue; your room; influential character; topic of your choice.

Cornell University

Ithaca, NY 14850

Cornell's reputation as a pressure cooker comes from its preprofessional attitude and "we try harder" mentality. Spans seven colleges—four private and three public— and tuition varies accordingly. Strong in engineering and architecture, world-famous in hotel administration. Easiest Ivy to get into.

Cornell has a long tradition for being the lone wolf among the Ivy League universities. So it should come as no surprise that Cornell has taken another huge step away from its Ivy League counterparts by announcing its intention to become the finest research university for undergraduate education in the nation. Cornell's president recently unveiled a $400 million, 10-year plan to improve undergraduate education by combining education and research and having all freshmen live in the same residential area. And the mixture of state and private, preprofessional, and liberal arts at one institution provides a diversity of students rare among America's colleges.

Aside from the great strides in undergraduate education, Cornell also has its stunning campus to lure students to upstate New York. Perched atop a hill that commands a view of both Ithaca and Cayuga lakes, the campus is breathtakingly scenic; or, as the saying goes, "Ithaca is gorges." Ravines, waterfalls, and parks border all sides of the school's campus. The Cornell Plantation, more than 3,000 acres of woodlands, natural trails, streams, and gorges, provides space for walking, picnicking, or contemplation.

At the undergraduate level, Cornell has four privately endowed colleges: architecture, art, and planning; arts and sciences; engineering; and hotel administration. "Architecture and hotel administration are two of the best programs in the country," confirms one senior. Cornell is also New York State's land grant university. Therefore, Cornell operates three other colleges under contract with New York State: agriculture and life sciences, human ecology, and the school of industrial and labor relations (ILR). Thirty-seven percent of the students in these state-assisted colleges are New York State residents who pick up their Ivy League degrees at an almost-public price (as tuition at these schools is slightly steeper than SUNY rates).

The College of Arts and Sciences boasts considerable strength in history, government, and just about all the natural and physical sciences. The English program has turned out a number of renowned writers, including Toni Morrison, Thomas Pynchon, and Richard Farina. Foreign languages, required for all arts and sciences students, are also strong, and the performing arts, mathematics, and most social science departments are considered good. Among the state-assisted units, the agriculture college is solid and a good bet for anyone hoping to make it into a veterinary school (there's one at Cornell with state support). The School of Hotel Administration is top-notch, and ILR is also well-regarded. Hotel's wine tasting course draws students from across the university. Human ecology is among the best in the nation in human service related disciplines, and the Department of Applied Economics and Management offers an undergraduate business major. The Johnson Museum of Art, designed by IM Pei, has been rated as one of the 10 best university museums in America. Students enjoy the $22 million theater arts center, designed specifically for undergraduates.

Student/faculty relations at Cornell are a mixed bag, but for the most part students do have a lot of respect for their professors. A junior muses, "It's always entertaining when your professor wrote the book for the class." First-year courses in the sciences and social sciences are generally large lectures, though many are taught by "charismatic profs" who try to remain accessible. The largest course on campus,

Website: www.cornell.edu
Location: Small city
Total Enrollment: 19,518
Undergraduates: 13,625
Male/Female: 51/49
SAT Ranges: V 630–730
  M 660–760
ACT Range: 28–32
Financial Aid: 49%
Expense: Pr $ $ $
Phi Beta Kappa: Yes
Applicants: 20,822
Accepted: 29%
Enrolled: 50%
Grad in 6 Years: 92%
Returning Freshmen: 96%
Academics: ✍ ✍ ✍ ✍ ✍
Social: ☎ ☎ ☎ ☎
Q of L: ★ ★ ★
Admissions: (607) 255-5241
Email Address:
  admissions@cornell.edu

Strongest Programs:
Biology
Physical Science/Math
English
Economics/Business
Architecture
Hotel Administration
Industrial and Labor Relations
Agriculture

Psych 101, packs in more than 1,000, but students report that scintillating lectures make it a well-loved rite of passage. Some undergrads complain that it is difficult to get into popular courses unless you are a major. "Students joke about the 'Big Red Tape' because it can be daunting trying to navigate Cornell's bureaucracy," explains one senior. The administration, however, is hoping that the decision to make undergraduates a priority will improve most of the problems in the lower-level courses. For now, first-year students do have access to senior faculty members through mandatory First-Year Writing Seminars.

The Fund for Educational Initiatives gives professors money to implement innovative approaches to undergraduate education, which have included a visual learning laboratory and a course on electronic music. Cornell was early among universities to add women's studies to the curriculum and continues to be an innovator, with programs in Asian American studies and by offering its students programs like its Sea Semester.*

Cornell academics are demanding and foster an intensity found on few campuses. "The easiest Ivy to get into; the toughest to get out of," quips one student. "Students at Cornell take college life from the work hard, play hard perspective." Another student adds, "While it's hard to ignore the prestige of attending an Ivy League school, most Cornellians seem to have checked their attitudes at the door."

"It's always entertaining when your professor wrote the book for the class."

Eighty-five percent of Cornell students ranked in the top tenth of their high school class, so those who were the class genius in high school should be prepared for a struggle to rise to the top. To cope with the anxieties that the high-powered atmosphere creates, the university has one of the best psychological counseling networks in the nation, including an alcohol-awareness program, peer sex counselors, personal-growth workshops, and EARS (Empathy, Assistance, and Referral Service).

The library system is superb. Cornell students have access to more than seven million volumes, 63,000 journals, and 1,000 networked resources in the 20 libraries comprising Cornell's library system. The resources are available to a wide range of students, faculty, staff, and, in some cases, the community. Also, the digital initiative has made many resources and collections available online. Within the beautiful underground Carl A. Kroch Library, students study in skylit atriums and reading rooms and move about the renowned Fiske Icelandic Collection and the Echols Collections, the finest Cambodian collection on display.

Cornell offers over 4,000 courses in a wide range of pursuits. A co-op program is available to engineering students, and Cornell-in-Washington, with its own dorm, is popular among students from all seven undergraduate colleges. Students looking to study abroad can choose from more than 200 programs and universities throughout the world, including those in Indonesia, Belgium, Ireland, and Nepal. Research opportunities are outstanding at Cornell, and students can take part in some of the most vital research happening in the nation. Recent research findings include solving the mystery of how Jupiter's rings are formed and discovering a new technology to make computer software less vulnerable to bugs.

Prospective students apply to one of the seven colleges or schools through the central admissions office, and admissions standards vary by school. City slickers, country folk, engineers, and those with an artsy flair all rub shoulders here. Fifty-five percent of Cornell's students are out-of-staters; another 8 percent are foreign. African Americans and Hispanics each account for 5 percent of the student body, and Asian Americans comprise 16 percent. Cornell offers many workshops and discussion groups aimed at increasing tolerance. The state-assisted schools draw a large number of in-staters, as well as students from New Jersey, Pennsylvania, and New England, while arts and sciences and engineering draw from the tristate metropolitan New

York City area, Pennsylvania, Massachusetts, and California. Whatever their origin, students seem self-motivated and studious. Upon graduation, 35 percent of Cornell students take jobs, and 34 percent continue to graduate and professional schools.

Cornell is need-blind in admissions and guarantees to meet the demonstrated need of all accepted applicants, but the proportion of outright grants—as opposed to loans that must be repaid—in the financial aid package varies depending on how eager the university is to get you to enroll. The Cornell Installment Plan (CIP) allows students or their parents to pay a year's or semester's tuition in monthly interest-free installments. The university also takes pride in its Cornell Tradition fellowship programs that recognize leadership and academic excellence, work and service, and research and discovery. Tradition students may receive up to $4,000 in loan replacement each year in addition to other financial rewards.

> **"The easiest Ivy to get into; the toughest to get out of."**

North Campus residence halls are the home of all freshmen. A few students are housed in two dorms on the edge of Collegetown, the blocks of apartments and houses within walking distance of the campus. There are dorms devoted to everything from ecology to music, and cultural houses include the International Living Center, Latino Living Center, Ujamaa Residential College, and Akwe:kon, a program house focusing on American Indian culture (the only facility of its kind in the nation). "Just make sure that you turn in your housing form on time," warns a senior psychology major. Also available are a small number of highly coveted suites—six large double rooms with kitchens and a common living area. Just under half of Cornell's students live in university housing, though many try their luck in Collegetown, where demand keeps the housing market tight and rents high. Cornell's food service is reputedly among the best in the nation. There are eight residential meal plan dining halls that function independently, so, one student enthuses, "the food is fantastic and incredibly diverse." Milk products and some meats come right from the agriculture school, and about twice a semester a cross-country gourmet team—the staff of a famous restaurant—prepares its specialties on campus.

Despite the intense academic atmosphere—or maybe because of it—Cornell social life beats most of the other Ivies hands down. Once the weekend arrives, local parties and ski slopes are filled with Cornell students who have managed to strike a balance between study and play. Collegetown bars offer good eats and drinks, but those under 21 are barred. "Social life can be found on or off campus. There is always something going on and I've always felt socially stimulated." With 28 percent of men and 22 percent of women pledging, fraternities and sororities also play a significant role in the social scene. Big events include Fun in the Sun (a day of friendly athletic competition), Dragon Day (architecture students build a dragon and parade it through campus), and Springfest (a concert on Libe Slope). Students celebrate the last day of classes—Slope Day—by hanging out at Libe Slope. There are also innumerable concerts and sporting events. In addition, there are more than 600 extracurricular clubs ranging from a tanning society to a society of women engineers.

> **"Most Cornellians seem to have checked their attitudes at the door."**

Hockey is unquestionably the dominant sport on campus (the chief goal being to defeat Harvard), and camping out for season tickets is an annual ritual. Cornell boasts the largest intramural program in the Ivy League; it includes more than a dozen sports, including 100 hockey teams organized around dorms, fraternities, and other organizations. The aforementioned "four seasons of Ithaca" can make walking to class across the vast and hilly campus challenging, but with the first snow of the winter, "traying" down Libe Slope becomes the sport of choice for hordes of fun-loving Cornellians. Ithaca boasts "wonderful outdoor enthusiast stores," says one student. It also hosts Greek Peak Mountain for nearby skiing,

Despite the intense academic atmosphere—or maybe because of it—Cornell social life beats most of the other Ivies hands down.

City slickers, country folk, engineers, and those with an artsy flair all rub shoulders here.

Cayuga Lake for boating and swimming, and lots of space for hiking and watching the clouds roll by.

One junior sums it up: "Cornell is a world unto itself. It has the academic rigor of an Ivy League university, but the size and diversity of a large state school. The city and surrounding geography is beautiful and unique, which fosters a love of this place that is unmatched." Like most other Ivy League universities, Cornell is a premiere research institution with a distinguished faculty and outstanding academics. What sets it apart is the university's willingness to stray from the traditional Ivy League path. Cornell University is a pioneer in the world of education, and students unafraid to blaze their own trails will feel at home here.

If You Apply To ➤   **Cornell:** Early decision: Nov. 1. Regular admissions: Jan. 1. Financial aid: Jan. 1. Housing: May 1. Campus interviews: not available. Alumni interviews: recommended, informational (varies by program). SATs or ACTs: required. SAT IIs: required (varies by program). Accepts Common Application and electronic application. Essay question: Common Application. Apply to individual programs or schools.

University of Dallas

Irving, TX 75062

Bulwark of academic traditionalism in Big D. Despite being a "university," UD has fewer than 1,200 undergraduates. Except for the Business Leaders of Tomorrow program, the curriculum is exclusively liberal arts. The only outpost of Roman Catholic education between Loyola of New Orleans and University of San Diego.

While many universities around the nation have reexamined their Eurocentric core curriculums, the University of Dallas—the best Roman Catholic college south of Washington, D.C.—remains proudly dedicated to fostering students in "the study of great deeds and works of Western civilization."

UD's 744-acre campus occupies a pastoral home in a Dallas suburb on top of "the closest thing this region has to a hill." Texas Stadium is right across the street. A major portion of the campus is situated around the Braniff Mall, a landscaped and lighted gathering place near the Braniff Memorial Tower, the school's landmark. The primary tone of the buildings is brown, and the architecture, as described by one student, is "post-1950s, done in brick, typical Catholic-institutional." While it may not be a picture-perfect school, it does have a beautiful chapel and a state-of-the-art science building. The Art Village has five buildings and each art major has private studio space.

At UD, students choose from more than 20 majors and 27 concentrations. English is the most popular major, followed by business leadership, biology, politics, and psychology. The business leadership program draws on the learning opportunities in the Dallas Metroplex to develop responsible and competent managers through classroom and industry experi-

"The environment provides a good balance of drive and priority."

ences. Political philosophy is also a popular major, although students claim that most courses tend to be slanted toward the conservative side. Students can also participate in an accelerated BA-to-MBA five-year plan. Premed students are well served by the biology and chemistry programs, and 80 percent of UD graduates go on to grad school.

Appropriately for a Roman Catholic school, much of the focus is on Rome, where most of the sophomore class treks every year. The unique and intense program focuses on the art and architecture of Rome, the philosophy of man, classical literature, Italian, and the development of Western civilization. The Rome semester is part of UD's four-semester Western Civilization core curriculum. Included in the core are philosophy, English, math, fine arts, science, American civilization, Western civilization, politics, and economics, as well as a serious foreign language requirement. Two theology courses (including Scripture and Western Theological Tradition) are also required of all students. Those inclined toward the sciences may take advantage of the John B. O'Hara Chemical Science Institute, which offers a hands-on nine-week summer program to prepare new students for independent research and earns them eight credits in chemistry. The classics program has been divided into two majors, classical philology and classics, and the new "Politics, Policy, and Leadership" major emphasizes the application of political principles in the world of domestic policy.

> **"The students here are diverse, intelligent, and well-rounded."**

Students report the academic pressure and the workload can be intense. "The courses are intellectually stimulating and academically challenging," says a junior. A classmate adds, "Students are supportive of each other. The environment provides a good balance of drive and priority." The university uses no teaching assistants, and professors are easy to get to know. "Most of the professors are well loved and well known," a senior says. An English major adds, "They endeavor not just to relay information to us but to shape the whole person." Seventy-eight percent of classes taken by freshmen have 25 or fewer students.

"The students here are diverse, intelligent, and well-rounded," a student says. Two-thirds of UD students are Roman Catholic, and many of them choose this school because of its religious affiliation. Fifty-five percent are from Texas, and 15 percent of the student body is Hispanic; Asian Americans account for 5 percent and 2 percent are African American. UDers tend to lean to the right politically, with topics such as abortion and homosexuality leading the discussions. "We are clearly a conservative school," admits one senior. "However, dialogue is open and students are respectful" of opposing viewpoints. UD offers various merit scholarships, averaging $9,435, but no athletic scholarships.

Tradition and religion govern conduct. Students under 21 who don't reside at home with their parents must live on campus in single-sex dorms, "where visitation regulations are relatively strict," one student reports. As for the dorms, they "aren't huge," but are "clean and updated," according to one student. The most popular dorms are Jerome (all-female) and Madonna (all-male). At Gregory, the dorm reserved for those who like to party, the goings-on are less than saintly. In addition to a spacious and comfortable dining hall with a wonderful view of North Dallas, there is the Rathskeller, which serves snacks and fast food (and great conversation). A car is a must for off-campus life because there is virtually no reliable public transportation in the Dallas/Fort Worth Metroplex.

> **"We are clearly a conservative school."**

The University of Dallas is unusual for a Texas school in that its entire population does not salivate at the sight of a football or basketball.

With no fraternities or sororities at UD, the student government sponsors most on-campus entertainment. Three free movies a week, dances, and visiting speakers are usually on the agenda. Church-related and religious activities provide fulfilling social outlets for a good number of students. "Social life here mostly takes place on campus or at the apartment complex across the street," says a freshman. Annual events include Mallapalooza, a spring music festival, and Groundhog, a party on Groundhog's Day weekend. Then there's Charity Week in the fall, when the junior class plans a week's worth of fund-raising events. Each year, students dread Sadie Hawkins Day and the annual Revenge of the Roommate dance—dark nights of the

The Rome semester is part of UD's four-semester Western Civilization core curriculum.

soul, each. The university can be vigorous in enforcing restrictive drinking rules and, as a result, it is difficult for a minor to drink at campus events. "There are very strict penalties for anyone who breaks the rules," a student warns.

Students describe Irving as "a suburb, just like any other," but the Metroplex offers almost unlimited possibilities, including a full agenda for bar-hopping on Lower Greenville Avenue, about 15 minutes away. The West End and Deep Ellum offer a taste of shopping and Dallas's alternative music scene. And for the more adventurous, Austin and San Antonio aren't too far away.

The University of Dallas is unusual for a Texas school in that its entire population does not salivate at the sight of a football or basketball. But baseball and men's basketball are competitive, as is women's soccer, which reached the NCAA Division III tournament three of the last four years. Intramural sports are well organized and very popular, with volleyball attracting the most players. Chess is also a favored activity.

UD appeals to those students who pride themselves on being the "philosopher kings of the 21st century," but whose roots go back to the Roman thinkers of an earlier era. The mix of religion and liberal arts can serve a certain breed of students well. In the words of one senior, "Come here to have fun, build sincere friendships, work hard, and graduate with a deep sense of your place in the Western cultural tradition."

Overlaps

Austin College, Trinity University, Baylor, Rice, Notre Dame, Southern Methodist

If You Apply To ➤

Dallas: Rolling admissions. Early action: Nov. 1, Dec. 1. Financial aid: Mar. 1. Housing: Apr. 1. Meets full demonstrated need of 20%. Campus interviews: recommended, informational. No alumni interviews. SATs or ACTs: required. SAT IIs: optional. Accepts the Common Application and electronic applications. Essay question: describe a character in fiction, a historical figure, or a creative work (as in art, music, science, etc.) that has had an influence on you, and explain that influence; favorite joke; your activities; why you want to attend UD and what you expect to gain.

Dartmouth College

6016 McNutt Hall, Hanover, NH 03755

The smallest Ivy and the one with the strongest emphasis on undergraduates. Traditionally the most conservative member of the Ivy League, it has been steered leftward in recent years. Ivy ties notwithstanding, Dartmouth has more in common with places like Colgate, Williams, and Middlebury. Great for those who like the outdoors.

Website: www.dartmouth.edu
Location: Rural
Total Enrollment: 5,561
Undergraduates: 3,991
Male/Female: 50/50
SAT Ranges: V 670–770
    M 680–780
ACT Range: 29–34
Financial Aid: 50%
Expense: Pr $ $ $
Phi Beta Kappa: Yes
Applicants: 12,756
Accepted: 17%
Enrolled: 49%

Unlike the other seven members of the Ivy League, which trace their roots to Puritan New Englanders or progressive Quaker colonists, Dartmouth College was founded in 1769 to educate Native Americans. The student body has always been the smallest in the Ancient Eight, and the school's focus on undergraduate education differentiates Dartmouth from its peers, though the college does offer graduate programs in engineering, business, and medicine. Dartmouth's campus is probably the most remote of the Ivies, and its winters may be the coldest, with the possible exception of those at Cornell. The Big Green compensates with the warmth of community, keeping sophomores on campus for the summer to build closeness, and using intensive language training and study abroad to emphasize the importance of global ties.

While tradition is revered at Dartmouth, the school also continues to grow, change, and evolve. That's the legacy of former president James Freedman, hired in the 1980s to "lead Dartmouth out of the sandbox" and make it a more hospitable place (especially for women and minorities) with a more scholarly feeling. Freedman said he wanted students "whose greatest pleasures may not come from the camaraderie of classmates, but from the lonely acts of writing poetry, or mastering the cello, or solving mathematical riddles, or translating Catullus." (The reference to the Latin poet Catullus may have been the president's little joke, as Catullus wrote erotic, sometimes obscene, verse on topics such as his passion for his mistress Lesbia, a boy named Juventius, and the sexual excesses of Julius Caesar.) Today's Dartmouth hardly resembles the college of yesteryear, whose rowdy Greek scene inspired the movie *Animal House*. The school still attracts plenty of hiking and skiing enthusiasts, and the most popular extracurricular organization is the Dartmouth Outing Club, the oldest collegiate outdoors club in the nation. But these days, students are just as likely to join a hip-hop dance group called Sheba, or to spend a vacation doing Dartmouth-sponsored community service in South America.

Set in the "small, Norman Rockwell town" of Hanover, New Hampshire, which is bisected by the Appalachian Trail, Dartmouth's picturesque campus is arranged around a quaint green. It's bounded by the impressive Baker Library at one end, and by the college-owned Hanover Inn at the other. Architectural styles range from Romanesque to postmodern, but the dominant theme is copper-topped Colonial frame. The nearest big city, Boston, is two hours away, but major artists like Itzhak Perlman routinely visit Dartmouth's Hopkins Center for the Creative and Performing Arts, adding a touch of culture. A spate of new construction has been completed recently, including the MacLean Engineering Sciences Center, Kemeny Hall (which houses the mathematics department), Haldeman Center, and the McLaughlin Cluster (a 342-bed undergraduate residence hall complex).

Dartmouth's status as a member of the Ivy League means academic excellence is a given. But that doesn't mean students have complete freedom when it comes to choosing courses. First-years must take a seminar that involves both independent research and small-group discussion; about 75 are offered each year. The seminars "ensure that every student's writing is up to par," while supplementing the usual introductory survey courses available in most disciplines, and offering a glimpse of the self-directed scholarship expected at the college level. Students must also demonstrate proficiency in at least one foreign language. And they must take three world cul-

> **"The whole campus is one giant, interactive, constant learning community."**

ture courses (one non-Western, one Western and one Culture and Identity), and 10 courses from various distribution areas: the arts; literature; systems and traditions of thought, meaning, and value; international or comparative studies; social analysis; quantitative and deductive science; natural or physical science and technology; or applied science. In addition, Dartmouth has a senior culminating activity—a thesis, public report, exhibition, seminar, production, or demonstration—which allows students to pull together work done in their major, with a creative and intellectual twist of their own. "The whole campus is one giant, interactive, constant learning community," says a senior.

Though Dartmouth students work hard, the climate is far from cutthroat. "There is a very cooperative learning atmosphere," an economics major says. "Competition is extremely rare," a senior adds. "Helping peers is common." The most popular majors are economics, government, history, psychology, and English. The languages are also well-regarded, and students benefit from the Intensive Language Model developed by Professor John Rassias. Computer science offerings are among the best in the nation, thanks in no small part to the late John Kemeny, the former Dartmouth president who coinvented time-sharing and the BASIC language. Indeed, computing is a way of life here; well before the Internet, all of Dartmouth's dorm rooms were networked, and students would "Blitzmail" each other to set up meetings, discussions, or meals. Now, a Voice over Internet Protocol phone system also lets students use their laptops as telephones, which means long-distance calls home are free.

(Continued)
Grad in 6 Years: 95%
Returning Freshmen: 98%
Academics: 🏛🏛🏛🏛🏛
Social: 🍷🍷🍷🍷🍷
Q of L: ★ ★ ★
Admissions: (603) 646-2875
Email Address: admissions .office@dartmouth.edu

Strongest Programs:
Biological Sciences
Computer Science
Engineering
Economics
Languages
Psychological and Brain
  Sciences

Dartmouth attracts outdoorsy, down-to-earth students who develop extremely strong ties to the school—and each other—during four years together in the hinterlands.

Professors get high marks at Dartmouth, perhaps because of the school's focus on undergraduates. The isolated location also helps; faculty make a conscious choice to teach here, leaving behind some of the distractions afflicting their peers at more urban schools. "Dartmouth is a rare school that is an undergraduate institution first," says a religion major. "Teaching is clearly a big part of that commitment to excellence." Sixty-four percent of classes taken by freshmen have 25 or fewer students. The Presidential Scholars Program offers one-on-one research assistantships with faculty, and the Senior Fellowship Program enables 10 to 12 students a year to pursue interdisciplinary research projects, and to pay no tuition for their final term. The Women in Science Project encourages female students to pursue courses and careers in science, math, and engineering, with mentors, speakers, and even research positions for first-years. The Montgomery Fellowships bring well-known politicians, writers, and others to campus for periods ranging from a few days to several months, while the Visionary in Residence program invites notable thinkers to campus to share their talents and insights.

"Helping peers is common."

The school's most notable eccentricity is the Dartmouth Plan, or "D Plan"—four 10-week terms a year, including one during the summer. Students must be on campus for three terms during the freshman and senior years, and also during the summer after the sophomore year, but otherwise, as long as they're on track to graduate, they can take off whenever they wish. More than half of the students use terms away for one of Dartmouth's 44 study abroad programs, where they may focus on the classics in Greece or the environment in Zimbabwe; other students pursue part-time jobs, internships, or independent travel. The college also participates in the Twelve College Exchange* and the Maritime Studies Program.*

Dartmouth went co-ed in 1972, and women now make up half of the student body. Ethnic minorities also comprise a substantial chunk, with African Americans making up 7 percent, Asian Americans 13 percent, and Hispanics 6 percent. Another 3 percent of the students are Native Americans. "Whether a student's passion is organic chemistry or organic farming—or both!—the students at Dartmouth love what they do," says a senior. "Everyone is incredibly dedicated and passionate." Admissions are need-blind, even for students who get in off the waitlist, and the school guarantees to meet each student's demonstrated need for all four years. The college has added about $4

"Dartmouth is a rare school that is an undergraduate institution first."

million in scholarship funds over the past few years, enabling students to borrow less for their education. It has also sold tax-exempt bonds through a state authority to help fund low-interest loans for students and their families. Dartmouth is trying to lure more middle-class students by offering bigger grants to those whose parents earn less than $60,000 a year. No merit or athletic scholarships are awarded; the Ivy League prohibits the latter.

Eighty-two percent of Dartmouth students live on campus in one of more than 30 dorms, which have been grouped into 11 clusters to help create a sense of community. "The dorms are spacious and comfortable and very well-maintained," says an environmental studies major. "You can live in a single, double, triple, or quad, with anywhere from one to three rooms of different shapes and sizes." First-year students may choose freshman-only housing, or dorms where all classes live together. Sophomores may get squeezed during the housing lottery, but because of the D Plan, people are always coming and going; it may be easier to find a new room or roommate than at schools on the semester system. The new residence halls will likely ease any housing crunches. Dining facilities are open until 2:30 a.m., for those needing sustenance during late-night study sessions. Seniors may move off campus, into group houses, and many choose to do so. Safety is not a big concern here. "Safety and Security [S&S] is a constant presence around the clock," a student says.

Dartmouth's Greek system attracts 37 percent of the men and women; it's "big, but not exclusive—all parties are open to everyone," says a junior. The Greeks have become less of a force in campus social life because of the Student Life Initiative, a steering committee that developed more-rigorous behavior standards. Parties and kegs must also be registered, and the houses where they're being held are subject to walk-throughs by college safety and security personnel. Alcohol policies are aimed at keeping booze away from those under 21. Indeed, recognizing that the nickname of its hometown has long been "Hangover," Dartmouth was one of the first schools to develop a counseling and educational program to combat alcohol abuse. And students say a campus bar called the Lone Pine Tavern is more likely to be the scene of Scrabble and chess tournaments than hardcore drinking games. Popular road trips include Montreal or Boston, for a dose of bright lights and the big city, or the White Mountains for camping. (Dartmouth has a 27,000-acre land grant in the northeast corner of New Hampshire, where cabins may be rented for five dollars a night.) Still, "the social life revolves almost 100 percent around the college itself," a senior says. "Very little occurs off campus."

"I was sold on Dartmouth because it is steeped in tradition," says one senior. Weekends include traditions such as a 75-foot-tall bonfire at homecoming, and Winter Carnival, which includes ski racing at the college's bowl, 20 minutes away, as well as snow-sculpture contests, and partiers from all over the Eastern seaboard. Spring brings mud as the snow slowly melts, and also Green Key weekend, which one student calls "an excuse to drink under the guise of community service." (The school is striving for 100 percent participation in community service by graduation, "and we're close," says one student. "I've traveled twice to rural Nicaragua on a Dartmouth-sponsored, student-organized crosscultural education and service program; we provide medical assistance to villagers and construct clinics, compostable latrines, and organic farms.") During the summer term, the entire sophomore class heads to the Connecticut River for "Tubestock," a day of lazy, floating fun in the sun.

Love of the outdoors at Dartmouth extends to varsity athletics. The women's ice hockey team is a perennial powerhouse, and 19 different teams have either won an Ivy/ECAC Championship, been nationally ranked, or qualified for an NCAA/national championship within the past five years. Hanover residents support the basketball teams, and few Dartmouth students miss a trip to Cambridge for the biennial Dartmouth–Harvard football game. The school's sports facility boasts a 2,100-seat arena, a 4,000-square-foot fitness center, and the only permanent three-glass-wall squash court in North America. The $3 million Scully-Fahey Field offers an Astroturf surface for soccer, field hockey, and other team sports. Three-quarters of the student body participate in 40 intramural sports. Nonathletes beware: Dartmouth does have a non-timed swimming test and a physical education requirement for graduation; you can fulfill the latter with classes such as fencing or ballet, or participation in a club or intramural sport.

"Dartmouth is really a place for a go-getter," advises one student. The college attracts outdoorsy, down-to-earth students who develop extremely strong ties to the school—and each other—during four years together in the hinterlands. It seems as if every other grad has a title like deputy assistant class secretary, and many return to Hanover when they retire, further cementing their bonds with the college, and driving local real-estate prices beyond the reach of most faculty members. You'll have to be made of hardy stock to survive the harsh New Hampshire winters. But once you defrost, you'll be rewarded with lifelong friends, and a solid grounding in the liberal arts, sciences, and technology.

Professors get high marks at Dartmouth, perhaps because of the school's focus on undergraduates.

"Dartmouth is really a place for a go-getter."

During the summer term, the entire sophomore class heads to the Connecticut River for "Tubestock," a day of lazy, floating fun in the sun.

Overlaps
Brown, Harvard, Princeton, Stanford, Yale

Dartmouth: Early decision: Nov. 1. Regular admissions: Jan. 1. Financial aid: Feb. 1. Housing: Jul. 1. Guarantees to meet full demonstrated need. Campus and alumni interviews: optional, evaluative. SATs or ACTs: required (including optional ACT Writing Test). SAT IIs: required (any two). Accepts the Common Application and prefers electronic applications. Essay question: significant experience, achievement, risk or ethical dilemma; issue of personal, local, national, or international concern; influential person, fictional character, historical figure, or creative work; what you would bring to diversity of college; topic of your choice.

Davidson College

P.O. Box 7156, Davidson, NC 28035

Has always been styled as "The Dartmouth of the South." Goes head-to-head with Washington and Lee (VA) for honors as the most selective liberal arts college below the Mason-Dixon Line. At just under 1,700 students, it is slightly bigger than Rhodes and Sewanee and slightly smaller than W&L.

Website: www.davidson.edu
Location: Small town
Total Enrollment: 1,683
Undergraduates: 1,678
Male/Female: 50/50
SAT Ranges: V 640–730
   M 640–710
ACT Range: 28–31
Financial Aid: 35%
Expense: Pr $ $
Phi Beta Kappa: Yes
Applicants: 4,258
Accepted: 27%
Enrolled: 40%
Grad in 6 Years: 92%
Returning Freshmen: 96%
Academics: ✐ ✐ ✐ ✐ ½
Social: ☎ ☎ ☎
Q of L: ★ ★ ★ ★
Admissions: (800) 768-0380
Email Address:
   admission@davidson.edu

Strongest Programs:
Biology
Psychology
English
Political Science
Theatre
Chemistry
International Studies
History

Davidson College boasts the Southern tradition and gentility of neighbors like Rhodes and Sewanee, with the athletic and academic prowess more common to Northern liberal arts powerhouses such as Dartmouth and Middlebury. Often overlooked because of its small size and sleepy location, Davidson is one of the most selective liberal arts colleges below the Mason-Dixon Line. The school's size, honor code, and strong interdisciplinary, international, and preprofessional programs distinguish Davidson from many of its contemporaries. "Davidson offers one of the best undergraduate experiences and is the liberal arts school of the South," says a senior economics major.

Located in a beautiful stretch of the North Carolina Piedmont, Davidson's wooded campus features Georgian and Greek Revival architecture. The central campus is designated as a National Arboretum, and college staff lovingly maintain a collection of the woody plants that thrive in the area. In fact, the arboretum serves as an outdoor laboratory for students, with markers identifying the varieties of trees and shrubs. Davidson retains its original quadrangle, which dates from 1837, plus two dorms and literary society halls built in the 1850s.

Davidson's honor code allows students to take exams independently and to feel comfortable leaving doors unlocked. "Because of an honor code that works, Davidson students are able to walk around campus feeling safe and can leave their belongings anywhere without worrying that they will be stolen," says a senior. Every entering freshman agrees to abide by the code, and all work submitted to professors is signed with the word "pledged." Core requirements

"Students are self-motivated and very hardworking."

include one course each in fine arts, literature, and history, and two courses each in religion and philosophy, natural sciences and math, and social sciences. Students must also take classes with significant writing and discussion in the first year to satisfy the composition requirement. Four physical education classes are required, as is proficiency in a foreign language equal to three semesters of work. Many requirements, including in-depth or comparative studies of another culture, may be met through the two-year interdisciplinary humanities program.

Davidson's academic climate is rigorous, but not grueling. "It's not competitive between students," says a psychology major, "but academics are tough." The most popular majors are history, biology, English, political science, and psychology. Professors are highly lauded for being friendly and accessible. "The professors are extremely well-prepared and top rate," a senior says. A classmate adds, "Their love of teaching shines through." Students also report that profs have ample office hours and are available for help outside of the classroom.

For those whose academic interests lie outside the mainstream, Davidson's Center for Interdisciplinary Studies allows students to develop and design their own majors, with faculty or on their own. Environmental studies majors may apply to the School for Field Studies to spend a month or a semester studying environmental issues in other countries or to work and conduct research at Biosphere 2. Aspiring diplomats can tap into the Dean Rusk Program for International Studies, named for the Davidson alumnus who served as Secretary of State to Presidents Kennedy and Johnson. The South Asia Studies program focuses on India, Pakistan, Bangladesh, Sri Lanka, Nepal, and Bhutan; study abroad is also available in countries from France, Germany, and England to Cyprus and Zambia, and more than 65 percent of the students go abroad. A 3–2 engineering program is available with five larger universities. A Sloan Foundation grant

"Housing is way above par."

helps Davidson integrate technology into the liberal arts, with courses such as From Petroleum to Penicillin, and Sex, Technology, and Morality. On campus, class size is restricted; you won't find a room other than the cafeteria with more than 50 students.

Although most Davidson students come from affluent Southern families, the college is attracting more students from the Northeast and nearly half hail from outside the Southeast. Many students are Presbyterian, and the school has ties to the Presbyterian Church. Six percent of the student body is African American, with 4 percent Hispanic and 2 percent Asian American. "Students are self-motivated and very hardworking," says one student. Hot-button issues range from "what to do with the empty frat houses to global poverty," according to one political science major. Davidson lures top students with merit scholarships, and its financial aid packages have no loans. One hundred eighty-five athletic scholarships are available.

Ninety-one percent of Davidson's students live on campus in co-ed or single-sex dorms. "Housing is way above par," a senior says. Freshmen are housed together in two five-story halls and they eat in Vail Commons, where the "food is great—all you can eat, and lots of options, though it's hard to be a vegan at Davidson." Upperclassmen may live in the dorms, off campus, or in college-owned cottages on the perimeter of campus, which hold about 10 students each. Davidson is the only college in the country that still provides complementary laundry service in the dorms—a throwback to its days as an all-male school when it was the only way to keep undergrads clean. Seniors get apartments with private bedrooms. Most upperclassmen take meals at one of the 11 eating clubs—six for men, four for women; one coed eating club has opened for the adventurous. These Greek-like groups have their own cooks and serve meals family style.

The eating clubs are the center of social life on campus, as is the Alvarez College Union. All but one are in Patterson Court, which freshmen are not allowed to enter for the first three weeks of school. The dues charged by these clubs cover meals, as well as parties and other campuswide events. The fraternities, which claim 40 percent of Davidson's men, are not much different from the eating clubs, and freshmen simply sign up for the group they want to join on Self-Selection Night, with no "rushing" allowed. There are no sororities on campus. And even if you don't join up, don't despair; Davidson requires that most parties—"at least two per weekend" at the eating clubs—be open to the entire community. Alcohol policies comply with North Carolina law; officially, no one under 21 can be served. "Policies are tied into the Honor Code, so they are enforced," a student says.

Davidson's five-day freshman orientation includes a regatta, scavenger hunt, and the Freshman Cake Race. The program helps introduce students to the cozy town of Davidson, which has coffee shops and cafes, nearby Lake Norman for sailing,

The eating clubs are the center of social life on campus, as is the Alvarez College Union.

Davidson's honor code allows students to take exams independently and to feel comfortable leaving doors unlocked.

Most Davidson students come from affluent Southern families.

swimming, and waterskiing, and plenty of volunteer opportunities. The equally quaint town of Cornelius is only a 10-minute drive from campus, so it's a common destination for dinner and a movie or a relaxed night out. When those diversions grow old, North Carolina's largest city, Charlotte, is just 20 miles away, with clubs and other attractions. A car definitely helps here, as Myrtle Beach and skiing are several hours from Davidson, in different directions. Intramural and club sports are varied and popular. The men's basketball squad brought home the 2005 Southern Conference championship.

> "Sure, we live in a bubble, but we are all genuinely happy to be here."

Despite its North Carolina location, Davidson has the look and feel of a New England liberal arts college and continues to attract top students to its charming neck of the woods. "Sure, we live in a bubble," admits a physics major, "but we are all genuinely happy to be here." From study abroad and independent research to a strawberries-and-champagne reception with the college president for graduating seniors, students here combine tradition with forward thinking, to make great memories, friends, and intellectual strides.

Overlaps

UNC–Chapel Hill, Duke, Wake Forest, Washington & Lee, University of Virginia, Vanderbilt

If You Apply To ➤

Davidson: Early decision: Nov. 15. Regular admission and second early decision: Jan. 2. Does not guarantee to meet full demonstrated need. Campus and alumni interviews: optional, informational. SATs or ACTs: required. SAT IIs: recommended. Accepts the Common Application. Essay question: varies year to year.

University of Dayton

Dayton, OH 45469-1323

Among a cohort of second-tier Midwest Roman Catholic institutions that includes Duquesne, Xavier (OH), U of St. Louis, and DePaul and Loyola of Chicago. Drawing cards include business, education, and the social sciences. The city of Dayton is not particularly enticing and UD's appeal is largely regional.

Website:
admission.udayton.edu
Location: City outskirts
Total Enrollment: 10,569
Undergraduates: 6,913
Male/Female: 50/50
SAT Ranges: V 520–620
M 540–650
ACT Range: 23–28
Financial Aid: 60%
Expense: Pr $
Phi Beta Kappa: No
Applicants: 8,675
Accepted: 80%
Enrolled: 29%
Grad in 6 Years: 79%
Returning Freshmen: 87%

Anyone who thinks college students of today subscribe to postmodern cynicism ought to take a peek at Dayton, where optimism and Christian charity are alive and well. "If you used one word to describe UD students, it would be friendly," a senior says. "Everyone on campus is very welcoming. We smile and say 'hi' to people we don't know and hold the doors open for each other." There's good reason for the cheery disposition: Applications recently hit a record level for the fourth year running.

Founded by the Society of Mary (Marianists), Dayton continues to emphasize that order's devotion to service. The majority of UD students volunteer their time in 30 different public service areas. Christmas on campus, where UD students "adopt" local elementary students for a night of crafts, games, and a visit with Santa, is one of the most student-involved activities. "We bring in about 1,000 inner-city Dayton school kids and walk them around campus which is transformed into a winter wonderland," says one student.

The parklike campus is on the southern boundary of the city, secluded from the traffic and bustle of downtown. The more historic buildings make up the central core of the campus and blend architectural charm with modern technological conveniences. A new $25 million fitness and recreation complex, dubbed the "RecPlex," opened in 2006. It houses classrooms, a climbing wall, basketball and volleyball courts, a juice bar, and many other sports-related facilities. A new 400-bed residence hall with amenities such as a post office and credit union is in the heart of campus. ArtStreet, a $9 million arts-centered living/learning complex, includes six two-story townhouses and five loft apartments above performance spaces, art studios, the campus radio station, and a recording studio.

"The academic climate is competitive," says a junior. "The professors push you to the next level to ensure you will have the competitive edge when you reach the workforce." UD students take full advantage of the strong offerings found in communication, mechanical engineering, psychology, accounting, and management—the most popular majors. Weaker offerings include theatre and physical education. Entrepreneurship is the fastest-growing major in the School of Business Administration and participating sophomores are given $3,000 loans to start their own businesses, with any profits going to charity. The program includes mentoring by local entrepreneurs and courses taught by entrepreneurs and PhD faculty.

Dayton's general education requirements include courses in five "domains of knowledge": arts, history, philosophy and religion, physical and life sciences, and social sciences. Students must also complete the humanities base and a thematic cluster. Faculty members in the College of Arts and Sciences have developed a 12-course core curriculum that satisfies the general education requirements through an interdisciplinary program that clumps mandatory classes into sequences pertinent to academic disciplines. The prestigious Berry Scholars program includes seminars, study abroad opportunities, service and leadership projects, and a major independent research project. Students with at least a 1300 combined SAT score or a 30 ACT score and who place in the top 10 percent of their graduating class or have a 3.7 GPA may join the University Honors Program, which provides guest speakers in small classes and requires an honors thesis project.

"If you used one word to describe UD students, it would be friendly."

The Interdepartmental Summer Study Abroad Program is a popular ticket to the world's most exciting cities, while the Immersion Program in Third World countries is much praised by participants. The University of Dayton Research Institute (UDRI) is one of the nation's leading university-based research organizations. All students purchase a notebook computer upon entering UD. Students speak enthusiastically about the quality of teaching, stating that professors are "passionate" about their courses. "The professors bend over backwards for the students in order to make sure they achieve," a communication major says. New students unsure of their majors can take advantage of First Year Experience, a structured program where students are required to meet with their advisors once a week.

"The students at UD are very welcoming," says a student. "It's a 'sweatshirt' campus where you don't have to dress to impress." A classmate adds, "There is a strong sense of community here." Two-thirds of Dayton's students are from Ohio, and minorities make up just 7 percent of the student population: 4 percent are African American, 2 percent are Hispanic, and 1 percent are Asian American. The Task Force on Women's Issues, the Office of Diverse Student Populations, and an updated sexual-harassment policy demonstrate UD's growing sensitivity to campus issues. More than 80 percent of incoming students rank in

"It's a 'sweatshirt' campus where you don't have to dress to impress."

the top half of their high school class, and the school is becoming more selective. Dayton's 91 athletic scholarships go to athletes in six men's sports and seven women's sports. There are also many merit awards, averaging $5,159.

Seventy-nine percent of students are campus residents; those who live off campus generally live adjacent to it. "All first-year students live in one of four traditional residence halls," a senior explains. "Sophomores opt to live in suites or apartments, while upperclass students live in university-owned houses, apartments or townhouses."

(Continued)
Academics: ✍ ✍ ✍
Social: ☎ ☎ ☎ ☎ ☎
Q of L: ★ ★ ★
Admissions: (937) 229-4411
Email Address:
admission@udayton.edu

Strongest Programs:
Education
Communication
Marketing
Finance
Psychology

UD students take full advantage of the strong offerings found in communication, mechanical engineering, psychology, accounting, and management.

Students say the dorms are well maintained yet somewhat outdated. Sophomores have an opportunity to live in Virginia Kettering, a residence hall whose amenities evoke luxurious apartments. The food in the dining halls that dot the campus is generally well-received; one dining hall is located in one of the first-year dorms, another in the sophomore complex, the third is centrally located in the student union, and a fourth is in Marianist Hall.

The student neighborhood (aka "the Ghetto") serves as a sort of continuous social center. A lit porchlight beckons party-seeking students to join the weekend festivities. Because the university owns most of the properties, a 24-hour campus security patrol keeps watch over the area. "You never really need to leave campus to have a memorable time," says a junior. The more adventurous weekend excursions are trips to Ohio State University, Ohio University, Indianapolis, and the restaurants, shops, and sports arenas in Cincinnati. But the best road trip is the Dayton-to-Daytona trip after spring finals, a 17-hour trek that draws loads of students each year. Partying on campus is commonplace and controlled, but parties have sized down due to the university's enforcement of the 21-year-old drinking age. "There is a 'no keg policy' that most students ignore," admits one student. Still, "There is a 'three strikes, you're out' policy regarding drinking," which students say makes them cautious. Greek organizations draw 15 percent of UD men and 19 percent of the women, with all chapters playing an active role in the community service and social life.

More important, though, are sports, particularly basketball. "The Xavier game is our biggest rivalry and attracts the attention of most students," says a student. The football team, which is Division I-AA, plays in the Pioneer Football League and has had 29 consecutive winning seasons. UD is in the Atlantic 10 for Division I athletics in all other sports. Women's soccer has won three of the last five regular season titles and four tournament crowns.

"You never really need to leave campus to have a memorable time."

In 2003, 2004 and 2005, the women's volleyball team won the Atlantic championship and played in the NCAA championships. When students aren't cheering, they can participate in an intramural program that offers 45 sports. Other activities in the city include a minor-league baseball team, an art institute, an aviation museum, and a symphony and ballet in the new Schuster Performing Arts Center. Two large shopping malls are also easily accessible.

The success of Dayton's attempts to provide its students with a high quality of life and a sense of cohesiveness is reflected in many of the students' comments about the terrific social life and family-like atmosphere among both students and faculty. As a midsize university where the undergraduates come first, Dayton has managed to maintain an exciting balance of personal attention, academic challenge, and all-American fun. "We're a community. We live together, learn together and socialize together," a senior says. "UD isn't just a school. It's a family!"

Overlaps

Miami (OH), Ohio University, Xavier (OH), Ohio State, Notre Dame, Marquette

If You Apply To ➤ **Dayton:** Rolling admissions: Jan. 1 (priority). Financial aid: Mar. 1. Housing: May 1. Campus interviews: recommended, informational. Alumni interviews: optional, informational. SATs or ACTs: required. SAT II: optional. Requires electronic applications. Essay question: personal statement on an achievement, experience, or risk and its impact on you.

Deep Springs College

Deep Springs, CA Mailing address: Dyer, NV 89010-9803

Picture 26 Ivy League–caliber men living and learning in a remote desert outpost—that's Deep Springs. DS occupies a handful of ranch-style buildings set on 50,000 acres on the arid border of Nevada and California. Most students transfer to highly selective colleges after two years.

If the thought of spending countless hours under the fluorescent lights of the classroom makes you grimace, you may consider getting your hands dirty at Deep Springs College. This two-year institution doubles as a working ranch. Bonding is easy here, and students enjoy a demanding and individualized education supplemented by the challenges and lessons of ranch life. Both, it seems, demand the same things: hard work, commitment, and pride in a job well done. Deep Springs College students are also rewarded for their efforts in other ways: tuition is free and so is room and board. Students pay only for books, travel, and personal items; the average cost of one year at Deep Springs is $500. Still, students do have a few complaints, says a sophomore: "Loneliness, saddle-chafing, and bad coffee."

Many of the men who work, study, and live at this college have shunned acceptance at Ivy League schools to embrace the rigors of a truly unique approach to learning. Deep Springs students tend to be of the academic Renaissance-man variety with wide-ranging interests in many fields. Almost all transfer to the Ivies or other prestigious universities after their two-year program, and 70 percent eventually earn a PhD or law degree.

California's White Mountains provide a stunning backdrop for the Deep Springs campus, set on a barren plain 5,200 feet above sea level, near the only water supply for miles around. The campus is an oasis-like cluster of trees and a lawn with eight recently renovated, ranch-style buildings that were built from scratch by the class of 1917. Deep Springs is 28 miles from the nearest town, a thriving metropolis known as Big Pine, population 950. The focal point of campus is the Main Building, a venerable ranch-style building with wide eaves that includes a computer room and offices. Dorm rooms are now housed in the new, spacious Student Residence. Faculty houses and the dining facilities are grouped around the circular lawn a few yards away, and the trappings of farm life surround the tiny settlement. The college has 170 acres under cultivation, mostly with alfalfa, and an assortment of barnyard animals. Once threatened with extinction (thanks to meager financial resources), the college launched a capital campaign that generated $18 million in six years. True to its practical spirit, the school used much of the money to enhance facilities and put itself back on track to a long future.

Founded in 1917 by an industrialist who made a fortune in the electric-power industry, Deep Springs today remains true to its charter "to combine taxing practical work, rigorous academics, and genuine self-government." Ideals of self-government, reflectiveness, frugality, and community activity have weathered more than 85 years of a grueling academic climate. "What is most remarkable about the academic climate is how passionate the students are," says a freshman. "Because we choose what classes are offered and work together with the professors to plan the syllabi, students are always interested and engaged." Academic learning is the primary activity here, but students are also required to perform 20 hours per week of labor, which can include everything from harvesting alfalfa to cooking dinner. When asked which are

> **"What is most remarkable about the academic climate is how passionate the students are."**

Sidebar

Website:
  www.deepsprings.edu
Location: Rural
Total Enrollment: 26
Undergraduates: 26
Male/Female: 100/0
SAT Ranges: V 700–800
  M 685–800
Financial Aid: N/A
Expense: Pr $
Phi Beta Kappa: No
Applicants: 160
Accepted: 11%
Enrolled: 86%
Grad in 6 Years: N/A
Returning Freshmen: 93%
Academics: ✑ ✑ ✑ ✑ ½
Social: ☎
Q of L: ★ ★ ★
Admissions: (760) 872-2000
Email Address:
  apcom@deepsprings.edu

Strongest Programs:
Humanities
Liberal Arts
Environmental Studies
Philosophy

the best majors, one wit exclaims, "Dairy is the most popular, but many students swear by irrigation."

Students' input carries a lot of weight at this school. The student body committees are an essential part of the self-governance pillar at DS. There are four committees: the student-run Applications Committee, which is made up of nine students, a faculty member, and a staff member, the Curriculum Committee, the Review and Reinvitations Committee, and the Communications Committee. They help choose the college's faculty and even elect two of their own to be full-voting members on the board of trustees. They play a determining role in admissions and curricular decisions. And they abide by a Spartan community code that bans all drugs, including alcohol, and forbids anyone to leave Deep Springs Valley (the 50 square miles of desert surrounding the campus) while classes are in session, except for medical visits and college business. Lest these rules sound unnecessarily strict, keep in mind that these are all decided on and enforced by the student body, not the administration.

> **"Deep Springers are brilliant, bitter, and achingly cool."**

Like almost everything else about it, Deep Springs has an unorthodox academic schedule: two summer terms of seven weeks each, and a fall and spring semester of 14 weeks each. Between seven and 10 classes are offered every term. The faculty consists of three "permanent" professors (they sign on for two years, but can stay for up to six), plus an average of three others who are hired on a temporary basis to teach for a term or two. The quality of particular academic areas varies as professors come and go. "Teaching quality changes as faculty change, which is frequently," says one student, "but, generally, the quality is excellent." Although the curriculum is altered yearly, students predict that the humanities will always remain superior. The students control the academic program and quickly replace courses—and faculty—that do not work out. "This year we had several incredible artists and poets teach," says one student. Currently, the only required courses are public speaking and composition.

> *Like almost everything else about it, Deep Springs has an unorthodox academic schedule: two summer terms of seven weeks each, and a fall and spring semester of 14 weeks each.*

Deep Springers aren't much for the latest conveniences, but computers have taken the campus by storm; whether provided by DS or brought from home, almost every student room has one, plus several in a common area. With class sizes ranging from two to 14, there is ample opportunity for close student/faculty interaction. Close living arrangements have fostered a kind of kinship between faculty and students. Students routinely visit their mentors in their homes, sometimes to confer on academic matters and sometimes to play soccer with their children.

Deep Springers can truly boast of being hand-picked to attend; of the approximately 200 applications received each year, only 11 or so students are accepted. Most DS students are from upper-middle-class families and typically rank in the top 3 percent of their high school class. "Deep Springers are brilliant, bitter, and achingly cool," quips a sophomore. Many Deep Springers are transplanted urbanites; the rest hail from points scattered across the nation or across the seas. Political leanings run the gamut, and there is diversity even among this small population: 76 percent of the student body is

> **"We try to understand what service means in a nuanced and original way."**

white, Asian Americans make up 20 percent, and Hispanics account for 4 percent. Local issues spark the most debate and include topics such as "the drinking policy, isolation, diversity, and the meaning of service," says a freshman.

Dorm selection and maintenance is entirely the responsibility of the students. "Dorms are gigantic and very comfortable. Some are full of old furniture and art. One is filled with Greek busts and another has a mural on the ceiling," says a student. "The nicest rooms I have ever seen; anywhere. Too nice for us actually—some students like to sleep out in the desert." Students all pitch in preparing the meals, from butchering the meat to milking the cows to washing the dishes. "It's always edible and often extraordinary," boasts a student. And what about security? "Unless a tractor runs over you, you're fine," says a freshman. A classmate adds, "Sometimes the bulls get loose."

They used to have one telephone line for the whole school; now they have six.

Social life can be a challenge. "The lack of women, alcohol, and nightlife can at times be frustrating," reports one student. And loneliness can be an issue. Still, "we're too busy with the three pillars—academics, labor, and self-governance—to have much of a social life." When the moon is full, students go out en masse in the middle of the night to frolic in the 700-foot-high Eureka Sand Dunes with Frisbees and skis. "We slide down the Eureka Valley sand dunes au naturel," says one student. As for "recent technological advancement," they used to have one telephone line for the whole school; now they have six. Perhaps the most popular social activity on campus is conversation over a cup of coffee in the dining hall, where the chatter is usually lively until the wee hours of the morning. Other common activities are road trips to nearby national parks, hikes in the nearby mountains, and horseback riding. The Turkey Bowl, the potato harvest, the two-on-two basketball tournament, and Sludgefest, an annual event involving cleaning out the reservoir, are only some of the time-honored Deep Springs traditions.

Critics of Deep Springs charge that DS cultivates arrogance and social backwardness among students who were too intellectual to be in the social mainstream during high school. They argue that students who come here are doomed to be misfits for life, citing a survey that shows many Deep Springers never marry. While that charge is debatable, even supporters of Deep Springs confess to a love-hate relationship with the college. Although the interpretations may vary, one common thread winds through the DS mission from application to graduation: training for a life of service to humanity.

"Unless a tractor runs over you, you're fine."

Perhaps more than any other school in the nation, Deep Springs is a community where students and faculty interact day-to-day on an intensely personal level. Though the financial commitment is small, the school demands an intense level of personal commitment. All must quickly learn how to get along in a community where the actions of each person affect everyone. "We are oriented towards serving humanity," says a freshman, "and we try to understand what service means in a nuanced and original way." Urban cowboys who dream of riding into the sunset are in for a rude awakening. For a select few, however, the camaraderie and soul-searching fostered in this tight-knit community can be mighty tempting—just stay clear of those bulls.

Overlaps

Harvard, Yale, University of Chicago, Columbia, Stanford, Swarthmore

If You Apply To ➤ **Deep Springs:** Regular admissions: Nov. 15. Campus interviews: required, evaluative. No alumni interviews. SATs: required. SAT IIs: optional. Essay questions: describe yourself; critical analysis of book or other work of art; and why Deep Springs?

University of Delaware

116 Hullihen Hall, Newark, DE 19716

Plenty of students dream of someday becoming Nittany Lions or Cavaliers–even Terrapins–but not many aspire to be Blue Hens. The challenge for UD is how to win its share of students without the name recognition that comes from big-time sports. Less than half the students are in-staters.

Aspiring engineers and educators, and practically everyone in between, can find something to delve into at the University of Delaware.

The University of Delaware is a public gem that boasts more than 124 solid academic programs, from engineering to education. Though lacking a big-time sports program, UD attracts its share of students who are looking for solid academics and a friendly atmosphere. It all adds up to "the small-school feel with the opportunities of a larger university," as one junior says.

Delaware's 1,000-acre campus has an attractive mix of Colonial and modern geometric buildings, set among one of the nation's oldest Dutch elm groves. The hub of the campus is a grassy green mall, flanked by classic Georgian buildings. Mechanical Hall has undergone a $4.6 million renovation and is now a climate-controlled art gallery and home to the Paul R.

"There is diversity, you just have to look below the surface to find it."

Jones collection. Hotel and restaurant management students benefit from a recently-opened Courtyard by Marriott right on campus, which doubles as a learning and research facility. A performing arts center is the newest campus addition.

Delaware's academic menu includes more than 125 majors, ranging from the liberal arts and sciences to more professional programs such as apparel design and fashion merchandising. To graduate, students must pass freshman English (critical reading and writing) and earn at least three credits of discovery-based or experiential learning, such as an internship, research, or study abroad. Other requirements vary by college; Delaware has seven colleges, and all except the College of Marine Studies award undergraduate degrees. Biological sciences and psychology are the most popular majors, followed by teacher education, English, and nursing. New majors include agriculture, finance, human services, marketing, natural resource management, sport management, and landscape design.

Engineering, especially chemical engineering, is one of UD's specialties, and the school benefits from the close proximity of DuPont, which developed Lycra spandex. The music department is another attraction, with a 350-member marching band and several faculty members holding impressive professional performance credits. UD created the nation's first study abroad program in 1923, and more than 70 study abroad programs are available on all seven continents. Journalism students can take a winter-term trip to Antarctica aboard a Russian icebreaker. Each year, about 600 UD students hold research apprenticeships with faculty members; in fact, the Carnegie Endowment Reinvention Center at SUNY–Stony Brook has called Delaware a "national model" for undergraduate research.

As UD has grown in popularity, academic standards have become more rigorous. Delaware routinely gets the highest number of nonresident applications for state-affiliated U.S. institutions. "In the last few years, Delaware has become more competitive and more expensive," says a senior. "Particularly, the out-of-state tuition has become a topic of dissent." About 500 new students enter the University Honors Program each year, which offers interdisciplinary colloquia, priority seating in "honors" sections of

"It's an awesome social life—lots of on–campus activities and off–campus parties."

regular courses, along with talented faculty, personal attention, and extracurricular and residence hall programming.

Forty percent of students at Delaware hail from the First State; many of the rest are from the Northeast. Minority enrollment has been increasing; 6 percent of the student body are African American, 4 percent are Hispanic, and 3 percent are Asian American. "There is diversity," says a senior, "you just have to look below the surface to find it." Merit scholarships average $4,400 each, and 200 athletic scholarships are offered. The university maintains a search program to ensure that deserving Delaware students are aware of the scholarships for which they are eligible.

Forty-seven percent of students live on campus, including all freshmen—except those commuting from home. After that, dorm housing is guaranteed and awarded by lottery, though many juniors and seniors move into off-campus apartments. Those who stay on campus find a range of accommodations: co-ed and single-sex halls with single and double rooms, as well as suites and apartment-style buildings. LIFE (Learning Integrated Freshman Experience) is a popular program that allows freshmen to live with others in the same major. Honors students also live together in designated residence halls. Campus dwellers must buy the meal plan, which offers "enough choice to accommodate pretty much every tastebud," says a senior.

When the weekend comes, Delaware students know how to let loose, though a ban on alcohol at campus parties has really taken things down a notch. "The three-strike policy works fairly well, especially in the freshman dorms," says a student. (First two strikes: fines and meetings. Third strike: suspension.) Fraternities attract 13 percent of the men and sororities 13 percent of the women. Greek groups often throw parties, but there are plenty of other options, from concerts and plays on campus to casual gatherings in friends' rooms or apartments. "It's an awesome social life—lots of on-campus activities and off-campus parties," says an English major.

Main Street, the heart of downtown Newark, "practically runs right through campus," one student says. "It's easy walking distance from anywhere, and there are tons of coffee shops, pizza places, restaurants, a movie theater, a bowling alley, bookstores, and shops—anything you could possibly want." For those seeking further excitement, New York, the Washington/Baltimore area, and Philadelphia are all within a two-hour drive. When the weather is warm, the beaches of Rehoboth and Dewey beckon, and in chilly months, the Pennsylvania ski slopes aren't too far. Mallstock is the annual spring bacchanal, bringing music and a carnival to the central campus green.

> "Now that our football team won the NCAA IAA championship, it is getting more popular."

Delaware's Blue Hens compete in Division I, and on Saturdays in the fall, watch out. "Football is big," says one student. "Now that our football team won the NCAA IAA championship, it is getting more popular. The cheerleaders are also very good." Tailgate picnics are popular before and after the game. The Blue Hens are also strong in field hockey, women's basketball, and men's lacrosse, and spectators also love to cheer for the soccer and women's crew squads. Delaware's sports center has space for 6,000 to cheer; intramural sports are popular.

Aspiring engineers and educators, and practically everyone in between, can find something to delve into at the University of Delaware. With a challenging and stimulating academic environment, an increasingly smart student body, a healthy social scene, and up-and-coming athletic teams, UD offers a blend of strengths that would make many schools envious—and leads to many happy Blue Hens.

Delaware's academic menu includes more than 125 majors, ranging from the liberal arts and sciences to more professional programs such as apparel design and fashion merchandising.

Overlaps

Penn State, University of Maryland, Rutgers, University of Connecticut, Boston University, James Madison

If You Apply To ➢

Delaware: Regular admissions: Jan. 15 (Dec. 15 for scholarships) Financial aid: Mar. 15. Housing: May 1. Does not guarantee to meet full demonstrated need. Campus and alumni interviews: optional, evaluative. SATs or ACTs: required. SAT IIs: recommended. Accepts the Common Application and electronic applications. Essay question: describe a life-changing experience; how your ethnic or cultural heritage has shaped your worldview; or use your imagination to surprise or entertain the admissions officers.

Denison University

Granville, OH 43023

Website: www.denison.edu
Location: Small town
Total Enrollment: 2,292
Undergraduates: 2,292
Male/Female: 44/56
SAT Ranges: V 570–660
  M 580–670
ACT Range: 25–29
Financial Aid: 44%
Expense: Pr $ $
Phi Beta Kappa: Yes
Applicants: 5,144
Accepted: 39%
Enrolled: 31%
Grad in 6 Years: 79%
Returning Freshmen: 91%
Academics: ✍ ✍ ✍ ½
Social: 🍸 🍸 🍸 🍸 🍸
Q of L: ★ ★ ★
Admissions: (740) 587-6276
Email Address:
   admissions@denison.edu

Strongest Programs:
Natural and Social Sciences
Philosophy
Theatre
Creative Writing
Geology
International Studies
Computer Science

Denison University, tucked into the "quaint, small, and beautiful" hamlet of Granville, draws "bright and laid-back, engaging and interesting, athletic, good-looking, generally privileged and ambitious students," says an immodest senior. Thanks to Denison's small size, there's ample opportunity to interact (and do research with) professors, and to form close relationships with peers, as everyone focuses on the liberal arts. "Our experience is a much more holistic one," an English major explains.

Denison's campus is set atop rolling hills in central Ohio. Huge maples shade the sloping walkways, which offer a panoramic view of the surrounding valley. And don't be surprised if you're reminded of New York's Central Park—Denison retained park architect Frederick Law Olmsted for its first master plan back in the early 1900s. The Georgian style of many buildings—redbrick with white columns—also evokes shades of New England and its private liberal arts colleges.

Denison's general education requirements are division-based and students take two first-year seminars, and then during their four years, two courses each in the fine arts, the sciences (one with lab), the social sciences, the humanities, and a foreign language. Some courses may be double-counted to fulfill oral communication and quantitative reasoning requirements. Students must also complete a course focusing on interdisciplinary and world issues. The environment is more collaborative than competitive. "Students tend to enjoy each other in class and aim for individual success," says a student.

Students say that some of the best majors are unique to Denison. The PPE major is effectively a triple major in philosophy, political science, and economics. The economics department is nationally ranked among undergraduate institutions. "Geology, astronomy, philosophy, and psychology offer a thinking person's alternative to the more technically demanding sciences," advises a senior. Denison's 350-acre biological reserve is a boon for environmental studies majors. The school

> **"Our experience is a much more holistic one."**

encourages students to pursue independent research or to work with faculty members on their projects, and gives 125 students summer stipends of $3300 plus housing and research support for supplies and travel each year. The Honors Program offers up to 50 courses per year, and also sponsors a "Chowder Hour," where students and faculty gather for an informal presentation while dining on a faculty member's culinary specialty.

Classes are small, and individual attention is the norm. "Professors here love teaching as much as research," a student says. "Their office doors are always open and they will drop their sandwich if you stop by during lunch in order to answer your academic or personal questions." When Granville gets too small, internships and off-campus studies are available through the Great Lakes Colleges Association* and the Associated Colleges of the Midwest.* Those considering a run for office may be interested in the Richard G. Lugar Program in American Politics and Public Service, which includes political science courses on campus and culminates in a House or Senate internship in Washington. (The Senator happens to be a Denison grad.) As a participant in the Denison Internship Program, formerly known as May Term, students select from more than 250 internships around the country.

Thirty-nine percent of Denison's population is homegrown, and 4 percent come from abroad. The Posse Program funds 10 full-tuition scholarships per class for multi-cultural student leaders from Chicago and Boston public high schools, which has helped boost diversity on campus. African Americans now constitute 6 percent of the student body, while Hispanics and Asian Americans combine for 6 percent. "The environment, social justice, and fair trade are the largest issues on campus. The percentage of liberal students on campus is relatively high despite the somewhat conservative and wealthy backgrounds of many students," explains a student. A senior agrees, "Social justice is a major concern for a lot of students at Denison. Students are always discussing current events and how they relate to what they are learning in class." Forty-nine percent of students receive merit scholarships, including the Wells, Dunbar, and Faculty Achievement awards, which cover full tuition. The average non-need-based award is more than $12,400, though there are no athletic scholarships.

> **"The percentage of liberal students on campus is relatively high despite the somewhat conservative and wealthy backgrounds of many students."**

Ninety-eight percent of Denison students live on campus; the only ones allowed to live elsewhere are those commuting from home. (One exception is the dozen Homesteaders, who live in three student-built solar-paneled cabins on a farm a mile away, and grow much of their own food.) Options range from singles to nine-person suites. Housing is guaranteed for four years, and the recent housing crunch has eased, thanks to the school's new apartment-style dorms. Crawford Hall, the freshman dorm, offers counseling, entertainment, and other assistance on the premises. Former fraternity houses now focus on common interests, such as service-learning (Morrow House) or honors courses (Gilpatrick House), while Stone Hall, Curtis East, and Smith Hall have been upgraded in recent years. Entrees in the two dining halls change daily.

When the weekend comes, "Denison does a wonderful job at providing a ton of events for students to choose from," says a psychology major. "The only problem I ran into was deciding which to attend." Options include movie screenings, comedy shows, and concerts at the student union, where guest artists have included Bela Fleck and the Flecktones, Edgar Meyer, Ani di Franco, Bobby McFerrin, George Clinton and the P-Funk All-Stars, and Joshua Bell, to name a few. The school also runs trips to the local mall and to Meijer, a Midwestern superstore similar to Wal-Mart, and the student government has even been known to hold fireworks shows. Alcohol policies are less punitive and more focused on ensuring safety. "Alcohol consumption is common and frequently high. Thankfully, everything at Denison is within walking distance, so it is far more likely to find drunken underclassmen nursing tacos or pizza in the student union than crumpled around a highway divider," says one student. Though the Greek

> **"Denison does a wonderful job at providing a ton of events for students to choose from."**

system is now nonresidential, 35 percent of the men and 40 percent of the women still join up. Students also look forward to two blowout parties each year—November's D-Day, which recently featured Third Eye Blind, and Festivus, on the last day of spring classes, headlined in the past by Dave Matthews Band, Guster, Rusted Root, and the Samples.

The tiny town of Granville "is a gorgeous slice of small town Americana," says one student. "Granville in spring is a sight to behold, with a seemingly endless supply of blooming, fragrant flowers," enthuses a senior. There are four churches on the corners of the town's main intersection, along with "two bars, a coffee shop, a bank, a greasy spoon restaurant, a library, and gift shops," says another student. Most stores and restaurants close by 8 p.m., though students appreciate the feeling of safety and security that results. The Denison Campus Association frequently sends

Ninety-eight percent of Denison students live on campus; the only ones allowed to live elsewhere are those commuting from home.

The economics department is nationally ranked among undergraduate institutions.

students into Granville and nearby Newark to provide tutoring, mentoring, and environmental cleanup. Popular road trips include Ohio University, Ohio State (in the state capital, Columbus), and Miami University in Oxford, Ohio. Pittsburgh, Cleveland, Cincinnati, and Dayton are also close by.

Denison students are enthusiastic supporters of the Big Red, and men's lacrosse games against Ohio Wesleyan always draw large crowds. The school competes in the North Coast Athletic Conference and has won the All-Sports Championship for nine years running. In 2005–06, 19 of 22 Denison sports programs finished in the upper half of the NCAC standings. In its history of participation in the NCAA, 39 students have received NCAA Post-Graduate Scholarship awards, the third largest number in Division III sports. Enthusiasm reaches a fever pitch during homecoming weekend, when the all-campus gala always includes a chocolate volcano.

Denison University is a school on the move, aiming to graduate independent thinkers who become active citizens of a democratic society. The school continues to value tradition—woe to the student who steps on the school seal in front of the chapel, for doing so will cause him to fail all his finals—while growing and evolving to emphasize academics and the life of the mind. A recent graduate says, "Denison is distinctive for its commitment to preparing its students to be life-long learners. The liberal arts tradition at Denison teaches students how to become analytical and critical thinkers—something valued in nearly any field of work."

Overlaps

Miami University (OH), College of Wooster, Kenyon, Ohio Wesleyan, Wittenberg, DePauw

If You Apply To ➤

Denison: Early decision: Nov. 1, Jan. 1. Regular admissions: Jan. 15. Financial aid: Jan. 15. Housing: Jun. 1. Does not guarantee to meet full demonstrated need. Campus and alumni interviews: recommended, evaluative. SATs or ACTs: optional. SAT IIs: optional. Accepts the Common Application and prefers electronic applications. Essay question: paragraph or two describing reasons for applying to Denison.

University of Denver

2199 South University Boulevard, Mary Reed Building, Denver, CO 80208

The only major midsized private university between Tulsa and the West Coast, DU's campus in residential Denver is pleasant but uninspiring. Brochures instead tout Rocky Mountain landscapes. DU remains a haven for ski bums and business majors.

Website: www.du.edu
Location: City outskirts
Total Enrollment: 7,439
Undergraduates: 4,431
Male/Female: 47/53
SAT Ranges: V 530–630
　M 530–640
ACT Range: 23–28
Financial Aid: 43%
Expense: Pr $ $
Phi Beta Kappa: Yes
Applicants: 4,038
Accepted: 82%
Enrolled: 27%

The oldest private university in the Rocky Mountain region, the University of Denver is where Secretary of State Condoleeza Rice earned her BA in political science at age 19 and later returned for a PhD in International Studies. Her mentor was Soviet specialist Joseph Korbel, father of former Secretary of State Madeline Albright. Thus, it's not surprising that DU boasts strong programs in political science, international studies, and public affairs. However, many students opt for DU's business program, and the campus location offers ample opportunities for networking, skiing, and taking in the beautiful Colorado landscape. "Academics are high, social life is awesome, friendships are easily formed, and it's four years of unforgettable memories," a junior says.

DU's 125-acre main campus is located in a comfortable residential neighborhood only eight miles from downtown Denver and an hour east of major ski areas. The north campus is home to several programs, including the Women's College, but a multiyear, $450 million project is underway to unite all university programs onto the same campus. Architectural styles vary, and include Collegiate Gothic, brick, limestone, Colorado sandstone, and copper. Nearby Mount Evans (14,264 feet) is home to the world's loftiest observatory, a DU facility available to both professors and students. The 2,000-seat Peter Barton Lacrosse Stadium opened in 2005 as the nation's first campus facility devoted to the sport. A building to house the School of Hotel, Restaurant, and Tourism Management also functions as a student-operated conference center.

DU is known for its business school—especially the hotel, restaurant, and tourism management offerings—and for its innovative core curriculum. Pre-professional programs are feeders for graduate schools in business, international studies, engineering, and the arts. Chemistry, atmospheric physics, music, psychology, and computer science have solid reputations.

"I really appreciated the orientation program and the first-year program."

Undergrads can opt for a five-year program toward a master's degree in business, international studies, or law. Students report that the academic climate can be demanding but is mostly relaxed. "Courses at DU are relatively rigorous and can be competitive," says a junior. Professors receive high marks for their intelligence and passion. "The professors are very helpful and easy to approach," says a sophomore.

All undergraduates must take courses in English, natural science, a foreign language, arts and humanities, social sciences, math and computer science, and creative expressions, in addition to core curriculum courses in communities and environments, self and identities, and change and community. A freshman orientation program brings new students onto campus a week early, during which they spend 10 hours with a small group of students and a professor discussing a collection of essays by prominent writers. Freshmen also take a four-credit, first-year seminar limited to 15 students. "I really appreciated the orientation program and the first-year program," says a junior. The Honors Program draws more than 400 students; in the past six years, the university has fielded a Rhodes Scholar, a Marshall Scholar, a Truman Scholar, and 12 Fulbright Scholars.

University rules stipulate that all core courses must be taught by senior faculty. "At first I thought, 'Who wants to take these science, art, and English classes?'" explains a business major. "But now that I've completed the core, I feel better about myself and my world knowledge. Now I can speak of Goya, Berlioz, and define my favorite artists with a knowledge of the period, styles, and works." All juniors and seniors have the chance to study abroad at no extra cost. Writing offerings have been

"Now that I've completed the core, I feel better about myself and my world knowledge."

expanded to include small classes focused on helping students write effectively within their field. Dual degree programs are available for students who want to accelerate their graduate studies by up to one year.

For the most part, students come from fairly affluent families. One student says DU is "a lot of rich kids who like to ski," but adds, "if you look hard enough you'll find impassioned students who want to make a difference." Fifty-five percent of the students are from Colorado; minorities account for 14 percent of the student body. Because it is one of the few private colleges in the West, DU is also among the most expensive in the region. In a nod to the diminishing importance of standardized test scores and GPAs, the university now conducts personal interviews of every candidate for undergraduate admissions, either in person or by phone. There are hundreds of merit scholarships averaging $8,489 and 191 athletic scholarships.

Students are required to live their first two years on campus in the residence halls. "The dorms are comfortable enough," says a sophomore. "They have a beautiful view of the mountains, and at sunset or during lightning storms, students often crowd around lounge windows to watch." The Johnson-McFarlane hall ("J-Mac") is supposed to be the best place for freshmen, though another student says that the Towers are a much quieter on-campus option. "Dorms are very well-maintained and

(Continued)

Grad in 6 Years: 69%
Returning Freshmen: 88%
Academics: 🖉 🖉 🖉
Social: ☎ ☎ ☎ ☎
Q of L: ★ ★ ★ ★
Admissions: (303) 871-2036
Email Address:
  admission@du.edu

Strongest Programs:
Biological Sciences
Business
Psychology
History
English
Political Science
Public Affairs
International Studies

The Honors Program draws more than 400 students; in the past six years, the university has fielded a Rhodes Scholar, a Marshall Scholar, a Truman Scholar, and 12 Fulbright Scholars.

provide a sense of closeness for activities," says a junior. Greeks can live and dine together in their houses. Most juniors and seniors opt for the decent quarters found within walking distance of campus.

With consistently beautiful sunny weather and great skiing, hiking, and camping less than an hour away in the Rockies, many DU students head for the hills on weekends. "During the winter weekends, the campus is empty because everyone is skiing," a senior says. Besides various ski areas, one can explore Estes Park, Mount Evans, and Echo Lake. Additionally, DU is not far from Moab, Albuquerque, and Las Vegas. Since Denver is not primarily a college town, many students with cars head for Boulder (home of the University of Colorado), about 30 miles away. For those staying home, the transit system makes it easy to get to downtown Denver. The options there are tremendous and include great local restaurants, bars, and stores, many of which cater to students. Students gripe about the lack of a student union. Twenty percent of the men and 19 percent of the women belong to a fraternity or sorority, and Greeks tend to dominate the social life. Drinking policies abound, and although DU enforces the law, students say drinking is prevalent. "The school has harsh rules against underage drinking," notes a sophomore, "yet such activities exist."

The students unite when the DU hockey team, a national powerhouse, skates out onto the ice, especially against archrival Colorado College. The team brought home the NCAA championship in 2004–2005. That same year, the skiing team also captured the conference championship, along with women's golf and men's lacrosse, basket-

"During the winter weekends, the campus is empty because everyone is skiing."

ball, and golf. Intramural and club sports are varied and popular. Each January, academics are put aside for the three-day Winter Carnival. Top administrators, professors, and students all pack off to Steamboat Springs, Crested Butte, or another ski area to catch some fresh powder and see who can ski the fastest, skate the best, or build the most artistic ice sculptures. In the spring, the whole campus turns out for the annual Chancellor's Barbecue. May Daze, held the first week of May, includes events on the lawn such as a musical festival.

Students like DU for its modest size and friendly atmosphere. And while there remain some moneyed students with attitude problems, plenty of others are more down-to-earth. "In my three years here I have found a wide variety of people," a senior says. As the school pushes for a more ethnically diverse student body and improves its curriculum and facilities, the University of Denver is striving to become better known for its intellectual rigor than its gorgeous setting in the Rocky Mountains.

Overlaps

University of Colorado at Boulder, Colorado College, Colorado State, Santa Clara, George Washington, University of Southern California

If You Apply To ➤ **DU:** Early action: Nov. 1. Regular admissions: Jan. 15. Financial aid: Mar. 1. Guarantees to meet demonstrated need of 10%. Campus and alumni interviews: required, evaluative. ACTs or SATs: required. SAT IIs: optional. Accepts the Common Application and electronic applications. Essay question: Open-ended question.

DePaul University

Chicago, IL 60604

Gets the nod over Loyola as the best Roman Catholic university in Chicago. DePaul's Lincoln Park setting is like New York's Greenwich Village without the headaches. About half of the student body is Catholic. Especially strong in business and the performing arts.

While there is no refuting that DePaul is the largest Roman Catholic university in the nation, students claim its diversity and liberal leanings set it apart from rival institutions. Based in the heart of Chicago, DePaul is a feeder to Chicago's business community. A spate of campus construction has transformed it from the "little school under the tracks" to Chicago's version of NYU.

DePaul has two residential campuses. The Lincoln Park campus, with its state-of-the-art library and new student center, is home to the College of Liberal Arts and Sciences, the School of Education, the Theatre School, and the School of Music, as well as residence halls and academic and recreational facilities. Lincoln Park itself is a fashionable Chicago neighborhood with century-old brownstone homes, theaters, cafes, parks, and shops. The Loop, or "vertical" campus, 20 minutes away by elevated train in downtown Chicago, houses the College of Law, the School for New Learning, the College of Commerce, and the School of Computer Science, Telecommunications, and Information Systems in four high-rise buildings. The DePaul Center, a $70 million teaching, learning, and research complex, is the cornerstone of this campus. A new residence hall shares space with Columbia College and Roosevelt University.

> "There are student organizations to represent all different faiths, ethnicities, and backgrounds."

DePaul's name is closely associated with Midwestern business and law, and undergraduates can find internships with local legal and commercial institutions. The School of Accountancy draws many majors and is reported to be the most challenging department in the College of Commerce, which has added majors in e-business and management information systems. Programs in music and theater are renowned, while the School of Computer Science, the School of Education, and several of the science departments (including biology, chemistry, and physics) have been rejuvenated, and a digital cinema program has been added. DePaul's academic climate is demanding but not overwhelming. "Some courses are difficult and rigorous, but others are easy-going," says an education/sociology major.

Classes are small—97 percent have fewer than 40 students—and professors teach at all levels. The administration appoints student representatives from each school and college to faculty promotion and tenure committees. Social activities bring undergraduates and faculty members together, and students receive a phone book with all the profs' home numbers. "Especially during freshman year, professors really take care to make sure that students become well acclimated to college life," says a communications and American studies double major. All freshmen take a course called Discover Chicago or its alternative, Explore Chicago. "This class is one of the best things about DePaul. It's really fun and valuable," says a senior. Other common core courses include composition and rhetoric as well as quantitative reasoning for freshmen, a sophomore seminar on multiculturalism in the U.S.; and a junior-year program in experiential learning. Students also complete a series of "learning domains," consisting of arts and literature, philosophical inquiry, religious dimensions, scientific inquiry, social science, and history. To earn a BA, students must take three foreign language courses. The highly selective honors program includes interdisciplinary courses, a modern language requirement, and a senior thesis. Nearly a quarter of students participate in study abroad programs.

> "Chicago is the ultimate college town. There is everything to do here."

DePaul's president is a priest, and priests teach some courses and hold (voluntary) Mass every day. In addition, the University Ministry hosts other religious services and leads programs to teach students about other faiths. Seventy-one percent of DePaul students hail from Illinois. Hispanics represent 13 percent of the student body, African Americans 10 percent, and Asian Americans 9 percent. DePaul hopes to boost those

Website: www.depaul.edu
Location: Urban
Total Enrollment: 23,149
Undergraduates: 14,893
Male/Female: 44/56
SAT Ranges: V 510–610
M 500–610
ACT Range: 22–26
Financial Aid: 68%
Expense: Pr $
Phi Beta Kappa: No
Applicants: 10,414
Accepted: 70%
Enrolled: 35%
Grad in 6 Years: 61%
Returning Freshmen: 84%
Academics: 🖋🖋🖋
Social: ☎ ☎
Q of L: ★ ★ ★
Admissions: (312) 362-8300
Email Address:
admission@depaul.edu

Strongest Programs:
Accounting
Computer Science
Finance
Music
Theatre
Communication
Psychology

On the sports scene, men's basketball is the headline story, beginning with the Midnight Madness of each fall's first practice in October.

figures by reaching out to disadvantaged inner-city students with high academic potential. "The students at DePaul are very diverse," says a sophomore. "There are student organizations to represent all different faiths, ethnicities, and backgrounds." DePaul has a reputation for being liberal, and it has become more liberal in recent years," says a junior. "The anti-Coke movement is huge. So is Fair Trade Coffee," says a senior. Athletic and merit scholarships are available to qualified students.

One student complains, "Weekends can be boring because many students go home." Traditionally, DePaul has been a commuter school, but more than 3,000 students live in university housing. Students find the dorms comfortable and well maintained, but they advise applying early to secure a bed, especially after sophomore year. "Housing is in high demand around here," says a junior. The Lincoln Park campus includes eight modern co-ed dorms, several townhouses and apartments; and a new residence hall that opened in fall 2006. At the Loop campus, a 1,700-student residence hall includes a rooftop garden, fitness center, and music, art, and study rooms.

"The campus and its students are friendly, open, and always inviting." Although students like campus housing, some find the food overpriced and limited. "I'm a vegan, and there is not a large or good variety," says a junior. While Chicago may have a high-crime reputation, students say campus security is visible, with officers patrolling in cars and on foot, emergency blue lights on campus, and dorms requiring students to swipe ID cards at two or three places before allowing entrance.

Fraternities and sororities draw just 3 percent of DePaul men and 2 percent of women. Not surprisingly, with the school's proximity to Chicago's clubs (especially on Rush Street), sporting events, and bars, most social life occurs off campus. In the warmer months, the beaches of Lake Michigan beckon downtown students, while the huge annual outdoor Fest concert attracts large crowds from both campuses. "Chicago is the ultimate college town," says a music major. "There is everything to do here." On campus, the alcohol policy forbids beer for underage students, but students say that enforcement doesn't always work.

On the sports scene, men's basketball is the headline story, beginning with the Midnight Madness of each fall's first practice in October. Recently, they were conference USA champions. The game against Notre Dame always draws a capacity crowd, though Loyola is DePaul's oldest rival. Women's basketball is also strong, as is DePaul's solid intramural program.

DePaul's student body has become more diverse while increasing in size, an admirable achievement. The administration credits the school's "increased academic reputation" for growth, but students say DePaul's popularity is due as much to the special bonds they feel with fellow Blue Demons. "DePaul University provides students a unique atmosphere in which to learn and grow," says a sophomore. "The campus and its students are friendly, open, and always inviting."

Overlaps

University of Illinois–Urbana Champaign, Loyola University, Marquette, Illinois State, Northern Illinois

If You Apply To > **DePaul:** Rolling admissions. Early action: Nov. 15. Regular admissions: Feb. 1 (priority). (Music/Theatre: Jan. 15) Financial aid: Mar. 1. Campus interviews: optional, evaluative. No alumni interviews. SATs or ACTs: required. SAT IIs: optional. Accepts electronic applications. Essay questions: who or what influenced your decision to apply to DePaul; impact of a recent failure and achievement.

DePauw University

315 South Locust Street, P.O. Box 37, Greencastle, IN 46135

DePauw is a solid Midwestern liberal arts institution in the mold of Illinois Wesleyan, Ohio Wesleyan, Denison, and Dickinson. Its Greek system is among the strongest in the nation and full of students destined for Indiana's business and governmental elite.

DePauw University offers a liberal arts education with an orientation toward experiential learning. Art history, creative writing, and music are solid, as are (more surprisingly) computer science and economics. Indeed, students here are career-oriented and happy to take advantage of the rigorous classwork and ample real world experiences. And with an undergraduate population of 2,300 students, close ties to classmates and faculty is a given.

DePauw is set amid the gently rolling hills of west-central Indiana. The lush green campus has a mix of older buildings and more modern redbrick structures, centered on a well-kept park with fountains and a reflecting pool. Newer additions include a $15 million art building, which provides studio space, a digital image library, classrooms, and galleries; a $36 million technology and physical science center, which offers 200,000 square feet of classrooms and labs; and a new set of residence life units collectively known as Rector Village.

DePauw's first-year program helps students transition to college by combining academically challenging coursework with co-curricular activities and programs. When they arrive, students are assigned to mentor groups with 10 to 12 peers, plus an upperclassman advisor and a faculty member who will teach their first-year seminar and serve as their academic advisor. By graduation, students must demonstrate competence in writing, quantitative reasoning, and oral communication. And they must fulfill distribution requirements in six areas: natural science and math, social and behavioral sciences, literature and the arts, historical and philosophical understanding, foreign language, and self-expression through performance and participation.

"I am aided by professors with years of experience."

Academically, the DePauw student body is as career-oriented as they come. Aspiring business leaders benefit from courses, speakers, and internships offered through the McDermond Center for Management and Entrepreneurship. Future reporters, editors, anchors, and producers will find a home in the Pulliam Center for Contemporary Media, which supplements DePauw's strong student-run newspaper, TV station, and radio stations. And DePauw's School of Music is also worth a mention, offering all students the chance to take lessons, join ensembles, and perform, in genres from orchestra to jazz to opera. For exceptionally motivated students, five programs of distinction offer the chance to focus on an area of interest, such as media, management, scientific research, or information technology. The latter program includes real-world work experience, both in IT departments at DePauw and with off-campus employers. Recent additions to the curriculum include courses in statistics and Middle Eastern studies. A new major in education studies qualifies students for a fifth-year program in teaching with a master's degree and certification. Interdisciplinary studies at DePauw range from long-established programs in Asian studies, women's studies, and black studies to new programs in biochemistry, conflict studies, film, Jewish studies, and courses focusing on Latin America, the Caribbean, and Europe.

True to its focus on experiential learning, the DePauw curriculum includes a January term, during which first-year students remain on campus for a focused,

Website: www.depauw.edu
Location: Small town
Total Enrollment: 2,397
Undergraduates: 2,345
Male/Female: 45/55
SAT Ranges: V 560–660
  M 570–670
ACT Range: 24–29
Financial Aid: 50%
Expense: Pr $ $
Phi Beta Kappa: Yes
Applicants: 3,445
Accepted: 66%
Enrolled: 26%
Grad in 6 Years: 79%
Returning Freshmen: 92%
Academics: 🏫 🏫 🏫 ½
Social: 🍺 🍺 🍺
Q of L: ★ ★
Admissions: (800) 447-2495
Email Address:
  admission@depauw.edu

Strongest Programs:
Chemistry and Biochemistry
Biology
Economics and Management
Communications and Theatre
English and Creative Writing

interdisciplinary course, while upperclassmen pursue independent study or off-campus study or service in the U.S. or abroad. Approximately 40 percent of DePauw students spend a semester off campus, and when the January term is included, that fraction rises to more than 80 percent. The school offers its own programs as well as those arranged by the Great Lakes Colleges Association.* Students say that DePauw's classes are rigorous but well-supported. "I am aided by professors with years of experience," says a junior, one of whom "actually came to see my theater production!"

One student describes her peers at DePauw as "intellectually curious, philanthropically minded, and socially active." The minority presence has grown, thanks in part to recruitment of Posse Foundation students from New York and Chicago. African Americans now account for 5 percent of the student body, with Hispanics adding 3 percent, and Asian Americans 2 percent. Nearly three-quarters of students receive a merit scholarship, and the average award is more than $12,000. Three-quarters of DePauw students volunteer with area churches and social service agencies, which can also help them qualify for scholarships.

Ninety-nine percent of DePauw students live in university housing, which is guaranteed for four years. Overcrowding is a thing of the past, since the school has opened 17 new duplexes and renovated another 46 apartments in recent years. The homey buildings have computer labs, common areas, and TV lounges; students recommend Humbert Hall for freshmen because of its hotel-like atmosphere. A whopping 72 percent of DePauw's men and 68 percent of the women go Greek.

"Fake IDs get confiscated quicker than you can take them out."

Perhaps because of the prevalence of Greeks on campus, and spurred by the disciplining of a sorority accused of purging its overweight members, these groups have worked hard to change the stereotypes of fraternities and sororities. They've devised a risk-management policy and instituted a community council to review conduct violations. In addition, rush is delayed until second semester so freshmen can first get their feet on the ground academically. Fraternities still maintain the old custom of having "house moms." Still, students say it's easy for underage drinkers to imbibe, especially at fraternity parties. Off campus, however, fake IDs "get confiscated quicker than you can take them out," one student warns.

The town of Greencastle has a movie theater, a bowling alley, and several pizza places, but it "lacks an atmosphere," says a recent grad. "It is fine for sustaining day-to-day living, but doesn't offer many alternatives to the university." In good weather, several state parks offer hiking trails and a lake for the sailing club; Indianapolis is only a 45-minute drive, and St. Louis, Chicago, and Cincinnati make for good road trips. Another cherished tradition is a takeoff on Indiana University's famed Little 500 bike race, itself a takeoff on the Indianapolis 500 auto race—teams of cyclists compete on a course that circles the heart of the DePauw campus.

Aside from Greek parties, social life at DePauw revolves around varsity athletics, especially the annual football game against Wabash College. The Wabash–DePauw rivalry is the oldest west of the Alleghenies, and the winner of each year's contest gets the much-cherished Monon Bell, hence the popular T-shirt: "Beat the bell out of Wabash." Several teams have recently played in NCAA Division III championship games, including women's soccer and golf, men's and women's cross-country and tennis, and men's golf. Swimming, track, baseball, and women's basketball are also strong. Intramural sports attract some 90 percent of students.

For a small school, DePauw offers a multitude of opportunities, balancing strong academics with a healthy dose of school spirit and a wealth of opportunities to lead—whether in one of the abundant extracurricular activities or by blazing a trail through study abroad.

Dickinson College

P.O. Box 1773, Carlisle, PA 17013

Dickinson occupies an historic setting in the foothills of central Pennsylvania. Known foremost for study abroad and foreign languages, Dickinson's curriculum combines liberal arts with a business program that picks up the international theme. Competes head-to-head with nearby Gettysburg.

Dickinson College won its charter just six days after the Treaty of Paris recognized the United States as a sovereign nation, and this small liberal arts school has been blazing trails ever since. Though it has occupied the same central Pennsylvania plot for more than 200 years, location is one of the few constants in Dickinson's history. The school asks students to challenge what is safe and comfortable, to meet the future with a voice that reflects America and engages the world. To that end, its more than 40 study abroad programs in 24 countries draw well over half of the student body. Now, administrators are focused on diversity, global education, and attracting the best and brightest academic talent. "This is a school where one can find abundant resources to develop their academic skills as well as a variety of people to develop friendships," says a biochemistry major.

Almost all of Dickinson's Georgian buildings are carved from gray limestone from the college's own quarry, which lends a certain architectural consistency. Even the three-foot stone wall that encloses much of the wooded 90-acre Dickinson yard is limestone. The campus is part of the historic district of Carlisle, an economically prosperous central-Pennsylvania county seat nestled in a fertile valley. The renovated Community Studies Center fosters interdisciplinary hands-on learning in the social sciences and humanities, including taped interviews, surveys, and videotapes produced by students and faculty engaged in field work.

Dickinson is best known for its workshop approach to science education, for its outstanding and comprehensive international education program, and for the depth of its foreign language program—a total of 13 languages are offered, including Arabic, Chinese, Japanese, Hebrew, Portuguese, and Italian. A 3–3 program with Penn State's Dickinson School of Law allows students to obtain undergraduate and JD degrees in six years. The international business and management major includes coursework in economics, history, and financial analysis, as well as internships and overseas education. Regardless of major, about half of all students complete internships, where they may learn about stock trading at a brokerage firm, assist a judge in a common pleas court, or work with the editorial staff of a magazine. The required First Year Seminar Program introduces new students to college-level study and reflection, with interdisciplinary courses such as Law, Justice and the Individual; and Physics or Philosophy? The Debate over "Intelligent Design." To help students understand how the liberal arts fit into the broader world, Dickinson requires distribution courses in the arts and humanities, social sciences, and laboratory sciences; writing; and quantitative reasoning. Required cross-cultural studies courses include comparative civilization, United States diversity, and a foreign language.

Website: www.dickinson.edu
Location: Small town
Total Enrollment: 2,292
Undergraduates: 2,292
Male/Female: 44/56
SAT Ranges: V 600–690
  M 600–680
ACT Range: 26-30
Financial Aid: 58%
Expense: Pr $ $ $
Phi Beta Kappa: Yes
Applicants: 5,294
Accepted: 43%
Enrolled: 27%
Grad in 6 Years: 84%
Returning Freshmen: 89%
Academics: ✑ ✑ ✑ ✑
Social: ☎ ☎ ☎
Q of L: ★ ★ ★
Admissions: (717) 245-1231
Email Address:
  admit@dickinson.edu

Strongest Programs:
Political Science
Foreign Languages
English
Biological Sciences
International Business and
  Management

Academics are demanding, but not cutthroat. "Courses are rigorous and often reading and writing intensive," says a sophomore, who describes students as "goal-oriented and hard working." Eighty-one percent of classes have 25 or fewer students, allowing freshmen easy access to their professors. "Professors here are very accessible and have great credentials and amazing experiences to bring to the classroom," says a French major.

Dickinson is trying to recruit more minority and international students, especially through partnerships with New York's Posse Foundation and the Philadelphia Futures Foundation. For now, though, "you have your wealthy preppy kids and your artsy kids and your scholarship academic kids and every combination," says an English major. One-third of the student body hails from the Keystone State, and many of the rest are white, middle-class or upper-middle-class, coming from elsewhere in the Northeast. What else do they have in common? According to one junior, "Students who attend Dickinson are very open-minded and tend to think globally." Perennial hot topics include sustainability and social justice. African Americans, Asian Americans, and Hispanics each make up 4 percent of the student body, and foreign students comprise 3 percent. Dickinson awards three types of merit scholarships worth an average of $11,000; the school does not offer athletic scholarships.

"Courses are rigorous and often reading and writing intensive."

Only seniors are allowed to live off campus, and Dickinson guarantees housing for four years, so 92 percent of students remain in the dorms. First-year students have their own halls and upperclass students live in traditional dorms, which are co-ed by floor, or in townhouses with eight-person suites. The college transformed an abandoned factory into a combination of art studios and loft-style apartments, with 118 beds for juniors and seniors, and also recently renovated Morgan Hall, the largest dorm on campus. All Greek houses are owned and maintained by the college, and special interest housing includes areas devoted to French, Spanish, volunteerism, cultural diversity, and the environment.

Most social life at Dickinson occurs on campus. Fraternities and sororities attract 19 percent of the men and 24 percent of the women, respectively, and though they throw open parties at their houses, "you will not find many large, raging, house parties, but if you know where to look, you can find some," says an international studies major.

"Students who attend Dickinson are very open-minded and tend to think globally."

At the Quarry, a former frat house, you can grab a cup of coffee, play some video games, or bust a move on the dance floor. When it comes to booze, Dickinson follows Pennsylvania state law, so you must be 21 to drink; monitors check IDs at parties. Kegs aren't permitted in any college housing, and four underage drinking incidents will get you expelled. "The policies are mostly effective," a student says. There's a big concert each semester, where The Roots and Guster have played, and Hub-All-Night includes "music, food, and lots of free stuff," says a political science and economy major. Each fall brings an arts festival, and a spring carnival gives students one last blast before finals. One English major concludes, "On campus, if you can't find something to do, you live in a closet."

Carlisle is 20 miles from the Pennsylvania state capital of Harrisburg, and has "every fast-food restaurant you could want, as well as a Wal-Mart," says an international studies major. "Other amenities are lacking," grumbles a sophomore. For those seeking more of a college town vibe, Amy's Thai and Pomfret Street Books are good bets. Big Brothers Big Sisters programs, the Alpha Phi Omega community service fraternity, and programs like Adopt-a-Grandparent help bring the school and community together. In the spring and early fall, Maryland and Delaware beaches beckon; they're just a two- to three-hour drive. Come winter, good skiing is a half-hour away. Nature lovers will enjoy hiking the Appalachian Trail, just 10 minutes away. For those craving urban stimulation, the best road trips are to Philadelphia, New York, and Washington, D.C. All are accessible by bus or train—a good thing, since first-years can't have cars.

Dickinson students get riled up for any match against top rival Franklin and Marshall; the schools battle it out each year for the Conestoga Wagon trophy. Dickinson also squares off with Gettysburg College each year for the Little Brown Bucket. Football, the women's basketball team, cross-country team and indoor and outdoor track and field squads have won Centennial Conference championships in the past two years. About three-quarters of the men and half the women take part in intramurals, where dodgeball, basketball, floor hockey, and soccer are most popular. The soccer, softball, baseball, and varsity football fields sit within a 30-acre park, and they're lighted for night games. There is an outdoor track, jogging trails, and an indoor rock climbing wall. Students may also organize club teams to compete with other schools in sports like ice hockey, where Dickinson doesn't field varsity squads.

> "On campus, if you can't find something to do, you live in a closet."

Dr. Benjamin Rush founded Dickinson more than two centuries ago, after signing the Declaration of Independence, and some things haven't changed since then. Seniors still share a champagne toast before graduation. And the steps of Old West, the first college building, are still used only twice a year—in the fall, at the convocation ceremony that welcomes new students, and in the spring, for commencement. But Dickinson continues to honor Rush's global vision, with its wealth of study abroad options and its demand that students cross the traditional borders of academic disciplines to grasp the interrelated nature of knowledge. "Five years ago, Dickinson was a good school," says one senior. "Today, Dickinson is a great school."

Overlaps

Gettysburg, Franklin and Marshall, Bucknell, Lafayette, Hamilton, Skidmore

If You Apply To ➤ | **Dickinson:** Early decision: Nov. 15, Jan. 15. Early action: Dec. 1. Regular admissions and financial aid: Feb. 1. Meets demonstrated need of 83%. Campus interviews: recommended, informational. Alumni interviews: optional, informational. SATs or ACTs: optional. SAT IIs: optional. Accepts the Common Application and electronic applications. Essay question: Common Application questions, and why Dickinson is a good match.

Drew University

Madison, NJ 07940-4063

From Drew's wooded perch in suburban Jersey, Manhattan is only a 30-minute train ride away. That means Wall Street and the UN, both frequent destinations for Drew interns. Drew is New Jersey's only prominent liberal arts college and one of the few in the greater New York City area.

Founded more than a century ago as a Methodist university, Drew University has grown into a place where an emphasis on hands-on learning, research, independent studies, and internships are just as important as performance in the classroom. And the changes keep coming. University president Robert Weisbuch came up with the idea of dropping the SAT and ACT requirement for entrance, and the response has been overwhelmingly positive. In addition, the university sends its students abroad for month-long educational ventures, promotes internships on Wall Street, and encourages theater and the arts to thrive.

Website: www.drew.edu
Location: Suburban
Total Enrollment: 2,627
Undergraduates: 1,561
Male/Female: 44/56
SAT Ranges: V 550–660
  M 540–650

The school occupies 186 acres of peaceful woodland in the upscale New York City suburb of Madison and is known as "The University in the Forest." Fifty-six campus buildings peek through splendid oak trees and boast classic and contemporary styles, a physical reflection of Drew's respect for both scholarly traditions and progressive education. The school is currently sprucing up the atmosphere through a reforestation project and the opening of the $20 million Dorothy Young Center for the Arts for the theater and studio arts departments.

Political science is Drew's strongest undergraduate department, and future politicos can take advantage of off-campus opportunities in Washington, D.C., London, Brussels, and the United Nations in New York City. Other popular majors include behavioral science, economics, English, and psychology. The Dana Research Institute for Scientists Emeriti offers opportunities for students to do research with distinguished retired industrial scientists, and recently won the lofty Merck Innovation Award for Undergraduate Science Education for fresh thinking and imaginative use of resources. Even more impressive is a program whereby students can earn a BA and MD from Drew and the University of Medicine and Dentistry of New Jersey/New Jersey Medical School in seven years. Future financiers can follow in the footsteps of the school's founder and take advantage of Drew's Wall Street Semester, an on-site study of the national and international finance communities. Other recent program additions include Pan-African studies and Chinese studies.

Drew's commitment to liberal arts education includes the lofty goal of universal computer literacy. In fact, tuition for full-time students includes a notebook computer and supporting software, which students take with them when they graduate. The school's campuswide fiber-optic network links all academic buildings and many residence halls.

General education requirements, which take up a third of each student's total program, involve coursework in natural and mathematical sciences, social sciences, humanities, and arts and literature. Students must also show competency in writing, and each first-year student enrolls in seminars limited to 16 people, 80 percent of which are taught by senior faculty. The theater arts department works closely with the Playwrights Theater of New Jersey (founded by faculty member Buzz McLaughlin) to produce plays that are written, directed, and designed by students. Drew has long been a proponent of study abroad programs, including the Drew International Seminar program, where students study another culture in-depth on campus, then spend three to four weeks in that country. "These seminars are a great way of learning firsthand about a country and experiencing a once-in-a-lifetime chance to put your education into practice," says one student.

"Teaching here is eclectic and usually very interesting and enjoyable."

Maintaining a rigorous study schedule is key, according to many upperclassmen. A cast of highly praised, interactive faculty who generate enthusiasm and ambition fuels the industrious grind of hard work. "The professors are extremely knowledgeable and experienced in their fields," says a junior, "but, more importantly, they are eager to share their knowledge with you." A classmate agrees: "Teaching here is eclectic and usually very interesting and enjoyable." Drew's library complex, a cluster of three buildings, contains more than 450,000 titles and offers ample study accommodations, though some students complain that some collections are outdated.

Fifty-seven percent of Drew's students are from New Jersey, and most attended public high schools. The school continues to work on increasing its racial diversity—7 percent are Asian American, 3 percent are African American, and Hispanics comprise 8 percent. And the differences extend past heritage: "My favorite thing about Drew is that although people are politically and socially active, they are more interested in listening to what others have to say than in projecting their

Eighty-five percent of the students live in university housing, which includes both single-sex and co-ed dorms, and six theme houses.

own opinions," says a senior. Merit awards average $11,802 although there are no athletic scholarships available.

Eighty-five percent of the students live in university housing, which includes both single-sex and co-ed dorms, and six theme houses. Themes have included Earth House, Umoja House, Womyn's Concerns, Asia Tree House, and Spirituality Home. Several housing options are available to upperclassmen, from dorm rooms of all sizes to suites and townhouses. A lottery gives housing preference to seniors and juniors, and most freshmen reside in dorms situated at the back of campus, which aren't the best. "The dorms are maintained," says a student, "but first-year housing is deplorable." Still, most students live on campus because housing prices in Madison are out of reach for collegians. Although the dorms have kitchenettes, everyone must buy the meal plan, which students describe as "not at all appetizing" and "poor."

"It's easy to get involved, make friends, and feel like you make a difference."

There is no Greek system and social life mostly takes place on campus. Officially, nobody under 21 is allowed to drink, but alcohol is said to be easy to come by. There is a 21-and-over pub on campus as well as two coffeehouses. On-campus social programming is extensive. "With more than 100 clubs on campus and the Student Activity Board, there is always a concert, play, or lecture," a psychology major says. New York City's Pennsylvania Station is less than an hour away by commuter train, and Philadelphia, the Jersey shore, and the Delaware River are close by.

The First Annual Picnic, held on the last day of classes and numbered like Super Bowls (FAP XVII), provides an opportunity to enjoy live music and food. On Multi-cultural Awareness Day, students are excused from one day of classes to celebrate cultural diversity by attending lectures, workshops, and social events. Drew also launched an initiative to give all students and staff opportunities to participate in diversity training.

The commuter town of Madison tends to get discouraging reviews as a college town by students who feel its wealthy residents don't take too kindly to Drewids, as they affectionately call themselves. However, Community Day—designed to bring students and residents together—has become an annual event, and approximately 50 percent of students volunteer in activities such as Mentors at Drew and The Honduras Project, in which a group of Drew students travel to Honduras to help at an orphanage. Madison does provide several unique shops and restaurants within walking distance of campus, though the town "is a nice

"With more than 100 clubs on campus and the Student Activity Board, there is always a concert, play, or lecture."

college town according to my parents, but not to students. Everything closes up pretty early." Nearby Morristown is more of a college place. The New Jersey Shakespeare Festival is in residence part of every year and offers both performances and internships. If that doesn't float your boat, NYC is a short train ride away.

Students used to seem more interested in intramural sports than in the school's Division III varsity teams, but interest has grown as the teams have become more successful. The $15 million athletic center is a 126,000-square-foot state-of-the-art facility that seats 4,000 and is used by varsity sports teams and intramural programs. The fencing team has done exceptionally well, too.

From a tight-knit campus to far-flung study abroad programs, Drew offers its small body of students a wide range of opportunities in a classic liberal arts structure. "It's easy to get involved, make friends, and feel like you make a difference," says a senior. Not too bad for a school in the forest.

Political science is Drew's strongest undergraduate department, and future politicos can take advantage of off-campus opportunities in Washington, D.C., London, Brussels, and the United Nations in New York City.

Overlaps

Rutgers, College of New Jersey, Muhlenberg, Boston University, NYU, Fordham

If You Apply To >

Drew: Early decision: Dec. 1, Jan. 15. Regular admissions: Feb. 15. Meets demonstrated need of 36%. Campus interviews: recommended, evaluative. Alumni interviews: optional, evaluative. SATs: optional. SAT II: optional. Accepts the Common Application and electronic applications. Essay question: graded paper for those not submitting SATs.

Drexel University

3141 Chestnut Street, Philadelphia, PA 19104

Drexel is a streetwise, no-nonsense technical university in the heart of Philadelphia. Go to school, work an internship, go to school again, work again—that's the Drexel way. Like Lehigh, Drexel also offers programs in business and arts and sciences, and its most distinctive offering is a College of Media Arts and Design.

Website: www.drexel.edu
Location: City center
Total Enrollment: 13,626
Undergraduates: 10,158
Male/Female: 59/41
SAT Ranges: V 530–630
    M 550–660
Financial Aid: 63%
Expense: Pr $
Phi Beta Kappa: Yes
Applicants: 12,093
Accepted: 82%
Enrolled: 25%
Grad in 6 Years: 60%
Returning Freshmen: 80%
Academics: ✐ ✐ ✐
Social: ☎ ☎
Q of L: ★ ★
Admissions: (215) 895-2400
Email Address:
    enroll@drexel.edu

Strongest Programs:
Engineering
Graphic Design
Architecture
Film and Video

For career-minded students who want to bypass the soul-searching of their liberal arts counterparts, Drexel University offers both solid academics and an innovative co-op education that combines high-tech academics with paying job opportunities. "If you want a good job, you go to Drexel and you do co-op." It's easy to see why Drexel University is nicknamed "the Ultimate Internship."

"Drexel's campus is impressive for its downtown Philadelphia location, with gardens and greenery on every block," says a student, "but the campus is woven tightly into the fabric of the city." The buildings are simple and made of brick; most are modern and in good condition. Sitting just west of the city center and right across the street from the University of Pennsylvania, the campus is condensed into about a four-block radius. Newer facilities include the Edmond Bossone Research Center, Ross Commons (a student center), and a new residence hall.

> **"If you want a good job, you go to Drexel and you do co-op."**

Cooperative education is the hallmark of the curriculum, which alternates periods of full-time study and full-time employment for four or five years, providing students with six to 18 months of moneymaking job experience before they graduate. And the co-op possibilities are unlimited: students can co-op virtually anywhere in this country, or in 11 foreign countries, and 98 percent of undergraduates choose this route. Freshman and senior years of the five-year programs are spent on campus, and the three intervening years (sophomore, prejunior, and junior) usually consist of six months of work and six months of school. A pre-cooperative education course covers such topics as skills assessment, ethics in the workplace, résumé writing, interviewing skills, and stress management. Each co-op student has the opportunity to earn from $7,000 to $30,000 while attending Drexel. And although some students complain that jobs can turn out to be six months of make-work, most enjoy making important contacts in their potential fields and learning while earning. "The courses are very rigorous," explains a junior. "The profs tend to move very fast to fit everything into the ten-week terms."

To accommodate the co-op students, Drexel operates year-round. Flexibility in requirements varies by college, but in the first year everyone must take freshman seminar, English composition, mathematics, and Cooperative Education 101; engineering majors must also complete the Drexel Engineering Curriculum, which integrates math, physics, chemistry, and engineering to make sure that even techies enter the workforce well-rounded and able to write as well as they can compute and design. Students enjoy the 700,000-volume library, which offers good hours and lots of room for studying. Professors receive high praise from most, and are noted for their accessibility and warmth. Says one student, "I have had exceptional teachers who go out of their way to ensure the success of their students."

Drexel's greatest strength is its engineering college, which churns out more than 1 percent of all the nation's engineering graduates, BS through PhD. The electrical and architectural engineering programs are particular standouts. The College of Arts and Sciences is well recognized for theoretical and atmospheric physics; chemistry is also recommended. The futuristic Center for Automated Technology complements the strong computer science program. Students mention that the biology and chemistry departments are weak, primarily due to lack of organization and foreign teachers, who are hard to comprehend. New programs include urban environmental studies, entertainment and arts management, and the Drexel College of Law, which will accept its first class this year.

The co-op program often undermines any sense of class unity, and can strain personal relationships. Activities that depend on some continuity of enrollment for success—music, drama, student government, athletics—suffer most.

Drexel students are "professional, experienced, bright, and ambitious," according to one junior. The student body is 52 percent Pennsylvanian, with another large chunk of students from adjacent New Jersey. The foreign student population is 7 percent, while Asian Americans and African Americans account for 21 percent of the student body. Thirty

"The profs tend to move very fast to fit everything into the ten-week terms."

percent of Drexel undergrads graduated in the top tenth of their high school class, and the student body tends to lean right politically. "This is a science and technology school full of conservative students who don't really have the time to worry about liberal issues," says a student. In addition to need-based financial aid, a wide range of athletic and merit scholarships (the latter in amounts up to $10,000 per year) is offered.

The co-op possibilities are unlimited: students can co-op virtually anywhere in this country, or in 11 foreign countries, and 98 percent of undergraduates choose this route.

Freshmen live in one of six co-ed residence halls, including a luxurious high-rise, but many upperclassmen reside in nearby apartments or the fraternities, which are frequently cheaper and more private than university housing. Overall, 25 percent of the students live in the dorms; another third commute to campus from home. The cafeteria offers adequate food and plenty of hamburgers and hot dogs, but it's far away from the dorms. While on-campus freshmen are forced to sign up for a meal plan, most upperclassmen make their own meals; the dorms have cooking facilities on each floor. If all else fails, nomadic food trucks park around campus providing quick lunches. Students are encouraged to use a shuttle bus between the library and dorm at night, and access to dorms, the library, and the physical education center is restricted to students with ID, so most feel safe on campus.

With so many students living off campus and the city of Philadelphia at their disposal, Drexel tends to be a bit deserted on weekends. A student notes, "In a single weekend, I may play paintball in the Poconos, swim at the Jersey shore, see an opera in Philadelphia, and go mountain biking in nearby Wissahickon Park." Friday-night flicks are cheap and popular with those who stay around, and dorms sponsor floor parties. The dozen or so fraternities also contribute to the party scene, especially freshman year, but a handful of smaller sororities has little impact. Still, Greek Week is well attended by members of both sexes, as is the spring Block Party, which attracts four or five bands. The

New programs include urban environmental studies, entertainment and arts management, and the Drexel College of Law, which will accept its first class this year.

"This is a science and technology school full of conservative students."

Greeks recruit 5 percent of the men and women. Drinking is "not a big deal to everyone," and campus policies are strict; dorms require those of age to sign in alcohol and limit the quantities they may bring in. "This policy encourages many students to explore non-alcohol related activities," says an education major.

The co-op program often undermines any sense of class unity, and can strain personal relationships. Activities that depend on some continuity of enrollment for success—music, drama, student government, athletics—suffer most. "It's hard to get people involved because of the amount of schoolwork and co-ops," says one woman. There is no football team, but men's basketball and soccer are strong. "Our biggest rivalry is our feud with Delaware," admits one frenzied student. "We delight

"We delight in sacrificing blue plastic chickens!"

in sacrificing blue plastic chickens!" Men's and women's swimming and women's volleyball also generate interest. An extensive intramural program serves all students, and joggers can head for the steps of the Philadelphia Art Museum, just like Rocky did in the movies. Students take full advantage of their urban location by frequenting clubs, restaurants, cultural attractions, and shopping malls in Philadelphia, easily accessible by public transportation.

Aspiring poets, musicians, and historians may find Drexel a bit confusing. But for future computer scientists, engineers, and other technically oriented minds, the university's unique approach to learning inside and outside the classroom could give your career a fantastic jumpstart. As one satisfied customer explains, "The terms are intense, the activities unlimited, but Drexel graduates are surely among the most capable and motivated individuals I have ever met. When I graduate, I will be prepared and proud of it."

Overlaps

Penn State, Temple, University of Delaware, University of Pittsburgh, Northeastern, Lehigh

---

If You Apply To ➤ | **Drexel:** Rolling admissions. Financial aid: Mar. 1. Does not guarantee to meet demonstrated need. Campus interviews: recommended, informational and evaluative. No alumni interviews. SATs or ACTs: required. SAT IIs: optional. Accepts the Common Application and electronic applications. Apply to particular schools or programs.

Duke University

2138 Campus Drive, Durham, NC 27708

What fun to be a Dukie—face painted blue, rocking Cameron Indoor Stadium as the Blue Devils score again. Duke is the most prestigious private university in the South—similar to Rice in selectivity, and academically competitive with the Ivies and Stanford. Duke is strong in engineering as well as the humanities. Offers public policy and economics rather than business.

Website: www.duke.edu
Location: Small city
Total Enrollment: 13,088
Undergraduates: 6,244
Male/Female: 52/48
SAT Ranges: V 670–760
  M 690–780
ACT Range: 29–33
Financial Aid: 40%
Expense: Pr $ $ $ $
Phi Beta Kappa: Yes
Applicants: 18,090
Accepted: 22%
Enrolled: 43%

Duke University is one of the few elite U.S. colleges where strong academics and championship-caliber sports teams manage to coexist. It might be south of the Mason-Dixon Line, and may seem a bit wet behind the ears compared to those ancient and prestigious Northeastern schools, but Duke is competing with them and winning its fair share of serious students as well as superb athletes. A rising star in the South, Duke is on even footing with the Ivies and Stanford.

Founded in 1838 as the Union Institute (later Trinity College), Duke University is young for a school of its stature. It sprouted up in 1924, thanks to a stack of tobacco-stained dollars called the Duke Endowment. Duke's campus in the lush North Carolina forest is divided into two main sections, West and East, and, with 8,300 acres of adjacent forest, offers enough open space to satisfy even the most diehard outdoors enthusiast. West Campus, the hub of the university, is laid out in spacious quadrangles and dominated by the impressive Gothic chapel, a symbol of the university's Methodist tradition. Constructed in the 1930s, West includes Collegiate Gothic residential and classroom quads, the administration building, Perkins

Library (with 4.2 million volumes, nearly 8.9 million manuscripts, and two million public documents), and the student union. East Campus, built in the 1920s, consists primarily of Georgian redbrick buildings. East and West are connected by shuttle buses, though many students enjoy the mile-or-so walk between them along wooded Campus Drive. Befitting its rising stature, Duke has become a construction zone. Recent projects include expansions to the library and nursing school, a student center plaza and commons, and the French Science Center, a $110 million teaching and research laboratory. Soon to begin is a $500 million investment in a new Central Campus that will house arts, language, and international programs as well as 1,500 student apartments.

Students opt for one of two undergraduate schools: the Pratt School of Engineering and Trinity College of Arts & Sciences (the latter resulted from a merger of the previously separate men's and women's liberal arts colleges). The school's engineering programs—particularly electrical and biomedical—are national standouts. Natural sciences, most notably ecology, biology, and neuroscience, are also first-rate. The proximity of the Medical Center enhances study in biochemistry and pharmacology. Economics attracts the most majors, followed by psychology, biology, political science, and public policy studies.

Duke's Sanford Institute of Public Policy offers an interdisciplinary major—unusual at the undergraduate level—that trains aspiring public servants in the machinations of the media, nonprofit organizations, government agencies, and other bodies that govern public life. Internships and apprenticeships are a big part of the program. A junior praises the department's "very impressive faculty." There is a new major available in statistical science, as well as a performance concentration in the existing music major. In addition, a new dance major is available to undergraduates—the first in the university's history. Certificate programs are available in media studies and in markets and management studies. Additionally, Duke offers more than 100 interdisciplinary courses, bolstered by the John Hope Franklin Center for International and Interdisciplinary Studies. Duke is looking to double the number of undergraduates participating in mentored research.

More than 40 percent of Duke students study abroad, and there are ample opportunities for those who want a break from campus life without leaving the country. The university recently launched DukeEngage, a program backed by a $30 million endowment that will make civic engagement an integral part of the undergraduate experience. It will support students willing to spend summers working on projects ranging from building schools in Kenya to working with Gulf Coast flood victims.

Trinity College's curriculum, part of the traditional undergraduate coursework known as Program I, requires courses in five general areas of knowledge: arts, literature, and performance; civilizations; social sciences; natural sciences; and quantitative studies. Students must also fulfill requirements in six modes of inquiry, including foreign language, writing, research, and ethical inquiry. All students also must complete three Small Group Learning Experiences: one seminar course during the freshman year—offered in topics such as Imagining Dinosaurs and The Psychology of Social Influence—and two more as upperclassmen. Students must finish 34 courses to graduate; those who wish to

"Getting A's is hard and teachers have high expectations."

explore subjects outside and between usual majors and minors may choose Program II, to which they are admitted after proposing a topic, question, or theme, for which they plan an individualized curriculum with faculty advisors and deans. "Students are given considerable freedom, and with it, responsibility," says a student.

When college counselors say Duke is hot, they're not referring to the temperatures in the South. Duke competes with the Ivies and a select few other colleges for top-notch students. Courses here are rigorous and the academic atmosphere has become

(Continued)

Grad in 6 Years: 94%
Returning Freshmen: 96%
Academics: 🖾 🖾 🖾 🖾 🖾
Social: ☎ ☎ ☎ ☎
Q of L: ★ ★ ★ ★
Admissions: (919) 684-3214
Email Address: undergrad-admissions@duke.edu

Strongest Programs:
Biology
Ecology
Neuroscience
Political Science
Public Policy
Economics
Literary/Cultural Studies
Romance Languages

The university recently launched DukeEngage, a program backed by a $30 million endowment that will make civic engagement an integral part of the undergraduate experience.

Duke's Southern gentility is reflected in campus attire, which is generally neatly pressed on guys and maybe a bit outfitty on women, in contrast to the thrown-together anti-status uniform of jeans and sweats that dominates on some other campuses.

more intense, particularly in the sciences and engineering. "Students are high achievers," a student says. "Getting A's is hard and teachers have high expectations." In recent years, the university has focused resources on undergraduate education and having senior professors teach more classes. The new president, lured from Yale, has made the expansion of interdisciplinary work—already part of the Duke culture—a priority for faculty and students alike. "Certainly not all professors desire out-of-classroom interaction, but students don't have to look far to find ones that do," a senior English major says. The nationally recognized FOCUS program offers groups of seminars with 15 or fewer students clustered around a single broad theme such as Biotechnology and Social Change, and Humanitarian Challenges at Home and Abroad. It is "an incredible opportunity to engage with the university's top professors," a senior says. Despite Duke's relatively large undergraduate population, 75 percent of courses have 25 or fewer students.

Only 12 percent of Duke students are from North Carolina, although a large fraction of the student body hails from the South, and the Northeastern corridor sends a fair-sized contingent. Despite the unmistakable air of wealth on campus, two-thirds come from public high schools. "Students here are driven and independent," a junior says. Eleven percent of students are African American, 7 percent are Hispanic, and the growing Asian American proportion has reached 20 percent. Students of different ethnicities and races tend to "self-segregate," students say, producing little tension but also little interaction. Overcoming these self-imposed barriers has been an ongoing quest for students and administrators, who conduct a diversity orientation program each year. There is also a student-run Center for Race Relations and a new Campus Culture Initiative, which seeks to evaluate and improve the way Duke educates its students about diversity and conflict resolution.

Duke's Southern gentility is reflected in campus attire, which is generally neatly pressed on guys and maybe a bit outfitty on women, in contrast to the thrown-together anti-status uniform of jeans and sweats that dominates on some other campuses. Duke is also a culturally active campus; theater groups thrive, and the Freewater Film Society shows classic movies each week. During the summer, Duke is home to the splendid American Dance Festival. Undergraduates use the school's cable television system to make and broadcast parodies of game shows and other entertainment.

"Students here are driven and independent."

Despite its imprimatur of wealth, Duke admits students without regard to financial need and guarantees to meet all accepted applicants' full demonstrated need for four years. More than 40 percent of students receive need-based financial aid; four percent of students receive merit scholarships averaging $24,016. There are a number of scholarships earmarked for outstanding African Americans. Like most other NCAA Division I universities, Duke hands out athletic scholarships annually, offering them in seven men's sports and eight women's sports.

Duke undergrads are required to live on campus, and each student is loosely affiliated with one of 60 "living groups," ranging in size from 14 to 250 students (in a nod to the college- or house-based living units at some Ivy League schools). Freshmen all reside in dorms on the East Campus led by a faculty member and his or her family. "The dorms look like castles on the outside and feel like Harry Potter," said a junior, who added that the new dorms "are like five-star hotels." Seniors can move off campus, but "the apartments vary in quality." The decision to have all freshman live on the dry East Campus was aimed at making it easier to adjust to academic life and insulating them from the wilder aspects of Duke's social scene, which attracted national attention following a scandal involving off-campus behavior of members of the men's lacrosse team. Sophomores move to West Campus, where there are also special interest dorms focused on themes such as women's

The new president, lured from Yale, has made the expansion of interdisciplinary work—already part of the Duke culture—a priority for faculty and students alike.

studies, the arts, languages, and community service. Students say the dining choices range from edible to excellent. "There are over 20 eateries to choose from," says a junior, who notes that off-campus restaurants—many that will deliver—are linked to the Duke meal plan. Unused "money" from prepaid meal cards is refunded at the end of the semester, an unusual and much-appreciated policy.

When college counselors say Duke is hot, they're not referring to the temperatures in the South. Duke competes with the Ivies and a select few other colleges for top-notch students.

Duke undergrads take both their studies and their play seriously. "Duke students are the type who will start a club if they are interested in something that nobody else is doing, work hard on a paper late into the night and then go out Thursday, Friday, and Saturday," says a public policy major. Students agree that most social life takes place on campus or in surrounding houses and apartments. "As a college town, it has several of the requisite bars and pizza places, but compared to a place like Madison, Ann Arbor, or Boston, or even Chapel Hill, it's pitiful," says a junior. Although it has been pushed away from the center of campus, "the Greek scene dominates," says a history major. Fraternities and sororities attract 29 percent of men and 42 percent of women, respectively. Fraternity parties are open to everyone, and the free shuttle bus service that connects the school's various dorm and apartment complexes runs until 4 a.m., making it easy to socialize in rooms or suites. There are two tightly regulated bars on campus for students

"The dorms look like castles on the outside and feel like Harry Potter."

who are of age to drink. "Alcohol, it seems, is quite easy to find," says a freshman, at least on West Campus. "The university struggles with the drinking culture," explains a student, "and continues to tinker with the policies."

During the basketball season, men's games always sell out, and the town is proud of its Durham Bulls, the local minor-league baseball team, which coined the term "bullpen." Popular road trips include nearby Chapel Hill, home of archrival UNC, and Raleigh, the state capital and home of North Carolina State University. In warm weather, the broad beaches on North Carolina's outer banks are two to three hours away, while winter ski slopes are three to four hours distant. The popular Oktoberfest and Springfest bring in live bands and vendors peddling local crafts and exotic foods each fall and spring, and concerts are being brought back to Cameron Indoor Stadium.

Relations between Duke and the city of Durham got some unjustified bad press during coverage of the lacrosse scandal. Durham is a working class city that has had its share of racial tensions but also boasts a vibrant African American middle class and good political leadership. The contrast between Duke's wealth and the economic depression afflicting many of Durham's residents is obvious. But Duke as an institution has been active in the community, especially in public schools, and hundreds of undergrads are involved in service learning, tutoring, and related activities. "Community service is huge," said one student. Downtown Durham is undergoing a revival, with old tobacco warehouses being converted into restaurants, stores, offices, and apartments. No one misses the irony of the fact that Durham, once known as the "City of Tobacco," now bills itself as the "City of Medicine." Durham is about 15 minutes from Research Triangle Park, the largest research center of its kind in the world. Duke, North Carolina State, and the University of North Carolina at Chapel Hill created the park for nonprofit, scientific, and sociological research. Many Silicon Valley technologies companies have East Coast outposts in the park, which has helped make the Raleigh–Durham area one of the most productive regions in the nation, with the highest percentage of PhDs per capita in the U.S.

Duke's official motto is *Eruditio et Religio* only to a few straitlaced administrators; everyone else knows it as "Eruditio et Basketballio," which translates more or less as "Go to hell, Carolina"—meaning UNC at Chapel Hill, Duke's archrival in the rough and tough Atlantic Coast Conference. At games, students get the best courtside seats, where they make life miserable for the visiting team. Their efforts paid off in 2001 when the Blue Devils won the national Division I men's basketball championship for the third time in a decade. "By far, basketball season brings out the best of student support," a senior says. Sports-crazed Blue Devils erect a temporary tent city—dubbed "Krzyzewskiville" after the surname of fabled coach Mike—to vie for the best seats. This is far from "roughing it"—students form groups to hold their places so that some fraction can go to class and keep their peers on track academically while those who hold down the fort check their email through wireless con-

> **"By far, basketball season brings out the best of student support."**

nections. "The line for the Duke–UNC men's game starts two to three months in advance," says one fan. The women's basketball team has become a dynasty in its own right and usually makes it to the Final Four. In 2006 the women's golf team won its fourth national championship in eight years. Football is, to put it kindly, rebuilding—a process that has been underway for about 40 years. Intramurals are big and operate on two levels, one for competitive types and one for strictly weekend athletes, which draw heavy participation from the Greeks and guys.

Meandering around Duke's up-to-date campus, you can see the latest technology, but also can hear the whisper of the Old South through those big old trees. "If you come here, there isn't a chance in the world that you won't fall in love with it, with its possibilities and opportunities and people and beauty," one student says. In addition to blending old and new, Duke also does an amazing job combining sports and academia, producing students who almost define the term "well-rounded." But this may be changing. Says a junior, "It's attracting better students, shifting the focus away from basketball and fraternities, and trying to create a more intellectual environment on campus."

Overlaps
UNC–Chapel Hill, University of Virginia, Washington University, Cornell, Penn, Northwestern

If You Apply To ➤

Duke: Early decision: Nov. 1. Regular admissions: Jan. 2. Financial aid and housing: Feb. 1. Guarantees to meet full demonstrated need. Campus interviews: optional, evaluative. Alumni interviews: recommended, evaluative. SATs or ACTs: required. SAT IIs: required. Accepts the Common Application and electronic applications. Essay question: how you responded to someone doing something wrong, most profound or surprising intellectual experience, or a matter of importance to you.

Earlham College

Richmond, IN 47374

BEST BUY

Earlham is a member of the proud circle of liberal colleges in the Midwest that includes Oberlin, Kenyon, Grinnell, and Beloit, to name just a few. Less than half the size of Oberlin and comparable to the other three, Earlham is distinctive for its Quaker orientation and international perspective.

Website: www.earlham.edu
Location: City outskirts
Total Enrollment: 1,272
Undergraduates: 1,178
Male/Female: 42/58
SAT Ranges: V 570–700
  M 530–650

Earlham is a study in contradictions—a top-notch liberal arts college in a conservative city that few could place on a map, and an institution that even in the 21st century remains true to the traditions of community, peace, and justice that are hallmarks of its Quaker heritage. Earlham's curriculum and programs engage students with the world by exposing them to classmates from 61 countries and offering more than 200 academic courses that incorporate an international perspective. A variety of study abroad programs offer close faculty involvement and a thoughtful focus on crosscultural perspectives.

Earlham's 800-acre campus sits in the small, quintessentially Midwestern city of Richmond, just a short distance from Cincinnati and Indianapolis. Georgian-style buildings dominate, surrounded by mature trees and plantings, while the Japanese gardens symbolize the college's long friendship and closeness with Japan. Recent additions to campus include a suite-style residence hall which accommodates 132 students. Each level includes spacious kitchen and dining areas, as well as study spaces, meeting rooms, and spacious porches.

> **"Professors expect students to expect a lot out of themselves."**

To graduate, students must complete general education requirements in the arts, analytical reasoning, wellness, scientific inquiry, foreign language, and, not surprisingly, diversity. Psychology is the most popular major, followed by politics, sociology/anthropology, art, and English. A wide range of interdisciplinary offerings includes such programs as peace and global studies, legal studies, Quaker studies, Latin American studies, and Japanese studies, a field in which Earlham is a national leader. Challenged to think and meet high academic expectations, students see themselves as capable and eager to learn. "More than that, professors expect students to expect a lot out of themselves," says a politics major.

Class discussion, rather than lecture, is the predominant learning style here. Earlham faculty members are selected for their excellence in teaching and their ability to cross disciplinary lines. "The quality of teaching at Earlham is one of its best attributes," states a freshman. "Not only are most professors very experienced and knowledgeable in their field, they are also very approachable and easy to talk to." While profs are available, class outlines demand that individuals "figure things out" by taking the initiative to take their work seriously.

About three-quarters of students eventually pursue postgraduate study, often after taking some time off for a job or to participate in volunteer or service programs. During their undergrad years, more than 70 percent of Earlham students participate in at least one off-campus study experience. Earlham offers study abroad programs in 25 countries, including Mexico, India, England, Spain, Martinique, Northern Ireland, France, and Japan. In a Border Studies program, students live with families in El Paso or Ciudad Juarez and take courses focusing on U.S.–Mexico border issues. Most programs are managed by the college; students first receive preparation for a multicultural experience, and most programs have an onsite director. The popular May term courses send students off campus with faculty for one-month intensive courses in various locations around the world.

Earlham may be small, but its student body is exceptionally diverse. Only 21 percent of the students are Hoosiers; 8 percent hail from abroad, representing 61 different countries. Another 7 percent are African American, with Hispanic students adding 3 percent and Asian Americans 2 percent. With a strong emphasis on conversation, the campus

> **"At a liberal school like Earlham, there are plenty of vegetarians and vegans, so the dining hall is always prepared for special needs."**

is full of well-intentioned activists blazing their own trails through life, albeit on "Earlham Time" (a tardy-favorable clock widely accepted in this laid-back climate). "There is also a huge feeling of community here, and students care more about their fellow students instead of being only concerned for one's self and one's own pursuits," says a peace and global studies major. Merit scholarships are available for qualified students; there are no athletic scholarships.

Eighty-eight percent of Earlham students live on campus in 8 residence halls. Single, double, and triple rooms are available in the two older dorms, which connect with the new Mills Hall. Along with wireless Internet connectivity, the hall features two- to four-bedroom suites sharing private baths, and, on each floor, a kitchen, study room, laundry, and, yes, TV lounges. Dorm space is reserved for first-years, and

(Continued)
ACT Range: 23–29
Financial Aid: 63%
Expense: Pr $ $
Phi Beta Kappa: Yes
Applicants: 1,554
Accepted: 70%
Enrolled: 30%
Grad in 6 Years: 71%
Returning Freshmen: 83%
Academics: ✐ ✐ ✐ ✐
Social: ☎ ☎ ☎
Q of L: ★ ★ ★ ★ ★
Admissions: (765) 983-1600
Email Address:
admission@earlham.edu

Strongest Programs:
Biology
English
Psychology
Politics
Sociology/Anthropology
Art

With a strong emphasis on conversation, the campus is full of well-intentioned activists blazing their own trails through life, albeit on "Earlham Time."

A wide range of interdisciplinary offerings includes such programs as peace and global studies, legal studies, Quaker studies, Latin American studies, and Japanese studies, a field in which Earlham is a national leader.

upperclassmen enter a lottery for the remaining rooms or petition to live together in small houses. All residence halls and academic buildings provide access to the campus computer network, and students may also tap into the extensive resources of Earlham's Lilly or Wildman libraries. Most students eat in the college dining hall, which offers an impressively diverse selection for special diets. "At a liberal school like Earlham, there are plenty of vegetarians and vegans, so the dining hall is always prepared for special needs," says a junior.

Quaker beliefs and Indiana's liquor laws prohibit alcohol on campus, making Earlham a dry campus, at least technically. While any college has its dissenters on alcohol policy making it more realistically a "damp campus," Earlham seems to embrace its policy well enough. "There are a significant number of students who adamantly support the policy," says a junior. "Alcohol is kept quiet and is a significantly less crucial aspect of EC's social scene than at other schools." The atmosphere this creates is very respectful of nondrinkers' decisions and avoids pressure.

With no fraternities or sororities at Earlham, gatherings and parties on weekends may be at a minimum and quiet when they do happen, but on-campus activities abound. Students enjoy improv comedy, a cappella music, equestrian programs, a lip synch competition, fall and spring festivals, concerts, and sports. "Very little of the social life of Earlham students takes place off campus," says a freshman, "simply

"Alcohol is kept quiet and is a significantly less crucial aspect of EC's social scene than at other schools."

because students don't need to seek outside sources of entertainment." Student organizations include numerous cultural, ethnic, and religious groups as well as left-of-center organizations such as Amnesty International and the Earlham Progressive Union. Apart from day trips to Cincinnati, Indianapolis, or Columbus, students stick with a laid-back social atmosphere of visiting with others or checking out one of the musicians, speakers, or other groups that Earlham brings to campus.

Richmond and the surrounding county offer standard American as well as Mexican restaurants, movie theaters, bowling alleys, roller-skating, and golf. Students fan out into the city racking up nearly 50,000 hours of volunteer service a year. Guaranteed to impress, outreach programs are truly getting students involved in their community and building a close relationship with the city. "Volunteerism is an important value of many Earlham students, and despite class work and other commitments, many students still make time to volunteer," says a freshman.

Basketball, flag football, racquetball, and soccer are especially popular intramurals. The school's 16 varsity teams include nearly a third of the student body and compete in Division III sports, including basketball, track, cross-country, baseball, volleyball, and tennis. Volleyball and men's soccer are among the school's strongest squads.

Although Earlham students are based in the Midwest, they graduate ready to take on the world, thanks to the school's cooperative, can-do spirit, international perspective, and caring student/faculty community.

Overlaps

Oberlin, Grinnell, Kenyon, College of Wooster, Macalester, Carleton

If You Apply To ➤

Earlham: Early decision: Dec. 1. Early action: Jan. 1. Regular admissions: Feb. 15. Financial aid: Mar. 1 (priority). Campus and alumni interviews: recommended, evaluative. SATs: required (SATs preferred). SAT IIs: optional. Accepts the Common Application and electronic applications. Essay question.

Eckerd College

4200 54th Avenue South, St. Petersburg, FL 33711

There are worse places to go to school than the shores of Tampa Bay. Eckerd's only direct competitor in Florida is Rollins, which has a business school but is otherwise similar. Marine science, environmental studies, and international studies are among Eckerd's biggest draws.

Attending Eckerd College demands a special sort of willpower. Why? In the words of an international business major: "We are right on the water, and it is like going to college in a resort." With free canoes, kayaks, boats, coolers, and tents always available for student use, it's a wonder anyone studies. But study they do, as administrators continue to lure capable students to Eckerd with small classes, skilled professors, renovated housing, and a reinvigorated social scene. "The standard of education and competition has improved dramatically," says a computer science major.

Founded in 1958 as Florida Presbyterian College and renamed 12 years later after a generous benefactor (of drug store fame), Eckerd considers itself nonsectarian. Still, the school maintains a formal "covenant" with the major Presbyterian denomination, from which it receives some funds. The lush, grassy campus is on the tip of a peninsula bounded by the Gulf of Mexico and Tampa Bay, with plenty of flowering bushes, trees, and small ponds—and it's not unusual to spot dolphins frolicking in the adjacent waters. Campus buildings are modern, and none are taller than three stories.

Freshmen arrive three weeks early for orientation and take a one-credit seminar on the skills required for college-level work. First-years also take a yearlong course called Western Heritage in a Global Context, which focuses on influential books, and students must meet composition, foreign language, information technology, oral communication, and quantitative skills requirements. Also required are one course in each of the four academic areas—arts, humanities, natural sciences, and social sciences—plus one course each in environmental and global perspectives. The capstone senior seminar, organized around the theme "Quest for Meaning," asks students to draw on what they've learned during college to find solutions to important issues. Popular departments include marine biology, business management, international business, environmental studies, and human development.

Watery subjects such as marine science are especially strong. "If you are only interested in swimming with dolphins, don't be a marine science major," says a junior. The program is "very rigorous." The college was granted a Phi Beta Kappa chapter in 2003, making it the

"We are right on the water, and it is like going to college in a resort."

youngest private college ever to receive the honor. Eckerd pioneered the 4–1–4 term schedule, in which students work on a single project for credit each January. Every student has a faculty mentor, and there are no graduate assistants at the blackboards. "Teaching is excellent," a junior says. "Professors are also available outside the classroom to help students." The academic climate is challenging but not competitive, students say. "Eckerd College is pretty laid-back," says a freshman, "but professors expect a lot out of you." A major in interdisciplinary art and a minor in film studies have been added to the curriculum. A Freeman Foundation grant funds significant coursework in the Chinese and Japanese languages.

While St. Petersburg isn't a college town (a senior says it "is closer to a retirement community" and closes up by 10 p.m.) a side benefit to the school's location is the Academy of Senior Professionals, a group of senior citizens who mentor undergrads. Academy members, who come from all walks of life, take classes with students, work

Website: www.eckerd.edu
Location: City outskirts
Total Enrollment: 1,845
Undergraduates: 1,845
Male/Female: 44/56
SAT Ranges: V 510–630
 M 500–620
ACT Range: 22–27
Financial Aid: 38%
Expense: Pr $ $
Phi Beta Kappa: Yes
Applicants: 2,774
Accepted: 72%
Enrolled: 27%
Grad in 6 Years: 58%
Returning Freshmen: 82%
Academics: ✍ ✍ ✍
Social: ☎ ☎ ☎
Q of L: ★ ★ ★ ★ ★
Admissions: (727) 864-8331
Email Address:
 admissions@eckerd.edu

Strongest Programs:
Marine Science
Biology
Psychology
International Relations
Creative Writing and Literature
Management and International
 Business
Environmental Studies

with professors on curriculum development, help students with career choices, and lead workshops in their areas of expertise. About half of Eckerd's students study abroad, in countries ranging from Austria and France to Bermuda and China. The school also has its own campus in London. Marine science programs include a Sea Semester* and the Eckerd College Search and Rescue, which performs more than 500 marine rescues annually and inspires a popular campus T-shirt that tells students to "GET LOST! Support Eckerd Search and Rescue."

Eckerd's president once referred to students as "intellectuals in sandals," says a junior. "I like the quote and it really works for the students." Another student says Eckerd is "just a bunch of laid-back, liberal people who work hard and love where we are at." About two-thirds of the student body hails from out of state, with a large contingent coming from the Northeast; 7 percent are foreign. Hispanics are the largest minority group at 4 percent, African Americans comprise 3 percent, and Asian Americans account for 2 percent. Ten athletic scholarships are awarded annually, and merit scholarships are available to qualified "barefoot and brainy" types.

Seventy-nine percent of students live in one of eight housing quads, separated from the rest of campus by the imaginatively named Dorm Drive. Rooms are fairly large and air-conditioned, and waterfront views and beach access are in-your-face—and free. Two trendy townhouse- and apartment-style residence halls provide suite living and other dorms have been renovated to add computer labs and kitchens in lounges. And how about the grub? "The food is edible and very diverse in all three of the dining facilities," says a junior.

"Eckerd College is pretty laid-back, but professors expect a lot out of you."

There are no Greek organizations at Eckerd, and a strict alcohol policy—no kegs on campus, no alcohol at university events—means wristbands at campus parties, even for those over 21. The policy has been relaxed a bit to allow students of drinking age to imbibe at the campus bar, the Triton Pub, and to drink in public areas of the dorms. Students say those who are underage still manage to get booze and consume it in their rooms, away from prying eyes. "The policies work," says a political science major, "but kids do drink." Off campus, it's next to impossible for underage students to be served at bars and restaurants, students say—though they do enjoy the new Baywalk shopping complex, about 15 minutes from campus, with a stadium-seating movie theater, bars, and restaurants.

On campus, students can partake in concerts, lectures, shows, and games arranged by the student activity board. At the Festival of Hope, seniors present their Quest for Meaning social work. The Kappa Karnival offers rides and games galore. Off campus, students can take in the nightclubs and bars of Latin-flavored Ybor City about 30 minutes away. Tampa and St. Pete also offer a Salvador Dali museum—which Eckerd students get into for free—and professional baseball, football, hockey, and soccer teams. Tempting road trips include Orlando's Walt Disney World and Islands of Adventure theme parks, Miami's South Beach, and that hub of debauchery on the delta, New Orleans.

"I love what I'm learning."

Eckerd doesn't have a football team, but popular intramurals include flag football, soccer, baseball, softball, and the assassin game, in which students try to shoot their peers with dart guns. Varsity teams compete in NCAA Division II, and the men's basketball squad recently competed in the final Elite Eight, and were NCAA conference champs in 2004. "Men's basketball is the only sport that attracts lots of fans and spectators," a senior says, and a night of Midnight Madness helps kick off the season. The co-ed sailing team has claimed several recent divisional and regional championships.

"I love what I'm learning," says a junior marine biology student. Eckerd is striving to add "experiential, service, and international learning" to the traditional classroom experience and attract a higher caliber of students. That mission, combined with new facilities and the fun to be had in the Florida sun, makes Eckerd a standout.

Overlaps

University of Tampa, University of Miami, Stetson, Rollins, College of Charleston, Florida State

Elon University

2700 Campus Box, Elon, NC 27244

A rapidly rising star among liberal arts colleges in the Southeast and an emerging name nationwide. With a welcoming environment and a supportive faculty, Elon is good at taking average students and turning them on to the life of the mind. Strong emphasis on global perspectives and hands-on learning in the classroom.

Elon University derives its name from the Hebrew word for "oak," which is fitting when you consider the many ways in which the school is growing. At each year's opening invocation, entering students are given an acorn. Four years later, they are presented with an oak sapling at commencement. It's a charming tradition and a reminder of how things grow and change. Indeed, it seems everything is changing here—from the name and the mascot, to the buildings, academic majors, and programs. With an emphasis on undergraduate research, group work, service learning, and study abroad, the university also provides its students with plenty of opportunities to grow—intellectually and socially.

Elon was founded in 1889 and occupies a 575-acre campus set in the woods of North Carolina's Piedmont region. It is arguably the most architecturally consistent campus in the nation. Buildings are Georgian-style brick with white trim, and newer buildings have been adapted to modern

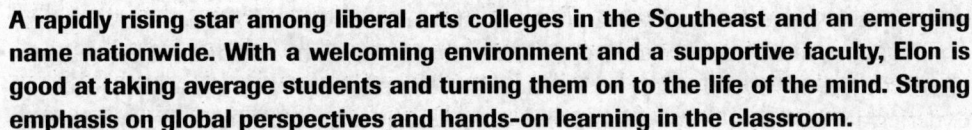

"The academic climate here is definitely under a state of transition, becoming much more competitive and demanding."

architectural lines while maintaining this classic collegiate feel. At the center of campus is Lake Mary Nell, home to an abundance of geese and ducks. Academic buildings are organized in two clusters: an arts and sciences quad near a fountain in the older section of the campus, and a newly constructed "academic village," complete with a colonnade and rotunda. The fountain area serves as an informal gathering spot where students and faculty come together weekly for bagels and coffee, a tradition since 1984.

All of the newer facilities have been designed to support Elon's highly interactive academic programs. The university offers over 40 undergraduate degrees, with strong programs in business, communications, psychology, education, and biology (which also happen to be the most popular). The School of Communications— where the curriculum was recently completely revamped—is nationally recognized and benefits from two ultramodern digital television stations that broadcast seven hours of live programming each week. The communications and business schools were recently accredited, and chemistry and education curricula have been revised as the university strives for accreditation in those areas. Majors in computer information systems and performing arts and design were added recently, along with minors in global information systems, art history, and multimedia authoring. Weaker academic areas include nonviolence studies and Asian-Pacific studies. Students agree the academic climate is rigorous but varies greatly depending on your

Website: www.elon.edu
Location: Small town
Total Enrollment: 4,796
Undergraduates: 4,622
Male/Female: 39/61
SAT Ranges: V 540–620
  M 550–630
ACT Range: 22–27
Financial Aid: 35%
Expense: Pr $
Phi Beta Kappa: No
Applicants: 8,063
Accepted: 41%
Enrolled: 37%
Grad in 6 Years: 74%
Returning Freshmen: 88%
Academics: ✐ ✐ ✐
Social: ☎ ☎ ☎ ☎
Q of L: ★ ★ ★ ★ ★
Admissions: (336) 278-3566
  or (800) 334-8448
Email Address:
  admissions@elon.edu

Strongest Programs:
Business
Communications
Performing Arts
Psychology
Political Science
Biology
Education

major. "The academic climate here is definitely under a state of transition, becoming much more competitive and demanding," a junior says.

Elon has an elaborate support system designed to ensure that first-year students don't fall through the cracks. Students begin general studies with a first-year course called The Global Experience, a seminar-style interdisciplinary class that investigates challenges facing the world. First-year orientations include Move-In Day, in which faculty members literally help students lug their belongings from their cars to their new rooms, and an optional experiential learning program that partners 120 freshmen with returning students for activities ranging from whitewater rafting to volunteer work. Elon 101 is taken by all first-years; students meet weekly in groups of no more than 16 during the first semester and discuss academic, social, and personal concerns with a faculty member and an upper-level student.

"Often, the most social place on campus during the semester is the library."

Elon has an elaborate support system designed to ensure that first-year students don't fall through the cracks.

Students must complete a core that includes English, mathematics, wellness, eight courses in liberal arts and sciences, three courses at the advanced level, an experiential learning component, and a general studies interdisciplinary seminar. The university places a big emphasis on service learning and service research. Ten percent of undergrads are engaged in research work with faculty. Nearly two-thirds study abroad, thanks to the 4–1–4 academic calendar. The Honors Program offers a series of demanding courses that focus on writing and critical thinking skills and the university offers a variety of prestigious fellows programs.

More than three-quarters of all classes have 25 or fewer students, and professors are highly praised. "The environment in class is cooperative, geared towards collective learning and application. I have developed some very meaningful relationships with the professors at the university," explains an economics major. Elon prides itself on attracting students who may not have been academic stars in high school (but who have leadership potential) and turning them on to the life of the mind. "Academics are always a priority, but it does not inhibit an Elon student's desire to be involved," a senior says. "Often, the most social place on campus during the semester is the library." Nearly 70 percent come from outside North Carolina. Eighty-nine percent of Elon's student body are Caucasian, 7 percent African American, 1 percent Hispanic, 1 percent Asian American and 1 percent foreign. Students can vie for hundreds of merit scholarships averaging $3,713, and there are 147 athletic scholarships available.

"I believe in working hard and playing hard, and at Elon, you can have a healthy balance."

Elon 101 is taken by all first-years; students meet weekly in groups of no more than 16 during the first semester and discuss academic, social, and personal concerns with a faculty member and an upper-level student.

Fifty-nine percent of students reside on campus and are required to for their first two years. "The rooms are large and spacious with big windows," says a senior. "Other universities that I visited had dorm rooms that made it feel like I was in a prison cell." Options include traditional residence halls, an academic village complex where students and faculty live and study together, and university-owned apartments. Campus dining gets mixed reviews, but most agree that there are a variety of options for the discerning palate.

When it's time to let off steam, students generally turn to the active Greek scene—which attracts 25 percent of the men and 44 percent of the women—or countless activities on campus. "There is always a lot going on. I like to have fun on the weekends and have rarely been bored here," a junior says. "I believe in working hard and playing hard, and at Elon, you can have a healthy balance." Students say that while alcohol is ever-present, there is little pressure to drink. Says a student, "Elon's alcohol policy focuses on education not prohibition. They encourage students to make wise decisions and optimize their time at Elon." Road trips to the beach (three hours), the mountains (one hour), and Washington, D.C. (five hours) are popular diversions.

The tiny town of Elon is virtually indistinguishable from the university, which even owns the two main restaurants. Students take an active role in the community through volunteer projects both "through certain classes as well as on their own time," says a student. Back on campus, popular events include the Festival of the Oaks, homecoming, and a weekly College Coffee, where students and faculty mingle over free breakfast and coffee.

And let's not forget the road trip to rival Furman University. "Six to eight chartered buses are filled and we tailgate at the university before the game," a student says. Elon competes in the Division I Southern Conference, and offers eight women's sports and seven men's. There's also an intramural program covering 20 sports, which almost half of the students participate in, and a successful club sports program that lets students compete with those at other schools.

Without a doubt, Elon University has come a long way in recent years. "Elon is truly an up-and-coming school. The leadership of the university is taking Elon to a new level of national prominence. "Our programs of study are getting better and better and the campus is growing by leaps and bounds," a communications major reports. By steadily ramping up its educational offerings, growing and improving its facilities, and upping its admissions standards, this quality liberal arts university is quickly outgrowing its local reputation and making a name for itself across the country.

If You Apply To ➤ **Elon:** Early decision: Nov. 1. Early action: Nov. 10. Regular admissions: Jan. 10. Financial aid: Mar. 15. Housing: Jun. 16. Campus and alumni interviews: optional, informational. SATs or ACTs: required. SAT IIs: optional. Accepts the Common Application and prefers electronic applications. Essay question: significant influence; impact of significant achievement; what do you expect to contribute?

Emerson College

120 Boylston Street, Boston, MA 02116-4624

Emerson is strategically located in the heart of Boston's theatre district and within walking distance of the city's major attractions. Media production-film heads the list of strong programs. With roughly 3,000 undergraduates, Emerson is a smaller alternative to neighboring giants Northeastern and Boston U.

Those who aspire to a career in Hollywood or Manhattan may want to start with a four-year stint in Boston. There they will find Emerson College, a small liberal arts school that offers strong programs in communications and the performing arts. Here, students take notes from professors who also happen to be working directors, producers, actors, and writers. It's an approach that helps talented students find their voices and prepare for the spotlight.

Founded in 1880, Emerson is located on Boston Common in the middle of the city's theatre district and features a mix of traditional and modern high-rise buildings. Nearly half of the campus facilities have been refurbished or newly built since 2002, and much of the surrounding city is accessible by foot, including the historic Freedom Trail and the Boston Public Garden. The historic 1,200-seat Cutler Majestic Theatre is the anchor of Emerson's campus. In 2003, the college completed restoration of this landmark to its original 1903 appearance and constructed an 11-story performance and production center for rehearsal space, a theatre design/technology center, costume shop, makeup lab, and television studio. Current construction

Website: www.emerson.edu
Location: Urban
Total Enrollment: 4,324
Undergraduates: 3,133
Male/Female: 45/55
SAT Ranges: V 580–670
  M 550–640
ACT Range: 24–29
Financial Aid: 70%
Expense: Pr $
Phi Beta Kappa: Yes
Applicants: 4,824
Accepted: 47%
Enrolled: 32%

(Continued)

Grad in 6 Years: 74%

Returning Freshmen: 88%

Academics: ✍ ✍ ✍

Social: ☎ ☎

Q of L: ★ ★ ★

Admissions: (617) 824-8600

Email Address:
admission@emerson.edu

Strongest Programs:
Media Production-Film
Writing, Literature, &
    Publishing
Performing Arts
Journalism
Marketing Communication

includes renovations to Boston's Paramount Theater complex to create additional student housing, a cinema, and performing arts space.

Emerson was founded with an emphasis on communication and performance, and the school still offers a plethora of strong programs in this vein. Undergraduates may choose from more than a dozen majors, including acting, broadcast journalism, print and multimedia journalism, film, communication sciences, television/radio, political communication, and theatre design/technology. General education requirements consist of a combination of interdisciplinary seminars and traditional courses. All students must take courses in three areas: foundations, which includes courses in writing, oral communications, and quantitative reasoning; perspectives, which includes courses in aesthetics, ethics and values, history, literature, and scientific, social, and psychological perspectives; and multicultural diversity, which includes classes in global and U.S. diversity. Interdisciplinary seminars of no more than 20 students stress the interrelationships between different communication fields; recent seminars include Minds and Machines; The City; Ways of Knowing: Philosophy and Literature; and Words, Imagination, Expression.

"It's not easy to turn in a short film or write a coherent 20-page creative fiction piece."

The most popular major is media production-film, followed closely by a major that combines writing, literature, and publishing. Performing arts, journalism, studio television production, and marketing communication are also popular. "It's not easy to turn in a short film or write a coherent 20-page creative fiction piece," says a senior. "It's definitely rigorous but it's fun, useful, and very hands-on." In visual and media arts, new production courses focus on animation, screenwriting, and digital media. The college also provides students with access to state-of-the-art equipment and facilities—including digital labs, audio post-production suites, sound mix studios, radio stations, an all-digital newsroom, and television studios—and the campus is home to the oldest noncommercial radio station in Boston.

For those seeking a spotlight and stage in a different setting, Emerson offers a semester-abroad program at Kasteel Well (the Netherlands), where students are housed in a restored 12th-century castle complete with moats, gardens, a gate house, and peacocks. Film students may attend a summer program in Prague, and each year about 200 students spend a semester at Emerson's Los Angeles Center. There, they can participate in internships with companies such as Interscope Records, CNN, Warner Bros., Dreamworks, and NBC. Back on campus, students may crossregister with nearby Suffolk University and the six-member Boston ProArts Consortium.

"Students are very liberal and loud about it."

Nearly two-thirds of all classes have fewer than 25 students, and professors receive high marks for their real-world experience. "The professors are all practitioners in their fields and are incredibly knowledgeable," says a senior. "I'm often in awe of their wisdom and past work," raves a sophomore. The library holds 200,000 volumes and other media, mostly related to communication and performing arts. Students may also take advantage of more than 700,000 volumes in the collections of 10 nearby academic and museum libraries.

The most popular major is media production-film, followed closely by a major that combines writing, literature, and publishing.

Emerson students are "creative, passionate, and artsy," according to a media major. Nearly two-thirds hail from outside of Massachusetts, and 76 percent come from public high schools. African Americans account for only 3 percent of the student body, Hispanics 5 percent, and Asian Americans another 4 percent. Hot campus issues include politics, government, and the war in Iraq. "Politics are huge here," a junior says. "Students are very liberal and loud about it." Emerson offers 204 merit scholarships to qualified applicants, ranging from $5,000 to a half-ride. There are no athletic scholarships.

Forty-eight percent of students live on campus, some in dorms with special theme floors, including the Writers' Block (cute, huh?), and the Digital Culture floor. Freshmen are guaranteed on-campus housing, and the new 14-story residence hall overlooking Boston Common is helping to create a residential feel to the campus. The current residence halls are "really nice" with no "cinderblocks or prison-like rooms," says one sophomore. Others agree that the dorms are spacious and well maintained. Campus dining is not only about eating but also, according to one sophomore, "the social mecca for kids on campus." The food rates well, with lots of options for vegans, vegetarians, and those with special diets. Students feel safe on campus. Each building requires an ID to enter, and public safety officers regularly patrol the streets outside the buildings.

The college also provides students with access to state-of-the-art equipment and facilities—including digital labs, audio post-production suites, sound mix studios, radio stations, an all-digital newsroom, and television studios.

"Social life is very diverse," says a marketing major, and includes plays and film shoots that take place on a regular basis. More than 60 student clubs, organizations and performance groups offer students ample opportunity for involvement, including two radio stations, six humor and literary journals, 10 performance troupes, and six production organizations. "People make tons of friends in organizations," a senior says. Party animals can be found among performing arts majors who "use substances to thrive," according to one student, and among members of the Greek scene, which attracts 5 percent of Emerson men and 4 percent of the women. Though the campus is considered "dry," parties at off-campus apartments make it possible for students to drink. However, "Alcohol is not a huge problem," says a sophomore.

"Emerson students live, study, work, and volunteer in almost every major neighborhood and area of the city."

When students tire of on-campus events, they can rush headlong into Boston, arguably the best college town in the nation. One student says, "There are so many colleges in the Boston area that this city is crawling with young people." Another gushes, "Emerson students live, study, work, and volunteer in almost every major neighborhood and area of the city." There are plenty of diversions, including museums, the Franklin Park Zoo, Freedom Trail, the Boston Symphony Orchestra, and major league baseball at Fenway Park. Back on campus, students enjoy poetry slams and comedy sketches. Popular festivities include EVVY Awards, the largest student production/organization in the country. There is also Hand-Me-Down Night (during which outgoing club officers "hand down" their positions to incoming officers), Greek Week, and the New Student Revue.

Emerson fields 13 Division III athletic teams, and the Lions compete as a member of the Eastern College Athletic Conference and the Great Northeast Athletic Conference. Men's basketball, women's soccer, men's tennis, women's softball, and the cross-country team are perennial GNAC conference finalists. The college has built a new athletic field and gym to support a growing intercollegiate presence. Emersonians also enjoy an active intramural program and take advantage of the 10,000-square-foot fitness center featuring state-of-the-art equipment, classes, and wellness workshops.

While you are not guaranteed to become the next Julia Roberts, Spike Lee, or Brad Pitt, the possibility for stardom exists at Emerson. And even if a lifestyle of fame is not for you, the excellent education, small classes, and attentive professors may teach you how to be the "star" of your own life.

Overlaps

New York University, Boston University, Northeastern, USC, Ithaca, Syracuse

If You Apply To > **Emerson:** Regular admissions: Jan. 5. Financial aid: Mar. 1. Housing: May 1. Meets demonstrated need of 83%. Campus interviews: optional, informational. No alumni interviews. SATs or ACTs: required. SAT IIs: recommended. Accepts electronic applications. Essay question: personal statement.

Emory University

200 Boisfeuillet Jones Center, Atlanta, GA 30322-1950

Often compared to Duke and Vanderbilt, Emory may be most similar to Wash U in St. Louis. Both have suburban locations in major cities and both tout business and premed as major draws. If the campus is uninspiring, the suburban Atlanta location is unbeatable.

Website: www.emory.edu
Location: Suburban
Total Enrollment: 10,024
Undergraduates: 6,378
Male/Female: 44/56
SAT Ranges: V 640–730
   M 660–740
ACT Range: 29–33
Financial Aid: 38%
Expense: Pr $ $ $
Phi Beta Kappa: Yes
Applicants: 12,011
Accepted: 36%
Enrolled: 29%
Grad in 6 Years: 89%
Returning Freshmen: 94%
Academics: 🖉 🖉 🖉 🖉
Social: ☎ ☎ ☎
Q of L: ★ ★ ★ ★
Admissions: (800) 727-6036
Email Address: admiss
   @learnlink.emory.edu

Strongest Programs:
Business
Psychology
Natural Sciences
Political Science
English
History
Art History

Emory University may lack the liberal arts prowess of the Northeastern schools with which it competes, but it's a favorite of preprofessional students from both U.S. coasts. They come for its size (big, but not too big), location, and reputation (increasingly prominent on the national stage). Though most students are clean-cut and career-oriented, a freshman says the population ranges "from preppy, to Northeast and very designer-oriented, to hippie, and everything in between." Regardless of how they're dressed, students are challenged, not coddled, in the classroom; they form study groups and work together to succeed. An atmosphere of friendliness and southern hospitality enhance the vibrant campus life.

Set on 631 acres of woods and rolling hills in the Druid Hills area of Decatur, Emory's campus spreads out from an academic quad of marble-covered, red-roofed buildings. Contemporary structures dot the periphery of the lush, green grounds. In recent years, Emory has gained a new performing arts center, new computer labs, and a new business school. Most recently, ten new sorority townhouses were completed.

Emory's distribution requirements aim to develop competence in writing, quantitative methods, a second language, and physical education, and include exposure to the humanities, social sciences, and the natural sciences. Other required coursework helps broaden students' perspectives on national, regional, and global history and culture. Finally, students take two seminars—one as freshmen (50 to 60 are available each term, limited to 18 students each) and one at an upper level. Entering freshmen seeking a smaller environment may want to consider Emory's two-year Oxford College, where 600 students earn associate's degrees in a "small town" atmosphere, and transfer to the main campus to finish up. Additionally, all freshmen participate in FAME (Freshmen Advising and Mentoring at Emory). The program brings together faculty, staff, and student leaders to mentor first-year students on the aspects of

> **"I went from being the smart kid in my high school to being in college classes composed entirely of these kinds of kids."**

college life. Emory also belongs to the Atlanta Regional Consortium for Higher Education, which lets students take courses at other area schools. The Center for International Programs Abroad (CIPA) offers study in more than 70 locations around the world. Participants earn Emory credit and Emory grades, and they can receive Emory financial aid, scholarships and grants.

Emory can be challenging, even for those accustomed to hard work. "My friends and I jokingly call ourselves 'the students formerly known as gifted,'" quips a sophomore. "I went from being the smart kid in my high school to being in college classes composed entirely of these kinds of kids." Just as Emory has invested in its physical plant, the school has spent lavishly to add star faculty members to key departments, such as Archbishop Desmond Tutu in the school of theology. "Emory has teachers who actually want to teach," says one student, and personal attention is nearly a given. Chemistry and biology benefit from physical proximity to the federal Centers for Disease Control, while many political science professors have ties to the Carter Center (named for the former president, who holds a town hall meeting on campus each year), and serve as regular guests on nearby CNN. The most popular majors are business, economics, political science, psychology, and biology. A 4–2 program allows students to earn a bachelor's degree at Emory, then a master's in engineering at Georgia Institute of Technology. Emory has received a significant portion of Nobel laureate Seamus Heaney's archive, and its Irish studies program is said to rival those of Notre Dame and Boston College.

Students agree that the typical Emory student is hardworking and balanced. "Emory has intelligent students that want to have fun and expand their lives instead of just being dorks and studying all day," says a senior. Only 20 percent of Emory students are Georgians, and a little over half are from the Southeast. New York, New Jersey, California, and Florida are also well represented. African Americans make up 9 percent of the student body, Asian Americans 16 percent, and Hispanics 3 percent. Politically, the campus is less conservative than many southern institutions. "Most people are very aware of what's going on in the world and bring humanitarian and political activism onto campus," a sociology major says. Merit scholarships worth an average of $17,985 are awarded annually; there are no athletic scholarships.

> "Emory has an amazing social life that combines on-campus fun with the best Atlanta has to offer."

Sixty-six percent of Emory students live on campus; freshmen and sophomores are required to do so. Lucky juniors, seniors, and graduate students may hang their hats in the one- to four-bedroom Clairmont Campus apartments, which boast private bedrooms with full-size beds, kitchens, and baths, and a washer-dryer in each unit. Clairmont residents also get an activity center with basketball, volleyball, and tennis courts, a heated, outdoor, Olympic-sized pool, and weight-training facilities. Housing is guaranteed for four years, and students can request to live in a building that is co-ed by floor, co-ed by room, or single sex. They can also request a specific roommate. "Emory has dining options for even the pickiest eaters," says a satisfied junior. In addition to the dining halls, there are small cafés, grills, and a food court on campus.

"Emory has an amazing social life that combines on-campus fun with the best Atlanta has to offer," says a junior. Fraternities and sororities attract 31 percent of Emory's men and 33 percent of the women so, of course, Greek parties are prevalent. Upperclassmen enjoy the Atlanta bar scene. Other options include college nights at local dance clubs and concerts organized by the Student Programming Committee, which recently brought Usher, Guster, and Dave Matthews Band to campus. Alcohol isn't allowed in the freshman dorms, though "if you want to drink, it is possible," a sophomore says. A very popular highlight of the social calendar is Dooley's Week, a spring festival in honor of Emory's enigmatic mascot, James W. Dooley, a skeleton who reportedly escaped from the biology lab almost 100 years ago. If Dooley walks into your class, the class is dismissed, and the week culminates with a costume ball in his honor. Freshman halls also have Songfest, a competition where residents make up spirit-filled song-and-dance routines.

> "Emory has dining options for even the pickiest eaters."

Popular road trips include: Stone Mountain for hiking or a picnic; the cities of Athens and Savannah for concerts, beaches, or big-time college sports; Florida and the Carolinas.

Atlanta also offers a multitude of diversions, from Braves baseball, Thrashers hockey, and Hawks basketball to plays at the Fox Theatre, exhibits at the High Museum of Art, and shopping at Underground Atlanta or the Lenox Mall, to which Emory provides a free shuttle every Saturday.

Emory doesn't field a varsity football team, though students show their spirit with T-shirts that proudly proclaim: "Emory Football: Still Undefeated (since 1836)." Recent national champs include the men's and women's tennis teams and women's

All freshmen participate in FAME (Freshmen Advising and Mentoring at Emory). The program brings together faculty, staff, and student leaders to mentor first-year students on the aspects of college life.

A very popular highlight of the social calendar is Dooley's Week, a spring festival in honor of Emory's enigmatic mascot, James W. Dooley, a skeleton who reportedly escaped from the biology lab almost 100 years ago.

swimming; men's tennis, men's swimming and diving, women's volleyball, and men's golf are competitive as well. Men's basketball competes in the University Athletic Association, against such academic powerhouses as the University of Chicago, Johns Hopkins, and Carnegie Mellon. Most students join at least one intramural sports team at either a competitive or a recreational level. Popular intramurals include flag football, volleyball, soccer, basketball, water polo, and ultimate Frisbee.

While many Southern schools suffer from a regional provincialism, that isn't true at Emory, which blends a focus on teaching and research to nurture creativity and graduate leaders who are highly sought-after in the working world—and by postgraduate law, medical, and business programs. "It is an awesome environment for academics and athletics and social life," says a junior. "You make some life-long friends."

If You Apply To ➤

Emory: Early decision: Nov. 1, Jan. 1. Regular admissions: Jan. 15. Financial aid: Apr. 1. Housing: May 30. Guarantees to meet full demonstrated need. No campus or alumni interviews. SATs or ACTs: required. SAT IIs: recommended. Accepts the Common Application and electronic applications. Essay question: half-page on a meaningful activity or work experience, and half-page on why Emory is a good match; and one to three pages on any topic of genuine personal interest.

Eugene Lang College—The New School for Liberal Arts

(formerly New School for Social Research)
65 West 11th Street, New York, NY 10011

Eugene Lang College is home to more than 800 street-savvy, freethinking students. New York City is the campus, and Lang offers little sense of community. In keeping with the New School's traditional ties to Europe, an internationalist perspective is predominant. Strong in the arts and humanities.

Website: www.newschool.edu
Location: Urban
Total Enrollment: 985
Undergraduates: 939
Male/Female: 32/68
SAT Ranges: V 568–673
M 530–620
ACT Range: 20–26
Financial Aid: 80%
Expense: Pr $ $
Phi Beta Kappa: No
Applicants: 1,244
Accepted: 61%
Enrolled: 19%
Grad in 6 Years: 50%
Returning Freshmen: 73%
Academics: ✍ ✍ ✍
Social: ☎
Q of L: ★ ★ ★
Admissions: (212) 229-5665

Students seeking a typical college experience—large lectures, rowdy football games, and rigid academic requirements—would do well to avoid Eugene Lang College. That's because Lang has no required courses, seminars instead of traditional lectures, and not a single varsity sport. Instead, this small, urban liberal arts college offers individualized academic programs, small classes, and a campus that reflects the quirky and kinetic atmosphere of Greenwich Village. "We don't want to become business leaders, but instead teachers, community organizers, thinkers, professors, and writers," a junior says. "Students here want to change the world—and I think in many ways we are."

Lang fits right in amid the brownstones and trendy boutiques of one of New York's most vibrant neighborhoods. The majority of Lang's classrooms and facilities are in a single five-story building between Fifth and Sixth Avenues on West 11th Street, although The New School occupies 16 buildings along Fifth Avenue. NYU and the excitement of Greenwich Village and Washington Square Park are just a few blocks away.

The New School was founded in 1919 by a band of progressive scholars that included John Dewey, Charles Beard, and Thorstein Veblen. A decade and a half later, it became a haven for European intellectuals fleeing Nazi persecution, and over the years it has been the teaching home of many notable thinkers, including Buckminster Fuller and Hannah Arendt. Created in 1978, the undergraduate college was renamed in the late '80s for Eugene Lang, a philanthropist who (surprise, surprise!) made a significant donation to the school.

The two most distinctive features of Lang College are the small classes—fewer than 16 students—and undergraduates designing their own path of study with no required courses. As freshmen, students choose from a broad-based menu of seminars, and as sophomores they select from 11 paths of study: cultural studies and media; literature; writing; the arts (including dance and theater); philosophy; psychology; science, technology and society; social and historical inquiry; urban studies; religious studies; and education studies. In their final year, students take on advanced "senior work" through a seminar or independent project that synthesizes their educational experience. The standard courseload is at least four seminars a semester, with topics such as From Standup to Shakespeare and the History of Jazz. All first-

"Students here want to change the world—and I think in many ways we are."

year students must take one year of writing and workshops focusing on nonacademic concerns and library research skills. Cooperation, not competition, is the norm. "The courses are rigorous in the way that they ask for mental, emotional, and intellectual energy from each student," a junior says. "Students are usually excited about what they are taking, so the rigor takes a back seat to the excitement involved in learning," another student adds.

Lang's top offerings include political and social theory, writing, history, literature, and literary theory. Its city location lends strength to the urban studies and education programs. Writing is highly praised, especially poetry, and theater is strong. The natural sciences and math are weak, though courses are offered through an arrangement with nearby Cooper Union. While introductory language courses are plentiful, upper-level language offerings are limited. And the college has beefed up its offerings on the history and literature of Third World and minority peoples, which were already better than those at most colleges. The professors at Lang are well versed and engaging, according to many students. "Getting lectured is rare, and I appreciate the way that class discussions are so well planned and thought out," says one student.

The main academic complaint is the limited range of seminars, but outside programs offer more variety. After their first year, students may enroll in a limited number of approved classes in other divisions of The New School. A joint BA/BFA with Parsons The New School for Design has proven very popular. There's also a BA/BFA program in jazz and a BA/MA in media studies with The New School for General Studies' communications department. A newer addition is the exchange program with Sarah Lawrence College, established to provide motivated students with additional academic opportunities. The New School's library is small, but students have access to the massive Bobst Library at nearby New York University.

Lang College attracts a disparate group of undergraduates, but most of them can be described as idealistic and independent. "We are nontraditional college students who relish in this difference and exciting uniqueness that sets us apart from conforming NYU students," a junior says. Some are slightly older than conventional college age and are used to looking after themselves.

"The courses are rigorous in the way that they ask for mental, emotional, and intellectual energy from each student."

Nine percent are African American or Hispanic, another 4 percent are Asian American, and 3 percent are foreign. A junior says "the freedom involved with concentrations, as opposed to majors, attracts the type of student that really wants to just learn things." Twenty-nine percent of Lang's students are from New York and many cite the school's location as one of its best features. "Whatever is desired can be found somewhere in New York City," says a junior. "It's a nice place to be if you want to party or be a stone-cold intellectual." Lang College admits students regardless of their finances and strives to meet the demonstrated need of those enrolled. However, the school does not guarantee to meet the demonstrated financial need of

(Continued)
Email Address:
lang@newschool.edu

Strongest Programs:
Writing
Fine Arts
Education Studies
Cultural Studies

The social network at Lang is quite small, and like many things, is left up to the student.

The main academic complaint is the limited range of seminars, but outside programs offer more variety.

all admits. A deferred-payment plan allows students to pay tuition in 10 installments, and there are various loan programs available. There are a handful of merit scholarships, but no athletic awards.

Dorm life at Lang engages one-third of the student body, though the rooms are in good shape. One student offers this assessment: "Union Square is comfortable and fun to live in. Loeb Hall is the newest and is mostly for freshman. Marlton Hall is in sort of a drab location…and is just old and generally uncomfortable." Off-campus dwellers live in apartments in the Village, if they can afford it, or in Brooklyn or elsewhere in the New York City area. Eighty percent of freshmen live on campus. A meal plan is available, but most students opt for the hundreds of delis, coffee shops, and restaurants that line Sixth Avenue.

The social network at Lang is quite small, and like many things, is left up to the student. "Our lack of campus kind of makes all activity 'off campus,' though we do have dances and club activities within the school facilities themselves," a junior

> **"Our lack of campus kind of makes all activity 'off campus,' though we do have dances and club activities within the school facilities themselves."**

says. The social activities found on campus generally involve intellectual pursuits such as poetry readings and open-mic nights, as well as typical college activities like the student newspaper and the literary magazine. Occasionally, students organize dances and parties, like the Spring Prom, a catered affair with live music that is "a satirical offshoot of the high school tradition." Students generally avoid drinking on campus, and when they do imbibe, alcohol is "far from the central focus of activity," says a junior. With no intramural or varsity sports, another junior says "the most popular sports including 'chopstick using' and 'car dodging.'"

Students relish the freedom and independence they have at Eugene Lang College. For a student who yearns for four years of "traditional" college experiences, Lang would likely be a disappointment. But for those desiring an intimate education in America's cultural capital, Lang offers all the stimulation of the city it calls home.

Overlaps

Sarah Lawrence, NYU, Bard, Hampshire, Fordham

If You Apply To ➤

Eugene Lang: Rolling admissions. Does not guarantee to meet full demonstrated need. No campus interviews. Alumni interviews: optional, informational. SATs or ACTs: required. SAT IIs: required. Students taking the ACT must take the Writing Test. Accepts Common Application and prefers electronic applications. Essay question.

The Evergreen State College

Olympia, WA 98505

There's no mistaking Evergreen for a typical public college. Never mind the way-out garb favored by its students. Evergreen's interdisciplinary, team-taught curriculum is truly unique. To find anything remotely like Evergreen, you'll need to go private and travel East to places like Hampshire or Sarah Lawrence.

Website: www.evergreen.edu
Location: City outskirts
Total Enrollment: 3,799
Undergraduates: 3,642

In "La Vie Boheme," the anthem of Jonathan Larson's rock opera *Rent*, one of the characters asks, "Anyone out of the mainstream / Is anyone in the mainstream?" At Evergreen State College, the answer has always been a vehement "No!" The school's unofficial motto is *Omnia Extares*, Latin for "Let it all hang out." Founded in 1967 as Washington State's experimental college, Evergreen lacks grades, majors, and even departments. This system may sound strange, but it works: Alumni include Matt Groening, creator of *The Simpsons*, *Futurama*, and *Life in Hell*. "The character of the school is openly artistic, earth-friendly, musically open, and a place to truly be an individual," says a freshman. "Students are free to explore and be whoever they want to be," agrees a sophomore.

Evergreen lies in a fir forest at the edge of the 90-mile-long Puget Sound. The peaceful, 1,000-acre campus includes a 24-acre organic plant and animal farm, as well as 3,300 feet of undeveloped beach. Most of Evergreen's buildings are boxy concrete-and-steel creations, though the Longhouse Education and Culture Center is designed in the Native American style typical of the Pacific Northwest. Seminar II, a five-building, 160,000-square-foot complex, opened in 2004. They are the first new academic buildings to be built on campus in more than 25 years. The complex incorporates "green" design concepts, construction methods, and materials, and houses classrooms, workshop spaces, lecture halls, art studios, and offices. Current projects include renovations to the library and campus housing.

> "Students are free to explore and be whoever they want to be."

At first glance, Evergreen's wide-open curriculum looks like Easy Street: It's based on five "planning units"—culture, text, and language; environmental studies; expressive arts; scientific inquiry and society; and politics, behavior and change— which means no required classes and few traditional exams to slog through at the end of each 10-week quarter. And instead of signing up for unrelated courses to fulfill distribution requirements, students enroll in a coordinated "program" team-taught by multiple professors. One recent program, Problems Without Solutions, looked at AIDS and homelessness from the perspective of political science, philosophy, anthropology, economics, statistics, and writing. Still, there is some structure at the college. Freshmen select an interdisciplinary core program, while upperclassmen concentrate in more specialized areas, often writing a thesis or fulfilling an Individual Learning Contract developed in partnership with a faculty sponsor.

Students praise Evergreen's environmental science major, which offers classes in ornithology, marine biology, and wetlands studies. To supplement their coursework, environmental scientists may also study marine animals while sailing in Puget Sound, spend seven weeks at a bird sanctuary in Oregon, or trek to the Grand Canyon or the tropical rainforests of Costa Rica. Various arts programs—dance, writing, visual arts, and media arts—also get high marks. And regardless of what they study, students warn that—while the integrated approach to learning may improve comprehension and deepen understanding—it likewise means a lot of work. "The courses can be rigorous," says a junior. A senior says that students can expect a "comfortable pace and high expectations."

Because Evergreen attracts many nontraditional students, and students who are older than the typical college freshman, administrators take advising and career counseling seriously. They've also asked faculty members to do more to help students adjust to life on campus. "It's not about teaching—it's more like guidance through a subject," a junior says. "If students want to understand the subjects, they must motivate to do more on their own." Another bonus: Because Evergreen doesn't award tenure, there's less pressure for professors to conduct research and publish their findings—and less to distract them from teaching undergraduates.

> "It's not about teaching—it's more like guidance through a subject."

"We have a lot of hippies, tree-huggers, and activists," a student says. Seventy-eight percent of Evergreen's students are Washington natives; African Americans and Native Americans each account for 5 percent of the student body, while Asian Americans and Hispanics add another 4 percent each. Hot topics on campus include

(Continued)
Male/Female: 45/55
SAT Ranges: V 530–650
   M 480–600
ACT Range: 21–27
Financial Aid: 56%
Expense: Pub $
Phi Beta Kappa: No
Applicants: 1,657
Accepted: 97%
Enrolled: 38%
Grad in 6 Years: 56%
Returning Freshmen: 70%
Academics: ✍ ✍ ✍
Social: ☎ ☎ ☎
Q of L: ★ ★ ★ ★
Admissions: (360) 867-6170
Email Address:
   admissions@evergreen.edu

Strongest Programs:
Environmental Science
Media Arts
Physical and Biological
   Sciences
Social Sciences
Computer Science
Humanities

Seminar II, a five-building, 160,000-square-foot complex, opened in 2004. The complex incorporates "green" design concepts, construction methods, and materials, and houses classrooms, workshop spaces, lecture halls, art studios, and offices.

gender and sexuality, U.S. foreign policy, and (surprise!) environmental issues. Evergreen awards merit scholarships each year, averaging $3,793 each, along with 33 athletic scholarships in four sports. Special admissions consideration is given to applicants 25 years of age and older, as well as Vietnam-era veterans and applicants whose parents are not college graduates.

Twenty-two percent of Evergreen students, mostly freshmen and sophomores, live on campus. The school's apartment complexes have single bedrooms, shared bathrooms—with bathtubs, not just shower stalls—and full kitchens, according to one student. "The dorms are comfortable and well maintained and give you the opportunity to meet new people," a sophomore says. Still, because campus housing can be expensive, most upperclass students live off campus. There's an efficient bus system to get nonresidents to class on time, though it helps to have a car. Evergreen's food service offers a wide variety of dishes, including vegetarian and vegan options; some choices are organic, too. That said, opinions on the quality of campus chow are mixed.

Not surprisingly, Evergreen lacks a Greek system, but students say the housing office organizes plenty of weekend events—including open-mic nights, soccer and other field games, performances, and parties. Nearby Olympia (the state capital) doesn't really qualify as a college town, but it is progressive and open-minded, with art walks through local galleries, coffee shops, clothing stores, co-ops, and lots of live music. Even better, "Evergreen is 10 minutes from theaters, parks, and recreation, an hour or so from skiing, hiking, or the beach, and about an hour from Seattle," says a student concentrating in sociology. The college offers all types of outdoor equipment for rent, from backpacks and skis to kayaks and sailboats. Its large College Activities Building houses a radio station and the student newspaper, along with space for student gatherings. Portland and the rugged Oregon coast (three to four hours away) provide other changes of scenery for students with wheels; everything is kept green and lush by the (interminable) rain, which stops in time for summer break and resumes by October.

"We have a lot of hippies, tree-huggers, and activists."

You may chuckle at Evergreen's mascot, an eight-foot clam named "Gooeyduck," for the large geoduck clams found in Puget Sound, but the school is getting more serious about organized sports. Its basketball, cross-country, soccer, and women's volleyball teams compete in the NAIA Region I. While those squads haven't brought home any titles yet, Evergreen has produced All-American players. A mere 8 percent of students participate in recreational or intramural sports, which include Frisbee, volleyball, skiing, tennis, basketball, and sailing.

Evergreen isn't for everyone; indeed, this remains one of the best choices for students who think they were born decades too late. Freed from requirements and grades, Greeners delight in exploring the connections between disparate disciplines at their own pace. Succeeding in that endeavor, however, requires an incredible ability to focus; while some students would find the task burdensome, students here welcome the challenge. "It's up to us to learn what to believe in, not some instructor telling us what it's going to be," says a satisfied math and computer science student.

Overlaps

Western Washington, University of Washington, Central Washington, Hampshire, Washington State, UC–Santa Cruz

If You Apply To > **Evergreen:** Rolling admissions: Mar. 1. Financial aid: Mar. 15. Housing: Jun. 1. Guarantees to meet full demonstrated need. Campus interviews: optional, informational. No alumni interviews. SATs or ACTs: required. SAT IIs: optional. Accepts electronic applications. Essay question: academic preparation and why you're ready for college-level studies at Evergreen, along with a description of educational and career goals, and how Evergreen will help you reach them.

Fairfield University

Fairfield, CT 06824

Fairfield is one of the up-and-coming schools in the Roman Catholic higher education scene. Undergraduate enrollment has grown in recent years. Strategic location near New York City is a major attraction. Lack of big-time sports keeps Fairfield from enjoying the visibility of Boston College/Holy Cross.

No doubt about it, Fairfield University is moving into the same class as older, more revered East Coast Jesuit institutions. The school provides a dynamic living and learning environment, combining solid academics, real-world opportunities in and outside of the classroom, and an abundance of community service projects. Extensive study abroad programs are available as well.

The physical beauty of the university's scenic, tree-lined campus just 90 minutes from Manhattan is a source of pride. The administration takes pains to preserve a lush atmosphere of sprawling lawns, ponds, and natural woodlands. Buildings are a blend of collegiate Gothic, Norman chateau, English manor, and modern. Students enjoy a 24-hour computer lab, Geographic Information Systems lab, wireless 125-person computer lab in the School of Nursing, and wireless area in the Barone Campus Center. The Kelley Administrative Center opened in 2006 and houses administrative departments. In addition, the university has opened six state-of-the-art "smart" classrooms, with four more on the way.

Students may have difficulty finding time to enjoy the beautiful facilities. A demanding class schedule requires everyone to complete the liberal arts core curriculum over four years, with two to five courses from each of five areas: math and natural sciences; history and social and behavioral sciences; philosophy, religious studies, and applied ethics; English and visual and performing arts; modern and classical languages. The core constitutes almost half of a student's total courseload. "This pretty much means that a student is taking at least 15 credits a semester. Being a major in the sciences increases that load," says a senior physics major who has taken up to 21 credits at a time. Fairfield's main academic strengths are business (accounting, finance, and management), art history, communication,

> **"Students always find ways around the policies and drink in dorms, in upperclass housing, and off campus."**

and religious studies. "Fairfield is mainly known by reputation as a liberal arts school, but I believe their biology program is underrated," says one biology major. Upperclassmen with the necessary academic standing can design their own majors.

Fairfield's academic climate is challenging. "Most courses are competitive," reports a senior. "The workload is challenging but fair." Recent additions to the curriculum include minors in information systems; operations management; classical studies; Irish studies, which has strong ties to the University of Galway; and Italian studies, with links to the the Florence University of the Arts. Engineering, nursing, and business curricula have been revamped, and a major in new media, television, and film has been added, as has a minor in Catholic studies. Engineering students may enroll in joint five-year programs with the Rensselaer Polytechnic Institute, Columbia University, the University of Connecticut, or Stevens Institute of Technology. MBA candidates can choose a concentration in e-business or minor in information systems or operations management. Since 1993, 39 Fairfield students have been awarded Fulbright Scholarships for studies abroad. About 8 percent are part of the four-year honors program. Students looking to travel abroad without committing a full semester can take a trip with one of several professors who lead educational summer tours for

Website: www.fairfield.edu
Location: Suburban
Total Enrollment: 5,173
Undergraduates: 3,688
Male/Female: 43/57
SAT Ranges: V 560–640
    M 580–660
ACT Range: 23–27
Financial Aid: 45%
Expense: Pr $ $
Phi Beta Kappa: Yes
Applicants: 8,035
Accepted: 61%
Enrolled: 19%
Grad in 6 Years: 81%
Returning Freshmen: 91%
Academics: ✍ ✍ ✍
Social: ☎ ☎ ☎ ☎
Q of L: ★ ★ ★ ★
Admissions: (203) 254-4100
Email Address:
    admis@mail.fairfield.edu

Strongest Programs:
Art History
Communication
Religious Studies
Accounting
Finance
Management
Marketing
Nursing

credit. About 225 students study abroad each year, through their choice of more than 100 programs in 50 nations. Sophomores can join the Ignatian Residential College; afterward, they can continue on to the Companions program, which includes cultural activities and mentoring.

Freshmen are introduced to Fairfield with a thorough orientation program. A formal academic convocation in the first week of classes includes a speaker chosen to reflect the school's Jesuit values. There are no teaching assistants at Fairfield; 74 percent of classes have 25 or fewer students, and none have more than 50. "The professors are very available and they take the time to get to know their students," a sociology major says. "They are enthusiastic about the subjects they teach."

The vast majority of Fairfield's students come from Roman Catholic families, and 25 percent are from Connecticut. Minority enrollment is small, with African Americans constituting 1 percent of the student body, Hispanics 4 percent, and Asian Americans another 3 percent. "The students at Fairfield are largely white, upper-middle class or upper-class, and from the Northeast," a senior says. "The diversity of opinion and ethnicity is not great." To help students with Fairfield's steep price, the school offers merit scholarships annually, averaging $11,762, as well as athletic scholarships in 19 sports.

"Fairfield seeks to help people realize that there is more to life than your marketing degree."

Fairfield's "comfortable and well-maintained" residence halls house 85 percent of the student body. The school recently built a new upperclassman apartment village, and four of the traditional halls have been renovated. "The dorms are nice but usually overcrowded for freshmen," one student says. Seniors are given the opportunity to experience living off campus and among the most popular options are the privately owned beach houses and apartments on Long Island Sound. Meal plan options are available to all students, and many say the food is constantly improving. "Although students gripe, we have a pretty vast selection," admits one sophomore.

Fairfield's proximity to the beaches of the Long Island Sound, a quick five-minute drive from campus, provides students with a scenic social space for everything from romantic retreats to lively parties. Still, students say much of the social life takes place on campus, where sponsored events range from dances to hanging out at the coffeehouse and on-campus pub to concerts with stars such as Blues Traveler and Third Eye Blind. Harvest Weekend at the end of October and Dogwoods Weekend at the end of April provide relief from the stress of studying. Road trips to New York (an hour by train) and Boston (two hours away) are popular. The college recently instituted a new program for alcohol misuse or abuse. "Students always find ways around the policies and drink in dorms, in upperclass housing, and off campus," one senior reports. Although Jesuits are very much in evidence and often live in the dorms, students say they do not hinder the social scene. The Campus Ministry draws a large following, with daily masses, retreats about three times a semester, and regular community service work, including two weeks of programs in the Caribbean and Latin America. But one student complains, "Our Jesuit identity needs to be expressed more clearly and something more must be done to excite students about the spiritual life."

"The workload is challenging but fair."

As for the surrounding area, some students say the quaint, wealthy Fairfield area can feel a bit "snobby," though there are plenty of shopping outlets and restaurants that fit college student budgets. Beach residents don't always approve of beach-apartment students and their activities. "Community members are not often comfortable participating in university events, and vice versa," a student says. Volunteerism abounds, much of it in the nearby city of Bridgeport. "The influence of Bridgeport on my Fairfield experience has been profound," says a student.

Athletics have come of age at Fairfield. Competitive teams include men's and women's soccer, men's and women's lacrosse, men's and women's tennis, volleyball, and softball. Men's and women's basketball both draw crowds, and the boisterous home-court fans, who come to games in full Fairfield regalia, have been dubbed the "Red Sea." The university moved varsity football and men's ice hockey into club sports as a budget-cutting move. Living up to the Jesuit motto of sound mind and body, about three-quarters of students play on one of 25 intramural teams, whose exploits are copiously chronicled in the campus newspaper. "Many students sign up and create teams with their friends, and have a great time playing during the weekends," one student says. The school also takes pride in its high graduation rate for athletes, regularly one of the highest rates in the country.

Fairfield University has combined several traditions to create a rich undergraduate experience, including close bonds with faculty, an emphasis on community involvement, challenging academics, and an emphasis on the holistic development of each student. "Fairfield seeks to help people realize that there is more to life than your marketing degree," a satisfied student says. "Students are taught to open their eyes and see the real world."

Overlaps

Boston College, Providence, Villanova, Loyola (MD), College of the Holy Cross, Fordham

If You Apply To >

Fairfield: Early action: Nov. 15. Regular admissions: Jan. 15. Financial aid: Feb. 15. Meets full demonstrated need of 26%. Campus and alumni interviews: optional, informational. SATs or ACTs: required. SAT IIs: optional. Accepts the Common Application and electronic applications. Essay question: Common Application; personal statement.

University of Florida

Gainesville, FL 32611

It should come as no surprise that UF is a world leader in citrus science. Throw in communications, engineering, and Latin American studies to the list of renowned programs. Among Deep South public universities, only the University of Georgia rivals UF in overall quality.

Set on 2,000 acres of rolling, heavily forested terrain in north-central Florida, the University of Florida is an athletic powerhouse, and administrators are working hard to gain the same level of national recognition for their academic offerings. The school is massive and continues to grow, and in this case, bigger does seem to be better. While some students certainly get lost in the shuffle, those who can navigate the bureaucratic red tape will find ample resources at their fingertips, including the world's largest collection of butterflies and moths and an $85 million Cancer and Genetics Research Complex.

UF's central campus has 21 buildings on the National Register of Historic Places. Most are collegiate Gothic in style—redbrick with white trim. They're augmented by more modern facilities, including a 173,000-square-foot complex for nursing, pharmacy, and the health professions; and an honors dorm complex, which offers suite-style living, a computer lab, classrooms, and a full-time honors staff in residence. UF's research capabilities and equipment are likewise impressive, and a boon to aspiring physicians. The school has one of the nation's few self-contained intensive care hyperbaric chambers for treatment of near-drowning victims, and a world-class, federally funded brain institute. Media types flock to the school's public TV

Website: www.ufl.edu
Location: City center
Total Enrollment: 43,211
Undergraduates: 31,742
Male/Female: 46/54
SAT Ranges: V 570–670
  M 590–690
ACT Range: 25–29
Financial Aid: 39%
Expense: Pub $
Phi Beta Kappa: Yes
Applicants: 21,151
Accepted: 57%
Enrolled: 60%
Grad in 6 Years: 79%
Returning Freshmen: 94%

and radio stations, and to its two commercial radio stations. Pugh Hall, scheduled for completion this year, will house the Bob Graham Center for Public Service and will train students in languages, culture, and other skills vital to careers in public service.

Academically, UF's strongest programs are those with a preprofessional bent, including engineering, tax law, and pharmacy. Popular majors include business administration, finance, elementary education, and advertising. Students also give high marks to the College of Journalism and Communications, the first in the nation to offer students an electronic newsroom. Students mention foreign languages and math as weaker, saying they rely too much on teaching assistants. Fine arts and music also suffer, probably because they're perceived as less helpful in the eventual job search. Students with weighted high school GPAs of 4.0 or higher and SAT scores of at least 1400 are invited into the Honors Program, where most classes are limited to 25 students. The program offers honors sections in standard academic subjects, and interdisciplinary courses such as masterworks of music, writing and love, and the history of rock and roll. Additionally, honors students are invited to live in the Honors Residential College. The University Scholars Program offers $2,500 stipends

> "The dorms are comfortable and clean and there is a great sense of community."

to 205 students each year for one-on-one research with a faculty member. Results must be published in the online Journal for Undergraduate Research or another peer-reviewed journal. Even if you don't qualify for those options, you should find something of interest, since only Ohio State and the University of Minnesota offer more degree programs on one campus than UF, among the five biggest universities in the nation.

To balance students' preprofessional coursework, UF's general education program requires credits in composition, math, humanities, social and behavioral sciences, and physical and biological sciences. Students must also take six credits in the humanities or social or physical sciences that focus on themes of internationalism or diversity. Internships abound, along with volunteer and leadership opportunities and foreign study in Latin America, Asia, the Middle East, and more than a dozen cities in Eastern and Western Europe. UF has recently added the following majors: computer science, landscape and nursery horticulture, astronomy, athletic training, and biology.

"The classes at UF are challenging," says a senior, "but not impossible." While UF offers programs in every conceivable discipline, like many super-sized schools, it also forces students to climb a mountain of bureaucracy to get the courses and credits they need. Occasionally, for example, lectures in the College of Business Administration have to be videotaped and rebroadcast on the campus cable network, so that everyone can see them. Still, administrators are working to fix these problems. Professors often have deep professional experience, and bring enthusiasm to their work, though students often find TAs behind the lectern. "Many of the lower division courses are taught by graduate students," says a student, "and while some are good instructors, others may be too engrossed in their own work to provide adequate teaching."

UF is one of Florida's 11 state universities, and 77 percent of students here hail from the Sunshine State. Despite the geographical homogeneity, they're an ethnically diverse bunch, with African Americans adding 8 percent of the study body,

> "Gainesville revolves around UF."

Asian Americans 7 percent, and Hispanics 11 percent. UF has established the Latino-Hispanic Cultural Center to serve the largest minority on campus, and many African American students belong to historically African American fraternities and sororities. The multicultural celebration known as People Awareness Week has grown into a popular campus event, but students say that, unfortunately, while there's no hostility between ethnic groups, there's also not much mixing. UF offers

(Continued)
Academics: 🖊🖊🖊🖊
Social: ☎☎☎☎
Q of L: ★★★★
Admissions: (352) 392-1365
Email Address:
www.admissions.ufl.edu

Strongest Programs:
Business
Engineering
Journalism/Communications
Citrus Science
Latin American studies

Academically, UF's strongest programs are those with a preprofessional bent, including engineering, tax law, and pharmacy.

more than 200 athletic scholarships and thousands of merit scholarships. National Merit Scholars automatically qualify if they list UF as their first choice by the required date.

Twenty-two percent of Florida's undergrads live on campus, and students say rooms are tough to come by if you're not a first-year student. "The dorms are comfortable and clean," says a senior, "and there is a great sense of community." Doubles, triples, and suites in the co-ed dorms are awarded by lottery, based on Social Security numbers, and there just isn't enough room for everyone. Dorm-dwellers buy the campus meal plan or use kitchens in their residence halls. Fifteen percent of UF's men and women go Greek; rush is held before classes start in the fall and again in the spring. The 22 traditional fraternities and 16 Panhellenic Council sororities have privately owned houses in Gainesville, which also offer meal service. Nine historically African American Greek groups and seven culturally based Greek organizations also recruit at various times during the year; they don't offer housing.

Students say Gainesville, a city of about 125,000 between the Atlantic Ocean and the Gulf of Mexico, is a great college town. "Gainesville revolves around UF," a student says. There are plenty of stores, restaurants, and bars, as well as a sports arena and the Center for Performing Arts, which brings in world-class symphony orchestras, Broadway plays, opera, and large-scale ballet productions. The university owns a nearby lake, which is "great for lazy Sundays" and more vigorous water sports, and there's a plethora of parks, forests, rivers, and streams for backpacking, camping, and canoeing. Beaches are also a popular destination.

Sports are a year-round obsession here, and the university recently pulled off an impressive feat by becoming the first Division I school to simultaneously hold national championships in basketball (two years in a row) and football. The annual homecoming extravaganza, known as "Gator Growl," is billed as the biggest student-run pep rally in the country. Other sports are not forgotten, though; the university has one of the top intercollegiate programs in the nation, with varsity competition for men and women in 16 sports, including nationally ranked

> **"UF is a school with an impressive history, and its students and alumni remain proud of that."**

teams in baseball, track, golf, tennis, gymnastics, volleyball, swimming, and diving. Intramural sports are popular, and for those who don't want to join a team, the 60,000-square-foot fitness park offers aerobics classes, martial arts, strength training equipment, and squash and racquetball courts.

For some students, Florida's sheer size is overwhelming. For others, it's a drawing card. "UF is a school with an impressive history," says a senior, "and its students and alumni remain proud of that." Combine great weather with nationally recognized programs in engineering and business, and nationally ranked athletic teams, and it's easy to see why Sunshine State natives clamor to study here.

> *Sports are a year-round obsession here and the university recently pulled off an impressive feat by becoming the first Division I school to simultaneously hold national championships in basketball and football.*

Overlaps

Florida State, University of Central Florida, University of Miami (FL), University of South Florida, Florida International

If You Apply To ➤ **Florida:** Early decision: Oct. 1. Rolling admissions: Jan. 17 (for fall semester). Financial aid: Mar. 15. Campus interviews: optional, evaluative. No alumni interviews. SATs or ACTs: required. SAT IIs: required for placement in some programs. Accepts the Common Application and prefers electronic applications; students apply to a particular school within UF. Essay question (choose two): meaningful activity, interest, experience, or achievement; how family history, culture, or environment has influenced you; qualities or unique characteristics.

Melbourne, FL 32901-6975

FIT is practically a branch of the nearby Kennedy Space Center, so aeronautics and aviation are popular specialties. The Atlantic Ocean is close at hand, also making the school an ideal spot for marine biology. With a total enrollment of about 2,300, FIT is the smallest of the major technical institutions in the Southeast.

Website: www.fit.edu
Location: Small city
Total Enrollment: 2,884
Undergraduates: 2,259
Male/Female: 69/31
SAT Ranges: V 510–630
   M 550–660
ACT Range: 22–29
Financial Aid: 64%
Expense: Pr $ $
Phi Beta Kappa: Yes
Applicants: 2,463
Accepted: 83%
Enrolled: 29%
Grad in 6 Years: 52%
Returning Freshmen: 78%
Academics: ✍ ✍ ✍
Social: ☎ ☎ ☎
Q of L: ★ ★ ★
Admissions: (321) 674-8030
Email Address:
   admission@fit.edu

Strongest Programs:
Computer Science
Marine Science
Electrical Engineering
Aerospace Engineering
Computer Engineering
Mechanical Engineering
Aviation Management

Students at the Florida Institute of Technology can explore the endless depths of the ocean or shoot for the stars. Located just 40 minutes from one of NASA's primary launch pads, FL Tech is a child of the nation's space program. The school's subtropical setting is perfect for scientific research and study in oceanography, meteorology, marine biology, and environmental science. It comes as no surprise that some of the most cutting-edge work in space and water-related sciences happens here. The combination of academic excellence and a convenient central Florida location—just an hour from the dizzying bustle of Walt Disney World—draws students to this unique and innovative school.

Founded in 1958 to meet the academic needs of engineers and scientists working at what is now the Kennedy Space Center, Florida Tech's contemporary 130-acre campus features more than 200 species of palm trees and botanical gardens in a tropical setting. Campus architecture ranges from modern to Georgian Gothic. Newer facilities include the F.W. Olin Physical Sciences Building, Charles Ruth Clemente Center for Sports & Recreation, and the Columbia Village Residence Hall complex.

If you're considering Florida Tech, the only independent technological university in the Southeast, make sure you have a strong background in math and science, especially chemistry and physics. Few students major in the less practical sciences. Though many students grouse that Florida Tech is too expensive for their tastes, those who plan their education well are able to get high-paying technical jobs as soon as they graduate. Seventy percent of students go into the workforce, and another 23 percent move on to graduate school. Prospective aviation students can major in aviation management, aviation meteorology, or aviation computer science, as well as aeronautics with or without a flight option. The flight school has a modern fleet of 30 airplanes and three flight-training devices, and the precision-flying team regularly wins titles. The most popular majors are aerospace engineering, aviation management, mechanical engineering, marine biology, and computer science. Newer programs include software engineering, aviation meteorology, business accounting, business information systems, environmental studies, forensic psychology, and mathematical sciences.

The academic climate at FIT is challenging. "The atmosphere is not disastrously competitive, students are more than willing to help each other out," says an senior. A junior reports, "The focus is on getting the grade rather than beating your classmates." Graduate teaching assistants are not overused. All majors offer co-op programs and senior independent research at the Indian River Lagoon or on the RV Delphinus, a 60-foot research boat the school owns. Recent marine research includes manatee preservation, beach erosion, and sea turtle studies. Everyone must take courses in communication, physical or life science, math, humanities, and social sciences, and be proficient in using computers.

> **"The focus is on getting the grade rather than beating your classmates."**

Fifty-four percent of Florida Tech students are out-of-staters, while 15 percent hail from outside the country. "It is a microcosm of intelligent people representing 97 countries," says a sophomore. "It's like traveling the world in four years." The student body is 59 percent white, yet political correctness and diversity-related issues seem to be taken in stride. African Americans comprise 4 percent of the student body and Hispanics 7 percent. "There is no 'typical' Florida Tech student," explains one junior. Florida Tech offers merit scholarships that average $8,640, and 118 athletic scholarships in nine sports. Incoming freshmen are welcomed with a weeklong orientation program highlighted by trips to Disney World and the beach, just three miles away. On campus, freshmen may take part in the University Experience Program, which helps first-years adapt to college life.

The growing selection of dorms at Florida Tech are modern, air-conditioned (whew!), and well maintained. Fifty-five percent of students make their home on campus, though some complain about the price of it. Still, the rooms are well received. "Residence Life takes a holistic approach to making every resident feel right at home," says a junior. Freshmen are required to live on campus in large double rooms. Four-student apartments are available to a small percentage of qualifying upperclassmen by lottery. Students who live off campus are drawn by cheap rent and not much else, because Melbourne

Though many students grouse that Florida Tech is too expensive for their tastes, those who plan their education well are able to get high-paying technical jobs as soon as they graduate.

"It is a microcosm of intelligent people representing 97 countries."

is "quiet and is not a typical college town," reports an aerospace engineering major. The meal plan is an open, unlimited arrangement, and students report the food is improving every year.

Most Florida Tech students who don't have cars choose bikes as their favorite mode of transportation. "If you don't have a car of your own it may be hard because public transportation is limited," cautions one student. Diversions can be found in Orlando (with Epcot, MGM Studios, and Animal Kingdom abutting Disney World) or at the Kennedy Space Center. Students also hit the road for other Sunshine State cities, including Tampa, Key West, Miami, Daytona, and St. Augustine. Watching space shots from campus with a trained eye and a cold brew is a treasured pastime. The campus bar, the Rat, is a popular hangout. Otherwise, campus social life is predictably hampered by the low male/female ratio. Fraternities and sororities are slowly becoming more popular at Florida Tech, claiming 16 percent of the men and 10 percent of the women. While the campus is officially dry, every frat party has beer that the underage eagerly guzzle, students say. Besides partying, students spend their downtime surfing, fishing, hanging out at the beach, shopping, or going for a "Sunday drive" (in the sky) with a flight school student. Every April, students brace

Students spend their downtime surfing, fishing, hanging out at the beach, shopping, or going for a "Sunday drive" (in the sky) with a flight school student.

"Residence Life takes a holistic approach to making every resident feel right at home."

for the invasion of other collegians on spring break. Techies also look forward to Greek Week and intramural sports competitions. Florida Tech is in Division II, and with so much water around, it's not surprising that crew is popular and awesome.

Whether it's surveying marine coral 50 feet below the sea or the sky 30,000 feet above, students at Florida Tech get hands-on experience that serves to sharpen the school's already specialized, high-quality academics. The administration continues to focus on capital improvements, sponsor cutting-edge research, and embrace diversity. And with beaches and amusements close at hand, students can have some real fun in the sun while they prepare for high-flying—or low-lying—careers.

Overlaps

University of Florida, Embry-Riddle, University of Central Florida, University of Miami, Georgia Tech, MIT

If You Apply To ➤

Florida Tech: Rolling admissions. Campus interviews: recommended, informational. Alumni interviews: optional, informational. SATs or ACTs: required. SAT IIs: optional. Accepts the Common Application and prefers electronic applications. No essay question.

A2500 University Center, Tallahassee, FL 32306-2400

With an assist from its football program, FSU's popularity has burgeoned in recent years. Not that there weren't some quality programs to begin with. The motion picture school is among the best around, and business and the arts are also strong. So long as the football team beats the hated Gators, all is well.

Website: www.fsu.edu
Location: City outskirts
Total Enrollment: 39,146
Undergraduates: 27,055
Male/Female: 43/57
SAT Ranges: V 530–620
 M 540–630
ACT Range: 23–27
Financial Aid: 32%
Expense: Pub $
Phi Beta Kappa: Yes
Applicants: 22,450
Accepted: 62%
Enrolled: 43%
Grad in 6 Years: 66%
Returning Freshmen: 86%
Academics: ✍ ✍ ✍
Social: ☎ ☎ ☎
Q of L: ★ ★ ★ ★
Admissions: (850) 644-6200
Email Address:
 admissions@admin.fsu.edu

Strongest Programs:
Accounting
Biological sciences
Meteorology
Oceanography
Criminology
Dance
Theatre
Film

Florida State University has long been synonymous with football, but students at this Sunshine State university enjoy success off the gridiron, too. Here, you could have a Nobel laureate for a professor, study in one of the finest science facilities in the Southeast, or network at the state capitol. The choices are plentiful at FSU, and the pace of life makes it possible to taste a little of everything: a wide array of academic choices, Florida sunshine, and plenty to do, from football to Tallahassee hang-outs.

FSU is located in the "Other Florida": the one with rolling hills, flowering dogwoods and azaleas, and a canopy of moss-draped oaks. Glistening Gulf of Mexico waters are only half an hour away. The main campus features collegiate Jacobean structures surrounded by plenty of shade trees, with some modern facilities sprinkled in. Situated on 450 compact acres, the campus is the smallest in the state university system—it's just a 10-minute walk from the main gate on the east side to the science complex on the west side. Bicycling and skating are popular forms of transportation, and a free shuttle bus circles campus for those without wheels. The most recent campus improvements include a medical school complex, new dining halls, and a student housing complex.

FSU has outstanding programs in music, drama, art, and dance; it's moving up fast in the sciences, thanks to strong faculty and some cutting-edge research. The College of Medicine, which focuses on serving the elderly and underserved communities, graduated its first class in 2005. Communications, statistics, and business (especially accounting) have strong reputations in the Southeast. The English department and the School of Motion Picture, Television, and Recording Arts have consistently won an impressive array of national and international awards. For gifted students, the honors program offers smaller classes, closer faculty contact, and has been revamped to include service and research components. Certain students can even earn their degrees in three years. Directed Individual Study courses offer undergraduates the chance to participate in independent research projects with faculty direction. Internships and political jobs abound for tomorrow's politicians, since the state capitol and Supreme Court are nearby.

"It is a place I can consider almost like my home."

Students report the academic climate is somewhat laid-back but that "the courses are rigorous." Freshmen can take advantage of the First-Year Experience (FYE), Living/Learning Communities, and Freshmen Interest Groups. FYE is an extended orientation that introduces students to campus organizations, events, and activities; students who enroll in FIGs attend classes with the same group of peers, who have similar academic interests.

Within FSU's liberal studies program, students must also complete six hours of multicultural understanding coursework focusing on diversity within the Western experience and crosscultural studies. Freshmen must take math and English, and may find a TA at the helm in these courses. But overall, faculty members do teach. "The quality of teaching is excellent," a senior says. For those with wanderlust, FSU offers extensive study abroad options. They include a branch campus in Panama, year-round programs in Italy, England, and Spain, and summer programs in Greece,

Vietnam, Switzerland, France, Costa Rica, Russia, the Czech Republic, Ireland, Japan, Belize, Brazil, and China. The university is making strides in the world of distance learning, allowing some students with an associate's degree to earn their bachelor's degree online. New degree programs include French and Francophone studies, biostatistics, and Middle Eastern studies.

Perhaps not surprisingly, FSU's student body has a distinctly Floridian flavor: in-staters comprise 87 percent of the group. Twelve percent are African American, 3 percent are Asian American, and 11 percent are Hispanic. There's little evidence of racial tension on the diverse campus. Seminoles are a mixture of friendly small-towners and city dwellers, and political tastes tend toward the conservative. Hot topics include voter registration, the environment, and student government concerns. Merit scholarships averaging $1,797 are available to qualified scholars and student athletes vie for 238 scholarships in a variety of sports.

Fourteen percent of FSU's undergrads live in the university dorms, all of which are air-conditioned (a must in Florida) and wired for Internet access. Students may opt for typically spacious older halls or newer ones that tend to be more cramped. The dorms get mixed reviews from students, and the number who can live in them is limited, so rooms are assigned on a first-come, first-served basis. Upperclassmen generally forsake the housing rat race and move into nearby apart-

> **"The quality of teaching is excellent."**

ments, houses, or trailers, where they take advantage of the city and campus bus systems to get to school. The dorms are equipped with kitchens; meal plans that offer "good but expensive" food are also available.

When they're not studying, plays, films, concerts, and dorm parties keep FSU students busy. "The social life is fine," a freshman says. Those with a valid ID can head for one of Tallahassee's bars or restaurants, which fall somewhere between "college hangout" and "real world." Generally, though, students give the area a thumbs-up. As for Greek life, 14 percent of the men and women join fraternities and sororities, which constitute another important segment of the social scene.

In sports, the big-time Seminole football team won two national titles in the '90s and, until recently, has dominated the Atlantic Coast Conference. Coach Bobby Bowden is the all-time leader in wins in major college football. Going to games is an integral part of the FSU social scene, especially when games are against FSU's two most hated rivals: the University of Florida and the University of Miami. FSU's baseball team also draws an enthusiastic following, as do women's basketball, volleyball, soccer, and softball teams.

Florida State remains a solid choice for those seeking knowledge under the blazing Florida sun. FSU students take pride in their school and what it has to offer. "It is a place I can consider almost like my home," says a business major.

> *Fourteen percent of FSU's undergrads live in the university dorms, all of which are air-conditioned (a must in Florida) and wired for Internet access.*

> *FSU has outstanding programs in music, drama, art, and dance; it's moving up fast in the sciences, thanks to strong faculty and some cutting-edge research.*

Overlaps

University of Florida, University of Central Florida, University of South Florida, University of Miami, Florida International, Florida Atlantic

---

If You Apply To ➤

Florida State: Regular admissions: Feb. 14. Guarantees to meet demonstrated need of 76%. No campus or alumni interviews. SATs or ACTs: required. SAT IIs: optional. Accepts the Common Application and prefers electronic applications. Essay question: personal experience; family history; personal qualities.

Fordham University

Rose Hill Campus: 441 East Fordham Road, Bronx, NY 10458
Lincoln Center Campus: 113 West 60th Street, New York, NY 10023

New York City's Fordham has climbed a few notches on the selectivity scale. There is no better location than Lincoln Center in Manhattan, where the performing arts programs are housed. The Bronx location is less appealing but better than the horror stories you may hear.

Website: www.fordham.edu
Location: Urban
Total Enrollment: 14,861
Undergraduates: 6,806
Male/Female: 42/58
SAT Ranges: V 560–660
  M 560–650
ACT Range: 24–28
Financial Aid: 62%
Expense: Pr $ $
Phi Beta Kappa: Yes
Applicants: 15,225
Accepted: 50%
Enrolled: 23%
Grad in 6 Years: 78%
Returning Freshmen: 90%
Academics: ✍ ✍ ✍
Social: ☎ ☎ ☎
Q of L: ★ ★ ★
Admissions:
  (800) FORDHAM
Email Address:
  enroll@fordham.edu

Strongest Programs:
Business
International Political Economy
Theatre
Psychology
English
Philosophy
Theology
History

The Jesuit tradition pervades all aspects of life at Fordham University, from the quality of teaching, to the emphasis on personal relationships, to the pursuit of both "wisdom and learning," which also happens to be the school's motto. Students benefit from three campuses: the suburban Marymount College in Tarrytown, New York; the gated Bronx community of Rose Hill; and the Lincoln Center facility, just a short subway ride away from the heart of midtown Manhattan. Small classes offer individual attention, and though a majority of students are from New York, there's plenty of variation in ethnic background and in students' political and social views. Fordham is "more diverse than Boston College, less funky than NYU," says a German and English double major. "We have an even mix of preppy, athletic types and independent, Manhattan types, the latter more so at the Lincoln Center campus."

The 85-acre Rose Hill campus is an oasis of trees, grass, and Gothic architecture; it's close to the New York Botanical Garden and Yankee Stadium and had cameo appearances in films such as *A Beautiful Mind*. Rose Hill is home to Fordham College, the largest liberal arts school at the university, as well as the undergraduate schools of business administration and liberal studies. The Lincoln Center campus benefits from its proximity to the Juilliard School, to the CBS and ABC television studios between Tenth and Eleventh Avenues, and to Lincoln Center itself, Manhattan's performing arts hub. This campus has its own undergraduate college, and also houses Fordham's law school and other graduate programs. Started as an alternative-style urban institution with no grades, the Lincoln Center campus has become more traditional over the years, and now uses the same 18-course core curriculum as Rose Hill.

> "Many students go to Manhattan weekly, if not more."

No matter where at Fordham you study, humanities are a good choice. Strengths at Rose Hill include history, philosophy, psychology, and economics, while at Lincoln Center, the forte is, appropriately, drama (music isn't bad, either). The BFA in dance is offered along with the Alvin Ailey American Dance Theater; students must be accepted by both Fordham and by the Ailey audition panel. Fordham's public radio station, WFUV, offers hands-on experience for aspiring deejays and radio journalists, and there are TV production studios in both Manhattan and Rose Hill, but so far, no university-sponsored station. Both colleges offer interdisciplinary majors, such as African American and Latin American studies, as well as 3–2 engineering programs with Columbia and Case Western Reserve, a 3–3 program with Fordham Law School, and a teacher-certification program. The latter may include a fifth year, culminating in a master's degree in education.

Popular and well-regarded majors include business, history, communications, psychology, and English language and literature. Helped by alumni connections, business students often obtain internships on Wall Street or elsewhere in the Manhattan financial community; some of these positions lead to jobs after graduation. The GLOBE Program in International Business includes an international internship or study abroad assignment, two courses with an international focus, and proficiency in a foreign language. Core requirements include English, social and natural sciences, philosophy, theology, history, math/computer science, fine arts, and foreign languages. "The best program at Fordham is the new American Catholic studies concentration," offers one senior. "It provides students with great professors, excellent graduate contacts, and a terrific education."

When it comes to teaching, "I have had amazing professors who speak on CNN," says a student majoring in international political economy. "The professors foster an environment that stimulates discussion and continuously challenges the students to think," adds a classmate. "Many of the lower level core courses are taught by adjunct professors, but I have never had a bad experience with a professor," says a sophomore.

"Fordham students constantly want to do 'magis,' the Jesuit term for 'more,'" says a student, "whether in class or in the community." Fifty-six percent of the students at Fordham are from New York state, and many of the rest are Roman Catholics from elsewhere on the East Coast, even though the school is independent of the church. The atmosphere is less intellectual than at nearby Columbia and NYU, but students are driven to do well academically and to get good jobs after graduation, says a senior. "Our Democrat and Republican groups go at it!" says a marketing major. African Americans comprise 5 percent of the student body; Asian American make up 6 percent, and Hispanics 11 percent. In the Jesuit tradition of "men and women for others," students travel as far away as Romania and Belize during vacations for intensive study and service projects. There are hundreds of merit scholarships worth an average of $8,276, and more than 100 athletic scholarships as well. Students willing to commute from home are eligible for a $5,000 Metro Grant.

> "More diverse than Boston College, less funky than NYU."

Fifty-nine percent of Fordham students live in the dorms, and they are guaranteed university housing for four years. Those lucky enough to snag rooms in the 20-story, 850-bed residence hall near Lincoln Center are saved from the borough's unscrupulous brokers and unconscionable rents. At Rose Hill, "Dorms are spacious, and community bathrooms are cleaned daily," says a senior. Students must sign in to gain admission to dorms other than their own, and Rose Hill has its own security personnel, but the Bronx "is very much an eye-opener for all the suburban white kids," one student quips. "It's a good experience if you have the common sense to keep a low profile among the locals." Those who venture off campus often choose to live in nearby Little Italy, says a senior.

Fordham's social life is an embarrassment of riches. "Most night-time socializing takes place off campus and in downtown Manhattan, which offers endless choices for things to do," says a marketing major. "Many students go to Manhattan weekly, if not more. On campus, there are also many activities going on." Fordham sponsors an intramurals program in Central Park, as well as movies and concerts on both campuses; recent guest

> "Fordham students constantly want to do 'magis,' the Jesuit term for 'more.'"

artists include Wyclef Jean, Busta Rhymes, and Mos Def. There's a "huge bar scene, both in the Bronx and in Manhattan," one student says—presumably aimed at students over 21, but this being New York, fake IDs are easy to come by, and sometimes they work. Off-campus parties also provide another opportunity to imbibe, if students are interested. The Rose Hill campus backs up against the Bronx Zoo and it's around the corner from Arthur Avenue, another Little Italy. Both provide welcome weekend diversions. Students look forward to homecoming, Fall Fest, Fordham Week, Spring Weekend, and Senior Week, as well as to the Columbia–Fordham football game and the Irish v. Italian rugby match. "The 10 o'clock scream," a Thursday night ritual in which everyone leans out their window and screams for one minute, is a favored stress reliever.

Fordham competes in Division I and the Atlantic 10 Conference, and its location on the Hudson River has also helped to produce the women's rowing Metropolitan champs.

No matter where at Fordham you study (Rose Hill or Lincoln Center), humanities are a good choice.

Fordham competes in Division I and the Atlantic 10 Conference, and its location on the Hudson River has also helped to produce the women's rowing Metropolitan champs. The marvelous Lombardi Memorial Athletic Center (named for Vince, an alumnus) supports 12 club sports and 24 intramurals, including ultimate Frisbee and tae kwon do, as well as "grandstand athletes," who root for the varsity basketball team. With a new head coach and strong recruiting classes, basketball is poised for increasing competitiveness.

Consistent with its Jesuit tradition, Fordham fancies itself a family. Indeed, students are often so happy with what they find that they preach the gospel to younger siblings, who obligingly follow them to one of the school's three campuses. Some things are changing at Fordham—including its admission and academic standards, which are slowly creeping up, and its national profile, which is also far higher than in years past. What hasn't changed is the idea that diversity and community can coexist, instilling confidence and pride in Fordham students and loyalty in the expanding alumni base.

If You Apply To ➤

Fordham: Early action: Nov. 1. Regular admissions: Jan. 15. Financial aid: Feb. 1. Meets demonstrated need of 21%. Campus and alumni interviews: optional, evaluative. SATs or ACTs: required. SAT IIs: recommended. Apply to particular school or program. Accepts the Common Application and electronic applications. Essay question: significant life experience; literary character with whom you identify; important political, social or cultural issue.

Franklin and Marshall College

637 College Avenue, Lancaster, PA 17604-3003

F&M is known for churning out hard-working preprofessional students. Faces tough competition from the likes of Bucknell, Gettysburg, Lafayette, and Dickinson for Pennsylvania-bound students. Known for natural sciences, business, and internships on Capitol Hill. Check out its witty website.

At Franklin and Marshall College, set in the serene hills of Pennsylvania's Amish country, you might come nose-to-nose with a horse and buggy, but you can still enjoy the perks of being in one of the country's 50 largest metro areas. While the city has modernized beautifully, parts of this historic town and many of its residents look much the same as they did when two acclaimed but struggling colleges decided to pool their resources. Marshall College (named for Chief Justice John Marshall) merged with Franklin College (started with a donation of 200 English pounds from Ben himself) in Lancaster. These days, F&M is trying to modernize too, particularly by bringing a more international bent to the curriculum.

F&M's 125-acre campus is surrounded by a quiet residential neighborhood shaded by majestic maple and oak trees. The campus itself is an arboretum and boasts 47 buildings of Gothic and Colonial architecture. The College Square complex appeals to students seeking a study respite. Recent additions

"The one-on-one interaction is priceless."

to the campus include the Life Sciences & Philosophy Building, a tennis center, and a 400-bed residential facility.

Although there are no required courses freshman year, nine out of 10 students enroll in First-Year Residential Seminars. Participating students live together in groups of 16 on co-ed freshman floors and study a major theme or concept within a discipline.

General education requirements include writing and language requirements, two "Foundation" courses, and distribution requirements in the arts, humanities, social sciences, sciences, and non-Western cultures. Collaborations are optional opportunities to get course credit for an experience that includes working with others. F&M has long been known for being strong in the natural sciences, and the school is now placing more emphasis on courses with a service-learning component. Majors have been added in dance and creative writing, along with a minor in international studies. All majors can be combined with the international studies concentration that requires students to study abroad and become proficient in a foreign language. About 40 percent of

"We are both laid-back and spirited, liberal and conservative, athletic and artsy."

F&M students have the chance to complete an independent research project for credit. A preprofessional college in line with Lafayette and Bucknell, F&M has an excellent reputation for preparing undergrads for medical school, law school, and other careers.

Students uniformly describe the coursework as strenuous and demanding, but say the environment remains more cooperative than competitive. "The courses are definitely challenging and there are some fantastically driven students," a junior says, "but I haven't encountered a cutthroat environment at all." Students say teaching is outstanding, and the relatively small student body and intimate class sizes help create a strong sense of community between students and professors. "Hats off to the teachers," cheers a psychology major. "The profs are the cream of the crop," adds a student. "The one-on-one interaction is priceless." F&M offers crossregistration with two other small Pennsylvania colleges—Dickinson and Gettysburg—and several domestic-exchange and cooperative-degree programs. In the summer, the college sends students to countries such as Japan and Russia, and nearly 25 percent study in locations around the world during the junior year. Others participate in the Sea Semester.*

"F&M students are driven and determined to be successful," says a junior. Eighty-three percent of students rank in the top quarter of their graduating high school class, and almost half rank in the top tenth. Only 32 percent hail from Pennsylvania. Asian American students comprise 4 percent of the student body, African Americans 3 percent, and Hispanics 3 percent. The student body has been diversifying, but remains fairly homogeneous, students say. An occasional political debate may waft through the murmurs of light social exchanges during dinner, but according to one student, the big issue on campus is the lack of issues

"Lancaster is actually a pretty cool town."

on campus. Fummers do, however, take an interest when it comes to extracurricular activities and social opportunities. The 115 clubs on campus attest to that, as does an unusually high level of participation in community service activities.

Approximately 50 Marshall and 75 Presidential scholars are named each year. Marshalls receive a $12,500 tuition grant, a Macintosh computer, and the chance to apply for up to $3,000 in research travel funds. Presidential scholars receive a $7,500 tuition grant. F&M also offers two Rouse scholarships worth full tuition, books, and fees, and about 40 Buchanan community service grants of $5,000 each. The school also offers merit-based financial aid to outstanding students of ethnic backgrounds that have been traditionally underrepresented in higher education. There are no athletic scholarships.

The college requires students to live in college-operated housing all four years and housing options include dorms, special interest housing, suites, and apartments. A faculty-led College House system, begun in 2005, is designed to increase the quality of the residential experience. Boarders eat most of their meals in the campus cafeteria under a flexible meal plan, but students are issued debit cards that they may use at a number of different food stops on campus. Campus security is described as "efficient and friendly" and students report feeling safe on campus.

(Continued)
Admissions: (717) 291-3953
Email Address:
admission@fandm.edu

Strongest Programs:
Biology
Chemistry
Geosciences
Physics
Business
Government
English

Nine out of 10 students enroll in First-Year Residential Seminars. Participating students live together in groups of 16 on co-ed freshman floors and study a major theme or concept within a discipline.

Reorganized by the college in 2004, the seven fraternities attract 35 percent of men, and two sororities attract 15 percent of women. They are integral to much of the nightlife, although the residence halls and special-interest groups offer a range of alternatives, including concerts and comedians. Ben's Underground, a popular student-run nightclub, and Hildy's, a tiny local bar, are favorite meeting places. "The social life is expansive," a junior says. "There are obviously Greek parties and house parties, but there are tons of movies, field trips, and programs to participate in, too." In recent years, the student-run and college-funded College Entertainment Committee has brought a number of popular musicians and bands to the campus.

Lancaster is a historical and well-to-do city of 60,000 people located in a larger metro area of more than 400,000. Lancaster offers a 16-screen cinema, scores of shops, a farmer's market, brick-and-cobblestone streets (with hitching posts for the Amish horses and buggies that are almost never used), and a plethora of quaint restaurants and cafes. Students have a measured, realistic appreciation of its urban amenities and rural ambiance. "Lancaster is actually a pretty cool town," muses one math major. "It's really artsy and has an awesome music scene. I was pleasantly surprised." The Amish culture draws the interest of some students, and many frequent the charming farmer's market to shop for handmade quilts. Those with a hankering for contemporary action take road trips to Philly, Baltimore, Washington, D.C., and New York City. The biggest annual event is Spring Arts, held the weekend before finals, which includes student air-band contests, live concerts, art exhibits, games, booths, and barbecues. Other highlights include the freshman Pajama Parade, the Sophomore Sensation, the Senior Surprise, International Day, Black Cultural Arts Weekend, Flapjack Fest (when professors serve pancakes to students), and Fum Follies, a faculty-produced play.

The college has a good selection of intramural sports, which include popular co-ed competitions, and about 40 percent of nonvarsity athletes take part. In addition to competing in Division I in wrestling, F&M boasts recent Centennial Conference championships in men's swimming, football, baseball, and women's swimming. Varsity squads are called the Diplomats, a name that is irresistibly abbreviated to "the Dips." The annual football game against Dickinson for the Conestoga Wagon trophy is always a crowd-pleaser.

"Franklin and Marshall may fly under the radar but we get a solid education from the professors while cultivating an idiosyncratic character that defies an easy label," boasts a student. "We are both laid-back and spirited, liberal and conservative, athletic and artsy; you get a great mix here." The school's true illustrious namesakes, who are spoofed on the admissions office website, would be proud.

If You Apply To ➤

F&M: Early decision: Nov. 15. Regular admissions and financial aid: Feb. 1. Does not guarantee to meet full demonstrated need. Campus interviews: recommended, evaluative. Alumni interviews: optional, evaluative. SATs or ACTs: optional (required for international students). SAT IIs: optional. Accepts the Common Application and electronic applications. Essay question: Common application personal essay.

Furman University

3300 Poinsett Highway, Greenville, SC 29613-5245

Furman's campus is beautiful, and the swans are definitely a nice touch. At just less than 3,000 total enrollment, Furman is nearly twice the size of Davidson and half the

size of Wake Forest. As befits its Baptist heritage, Furman is a conservative place and still a largely regional institution.

Furman University has been called the "Country Club of the South." And if you're Southern, white, Christian, and conservative, you're likely to love it here. As a political science major says, "One of the biggest issues on campus is the distribution of condoms. Students here tend to be sheltered and ignorant of real-world issues. They are very image-conscious and our gym stays busier than most others." Beyond the lush lawns lie small classes led by caring faculty, and plenty of opportunities for independent research, all part of what administrators call "engaged" or "active" learning.

Furman's 750-acre campus is one of the country's most beautiful, with tree-lined malls, fountains, a formal rose garden and Japanese garden, and a 30-acre lake filled with swans and ducks. Flowering shrubs dot the well-kept lawns, which surround buildings in the classical revival, Colonial Williamsburg, and modern architectural styles. Many have porches, pediments, and other Southern touches, such as hand-made Virginia brick. The library has undergone renovations that doubled its size; the facility now includes a computer lab, 24-hour study area, study rooms, multimedia areas, and more than 400,000 volumes.

Furman considers itself a new type of liberal arts institution, in that it prepares students for life in a rapidly changing society by placing an emphasis on experiential learning outside of the classroom. While this is hardly novel, few other schools make it their core emphasis. As such, general education requirements include freshman composition, four other humanities courses, one to three courses in math, two courses each in natural sciences and social sciences, and one course each in fine arts, health and exercise science, and the Asian-African program. Students must also achieve foreign language proficiency and attend nine events a year from the Cultural Life Program. High marks go to the departments of psychology, chemistry, and music, but students reserve their most vocal praise for political science. "The department promotes the most involvement, through debate clubs, opportunities for foreign travel, and summer internships in D.C.," says one student. The academic calendar enables students to focus on topics in greater depth, with three courses each during the fall and spring terms, and two courses during the eight-week winter term.

> "Students, in general, care more about GPA than enjoying life."

Perhaps because of the calendar, students say Furman's academic climate is intense. "The climate is extremely competitive," says one student. "Students, in general, care more about GPA than enjoying life." Seventy-six percent of classes taken by freshman have fewer than 25 students, and none has more than 50, helping students get to know faculty members well. "Classes are small and interactive, and professors are nurturing and passionate," says a communication major. The Furman Advantage program helps fund research fellowships and teaching assistantships for more than 120 students a year. Furman also typically sends one of the largest student delegations to the annual National Conference of Undergraduate Research. More than 250 students study abroad each year, through one of 13 Furman-sponsored programs on five continents, including a special exchange with Japan's Kansai-Gaidai University. Furman also belongs to the Associated Colleges of the South.* Entering freshmen have the opportunity to travel in small groups to an island off the coast of Charleston, the mountains of NC, or even China during the summer before they enroll.

Furman broke with the South Carolina Baptist Convention in 1992 after 166 years, but it remains in South Carolina, where religion ranks second only to football. Furman is trying to diversify, but those efforts have been slow to bear fruit. African Americans make up 7 percent of the student body, and Hispanics and Asian Americans add 1 and 2 percent, respectively. "There are too many middle- to upper-class

Website: www.furman.edu
Location: City outskirts
Total Enrollment: 2,723
Undergraduates: 2,679
Male/Female: 44/56
SAT Ranges: V 600–700
  M 600–690
ACT Range: 25–30
Financial Aid: 44%
Expense: Pr $ $
Phi Beta Kappa: Yes
Applicants: 4,007
Accepted: 53%
Enrolled: 33%
Grad in 6 Years: 84%
Returning Freshmen: 93%
Academics: 🖋🖋🖋½
Social: ☎ ☎ ☎
Q of L: ★ ★ ★
Admissions: (864) 294-2034
Email Address:
  admissions@furman.edu

Strongest Programs:
Chemistry
Business
Psychology
Political Science
Music
Health and Exercise Science
History

More than 250 students study abroad each year, through one of 13 Furman-sponsored programs on five continents, including a special exchange with Japan's Kansai-Gaidai University.

white people," observes one sophomore. There's also a lot of competition for leadership positions among this resumé-conscious bunch. Each year, Furman awards a number of merit scholarships (averaging $10,830), plus 251 athletic scholarships.

Ninety percent of students live on campus, as Furman has a four-year residency requirement. "The dorms are quite comfortable and well-maintained, and the housing lottery is fair." Furman is a dry campus and the policy is strictly enforced in freshman and sophomore dorms, where underage students shouldn't be imbibing anyway. The atmosphere is more relaxed in North Village, a newer, university-owned apartment complex of 10 buildings. All dorms are equipped with telephone, cable TV, and Internet access, and students enjoy the camaraderie that results from a residential campus. Meal plan credits can be used in the dining hall or food court, which always offers student favorites like hamburgers and hot dogs. "There is always something to eat. Whether or not you'll like it is the question," quips a freshman political science major. Campus security helps provide a relatively safe environment. "Security is so tight that it is annoying," grumbles one student.

"Classes are small and interactive, and professors are nurturing and passionate."

When the weekend comes, Furman's Student Activities Board sponsors "free movies, weekend trips, restaurant deals, huge concerts, and basically always something to do," says a communication major. Fraternities claim 35 percent of the men and sororities 40 percent of the women, and off-campus Greek parties draw crowds. "I am increasingly becoming a fan of Greenville," says a political science major. The Peace Center for the Performing Arts, located downtown, brings in touring casts of Broadway shows and other top-rated acts. More than 60 percent of Furman's students devote spare time to the Collegiate Educational Service Corps, which provides volunteers to 85 community agencies and organizes the annual May Day–Play Day carnival, converting the campus into a playground for underprivileged kids. However, one freshman remarks, "A good amount of students volunteer, yes, but most are content to simply live out their privileged lives in the peace of the Furman bubble." The best road trips are to the mountains of Asheville (only 45 minutes away), Atlanta (for the big city and shopping, about two hours), and Charleston or Myrtle Beach (four hours).

Furman's athletic teams compete in the Southern Conference, and in recent years, the men's rugby and women's tennis and golf teams have brought home conference titles. Furman also fields a handball team, which was the runner-up in the national collegiate tournament in a previous year. Students happily yell out the school's tongue-in-cheek cheer ("F.U. one time, F.U. two times, F.U. three times, F.U. all the time!") during football games against archrivals Wofford and Georgia Southern. Seventy percent of the student body competes for the coveted All Sports Trophy by participating in intramurals, which range from flag football to horseshoes.

"There is always something to eat."

Furman may call itself a university, but its educational approach is closer to that of a liberal arts college, emphasizing problem solving, projects, and experience-based learning. More than a decade after severing its religious ties, the school continues to evolve, drawing more academically capable students from increasingly diverse backgrounds.

If You Apply To ➤

Furman: Early decision: Nov. 15. Regular admissions, financial aid, and housing: Jan. 15. Meets demonstrated need of 45%. Campus and alumni interviews: optional, informational. SATs or ACTs: required. SAT IIs: optional. Accepts the Common Application and prefers electronic applications. Essay question.

George Mason University

4400 University Drive, Fairfax, VA 22030-4444

Located in one of the richest suburbs in America, GMU is poised to become a major university. Though still mainly a commuter school, campus housing continues to grow. The presence of prominent conservatives such as Walter Williams has added cachet to economics and public policy.

Located in the middle of greater Washington, D.C.'s budding high-tech corridor, George Mason University is a leading center of conservative political and economic thought. The urban campus and symbiotic relationship with the surrounding region contrast starkly with Virginia's two other major universities, which have held classes for a hundred years in the relative isolation of Charlottesville and Blacksburg. Mason has grown by leaps and bounds for most of the past two decades and now boasts more than 30,000 students pursuing majors in nearly 150 degree programs.

Founded as a sleepy outpost of the University of Virginia, GMU sits on a 583-acre wooded campus in the Washington, D.C., suburb of Fairfax, Virginia. Campus architecture is modern and nondescript; most structures were erected after the mid-1970s. GMU's 10,000-seat arena, the Patriot Center, hosts both sporting and entertainment events. In addition, an aquatic and fitness center, featuring two pools, a whirlpool, and co-ed saunas, was recently completed. And although GMU's campus doesn't have the Colonial ambiance or tradition of William and Mary or UVA, its namesake does have the same Old Virginia credentials. George Mason drafted Virginia's influential Declaration of Rights in 1776, and he later opposed ratification of the federal Constitution because there was no Bill of Rights attached.

Mason's general education requirements stipulate that all students take the equivalent of two courses in English composition, humanities, social sciences, and math and sciences. Students who prefer to find their own way can design a major under the Bachelor of Individualized Study program. The academic climate is intense but manageable.

> **"Mason has Pulitzer and Nobel Prize winners who teach freshman classes."**

"Courses run the gamut from relaxed, entry-level electives to rigorous, upper-level courses," says a senior. Professors are praised for their knowledge and Mason has the distinction of being the only university in Virginia to have two Nobel Prize winners on its faculty. "Mason has Pulitzer and Nobel Prize winners who teach freshman classes," marvels one student. If they do fall behind or need guidance, academic counseling is likely to put them back on course. Advisors "have been so helpful and really supportive," says a marketing major.

The most recent additions to the curriculum include degree programs in classical studies, international transactions, computational sciences, public policy, and urban systems engineering. Another option is the New Century College degree program, which teams small groups of faculty and undergraduates on projects that can be easily connected to the world outside GMU. Though it is growing up fast, Mason's youth shows in a number of ways. First, programs taken for granted at more established universities are just hitting their stride here. Next, GMU's relatively small endowment means almost constant tuition increases. Last, some of the school's facilities are just plain inadequate for its more than 20,000 students. The library, for example, has fewer than 700,000 volumes, though it now subscribes to more than 300 online databases and allows students to borrow books from all eight members of the Washington Research Library Consortium.

Website: www.gmu.edu
Location: Suburban
Total Enrollment: 29,728
Undergraduates: 17,525
Male/Female: 44/56
SAT Ranges: V 480–580
  M 490–590
ACT Range: 19–23
Financial Aid: 38%
Expense: Pub $ $ $
Phi Beta Kappa: No
Applicants: 28,432
Accepted: 52%
Enrolled: 24%
Grad in 6 Years: 50%
Returning Freshmen: 76%
Academics: ✍ ✍ ✍
Social: ☎ ☎
Q of L: ★ ★
Admissions: (703) 993-2400
Email Address:
  admissions@gmu.edu

Strongest Programs:
Economics
Engineering
Public Policy
Nursing
Government
English
Communications
Information Technology

The lack of resources in the library may present less of a problem for GMU's career-focused students, who seem to like learning on the job: 70 percent enter the working world after graduation, and just 20 percent proceed to graduate and professional schools. Psychology tops the list of popular majors, and economics—which boasts its own Nobel laureate—is probably the strongest department. Other well-regarded majors include computer science, nursing, engineering, and English; not surprisingly, given the school's location, the public policy department also receives accolades. The drama department, once a weak sister, is now part of the Institute of the Arts, created to make the arts an intrinsic part of every student's GMU experience. The institute includes a professional theater company, which hosts actors and playwrights in residence.

Eighty-six percent of GMU students are homegrown, and Mason's student body is fairly diverse, likely due to the diversity of the surrounding area. Minorities make up 28 percent of the student population—9 percent African American, 6 percent Hispanic, and 13 percent Asian American. More than 130 nations are represented as well. Students are politically aware and tend to lean rightward. That said, racial tensions haven't been a problem, perhaps thanks to the four-year-old Stop, Look, and Learn program. The program attempts to increase campus discussion on prejudice, discrimination, and harassment. Athletic and merit scholarships are available to those who qualify.

George Mason has traditionally been a commuter school, but there is on-campus housing and the residential population continues to grow. Currently, 7,000 students live on campus and another several thousand live around campus in university-sponsored housing. The administration admits that room and board costs are inflated because the university's entire housing stock dates from 1978 or later, which means the buildings are modern and air-conditioned—but still being paid for. And though the dorms are comfortable and well maintained, there's still a lot of building to do. "The demand and production of housing has skyrocketed, and more and more students are moving on campus," says a student. Freshmen live together in Presidents Park, while other students get rooms on a first-come, first-served basis based on class status. Those looking for an active social life should definitely consider a stint in the dorms, particularly in Presidents Park or the Freshman Center, but freshman dorms are dry, and you can get the boot if you're caught having a party with alcohol.

"Courses run the gamut from relaxed entry-level electives to rigorous upper-level courses."

GMU's University Center, with its food court, movie theater, classrooms, computer labs, and study areas, has become the center of on-campus social life. The center is a convenience and a lure for students who commute to school and have gaps between classes. On the weekends, students find a predictable assortment of malls and shopping centers in Fairfax, just southwest of D.C., but off-campus parties and the sights and sounds of downtown Washington, Georgetown, and Old Town Alexandria beckon when the sun goes down. Best of all, these are only a short commute away via a free shuttle bus to the subway. Those searching for a more lively collegiate scene take road trips to other local schools, including James Madison and UVA.

With barely a generation of history under its belt, Mason is notably lacking in traditions and annual events: "Come here and invent one!" a student urges. Patriots Day and Mason Day are the two major bashes, in addition to homecoming, Greek Week, and International Week. GMU competes in Division I, and the basketball team became a player on the national scene when it advanced to the NCAA Final Four in 2006. Any game against James Madison University draws a big crowd. Other successful teams include women's soccer, men's and women's track, and women's volleyball. Intramurals are catching on now that many games are held in the Patriot Center.

Overlaps

Virginia Tech, James Madison, Mary Washington, University of Virginia, University of Maryland

The name of George Mason may not have the cachet of George Washington, James Madison, or the other luminaries of Virginia history who have had universities named for them, but with improving academics, a growing and improving physical campus, and the rich cultural and economic resources of Washington, D.C., Mason's namesake looks like it's set to follow in those other schools' fine footsteps.

George Washington University

Washington, D.C. 20052

Ten years ago, GW was a backup school with an 80 percent acceptance rate maligned for its lack of identity. But the allure of Washington, D.C.'s opportunities coupled with an intellectually stimulating educational environment has proved to be a strong drawing card, and GW now accepts less than half of those who apply. The nation's leader in internships per capita.

Like Washington, D.C., itself, George Washington University draws students from all over America—and around the world. Upon arrival, they find a bustling campus in the heart of D.C., enriched with cultural and intellectual opportunities. Students have easy access to the Smithsonian Institution museums, the Folger Shakespeare Library, the Library of Congress, and other national treasures. GW also offers top political officials as guest speakers and visiting professors. Since Congress chartered GW in 1821, perhaps it's not surprising that the school has learned well from nearby government agencies how to create red tape. "The bureaucracy is annoying," a junior laments.

Today, George Washington University is comprised of two campuses—the main, older campus in the Foggy Bottom neighborhood on Pennsylvania Avenue near the White House, and the new Mount Vernon campus three miles away with five residence halls and some classroom buildings. The Foggy Bottom campus has a mix of renovated federal row houses and modern buildings and is virtually indistinguishable from the rest of the neighborhood, while the wooded Mount Vernon campus spans 26 acres near Georgetown, and also includes athletic fields, tennis courts, and an outdoor pool. Formerly a women's college, Mount Vernon now permits all GW students to take classes and attend activities, though certain programs and academic initiatives are geared toward women. At the Mount Vernon campus, three new science laboratories and a 379-bed residence hall have opened.

Aside from the Elliott School, freshmen may enroll in the School of the Engineering and Applied Science, the School of Business and Public Management, the School of Media and Public Affairs, and the Columbian College of Arts and Sciences, which is the largest undergraduate division. During freshman year, all undergraduates take English composition. Other requirements vary by school. To graduate, students fulfill requirements covering seven areas of knowledge: literacy, quantitative and logical reasoning, natural sciences, social and behavioral sciences, creative and performing arts, humanities, and foreign languages and cultures. The new University Writing Program

Website: www.gwu.edu
Location: Urban
Total Enrollment: 15,489
Undergraduates: 9,660
Male/Female: 44/56
SAT Ranges: V 600–700
  M 600–690
ACT Range: 25–29
Financial Aid: 43%
Expense: Pr $ $ $ $
Phi Beta Kappa: Yes
Applicants: 19,406
Accepted: 38%
Enrolled: 33%
Grad in 6 Years: 78%
Returning Freshmen: 92%
Academics: ✍ ✍ ✍ ½
Social: ☎ ☎ ☎
Q of L: ★ ★ ★
Admissions: (202) 994-6040
Email Address:
  gwadm@gwu.edu

Strongest Programs:
Astronomy
Biology
Chemistry

requires that students take University Writing 20 as freshmen and two additional Writing in the Discipline courses during their remaining undergraduate years.

For highly motivated and capable undergraduates seeking a challenge, GW's honors program offers special seminars, independent study, and a university symposium on both campuses. The intensive Enosinian Scholars Program culminates with a written thesis and oral examination. The School of Engineering also offers an honors program in which students work with professors on research projects; a team recently collaborated with America Online to create a wireless technology lab. GW's political communications major, which combines political science, journalism, and electronic media courses, is one of the few undergraduate programs of its kind and benefits from its Washington location. "GW is famed for its international affairs, communication, and political science programs," says a junior.

> "GW is famed for its international affairs, communication, and political science programs."

Many GW students describe the academic climate as competitive and rigorous. "Generally, the courses are challenging," says one student, "but professors understand that students have a number of outside commitments such as jobs and internships." Sixty-five percent of the classes taken by freshmen have 25 students or less; professors handle lectures and seminars, and TAs facilitate discussion or labs. "The quality of teaching has been hit or miss," says a sophomore. "I have had some really inspirational professors and some really bad ones." Almost half of GW's faculty members divide their time between the halls of academia and the corridors of power, with many holding high-level government positions. "Avoid the 'super-profs,'" says a junior history major. "They tend to cancel classes more for things like an appearance on CNN." Then again, those connections allow students tremendous access to the bigwigs of D.C.—prominent figures such as President Bush, Virginia Governor Mark Warner, and Larry King have all addressed GW students.

The School of Engineering also offers an honors program in which students work with professors on research projects; a team recently collaborated with America Online to create a wireless technology lab.

Given GW's location and its improved academic reputation, students "are from diverse economic, religious, racial, ethnic, and social backgrounds," says a freshman. A psychology major adds, "Since GW in is the center of Washington, D.C., most people are educated, young professionals and accepting of everyone." Six percent of the students are African American, 5 percent are Hispanic, 9 percent are Asian American, and 4 percent hail from foreign countries. As you might expect, political issues important on the national stage are also important here. "Students do know what's going on and are active for causes," says a political science major. "The school is seen as liberal, but College Republicans is very active." Merit scholarships worth an average of $19,290 are awarded annually and student-athletes vie for 116 awards. The university has also adopted a fixed-rate tuition plan that guarantees fees will not increase for up to five years of full-time undergraduate study. Students who receive need-based grant aid for their first year are guaranteed the same amount of aid for four years.

> "Since GW in is the center of Washington, D.C., most people are educated, young professionals and accepting of everyone."

Sixty-seven percent of GW students live in campus housing, where freshmen and sophomores are guaranteed housing. Dorms are "primarily converted hotels and apartment buildings, so housing is very plush," says a junior. "We have the best dorms in the country," boasts another. Those who move off-campus typically find group houses in Foggy Bottom or go to nearby neighborhoods like Dupont Circle and Georgetown, just a short walk from campus. Some also choose the Maryland or Virginia suburbs, where housing stock is newer, a little more affordable, and just a short subway ride away. Many freshmen are assigned to suites with up to four roommates in Thurston Hall, the biggest and rowdiest dorm on campus. They may also choose one of 24 Living and Learning Communities, groups of students who share

Many freshmen are assigned to suites with up to four roommates in Thurston Hall, the biggest and rowdiest dorm on campus.

similar interests. These groups have gone to the Kennedy Space Center for a rocket launch and to New York City to tour the United Nations. Despite the university's urban location, students say the campus is safe and security is visible and active.

Sixteen percent of GW men and 13 percent of the women go Greek, and students say it does not dominate the social scene. "There are over 350 student organizations on campus so there is always something going on," a senior says. Although underage drinking does occur, a D.C. police crackdown has made it extremely difficult for those under 21 to be served at off-campus restaurants and pubs. Major annual events include the Fall Fest and Spring Fling carnivals, with free food **"We have the best dorms in the country."** and such nationally known entertainment as The Roots and Busta Rhymes. Popular road trips include the beaches of Ocean City, Maryland, and Virginia Beach, Virginia. Philadelphia and New York City are easily accessible by bus or train, a boon because most GW students don't have cars.

GW doesn't field a football team, but its men's and women's basketball teams, men's soccer squad, and baseball team have won Atlantic 10 conference championships in recent years. The gymnastics squad is also strong, and the men's and women's rowing teams compete on the Potomac River, right in GW's backyard. Approximately 20 percent of undergraduates participate in the intramural sports program, which offers more than 30 events throughout the year. The school's unofficial mascot is the hippopotamus.

A popular GW T-shirt proclaims: "Something Happens Here." Something certainly has happened on both of the school's campuses in the past five years, says a junior. "GWU has become a much more well known institution," a sophomore says. "Standards are higher and admission criteria is much more competitive." For students interested in urban living in the heart of the nation's political establishment, GW may fit the bill.

Overlaps

Boston University, NYU, Georgetown, American, University of Maryland, Boston College

---

If You Apply To ➤

GW: Early decision: Nov. 10, Jan. 10. Financial aid: Jan. 31. Housing: May 1. Campus and alumni interviews: optional, evaluative. SATs or ACTs: required. SAT IIs: recommended. Accepts the Common Application and prefers electronic applications. Essay question: why GW; media and public affairs applicants must answer one concerning political communication, electronic media, or journalism. Freshmen must answer about an experience or design their own monument.

Georgetown University

37th and O Streets, NW, Washington, D.C. 20057

For anyone who wants to be a master of the political universe, this is the place. Strong international and intercultural culture. In the excitement of studying in D.C., students may pay little attention to the Jesuit affiliation, which adds a conservative tinge to the campus.

As the most selective of the nation's Roman Catholic schools, Georgetown University offers students unparalleled access to Washington, D.C.'s corridors of power. Aspiring politicos benefit from the university's emphasis on public policy, international business, and foreign service. The national spotlight shines brightly on this elite institution, drawing dynamic students and athletes from around the world.

From its scenic location just blocks from the Potomac River, Georgetown affords its students an excellent vantage point from which to survey the world. The 104-acre

Website:
www.georgetown.edu
Location: Center city
Total Enrollment: 13,652
Undergraduates: 6,250
Male/Female: 46/54

(Continued)

SAT Ranges: V 640–750
   M 650–740

ACT Range: 27–32

Financial Aid: 42%

Expense: Pr $ $ $

Phi Beta Kappa: Yes

Applicants: 15,285

Accepted: 21%

Enrolled: 47%

Grad in 6 Years: 93%

Returning Freshmen: 98%

Academics: ✍ ✍ ✍ ✍ ½

Social: ☎ ☎ ☎ ☎

Q of L: ★ ★ ★ ★

Admissions: (202) 687-3600

Email Address: N/A

Strongest Programs:
Government
Chemistry
Philosophy
Business
International Relations
Diplomatic History
International Economics

There are no special academic requirements for the freshman year, but about 30 Georgetown College freshmen are accepted annually into the liberal arts colloquium.

campus reflects the history and growth of the nation's oldest Jesuit university. The Federal style of Old North, home of the school of business administration, which once housed guests such as George Washington and Lafayette, contrasts with the towers of the Flemish Romanesque–style Healy Hall, a post–Civil War landmark on the National Register of Historic Places. The Royden B. Davis Performing Arts Center—home to the theater program—opened recently.

Although Georgetown is a Roman Catholic university (founded in 1789 by the Society of Jesus), the religious atmosphere is by no means oppressive. Just more than half of the undergraduates are Roman Catholic, but all major faiths are respected and practiced on campus. That's partially due to the pronounced international influence here. International relations, diplomatic history, and international economics are among the hottest programs, as evidenced by former Secretary of State Madeline Albright's return to the School of Foreign Service. Through its broad liberal arts curriculum, GU focuses on developing the intellectual prowess and moral rigor its students will need in future national and international leadership roles. The curriculum has a strong multidisciplinary and intercultural slant, and students can choose among several programs abroad to round out their classroom experiences.

> **"Students at Georgetown are unique because one moment they're doing a keg stand and the next they're talking about social reform."**

Would-be Hoyas may apply to one of four undergraduate schools: Georgetown College for liberal arts, the School of Nursing and Health Studies, McDonough School of Business, and the Walsh School of Foreign Service (SFS), which gives future diplomats, journalists, and others a strong grounding in the social sciences. Prospective freshmen must declare intended majors on their applications, and their secondary school records are judged accordingly. This means, among other things, intense competition within the college for the limited number of spaces in Georgetown's popular premed program.

Georgetown's liberal arts program is also very strong: The most popular programs include international relations, government, English, finance, and nursing. Of course, the theology department is also strong. The School of Foreign Service stands out for its international economics, regional and comparative studies, and diplomatic history offerings. SFS also offers several five-year undergraduate and graduate degree programs in conjunction with the Graduate School of Arts and Sciences. The business school balances liberal arts with professional training, which translates into strong offerings in international and intercultural business as well as an emphasis on ethical and public policy issues. The School of Nursing and Health Studies runs an integrated program combining the liberal arts and humanities with professional nursing theory and practice, and offers majors in nursing and health studies. The Faculty of Languages and Linguistics, the only undergraduate program of its kind nationwide, grants degrees in nine languages, as well as degrees in linguistics and comparative literature. There are few weak spots in the university's programs, and "the appeal of a department is mainly due to the professors," says a student.

All students must complete requirements in humanities and writing, philosophy, and theology; other requirements are specific to each school. That GU views most subjects through an international lens is evidenced by the 38 percent of students who study abroad. University-sponsored study programs in 90 countries—in Asia, Latin America, Poland, Israel, France, Germany, and at the university's villas in Florence, Italy, and Alanya, Turkey—attract the culturally curious. First-years read the same novel during the summer and the author visits campus during the first few weeks for a daylong seminar. There are no special academic

> **"Washington is an ideal place to spend your college years."**

requirements for the freshman year, but about 30 Georgetown College freshmen are accepted annually into the liberal arts colloquium.

Georgetown likes to boast about its faculty, and well it should. "For the most part, I have had phenomenal teachers," says a senior, "although it varies by class and department." The academic climate can be rigorous, but students are quick to point out that competition is friendly and community-oriented, with study groups the norm. "Overall, people are intense about school," says a nursing major, "but want their peers to do well, too."

"Students at Georgetown are unique because one moment they're doing a keg stand and the next they're talking about social reform," a senior says. The GU community includes students from all over the United States, and 7 percent of the student body comes from abroad. African Americans make up 7 percent of undergrads, Hispanics 6 percent, and Asian Americans comprise 9 percent. Almost half the students come from private or parochial schools. A student committee works with the vice president for student affairs to improve race relations and develop strategies for improving inclusiveness and sensitivity to issues of multiculturalism. About two-thirds of graduates move directly into the job market after graduation, and 29 percent of the graduating class heads straight to graduate or professional school. Georgetown offers no academic merit scholarships, but it does guarantee to meet the full demonstrated need of every admit, and 104 athletic scholarships draw male and female athletes of all stripes.

University-owned dorms, townhouses, and apartments accommodate 72 percent of undergrads, and "with the recent completion of the Southwest Quadrangle, campus housing offerings have drastically improved," says a history and government major. All dorms are co-ed, and some have more activities and community than others. "All have great amenities like Ethernet and landscaping," says one student. Two dining halls serve "edible" food, but "not diverse at all," says a student. GU students feel relatively safe on campus, thanks to the school's ever-present Department of Public Safety and its walking and riding after-dark escort services.

"The appeal of a department is mainly due to the professors."

Jesuits, who know a thing or two about secret societies, frown upon fraternities and sororities at their colleges, and so there are none at Georgetown. The university's strict enforcement of the 21-year-old drinking age has led to a somewhat decentralized social life, which is not necessarily a bad thing. Alcohol is forbidden in undergrad dorms, and all parties must be registered. The dozens of bars, nightclubs, and restaurants in Georgetown—Martin's Tavern and the Tombs are always popular—are a big draw for students who are legal, but they can get pricey. The Hoyas, a campus pub in the spectacular student activity center, is a more affordable alternative. Popular annual formals such as the Diplomatic and the Blue/Gray Ball force students to dress up and pair off. "There is always something going on that you don't want to miss," says a sophomore.

Washington offers unsurpassed cultural resources, ranging from the museums of the Smithsonian to the Kennedy Center. "Washington is an ideal place to spend your college years," says a student. "The city has everything students could want, including culture, shopping, museums, monuments, social life, and the clean and convenient Metro for transportation." Given the absence of on-campus parking, a car is probably more trouble than it's worth. Indeed, roadtrips "are not popular," a student says.

Should you notice the hills begin to tremble with a deep, resounding, primitive chant—"Hoya...Saxa...Hoya...Saxa"—don't worry; it's probably just another Georgetown basketball game. Hoya is derived from the Greek and Latin phrase *hoya saxa*, which means, "What rocks!" Some say it originated in a cheer referring to the

The Walsh School of Foreign Service (SFS) gives future diplomats, journalists, and others a strong grounding in the social sciences.

That GU views most subjects through an international lens is evidenced by the 38 percent of students who study abroad.

stones that formed the school's outer walls. The Hoya men's basketball team has a long history of prominence; in 2005–06 it advanced to the NCAA Sweet 16. The Hoyas are especially competitive in men's and women's lacrosse and sailing; the women's lacrosse team has won six straight Big East Championships. The thrill of victory in intramural competition at the superb underground Yates Memorial Field House is not to be missed, either.

For anyone interested in discovering the world, Georgetown offers an outstanding menu of choices in one of the nation's most dynamic cities. Professors truly pay attention to their undergrads and the diverse students who are "hard-working, diligent, caring individuals," says one sophomore. "Georgetown is a place where students of all backgrounds, all traditions, and all faiths come together for a common purpose of educating each other and making an impact on the world."

> ## Overlaps
> **Boston College, University of Pennsylvania, NYU, Duke, George Washington, University of Virginia**

> ## If You Apply To ➤
> **Georgetown:** Early action: Nov. 1. Regular admissions: Jan. 10. Financial aid: Feb. 1. Campus interviews: not available. Alumni interviews: required, evaluative. SATs or ACTs: required. SAT IIs: recommended. Apply to particular schools or programs. Essay question: personal statement plus one additional question for each school.

University of Georgia

212 Terrell Hall, Athens, GA 30602-1633

What a difference free tuition makes. Top Georgia students now choose UGA over highly selective private institutions. Business and social and natural sciences head the list of strong and sought-after programs. The college town of Athens boasts great nightlife and is within easy reach of Atlanta.

College-aged Georgians hit the jackpot when the state began using lottery receipts to fund the Hope Scholarship program. The program covers tuition, books, and most fees at the University of Georgia for all four years for students who finish high school in the state with a B average and maintain that average in college. In fact, the scholarship has made it much tougher to get into to UGA, which not long ago was known mostly for its dynamite football team and raucous parties. The university has moved aggressively to challenge its new and brainier breed of students, but some things haven't changed. "While we are academically competitive, we are not pretentious," a senior says. "We are very welcoming."

Founded in 1785, Georgia was the nation's first state-chartered university. Its attractive 706-acre campus is dotted with greenery and wooded walks. The older north campus houses the administrative offices and law school, and features 19th-century architecture and landscaping. The southern end of campus has more modern buildings and residence halls. A $40 million biomedical and health sciences facility opened in 2006 and provides 200,000 square feet for faculty research.

UGA's core curriculum includes nine to 10 credit hours of essential skills and four to five hours of institutional options selected with an academic advisor. Also

> **Website:** www.uga.edu
> **Location:** Small city
> **Total Enrollment:** 28,412
> **Undergraduates:** 22,564
> **Male/Female:** 44/56
> **SAT Ranges:** V 560–660
> M 570–670
> **ACT Range:** 24–28
> **Financial Aid:** 25%
> **Expense:** Pub $
> **Phi Beta Kappa:** Yes
> **Applicants:** 12,326
> **Accepted:** 65%
> **Enrolled:** 59%
> **Grad in 6 Years:** 74%
> **Returning Freshmen:** 93%
> **Academics:** ✑ ✑ ✑
> **Social:** 🐿 🐿 🐿 🐿 🐿
> **Q of L:** ★★★
> **Admissions:** (706) 542-8776

> **"While we are academically competitive, we are not pretentious."**

required are six hours in arts and humanities, 10 to 11 hours in science, math, and technology, 12 hours in social sciences, and at least 18 hours in the major field. Finally, students must demonstrate proficiency in written and oral communication, computer skills, and reading comprehension, and pass an exam on the U.S. and Georgia constitutions. Before the semester starts, freshmen may spend a month on campus to learn their way around, meet new friends, and even earn six hours of credit. Don't want to stay for an entire month? Choose the Dawg Camp weekend retreat, which promotes networking and leadership skills, and includes programs on time and stress management, diversity, and "What It Means to Be a Georgia Bulldog." During the year, freshman seminars allow first-year students to study under a senior faculty member in a small, personalized setting while earning an hour of academic credit.

Students give high marks to UGA's Terry College of Business, Grady College of Journalism, the School of Public and International Affairs, and the colleges of education and of agriculture and environmental sciences. Newer majors include water and soil resources, environmental resource science, dance, avian biology, and environmental chemistry. "I would say the academic climate is pretty competitive," says a senior. "I think it has become more competitive since I started here four years ago." As you might expect, large lecture classes are common. But "teaching assistants do just that," says a senior. "They assist the professor in break-out sessions, and the monitoring and grading of exams." A computerized registration system helps students sign up for courses; first pick usually goes to the 2,500 honors students and to UGA's varsity athletes, with everyone else prioritized by class standing. (That's become more important as recent cuts in state funding leave hundreds of

"The dining halls are incredible!"

faculty and staff positions unfilled. Classes are larger, some programs have stopped enrolling new students entirely, and students report more difficulty getting into the classes they want.) The Center for Undergraduate Research Opportunities allows students to conduct a research or service project, write a thesis, or develop a creative work with close faculty supervision. About 25 percent of undergrads study abroad, through 90 programs in more than 30 countries. UGA operates year-round residential study abroad programs of its own in Oxford, England; Avingon, France; and Cortona, Italy; and the school has an ecological research center in Costa Rica.

Eighty-five percent of UGA students are Georgian, and 84 percent graduated in the top quarter of their high school class. "Students tend to be either suburban kids from Atlanta or from more affluent South Georgia rural backgrounds," says a broadcast news major. African Americans and Asian Americans each make up nearly 5 percent of the student body, and Hispanics add another 2 percent. Politically, the school tends to lean right, but as a senior notes, "For the most part, outside of election time, I would say a majority of students steer away from political debate." Merit scholarships are available, worth $1,870 on average, and UGA also gives out 418 athletic scholarships. As many as 100 top undergraduates are named Foundation Fellows, netting a full scholarship plus stipends for international travel and research.

Twenty-seven percent of Bulldogs live in the dorms, and freshmen are required to do so. Rooms typify "stereotypical freshman living," says a student: "small rooms and community bathrooms." After freshman year, it's much harder to get a room, students say. Most keep their meal cards even after moving off campus. "The dining halls are incredible!" says a student. There are four campus dining halls—each with its own specialty cuisine—and a snack bar, and the excellence of UGA's food service program has been recognized with the Ivy Award.

When the weekend comes, students know how to have a good time. "Most student GPAs suffer because of the social aspect of UGA," says a junior. There are more than 500 student organizations; sororities and fraternities attract 18 percent of the men and 23 percent of the women, respectively. More significant than Greek life is the funky mix of shops, restaurants and clubs, and various cultural events found in downtown Athens, a 10-minute walk from most residence halls. "Athens is the ultimate college town, with an amazing music scene. It gave Widespread, R.E.M., John

(Continued)
Email Address: undergrad @admissions.uga.edu

Strongest Programs:
Life Sciences
Ecology and Environmental Studies
Agriculture
International and Public Affairs
Business
Education
Journalism/Mass Communication

Students give high marks to UGA's Terry College of Business, Grady College of Journalism, the School of Public and International Affairs, and the colleges of education and of agriculture and environmental sciences.

Athens residents worship UGA's perennially fierce football team, which earned another bowl appearance (and victory) during the 2006–07 season.

Mayer, and many others their start," says a psychology major. Many clubs in Athens cater to UGA students and admit those under 21, as long as they get a stamp saying

"Athens is the ultimate college town, with an amazing music scene."

they can't drink. Alcohol is prohibited in the dorms, but as at most schools, the determined manage to imbibe anyway. Athens is also "far enough away from Atlanta to maintain the college community, yet close enough to provide an escape," a senior says. Other popular road trips include the Florida and Carolina beaches, and anywhere the Bulldogs are playing on a fall Saturday.

Indeed, Athens residents worship UGA's perennially fierce football team, which earned another bowl appearance (and victory) during the 2006–07 season. The Georgia–Florida rivalry is particularly notable: "The game takes place on neutral ground in Jacksonville, and since it coincides with fall break, 90 percent of the students go," says an accounting major. The women's gymnastics team won its seventh national championship in 2006, and the men's baseball team earned a spot in the College World Series. Women's softball and men's golf are also competitive. For the weekend warrior, there are intramural teams in 15 sports and the Oconee River Greenway, a 13-mile paved trail.

UGA's sheer size means you could coast through four years here as nothing more than a number. But with a little effort, that doesn't have to happen. Freshman seminars, research projects, study abroad, and honors courses offer the opportunity to graduate with a solid background in any number of areas and fond memories of Saturdays spent cheering on the Bulldogs—along with 92,000 of your closest friends. "UGA is a land of opportunity," a senior says. "Of course you have academic opportunity, but the relationships you form here are more valuable than anything you learn from a book."

Overlaps

Georgia Tech, Emory, Auburn, Georgia Southern, UNC–Chapel Hill, Clemson

If You Apply To ➢

Georgia: Early action: Oct. 15. Regular admissions: Jan. 15. Meets demonstrated need of 31%. No campus or alumni interviews. SATs or ACTs: required. SAT IIs: recommended. Accepts electronic applications. Essay questions: How you would contribute to and/or benefit from a broadly diverse learning environment; personal experience that showed integrity and/or maturity; how you will make positive contributions to UGA; experience though which you gained respect for intellectual, social, or cultural differences; how you grew through an intellectual or creative opportunity.

Georgia Institute of Technology

Atlanta, GA 30332-0320

As the South's premier technically-oriented university, Ma Tech does not coddle her young. That means surviving in downtown Atlanta and fighting through a wall of graduate students to talk with your professors. Architecture and big-time sports supplement the engineering focus.

Website: www.gatech.edu
Location: City center
Total Enrollment: 15,281
Undergraduates: 10,951
Male/Female: 72/28
SAT Ranges: V 600–700
     M 650–740

If you're looking for lazy days on the college green and hard-partying weekends, look elsewhere. You won't find those at Georgia Institute of Technology, the South's premier tech university. What you will find are challenging courses that prepare you for a high-paying job as an engineer, architect, or computer scientist. "Tech is tough," reasons a graduate student. "You have to want to be here." Still, even those who want to be here are happy to finally arrive at graduation day. What makes Tech a special place? "The fact that I survived it and got out with a degree," says a computer science major, only partially joking.

Located just off the interstate in Georgia's capital city, Tech's 330-acre campus includes seven residence halls, an aquatic center, a sports performance complex, and an amphitheater. Reflective of the history of Georgia Tech, the building styles include the Georgian Revival and Collegiate Gothic of the historic Hill District (listed on the National Register of Historic Places) and surrounding area, the International Style buildings constructed from the 1940s into the 1960s, the modernist structures of the 1970s and '80s, the postmodern facilities of the '90s, and the recently constructed high-tech facilities. All these styles coexist comfortably on a tree-filled and landscaped campus seen as a green oasis in the midst of a dense urban environment. The ambitious Campus Master Plan resulted in the opening of more than two million square feet of new and renovated space at a cost of nearly $500 million. Current construction includes a molecular science and engineering building, the Christopher Klaus Advanced Computing Building, and the expanded Fifth Street Bridge.

Courses at Tech are "extremely rigorous," says a senior, at least in the sciences and engineering. "Grading on a curve creates hypercompetitive situations because your absolute grade is largely irrelevant—you just have to do better than most of the others." Strong programs include math and computer science ("It's hard to have a life and be a CS student," one major quips) as well as most types of engineering, especially electrical, computer, and mechanical. Tech also offers materials, ceramic, chemical, and nuclear engineering. Tech has plenty of liberal arts courses, but a grad student says history, philosophy, and English aren't the reasons why most students enroll. "International affairs, while it has some interesting classes, seems to be a haven for people that can't hack the engineering stuff," another student adds. The newest undergraduate offering is a multidisciplinary degree in computational media.

> "Tech is tough. You have to want to be here."

Aside from the technical fare, Tech's management college is increasingly popular, and its school of architecture has done pioneering work in historic preservation and energy conservation. Among the architecture program's alumni is Michael Arad, whose winning design for the September 11 memorial in lower Manhattan was selected from a field of more than 5,200. The prelaw certificate is a boon to aspiring patent attorneys, as is the minor in law, science, and technology. As most courses need computers, the school requires students to bring their own desktops or laptops.

Regardless of major, students must complete nine semester hours of social sciences, eight hours each of math and science, six hours each of English and humanities and fine arts, three hours each of computer science and U.S. or Georgia history, and two hours of wellness. As students move from those core and required courses to upper-level options within their majors, the quality of teaching improves. "It's absolutely horrible for things like freshman math classes," says a computer science major. "You're typically taught by TAs (grad students), maybe half of whom have only the slightest grasp of English. Things get better as you progress and get to know professors." That's because those professors are indeed exceptional; some have worked on projects such as

> "Atlanta is not a college town. However, it is the best thing going in Georgia."

the Star Wars missile-defense system and the space shuttle. Still, classes are big—12 percent of those taken by undergraduates have more than 100 students—and the problem has been getting worse rather than better. "Students are generally stressed and tired," sighs a grad student. "Not working hard is not an option here."

In fact, Tech's demanding workload means it's common to spend five years getting your degree. There's also the frustrating course selection process: "Sleep through your registration time ticket, and you may blow your semester because you won't get into anything," warns a senior. Also contributing to delayed graduation dates is the popular co-op program, through which more than 3,000 students earn money

(Continued)
ACT Range: 26–30
Financial Aid: 30%
Expense: Pub $
Phi Beta Kappa: No
Applicants: 9,172
Accepted: 67%
Enrolled: 39%
Grad in 6 Years: 76%
Returning Freshmen: 92%
Academics: 🖋🖋🖋🖋
Social: ☎ ☎
Q of L: ★ ★
Admissions: (404) 894-4154
Email Address:
  admission@gatech.edu

Strongest Programs:
Engineering
Computer Science
Architecture

for their education while gaining on-the-job experience. Those eager to experience another culture or environment can tap into the courses and resources of the Atlanta Regional Consortium for Higher Education,* or choose Tech study abroad programs in Paris, London, Australia, or England's Oxford University.

Two-thirds of Georgia Tech's students come from Georgia, and most are too focused on school or their co-op jobs to care about politics, causes, or any of the issues that get their peers riled up on nearby campuses. "There are a lot of left-brain types here—high on the introspection and thinking, low on the social skills," says a senior. And though they may be united in their pursuit of technical expertise, the campus is hardly homogenous: African Americans account for 8 percent of the student body, Hispanics 4 percent, and Asian Americans 15 percent. To limit burgeoning enrollment, out-of-state applicants must meet slightly higher criteria than their Georgia counterparts. The university awards merit scholarships to 44 percent of students, with awards averaging $3,041. There are also 395 athletic scholarships available to student athletes.

Sixty-one percent of students live in the dorms, where freshman are guaranteed a room. A senior says that despite the new construction and renovation that took place before 1996, when Tech was transformed into the Olympic Village, the quality of residence halls varies widely. "Some dorms are new, apartment-style, and nice," the student says. "Others are foul dungeons." Many dorms have full kitchen facilities, though, and while most halls are single-sex, visitation rules are lenient. Off-campus housing is generally comfortable, but parts of the surrounding neighborhood are unsavory. "Far too many cars are broken into or stolen," says one student. "There's usually a couple of armed robberies (at least) a semester." The campus dining halls offer "little variety and less quality," according to another student.

"I love a good challenge, and Tech is perfect for that."

Fortunately, even if mystery meat is on the day's menu at Tech, the school is smack-dab in the middle of "Hot-Lanta," with its endless supply of clubs, bars, movie theaters, restaurants, shopping, and museums, both in midtown Atlanta and the Buckhead district. "Atlanta is not a college town," reasons a computer science major. "However, it is the best thing going in Georgia," with friendly, young residents, good cultural activities, beautiful green spaces, and a booming economy. The city also offers plenty of community service opportunities. Fraternities draw 21 percent of Tech's men and sororities draw 24 percent of the women, and members may live in their chapter houses. Alcohol flows freely at frat parties, but otherwise, students say, Tech's policies against open containers and underage drinking are strictly enforced. "There's not much in the way of social life here outside of the frats," says a senior. "You have your group of friends and you do your own thing." The best road trips include Florida's beaches, which are a half-day's drive, and Athens, Georgia, for basketball or football games against the University of Georgia.

Tech's varsity sports have become as big-time as any in the South, and when the weekend comes, students throw off their lab coats and pocket protectors and become wild members of the "Rambling Wreck from Georgia Tech." Recent championship teams include men's basketball, football, golf, volleyball, softball, and women's tennis. Among Tech's many other traditions are "stealing the T," in which students try to remove the huge yellow letter "T" from the tower on the administration building and return it to the school by presenting it to a member of the faculty or administration. The addition of alarms, motion sensors, and heat sensors on the T has made the task more difficult, but "certainly not impossible for a Georgia Tech engineer," says an electrical engineering major. And then there's the Mini 500, a 15-lap tricycle race around a parking garage with three pit stops, a tire change, and a driver rotation.

Forget fitting the mold; the engineers of Georgia Tech are proud to say they make it. Self-direction, ambition, and motivation will take you far here, as will dexterity with a graphing calculator and a fondness for highly complex software algorithms. And despite their complaints about the workload, the social life (or lack thereof), the safety of their surrounding neighborhood, and the impact of budget cuts, Tech students do have a soft spot for their school. Says one student, "I love a good challenge, and Tech is perfect for that."

If You Apply To ➣

Georgia Tech: Regular admissions: Jan. 15. Financial aid: May 1. Housing: May. 1. Does not guarantee to meet demonstrated need. No campus or alumni interviews. SATs or ACTs: required. SAT IIs: optional. Accepts the Common Application and electronic applications. Essay question: the personal experience that gave you the feeling of greatest achievement or satisfaction because of the challenges you met. Looks for high math and science aptitude.

Gettysburg College

300 North Washington Street, Gettysburg, PA 17325-1484

The college by the battlefield is strong in U.S. history—that's a given. The natural sciences and business are also popular, and political science majors enjoy good connections in D.C. and New York City. Students can also take courses down the road at Dickinson and Franklin and Marshall.

Mention the word "Gettysburg," and patriotic heart palpitations and echoes of the "Battle Hymn of the Republic" are likely to result. Whether the reference is to the Pennsylvania town steeped in Civil War history or the small, high-caliber college located in the famed battlefield's backyard, a certain pride and reverence are immediately evident. This feeling is not lost on students at Gettysburg College, who come to southeastern Pennsylvania to acquaint themselves with American history while gearing up for the future.

Situated in the midst of gently rolling hills, Gettysburg's 200-acre campus is "a historical treasure," an eclectic assemblage of Georgian, Greek, Romanesque, Gothic Revival, and modern architecture, plus several styles not easily categorized. One campus building—Penn Hall—was actually used as a hospital during the Battle of Gettysburg. Rumor has it that ghostly soldiers can still be seen walking the grounds. The 86,000-square-foot Science Center provides a home and state-of-the-art equipment for the sciences, and three new athletic fields have been added to the existing athletic facilities.

Indoors, the English department, home of the *Gettysburg Review*, is among the strongest at Gettysburg, as are the natural sciences, which are well endowed with state-of-the-art equipment. The fine psychology department offers opportunities for students to participate in faculty research. The management major is the most popular. Also popular, of course, is the excellent history department, which **"Gettysburg is in a league of its own."** is bolstered by the school's nationally recognized and prestigious Civil War Institute. The library system boasts nearly 400,000 volumes, a library/learning resource center, and an online computer catalog search. "The academic climate is definitely competitive and the courses can be rigorous," says a senior.

The small class sizes make for close student/faculty relationships. "Professors here are engaging and have the drive to get us thinking critically," says an English major. "I have grown so much as a student greatly due to their dedication." The academic

Website: www.gettysburg.edu
Location: Small town
Total Enrollment: 2,463
Undergraduates: 2,463
Male/Female: 48/52
SAT Ranges: V 600–670
  M 600–670
ACT Range: N/A
Financial Aid: 56%
Expense: Pr $ $ $ $
Phi Beta Kappa: Yes
Applicants: 5,313
Accepted: 41%
Enrolled: 34%
Grad in 6 Years: 77%
Returning Freshmen: 92%
Academics: ✐ ✐ ✐½
Social: ☎ ☎ ☎
Q of L: ★ ★ ★
Admissions: (800) 431-0803
Email Address:
  admiss@gettysburg.edu

Strongest Programs:
English
History

(Continued)
Psychology
Natural Sciences
Business
Political Science

honor code contributes to the atmosphere of community and mutual trust. The popular first-year seminars explore topics such as "Why Do People Dance?"; participants live in the same residence hall and are in the same first-year residential college program. Another popular program is the Area Studies Symposium, which focuses each year on a different region of the world and offers lectures and films for the whole campus, in addition to academic credit for participating students. There are disciplinary programs such as environmental studies, Latin American studies, and biochemistry and molecular biology. Students may pursue a music in performance degree within the Sunderman Conservatory of Music; there are also new degrees in globalization studies and African studies.

Gettysburg sponsors a Washington semester with American University, a United Nations semester through Drew University in New Jersey, and cooperative dual-degree programs in engineering and forestry. Most departments offer structured internships, and the chemistry department offers a summer cooperative research program between students and professors in which most majors participate and work on a joint publication. Through the Central Pennsylvania Consortium, students may take courses at two nearby colleges—Dickinson and Franklin and Marshall. Outstanding seniors may participate in the Senior Scholar's Seminar, with independent study on a major contemporary issue, but all students have a chance to do independent work and/or design their own majors. Study abroad programs are global and popular with more than half of the students taking part during their college career.

About a quarter of Gettysburg's students earn varsity letters, and the college boasts a strong athletic program. In fact, school teams have brought home 74 conference titles in the past 13 years.

"Professors here are engaging and have the drive to get us thinking critically."

The Gettysburg student body is 89 percent Caucasian and mostly middle- to upper-middle-class. A math major says Gettysburg students are "involved, inquisitive, and actively engaged." African American, Asian American, and Hispanic enrollment accounts for just 8 percent of the student body. Students are so interested in public service that the school set up a Center for Public Service to direct their community activities. Seventy percent of the students come from public high school and 53 percent were in the top tenth of their high school class. No athletic scholarships are available, but academic scholarships average nearly $10,000.

Campus housing is guaranteed all four years, and students can choose from apartment-style residence halls, special interest halls, and the Quarry Suites. The top scholars in each class get first crack at the best rooms. Student rooms have been added in renovated historical properties (some reputed to be haunted) on campus, and there are more options for interesting housing and suite living. "The rooms are pretty nice and fairly large," a student says. Off-campus apartments lure 6 percent of the student body while freshmen are required to remain in the residence halls. There are a variety of dining options, including the ever popular Cafe 101, the campus snack bar and grill room where many students take their regular meals. Kitchens are also available in the residences for upperclassmen. In a word, "Dining facilities are fantastic!" says a French major.

The Area Studies Symposium, which focuses each year on a different region of the world and offers lectures and films for the whole campus, in addition to academic credit for participating students.

Social life at the 'Burg involves the Greek system and other activities. Forty percent of the men belong to the dozen fraternities; the seven sororities draw 26 percent of the women. Greek parties are open and attract crowds eager to dance the night away, although students insist they're not the only source of fun on campus. A Student Activities Committee provides alternative social events, including concerts, comedians, bus trips to Georgetown, movies, and campus coffeehouses. "Students spend their weekends doing community service, going to sporting events, hanging out in town, going to campus special events, or just hanging out in their dorms," says a senior. Officially the campus is dry, but like many such campuses, drinking can be done,

"The academic climate is definitely competitive and the courses can be rigorous."

albeit carefully, students report. The orchards and rolling countryside surrounding the campus are peaceful and scenic, and there is a small ski slope nearby. Students also get free passes to the historic attractions in town. Many participate in the November 19 Fortenbaugh Lecture by noted historians commemorating the Gettysburg Address and in the yearly wreath-laying ceremony in front of the Eisenhower Admissions Office to commemorate the general's birthday. Tourist season is a common complaint among students. But those who want to escape can do so—the campus is within an hour and a half of Washington, D.C., and considerably closer to Baltimore, where students enjoy the scenic Inner Harbor area.

About a quarter of Gettysburg's students earn varsity letters, and the college boasts a strong athletic program. In fact, school teams have brought home 74 conference titles in the past 13 years. The annual football game against Dickinson draws a good turnout, and the Little Brown Bucket, mahogany with silver handles, is passed to the team that wins. Both track and swimming frequently produce All-Americans. The college is a 10-time winner of the President's Cup, awarded to the top overall athletic program in the Centennial Conference.

At Gettysburg, students stay true to their slogan: "Work Hard, Play Smart." "Gettysburg is in a league of its own," a music/political science double major says. Students wanting personal attention from professors, solid academics, and an area rich with history might consider getting their education with a Gettysburg address.

If You Apply To ➤

Gettysburg: Regular admissions and financial aid: Feb. 15. Guarantees to meet full demonstrated need. Campus interviews: recommended, evaluative. No alumni interviews. SATs or ACTs: required. SAT IIs: optional. Accepts Common Application and electronic applications. Essay question: describe how you have made a difference to your school or community.

Gordon College

255 Grapevine Road, Wenham, MA 01984

Gordon is the most prominent evangelical Christian college in New England and competes nationally with Wheaton (IL) and Calvin without all those strict rules. Not quite in the Boston area, but close enough.

Evangelical Christian values color almost all aspects of life at this New England college, where faith and spirituality are not only character traits, but required courses of study as well. The college, founded as a missionary training school, is always evolving, sharpening its offerings across the board, from neuroscience to music education, and looking to increase its diversity. The students revel in the atmosphere. "Since we are a Christian college, our beliefs shine through in our character," says a senior. "Gordon kids are trustworthy, loyal, honest, dedicated, and warm."

Gordon is located on Massachusetts's scenic North Shore, three miles from the Atlantic Coast and 25 miles from Boston. The campus sits on hundreds of forested acres with five lakes. Academic buildings and dorms are clustered in one small section. Most structures are Georgian-influenced traditional redbrick, except for the old stone mansion that houses administration and faculty offices. The college opened a 162-student residence hall, along with an athletic complex with facilities for tennis, soccer, lacrosse, and track and field. Many college offices recently moved to a nearby corporate park, alleviating a space crunch.

Website: www.gordon.edu
Location: Suburban
Total Enrollment: 1,645
Undergraduates: 1,553
Male/Female: 37/63
SAT Ranges: V 550–670
  M 540–650
ACT Range: 23–29
Financial Aid: 87%
Expense: Pr $
Phi Beta Kappa: No
Applicants: 1,098
Accepted: 84%
Enrolled: 45%

For outdoorsy types, Gordon's setting on rugged Cape Ann is ideal, and it attracts its share of tourists.

Because religious commitment is seen as an enhancement to, not a threat against, serious academic inquiry, Gordon's core curriculum includes 46 hours of instruction distributed among religion, the fine arts, humanities, social and behavioral sciences, natural sciences, math, and computer science. Freshmen also take a first-year Christianity, Character,

"Since we are a Christian college, our beliefs shine through in our character."

and Culture seminar to help them learn how to integrate faith into their academic experience. The City Scholars Program provides inner-city students with mentoring and support both on campus and at home. The most popular major is English, followed by psychology, business, communications, history, and Biblical studies. Finance was recently added as a major—a rarity at small Christian colleges—and theatre was spun off from communications to become its own major. Some students complain that some of the science departments, such as chemistry and physics, lack modern equipment, and administrators admit this limits the programs' strength.

Gordon's faculty receives high marks; a junior psychology major calls the quality of teaching "outstanding." "They all certainly know what they're talking about," says a freshman, "although some are better communicators than others." Off-campus opportunities include stints in Washington, D.C., for aspiring politicos, in Michigan for environmentalists, in Los Angeles for filmmakers, and trips abroad through the Christian College Consortium.* The college also has its own programs in Orvieto, Italy, focused on the country's history, art, and language; Aix-En-Provence in France; and at Oxford University in England; and partners with programs elsewhere around the world, from Africa to Israel.

Hard-working Christians come to Gordon from all over the United States. "Students at Gordon are characterized by their global mindset and concern for social justice," observes a student. Still, the global concern doesn't translate into ethnic diversity on campus. "Most of the students are white, middle-class New Englanders," says a sophomore. African Americans constitute 1 percent of the campus, while Asian Americans, international students, and Hispanics each comprise 2 percent. Gordon students face the same campus issues as their peers at most colleges—homosexuality and tolerance, for example—but are required to sign a Statement of Faith promising acceptance of racial and gender equality, and moderation in behavior. "The

"Students at Gordon are characterized by their global mindset and concern for social justice."

student body is mostly split between Democrat and Republican," says a Bible studies and English major, "which is very unusual in Massachusetts." Merit scholarships worth an average of $12,000 are offered, but there are no athletic scholarships.

Eighty-eight percent of Gordon students live in the co-ed dorms, where men and women live in separate wings of the same buildings—separated by a lobby, a lounge, and a laundry room. Persons of the opposite sex may traverse these barriers only at specified times. That policy draws some complaints from students, and the Student Council has been trying to extend the hours. Permission to move off campus is granted only after the dorms are filled. "Most students choose to live on campus because the community and living experience at Gordon is so valuable," says a sophomore.

Social life at Gordon is "fairly quiet. The academic rigors keep everyone pretty busy." Another student agrees, but adds, "The social life at Gordon is well-balanced by students with their academic life." As drinking and smoking are forbidden on campus (and may result in suspension or expulsion), students focus on other activities. "There is always something to do on campus," a junior reports, "whether it be going to one of the varsity games, seeing a play, or going to a dance or talent show." Those who are 21 or older may drink off campus, but are expected to do so responsibly. Other options include weekend excursions to Boston (25 miles away by a five-minute walk to the T, the city's public transit system), the beach, church-related functions, movies, and an occasional square dance.

Everyone looks forward to homecoming, the Winter Ball formal (held at the Danversport Yacht Club), and the Last Blast spring party. Each year, the most popular guys in each class face off in the "hilarious" Golden Goose talent show, where the winner is crowned "Mr. Gordon." Gordon has no Greek system—so drunken toga parties are out. For outdoorsy types, Gordon's setting on rugged Cape Ann is ideal, and it attracts its share of tourists. The campus has cross-country ski trails and ponds for swimming, canoeing, and skating. The ocean is a quick bike ride away, nice beaches are available on Cape Cod and in Maine, and students frequently ski New Hampshire's nearby White Mountains. Having a car is essential. Volunteering through prison ministry and in soup kitchens and local churches is popular, and missionary road trips take students to Tennessee, Florida, and Washington, D.C.

> **"Most students choose to live on campus because the community and living experience at Gordon is so valuable."**

Gordon competes in NCAA Division III athletics and "a good portion of the student body comes out to the games" when the opponent is rival Endicott College, says a history major. Another adds: "The basketball games are standing-room-only when we play them." The men's and women's lacrosse teams brought home Commonwealth Coast Conference championships in recent years with women winning in 2005. Other popular sports are men's and women's soccer. Intramural sports draw a lot of players.

For many students, Gordon's combination of Christian values, strong academics, and relaxed setting is a winning one. "Gordon wants to graduate men and women of academic excellence and high Christian character." And as the school tries to attract students of different backgrounds, those here already find it welcoming. "I don't need to be anything I'm not. I have the respect and acceptance of my peers," says another junior. "Gordon is a Christian community where students from all different backgrounds can come together and learn from each other with freedom within a framework of faith."

Gordon students face the same campus issues as their peers at most colleges—homosexuality and tolerance, for example—but are required to sign a Statement of Faith promising acceptance of racial and gender equality, and moderation in behavior.

Overlaps
Wheaton (IL), Houghton, Messiah, Eastern, Calvin

If You Apply To ➤ **Gordon:** Early decision: Nov. 15. Regular admissions and financial aid: Mar. 1. Does not guarantee to meet demonstrated need. Campus interviews: required, evaluative. No alumni interviews. SATs: required. SAT IIs: recommended. Accepts the Common Application and electronic applications. Essay question: do you consider yourself a Christian; why you are interested in Gordon; response to a meaningful educational experience or achievement, or a pressing issue in American life.

Goucher College

Baltimore, Maryland 21204

This is not your grandmother's Goucher. Once a staid women's college, Goucher has added men and a more progressive ambience, similar to places like Skidmore and Sarah Lawrence. Strategically located near Baltimore and not far from D.C., Goucher offers an excellent internship program.

Goucher is the kind of place where a student starts off with a dance class, then dashes to a lab to use a nuclear magnetic resonance spectrometer, and finally wraps up the afternoon chatting with a professor about studying abroad in Ghana. The school's

Website: www.goucher.edu
Location: Suburban

mission is to prepare students for a life of inquiry, creativity, and critical and analytical thinking. According to students, the atmosphere is academically challenging and competitive but not uptight. A sophomore says it succinctly: "We're a family."

Goucher, a former woman's college that went co-ed in 1987, has a long-standing history of excellence. Phi Beta Kappa established a chapter on campus only 20 years after the college was founded, and the college ranks among the nation's top 50 liberal arts colleges in turning out students destined for PhDs in the sciences. Set on 287 landscaped acres in the suburbs of Baltimore, Goucher's wooded campus features lush lawns, stately fieldstone buildings (the fieldstone is mined from local quarries), and rare trees and shrubs from all corners of the globe. A new residence hall opened in 2005 and among its other features, it incorporates two faculty apartments designed for use by visiting scholars.

A rigorous general education program, adopted in 2006, forms the foundation of every Goucher student's education. The core curriculum requires a first-year colloquium (Frontiers), one course in each of the humanities, social sciences, and mathematics, a lecture/lab course in the natural sciences, computer proficiency, a writing class, completion of the intermediate level of a foreign language, two physical education courses, and a one-credit Connections course that covers the "hot topic issues of college," says a sophomore. "This class was helpful because I found a mentor on campus in the residence life department and now I always have someone to talk to about issues." Of Goucher's offerings, the science departments are arguably the strongest, with a nuclear magnetic resonance spectrometer and scientific visualization lab available for student use. Other facilities include dedicated research space, a greenhouse, and an observatory with a six-inch refractor telescope. The dance department is especially strong. Administrators acknowledge that psychology and communications, ironically two of the most popular majors, are the weakest programs due to their large size and lack of full-time faculty. Administrators have recently revamped the programs and predict that they will emerge among the strongest majors.

> "Goucher has small classrooms and the professors always keep their office doors open if you need help."

An honors program has been dropped, but an international scholars program has been added. There's also a German minor offered through Loyola College (MD), and future engineers can take advantage of the 3–2 program offered in conjunction with the Whiting School of Engineering at Johns Hopkins University. A curriculum change has transformed the Intercultural studies major into the new International Portfolio Program and there is a new peace studies major. Goucher students may take courses at nearby Johns Hopkins and seven smaller area colleges. The campus library houses 280,000 volumes and draws complaints from some students, mainly because it closes at 6 p.m. on Saturdays. However, Goucher students have free access to the libraries at Hopkins and other nearby schools.

> "A lot of students hold forums, discussions, and workshops to talk about the issues."

Faculty members here devote most of their time and energy to undergraduate teaching and have a good rapport with students. "Professors are eager to help you learn and the one-on-one [interaction] really improves grades and general attitude," says one student. Each freshman has a faculty advisor to assist with the academic and overall adjustment to college life, which is made easier by Goucher's trademark small classes and individual instruction. "Goucher has small classrooms and the professors always keep their office doors open if you need help," says one sophomore.

In addition to their academic work, all Goucher students are required to do a three-credit internship or off-campus experience related to their major. Popular choices include congressional offices, museums, law firms, and newspapers. Another

Study abroad is no longer just an option: It's mandatory. Students are required to take a three-week course abroad, but may elect to stay longer.

option is the three-week-long Public Policy Seminar in Washington, D.C., where students meet informally with political luminaries. Goucher also hosts College Summit, a nonprofit organization for economically challenged high school students. It offers summer mentoring workshops where members of the campus community serve as writing coaches. Study abroad is no longer just an option: It's mandatory. Students are required to take a three-week course abroad, but may elect to stay longer. The college has pledged to provide every student with a $1,200 voucher to offset the cost.

"It is difficult to understand what the words 'eclectic,' 'unique,' and 'individual' mean until you have been here for 24 hours!"

Thirty-one percent of Goucher's students are homegrown, and most of the rest hail from Pennsylvania, Virginia, New York, and New Jersey. African Americans make up 4 percent of the student body, Hispanics 3 percent, Asian Americans 3 percent, and foreign students 1 percent. The administration hosts "campus conversations" with students, faculty, and the college president to discuss various issues. Newer, more intensive programs such as the Study Circle on Diversity and the peace studies dialogue sessions have also been implemented to address diversity issues. "A lot of students hold forums, discussions, and workshops to talk about the issues," says one sophomore. Goucher offers merit scholarships worth an average $8,825, for those who are qualified, some providing full tuition, room, and board each year.

Eighty-one percent of students live on campus and this figure promises to increase as students fill the new residence hall. Freshmen double up in spacious rooms, while upperclassmen select housing through lotteries; the available singles usually go to juniors and seniors, though a lucky sophomore may occasionally get one. "Housekeeping is friendly and most students get the rooms that they want. It's a very family-like atmosphere," says a sophomore. Campus dining options include vegan and vegetarian fare. One satisfied sophomore claims it's the "best homemade campus food" anywhere.

The social life on campus is improving, though it is often up to students to make their own fun. One student says, "Most students go off campus at least one night a weekend," so access to a car is a virtual necessity because many students travel to nearby universities (Loyola and Towson State) or Baltimore's Inner Harbor for entertainment. The CollTown Network, however, provides transportation to nearly 20 colleges in the area. Students who are of age frequent restaurants and bars in Towson, the small but bustling college town a five-minute walk away. Goucher has no sororities or fraternities, but the close-knit housing units hold periodic events, and the college hosts weekend movies, concerts, and lectures. Major annual social events include Rocktoberfest, Spring Fling, and the Blind Date Ball each fall. Biggest of all is GIG, Get-into-Goucher Day, when classes are canceled and the whole campus celebrates. Popular road trips include Ocean City, New York, Philadelphia, and Washington, D.C.

As Goucher was a women's college for so long, women's athletics are more highly developed than those at many co-ed schools. The women's lacrosse team is popular, along with men's basketball and lacrosse. The genteel sport of horseback riding is popular, thanks to the indoor equestrian ring, stables, and beautiful wooded campus trails. Goucher also has several tennis courts, a driving range, practice fields, a swimming pool, and saunas.

Goucher is far from a stagnant place. Indeed, it is constantly rethinking its mission and redirecting its resources to broaden student experiences. With self-designed interdisciplinary majors, an emphasis on experiential learning, and several partnerships with other top schools, Goucher students find themselves sampling from a buffet of options. Says one student, "It is difficult to understand what the words 'eclectic,' 'unique,' and 'individual' mean until you have been here for 24 hours!"

Goucher students may take courses at nearby Johns Hopkins and seven smaller area colleges.

Freshmen double up in spacious rooms, while upperclassmen select housing through lotteries; the available singles usually go to juniors and seniors, though a lucky sophomore may occasionally get one.

Overlaps
American, Skidmore, Clark, Dickinson, Towson, Wheaton (MA)

Grinnell College

Grinnell, IA 50112

Iowa cornfields provide a surreal backdrop for Grinnell's funky, progressive, and talented student body. At just less than 1,500 students, Grinnell is half the size of Oberlin. That translates into tiny classes and tutorials of 13 students or fewer. Second only to Carleton as the best liberal arts college in the Midwest. Grinnell's biggest challenge is simply getting prospective students to the campus.

Website: www.grinnell.edu
Location: Small town
Total Enrollment: 1,541
Undergraduates: 1,541
Male/Female: 46/54
SAT Ranges: V 640–750
  M 640–730
ACT Range: 29–33
Financial Aid: 55%
Expense: Pr $ $
Phi Beta Kappa: Yes
Applicants: 3,121
Accepted: 45%
Enrolled: 28%
Grad in 6 Years: 88%
Returning Freshmen: 92%
Academics: ✍ ✍ ✍ ✍ ½
Social: ☎ ☎
Q of L: ★ ★ ★
Admissions: (800) 247-0113
Email Address:
askgrin@grinnell.edu

Strongest Programs:
Foreign Languages
Biology
Chemistry
History

"Go West, young man, go West," Horace Greeley said to Josiah B. Grinnell in 1846. The result of Grinnell's wanderings into the rural cornfields, 55 miles from Des Moines and 60 miles from Iowa City, is the remarkable college that bears his name. Despite its physical isolation, Grinnell is a powerhouse on the national scene. Ever progressive, it was the first college west of the Mississippi to admit African Americans and women, and the first in the country to establish an undergraduate political science department. It was once a stop on the Underground Railroad, and its graduates include Harry Hopkins, architect of the New Deal, and Robert Noyce, inventor of the integrated circuit, two people who did as much as anyone to change the face of American society in the 20th century.

The school's 108-acre campus is an attractive blend of collegiate Gothic and modern Bauhaus academic buildings and Prairie-style houses. (Architecture buffs should take note of the dazzling Louis Sullivan bank facade just off campus.) The Noyce Science Center, a technological showpiece, recently underwent a $15.3 million renovation (Phase II is expected to be completed in 2008), and a 75,000-square-foot addition to the Fine Arts Center—including gallery, studio, performing, and rehearsal space—has opened.

True to its liberal arts focus, Grinnell mandates a first-semester writing tutorial, modeled after Oxford University's program, but doesn't require anything else. The more than 30 tutorials, limited to 12 students each, help enhance critical thinking, research, writing, and discussion skills, and allow first-year students to work individually with professors. When it comes to declaring a major, students determine their own course of study with help from faculty. Strong departments include the natural sciences and foreign languages,

"With the best equipment and graduate-level research at the undergraduate level, Grinnell's science division is a popular choice."

including German and Russian, bolstered by an influx of research grants, including one from the National Science Foundation. "Due to the college's enormous endowment, the sciences are top-notch," offers one student. "With the best equipment and graduate-level research at the undergraduate level, Grinnell's science division is a popular choice." The chemistry department draws majors with independent research projects, and English, anthropology, sociology, and economics are popular, too.

Grinnell's admissions standards are high—93 percent of students were in the top quarter of their high school class—and 30 percent of graduates move on to graduate and professional schools. Students who don't mind studying, even on weekends, will be happiest here. "The academics are definitely rigorous," says a senior, "but for the most part, we really aren't competitive." During finals, perhaps to help ease the stress, costumed superheroes run around the library giving out candy, says a sociology major. Teaching is the top priority for Grinnell faculty members, and because the college awards no graduate degrees, there are no teaching assistants. "I have loved just about every prof I've had so far," cheers one student. "Having great professors really makes class enjoyable even if the subject matter is difficult or dry."

Academic advising is also well regarded, as students are assisted by the professor who leads their first-year tutorial, and then choose another faculty member in their major discipline. "Tutorials are fun, interesting, and a great introduction to the academic possibilities that Grinnell has to offer," one student says. It's rare to find classes with more than 50 students, and 94 percent of classes have 30 or fewer. When the urge to travel intrudes, students may study abroad in more than 100 locations, through the Associated Col-

"We're a bunch of liberal students from all over the world."

leges of the Midwest* consortium and Grinnell-in-London. Fifty percent of students spend some time away from campus, and financial aid extends to study abroad, administrators say, and there are opportunities for research across disciplines. Co-ops in architecture, business, law, and medicine, and 3–2 engineering programs are also available.

Grinnell is a bit of Greenwich Village in corn country. Despite the rural environment, the college attracts an urban clientele, especially from the Chicago area. Only 11 percent are from Iowa, and 10 percent are foreign. A senior describes Grinnell students like this: "We're a bunch of liberal students from all over the world who spend more time talking with friends about activism, politics, or literature than we do deciding what to wear in the morning." The student body is 4 percent Hispanic, 6 percent Asian American, and 4 percent African American. Women's rights, gay rights, labor rights, human rights, globalization, the environment, and groups such as PAFA (the Politically Active Feminist Alliance), GEAR (Grinnell Escalating AIDS Response), and Fearless (formed to combat gender-based violence) set the tone. Grinnell benefits from a hefty endowment, not surprising when you consider the fact that Warren Buffett is a trustee and Intel cofounder Robert Noyce is an alumnus. All told, the endowment funds 50 percent of the college's operating budget. "Funding is never a problem here and students are always benefiting from the college's vast wealth," boasts one student. Merit awards averaging $10,564 are handed out annually, but there are no athletic awards.

The college guarantees four years of campus housing and 87 percent of students take advantage of the dorms, each of which has kitchen facilities, cable television, and a computer room. All but two dorms are co-ed, and after freshman year, students participate in a room draw, which can be stressful but usually works out. "You really can't go wrong wherever you live," a student says. Students who move off campus, mostly seniors, live just across the street. Meal plans range from full board to just dinner and special family-style dinners are served every other Wednesday. Students are betting that the Rosenfield Campus Center will dramatically improve dining options and quality. Despite being located in the middle of Iowa, students say they are glad to have campus security available. "There isn't much to worry about," says a student.

That town, Grinnell (population 9,100), is "a quaint, rural town that students find to be very friendly and laid-back. It holds a variety of stores and facilities that, unfortunately, many students do not discover because there is so much to do on campus." Community service helps bridge the town/gown gap, with some students serving as tutors and student teachers at the local high school and others participating

Strong departments include the natural sciences and foreign languages, including German and Russian, bolstered by an influx of research grants, including one from the National Science Foundation.

in mentoring programs and community meals, among other projects. Outdoor recreation is popular, and nearby Rock Creek State Park lends itself to biking, running, camping, kayaking, and cross-country skiing, as well as other pursuits sponsored by the Grinnell Outdoor Recreation Program, or GORP. There are a few bars and pizza joints downtown, but for those craving bright lights, Iowa City and Des Moines are within an hour's drive, and the college runs a shuttle service to them. Chicago and Minneapolis are each about four hours distant.

With no fraternities or sororities, intramurals and all-campus parties revolve mainly around the dorms. Each dorm periodically sponsors a party using wordplay from its name in the title. For instance, Mary B. James Hall puts on the Mary-Be-James party, for which everyone comes in drag. As for alcohol, a senior reports, "Grinnell is not a dry campus, but there are no bars on campus and there is no peer pressure to drink or culture of problematic drinking. There are liberal limits on how much alcohol can be served at parties and these events always have trained servers who check ID." Nondrinkers need not sit home, however. Grinnell's social groups and activities range from the Society for Creative Anachronism and the Black Cultural Center to improvisational workshops, poetry readings, symposia, concerts, and movies. Highlights of the campus calendar include semiformal Winter and Spring Waltzes, where "most people wear formals and look very nice, not a common occurrence at a school where comfort is the usual standard and women rarely wear makeup," notes one student. At Disco, "everyone dresses up in clothes from the '70s and dances all night." Other noteworthy events include a band/tie-dyeing fest called Alice in Wonderland, Titular Head (a festival of five-minute student films), and Pipe Cleaner Day (May 5 generally brings upwards of 20,000 of the sculptable wires to campus).

> **"We like being in the middle of Iowa, we like that you've probably never heard of us, we love that you won't come here because you want a big name."**

The Grinnell Pioneers compete in the Division III athletic conference, and the men's basketball team has won national attention for an unusual run-and-gun offense that uses waves of five players like hockey shifts in an effort to wear down opponents. Several teams have brought home championships recently, including men's cross-country, tennis, and men's and women's swimming, and women's soccer. One psychology major explains, "Because almost 40 percent of the student body participate in intercollegiate athletics, the intramural program isn't as extensive as it would be in a larger school."

Grinnell wouldn't put a grin on every prospective college student's face. "Most of us don't apologize for what at first turns people off about Grinnell," explains a senior. "We like being in the middle of Iowa, we like that you've probably never heard of us, we love that you won't come here because you want a big name." But there's no denying that Grinnell—a first-rate liberal arts college in the cornfields—is a real gem of a school.

Overlaps

Carleton, Macalester, Washington University (MO), Oberlin, University of Chicago, Kenyon

If You Apply To ➤ **Grinnell:** Early decision: Nov. 20. Regular admissions: Jan. 20. Guarantees to meet full demonstrated need. Campus interviews: recommended, informational. No alumni interviews. SATs or ACTs: required. SAT IIs: optional. Accepts the Common Application and electronic applications. Essay question: Common Application.

Guilford College

5800 West Friendly Avenue, Greensboro, NC 27410

Guilford is one of the few schools of Quaker heritage in the South. Emphasizes a collaborative approach and is among the most liberal institutions below the Mason-Dixon line. A kindred spirit to Earlham in Indiana. Guilford's signature program is justice and policy studies.

If your idea of a rousing road trip is protesting in Washington, D.C., you'll likely find plenty of kindred spirits at Guilford College. This generally left-leaning campus loves to debate just about any issue and get involved in the world around them. There's none of that college "bubble" that envelopes many other colleges. Instead, enrollment numbers are going up and the student body is becoming ever more diverse. "Intelligent, socially active, liberal students form the backbone of the school," says one senior, "but the Conservative Club is thriving." Founded in 1837 by the Religious Society of Friends (Quakers), Guilford is sticking to its original principles of inclusiveness as it constantly encourages its students to broaden their minds.

Located on 340 wooded acres in northwest Greensboro, Guilford's redbrick buildings are mainly in the Georgian style. The school is the only liberal arts college in the Southeast with Quaker roots, as well as the oldest coeducational institution in the South and the third oldest in the nation. During the Civil War, Guilford was one of a few Southern colleges that remained open—perhaps because it was also an embarkation point on the Underground Railroad. Newer additions to the campus include three student housing buildings and a community center.

In addition to their majors, Guilford students fulfill general education requirements in three areas: Foundations, Explorations, and Capstone. All students must also demonstrate quantitative literacy. Foundations consists of four skills and perspectives courses. The college has five areas of study—arts, business and policy, humanities, natural sciences and math, and social science—and the first set of Explorations courses provides academic breadth outside the area covered by a student's major and concentration. The second set of Explorations courses consists of three critical perspectives classes, one each from the categories of intercultural, social justice and environmental responsibility, and U.S. diversity. During the senior year, students take an interdisciplinary studies course to meet the Capstone requirement.

Guilford's academic climate is "far from competitive," according to an English major. "Classes aren't easy, but they're not impossible." Students say Guilford's best programs include physics, religious studies, peace and conflict studies (owing to the Quaker influence), and political science. "The English and art departments are really good," offers one freshman. Guilford believes that experiential learning adds immeasurably to classroom work, so the college offers study abroad from

"This is the place to find your niche!"

China and Japan to Mexico, Germany, and France. In the summer, students may participate in a five-week seminar that includes hiking, camping, and geological and biological research in the Grand Canyon, or in a seminar on the East African rift, which includes a three-week trip to Africa.

Despite its small size, Guilford gives students the tools needed for groundbreaking work. Physics majors get professional-grade optics and robotics equipment, and students working on complex geology and chemistry projects have access to the Scientific Computation and Visualization Facility, with more than 20 Unix workstations for information-based modeling. Perhaps not surprisingly, Guilford publishes both the *Journal of Undergraduate Mathematics* and the *Journal of Undergraduate Research in*

Website: www.guilford.edu
Location: City outskirts
Total Enrollment: 2,682
Undergraduates: 2,682
Male/Female: 38/62
SAT Ranges: V 520–640
   M 500–600
ACT Range: 20–25
Financial Aid: 44%
Expense: Pr $
Phi Beta Kappa: No
Applicants: 2,492
Accepted: 58%
Enrolled: 26%
Grad in 6 Years: 68%
Returning Freshmen: 72%
Academics: ✍ ✍ ✍
Social: ☎ ☎
Q of L: ★ ★ ★ ★
Admissions: (800) 992-7759
Email Address:
   admission@guilford.edu

Strongest Programs:
Psychology
Political Science
English
Business Management
Justice and Policy Studies
Education Studies

A popular Guilford mantra is "how are you going to change the world?" And with students who'd rather get involved than sit back and watch, you can expect some pretty passionate answers to that question.

Physics. Wireless Internet connections are available in the Hege Library, where students may check out laptop computers from the circulation desk. Students call professors by their first names and "for the most part, the quality of teaching I have received has been very high," says one student.

Guilford students come from 40 states, 25 countries, and a range of socioeconomic backgrounds; though most are liberal, there are a rising number of conservative students. Students "tend to be sympathetic to the little man," says a sophomore. African Americans comprise 23 percent of the student body, Hispanics 2 percent, and Asian Americans 1 percent. "There are two major groups of students at Guilford," says a freshman, "those that came here because it's small, liberal, and Quaker and those who were recruited for sports teams." Thirty-four percent of students receive merit scholarships worth an average of $7,006; there are no athletic scholarships.

Seventy-five percent of students live in Guilford's dorms, which feature newer furniture and carpeting, big-screen TVs, remodeled kitchens, and upgraded heating and air-conditioning systems. "All are a little gross, but not too bad," says one student. On-campus apartments for juniors and seniors are comparable in price to off-campus digs—a good thing, since getting permission to move off campus is tough. Bryan is the party dorm, and English (for men) and Shore (for women) are the single-sex quiet dorms. The female residents of Mary Hobbs, a co-op dorm built in 1907, do their own housekeeping in exchange for cheaper rent. The college is also piloting a "Community Agreements" project to strengthen community life in a first- and second-year residence hall. Food options include vegan and ethnic dishes, although "the food isn't always like mother used to make," says a student.

"It gives me the tools I need to make a change in the world when I leave."

Guilford believes that experiential learning adds immeasurably to classroom work, so the college offers study abroad from China and Japan to Mexico, Germany, and France.

Guilford's social life revolves around various clubs and organizations, ranging from the Entrepreneur's Network and Strategic Games Society to Hillel and the African American Cultural Society. "This is the place to find your niche!" says an economics major. No alcohol is allowed at college functions, but it remains pretty easy for underage students to drink—despite efforts to impose fines on those who are caught. Serendipity, a celebration of spring with games, mud wrestling, streakers, famed musicians such as the Violent Femmes, and "a sense of mass disorientation," is a cherished tradition. "If you love to drink and be loud, and throw chairs and trash cans off of balconies and buildings, then you will have a great time," says one student.

Beyond the campus gates, students find all of the essentials—Wal-Mart, some clubs in downtown Greensboro (only 10 minutes away), the ethnic restaurants of Tate Street, and the college's Quaker Village, which has a $1 movie theater, pool hall, and a Starbucks. Getting off campus and back will be easier now that Guilford plans a free shuttle service, also stopping at some of the five other schools in the area, and running until 2 a.m. on weekends. Popular road trips include UNC–Chapel Hill (one hour), Asheville and the mountains (three-and-a-half hours), and the famous Outer Banks beaches (four hours).

Guilford's 16 athletic teams compete as the "Fighting Quakers," and students love the oxymoron, as in their cheer: "Fight, fight, inner light! Kill, Quakers, kill!" Students root for the football team in the annual Soup Bowl against Greensboro College, while

Overlaps

UNC–Greensboro, Appalachian State, UNC–Chapel Hill, Elon, Earlham, UNC–Asheville

"Intelligent, socially active, liberal students form the backbone of the school."

the men's golf team brought home the NCAA Division III national championship in 2002 and 2005. The women's basketball team and the women's rugby team are strong, too. Because of Guilford's emphasis on developing the whole person, physically, mentally, and spiritually, students are encouraged to participate in school-sponsored outdoor adventures, such as a ropes course, sailing, and white-water rafting.

A popular Guilford mantra is "how are you going to change the world?" And with students who'd rather get involved than sit back and watch, you can expect some pretty passionate answers to that question. It all goes back to Guilford's traditional Quaker goal of "educating individuals not only to live, but to live well." As one student explains it, the college "supports me while allowing me to grow as a person." She then adds, in typical Guilford-speak: "It gives me the tools I need to make a change in the world when I leave."

<table>
<tr><td>If You Apply To ➤</td><td>Guilford: Early action: Jan. 15. Regular admissions: Feb. 15. Financial aid and housing: May 1. Does not guarantee to meet demonstrated need. Campus interviews: recommended, informational. Alumni interviews: optional, informational. SATs or ACTs: required; personal portfolio or presentation may be substituted. SAT IIs: optional. Accepts the Common Application and prefers electronic applications. Essay question.</td></tr>
</table>

Gustavus Adolphus College

800 West College Avenue, St. Peter, MN 56082

A touch of Scandinavia in southern Minnesota, GA is a guardian of the tried and true in Lutheran education. With Minnesotans comprising three-quarters of the students, GA is less national than cross-state rival St. Olaf. Extensive distribution requirements include exploring values and moral reasoning.

Gustavus Adolphus College is named for Sweden's King Gustav II Adolph (1594–1632), who is credited with making Sweden a major European power and defending Lutheranism against the Roman Catholics. While the king's battle victories earned him the title Lion of the North, he was also an advocate of education and culture. Save for the women now attending classes, King Gustav would probably feel at home at the college that bears his name, where a not-so-subtle Swedish influence pervades everything from the buildings to the curriculum. Comfy and spacious new dorms, with free laundry, help students feel at home right away. And the sidewalk running through the middle of campus is nicknamed the Hello Walk, because it's a tradition for students to greet one another as they pass—whether they know each other or not.

The 330-acre GA campus is about 65 miles southwest of the Twin Cities. Not surprisingly, the prevailing architectural theme is Scandinavian, with mostly modern and semimodern brown brick buildings. Highlights include the 113-year-old Old Main, which recently underwent renovation and restoration, and the centrally located Christ Chapel, with spires and shafts resembling a crown. Thirty bronze works by sculptor-in-residence Paul Granlund are strategically placed, and the 130-acre Linnaeus Arboretum and Interpretive Center offers plant study and retreats. The new Southwest Hall houses 190 students in apartment- and suite-style configurations.

In the classroom, students find an academic smorgasbord, as GA aims to offer an education both "interdisciplinary and international in perspective." There are interdisciplinary programs in Scandinavian studies, environmental studies, women's studies, and materials science, and if neither those nor traditional departments suffice, students may design their own courses of study. While the social sciences—psychology, economics, communication studies—are among the most popular majors, students give the highest marks to GA's science and premed programs and its offerings in music and education. Students also say weak departments are hard to

Website: www.gustavus.edu
Location: Small city
Total Enrollment: 2,546
Undergraduates: 2,546
Male/Female: 43/57
SAT Ranges: V 570–670
  M 565–675
ACT Range: 23–28
Financial Aid: 68%
Expense: Pr $
Phi Beta Kappa: Yes
Applicants: 2,689
Accepted: 79%
Enrolled: 33%
Grad in 6 Years: 82%
Returning Freshmen: 89%
Academics: ✍ ✍ ✍
Social: ☎ ☎ ☎
Q of L: ★ ★ ★
Admissions: (507) 933-7676
Email Address:
  admission@gustavus.edu

Strongest Programs:
Biology
Biochemistry

find, although some programs do enroll only a small number of students. Outside the classroom, learning opportunities come from several internationally renowned meetings, such as the Nobel Conference, which brings Nobel laureates and other experts to campus for two days each October.

To fulfill core requirements, Gustavus students have two options. Curriculum I includes 12 courses from seven areas of knowledge, plus a first-term seminar covering critical thinking, writing, speaking, and recognizing and exploring values. Curriculum II is an integrated 12-course sequence focused on related classic works from various disciplines. Sixty students may select this option on a first-come, first-served basis. In addition to the core courses, students must satisfy a "writing across the curriculum" requirement, with three courses that have a substantial amount of writing, and the first-term Values in Writing seminar, which explores questions of value while emphasizing critical thinking, writing, and speaking.

> "Gustavus has some of the nicest freshman housing I've seen."

Overall, academics at Gustavus are rigorous, but study groups are common, and students don't compete for grades. "There is a high level of academic excellence," says a student, "with rigorous coursework." Undergraduate research is a hallmark—despite the library's paltry 250,000 volumes—and Gustavus Adolphus students recently presented 42 papers at the National Conference on Undergraduate Research, the third-highest total in the nation. That's not as surprising when you consider that each year, about 40 freshmen are selected for the Partners in Scholarship program, which matches them with faculty research mentors and gives them renewable grants of $9,000 annually. For the professionally minded, Gustavus offers 3–2 engineering programs with the University of Minnesota and Minnesota State in Mankato.

During the January term, when winter winds force almost everyone indoors, Gustavus students (known as Gusties) may take concentrated courses on campus or pursue travel and co-op opportunities. The school sponsors study abroad programs at five colleges and universities in—surprise, surprise—Sweden, as well as in non-Scandinavian haunts such as Japan, India, Malaysia, Australia, Russia, the Netherlands, and Scotland, and about half of the students participate. Back on campus, students find faculty members knowledgeable and friendly. "They care about how their students are doing both in and out of class," says a senior. "They want you to succeed." Profs even serve up a free meal for students during Midnight Express, which precedes final exams.

For all its good points, though, this liberal arts college is hardly a model of diversity. In fact, the population is more reminiscent of Garrison Keillor's Lake Wobegon: 91 percent of students are white, 77 percent are Minnesotan, and more than half are Lutheran. Gusties "generally have a strong sense of acceptance, community, and care for the college and the world," a religion major says. The school is trying to boost diversity, though African Americans and Hispanics account for only 1 and 2 percent of the student body, respectively, and Asian Americans add only 4 percent. Politically, the campus has its fair share of both conservatives and liberals, and students debate everything from campus issues to topics of global concern. Merit scholarships averaging $8,252 are awarded annually but there are no athletic scholarships.

> "There is a high level of academic excellence."

While the social sciences—psychology, economics, communication studies—are among the most popular majors, students give the highest marks to GA's science and premed programs and its offerings in music and education.

Eighty-five percent of Gusties live in the dorms, and why not? Housing is guaranteed for four years, and all residences are smoke-free; even the halls without suites or apartments have kitchenettes, 24-hour computer labs, and foosball tables. "Gustavus has some of the nicest freshman housing I've seen," says an English major. Substance-free floors are available, as is the Crossroads International House, for students interested in languages and contemporary global issues. Norelius is exclusively for freshmen and sophomores, which helps students form friendships by encouraging group activities. Two dorms with single rooms and apartment-style suites, and some college-owned houses, are exclusively for upperclassmen, who get priority at room draw. Juniors and seniors (and all students over 21) may also request permission to live off campus. Students rave about the *a la carte* meal plan, and especially about the Marketplace dining hall, which is open from 7 a.m. until 11 p.m. daily. Students say security is good. "I feel very safe on campus," a senior says.

Most students stick around campus for fun, says a student. Twenty percent of the men and 17 percent of the women go Greek, but GA's social life does not revolve around fraternities and sororities. In fact, service projects are far more important, with about 70 percent of students participating, and giving 15,000 hours of service each semester. Projects include working with children, the elderly, and the local animal shelter, as well as with Habitat for Humanity. The Campus Activities Board sponsors more than 50 events each week, such as concerts, dances, and outdoor movies (in the warmer months). While the town of St. Peter has coffee shops and bowling, the college offers periodic trips to Mankato, 10 miles away, and to the Twin Cities, for "real" shopping at the Mall of America or for a professional baseball, basketball, or hockey game. Because students 21 and older may drink in their rooms—with the door closed—underage students can get alcohol if they want it, but

"On campus, you feel like you belong."

students say drinking isn't a popular pastime here. GA's many musical ensembles all perform together at the Christmas in Christ Chapel concert. The chapel holds 1,500 people and there are a total of five performances, all of which usually sell out.

When it comes to athletics, "any time, any sport—if St. Olaf is in town, the event is packed," says a lusty Gustie fan. GA competes in Division III, and men's golf, men's and women's tennis, swimming, and Nordic skiing are competitive, as are men's basketball and women's cross-country. The fitness center features treadmills, stair-climbers, Nautilus equipment, and a pristine weight room. Approximately three-quarters of Gusties participate in intramurals.

The Gustavus Adolphus campus may be gorgeous in the spring and fall and too cold in the winter, but it's warmhearted all year long. Small classes, one-on-one academic attention, a plethora of research opportunities, and an active campus social life go a long way toward making St. Peter, Minnesota, seem a lot less isolated. "On campus," says another student, "you feel like you belong." Hello!

Overlaps

St. Olaf, University of St. Thomas, Luther, University of Minnesota, College of St. Benedict, Carleton

If You Apply To ➤ **Gustavus Adolphus:** Early action: Nov. 1. Rolling admissions and financial aid: Apr. 15. Housing: May 1. Meets full demonstrated need of 76%. Campus interviews: recommended, evaluative. No alumni interviews. SATs or ACTs: optional (recommended). SAT IIs: optional. Accepts the Common Application and electronic applications. Essay question: significant experience; what you hope to gain from college; personal goals; or submit a writing sample from one of your classes.

Hamilton College

198 College Hill Road, Clinton, NY 13323

Hamilton is part of the network of elite, rural, Northeastern liberal arts colleges that extends from Colby in Maine through Middlebury and Williams to Colgate, about half-an-hour's drive to Hamilton's south. Hamilton is on the small side of this group and emphasizes close contact with faculty and a senior project requirement.

Back in 1978, Hamilton College seemed to have everything: money, prestige, and academic excellence. Everything, that is, except women. So, after 166 years of bachelorhood, Hamilton walked down the aisle with nearby Kirkland College, the artsy women's college founded under its auspices a decade before. The arrangement took some getting used to, but ultimately, it's been a success. Hamilton's student body is now evenly split between men and women, and students of both sexes benefit from the mix of old-boy tradition and right-brain flair.

Set on a picturesque hilltop overlooking the tiny town of Clinton, the old Hamilton campus features collegiate Victorian architecture rendered in rich, warm brownstone. In fact, the only facility interrupting the rhythmic beauty of campus is the eyesore housing the library. By contrast, the adjacent Kirkland campus consists mostly of boxy concrete structures of a 1960s "brutalist" vintage, otherwise described as "faux IM Pei." Straddling the ravine that divides the campuses and joining them literally and figuratively is a student activities building

"The students are very well rounded."

with a diner, lounges, and areas for student and faculty relaxation. Surrounding the campuses are more than 1,200 college-owned acres of woodlands, open fields, and glens, with trails for hiking or cross-country skiing. The college has recently added a new science building as well as a new fitness and dance center.

In the classroom, Hamilton is pure liberal arts. Government, economics, psychology, English, and math are the most popular majors. "Many avoid history because the professors are notoriously difficult. You won't get an A in those classes," predicts a sophomore government major. Another student agrees, but qualifies, "History, English, government, and science are some of the best departments because the professors are renowned and the students love them. They come out of those courses knowing 100 percent of what they need to know." Hamilton's Arthur Levitt Public Affairs Center, named for the former New York State comptroller, is a working think tank where students can pursue research for local, regional, and state social service groups and government agencies. The natural sciences are strong, bolstered by a new $56 million science center. A grant from the National Science Foundation helps send geoscience students to Antarctica for research each year, while Hamilton's rocky terrain provides fertile ground for those who remain. The administration has taken steps to strengthen the communication department with a new major and the establishment of an Oral Communication Center.

A major curriculum reform eliminates distribution requirements and features a series of proseminars—classes of no more than 16 that require intensive interaction—that emphasize writing, speaking, and discussion. The required sophomore program stresses interdisciplinary learning and culminates in an integrative project with public presentation. In fact, Hamilton is among a handful of schools requiring all students to undertake a senior program in their area of concentration. Another interesting development is the administration's decision to let students choose which standardized tests to submit with their application for admission, including SAT I, the ACT, three SAT IIs, or three Advanced Placement (AP) or International Baccalaureate (IB) exams. Students may also select more than one type of test, so long as their portfolio includes an English test and a quantitative test, plus one other exam.

Academically, students say that the emphasis is on performance, allowing students to remain laid-back rather than competitive. You'll always find a professor at the lectern, not a teaching assistant. "Teachers

"Dorms are clean and cozy."

are excited and passionate about their subject matter and, overall, very good," says one junior. Even with the intimate size of most classes, students report few problems getting needed courses, once they declare a major. A chemistry major says, "If you're in chemistry and you want to do research,

In the classroom, Hamilton is pure liberal arts.

you are almost guaranteed to find a faculty member willing to take you on." When the town of Clinton gets claustrophobic, students can spend a semester or a year in France, Spain, or China, or take a term in Washington, D.C., or New York; about 40 percent of graduates have studied abroad. There are 3–2 programs in engineering with Columbia, Rensselaer Polytechnic Institute, and Washington University (MO), and a 3–3 program in law with Columbia.

About one-third of Hamilton students are New York residents and 74 percent were in the top 10 percent of their high school class. "The students are very well rounded," says a student. "They are hard-working, athletic, and interested in getting involved in clubs and the community." African Americans and Hispanics each comprise 4 percent of the student body, and Asian Americans add 7 percent. Hamilton offers a few merit scholarships of up to half tuition but there are no athletic scholarships, as the school competes in Division III.

Ninety-eight percent of the students reside on campus. Options range from old fraternity houses renovated and turned into dorms to stately mansions with posh amenities, and newer apartments that accommodate three to four students each. All halls are co-ed and include residents from all four classes. "Dorms are clean and cozy," says a French major. Smart students become resident assistants, which guarantees them a single and lets them bypass the hated lottery. The Hamilton side of campus is the place for party animals; Dunham gets cheers for being social, but some students liken it to a "dungeon." The Kirkland dorms have a more mellow reputation. Students may also choose to live in a co-ed cooperative house, or in houses that are substance-free or quiet. And how about the grub? "There is plenty of variety," says a sophomore. "You will never go hungry."

Social life at Hamilton ranges from the campus pub, which occupies an old barn, to programming arranged by the campus activities board, such as comedy shows, a casino night, and an award-winning acoustic coffeehouse series featuring Ellis Paul, The Shins, Guster, Ben Folds, Joshua Redman, Jason Mraz, Talib Kweli, **"Make sure you enjoy the snow."** Keller Williams, SouLive, and Rahzel. There's also the popular Greek system, which draws 20 percent of the men and 15 percent of the women, even though fraternities may not have on-campus houses. Clinton itself is "a very small town with one stop light. I would say more but that's really all you need to know." Because there's relatively little to do, much of the social life does revolve around alcohol, one student says. The nearest small city, Utica, is only 10 minutes away by car. Other popular road trips include Syracuse and New York City, while Boston, Toronto, and Montreal are each less than five hours away.

In athletics, Hamilton College offers 28 intercollegiate varsity sports and is a member of both the prestigious New England Small College Athletic Conference and the Liberty League. Since 2003, the men's basketball, men's and women's soccer, men's lacrosse, men's golf, and men's and women's swimming and diving programs competed in the NCAA championships. In 2005–06, three Hamilton athletes earned All-American recognition, including a diver who won two national titles. Even some school traditions are athletically minded. Class and Charter Day marks the last day of classes in the spring, with ceremonies, a picnic, and even a triathlon.

Hamilton students are, by necessity, hearty. They're used to the cold and the snow—and perhaps that's what leads to the strong sense of community evident on campus. "Make sure you enjoy the snow; the winters are long, and if you need sunny days to be happy, this is not the place for you." So much, indeed, for the battle of the sexes. As the Hamilton-Kirkland union proves, women may be from Venus and men from Mars, but with time, both sexes can find common ground.

Hamilton lets students choose which standardized tests to submit with their application for admission, including SAT I, the ACT, three SAT IIs, or three Advanced Placement (AP) or International Baccalaureate (IB) exams.

Overlaps
Colgate, Bates, Colby, Amherst, Williams, Cornell

Hampden-Sydney College

P.O. Box 667, Hampden-Sydney, VA 23943

The last bastion of the Southern gentleman and one of two all-male colleges in the nation. Feeder school to the economic establishment in Richmond. Picturesque rural setting evokes the old South.

Website: www.hsc.edu
Location: Rural
Total Enrollment: 1,060
Undergraduates: 1,060
Male/Female: 100/0
SAT Ranges: V 520–630
　M 530–640
ACT Range: 21–27
Financial Aid: 47%
Expense: Pr $
Phi Beta Kappa: Yes
Applicants: 1,374
Accepted: 67%
Enrolled: 35%
Grad in 6 Years: 65%
Returning Freshmen: 83%
Academics: ✍ ✍ ✍
Social: 🐎 🐎 🐎 🐎
Q of L: ★ ★ ★ ★
Admissions: (434) 223-6120
Email Address:
　hsapp@hsc.edu

Strongest Programs:
Economics
Biology
Chemistry
Physics
Rhetoric
History
Political Science

Perhaps a bit of an anachronism in a society increasingly focused on diversity, the all-male Hampden-Sydney College still aims to expose its small student body to a broad liberal arts education, which is entirely focused on undergraduate success. H-SC is one of only two all-male colleges without a coordinate women's college in the nation. Tradition reigns here and students like to call themselves "Southern gentlemen." Of course, there's plenty of not always gentlemanly fun to be had when you have more than 1,000 guys together.

Hampden-Sydney's 660-acre campus, surrounded by farmland and woods, features mainly redbrick buildings in the Federal style. The Johns Auditorium, originally completed in 1951, was recently renovated and a new campus library is currently under construction. The nearby town of Farmville, population 6,600 and home to Longwood College, offers restaurants, stores, and a movie theater; it's just five miles from H-SC, but one student describes the town as "a black hole inside a time warp."

Hampden-Sydney's most popular major is economics, which may help explain why more than half of the school's alumni have pursued business careers. The department offers several concentrations, including managerial and mathematical concepts; instruction is "intensive," a junior says. History, political science, psychology and religion are also popular, and the rhetoric program has earned international recognition, administrators say. The Wilson Center for Leadership in the Public Interest puts a public service focus on the study of political science, preparing students for government work and garnering high marks in return. The school's small size offers many opportunities to work

"When you come here, teachers challenge students to think deeper."

closely with professors, but has some academic drawbacks, including few computer courses and fewer than 30 majors. The fine arts program, with concentrations in music, theater, and visual arts, has begun emphasizing performance rather than the study of these disciplines, but it suffers from a lack of facilities and student interest.

Students at Hampden-Sydney say there are no free passes when it comes to classwork. "The courses and curriculum are challenging," says a student, "but professors are very willing to go out of their way to help students any way they can." To graduate, students must demonstrate proficiency in rhetoric and a foreign language, along with completing seven humanities courses, three in the social sciences, and four in the natural sciences and mathematics. All freshmen have a special advising program, and 60 percent take freshman seminars. Classes are small; 97 percent have fewer than 25 students, and less than 4 percent have more than 50. "Most courses require strict attendance," a history major says—after all, with a class of 12 or 14 students, your presence or absence will certainly be felt. "It is not a cake walk," says a managerial economics major. "When you come here, teachers challenge students to think deeper." Most H-SC professors live on campus and encourage students to drop by their offices often. Some even make house calls to find out why a student missed class. "I have been invited to numerous dinners at professors' homes, and professors encourage us to contact them whenever we have a question, even if that means at nine o'clock on a Wednesday night," says a junior.

One thing you won't see a lot of at H-SC are students with long hair, body piercing, or much else that would not fit into a clean-cut profile. "Students are wealthy and Southern and very proud of it," quips a senior. Sixty-seven percent of students are state residents, and 92 percent are white. African Americans make up 4 percent of the student body, Hispanics constitute 1 percent, and Asian Americans and Native Americans 1 percent each. "Political correctness is not an issue," says a political science major. The college has, however,

The Wilson Center for Leadership in the Public Interest puts a public service focus on the study of political science, preparing students for government work and garnering high marks in return.

"Students are wealthy and Southern and very proud of it."

added a director of intercultural affairs to help increase tolerance for diversity. As a Division III school, Hampden-Sydney offers no athletic scholarships. There are hundreds of merit awards worth an average of $16,595.

Ninety-five percent of students live on campus, as housing is guaranteed for four years, and H-SC is renovating older residence halls with mostly single rooms to offer more apartment-style living. "The rooms—not all, however—are spacious and bright," says a psychology and music major. "There can be some trouble getting the room that a student wants, because everyone wants basically the same rooms." All rooms have Internet connections and cable. Cushing Hall, built in 1824, is the dorm of choice for first-year students, with "big rooms, excellent parties, and at least three ghosts."

Students praise the close-knit atmosphere fostered by Hampden-Sydney's all-male status. "Brotherhood amongst your peers" is what makes the place special, says a senior. One student reports that showering during the week is really optional: "We clean up on Fridays before the girls come." Men seeking members of the opposite sex can find them at four all-female schools nearby—Sweet Briar, Hollins, Mary Baldwin, and Randolph College. Those who make a love connection will be glad to know that H-SC's dorms have 24-hour visitation.

Hampden-Sydney's social nexus is the Circle, the site of 11 of the school's 12 fraternities, which claim a third of the students. "On the weekends, everyone goes to 'frat circle' where all the fraternities have houses on campus," says a student. Students under 21 can't drink, but where there's a will, there's a way. "To be honest, a three-year-old could drink here," reports one student. But campus security will crack down if students get wild. The annual spring Greek Week brings out the Animal House aspect of Hampden-Sydney's budding gentlemen. Homecoming and various music festivals are also eagerly anticipated.

Perhaps because of all that testosterone on campus, Hampden-Sydney men are competitive, and that spells excellence in athletics.

Despite its lack of bright lights ("It sucks," says a particularly blunt senior), nearby Farmville does provide numerous community service and outreach opportunities. A campus volunteer group called Good Men, Good Citizens spearheads projects such as tutoring, highway clean-up, and Habitat for Humanity home-

"We clean up on Fridays before the girls come."

building. When rural Virginia gets too insular, H-SC students can be found on road trips to the University of Virginia and James Madison University, Virginia's beaches, or Washington, D.C. The ski slopes of Wintergreen are within three hours' drive.

Perhaps because of all that testosterone on campus, Hampden-Sydney men are competitive, and that spells excellence in athletics. Football is big; students attend games in coat and tie, and H-SC's football rivalry with Randolph-Macon (not the famed women's college!) is the oldest in the South. At the annual pregame bonfire, the college rallies to sing songs and hear student and faculty leaders vilify the enemy and extol "the garnet and gray." Basketball, baseball, soccer, and lacrosse are also competitive. For weekend warriors, intramurals are available in several sports, including football, baseball, soccer, and lacrosse.

So it's largely conservative and a bit homogenous. But Hampden-Sydney offers more than just a flat demographic profile. Two centuries of tradition and a tight-knit student body make for a rich undergraduate experience. "I like the small size because I don't feel like a number," a junior says. "And I have lived all my life in the big city. I kinda like being in the country." Of course, it's easy to counter all that profundity with some of the school's more popular sayings, such as this gem: "We don't need girls. We're doing just fine with yours."

Overlaps

University of Virginia, Virginia Tech, James Madison, Longwood

If You Apply To ➤ **Hampden-Sydney:** Early decision: Nov. 15. Early action: Jan. 15. Regular admissions: Mar. 1. Financial aid: May 1. Does not guarantee to meet demonstrated need. Campus interviews: recommended, evaluative. Alumni interviews: optional, informational. SATs or ACTs: required. (SATs preferred.) SAT IIs: recommended. Accepts the Common Application and electronic applications. Essay question: a prominent person you would interview; significant experience or achievement; why you have saved a personal object or item; experience with those of different race, background, or culture.

Hampshire College

P.O. Box 5001, Amherst, MA 01002-5001

Part of a posse of nonconformist colleges that includes Bard, Bennington, New School University, and Sarah Lawrence. Instead of conventional majors, students complete self-designed interdisciplinary concentrations and independent projects. Gains breadth and resources from the Five College Consortium.

Website: www.hampshire.edu
Location: City outskirts
Total Enrollment: 1,448
Undergraduates: 1,448
Male/Female: 41/59
SAT Ranges: V 610–700
  M 540–660
ACT Range: 26-29
Financial Aid: 70%
Expense: Pr $ $ $ $
Phi Beta Kappa: No
Applicants: 2,553
Accepted: 55%
Enrolled: 29%
Grad in 6 Years: 63%
Returning Freshmen: 79%
Academics: 🐌🐌🐌🐌
Social: 🐿🐿🐿
Q of L: ★★★
Admissions: (413) 559-5471
Email Address:
  admissions@hampshire.edu

Passion reigns at Hampshire College. It's found in just about everything students do—from devising their own courses to starting new clubs to debating the most current social issues. There's no one way to do things at Hampshire, and the students revel in the freedom they have to direct the path of their educations. "They love what they are studying because they get to choose what they are studying," says a junior studying sustainable agricultural methods. Without the yoke of traditional majors and the nail-biting stress of regular grades, Hampshire offers a virtually boundary-free exercise in intellectual nirvana.

Located in the Pioneer Valley of western Massachusetts, Hampshire's 800-acre campus sits amid former orchards, farmland, and forest. Buildings are eclectic and contemporary, and the school is most proud of its bio-shelter, arts village, and multisports and multimedia centers. Two nationally known museums—the National Yiddish Book Center and the Eric Carle Museum of Picturebook Art—are located right on campus.

Instead of grades, Hampshire professors hand out "narrative evaluations," which consist of written evaluations and critiques. Degrees are obtained by passing a series of examinations—not tests, but portfolios of academic work, evaluations, and students' self-reflections on their academic development. The first hurdle, known

"We Hampshire students will be the ones at a party talking about our exciting work."

as Division I, begins with a course in each of five multidisciplinary schools: natural science; social science; cognitive science; interdisciplinary arts; and humanities, arts, and cultural studies; and other coursework.

The second hurdle, Division II, is each student's "concentration"—the rough equivalent of a major elsewhere. Unlike a major, the requirements of a concentration are unique to each student, emerging from regular discussions with two faculty members and include courses, independent study, and fieldwork or internships. Division III, or "advanced study," begins in the fourth year. Students are asked to complete a sizable independent study project centered on a specific topic, question, or idea, much like a master's thesis. Because of the division system, there are as many curricula at Hampshire as there are students; each individual must devise a viable, coherent program specific to himself or herself. Not surprisingly, competition is virtually nonexistent, and the academic climate is "challenging, empowering, liberating, in the sense that it allows students to study what they love," says a philosophy major. The common denominator is a heavy workload, an emphasis on self-initiated study, close contact with faculty advisors, and the assumption that students will eventually function as do graduate students at other institutions. "We Hampshire students will be the ones at a party talking about our exciting work," says a sophomore. (One popular campus T-shirt says Hampshire is "The Undergraduate Graduate School.")

Given the emphasis on close working relationships with faculty and those "narrative evaluations," the importance of qualified, attentive faculty is not to be underestimated. Students at Hampshire heap praise on their professors. "The mentoring relationships I have formed with my professors will last well after I graduate," raves a literature major. The Hampshire academic year has fall and spring semesters, each four months long; an optional January term; and internships and other real-world experience are

"Hampshire is where all the wacky and wonderful people reside."

encouraged during all three. Despite Hampshire's entrepreneurial nature, a large percentage of grads do go on to graduate school, and many Hampshire students begin their own businesses.

Hampshire's flexibility is ideal for artists, and the departments of film and photography are dazzling, which is also the reason they are overcrowded. Communications, creative writing, and environmental studies are also good bets, and Hampshire was the first college in the nation to offer an undergraduate program in cognitive science. A popular program called Invention, Innovation, and Creativity exposes students to the independent reasoning and thinking essential to the process of inventing. "Culture, Brain, and Development" emerged three years ago with funding from the Foundation for Psychosocial Research. Students also have access to selected courses at sister schools in the Five College Consortium.*

Although the school's library is a quiet and pleasant place to study, it has only 114,000 volumes. Still, if you count the library resources at all five institutions, students have ready access to more than eight million volumes. There is no extra cost to use the other schools' facilities or the buses that link them. And use them they do—Hampshire students take 1,200 classes per year at the other schools.

Hampshire draws students from across the country who tend to be "free-loving, innovative-thinking scholars who challenge institutions and bureaucracy," says an English major. Another student reports that Hampshire students are "progressive, passionate, critical people who like barefoot picnics at sunset." For a school that's so focused on social issues, the minority community is relatively small—3 percent of students are African American, 5 percent Hispanic, and 4 percent Asian American—and most students would like to see these numbers rise. Another 3 percent are international. The school is "100 percent politicized in every way," says a philosophy major. "'Political correctness' doesn't even begin to describe it." Thirty-two merit scholarships are available to qualified students.

Ninety-three percent of students live on campus. First-year students live in co-ed dorms, about 25 percent in double rooms. Many single rooms are available for older

(Continued)
Strongest Programs:
Film and Television
Environmental Studies
Cognitive Science
Creative Writing
Natural Science
History
Human Health
Agriculture

Hampshire is no place for competitive jocks, since many sports are co-ed and primarily for entertainment (there never was a football team here).

students who may move to one of more than 100 "mods"—apartments in which groups of four to 10 students share the responsibility for cleaning, cooking, and maintaining their space. "The dorms and apartments were quickly and cheaply built, and have problems," warns a senior. Special quarters are available for nonsmokers, vegetarians, and others with special preferences.

On weekends, some students head for Boston, New York, Hartford, or, in season, the ski trails of Vermont and New Hampshire. But there are plenty of cultural resources within the Five College area, and the free buses to Amherst (the ultimate college town), Northampton, and South Hadley (all within 10 miles) are always crowded. From edgy record stores to ethnic restaurants and boutiques, the area abounds with diversions. The annual Spring Jam brings live bands to campus, and throughout the year there's almost always a party going on, including the drag ball and the much-anticipated Halloween bash—an intense, all-campus blowout complete with fireworks. A tradition called "Div Free Bell" celebrates the completion of Division III requirements—and graduation—with soon-to-be alumni ringing a bell outside the library, surrounded by friends.

"This is the only place where I could see myself being happy."

Hampshire is no place for competitive jocks, since many sports are co-ed and primarily for entertainment (there never was a football team here). The school offers paid instructors in a handful of sports, but most students organize their own clubs (men's and women's soccer, basketball, and fencing are the biggies, and there's also the competitive Red Scare Ultimate Frisbee Team) and intramural teams. The outdoors program offers mountain biking, cross-country skiing, and kayaking; equipment may be borrowed for free. The school also has its own climbing wall and cave, a gym with solar-heated pool, and a co-ed sauna.

"Hampshire is where all the wacky and wonderful people reside," says a sophomore. There's a niche for every type of student, and even those pigeon holes are blown apart quite regularly. When you can make up your own education, and do it with great faculty and the option of studying at several other top-notch schools, there's little Hampshire students can't accomplish. "You set limits with yourself, and then push those limits and strive even higher," says one student, who adds, "This is the only place where I could see myself being happy."

Hartwick College

Oneonta, NY 13820

Hartwick is known for its cozy atmosphere and ability to take good care of students. Combines arts and sciences with a nursing program. The campus is beautiful, but small-town upstate New York has proven to be a hard sell in recent years.

Website: www.hartwick.edu
Location: Small city

Hartwick College has strived in recent years to transform itself and has chosen to better focus on its ever-improving liberal arts profile. Hartwick emphasizes study abroad—especially for freshmen—crystallizing the school's philosophy that learning isn't about rote memorization, it's about creating knowledge and developing skills. The students here take full advantage of what's offered and feel at home on this close-knit campus. "The atmosphere here is much more welcoming than at other schools," says a senior.

Hartwick's campus has a New England feel with its ivy-covered, redbrick buildings and white cupolas, gables, and trim. The campus setting on the Oyaron Hill, overlooking the city and the Susquehanna Valley, provides a breathtaking view, though the steepness of the campus may have some wishing for the legs of a mountain goat. Campus facilites include a tissue culture lab, electron microscopes, a greenhouse, an herbarium, a cold room, a biotechnology "clean lab," and a graphics imaging lab.

Hartwick's liberal arts and sciences framework ensures that its students are exposed to what one administrator terms "a broad swath of human knowledge." The most popular major is business administration, followed by nursing, psychology, biology, and political science. Students are enthusiastic about political science and life sciences, which

"The atmosphere here is much more welcoming than at other schools."

"attract many students due to the diverse faculty and variety of courses," says a junior. Art and music also are praised, while modern and classical language and sociology are works in progress. Student/faculty collaborations are the norm, as is the emphasis on learning through real-world experiences, from working in a Jamaican hospital to interning with the New York Mets.

Hartwick's general education program is divided into five areas: continuity (Western tradition), interdependence, science and technology, critical thinking and effective communication, and choices. Among the voluminous requirements are two Great Books courses, a course in Western and non-Western culture, foreign language, two decision-making seminars, a course in the creative or performing arts, and a senior research thesis. Another example of Hartwick's academic enrichment is the honors program, which provides students with the opportunity to design and carry out a coherent program of study characterized by challenges exceeding those offered in typical coursework required for graduation. The course offerings are necessarily limited by Hartwick's small size, but the Individual Student Program (ISP) enables students to create their own major dealing with a particular interest. Students may take courses at the nearby State University College at Oneonta (SUCO).

"Courses are fairly rigorous because the majority of professors have high academic standards," says one student. "Many opportunities are available for students to become involved with research in their fields." Hartwick offers other unconventional learning options, many of them in off-campus locations. Students have traveled to all parts of the world while pursuing their Hartwick education—first-year students are especially encouraged to leave the "Bubble on the Hill," as some call Hartwick. The Awakening Challenge, completed by freshmen as part of orientation, is also an option for management majors who want to test their leadership skills. The four-week January term is also a favorite time to explore the world beyond Oneonta.

The faculty wins universal praise from the students. "Because of the extremely low student-teacher ratio, students receive a much individual attention from professors and have a fair amount of resources outside the classroom," says a student. Private tutoring and help sessions are offered, along with an innovative freshman early-warning program that identifies struggling students early and offers counseling. "Many students do research alongside professors," says a biology major.

Hartwick has traditionally attracted a somewhat less academically oriented student body than most of the colleges with which it competes, but it has improved its academic position in recent years, thanks to a focused recruitment program. "We have the granola kids, the hipsters, the preps, the frat boys, the jocks, and everything in between," says a junior. Twenty-four percent of students come from the top tenth of

(Continued)
Total Enrollment: 1,463
Undergraduates: 1,463
Male/Female: 44/56
SAT Ranges: V 520–620
  M 500–620
Financial Aid: 76%
Expense: Pr $ $
Phi Beta Kappa: No
Applicants: 2,211
Accepted: 87%
Enrolled: 21%
Grad in 6 Years: 54%
Returning Freshmen: 76%
Academics: ✍ ✍ ✍
Social: ☎ ☎ ☎ ☎
Q of L: ★ ★ ★
Admissions:
  (607) HARTWICK
Email Address:
  admissions@hartwick.edu

Strongest Programs:
Business Administration
Political Science
Psychology
Nursing
Elementary Education
Biology
Anthropology

The course offerings are necessarily limited by Hartwick's small size, but the Individual Student Program (ISP) enables students to create their own major dealing with a particular interest.

their class. Sixty-one percent are from New York State, especially upstate, and most of the rest come from New England or the mid-Atlantic States. Hartwick has worked hard to improve the diversity of its student body: 5 percent are African American, 4 percent Hispanic, and 1 percent Asian American. Hartwick college awards merit scholarships as well as 15 athletic scholarships for Division I men's soccer and women's water polo.

"Many students do research alongside professors."

In general, the dorms receive mixed reviews, but the administration is hoping for more positive reactions because all residence halls have been recently renovated and improved. Upperclassmen covet a place in one of the four townhouses described by one as "the yuppie version of on-campus living." Freshmen, sophomores, and juniors are required to live on campus, though the latter may move into one of the fraternity or special-interest houses. Each dorm has designated quiet hours, though they may not always be observed. Hartwick's 920-acre environmental campus, Pine Lake, has cabins that are heated by pellet stoves and a lodge where environmentally inclined students can live in rustic style. One student says that it is "very selective, but well worth the application process." On-campus dining has been improving and students say their menu options are expanding.

Hartwick is nationally ranked at the Division I level in soccer; the remainder of the teams compete in Division III.

From campus it's only a short walk, bike ride, or bus ride downhill into the small city of Oneonta, with its tantalizing profusion of bars. But the underage Hartwick students usually don't get past the front doors of these taverns, and the administration is tough about enforcement on campus. "There is not a lot of drinking on campus, but many people drink off campus," says a sophomore. Tamer entertainment includes Sunday night movies as well as occasional lecturers and comedians, and just hanging out at the student union. The Greek system attracts 4 percent of the men and 7 percent of the women. Popular campuswide bashes include a Last Day of Classes party, the Holiday Ball, and Winter and Spring Weekends, the latter of which features the notorious "Wick Wars," a schoolwide sports competition. There's also the Breakfast of Champions before final exams when professors and administrators serve students breakfast between 11 p.m. and 2 a.m. Walking to class each day provides great hill workouts for your ski legs, and skiing is popular throughout the region. Another nice diversion is Pine Lake, which offers cross-country trails, swimming, boating, and fishing. Nearby Cooperstown offers entertainment for baseball and history buffs.

"We have the granola kids, the hipsters, the preps, the frat boys, the jocks, and everything in between."

Oneonta is big on soccer, and Hartwick fits that bill. The Mayor's Cup Soccer Tournament weekend is a big event. Hartwick is nationally ranked at the Division I level in soccer; the remainder of the teams compete in Division III. The women's water polo team has been a real success story by recently winning the Eastern Collegiate Conference Tournament championship. The women's field hockey team reached the NCAA semifinals, and conference championships have been won by men's basketball, swimming, and baseball, and by the women's basketball, soccer, and lacrosse teams. Intramurals are popular as well.

Change is good, the sages say, and the folks at Hartwick would definitely agree. By refocusing its efforts on recruiting higher-caliber students and emphasizing top-notch experiential learning, Hartwick is shedding its image as a party school. Even some of the T-shirts sold on campus broadcast the students' attitudes about their education: one simply says "Smartwick."

Overlaps

SUNY–Oneonta, Hobart & William Smith, Ithaca, Union, University of Vermont, Elmira

<table>
<tr><td>If You Apply To ➤</td><td>Hartwick: Early decision: Nov. 15, Jan. 15. Regular admissions: Feb. 15. Campus interviews: recommended, informational. Alumni interviews: optional, informational. SATs or ACTs: optional. SAT IIs: optional. Accepts the Common Application and electronic applications. Essay question.</td></tr>
</table>

Harvard University

Byerly Hall, 8 Garden Street, Cambridge, MA 02138

An acceptance here is the gold standard of American education. Gets periodic slings and arrows for not paying enough attention to undergraduates, some of which is carping from people who didn't get in. It takes moxie to keep your self-image in the midst of all those geniuses, but most Harvard admits can handle it.

Over the past 350-plus years, the name Harvard has become synonymous with excellence, prestige, and achievement. At this point, Harvard University is the benchmark against which all other colleges are compared. It attracts the best students, the most academically accomplished faculty, and the most lavish donors of any institution of higher education nationwide. Sure, some academic departments at Hah-vahd are smaller than others, but all have faculty members who have made a name for themselves, many of whom have written the standard texts in their fields. Olympic athletes, concert pianists, and Rhodes Scholars blend in nicely here, ready to embrace the challenges and rewards only Harvard's quintessential Ivy League milieu can offer.

Spiritually as well as geographically, the campus centers on the famed Harvard Yard, a classic quadrangle of Georgian brick buildings whose walls seem to echo with the voices of William James, Henry Adams, and other intellectual greats who trod its shaded paths in centuries past. Beyond the yard's wrought-iron gates, the campus is an architectural mix, ranging from the modern ziggurat of the science center to the white towers of college-owned houses along the Charles River. Loker Commons, a student center beneath the new Annenberg freshman dining hall, provides a place for students to meet and philosophize over gourmet coffee or burritos of epic proportions. The Barker Center for humanities has emerged from the shell of the Union and the old freshman dining hall. The Maxwell Dworkin building, housing the computer science and engineering departments, has also been completed.

Harvard's state-of-the-art physical facilities are surpassed only by the unparalleled brilliance of its faculty. Under its "star" system, Harvard grants tenure only to scholars who have already made it—usually someplace else—and then gives them free rein for research. It seems like every time you turn around, a Harvard professor is winning a Nobel Prize or being interviewed on CNN; every four years, half the government and econ departments move to Washington to hash out national policy. But one of Harvard's finest qualities is also one of its biggest problems. "You can have unlimited contact with professors, but it must be on your initiative," notes a biology major. "This is not a small liberal arts college where people will reach out to you." That's not to say profs are completely uncaring. Most teach at least one undergraduate course per semester, and even the luminaries occasionally conduct small undergraduate seminars (including those reserved for freshmen, which can be taken pass/fail). Harvard also sponsors a faculty dining program, encouraging professors to eat at the various residential houses and chew over ideas as well as lamb chops.

Website: www.fas.harvard.edu
Location: City outskirts
Total Enrollment: 18,639
Undergraduates: 6,649
Male/Female: 51/49
SAT Ranges: V 700–790
  M 700–790
ACT Range: 30–34
Financial Aid: 50%
Expense: Pr $ $ $
Phi Beta Kappa: Yes
Applicants: 22,796
Accepted: 9%
Enrolled: 78%
Grad in 6 Years: 98%
Returning Freshmen: 97%
Academics: ✑ ✑ ✑ ✑ ✑
Social: ☎ ☎ ☎
Q of L: ★ ★ ★ ★
Admissions: (617) 495-1551
Email Address:
  college@fas.harvard.edu

Strongest Programs:
Economics
Biology
Social Studies
Government
English
African American Studies
East Asian Studies
Anthropology
Music
History of Science

Harvard's best-known departments tend to be its largest; economics, government, biology, English, and biochemistry account for a large chunk of majors. But many smaller departments are gems as well: East Asian studies is easily tops in the nation. And under the leadership of Henry Louis Gates, the African American studies department has assembled the most high-powered group of black intellectuals in American higher education. Smaller, interdisciplinary honors majors, to which students apply for admission, boast solid instruction and happy undergraduates, too. These programs—social studies, history and science, history and literature, and folklore and mythology—are the only majors that require a senior thesis, although many students elect to do one in other departments.

> "You can have unlimited contact with professors, but it must be on your initiative."

Harvard's visual and environmental studies major serves filmmakers, studio artists, and urban planners, and concentrations in women's studies and environmental sciences have been well received. Students can also petition for individualized majors, typically during the sophomore year. All students must choose some sort of major at the end of their freshman year, a year earlier than most schools. The field of concentration can be changed later, but Harvard expects its students to hit the ground running. Regardless of the department, students uniformly complain about the overuse of teaching fellows (graduate students) for introductory courses in mathematics and the languages. TFs aren't all bad, though, says a junior: "They can give good advice, having just been in our position." Besides, it's easier to ask "dumb questions" of mere mortals than of the demigod-like professors.

Back in the mid-1970s, Harvard helped launch the current curriculum reform movement. The core curriculum that emerged ranks as perhaps the most exciting collection of academic offerings in all of American higher education. The best and brightest freshmen can apply for advanced standing if they have enough Advanced Placement credits. And should you not find a class you are looking for, admittedly highly unlikely, Harvard offers crossregistration with several of its graduate schools and the Massachusetts Institute of Technology.

In formal terms, the core requires students to select eight courses, or a quarter of their program, from a list of offerings in six different "modes of inquiry": foreign cultures, historical studies, literature and arts, moral reasoning, sciences, and social analysis. For the three or four most popular courses, enrollment is limited by the number of seats in the various large auditoriums on campus; sometimes places in these lectures are determined by lottery. Freshmen also face quantitative reasoning and foreign language requirements, as well as a semester of Expository Writing (Expos), taught mainly by preceptors.

> "Yes, despite what you may think, Harvard people have parties and Harvard people date!"

For many students, the most rewarding form of instruction is the sophomore and junior tutorial, a small-group directed study in a student's field of concentration that is required in most departments within the humanities and social sciences. Teaching of the tutorials is split between professors and graduate students, and the weight of each party's responsibility varies with the subject and the professor. Juniors and seniors seek out professors with whom they want to work.

The oft-made claim that "the hardest thing about Harvard is getting in" is right on target. Flunking out takes real effort. Once on campus, the possibilities are endless for those who are motivated. Then again, Harvard can feel uncaring and antisocial. While it offers unparalleled resources—including fellow students—brilliant overachievers who desire the occasional ego stroke might be better off at a small liberal arts college. Although most students feel little competition, the academic climate is still intense. "The courses at Harvard are very demanding," says a social studies major.

Beware: It is only the most motivated and dedicated student who can take full advantage of the Harvard experience. Others who attempt to drink from the school's perennially overflowing cup of knowledge may find themselves drowning in its depths.

"If you choose to be competitive, you'll find the competition can be cutthroat." Sooner or later, all roads lead to Widener Library, where incredible facilities lie in wait (and where snow-covered steps make prime sledding runs in the winter).

Harvard does have one thing its $29 billion endowment can't buy: a diverse, high-powered, ambitious, and exciting student body. You will meet smooth-talking government majors who appear to have begun their senatorial campaigns in kindergarten. You will meet flamboyant fine arts majors who have cultivated an affected accent all their own. You will sample the intensity of Harvard's extracurricular scene, where more than 6,600 of the world's sharpest undergrads compete for leadership positions in a luminous galaxy of extracurricular opportunities. "Most of the social life takes place on campus, and there are a million things to do," says a history/government double major. "Yes, despite what you may think, Harvard people have parties and Harvard people date!" Stressed-out students can count on help from a variety of quarters, including the various deans' offices, the Bureau of Study Counsel, the Office of Career Services ("dedicated to working with Harvard students and alums for the rest of their lives," claims a senior), and counselors associated with each residential house. All students participate in weeklong orientation, and the First-Year Urban and Outdoor Programs help students acquaint themselves with one another and the Boston area.

No one can tell you exactly what it takes to gain admission to Harvard (and if anyone tries, apply a large grain of salt), but here's a hint: 96 percent of the current student body ranked in the top tenth of their high school class and two-thirds went to public high school. Though there are a few old-money types who probably spit up their baby food on a Harvard sweatshirt, their numbers are smaller than one might imagine on this liberal campus. (Many enter as sophomores when no one is looking.) Undergrads come from all 50 states and scores of foreign countries, although the student body is weighted toward the Northeast. Minority groups account for nearly a third of the enrollment. There are no merit or athletic scholarships to ease the pain of Harvard's hefty tuition, but a generous financial aid policy recently added $2,000 annually to every student aid package. Under a new financial aid policy, students whose families earn less than $60,000 a year will receive full scholarships.

"Freshmen have amazing rooms and upperclass houses are great."

In the past, female students benefited from "dual citizenship" in both Harvard and Radcliffe colleges, receiving degrees ratified by the presidents of both colleges. However, Radcliffe has long since been phased out as a separate institution; everyone is now considered a Harvard student, though students can still take advantage of Radcliffe's network of professional women, researchers, and alumnae.

Every first-year class lives and eats as a single unit in Harvard Yard, a privilege made more enticing by recent renovation of all the freshman dorms. Freshmen now eat in Annenberg Hall, the new name for beautifully renovated Memorial Hall. For their last three years, students live in one of 12 residential houses, built around their own courtyards with their own dining halls and libraries. All the houses are co-ed, and each holds between 300 and 500 students. Designed as learning communities, the upperclass houses come equipped with a complement of resident tutors, affiliated faculty members, and special facilities—from art studios to squash courts. Each house has a student council, which plans programs and parties and arranges the fielding of intramural teams. Students are now randomly assigned (with up to 15 friends) to one of the houses, but some houses still retain a personality from the days of old when each stood for a particular ideology, interest, or economic class. "Harvard housing is beautiful," says a history of science major. "Freshmen have amazing rooms and upperclass houses are great."

The nine houses along the Charles River feature suites of rooms, while the three houses at the Radcliffe Quad, a half-mile away, offer a mixture of suites and single

Harvard's visual and environmental studies major serves filmmakers, studio artists, and urban planners, and concentrations in women's studies and environmental sciences have been well received.

rooms. Some students value the greater privacy of the Quad houses' singles; others consider it equivalent to a Siberian exile, especially during harsh Cambridge winters. The older dorms provide spacious wood-paneled rooms, working fireplaces, and the gentle reminders of Harvard's rich traditions. Most rooms are also wired for direct Internet access. With all these features and amenities, it's no wonder few students move off campus.

Socializing at Harvard tends to occur on campus and in small groups. "It's certainly normal to spend Friday and Saturday nights studying," says a philosophy major. With the exception of the annual all-school Freshman Mixer and the annual theme festivals each house throws, parties tend to be private affairs in individual dorm rooms. A new student-run website provides information and commentary on parties for those in the dark. Though Harvard does enforce the drinking age at university events, in individual houses, it's up to the resident tutors. For some, the key to happiness in Harvard's high-powered environment is finding a niche, a comfortable academic or extracurricular circle around which to build your life. Outside activities include about 80 plays performed annually, two newspapers and several journals, and plenty of community service projects.

"I gauge myself by how many allusions in the *New Yorker* I understand."

The possibilities of Harvard's social life are increased tenfold by Cambridge and Boston, where there are many places to have fun. Harvard Square itself is a legendary gathering place for tourists, shoppers, bearded intellectuals, and coffeehouse denizens. The American Repertory Theater, transplanted from Yale in the mid-1980s, offers a season of professional productions and nearly as professional student shows. Cambridge also enjoys an exceptional selection of new and used bookstores, including the Starr Bookshop (behind the Lampoon building), McIntyre & Moore, Grolier Books, and, of course, the Harvard Bookstore and the mammoth Harvard Co-op, known universally as "the Coop." Boston itself features Faneuil Hall, the Red Sox, the Celtics, and 52 other colleges. "Cambridge/Boston is the ultimate college town," says an English major. "Everything is geared toward the students."

Harvard fields more varsity teams—41—than any other university. The athletic facilities are across the river from the campus, and their incredible offerings often go unnoticed by students buried in the books. Both the men's and women's squash and crew teams are perennial national powers, and the men's ice hockey team draws a crowd of a few dedicated fans. The women's lacrosse team is strong, as are tennis, swimming, and sailing. As for football, the team has been doing better in recent years, but the season always boils down to the Yale game, memorable as much for the antics of the spectators and marching band as for the fumbles of the players. Intramural sports teams are divided up by house, and each fall, league champs play teams from Yale the weekend of the game. Another fall highlight is the annual Head of the Charles crew race, the largest event of its kind in the world, where as many as 200,000 people gather to watch the racing shells glide by.

Nowhere but Harvard does the identity of a school—its history, its presence, its pretense—intrude so much into the details of undergraduate life. Admission here opens the door to a world of intellectual wonder, academic challenges, and faculty minds unmatched in the United States—but then drops students on the threshold. "I have quickly gained exposure to major theories in literature, psychology, anthropology, social sciences, and evolutionary biology," says a junior. "I gauge myself by how many allusions in the *New Yorker* I understand." That's the way Harvard is; what other kind of place could produce statesmen John Quincy Adams and John F. Kennedy, pioneers W. E. B. DuBois and Helen Keller, and artists T. S. Eliot and Leonard Bernstein? Even its dropouts are movers and shakers (witness Bill Gates). But beware: It is only the most motivated and dedicated student who can take full advantage of

Overlaps

Princeton, Yale, Stanford, MIT, Brown

the Harvard experience. Others who attempt to drink from the school's perennially overflowing cup of knowledge may find themselves drowning in its depths.

If You Apply To ➤

Harvard: Regular admissions: Jan. 1. Financial aid: Feb. 1. Guarantees to meet full demonstrated need. Campus interviews: optional, evaluative. Alumni interviews: required, evaluative. SATs or ACTs: required. SAT IIs: required (any three). Accepts the Common Application and electronic applications. Essay question: uses Common Application questions.

Harvey Mudd College: See page 139.

Haverford College

Haverford, PA 19041-1392

Quietly prestigious college of Quaker heritage. With an enrollment of about 1,100, Haverford is half the size of some competitors but benefits from its relationship with nearby Bryn Mawr. Close cousin to nearby Swarthmore but not quite as far left politically. Exceptionally strong sense of community.

An overarching honor code covering everything from the classroom to the dorm room defines student life at Haverford College. Students schedule their own final exams, take unproctored tests, and police underage drinking on their own. "The honor code, in some respects, is a self-selecting system which draws many students to Haverford. For this reason, nearly all students who come here share common values of trust, concern, and respect for others as well as academic integrity," says a junior econ major. Haverford may be smaller and less well-known than some of its peers, but it holds its own against the finest liberal arts colleges in the country, especially for students who want to work hard and play hard. The competitive edge doesn't stop at academics. Sixty-five percent of Haverford students were varsity athletes in high school, and they take many Haverford teams to championships.

Founded under Quaker auspices in 1833, Haverford functions much like a family. The campus consists of 204 acres just off Philadelphia's Main Line railroad, and resembles a peaceful, well-ordered summer camp. The densely wooded campus has an arboretum, duck pond, nature trails, and more than 400 species of shrubs and trees. Architectural styles range from 19th- and early-20th-century stone buildings to a sprinkling of modern structures here and there. The combination enhances the sense of a balanced community, bringing together two traditional Quaker philosophies: development of the intellect and appreciation of nature. The 188,000-square-foot, $50 million Koshland Center for Integrated Natural Sciences (for the departments of astronomy, biology, chemistry, physics, math, and more) provides a place for interdisciplinary teaching, learning, and research.

Haverford's curriculum reflects commitment to the liberal arts. Biology, English, history, political science, and philosophy are among the most popular majors. There are nine areas of concentration—which are different from minors—that are attached to certain majors, including peace and conflict studies and mathematical economics. Unusual offerings include a philosophy seminar called "From Zen Buddhism to

Website: www.haverford.edu
Location: Suburban
Total Enrollment: 1,168
Undergraduates: 1,168
Male/Female: 47/53
SAT Ranges: V 640–760
  M 650–740
Financial Aid: 61%
Expense: Pr $ $ $
Phi Beta Kappa: Yes
Applicants: 3,351
Accepted: 26%
Enrolled: 36%
Grad in 6 Years: 92%
Returning Freshmen: 99%
Academics: 🖉 🖉 🖉 🖉 🖉
Social: ☎ ☎ ☎
Q of L: ★ ★ ★ ★ ★
Admissions: (610) 896-1350
Email Address:
  admission@haverford.edu

Strongest Programs:
Biological and Physical
  Sciences
English

Contemporary Public Black Intellectuals," and a course called "Happiness, Virtue, and the Good Life." Comparative literature and East Asian studies are newer, growing majors. Haverford's general education requirements call for taking three courses in each of the three divisions: social sciences, natural sciences, and humanities. One of these nine courses must fulfill a quantitative reasoning requirement, and every student must also study a foreign language for a year and complete a semester each of coursework in freshman writing and social justice.

The bicollege system with Bryn Mawr College, a nearby women's school, allows Haverford students to major in subjects such as art history, growth and structure of cities, and environmental studies. Also, by combining resources with Bryn Mawr and nearby Swarthmore, Haverford offers students an extensive language program including, most recently, Arabic. The unique relationship between Bryn Mawr and Haverford dates to the days when Haverford was all-male, more than 25 years ago. However, students at each institution can still take courses, use the facilities, eat, and even live in the dormitories of the other. Haverford and Bryn Mawr students cooperate on a weekly newspaper, radio station, orchestra, and other clubs and sports, and a free shuttle bus connects the campuses. Crossregistration is also available at Swarthmore and the University of Pennsylvania. Study abroad programs in one of 36 countries attract nearly 50 percent of juniors.

Since there are no graduate students at Haverford, undergraduates often help professors with research, and several publish papers each year. In fact, Haverford's biggest strength may be its faculty members, 60 percent of whom live on campus and are said to be very accessible. A junior political science student notes that one professor personalized a 90-student class by dividing it up into groups of 10 and having each group over for a pre-exam dessert. "The professors are only here for us. They really enjoy teaching as well as involving us in research." Perhaps because of the intense classroom interaction, the workload is sizable, although students say they don't worry about each other's grades and try to squeeze in nonscholarly pursuits, too. "Everyone always has work to do but we're willing to drop it and throw a Frisbee," a political science student says. Advising is ever-present: freshmen are matched with professors who work with them from their arrival until they declare majors two years later, while upperclass "Customs people" are resources and mentors for living groups of eight to 16 first-year students.

One of Haverford's most distinctive features is the honor code that governs all aspects of campus life. "I can take my final exam at 3 a.m. on Founder's Green" says a junior, who says the honor code "means we look out for ourselves." The

"The honor code, in some respects, is a self-selecting system which draws many students to Haverford."

code, administered by students and debated and re-ratified each year at a meeting called Plenary, helps instill the values of "integrity, honesty, and concern for others." While the social honor code encourages students to "voice virtually any opinion so long as it is expressed rationally," this can also mean self-censorship, says a philosophy major. "Sometimes you feel like you are walking on eggshells to avoid offending anyone," the student says. In good Quaker tradition, decisions are made by consensus rather than formal voting, and students play a large role in college policy.

Only 15 percent of the Haverford's students hail from Pennsylvania, but a large percentage are East Coasters nonetheless. Approximately 13 percent of students are Asian American, 7 percent are Hispanic, and 8 percent are African American. Though the college is nonsectarian, the Quaker influence lives on in the form of an optional meeting each week. "Although students tend to be very well informed on political issues and have great concern about these issues, few are very involved in political activism," says a philosophy major.

Haverford's residence halls are spacious and well maintained, and 64 percent of available rooms are singles—even for freshmen—so it's not surprising that 99 percent of all students live on campus.

Haverford's residence halls are spacious and well maintained, and 64 percent of available rooms are singles—even for freshmen—so it's not surprising that 99 percent of all students live on campus. All dorms are co-ed, but students may request single-sex floors. Freshmen are guaranteed housing, and even sophomores, who draw last in the lottery, can usually get decent rooms. The extremely popular, school-owned Haverford College Apartments sit on the edge of campus. These include one- and two-bedroom units, each with a living room, kitchen, and bathroom. Upperclassmen in the apartments may cook for themselves, but all others living on campus (and all freshmen regardless of where they live) must buy the meal plan, which includes weekend board. "Some of the dorms are incredible. The apartments are the best freshman housing anywhere," says an economics student. Crime is virtually nonexistent, owing to the school's location in the ritzy Philadelphia suburbs. "There is little to worry about coming from the surrounding area," explains a philosophy major: "There is a very safe and trusting relationship between students and security."

While the community spirit at Haverford works well for academics and personal development, it doesn't always carry over to the social scene. Without fraternities and sororities, Haverford and Bryn Mawr hold joint campus parties. These alcohol-soaked affairs can get tiresome after freshman and sophomore years, which is why students

"Everyone always has work to do but we're willing to drop it and throw a Frisbee."

tend to spend at least part of their junior year abroad. The alcohol policy respects the law of the commonwealth of Pennsylvania—no drinking if you're under 21— and is connected, as well, to the honor code. "Haverford is very unique in that the drug and alcohol policy is set and administered by the student body," explains a junior. For nondrinkers, there are frequently free movies, concerts, and other activities on campus. Other traditional events include the weekend-long pre-exams Haverfest—Haverford's approximation of Woodstock—as well as the winter Snowball dance and Taste the Rainbow drag ball. Life in the close-knit, introspective environment that is Haverford can get stifling, but there are easy escapes: downtown Philadelphia is 20 minutes away by train. New York City, Washington, D.C., the New Jersey beaches, Pocono ski areas, and Atlantic City are only a couple hours away by car or train. Many students participate in the Eighth Dimension, which coordinates volunteer opportunities.

Soccer, a sport in which Haverford played in the first intercollegiate game more than 100 years ago, is the most popular sport for both men and women. Track and cross-country are also strong, with the men's and women's teams winning conference championships many times, along with recent league titles in softball and volleyball. In 2005, Haverford opened the Gardner Integrated Athletic Center, a 102,000-square-foot green building, to serve the athletic and fitness aspirations of the entire campus. Every year, Haverford and archrival Swarthmore vie for the Hood Trophy, awarded to the school that wins the most varsity contests between the two. "The Haverford/Swarthmore men's basketball and lacrosse games are the two biggest athletic events of the year," says one student. Haverford also boasts the number-one varsity college cricket team in the country because, well, it's the only school that has one! Intramural sports are popular, especially because partici-

"Some of the dorms are incredible."

pation counts toward the six quarters of athletic credit Haverford requires during the freshman and sophomore years. In spite of all the rivalries, these Quakers have struggled to justify their peace-loving heritage with the desire to bash opponents' brains out on the court or the field. For now, students root for the Black Squirrels, and cheer with this gem: "Fight, fight, inner light—kill, Quakers, kill!"

Haverford's student body may be beyond the norm in regards to personal values but the down side of the honor code is that, "students are challenged to meet an ideal set before them of creating the best community possible. For this reason, students are constantly criticizing themselves and the community as a whole to find ways of solving the problems facing them." See? Mom was right. With freedom comes responsibility.

If You Apply To ➤

Haverford: Early decision: Nov. 15. Regular admissions: Jan. 15. Financial aid: Jan. 31. Guarantees to meet full demonstrated need. Campus interviews: recommended (strongly recommended for those living less than 150 miles from campus), informational. Alumni interviews: recommended, informational. SATs or ACTs: required. SAT IIs: required (two of the student's choice). Accepts the Common Application and electronic applications. Essay question: personal statement that reveals something other application materials have not covered, and how the honor code would change you or help you grow.

University of Hawaii at Manoa

2530 Dole Street, Room C200, Honolulu, HI 96822

Who wouldn't want to go to Hawaii for college? To make it work, aim for one of UH's specialties, such as Asian studies, marine science, and travel industry management. Bear in mind the measly 55 percent graduation rate. Too many luaus and not enough studying can be a bad combination.

Website: www.hawaii.edu
Location: Center city
Total Enrollment: 14,475
Undergraduates: 11,525
Male/Female: 45/55
SAT Ranges: V 480–580
  M 520–620
ACT Range: 21–25
Financial Aid: 32%
Expense: Pub $
Phi Beta Kappa: Yes
Applicants: 6,896
Accepted: 68%
Enrolled: 43%
Grad in 6 Years: 55%
Returning Freshmen: 75%
Academics: ✍ ✍
Social: ☎ ☎
Q of L: ★ ★ ★
Admissions: (808) 956-8975
Email Address:
  ar-info@hawaii.edu

Strongest Programs:
Astronomy
Asian and Pacific Area Studies
Languages
Travel Industry Management

One of the goals of the University of Hawaii at Manoa is to "serve as a bridge between East and West." This multiculturalism is evident in everything from course offerings to the student body. And while you may be thinking about surfing as much as studying, don't be fooled: It will take more than a great tan and the ability to catch a wave to earn your degree here.

The UH campus occupies 300 acres in the Manoa Valley, a residential Honolulu neighborhood. The architecture is regionally eclectic, mirroring historical and modern Asian Pacific motifs, and is enhanced by extensive subtropical landscaping. "There are many plants and trees that make our campus more environmentally friendly," says a sophomore.

UH offers bachelor's degrees in 88 fields. Among the best are astronomy, Asian and Pacific area studies, languages and the arts, ethnomusicology, and tropical agriculture. It should come as no surprise that marine and ocean-related programs are also first-rate. The university also takes pride in its programs in engineering, geology and geophysics, international business, political science, and travel industry management. Both medical and law schools are gaining reputations for excellence. UH also offers a BA degree in information and computer science. Beyond these few specialties, programs are adequate but hardly worth four years of trans-Pacific flights for students from the mainland. Students describe the academic climate as "fairly competitive" and somewhat laid-back. Despite the relaxed atmosphere, core requirements are extensive. All students must take a semester in expository writing and math, two courses in world civilization, two years of a foreign language or Hawaiian, and three courses each in the humanities, social sciences, and natural sciences. Freshman seminar classes offer small-group learning in a variety of subjects. Desirable classes and times are said to be difficult to get into for freshmen and sophomores. You may need to talk to profs, one student advises. Another problem seems to be that certain classes are only offered one semester a year. The academic advising is described as "good" but requires that students take the initiative to seek guidance.

> **"Hawaii is a unique place where diversity is recognized and accepted."**

Hawaii stands out among major American universities in that 64 percent of the students are of Asian descent. Caucasians account for 26 percent, African Americans 1 percent, and Hispanics 3 percent. The different groups seem to get along well, according to students. "Hawaii is a unique place where diversity is recognized and accepted. There are many mixed-race students and many interracial couples," a senior says. Hot campus issues include gay rights, campus parking, and Hawaiian sovereignty. Especially promising students can compete for merit scholarships and athletes receive grants-in-aid.

(Continued)
English as a Second Language
Ethnomusicology
Tropical Agriculture
Geosciences

Only 15 percent of students live in campus housing, which is parceled out by a priority system that gives preference to those who are from across the sea. Students recommend the four towers, Ilima, Lehua, Lokelani, and Mokihana; the rooms are small and the hallways are happening. "Students never know what to expect," one student explains. If you're thinking about off-campus housing, take note: Housing in Honolulu is scarce and expensive. Once you are accepted into housing, continuous residency is not that difficult to obtain. Cafeterias are located throughout the campus and serve "edible" fare.

UH offers bachelor's degrees in 88 fields. Among the best are astronomy, Asian and Pacific area studies, languages and the arts, ethnomusicology, and tropical agriculture.

Because of all the commuters, UH is pretty sedate after dark. "On the whole, UH seems to be an academically focused campus, meaning that school is for on campus and socializing is for off campus," a psychology major says. Many students hit nearby dance clubs or movies, or else head for home. Less than 1 percent of the men and women join the tiny Greek system. Drinking is not allowed in the dorms. A couple of local hangouts provide an escape, and the campus pub, Manoa

"Many do not recognize the high quality of education possible through choosing challenging courses and instructors who urge achievement and high-quality work."

About the only things that generate excitement on campus are the athletic teams, the Rainbow Warriors, with football, volleyball, basketball, baseball, and swimming among the top draws.

Garden, is also an option. Lest anyone forget, some of the world's most beautiful resorts—Diamond Head and all the rest—are less than a 20-minute drive away. Waikiki Beach? Within two miles' reach. And round-trip airfare to the neighboring islands—including Maui, Kauai, and the Big Island—is not unreasonable.

About the only things that generate excitement on campus are the athletic teams, the Rainbow Warriors, with football, volleyball, basketball, baseball, and swimming among the top draws. The Rainbow women's teams are also well-supported, especially the championship volleyball team. The homecoming dance is one of the most popular events of the year. But what students really look forward to is Kanikapila, a festival of Hawaiian music, dance, and culture. Don Ho, eat your heart out.

Students seeking warm weather and great surfing won't be disappointed, but mainlanders should think twice about making the leap to UH unless they are set on one of the university's specialized programs. It's up to you, one student says, to get the best out of Hawaii. "Many do not recognize the high quality of education possible through choosing challenging courses and instructors who urge achievement and high-quality work." And if you can catch a few waves in the process, so much the better.

Overlaps

Hawaii Pacific, UH–Hilo, University of Washington, UCLA, University of Southern California

If You Apply To ➤

UH: Rolling admissions: May 1. Financial aid: Mar. 1. Housing: May 1. Meets demonstrated need of 25%. Campus interviews: optional, informational. No alumni interviews. SATs or ACTs: required. Achievement tests: optional. Prefers electronic applications. No essay question.

Hendrix College

1600 Washington Avenue, Conway, AR 72032

Hendrix is in the same class of mid-South liberal arts colleges as Millsaps and Rhodes. The smallest and most progressive of the three, Hendrix has a strong emphasis on international awareness. Small-town Arkansas is a tough sell, and the college accepts the vast majority of students who apply.

Website: www.hendrix.edu
Location: Small town
Total Enrollment: 1,094
Undergraduates: 1,088
Male/Female: 44/56
SAT Ranges: V 560–690
  M 550–690
ACT Range: 25–31
Financial Aid: 56%
Expense: Pr $
Phi Beta Kappa: Yes
Applicants: 1,310
Accepted: 84%
Enrolled: 32%
Grad in 6 Years: 66%
Returning Freshmen: 87%
Academics: ✍ ✍ ✍
Social: ☎ ☎ ☎ ☎
Q of L: ★★★★
Admissions: (501) 450-1362
Email Address:
  adm@hendrix.edu

Strongest Programs:
Biology
Chemistry
History
Religion
Psychology

For a school in the heart of the Bible Belt, Hendrix College is surprisingly liberal. Academics are demanding but students are laid-back—even radical—in their political and social views. Ironically, healthy dialogue about tough issues such as gay rights, the environment, and capital punishment draws students together. "Students are more liberal and very open-minded. People want to learn about each other and their ideas," says a junior.

Hendrix's compact and comfortable campus stretches for 160 acres between the Ouachita and the Ozark mountains. College land boasts more than 80 varieties of trees and shrubs, and more than 10,000 budding flowers each spring. The main campus—with its own lily pool, fountain, and gazebo—occupies about one-fourth of the total acreage. The redbrick buildings are a mix of old and new, and a pedestrian overpass connects the main campus to the college's athletic facilities and a wooded fitness trail. The campus is undergoing a building boom. Hendrix's commitment to the arts is evident in the three-building complex that houses the Department of Art, which includes separate buildings for photography and art history, two-dimensional arts, and three-dimensional arts. There is a new soccer field, an all-weather track, and the college has broken ground on a 95,000-square-foot wellness and athletic center.

> **"I chose to attend Hendrix for its exceptional ability to prepare students for medical school."**

Hendrix is strong in many areas, but natural and social sciences are definitely the school's forte. Thirty-five percent of students major in biology or psychology, the school's two most popular majors. Students also give high marks to English, history, religion, philosophy, and politics. "I chose to attend Hendrix for its exceptional ability to prepare students for medical school," says a senior. Doing well at Hendrix means keeping up with the workload. "The courses are extremely rigorous," says a sophomore.

At Hendrix, undergraduate research takes priority, especially within the sciences, and students get the chance to present original papers at regional and national symposia. Hendrix offers exchange programs in Austria and England, sends aspiring ecologists to Costa Rica for two weeks and budding archaeologists to Israel, and allows other students to get course credit for internships at U.S. embassies and organizations such as the National Institutes of Health and Agency for International Development. Hendrix-in-London sends 14 students and a professor abroad for a semester. About 40 percent of Hendrix students participate in study abroad. The school is also a member of the Associated Colleges of the South* consortium, and offers five-year programs with Columbia, Vanderbilt, and Washington University in St. Louis for aspiring engineers.

Hendrix freshmen participate in a six-day orientation program, which includes a two-day, off-campus trip emphasizing outdoor experiences, urban exposure, or volunteer service. Professors from many departments teach the course, which considers the lasting importance and global influence of Western and non-Western traditions. Students are also required to complete one course from a category called challenges of the contemporary world, and to complete seven courses across six learning domains—expressive arts, historical perspectives, literary studies, natural science inquiry (two courses, one a lab), social and behavioral analysis, and values, beliefs, and ethics. Hendrix also expects students to take advantage of at least three experiential learning opportunities from artistic creativity, global awareness, professional and leadership development, service to the world, undergraduate research, and special projects. New majors in chemical physics and allied health have been added.

"Students at Hendrix are driven to make a difference," says a junior. "We come to work and we leave to help." Fifty-five percent of Hendrix students are from Arkansas. African Americans and Hispanics comprise 4 percent each and Asian Americans make up 3 percent. Students celebrate diversity unrelated to skin color, with events such as the Miss Hendrix drag show and pageant, where proceeds are given to charity. "Hendrix students are very passionate and highly involved in activist movements," says an English major. "Hendrix is liberal, but levels of participation vary." The college offers a variety of merit scholarships averaging $6,653 to academically qualified students, but there are no athletic scholarships.

> **"Students at Hendrix are driven to make a difference."**

All but one of Hendrix's dorms are single-sex, and freshmen are required to live on campus, which students say adds to the sense of community. Eighty-five percent of students live on campus, but some seniors get permission to move into nearby college-owned apartments. "It is not very difficult getting a room, but getting the room you want may be more trouble," says a junior. A sophomore adds that dorms are "comfortable and people stay there because they like it." Students generally praise the dining options. "It's institutional food but there are a lot of choices," reports an anthropology major.

The two Fs that dominate social life at most Southern schools—football and fraternities—are conspicuously absent at Hendrix. Students are proud of their independence; the annual Hendrix Olympics allows them to celebrate the absence of Alphas, Betas, and Gammas from campus. Other major affairs include the Toga Party, Oktoberfest, and Beach Bash, as well as the annual Toad Suck Daze, a rollicking carnival that features bluegrass music. Last but not least is the Shirttail Serenade, in which first-year men and women from each dorm croon out a song-and-dance routine in their shirts, ties, shoes, and socks for classmates. Judges rate each performance on the basis of singing, creativity, legs, and so on. "We love these traditions!" cheers a student.

Conway offers shops and restaurants, but the 30-minute ride to Little Rock is the preferred destination. Most social life takes place on campus, or on the campuses of two other nearby colleges. "Conway can be fun," a student says, "but you have to search for things to do." Faulkner County, where the school is located, is officially "dry," so students must travel to buy booze—or find older peers to help out. School policies prohibit underage drinking, but "alcohol can be taken anywhere by anyone," says one student. Other popular road trips are Memphis (two hours by car) and Dallas and Oklahoma City (each a five-hour drive) for concerts and the like. For those who stay in town, the Volunteer Activities Center coordinates participation in projects on Service Saturdays.

> **"Hendrix students are very passionate and highly involved in activist movements."**

In the absence of a football team to cheer for, basketball, soccer, baseball, and softball are hot sports on campus. Men's lacrosse and women's field hockey were also added recently. Rhodes College is the chief rival. For outdoor buffs, the college sponsors trips around Arkansas for canoeing, biking, rock climbing, and spelunking.

Musician Jimi Hendrix—whose mug inevitably adorns a new campus T-shirt each year—once asked listeners, "Are you experienced?" After four years at Hendrix

Hendrix also expects students to take advantage of at least three experiential learning opportunities from artistic creativity, global awareness, professional and leadership development, service to the world, undergraduate research, and special projects.

For outdoor buffs, the college sponsors trips around Arkansas for canoeing, biking, rock climbing, and spelunking.

Overlaps

University of Arkansas–Fayetteville, University of Central Arkansas, Rhodes, Louisiana State, University of Arkansas–Little Rock, Millsaps

College, with small classes, an emphasis on research, and a laid-back atmosphere in which to test their beliefs and boundaries, students here can likely answer, "Yes!"

If You Apply To > Hendrix: Rolling admissions. Financial aid: Feb. 15 (priority). Housing: May 1 (priority). Does not guarantee to meet full demonstrated need. Campus interviews: recommended, informational. Alumni interviews: optional, informational. SATs or ACTs: required. SAT IIs: optional. Accepts the Common Application and electronic applications. Essay question.

Hiram College

P.O. Box 96, Hiram, OH 44234

At only 1,239 students, Hiram is the smallest of the prominent Ohio liberal arts colleges. Less nationally known than Wooster or Denison, Hiram attracts the vast majority of its students from in-state. Many classes are taught in seminar format, and an extensive core curriculum ensures a broad education.

Website: www.hiram.edu
Location: Rural
Total Enrollment: 1,239
Undergraduates: 1,205
Male/Female: 44/56
SAT Ranges: V 490–610
   M 490–600
ACT Range: 20–25
Financial Aid: 93%
Expense: Pr $
Phi Beta Kappa: Yes
Applicants: 1,058
Accepted: 88%
Enrolled: 32%
Grad in 6 Years: 64%
Returning Freshmen: 78%
Academics: ✍ ✍ ✍
Social: ☎ ☎ ☎
Q of L: ★ ★ ★
Admissions: (800) 362-5280
Email Address:
   admission@hiram.edu

Strongest Programs:
Biology
Chemistry
English
History
Psychobiology
Environmental studies
Communications

Hiram College's secluded campus breeds a strong sense of community among its students. "Hiram has a unique character," extols one junior. "Because of its small size, students get noticed, not lost in the shuffle." But for all their emphasis on closeness, Hiram students are hardly homebodies—more than 50 percent study abroad in locales ranging from Europe to Australia to Costa Rica. The school's flexible schedule makes it even easier to split campus for a while. Clearly, these Hiram Dawgs are loyal to their school but are always willing to learn new tricks.

Set on a charming hilltop campus that occupies the second-highest spot in Ohio, Hiram is blessed with an abundance of flowers and trees as well as a nice view of the valley below. The prevailing architectural motif is New England brick, and many Hiram buildings are restored 19th-century homes. The Les & Kathy Coleman Sports, Recreation, and Fitness Center houses a competition gymnasium; two multipurpose field-houses that feature tennis, volleyball, and basketball courts; a pool; an indoor track; and facilities for fitness training. Hiram's bio majors work at a 260-acre, college-owned ecology field study station a mile away, with a specialized lab, a 70-acre beech and maple forest, artificial river, and numerous plant and animal species.

The Hiram Plan allows students to cover a breadth of material in three courses during each semester's longer 12-week session, and to focus on a seminar-style class during the additional three-week term. Even nonseminars are small, though; 95 percent of Hiram's courses have 25 or fewer students, allowing for an impressive degree of faculty accessibility. "They are involved and are fun to talk with," says a freshman.

Hiram's core curriculum is extensive. Students are required to select courses from at least six different academic disciplines that fall under two categories, "Ways of Knowing" and "Ways of Developing Responsible Citizenship." Through their studies, students learn various methods for acquiring knowledge and understanding about human beings and the world while exploring what it means to be socially responsible citizens. In addition, Hiram

"Because of its small size, students get noticed, not lost in the shuffle."

has the Freshman Colloquium, a writing and speaking skills seminar, and an upper-division interdisciplinary requirement. First-year students are also enrolled in a seminar with a focus on Western intellectual traditions and an emphasis on writing. Though the academic climate can be challenging, "there are a lot of resources to help students on campus," says a freshman, including tutors, the writing center, and help from peer or profs. The sciences, especially chemistry, are strong, and the music program has been bolstered by the addition of a student-created marching band. Recent additions to the curriculum include an accelerated biomedical humanities program, in which students prepare for medical or veterinary schools in three years; accounting; and a nursing degree.

Hiram offers several unusual summer opportunities, most notably the Northwoods Station up in the wilds of northern Michigan, where students choose courses ranging from photography to botany and geology to writing. And Hiram is the only affiliate college of the Shoals Marine Lab, run by Cornell University and the University of New Hampshire, which offers summer study in marine science, ecology, coastal and oceanic law, and underwater archaeology.

Hiram goes to great lengths to offer outstanding travel abroad programs. Trips led by professors make it to all corners of the globe and all participating students get academic credit. Students can also study at Hiram's Rome affiliate, John Cabot International University, and transfer their credits. Hiram's unique academic calendar allows ample opportunity for off-campus endeavors of all types, including the Washington Semester* at American University, which Hiram helped found.

The Hiram Plan allows students to cover a breadth of material in three courses during each semester's longer 12-week session, and to focus on a seminar-style class during the additional three-week term.

Seventy-eight percent of Hiram students are in-staters, and many of the rest hail from New York and Pennsylvania, though the administration is working to broaden the college's geographic base. Minority students are present, too, with African Americans constituting 11 percent and Asian Americans

"There are a lot of resources to help students on campus."

and Hispanics together comprising another 2 percent. Hiram heads off race-related conflict with a dorm program called Dialogue in Black and White that encourages open discussion on multicultural issues. There's also a one-credit course that has as its final project the creation of a plan of action on campus race relations. In addition to need-based financial aid, Hiram awards merit scholarships worth an average of $6,800. Hiram lures good students with irresistible financial aid. "My financial aid package at Hiram is unbelievable," says a junior. "After applying to larger, cheaper public universities, I came to realize it would actually cost me less to attend Hiram."

Almost all Hiram students—88 percent—live on campus, and everyone who wants a room gets one. "Ours are much better than some I have seen at other schools," says a junior. Community lounges in each hall boast big-screen TVs and computer labs. Most halls are co-ed, and students choose between 24-hour quiet, 24-hour noise, or a happy medium. Upperclassmen who like their location can stay in the same room year after year. Most students live in two-person suites; the popular (and larger) triples and quads are scarcer and thus harder to get.

Hiram is hardly a mecca for budding athletic superstars, but it does have a decent Division III sports program, including a volleyball team that has gone to the NCAA Tournament the past two years.

When the weekend rolls around, don't expect to find all Hiram students gathered around a keg. There are dorms designated as totally dry, and the college has cracked down on underage drinking. Still, students admit that it's easy for underage students to get alcohol if they really want to drink. Hiram is a bit isolated, and there are few distractions in town, so students must make their own fun. Typically, that means hanging out in each other's rooms, or if they're 21, at the on-campus pub that serves pizza (and features karaoke on Tuesdays). The Student Programming Board plans concerts, comedians, speakers, movies, and both formal and informal dances. "We always have tons of things happening on campus," says one student, noting the college also sponsors trips to events. Cleveland's Jacobs Field is

"My financial aid package at Hiram is unbelievable."

a short road trip away. Cleveland's Rock and Roll Hall of Fame can get students rockin' all year-round, and Cedar Point amusement parks get them rollin' in good weather, along with Geauga Lake. Sometimes the college offers free tickets to concerts, plays, and ballets in town. Every semester also brings a surprise Campus Day, when classes are cancelled and a slew of activities are planned. Other diversions include an excellent golf course three miles away, a college-owned cross-country ski trail, and good downhill slopes about an hour distant.

Hiram is hardly a mecca for budding athletic superstars, but it does have a decent Division III sports program, including a volleyball team that has gone to the NCAA Tournament the past two years. Football, baseball, and soccer are among the most popular men's teams, while soccer and softball attract women. Intramurals, including soccer, floor hockey, basketball, and dodgeball, are popular.

Those looking for a school where anonymity will be ensured need not apply. People here are so close that they share an equivalent of the secret handshake. "Everyone smiles at you as you pass—faculty, staff, a senior football player, a freshman chemistry major, the lady that vacuums in the morning, the gardener," says a junior. Indeed, those seeking a friendly, all-American institution with a touch of internationalism might want to give Hiram a look.

Overlaps

College of Wooster, John Carroll, Mount Union, Miami University (OH), Wittenberg

If You Apply To >

Hiram: Rolling admissions. Early action and regular admissions: Apr. 15. Does not guarantee to meet demonstrated need. Campus interviews: recommended (required for scholarship consideration), evaluative. Alumni interviews: optional, informational. SATs or ACTs: required. SAT IIs: optional. Accepts the Common Application and electronic applications. Essay question: last book you read; government spending; space exploration a good thing or not; most significant invention ever; your choice.

Hobart and William Smith Colleges

Geneva, NY 14456

Coordinate single-sex colleges overlooking one of New York's Finger Lakes. Because of the two-college system, relations between the sexes are more traditional than at most schools. Greek life and the men's lacrosse team set the pace of the social scene. Geneva is an old industrial town.

Website: www.hws.edu
Location: Small city
Total Enrollment: 1,847
Undergraduates: 1,839
Male/Female: 46/54
SAT Ranges: V 540–630
  M 545–640
Financial Aid: 62%
Expense: Pr $ $ $ $
Phi Beta Kappa: Yes
Applicants: 3,266
Accepted: 63%
Enrolled: 24%
Grad in 6 Years: 72%
Returning Freshmen: 86%
Academics: ✍ ✍ ✍
Social: ☎ ☎ ☎
Q of L: ★ ★ ★
Admissions: (315) 781-3472

At Hobart College, men can be men, while next door at William Smith College, women rule the roost. Each school has its own dean, admissions office, and student government, but their "coordinate system" means students eat together, study together, and even live together in co-ed residence halls. If you can tolerate the chilly winters and relative isolation of upstate New York, you'll be rewarded with small classes, caring faculty, and a place where tradition still matters. Alumni are loyal and devoted.

Hobart College was founded in 1822 by Episcopal Bishop John Henry Hobart, who conceived it to be an outpost for civilized and learned behavior. In 1908, William Smith College opened. The college bears the name of a wealthy businessman who wanted to introduce women to opportunities that were largely unrecognized at the time. The H-WS campus stretches for 200 tree-lined acres, and includes a forest, farmland, and a wildlife preserve. Architectural styles range from Colonial to postmodern, with stately Greek Revival mansions and ivy-clad brick residences and classrooms. Stern Hall houses the departments of economics, political science, anthropology, sociology, and Asian languages and cultures. There's a boathouse on the shores of Seneca Lake for the nationally ranked sailing team. Two new residence halls opened in 2005.

The innovative H-WS curriculum has no distribution requirements. Instead, students take an interdisciplinary seminar in the first year, capped at 12 students, and focused on writing and critical thinking. Students must also complete a major and minor or a double major, one from a traditional department and one from an interdisciplinary program. Newer majors include international relations, European studies, and media and society, and the long list of minors includes the new Sacred in Crosscultural Perspective, men's studies, The Good Society, and aesthetics. "Some classes are more laid-back but the majority are competitive and every course is challenging. Professors have different ways of encouraging students to do their best work," explains one bioethics major.

Small classes are the norm here; 95 percent of all classes have 50 students or fewer, and more than two-thirds have 25 or fewer. "Freshmen are always taught by full professors." Another senior says, "Professors go far beyond the basics and extend their knowledge to students in a way that engages passion." The "Finger Lakes Institute" at the college gives students wide opportunities to work in various fields of scientific inquiry, as well as public policy. H-WS students may take a term away from campus in 28 cities from Copenhagen, Rome, and Edinburgh to Sao Paolo, Taipei, and even New York and Los Angeles. Each year, about 30 seniors elect to "do Honors," producing a research or critical paper or an equivalent creative work, and then taking written and oral exams on their projects.

> "Freshmen are always taught by full professors."

Hobart and William Smith also participate in the Venture Program* and provide independent research opportunities; such work helped chemistry major Julia James become the colleges' first Rhodes Scholar in 2004.

H-WS has a way to go when it comes to diversity, whether geographic or ethnic in nature. New Yorkers make up 47 percent of the H-WS student body, and African Americans, Hispanics, and Asian Americans each constitute 4 percent of the total. Still, the departments of women's, African American, and third-world cultural studies are small but flourishing. Politically, the colleges lean liberal, but "everyone is respectful of each other's opinions." "Students at H-WS are motivated, determined, and constantly engaged in stimulating conversation. They continue their interests in academics beyond the classroom," says a senior.

Ninety percent of H-WS students live on campus, where first-years may opt for single-sex or co-ed dorms, and "a quarter of the people on your floor will be in your first-year seminar," says a sophomore. Favorites include Geneva (Hobart) and Hirshon (William Smith). Durfee, Bartlett, and Hale halls, known as "Miniquad" and formerly the most avoided living spaces on campus, are again popular among Hobart men. Sophomores, who may lose out in the "wacky" housing lottery, typically live in a large co-ed complex known as J-P-R (for Jackson, Potter, and Rees halls). Juniors and seniors may choose Victorian houses with high ceilings and wood floors, or the more modern "Village at Odell's Pond," where townhouses have four to five bedrooms and two bathrooms each. Housing is guaranteed for four years, and only seniors are allowed to live off campus, though few choose to do so.

Five Hobart fraternities claim 15 percent of the men, who aren't permitted to pledge until sophomore year, but there are no sororities at William Smith. For men who'd rather not join, Bampton House (the men's honors house) and McDaniel's House are good bets. Women may choose smaller residence halls, like Blackwell and Miller houses, which contribute to a feeling of community without rigid structure. William Smith has also retained a number of traditions typical of women's colleges, such as Moving Up Day, in which seniors symbolically hand over their leadership role to juniors. (Hobart, not to be left out, has a similar event called Charter Day.) Social life includes Greek parties and bashes at off-campus houses, while the Campus Activity Board plans movies, concerts, dances, plays, and other events each weekend. There's a "zero tolerance" policy for underage drinking, and students caught with booze must attend alcohol awareness classes and may face social probation.

(Continued)
Email Address:
admissions@hws.edu

Strongest Programs:
Creative Writing
Environmental Studies
Architectural Studies
Biology
Political Science
History
Economics
Elementary, Secondary, and
    Special Education

The old industrial city of Geneva "is a decent college town, but not the best."

The innovative H-WS curriculum has no distribution requirements. Instead, students take an interdisciplinary seminar in the first year, capped at 12 students, and focused on writing and critical thinking.

The old industrial city of Geneva "is a decent college town, but not the best" says a senior. But features that get rave reviews are Seneca Lake and the Smith Opera House, which "offers unique events, free or discounted to students, that you might not expect in a small, rural area," praises one junior. Other notable perks include

"Everyone is respectful of each other's opinions."

movie theaters, restaurants and bars, a state park, an outlet mall, and a bowling alley—if not right in town, then not far, as long as you've got a car. Popular road trips include Rochester, Ithaca, and Syracuse, all about 45 minutes away; more intrepid (or bored) souls trek as far as New York City, Washington, D.C., and Toronto, Canada. Most of the campus takes part in Celebrate Service, Celebrate Geneva, which brings students and professors together for a day of service in the area. Students look forward to the two-day Folk Fest, a music-, food-, and craft-filled party held in the fall. On weekends, the H-WS outdoor recreation program takes students hiking, camping, and skiing in the Adirondacks and surrounding areas.

Sports are the most popular diversion from studying here, whether you're a spectator or participant. H-WS teams compete in Division III, except for men's lacrosse, which won 16 straight NCAA Division III championships before joining Division I. The William Smith field hockey team has won national titles in past years, and the WS soccer and lacrosse squads are likewise strong. Both men and women excel on the water, whether in sailing or crew.

Hobart and William Smith Colleges has been overshadowed by some other Northeastern schools with similar names—a popular campus T-shirt explains, "Not Williams...Not Smith...William Smith!"—but students say, at last, that's changing. If you're unsure about a single-sex institution, the H-WS "coordinate system" may offer the best of both worlds.

Overlaps

Union, St. Lawrence, Colby, Hamilton, Skidmore

If You Apply To >

Hobart and William Smith: Early decision: Nov. 15, Jan. 1. Regular admissions: Feb. 1. Financial aid: Feb. 15. Meets demonstrated need of 75%. Campus and alumni interviews: recommended, evaluative. SATs or ACTs: required. SAT IIs: optional. Accepts the Common Application and electronic applications. Essay question: uses question from the Common Application.

Hofstra University

Hempstead, NY 11549

Like nearby Adelphi, offers combination of suburban setting with ready access to the Big Apple. Hofstra has outgrown its commuter-school origins and offers a broad range of academic programs. Greek system an option but does not dominate the social scene. Well-known as a lacrosse powerhouse.

Website: www.hofstra.edu
Location: City
Total Enrollment: 10,131
Undergraduates: 7,960
Male/Female: 47/53
SAT Ranges: V 520–620
  M 540–620
ACT Range: 21–26

Although it sits within easy striking distance of Manhattan, Hofstra University occupies one of the prettiest campuses you'll find anywhere. Its parklike setting is not only an accredited museum and arboretum but also home to the school's blossoming professional offerings. Whatever their field, Hofstra students enjoy study abroad options, research opportunities, and first-year programs that help harried freshman get off to a good start.

Founded in 1935 with one building—the Dutch Colonial mansion left in trust by Kate and William Hofstra—on their 15-acre estate, the campus is now home to 113 buildings on 240 acres. The suburban campus offers a park-like environment with a variety of architecture, from ivy-covered stone buildings to modern facilities with sleek angles and electronic signage, which surround open green quads, where students gather, play sports, or enjoy outdoor classes when weather permits. The campus is especially beautiful in the spring, when Hofstra's 100,000 tulips, a tribute to the Dutch heritage of Hofstra's founders, are in bloom. The most recent campus projects include Hagedorn Hall, opened in 2004, and a new academic building that houses a 230-seat theater, an orchestra rehearsal room, music practice rooms, seminar rooms, and offices. A new residence hall is currently in the planning stages.

With more than 130 academic programs for undergraduates, Hofstra offers students plenty of academic options. Accounting, biology, drama, and international business are especially strong, and the most popular majors include psychology and education. "The best seem to be the schools of business and communications," reports a radio major, "because of the facilities and resources." First Year Connections offers new students a combined social and academic experience centered on small seminars (limited to 15 students) taught by senior faculty; recent seminars include "New York Raw," "Imagining Iraq," and "Design and the Human Body." Freshmen may also enroll in clusters of thematically related courses. The Honors College offers approximately 600 qualified students a multidisciplinary program that allows them to graduate

> **"Most professors focus on learning versus grades, which is great."**

with both a bachelor's degree and Honors College designation. They may also take advantage of faculty mentors and special housing. Those with wanderlust can take advantage of study abroad options around the world, including programs in England, Italy, Argentina, Japan, France, Jamaica, and South Korea. Hofstra is home to one of the largest simulated trading rooms in the New York area, boasting 34 dual-panel Bloomberg terminals. The Department of Dance and Drama has turned out such luminaries as Francis Ford Coppola, the late Madeline Kahn, and *Everybody Loves Raymond* creator Phil Rosenthal.

Regardless of what major they choose, all undergrad students must complete distribution requirements, including coursework in humanities, natural sciences and mathematics/computer science, social sciences, crosscultural studies, and interdisciplinary studies. Students must also pass the English Proficiency Exam. "The academic climate is fairly laid-back," says a senior, "and most professors focus on learning versus grades, which is great." Seventy-nine percent of classes have 25 or fewer students, giving students plenty of access to professors. "Most professors are really helpful and understanding," a student says.

Sixty-nine percent of the student body hails from New York, and 47 percent ranked in the top quarter of their high school class. "A large percentage of the students are too preoccupied with the mall or tanning," gripes one student. "It would be nice if some of the students put as much energy into school as they do getting ready in the morning." African Americans account for 9 percent of the student body, Hispanics 8 percent, and Asian Americans 5 percent. "Hofstra is generally a liberal campus," says a student, and hot topics include gay rights, women's rights, and abortion. The university offers merit scholarships worth an average of $8,050 and 140 athletic scholarships in a variety of men's and women's sports.

Seventy-three percent of students live in university housing, which is described as "really good or really bad, depending on where you live." A public relations major explains, "The suite-style buildings are better maintained than the tower buildings." Although the housing restrictions leave many students

> **"Hofstra is generally a liberal campus."**

grumbling, the tasty cafeteria fare draws quick praise. "Most of the food is really good," says one student. "My only complaint is that it's expensive. Very expensive." Security is "moderate," and students generally feel safe on campus. "Campus secu-

(Continued)

Financial Aid: 59%

Expense: Pr $

Phi Beta Kappa: Yes

Applicants: 15,981

Accepted: 62%

Enrolled: 19%

Grad in 6 Years: 55%

Returning Freshmen: 78%

Academics: ✍ ✍ ✍

Social: ☎ ☎ ☎

Q of L: ★ ★ ★

Admissions: (516) 463-6710

Email Address:
admitme@hofstra.edu

Strongest Programs:
Accounting
Biology
Creative Writing
International Business
Journalism
Music
Audio/Video/Film
Drama and Dance

The Department of Dance and Drama has turned out such luminaries as Francis Ford Coppola, the late Madeline Kahn, and **Everybody Loves Raymond** *creator Phil Rosenthal.*

First Year Connections offers new students a combined social and academic experience centered on small seminars (limited to 15 students) taught by senior faculty.

rity has greatly increased during my time here," says a senior. "Students feel relatively safe on campus." Unfortunately, the surrounding neighborhoods don't foster this same sense of safety.

Hofstra's social scene offers a liberal mix of campus events, off-campus bars, and "small hangouts in people's rooms," says one student. "Being involved is absolutely crucial to having a social life on campus," explains a sophomore. "This is where Hofstra excels, though, offering an incredible number of clubs, in-dorm activities, and departmental organizations." Six percent of the men and 7 percent of the women go Greek, but they don't dominate the social landscape. Alcohol policies include severe consequences for underage drinkers, "but most students assume the

"Campus security has greatly increased during my time here."

risk and consume anyway," says a junior. Festivals include Greek Week, and "the Irish, Italian, and Dutch festivals are really fun events," a student says. When students tire of on-campus activities, they trek into New York City (40 minutes away) for more urban activities.

Although "the surrounding town has plenty of Targets, Wal-Marts, and supermarkets, Hempstead is definitely not a college town," says a sophomore. Although Long Island has "absolutely everything," the area surrounding Hofstra is "a very bad neighborhood," according to a freshman. Nevertheless, some students get involved in the community through service clubs or organizations on campus.

Hofstra teams compete in Division I and the men's and women's soccer teams and women's softball squad have captured conference championships in recent years. Other competitive teams include men's and women's basketball and wrestling. The men's and women's lacrosse teams are perennial powerhouses ranked in the top 20 nationally. Intramurals and recreational sports attract 35 percent of undergrads, and students take advantage of the Recreation Center, which features daily fitness classes, a multipurpose gym, and a fully equipped weight room.

Without a doubt, Hofstra succeeds at making the transition from high school to college less overwhelming. "What makes this school great is the feeling of a small school within a big school," says a Spanish and secondary education major. "There are plenty of things to do on campus, and even with such a large student body, you get to know so many people." By offering solid academics and a bevy of programs aimed at first-year students, the university seeks to put its students in a New York state of mind.

Overlaps
NYU, Fordham, SUNY–Stony Brook, SUNY–Binghamton, Boston University, St. John's

If You Apply To ➤

Hofstra: Early action: Nov. 15, Dec. 15. Financial aid: Mar. 1. Campus interviews: recommended, informational. Alumni interviews: optional, informational. SATs or ACTs: required. SAT IIs: recommended. Accepts the Common Application and prefers electronic applications. Essay question: meaningful event; describe your educational and personal goals; discuss a work's influence; personal statement.

Hollins University

(formerly Hollins College)
P.O. Box 9707, Roanoke, VA 24020

Hollins is among a trio of western Virginia women's colleges—along with Sweet Briar and Randolph College. Hollins is in the biggest city of the three, Roanoke, and has long been noted for its program in creative writing. Social life often depends on road trips to Washington and Lee, and Virginia Tech.

Traditions rule at Hollins University, a private, single-sex college on a lush 475-acre campus in the Virginia mountains. Each fall, students and staff hike up Tinker Mountain, for skits, a picnic lunch, and a bird's-eye view of the changing foliage. There's a secret society known as Fraeya, since the school doesn't have sororities. And there are boys on campus, too, since Hollins offers co-ed graduate programs in children's literature, creative writing, dance, screenwriting and film studies, liberal studies, and teaching. "Many would say that students come to Hollins for one of two things: English or equestrian riding," says a freshman. "Many would be wrong, though—students are now choosing degrees in biology, sociology, women's studies, business, and international studies."

Described by the New York Times as "achingly picturesque," the neoclassical redbrick buildings at Hollins date back to the mid-19th century. There are some modern structures, too, such as the Wetherill Visual Arts Center, which opened in 2004. The building offers space for programs in studio art, art history, and film and photography. The center also houses the Eleanor D. Wilson Museum, which is a partner institution of the Virginia Museum of Fine Arts. New space for the English and creative writing department opened last year.

General education requirements at Hollins stress breadth and depth across the curriculum. The skills component teaches students to write successfully, reason quantitatively, express themselves effectively, research astutely, and be adept technologically. The perspectives component includes seven areas of knowledge that help to explain how we view and understand the

> "Students are now choosing degrees in biology, sociology, women's studies, business, and international studies."

world: Aesthetic Analysis, Creative Expression, Ancient and Medieval Worlds, Modern and Contemporary Worlds, Scientific Inquiry, Social and Cultural Diversities, and Global Systems and Languages. Two terms of physical education are also mandatory, as is Orientation Week, which includes academic programming, a day of community service, and plenty of time to form friendships with new classmates.

Motivated students are encouraged to design their majors, while the technically inclined may enroll in a combined-degree program in engineering. Additionally, because Hollins belongs to the Seven-College Exchange,* its students may crossregister at any of the other six institutions. Those with wanderlust may spend semesters at Hollins' campuses in England and France or programs in Argentina, Spain, Japan, Mexico, Ghana, Italy, Ireland, Greece, and the School for Field Studies; 45 percent of all students spend some time abroad, and a service-learning program takes altruistic students to Appalachia and Jamaica every year. The annual Science Seminar features undergrad research in biology, chemistry, psychology, physics, and math, while the Hollins Leadership Institute is the only college program in the nation that promotes personal and professional growth with videotaped performance reviews, senior mentoring, intensive communication skills groups, and board governance experience. The shorter January term offers a break for on-campus projects, travel, or internships, and alumnae help arrange housing in Washington and other cities.

Academics are a top priority at Hollins, though competition is rare. "Courses are, overall, challenging and rewarding," says a senior. Communication studies, psychology, and studio art are some of Hollins's most popular majors; a new major in environmental science is designed to be interdiscipli-

> "Courses are, overall, challenging and rewarding."

nary. Artists benefit from extensive internship opportunities at Christie's, Sotheby's, and various parts of the Smithsonian Institution, while the Hollins Repertory Dance Company, closely affiliated with the American Dance Festival, has been selected to perform at the Kennedy Center and at Aaron Davis Hall, Harlem's

Website: www.hollins.edu
Location: City outskirts
Total Enrollment: 1,123
Undergraduates: 848
Male/Female: 0/100
SAT Ranges: V 530–640
  M 490–590
ACT Range: 22–27
Financial Aid: 61%
Expense: Pr $
Phi Beta Kappa: Yes
Applicants: 686
Accepted: 86%
Enrolled: 31%
Grad in 6 Years: 62%
Returning Freshmen: 79%
Academics: ✍ ✍ ✍
Social: ☎ ☎ ☎
Q of L: ★ ★ ★ ★
Admissions: (800) 456-9595
  or (540) 362-6401
Email Address:
  huadm@hollins.edu

Strongest Programs:
English/Creative Writing
Visual and Performing Arts
Psychology
Communication Studies
Biology
Political Science

As the number of women's colleges continues to dwindle, Hollins remains committed to offering single-sex education and close-knit community.

premier performing arts center. Classes are small—95 percent have 30 or fewer students, and none have more than 50.

Just half of the students at Hollins are Virginians; African Americans make up 8 percent of the total, while Asian Americans and Hispanics add 1 and 3 percent, respectively. "We have a very active gay and lesbian club called Outloud, and also an active women's studies department, that put together a lot of rallies and events," says a freshman. "Our campus has a definite liberal leaning, although there is a small, committed Republican group." Hollins hands out $17 million annually in financial aid and scholarships. There are no athletic scholarships, although selected students have a chance to win full scholarships.

Eighty percent of Hollins students live in the dorms; they have to, unless they're married, older than 23, or living at home in Roanoke with their parents. "The dorms are beautiful with great views of the campus," says a history major, "but could definitely use a lot of renovation—they are old!" Freshmen live in Randolph and Tinker; upperclassmen in Main, West, and East, which have 10-foot ceilings and hardwood floors. Singles are only available to upperclassmen, and there is a fee. Some rooms in Main, West, and East also boast brass doorknobs, walk-in closets, and even fireplaces. Students chow down in Moody or at the campus coffee shop, which is open late. Food is "pretty good, but it does get repetitive after a few months," one student says.

Hollins shuns sororities, but sporadic student efforts to bring them to campus draw lively debate. To fill the gap, the school organizes mixers, concerts, dances, and second-run movies each weekend. There's also a free shuttle to help students get around Roanoke, a city with "malls, shopping centers, and movie and live theaters," says a freshman. "It has a nice historic downtown with an organic farmer's market, but it is not densely urban like New York or Washington, D.C." As a result, road-tripping remains the preferred social option—to Hampden-Sydney College, Virginia Tech, the University of Virginia, or Washington and Lee.

> **"The friendliness of staff, faculty, and students is infectious."**

The Hollins Outdoor Program offers hiking, spelunking, and other activities in the beautiful Shenandoah Valley and Blue Ridge Mountains. The on-campus stable complements the school's equestrian program, which since 1998 has been champion, reserve champion, and third in national competitions sponsored by the Intercollegiate Horse Show Association. Women's tennis placed second in the Old Dominion Athletic Conference in 2004, and the swim team has won national Division III championships. In addition to the varsity lacrosse, golf, soccer, basketball, swimming, and tennis teams, fencing and field hockey are available as club sports.

As the number of women's colleges continues to dwindle, Hollins remains committed to offering single-sex education and a close-knit community. Students leave with confidence, critical-thinking skills, and intellectual depth, thanks to a solid grounding in the liberal arts. And the school's Southern heritage doesn't hurt, either. "The friendliness of staff, faculty, and students is infectious," says a business and communications major.

Overlaps

Sweet Briar, University of Mary Washington, James Madison, Randolph College, Roanoke, Virginia Commonwealth

If You Apply To ➤

Hollins: Rolling admissions. Financial aid: Feb. 15. Housing: May 1. Meets demonstrated need of 21%. Campus interviews: recommended, informational. Alumnae interviews: optional, informational. SAT or ACT: required. SAT IIs: optional. Accepts the Common Application and electronic applications. Essay question: significant experience, achievement, or risk you have taken; ethical dilemma you have faced, and its impact; or why your volunteer service is important, and how you'd like to serve in the future. Also, how you enjoy expressing your creativity, or have used it to solve problems; and a woman you admire outside your family, and why.

College of the Holy Cross

Worcester, MA 01610

A tight-knit Roman Catholic community steeped in church and tradition. Many students are the second or third generation to attend. Set high on a hill above gritty Worcester, an hour from Boston. Sports teams compete with (and occasionally beat) schools 10 times HC's size.

Students at Holy Cross, a Roman Catholic college in the heart of New England, are devoted to the Jesuit tradition of becoming "men and women for others." Students on "the hill" are driven to do something for their college or community, whether it's a football player becoming a Big Brother or an off-campus senior becoming a student government senator. Peers and professors alike offer support and spiritual guidance, and bonds forged in the lab or on the field are strengthened through activities like SPUD (Student Programs for Urban Development), which provides community service opportunities. The classroom focus is critical thinking and writing, but the school's proximity to nine other colleges in the Boston area means Crusaders focus on their social lives, too.

Located on one of the seven hills overlooking the industrial city of Worcester, the 174-acre Holy Cross campus is a registered arboretum. The school's landscaping has won a half-dozen national awards, including two first-place prizes, as the best-designed and -planted campus in the nation. Architectural styles range from classical to modern. The college recently completed a 244-bed, apartment-style residence hall and covered parking facility. There is also a new 1,350-seat, natural-turf soccer facility and, in conjunction with a professional minor league baseball team, a 3,500-seat baseball stadium.

Holy Cross offers small classes—71 percent of those taken by freshmen have 25 or fewer students—which help faculty members keep in touch with undergraduates. Crusaders come from a variety of backgrounds and hometowns, though most are Irish Catholics from the Northeast who want to learn, succeed, and help others. Students say the climate is challenging and the classes intense. "Holy Cross is very competitive," says a junior. "This is evident in the amount of time students spend doing work and in the library, as well as from the noticeable desire to outperform their peers." Even so, most students agree that the competition is healthy and rarely cutthroat. "Students want to succeed, but also want to see their classmates succeed," a student says. Professors earn high marks for their knowledge and compassion. "My professors are teachers, role models, mentors, and friends," a math major says. "All the professors that I have had are extremely gifted and genuinely care about their students," adds a junior.

HC's premed program boasts that it has twice the number of students accepted to medical school as the national average. Students also give high marks to the English, history, and economics and accounting programs. As might be expected, philosophy and religious studies are strong, and concentrations in Latin American studies and peace and conflict studies are popular, as these are the disciplines central to Jesuit missionary work.

"Students want to succeed, but also want to see their classmates succeed."

Community-based learning courses include two to two-and-a-half hours of weekly service with local volunteer, education, or health organizations, in addition to time in the classroom. Computer science is now available as an undergraduate major, and there's an environmental science concentration.

For more than a decade, a quarter of Holy Cross's first-year students have enrolled in the First Year Program, which attempts to answer a question adapted

---

Website: www.holycross.edu
Location: City outskirts
Total Enrollment: 2,777
Undergraduates: 2,777
Male/Female: 45/55
SAT Ranges: V 620–640
  M 580–680
Financial Aid: 53%
Expense: Pr $ $ $
Phi Beta Kappa: Yes
Applicants: 4,744
Accepted: 48%
Enrolled: 32%
Grad in 6 Years: 90%
Returning Freshmen: 93%
Academics: ✍ ✍ ✍ ✍
Social: 🐻 🐻 🐻 🐻
Q of L: ★ ★ ★ ★
Admissions: (508) 793-2443
Email Address:
  admissions@holycross.edu

Strongest Programs:
Biology/Premed
History
Economics
English

Holy Cross is keeping the faith—its emphasis on Catholicism and the Jesuit tradition, that is—even as administrators place a renewed emphasis on academics.

from Tolstoy: "How then shall we live?" The FYP includes seminars, each limited to 15 students, shared readings, co-curricular events, and a common dorm. Aside from the FYP, Holy Cross's general education requirements comprise 12 courses in 10 areas: modern or classical languages, social sciences, arts, literature, religion, philosophy, history, cross-cultural studies, science, and mathematics. Ideas and thinking are the focus rather than preparation for a specific vocation.

Holy Cross is part of the Worcester Consortium,* which offers registration privileges at the region's most prestigious colleges and universities. Aspiring teachers will find education courses and student-teaching opportunities at local primary and secondary schools and a teacher certification program accredited by the Massachusetts Department of Education. Would-be engineers can choose Holy Cross's 3–2 dual-degree programs with Columbia or Dartmouth; a partnership with nearby Clark offers a BA and MBA, or a BA and MA in finance, in five years. Off-campus opportunities include academic internships in the community, 17 study abroad programs, and the Washington Semester.* "Students who participate in study abroad and the First Year Program usually speak highly of the programs," a senior says. HC's honors program enables a small number of juniors and seniors to enroll in exclusive courses and thesis-writing seminars, while the Fenwick Scholars program helps students design and carry out independent projects.

"My professors are teachers, role models, mentors, and friends."

"Students here are dedicated, hardworking, goal-oriented, active, and involved," says a history and economics double major. The religious influence at Holy Cross is somewhat greater than at other Jesuit schools—most students are Roman Catholic, one-third are in-staters, and just more than half attended public high school—but daily mass is not required. The chaplain's office does offer an optional five-day silent retreat four times a year, in which student volunteers follow the spiritual exercises of Jesuit founder St. Ignatius Loyola. African Americans make up 4 percent of the student body, Asian Americans comprise 4 percent, and Hispanics account for 5 percent. Hot political issues include social justice, campus diversity, and homosexuality. Merit scholarships are awarded annually, with an average payout of $15,517. Student athletes may vie for 185 athletic scholarships in six sports.

Eighty-eight percent of Holy Cross students live in the residence halls, where freshmen and sophomores have double rooms, and juniors and seniors may opt for two- and three-bedroom suites—with living rooms and bathrooms, but no kitchens. Floors are single-sex; buildings are co-ed. Most first-years live on "Easy Street," the row of five dorms (Healy, Leahy, Hanselman, Clark, and Mulledy) on the college's central hill next to the Hogan Campus Center. Wheeler, Loyola, Alumni, and Carlin house mostly upperclassmen. "The dorms are very comfortable and the housekeeping staff does a great job keeping them clean and well maintained," says one student. Because there are no Greek organizations, dorm life takes center stage. Each dorm has its own T-shirt, and they compete against each other for prizes in athletic and other contests, says a philosophy major. Doors have combination locks, which means no worries about forgetting your keys when you walk to the shower. Students say they feel safe on campus. "Campus security is always present," a junior observes.

While Worcester isn't Boston, the town doesn't deserve its lousy reputation, students say. An hour from Beantown, Cape Cod's beaches, and the ski slopes of the White Mountains, Worcester "is a great town if you give it a chance," says a student. A school shuttle service takes students to the orchestra, the Worcester Centrum for athletic events and rock concerts, and the town's museum. Nightlife is good, especially with so many other schools nearby.

"Volunteering is one way students live out the Holy Cross mission."

"For underclassmen, most gatherings are in the dorm rooms or at a nearby house," explains one student. "Upperclassmen frequent a few of the bars in Worcester." HC abides by Massachusetts liquor law: Students under 21 can't drink at the campus pub. If they're caught with alcohol, they're put on probation and parents are notified. Still, as with most schools, students who seek to imbibe can find booze, regardless of their age. Since students are discouraged from having cars, most take advantage of concerts, hypnotists, comedians, and other events organized by the Campus Activities Board. The college also organizes trips to New York City and Providence, Rhode Island, and there are countless service opportunities. "Volunteering is one way students live out the Holy Cross mission," a student says.

Tradition is big at Holy Cross, from Alumni Weekend to HC by the Sea (a week in Cape Cod at the end of the year). Midnight breakfasts provide sustenance as students cram for finals, while the 100 Days weekend begins the senior class countdown to graduation. Spring Weekend brings well-known performers. Would-be matchmakers can set up their roommates on dates at the "Opportunity Knocks" dance. And of course, given the high percentage of Irish Catholic students, St. Patrick's Day is an occasion for celebration. Students cheer with religious zeal when the Crusaders battle Boston College at HC's football stadium, which holds 23,500 screaming fans. Recent Patriot League champs include women's lacrosse and women's basketball. The men's ice hockey team captured the American Hockey League championship in 2006. More than one-third of HC students participate in intramural sports,

Holy Cross is keeping the faith—its emphasis on Catholicism and the Jesuit tradition, that is—even as administrators place a renewed emphasis on academics. New faculty have come to campus, allowing HC to offer more courses while keeping class sizes small. "The admissions pitch is that Holy Cross is small, liberal arts, Jesuit, and exclusively undergraduate," says a senior. "Holy Cross is more than an institution for education. It's the people that make the school what it is."

> ## Overlaps
> Boston College, Providence, Villanova, Fairfield, Georgetown, Tufts

> **If You Apply To ➤**
>
> **Holy Cross:** Early decision: Dec. 15. Regular admissions: Jan. 15. Financial aid: Feb. 1. Guarantees to meet full demonstrated need. Campus and alumni interviews: recommended, evaluative. SATs or ACTs: optional. SAT IIs: optional. Accepts the Common Application and electronic applications. Essay question: personal reflection; prominent personality traits; where would you travel and why?

Hood College

401 Rosemont Avenue, Frederick, MD 21701

The newly co-ed Hood faces a challenging task in building a co-ed environment. A major asset: Hood's strategic location, one hour from D.C. and Baltimore. Hood's distinctive core curriculum stresses thematic study. Alumni children and grandchildren get a great deal!

Newly co-ed Hood College has given new meaning to the term "legacy" in college admissions. This small liberal arts college lets the children and grandchildren of alumni pay the same tuition for their freshmen year as their elders. If Granny was Hood Class of '50, that means the tab is only $500! The new tuition policy is only one of the creative steps that Hood has been taking to reinvent itself in the wake of declining enrollment and serious financial problems in the late 1990s. The big change came in 2003 when Hood began admitting men as regular residential students (males had been commuting students since 1971). Now members of both sexes

Website: www.hood.edu
Location: Small city
Total Enrollment: 2,248
Undergraduates: 1,274
Male/Female: 29/71
SAT Ranges: V 490–600
  M 490–600

have the opportunity to partake of Hood's traditional mix of professional and liberal arts offerings on a campus strategically located in an historic Civil War setting.

Hood was founded as a women's college in 1893. Its strikingly beautiful, 50-acre campus, which features redbrick buildings and lush, tree-shaded lawns, sits in the Civil War town of Frederick. Hood is within an hour and a half of nearly 30 colleges, and within minutes of a major National Cancer Institute research complex, high-tech firms, small and large businesses, and both Washington, D.C., and Baltimore. On campus, technology programs, which are already important, now get a further boost with the new Hodson Science and Technology Center.

Students see their school's biggest strength in its people: students, staff, and faculty. "There are so many cultures and ethnicities and traditions to be shared," says one junior. "I love living here. I'm having the time of my life." All incoming students have the opportunity to participate in the First Year Advantage (FYA) program, which introduces participants to a plethora of resources via workshops, leadership programs, athletic and social events, and community interaction. Sophomore Experience helps students pick a major and plan a career. The honor system also is an important part of a Hood education. The academic honor code permits unproctored exams and self-scheduled finals; the social code allows for self-governed residence halls where students actively participate in hall politics.

"There are so many cultures and ethnicities and traditions to be shared."

Hood's required core curriculum is divided into three parts. Foundation courses include English, foreign language, computation, physical education, and fitness. Methods of Inquiry offers courses that acquaint students with scientific thought, historical and social/behavioral analysis, and philosophy. The Civilization section requires coursework in modern technology and Western and non-Western civilization at the junior/senior level. Even with these comprehensive requirements, there is still a great deal of flexibility; creative interdepartmental majors are often approved. New programs include a major in German, and minors in theater arts, women's studies, and archeology.

Hood's major strength lies in the sciences, especially the biology department, with its special emphases on molecular biology, marine biology, and environmental science and policy. A semester-long coastal studies program takes students along the East Coast on a biological educational mission. Education, especially early childhood, is a program of note, as are psychology, communications, English, and management. The graduate school of mostly commuting students is as large as the undergraduate program.

Students say the learning environment is more rigorous than competitive. "The professors here are amazing," raves one mathematics and computer science major. In general, Hood students are "interested in their education and are serious and hard-working," says a sophomore. Hood students praise the competence and accessibility of the faculty. "The teachers want their students to succeed and are very accessible when students need help," says a math major. Only labs are taught by graduate assistants, and 90 percent of freshmen classes have 25 or fewer students. A computer network links every dorm room and academic building to the campuswide information system, and students have 24-hour Internet access.

Hood's major strength lies in the sciences, especially the biology department, with its special emphases on molecular biology, marine biology, and environmental science and policy.

If you really want to stimulate the brain cells, the four-year honors program features team-taught courses and a sophomore-year seminar on the ethics of social and individual responsibility with student involvement in a community service project. One-tenth of students complete internships that include overseas jobs for language and business majors and legislative and cultural positions in Washington, D.C. With the outstanding resources of the Catherine Filene Shouse Career Center (including a national electronic listing for resumés), students

"I love living here. I'm having the time of my life."

have a leg up on their next step in life—81 percent of graduates go straight into jobs after graduation; 37 percent enroll in graduate or professional schools. A unique French/German major is available to Hood students at the college's center at the University of Strasbourg in France. Other study abroad destinations are Korea, Japan, South America, and South Africa.

The Hood student body is 74 percent Caucasian, and one student says that characterizing her classmates is difficult because "we all come from such different financial, cultural, ethnic, and personal backgrounds. The only thing I can say for sure is that we come here to learn." Twelve percent of the student body are African American, while Hispanics and Asian Americans each make up 2 percent. Five percent are from foreign countries. Students say major issues on campus include diversity, sexuality, and alcohol. Hood provides more than 200 merit scholarships worth an average of $14,337. And then, of course, there's that unbeatable deal for legacies.

Hood's residence halls are well liked, with good-sized, air-conditioned rooms. One of the dorms, Shriner Hall, has been renovated. The lottery system is based on seniority, and 53 percent of students live on campus. Freshmen can expect to be assigned to doubles (seniors and juniors can compete for singles), and language majors may choose to live in French-, Spanish-, or German-language houses.

Social life among the students is centered on the dorms, as each has its own personality as well as its own house council, rules, and social activities. Students report that there are parties every weekend, along with movies, dances, or other forms of entertainment. The Whitaker Campus Center, with its pool tables, snack bar, bookstore, and meeting rooms, offers a great gathering place for residents and commuters 24 hours a day. "If you don't like to stay on campus, there are restaurants, bars, clubs, malls, and coffeehouses within 10 minutes of the college by car," explains an English major. About an hour in the car brings students to the multiple diversions in Baltimore and Washington, D.C. Campus alcohol policies have been tightened and follow state law and the honor code, but drinking is generally not a big deal at Hood. "At parties and events, you have to show ID to get alcohol," one senior says. "However, in the dorms at other times, it is possible for underage students to get alcohol from those 'of age.'" Students also frequent scenic Frederick, which is described as small, safe, and beautiful.

> **"If you don't like to stay on campus, there are restaurants, bars, clubs, malls, and coffeehouses within 10 minutes of the college by car."**

With a 100-year history, Hood is rife with traditions. Some of the most important ones include Class Ring dinner and formal, a performance of Handel's Messiah, and Spring Parties, a weekend of carnival activities and dances. In sports, tennis, swimming, cross-country and basketball are tops. Students also brag that Hood has never lost a football game since 1893 (or, they might add, played one).

While Hood has undergone the major change from a women's college to a co-ed institution, its mission remains the same: to prepare students to face the challenges of a fast-changing society and professional environment. Century-old traditions are sure to remain, while new ones are sure to be made.

A semester-long coastal studies program takes students along the East Coast on a biological educational mission.

Overlaps

Salisbury University, University of Maryland–College Park, McDaniel College, Towson, Mount St. Mary's University

If You Apply To ➤ **Hood:** Early action: Oct. 1, Nov. 1. Regular admissions: Feb. 15. Financial aid: Feb. 15. Housing: May 1. Guarantees to meet demonstrated need of 50%. Campus and alumni interviews: recommended, informational. SATs or ACTs: required. SAT IIs: optional. Accepts the Common Application and electronic applications.

Hope College

P.O. Box 9000, Holland, MI 49422

Hope has an in-between size—bigger than most small colleges but smaller than a university. It's Evangelical in orientation, but less than a fourth of the students are members of the Reformed Church in America. In addition to the liberal arts, Hope offers education, engineering, and nursing, and makes undergraduate research a priority.

Website: www.hope.edu
Location: Small city
Total Enrollment: 3,141
Undergraduates: 3,141
Male/Female: 39/61
SAT Ranges: V 550–680
  M 560–680
ACT Range: 23–29
Financial Aid: 55%
Expense: Pr $
Phi Beta Kappa: Yes
Applicants: 2,674
Accepted: 77%
Enrolled: 37%
Grad in 6 Years: 75%
Returning Freshmen: 87%
Academics: ✐ ✐ ✐
Social: ☎ ☎ ☎
Q of L: ★ ★ ★
Admissions: (616) 395-7850
Email Address:
  admissions@hope.edu

Strongest Programs:
Biology
Chemistry
Dance
Education
English
Political Science
Psychology
Religion

Each fall since 1897, Hope College freshmen have spent three grueling hours engaged in "the Pull," an epic tug-of-war against the sophomores, who stand assembled on the opposite end of a 650-pound rope across the 15-foot-wide Black River. This well-known annual tradition evokes the daily struggle Hope students face: maintaining their faith in a world eager to challenge it at every turn. The heritage of Hope's Dutch founders remains strong and visible on campus, but you don't have to be a member of the Reformed Church in America to appreciate this conservative Christian college. "It is a very happy, energetic, caring place to be," a senior says.

The college, founded in 1866, is situated on six blocks near downtown Holland, the tulip capital of the nation (population 60,000), and a short bike ride from the shores of Lake Michigan. There's a lush pine grove in the center of campus, which features an eclectic array of buildings in architectural styles ranging from 19th-century Flemish to modern. The college has undergone a bit of a building boom, opening $70 million in new facilities in the last few years, including the Martha Miller Center for Global Communication, a new science center, an addition to Cook Residence Hall, and the DeVos Fieldhouse.

Among Hope's academic offerings, the sciences (especially biology and chemistry) stand out, with excellent laboratory facilities and faculty who are eager to involve students in their funded research. During the school year, undergraduates often conduct advanced experiments and even publish papers; come summer, more than 75 biology, chemistry, mathematics, computer science, and physics and engineering majors participate in research full-time. Not surprisingly, many science majors go on to medical and engineering schools and PhD programs. For those otherwise inclined, Hope's offerings in political science, psychology, dance, English, and business administration are solid, too. Hope's Department of Communication is one of the Speech Communication Association's two nationwide Programs of Excellence. And Hope is one of only 14 colleges and universities in the nation with accredited programs in art, dance, music, and theater. The First-Year seminar and GEMS (General Education in Math and Science) courses are two new fully implemented courses. The bachelor's-level nursing program, offered with Calvin College since 1982, is now an independent Hope-only degree.

"The courses are demanding and will challenge you," says a sophomore, "but they will also help you more than you will realize." Most Hope students select a major from one of the college's 39 fields, although the truly adventurous may design their own composite major. Hope's general education program, designed around the themes "knowing how" and "knowing about," includes a first-year seminar, which provides

"It is a very happy, energetic, caring place to be."

"an intellectual transition into Hope." Courses in expository writing, health dynamics, math and natural science, foreign language, religious studies, social sciences, the arts, and cultural heritage are also required; some must have a focus on cultural diversity. Students also complete a senior seminar, and a two-credit freshman seminar linked with academic advising. "The quality of teaching is off the charts," raves a student. "Professors are willing to go out of their way for any student."

Hope offers off-campus programs through the Great Lakes Colleges Association,* including semesters at other U.S. colleges and options combining classes and internships. Students may also study abroad in Austria, England, Greece, Japan, or Israel. The modern and classical language departments offer students proficient in a second language the chance to use their skills in volunteer work and research with faculty members, while the Visiting Writers Series gives students an opportunity to interact with noteworthy authors. A new interdisciplinary leadership program has recently been introduced.

Less than a fourth of Hope's students belong to the Reformed Church in America, and the student body is "outgoing, friendly, involved, and diligent," says a junior. Thrice-weekly chapel is voluntary, but administrators say it's typically filled to capacity. Sixty-seven percent of students hail from Michigan; African Americans, Hispanics, Asian Americans and Native Americans each make up 2 percent of the student body. Students say conservatism abounds and debates over issues such as homosexuality and politics can become heated. One sophomore says campus politics are generally split between the "party scene and chapel goers." Merit scholarships averaging $6,100 are available to qualified students but there are no athletic scholarships.

Seventy-seven percent of Hope students live in university-sponsored housing. Hope's housing options include on-campus apartments, small houses called cottages, and traditional dorms, arranged in freshman clusters or co-ed by suite. First-year students are assigned dorms and roommates; upperclassmen get first pick in the annual lottery. Students say dorms are comfortable and well maintained. Only seniors and married students may live off-campus. Students **"The quality of teaching is off the charts."** complain about the lack of parking, but there are few negative comments about campus security. On-campus students eat in one of two large dining halls where the fare—especially homemade bread and desserts—is tasty.

Most of Hope's social life takes place on campus, for those who stay around on weekends. The Social Activities Committee brings in comedians, bands, and hypnotists, shows movies in campus auditoriums, and plans the Spring Festival carnival, Winter Fantasia dance, and May Day celebration. "'Good clean fun' is their motto," says a student, "and that's exactly what it is." Seven fraternities and six sororities, all local organizations, claim 26 percent of the men and 28 percent of the women. While there are no echoes of *Animal House* on campus, some students do imbibe. Students caught drinking must perform community service, although plenty of students do that anyway, through activities like charity walks in Holland, a "very quaint and simple town" with "great downtown shopping" and a "simple, slow-paced lifestyle." Holland is also the site of spring's Tulip Time, one of the largest U.S. flower festivals. When Hope's cozy campus and the quaint town of Holland get too close for comfort, students find relief at the beaches of Lake Michigan or drive 30 minutes to Grand Rapids, which offers some large-city amenities and good weekend rental deals at the ski slopes. Chicago and Detroit are other typical destinations for those trying to hit the road.

On the field and on the court, Hope's Flying Dutchmen are fearless and talented Division III competitors. Hope teams won seven conference championships during the 2005–06 season, including men's golf, men's and women's soccer, volleyball, women's basketball, baseball, and softball. The college also won the Commissioner's Cup of the Michigan Intercollegiate Athletic Association for a record 28th time; the trophy recognizes the conference's best cumulative sports program for men and women. Especially important are any competition against Calvin (a century-old rivalry) and football versus Albion and Kalamazoo. Sixty percent of students take part in 17 intramural sports, which range from soccer and softball to innertube water polo.

Hope's housing options include on-campus apartments, small houses called cottages, and traditional dorms, arranged in freshman clusters or co-ed by suite.

Overlaps

University of Michigan, Michigan State, Grand Valley State, Western Michigan, Calvin, Albion

Hope's academic and athletic programs continue to grow and prosper, helped out by an array of new facilities. For those seeking an institution with traditional Christian roots and an emphasis on undergraduates, Hope may be worth a look.

Houghton College

Houghton, NY 14744

The mid-Atlantic's premier evangelical Christian college. Women outnumber men nearly two to one and enjoy perks such as a 386-acre horseback-riding facility. All students are required to take a Biblical literature class, and most go to chapel three times a week.

Website: www.houghton.edu
Location: Rural
Total Enrollment: 1,350
Undergraduates: 1,337
Male/Female: 40/60
SAT Ranges: V 530–660
  M 520–630
ACT Range: 24–28
Financial Aid: 76%
Expense: Pr $
Phi Beta Kappa: No
Applicants: 1,009
Accepted: 90%
Enrolled: 35%
Grad in 6 Years: 65%
Returning Freshmen: 84%
Academics: 🎓 🎓 🎓
Social: 🐧 🐧 🐧
Q of L: ★★★★
Admissions: (800) 777-2556
Email Address:
  admission@houghton.edu

Strongest Programs:
Music
Biology
Bible
English
Education

Located in the bucolic New York town that shares its name, Houghton College offers a solid, growing academic program and strong athletic teams, while remaining committed to its core mission as a Christian liberal arts school. Run by the Wesleyan Church of America, Houghton celebrates its Christian heritage and tries to ensure that students do the same. Applicants must explain in their essays how Jesus Christ became personal to them and why they want to go to a Christian college, and thrice-weekly chapel attendance is a must. These strict mandates help create true community on campus. "I feel loved, supported, and honestly blessed to be at such a special place with such a warm, loving community of friends and mentors," says a sophomore.

Houghton's scenic hilltop campus covers 1,300 acres of rural beauty, surrounded by vast expanses of western New York countryside. The academic buildings are a mix of area fieldstone and brick with ivy-covered walls. Students can go online with the laptops they get at matriculation; through Houghton's Educational Technology Initiative, the computer's price is included in the tuition bill. Current construction includes a third-floor addition to the library.

Students say Houghton's academic climate is rigorous, but not overwhelmingly so. "The climate helps me to stay on my toes and take my studies seriously," says a junior, "yet it is still laid-back enough to be fun." The school's most popular programs are education and biology, but music, psychology, and business also draw crowds. Unusual minors

"I feel loved, supported, and honestly blessed to be at such a special place."

such as equestrian studies—which takes advantage of Houghton's 386-acre riding facility—are available and the school continues to make new majors and courses available. Across departments, the faculty is lauded. "Our professors are one of our best features and a big reason why I am at this school," says a psychology major.

Houghton students must complete general education requirements known as Integrative Studies, designed to provide a context and framework for the entire educational program. Freshmen must take Biblical Literature, Principles of Writing, and a course titled FYI (First-Year Introduction), aimed at easing the transition to college. The First-Year Honors Program allows about 25 students to spend the spring semester of their freshman year in London studying under the supervision of two Houghton professors. According to the administration, all physics students are involved in ongoing research with their professors, providing a model that will be utilized in chemistry and biology as well.

More than 40 percent of Houghton students have taken advantage of off-campus study, with programs in Tanzania, Australia, London, and many locales in between. Students who want to get away within the U.S. can spend a semester at any Christian College Consortium* member school or participate in the American Studies program in Washington, D.C., sponsored by the Council for Christian Colleges and Universities.* The Oregon Extension program lets 30 Houghton students spend the fall studying in the Cascade Mountains while Houghton's extension campus in Buffalo offers internships and provides living quarters for students completing student teaching assignments. A 3–2 engineering program with Clarkson University (NY) is also available.

Sixty percent of those who enroll at Houghton are from New York. The minority community is tiny, accounting for only 6 percent of the student body. Hot-button issues tend to center on campus issues rather than those of global concern. For example, the increasing use of technology sparked a campus debate when the school's board blocked access to some Internet sites with sexually explicit materials or references. "There were students on both sides of the censorship issue, and eventually some of the restrictions were loosened," says a junior. Forty-four percent of undergrads receive merit scholarships worth an average of $3,798, and student athletes vie for 92 awards in a variety of sports.

"The climate helps me to stay on my toes and take my studies seriously."

Houghton's single-sex dorms, 16 townhouses, and 36 apartments house 90 percent of students and "aren't plush but are certainly comfortable," according to a sophomore. Students are required to live in the residence halls as freshmen and sophomores. There, in-room visitation is only allowed at weekly "open houses," but dorm lounges are open daily to members of the opposite sex. After their first two years, some students move off campus, but many opt for college-approved townhouses where regulations are self-imposed. This isn't exactly surprising, considering the other rules students voluntarily obey here, including abstention from tobacco, alcohol, drugs, and swearing; and optional Sunday church and Tuesday prayer meetings.

Houghton's boondocks village is truly small, lacking even a traffic light, says a junior. "The town of Houghton is pretty much just the college, a diner, two Laundromats, and a gas station," a junior adds. Students are involved in service projects such as Big Brothers Big Sisters and nursing home visitation, because the area surrounding the college is one of the poorest in New York State. Social life consists of on-campus movies, coffeehouses, concerts, and picnics. Because the student body is predominately female, dating can be a challenge. "You make your own fun at Houghton or you go off campus—usually Buffalo, Rochester, Olean, or Latchworth State Park," says a political science major. The college even has its own ski trails.

"We are distinctly evangelical Christians, and it influences the friendly atmosphere of the campus."

The town of Houghton is dry and college policy forbids alcohol. "Although there are isolated violations, the policies are typically very effective," says a chemistry major. Students eagerly anticipate annual celebrations for homecoming, Christian Life Emphasis Week, and the Christmastime Madrigal Banquets.

Soccer is the spectator sport of choice at Houghton, especially since there is no football team. The women's squad has captured four consecutive division titles and the men's team is solid, too. The women's basketball and volleyball teams also advanced to their respective national tournaments recently. The school's sports facilities have undergone extensive renovation and now include new tennis courts, an all-weather track, and lighted soccer and field hockey fields—nearly 50 percent of students participate in intramural sports.

Students don't come to Houghton for the surrounding town, which is 30 minutes by car from the nearest mall, or for the weather, which can be brutal once winter sets in. But they do come, and for good reason, says a sophomore: There's little to distract them from their studies, their campus's natural beauty, and their connection to God. "We are distinctly evangelical Christians," says a senior, "and it influences the friendly atmosphere of the campus."

If You Apply To ➤ **Houghton:** Rolling admissions. Does not guarantee to meet full demonstrated need. Campus interviews: recommended, informational. No alumni interviews. SATs or ACTs: required. SAT IIs: optional. Music majors apply directly to music program. Accepts electronic applications. Essay question: when and how Christ became personal; how you are cultivating spiritual growth; why Houghton; opinion of the college's policies on drugs, alcohol, and tobacco; how a liberal arts college will contribute to your goals.

Howard University

2400 Sixth Street NW, Washington, D.C. 20059

The flagship university of Black America and the first to integrate the Black experience into all areas of study. Strategically located in D.C., Howard depends on Congress for much of its funding. Preprofessional programs such as nursing, business, and architecture are among the most popular.

Website: www.howard.edu
Location: City center
Total Enrollment: 10,522
Undergraduates: 6,982
Male/Female: 36/64
SAT Ranges: V 440–680
  M 430–680
ACT Range: 20–29
Financial Aid: 43%
Expense: Pr $
Phi Beta Kappa: Yes
Applicants: 7,488
Accepted: 56%
Enrolled: 33%
Grad in 6 Years: 57%
Returning Freshmen: 85%
Academics: ✍ ✍
Social: ☎ ☎ ☎
Q of L: ★ ★ ★
Admissions: (202) 806-2763
Email Address:
  admission@howard.edu

Contrary to the advice of early black leaders such as Booker T. Washington, who argued in favor of technical training, Howard has promoted the liberal arts since its inception. This focus has served the school well; Howard's law school counts the late Supreme Court Justice Thurgood Marshall among its alumni, and Nobel Prize–winning author Toni Morrison went here, too. In recent years, Howard has strengthened its financial position and has begun implementing a strategic plan structured around "Leadership for America and the Global Community." The four-part plan focuses on strengthening academic programs and services, promoting excellence in teaching and research, increasing private support, and enhancing national and community service.

Founded in 1866 by General Oliver Howard primarily to educate freed slaves, the university now operates five campuses and serves nearly 11,000 students. The 89-acre main campus houses most classrooms, dorms, and administrative offices, as well as the university center, the Founders, and undergraduate and medical and dental libraries. The Howard Law Center, with a new library, is on the west campus near Rock Creek Park; the Divinity School is on a 22-acre site in northeast Washington; and there's also a 108-acre campus in suburban Beltsville, Maryland, and a **"Come to Howard ready to study."** campus in Silver Spring. Architecturally, the main campus is a blend of old and new, with numerous sculptures and murals created by Jacob Lawrence, Richard Hunt, Elizabeth Catlett, and the late Romare Bearden. The campus is an easy bus ride from the attractions of the nation's capital; and auditoriums, office spaces, classrooms, galleries, and computer labs across campus have undergone large-scale renovation in recent years; the physics, chemistry, and fine arts facilities have also been completely upgraded.

The school has excellent programs in business, computer sciences, and psychology. Other intriguing academic options are accelerated programs for a BS on the way to a medical or dental degree, coursework in the institute of jazz studies, programs in zoology and engineering (especially electrical engineering), and programs in communication science and disorders. The most popular major is biology, followed by political science, psychology, information sciences, and radio and television. Newer programs in ancient Mediterranean and international studies have become robust.

All students must complete general education requirements, which vary by school or college but uniformly encompass 18 credits in science, social sciences, humanities, computer literacy, math, languages, and one Afro American studies course. Freshman seminars and various other special programs for first-year students are available in the undergraduate schools, such as communication, engineering, and arts and sciences. Seniors in arts and sciences must weather a comprehensive exam to graduate.

In general, students say that the workload at Howard is demanding. "Some courses are more rigorous than others. But overall this school is tough," says a junior. Another student adds, "Come to Howard ready to study." Most students agree that professors are ready and willing to help

"It's a very competitive school, from grades to fashion."

when asked, though academic advising is not Howard's strength. "Sometimes you may get professors who do not know how to break down anything," explains a psychology major. "Then it is your job to talk up and ask questions. You must ask questions because a closed mouth does not get fed!" Students who need a break from the academic scene seek out internships in town or across the country. Many also study abroad at one of the more than 200 institutions in 36 countries where Howard grants credit.

Eighty-seven percent of Howard students are African American, and 11 percent hail from foreign countries. Most come from decidedly middle-class backgrounds. Although Howard seems to be a very cohesive community, career-minded and highly motivated men and women fit in best, students say, and most are politically liberal. "It's a very competitive school, from grades to fashion," says a junior. Hot issues include women's empowerment, student government, and fraternities and sororities. Fraternities and sororities do not have their own housing or dining facilities, and only 2 percent of men and 3 percent of women go Greek. Howard awards more than 100 athletic scholarships in a variety of sports, and more than 1,500 merit scholarships. A deferred-payment plan also allows families to pay each semester's tuition in three installments. But even with financial aid, costs are steep; President Swygert hopes that will change as he encourages more alumni to give back to their alma mater.

Interestingly, Howard is one of a handful of universities in the nation supported partly by federal subsidies; these days, the school gets about 55 percent of its budget from Congress. Bethune Hall, a $14 million housing complex, has helped ease the space crunch, but less than half of Howard's students are accommodated on campus. "It is quite difficult to get a room at times," says a sophomore economics major. "Housing at Howard is average in regards to availability, maintenance, and comfort," says one student. Freshmen get room assignments, while upperclassmen take their chances in a lottery. The halls are co-ed, and the 11 residential computer labs have more than 200 state-of-the-art machines for student use. Many students live off campus purely to avoid the mandatory meal plan. Still, the administration is doing its best to bring students back, and Drew, Meridian Hill, Baldwin, Carver, Truth, and Crandall halls have recently gotten facelifts.

Weekends bring an assortment of social happenings to campus, many of which take place in the student center. On-campus parties and sports events are always big draws, but the bars of Georgetown and Adams Morgan, the restaurants and clubs in

(Continued)
Strongest Programs:
Biology
Psychology
Business
History
Communications

Students who need a break from the academic scene seek out internships in town or across the country. Many also study abroad at one of the more than 200 institutions in 36 countries where Howard grants credit.

Athletics are an important presence on campus. The highlight of the season is always the grudge match with Hampton University.

the "New U" Street corridor, and the MCI Center arena (home to the NBA's Wizards and NHL's Capitals)—most accessible by public transit—also beckon. Though small in numbers, the Greeks are "an integral part of the university."

Athletics are also an important presence on campus, particularly varsity basketball, soccer, football, and volleyball, and the highlight of the season is always the grudge match with Hampton University to decide which school is the "true HU." The women's cross-country team captured the MEAC championship recently. Students list Howard's homecoming as one of the best annual events, along with various Greekfests, concerts, and talent shows that alumni, current students, and members of the community enjoy together.

Among America's historically black colleges and universities, Howard stands out as the standard-bearer, a longtime center of excellence and leadership. Its scholarship and collections of artworks, rare books, manuscripts, and photographs are a repository of the African American experience and offer students unique educational opportunities.

University of Illinois at Urbana-Champaign

901 West Illinois, Urbana, IL 61801

Half a step behind Michigan and neck-and-neck with Wisconsin among top Midwestern public universities. U of I's strengths include business, communications, engineering, architecture, and the natural sciences. Nearly 90 percent of the student body hails from in-state.

Website: www.uiuc.edu
Location: Small city
Total Enrollment: 41,938
Undergraduates: 30,909
Male/Female: 53/47
SAT Ranges: V 560–670
  M 620–730
ACT Range: 26–31
Financial Aid: 39%
Expense: Pub $ $ $ $
Phi Beta Kappa: Yes
Applicants: 18,987
Accepted: 75%
Enrolled: 53%
Grad in 6 Years: 83%
Returning Freshmen: 93%
Academics: ✑ ✑ ✑ ✑ ✑
Social: ☎ ☎ ☎
Q of L: ★ ★ ★

Like many of its Midwestern neighbors, the University of Illinois has its roots in agriculture. The Morrow Plots, the oldest experimental fields in the nation, still stand in the middle of campus—and when the wind blows the wrong way, students are not-so-subtly reminded of their heritage as a farm school. Like most big, public universities, U of I has a smorgasbord of choices, and with a strong Greek system and 1,200 clubs, social activities are more than plentiful. Homecoming weekend was invented at the University of Illinois, and whether cheering for the Illini, pledging one of 91 Greek houses, or celebrating Moms', Dads', or Siblings' Weekends, students here stir up

"Everyone seems laid-back until exam time—the entire campus feels anxious and tense."

a vibrant mix of school spirit and good times. This may look and feel like a laid-back Midwestern campus, but students work hard for the degrees they receive, especially in prestigious departments such as engineering and business administration.

Befitting the oldest land grant institution, the Illinois campus was built in farm country between the twin cities of Champaign and Urbana. The park-like campus was designed along a mile-long axis where trees and walkways separate stately white-columned Georgian structures made of brick. Physically challenged students tend to appreciate the campus because it is flat and well equipped with ramps and widened doorways. A new 250,000-square-foot computer science center and new physical education center have recently opened, as has an incubator facility, Enterprise Works, and the Technology Commercialization Laboratory, which provides faculty and students the opportunity to benefit from the commercialization of their research.

Illinois has eight undergraduate colleges and more than 150 undergraduate programs; if nothing strikes your fancy, you may design your own. Requirements include composition, quantitative reasoning, proficiency in a foreign language, and six hours each of cultural studies, natural sciences and technology, humanities and arts, and social and behavioral sciences. Engineering, business, education, and the sciences—especially agriculture and veterinary medicine—get high marks from students and lots of resources from administrators. "It's pretty competitive here," says a senior. "Everyone seems laid-back until exam time—the entire campus feels anxious and tense."

Partially because of its size, Illinois can afford to support excellent programs across the university, including the expansion of undergraduate minors campuswide. For a huge university, registration can be relatively painless, thanks to an online system allowing course selection from one's own computer. Nevertheless, freshmen and sophomores, who register last, may have trouble getting into certain general education classes, such as foreign languages. Professors and academic advisors can usually help if classes you need are full, and students appreciate their dedication. "The professors don't coddle the students. If you don't understand, you need to ask," explains a senior.

> "In Champaign, the ticket for underage drinking is around $300, so don't risk it!"

The impressive Illinois library system, the largest public university collection of its kind worldwide, makes it easier to keep up with class work. Aside from engineering and business, other notable programs at Illinois include the Beckman Institute for Advanced Science and Technology, an interdisciplinary center designed to bring biological and physical sciences together to pursue new insights in human and artificial intelligence. The National Center for Supercomputing Applications at Illinois developed Mosaic, the predecessor to Netscape's Navigator World Wide Web browser. The undergraduate honors program includes faculty mentoring, intensive seminars, advanced sections of regular courses, and access to special resources. More than a quarter of undergraduates travel and study abroad each year, roaming 45 countries around the globe, while the Ronald E. McNair Scholars Program helps fund independent, original research by minority, low-income, and first-generation college students who are completing bachelor's degrees.

Illinois has its share of stellar faculty, including 12 Nobel Laureates, 8 National Medal of Science winners, and 38 members of the National Academy of Sciences. Professors are terrific, says a student, providing "their time isn't completely consumed by research." Even freshmen stuck in large lectures (750 seats) will find some personal attention in the associated discussion sections, led by graduate teaching assistants. (Although many students note that due to budget cutbacks, even these classes are getting larger.) Freshmen Discovery Courses, seminars limited to 19 students, enable first-year students to interact closely with full professors. First-semester freshmen can ease into the rigors of college-level work in one of the Learning Communities.

Eighty-four percent of Illinois undergrads are homegrown. But since Illinois stretches from the wealthy north suburbs of sophisticated Chicago to the unspoiled rural hills bordering Kentucky and encompasses classic farm towns as well as factory towns, students do come from multiple backgrounds and fit less into the stereotypical "Midwest" mold than one might think. African Americans comprise 7 percent of the student body and Hispanics 6 percent. Asian Americans account for 13 percent, thanks in large part to the administration's efforts to attract high-achieving students via the President's Award. International students account for 5 percent of students. The Illini basketball team won its fourth Big Ten championships in six seasons, and

(Continued)
Admissions: (217) 333-0302
Email Address:
   ugradadmissions@uiuc.edu

Strongest Programs:
Accounting
Engineering
Agricultural Economics
Architecture
Business Administration/
Business Management
Psychology

The impressive Illinois library system, the largest public university collection of its kind worldwide, makes it easier to keep up with class work.

Many sophomores live in fraternity or sorority houses; Illinois claims to have the largest Greek system anywhere, with more than 91 chapters drawing 20 percent of men and 22 percent of women.

the men's tennis team is ranked as one of the first in the nation. The trustees recently ended a long-running controversy (and 81-year old tradition) of having the mascot Chief Illiniwek dance at athletic events, a practice that led the NCAA to ban Illini teams from playing host to postseason sporting events. The intramural program is extensive with available facilities that include 16 full-length basketball courts, five pools, 19 handball/racquetball courts, a skating rink, a baseball stadium, and the $5.1 million Atkins Tennis Center, with six indoor and eight outdoor courts. Illinois has a strong athletic program for students with disabilities, including wheelchair basketball, which Illinois invented.

Thirty-nine percent of students live in the U of I's 22 co-ed and single-sex residence halls, which range in size from 51 to 660 beds and are arranged in quadrangle-like groups. Some dorms are quite a hike from classrooms, veterans warn. Daniels Residence has been renovated. All bedrooms have fast Internet connections, and many residence halls house living/learning programs, such as WIMSE (Women in Math, Science, and Engineering) and Unit One (academic support and educationally focused programming). Each residence hall is a mini-neighborhood, with dining halls, darkrooms, libraries, music practice rooms, computers, and lounges creating a sense of community. But students warn that housing is getting crowded. The food is good when the chefs keep things interesting, and campus security maintains a visible presence.

Many sophomores live in fraternity or sorority houses; Illinois claims to have the largest Greek system anywhere, with more than 91 chapters drawing 20 percent of men and 22 percent of women. Illinois attracts many socially oriented students who love parties and intramural sports, which may be why the Greek influence is particularly strong. Independents don't have to suffer boredom, though, as there are also more than 1,200 registered student clubs and organizations ranging from the rugby team to ethnic advocacy groups. On most weekends, the Illini Union showcases bands, comedians, and hypnotists in its central cafe. The impressive Krannert Center for the Performing Arts, with four theaters and more than 350 annual performances, serves as the area's cultural center, while Assembly Hall hosts national touring acts, including popular musicians such as Pearl Jam, James Styx, Kenny Chesney, and James Taylor. Students get a discount at both facilities. Chicago and the shores of Lake Michigan beckon when the weather warms up, and Mardi Gras makes for a good road trip in the dead of winter. Though drinking is prohibited in the dorms, the campus policies regarding alcohol are a "token gesture," a business major says. "You only need to be 19 to get into bars, but 21 to drink. In Champaign, the ticket for underage drinking is around $300, so don't risk it!" For those who itch for the stimulation of a big city, the campus is just about equidistant from Chicago, Indianapolis, and St. Louis.

"We have a great reputation and it only grows stronger and stronger."

In 2006, the Illini basketball team won its fourth Big Ten championship in six seasons, and the men's tennis team is ranked as one of the first in the nation. The intramural program is extensive with available facilities that include 16 full-length basketball courts, five pools, 19 handball/racquetball courts, a skating rink, a baseball stadium, and the $5.1 million Atkins Tennis Center, with six indoor and eight outdoor courts. Illinois has a strong athletic program for students with disabilities, including wheelchair basketball, which Illinois invented.

While the University of Illinois may seem mammoth to some students, don't be scared off by this giant institution. Academic and social opportunities are incredibly diverse and classroom sizes, while growing, are supplemented by smaller group discussions. The breadth of the programs offered, combined with an active campus life makes for a well-rounded college experience, students say. "We have a great reputation and it only grows stronger and stronger."

Illinois Institute of Technology

10 West 33rd Street, Chicago, IL 60616

Forget about cheerleaders, homecoming games, and other trappings of college life. IIT is about learning technology, getting a degree, and landing a job. IIT is all engineering with a little bit of architecture thrown in for good measure. If your goal is a technical job in the Chicago area, this is your place.

Engineers unite at the Illinois Institute of Technology, where classwork and real-world experience promise to propel students to the top of their fields. After all, when you're taught by Nobel laureates, engaged in comprehensive undergraduate research, and able to take advantage of state-of-the-art labs, you're nearly guaranteed a high-paying job after graduation. Students here burn the midnight oil, but frequently escape to downtown Chicago for much-deserved fun and culture.

IIT's home is an urban, 120-acre campus designed by Ludwig Mies van der Rohe, the influential 20th-century architect who directed the architecture school for 20 years. Founded in 1890, the school is just three miles south of Chicago's Loop, one mile west of Lake Michigan. Miesian-style buildings are adorned by trees and grassy open parks. Comiskey Park, home of the White Sox, is located directly across from the campus. The S.R. Crown Hall, home of IIT's College of Architecture, is considered a landmark. In fact, major campus renewal is currently in progress, with a technology business center in the works.

Engineering sets the tone at IIT. All engineering departments are outstanding. Architecture is the most popular major, followed by computer science, biomedical engineering, aerospace engineering, and computer engineering. The sciences, physics in particular, are first-rate; high-energy physicist and Nobel laureate Leon Lederman teaches freshman—yes, freshman—physics. Computer literacy is demanded of all students. In addition, all freshmen take an introduction to the professions seminar, which includes discussion of innovation, ethics, teamwork, communication, and leadership.

Multidisciplinary, group-based learning is big at IIT. Every student must complete two semester-long interprofession projects that sharpen real-world skills. The architecture curriculum emphasizes a team approach that mixes third- through fifth-year students under the supervision of a master professor. Guided by an academic reorganization, the physical sciences have been bolstered, grouped together with career-oriented fields such as psychology, political science, and computer information systems. Newer academic options include majors in business administration, humanities, and dual admissions programs in pharmacy, optometry, and osteopathic medicine.

> **"You always have a lot of work to do, especially with homework and labs."**

Along with humanities and social science courses, students must fulfill general education requirements that include mathematics, computer science, natural science, and engineering; writing is emphasized across the curriculum. IIT's academic

Website: www.iit.edu
Location: Urban
Total Enrollment: 6,469
Undergraduates: 2,216
Male/Female: 74/26
SAT Ranges: V 560–660
  M 620–720
ACT Range: 25–30
Financial Aid: 58%
Expense: Pr $
Phi Beta Kappa: No
Applicants: 2,514
Accepted: 63%
Enrolled: 26%
Grad in 6 Years: 67%
Returning Freshmen: 81%
Academics: ✎ ✎ ✎ ½
Social: ☎ ☎
Q of L: ★ ★
Admissions: (312) 567-3025
Email Address:
  admission@iit.edu

Strongest Programs:
Electrical Engineering
Chemical Engineering
Mechanical Engineering
Aerospace Engineering
Architecture

climate is pretty unforgiving, students say. Both the workload and the competition are fierce. "You always have a lot of work to do, especially with homework and labs," says an electrical engineering major. Adds another student: "The courses usually require three hours of time outside class for every lecture hour." Professors always teach their own classes at IIT, while TAs are available for labs and extra help. Most students praise the faculty for their knowledge and tendency to offer as much help as is needed. More than two-thirds of the classes have 25 or fewer students, and 96 percent have 50 or fewer.

In addition to meeting outside of class to go over problem sets or for career direction, IIT students and professors often work side-by-side on research projects. Engineering students have the use of sophisticated labs, and independent research labs in Chicago are also available. The five-year co-op program—another possibility for hands-on experience—helps lead IIT grads into high-paying jobs after graduation. There are study abroad programs that include France, Spain, Scotland, and Germany.

"The students here are ambitious, dedicated, and overachieving," reports a business administration major. "Most students here can be considered nerds." Sixty-nine percent of IIT students graduated in the top quarter of their high school class. Out-of-state students account for 41 percent of the undergraduate population, and 15 percent are from foreign countries. African Americans and Hispanics together constitute 12 percent of the student body, and Asian American students constitute another 14 percent. Students say International Fest is one of the year's most popular events, and IIT offers a multitude of cultural awareness workshops and sponsors "awareness" weeks and months on

> **"The students here are ambitious, dedicated, and overachieving."**

different topics to help avert potential problems. A three-day workshop on topics including race relations, international diversity, homophobia, and sexism is required of all new students. Politics and political correctness don't really stir up campus because "we are so conservative and diverse," says an aerospace engineering major. IIT offers merit scholarships (worth an average of $10,259) and athletic scholarships in a number of sports. IIT's ROTC program has grown and matured into one of the finest in the nation and even hosts a popular annual formal ball.

As befits the school's urban location, a large chunk of students commute. The 54 percent of students who live in residence halls report that rooms are "kind of small," but comfortable. Six of the seven dorms are co-ed, with one hall for women only. The McCormick Student Village is popular, and South and North are said to be the nicest dorms. Fowler has the biggest rooms, but no air-conditioning; the rest of the dorms have A/C. Some students live in apartments in the area or on Chicago's North Side; others inhabit one of the eight fraternities, which claim 17 percent of the men. Sororities attract 17 percent of the women, and many students say the social aspect of Greek life is a welcome addition to campus. The dining hall has several meal plans and a special vegetarian menu. Breakfast and lunch can also be eaten in the cafeteria at the student union, while the campus pub serves lunch and dinner. Engineers and architects—notorious late-night studiers—have to hit the library early, since it closes at 10 p.m. Though students tend to feel safe on campus, the surrounding area is a different story. "It is difficult to get to a doctor, pharmacy, or grocery store and feel safe unless you have a car," says a student.

IIT's six-block campus is contiguous to Chicago's "Gap" community, where historic but rundown homes are being rehabilitated to form one of the city's hottest new urban residential areas. Most students love exploring Chicago; the city skyline is beautiful and a veritable museum, with buildings designed by the likes of Frank Lloyd Wright, Louis Sullivan, and, of course, van der Rohe. "Chicago is awesome—there is always something to do," a sophomore says, but adds that it can be a bit pricey. Thus, the university provides free shuttle bus service to downtown on weekends. Lake Michigan is within jogging distance, and Chinatown is a walk away for lunch or dinner.

Students who stick around campus on weekends must work hard to find social events. "Social life on campus is the worst part of the school," says a sophomore. The Union Board offers movies, concerts, and comedians, and the Bog brings in bands on Thursdays and Saturdays. Students can also plan events like a formal on the Odyssey, a sightseeing boat, or an outing to the Chicago Symphony. The eight-day Winter Festival and the Spring Formal are other popular annual events. Returning sophomores are invited to a weekend retreat in Lake Geneva, Wisconsin, for some bonding time and to celebrate making it through their first year. As for alcohol, the school follows the 21-year-old law and students say it generally works.

> "Chicago is awesome—there is always something to do."

In sports-crazy Chicago, IIT athletic teams are not much of a draw. Students praise the men's baseball and swimming teams along with women's volleyball, which compete in the NAIA Division I. Men's and women's soccer teams were recently started. The intramural program is strong, but students hate the fact that the facilities close at 5 p.m. on weekends. The Olympics occur every year at IIT when Greek Week and Sports Fest kick off, featuring Olympic-type competition for all students.

Shipping off to Chi-town to take on the mammoth workload at IIT means hitting the books for hours upon hours and a fair share of all-nighters. But the payoff is undeniable. One student says bluntly, "This school is for people who want to make a lot of money after college." Indeed, students who take advantage of this small school's ever-improving engineering departments are likely to have their pick of careers after graduation. And with the innumerable diversions offered in the Windy City, students at IIT revel in the best of two worlds: a challenging academic climate and a great city in which to let off all that steam.

Overlaps

University of Illinois–Chicago, University of Illinois–Urbana-Champaign, Loyola University (IL), DePaul, Purdue, Northwestern

If You Apply To ➤

IIT: Rolling admissions. Housing: Jun. 1. Guarantees to meet full demonstrated need. Campus interviews: recommended, informational. No alumni interviews. SATs or ACTs: required. SAT IIs: optional. Prefers electronic applications. Essay question: Influential person; significant personal experience; prominent figure you would interview; most important modern invention.

Illinois Wesleyan University

Bloomington, IL 61702-2900

IWU is an up-and-coming small college with a new emphasis on creativity and the spirit of inquiry. The curriculum is basic liberal arts with additional divisions devoted to fine arts and nursing. An optional three-week term in May allows students to travel or explore an interest.

Illinois Wesleyan University has its sights set on a special breed of student—the kind who isn't afraid to be many things at once. Its new mission statement highlights the school's goal of becoming the ideal liberal arts institution by providing unique opportunities, fostering creativity, and preparing students for a life in a global society. Students here are very much encouraged to pursue multiple passions and IWU is a mecca for students who have preprofessional interests, especially those with unusual pairings like management and music.

Website: www.iwu.edu
Location: Small town
Total Enrollment: 2,140
Undergraduates: 2,140
Male/Female: 43/57

(Continued)

SAT Ranges: V 600–690
  M 590–690
ACT Range: 26–31
Financial Aid: 54%
Expense: Pr $ $
Phi Beta Kappa: Yes
Applicants: 2,990
Accepted: 57%
Enrolled: 36%
Grad in 6 Years: 82%
Returning Freshmen: 92%
Academics: 🖉 🖉 🖉 ½
Social: ☎ ☎ ☎
Q of L: ★ ★ ★ ★
Admissions: (800) 332-2498
Email Address:
  iwuadmit@iwu.edu

Strongest Programs:
Accounting
Biology
Computer Science
International Studies
Political Science
Philosophy
Economics
Psychology

The business administration department offers a Portfolio Management course, in which students buy and sell orders overseen by a Client Board composed of University Trustees.

Founded in 1850, IWU occupies an 80-acre campus site in a north side residential district of Bloomington. The heart of the campus is the central quadrangle, and tree-lined walkways connect buildings that range in style from gray stone Gothic to ultra-modern steel and glass. The College of Fine Arts houses the three separate schools of music, art, and drama; music is the standout, having turned out such talents as opera star Dawn Upshaw. Among the top-notch programs in the College of Liberal Arts are biology, English, chemistry, and math. In addition to the usual fall and spring semesters, IWU has an optional three-week May term. The courses during this term must have one of five features: curricular experimentation, nontraditional approaches to traditional subject matter, student/faculty collaboration, crossing of disciplinary boundaries, or experiential learning through travel, service, or internships. Over half the students take a May term class, and about one-fourth take off-campus travel courses. The university's study abroad program offers students the opportunity to travel to countries such as England, Denmark, and Japan. The business administration department offers a Portfolio Management course, in which students buy and sell orders overseen by a Client Board composed of University Trustees. IWU hosts an annual student research conference that attracts people from all disciplines.

> "The academic climate is very competitive and it really forces you to raise the bar and work harder."

Illinois Wesleyan's general education requirements emphasize critical thinking, imagination, intellectual independence, social awareness, and sensitivity to others. All first-year students must take a Gateway Colloquium, a topic-based, seminar-style class of 15 that stresses critical reading, writing, discussion, and analytical skills and introduces students to the intellectual life of the university. Some of the topics include Jesus at the Movies, Are We What We Eat?, The Mommy Wars, and Jewish Humor. Students have differing views regarding which programs are weaker, but most agree that specific professors are avoided rather than departments. Newer programs include majors in environmental studies, Greek, and Roman studies. Students do jockey for high grades, especially since the institution of a plus/minus grading system. "The academic climate is very competitive and it really forces you to raise the bar and work harder," says one student. Professors are lauded for their knowledge and accessibility. "They are brilliant scholars that actually care about teaching," says a junior.

Students at IWU are mostly the homegrown variety, with 85 percent hailing from Illinois. "Students at this college are very driven and motivated by some passion," a senior says, "whether that is academic, athletic, or service." Although IWU began admitting African American students in 1867, the campus is still predominantly white. African Americans account for only 4 percent of the student body, Hispanics 3 percent, and Asian Americans 3 percent. A multicultural task force has been formed to address the issue of diversity. The university has placed great emphasis on educating students about sexual harassment. Part of freshman orientation is spent role-playing harassment situations. Active participation in groups like Circle K, the Alpha Phi Omega service fraternity, and Habitat for Humanity provides evidence for the social consciousness of the IWU campus. Merit scholarships worth an average of $8,691 are available to qualified students; there are no athletic scholarships.

> "Our dorms rock. All are modern with good amenities available for everyone."

Housing is guaranteed for four years, and 75 percent of the students live in the dorms, which receive stellar marks from residents. "Our dorms rock," says a student. "All are modern with good amenities available for everyone." "Room lotteries can get hairy," warns another student. "But you always get a room." Students must be 21 to live off campus. The food gets high marks as well. "The food is generally good but the menu eventually gets old," says a psychology major. Most students say campus security is good, though common sense must be exercised.

Thirty-three percent of the men and 25 percent of the women go Greek because fraternities and sororities are the focus of IWU's social life. Non-Greeks also use the system for social life, but if partying isn't your thing, there are plenty of other options. "Our student center always has free stuff going on like comedians, hypnotists, movies, and bands every Friday and Saturday night," says one senior. The alcohol policy allows drinking on campus for those over the age of 21, and students admit that the policies don't always stop underage drinkers. Each fall during homecoming, the fraternities and residence halls compete in the Titan Games to get appropri-

> "Bloomington has the highest restaurant-per-capita ratio of any community."

ately psyched. Other annual festivities include the Far Left Carnival, the Gospel Festival, and Earthapalooza (on Earth Day). The Student Senate also sponsors guest speakers; Spike Lee, Bonnie Blair, and Maya Angelou have addressed audiences in recent years.

Thanks to the proximity of Illinois State University in nearby Normal, IWU offers more than the typical small college town atmosphere. The total area school population of about 25,000 helps to offer students at tiny IWU "the best of both worlds," says a senior. An accounting major says, "Bloomington has the highest restaurant-per-capita ratio of any community. The students love this." The best road trips are to Peoria or Urbana-Champaign (home of the University of Illinois), or to Chicago or St. Louis, each two-and-a-half hours away.

In the IWU arena, baseball and football are well and good, but basketball really gets students going; the men's team reached the Final Four in Division III four times within the past decade, winning the championship once and placing third three times. In addition, the Titans have won CCIW conference championships in softball, baseball, women's basketball, football, and men's and women's golf. The Fort Natatorium houses a whopping 14-lane swimming pool, and the swim team had its share of stars along with the track team. Intramural sports include volleyball, badminton, and co-ed inner-tube water polo. More than half of IWU men and 25 percent of IWU women participate.

One of the Midwest's better-kept secrets, Illinois Wesleyan is at once cozy and diverse, loaded with opportunities for ambitious students with traditional or offbeat interests. As one junior advises, "The school provides a multitude of paths down which one can travel and it is up to the student to decide which path to take. IWU allows you to become who you want to be, but only if you let it."

All first-year students must take a Gateway Colloquium, a topic-based, seminar-style class of 15 that stresses critical reading, writing, discussion, and analytical skills and introduces students to the intellectual life of the university.

Overlaps

University of Illinois, Augustana, Northwestern, Washington University (MO), Loyola, Bradley

If You Apply To ➤ | **Illinois Wesleyan:** Rolling admissions: Dec. 15. Financial aid: Mar. 1. Campus interviews: recommended, evaluative. No alumni interviews. SATs or ACTs: required. SAT IIs: optional. Accepts the Common Application and prefers electronic applications. Essay question: significant quote; what diversity means to you; what you would do with a year of service; or topic of your choice.

Indiana University

300 North Jordan Avenue, Bloomington, IN 47405

Though men's basketball is IU's most famous program, it may not be its best. That distinction could easily go to the world-renowned music school or to the distinguished foreign language program. IU enrolls three times as many out-of-staters as University of Illinois.

Website: www.indiana.edu
Location: Small city
Total Enrollment: 38,263
Undergraduates: 29,299
Male/Female: 48/52
SAT Ranges: V 490–610
   M 510-630
ACT Range: 22–27
Financial Aid: 51%
Expense: Pub $ $ $
Phi Beta Kappa: Yes
Applicants: 24,169
Accepted: 80%
Enrolled: 38%
Grad in 6 Years: 71%
Returning Freshmen: 88%
Academics: 🖎 🖎 🖎 🖎
Social: ☎ ☎ ☎ ☎
Q of L: ★ ★ ★ ★
Admissions: (812) 855-0661
Email Address:
   iuadmit@indiana.edu

Strongest Programs:
Accounting
Business
Chemistry
Journalism/Communications
Languages
Music
Optometry

With more than 38,000 students on its enormous campus, Indiana University is the prototype of the large Midwestern school. With strong academics, a thriving social scene, and some of the best sports teams around, this top-notch public institution is a testament to Hoosier determination.

Located in southern Indiana's gently rolling hills, the 1,900-acre campus boasts architecture from Italianate brick to collegiate Gothic limestone to the distinctive style of world-famous architect IM Pei. Other unique campus features include fountains, gargoyles, an arboretum of more than 450 trees and shrubs surrounding two reflecting pools, a limestone gazebo, and the Jordan River, a pretty creek that runs alongside a shaded path. Recent campus additions include the 117,000-square-foot Theatre/Neal-Marshall Education Center, a new parking garage, and an addition to the Psychology Building.

IU's 18 schools and colleges offer many majors and minors, crossdisciplinary study, an individually designed curriculum, intense honors and research programs, and yearlong study in 37 countries (and 17 languages). The highly touted business school, with its respected international studies component, is second only to arts and sciences in popularity. The internationally known Kinsey Institute for the Study of Human Sexual Behavior is housed on IU's campus, and the music school is tops in its field, setting the, ahem, tone for much of the campus. Students don't complain about many departmental weaknesses but note that large introductory lectures, especially in the sciences, are a hazard of IU's size. Indiana prides itself on its liberal arts education—freshmen are admitted not to preprofessional schools but to the "university division." Majors are declared after one or two years, and the university discourages premature specialization. IU's communications and culture department advances the study of communication as a cultural practice, while the Environmental Science Joint Program is an undergraduate degree program that specifically considers the environment as a scientific entity.

> "With 4,000 different courses per semester, a variety of intensity levels exist."

General education requirements vary from school to school but usually include math, science, arts and humanities, social and behavioral sciences, English and writing, culture, and a foreign language. Students describe the academic climate as rigorous but not cutthroat. "With 4,000 different courses per semester, a variety of intensity levels exist," says a marketing major. "There is a balance with room for both competitive overachievers and laid-back, carefree individuals." Students say they regularly share ideas with each other, and group projects are commonplace. Faculty members bring their research results directly to students, and some profs bring undergrads into their labs to assist with ongoing projects. Students say the quality of teaching is excellent. "The professors here are remarkable," says an art history/telecommunications major. As for advising, many students seem surprised by the personal attention they receive at such a large university, and they soon learn that many available resources are helpful to those students who seek them out. Some students, though, complain of confusing bureaucracies and parking problems.

Sixty-six percent of IU students are from in-state, while the remainder hail from every state and more than 100 foreign countries. Out-of-staters face much more rigorous minimum admissions standards, including rank in the top quarter of their high school class and SAT scores in the 1050 to 1100 range. African Americans comprise 4 percent of the student body, Hispanics 2 percent, and Asian Americans 4 percent. By and large, students do not seem to dwell on political or social issues. The school's rolling admissions system enables students to know their fate only a month after their application is filed. And while IU does not guarantee to meet the full demonstrated need of every student, it admits on a need-blind basis

> "Bloomington is a great small college town."

The school's rolling admissions system enables students to know their fate only a month after their application is filed.

and offers the Early Approximate Student Eligibility (EASE) program to help prospective freshmen gauge how much financial aid they will get. Merit scholarships are awarded to qualified students; applicants must be in the top 10 percent of their graduating class and have a minimum combined SAT score of 1200. Athletic scholarships are available for qualified jocks.

Housing ranges from Gothic quads (co-ed by building) to 13-floor high-rises (co-ed by floor or unit, except for one all-women dorm), and halls are considered "clean and comfortable." One student explains the housing situation this way: "All dorms have laundry facilities, cafeterias, computer clusters, and undergraduate advisors, and some even have special amenities like language-speaking floors." A junior adds, "There is no trouble getting a room, but preference of dorm may be harder." Academic floors (requiring a GPA of 3.1 or better) are popular with more serious students who are not interested in intense nightlife. Housing is guaranteed to all incoming freshmen, and those who stay in the university housing system won't ever face rent increases. Based on results from a student survey, some dining halls have been modernized to resemble mall food courts with outlets offering international and healthful menus sprinkled among the fast-food options. Alcohol is prohibited in the dorms, which may help explain why 65 percent of the student body lives off campus. Most off-campus residents choose apartments or small houses with big front porches within walking distance of the campus or of the IU bus system.

Although campus organizations host numerous events, the most active on-campus groups, in terms of social life, seem to be the Greeks. About 16 percent of IU men and 18 percent of IU women are in the Greek system, and membership is a status symbol. Some complain of a polarized atmosphere. "There is a large separation between the Greek community and the rest of the social body," says a senior. Every fall there is a 36-hour Dance Marathon to raise money for Riley's Children's Hospital in Indianapolis. The Little 500 bike race, which was modeled after the Indianapolis 500, is one of the most highly attended events of the year at Indiana. With concerts, ballets, recitals, and festivals right on campus, students are not lacking for things to keep them busy. The IU student union is the largest in the nation, and the range of extracurricular organizations is also impressive. The Office of Diversity Programs, Committee on Multicultural Understanding, and Students Organized Against Racism are a few more ways students can make a difference on campus.

> **"All dorms have laundry facilities, cafeterias, computer clusters, and undergraduate advisors."**

"Bloomington is a great small college town," says one senior. There are many excellent bars, shops, and restaurants, including one of the few Tibetan restaurants in the country. Locally, the area offers some impressive rock quarries (often used as illegal but refreshing swimming pools), miles of public forests, and three nearby lakes. Spelunkers will find heaven down below in the many nearby caves. Chicago, Cincinnati, Indianapolis, St. Louis, and even New Orleans are popular road trips.

Intramurals pale in comparison with varsity athletics here; basketball is an established religion in the state of Indiana. Although students and faculty are all eligible for tickets, they've got to get requests in early—and even those lucky enough to get tickets don't count on going to more than a quarter of home games. The Indiana Hoosiers soccer team won the 2003 and 2004 NCAA championships. In recent years, women's golf and women's tennis have both claimed at least a share of the Big Ten championship, and even the football team is beginning to draw red and white crowds. Recently, women's water polo has attained varsity status. Purdue is IU's traditional athletic rival, and teams play for the Old Oaken Bucket, found on a farm in southern Indiana in 1925 and alleged to have been used during the Civil War.

IU is a huge university, but it seems as if IU has it all. Students graduate proud of their school, with a degree that is highly respected in Indiana and outside the state.

IU's communications and culture department advances the study of communication as a cultural practice, while the Environmental Science Joint Program is an undergraduate degree program that specifically considers the environment as a scientific entity.

Overlaps

Purdue, Ball State, University of Illinois, University of Iowa, Miami University (OH)

International Colleges and Universities

Do you thrive on new experiences? Like to meet new people? Want to learn about different cultures? You can do all that at a college or university in the United States, but if you really want to jump in with both feet, think about attending a school in a foreign country. This section highlights the opportunities available in Canada and Great Britain, by far the most common destinations outside the U.S. for degree-seeking undergraduates.

The absence of a language barrier is the most obvious reason why Canada and Britain are the preferred destinations for study abroad. Plenty of students do a junior-year-abroad where the language is Spanish or Swahili, but only a handful can realistically expect to earn an entire degree in a foreign tongue. A smattering of American universities do exist in places ranging from Paris to Cairo, but most are small and the majority of their enrollment is students from other countries seeking an American-style education. If you're willing to venture halfway around the world, Australia is an English-speaking destination that might be worth a look for its combination of beautiful scenery and bargain-basement tuition.

Look for coverage of Australian institutions in a future edition of the *Fiske Guide*. The following sections examine Canada and Britain in more detail, followed by full-length articles on selected institutions.

Canadian Colleges and Universities

Horace Greeley told ambitious young men of his generation to "go West." Today his admonition to young men and women seeking a quality college education at a fraction of the usual cost would probably be to "go North"—to Canada. A growing number of American students are discovering the educational riches that lie just above their northern border in this huge land of 30,000,000 people that is known for its rugged mountains, bicultural politics, spirited ice hockey, and cold ale. What's drawing them is easy to discern.

The top Canadian universities are the academic equals of most flagship public universities and many leading privates in the United States, but the expense of a bachelor's degree is far lower, even taking travel into account. Canadian campuses and the cities in which they are located are safe places and, unless one opts for a French course of study, there are no language and few cultural barriers. Canadian schools are strong on international exchange programs, and their degrees carry weight with U.S. graduate schools.

Canada has 90 institutions of higher learning, ranging from internationally recognized research universities to the small undergraduate teaching institutions in the country's more rural areas; the country ranks second after the U.S. in the percentage of citizens attending university. Most of the larger universities are located in highly urban centers, but some are situated in smaller towns where they dominate the life of the community. Most are almost literally next door to the United States, within 100 miles of the Canada–U.S. border. In this guide, we feature four of Canada's strongest universities: the University of British Columbia, McGill University, Queen's University, and the University of Toronto.

Institutions of higher learning in Canada were established from the earliest days of French settlement in the mid-17th century, making them some of the oldest in North America. The precursors to the public universities in Canada were the small, elite, denominational colleges that sprang up in Quebec, in the Maritimes, and later in Ontario. A few private denominational colleges and universities still exist in Canada, but most have been subsumed into affiliations or associations with the larger universities. Education in Canada, including university education, became the exclusive jurisdiction of provincial governments. As the Canadian West was developed, the large Western provinces of British Columbia, Alberta, Manitoba, and Saskatchewan set up provincially chartered universities similar to land grant colleges in the U.S.

One of the key differences between Canadian and U.S. universities is that Canadian universities (and this is what they are, not "colleges") are primarily funded from public monies. Despite steady tuition increases in the past five years, the average Canadian student still only pays on average about $2,500 in Canadian dollars, or U.S. $1,700. Although non-Canadians may be charged up to six times the domestic rate, most costs are still lower than out-of-state tuition in the U.S. Tuition at the four universities described below ranges from U.S. $5,500 to $9,200.

Canadians have come to expect easy and affordable access to a uniformly high quality of education whether they live in Halifax or Vancouver. After diminishing government funding in the past several years, a now booming economy, a large government surplus, new federal initiatives, grants for innovations and scholarships, and the universities' own aggressive fund-raising campaigns bode well for the continued growth and quality of Canadian higher education in the immediate future.

Federal and provincial loans and grants that are readily available to Canadian students are generally not available to students from the U.S. and other countries. However, the majority of universities with competitive admissions, particularly those featured in the *Fiske Guide*, offer merit-based awards and scholarships to students of all nationalities. American students who attend leading Canadian schools can apply their U.S. student assistance funds, including Stafford Loans and Pell Grants, as well as the recently implemented HOPE Scholarship and Lifetime Learning tax credits.

The requirements for obtaining a degree are set by each institution, as are the admission requirements and prerequisites. Unlike the U.S., Canada does not offer nor require its own students to take a Canadian college entrance test. Some Canadian universities admitting students from the United States will require SAT or ACT scores along with high school marks from academic subjects in the last two or three years of high school. In general, top universities are about as selective as their American counterparts.

Application fees vary by institution, as do deadlines. Canadian universities are aware of the May 1 deadline operative in the U.S. and they try to accommodate. Applications to the University of Toronto and Queen's University in Ontario are handled centrally through the Ontario Universities' Application Service. McGill handles applications directly and accepts both Web-based and paper applications. British Columbia has its own application; it can be mailed, but students are encouraged to apply online. Canadian universities differ widely in the amount of credit and/or advanced standing they offer for Advanced Placement Examinations or International Baccalaureate Higher Level Examinations.

The following admission requirements apply to applicants from an American school system. The University of British Columbia bases admission decisions on the average on eight full-year academic courses over the last two years of high school, and there are also specific program requirements for students entering the science-based faculties. SAT test results are not required, but if students submit them, the results can be helpful in the evaluation process. McGill bases its assessment of American high school graduates on the overall record of marks in academic subjects during the final three years of high school, class standing, and results obtained in SAT I and SAT II and/or ACT tests. Queen's wants applicants with a minimum score of 1200 on SAT I (with at least 580 in the verbal section and 520 in the mathematical one) and looks at class rank. There are also program-specific requirements for programs where mathematics and/or biology, chemistry, and physics are a requirement. Toronto's Arts and Science faculties want a high grade point average and good scores on the SAT I and on three SAT II subject tests. ACT and CEEB Advanced Placement Examination scores are also considered.

It is hard to beat Canadian universities for the quality of student life. Although many students commute, most of the universities in Canada offer on-campus housing; some even guarantee campus housing for first-year students. Universities offer active intramural and intercollegiate sports programs for both men and women, and the usual student clubs, newspapers, and radio stations provide students with opportunities to get involved and develop friendships. As in the United States, student-run organizations are active participants in university life, with leaders serving on university committees and lobbying on issues ranging from creating more bicycle paths to keeping tuition low. Few Canadian campuses are troubled by issues of student safety or rowdiness. In the larger urban centers, Canadian campuses reflect the rich diversity of Canada's cultural mosaic, and most encourage their students to gain international experience by spending a term or a full year abroad.

Americans wondering about the currency of a Canadian degree in the U.S. should be reassured that top American and multinational countries—the likes of Archer Daniels Midland, Chase Manhattan, IBM, Microsoft, Nortel Networks, and Solomon Smith Barney—actively recruit on Canadian campuses, as do American graduate schools. According to the Institute of International Education in New York, more than 7,000 Canadians are currently enrolled in graduate schools in the United States.

The one thing that is different for U.S. and other international students intending to study in Canada is that they will have to obtain a Student Authorization, equivalent to a visa, from Canadian immigration authorities. Getting a Student Authorization is fairly straightforward for American citizens, but this slight bureaucratic hurdle is a reminder that Canada, for all of its similarities in language and culture with the United States, is still another country. For many American students who have chosen to study in Canada, this is part of the draw—they get to enjoy all the excitement of studying abroad in a foreign country with few of the cultural and none of the linguistic barriers to overcome. One of the most pleasant differences American students soon discover is how far their American dollar goes in Canada with the favorable exchange rate.

The Association of Universities and Colleges of Canada has a website at www.aucc.ca. Another source of scholastic information is the website of the Canadian Embassy in Washington, D.C., at www.canadianembassy.org/studyincanada.

Canadian universities are currently playing host to about 3,000 American students on their campuses and, as a result of funding cutbacks and internationalization policies in the early 1990s, they have become increasingly active in recruiting students from south of the border. This is but one more reason why it makes sense for more young Americans to check out the "Canadian option." Canada, eh?

University of British Columbia

Vancouver, British Columbia, Canada V6T 1Z1

Natural beauty is the first thing that draws Americans to Vancouver—and Canada's premier western university. A similar scale to places like University of Washington but with two major differences—no big-time sports to unite the campus, and limited dorm life.

Website: www.ubc.ca
Location: Suburban
Total Enrollment: 42,427
Undergraduates: 19,961
Male/Female: 45/55
SAT Ranges: N/A
ACT Range: N/A
Financial Aid: N/A
Expense: Pub $
Phi Beta Kappa: No
Applicants: 17,001
Accepted: 57%
Enrolled: 49%
Grad in 6 Years: 80%
Returning Freshmen: 91%
Academics: ✍ ✍ ✍ ✍ ½
Social: ☎ ☎
Q of L: ★ ★ ★
Admissions: (604) 822-8999
Email Address: international .reception@ubc.ca

What do two prime ministers of Canada, three provincial premiers, an astronaut, a world-renowned opera singer, and a Nobel Prize winner have in common? They are all graduates of the University of British Columbia. Founded in 1908, UBC offers students hundreds of solid programs such as business, science, engineering, the social sciences, and fine arts, as well as ready access to beaches and mountains and a diploma with instant name recognition. Though the massive campus can sometimes feel isolating, students are nevertheless happy to be here in such illustrious company.

Located just 25 minutes from downtown Vancouver, UBC's striking Point Grey campus covers a peninsula that borders the Pacific Ocean and is bounded by an old-growth forest. Mountains—perfect for skiing—loom in the distance. Architectural styles are a mix of Gothic and modern, and students can enjoy a leisurely stroll through the university's botanical gardens. Recent campus additions include the Fred Kaiser Building (the central hub of engineering), the UBC Life Sciences Centre, and an Aquatic Ecosystems Research Laboratory. The university also has a new, smaller campus—UBC Okanagan—located in Kelowna, in the Okanagan Valley.

Strong programs include microbiology, international relations, economics, and business administration. Asian studies is highly regarded, and music majors benefit from the Chan Centre for the Performing Arts. Other popular majors include psychology and computer science. The administration admits that some home economics and agricultural science courses could be strengthened, and one student grumbles about his 8 a.m. philosophy lecture: "Who can focus on the big questions at that time of the morning?"

Freshmen benefit from a wide array of first-year programs, including Imagine UBC and Create UBC Okanagan, a first-day orientation. Arts One and the Coordinated

Arts programs offers enriched, integrated approaches to broad interdisciplinary themes in arts and humanities. Qualified students can take advantage of Science One, featuring team-taught courses in biology, chemistry, math, and physics. Student exchange programs are available through 150 partner universities in 30 different countries, and co-op programs in engineering, science, arts, commerce, and forestry give students an opportunity to earn while they learn. In addition, honors and double-honors programs are available to super-brains and budding geniuses. The School of Human Kinetics has revised its undergraduate program, allowing students to choose from one of three majors at the end of their second year; choices include kinesiology

"A lot is expected of the students, but if you put in the effort you will do well."

and health science, physical and health education, and interdisciplinary studies in human kinetics. A major in United States studies has also been added. Additionally, applicants to undergrad programs following the U.S. curriculum are now required to submit test scores for the SAT (minimum 1700) or ACT (minimum 26).

The academic climate is exactly what you would expect from a university of UBC's international stature. "A lot is expected of the students," says a biology major, "but if you put in the effort you will do well." Most classes have fewer than 50 students, while larger lectures are supplemented with smaller labs and discussion groups. Overall, the faculty receives good marks. "The professors are extremely intelligent people who are truly dedicated to their discipline," says a junior. Academic advising is a mixed bag, with some students complaining that finding a knowledgeable advisor can be time-consuming.

With 42,000 students attending the Vancouver campus, it's no surprise that UBC's student population is a melting pot. "There is a huge diversity here that many smaller schools may lack," says a sophomore. The typical UBC student is bright, hardworking, and gregarious. A history major divides his classmates into two categories: commuters "who come in their fancy new cars with cell phones" and "those who live on campus and enjoy the community spirit." Minorities are well represented on campus (Asians make up the largest contingency), and the university encourages diversity through a series of special programs and active recruiting. Hot political issues include gay and lesbian rights, abortion, and sexual harassment. UBC offers 75 merit scholarships to qualified students and athletic scholarships in many varsity sports.

UBC Vancouver recently opened a new residence hall, adding more than 550 beds to student housing. An additional 1,100 beds will open in late 2007. A space is guaranteed to new first-year students who come from beyond local areas. The rest must fend for themselves against Vancouver's pricey rental market or commute from home. On-campus options include co-ed complexes (primarily for freshmen), university apartments, and family units for upperclassmen. Theme houses are another

"If you don't get involved in the nonacademics, you'll graduate with only half an education."

alternative, and offer like-minded students the opportunity to mingle, including three international houses—Korea University/UBC House, Tec de Monterrey/UBC House, and Ritsumeiken House. Hungry students will find a wide variety of meal options, including "Japanese, Lebanese, Italian, and vegetarian" plates, according to one student

On such a large campus, isolation is a real threat. "You need to get in touch with other students quickly when you get here or you could feel lost on such a big campus," says a freshman. A history major offers another point of view, "One of the biggest complaints is that UBC is too big. I totally disagree with this. I think that the school's biggest strength is its size; there are so many opportunities here." Social life happens mostly on campus but largely "depends on the crowd you hang with,"

Strong programs include microbiology, international relations, economics, and business administration.

Minorities are well represented on campus (Asians make up the largest contingency), and the university encourages diversity through a series of special programs and active recruiting.

according to one student. For partying types, there are the requisite beer bashes, courtesy of UBC's small but active Greek scene—one of the few places where under-age drinkers may sneak a sip of booze. Alternatives include university-sponsored events, such as movies and guest speakers. Popular campus events include Storm the Wall, long-boat racing, and the Arts County Fair.

Vancouver offers students countless opportunities, though one health science major says, "The town is way too rich and posh. Food and activities are expensive and places for rent are pricey. Beautiful place, but not student oriented." Still, other students give the town higher marks, "Vancouver is one of the most livable cities in the world and UBC is located in the nicest, most beautiful part—it's not too hard to imagine what a pleasure it is to go to school with that surrounding you." Beautiful weather draws students outdoors and to nearby beaches and mountains for in-line skating, snowboarding, and swimming. Vancouver will host the 2010 Winter Olympics, giving sports fans and Human Kinetics majors plenty to cheer about. Varsity and intramural competition are favorite pastimes; popular sports include soccer, basketball, hockey, volleyball, and skiing. The swimming teams are perennial championship winners (eighth consecutive year), and other recent championship teams include women's golf, field hockey, and rowing. Additionally, UBC varsity sports have produced 213 Olympic participants.

> "I think that the school's biggest strength is its size; there are so many opportunities here."

"If you don't get involved in the nonacademics, you'll graduate with only half an education. Academics aren't everything," admits a junior. Indeed, spending four years at this mammoth university can be isolating for the shy student. But for those willing to take control of their social lives, UBC offers an impressive academic milieu.

If You Apply To ➢

British Columbia: Regular admission: Feb. 28. Housing: May 1. Guarantees to meet demonstrated need of Canadian citizens and permanent residents. No campus or alumni interviews. SATs or ACTs: recommended (required for those following the U.S. curriculum). SAT IIs: optional. Essay question.

McGill University

Montreal, Quebec, Canada H3A 2T5

BEST BUY

The Canadian university best-known south of the border. Though instruction is in English, McGill is located in French-speaking Montreal. Individualism is encouraged, and there's a strong international flavor.

Website: www.mcgill.ca
Location: City center
Total Enrollment: 30,333
Undergraduates: 21,335
Male/Female: 40/60
SAT Ranges: V 620–720
 M 650–720
ACT Range: N/A
Financial Aid: 28%

Strong preprofessional programs and a diverse student body are just two of the drawing cards of McGill University. With such notable alumni as singer Leonard Cohen, musician Burt Bacharach, astronaut Julie Payette, and actor William Shatner (*Star Trek*'s Captain Kirk), it's easy to see why enterprising men and women from around the world flock to McGill University. But beware: this is not a cookie-cutter school, and fitting in actually seems to be discouraged. "McGill is a university where people are allowed to become individuals," a senior says. "Difference and creativity are celebrated here."

Montreal's climate alternates between hot summers and freezing winters. A junior describes McGill's 88-acre main campus as "an oasis in the heart of the city."

Located in downtown Montreal amidst the hustle and bustle, the campus provides students with ample greenspace and a welcome respite from the decidedly urban atmosphere of the city. Campus buildings range from "Gothic-like" structures with vines growing up the sides to more modern structures. Trees and greenery dot the campus landscape, and the sprawling recreation trails of Mount Royal rise to its immediate north. New construction includes a $71 million life sciences complex that will serve as the centerpiece of the largest research complex of its kind in Eastern Canada. The Schulich School of Music offers an ultra-modern symphony and multimedia hall that functions as a recording studio, performance venue, and research studio. A short drive west of downtown, the Macdonald Campus occupies 1,600 acres of woods and fields on the shores of Lac St-Louis, providing unique opportunities for fieldwork and research.

"Difference and creativity are celebrated here."

Though the most popular majors are management, education, and psychology, there is no denying that the university's strengths lie in preprofessional programs such as medicine, law, and engineering. The sciences receive uniform praise, as does the School of Environment, where environment-related courses are offered. For those who want to escape Montreal's brutal winters, there are internships; field studies in Barbados, Africa, and the Smithsonian Tropical Research Institute in Panama; exchange programs with more than 500 partner universities around the world; and study abroad options via the Canadian University Study Abroad Program (CUSAP).

To fulfill the university's general education requirements, students must first choose which discipline (or faculty) to enter. A senior says, "It is important to consider the university on the basis of which faculty you would be interested in, because they vary greatly and operate almost as independent units." On average, students must earn 120 credits to graduate with a four-year degree. Freshmen must accumulate six to 12 credits in three of four disciplines, including languages, math and science, social sciences, and humanities, and declare a major before their sophomore year. Upon entering their major, students have a menu of course options that includes honors programs and double majors. A double-degree interdisciplinary program allows students to combine a bachelor of arts program with one in the sciences. Several programs help freshmen with the transition to college, and some are tailored to international students, which includes those from the United States.

Regardless of the major, students can expect classes to be demanding. "Profs at McGill have very high standards for their students," a senior says. "Good time management is essential in handling the load." Classes tend to be large—especially for freshmen, who have more than 100 students in two-thirds of their classes—and students must be willing to seek out professors and advisors. Professors receive high marks for their knowledge and accessibility outside

"Good time management is essential in handling the load."

of class. "Some teachers have different lecture styles, but they're all very knowledgeable on their subject and encourage students to ask questions," an English major says. Academic advising is a bureaucratic tangle, but the new Principal's Task Force on Student Life and Learning promises to improve accessibility and consistency of academic advising. "Career and placement service runs lots of programs, so all you have to do is sign up," says one student. McGill's 6.1 million library holdings are reportedly adequate, though "not outstanding."

McGill students are a diverse lot—more than 140 countries are represented here—and the only common thread among students seems to be their fierce independence. "McGill expects its students to be mature and independent, but a complaint that some students have is that McGill does not communicate and advise students as much as they should," says one senior. Qualified students are eligible for

(Continued)

Expense: Pub $
Phi Beta Kappa: No
Applicants: 18,963
Accepted: 56%
Enrolled: 45%
Grad in 6 Years: 83%
Returning Freshmen: 92%
Academics: 🖊 🖊 🖊 🖊 ½
Social: ☎ ☎ ☎ ☎
Q of L: ★ ★ ★ ★
Admissions: (514) 398-3910
Email Address:
admissions@mcgill.ca

Strongest Programs:
Medicine
Law
Engineering
Management
Environmental Studies
Music

The Schulich School of Music offers an ultra-modern symphony and multimedia hall that functions as a recording studio, performance venue, and research studio.

merit scholarships, but there are no awards for athletes. There is also a work-study program for those in need of financial assistance.

The university's six traditional and 14 alternative residence halls house approximately 2,300 undergrads in dorms, apartments, and shared facilities houses. Dorms run the gamut but one recent acquisition is, according to a senior, a "four-star hotel, turned into a six-star dorm." Party animals will feel free to crank up the stereo in Molson or McConnell, while bookworms might be better suited for Gardener. Douglas denizens enjoy their hall's quaint charm, and women who want to skip the co-ed scene can find a room in Royal Victoria College, an all-female dorm. "The food service is quite diverse, ranging from pizza to sushi," says a senior. Off-campus apartments are a popular alternative for upperclassmen, who take advantage of

"There are always free concerts and festivals all over the city throughout the year."

Montreal's clean, affordable housing. Despite its urban location, the McGill campus is safe and security is considered more than adequate. "There are student organizations like 'Walksafe' and 'Drivesafe' that will walk or drive students to their residences at night regardless of where they are or where they are going," reports one student.

"McGill students work hard and play hard!" says one student. Though there are "considerable on-campus social activities, with many clubs and associations," many students venture off campus into Montreal for fun and adventure. "A cultural epicenter, Montreal is home to some of the world's best museums, galleries, restaurants, shops, and music," a senior says. "There are always free concerts and festivals all over the city throughout the year." Drinking is a popular pastime, but underage drinkers are few and far between since the legal age in Quebec is 18. "McGill treats its students as mature, educated adults and offers them the decision to choose whether or not to drink. Their decision is respected," reports a senior English major. Well-attended campus events include homecoming, Winter Carnival, and Frosh Week activities. Popular roadtrips include New York City, Ottawa, and Toronto. Ski slopes are less than an hour away.

Rugby, ice hockey and women's track are the most popular varsity sports, and all have captured recent championships. According to one student, "The McGill–Harvard rugby match is a must-watch." Intramurals offer would-be jocks an opportunity to blow off steam after classes and on weekends, with soccer and ice hockey attracting the most interest.

Large and tough classes, brutal winters, and mountains of red tape are undeniably part of the McGill experience. "Students here know that it's a competitive school, and, so, know much is expected of them," a senior says. Nevertheless, most denizens seem happy. "The students who go to McGill are very invested in their academic life and are proud of their school," another student says.

If You Apply To ➤ **McGill:** Rolling admissions. Regular admissions: Jan. 15. Does not guarantee to meet full demonstrated need. Campus and alumni interviews: optional, informational. SATs or ACTs: required. SAT IIs: required (three with old SAT I, depending on program; two with new SAT I). Prefers electronic applications. No essay.

Queen's University

Kingston, Ontario, Canada K7L 3N6

With "only" 18,131 undergraduates, Queens is the smallest of the major Canadian universities. It is also the only one set in a metropolitan area of modest size. Engineering is perhaps its strongest area, with business a close second. Toronto and Montreal are both about three hours away.

Students at Queen's University approach work and play with equal zeal and enjoy a potent mix of school spirit and intellectual drive. Success requires energy and a willingness to get into the thick of things. "People who aren't interested in being a part of the school community are better off at a school that isn't such a big family," warns a sophomore. Solid academics, a pervasive school spirit, and longstanding traditions make life at this storied university unique—and demanding. "Getting into Queen's is just the first challenge," says a senior. "Succeeding at Queen's is another battle."

The 161-acre Queen's campus is located on the north shore of Lake Ontario, just minutes from the heart of Kingston, Ontario ("the limestone city") and directly between Montreal and Toronto. "Almost all buildings are constructed using limestone," explains a senior. Historically significant buildings have been maintained and "there are some modern buildings with a lot of glass to provide a bright and welcoming atmosphere." Ample greenery and open spaces provide students a place to stretch out under the sky and hit the books. A $57 million chemistry building features a 250-seat multimedia lecture theater and environmentally friendly labs that consume less electricity and produce less waste.

Established in 1841 by Royal Charter of Queen Victoria, Queen's University offers undergraduate degrees in a variety of faculties, including arts, science, engineering, commerce, education, music, nursing science, and fine art. Academics are unilaterally solid, but the most demanding are engineering and commerce. The bachelor of commerce program was the first of its kind in Canada and provides students with an internationally focused liberal business education, enhanced by leadership modules and the integration of technology. The School of Computing offers Bachelor of Computer degrees in biomedical computing, cognitive science, and software design, as well as BA and BS degrees.

> **"Queen's is fairly competitive, but not bloodthirsty."**

General education requirements vary by program, but all students can expect to complete a rigorous series of core and elective courses. Students participating in the Queen's International Study Centre are whisked away to the university's England campus, where they enjoy small classes and integrated field studies while residing in a 15th-century castle. In addition, there are exchange programs with universities around the world.

The academic climate is challenging and competitive, which comes as no surprise to students. "Queen's is fairly competitive," says a junior, "but not bloodthirsty." The general consensus among struggling students is that A's are hard to come by. "After working your butt off and reading stacks of textbooks, your grades pale in comparison to the marks of students at other universities," gripes a biology major. Classes tend to be large for freshmen and sophomores, but dwindle in size as one approaches graduation. The majority of classes are taught by full professors who receive praise for their accessibility and intelligence. "The teachers I have had have been thorough, challenging, and concerned about my success," says a junior. Office hours and special "wine and cheese" functions give students ample opportunity to

Website: www.queensu.ca
Location: City center
Total Enrollment: 18,649
Undergraduates: 15,483
Male/Female: 44/56
SAT Ranges: N/A
ACT Range: N/A
Financial Aid: N/A
Expense: Pub $
Phi Beta Kappa: No
Applicants: 39,135
Accepted: 34%
Enrolled: 24%
Grad in 6 Years: 95%
Returning Freshmen: 96%
Academics: ✏️ ✏️ ✏️ ✏️ ½
Social: 🍷 🍷 🍷 🍷
Q of L: ★ ★ ★ ★
Admissions: (613) 533-2218
Email Address:
admissn@post.queensu.ca

Strongest Programs:
Engineering
Commerce
Music

Students participating in the Queen's International Study Centre are whisked away to the university's England campus, where they enjoy small classes and integrated field studies while residing in a 15th-century castle.

mingle with faculty. Students report that there is little trouble getting into desired classes, and "there is lots of counseling available for students who need it."

Queen's students are an industrious, intelligent group, and most are used to academic success. "The Queen's student seems to take on the leader role and is more likely to succeed," asserts one student. School spirit runs high and campus issues include rising tuition fees—and determining just who is responsible for the cost. Students come from every Canadian province and 80 countries, and a sociology major says that "Queen's is very PC and inclusive, regardless of gender, race, religion, or sexual orientation." A large percentage of the student body is active in extracurriculars, and school spirit is a must. Though there are no athletic scholarships, more than 800 merit awards of $1,000 to $12,000 are handed out annually. "I have had great help through scholarships and financial aid," relates a senior. "There is quite a lot of money for you. You just have to go after it."

> "The Queen's student seems to take on the leader role and is more likely to succeed."

Twenty-five percent of the student body live in one of 11 residence halls, and all freshmen are guaranteed a place to hang their hat. "The residences are very comfortable and the custodial staff is in every day, becoming your parents away from home," says a junior. Co-ed and single-sex dorms are available. A mandatory meal plan gives freshmen a wide variety of foods to choose from, including pasta, salad, pizza, and a soup-and-salad bar. The surrounding city also offers a plethora of dining options. "Kingston is known in our house as the 'city of restaurants,'" says a student. "They are everywhere." After freshman year, most students pack their bags and head off campus to the "student village," where comfortable apartments are available. In fact, 80 percent of Queen's students live within a 15-minute walk of campus. Though always a concern, safety is practically a nonissue on campus. Students report that they feel quite safe and that security is more than adequate.

Make no mistake about it, Queen's students know how to have a good time. "Social life is huge at Queen's," says a student. Adds another, "Campus pubs and city pubs have both found their niche." On Thursday nights, students flock to campus bars such as Afie's for a drink or two (or three), while Saturday nights are reserved for city bars and nightclubs. The legal drinking age is 19, and kiddies will have a tough time skirting the law. "The bouncers in Kingston actually have a couple of brain cells and can spot a fake ID from 90 kilometers away," says a senior.

> "Extracurricular activities are a must, not an option!"

Nonalcoholic alternatives include school-sponsored movies and extracurricular clubs (there are more than 220!). "Extracurricular activities are a must, not an option!" says one student. Frosh Week is a favorite event with "cheers that even the most blasé of students will be shouting out with pride by the end of the week." The school is steeped in Scottish tradition, and it's normal to see kilt-wearing bandsmen at important campus events.

Once the capital of Canada, Kingston is described as "very much a university town." There are several universities in the area (including the Royal Military College), and downtown provides students with places to shop. "Kingston itself has several clubs, three malls, a number of museums, numerous gyms, and three or four movie theaters," says a student. The city's relative isolation makes it the favored stomping ground for students without wheels. Town/gown relations are good, and students are very active in the community. Toronto and Montreal (less than three hours away) are popular roadtrips.

With 41 varsity teams, Queen's athletic program is not only the largest in Canada, but also ties with Harvard University for the largest program in North America. Popular sports include men's and women's rugby, women's squash, rowing, golf, and women's lacrosse. The annual "kill McGill" football game against rival

With 41 varsity teams, Queen's athletic program is not only the largest in Canada, but also ties with Harvard University for the largest program in North America.

Overlaps

University of Toronto, University of Western Ontario, McGill University, University of Waterloo, University of Ottawa

McGill University draws pigskin-crazed students from every corner of campus; homecoming is reputed to be a raucous affair featuring "alumni from the 1920s parading around the football field during halftime." Intramural competition is fierce, too, and nearly every student is involved on some level. A student says, "There is so much school spirit, sometimes it makes you sick."

Life at Queen's University is one of extremes. "Students who are able to balance work and pleasure fit in best here," says a junior. The pressure to succeed can be tough and expectations are high. But for those who pull it off, the rewards are well worth the effort. "Queen's has its own culture," says a student. "Don't be afraid to engage it."

If You Apply To ➤ **Queens:** Regular admissions: Feb. 28. Financial aid: Mar. 15. Does not guarantee to meet demonstrated need. Campus interviews: optional, informational. No alumni interviews. SATs: required. SAT IIs: required (for engineering candidates only). No essay question.

University of Toronto

Toronto, Ontario, Canada M5S 1A3

U of T is the largest institution in the *Fiske Guide* and one of the biggest in the world. It is also, for most readers, in a foreign country. If ever there were a place where go-getterism is a necessity, this is it. In the absence of American-style school spirit, U of T students cut loose to find their fun in the city of Toronto.

Students at the University of Toronto avoid getting lost in the shuffle by taking part in a unique residential college system that allows them to model their educational experience after their own personalities. Each college has a distinct character and appeal, yet blends seamlessly into the university's overall academic milieu. And when it comes to academics, the U of T delivers, says a senior: "The students were likely at the top of their class in high school, and are very competitive—more so than at Queen's or York universities."

Indeed, the University of Toronto is so large that it spans three campuses. The St. George campus, in downtown Toronto, features Gothic architecture and historic buildings. The suburban campuses, in Mississauga and Scarborough, feature more modern structures. Newer facilities include the Centre for Biological Timing and Cognition, where researchers will study how sleep cycles affect learning and physical and mental health. There are also 274 additional beds for students who want to live on campus, thanks to the opening of the Morrison Residence at University College.

Students apply directly to one of Toronto's nine colleges, seven of which are on the St. George campus. U.S. citizens who are admitted typically have a combined SAT I score of at least 1150, or a minimum score of 1300 for most engineering programs. Students must also take at least three SAT II tests, scoring 500 or higher on each, while those applying to engineering, business or science programs are also strongly advised to complete Advanced Placement Calculus AB or BC, or an International Baccalaureate program in math. The most popular majors include business (known as commerce here), computer science, engineering, English literature, and forensic science, and students also give high marks to sociology and anthropology.

> "I feel like our student population is mature, aware of current events, and ambitious."

Website: www.utoronto.ca
Location: City center
Total Enrollment: 59,968
Undergraduates: 49,881
Male/Female: 44/56
SAT Ranges: N/A
ACT Range: N/A
Financial Aid: N/A
Expense: Pub $
Phi Beta Kappa: Yes
Applicants: 59,055
Accepted: 66%
Enrolled: 30%
Grad in 6 Years: 74%
Returning Freshmen: 94%
Academics: ✑ ✑ ✑ ✑ ½
Social: ☎ ☎ ☎
Q of L: ★ ★ ★
Admissions: (416) 978-2190
Email Address: admissions .help@utoronto.ca

Strongest Programs:
Arts

First-Year Seminars, capped at about 25 students each, are "a good transition from high school to university," says a senior, and give incoming freshmen the chance to learn from leading faculty members in a less-intimidating environment. In addition, the Trinity One and Vic One programs each offer 25 to 50 highly capable students the opportunity to develop their critical thinking, speaking, and writing skills, while fostering close relationships with fellow students, instructors, guest lecturers, and visiting scholars. But the U of T's size can become a problem for juniors, says a physical anthropology major: "Some of the classes are huge because the department cannot afford to hire more professors."

Courses require a great deal of reading outside the classroom and are typically demanding. "The school is very competitive and most courses are quite rigorous," says a management major. A sophomore explains that competition is generally genial and students are more apt to work together: "Students are all willing to share notes and group study sessions are popular." Large lecture classes are accompanied by smaller tutorials, facilitating personal attention, and professors get high marks for their teaching skills and their smarts. "The quality of teaching, overall, has been very good, with only a few questionable professors—and that's due to the personality type, more than their knowledge."

> "I have met many people here from countries I'd only read about."

"The students here seem more 'put together' and composed than other universities," says a student. "I feel like our student population is mature, aware of current events, and ambitious." Only 8 percent of those enrolled are "foreign"—from outside Canada, that is—and the most significant issue on campus is rising tuition and fees, blamed on the provincial and national governments. The Ontario Public Interest Research Group and Amnesty International attract sizable followings, and students turn out in droves to celebrate PRIDE, the largest gay pride event in North America, along with Frosh Week, which includes wacky fun such as bed races between the colleges. (Presumably, these involve carrying new friends from place to place on a bed, not racing to see who can be first to hop under the covers with the guy or gal from down the hall.) There are no athletic scholarships.

Students apply directly to one of Toronto's nine colleges, seven of which are on the St. George campus.

Only 10 percent of students at U of T live in campus housing; while first-years are guaranteed rooms, more than half of the students commute from off-campus apartments or their parents' homes. "There are new buildings, and historic old buildings that are beautiful," says an anthropology major. "They are cozy and comfortable, and pretty well-maintained." All of the dorms are affiliated with one of the nine undergraduate colleges, which act as "local neighborhoods." Campus security is good, thanks in part to a walking-escort service that operates after dark.

Social life here revolves around the very energetic city in which the school is located, says a sociology and English major. Toronto boasts great culture, super shopping, a clean and safe nightlife district—and the picturesque shores of Lake Ontario, lovely in warmer weather. "Toronto has a huge club scene, and U of T pub nights happen in the city," the student

> "The school is very competitive and most courses are quite rigorous."

explains. The legal drinking age here is 19; students who are of age may have alcohol in their rooms, but not in common spaces, and anyone caught violating local laws or the open-container policy is reported to the dean of the residence. Those wishing to party alcohol-free will find plenty of school-sponsored events, such as movies, guest speakers, and countless clubs. There are no fraternities or sororities.

All of the dorms are affiliated with one of the nine undergraduate colleges, which act as "local neighborhoods."

Sports are not a focus of campus life at Toronto, though hockey, volleyball, and basketball are among the varsity teams that draw something of a following, especially when the opponent is Queen's University or the University of Western Ontario. The intramural program, however, is another story. It's the largest in Canada, involving 8,500 students across 57 leagues and 56 championship tournaments each year. Residence halls and groups of friends compete in everything from badminton to indoor cricket, inner tube water polo, squash, triathlon and ultimate Frisbee. Students can also be found cheering the city's many professional teams, including the Blue Jays (baseball), the Raptors (basketball) and the Maple Leafs (hockey).

Toronto's biggest liability, its sheer and sometimes overwhelming size, may also be its biggest asset, students say—as long as they learn to speak up, and proactively take advantage of all of the school's resources. "I have met many people here from countries I'd only read about, and have learned a lot about different cultural beliefs and practices," says a junior. "There are lots of people from all over the globe," adds a senior. "You learn about what is happening not only in your country, but in others as well."

If You Apply To ➤ **University of Toronto:** Rolling admissions: March 1. Guarantees to meet demonstrated need (for Canadian residents). No campus or alumni interviews. SATs or ACTs: required. SAT IIs: required. Accepts electronic applications but not the Common Application. No essay question.

British Colleges and Universities

If going to college in Canada takes nerve, you'll need even more moxie to venture overseas. But give it some thought. At the most popular overseas destination, Great Britain, about 2,300 Americans are currently enrolled in undergraduate degree programs, and another 30,000 per year are doing semester- or yearlong study abroad. Studying in Britain is not as cheap as in Canada—count on a total bill of $20,000 each year or more, not including travel—but Britain offers a richer international experience and a sense of history that cannot be had on this side of the Atlantic.

Britain makes the most sense for students interested in subjects such as English, history, foreign language, and anything related to international studies. If medieval history is your passion, why not go to school where the remains of that long-ago world still dot the landscape? If you're looking for a career in international business, consider a country where the global village has been a way of life. Though Britain is an English-speaking country, it offers far better instruction in European languages than you can get in the U.S. No matter what your academic interests, your classes will include a cross section of nationalities that would be the envy of any North American institution. Most importantly, study in Britain has the potential to be a life-changing experience that will broaden your horizons and deepen your understanding of the world.

If you do consider Britain, there is one important wrinkle that may come as a surprise. You may picture yourself in England, the most populous region of Great Britain that includes London as well as fabled universities such as Oxford and Cambridge. But here's the rub: The English have a different system of higher education than that of the U.S. that makes degree study impractical in many cases. In England, undergraduate degrees are completed in three years, not four, and students are generally assumed to have completed 13 years of schooling rather than 12. As a result, the most selective English universities are reluctant to admit American high school graduates—some refuse to admit any—and the students who do get in will find themselves thrust into a world more appropriate for juniors and seniors in college.

The answer? Look to Scotland, England's less populous neighbor, where universities offer four-year degrees that are much better suited to the needs of American high school graduates. For those who are hazy on their geography, Scotland lies north of England. Together with Wales, they make up the island of Great Britain. Throw in Northern Ireland and the moniker changes to the United Kingdom. Scotland, which was an independent nation until 1706, has an illustrious intellectual history and has produced the likes of David Hume, Adam Smith, Rudyard Kipling, Robert Louis Stevenson, J. K. Rowling, and the world's most famous ogre, Shrek.

Most people assume that the American version of higher education was imported from England. Not so. The American-style liberal arts institution came directly from Scotland in the person of John Witherspoon, a graduate of the University of Edinburgh who was lured to the U.S. in 1768 to head Princeton University. With the model of his alma mater in mind, Witherspoon transformed Princeton from a small-time school for ministers into a broad-based institution that taught philosophy, history, geography, science, mathematics, and theology. In the process, he became the most influential educator of his time and charted the path on which the American university system has continued, more or less, to this day.

Though the American and Scottish systems share the same heritage, they are not identical. Any American who is considering Britain—whether in England or Scotland—should be aware of the differences that lie "across the pond." The most important: while American universities generally encourage students to sample a variety of fields before choosing a major, British institutions expect students to know their major before they set foot on campus, in part because British students take their general education courses in high school. As a result, college coursework focuses almost exclusively on the student's major field, and American-style distribution requirements are all but unheard of. For students who want to get out of those nasty math or foreign language requirements, a British institution will probably fill the bill. But anybody who wants to change majors after a year or two may encounter difficulty.

Students in Britain take only two or three courses at a time, all of which are generally related to their major. This may sound like easy street to harried American students who are expected to take four or five courses per semester. But because British students tend to take courses only in subjects that seriously interest them, all classes are taught at a high level, even introductory ones. Though students in Britain get fewer hours in class, they are expected to put in more hours of study per course outside of class.

Education aside, American students will need to adjust to a different tenor of life in Britain. To put it bluntly, don't expect the comforts of home. There are no posh food courts or fitness facilities that look like a ritzy health club. The student body will not come out on a Saturday afternoon for the big game—there are no big games. Dorms are generally the domain of first-year students; expect to find a "flat" (apartment) for your upper-class years. Generally speaking, there will be no campus "bubble" to cloister you away from the real world—a good thing if you want an authentic sample of life in your new home.

Before we go further, here's a word to moms and dads. We know that you get queasy at the thought of sending your little cherub across a 3,000-mile ocean. Take a chill pill. Safety fears are based on illusions, not facts. Even in the age of terrorism, a flight to Britain is quicker and safer than driving 10 hours to get to First Choice U. Once you're in Britain, the cities are at least as safe as those in the U.S., and the small towns have a crime rate roughly equivalent to that of the town of Mayberry on The Andy Griffith Show. The best part for parents: You'll need to visit at least once (and preferably several times).

As noted earlier, the price tag for all this crosscultural enrichment is about one-third less than that of a selective private institution in the U.S., not including travel. The dark lining to the silver cloud is that academic scholarships are scarce and institutional financial aid is all but nonexistent. Federal aid such as Stafford loans and Pell Grants can be transported in some cases, but most families will find themselves paying the full freight. And because exchange rates fluctuate, the bill can vary significantly depending on whether the dollar is weak or strong. For a searchable database of the few scholarships available for study in Great Britain, visit the British Council at www.britishcouncil.org/usa.

We said earlier that English universities are often not the best choice. The University of Cambridge (www.cam.ac.uk) is particularly blunt about "the possible mismatch between the broad liberal arts curriculum of the North American high school and the specialist emphasis of British degree courses." In a recent year, Cambridge accepted only three students from U.S. high schools. The University of Oxford (www.ox.ac.uk) does offer a glimmer of hope for a select few superachievers; Oxford considers American students who graduate in the top 2 percent of their class, and in a recent year it enrolled about 30. Even so, the odds of admission to Oxford are lower than at any college in the U.S., including Harvard. The vast majority of American undergraduates at both Oxford and Cambridge are there for a second bachelor's degree after earning one from an American institution. Students with their hearts set on the Oxbridge institutions should consider them for graduate school, where both do admit Americans in significant numbers.

Students will get a similar story at the third-most recognized name in English higher education, the London School of Economics (www.lse.ac.uk), which enrolls about 3,600 undergraduates. The LSE says it will not normally consider U.S. students until they have a year of higher education under their belts. Less selective English institutions are more receptive to Americans, but only those who feel certain of what they would like to study should apply. If you are in this category, there is one potential benefit to an English degree: the three-year degree program will save you a year of tuition bills.

Most students would be better served by looking at a place like Scotland's University of St. Andrews, which has a decades-long history of catering to American undergraduates. About a sixth of the U.S. students who earn a bachelor's degree in Great Britain do so at St. Andrews (www.st-and.ac.uk), a highly selective institution that considers applicants with a GPA of approximately 3.3 or higher, a combined SAT score of at least 1300, or an ACT of 29 or better. A less selective option is the University of Aberdeen, a Scottish university that has stepped up its North American recruitment efforts. St. Andrews and Aberdeen are profiled with full-length write-ups below.

Other notable Scottish institutions include the University of Edinburgh (www.edinburgh.ac.uk), which is set in one of Europe's most vibrant and interesting cities. With an undergraduate population of about 16,000, Edinburgh has roughly 100 degree students from the United States. Adam Smith's old haunt, the University of Glasgow (www.glasgow.ac.uk), is an option of similar size. Glasgow is less international than Edinburgh or St. Andrews, but is believed by some to provide a more authentic Scottish experience. For more Scottish or English institutions, head to www.britishcouncil.org/usa.

British universities evaluate applicants largely on the basis of academic credentials, with emphasis on demonstrated ability in their field of study. British institutions put much less weight on extracurricular activities and personal qualities than those in the U.S. and they tend to prefer the SAT over the ACT, though many will accept either. Scores from SAT II or AP tests may also be required. For the application essay, the British usually ask about commitment to your intended major and why you want to study it. They often view American-style personal essays as fluff.

Students applying to U.K. institutions should generally use the Universities and Colleges Admissions Service (UCAS, www.ucas.ac.uk), which functions like the Common Application group in the U.S. The UCAS form asks you to list all your courses, and the grades you received in them, as well as your SAT and/or ACT scores. It also includes an essay and a letter of recommendation. Most institutions will accept applications through the spring, though we recommend that you apply by the deadline for British students, January 15. The deadline for applying to Oxford and Cambridge, or to apply to any program in medicine, is October 15 for entrance the following fall. Many institutions have rolling admissions, another reason to apply early. One note on terminology: In Britain, a program of study is called a "course." The British word for what we call a course is "module."

College in Britain is not for the faint of heart, but those with the initiative to go will be richly rewarded. After college in Britain, students will have the skills and savvy to succeed almost anywhere in the world.

University of Aberdeen

Aberdeen, Scotland AB24 3FX

Aberdeen is a leading Scottish institution that has stepped up its efforts to recruit Americans. Major attractions include engineering, life sciences, and anything related to Europe. Located on Scotland's picturesque east coast, the city of Aberdeen combines charm with the bustle of a small city.

The University of Aberdeen was founded in 1495—three years after a certain well-known explorer sailed from Spain to the New World. Students seeking the flavor of old Europe will not be disappointed; with plenty of cobblestone streets and buildings made of ancient stone, the university has a distinctly medieval aura. But Aberdeen is more than just a history theme park. With a full-time U.S. recruiter and exchange programs with 14 North American universities, Aberdeen is one of the few British universities with a critical mass of Americans. And why not? It offers top-notch academics, a curriculum that is unusually flexible by U.K. standards, and a slice of life far richer than any U.S. institution can muster. "I love the diversity at the University of Aberdeen," says a history major, "It is great to be surrounded by people from other countries and to hear their points of view."

Aberdeen is perched at a latitude roughly the same as Juneau, Alaska, but because of the Gulf Stream, winter temperatures are generally milder than those on the East Coast of the United States. December days are short in winter, but sky-gazers are

Website:
   www.abdn.ac.uk/sras
Location: Urban
Total Enrollment: 13,900
Undergraduates: 10,260
Male/Female: 45/55
SAT Ranges: 1800
   (combined)
ACT Range: 27 or above
Financial Aid: N/A
Expense: Pub $
Phi Beta Kappa: No
Applicants: N/A

(Continued)

Accepted: N/A

Enrolled: N/A

Grad in 6 Years: N/A

Returning Freshmen: N/A

Academics: ✑ ✑ ✑ ½

Social: ☎ ☎ ☎

Q of L: ★ ★ ★ ★

Admissions:

(+44) (0) 1224 272090

Email Address:

sras@abdn.ac.uk

Strongest Programs:

Divinity

Biological Sciences

Engineering

English

French

Geography

History

Law

Medicine

Politics and International
    Relations

Aberdeen is one of the few British universities with a critical mass of Americans.

often treated to glimpses of the fabled Northern Lights. Most university buildings are concentrated in a quiet enclave known as "Old Aberdeen." The campus is crowned—literally—by a 16th-century tower in the shape of an imperial crown. Lightly traveled streets pass through the campus, and the multitude of green lawns and picturesque courtyards are ideal for lounging on sunny days.

Aberdeen is divided into three Colleges: College of Arts and Social Sciences, College of Life Sciences and Medicine, and the College of Physical Sciences. Popular majors include English, French, geography, history, and a joint politics/international relations major. An interdisciplinary program in physical sciences mixes scientific study with writing, communication, and computer and Internet-related skills. Engineer-

> "It is great to be surrounded by people from other countries and to hear their points of view."

ing is also strong, especially programs related to the oil industry. (The City of Aberdeen is headquarters for the thriving oil extraction business in the North Sea.) The life sciences are a major draw—with zoology being the most popular of them—and the university offers unique programs in marine resource management and tropical environmental science.

Aberdeen is a medium-sized university by U.S. standards, and courses in the first two years generally consist of lectures supplemented by smaller weekly discussion sections. "One of the finest points about Aberdeen is that you are always lectured to by full professors starting in the first year," declares an English major. "I have found my professors very approachable, friendly, and helpful," chirps a history major. Unlike in the U.S., professors typically team-teach introductory "modules," with each covering the topics that are his or her specialty. Perhaps the biggest academic difference from stateside institutions is that students generally take only three subjects at a time in the first two years, though extensive reading and research outside of class is generally taken for granted. "Often we are expected to come to class prepared to discuss certain topics, but given no minimum reading assignment—the professor gives out a list of selected readings from which we can choose," says a history major. As at most U.K. institutions, grades are typically determined by end-of-the-term evaluations with few intermediate assignments. At the end of their second year, students must typically pass exams in order to advance to "honors level," the equivalent of the junior and senior years of college in the States. These advanced students typically take two yearlong classes that may meet as many as four times per week with both lectures and interactive instruction. Upper-level scientists typically spend long hours in the lab. One nice feature: There is generally no limit to the number of students

> "It is such a carefree atmosphere with that special touch of Scottish tradition."

who can sign up for a particular course, thereby giving students the freedom to sign up for anything that strikes their fancy. "The university has a relaxed climate," says a junior, who adds the "emphasis is on self-study with strong background support."

Aberdeen has about 225 degree students from the States. Overall, 81 percent of the students hail from the U.K. while the remaining 19 percent arrive from overseas. In fact, the University currently welcomes students from 120 different countries. "Wandering around campus, you are bound to hear at least three different languages being spoken in a day," reports a first-year student, adding, "The people here are extremely friendly and outgoing." The political climate on campus is relatively subdued—several students describe the tenor as "conservative"—though there were protests regarding Iraq and its aftermath. Upon their arrival at the university, students partake of Freshers Week. No classes are held, and various student organizations sponsor informational meetings. Aberdeen is a selective institution for U.S. students, though less so than University of St. Andrews. The bill for an Aberdeen education depends on your course of study. Students in the humanities and social sciences can expect to fork over in the neighborhood of $20,000 for all expenses except travel, depending on the exchange rate, while scientists will pay closer to $25,000. No scholarships are available for U.S. students, but those eligible for federal student loans can use them for study at Aberdeen.

Housing at Aberdeen is varied. "Accommodations themselves are basic," notes one student, "but you get what you pay for." A majority of the international students (along with first-years from the U.K.) live in Hillhead Halls of Residence, an complex of houses and flats that is about a 15-minute walk from the campus. Students may elect catered rooms (two meals per day) or self-catering, wherein they cook their own food with kitchen facilities generally located down the hall from the rooms. Many students choose to move off campus after their first year, and a variety of housing options are available near the campus. Only a few students own cars, as the university is within easy walking distance of the city center and the North Sea and is on a regular bus route. Parking permits are made available to those students who require them.

One staple of American college life will not be necessary at Aberdeen: a fake ID. The drinking age in Britain is 18, so social life at Aberdeen revolves around legal consumption rather than drinking on the sly. "Most people go out on most nights if only for a pint at the pub," reports an exchange student, "More often than not, one pint turns into a whole night out, but that's tricks." Aberdeen offers a varied nightlife with clubs and bars to suit all tastes. You may find yourself doing what Britons call "the pub crawl," which means sampling the refreshment of several pubs before heading home in the wee hours. For those who find themselves inebriated in a distant part of the city, the university has a standing deal with a local cab company to take home any student who needs a ride with the fare put on the student's university bill.

> **"One of the finest points about Aberdeen is that you are always lectured to by full professors starting in the first year."**

Campus social events consist mainly of periodic formal balls, to which the men wear kilts and the women wear evening gowns. Perhaps the biggest campus event of the year is the Torcher's Parade, which is held every spring and features floats made by various student organizations. Sports are mainly for playing rather than watching at Aberdeen. Individual sports rather than team intramurals are the staple of weekend warriors, and students can purchase passes for various athletic facilities depending on their interests. Among the most successful of Aberdeen's intercollegiate teams are field hockey, football (known in the Colonies as soccer), crew, table tennis and, of course, golf.

Aberdeen is a city of about 220,000, and is described by one student as "a fantastic college town!" The city has plenty of old world charm, and outdoorsy types will love the dramatic scenery that is everywhere in northeast Scotland. Picturesque cliffs overlooking the North Sea are within an easy bus or train ride. Fifteen miles south of Aberdeen is breathtaking Dunnottar Castle, a 14th-century ruin set high on a rocky outcrop that was the set for Mel Gibson's film rendition of *Hamlet*. Within a half-an-hour ride inland is the edge of the legendary Scottish Highlands. Famous castles are in all directions, including the royal family's summer hideaway, Balmoral. For the Scottish version of the big city, Glasgow and Edinburgh are close by and two hours on a plane will get you to most places in Western Europe.

Though Aberdeen may lack some of the conveniences of home, most Americans are happy they came. "Between classes on a sunny day, students will buy something from the bakery and sit on the grass in the midst of 500-year-old buildings and cobblestone streets. It is such a carefree atmosphere with that special touch of Scottish tradition," says a satisfied history major. If you're the kind of person who likes to meet new people and learn about different cultures, take a chance. You, too, may thrive on the Aberdeen air.

Aberdeen is perched at a latitude roughly the same as Juneau, Alaska, but because of the Gulf Stream, winter temperatures are generally milder than those on the East Coast of the United States.

Overlaps
Dundee, Edinburgh, Glasgow, St. Andrews, Stirling

University of St. Andrews

St. Andrews, Scotland KY16 9AJ

St. Andrews is the most popular destination for Americans who want to study in Britain. Often compared to Princeton, St. Andrews is small by British standards and middle-sized by American ones. Major drawing cards include English, international relations, medieval history, and modern languages.

Website: www.st-and.ac.uk
Location: Small town
Total Enrollment: 6,512
Undergraduates: 5,508
Male/Female: 48/52
SAT Ranges: N/A
ACT Range: N/A
Financial Aid: N/A
Expense: Pub $
Phi Beta Kappa: No
Applicants: N/A
Accepted: N/A
Enrolled: N/A
Grad in 6 Years: N/A
Returning Freshmen: N/A
Academics: ✑ ✑ ✑ ✑ ½
Social: ☎ ☎ ☎
Q of L: ★ ★ ★
Admissions:
  (+44)1334 463326
Email Address:
  intoff@st-and.ac.uk

Strongest Programs:
Art history
Medieval history
English
International relations
Modern languages
Physics
Psychology

Harvard likes to brag about the fact that it was founded back in 1636. Think that's old? Try 1413—the date Pope Benedict XIII issued a Papal Bull recognizing the University of St. Andrews. Set in an ancient seaside town on the eastern shore of Scotland, St. Andrews is an ideal spot for adventuresome Americans who want a world-class education and an introduction to life outside North America. With 550 U.S. students in degree programs and another 250 in residence each year for study abroad, St. Andrews has the highest proportion of Americans of any major university outside North America. Yet success here requires a go-getter mentality. Support services are available, but students on this side of the Atlantic are accustomed to being treated like adults. "They don't hold your hand—you're going to be dropped in and it's sink or swim," says one U.S. student.

St. Andrews is the only institution in the *Fiske Guide* that saw a heretic burned at the stake beneath its most prominent landmark. In 1528, a Protestant reformer named Patrick Hamilton was the victim of a botched burning imposed by the local archbishop that dragged on through six hours and several relightings. Hamilton's initials are carved in the cobblestones where he died, and legend has it that any student who steps on them will fail exams. The spot is at the base of St. Salvator's Tower, which dates from 1450. An adjoining quadrangle is the hub of the university, but most academic buildings are interspersed through the town's narrow medieval streets. Town and university alike are made of ancient stone, and the ruins of a 14th-century castle and cathedral are adjacent. Narrow alleys, called "wynds" by the Scots, lead to secluded gardens and courtyards that add to the old world aura.

Though more flexible than most British universities, St. Andrews offers less latitude to explore a variety of subjects than would be found in a U.S. institution. "Most U.K. students have a very good idea of what they want to study. This is not a university

"My four best friends are Russian, Scandinavian, English, and Irish/Slovakian."

for an undecided major," says a senior who is studying international relations and French. Students typically take three yearlong classes in each of their first two years—the equivalent of introductory survey courses in the States. These "modules" generally consist of three lectures per week with 100 or more students and a tutorial with 10 to 20. With satisfactory progress, students move on to honors level courses for the final two years, which are generally taught in seminar format. Fewer courses means less time in class and more emphasis on outside reading—"much more than in the States" says a UCLA study abroad student. Instead of a textbook or two, most modules have sizable reading lists that students must navigate largely on their own. Modules typically end with papers or exams that account for most of the grade, and there is a full week without classes prior to exams. More so than in the U.S., the onus is on the students to keep current with their work and seek help when necessary. Nevertheless, the faculty gets high marks and students like the fact that professors are often internationally known in their fields. "The faculty is very accessible and friendly," says a fourth-year student. "The academic counseling has been far superior to my counseling back home," notes a junior-year-abroad student.

St. Andrews is the smallest of the leading British universities and among the few located in a small town. It is a liberal arts university with additional programs in divinity, medicine, and business. Signature offerings include art history, medieval history, English, international relations (IR), modern languages, physics, and psychology. IR is the biggest draw for Americans, and St. Andrews is a world leader in the study of international terrorism. "I thought doing international relations in the U.S. would be a bit silly," says one American. Standards in foreign language are higher than in the U.S.—an opportunity but also a challenge. "I got close to 700 on my SAT IIs in French and I was completely lost," says a second-year student. As at many U.S. universities, natural science students tend to work the hardest. Premed is a killer, and most scientists will have spent hour upon hour in the lab by the time they get to honors level.

Though more flexible than most British universities, St. Andrews offers less latitude to explore a variety of subjects than would be found in a U.S. institution.

"We don't live on a campus—we live in a town. It is more like real life."

While there are few weak majors, several students report that access to information technology has been less than they expected. Computer labs can be booked solid near the end of the term, and students are better off bringing their own machines.

St. Andrews is one of the world's most international universities—25 percent of the students come from outside the U.K. It is an oft-noted fact that more English students than Scottish are enrolled, and sizable contingents also come from Scandinavia, eastern Europe, Asia, and the Middle East. "My four best friends are Russian, Scandinavian, English, and Irish/Slovakian," says one American who attended English boarding school. The whole seems to mesh reasonably well, though one student notes that class differences are sometimes apparent. The English tend to be from upper-middle-class backgrounds and help give the campus a more conservative tenor than some Americans might expect.

In the U.K., St. Andrews is seen as a Scottish alternative to Oxford and Cambridge, inviting enough to attract the likes of Prince William, the heartthrob son of Prince Charles and Princess Diana. Admission for U.S. students is highly competitive—on par with places like Cornell, Tufts, or Emory. A high proportion of the U.S. students at St. Andrews attended private schools. The Americans who thrive at St. Andrews tend to have international experience, though not all do. Many students are "diplobrats" whose parents have worked in international organizations such as the World Bank or the State Department. (The Scots generally give Americans a warm welcome, despite huge reservations about U.S. foreign policy.) Tuition varies depending on the course, but the total bill for a year at St. Andrews is likely to be about $25,000, depending on the exchange rate.

Housing is guaranteed for first-year students, and students can request a single or shared room. They also have the option of the university meal plan or "self-catering," in which they use kitchens in the dorms to prepare their food. Reviews of the food are mixed at best. "British cuisine is not world renowned. There is good reason for this," says one student. Nor should students expect the glitzy food courts or all-you-can-eat service typical in the States. Meals are served at specified times with limited portions. On the plus side, dorm life includes once-a-week maid service. After their first year, students generally move to one of the many apartments ("flats") near the university. "There is a bit of a scramble for flats in February, but most students

St. Andrews is one of the world's most international universities—25 percent of the students come from outside the U.K.

are able to find good accommodations," reports an international relations and French major. Because there is no boundary between town and university, students are less sheltered from the real world than they might be on a typical U.S. campus. "This is a grown-up place," declares one student, "We don't live on a campus—we live in a town. It is more like real life."

A peek inside the student union reveals something never seen on a U.S. campus—a fully equipped bar with everything from vodka to vermouth (not to mention scotch). The drinking age is 18 in Britain, and though there may not be more alcohol than at an American institution, it is certainly more out in the open. There are 28 pubs in St. Andrews, a town of about 18,000, and imbibing in them is the main form of social life. Black-tie balls are also a staple, as are ceilidhs (pronounced "kaylees"), which feature traditional Scottish dancing that resembles a square dance. Soccer ("football") is the national sport, and students congregate to watch pro teams in the pubs or on the big-screen TV at the union. University sports are not

"The faculty is very accessible and friendly."

nearly as big in the U.K. as in the U.S. and draw few spectators. "Interhall" competitions in soccer or rugby are the equivalent of intramurals. There is a pay-per-use sports center with facilities ranging from basketball to squash. Golf enthusiasts know that the game was invented in St. Andrews in the 1500s, and the fabled Old Course is a short walk from campus. (Students can get a yearlong pass to play at the courses in town for about $150, though Old Course players must have a specified handicap rating.)

The nearest road-trip destination is the medium-sized city of Dundee, about 30 minutes away, which offers a mall, movie theaters, and a McDonald's. (Subway is the only American fast-food joint to crack the St. Andrews market—so far.) Scotland's two largest cities, Edinburgh and Glasgow, are about an hour away, and for outdoorsy types, the legendary Scottish Highlands are within easy reach.

As befitting a 600-year-old institution, St. Andrews is rife with tradition. At dawn on May 1, hundreds of sleep-deprived and/or drunk students take an icy dip in the North Sea as part of May Morning, a tradition with roots in pagan times. The all-male Kate Kennedy Club also goes back centuries and was created in tongue-in-cheek honor of the daughter of the university's founder, a lass who was supposedly liberal in her sexual favors for St. Andrews students. The KK Club sponsors numerous social events throughout the year, most notably Raisin Weekend, which culminates in a massive shaving cream fight in St. Salvator's Quad.

St. Andrews is not the U.S., and those who come here must be ready to adjust to a different way of life (not to mention constant rain and winter nights that begin with sunset at 3:30 p.m.). But this is a small price to pay for Scotland in all its ancient glory. Mothers may tremble at the thought of their offspring going overseas for college, but the U.K. is not that much further from the East Coast than Berkeley or Pomona. The education from merely living in St. Andrews will far exceed anything learned in the classroom. U.S. institutions may trumpet their diversity, but nothing stateside compares to the richness of living abroad among the best and brightest from all corners of the globe. St. Andrews delivers it all against a hauntingly beautiful backdrop that will remain forever etched in the minds of all who come here.

Overlaps

N/A

If You Apply To ➤ **St. Andrews:** Rolling admissions: May 31. Interviews: not available. SATs or ACTs recommended. Essay: reasons for wishing to study at St. Andrews.

University of Iowa

107 Calvin Hall, Iowa City, IA 52242-1396

A bargain compared to other Big Ten schools such as Michigan, Wisconsin, and Illinois. Iowa is world-famous for its creative writing program and Writers' Workshop. Other areas of strength include health sciences, social and behavioral sciences, and space physics. Future scientists should check out the Research Scholars Program.

At first glance, one might dismiss Iowa as a standard-issue Midwestern state U. But beneath the bland exterior of fields and corn lies one of the most dynamic schools in the country—and one of the best values to boot. Iowa is known for breeding stellar nurses, future doctors and of course, wrestlers. Iowa has long been a major player in the creative worlds, particularly writing, and its small-town atmosphere is just one of many reasons students nationwide flock to this "budget Ivy League."

The 1,880-acre campus, located in the rolling hills of the Iowa River valley, is bisected by the Iowa River and merges with downtown Iowa City. Among the 90 primary buildings is Old Capitol, the first capitol of Iowa, a national historic landmark and the symbol of the university. The primary architectural styles of the campus buildings are Greek Revival and modern. The face of the campus is changing, with many new buildings recently opened and several more on the way. The biggest **"Residence life here is fabulous!"** include a $47 million medical education and biomedical research facility, a new art and art history building, a journalism and mass communications building, a center for the Honors Program, and a planned $3 million addition to the historic Dey House, which will serve as additional space for the Iowa Writers' Workshop.

Iowa has a long tradition in creative arts. It was one of the first universities to award graduate degrees for creative work and is also the home of the famed Writers' Workshop. The school also prides itself on its International Writing Program, which brings a wide array of prominent authors to the campus. "The English department is stellar," raves one English major. "It's possibly the best in the country—at least for creative writing." Iowa's on-campus hospital is one of the largest teaching hospitals in the United States. Undergraduates benefit from the strong programs in health professions such as physician's assistant and medical technician. Iowa is also strong in the social and behavioral sciences, space physics, and paleontology. Combined degree programs permit students to earn degrees in liberal arts and their choice of business, engineering, nursing, or medicine. Among the newest bachelor's degree programs are performing arts, entrepreneurship, and international studies, which includes a scholarship for study abroad. The University Honors Program provides special academic, cultural, and social opportunities to undergraduates who maintain a cumulative grade point average of 3.3 or higher. Iowa's study abroad program gives students a choice of more than 40 different countries. Agriculture, veterinary medicine, forestry, architecture, and animal science are not offered at Iowa but are taught at its sister institution, Iowa State.

Each of the three undergraduate colleges has its own general education requirements. Liberal arts students must take courses in rhetoric, natural science, social sciences, foreign language, historical perspectives, humanities, and quantitative or formal reasoning. Also required are general education courses in the areas of cultural diversity, foreign civilization and culture, and physical education. Most classes have fewer than 50 students, but "freshmen do tend to spend a majority of their time in large lectures," says a senior. The University of Iowa's Four-Year Graduation Plan guarantees that students who fulfill certain requirements will not have their

Website: www.uiowa.edu
Location: Small city
Total Enrollment: 28,426
Undergraduates: 18,130
Male/Female: 47/53
SAT Ranges: V 520–650
   M 540–660
ACT Range: 22–27
Financial Aid: 51%
Expense: Pub $ $
Phi Beta Kappa: Yes
Applicants: 13,241
Accepted: 84%
Enrolled: 35%
Grad in 6 Years: 66%
Returning Freshmen: 84%
Academics: ✍ ✍ ✍ ✍
Social: ☎ ☎ ☎
Q of L: ★ ★ ★
Admissions: (800) 553-IOWA
Email Address:
   admissions@uiowa.edu

Strongest Programs:
Creative Writing
Theater Arts
Music
Printmaking
Business
Social and Behavioral
   Sciences
Space Physics
Paleontology

graduation delayed by unavailability of a needed course. Students say the academic climate is challenging, and they give most professors high marks. "Most teachers are incredibly knowledgeable and those in small sections really make sure the students comprehend the material," says a junior.

"I think that the students here are more open-minded than at our closest rival," says a pre-med student. "I think this is because of the strong artistic and performing culture present here." Sixty-one percent of the undergraduates hail from Iowa, with most of the rest coming from contiguous states, especially Illinois. African Americans, Hispanics, Asian Americans, and Native Americans account for only 9 percent of the student body, but as the administration points out, the state of Iowa has only a 4 percent minority population. Students say the campus is extremely tolerant and a community atmosphere is fostered in and out of the classroom. Besides the 436 athletic scholarships available, there are more than one thousand academic scholarships, averaging $1,950. The Roy J. Carver Scholarships are awarded to 82 students who have overcome social or psychological barriers; it is one of the first awards to be bestowed on a former homeless man.

> "There are two university theaters right on campus and many affordable cultural events take place at Hancher Auditorium."

Students say that campus residence halls are very sociable and therefore not very quiet. Still, "Residence life here is fabulous!" cheers a freshman. All are co-ed by floor or wing. Students can choose to live in one of eight "learning communities," such as women in science and engineering, or performing arts. There are many praises for the custodial staff, which keeps the buildings clean. Only 27 percent of the students live in university housing, and more than half live in apartments or houses adjacent to the campus. Many students move off campus after their freshman year. The "very nice" dining halls are "set up like food courts, with numerous options for varying ethnic and special taste backgrounds," says a senior. The student union includes a pastry and coffee shop, two cafeterias, and the State Room Restaurant.

Seven percent of the men and 12 percent of women belong to fraternities and sororities, and these groups tend to play less of a role in the social life than they do elsewhere. "There always seems to be something to do," a student says. Football, basketball, and wrestling events are especially popular on campus. On weekends, students often venture to the downtown area, across the street from campus, which "is built with the college student in mind," a student says. "There are two university theaters right on campus and many affordable cultural events take place at Hancher Auditorium. The Union Bar and Grill, Mickey's, Sports Column, and George's are all popular hangouts with students. The

> "The English department is stellar. It's possibly the best in the country —at least for creative writing."

school officially follows the state policy regarding alcohol; the legal drinking age is 21. "Drinking will always happen," says a student, "but I think the policies really teach up-and-coming students what may happen to them if they make poor choices." For a change of scene, Chicago, Kansas City, or St. Louis are all within six hours by car, a short road trip by Midwestern standards. Riverfest, held at the Iowa Memorial Union and on the banks of the Iowa River, is a weeklong, all-campus event celebrating the long-awaited spring. "It is a unique tradition that brings the campus and the community together," says a sophomore. Students also look forward to the annual Iowa City Jazz Festival, homecoming, Dance Marathon, and Big Ten football, especially the game against Iowa State.

Now that Iowa's football team has become a national power and regularly appears in New Year's Day bowl games, that big football stadium is getting an $87 million facelift, including seats that are two inches wider and lots of new concession stands. Hawkeye fans are serious about their team. "The whole town is basked in black

Combined degree programs permit students to earn degrees in liberal arts and their choice of business, engineering, nursing, or medicine.

The school also prides itself on its International Writing Program, which brings a wide array of prominent authors to the campus.

The University of Iowa's Four-Year Graduation Plan guarantees that students who fulfill certain requirements will not have their graduation delayed by unavailability of a needed course.

Overlaps

Iowa State, Northern Iowa, University of Illinois, Indiana University, University of Wisconsin and gold," a freshman says. Other traditional powerhouses include the Hawkeye wrestling, basketball, and softball teams.

Much more than a campus among the cornfields, Iowa boasts a beautiful university that is ever evolving. "The University of Iowa is continually renovating and improving its facilities to stay modern and keep up with technology," says a sophomore. The scope of its academic programs is broad and social activities abound—especially when it comes to rooting for their Hawkeyes.

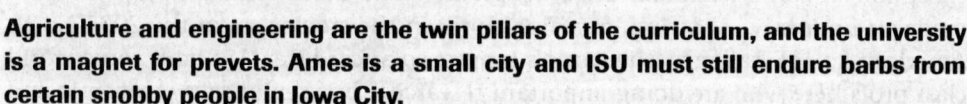

If You Apply To ➤ | **Iowa:** Rolling admissions. Regular admissions: Apr. 1. Does not guarantee to meet demonstrated need. Campus interviews: optional, informational. No alumni interviews. SATs or ACTs: required. SAT IIs: optional. Prefers electronic applications. No essay question.

Iowa State University

100 Alumni Hall, Ames, IA 50011

Agriculture and engineering are the twin pillars of the curriculum, and the university is a magnet for prevets. Ames is a small city and ISU must still endure barbs from certain snobby people in Iowa City.

Love for Iowa State University runs as deep as its Midwestern roots. Strong programs in engineering, business, and agriculture attract students from around the globe. The close-knit, small-town atmosphere fostered at this school of more than 25,000 keeps them here.

The university has lavished attention on its park-like campus, located on a 1,984-acre tract in the middle of Ames, population 50,000. The campus, which boasts a combination of dignified old buildings and award-winning new ones, is a landmark of landscape design with numerous shady quadrangles with floral plantings and artwork that create a garden-like quality. History and tradition prevail, from the campanile, which serenades the campus with its carillon bells, to the huge public art collection including sculptures by Danish artist Christian Petersen. The newer Howe Hall, home to aerospace engineering, boasts a virtual reality application center and a six-sided virtual reality cave. Along with Hoover Hall, the complex provides the teaching and researching home for the College of Engineering. Other additions to campus include the Gerdin Building for the College of Business and the 4-H extension building for the College of Agriculture. Much of the campus is closed to cars, largely for the benefit of walking, bicycling, and in-line skating students, as well as the swans (named Sir Lancelot and Lady Elaine) and the ducks that reside on Lake LaVerne. The CyRide fare-free bus system delivers students around campus and the city.

When Iowa State opened in 1869 as a land grant university, agriculture and engineering ruled the academic roost. These days, though, the liberal arts are just as popular, and the College of Liberal Arts and Sciences is the largest of ISU's seven colleges. Among the university's 100-plus majors, the College of Agriculture still fields outstanding programs in animal science, turf grass management, agribusiness, and agronomy. Other colleges include business, design, veterinary medicine, human sciences (formerly family consumer sciences and education), and the graduate college. Newer programs include a BS/MS in diet and exercise, a BS in software engineering, a

Website: www.iastate.edu
Location: Small city
Total Enrollment: 25,462
Undergraduates: 20,440
Male/Female: 56/44
SAT Ranges: V 530–650
  M 550–690
ACT Range: 22–28
Financial Aid: 57%
Expense: Pub $ $
Phi Beta Kappa: Yes
Applicants: 9,172
Accepted: 90%
Enrolled: 46%
Grad in 6 Years: 67%
Returning Freshmen: 85%
Academics: ✐ ✐ ✐
Social: ☎ ☎ ☎
Q of L: ★ ★ ★
Admissions: (515) 294-5836
Email Address:
  admissions@iastate.edu

Strongest Programs:
Exercise Science
Apparel Design
Chemistry

BA in criminology and criminal justice, and a minor in engineering studies. Students find the academic climate competitive, but that can vary by program.

All undergraduates must take two semesters of English composition their freshman year and demonstrate proficiency in English prior to graduation. Other general education requirements, which vary by college, focus on gaining breadth in the natural and social sciences, but everyone takes a half-credit course on the use of the library and must satisfy a three-credit requirement in diversity. Students enjoy both wired and wireless access to link their personal computers or laptops with the university system. Students can also use the AccessPlus system of electronic kiosks sprinkled around campus to check the status of their university bill or financial aid package, print an unofficial transcript, or get their current schedule. Iowa's highly touted learning communities offer transfer students and freshmen the opportunity to join a group of other newcomers who share similar academic interests in taking a common set of classes together or living together on the same residence hall floor.

"We have some world class profs here who are doing important research."

An honors program enrolls 400 outstanding freshmen each year, many of whom live in honors housing. Iowa State ranks in the top 20 for sending students overseas; popular destinations include Italy, the United Kingdom, Spain, Australia, and Ireland. Despite the university's size, professors teach most classes, with the exception of some freshman English options. Iowa State's student-built solar car made an impressive showing, finishing third, in the stock class of the 2005 North American Solar Challenge, the world's longest solar car race. "We have some world class profs here who are doing important research," one senior says. Academic and career counseling draw praise, too, and advisors are "always readily available" to help students.

Among the university's 100-plus majors, the College of Agriculture still fields outstanding programs in animal science, turf grass management, agribusiness, and agronomy.

Seventy-seven percent of ISU's students are Iowans, though all 50 states and more than 100 countries are represented in the student body. Foreign students comprise 8 percent of the student body. Iowa State was the first co-ed land grant institution, but attracting minorities has proven more difficult: minority students account for 8 percent of the student body. Hispanics comprise 2 percent, Asian Americans 3 percent, and African Americans 3 percent. "We have a lot of farmers and small-town Iowans," says a senior. "This is a pretty white campus." To help remedy this situation, ISU launched a $25 million campaign aimed at increasing the number of scholarships available for minority students, student athletes, and student leaders. In addition to need-based financial aid and a variety of athletic scholarships, many merit awards are available.

Eight thousand students live in on-campus residence halls and apartments and another 2,600 live in Greek housing. Eaton Residence Hall offers suite-style rooms. Single-sex and co-ed dorms are available, and rooms are said to be comfortable and well maintained. One student indicates that the janitors throw a picnic every spring. "The dorms are sterile to begin with but the traditions are very strong here and students transform them into home with their own personal touch," explains a junior. Special floors are available for international students, tee-totalers, and particularly studious undergraduates; separate housing is available for married students and students with dependent children. The college boasts dining at the Union Drive Marketplace.

"We have a lot of farmers and small-town Iowans."

Eight thousand students live in on-campus residence halls and apartments and another 2,600 live in Greek housing.

Iowa State is not simply located in Ames—in many respects it is Ames, but whether or not it qualifies as a college town depends on who you ask. Des Moines, the state capital, is about 30 minutes away, and Iowa City, Minneapolis, and Chicago are other easy and enjoyable road trips. Socializing tends to stay on campus, with big-name bands playing at Hilton Coliseum and parties always rocking.

There are also 500 student organizations that cater to just about any interest. The campus is officially dry but, according to one sophomore, many older students will buy alcohol for minors. The big event every spring is a weeklong campus festival called VEISHEA (an acronym for ISU's original five colleges), which features parades, exhibitions, food, and a fun-run. Another tradition is campaniling, where students must kiss under the campanile at the stroke of midnight to be considered "true" co-eds. And students have learned not to walk over the Zodiac sign in the Memorial Union—it brings bad luck.

In sports, basketball is king; the men's and women's teams are usual invitees to the NCAA tournament. The university boasts college wrestling's only four-time undefeated champion and an Olympic Gold Medal winner as the head coach. A majority of students participate in one of the largest intramural sports programs in the nation. The Hawkeye rivalry is one of the strongest in the nation.

From that first class of 28 men and two women in 1869, Iowa State has taken to heart Abraham Lincoln's land grant ideal; to open higher education to all, to teach practical courses, and to share that knowledge beyond the borders of the school. According to one junior, it's this dynamic combination that draws "hardworking, kind students" from near and far.

Overlaps

University of Iowa, University of Minnesota, University of Illinois, Purdue, Des Moines Community College, University of Northern Iowa

If You Apply To ➤

Iowa State: Rolling admissions. Admissions: Jul.1. Financial aid: Mar.1 (priority). Meets demonstrated need of 56%. SATs or ACTs: required. SAT IIs: optional. Accepts Common Application and electronic applications. No essay question.

Ithaca College

100 Job Hall, Ithaca, NY 14850-7020

Students looking at Ithaca College also apply to Boston University, Syracuse, and NYU. The common thread? Outstanding programs in the arts and media. Students also clamor to get into physical therapy. Crosstown neighbor Cornell adds curricular and social opportunities.

Over South Hill, at the center of the upstate New York Finger Lakes region, sits Ithaca College. The school has a close-knit community, along with strong programs in music, theater, communications, and health sciences, such as physical and occupational therapy. With a "gorges" campus and a size that allows for easy friendships with peers and professors alike, Ithaca draws students from all over the U.S.—and 69 other countries. IC may be overshadowed on the national scene by its Ivy League neighbor, Cornell University. But the college's focus on undergraduate education and "hands-on" learning differentiates IC from its larger rival down the road, and helps prepare students for the rigors of life outside institutional walls.

Ithaca's campus, midway between Syracuse and Binghamton, boasts breathtaking views of Cornell, as well as of the city of Ithaca and Cayuga Lake. None of the streamlined, modern campus buildings on the 757-acre plot are more than a few decades old, since the college did not move to its present location until the 1960s. The surrounding area is dotted with forests, waterfalls, and rolling hills, and of course, those ever-present gorges. In fact, author Tom Wolfe dubbed the college "the emerald eminence at the fingertip of Lake Cayuga." Current construction includes a new $14 million

Website: www.ithaca.edu
Location: Small town
Total Enrollment: 6,412
Undergraduates: 5,942
Male/Female: 44/56
SAT Ranges: V 540–640
  M 540–640
Financial Aid: 70%
Expense: Pr $
Phi Beta Kappa: No
Applicants: 10,421
Accepted: 71%
Enrolled: 21%
Grad in 6 Years: 73%
Returning Freshmen: 86%

(Continued)

Academics: ✍ ✍ ✍
Social: 🐿 🐿 🐿 🐿 🐿
Q of L: ★ ★ ★
Admissions: (607) 274-3124
Email Address:
  admission@ithaca.edu

Strongest Programs:
Music
Physical Therapy
Theatre
Communications
Natural Sciences
Psychology

The Student Activities Board, IC After Dark, and the Bureau of Concerts bring in comedians and guest speakers, show movies, and organize karaoke and open-mic nights.

Students in the School of Health Sciences and Human Performance boast of a nearly 100 percent job-placement rate at graduation.

building for the business school designed to U.S. Green Building Council standards. Future construction will include the $25 million Athletics and Events Center.

Ithaca has five schools—music, communications, business, health sciences and human performance, and humanities and sciences—and a division of interdisciplinary and international studies. Together, they offer more than 100 majors and more than 50 minors. The newest programs include Italian studies, environmental science, biochemistry, and biology. Additionally, the School of Health Sciences and Human Performance now offers a BS in outdoor adventure leadership. While all students must demonstrate proficiency in writing and statistical analysis, only the 2,500 enrolled in humanities and sciences must complete the general education curriculum, which focuses on human communities—how they form and function, and how their members express meaning. The natural sciences benefit from a $23 million facility.

"Every freshman should take a freshman seminar."

Although they praise Ithaca's programs in writing and sociology, students reserve their highest acclaim for the school's preprofessional programs—and for its conservatory of music, which dates to 1892, when the college was founded. "These programs have the most prestige and strongest applicant pool," a freshman says. The conservatory takes students by audition only, making it a destination for already-accomplished musicians and composers. Students in the School of Health Sciences and Human Performance boast of a nearly 100 percent job-placement rate at graduation. The Park School of Communications has a $12 million facility, housing programs in radio and TV production (the most popular major), photography, and cinema. Ithaca's business school has grown rapidly; students choose one of six concentrations after their first or second year.

To ease the transition to college, all students read the same book before arriving, then discuss it the day after Convocation. Some students also come to campus early for "Community Plunge," a service-learning experience. Students enrolled in the School of Humanities and Sciences may also take a freshman seminar, which meets for four hours a week instead of the usual three. These classes have about 15 students, and focus on topics such as Growing Up With Television or The Relevance of History. Professors and students together decide how to use the "extra" hour, covering themes such as personal, social, and academic responsibility. "Every freshman should take a freshman seminar," says a sophomore. "They are a great way to get to know people." The Honors Program in the School of Humanities and Sciences offers special, intensive seminars and an array of out-of-class activities to qualified students.

"Ithaca is a pretty competitive school," says a physical therapy major. A sophomore adds, "Even though the work is challenging, the teachers and students work together well to create a comfortable environment." For students tired of Ithaca's small size,

"Even though the work is challenging, the teachers and students work together well to create a comfortable environment."

crossregistration is available at Cornell and Wells College. Those seeking respite from the harsh winters may pursue internships in Washington, D.C., or Los Angeles—or a one-semester "Walkabout" in Australia, which includes stints at three different universities Down Under. In all, Ithaca offers study abroad in more than 50 countries, from Japan to Chile and Italy to Zimbabwe.

Students at Ithaca tend to be "liberal" and "artsy," and almost half hail from New York State. Many of the rest come from elsewhere in New England; 3 percent hail from other nations. African Americans and Asian Americans each comprise 3 percent of the student body and Hispanics add another 3 percent; students recognize and lament the lack of ethnic diversity on campus, and say administrators are working hard to address it. "We are a very political campus," a student says. "People speak their minds here." While tuition continues to rise, students may vie for hundreds of merit scholarships each year worth an average of $9,582. There are no athletic awards.

Seventy percent of Ithaca's students live on campus, thanks partly to the College Circle Apartments, which have full kitchens and space for 630 upperclassmen in units accommodating two to six people each. Since everyone is guaranteed space in the residence halls, first-years may find themselves in triples that used to be doubles, or packed into a common-room lounge. The situation usually sorts itself out by Thanksgiving, students report. Dorm dwellers can eat in one of three dining halls, each with a different daily menu. Seniors and some juniors also rent off-campus houses with friends from sports teams or theater troupes. These take the place of Greek houses when it comes to throwing weekend parties, since Ithaca recognizes only academic fraternities or sororities, not social organizations.

"There are many opportunities for on-campus social activities," a student says, "though most students hit parties off campus." Ithaca—described by one student as an "eclectic oasis in central New York"—still is relatively isolated, and long, cold winters mean students often make their own fun. The Student Activities Board, IC After Dark, and the Bureau of Concerts bring in comedians and guest speakers, show movies, and organize karaoke and open-mic nights. Bars and clubs in Ithaca are 18 to enter, 21 to drink, and there's a mall, bowling alley, movie theaters, and go-carting in town. "There is a lot of drinking," admits one student, "but not a lot of pressure to drink." Each fall brings Applefest, which celebrates the harvest, and

> "It's a larger school that still feels like home."

in the winter, Chilifest helps students warm up. Cornell's fraternity houses make up for the lack of Greeks at Ithaca, and the area's hilly terrain and proximity to the Shawangunk Mountains provide opportunities for hiking, biking, sledding, and skiing. Popular road trips include Syracuse and Binghamton (each an hour away, with plenty of malls), and Philadelphia, Washington, D.C., New York City, or Canada, less than six hours' drive.

Ithaca competes in Division III athletics. The biggest annual tradition is the "Cortaca Jug" football game, pitting the school against rival SUNY–Cortland. (The winner gets the jug-shaped trophy, and both schools get bragging rights for playing in the only Division III football game televised on ESPN.) Women's softball and men's baseball are recent winners, and women's crew and gymnastics are solid, too. There are two levels of intramural competition at Ithaca, depending on just how competitive you want to be, and popular options include soccer, flag football, basketball, and floor hockey. Students may also train to referee these matches—and get paid for doing so. Ithaca also offers half-credit physical education courses in badminton, scuba diving, and cross-country skiing.

If you can endure the harsh winters, you'll appreciate the small size and personal attention characteristic of Ithaca College. These qualities help it stand out among the many Northeastern schools with excellent communications programs, such as Boston University, Emerson College, and nearby Syracuse. Ithaca also boasts superior options in the health sciences, music, and theater. While it may share a city with Cornell University, Ithaca has made a reputation all its own. "It's a larger school that still feels like home," says a senior. "You meet many friendly people and you never feel like just a number."

> *The biggest annual tradition is the "Cortaca Jug" football game, pitting the school against rival SUNY–Cortland. (The winner gets the jug-shaped trophy, and both schools get bragging rights for playing in the only Division III football game televised on ESPN.)*

Overlaps

Syracuse, NYU, Boston University, Northeastern, Penn State, University of Delaware

If You Apply To ➤ **Ithaca:** Rolling admissions. Early decision: Nov. 1. Regular admissions and financial aid: Feb. 1. Guarantees to meet full demonstrated need of 52%. Campus interviews: recommended, informational. Alumni interviews: optional, informational. SATs or ACTs: required. SAT IIs: optional. Apply to particular school or program. Accepts the Common Application and prefers electronic applications. Essay question: important personal, local, national, or international issue.

James Madison University

Harrisonburg, VA 22807

JMU has carved out a comfortable niche among Virginia's superb public universities. More undergraduates than UVA and more than double the number at William and Mary. Strong in preprofessional fields such as business, health professions, and education.

Website: www.jmu.edu
Location: Small town
Total Enrollment: 16,938
Undergraduates: 15,287
Male/Female: 39/61
SAT Ranges: V 530–620
   M 540–630
ACT range: 21–26
Financial Aid: 30%
Expense: Pub $ $ $
Phi Beta Kappa: No
Applicants: 16,388
Accepted: 68%
Enrolled: 34%
Grad in 6 Years: 80%
Returning Freshmen: 91%
Academics: ✐ ✐ ✐
Social: ☎ ☎ ☎ ☎
Q of L: ★ ★ ★ ★
Admissions: (540) 568-5681
Email Address:
   admissions@jmu.edu

Strongest Programs:
Business
Communication Sciences and
   Disorders
Integrated Science and
   Technology
Health Sciences
Education
Psychology

James Madison University means business. Its business programs continue to garner national attention and attract top-notch students from coast to coast. The university has been growing at a phenomenal rate, causing some to feel growing pains. But an emphasis on undergraduate teaching, close student/faculty interaction, and a warm and welcoming climate are business as usual at JMU, so students have plenty of things to cheer about.

JMU is in the heart of the Shenandoah Valley, two hours from Washington, D.C., and Richmond, Virginia. Three types of architecture make up the campus. The buildings on Front campus have red-tile roofs and are constructed of a distinctive limestone block known as bluestone. Back campus has more modern, redbrick structures. The recently constructed College of Integrated Science and Technology campus features modern beige buildings. The university straddles Interstate 81, an outlet to several major East Coast cities. Thanks to a $100 million bond issued by the state, a new physics and chemistry building opened in 2005.

James Madison University is recognized nationally for its programs within the business major, while social sciences and education are also strong. The most popular majors at JMU are psychology, interdisciplinary liberal studies, biology, media arts and design, and marketing. Undergraduates in the biology department have even employed recombinant DNA technology to help develop organisms that produce biodegradable plastics. Also worth noting is the geology and geography depart-

> **"We have been referred to as the door-holding university because people are always willing to wait a little longer in order to hold the door."**

ments' summer geology field camp for undergraduates. Although the math department has been cited as weak, its development of a mathematical modeling laboratory is used by select undergrads to solve real-world applied math problems. Programs in the health professions, integrated science, and computer science have all seen recent expansion, and the university has added programs in biotechnology and nanotechnology.

The General Education Program requires each student to take courses in several clusters, including Skills for the 21st Century, Arts and Humanities, the Natural World, Social and Cultural Processes, and Individuals in the Human Community. The idea is to offer students a basis for lifelong learning by challenging them to become active in their own education and to explore the foundations of knowledge. Freshmen are offered a variety of programs to help smooth their transition into the university. Freshmen Adventures, held before classes begin, gives students an opportunity to meet while hiking and mountain climbing in the Shenandoah Mountains.

With undergraduates far outnumbering grad students, JMU's main mission is undergraduate teaching. "Most professors are very hands-on and active with their students," a junior says. But, students say faculty advising can be hit or miss, though if you are willing to find the time, weekly meetings with these gurus are possible. Those looking for a more intense intellectual experience can check out the honors program, which offers small classes and opportunities for independent study. Many upper-level programs encourage undergraduate participation with faculty research, another plus of this school. Now that Madison's enrollment exceeds 15,000, parking—or rather the lack of space for parking—is a constant complaint, and registering for classes can be torturous. "Registering can be hard because classes fill up fast," says a junior. Students may, however, opt for a semester in London, Paris, Florence, Salamanca, or Martinique.

You won't find a lot of diversity among fellow students at JMU. Eighty percent attended public high school, and more than 70 percent are from Virginia. In fact, there's a general effort to keep out-of-state enrollment below 30 percent. African Americans account for just 3 percent of JMU's student body, and Hispanics and Asian Americans combine for another 7 percent. Students are friendly and comfortable with each other. "We have been referred to as the door-holding university because people are always willing to wait a little longer in order to hold the door," a sophomore notes. As for political involvement, a finance major says, "You will have your occasional small group demonstrate about something." A group called Orange Band tries to get people talking about current issues. JMU offers merit scholarships averaging around $2,000 and 192 athletic scholarships in four men's sports and seven women's sports.

"I loved living on campus my first two years, and I miss it now."

Thirty-eight percent of the students live in the dorms, which run the gamut from the old high-ceiling variety to newer, air-conditioned rooms that come complete with carpet and a fitness center in the building. "I loved living on campus my first two years, and I miss it now," says a junior. Freshmen are guaranteed a room, but the school's enrollment growth means some students wind up in triples, and even sophomores chance being placed on a waiting list. Most upperclassmen opt to move off campus. Students rave about the meal plan, which has a growing number of options to choose from, including a salad bar and low-calorie meals. "So delicious!" cheers a senior.

Much of James Madison's social life is off campus, including in nearby apartment complexes. The school does not allow underage drinking on campus, but a junior says, "Everyone drinks regardless. This is a huge party school." And a sophomore says, "If you wish to participate in parties off campus, that is always an option, but if not, there are plenty of things going on on campus," such as a movies and concerts. The Greek system attracts 9 percent of the men and 11 percent of the women. Greeks and independents alike participate in JMU's many annual rites, including homecoming and Christmas on the Quad. As for road trips, the favorite destination seems to be the University of Virginia, almost an hour's drive to the south. Equally enticing, however, are the many natural delights of the Shenandoah Valley, including hiking, camping, and even skiing, all nearby. Most students find local Harrisonburg a friendly Southern town, though it doesn't necessarily embrace the college.

Sports fans here are known as the Electric Zoo and are enthusiastic about their teams, which are all Division I. Recently, however, the school eliminated eight sports, including men's cross country and track, to comply with Title IX mandates. More than two-thirds of students participate in the intramural program's many offerings.

Though JMU still has a ways to go before establishing itself as a front-rank national university, it is making considerable progress. The school is growing, but not outgrowing its Southern charm. "The students here are very friendly and outgoing. I chose JMU because when I came to take a tour, the environment was just great," a satisfied junior says.

Those looking for a more intense intellectual experience can check out the honors program, which offers small classes and opportunities for independent study.

Freshmen are guaranteed a room, but the school's enrollment growth means some students wind up in triples, and even sophomores chance being placed on a waiting list. Most upperclassmen opt to move off campus.

Overlaps

University of Virginia, Virginia Tech, College of William and Mary, University of Maryland, Mary Washington

Johns Hopkins University

3400 North Charles Street, Baltimore, MD 21218

The Hop's reputation as a premed factory can be misleading. It's apt, but Hopkins also has fine programs in international studies (with D.C. close at hand) and a variety of liberal arts fields. The national powerhouse lacrosse team is a major rallying point.

Website: www.jhu.edu
Location: Urban
Total Enrollment: 5,436
Undergraduates: 4,290
Male/Female: 54/46
SAT Ranges: V 630–740
   M 660–760
ACT Range: 28–32
Financial Aid: 49%
Expense: Pr $ $ $ $
Phi Beta Kappa: Yes
Applicants: 11,278
Accepted: 35%
Enrolled: 30%
Grad in 6 Years: 90%
Returning Freshmen: 96%
Academics: 🎓 🎓 🎓 🎓 🎓
Social: ☎ ☎ ☎
Q of L: ★ ★ ★
Admissions: (410) 516-8171
Email Address:
   gotojhu@jhu.edu

Strongest Programs:
Biomedical Engineering
Biology
International Studies
English/Writing
History
Art History
Philosophy

The typical Johns Hopkins University student is fueled by the pursuit of academic excellence and a need to achieve. This midsize Baltimore school has garnered widespread acclaim for its acclaimed professors, incredible resources, and unparalleled research opportunities. Though the university has built a reputation on its top-notch premed program, the administration has been working for a number of years to make it clear that Johns Hopkins has plenty to offer those whose interests are decidedly non-medical. Students who attend this elite university know they are at the top of the game, and they burn the midnight oil to stay there. "While the courses are rigorous, I feel that if I manage my time well, I'm fine with the demanding workload," says a classics major.

The arts and sciences and engineering schools are on the picturesque 140-acre Homewood campus, just three miles north of Baltimore's revitalized Inner Harbor. Tree-lined quadrangles, open lawns, and playing fields make for an idyllic setting on the edge of a major urban center. The architecture on this woody urban campus is mainly Georgian redbrick, with several recently built, more modern structures scattered throughout. Hodson Hall is home to high-tech classrooms, and a new chemistry building houses several research programs. Charles Commons, a new residence hall for upperclassmen, offers extensive space for student activities. Decker Quadrangle will open in mid-2007 and includes a 28,000-square-foot visitor center, an interdisciplinary computational sciences building, and a 604-space underground parking garage.

As much as some try to deny it, premeds dominate the campus. Biomedical engineering tops the list of most popular majors (followed by international studies, public health studies, biology, and neuroscience). Excellent research opportunities abound on the Homewood campus and for those who hop the cross-town shuttle to the medical campus. "Many [students] are involved in some form of research," says a junior biomedical engineering student. "That's a big part of Hopkins." Medicine at JHU plays such a major role in campus life that students sometimes fear it "overshadows the vibrant undergrad life that exists at Homewood." But administrators say the school is paying more and more attention to the undergraduate programs. Biomedical engineers also enjoy a strong department, which now includes options such as biomolecular engineering, biomechanics, and biomaterials engineering. Many say it's the hardest department at the school, thanks to its intense workload. But that doesn't deter a Hopkins student. The school has initiated a new major in environmental engineering, making it one of 40 institutions in the country to offer a bachelor's of science in the field. Also new is a major in Jewish studies.

Whether it's science or the humanities, undergraduates praise the talents and accessibility of their professors. Many are top experts in their fields and always willing to help. Although Johns Hopkins is a firm supporter of traditional scholarship, there are no university-wide requirements other than a four-course writing component. Each major has its own distribution requirements, and there are several creative seminar offerings for freshmen. Students can receive a BA in creative writing through the Writing Seminars program, where they study with authors and playwrights such as Alice McDermott and Stephen Dixon. The Humanities Center espouses a casual, interdisciplinary approach, and with maximum curriculum flexibility allowed them, undergraduates are free to range as broadly or focus as specifically as they want. Students can get a dual degree in music performance with the university's Peabody Conservatory. There also are broad "area majors," such as social sciences and behavioral sciences, humanistic studies, or natural sciences, and students can choose from a cluster of related disciplines to design their own program. Even the strictly structured engineering course plan stresses the importance of interdisciplinary and interdepartmental exposure.

> "I feel that if I manage my time well, I'm fine with the demanding workload."

Students also benefit from the well-developed graduate side of Johns Hopkins. The International Studies Program, for example, is enriched by its offerings at the university's Bologna Center in Italy, Nanjing Center in China, and at its Nitze School of Advanced International Studies in nearby Washington, D.C. Undergraduate research is also a hallmark of a Hopkins experience, with 70 percent of students having at least one research experience. The provost awards 60 grants of up to $2,500 each for undergraduate research.

Academically, Hopkins is one of the toughest schools in the country, and the workload reflects that. "There are very few slackers here," one student says. Although grade inflation is nonexistent, an international studies major declares, "It is not a cutthroat atmosphere." And a biology major says, "We always study in groups." There is a premajor faculty advisor for freshmen, and arts and sciences students are encouraged to wait until at least their sophomore year to declare a major. Freshmen may take their first semester on a pass/fail basis to ease them into the academic rigor of the place. After this "honeymoon" period, students buckle down to a Herculean workload. They get some relief from the optional January intersession, during which students can take courses or pursue independent study for one or two credits.

> "There are very few slackers here."

Hopkins students are remarkably talented and motivated, with 78 percent from the top tenth of their high school class. The image Johns Hopkins students once had as antisocial bookworms is giving way to more focus on social life. "Students are very academically driven but also know how to have a good time when the work is done," says a junior. Twenty-two percent of students are Asian American, 5 percent African American, and 6 percent Hispanic. "I think the average student is apathetic," says a senior. "It's probably because they're too concerned with schoolwork." Geographically, most students come from the mid-Atlantic states and New England.

At nearly $2 billion, Hopkins's endowment is among the top 25 in the country, and it strives to meet the full demonstrated financial need of nearly every admit. Hopkins generously rewards the extraordinarily talented with 18 hefty Hodson Trust scholarships worth $20,000 annually, regardless of need, and renewable annually for those who keep a 3.0 GPA. The Hodson Success Scholarship, based on need, replaces the loans in the aid packages of selected students from underrepresented minority groups. Twenty-four athletic scholarships are also awarded in women's and men's lacrosse, where Hopkins is a perennial national powerhouse. Word is that non-premeds are looked on with particular favor by the financial aid office.

Students say Hopkins' social scene is off campus, but there are also more than 200 clubs and student organizations, including several a cappella groups.

Academically, Hopkins is one of the toughest schools in the country, and the workload reflects that.

Freshmen and sophomores are required to live on campus in either single-sex or co-ed-by-floor dormitories. "We only have two older dorms that leave a little something to be desired, but they make up for it in the very social atmosphere," says a junior. Upperclassmen often choose to scope out the row houses and apartment buildings that surround JHU, but 25 percent are now guaranteed housing in the new Charles Commons or one of six other residence halls or university-owned "luxury" apartments.

Students say Hopkins' social scene is off campus, but there are also more than 200 clubs and student organizations, including several a cappella groups. Rowdy dorm parties and all-campus events are few and far between, but fraternity parties can be found on the weekends, though only 21 percent of the men and 22 percent of the women belong. The 21-year-old drinking limit is officially enforced, with violators facing disciplinary action. "Students can get alcohol at frat parties, but sometimes it's more difficult at local bars," one senior says.

"Baltimore is not wild and crazy, but it grows on you."

The biggest and most popular undergraduate social event of the year is the annual student-organized Spring Fair, which draws crowds from the surrounding communities as well. Downtown Baltimore and the famed Inner Harbor are not too distant, and some of the city's best attractions, such as the art museum, Wyman Park, and the funky Hampden neighborhood are right near campus. Trendy Baltimore hotspots—like Canton, Fells Point, and Little Italy—draw big crowds. Students also head downtown for plays, the symphony, films, clubs, restaurants, the zoo, and major league sports; Camden Yards, home of baseball's Orioles, is the most commodious park in the country. "Baltimore is not wild and crazy, but it grows on you," says a junior. Of course, the urban reality means parking can be hard on campus, some students say. Those with wheels can take off for Annapolis, less than an hour away. Washington, D.C., only an hour's train ride, beckons with tourist activities and "a great night life." In the warmer months, a trek out to the Delaware and Maryland beaches takes the mind off the books.

When the stellar lacrosse team—a national champion in 2005—takes the road against opponents, students often take advantage of the opportunity to road trip with them. Undergrads come together—even leaving the library at times—to cheer on their nationally acclaimed Division I Blue Jays and release some study tension. The rest of the Hopkins athletic program competes in Division III. The football, men's soccer, and tennis teams were Centennial Conference Champions in 2004–05.

With one of the world's premier medical and science programs as well as first-rate programs in areas as diverse as writing, international studies, and philosophy, and an increasing emphasis on the undergraduate experience, Johns Hopkins University is clearly among the best schools in the country. "Hopkins attracts incredible, interesting people from all over and offers fantastic career-building experiences to undergrads," says a junior. For those seeking top-notch professors, incredible resources, and unparalleled research opportunities, Johns Hopkins is hard to beat. Students truly take pride in the fact that they belong to the cream of the academic crop—premed or not.

Overlaps

Harvard, Yale, Cornell, Penn, Duke, Princeton

If You Apply To ➤ **Johns Hopkins:** Early decision: Nov. 15. Regular admissions: Jan. 1. Financial aid: Feb. 1. Housing: May 26. Does not guarantee to meet full demonstrated need. Campus interviews: recommended, informational. Alumni interviews: optional, informational. SATs or ACTs: required. SAT IIs: recommended. Biomedical engineering students must apply to that program. Accepts the Common Application and prefers electronic applications. Essay question: how you would have a day's adventure with $10. For engineering applicants: projects you enjoy, problems you enjoy solving, areas of engineering that interest you.

Kalamazoo College

1200 Academy Street, Kalamazoo, MI 49006-3295

K-zoo is a small liberal arts school that opens up the world to its students—literally. A whopping 85 percent of Kalamazoo Hornets study abroad. And if you need an extra boost to round out that resume, there is an extensive internship program.

Kalamazoo College is a small school in a small city in America's heartland. But college subsidies enable the majority of students to go abroad during their years here, making the school a launching pad to the world. In addition to international education, the school's K-Plan emphasizes teaching, internships (80 percent of students have at least one), and independent research (as seniors, all students complete a yearlong project, with one-on-one faculty supervision). Students are exposed to a demanding academic schedule and high expectations from faculty. "If you are looking for a party school," warns a student, "Kalamazoo College is not the place to come."

Life on Kalamazoo's wooded, 60-acre campus centers on the Quad, a green lawn where students ponder their destinies and play ultimate Frisbee with equal ease. With its rolling hills, Georgian architecture, and cobblestone streets, the campus has the quaint look more typical of historic New England than of nearby Kalamazoo, which, with surrounding communities, has 225,000 residents. Recent campus construction includes a $20 million expansion to the library.

Kalamazoo operates on the quarter system, and students must spend their entire first year on campus. Still, many freshmen begin the year with a "land/sea adventure," three weeks of climbing, rappelling, canoeing, and backpacking in the mountains of Canada. By the end, they're convinced they can survive anything, including the rigors of a Kalamazoo education (and the long Michigan winters). Once safely back on campus, they begin to fulfill distribution requirements, including three courses each in language, cultures, social sciences, and literature and fine arts, and two courses each in philosophy or religion and in math, natural sciences, or computer science. Also required are a first-year writing seminar and a quantitative reasoning course, as well as a senior individualized project—an internship, artistic work, directed research, student teaching, or a traditional thesis, basically anything that caps off each student's education in some meaningful way.

> **"K is full of all the kids that were presidents of all the associations and clubs at their high schools."**

After their freshman year, most of K-zoo's undergrads meet life's challenges with suitcase in hand, studying wherever their heart takes them, for the regular tuition price. A variety of off-campus programs are available, including those offered by the Great Lakes College Association.* Kalamazoo's Center for Experiential Education is another resource for information on careers, internships, and study abroad.

Kalamazoo aims to prepare students for real life by helping them synthesize the liberal arts education they receive on campus with their experiences abroad. This has filled the school with overachievers, says a freshman. "K is full of all the kids that were presidents of all the associations and clubs at their high schools." But students say they knew what to expect when they enrolled and a religion major admits to spending 70 percent of her waking hours studying. "Students are very academic and take pride in working hard and meeting the expectations they have for themselves," says one student. The natural sciences are exceptionally good and students heap praise on the psychology and languages departments. "The science program lays the foundation for students who want to continue their education and obtain a higher-level degree in the sciences," one student says. Professors give students lots

Website: www.kzoo.edu
Location: City outskirts
Total Enrollment: 1,345
Undergraduates: 1,345
Male/Female: 43/57
SAT Ranges: V 610–710
　M 600–690
ACT Range: 26–31
Financial Aid: 50%
Expense: Pr $ $
Phi Beta Kappa: Yes
Applicants: 1,800
Accepted: 68%
Enrolled: 32%
Grad in 6 Years: 86%
Returning Freshmen: 92%
Academics: 🖋 🖋 🖋
Social: ☎ ☎ ☎
Q of L: ★ ★ ★ ★
Admissions: (800) 253-3602
Email Address:
　admission@kzoo.edu

Strongest Programs:
Biology/Health Science
Psychology
English
Economics

of individual attention and are rewarded with some of Michigan's highest faculty salaries. "On a scale of one to 10 of teacher quality, I would give mine a nine," says a religion major. "The majority of them have truly inspired me and been there for me when I needed help."

Founded in 1833 and formerly associated with the American Baptist Churches, Kalamazoo is the oldest college in Michigan. Thirty percent of students come from out of state, and minorities comprise 9 percent of the student body—2 percent African American, 2 percent Hispanic, and 5 percent Asian American. Many K-zoo students crave more diversity on campus.

"It is a lot of fun to explore downtown."

The administration says it is continuing efforts to educate students on intercultural understanding. Other hot topics are women's rights and same-sex issues, students say, although debate is far from rancorous. Merit scholarships are available to qualified students.

Seventy-five percent of students live on campus. With 200 to 300 students away each quarter thanks to the K-Plan, a certain instability pervades all activities, from athletics to student government to living groups in the co-ed residence halls, where suites hold one to six students. "When the juniors come back from study abroad in the spring, students that aren't seniors are given the option of living off campus to make room for returning students," says a religion major. Dorms aren't divided by class standing, but Trowbridge is a good choice for freshmen. Those who return from abroad with bags of dirty laundry will be glad to know that dorm washers and dryers are free. While there are no Greek organizations at K-zoo, theme houses offer a more community-oriented atmosphere, including family-style dinners. But the fare in the central dining hall earns lackluster reviews. "The food here is mediocre," says a freshman. "I've learned to be pretty creative and make the most of it." Dorms require key cards for entry and students report feeling safe on campus.

The city of Kalamazoo "is a huge suburb," says one student. "If you make time for yourself it is a lot of fun to explore downtown," says another. Students volunteer at a variety of local churches and schools. On campus, students look forward to a casino night called Monte Carlo, homecoming, Spring Fling, and the Day of Gracious Living, a spring day where, without prior warning, classes are canceled and

"If you are looking for a party school, Kalamazoo College is not the place to come."

students relax by taking day trips or helping beautify the campus. (One popular T-shirt: "The end of learning is gracious living.") Off campus, Kalamazoo offers the typical collection of restaurants, theaters, and bars; K-zoo students also benefit from the physical proximity of colleges such as Western Michigan, where they may use the library or attend cultural events, as the college lacks a central place for students to hang out. Students also appreciate the city's proximity to Lake Michigan's beaches and Chicago's urban playground.

For those who equate college with big-time athletics, Kalamazoo has something to offer—even if it's not nationally televised games or tens of thousands of screaming fans. The Kalamazoo Hornets have a long-standing rivalry with Hope College, culminating in the football teams' annual competition for the "infamous wooden shoes," where the Hornets are cheered on by fans known as "the stingers." Kalamazoo also has an outstanding men's tennis team, which has won conference titles for the past 65 consecutive years. The K-Zoo men's swimming squad also finished 11th in the nation last year.

Kalamazoo is best suited for those "looking for a place where everyone is really enthusiastic about learning and thrives in that kind of environment," says one student. A classmate adds that it can be tough for students to "find a balance between the academic life and recreation." Despite the isolating nature of the program, Kalamazoo offers students a truly global education.

Overlaps

University of Michigan, Michigan State, Beloit, Grinnell, Albion, College of Wooster

If You
Apply
To ➤

Kalamazoo: Early action: Jan. 15. Early decision: Nov. 15. Regular admissions and financial aid: Feb. 15. Does not guarantee to meet demonstrated need. Campus interviews: recommended, evaluative. No alumni interviews. SATs or ACTs: required. Accepts the Common Application and requires electronic applications. Essay question.

University of Kansas

1502 Iowa Street, Lawrence, KS 66045

Memo to out-of-staters: Lawrence is not flat as a pancake and does not resemble Dorothy's home in The Wizard of Oz. The University of Kansas has a gorgeous campus and is one of the premier college bargains in the United States. Strong programs in a full slate of professional schools.

With solid academics, outstanding extracurricular programs, winning athletics and a stellar social life, the University of Kansas is one of higher education's best buys. It might be in the center of the conservative Midwest, but KU itself is an oasis of progressive activism and tolerance. And those students who are extremely dedicated can plunge into a great honors program the school provides to court them in its efforts to raise its academic profile.

The 1,000-acre campus is set atop Mount Oread ridge—once a lookout point for pioneer wagon trains—and spread out on rolling green hills overlooking valleys. Let's face it—Kansas isn't the first place that comes to many people's minds when asked to think of the most beautiful scenery. Never mind that the wooded and hilly Lawrence campus is one of the loveliest in the United States. Many of the buildings are made of indigenous Kansas limestone. But the real beauty of the campus lies in its landscape, particularly the breathtaking foliage that appears each autumn. There are nearly as many trees on campus—19,000 at last count—as there are undergrads. Recent construction includes a scholarship hall for women, a new humanities center, a ballpark for women's softball, and a Structural Biology Center. The Dole Institute of Politics—home to the world's largest Congressional archives and a World Trade Center memorial—has a new multi-media newsroom.

KU applicants apply to the individual school of their choice. Freshmen may apply to the School of Architecture, School of Engineering, College of Fine Arts or the College of Liberal Arts and Sciences. Those not admitted to one of the professional schools will automatically be considered for admission to the College of Liberal Arts and Sciences, where nearly 80 percent of the undergraduate population is enrolled. Students in the other professional schools spend their first two years completing the liberal arts requirements. The general education curriculum is intended to expose students to the foundations of the humanities, sciences, and social sciences. It also includes math, English, and oral communication, and requires courses in both Western and non-Western civilization. Foreign language and laboratory science courses are also required for all BA candidates.

"Journalism and Kansas are synonymous."

Of the 14 graduate and professional schools, those most noted for undergraduate programs are architecture and urban design, allied health, fine arts, social welfare, pharmacy, nursing, education, business, and engineering. The journalism, architecture, and business programs receive rave reviews from students. "Journalism and Kansas are synonymous," says a freshman. The lower-rated programs include math

Website: www.ku.edu
Location: Small city
Total Enrollment: 25,590
Undergraduates: 21,343
Male/Female: 49/51
ACT Range: 21–27
Financial Aid: 36%
Expense: Pub $ $ $
Phi Beta Kappa: Yes
Applicants: 10,442
Accepted: 69%
Enrolled: 59%
Grad in 6 Years: 57%
Returning Freshmen: 83%
Academics: ✍ ✍ ✍ ✍
Social: ☎ ☎ ☎ ☎
Q of L: ★ ★ ★ ★
Admissions: (785) 864-3911
Email Address: adm@ku.edu

Strongest Programs:
Architecture
Business and Economics
Engineering
Environmental Studies
Nursing and Pharmacy
Social Welfare
Premed Studies
Spanish and Portuguese

and Western civilization. The computer science department has been merged into the electrical engineering department, and the oft-maligned physical education program has been improved and reorganized under the Department of Health, Sport, and Exercise Sciences. Newer majors include Jewish studies, leadership studies, peace and conflict studies, and public service and civic leadership. On the Edwards' campus in Overland Park, students now can major in literature, language, and writing; molecular biosciences; public administration; or social work.

Students describe the academic atmosphere as challenging but laid-back. "You're rarely competing with your peers," notes a student. Classes can be large, and "large auditorium classes are taught by professors and smaller classes are taught by TAs," a sophomore reports. Twenty percent of courses taken by freshmen are taught by graduate students, but select freshmen can apply for the University Honors Program, which provides academically motivated students with honors courses, special advising, and opportunities for scholarships and research grants. In addition, freshmen in the top 20 percent of their high school class and with high ACT (composite score of 28 to 30) or equivalent SAT scores should definitely look into the Mount Oread Scholars program, which offers networking and special courses. Other options for undergraduates include independent study or the more than 75 study abroad programs in 52 countries, including Brazil, France, Germany, and Ghana. Kansas provides several area study programs supported by language instruction in more than 20 languages. The top-ranked Latin American, Spanish, and Portuguese studies programs, which benefit from an exchange with the University of Costa Rica, are three good examples. Undergraduates at KU may receive research awards to work with faculty members in publishing papers and poetry.

"We are a blue dot in the middle of a red state."

Seventy-three percent of the students are from Kansas, and most of the rest are fellow Midwesterners (many from Chicago). African Americans, Hispanics, Native Americans, and Asian Americans combine for only 12 percent of the students. "Our students have passion and pride," says a freshman, and passions are stirred over political topics including homelessness, genocide, and diversity. "We are a blue dot in the middle of a red state," quips a linguistics major. Out-of-staters must have a 2.5 high school core-curriculum GPA, be in the top third of their graduating class, or a 24 on the ACT to get in. KU gives out about 1,500 academic merit awards ranging from $1,000 to a full ride. More than 200 athletic scholarships are available to students.

Only 13 percent of Kansas students live in university housing, and both co-ed and single-sex dorms are available. Overall, students are pretty happy with the housing, much of which is being spruced up. "The recently remodeled dorms are very nice. The older dorms need to be refurbished," says a graphic design major. Students with 2.5 GPAs can live in one of the 11 scholarship halls where 50 men or women live in a cooperative-type arrangement, resulting in "a unique, close-knit, communal environment," says one participant. Groups of 20 students can live in thematic learning communities on their residence hall floors—past themes include aerospace engineering and the meaning of film. A vast majority of KU students live off campus in Lawrence apartments, which are considered expensive only by Kansas standards. A dining complex called Mrs. E's provides extended-hour access to food-court-style meals for 2,500 residence-hall occupants. "It has an amazing selection for every meal," says a student.

The university's bus system is run entirely by students and is much appreciated by tenderfeet, especially because that great big hill seems to double in size during the cold, windy winters. Lawrence, with its myriad boutiques, restaurants, and bars, receives rave reviews from students. "Lawrence is a great town for college students," a junior says. "There are a great variety of restaurants, coffeeshops, bars, and shopping. There are many ways to get involved with the community." City slickers can trek off to Topeka, the state capital, or to Kansas City, each less than an hour's drive. The KC airport makes for easy long-distance transportation, and the area is also served by Amtrak.

The Greek system, which attracts 14 percent of the men and 17 percent of the women, tends to be a major force in on-campus social life, though tension does exist between Greeks and independents. Sorority rush is completely dry, but rumor has it that the frats are a little more lenient when it comes to alcohol. "The social life here is a big part of the KU experience," says a junior. More than 400 organized groups keep things lively; other extracurricular activities include movies, poetry readings, and concerts. Scholarship halls, dorms, and other student groups sponsor large campus parties and events, but most of the social life takes place off campus.

"Lawrence is a great town for college students."

KU varsity teams—the only ones in the nation that carry the name Jayhawks—compete in the tough Big 12 Conference. The basketball team is legendary, and James Naismith, who invented basketball, was KU's first coach—and the only one with a losing record. Beloved coach Roy Williams has departed for his native North Carolina, but KU landed Bill Self, one of the top young coaches in America, and boasts that a new era for the team is underway.

The school year kicks off with Hawk Week, the official welcome for new students. The traditional "Rock Chalk Jayhawk" KU cheer is enough to bring a pang of nostalgia to the heart of even the most grizzled Kansas alum. To demonstrate their loyalty to the Jayhawks, thousands of students show up for the first basketball practice of the season at 12:01 a.m. This nocturnal tradition is lovingly labeled "Late Night in the Phog"—an allusion to the late, great coach Phog Allen. "Our 'Border War' rivalry with Mizzou is notable," says a student, and the winner of the annual football game takes possession of an "alumni spirit drum." Favorite road trips are determined by where the basketball team is playing. The women's basketball, softball, and volleyball teams are also worth watching.

With Kansas' huge number of high-ranking academic programs, its national reputation (the nonbasketball one) has certainly improved. "KU is has an incredible community feel that is hard to top," a history major says. With relatively low tuition, its academic value is also hard to top.

Undergraduates at KU may receive research awards to work with faculty members in publishing papers and poetry.

Overlaps

Kansas State, University of Texas, University of Oklahoma, University of Colorado, University of Missouri, University of Arizona

If You Apply To ➤

Kansas: Rolling admission. Financial aid: Mar. 1. Housing: Feb. 15 (priority). No campus or alumni interviews. SATs or ACTs: required. SAT IIs: optional. Apply to particular school or program. Prefers electronic applications. Essay question: choose from three topics (only for scholarship consideration).

University of Kentucky

100 Funkhouser Building, Lexington, KY 40506-0032

The state of Kentucky is better known for horses and hoops than higher education, but the University of Kentucky is working to change that. The basketball team is still a championship contender, but so too are programs in business, engineering, and health fields.

Website: www.uky.edu
Location: City center
Total Enrollment: 25,672
Undergraduates: 18,702
Male/Female: 44/56
SAT Ranges: V 510–630
   M 520–640
ACT Range: 21–27
Financial Aid: 40%
Expense: Pub $ $ $
Phi Beta Kappa: Yes
Applicants: 10,516
Accepted: 80%
Enrolled: 47%
Grad in 6 Years: 60%
Returning Freshmen: 79%
Academics: ✐ ✐ ✐
Social: ☎ ☎ ☎ ☎
Q of L: ★ ★ ★
Admissions: (859) 257-2000/
   (866) 900-4685
Email Address:
   admissio@uky.edu

Strongest Programs:
Business
Premed
Predentistry
Nursing
Engineering
Chemistry

The chemistry department has turned out three National Science Foundation fellowship winners, a feat matched only by Harvard, Cornell, Rice, Princeton, and the Massachusetts Institute of Technology.

You probably know that the University of Kentucky Wildcats are a perennial force in the NCAA postseason basketball tournament. What you may not know is that the University of Kentucky's excellence stretches beyond its winning athletic teams—into outstanding medical and premedical programs, scientific research involving both professors and students, and a social calendar packed so full of Southern tradition that it would make even the most composed debutante's head spin.

The University of Kentucky campus, home to a community college as well as major public research university, contains a mixture of old and new, modern and traditional buildings that date back to the late 1890s. The campus buildings indicate a transition beginning with the original redbrick structures to designs using contemporary glass and concrete as one moves south following the path of development. Most visitors would agree that the grounds are well-maintained, organized around the comfortable park-like spaces influenced by Frederick Law Olmsted's design. The campus contains a vast amount of mature trees and lawns set in a natural arrangement of open spaces, typical of the great land grant universities. Of course, UK's location in the heart of one of the finest horse-breeding areas in the world makes it a natural place for the Gluck Equine Research Center, a headquarters for research into horse diseases. The William T. Young Library is ranked 30th among public research libraries by the Association of Research Libraries.

Students sing the praises of many departments at UK, but several unique programs stand out. The Lexington campus is home to the Gaines Center of the Humanities, which is unusual in its study of public higher education. Lexington also hosts the Patterson School of International Diplomacy, one of the smallest yet most respected schools of its type in the country.

"UK students are typically self-assured, slightly competitive, and outgoing."

The chemistry department has turned out three National Science Foundation fellowship winners, a feat matched only by Harvard, Cornell, Rice, Princeton, and the Massachusetts Institute of Technology. Weaker areas include lower-level "monster" science classes, which one student describes as "extremely large and not at all personalized." Undergrads complain about trouble getting into courses they need, especially entry-level offerings. According to a marketing major, students "have difficulty if they are freshmen, because most of them have to take the same classes, and sometimes they don't get the right times—or the classes at all." It's hard to complete the engineering, health, business, and architecture programs in four years, students say. Term-time internships, known as co-ops, also complicate, but enliven, the picture.

Students praise UK's professors. "My professors have always shown a genuine concern for my grades," says a sophomore. TAs and full professors teach about the same number of freshman classes. The Central Advising Service, or CAS, is helping to improve the quality of academic guidance, though one student notes, "I do wish I had more assistance from my advisors on which classes to take."

To graduate, all students must take mathematics and a foreign language, as well as written and oral communication classes and a statistics, calculus, or logic course. The core program, called University Studies, also requires exposure to natural and social sciences, humanities, an introduction to cross-disciplinary education, and experience with non-Western ways of thinking. Additionally, all freshmen are encouraged to take an academic orientation class called UK101, designed to help them adjust to college life. The academic climate is laid-back, but students shouldn't expect easy A's. "When it comes to study time and class work, the students are always competing with themselves to earn the best grades they can," explains a junior.

For upperclassmen, UK offers a number of joint programs with other colleges and universities, including Transylvania, Centre, and Georgetown (in Kentucky). There's also a cooperative program with the Army and Air Force ROTC. Students studying prevet at UK will find coveted slots reserved for them at Auburn and Tuskegee in the advanced veterinary medicine program, at in-state tuition rates. UK is a member of the Academic Common Market, which provides students in 15 states the opportunity to pay in-state tuition at any of these states' schools if they want to enroll in a program not offered in their home state.

"UK students are typically self-assured, slightly competitive, and outgoing," says a psychology major (perhaps practicing analysis for her future career). The UK student body hails from all 50 counties in Kentucky, with 17 percent from out of state and 4 percent from foreign countries. The student body is predominantly white; African Americans account for 5 percent of students, and Hispanics **"In Kentucky, basketball is like a second religion."** and Asian Americans combine for 3 percent. Despite these small numbers, students say diversity is valued. "Respectfulness is an issue," says one student, "but Southern hospitality abounds." The university aims to be an "inclusive learning community," achieving academic excellence by working toward "social responsibility and community building, with particular focus on equity, fairness, and safety for each person," among other initiatives. Merit scholarships are offered to qualified students.

Kentucky's dorms are clean and convenient, as well as a great way to meet people, students say, though there's quite a range of what amenities you may get. Dorms are located on three parts of the campus—north, central, and south. North campus housing is old, but the halls are small, so they afford a chance to form close relationships. They're also within a short walking distance of classrooms, the student center, and the bookstore. South campus offers newer dorms with small rooms and air-conditioning, while central campus offers the biggest rooms. Recommended for freshmen: Kirwan-Blanding Complex, since "everything seems to happen there." Getting a room is not a problem as long as you apply by the deadline. Also, since students are not required to live on campus, only 29 percent do.

Students say that while Lexington is a great place to go to school, it's not a typical college town. "Lexington is almost 250,000 people strong," an upperclassman explains. "It's small enough to drive across town easily, but large enough not to see everyone you know when you go to Wal-Mart." Despite the lack of diversity on campus, Lexington abounds with a multitude of ethnic eateries, as well as theaters, shopping malls, and nightspots. On campus, students enjoy movies, presentations, seminars, and athletic events, the most **"My professors have always shown a genuine concern for my grades."** popular being basketball games at the legendary Rupp Arena. Other campus activities include the Little Kentucky Derby, a weeklong student-run festival that features a balloon race and concerts. Among the highlights of any student's career at UK are two one-month periods—one in the fall, one in the spring—when students spend afternoons at Keeneland Race Track enjoying the tradition of Kentucky horse racing.

About 15 percent of Kentucky men and 19 percent of the women go Greek, but fraternities and sororities offer the great majority of on-campus activities, as well as opportunities for volunteer work in the community. The university has a strict no-alcohol-on-campus policy, but it doesn't tend to affect students with fake IDs. When it's time for a road trip, UK students head to Cincinnati or Louisville (one hour away), or to Atlanta or Chicago (six hours)—that is, if they're not taking leisurely Sunday drives through nearby Blue Grass country. And the best road-trip destinations are anywhere there's a steamy, noisy gym and a basketball team ready to play UK's always-strong Wildcats. Home games at Lexington's Rupp Arena—what one student calls "a magical experience"—are consistently packed.

"In Kentucky, basketball is like a second religion," agrees another true-blue Wildcat fan. Although screaming yourself hoarse for five guys hitting the hardwood may not be as genteel as cheering while sipping a mint julep at the track, for many

Kentucky's dorms are clean and convenient, as well as a great way to meet people, students say, though there's quite a range of what amenities you may get.

Overlaps
University of Louisville, Miami University (OH), Indiana, Ohio State, University of Tennessee

students, the mix of collegiate craziness and old world Southern hospitality found in Lexington is just what they want.

If You Apply To ➤

Kentucky: Rolling admissions: Feb. 15. Financial aid: Feb. 15. Does not guarantee to meet full demonstrated need. Campus and alumni interviews: optional, informational. SATs or ACTs: required. SAT IIs: optional. No essay question.

Kenyon College

Ransom Hall, Gambier, OH 43022-9623

Kenyon is a vintage liberal arts college plunked down in the middle of the Ohio countryside. More mainstream than Oberlin, more serious than Denison, and more selective than Wooster, Kenyon is best-known for English and a small but distinguished drama program.

Website: www.kenyon.edu
Location: Rural
Total Enrollment: 1,646
Undergraduates: 1,646
Male/Female: 48/52
SAT Ranges: V 620–730
  M 610–700
ACT Range: 27–32
Financial Aid: 43%
Expense: Pr $ $ $ $
Phi Beta Kappa: Yes
Applicants: 4,368
Accepted: 32%
Enrolled: 32%
Grad in 6 Years: 88%
Returning Freshmen: 92%
Academics: ✍ ✍ ✍ ✍
Social: ☎ ☎ ☎
Q of L: ★ ★ ★
Admissions: (740) 427-5776
Email Address:
  admissions@kenyon.edu

Strongest Programs:
English
History
Political Science
Biology
Drama
Psychology
Economics

Kenyon College maintains a pure liberal arts and sciences emphasis that's less and less common in the world of higher education. Students here are proud of what sets Kenyon apart from other liberal arts colleges. "The one thing that unites us all is that we are passionate about something," explains one student. "Whether it be drama, physics, writing, activism—Kenyon students care!"

The oldest private college in Ohio, Kenyon's 1,000-acre campus sits on a hillside overlooking a scenic view of river, woods, and fields in a secluded villa of roughly 600 residents. The college's oldest building, Old Kenyon, dates from 1826 and is said to be the first collegiate Gothic building in America, and the campus is on the National Register of Historic Places. The Brown Family Environmental Center includes a butterfly garden and extensive perennial gardens planted with community donations. The sleek, modern Kenyon Athletic Center offers 263,000 square feet of fitness and recreational space, including a basketball/volleyball arena, 22-lane swimming pool, 200-meter track, four racquetball courts, and a 120-seat theater.

Kenyon's focus on liberal arts makes for a challenging, but largely noncompetitive, learning environment. "Students are very competitive with themselves, but not with each other," says a student. In fact, at Kenyon, it's hard to find a weak department—especially since there are so many opportunities for independent study. "It's hard to say there are bad departments, only apathetic and unimaginative students," says a religious studies major. English, a nationally renowned subject at Kenyon since the 1930s, is the most popular major, and it, along with the drama department (which turned out Paul Newman), sets the tone of campus life. Kenyon is, after all, the home of *The Kenyon Review*, a prestigious literary quarterly, and a school about which alum E. L. Doctorow has said, "Poetry is what we did at Kenyon, the way at Ohio State they played football."

"Whether it be drama, physics, writing, activism—Kenyon students care!" Political science is said to be solid, drawing undecided majors with its introductory class, Quest for Justice. The Integrated Program in Humane Studies, which incorporates English, history, political science, and art history, is also popular.

The hallmark of Kenyon's academic philosophy is an almost fanatical devotion to the liberal arts and sciences. "Academic life at Kenyon is rooted in three strong tenets," an administrator explains. "That students thrive when they can work closely with their professors; that they can best explore their own potential when they have enough flexibility to experiment; and that they learn most productively in an atmosphere of cooperation." A unique program is a farming independent study, which places students on nearby farms for field work each week, comments a student. Preprofessional content opportunities include 3–2 engineering programs with several universities, and with high acceptance rates to graduate programs in law, business, and medicine. On campus Summer Science research scholarships provide opportunities for collaborative research for aspiring scientists and doctors. The Career Development Center helps sort out grad schools and employment opportunities, both summer and postgraduation.

(Continued)
Modern Languages and
Literature

While there is no core curriculum at Kenyon, all students must have proficiency in a second language. Students must also complete requirements in quantitative reasoning. Change is constant at Kenyon, including changes in curriculum. American studies, which used to be a concentration, has now been designated as a major. A bevy of academic counselors, including upperclassmen and professors, help ensure that freshmen stay on the right track. About 20 percent of juniors are invited by their departments to read for honors, and about 15 percent graduate with departmental honors. The culmination of each student's coursework at Kenyon is the senior exercise, which may take the form of a comprehensive examination, an integrative paper, a research project, a performance, or some combination of these.

Classes are small at Kenyon—the average class size is 14—and even the larger introductory courses use a two-part format in which students meet for lectures one week and split up for discussion sections with the professor the next. "Our professors are unbelievably accessible, making it impossible to fall through the cracks, easy to get extra help, and rewarding to do your best," says a senior. Many profs live close to campus, which enhances the close-knit environment.

Students here are "well-rounded, passionate, and eager to learn," says a senior. Nineteen percent of Kenyon students are Ohioans, and one-third hail from New England and Mid-Atlantic states. "Kenyon students are preppy, rich, classy, liberal, well-educated, and high class," says a freshman. Students also have a global mindset. "Most

> **"You are going to work harder than you think and this is definitely not a party school."**

read the *New York Times* daily, and everyone seems to know what is going on," a senior says. African Americans account for 3 percent of the student body, Hispanics 3 percent, and Asian Americans 4 percent. The lack of diversity is noticed on campus. Kenyon awards merit scholarships ranging from $6,000 to $20,000.

Ninety-seven percent of the students live on campus, with housing guaranteed for four years. Freshmen start in five dorms at the north end of campus, and most move south to recently remodeled housing the next year. Although renovations and expansions are always in the works, they are currently having some trouble keeping up with demand. Rooms are selected via a recently modified but still sometimes harrowing housing lottery, and upperclassmen typically try to get into one the historic dorms. Housing options include a variety of spaces, from three-, four-, and six-person apartments to singles, suites, and traditional double rooms. Most dorms are co-ed. Rather than their own houses, fraternities occupy sections of the south-campus dorms, making that area the center of the party scene. "Frats have parties but these are definitely not your typical frats" says one student. Everyone, including those in the apartments with kitchens, must buy college chow; dining halls operate on each end of the campus, though only one is open on weekends. A unique aspect of the meals at Kenyon is a new service that uses foods grown by local farms in the dining halls.

The school's Greek system draws 27 percent of the men but only 9 percent of the women, and the frats throw lively parties that are open to all. Like most campuses, Kenyon is slowly moving away from the Animal House paradigm of social life. "You

The flagship sport is definitely swimming. Kenyon's swimming and diving teams dominate Division III competition, with the men's team having won a record 27 times in a row and the women capturing their 20th title.

are going to work harder than you think and this is definitely not a party school," says one freshman. But that does not mean there aren't social opportunities. "Every student performance—sports games, public presentations, music recitals, art shows—is incredibly well-attended," says a senior. Gambier is a small town, with a couple of bars and no movie theaters, but there are a few more options 10 minutes away in Mount Vernon, to which the college runs a daytime shuttle bus. On-campus events and college-sponsored activities are growing more popular to help keep boredom at bay. With its deli, market, coffeehouse, inn, restaurant, bank, and post office, Gambier is at least quaint. Students enjoy buying real maple syrup, fresh bread, and cheese from Amish farmers with stands on the main street on Saturdays.

Kenyon remains defined by its traditions, the most hallowed of which is renewed each year as incoming freshmen sing college songs to the rest of the community from the steps of Rosse Hall. Departing seniors sing the same songs at graduation. On Matriculation Day each October, after a formal ceremony, freshmen sign a book that contains the signatures of virtually every Kenyon student since the early 1800s. Other major events include homecoming and the Summer Send-Off. To break February's icy cold, the school holds a formal ball called Philander's Phling, remembering founder Philander Chase; an alum donates money for the dance. There are two small ski areas near campus, but for those seeking adventure further from home, Columbus and Ohio State University are a 45-minute drive south. The adventurous sometimes take road trips to Cleveland (home of the Rock and Roll Hall of Fame), Cincinnati, Chicago, or even Canada.

"Kenyon's rural setting is one of its greatest advantages."

Kenyon's varsity teams are known as the Lords and the Ladies. Women's tennis and men's soccer are competitive, but the flagship sport is definitely swimming. Kenyon's swimming and diving teams dominate Division III competition, with the men's team having won a record 27 times in a row and the women capturing their 20th title. Kenyon was instrumental in establishing the North Coast Athletic Conference, which includes a number of academically strong Midwestern schools, including longtime rival Denison. A junior cites the annual hockey game versus Denison, when "both teams have to drive to Newark and a surprising number of fans from both colleges attend." The school is second in the nation in NCAA Division III postgraduate scholarships. Soccer games against Ohio Wesleyan draw large crowds. Clubs sponsor everything from ultimate Frisbee to water polo.

Rituals and traditions of the past are still alive and well at Kenyon. English is the king of the castle, but other programs are also well-respected. Kenyon students are liberal, global thinkers who are as devoted to one another as their studies. While Gambier might be a bit of a culture shock for urbanites, one student says, "Kenyon's rural setting is one of its greatest advantages."

Overlaps

Oberlin, Denison, Carleton, Grinnell, Middlebury, Vassar

If You Apply To ➤ **Kenyon:** Early decision: Nov. 15, Jan. 15. Regular admissions: Jan. 15. Guarantees to meet demonstrated need of 54%. Campus interviews: recommended, evaluative. Alumni interviews: optional, evaluative. SATs or ACTs: required. SAT IIs: optional. Accepts the Common Application and prefers electronic applications. Essay question: significant educational experience; personal statement.

Knox College

2 East South Street, Galesburg, IL 61401

This friendly and progressive Illinois college was among the first in the nation to admit African Americans and women. Offers close interaction with faculty and pure liberal arts. More mainstream than Beloit and Grinnell and just over half as big as Illinois Wesleyan.

With the unconventional Prairie Fire as its mascot, Knox College has long made a name for itself by breaking away from the conventions of the day. Founded by abolitionists in 1837 as the Knox Manual Labor College, this liberal arts college has a tradition of debate that extends beyond the Lincoln-Douglas event that occurred there in 1858. And through a warm and supportive academic community, the college continues to foster a strong sense of individualism.

Located in the heart of the Midwest—almost midway between Chicago and St. Louis—the 82-acre campus has spacious, tree-lined lawns and a dynamic mixture of architecture that reflects the 145-year span of construction dates of existing buildings. Old Main, constructed in 1857, is a National Historic Landmark and the only building remaining from the 1858 Lincoln–Douglas debates.

Students say the academic relationships at Knox are infused with a spirit of cooperation and equality. Beyond the classroom, students, faculty, and administrators make decisions on boards together, each with identical voting power. First-year students confront the core issues of liberal education in Preceptorial, a one-term seminar examining questions of ethics and truth through multidisciplinary reading and critical writing. But while many schools have small, intense classes for first-year students, Knox takes things a bit farther by mandating an advanced preceptorial for seniors. This class connects their expertise in their major to a broad topic.

The general education curriculum requires students to take one course each in the arts, the natural and social sciences, and the humanities, as well as courses in writing, speaking, mathematics, information technology, foreign language, and human diversity. Knox also boasts the Ford Foundation Research Fellowship Program, which was created in the mid-1980s to encourage students to consider careers in college teach-

> "Students here are adventurous, curious, intelligent, laid-back, creative, and involved."

ing and research. Ford Fellows work with selected faculty mentors to design and carry out a research project in an area of interest. Through this permanently endowed foundation, Knox is able to offer stipends for summer research to a full one-fifth of the junior class. Moves like these have helped Knox earn a national reputation for its independent undergraduate research. More than 85 percent of students do some type of independent study.

Strong departments include creative writing, math, psychology, political science, and the natural sciences, with biology attracting lots of research grant money. The school's literary journal, *Catch*, has won national awards. Students can take part in the Chicago Semester in the Arts, and dramatists also benefit from several theaters, including one with a revolving stage. Study abroad options include programs in more than 20 countries, and the college is a member of the Associated Colleges of the Midwest Consortium.*

Knox operates on an honor system that allows students to take tests unproctored in any public area, but few students would even think of cheating. "There's not much competition among students," says a senior, "yet people push themselves and each other to do well." Where faculty is concerned, students offer uniformly glowing

Website: www.knox.edu
Location: Small city
Total Enrollment: 1,163
Undergraduates: 1,163
Male/Female: 45/55
SAT Ranges: V 570–700
  M 560–670
ACT Range: 25–30
Financial Aid: 69%
Expense: Pr $ $
Phi Beta Kappa: Yes
Applicants: 1,868
Accepted: 72%
Enrolled: 19%
Grad in 6 Years: 79%
Returning Freshmen: 89%
Academics: ✍ ✍ ✍
Social: ☎ ☎ ☎
Q of L: ★ ★ ★
Admissions: (309) 341-7100
Email Address:
  admission@knox.edu

Strongest Programs:
Biology
Creative Writing
Music
Math
Political Science
Psychology

reviews for their performance in the classroom and availability outside of it. "Professors are very passionate and knowledgeable," a student says. Eighty-five percent of classes have fewer than 25 students. Knox's trimester system packs a great deal of studying into a short period, but students are only required to take three courses per term.

Students praise Knox's student advising system. "Academic advising has been excellent for me, and advisors are always available to discuss anything ranging from future plans to personal problems," says a junior. An early identification of premed freshmen guarantees six students admission to Rush Medical College in Chicago if they maintain a four-year B average. Knox also offers 3–2 or 3–4 programs in engineering, nursing, medical technology, law, and architecture.

It takes a minor miracle for students to obtain permission to move off campus, which has become a common complaint among juniors and seniors.

"Students here are adventurous, curious, intelligent, laid-back, creative, and involved," says a senior. Forty-seven percent of students are from Illinois, and 8 percent of students come from foreign countries. Students of color make up 15 percent of the student body (5 percent African American, 5 percent Asian American, and 5 percent Hispanic) and maintain an active profile on campus.

"The dorms are spacious, well-maintained, and come in a variety of shapes and layouts."

"There is no typical student at Knox. Students come here with vastly different backgrounds, political leanings, interests, and goals," says a junior. While Knox is not a terribly politically active school, there appears to be a commitment to diversity across campus. One of the most popular forms of activism is "chalking," where students write messages in chalk on campus walkways. Most students went to public high school, and 64 percent graduated in the top quarter of their class. Merit scholarships are available, but athletic scholarships are not.

Housing is not a problem on the Knox campus. Renovations have improved housing for most students, though some students complain that most rooms are not air-conditioned. "The dorms are spacious, well-maintained, and come in a variety of shapes and layouts," says a senior. Co-ed living arrangements are available, although most freshmen live in single-sex suites with one or two upperclassmen as residential advisors. Students suggest freshman women would be happiest in Post Hall, while men should try to live anywhere in Old Quad. Older students may band together with friends or form a special-interest or theme suite. Twenty-eight percent of men belong to fraternities, while 17 percent of women join sororities. It takes a minor miracle for students to obtain permission to move off campus, which has become a common complaint among juniors and seniors. Food service, as at many colleges, "gets a bit predictable," says a senior. Security is "very visible on campus, driving around in their 'chariots of justice,' or golf carts," says a student.

Galesburg is a small Midwestern railroad town, and some students say they've had trouble adjusting to the sounds of locomotives.

Galesburg is a small Midwestern railroad town, and some students say they've had trouble adjusting to the sounds of locomotives. At one time this city of about 35,000 was a center of abolitionism, and the honorary degree that the college bestowed on then presidential candidate Abraham Lincoln was his first formal title. A political science major says, "Galesburg is a small town with a wonderful history. It has several aspects of a college town, but students usually do not look off campus for things to

"Students are going to drink, but administration will enforce disciplinary action if caught."

do." Nearby Lake Storey offers boating, water slides, and nature trails, and students looking for more excitement can travel to Peoria, about 40 miles away. Slightly farther away, Chicago is about 140 miles to the northeast. Weekends are filled with dances, campus activities, and fraternity parties. The alcohol policy is strict, but as one freshman notes, "Students are going to drink, but administration will enforce disciplinary action if caught. They are not to be reckoned with." One of the best all-time traditions is Flunk Day. At 5:30 on a spring morning, Old Main's bell rings, whistles blow, and classes are canceled to make way for dunk tanks and Jell-O pits.

Athletics generate a reasonable degree of enthusiasm. Both the men's and women's golf teams are strong—the men have won 17 of the last 19 conference championships. Every fall, the football team endures lots of hard Knox against archrival Monmouth to bring home the highly prized Bronze Turkey Award, a throwback to the time when the game was played on Thanksgiving Day.

Knox may not be a well-known school, but students here have little else to complain about. Academics are the priority and students are encouraged to be individuals, but the close-knit atmosphere helps them form strong connections with different types of students and down-to-earth professors. Says a student, "There is a 'freedom to flourish' at Knox. My opportunities are limitless."

If You Apply To ➤

Knox: Early action: Dec. 1. Regular admissions: Feb. 1. Financial aid: Mar. 1. Meets demonstrated need of 47%. Campus interview: recommended, evaluative. No alumni interviews. SATs or ACTs: optional. SAT IIs: optional. Accepts the Common Application and prefers electronic applications. Essay question: Common Application.

Lafayette College

Easton, PA 18020

Geographically close to Lehigh, but closer kin to Colgate and Hamilton. Does offer engineering, as do Swarthmore, Trinity (CT), Bucknell, and Union. Attracts relatively conservative, athletic students who work hard and play hard. A recent spate of building shows Lafayette's financial health.

Lafayette College has become one of the small elite liberal arts colleges with a huge presence abroad. Lafayette is a national leader in undergraduate faculty-mentored research and is ranked among the top colleges in study abroad participation. Its commitment to excellence in academics, and fostering an accepting environment shows in its new first-year program, which examines life in America after Sept. 11 and asks students to participate in a yearlong discussion of the country's values. The liberal arts curriculum mixes nicely with engineering in a small college atmosphere. Administrators have cracked down on the party-school reputation. Applications have increased, signalling that the world outside the once-industrial town Lafayette calls home is beginning to notice.

Lafayette is situated on a stately hill in Easton, Pennsylvania, just one and a half hours west of New York City and even closer to Philadelphia. The campus has an eclectic blend of architectural styles and more than 125 species of trees. The main library holds more than 500,000 volumes, with a 24-hour study area and online access to the card catalog from the comfort of dorm rooms. A $22 million expansion of the Skillman library has added 30,000 square feet of open learning space—a common theme in many Lafayette buildings. The 90,000-square-foot Acopian Engineering Center stays open all night and weekends, and features lots of open workspaces and glass walls to build a sense of shared purpose—and allow for extra mingling. There is also a new $28 million residential complex, complete with parking garage.

Lafayette's engineering, economics, and government and law programs are among the most popular majors, followed by English and psychology. "Even though many of the courses can be rigorous, [students] always look out for one another rather than compete against each other," says a senior. The standard course

Website: www.lafayette.edu
Location: City outskirts
Total Enrollment: 2,281
Undergraduates: 2,281
Male/Female: 52/48
SAT Ranges: V 580–670
  M 600–700
ACT Range: 25–30
Financial Aid: 50%
Expense: Pr $ $ $
Phi Beta Kappa: Yes
Applicants: 5,728
Accepted: 37%
Enrolled: 28%
Grad in 6 Years: 87%
Returning Freshmen: 96%
Academics: ✍ ✍ ✍ ✍
Social: 🍷 🍷 🍷 🍷 🍷
Q of L: ★ ★ ★
Admissions: (610) 330-5100
Email Address:
  admissions@lafayette.edu

load is four classes a semester, and all first-year students take an interdisciplinary seminar designed to engage them as thinkers, speakers, and writers. New student orientation is the beginning of a yearlong exploration of America's identity and core values in a post-9/11 society. The main text is Art Spiegelman's graphic diary, *In the Shadow of No Towers*. In the second semester of sophomore year, students take a second seminar to promote scientific and technological literacy. All students must take an intensive writing course, as well as four units in both math and natural sciences and humanities and social sciences. Students working for a BA must meet a foreign culture requirement through foreign language classwork or study abroad, or complete a group of courses providing intensive exposure to a specific foreign culture. Lafayette was noted by the Institute of International Education as having the high-

> **"Students here are generally very kind. There's a lot of door holding and smiling."**

est study abroad participation among liberal arts colleges. The EXCEL program pays students who take research positions with faculty. Marquis Scholars benefit from a free interim trip abroad and special trips like a day in New York.

Engineering students, too, may explore a foreign culture through an unusual arrangement with the Free University of Brussels, which allows them to study abroad while maintaining normal progress toward their degrees. There is also crossregistration available with other schools in the Lehigh Valley Association of Independent Colleges.* Most students are happy back on their own campus, though. "I feel like every college likes to boast that its professors give out their home numbers and invite their students over for dinner," says a senior. "Here, our president even knows our names, teaches classes, and routinely invites groups of students over for dinner with his family. His passion for teaching is representative of the college's attitude as a whole."

Students are described as "intelligent and hardworking," and 69 percent of Lafayette students are from out of state. "Students here are generally very kind," says a student. "There's a lot of door holding and smiling." Students tend to shy away from politics, although there are plenty of student groups for everything from the environment to getting out the vote. African Americans account for 5 percent of the student body; Hispanics and Asian Americans combine for another 5 percent. Hundreds of merit scholarships worth an average of $16,000 each are available to qualified students; there are six athletic scholarships available in men's and women's basketball.

Ninety-six percent of students live on campus, and housing is guaranteed for all four years. Possibilities include Greek houses as well as independent dormitories and college-owned apartments with a variety of living and eating arrangements. The school is busily renovating and modernizing some of the older dorms. Students

> **"Here, our president even knows our names."**

say the dorms are air-conditioned and comfortable overall. Experienced students recommend Keefe, South, Kirby or PT Farinon/Conway. A 60-person residence hall has special-interest floors organized around themes such as science and technology. There's a lottery system that determines which dorm a student will live in, but most get into the dorm of their choice. Ruef and South College are more social, while Watson Hall and Kirby House are quieter, students say. Most upperclassmen, including women and non-Greek males, join meal plans at fraternities or the social dorms. "The food is good enough but it gets repetitive by the end of the semester," says one student.

Greek life attracts 25 percent of the men and 45 percent of the women. For students with cars, or those willing to hop a bus or train, the bright lights of Philadelphia, New York, and Atlantic City beckon on weekends; for a change of pace, there is also hiking the Appalachian Trail. Another big "draw" is touring the nearby Binney &

The football team brought home the league football championship in 2005 and 2006. The annual football game against nearby Lehigh is intense.

Smith factory where Crayola crayons are made. The Lafayette Activities Forum plans plays, movies, concerts, and coffeehouses for those who wish to stay on campus. Parties are BYOB, and all sororities and some fraternities are dry. The arts program brings a range of performers to campus. Blue-collar Easton "gets a bad rap, but is really nice," says a senior. "A lot of restaurants on the hill, lots of nice shops, great health and beauty salons. I really like Easton." From College Hill and Downtown to the South Side, the city offers plenty of opportunities for volunteer work in schools, prisons, rehabilitation centers, hospitals, and environmental sites, under the auspices of Lafayette's Community Outreach Center.

Sports add much flavor to the Lafayette experience. The football team brought home the league football championship in 2005 and 2006. The annual football game against nearby Lehigh is intense—students claim it's the oldest rivalry in the U.S. Men's and women's basketball and indoor track and field are also strong, as are football, men's track, and soccer. All Leopard varsity teams compete in Division I except for football, which is I-AA. For those not up to varsity level, there is an extensive intramural program, buoyed by the state-of-the-art, $35 million Kirby Sports Center. The most important nonathletic campus event of the year is All-College Day, a spring festival with beach balls, bathing suits, bands, and the like.

> "Keep an open mind, get involved, and follow your interests and dreams!"

Students looking for tradition and close contact with professors—and who aren't afraid of some serious study—should take a look. "There's more to college life than drinking," advises one senior. "Keep an open mind, get involved, and follow your interests and dreams!"

Overlaps

Bucknell, Lehigh, Colgate, Villanova, University of Pennsylvania, Boston College

If You Apply To ➤ **Lafayette:** Early decision and regular admissions: Jan. 1. Financial aid: Feb. 1. Housing: Jun. 1. Campus interviews: recommended, informational. Alumni interviews: optional, informational. SATs or ACTs: required (SATs preferred). Accepts the Common Application and electronic applications. Essay question.

Lake Forest College

555 Sheridan Road, Lake Forest, IL 60045

The only small, selective private college in the Chicago area. The college generally attracts middle-of-the-road and conservative students. In the exclusive town of Lake Forest, students can baby-sit for corporate CEOs at night and get internships at their corporations during the day.

Want the bright lights and big shoulders of Chicago, along with the secluded setting of suburbia? Then consider Lake Forest College, which students call the "Enchanted Forest." Its lush, green campus is just 30 miles north of the downtown Loop—but a world away when it comes to safety, security, and the feeling of community. This small liberal arts college offers excellent programs in business, communications, and psychology, along with abundant opportunities for study abroad and professional internships at Chicagoland companies such as Avon, Motorola, and Kraft Foods. Academic improvements at Lake Forest are drawing attention; applications are up significantly since 2000, and 54 percent of the most recent freshman class ranked in the top 25 percent of their high school classes. "Students are friendly,

Website: www.lakeforest.edu
Location: Suburban
Total Enrollment: 1,378
Undergraduates: 1,375
Male/Female: 42/58
SAT Ranges: V 540–640
  M 530–650
ACT Range: 23–28
Financial Aid: 75%

motivated, and actively involved in campus life," says a senior. What's more, the school is shedding its image as a haven for spoiled rich kids, says a senior, in favor of "increasingly challenging academics, more student engagement, and more responsible students."

With its mixture of century-old Gothic and modern glass structures, Lake Forest's 107-acre campus is storybook beautiful. Located on Chicago's North Shore, in a wealthy, quiet city of 25,000, the campus has three parts: North, Middle, and South. Each has a mix of residence halls and academic buildings. The $18 million Donnelley and Lee Library offers a 24-hour cyber cafe and computer lab, wireless access to the campus network, and state-of-the-art "smart classrooms," along with space where students can work collaboratively on projects. The Mohr Student Center opened recently and offers a central gathering place for students outside of the classroom.

> **"The courses may be hard, but they are extremely beneficial and informative."**

General education requirements at Chicago's national liberal arts college include the First-Year Studies Program, "very small classes designed to help freshmen integrate into the college," says an English major. "These courses are writing intensive and offer a variety of opportunities, including trips to Chicago for plays and museum visits." Students also complete two credits in each of three liberal arts areas (humanities, social sciences and natural sciences, and math), two cultural-diversity courses, and a senior-studies capstone course. "The courses may be hard, but they are extremely beneficial and informative," says one business major. "The teachers will make sure you reach your full potential." Those seeking academic autonomy benefit from the Independent Scholar program, which allows undergrads to create their own majors outside the boundaries of traditional disciplines. Each year, about 35 students become Richter Apprentice Scholars. They participate in an interdisciplinary seminar and a 10-week paid research assistantship before their sophomore year. As upperclassmen, Richter Scholars live and work together and take part in a weekly student/faculty colloquium designed to help them learn about—and pursue—careers in academe and primary research.

Students say Lake Forest's best and most popular departments include business, economics, psychology, and the sciences. Those majoring in the latter benefit from the Student/Faculty Science Research Center, which provides lab and office space for students and faculty alike. In addition, "the communications department is excellent already and improving every year," according to a sophomore. A senior adds, "Education and biology are the most challenging majors," causing some students to avoid those departments in favor of less demanding coursework. Students who don't like what's offered at Lake Forest can create their own classes, provided they find professors to teach them. Newer offerings include a major in theater and the Beijing Program of Asian Studies, which aims to expose students to Chinese language, culture, history, and society through classroom instruction and internships.

> **"Teaching is Lake Forest College's greatest asset."**

Indeed, while Foresters like the school's small class sizes, flexible academic guidelines, and large doses of individual attention, nothing seems to compare to the quality of the faculty. "Teaching is Lake Forest College's greatest asset," says a senior. "All professors are experts in their fields and truly care about students' academic and personal development." The Career Advancement Center is run by the same person who oversees admissions, meaning that the person who brings you to Lake Forest is also looking out for you as you graduate and move into the workforce. The center offers symposia, workshops, and resume clinics, and students benefit from the college's proximity to downtown Chicago, just an hour away by train. Many pursue term-time internships in the city's business district, known as the Loop, or at nonprofits and other organizations in surrounding communities.

Those seeking academic autonomy benefit from the Independent Scholar program, which allows undergrads to create their own majors outside the boundaries of traditional disciplines.

Study abroad is also integral to the Lake Forest experience. The college's Greece and Turkey program uses archeological sites and museums to study the ancient Aegean world, while international internship programs in Paris and Santiago allow aspiring international executives to hone their language and financial skills. If those choices aren't sufficient, Lake Forest also belongs to the Associated Colleges of the Midwest,* which offers programs in Russia, Zimbabwe, Japan, India, and central Europe, among other locales.

Forty-one percent of Lake Forest students hail from the Land of Lincoln. "Some people call LFC a 'boarding school for adults,' but I think we're much more heterogeneous than that," says a sophomore. African Americans make up 5 percent of the student body, and Hispanics and Asian Americans combine for another 9 percent. Political issues include diversity, campus alcohol policies, and homosexuality. The student body is "mostly liberal to moderate, but conservatives have a loud presence," according to one senior. Generous financial aid packages are helping to bring more students of color and those from less-advantaged backgrounds to LFC. LFC awards numerous merit scholarships averaging $9,500 but does not offer athletic scholarships.

Eighty-one percent of students live in the dorms; all rooms are wired for cable and the campus computer network, and there are also computer labs on-site. Newer halls have cafes, study and social lounges (the latter with pool tables and big-screen TVs), exercise equipment, and kitchen areas. "Most people only share a bathroom with two or three other people," explains a student. Nollen and Deerpath have been renovated and given central air-conditioning; the latter "has themed suites for five or six people," says a senior. "There is a place for everyone to belong. All of the residence halls are well maintained and have a homelike feel, which is very important when making the transition from living at home to living at college," adds a music and education major. Freshmen are assigned rooms by the Dean of Students, while upperclassmen participate in a lottery based on seniority. Everybody eats in the central dining hall, where food is prepared to order at pizza, pasta, and stir-fry stations, and helpings are unlimited. There's also a prepared-foods section called "Outta Here," for students whose schedule requires meals on the go. "The food is fantastic," cheers a senior.

"All of the residence halls are well maintained and have a homelike feel."

LFC offers an active social life, with three fraternities and five sororities. They attract 19 percent of the men and 15 percent of the women, and their parties are open to all. "Greek life plays a prominent role," says one student. The Campus Entertainment Committee books movies, comedians, and big-name bands, such as Jurassic 5 and Black Eyed Peas. The Garrick Players also put on about four productions per semester. A crackdown on underage drinking and alcohol abuse is shifting the focus of LFC social life away from booze-soaked bashes. "Underage drinking is prevalent, but the judicial process is decreasing its occurrence," a senior says. Everyone enjoys the tradition of homecoming, as well as the semi-formal Winter Ball, the annual Spring Fling, and the Drag Show lip-synch contest planned by PRIDE.

The city of Lake Forest is "a bedroom community in an affluent city where the town closes early," says a student. There's a commuter train station five minutes away and the Lake Michigan beach is just as close. LFC also offers a weekend shuttle service to local malls and movie theaters, though a car is helpful. Seven student organizations are devoted to community service, and sororities and fraternities sponsor blood drives, bake sales, and car washes. One program sends students to the Appalachian Mountains in Virginia and Tennessee every spring break to help local townspeople repair substandard housing.

Varsity athletics aren't a big deal at Lake Forest; students are more likely to cheer for the intramural cricket club or the ultimate Frisbee team. Still, in 2005, 13 of the school's 17 varsity sports teams qualified for their national or conference postseason

The Campus Entertainment Committee books movies, comedians, and big-name bands, such as Jurassic 5 and Black Eyed Peas.

The college's Greece and Turkey program uses archeological sites and museums to study the ancient Aegean world, while international internship programs in Paris and Santiago allow aspiring international executives to hone their language and financial skills.

championship. The handball team claimed first place in the HACN tournament, women's soccer won a Midwest Conference championship, and softball won its fourth consecutive championship. Club and intramural sports programs attract a good number of students; ice hockey and indoor soccer are especially popular.

Although Lake Forest still has its fair share of "preppy" types, students are quick to point out that the college is taking steps to increase diversity, curb the excessive partying, and encourage school pride. "The administration is more attuned to student issues," says an English major. "Lake Forest takes very good care of its students by providing challenging academics, comfortable living, lots of entertainment and a small town feel with access to a big city," concludes a senior. "What more could anyone want?"

If You Apply To ➤

Lake Forest: Early decision: Dec. 1. Early action: Dec. 1. Regular admissions: Feb. 15. Campus interviews: recommended, evaluative. No alumni interviews. SATs or ACTs: required. SAT IIs: optional. Accepts the Common Application and electronic applications. Essay question: Personal statement. Applicants must also submit a graded paper.

Lawrence University

706 East College Avenue, Appleton, WI 54911

One of two small colleges in the nation that combines the liberal arts with a first-rate music conservatory. (Oberlin is the other.) Occupying a scenic bluff in northeastern Wisconsin, Lawrence is comparable to Beloit in size but not quite as nonconformist.

Website: www.lawrence.edu
Location: Center city
Total Enrollment: 1,405
Undergraduates: 1,405
Male/Female: 47/53
SAT Ranges: V 590–700
   M 600–690
ACT Range: 25–30
Financial Aid: 61%
Expense: Pr $ $
Phi Beta Kappa: Yes
Applicants: 2,060
Accepted: 68%
Enrolled: 20%
Grad in 6 Years: 73%
Returning Freshmen: 87%
Academics: ✍ ✍ ✍ ✍
Social: ☎ ☎ ☎
Q of L: ★ ★ ★
Admissions: (920) 832-6500
Email Address:
   excel@lawrence.edu

Lawrence University is an unpretentious school that can appeal to both the left and right side of students' brains. For those with an analytical bent, there is Lawrence's uncommon laser physics program. More creative types can take advantage of the school's renowned Conservatory of Music.

Lawrence's campus is on a wooded bluff above the Fox River, perfect for long walks, jogging, or simply meditating underneath the trees. It was chosen in 1847 by one of Appleton's earliest settlers. The pristine 84-acre campus reflects several architectural styles of the past 150 years, including classical revival, 1920s Georgian-inspired, and 1950s and 1960s institutional, unified by their limestone color. The award-winning Wriston Art Center and the Conservatory's Ruth Harwood Shattuck Hall of Music (both designed by Lawrence graduates) bring contemporary architectural touches to the campus.

The second coeducational college established in the nation, Lawrence was founded to educate German immigrants and Native Americans. While coeducation was shocking, innovators at Lawrence didn't stop there. More than 50 years ago, administrators introduced the Freshman Studies program, a required two-term course focusing primarily on the great works of art, music, and literature of both Western and non-Western origin. These days, general education requirements at Lawrence include freshman studies; distribution requirements; and diversity, foreign language, and writing-intensive courses.

At the school's Conservatory of Music, the instrument collection includes an 1815 Broadwood piano identical to Beethoven's Broadwood, and a Guarneri violin. There are first-rate jazz ensembles along with classical and world music programs; it offers a bachelor's degree in music within its liberal arts environment. "Music is the unifying theme at Lawrence," says one student. "Almost everybody plays it or studies it or likes to listen to it and talk about it." The most popular major is political science, followed by music performance, English, biology, and psychology. Students are encouraged to spend at least one term of their college career off campus. Lawrence is known for its London Study Center, which allows students **"Music is the unifying theme at Lawrence."** to take classes "across the pond" while taking advantage of the city's many cultural activities. Other off-campus programs involve the Kurgan Technical Pedagogical Institute in Russia, Waseda University in Japan, and the Ecoles des Beaux Arts in France. Programs in marine biology research are held in the Cayman Islands. In all, more than 40 off-campus programs are available.

(Continued)
Strongest Programs:
Music
History
Biology
Physics
Psychology
English

Back on campus, Lawrence students appreciate their professors' expertise and experience. "The professors are extremely knowledgeable about their fields and are very available to the students," says a freshman. Furthermore, "Freshmen are always taught by professors—there are no TAs," according to a senior. Because of the three-term calendar, the academic climate is intimate and intense. "Classes are challenging and the students are motivated to do well," says a student, "but everyone is very supportive of each other."

Most of Lawrence's students hail from Wisconsin or elsewhere in the Midwest. Three-quarters attended public high school, and 72 percent graduated in the top quarter of their class. The student body is described as a diverse group of individuals. "It's not unusual to find a classical violinist who majors in physics or a chemistry major who also minors in studio art," notes one junior. Asian Americans and Hispanics each make up 3 percent of the student body, and African Americans comprise another 2 percent. There is a sizable international population representing more than 40 countries. The political climate on campus is somewhat liberal—there's a leftist newspaper—and hot topics include same-sex marriage and the war in Iraq. There are no athletic scholarships, but brainy types vie for merit scholarships worth an average $9,270.

Programs in marine biology research are held in the Cayman Islands. In all, more than 40 off-campus programs are available.

The dorms at Lawrence are well populated; all but 3 percent of students live on campus. That's because you have to get permission to move off, which is no easy feat. Greek participation has dropped, to include 23 percent of men and 10 percent of women. One dorm is reserved for upperclassmen, and 15 small university- **"There's so much going on around campus on the weekends, students don't want to leave."** owned houses are devoted to special-interest groups. All halls are co-ed, by room or by floor, and all have laundry facilities, kitchens, televisions, Internet links, and lounges. Students praise the variety of housing choices. They report the older halls are more elegant, but the newer ones are more practical, with extra storage space and other amenities. "The dorms are pretty good but a little outdated," reports one sophomore. On-campus students have a choice of meal plans and eat in one of the two dining halls, but the meals leave some grumbling. "The food is one of the biggest complaints here," says a senior. "The food is mostly okay," says a classmate, "it's just boring."

Social life at Lawrence stays almost entirely on campus. "There's so much going on around campus on the weekends, students don't want to leave," a sophomore says. Alcohol policies aren't enforced, according to many students. Although it's almost impossible for underage students to be served at the on-campus bar, the story is different at private parties and dorm rooms. Fraternity parties, room parties, conservatory concerts, film series, coffeehouses, and art openings keep students busy. "There is usually enough going on on campus to keep everyone busy enough," says a junior. Those over 21 have numerous bars to pick from. The school radio station also broadcasts a 50-hour trivia contest in January, in which each hall has its own team,

Lawrence is known for its London Study Center, which allows students to take classes "across the pond" while taking advantage of the city's many cultural activities.

and students stay up for the entire weekend answering offbeat questions. Octoberfest is also a big weekend event, held in conjunction with the city of Appleton, which draws people in from nearby cities. A Shack-A-Thon has become a popular way to raise money for Habitat for Humanity.

There are good relations between Lawrence and Appleton. "Appleton is actually really cool," says a student. "Coming from Boston, I thought I would feel really isolated [but] I haven't experienced that." The nearest grocery store is a five-minute drive away, as is the nearest theater, and many students see a car as a necessity. Volunteerism is popular, and students regularly take part in activities such as tutoring

> **"Appleton is actually really cool."**

at local schools. The best road trips are Milwaukee (two hours), Green Bay (half an hour), and Chicago (four hours). There are also weekend seminars at Bjorklunden, the college's 425-acre estate on the shores of Lake Michigan.

And what would a Midwest fall Saturday be without football? The Lawrence team draws good crowds almost every weekend. The men's basketball team has brought home the conference championship for three consecutive years and made it to the NCAA Elite 8 in 2004. The sparkling recreation center helps students fend off midwinter blues. Participating in a rousing game of intramural broomball, which is ice hockey played on shoes with brooms as sticks and kickballs as pucks, is a must for students, even if all you do is watch. Ultimate Frisbee is another popular intramural program. "There are lots of fun intramural and club sports open to everyone," says a sophomore, "though not everyone participates."

With its outstanding liberal arts curriculum, knowledgeable and caring faculty, an administration that treats students like adults, and a charming country setting, Lawrence University is easily one of the best little-known schools in the country. "The people are just so nice and warm and welcoming here," raves a sophomore. And for students with a musical ear, Lawrence's symphony of offerings strikes just the right chord.

Overlaps

University of Wisconsin–Madison, St. Olaf, Macalester, Grinnell, Oberlin, Carleton

---

If You Apply To ➤

Lawrence: Early decision: Nov. 15. Early action: Dec. 1. Regular admissions: Jan. 15. Financial aid: Mar. 15. Guarantees to meet full demonstrated need. Campus interviews: recommended, informational. Alumni interviews: optional, informational. SATs or ACTs: optional. SAT IIs: optional. Music applicants must audition. Accepts the Common Application and prefers electronic applications. Essay question: important issue; character, historical figure or creative work that has influenced you; significant experience, achievement or person; topic of your choice.

Lehigh University

27 Memorial Drive West, Bethlehem, PA 18015

Built on the powerful combination of business and engineering, Lehigh occupies a middle ground between the techie havens, such as Rensselaer and Drexel, and the liberal arts/engineering institutions such as Bucknell and Union. By graduation, students are primed for the global job market.

Website: www.lehigh.edu
Location: City
Total Enrollment: 6,743
Undergraduates: 4,679

From the College of Arts and Sciences (the school with the highest enrollment and also the most selective) to the College of Business and Economics, Lehigh University combines the academic resources of a large research university with the collegial atmosphere of a much smaller institution. "Lehigh perfectly combines competitive academics with a great social life," says a junior.

Grand old oaks shade the buildings on Lehigh's 1,600-acre campus, which is tucked into the side of an eastern Pennsylvania mountain. Architectural styles range from ivy-covered collegiate Gothic to modern glass and steel. In an apt symbol of Lehigh's efforts to link tradition with what it takes to be part of a global workforce, the 1878 collegiate Gothic Linderman Library in the center of campus has been completely gutted and rebuilt with attention to computer access, group study areas, and a cafe. The Murray H. Goodman Campus provides first-class practice and playing facilities for Lehigh's sports teams, including a 16,000-seat stadium, a 5,600-seat basketball arena, and fields and facilities for many of the school's 25 Division I varsity sports teams.

Because much of Lehigh's reputation rests on its consistently strong engineering program, the school has invested millions of dollars to enhance critical academic programs such as optical technologies, nanotechnology, bioscience, biotechnology, and optoelectronics. Distribution requirements are divided into four domains—the mathematical sciences, the natural sciences, the social sciences, and the arts and humanities—and there are writing and foreign language requirements as well. Additionally, some degrees include a mandatory internship. Special degree options include a BA/MD seven-year program with Drexel University College of Medicine, a BA/DMD seven-year dental program with the University of Pennsylvania, a BA/OD seven-year program with SUNY State College of Optometry, and a five-year arts and engineering program, leading to BA and BS degrees. Co-ops allow students to spend eight months working for a major-related company—and getting paid to do so—while still graduating in four years.

Lehigh is big on connecting traditionally separate disciplines and prides itself on offering innovative, special programs such as Integrated Business and Engineering (IBE), Integrated Product Development (IPD), bioengineering, Information and Systems Engineering and Leadership Program (ISELP), computer science and business, environmental engineering, environmental studies, design arts, applied life science, and minors in business, ethics, and engineering. IPD—the acronym stands for Integrated Product Develop-ment—brings engineering, busi-ness, and arts students together to

"Lehigh perfectly combines competitive academics with a great social life."

design and make products for sponsoring companies. "Both the computer science and business programs are exceptional," says a senior, "but they are intense and there is not much flexibility built into those curriculums."

Lehigh also offers more than 50 study abroad options in 30 countries, with exchange programs in nine countries and winter and summer faculty-led programs in 13 countries. And every summer, the Martindale Scholars Program sends a dozen select juniors to another country, where they interview business and political leaders and write reports on the country's economy for a journal.

The workload at Lehigh is heavy; students are ambitious, and many pursue double majors. The burden is lightened somewhat by exceptional teaching. "Professors understand their students, challenge them in the classroom, and encourage them to incorporate class material into their daily lives," a junior says. Except for first-year English classes, professors teach all courses, with teaching assistants handling weekly recitation sections. Eighty-two percent of courses have fewer than 30 students and courses aren't hard to get into. Through the President's Scholars program, top Lehigh students who meet certain requirements and graduate with a 3.75 GPA or better are eligible for a fifth year of study tuition-free.

Lehigh is a wired campus in every sense of the word. A high-speed, fiber-optic campuswide network links classroom and administrative buildings, the libraries, and the computing center, and all student residences, including fraternities and sororities. There are more than 30 computing sites across campus with approximately 572 computers available for general student use. Students enjoy the convenience of

(Continued)
Male/Female: 54/46
SAT Ranges: V 610–690
  M 650–740
Financial Aid: 57%
Expense: Pr $ $ $
Phi Beta Kappa: Yes
Applicants: 10,685
Accepted: 39%
Enrolled: 28%
Grad in 6 Years: 84%
Returning Freshmen: 95%
Academics: 🐛 🐛 🐛 🐛
Social: 🍷 🍷 🍷 🍷
Q of L: ★ ★
Admissions: (610) 758-3100
Email Address:
  admissions@lehigh.edu

Strongest Programs:
Art and Architecture
Engineering
International Relations
Finance
Earth and Environmental
  Sciences
Architecture

online class registration, free high-speed Internet access in all rooms of campus buildings (including dorms), widespread wireless access, and laptops available for loan from libraries.

The Fairchild-Martindale and Linderman libraries house more than one million volumes. Lehigh has a full range of electronic indexes, reference works, full-text databases, and image databases, all of which are accessible to Lehigh students on or off campus, including students' rooms. These electronic resources are continuously being expanded and updated.

"Lehigh students are outgoing, determined, spirited, and enjoy a good time after their work," a student says. "Most come from affluent families." Less than 30 percent of Lehigh's students come from Pennsylvania, and many others hail from other Northeastern states; three-quarters are white, and the majority graduated from public schools. However, racial and ethnic diversity is improving—African Americans and Hispanics now constitute 5 percent of the student body, and Asian Americans add 6 percent. In addition, special interest housing such as the UMOJA

"Lehigh students are outgoing, determined, spirited, and enjoy a good time after their work."

House, established to encourage a sense of unity and pride among students of diverse backgrounds, exists to promote positive cultural exchanges. Nearly 400 merit scholarships are awarded annually, and student-athletes vie for 37 athletic scholarships. When it comes to politics, "we're very lethargic," says a senior.

About 69 percent of Lehigh students live on campus; first- and second-year students are required to do so. "The dorms are fine," offers one economics major, "nothing great and nothing terrible." Many upperclassmen choose to live in apartment-style dorms, Greek houses, and off-campus apartments. Other options include Campus Square, a residential and commercial complex that houses 250 upper-class students. Lehigh's dining service has been honored with the Ivy Award, given by the restaurant industry to first-class restaurants as well as to educational institutions. Students' biggest complaint is the walk uphill to the dorms, which makes gaining the "Freshman 15" near impossible.

An active Greek scene (33 percent of the men join fraternities and 39 percent of the women belong to sororities) fuels the campus social life. "Students tend to be outgoing and sociable," says one student, "with lots of frat parties and art/music events." No one under 21 is allowed to drink alcohol, and "these rules are seriously enforced," a junior says. There are 140 clubs and organizations on campus, and they sponsor concerts, comedians, performances, and other diversions, often at the striking Zoellner Arts Center, which hosts more than 75 performances per year, or the 5,600-seat Stabler Arena. Historic Bethlehem is five minutes from campus, and stu-

"There is a sense of community here that is felt from the moment you arrive on campus."

dents volunteered more than 75,000 hours last year both locally and nationally with a variety of charities, including the Boys and Girls Club and America Reads. In addition, more than 3,500 (or 75 percent) of students participate in intramural sports and club programs at Lehigh.

The bustling campus has helped revive Bethlehem, a once-great steel town in the heart of the Lehigh Valley, where some 27,000 students are enrolled in academic institutions. "Bethlehem is ever-improving," says a student. The town has drawn a number of emerging information-technology companies, many with ties to Lehigh professors and the university's mountaintop research park. Lehigh has also joined a $400 million redevelopment effort that may include a Smithsonian Institute Museum of Industry, a multiplex theater, and a large new hotel and conference center.

In early August the town hosts Musikfest, a 10-day music festival that attracts more than one million people and showcases nearly every musical style with hundreds of acts from throughout the country, international foods, arts and crafts, and free concerts. Shortly thereafter in September, Celtic Classic is underway, featuring the best of the Emerald Isle against a backdrop of autumn foliage. For those with wheels, Philadelphia is 50 miles to the south and New York City is 75 miles to the east. Skiers will appreciate the close proximity of the Poconos in the winter, while sun worshippers can enjoy the nearby Jersey shore in the early fall and late spring.

Men's basketball, softball, and men's soccer have recently earned invitations to NCAA tournaments, and swimming has won league championships. While Lehigh's Division I varsity wrestling team has been nationally strong for years, the biggest deal is still the annual sell-out football game against Lafayette. Lehigh is a Division I-AA powerhouse and its rivalry with Lafayette is college football's most-played rivalry. "I hate football," says one Mountain Hawk (the Lehigh mascot), "but I go to this game." Women's swimming is also strong, bringing home two championships recently. Even weekend warriors will find something to cheer about in the Welch Fitness Center's weight room in Taylor Gym—two pools, a climbing wall, and racquetball and squash courts.

Lehigh students proudly juggle rigorous classes and a packed extracurricular calendar. They give college life more than the old college try—and expect to succeed. Says one student, "Lehigh is a school with lots of tradition and pride. There is a sense of community here that is felt from the moment you arrive on campus. Alumni have an incredible allegiance to the school and once you are connected here, you are connected for life."

If You Apply To ➤ **Lehigh:** Early decision I: Nov. 15. Early decision II: Jan. 1. Regular admissions: Jan. 1. Financial aid: Feb. 1. Does not guarantee to meet demonstrated need. Campus interviews: optional, informational. No alumni interviews. ACTs or SATs: required. SAT IIs: recommended. Apply to particular colleges. Accepts the Common Application and electronic applications. Essay question: why Lehigh is a good match for you; respond to one of six topics.

Lewis & Clark College

Portland, OR 97219

The West Coast's leader in international and study abroad programs. Politically liberal, but not so far out as crosstown neighbor Reed. Portfolio Path to admission allows students to finesse standardized tests. With Mount Hood visible in the distance (sometimes), there is a wealth of outdoor possibilities.

The 19th-century explorers Lewis and Clark struck out from Middle America to find where the trail ended, and their travels took them to Portland, a lush, green paradise by the Willamette River. The college that bears the explorers' names encourages students to explore, too. Since 1962, more than 9,000 students and 200 faculty members have traveled to 66 different countries as part of one of the oldest off-campus student programs in the United States.

Lest students become too enchanted overseas, Lewis & Clark lures them back with a gorgeous campus perched atop fir-covered bluffs overlooking the river. The campus is an old estate, complete with elaborate gardens, fountains, and pools, where cement is almost nonexistent and the roads are paved with cobblestones. Lewis & Clark demonstrated its commitment to the environment by signing the Talloires Declaration, an agreement with other colleges to promote sustainable development.

Website: www.lclark.edu
Location: Urban
Total Enrollment: 2,853
Undergraduates: 1,940
Male/Female: 40/60
SAT Ranges: V 610–700
M 590–680
ACT Range: 26–30
Financial Aid: 60%
Expense: Pr $ $
Phi Beta Kappa: Yes

(Continued)
Applicants: 4,196
Accepted: 59%
Enrolled: 20%
Grad in 6 Years: 67%
Returning Freshmen: 86%
Academics: ✍ ✍ ✍
Social: ☎ ☎ ☎
Q of L: ★ ★ ★
Admissions: (503) 768-7040
Email Address:
admissions@lclark.edu

Strongest Programs:
Psychology
International Affairs
Foreign Languages
Biology and Environmental
    Sciences
English

The John S. Rogers Science Research Program teams students and faculty on research projects ranging from the adhesive power of geckos to molecular science.

Along those lines, the 50,000-square-foot John R. Howard Hall was built with the environment in mind, and received a gold certification from the U.S. Green Building Council. Howard Hall joins four other energy-saving, environmentally friendly buildings, including three residence halls and Wood Hall.

Lewis & Clark requires that all students achieve competency in a foreign language and international study; more than half of the students fulfill these requirements by studying overseas for a semester or more. Students may travel to countries including Australia, China, Colombia, Ecuador, Japan, Kenya, Germany, France, and Scotland, and may also study in a number of American cities—some study in two or three countries. The freshmen Inventing America class has been replaced with a program called Exploration & Discovery. In addition to the international studies requirement, students must complete courses in scientific and quantitative reasoning, creative arts, foreign languages, and physical education.

> **"Courses force students to think outside the box and form their own ideas."**

Not surprisingly, one of the most popular majors at Lewis & Clark is international affairs; others include psychology and biology. Minors in computer science, dance, and classical studies that cover the culture and history of ancient Greek and Roman forms are offered, and honors programs are available in all majors; 3–2 programs in engineering are also offered. The John S. Rogers Science Research Program teams students and faculty on research projects ranging from the adhesive power of geckos to molecular science.

Lewis & Clark offers a Portfolio Path to admission, where students present a package representing their talents and interests in lieu of standardized test scores. In addition to the essays and other items required of students who send in test scores, PP students supply three teacher recommendations and graded samples of high school work, such as essays, lab reports, or samples of art or music. Some students who use this approach feel standardized tests don't do them justice, while others have "incredible test scores." The key to a good portfolio is a "well-rounded approach," administrators say. "The more creative, the better, but be sure it's not solely artwork or writing samples."

Given this open attitude, it's not surprising that L&C students are given flexible deadlines yet are expected to follow through with a challenging workload. "L&C's academics are quite tough," says a sophomore. "Courses force students to think outside the box and form their own ideas." Freshmen and graduating seniors get priority in the registration process, helping ensure graduation in four years for those who declare majors early and plan a way to fit in all of the requirements. Professors get high marks for being knowledgeable and passionate. "They are willing to meet with students outside of class," says an art major. "This allows students to take advantage of their professor's knowledge."

"L&C is known as a hippie school," admits a sophomore. "It is a liberal campus, which is why we have a lot of free spirits." Lewis & Clark tends to attract West Coasters (almost half of the student body) who are seeking an emphasis on the liberal arts; it is also a haven for well-off Easterners who see L&C as a fresh contrast to the typical prep school or fancy suburban high school scene. The campus is politically active and predominantly left-leaning. The student body is 5 percent Asian American, 1 percent African American, and 4 percent Hispanic. Students pour their energies into community service and political activism via the numerous campus organizations and activities. "The students here are interested in making a difference," one sophomore says. "There is a high participation rate in activities and rallies here." Merit scholarships averaging $7,913 are available for qualified students, but student-athletes must go elsewhere for funding.

> **"It is a liberal campus, which is why we have a lot of free spirits."**

Lewis & Clark's residency requirement keeps students on campus their first two years; 80 percent live on campus all four years. Owing to the college's hilltop location, lucky dorm residents have views of Mount St. Helens, Mount Hood, or the Portland skyline—at least when it's not raining. Students involved in performing arts, foreign languages, outdoor pursuits, and other programs can live in theme wings. Dorms are unique and convenient. Despite L&C's location in a residential section of Portland, safety is a priority—residence halls have card-swipe entry systems and door alarms, and campus security has officers on duty 24 hours. Dining halls cater to the different diets and "the variety is great," says a freshman.

Fun-seekers at Lewis & Clark rely primarily on SOFA (Students Organized for Activities) for on-campus movies, contests, dances, and talent shows. Yearly events include Casino Night in February, the Naked Mile in the spring, and the homecoming dance in the fall. On the weekends, College Outdoors sponsors trips to Mount Hood (great skiing, about an hour distant) or the coastal beaches (an hour and a half). Seattle and Vancouver, B.C., three- and six-hours' drive, are favorite road trips, as are San Francisco and Las Vegas when there's more time. Despite the famous rains of the Pacific Northwest, the campus is officially dry. Per Oregon law, no one under 21 may drink alcohol. "You can get in trouble if you get caught," says a sophomore, "but the repercussions aren't that severe." The neighborhood immediately surrounding the college is pleasant, affluent suburbia, which means few stores,

"L&C's academics are quite tough."

restaurants, or bars. The activity of downtown Portland—mostly on Hawthorne Boulevard in the southeast section, and in the Pearl District or on 23rd Street in the northwest quadrant—is 15 minutes away on the city's public transit system or the campus shuttle service, the Pioneer Express.

As might be expected at a school in the outdoorsy Northwest, Lewis & Clark has a well-organized intramural program and plenty of outdoors activities. Men's basketball, women's crew, and volleyball are popular.

Students at Lewis & Clark College enjoy a rather laid-back atmosphere. Many are outdoor enthusiasts who are also unafraid to champion social causes. Like the school's namesakes, students are knowledge-seeking pioneers—ones who would have made Lewis and Clark, the explorers, proud. "Students here look outside their lives and experiences in order to find something greater," says a senior. "We constantly question, and search for the answers."

Fun-seekers at Lewis & Clark rely primarily on SOFA (Students Organized for Activities) for on-campus movies, contests, dances, and talent shows.

Overlaps

University of Puget Sound, Willamette, University of Oregon, Whitman, Reed, UC–Santa Cruz

If You Apply To ➤
Lewis & Clark: Early action: Nov. 15. Regular admissions: Feb. 1. Financial aid: Mar. 1. Housing: May 26. Guarantees to meet demonstrated need of 34%. Campus and alumni interviews: optional, informational. SATs or ACTs: required (except Portfolio Path). SAT IIs: optional. Accepts the Common Application and prefers electronic applications. Essay question: Common Application.

Louisiana State University

110 Thomas Boyd Hall, Baton Rouge, LA 70803

In the state that invented Mardi Gras, students come to LSU for a great time and a good education. The latter can be had in business, engineering, and life science fields. Administrators are trying to make LSU a more serious place with higher admission standards and less underage drinking.

Website: www.lsu.edu
Location: Urban
Total Enrollment: 31,264
Undergraduates: 23,633
Male/Female: 48/52
SAT Ranges: V 520–630
   M 540–660
ACT Range: 22–27
Financial Aid: 35%
Expense: Pub $
Phi Beta Kappa: Yes
Applicants: 10,825
Accepted: 73%
Enrolled: 63%
Grad in 6 Years: 59%
Returning Freshmen: 83%
Academics: ✍ ✍
Social: 🍷🍷🍷🍷🍷
Q of L: ★ ★ ★
Admissions: (225) 578-1175
Email Address:
   admissions@lsu.edu

Strongest Programs:
Biology
Chemistry/Chemical
   Engineering
English
French Studies
Geography and Anthropology
Geology and Geophysics
Mass Communication
Physics and Astronomy

Given its location and history as a sea grant college, LSU's offerings in coastal studies and coastal ecology are notable.

From abundant azaleas and Japanese magnolias, to the smell of Cajun cuisine, the sororities' antebellum mansions, and the "huge and legendary" rivalries with Ole Miss and Notre Dame, few schools evoke the spirit of the South like Louisiana State University in Baton Rouge. Though the worst of Hurricane Katrina's wrath was reserved for New Orleans (80 miles to the south), the university was directly impacted by the storm. Fortunately, things are returning to normal around the LSU campus. Enrollment has remained steady and students are again setting their sights on the future. "It is easy to get distracted from your work at such a fun school," sighs a freshman. "You must be disciplined and focused."

LSU sits on a 650-acre plateau along the banks of the Mississippi River on the grounds of a former plantation. Most of the 250 buildings are Italian Renaissance in style, with tan stucco walls and red-tile roofs. Lakes and huge oak trees dot the land-

"It is easy to get distracted from your work at such a fun school."

scape, helping to diffuse the strong sun and temper Louisiana's legendary humidity. The Manship School of Mass Communication has been renovated, and Cox Communications, headquartered in Atlanta, funded a new academic center for student athletes. There are also new facilities for the School of the Coast and the Environment, tied to LSU's history as one of the nation's 25 sea grant colleges.

For years, LSU was an open-admissions university for state residents. Standards have gone up as administrators work to make the school a national flagship university. Before enrolling, freshmen must complete 17.5 high school units in designated academic areas, including computer studies and foreign language. For out-of-state applicants, grades and test scores are weighed equally. Once on campus, students must complete a broad core curriculum, with six hours of coursework in each of three disciplines (English composition, analytical reasoning, and social sciences), nine hours in the humanities, eight to nine hours in the natural sciences, and three hours in the arts. "The academic climate is competitive," says a freshman. "I find most of my courses to be pretty challenging."

LSU students tend to be preprofessional and focused on practical courses of study, which will help them get into graduate school or find jobs after graduation. To that end, popular majors include biological sciences and psychology—both typical for premeds—as well as general studies, mass communication, and political science. Students also give high marks to LSU's programs in math and engineering, and say the nursing major is well-regarded, too. Given its location and history as a sea grant college, LSU's offerings in coastal studies and coastal ecology are notable as well. As for the professors, "every department has a teacher or two not up to par, but on the whole, the quality of teaching is great," says an electrical engineering major. "We have good professors that challenge us," says a junior.

Eighty-five percent of LSU Tigers are Louisiana natives, but that's where the similarities end. "We have different races, religions, and ethnicities here," says a junior. "We are very welcoming, hard-working, and we definitely have fun, especially during football season." African Americans make up 9 percent of the student body, Asian Americans 3 percent, and Hispanics 3 percent, though

"Dorms are a great idea for freshmen."

"many times, I find that the different types of crowds do not really interact," says a freshman. Thousands of merit scholarships are available, worth an average of $3,607 each. LSU also hands out 481 athletic scholarships each year, in eight men's and 10 women's sports.

Twenty-three percent of the students live on campus. "Dorms are a great idea for freshmen," one says. "Although you usually have to share a bathroom, they are well-maintained." All dorms are single-sex with visiting hours, one reason why upper-classmen may prefer off-campus digs. Campus dining halls offer a variety of dishes, and students say most are tasty. Despite Baton Rouge's high crime rate, students report feeling safe on campus. LSU has 70 full-time officers and a transit system so that students don't have to walk alone after dark.

Social life at LSU "is never-ending," cheers a dental hygiene major. "There is always an event to attend, or a place to hang out." Ten percent of the men and 17 percent of the women go Greek, and while doing so "is always a good way to meet people, it can be expensive," says a finance major. When it comes to drinking, you must be 21, though all bets are off on game days. "During football games, alcohol is very prevalent on the campus, and it's easy for underage people to drink," says a sophomore. Everyone looks forward to homecoming and to annual festivals such as Groovin' on the Grounds, "where big-name bands come play for the students, with lots of food and games—all free!" says a junior. Road trips to the Florida beaches are common during spring break.

Tiger football is king in Baton Rouge, having won a national championship in 2003, but the men's and women's track teams deserve equal praise, since they also brought home titles that year. The baseball team made it to the 2004 College World Series, and women's basketball is a perennial powerhouse. When the Tigers are on the road, campus tends to empty out, as students follow the team (and the fun) to Oxford, Mississippi (home of Ole Miss) or South Bend, Indiana (Notre Dame). Intramural soccer, football, and softball are popular, and even those who don't play may work out at the recreation center. It features a pool; courts for basketball, racquetball, and tennis; an indoor track; and two weight rooms, one for women only.

"I find most of my courses to be pretty challenging."

At LSU, the trees and traditions date back more than 100 years. Change may be slow in coming, but it's on the way, as the Chancellor's Flagship Agenda works to increase both class sizes and admission standards. And though it may be a while before the school's academic profile matches its athletic prowess, that's what administrators are shooting for. In the meantime, students are happy to *laissez les bons temps roulez!*

LSU students tend to be preprofessional and focused on practical courses of study, which will help them get into graduate school or find jobs after graduation.

Overlaps

Tulane, University of Texas, Texas A&M, University of Georgia, Florida State, University of Florida

If You Apply To ➤ **LSU:** Rolling admissions. Guarantees to meet full demonstrated need. Campus interviews: optional, evaluative. No alumni interviews. SATs or ACTs: required (including ACT Writing Test). SAT IIs: Optional. Prefers electronic applications. Essay question: only required for Honors College and scholarship candidates.

Loyola Marymount University

I LMU Drive, Suite 100, Los Angeles, CA 90045

There are four Loyolas and five Marymounts in American higher education, but only one has an academic dean who used to be a Hollywood producer. LMU is a Roman Catholic university known for its strategic L.A. location and strong programs in film and television, business, and communications.

At Loyola Marymount University, students are treated to perfect weather year-round, a vast array of internship opportunities, and a stellar academic lineup that includes solid programs in television and film, engineering, and business. What's more, LMU has the distinction of being the only Roman Catholic university in Los Angeles.

Website: www.lmu.edu
Location: Suburban
Total Enrollment: 8,582

(Continued)
Undergraduates: 5,419
Male/Female: 41/59
SAT Ranges: V 530–630
  M 540–640
ACT Range: 22–27
Financial Aid: 41%
Expense: Pr $ $ $ $
Phi Beta Kappa: No
Applicants: 7,733
Accepted: 56%
Enrolled: 31%
Grad in 6 Years: 72%
Returning Freshmen: 89%
Academics: ✍ ✍ ✍
Social: ☎ ☎ ☎
Q of L: ★ ★ ★
Admissions:
  (800) LMU-INFO
Email Address:
  admissions@lmu.edu

Strongest Programs:
Film Production
Accounting
Engineering
Psychology
Political Science
Theological Studies
Communication Studies

LMU's varsity teams compete in Division I and field a number of competitive teams, including men's and women's water polo, women's basketball, and men's soccer.

Established in 1911, LMU's 152-acre Westchester campus is located on a bluff overlooking the Pacific Ocean and Marina del Rey in a peaceful residential neighborhood of Los Angeles. As part of an ambitious ten-year construction program, the university recently opened two new apartment buildings and a traditional residence hall. Also, a full service diner was opened and the softball and tennis stadiums were completely reconstructed.

LMU offers more than 40 majors in five colleges. Although the general education requirements (known as the "core curriculum") differ slightly for each college, all are designed to encourage intellectual breadth. Students must take courses in American cultures; college writing; communication or critical thinking; critical and creative arts; history; literature; math, science and technology; philosophy; social sciences; and theological studies. "The nice thing about LMU's academics is that the core requirements encourage students to be open to various studies," says a freshman, "from science to theology to philosophy." Freshmen may take part in a number of programs designed to support first-year students, including an honors program. "The first-year program begins helping freshmen before they even arrive on campus," says one student. "The program seeks to integrate the freshmen so their first year will be successful."

The most popular programs include communication studies, management, marketing, biology, and psychology, and these are also among the university's strongest. Other solid programs include engineering, political science, and film production. Students in the School of Film and Television have access to a number of resources, including a student-run production office, a television stage, and a film soundstage with a professional "green bed" (for those cool CGI effects!), while those in the College of Science and Engineering take part in national competitions to design steel bridges and race concrete canoes. Thanks to its hip Los Angeles locale, LMU offers a plethora of internships to experience-hungry students, including stints at Disney, MTV, and Warner Brothers. Study abroad options include Germany, Africa, India, and New Zealand.

"The first-year program begins helping freshmen before they even arrive on campus."

Like nearby Tinseltown, LMU manages to be both competitive and laid-back. "Depending on the course, a lot of work might be required or barely any homework could be given," says a freshman. A theology major adds, "Most of the intro classes seem more laid back and the upper division classes are more rigorous." Ninety-six percent of classes have fifty or fewer students and students say teacher-student interaction is a given. "Professors are knowledgeable, enthusiastic, and very helpful outside of class," says a junior. Another student adds, "All of them encourage students to interact and to get to know them better."

LMU students hail from all fifty states and seventy foreign countries; 76 percent come from California. "In general, they are friendly and open," says a student. "There is an optimistic atmosphere on this campus because students want to be here." African Americans comprise 8 percent of the student body, Hispanics 20 percent, and Asian Americans 13 percent. Student activism is alive and well on campus and diversity is a perpetually hot issue, as is the lack of parking. Merit scholarships are available for qualified students, and student athletes vie for 130 athletic scholarships.

Sixty percent of LMU students live on campus. "The dorms—especially the upperclassmen dorms—are beautiful," gushes a sophomore. "Even the freshmen dorms are well-maintained and large." A number of themed living communities are available, including those dedicated to social action, substance-free living, and multicultural living. The university offers a variety of meal plan options and dining facilities and "the food is good," according to a theology major. Students describe campus security as good, too, and "Public Safety does an exceptional job of monitoring the campus," a freshman says.

The social life at LMU takes place "both on and off campus," says one student. "On campus, students will hang out in dorms or apartments and order Domino's." Student service organizations and clubs frequently host activities for the student body and Greek life influences the scene, too. Alcohol is readily available, but students say there is little pressure to drink. "If students want to drink, they can find alcohol," says a freshman. "Those that don't want to, don't have to." The university's Jesuit heritage promotes a social atmosphere that "motivates students to improve themselves by helping others," a student says. "Whether it's Greek life,

Thanks to its hip Los Angeles locale, LMU offers a plethora of internships to experience-hungry students, including stints at Disney, MTV, and Warner Brothers.

"There is an optimistic atmosphere on this campus because students want to be here."

service organizations, or intramurals, students have a number of possibilities." The city of Westchester is "definitely not a college town," groans a sophomore. Thankfully, Los Angeles is only a car or bus ride away, and the beach is within walking distance. "Since L.A. is a big city, there are plenty of places for a college student to eat, shop, and find entertainment." Popular road trips include San Diego, Santa Barbara, Las Vegas, and Mexico.

Back on campus, LMU's varsity teams compete in Division I and field a number of competitive teams, including men's and women's water polo, women's basketball, and men's soccer. The rivalry with Pepperdine always draws a huge crowd and "the annual pep rally—Madness at Midnight—is a pretty big event, especially for the basketball teams," says a student. Intramurals are popular, too, and include softball, basketball, soccer, volleyball, and ultimate Frisbee.

With its dynamic mix of solid academics, Jesuit tradition, and thriving social life, LMU offers students substance and style. "We're very friendly with a gorgeous campus," says a student. Whether you're a budding scientist or a future filmmaker, Loyola Marymount University may be worth a look.

Overlaps
University of Southern California, UCLA, UC–Santa Barbara, UC–San Diego, UC–Irvine

If You Apply To ➤

LMU: Financial aid: Mar. 1. Housing: May 1. Does not guarantee to meet demonstrated need. Campus interviews: optional, informational. No alumni interviews. SATs or ACTs: required. SAT II: optional. Accepts the Common Application and electronic applications. Essay question.

Loyola University New Orleans

6363 St. Charles Ave., Box 89, New Orleans, LA 70118

Of the four Loyolas in the nation, this is the only one where you can go to Mardi Gras and still get up in time for class. New Orleans is an ideal setting for this Roman Catholic university with strengths in business, communications, and the arts. More progressive than any other Deep South location.

In the aftermath of Hurricane Katrina, Loyola University's president calls upon students to "Be part of the resurrection." It's a clever turn of phrase, and one that sums up the faith and resilient character of this Jesuit liberal arts school. Signs of recovery are everywhere, from the restored residence halls to the array of activities and festivities created to help students reconnect and reestablish themselves with the Loyola community. It appears to be working; nearly 90 percent of the university's undergraduates

Website: www.loyno.edu
Location: Urban
Total Enrollment: 4,604
Undergraduates: 2,991
Male/Female: 41/59

SAT Ranges: V 560–690
  M 540–650
ACT Range: 23–28
Financial Aid: 52%
Expense: Pr $
Phi Beta Kappa: No
Applicants: 3,021
Accepted: 58%
Enrolled: 30%
Grad in 6 Years: 63%
Returning Freshmen: N/A
Academics: ✍ ✍ ✍
Social: ☎ ☎ ☎
Q of L: ★ ★ ★
Admissions: (504) 865-3240
Email Address:
  admit@loyno.edu

Strongest Programs:
Music
Music Industry
Advertising/Public Relations
Journalism
Biological Sciences/Premed
International Business
English/Creative Writing
Psychology

have returned to complete their education at Loyola directly following Hurricane Katrina. According to one communications major, the university is continuing to do what it does best: "encourage social justice and educate the whole person."

The school's attractive and well-kept 20-acre main campus, in the University section of Uptown New Orleans, mixes Tudor, Gothic, and modern structures. It overlooks acres of Audubon Park and, beyond, the mighty Mississippi River. Two blocks up St. Charles Avenue, Loyola's Broadway campus has an additional four acres. The J. Edgar and Louise S. Monroe Library houses 500,000 volumes, the Lindy Boggs National Center for Community Literacy, and an art gallery. There are multimedia classrooms and wireless Internet connections throughout the campus. Although Katrina left the university with $5 million in damages, the entire campus has been restored and residence halls have been opened to students.

Loyola offers comprehensive undergraduate degree programs in the College of Humanities & Natural Sciences and in the College of Social Sciences. The School of Mass Communications wins points with students. Also in demand is the international business program in the College of Business and virtually any major in the College of Music and Fine Arts. Newer programs include a multitrack music industry program, designed to serve both creative and business-minded students, a minor in jazz studies, and courses in music management and finance. A major and minor in forensic science are now part of the chemistry department.

There are no teaching assistants, and 91 percent of professors hold doctorate or terminal degrees in their field. "I have been taught by people who are at the top of their respective fields," a junior says. In addition, "Industry leaders and local executives are hired to give students the opportunity to experience and learn firsthand practical strategies and techniques," says a senior. Students take 24 credits of introductory courses and 24 more of advanced courses in English, history, math, philosophy, religious studies, and natural science. "The courses are a healthy challenge," says a political science major, "but very educational." First-year students participate in a three-day orientation followed by a comprehensive first-year experience that includes a common reading program, a series of lectures and panel discussions, educational excursions, exhibits, and service learning projects coordinated around a common academic theme. An Executive Mentoring program lets freshman business students meet regularly with local business leaders to discuss their career and personal development. A peer mentoring program helps new students adjust to college life during their initial semester. Loyolans also benefit from the New Orleans Consortium, with crossregistration and library access at other schools in the area. Study abroad programs take students to Belgium, Mexico, Spain, Berlin, Germany, Japan, and the Netherlands. Summer programs are offered in the Bahamas, Belgium, London, Mexico, Paris, Spain, and Europe.

A marketing major describes Loyola's student body as an "intellectual microcosm." Forty-two percent of Loyola students are Louisiana natives and many of the remaining students are from the Southeast. Religion—specifically Roman Catholicism—has a significant influence on campus. Daily mass is voluntary, but many students attend. Hispanics constitute 12 percent of the student body, African Americans 11 percent, and Asian Americans 4 percent. Recent hot-button issues at Loyola have included post-Katrina recovery, poverty, the 2008 presidential election, the war in Iraq, and that perennial student gripe, campus parking. Loyola awards merit scholarships each year and the school's first athletic scholarships, three male and three female basketball awards, were recently established.

Fraternities and sororities are rarities at Jesuit schools, but are popular at Loyola, with 20 percent of the men and 23 percent of women choosing to belong.

"I have been taught by people who are at the top of their respective fields."

Most Loyola students commute from home or off-campus apartments. Campus dining is good and students may choose from an array of culinary delights, including "a vegan bar, a salad bar, sandwiches, a grill with burgers and fries, and a bakery," says a history major. As for security, the campus is well lit and students feel safe but acknowledge that surrounding areas are still in various states of disrepair.

Students volunteer their sweat equity with the Loyola University Community Action Program, a coalition of 11 organizations that provides community service opportunities. With the help of a new service learning office, about 500 students make service learning part of their studies. Fraternities and sororities are rarities at Jesuit schools, but are popular at Loyola, with 20 percent of the men and 23 percent of women choosing to belong. Major campuswide social events include the Gator Crawl, Wolves on the Prowl community service day, Loyolapalooza spring music festival, and the Fr. Carter lecture series. February brings Mardi Gras, of course—the school shuts down that week. As for underage drinking, Louisiana law requires that you be at least 21 to buy alcohol, but only 18 to consume it in a private residence. While Loyola maintains

> "Overall, Loyola is an awesome small Jesuit college with a lot to offer."

that dorms are private residences, the school has also established a Coalition to Reduce Underage Drinking. But as one student says: "We live in New Orleans. They can't be too strict."

Loyola fields baseball, basketball, women's soccer and volleyball teams in the Gulf Coast Athletic Conference, NAIA Division II. Basketball, men's flag football (no real pigskins at Loyola), and women's volleyball are popular pastimes.

Students at Loyola know how to pull together and draw strength from their faith. Whether they're working closely with caring professors or relaxing with friends amid the Big Easy's boundless energy, students are satisfied with their choice. "Overall, Loyola is an awesome small Jesuit college with a lot to offer as a community and academic institution," a sophomore says.

An Executive Mentoring program lets freshman business students meet regularly with local business leaders to discuss their career and personal development.

Overlaps
Louisiana State, Tulane, University of New Orleans, Xavier, Spring Hill, St. Louis

If You Apply To ➤ **Loyola:** Rolling admissions. Financial aid: Mar. 1 (priority). Housing: May 1. Meets demonstrated need of 81%. Campus interviews: recommended, informational. Alumni interviews: optional, informational. Audition required for admission to the College of Music and Fine Arts. Portfolio required for admission to the Visual Arts Program. SATs or ACTs: required. SAT IIs: optional (writing is for placement only). Accepts the Common Application and electronic applications. Essay question.

Macalester College

1600 Grand Avenue, St. Paul, MN 55105

Former U.N. Secretary-General Kofi Annan, '61, typifies one of Mac's hallmarks: an internationalist view of the world. Carleton has a slightly bigger national reputation, but Mac has St. Paul, a progressive capital city. Mac is the only leading Midwestern liberal arts college in an urban setting.

Macalester College is an international island in the heart of the Great Plains. Liberal does not only describe its curriculum; it also describes its politics. Students here get riled up over all sorts of issues with local, national, or international import—from sweatshops and fair trade to the war in Iraq and gay rights. With its Scottish roots and international focus, Mac is worth a look.

Macalester is located in a friendly, family-oriented neighborhood in St. Paul, Minnesota, one mile from the Mississippi River, which divides St. Paul from Minneapolis. Summit Avenue, a tree-lined street with the longest, best-preserved stretch

Website: www.macalester.edu
Location: City outskirts
Total Enrollment: 1,827
Undergraduates: 1,827
Male/Female: 42/58
SAT Ranges: V 630–740
M 630–710

(Continued)
ACT Range: 28–32
Financial Aid: 69%
Expense: Pr $ $
Phi Beta Kappa: Yes
Applicants: 4,317
Accepted: 44%
Enrolled: 26%
Grad in 6 Years: 85%
Returning Freshmen: 93%
Academics: 🖉 🖉 🖉 🖉 ½
Social: ☎ ☎ ☎
Q of L: ★ ★ ★ ★
Admissions: (651) 696-6357
Email Address:
   admissions@macalester.edu

Strongest Programs:
Biology
Chemistry
Economics
Classics
International Studies
Math
Computer Sciences

Mac emphasizes collaboration and working together to handle the challenging workload, and students say most pressure to do well comes from within.

While Mac students may harbor radical political and social viewpoints, they live in traditional residences.

of Victorian homes in the nation, forms the campus's northern boundary. The self-contained, 53-acre campus is arranged around 115-year-old Old Main, a splendid Victorian structure listed on the National Register of Historic Places. The unifying theme is redbrick, the better to set off the octagonal Weyerhauser Chapel, constructed of black glass. In the past decade, Mac has spent more than $98 million to improve campus facilities.

Mac's general requirements include two courses in social sciences and two in natural sciences and math, plus one to two courses in fine arts and in humanities. Two courses must address cultural diversity, in the U.S. and internationally. Every student also completes a capstone experience during their senior year, such as an independent research project, performance, artistic work, or original work.

"Hippies hang out with kids in pink polo shirts."

Mac's academic strengths include economics, chemistry, and biology; the school's impressive science facilities include an observatory, an animal operant chamber, and labs for electronic instrumentation and laser spectroscopy.

Mac emphasizes collaboration and working together to handle the challenging workload, and students say most pressure to do well comes from within. "Courses are rigorous and challenging," says a junior, "but students do not compete with each other." Teaching is paramount, with professors often having students over for dinner or taking their students for drinks at a local watering hole. "The teaching at Macalester is second to none," says a senior. "Professors always have your best interest at heart and are always challenging you academically." A geography major adds, "They work hard to be good teachers and get to know their students as individuals." Mac attracts the best and the brightest, which might account for the competitive appearance of some programs. More than 100 students do stipend-supported research with Mac professors each summer, and since a number of faculty members play intramurals, students may find professors dishing off passes on the basketball court. Fifty percent of students participate in study abroad through approved independent programs or the Associated Colleges of the Midwest.* Before graduation, 60 percent of students complete an internship at a Twin Cities business, law firm, hospital, financial institution, government agency, or non-profit organization.

Macalester students are an open-minded, friendly bunch, more politically and socially progressive than their peers at similar institutions. "Beyond that, there's no 'Macalester type,'" says a student. "Hippies hang out with kids in pink polo shirts." Twenty-one percent of Macalester students hail from Minnesota, and the rest come from every state, the District of Columbia, and dozens of other countries. Despite

"Each day is an adventure in global cuisine."

Mac's small size, the student body is 4 percent African American, 4 percent Hispanic, and 7 percent Asian American—and a whopping 12 percent of the student body are international. Political debate is lively. "Mac students are activists," says a senior. "Whether working on political campaigns, raising awareness for AIDS victims in South Africa, or voicing discontent with college policies, Macalester students make their opinions known." Says another student: "Everyone is left of liberal, except the economics majors." Merit scholarships worth an average of $4,495 each are available to brainy types; there are no athletic scholarships.

While Mac students may harbor radical political and social viewpoints, they live in traditional residences, with double rooms for the first two years, when they're required to live on campus, and suites for upperclassmen. Single-sex floors are guaranteed to those who want them; many students have single rooms, and some live in language houses. Residents of the kosher house prepare their own meals, while the opening of the Campus Center has vastly improved chow elsewhere on campus. "Cafe Mac is one of the reasons I came here," says a student. "Each day is an adventure in global cuisine." Seventy-five percent of students remain in college-owned digs, and more say they'd stay, if only they could get rooms. Still, off-campus living is plentiful. Nearby neighborhoods are welcoming to students, with low-cost houses and apartments and families eager to have them move in. Food is unanimously praised for its tastiness and variety. "It's the best food I've ever had at a college campus," says an international studies major.

Without Greek organizations and given the proximity of a major metropolitan area, most of Mac's social life takes place off campus, although there are plenty of events in the "Macalester bubble" for those loath to leave. Minneapolis and St. Paul are close, with their bookstores, coffee shops, restaurants, bars, and movie theaters, plus dance and jazz clubs and professional sports teams. The Mall of America is also nearby, though Mac students tend to tire of it quickly, and the Twin Cities' public transportation isn't the greatest. Still, with about a dozen colleges and universities in town, there's plenty to do and the Twin Cities are an excellent place to live. For those with wheels, the best road trips include Chicago, Madison, and Duluth—and Bemidji, Minnesota, "to see Babe the Blue Ox," reports a senior.

Popular events include Spring Fest and the annual Brain Bowl football game against in-state rival Carleton, "the only school we can't chant 'We are smarter than you!' to," says a communications major. Although men's and women's soccer frequently bring home conference championships—joining the debate team in the winner's circle—the math and computer programming teams are the ones typically competing internationally. The Club Sports Program provides opportunities for students to participate in a variety of sports and recreational activities including crew, rugby, and hockey. More than half of the student body participates. "Although the student body seems to take pride in being much more focused on academia, whenever the weather's nice students dot the lawns playing Frisbee, soccer, or cricket—yes, cricket!" says one student.

> "The teaching at Macalester is second to none."

Macalester provides an atmosphere of high-powered scholarship and success, pairing academic rigor with global perspective. As the school's story travels, the skill and diversity of the student body is rising. "I love its multiculturalism, diversity, small-but-lovely campus, wonderful location, and, above all, its academic excellence," raves an economics student.

Although men's and women's soccer frequently bring home conference championships—joining the debate team in the winner's circle—the math and computer programming teams are the ones typically competing internationally.

Overlaps

Carleton, Grinnell, Oberlin, Wesleyan, University of Chicago, Vassar

If You Apply To ➤ **Macalester:** Early decision: Nov. 15, Jan. 3. Regular admissions: Jan. 15. Financial aid: Feb. 8. Housing: May 1. Guarantees to meet full demonstrated need. Campus interviews: recommended, evaluative. Alumni interviews: optional, evaluative. SATs or ACTs: required. SAT IIs: optional. Accepts the Common Application and electronic applications. Essay question: why Macalester—what you can add to the community; and Common Application essay.

University of Maine–Orono

Orono, ME 04469

A sleeper choice for out-of-staters amid better-known public universities such as UMass, UNH, and UVM. Not coincidentally, Maine is the least expensive of the four. A popular marine sciences program flourishes here, as do forestry and a range of preprofessional programs.

At the University of Maine–Orono, more than 8,500 undergraduates help themselves to a range of strong academic programs at a reasonable cost. UMaine is not only the state's only land grant university, it's also the only sea grant as well, and attracts top students to its marine sciences program. UMaine has become more academically competitive over the past several years. "It is becoming harder to get in because the school is well-known and very academic. A lot of the classes are competitive and have a large workload," explains a junior social work major.

Situated on an island between the Stillwater and Penobscot rivers, UMaine's campus is 660 acres, centered on a large, tree-shaded grass mall. Architectural themes at this flagship of the state university system range from English academic to contemporary. The Advanced Manufacturing Center is a new 30,000-square-foot engineering building, and Lord Hall is being renovated to house art facilities.

UMaine's five undergraduate colleges are education and human development; business, public policy, and health; engineering; liberal arts and sciences; and natural sciences, forestry, and agriculture. The growing Honors College, which has arisen out of one of the oldest honors programs in the nation, now enrolls about 700 students. Specific general education requirements vary from college to college, though all students must demonstrate writing proficiency and take two physical or biological science courses, 18 credits in human value and social context, six credits in math (including statistics and computer science), and at least one ethics course. A capstone experience in the major is also mandatory.

> "It is becoming harder to get in because the school is well-known and very academic."

The engineering programs are widely viewed as the most demanding on campus. Other best bets include new media, business, marine science, and the health professions. The interdisciplinary Institute for Quaternary Studies collaborates with other research centers around the world in focusing on the Quaternary period, a time of glacial and interglacial cycles leading up to the present. Research is a key part of an undergraduate education at UMaine, and is woven into many areas of the curriculum.

The university library, the state's largest, is the regional depository for American and Canadian government documents and houses some of alumnus Stephen King's papers. Former U.S. Senator William S. Cohen, a UMaine faculty member before he became defense secretary in the Clinton administration, donated his personal papers to the university as well. The papers, which chronicle Cohen's 24-year congressional career, will be used to develop a nonpartisan center on international policy and commerce focused on teaching, research, and public service, and named in Cohen's honor. They are also available to scholars interested in Cohen's career.

UMaine's Academic and Career Exploration program lets students work with professionals in different areas before declaring their degree choices. "It's becoming harder and harder to get in," a freshman says. "The classes are rigorous and require the student to work hard." Students report that the quality of teaching is high. "The teachers are very personable and really care about your performance," says a senior.

Outside the classroom, internships and co-ops are available in most fields, and there's a Semester-by-the-Sea and a Lobster Institute for nautical types. Juniors who want a reprieve from Maine's frigid winters can head for Brazil, while the heartier types choose Canada, Scandinavia, and Ireland. Most students, however, are immune to the weather, since 85 percent are from Maine and many of the rest hail from other parts of New England. The campus is 93 percent white but viewpoints are diverse, students say. "It's a fairly liberal campus," a senior says. "People are respectful of differing opinions." Merit scholarships offer an average of $5,272 a year for qualified students, and 298 athletic scholarships are available.

> "Dorms are a great way to meet new people."

Forty-three percent of UMaine students live on campus; the remainder seek shelter in Orono, nearby Bangor, or the sparsely populated area in between. Dorms are co-ed; some have gyms, computer labs, or apartment-style suites. "Dorms are a great

Outside the classroom, internships and co-ops are available in most fields, and there's a Semester-by-the-Sea and a Lobster Institute for nautical types.

way to meet new people," a business major says. "They are very well maintained." Some housing or wings are set aside for specific majors. Dining hall food is pretty good and catered to special tastes, but a sophomore warns "the food is the same each day, (with) little variety." Greeks can eat in their chapter houses.

Despite UMaine's relatively isolated location, the campus pulses with social life; nearly 240 student organizations plan plays, carnival nights, concerts, and comedy hours, with a different activity offered each night. "Our students are really support-ive of all sports and clubs," a junior says. Partiers find their niche off campus, at bars, clubs, and house parties. "If students don't make alcohol consumption obvi-ous, they get away with it," a sophomore says. "RAs aren't looking to get people in trouble." Come spring, students go all out for April's Bumstock Weekend, a three-day event featuring bands playing outdoors from dawn till dusk. Maine Day features a parade, cookout, and campus-wide cleanup.

The midsized town of Orono—described by one sophomore as a "great college town"—offers a few bars, a theater, and some other hangouts. Buses to Bangor, a fair-sized city 10 minutes away, run every 15 to 20 minutes. A car is helpful, although there are gripes about parking. UMaine students tend to be outdoor enthusiasts, and popular road trips include Acadia National Park, skiing at Sugarloaf USA, L.L. Bean's 24-hour store

> "Our students are really supportive of all sports and clubs."

in Freeport, and the real-life Mt. Katahdin, which appears on Bean's logo. More urban types enjoy Bar Harbor, Boston, or Montreal, just four hours away (and with a lower drinking age and cheaper drinks).

Hockey reigns here, especially when played against New Hampshire, Boston University, or Boston College, and the Black Bears are perennial champions. Women's ice hockey and cross-country have claimed the America East championship. The new Shawn Walsh Hockey Center houses the ice hockey teams and athletic facilities, and the new Recreation Center includes a pool and elevated jogging track. The popular intramural program covers a range of sports from swimming and wrestling to hoopball (golf with a basketball) and broomball (ice hockey with a dodgeball and a broom, played with shoes instead of skates).

UMaine is a big school with a small-school atmosphere. Combine the state's natural beauty with an increased emphasis on top-quality facilities and more inti-mate student/faculty interaction, and it's no surprise that this campus draws more die-hard "Maine-iaks" each year.

The university library, the state's largest, is the regional depository for American and Canadian government documents and houses some of alumnus Stephen King's papers.

Overlaps
University of Southern Maine, University of New Hampshire, University of Vermont, University of Maine–Farmington, University of Massachusetts, Northeastern

If You Apply To ➤

Maine: Rolling admissions. Early action: Dec. 15. Financial aid: Mar. 1. Housing: Jun. 1. Meets demonstrated need of 26%. Campus interviews: optional, evaluative. No alumni interviews. SATs or ACTs: required. Accepts Common Application and prefers electronic applications. Essay question: personal statement on academic goals and objectives or essay of student's choice.

Manhattanville College

2900 Purchase Street, Purchase, NY 10577

Though co-ed for nearly 40 years, Manhattanville is still almost 70 percent female. Strong programs include art, education, and psychology. Among the few small colleges in the NYC area, Manhattanville is a quick train ride from the city. Portfolio system emphasizes competency rather than rote learning.

Website:
www.manhattanville.edu

Location: Suburban

Total Enrollment: 2,677

Undergraduates: 1,651

Male/Female: 31/69

SAT Ranges: V 490–610
M 490–610

ACT Range: 19–24

Financial Aid: 79%

Expense: Pr $

Phi Beta Kappa: No

Applicants: 3,233

Accepted: 59%

Enrolled: 27%

Grad in 6 Years: 58%

Returning Freshmen: 74%

Academics: ✍ ✍ ✍

Social: ☎ ☎ ☎

Q of L: ★ ★ ★

Admissions: (914) 323-5124

Email Address:
admissions@mville.edu

Strongest Programs:
Management
Art
Psychology
Education

Fifty student-run organizations help fulfill the interests of the student body, but off-campus bars still draw many students—whether of age or not.

Manhattanville sees its mission as "educating students to become ethically and socially responsible leaders for the global community." The Portfolio System, Manhattanville's distinct approach to undergraduate education, requires students to create a body of work reflecting their entire college career. However, that's only one way Manhattanville encourages individuality and personal growth. Personal attention is another. "I like how our president is involved in everything and gets to know everyone," a junior says.

Manhattanville College, which began as a Roman Catholic academy for girls on Houston Street in New York City, pulled up stakes in the 1950s for a 125-acre estate in Purchase, New York. This estate is located in wealthy Westchester County, near the town of White Plains—home to several major corporations, but just 28 miles from the excitement of the Big Apple. The focal point of the campus, which was designed by Central Park architect Frederick Law Olmsted, is Reid Hall, a 19th-century replica of a Norman castle.

> **"I like how our president is involved in everything and gets to know everyone."**

M-ville's distribution requirements include courses in five areas: humanities, social sciences, fine arts, mathematics and sciences, and languages. Under the Portfolio System, students must craft a freshman assessment essay, a study plan and program evaluation, specific examples of work in writing and research, and a resume. Freshmen complete the Preceptorial, a two-semester introduction to college-level work, as well as a library and information studies course, and are required to participate in free weekend trips into New York City. Manhattanville's strongest offerings include art and design (enhanced by the proximity of New York City's many museums and galleries), music, and education, while business administration/management, psychology, and political science are also popular. Manhattanville's School of Education, which offers two five-year master's programs, boasts a near-perfect passage rate for the New York State Teaching Exam. The languages attract the fewest majors. Math and science students enjoy facilities that are independently ranked among the top 100 wired colleges in the country. Students may design their own major, and those studying psychology, biology, or chemistry can conduct research with faculty. Career Services, which offers internship opportunities at over 350 locations in the New York metro area and beyond, is "phenomenal," securing placements at places such as MTV, the Metropolitan Museum of Art, Fox News, U.S. Senate offices, MasterCard, PepsiCo, and the Westchester County Board of Legislators.

The academic climate at Manhattanville can be challenging or laid-back, depending on the major. "Each course requires a student to give his or her full effort," says one student. The low student/faculty ratio and the quality of teaching get high marks. Ninety-three percent of freshmen classes have 25 or fewer students. "The teachers are very knowledgeable and enjoy teaching," a freshman says. The Board of Trustees scholarships offer qualified students an Honors Preceptorial. An Honors Seminar and honors programs within majors are also available. The college also offers dual degree

> **"Since the college is small, it just feels like all the students belong to one big family."**

programs with New York Medical College (MS in physical therapy or MS in speech language pathology) and Polytechnic University (MS in computer science or MS in information technology). There is also a master's degree in publishing with Pace University and a master's in social work with Fordham, and Manhattanville is striving to add even more such programs. The college has exchange programs with Mills College and with American University's World Capitals Program, plus study abroad options in England, France, Germany, Ireland, Italy, Japan, Mexico, and Spain.

Fifty-three countries and 37 states are represented in Manhattanville's student body, and females outnumber males more than two to one. "Since the college is small, it just feels like all the students belong to one big family," says a senior. Thirty-three percent of undergraduates come from outside of New York, including 11 percent from overseas. Hispanics comprise the largest minority group at 18 percent, followed by African Americans at 6 percent and Asian Americans at 3 percent. The college does not guarantee to meet the need of every admit, but there are hundreds of merit scholarships worth an average of $8,496.

Students may design their own major, and those studying psychology, biology, or chemistry can conduct research with faculty.

Eighty percent of Manhattanville's students live on campus in one of four dorms, which have lounges, communal kitchens, laundry rooms, cable TV, and Internet access. Freshmen are assigned rooms that are "spacious and very comfortable" according to a student, while upperclassmen enter a lottery—and complain they never get what they want. Campus dwellers can choose 15- or 19-meal-a-week plans, and can also use their meal cards at The Pub (a deli-type eatery), the convenience store, and vending machines. Students can also order "room service" three times a semester. The dining hall has been renovated and offers fresh-baked goods and a well-stocked salad bar.

Manhattanville's hometown, Purchase, "is not a college town. It is an affluent area with mansions all over. It's gorgeous," a junior says. The city of White Plains is five minutes away, and New York City is 45 minutes away. With increasing numbers of male students enrolling, the campus social scene seems to be picking up, and with no fraternities or sororities, off-campus

"Each course requires a student to give his or her full effort."

parties are usually open to all. The student programming board is working to improve the social life, with weekend events such as dinners, formals in the castle, parties, comedy and talent shows, plays, and concerts. The student center has a movie theater. Fifty student-run organizations help fulfill the interests of the student body, but off-campus bars still draw many students—whether of age or not. On-campus alcohol policies are said to be strict. Road trips include Rye Beach in the warmer months and upstate New York or Vermont for skiing in the winter. The college offers a free Valiant Express bus service that runs into NYC, and to nearby venues. Every spring, students look forward to Quad Jam, "an all-day, all-night concert and carnival and party." There's also a Fall Jam and midnight brunches during finals served by faculty and staff.

Every spring, students look forward to Quad Jam, "an all-day, all-night concert and carnival and party."

Manhattanville's president has invested heavily in athletics as a way of making the school better known and attracting more males. It now offers 18 NCAA Division III sports. The men's and women's ice hockey teams and softball are strong. There's an intramural program, and weekend warriors and letter-winners alike applaud the college's gym, fitness center, swimming pool, tennis courts, and athletic fields.

Manhattanville's size can be both an asset and an annoyance, say students. The familial atmosphere can get claustrophobic at times, but for those wishing to be part of a close but growing community where values matter, Manhattanville may be worth a look. "It's a nice, close-knit community, small enough to be familiar not only with people but with your professors," a sophomore says.

Overlaps
Fordham, Manhattan College, NYU, Pace, Hofstra

If You Apply To ➢

Manhattanville: Early decision: Dec. 1. Regular admissions, financial aid: Mar. 1. Housing: July 15. Meets demonstrated need of 20%. Campus interviews: recommended, informational. Alumni interviews: optional, informational. SATs or ACTs: required. SAT IIs: optional. Accepts the Common Application and electronic application. Personal essay on your goals and objectives for next four years.

Marlboro College

Marlboro, VT 05344

Marlboro is a hilltop home to a few hundred nonconformist souls. Each develops a plan of concentration that culminates in a senior project. One of the few colleges in the country that is governed in town meeting style, where student votes carry equal weight of those from faculty.

Marlboro College is only a half-century old, but it is known far and wide as an innovator in liberal arts education. It was founded on the principles of independent and in-depth study just after World War II, when returning GIs renovated an old barn as the college's first building while living in Quonset huts. And today's Marlboro students are just as trail-blazing; they prepare for the future by digging into self-developed Plans of Concentration, including one-on-one tutorials and a thesis or project judged by visiting outside examiners from "the best Eastern colleges and universities." With less than 300 undergrads and fewer than 50 faculty members, Marlboro is its own little world, which students enter as novices and leave as pros.

Positioned atop a small mountain, surrounded by maples and pines and with a gorgeous view of southern Vermont, Marlboro's physical beauty is striking. Buildings are adapted from barns, sheds, and houses that stood on three old farms that today make up the 350-acre campus. Among the renovated structures, many with passive solar heating, are nine dormitories, a library, a science building, art studios and music practice rooms, a 350-seat theater, and a campus center. Above the science building is the college's astronomical observatory. The school has expanded the library and opened the Serkin Building, a performing arts and music building. While some schools see growth as a sign of success, Marlboro intends to remain one of the nation's smallest liberal arts institutions. Administrators believe the size stimulates dynamic relationships between students and faculty, making learning happen both inside and outside the classroom. This isn't a place where students can fade into the background: the institution relies on everyone to share their talents and skills. The same philosophy applies to the college's Graduate Center; its programs are as innovative as the college's heritage.

The cornerstone of a Marlboro education is the Plan of Concentration, which each undergraduate student develops independently. Juniors and seniors "on plan" take most coursework in one-on-one tutorials with the faculty sponsors. Seniors present their thesis or project to their sponsors, who are backed up by outside examiners, experts in the student's field unaffiliated with the college. The administration boasts that by bringing in these outsiders for two- to three-hour oral examinations of its seniors, Marlboro has created its own accountability system, ensuring that neither students nor faculty at this isolated institution are cut off from the most current academic thinking. Faculty members often find the exams as stressful as the students, as it means outsiders are judging their teaching. The only other requirement is the Clear Writing course, usually completed by the end of the third semester, with a 21-page portfolio reviewed by faculty. Marlboro offers solid instruction in literature, writing, social sciences, and fine arts. Administrators and students alike praise the World Studies program, which provides an eight-month professional internship and/or study experience abroad; 30 percent of students take part, and 40 percent do some type of study abroad.

Marlboro's flexibility should not be confused with academic flabbiness. Grades are an integral part of the evaluation process, professors are stingy with As, and

> **"Marlboro is incredibly demanding academically, yet very collaborative."**

most students work hard. "Marlboro is incredibly demanding academically," says a freshman, "yet very collaborative." Students give most profs high marks and appreciate the low student/teacher ratio: "Everything at Marlboro screams 'interactive learning,'" says one student. "The professors are fantastic. Most are always available, enthusiastic about the students education (rather than their own interests), and passionate about what they teach!" exclaims one senior political philosophy major.

In the old independent Yankee spirit, Marlboro's library operates on the honor system, where students sign out their own books 24 hours a day. It is the same for the computer center, and science and humanities buildings. The school has only one security guard. "I feel really safe here," says one freshman. "Everyone is out at night and nobody's worried about assault—except maybe by bears." Indeed, the college operates on a New England town meeting style of government involving students, faculty, and the staff and their spouses in every aspect of policymaking. Students can veto the faculty members on hiring and retention decisions, and it takes a two-thirds vote of the faculty to override them.

> "Everything at Marlboro screams 'interactive learning.'"

Marlboro remains a liberal-leaning campus, but students are quick to point out that campus politics don't dominate the scene. Minority enrollment continues to be low, in spite of the administration's program to recruit and support poor rural Vermonters. Hispanics make up 2 percent of students, Asian Americans add 1 percent and African Americans and Native Americans each another 1 percent. The college offers merit scholarships but no athletic scholarships—probably because there are no varsity sports.

Marlboro remains a liberal-leaning campus, but students are quick to point out that campus politics don't dominate the scene.

The dorms house 75 percent of Marlboro students and are mostly co-ed; students say they have a "rustic" appeal. "The rooms are cozy and friendly, not sterile like some dorms," says one dorm-dweller. Housing is based on credits, so freshmen have triples, sophomores have doubles, and upperclassmen have singles. There is one dining hall where "the entire community eats together." Food gets mixed reviews, but students agree it is adequate for putting on the "freshman 15." A small percentage of students live off campus in Brattleboro, 20 minutes away, and shuttle to and from campus in a school van.

The cornerstone of a Marlboro education is the Plan of Concentration, which each undergraduate student develops independently.

Students agree that the town of Marlboro, highlighted by a post office and general store, isn't much to write home about. Most head to Brattleboro for its restaurants, bookstores, and coffee shops. As might be expected, Marlboro has no Greek organizations; a staff member was recently hired to coordinate the planning of student activities like poetry readings, trips to Boston and New York City, vans to local movie theaters, and pumpkin-carving contests. Snowball fights by the library are a big draw in winter. On Community Work Day, students and faculty skip class and work together to improve the campus through various manual labor projects. The annual Cabaret and Halloween parties are unofficial costume contests showcasing student creativity.

Marlboro offers solid instruction in literature, writing, social sciences and fine arts.

The school mascot, the Fighting Dead Trees, is emblazoned on the shirts of the ever-popular co-ed soccer team, and broomball, a variation of ice hockey played using shoes and brooms instead of ice skates and sticks and a kickball instead of a puck, is also always popular for athletes and spectators. The intramural program, which has a 40 percent participation rate, also offers dodgeball, ultimate Frisbee and volleyball, among other sports. The "incredibly dynamic" outing club ensures plenty of opportunities to enjoy the local wilderness, including hiking and cross-country skiing on runs that radiate from the center of campus. Excellent downhill skiing is only a few minutes' drive away.

> "Everyone is out at night and nobody's worried about assault—except maybe by bears."

This iconoclastic school continues to push the academic envelope and remains proud of doing—and being—the unexpected. And that suits its students just fine. Says

Overlaps

Hampshire, Bennington, Bard, Reed, Sarah Lawrence, College of the Atlantic

a sophomore, "We tend to have students here who are a little left of center, very interested in the details, intelligent, engaged, interested in hands-on learning, passionate, and who will always go the extra mile to learn everything about their subject."

If You Apply To ➤

Marlboro: Early decision: Nov. 15. Early action: Jan. 15. Regular admissions: Feb. 15. Financial aid: Mar. 1. Housing: May 1. Does not guarantee to meet full demonstrated need. Campus interviews: recommended, evaluative. No alumni interviews. SATs or ACTs: required. SAT IIs: optional. Accepts the Common Application and prefers electronic applications. Essay question. Encourages "nontraditional" students.

Marquette University

Milwaukee, WI 53201-1881

Marquette is an old-line Roman Catholic university along the lines of St. Louis University and Loyola of Chicago. Milwaukee is not a selling point, and the university's clientele is mainly from the southern Wisconsin/northern Illinois corridor. About 80 percent of the students are Catholic.

Website: www.marquette.edu
Location: Urban
Total Enrollment: 10,997
Undergraduates: 7,413
Male/Female: 46/54
SAT Ranges: V 540–650
   M 540–660
ACT Range: 24–29
Financial Aid: 58%
Expense: Pub $ $ $ $
Phi Beta Kappa: Yes
Applicants: 10,348
Accepted: 70%
Enrolled: 25%
Grad in 6 Years: 80%
Returning Freshmen: 90%
Academics: 🖉 🖉 🖉
Social: ☎ ☎ ☎
Q of L: ★ ★ ★
Admissions: (800) 222-6544
Email Address:
   admissions@marquette.edu

Strongest Programs:
Humanities
Dentistry
Physical Therapy
Biomedical Engineering
Nursing
Journalism

At Marquette University, students practice what they preach. The college experience at this Roman Catholic institution includes an emphasis on civic responsibility, community service, and personal growth. "Marquette students really are a community. Everyone supports everyone else," says a senior. Innovative programs combine classroom theory with volunteer opportunities in Milwaukee and beyond.

Marquette occupies 80 acres of "concrete with interludes of grass and trees" just a few blocks from the heart of downtown Milwaukee. While offering the advantages of an urban setting, its campus does have plenty of open spaces suitable for everything from throwing a Frisbee to throwing a barbecue. Although most of the buildings are relatively modern, the campus is the site of the oldest building in the Western Hemisphere, the St. Joan of Arc Chapel, which was built in France in 1400 and later transported to Wisconsin. New buildings include the $31 million Al McGuire Center athletic facility, $55 million John P. Raynor, S.J., Library, and $12.8 million, 1,100-vehicle parking garage. The $30 million School of Dentistry has also opened, along with 61 new student apartments.

> "Volunteer work is huge at Marquette."

The 36-hour general education core curriculum is comprised of nine "knowledge areas:" diverse cultures, human nature and ethics, histories of cultures and societies, individual and social behavior, theology, literature/performing arts, mathematical reasoning, rhetoric, and science and nature. In addition to myriad study abroad programs, the university is the proud owner of the Les Aspin Center for Government in Washington, D.C., which allows students to take courses while participating in an internship with a federal government agency. Closer to home, students intern with local and state government agencies. The Honors Program, which offers small classes, admits 100 qualified freshmen each year. Each year, more than 1,800 students enroll in service learning courses and participate in service opportunities in 111 community settings.

Through an affiliation with the Milwaukee Institute of Art and Design, two art minors (studio art and art history) are available. Marquette has its own art museum and an active theater program. The foreign languages and literature departments are currently shifting focus and consequently are not as strong as other offerings.

(Continued)
Business

Sixty percent of classes have fewer than 25 students, but students usually manage to get into the ones they want. The administration encourages students to "put our beliefs into practice" through volunteer activity, which serves the elderly, the sick, and the poor in the Milwaukee area and elsewhere. "Volunteer work is huge at Marquette and there are departments that work to get anyone involved who wants to be involved," a junior says. Administrators claim students contribute more than 100,000 hours of community service each year. Student religious organizations are active, and weekly Masses are held in the dorms by the resident priest. Roman Catholics understandably predominate in the student body, but religious practice is left to the individual. The academic climate is described as challenging and competitive. A senior says, "I feel that all students work hard to get the grades that they do." Freshmen take a mandatory rhetoric course after reading the same book over the summer, and a first-semester seminar attracts about a quarter of entering students. The Freshman Frontier program offers admission and intensive assistance to students "who did not reach full academic potential in high school."

> **"Although Marquette is located in an urban environment, public safety does an absolutely outstanding job."**

Although Marquette actively recruits in 35 or so states and several U.S. territories, most of the student body is from the Midwest, 46 percent from Wisconsin itself. In general, Marquette boasts a friendly collection of traditional, middle-class students. "Activism is pretty high on both sides," a junior says. "It is a very conservative student body, but the liberals protest more issues." African Americans make up 5 percent of the student body and Hispanics and Asian Americans each add another 4 percent. Marquette offers a very successful Educational Opportunity Program, which enables low-income, disadvantaged students, most of whom are minorities, to have the advantage of a college education. Merit scholarships averaging $7,953 are available, as are 152 athletic scholarships.

Fifty-two percent of Marquette students make their home on campus; all but two residence halls are co-ed, and there are 440 apartments (which come with a separate electric bill). Residency is required for freshmen and sophomores, but by junior year an overwhelming majority of students choose to move off campus, though housing is guaranteed for all undergraduate students through a lottery. "I enjoyed both my residence halls and loved the communities that were formed there," one student says. Within the dorms are five different living/learning communities, including ones for freshman honors students, engineering majors, and nursing students. The university works hard to keep the campus safe. "Although Marquette is located in an urban environment, public safety does an absolutely outstanding job ensuring that students are safe and know how to remain safe in the area," a senior says.

Students don't characterize Milwaukee as a college town, but still say there are many good things about being there. For one, students say they have plenty of opportunities for community service. An old advertising slogan once claimed that "Milwaukee Means Beer," and few Marquette students would disagree. Marquette is stricter than most universities in enforcing the drinking age, but getting served off campus is not as difficult, and most students happily declare alcohol is there for the getting. There are fraternities and sororities, but they attract only 5 percent of the men and 7 percent of the women. Another well-loved tradition is the Miracle on Central Mall, the annual lighting of the campus Christmas tree and accompanying Mass.

> **"It is always easy to meet new people and become friends quickly."**

As the school grows, varsity sports are gaining a higher profile. Men's basketball and cross-country are competitive, along with women's cross-country, basketball, volleyball, and soccer. Seventy-five percent of students participate in at least one of the 39 intramural sports. Sports fans will be impressed with Milwaukee's Bradley Center, close to campus and home to Marquette basketball and the NBA's Milwaukee Bucks. Nature lovers can head to Lake Michigan, a 40-minute walk from campus, or to Kettle Moraine, a glaciated region ideal for hiking and cross-country skiing. Chicago is only 95 miles away.

Still, no matter how dynamic Marquette's athletic teams or how impressive the facilities, it's the familial atmosphere that makes Marquette what it is. "Students here are very friendly," a junior says. "It is always easy to meet new people and become friends quickly."

If You Apply To ➢

Marquette: Regular admissions: Dec. 1. Financial aid: Mar. 1. Meets demonstrated need of 32%. Campus and alumni interviews: optional, informational. SATs or ACTs: required. SAT IIs: optional. Accepts the Common Application and electronic applications. Essay questions: pick a Hollywood actor to play you in a movie about your life; greatest challenge your generation will face; risk you have taken that has helped you.

University of Maryland–Baltimore County

Baltimore, MD 21250

A mid-size public university with the feel of a private. Strategically located in a suburban setting between Washington, D.C., and Baltimore, UMBC invests heavily in learning communities and other efforts to ensure that its undergraduates thrive. Nationally known for selective Meyerhoff Scholars Program and a chess team that routinely bests its Ivy League competition.

Website: www.umbc.edu
Location: Suburban
Total Enrollment: 11,650
Undergraduates: 9,406
Male/Female: 54/46
SAT Ranges: V 540–650
 M 580–670
ACT Range: 22–28
Financial Aid: 46%
Expense: Pub $ $ $ $
Phi Beta Kappa: Yes
Applicants: 5,229
Accepted: 71%
Enrolled: 38%
Grad in 6 Years: 56%
Returning Freshmen: 81%
Academics: ✑ ✑ ✑
Social: ☎ ☎ ☎
Q of L: ★ ★ ★
Admissions:
 (800) UMBC-4U2
Email Address:
 admissions@umbc.edu

At the University of Maryland–Baltimore County, you can be king or queen of your academic world. Students here are given access to academic and social resources usually reserved for those attending mammoth public institutions or pricy private colleges. In addition to solid programs in the sciences and humanities, UMBC offers students a seemingly endless menu of social options. "There are so many ways to get involved academically and socially," says a student, "whether it's through clubs, research, internships, or service learning." What's more, the school fields a killer chess team that regularly mops the floor with the competition. UMBC encourages exploration and expects students to support one another and the community at large. "It is a tough school," admits one student. "Instead of fighting each other to be the best, we help each other so that we all succeed." It's your move.

UMBC's 500-acre suburban campus is located within the D.C.–Baltimore corridor, offering students access to an array of cultural attractions including restaurants, art galleries, specialty shops, and museums. In the past decade, the university has invested more than $300 million in new facilities. The Information Technology and Engineering building offers state-of-the-art amenities and the Chemistry and Biochemistry Building has undergone extensive renovations within the past two years.

UMBC's most popular programs are also its strongest, including information systems, psychology, biological sciences, computer science, and visual arts. "The computer science program is notorious for late-night, near-impossible assignments that only the biggest nerds can finish," a senior says. The interdisciplinary studies major gives students a chance to create individualized majors drawing on a wide range of disciplines; past majors include biomechanics, criminal justice, medical illustration, and intercultural conflict resolution. The Brown Center for Entrepreneurship sponsors programs and courses to inspire entrepreneurial thinking among students and faculty, while budding researchers may compete for undergraduate research awards through the Provost's Office and via Undergraduate Research and Creative Achievement Day (URCAD). The highly selective Meyerhoff Scholars Program was founded to address the shortage of African Americans in the sciences and engineering; today, it is one of the nation's top producers of African American science, engineering, and math undergrads who matriculate into PhD programs.

All students must complete general foundation requirements, which include three courses in the arts and humanities, three social science courses, one math course, and two biological/physical science courses. In addition, students must complete a foreign language requirement. "UMBC is extremely focused on academics," says a junior. "The courses can be very rigorous," adds a sophomore, "especially in the technical fields." The university offers a number of programs designed to help freshmen ease into college life. First-Year Academic Seminars allow students to

"Instead of fighting each other to be the best, we help each other so that we all succeed."

partner with faculty members to explore course material in an intimate, active learning environment. Students focus on creative and critical thinking, written and oral communications, and take part in faculty and peer critiques. The Small Study Groups program encourages freshmen and sophomores to form study groups and Campus Connect provides additional mentoring to undecided freshmen who appear at risk for not completing their degrees. In the classroom, students receive ample attention from professors. "I feel like I get a lot of personal attention from my professors," says a student. "My largest class had around 80 students and my smallest class was around eight students. In both cases, my professors took the time to learn all our names and made sure we understood the material." Sixty-five percent of classes have 25 or fewer students.

Fifty-eight percent of students graduated in the top quarter of their high school class and 88 percent hail from Maryland. "UMBC has the athletic teams, the Greek lifers, the nerds, the dancers, the actors, the singers, and the independents," says a student, but "the typical UMBC student is rather intelligent, self-motivated, somewhat introverted, and satisfied with a small, close group of friends." African Americans account for 14 percent of the student body, Hispanics 4 percent, Asian Americans 20 percent, and Native Americans 4 percent. Students say political activism is largely confined to political organizations on campus. "The students have more going on in their lives than to bother with what's going on in the White House," says a computer science major. UMBC awards merit scholarships worth an average of $9,000 and approximately 150 athletic scholarships.

The Brown Center for Entrepreneurship sponsors programs and courses to inspire entrepreneurial thinking among students and faculty.

Although 72 percent of freshmen live on campus, only 39 percent of the student body resides in university housing. "I love residential life," says a senior. "Even the lowest quality freshmen dorms have their own private bathrooms and brand new furniture." Living/Learning Communities connect students with similar interests and house them together in themed residence halls, which include the Center for Women and Information Technology, Emergency Health Services, Honors College, and Exploratory Learners. Campus dining options include a din-

"The students have more going on in their lives than to bother with what's going on in the White House."

ing hall ("Each meal is all-you-can-eat and there is a huge variety of food!") and the Commons, which offers a variety of fare, including Italian, Chinese, and Mexican cuisines. Campus security is described as "good" and "UMBC actually has its own police force located on campus," says a junior. In addition, students must swipe their student ID to get into residence halls and emergency call boxes are stationed throughout the campus.

"The social life at UMBC is very mellow in comparison to other schools," an information systems major says. "UMBC is hardly a party school." Instead, students flock to campus clubs and service organizations in search of a good time. A student says, "Most social groups are formed through the organizations on campus—volunteer, cultural, religious, academic, and others." Only 4 percent of the men and women go Greek and you're unlikely to find any alcohol-fueled toga parties here. "UMBC is not a place where alcohol and partying dictates who is cool or not," says a junior. "For this reason, alcohol abuse is not prevalent among the students." National acts—including comedians and rock stars—have been known to make an appearance at Quadmania, much to the students' delight. "This year we had Lewis Black as the main standup act and My Chemical Romance rocked to a sold-out crowd," a senior cheers. Block Party is "another carnival where all the residential students can win prizes and play games," gushes a student. "Last year I got to hit my Community Director in the face with a pie at one of the booths!" Homecoming always draws big crowds and popular roadtrips include treks into Baltimore (10 minutes away) and Washington, D.C. (40 minutes away).

The UMBC Retrievers compete in Division I and field a number of competitive teams, including men's soccer, volleyball, men's and women's track and field, men's and women's lacrosse, and men's and women's tennis. In 2004–05, the men's swimming and diving team won their eighth consecutive league championship. UMBC is a perennial

> **"UMBC is not a place where alcohol and partying dictates who is cool or not."**

collegiate chess powerhouse (and regularly makes the Final Four) and the university lures talented players with a bevy of scholarships. Intramurals are strong, too, and offer more than 15 sports and activities each semester. "There are intramurals and club teams all year-round," says a student. "Football is especially popular for the guys and there are co-ed teams for the ladies, too." The Retriever Activities Center offers students 18,000 square feet of fitness and recreation space, including a gymnasium, weight room, fitness studio, indoor pool, and tennis courts.

"I would say that UMBC is a school just coming into its own," says a junior. Unlike the gargantuan University of Maryland at College Park, UMBC capitalizes on its small size by providing students with solid academics and ample resources on a manageable scale. Says one student, "I love UMBC because it has all the opportunities of a huge university but still maintains its close community feel."

Overlaps

University of Maryland–College Park, Towson, Penn State, Johns Hopkins, University of Delaware, Salisbury

If You Apply To > **UMBC:** Early action: Nov. 1. Regular admissions: Feb. 1. Campus interviews: optional, informational. No alumni interviews. SATs: required. SAT IIs: optional. Accepts the Common Application and electronic applications. Essay question.

University of Maryland–College Park

College Park, MD 20742

The name says Maryland, but the location says Washington, D.C. Students in College Park can jump on the Metro just as they do at Georgetown or American. Maryland is nothing if not big, and savvy students will look to programs such as the honors program and living/learning communities for some personal attention.

For good luck on exams, University of Maryland students rub the nose of Testudo, the school's terrapin mascot. But even without touching the revered statue, most students here feel lucky to be at a school with so many courses, such a diverse student body, and state-of-the-art research programs and institutes. "The students here are very diverse, both ethnically and interest-wise," a student says. "They are very open to new ideas, new people and new things."

Maryland's 1,200-acre campus embraces an array of architectural styles, including the Georgian brick buildings ringing the oak-lined mall at the heart of the campus. Students now have their own athletic arena—the Comcast Center. Other new facilities include the Riggs Alumni Center, Kim Engineering Building, 700 beds of new housing, the renovated Stamp Student Union, a new Bioscience Research Building, and renovations to the Chemical/Nuclear Engineering Building.

Students select a major from within one of 12 schools and colleges; Maryland has earned a strong reputation for its engineering, physics, and computer science departments, as well as the Robert H. Smith School of Business and Philip Merrill College of Journalism. General education requirements comprise a third of every undergrad's courseload. These include classes in writing, math, the sciences, humanities and the arts, and social sciences and history, as well as two upper-level classes outside the major and one course focusing on cultural diversity.

"The teachers expect a lot of students, but they also give a lot back."

For students at the extremes of the academic spectrum, there are several honors programs and an intensive educational development and tutoring program. Students participating in individual studies combine established majors and create their own programs; other options include internships in nearby Washington, D.C., and Baltimore, and study abroad in locations such as Costa Rica, Israel, and Sweden. Several new offerings include a major in international business and minors in art history, philosophy, sports commerce and culture, and music performance.

Maryland's coursework "is rigorous, but not overwhelming," says a cell biology major. Lower-level courses tend to be large and impersonal ("easy to hide in, even easier to skip"), but the corresponding weekly discussion sections led by teaching assistants offer personal attention. The situation improves by junior year, when classes of 20 to 40 students become the norm. "The teachers expect a lot of students, but they also give a lot back," says an elementary education major. "Courses do require a lot of reading." Faculty members are "knowledgeable and enthusiastic," but, as might be expected given Maryland's size, students

"The incoming freshman class improves every year."

must seek them out for extra help, feedback, or assistance. The university is putting more emphasis on helping students make timely progress toward their degree. A two-day orientation, seminars and course clusters are offered for freshmen.

Seventy-five percent of students are Maryland natives, and New York and New Jersey are also well represented. "Students are very qualified and intelligent individuals. The incoming freshman class improves every year," says a senior. Diversity is more than just a buzzword: 13 percent of students are African American, 6 percent are Hispanic, and another 14 percent are Asian American. Big issues on campus range from the war in Iraq and gay and lesbian rights to tuition hikes. "The activism is great," a sophomore says. "Being so close to Washington, D.C., has a great impact." Students from families earning less than $21,000 a year receive all of their financial aid in the form of grants, with no loans required.

Forty-three percent of students—and more than 90 percent of freshmen—live on campus in single-sex or co-ed dorms; freshmen are guaranteed housing, and while many juniors and seniors seek off-campus accommodations, those who stay on campus all four years will find their digs improve as they gain seniority. (Upperclassmen

Website: www.umd.edu
Location: Suburban
Total Enrollment: 29,645
Undergraduates: 23,162
Male/Female: 51/49
SAT Ranges: V 580–670
  M 600–700
Financial Aid: 37%
Expense: Pub $ $ $
Phi Beta Kappa: Yes
Applicants: 22,463
Accepted: 49%
Enrolled: 38%
Grad in 6 Years: 76%
Returning Freshmen: 93%
Academics: ✍ ✍ ✍
Social: 🍺 🍺 🍺
Q of L: ★ ★ ★
Admissions: (301) 314-8385
Email Address:
  um-admit@uga.umd.edu

Strongest Programs:
Business
Management Information
  Systems
Engineering
Journalism
Computer Science
Government and Politics
Physics

A few bucks and a few minutes on the Metro (Washington's subway system) brings Terrapins into downtown D.C. at a hare's pace.

also have the option of on-campus apartments and suites.) "Housing is on a par with other schools. Many dorms are not air-conditioned, but are clean and well maintained," explains one senior. Decisions on financial aid and housing are affected by acceptance date, so the earlier you apply, the better off you'll be. Freshmen generally live in high-rises or low-rises; South Campus (more relaxing than the louder North Campus) features air-conditioning, carpeting, and new furniture. Safety features include triple locks on dorm-room doors, blue-light emergency phones, and walking and riding escort services to transport students after dark. "The area has its rough spots but it is constantly becoming safer," a sophomore says, noting the campus itself tends to be "extremely safe."

> "Being so close to Washington, D.C., has a great impact."

Social life at Maryland revolves around nonalcoholic events such as concerts, movies, and speakers, as well as around the traditional fraternity parties and football and basketball games. The university's reputation as a haven for those who prefer partying to studying is changing as students with better credentials apply, but there is still always something happening in the dorms, at local pubs, and in nearby Baltimore and Washington, D.C. On campus, only students over 21 may drink, in accordance with state law. "They say no tolerance and in recent years, there have been crackdowns. As a result, policies have increased in effectiveness," says one student. Nine percent of the men and 11 percent of women go Greek, but they don't dominate the tone of campus life.

Despite College Park's highly social atmosphere, Maryland's suburban campus can feel too small. A few bucks and a few minutes on the Metro (Washington's subway system) brings Terrapins into downtown D.C. at a hare's pace. Back on campus, Art Attack is a favorite annual event in which local artists share their crafts and national touring artists perform an evening concert. Other popular events include Maryland Day and homecoming.

Terrapin basketball fans are unsinkable and not always civilized, turning out *en masse* to cheer against Duke. "Students here have a lot of school spirit," a junior says. "Terrapin pride runs rampant around here." The basketball program looked impressive by slam-dunking the ACC championships in 2004, as did the men's lacrosse and women's volleyball teams. The men's football and soccer teams have also been successful, as have women's lacrosse and swimming. A handful of intramural sports are offered, along with an array of club-level sports.

The University of Maryland's overwhelming size is both a blessing and a curse for the increasingly capable undergraduates here. On one hand, "the diversity of the student body and the opportunities afforded are infinite," a sophomore says. On the other, largeness can translate into crowded dorms, big classes, parking problems, and hassles everywhere. Still, a junior insists, getting lost in the shuffle "is totally avoidable. Talk to professors, join clubs, attend campus meetings, and people will know you."

Overlaps

Cornell, NYU, University of Maryland–Baltimore County

If You Apply To ➤ **Maryland:** Rolling admissions. Early action: Dec. 1. Regular admissions: Jan. 20. Financial aid: Feb. 15. Housing: May 1. Campus interviews: optional, informational. No alumni interviews. SATs or ACTs: required. SAT IIs: optional. Accepts electronic applications. Essay question: how you will enrich the college community; if you could do anything without failing, what would it be; if you could meet any figure, who would it be and why. Students who do not meet academic standards may submit additional information for consideration.

University of Mary Washington

(formerly Mary Washington College)
1301 College Avenue, Fredericksburg, VA 22401

Mary Washington could easily be mistaken for one of Virginia's elite private colleges. It offers just as much history and tradition—for a much lower price. Once a women's college, it is still about two-thirds female. On the selectivity chart, UMW ranks behind only UVA and William and Mary among Virginia public universities.

Strolling among the university's elegant buildings of redbrick and white columns has led more than one pleased parent to declare, "Now this is what a college should look like." Indeed, for an aura of history and tradition, few schools stack up to this small college in Fredericksburg, a site of Civil War action and the boyhood town of George Washington. The University of Mary Washington campus features classical Jeffersonian buildings, sweeping lawns, brick walkways, and breathtaking foliage. If the campus architecture puts some people in mind of the University of Virginia, it's no accident: UMW (formerly Mary Washington College) was the all-female branch of that august institution before going co-ed in 1970 and cutting its ties in 1972. Of course, Mary Washington is not just beautiful, but smart, too. It has gained a reputation as one of the premium public liberal arts colleges in the country and continues to attract bright students from around the globe.

Mary Washington's core curriculum emphasizes a strong liberal arts focus. Students select courses to meet specific goals in the arts, literature, natural and social sciences, and mathematics. English composition and foreign language competency are also required. Students must complete general education requirements in five areas: writing, speaking, race and gender, global awareness, and environmental awareness. As for majors, business administration, English, and psychology are the most popular. "The English department has a great reputation," notes one student, "just recently having a teacher win the Nobel Prize in literature." Also very strong is a unique program in historic preservation. Among the sciences, biology is the clear favorite. The administration recently added an anthropology major, a new freshman advising system, and 10 freshman seminar courses capped at 15 students per class. Courses in Arabic are also now available.

Students are encouraged to take on research projects of their own design, and science students are eligible for a 10-week summer research program. Several departments offer grants for work abroad or in the U.S., and many students study abroad during their junior year. The college's location, roughly an hour from both Washington, D.C., and the state capital, Richmond, is a handy asset for the

"Now this is what a college should look like."

approximately 350 budding politicos who seek internships every year. The close ties between students and faculty are a great source of pride at Mary Washington. Classes rarely have more than 25 students, and the recent hiring of additional faculty may allow for even smaller classes. "The teachers are very dedicated and willing to put aside time for individual students' needs," says a freshman. While students find their courses demanding, "there is generally a supportive, non-combative atmosphere when it comes to academics," observes a junior.

"As a classmate of mine says, it is a 'sea of whiteness' here," quips one student. Seventy-seven percent of the student body is from Virginia. African Americans make up 5 percent of the student population, Hispanics 4 percent, and Asian Americans 4 percent. "UMW students are mostly down-to-earth and pleasant," adds a student, "and have a real penchant for learning and acceptance." A large majority continue

Website: www.umw.edu
Location: Small city
Total Enrollment: 4,734
Undergraduates: 3,952
Male/Female: 33/67
SAT Ranges: V 580–670
   M 560–640
ACT Range: 25-29
Financial Aid: 51%
Expense: Pub $ $ $
Phi Beta Kappa: Yes
Applicants: 4,635
Accepted: 64%
Enrolled: 20%
Grad in 6 Years: 74%
Returning Freshmen: 86%
Academics: ✍ ✍ ✍ ½
Social: ☎ ☎ ☎
Q of L: ★ ★ ★ ★
Admissions: (800) 468-5614
Email Address:
   admit@umw.edu

Strongest Programs:
Historic Preservation
Psychology
English
Biology
International Affairs
History
Political Science
Business Administration

on to jobs after graduation, rather than graduate school. Politically, "there seem to be a lot of students claiming to be moderates," says a freshman. "I call most of them apathetic." Twelve percent of undergraduates receive merit scholarships averaging $1,200, but there are no athletic scholarships.

An unusually strong sense of community characterizes everything from academics to dorm life at UMW. The 18 residence halls offer a variety of living arrangements, including suites, singles, doubles, and even quads. "The newer dorms are soulless and efficient and don't have much extra room or charm," a psychology major says, "although the older dorms are very spacious and pleasant to live in." Although

"There is generally a supportive, non-combative atmosphere when it comes to academics."

many upperclassmen move to private housing, most stay within a mile of campus. All told, 64 percent of students live in university housing. As for campus dining, "The secret to finding a well-balanced and tasty meal is to mix-and-match from the three rooms in Seacobeck, the main dining hall. Each room features a different salad bar," explains an English and business administration major. Security is described as good and students report feeling safe on campus.

Small and friendly, nearby Fredericksburg is a "quaint, historic town." While it lacks some of the nightlife of a larger community, there are historic homes to visit, museums with Civil War exhibits, and shops and restaurants that offer discounts to students. For dance clubs and bars, students drive to Richmond or D.C. Also an hour's drive away is the scenery of the Chesapeake Bay, due east, and the Blue Ridge Mountains, due west. On campus, student organizations offer special events each Friday, such as a student film festival or a Mardi Gras celebration. Although there are parties both on and off campus on any given weekend, alcohol does not dominate the social scene, and there are no fraternities or sororities. "Many students pack and go home on the weekends, leaving the others bored out of their minds," says one freshman. "This campus is definitely not geared for partying," adds a junior.

Mary Washington students take an uncommon interest in traditions. Several annual outdoor parties, including Grill on the Hill and Weststock, never fail to attract a large crowd. All third-year students brace themselves for Junior Ring Week, during

"'Intramural Champions' is the most coveted T-shirt on campus."

which they are the victims of practical jokes prior to receiving their rings from the school's president. Another tradition is Devil-Goat Day, an all-day competition pitting odd- and even-year classes against each other in events such as sumo wrestling, jousting, and the Velcro wall. Homecoming is observed with the usual round of sporting events (particularly soccer), dances, and dinners. The Multicultural Fair is also popular.

Mary Washington doesn't have a football team, but other sports are alive and well. Since the inception of the Capital Athletic Conference, UMW has won more conference championships than all other conference members combined. In 2004–05, it claimed eight of 18 conference championships. UMW teams advancing to NCAA tournaments in the past two years include women's soccer, men's and women's basketball, men's and women's swimming, baseball, and men's and women's tennis. Nonvarsity types also use the 76-acre sports and field complex, complete with an Olympic-size pool, for a variety of intramural and club sports, including men's and women's rugby and crew. Flag football, soccer, and basketball are popular intramurals. "'Intramural Champions' is the most coveted T-shirt on campus," says a senior.

Nestled midway between Washington, D.C., and Richmond, the University of Mary Washington offers a first-rate liberal arts education. It has the feel of a private school with a public school price tag and is an option that should be explored, says a happy junior. "I feel that [UMW] is overlooked by many due to its smaller size. But I feel if people took a closer look, they would see its real beauty and excellence."

Overlaps

College of William and Mary, University of Virginia, James Madison, George Mason, University of Richmond, Christopher Newport

University of Massachusetts–Amherst

Amherst, MA 01003

A liberal mecca in cosmopolitan and scenic western Massachusetts. UMass boasts strong study abroad programs and an international flavor. Science and engineering are also strong. Ready access to privates Amherst, Hampshire, Mount Holyoke, and Smith via the Five College Consortium.

The University of Massachusetts at Amherst, a leading land grant university with more than a century of tradition, offers students a dizzying array of majors and extracurricular options and the chance to take courses at nearby private colleges that are among the best anywhere. Students can live in one of the nation's top college towns, take advantage of an extensive research program and strong honors program, and enjoy an endless supply of social opportunities—without emptying their wallets.

UMass's sprawling 1,463-acre campus is centered on a pond full of ducks and swans, while architectural styles range from Colonial to modern. The school is located on the outskirts of Amherst, a city that combines the energy of a bustling cosmopolitan center with the quaintness of an old New England town. Students agree that Amherst caters to college life. Recent campus additions include on-campus apartments for undergrads and an extensive transformation of the library's main floor, which is now known as the Learning Commons.

UMass offers 89 undergraduate majors, and among them management and engineering are top-ranked. The English department is notable, and political science and creative writing also draw praise. Students report little difficulty getting into courses they want or are required to take, but some regard math, computer science, and the natural sciences as especially tough. Engineering students may face a bit of a challenge in finishing in four years; they are required to take 130 credit hours while most programs require 120 credit hours.

All undergraduates must demonstrate math proficiency and complete two courses in writing; six Social World courses, including literature, arts/liberal arts, historical studies, and social and behavioral sciences; three courses in biological and physical science; and a course in analytic reasoning. The writing requirement includes a freshman course taught in sections of 24 or fewer. The school's honors program, Commonwealth College, offers qualified students special courses and sponsors interdisciplinary seminars, student gatherings, service projects, and a housing option. Students seeking to stand out from the "Masses" might consider the interdisciplinary major in social thought and political economy, or the bachelor's degree in Individual Concentration program, a design-it-yourself major. About 17 percent of students study abroad at some point, and a growing number of service learning courses provide opportunities for internships and community service. The Center for Student Business offers one of the most unique programs at UMass, allowing students to staff and manage nine campus businesses and learn how to work with others and resolve conflicts professionally.

Website: www.umass.edu
Location: Small town
Total Enrollment: 20,197
Undergraduates: 17,939
Male/Female: 50/50
SAT Ranges: V 510–620
 M 520–630
Financial Aid: 51%
Expense: Pub $ $ $ $
Phi Beta Kappa: Yes
Applicants: 20,207
Accepted: 80%
Enrolled: 27%
Grad in 6 Years: 66%
Returning Freshmen: 84%
Academics: ✍ ✍ ✍ ½
Social: ☎ ☎ ☎ ☎
Q of L: ★ ★ ★
Admissions: (413) 545-0222
Email Address: mail @admissions.umass.edu

Strongest Programs:
Linguistics
Business
Psychology
Computer Science
Chemical Engineering
Electrical and Computer
 Engineering

UMass's intellectual and political climate is extraordinarily fertile for a state university, perhaps in part because of its membership in the Five College Consortium.* This special alliance allows students to attend UMass and take courses at the other four consortium schools: Amherst College, Smith, Hampshire, and Mount Holyoke. Students say that generally, the quality of teaching at UMass is excellent. "The classes are competitive and well set up for learning," a computer science major says. Full professors teach most courses, and some of the larger ones are broken down into smaller sections with graduate-level teaching assistants. "The quality of teaching is phenomenal," cheers a junior. Academic and career counseling receive mixed reviews, and it is usually up to students to pursue career help.

UMass has the sixth-largest residence-hall system in the country. Sixty-two percent of students are housed among five different neighborhoods.

"The classes are competitive and well set up for learning."

The majority of students are white public school graduates from Massachusetts. Out-of-state enrollment is capped at 25 percent of the student body, but applicants from other New England states are treated as Massachusetts residents for admission purposes if their own state schools don't offer the programs they want. Four percent of undergraduates are African American, while 7 percent are Asian American and another 3 percent are Hispanic. The university has established cultural centers on campus providing activities and support for students from different backgrounds, but affirmative action is still an issue, students report. "There are always rallies about better programs and aid for minorities," says a senior, but a "diversity action plan" introduced in 2005 seeks to improve first-year advising, increase diversity in the curriculum, and add minority faculty.

About 17 percent of students study abroad at some point, and a growing number of service learning courses provide opportunities for internships and community service.

UMass has the sixth-largest residence-hall system in the country. Sixty-two percent of students are housed among five different neighborhoods. Freshmen can choose single-sex or co-ed living and also submit a list of their preferred living areas, but they're required to live on campus through sophomore year. About half of the freshmen end up in the Southwest Area, a "huge, citylike complex" with five high-rise towers and 11 low-rise residence halls. Approximately 4,000 first-year students participate in Epoch, a yearlong residential program, or one of the university's other Living and Learning communities. Upperclassmen tend to move off campus.

UMass offers "a vast social life," says a student, with noisy dorms, overflowing frat houses, and frequent off-campus parties. "Social life is hot!" a senior raves. Both on campus and off, alcohol policies are strict and well-enforced; underage drinking is many times confined to students' rooms, if they can get away with it. First-time underage offenders are sent to alcohol-education programs. The school is losing its reputation as "ZooMass." Six percent of the men and 5 percent of the women belong to one of the nearly two dozen fraternities and sororities, but they are somewhat out of the mainstream. A free public transportation system allows maximum mobility—not only among the Five Colleges, but also to nearby towns, which are graced with a number of exceptional bookshops. The annual Spring Concert around the pond is a daylong event where musicians such as U2, Bob Dylan, and the Beastie Boys have performed.

Varsity sports are popular, and UMass has been a model for achieving gender equity in athletics.

"Social life is hot!"

Settled in the Pioneer Valley and surrounded by the Berkshire foothills, Amherst is close to good skiing, hiking, and canoeing areas. It's also 90 miles west of Boston, 150 miles north of New York City, and 25 miles south of Vermont and New Hampshire, making a car very useful (and very expensive if you get too many tickets from overzealous campus cops, students say).

Varsity sports are popular, and UMass has been a model for achieving gender equity in athletics. "Midnight Madness," the first men's basketball practice of the season, held annually at midnight with everyone invited, is a hot campus ticket. The men's lacrosse, skiing, and swimming squads have won conference championships in recent years, as have women's swimming, skiing, and softball. Two on-campus gyms offer facilities for the recreational athlete, and two Olympic-size skating rinks mark the recent reintroduction of intercollegiate hockey to the university.

UMass is big enough to offer a vast number of academic and extracurricular opportunities, though at times it can feel impersonal and overwhelming. But with special residential programs that group students with similar languages, cultures, and lifestyles, many students will easily find a home in Amherst.

If You Apply To ➤ **UMass:** Regular admissions: Jan. 15. Financial aid: March 1. Meets demonstrated need of 27%. Campus and alumni interviews: optional, informational. SATs or ACTs: required. SAT IIs: optional. Accepts the Common Application and electronic applications. Essay question: personal circumstances or academic experiences.

Massachusetts Institute of Technology

Room 3-108, 77 Massachusetts Avenue, Cambridge, MA 02139

If you're a science genius, come to MIT to find out how little you really know. No other school makes such a massive assault on the ego (with little in the way of support to help you pick up the pieces). Technology is a given, but MIT also prides itself on leading programs in economics, political science, and management.

MIT is, in a word, excellence. With a student body that averaged near-perfect scores on the math portion of their SATs with verbal scores not far behind, this is a place that restores faith in the American educational system. Engineering, science, and math are MIT's specialties, but students come here to learn about everything—and learn they certainly do.

MIT is located on 168 acres that extend more than a mile along the Cambridge side of the Charles River basin facing historic Beacon Hill and the central sections of Boston. The main campus of neoclassical architecture carved from limestone was designed by Welles Bosworth and constructed between 1913 and 1920. Since then, more modern designs in brick and glass have been added. The buildings have a utilitarian aura; most are even known by number instead of by name. Athletic playing fields, recreational buildings, dorms, and dining halls are closely arranged on the campus and provide a sense of unity. Sculptures and murals, including the works of Alexander Calder, Henry Moore, and Louise Nevelson, are found throughout the campus. The new State Center for Computer, Information, and Intelligence Sciences was designed by Frank Gehry.

Originally called Boston Tech and now frequently referred to as "the Tute," MIT stresses science and engineering studies with a "concern for human values and social goals." Every science and engineering department is superb. The biology department is a leader in medical technology and the search for designer genes. Nevertheless, pure sciences tend to play second fiddle to the engineering fields that, along with computer science, draw the bulk of the majors. Electrical engineering and computer science are almost universally credited as tops in the nation. Students in these two areas may pursue a five-year-degree option, where they can obtain a professional master's degree upon completion of their studies. Biomedical, chemical, and mechanical engineering; physics; and the aeronautics department are also highly praised programs. The most popular majors include electrical engineering and computer science, management science, mechanical engineering, and biology. The humanities are strong here as well, though not on par with science- and technology-oriented programs.

Website: http://web.mit.edu
Location: Urban
Total Enrollment: 10,206
Undergraduates: 4,066
Male/Female: 57/43
SAT Ranges: V 690–770
  M 740–800
ACT Range: 31–34
Financial Aid: 63%
Expense: Pr $ $ $
Phi Beta Kappa: Yes
Applicants: 10,440
Accepted: 14%
Enrolled: 67%
Grad in 6 Years: 94%
Returning Freshmen: 98%
Academics: 🎓 🎓 🎓 🎓 🎓
Social: ☎ ☎ ☎
Q of L: ★ ★ ★
Admissions: (617) 253-4791
Email Address:
  admissions@mit.edu

Strongest Programs:
Engineering
Computer Science
Management Science

MIT has always attracted top professors in a broad range of fields, including such luminaries as linguist Noam Chomsky. Political science, management, urban studies, linguistics, graphics for modern art, and holography—and just about anything else that can be linked to a computer—are strong, and the minority who major in these subjects receive enough personal attention to make any college student envious. "Some professors really know how to engage the interest of the student," says a senior.

MIT is tops in technology, but also strong in the social sciences. The administration worries that engineers of the future will need first-rate technical skills coupled with a good understanding of technology's social context and marketplace. As one dean put it, "Too many MIT graduates end up working for too many Princeton and Harvard graduates." Hence, faculty approved a new minor in management "in response to employers seeking graduates who are better prepared for today's increasingly complex responsibilities." New majors include biological engineering, mechanical and ocean engineering, chemical engineering, and comparative media studies, and a program that integrates science and global environmental issues. Faculty are reviewing the general education program, which requires undergraduates to take at least eight HASS (Humanities, Arts, and Social Sciences) courses. Perhaps to help ensure that they will be able to make their future discoveries known, students must take four communication-intensive subjects. Of course, there is also the intensive science requirement, which includes six math/science courses and two restricted electives in science, as well as a lab or two. There's also an eight-credit physical education requirement and a swimming test.

"MIT is intense and will take you for quite a ride."

One of MIT's most successful innovations is the Undergraduate Research Opportunities Program (UROP), a year-round program that facilitates student/faculty research projects. Considered one of the best programs of its kind in the nation, it allows students to earn course credit or stipends for doing research. Flexible, small-group alternatives for freshmen include five learning communities and the Experimental Study Group, which allows a self-paced course of study based on tutorials instead of a traditional lecture format. Not only can these offer support, but they can also engage geniuses who excel on exams without attending the lectures. Many students have access to world-renowned professors and the Nobel Prize winners who carry lighter teaching loads to allow them time for research and interaction with students. Faculty advising is "pretty good for freshmen," one student says, but after that, "it's as good as you make it." The vast library system contains more than two million volumes, including some one-of-a-kind manuscripts on the history of science and technology. One library is even open 24 hours a day, and "some students spend the majority of their time (awake or asleep) there," one student reports.

Hockey is popular, and even more popular is the extensive, well-organized intramural program (60 percent participate), with sports ranging from Ping-Pong, billiards, and bowling to the more traditional basketball and volleyball.

A pass/no record grading system helps freshmen adjust to "MIT brainstretching": in the first semester, freshmen receive grades of P, D, or F in all subjects they take. P means a C-or-better performance; Ds or Fs do not receive credit or appear on the permanent record. In the second semester, the Ps are replaced by A, B or C; Ds and Fs do not receive credit and are only noted internally. Grades or not, most MIT students set themselves a breathtaking pace. "MIT is intense and will take you for quite a ride," a biology/premed student says. "The courses demand your full attention and a lot of extra work," another says. Some relief from "tooling" (that is, studying) is found through the optional January Independent Activities Period, which offers noncredit seminars, workshops, and activities in fields outside the regular curriculum, as well as for-credit subjects. Participation in the engineering co-op program, junior year abroad (including a major new program at Cambridge University in England), or crossregistration at all-female Wellesley College are other

"The average MIT student can be characterized as having a passion and singular drive for what they really want in life."

helpful ways to get young noses away from the grindstone. For those with foreign language skills, summer internships abroad offer unique opportunities. Academic and psychological counseling are well thought of by students. A student-run hotline provides all-night peer counseling. Many upperclassmen return to school at least two weeks before classes start "to help integrate the freshmen."

While MIT somewhat justly earned an image as a "conservative, rich, white boys' school" in the past, there is certainly enough racial if not gender variety to beat the rap today. African Americans account for 6 percent of the student body, Hispanics 11 percent, and Asian Americans a hefty 27 percent. If anything, women may feel "a different tone, as the campus is nearly three-fifths male." Ninety-seven percent of students come from the top tenth of their high school class, and average SAT and ACT scores are simply mind-boggling. "The average MIT student can be characterized as having a passion and singular drive for what they really want in life," offers a chemical engineering major. MIT helps financially needy students pay the superhefty tuition bill by guaranteeing to meet all demonstrated need.

Ninety-four percent of undergraduates live on campus, and all freshmen are required to live in the dorms. Guaranteed housing is either single-sex or co-ed; the dorms are in the middle of campus, and most of the fraternities and living groups are a mile or less away across the Charles. Meal plans can be mandatory for dorms that don't have kitchens, or optional for equipped quarters. Frat-types feast on spreads prepared by their full-time cooks, and the Kosher Kitchen provides some refuge for others.

MIT's social scene is varied. There are campus movies and lectures if one can escape the ubiquitous workload worming its way into the uneasy consciousness of a techie's every waking hour. And there's the great city of Boston, with its many restaurants, clubs, parks, shopping opportunities, and more than 50 other colleges.

On-campus dances, parties, and dorm activities keep other students busy. Most on-campus drinking for over-21 students is relaxed and accepted, "as long as the alcohol does not result in unlawful behavior or cause any problems," a student explains. For those with the urge to roam, the multifaceted greater Boston metropolis sits only a few subway stops away. MIT's alcohol-prevention program is considered a national model, and drug-prevention initiatives are also comprehensive.

When the MIT megabrains take a break, practical jokes, or "hacks" (described by one student as "practical jokes with technical merit"), are sure to follow. In past years, popular hacks have included disguising the dome of the main academic building as a giant breast, unscrewing and reversing all the chairs in a 500-seat lecture hall, and, of course, welding shut Harvard's gates. Hacking can also involve Harry Potter–style late-night explorations by students in the tunnels and shafts that run through restricted parts of the campus, a practice that's definitely frowned upon by the school.

When not studying or hacking, these engineering jocks often turn into real jocks. Athletic accomplishments in the past few years include seven consecutive NEWMAC championships in men's tennis and cross-country; men's soccer national quarter finals; and championships for men's water polo, men's and women's track and field, women's fencing, and women's volleyball. Hockey is popular, and even more popular is the extensive, well-organized intramural program (60 percent participate), with sports ranging from Ping-Pong, billiards, and bowling to the more traditional basketball and

"Some professors really know how to engage the interest of the student."

volleyball. Everyone has access to MIT's extensive athletic facilities. Supposedly, there are more clubs and organizations at MIT than at any other school in the country, and a sampling of the offerings explains why. The Rocket Society, the Guild of Bell Ringers, and a singing group called the Corollaries are only a few of the diverse interests on this campus.

Every science and engineering department is superb. The biology department is a leader in medical technology and the search for designer genes. Nevertheless, pure sciences tend to play second fiddle to the engineering fields that, along with computer science, draw the bulk of the majors.

When the MIT megabrains take a break, practical jokes, or "hacks" (described by one student as "practical jokes with technical merit"), are sure to follow.

Though students often wonder what life at a so-called typical college would have been like, chances of survival and even satisfaction at MIT are excellent. Students are able to comprehend the incredible experience of attending one of the nation's leading academic powerhouses. A biology major puts it bluntly: "It will take you right up to what you think your limits are, and then MIT will shatter them and make you realize how great your potential is."

If You Apply To ➤ **MIT:** Early action: Nov. 1. Regular admissions: Jan. 1. Financial aid: Mar. 1. Guarantees to meet demonstrated need. No campus interviews. Alumni interviews: recommended, evaluative. SATs or ACTs: required. SAT IIs: required (math, science). Accepts electronic applications. Essay question: reaction to reaction to a disappointment or difficulty; how has your world shaped your dreams. Looks for aptitude in math and science.

McGill University: See page 332.

University of Miami

P.O. Box 248025, Coral Gables, FL 33124-4616

Football is the main reason UM is on the map, but it isn't the only reason. Renowned programs in marine science and music are big draws; business is also strong. Housing takes the form of a distinctive residential college system that offers living/learning opportunities.

Website: www.miami.edu
Location: Suburban
Total Enrollment: 15,674
Undergraduates: 10,132
Male/Female: 43/57
SAT Ranges: V 570–670
  M 590–690
ACT Range: 26–30
Financial Aid: 87%
Expense: Pr $ $
Phi Beta Kappa: Yes
Applicants: 18,810
Accepted: 46%
Enrolled: 26%
Grad in 6 Years: 71%
Returning Freshmen: 89%
Academics: ✍ ✍ ✍
Social: ☎ ☎ ☎ ☎
Q of L: ★ ★ ★
Admissions: (305) 284-4323
Email Address:
  admission@miami.edu

Year-round sunshine and the colorful Miami culture could make even the most dedicated students forget why they are at college. But at the University of Miami, students can have their fun and get a solid education at the same time. Solid academics and sunshine create a heady mix that attracts students from far and wide.

Twenty minutes from Key Biscayne and Miami's beaches, and 10 minutes from downtown Miami, the university's 260-acre campus is located in tranquil suburbia. With its own lake in the middle of the campus (and on the cover of most brochures), the campus is architecturally varied, from postwar, international-style structures to modern buildings, most with open-air breezeways to let in the warm, salty winds. The state-of-the-art Frances L. Wolfson School of Communication building, dedicated in 2001, has been joined by the more recent addition of the Jorge M. Perez Architecture Center, designed by noted architect and town planner Leon Krier. Finishing touches have been applied to the Frost School of Music's new library and technology center.

Miami has one of the nation's top programs in marine biology, and was the first university to offer a degree in music engineering. Miami's main strengths are in the preprofessional and professional areas. The school also boasts a unique program in jazz, and students recommend any of the strong premed offerings. Chemistry majors have access to a nuclear magnetic resonance spectrometer, an essential tool for modern chemistry. The school's seven-year medical program for outstanding students, and dual degree (grad/undergrad) programs in law, marine science, business, physical therapy, biomedical engineering, and medicine receive high marks. Women's studies and philosophy are said to be weaker. Business management is the most popular major, followed by visual and performing arts, biology, health professions, and engineering.

"The coursework is challenging," says a student, "especially if you have a scientific or engineering type major." Students give professors high marks for knowledge and accessibility. Full professors teach most courses. "The quality of teaching I have received is second to none," a sophomore says. "I have received that personal attention I wanted from my classes and professors." Though many programs have a pre-professional bent, students at Miami receive a broad liberal arts education. Distribution requirements vary from school to school, but general education require-

> **"The coursework is challenging, especially if you have a scientific or engineering type major."**

ments include proficiency in English composition, mathematics, and writing across the curriculum (courses that involve a substantial amount of writing). In addition, a certain number of credits must be earned in each of three areas of knowledge: natural sciences, social sciences, and arts and humanities. Students looking for a change of pace can take advantage of Miami's summer semester program in the Caribbean. In addition, the study abroad program offers more than 90 options in 37 countries such as Australia, Israel, France, Japan, the Netherlands, and Argentina.

Highly motivated students in any field can apply to the school's comprehensive honors program, which enrolls students who were in the top tenth of their high school class and have a combined SAT score of at least 1360. About 15 percent of each freshman class enrolls. The stereotypical beach bum who drops by for a couple of classes in the morning, spends the rest of the day at the shore, and almost never sees the inside of the library need not apply to UM. "The academic profile of UM students has risen so sharply over the last 10 years. Our standards are too rigorous for students to pull this off," asserts the administration.

Fifty-six percent of UM's students come from out of state, mostly from the Northeast, Ohio, and the Chicago area. UM is unique among universities of its caliber in the incredible diversity of its student body; Hispanics account for a substantial 27 percent of the total, African Americans 10 percent, and Asian Americans 8 percent. Seven percent are foreign-born. Students say that diversity is one of UM's best assets. "We have students from all 50 states and more than 110 foreign countries," says an English/political science major. "The most wonderful experience was hearing three different languages spoken by groups of students on my way from the dorm to the classroom." The school's large number of Hispanics is traceable to the influx of Cuban and other Caribbean refugees into southern Florida, and at times it seems that Spanish is the mother tongue on campus. The Latin influence mixes colorfully with

Miami offers a distinctive system of five co-ed residential colleges, modeled after those at Yale University. Each college is directed by a master—a senior faculty member who organizes seminars, concerts, lectures, social events, and the monthly community dinner.

> **"The most wonderful experience was hearing three different languages spoken by groups of students on my way from the dorm to the classroom."**

that of the New Yorkers who make up 8 percent of the freshman class, many of whom view their time at UM as a welcome vacation from the harsh weather. The one characteristic everyone seems to share is the hope of getting high-paying jobs after graduation. Nearly 2,000 merit scholarships and athletic awards in several sports ease the school's hefty price tag for qualified students.

Miami offers a distinctive system of five co-ed residential colleges, modeled after those at Yale University. Each college is directed by a master—a senior faculty member who organizes seminars, concerts, lectures, social events, and the monthly community dinner. Faculty members often host study breaks in their homes and provide guest speakers from all walks of life to discuss current issues. Generally, students give the dorms average marks; more than 40 percent of students live on campus, and others bunk in off-campus apartments or Greek houses. Still, all dorm housing at Miami is co-ed, and students can choose apartment-style housing when

Miami has one of the nation's top programs in marine biology, and was the first university to offer a degree in music engineering.

they tire of the residential colleges' closeness. Scrounging up grub on campus is easy; the residential colleges have their own cafeterias with a variety of plans, from five to 20 meals, and there is a kosher alternative.

On the weekends, Miami students frequent nearby bars or the campus Rat, home also to the popular Fifth Quarter postgame parties. On-campus alcohol policies are relatively strict for underage students, sometimes including parental notification for offenses. Fraternities still manage to thrive, accounting for 12 percent of the men and providing a space for most of the underage drinking at UM (although not during rush, which is dry). The sororities, with no housing of their own, attract 14 percent of the women. Many of the fraternities and sororities are small (averaging about 30 members), but they often join forces in throwing parties.

"The majority of students at Miami fit into the trendy, clubbing profile."

If keg parties aren't your scene, though, UM offers a plethora of other social opportunities. Those who shun sand between their toes bike down to the boutiques in Coconut Grove, Bayside, or South Beach, or attend on-campus events, such as International Week and Sportsfest. "The majority of students at Miami fit into the trendy, clubbing profile," explains a psychology major. "They like trendy up-to-date clothes and going to clubs in order to see and be seen." Public transportation runs in front of the residential colleges, but most students recommend a car in order to get "the full Florida effect." Parking can be a problem, though, says a senior: "If you want a parking space, you need to get to school by 8:00 a.m., and parking decals are way too expensive." The best road trips are Key West, Key Largo, and, of course, UM football games, especially those against Florida State.

Miami joined the Atlantic Coast Conference in 2004, creating quite a storm. The turbulence has continued as the football team has struggled of late, resulting in coach Larry Coker's dismissal. The team hopes to regain its foothold under the helm of new head coach Randy Shannon. Men's baseball, basketball, and tennis and women's track and field are strong and have captured recent championships. The $14 million Wellness Center, with juice bar and spa, and the annual intramural Sportsfest also draw crowds.

It's hard to imagine a school in the Sunshine State without a generous allotment of fun, and UM is no exception. Life really is a beach for students at the University of Miami, although its days as a beach-bum hideout are long over. These days, UM students are just as likely to search long and hard for the perfect instrumental phrase or mathematical proof as they are to scope out the perfect wave. "It is the whole college experience at U of M," says one sophomore. "It is so fulfilling and rewarding."

If You Apply To ➤ **Miami:** Early decision and early action: Nov. 1. Regular admissions: Feb. 1. Does not guarantee to meet demonstrated need. SATs or ACTs: required. SAT IIs: optional; math and science required for dual degree Honors Program in Medicine. Apply to particular schools or programs. Accepts the Common Application and electronic applications. Essay question: experience or achievement that is special to you; personal, local, or national concern important to you; person who has had significant influence on you; the role of academic integrity.

Miami University (OH)

301 S. Campus Avenue, Oxford, OH 45056

Rather than disappear into the black hole of Ohio State, top students in the Buckeye state come here to feel as if they are going to an elite private university. MU has a

niche like William and Mary's in Virginia—though MU is twice as big. Miami's top draw is business, and its tenor is conservative.

This Miami is about 1,000 miles from South Beach, but that doesn't mean it's without sizzle. The academic kind, of course. Miami University is actually tucked into a corner of Ohio and is gaining national recognition as an excellent state university that has the true look and feel of a private, with a picture-perfect campus and high-caliber student body.

The university is staked out on 2,000 wooded acres in the center of an urban triangle of approximately three million people, encompassing Cincinnati and Dayton, Ohio, and Richmond, Indiana. The campus is dressed in the modified Georgian style of the Colonial American period and it remains as impeccably groomed as its nattily attired students. Recent additions to campus include on-campus apartment-style housing and renovations to McGuffey Hall. Recently completed construction includes the new Ice Arena, a parking garage, and a psychology building. "The amount of construction and expansion at Miami is amazing," a student says.

Miami University was founded in 1809 to provide a classical liberal education, and has never strayed from its central commitment to liberal arts. All undergraduates must complete the Miami Plan for Liberal Education, which provides them with a background in fine arts, humanities, social sciences, natural sciences, and formal reasoning. Popular majors include marketing, zoology, finance, psychology, and political science. For those with an inclination toward forestry or the paper industry, the university's unique pulp and paper science technology degree is one of only eight in the nation for undergraduates.

University requirements, or foundation courses, provide for a broad education, and all undergraduates must complete foundation courses in English composition; fine arts; humanities; social sciences and world cultures; biological and physical sciences; and mathematics, formal reasoning, or technology. Additional requirements include 12 credits of an advanced liberal education focus consisting of nine credits of thematic and sequential study outside of the student's major, and three credits of the Senior Capstone Experience, which ties in liberal education with the specialized knowledge of their major. Interdisciplinary studies, economics, music, and botany are well regarded.

The academic atmosphere at Miami is competitive but not cutthroat. "The courses are challenging [but] the climate is cooperative," says a junior. A recent graduate adds, "The people in my classes were high achievers and therefore, there was a high level of competition." Professors are lauded for their knowledge and willingness to help. "Instruction is clear

"The amount of construction and expansion at Miami is amazing."

and extra help is always available," a student says. Sixty-six percent of classes have fewer than 25 students, and most are taught by full professors, though graduate students do pop up. Many students complain that it's increasingly difficult to get into required classes—especially foundation courses—and that this can complicate the task of graduating in four years. "I've had to force-add classes or wait to take them in the summer," gripes a speech communications major.

About 1,000 students study abroad each year. The Dolibois European Center in Luxembourg offers a semester or yearlong program in the liberal arts and an opportunity to live with a foreign family. Other exchange opportunities include universities in Denmark, Japan, Mexico, Austria, and England, as well as summer programs in France, Italy, Germany, Russia, Mexico, Luxembourg, the Caribbean, and Asia. Undergraduate research gets a lot of attention at Miami—the competitive Undergraduate Research Scholars program gives 100 students a stipend, free tuition, and an expense account to complete a 10-week academic project. "Honors and

Website: www.muohio.edu
Location: Rural
Total Enrollment: 15,343
Undergraduates: 14,302
Male/Female: 47/53
SAT Ranges: V 560–650
  M 580–670
ACT Range: 25-29
Financial Aid: 40%
Expense: Pub $ $ $ $
Phi Beta Kappa: Yes
Applicants: 15,579
Accepted: 69%
Enrolled: 29%
Grad in 6 Years: 80%
Returning Freshmen: 89%
Academics: ✍ ✍ ✍ ✍ ½
Social: ☎ ☎ ☎
Q of L: ★ ★ ★
Admissions: (513) 529-2531
Email Address:
  admission@muohio.edu

Strongest Programs:
Economics
Accountancy
Music
Chemistry
Botany
Microbiology
International Studies
Architecture

Many students complain that it's increasingly difficult to get into required classes—especially foundation courses—and that this can complicate the task of graduating in four years.

scholars programs are a great way to supercharge your educational experience," offers one junior.

"Miami is made up of a lot of preps who are well-dressed and have money to spend," says one student. Another adds that "a lot come from upper-middle-class families where Mom and Dad pay for everything." Three percent of the student body is African American, 2 percent Hispanic, and 3 percent Asian American. Seventy-two percent of students are from Ohio, and the campus has a reputation for conservatism. In recent years, Miami has begun an effort to attract more students of color with programs such as the Minority Professional Leadership Program. In another effort to foster better race relations, the university holds a series of forums designed to identify and solve issues within the African American community. Thousands of merit scholarships and 364 athletic scholarships are awarded annually.

> "Honors and scholars programs are a great way to supercharge your educational experience."

Forty-four percent of the student body call the campus home. A few of the dorms remain single-sex and are accompanied by visitation rules, and one student describes the rooms as "the cleanest dorms I've ever seen." Most upperclassmen find good and cheap off-campus housing by their senior year, but remain very involved on campus. Campus security is said to be good, and students report that they feel safe thanks to emergency call boxes, well-lit walkways, and key-card entrances.

Miami has a lively on- and off-campus social life, although students complain that social restrictions on campus are on the rise. "The campus is very strict about alcohol violations," explains one student. "There is no tolerance for minors drinking on campus." A lot of socializing takes place in the restaurants, bars, and clubs of Oxford. Twenty-two percent of the men and 25 percent of the women belong to fraternities or sororities, respectively. In fact, Miami is known as the "mother of fraternities" because several began here. Despite the hard-partying reputation of the Greeks, Miami boasts one of the best graduation rates in the nation among public institutions.

The town of Oxford "is as good as it gets." One student explains: "The students comprise two-thirds of its population, and most of its shops cater to the students. [Oxford] is a wonderful and distinct college town." For students who crave bigger fish to fry, Cincinnati is about 35 miles away. The well-organized intramural program provides teams for just about every sport (including the ever-popular korfball and broomball), and the Student Recreational Sports Facility is well used. Other annual events include Make a Difference Day in cooperation with Oxford, homecoming, and continued rivalries with Ohio University.

The Redhawks field a number of competitive athletic teams. Swimming, tennis, and volleyball are among Miami's best varsity sports, but football (2004 MAC Bowl champions) and men's basketball (2005 conference champs) reign as the most popular spectator sports. For women, soccer, basketball, and swimming teams do well (double-digit MAC champions; more than any other MAC school). Ice hockey, which is one of the top 10 programs in the country, is also extremely popular, and the women's precision ice-skating team is the only one at collegiate level in the nation. For cycling enthusiasts, the annual 20/20 Bike Race is one of the largest collegiate events of its kind in the country.

> "The campus is very strict about alcohol violations."

Miami University of Ohio, with its strong emphasis on liberal arts and its opportunities for research, travel abroad, and leadership, is looked upon as one of the rising stars among state universities. It recently broke with tradition and pegs its tuition at the out-of-state level—with automatic discounts for Ohio residents. The school effectively combines a wide range of academic programs with the personal attention ordinarily found only at much smaller institutions.

Swimming, tennis, and volleyball are among Miami's best varsity sports, but football (2004 MAC Bowl champions) and men's basketball (2005 conference champs) reign as the most popular spectator sports.

Overlaps

Ohio State University, Ohio University, Indiana University, University of Dayton, University of Michigan, Notre Dame

University of Michigan

1220 Student Activities Building, Ann Arbor, MI 48109-1316

The most interesting mass of humanity east of UC–Berkeley. UM is among the nation's best in most subjects, but undergraduates must elbow their way to the front to get the full benefit. Superb honors and living/learning programs are the best bet for highly motivated students. Costly for out-of-staters.

One of the nation's elite public universities, Michigan offers an excellent faculty, dynamite athletics, an endless number of special programs, and the most interesting collection of students east of Berkeley. "Michigan is a special place because it has a deep history and reputation," says a senior. "It is an excellent school and no matter what degree you have, it is respected."

Situated on 3,129 acres, Michigan's campus is so extensive that newcomers may want to come equipped with maps and a GPS system to find their way to class. The university is divided into two main campuses. Central Campus, the heart of the university, houses most of Michigan's 19 schools and colleges. North Campus, which is two miles northeast of Central, is home to the College of Engineering, School of Music, School of Art and Design, College of Architecture and Urban Planning, and the new Media Union. Other campus areas include the Medical Center complex containing seven hospitals and 15 outpatient facilities, and South Campus, featuring state-of-the art athletic facilities. Architecturally, the main drag of campus features a wide range of styles, from the classical Angell Hall to the Gothic Law Quad.

> "It is an excellent school and no matter what degree you have, it is respected."

Academically, students describe the courses as challenging and rigorous but not cutthroat competitive. "Although some students are overly ambitious, most are willing to share their notes and study together," says a senior. The university ranks among the best in the nation in many fields of study, mainly because it attracts some of the biggest names in academia to teach and research in Ann Arbor. The College of Literature, Science, and the Arts is the largest school at Michigan. The College of Engineering and School of Business are well respected, and the university's programs in health-related fields are also top-notch. Students report that professors are "knowledgeable." One student says, "The professors here are intelligent and seem to enjoy teaching." Students claim there is excellent academic and career advising available, but only for those who seek it. The administration, however, notes that the advising office, which registers nearly 12,000 clients each year, offers individually tailored services and workshops. The Career Planning and Placement Office processes about 120,000 transactions each year, provides individual and group career counseling/planning and individual job placement, and works with 950 companies annually in recruiting UM graduating students.

Michigan's special academic programs seek to offer the best of both worlds—personalized attention and a large university setting. Approximately 627 active

Website: www.umich.edu
Location: Suburban
Total Enrollment: 36,983
Undergraduates: 24,361
Male/Female: 49/51
SAT Ranges: V 590–690
  M 630–730
ACT Range: 26–31
Financial Aid: 40%
Expense: Pub $ $ $ $
Phi Beta Kappa: Yes
Applicants: 23,114
Accepted: 59%
Enrolled: 59%
Grad in 6 Years: 87%
Returning Freshmen: 96%
Academics: ✍ ✍ ✍ ✍ ✍
Social: ☎ ☎ ☎
Q of L: ★ ★ ★
Admissions: (734) 764-7433
Email Address:
  ugadmiss@umich.edu

Strongest Programs:
Premed
Engineering
Art and Design
Architecture
Music
Business

degree programs, including about 226 undergraduate majors as well as individualized concentrations, are offered, mainly through the College of Literature, Science, and the Arts. Special programs include double majors, accelerated programs, independent study, field study, and internships. The Screen Arts & Cultures major balances studies and production, with studies occupying approximately two-thirds of a student's coursework and the one-third devoted to creative, hands-on projects. In addition, students can choose from several small interdisciplinary programs. The instructors live and teach in the residential hall, the Residential College, and the Lloyd Hall Scholars Program. The Comprehensive Studies Program allows students to become part of a community of scholars who work in programs designed to best realize an individual student's potential.

> *The Comprehensive Studies Program allows students to become part of a community of scholars.*

The University of Michigan's honors program, considered to be one of the best in the nation, offers qualified students special honors courses, opportunities to participate in individual research or collaborative research, seminars, and special academic advisors. A preferred admissions program guarantees 150 top high school students admission to Michigan's professional programs in dentistry, biomedical engineering, social work, architecture, or pharmacy, provided they make satisfactory progress during their first years. The Undergraduate Research Opportunities Program enables students to work outside the classroom with a small group of students and a faculty member of their choice. The most popular majors at University of Michigan are psychology, business administration, mechanical engineering, political science, and English, but students say the statistics department needs improvement. Michigan also offers a number of foreign language majors not found many other places, including Arabic, Armenian, Persian, Turkish, and Islamic studies. Freshmen are no longer admitted directly to the School of Natural Resources.

> *Michigan's special academic programs seek to offer the best of both worlds—personalized attention and a large university setting.*

No courses are required of all freshmen at Michigan, but all students are required to complete some coursework in English (including composition), foreign languages, natural sciences, social sciences, and humanities. Students in the College of Literature, Science, and the Arts must also take courses in quantitative reasoning and race or ethnicity. In addition, the university offers a series of seminars designed specifically for freshmen and sophomores, which are taught by tenured and tenure-track faculty. Off-campus opportunities abound at UM. Students have the chance to visit and study abroad in more than 30 different countries, including Australia, China, Costa Rica, Finland, France, Greece, India, Ireland, Japan, Russia, Sweden, and Turkey. Some specific programs include a year abroad in a French or German university, a business program in Paris, summer internships in selected majors, and special trips organized by individual departments.

> **"It's a great city with something for everyone. There are coffeehouses, bars, sporting events, movie theaters, and a lot more."**

The University of Michigan's admissions office sifts through some of the best students in the country, with 90 percent of the students in the top tenth of their high school class. Sixty-eight percent hail from Michigan. The student body is remarkably diverse for a state university. In fact, Michigan's Program on Intergroup Relations, Conflict, and Community was recognized by former President Clinton's Initiative on Race as one of 14 "promising practices" that successfully bridge racial divides in communities across America. Minorities now comprise more than one-fourth of UM's total enrollment, an all-time high. African Americans and Hispanics combined make up 13 percent of the student body, and Asian Americans make up another 13 percent. There is a large and well-organized Jewish community at Michigan, and gays and lesbians are also organized and prominent. While the student body is more conservative today than it was a decade ago, it is still "most noticeably liberal," says a history major, and political issues flare up from time to time on cam-

> *No courses are required of all freshmen at Michigan, but all students are required to complete some coursework in English (including composition), foreign languages, natural sciences, social sciences, and humanities.*

pus. Michigan really socks it to out-of-staters with a hefty surcharge. However, the university guarantees to meet the demonstrated financial need of all admitted Michigan residents. Students can also vie for thousands of merit scholarships, as well as 415 athletic scholarships for men and women.

Dormitories at UM traditionally have well defined personalities. Sixties-inspired types and "eccentrics" find the East Quad the "most open-minded dorms" (the Residential College is here). The Hill dorms are "more sedate." For those seeking alternative housing arrangements, a plethora of special-interest housing is available, including substance-free residence halls. On-campus housing is comfortable and well maintained. "The dorms are a tad small but livable with a little bit of work," says a history major. Housing is guaranteed for all incoming freshmen, leaving many upperclassmen to play the lottery. For the student

> **"We are ranked high enough to be known for our academic success, but we still have a reputation for having a good time."**

who wants to live off campus, the UM housing office provides information, listings, and advice for finding suitable accommodations. Other alternatives include fraternity and sorority houses, and a large number of college- and privately owned co-ops.

Detroit is a little less than an hour away, but most students become quite fond of the picturesque town of Ann Arbor. "It's a great city with something for everyone," says a political science major. "There are coffeehouses, bars, sporting events, movie theaters, and a lot more." A surprising variety of visual and performing arts are offered in town and on campus. Underage drinking is not allowed, and a senior has a stern warning for any potential schemers: "Your fake ID will be taken. Plan on it. Do not be surprised, no matter how good it is." An annual art fair held in Ann Arbor draws craftspeople from throughout the nation and Canada. Many lakes and swimming holes lie only a short drive away and seem to keep the large summer-term population happy. As one junior says, "We are ranked high enough to be known for our academic success, but we still have a reputation for having a good time." Michigan winters, though, are known for being cold and brutal. Sixteen percent of the men and

> **"You shouldn't be allowed to graduate if you haven't gone to a hockey game."**

12 percent of the women go Greek, though these groups are the bane of campus liberals. Many students also volunteer in the community. One senior explains, "Most students get involved, especially if it has something to do with helping kids."

Football overshadows nearly everything each fall as students gather to cheer "Go Blue." Attending football games is an integral part of the UM experience, students say, and "you shouldn't be allowed to graduate if you haven't gone to a hockey game," quips a sophomore. Swimming, track and field, gymnastics, and basketball are also strong. Intramurals, which were invented at the University of Michigan, provide students with a more casual form of athletics.

The University of Michigan strives to offer its students a delicate balance between academics, athletics, and social activities. On one hand, this is American college as it's characterized in movies like Animal House—football and fraternities. But it's also a world-class university with a fine faculty and top-rated programs, intent on making America competitive in the 21st century. For assertive students who crave spirit and action as well as outstanding academics, Michigan is an excellent choice.

Intramurals, which were invented at the University of Michigan, provide students with a more casual form of athletics.

Overlaps

Michigan State, Northwestern, University of Pennsylvania, Cornell University, University of Wisconsin, Washington University (St. Louis)

If You Apply To ➤

Michigan: Rolling admissions. Regular admissions: May 1. Does not guarantee to meet full demonstrated need. Campus and alumni interviews: optional, informational. SATs or ACTs: required. SAT IIs: required for home-schooled students (English, math, science, foreign language, and social studies). Essay question: personal statement. Apply to particular school or program. Policies and deadlines vary by school.

Michigan State University

250 Administration Building, East Lansing, MI 48824-0590

Most people don't realize that Michigan State is significantly bigger than University of Michigan. (Classes via videotape don't help the situation.) Students can find a niche in strong preprofessional programs such as hotel and restaurant management, prevet, business, and engineering.

Website: www.msu.edu
Location: City outskirts
Total Enrollment: 45,166
Undergraduates: 35,678
Male/Female: 46/54
SAT Ranges: V 490–620
    M 520–650
ACT Range: 22–27
Financial Aid: 40%
Expense: Pub $ $ $ $
Phi Beta Kappa: Yes
Applicants: 21,844
Accepted: 76%
Enrolled: 43%
Grad in 6 Years: 71%
Returning Freshmen: 91%
Academics: ✑ ✑ ✑
Social: 🐿 🐿 🐿 🐿
Q of L: ★ ★ ★
Admissions: (517) 355-8332
Email Address:
    admis@mus.edu

Strongest Programs:
Social Sciences
Natural Sciences
Business
Engineering
Packaging
Criminal Justice
Education

Michigan State's roots are agricultural—the school became the state's first land grant institution in 1862—and future farmers and veterinarians still flourish here. So do those with wanderlust, thanks to study abroad programs on each of the world's seven continents. MSU's programs in natural sciences and multidisciplinary social sciences offer students the feel of a small, liberal arts college and the resources of a large research university. "Resources here abound," says a senior, "and that alone is a very worthwhile trait, giving students nearly limitless opportunities to succeed."

The heart of the MSU campus, north of the Red Cedar River, boasts ivy-covered brick buildings, some of which pre-date the Civil War, and are listed on the National Register of Historic Places. This area houses five colleges plus the MSU Union and 10 residence halls. Across the river are the medical complex, newer dorms, and two 18-hole golf courses. On the southernmost part of campus are University Farms, where researchers find ways to grow fatter hogs and cows that produce more milk. The newest addition to campus is the 200,000-square-foot Biomedical and Physical Sciences Building, with six floors of labs and another four of offices. The building connects MSU's Chemistry and Biochemistry buildings, and is near the Plant Biology Laboratories and the National Superconducting Cyloctron Lab. Eighty-two-year-old Spartan Stadium also got a $65 million facelift recently.

"Resources here abound."

Michigan State students tend to be preprofessional and clear about their interests; the premed and prevet programs are strong, and the most popular majors include business, communications and journalism, education, various fields in the social sciences, and engineering. More unusual options include museum studies, supply-chain management, and hospitality management; students in the latter program get real-world experience by staffing the university hotel. To graduate, all students must satisfy university requirements in math and writing, complete a major, and take a minimum of 26 credits in the integrative studies program, which includes arts and humanities; social, behavioral and economic sciences; and biological and physical sciences. There is a strong international component as well, with more than 200 study abroad programs in 60 countries.

The climate at MSU gets tougher as students advance through their majors, says a junior. A classmate adds, "The atmosphere is laid-back as a whole but there are several specialized colleges and programs that are very competitive." Online course selection means no long lines at registration, and advisors can help provide overrides for classes that are technically full. Most lectures are given by professors; labs and smaller recitation sections are led by TAs. "As a freshman, the majority of my classes were taught by professors," says an accounting major. "Not only were they great at teaching, but they worked closely with you to ensure your individual success."

"Aware, alert, and smiling, Spartans have a wide range of interests," says a senior. Whether they come from gritty Motor City or pretty Traverse City, or from somewhere outside the Midwest—a junior says every U.S. state and over 125 countries are represented on campus—Michigan State students care about the world around them. Indeed, more than 2,000 alumni have served in the Peace Corp during the school's 44-year partnership with the agency, a milestone reached by only four other universities. There's a balance between conservative and liberal factions on campus, says a premed student. African Americans make up 9 percent of the student body, Asian Americans add 5 percent, and Hispanics constitute 3 percent. Scholarships are offered in 25 Division I sports, and thousands of students also receive grants and awards based on academic merit.

"As a freshman, the majority of my classes were taught by professors."

Forty-two percent of MSU students—and 96 percent of freshmen—live in Michigan State's dorms, which one junior describes as "spacious, clean, and welcoming." Those seeking the "traditional" college experience can bunk in one of three huge living/learning complexes, each with about four residence halls plus libraries, faculty offices, classrooms, cafeterias, and recreation areas. There are also residential colleges housing less than 1,000 students each: Students in James Madison focus on the social sciences, while those in Lyman Briggs study the natural sciences and math. An honors college allows the brightest freshmen to live together, if they wish, and assigns them a special advisor. Other living/learning programs, known by their catchy acronyms, include RISE (focus on the environment), ROIAL (arts and letters), ROSES (science and engineering), and STAR (Support Teamwork Achievement Resources). All 20 of the school's residence halls are to be updated by 2020. MSU's dining services dish out 32,000 meals a day at 16 cafeterias, where salad, soup, and dessert bars help satisfy those who crave variety.

Safety is not a huge issue here; green-light emergency telephones are sprinkled throughout the campus, walking escorts are available for those who stay late at the library, and Lansing's bus system offers cheaper night-owl rates for those living father away. Still, parking places are in chronically short supply, and students complain about the tickets they receive as a result. Once they've earned 28 credits, which can be as soon as the spring of the first year, students may move off campus. Many do so because the city of East Lansing, just outside Michigan's capital, offers all the positive aspects of a large urban area, along with the safety and community feel of a much smaller town. Indeed, the population of the area more than doubles when school is in session.

Nine percent of MSU students go Greek; as at most schools, those under 21 may not drink alcohol, but Michigan State gives the policy teeth by permitting police to breathalyze any student suspected of being under the influence. Other weekend alternatives include bands, dances, and comedians brought in by the Student Activities Board. Movies that have left

"Aware, alert, and smiling, Spartans have a wide range of interests."

the theaters but haven't yet hit the video store are also shown in Wells Hall—free for campus dwellers, and $1 or $2 for those who live off campus. "MSU has a booming social scene," confirms a junior. Big-name performers who've come to campus recently include comedian Dave Chapelle and musicians Gavin DeGraw, Ani DiFranco, and John Mayer; Grammy winner Kanye West also made an appearance in 2005.

Weekends are dominated by Big Ten athletic competitions, with the Michigan–MSU rivalry especially fierce. "Our large campus is filled from end to end with individuals sporting green and white; alcohol-free tailgating is also available," says a junior. "Seeing 150,000 people in a space that usually has about 60,000 is quite an experience." In 2004–05, the women's basketball team brought home a Big Ten championship and advanced to the NCAA championship game, while the men's basketball team made it to the Final Four. MSU's marching band is a national award winner as well. After 63 years of standing guard at Kalamazoo Street and Red Cedar Road, the school's mascot, affectionately known as "Sparty," is being moved indoors to protect

The newest addition is the 200,000-square-foot Biomedical and Physical Sciences Building, with six floors of labs and another four of offices. Eighty-two-year-old Spartan Stadium also got a $65 million facelift recently.

All 20 of the school's residence halls are to be updated by 2020.

In 2004–05, the women's basketball team brought home a Big Ten championship and advanced to the NCAA championship game, while the men's basketball team made it to the Final Four.

Overlaps
University of Michigan, Western Michigan, Central Michigan, Grand Valley State, Eastern Michigan, Wayne State

him from the elements (and sneaky Wolverines). However, students will still be able to paint "the Rock," a large boulder donated in the 1960s, to advertise campus events, birthdays, anniversaries, and the like.

While 90 percent of MSU students are from the state of Michigan, they're far from a homogeneous lot. Future farmers, physicians, and financiers happily coexist here, in a "diverse, friendly, and expressive" bunch. And despite the university's size, "the majority of people are incredibly nice, outgoing, and laid-back," says a food-science major. "We are more willing to enjoy life and try new things than our counterparts at the University of Michigan."

<table>
<tr><td>**If You Apply To** ➤</td><td>**MSU:** Early action: Oct. 1. Rolling admissions. Housing: May 1 for fall; rolling for other semesters. Financial aid: June 30. Does not guarantee to meet demonstrated need. Campus interviews: optional, informational. No alumni interviews. SATs or ACTs: required. SAT IIs: optional. Accepts the Common Application and electronic applications; prefers electronic applications to paper. Essay question: personal statement.</td></tr>
</table>

Middlebury College

Middlebury, VT 05753

One of the small liberal arts colleges where applications have surged most significantly in recent years. Students are drawn to the beauty of Middlebury's Green Mountain location and strong programs in hot areas such as international studies and environmental science. Known worldwide for its summer foreign language programs.

Website: www.middlebury.edu
Location: Small town
Total Enrollment: 2,415
Undergraduates: 2,415
Male/Female: 49/51
SAT Ranges: V 620–740
    M 640–730
ACT Range: 27–32
Financial Aid: 41%
Expense: Pr $ $ $ $
Phi Beta Kappa: Yes
Applicants: 5,254
Accepted: 24%
Enrolled: 46%
Grad in 6 Years: 90%
Returning Freshmen: 94%
Academics: 🐝 🐝 🐝 🐝 ½
Social: ☎ ☎ ☎
Q of L: ★ ★ ★
Admissions: (802) 443-3000
Email Address: admissions @middlebury.edu

Middlebury College's nickname—"Club Midd"—may bring to mind a resort, but this school's rigorous workload means four years here is far from a vacation. The campus, with its picturesque sunsets, excellent skiing, and rural Vermont charm is a paradise for those interested in environmental studies, second and third languages, and a tight-knit community where highly motivated and intelligent students and faculty truly care about each other. "Be prepared to work," warns a senior, "but also to have an amazing time where every year gets better."

The college's 350-acre main campus overlooks the village of Middlebury, Vermont, which a junior calls a "small, quaint Vermont town of 6,000 people and five stoplights." The 1,800-acre mountain campus, site of the Bread Loaf School of English, the Bread Loaf Writers' Conference, and the college's Snow Bowl, is nearby. Old Stone Row cuts across the campus, where buildings with simple lines and rectangular shapes evoke the mills of early New England. (Middlebury was founded in 1800.) Academic halls and dormitories of marble and limestone sit in quadrangles and feature views of the Adirondacks and Green Mountains.

Between June and August, Middlebury banishes English from its campus and hundreds of students live, learn, and, hopefully, think only in their chosen language, some wearing T-shirts that proclaim "No English spoken here." The language departments continue their excellent instruction during the school year; especially notable are German, Chinese, and Japanese. Spanish, however, is less than stellar. "I was disappointed by the classes I took in the Spanish and Italian departments," says a senior. A classmate adds, "The other language departments are really amazing." Although there is no foreign language requirement, just about everyone studies another language, if

> **"Be prepared to work, but also to have an amazing time where every year gets better."**

only to take advantage of Middlebury's campuses in France, Germany, Italy, Spain, Russia, Mexico, Brazil, Argentina, Chile, China, and Uruguay. In fact, almost half of the departments on campus are affiliated with the international studies major. The school is also a member of the Maritime Studies Program* and there are 90 college-approved programs in total; about 60 percent of juniors take advantage of them. Other highly touted Middlebury departments include English (one of the school's most popular majors, bolstered by its connections to the famed Bread Loaf Writer's Conference), economics, and psychology.

"Middlebury is not competitive," reports one history major, "but it is academically challenging and most courses are fairly rigorous." Midd kids must take a discussion-based, writing-intensive First-Year Seminar with only 15 to 18 students; the instructor serves as advisor to those enrolled until they declare a major. By the end of sophomore year, students must complete a second writing-intensive course. In addition to a 10- to 16-credit major, students must also satisfy distribution requirements in seven of eight academic areas: literature, the arts, philosophical and religious studies, history, physical and life sciences, deductive reasoning and analytical processes, social analysis, and foreign language. Students also take four cultures and civilizations classes, and two noncredit courses in physical education. With all of these requirements, it's no wonder students and faculty become close. "Overall, the quality has been phenomenal," says an English major. "Professors are enthusiastic, accessible, and always willing to go the extra mile for their students."

Eighty-four percent of Middlebury's students graduated in the top tenth of their high school class, and students of color constitute 15 percent of the student body: 7 percent Asian American, 5 percent Hispanic, and 3 percent African American. International students account for 9 percent. The campus leans left politically and hot-button issues include race relations and environmental causes. The school's partnership with New York City's Posse Foundation brings 10

"While at some schools it might be considered 'dorky' to work hard, everyone here really values hard work."

inner-city students to the school each year. Lack of parking, the workload, the isolation of the campus, and the cold weather are familiar complaints. "While at some schools it might be considered 'dorky' to work hard, everyone here really values hard work," says a senior.

Few Middlebury students live off campus (2 percent), since tuition includes guaranteed housing for four years. A variety of "palatial" co-ed dorms offer suites, augmented by college-owned group houses, the Environmental House (where residents cook all of their own food), academic interest houses, and more "standard" situations. A substance-free social house now offers living space. Rooms for upperclassmen are distributed by a lottery based on seniority, a system that a sophomore calls "a confusing and elaborate process." It's easy to get a single room, even as a sophomore, students report. The meal plan is served at five dining halls, which get high marks for both their decor and their victuals. "The dining halls are all very different, and you can find the daily menus online," a sophomore reports.

Students at Middlebury play as hard on the weekends as they work during the week. Most stay on campus for school-sponsored dances, plays, dance performances, or parties at the Greek-like co-ed social houses, which draw 9 percent of students. Kegs are prohibited in the dorms but permitted at parties, which must be registered and also offer nonalcoholic drinks and snacks. Still, "Drinking is widespread here," a senior says, "so I guess policies to prevent it don't work."

Off campus, Middlebury is "the quintessential New England town, straight out of Norman Rockwell" that is "intrinsically linked to the school." It has necessities such as fast food, grocery stores, drug stores, hardware stores, and clothing shops, but the administration regularly brings in culture and entertainment. February can

Between June and August, Middlebury banishes English from its campus and hundreds of students live, learn, and, hopefully, think only in their chosen language, some wearing T-shirts that proclaim "No English spoken here."

be grim because the snow here comes early and stays late, so road trips are popular. The progressive city of Burlington is 45 minutes away, while Montreal is a three-hour drive, Boston four, and New York City five. Middlebury's own Snow Bowl ($100 for the season) and proximity to most Vermont ski slopes make this a paradise for ski fanatics, a breed Middlebury attracts in predictably large numbers. "We love the outdoors, and spend a lot of time outdoors, no matter what the weather is like," a sophomore says.

Middlebury athletics draw rabid fans, especially when cheering on the powerful ice hockey teams—men's and women's—that compete in Division III against archrival Norwich, and whose members all engage in tutoring in local schools. The college has won more than two dozen national titles since 1995—including men's and women's ice hockey in 2005–06—and offers 30 varsity sports. Students are also active in the large intramural program, with soccer, hockey, football, basketball, and softball drawing the most interest. Perhaps the

"Community service is a big part of who we are here at Midd."

biggest outdoor activity of all is the three-day Winter Carnival, an annual extravaganza including parties, cultural events, an all-school formal, sporting competitions, snow sculpture, and ice-skating at an outdoor rink. Townspeople support the school's hockey games, and students—including members of the hockey team—give back through volunteer work with children, women, the elderly, and local schools. "Community service is a big part of who we are here at Midd," says a sophomore.

Students have noticed physical changes at Middlebury over the past few years, with more in the works, and have seen the school grow more rigorous and competitive. But some things have remained the same—namely, the combination of "excellent academics with endless extracurricular opportunities," says a sophomore. "We're a small liberal arts college [that is] big on individuality and character," a sophomore says. "Whatever your character, you'll find your niche."

Overlaps

Dartmouth, Williams, Yale, Harvard, Amherst, Brown

If You Apply To ➤

Middlebury: Regular admissions: Jan. 1. Guarantees to meet full demonstrated need. Campus and alumni interviews: optional, evaluative. SATs or ACTs: optional (choose one, or three SAT IIs). Also accepts AP or IB exams in lieu of SAT II. Accepts the Common Application and electronic applications. Essay question and copy of recent graded essay.

Mills College

5000 MacArthur Boulevard, Oakland, CA 94613

One of two major women's colleges on the West Coast. Mills has the Oakland area to fall back on, including UC–Berkeley, where students can take classes. Mills is strongest in the arts and math, though it does offer smaller programs in preprofessional areas such as communications and business economics.

Website: www.mills.edu
Location: Small city
Total Enrollment: 1,410
Undergraduates: 927
Male/Female: 0/100

At Mills College, women receive more than just a liberal arts education. Students are expected to graduate with a deeper understanding of social issues and a broad knowledge base to help ensure they will be technologically savvy and artistically aware of the world around them. The school's dedication to equality is obvious: Mills was the first women's college in the West to award bachelor's degrees and also the first to offer a computer science major. Today, this small bastion of high learning continues to provide ambitious women with a stellar education and a host of opportunites.

The school's fascinating history started in 1852, when it began as a young ladies' seminary serving the children of California gold rush adventurers who were determined to see their daughters raised in an atmosphere of gentility. Now, the combination of student diversity and educational opportunity helps guarantee that no one will graduate without having her horizons well expanded. Mills is a place where issues are debated and analyzed in the classroom, even if some students perceive an apathetic atmosphere on campus with regard to social, political, and racial concerns. To foster the college's relationship with the surrounding community, the James Irvine Foundation has awarded the college a three-year grant to support the "Multicultural Engagement: Mills and Oakland" initiative. The enclosed, park-like 135-acre campus boasts both historic and modern architecture set among rolling meadows, woods, and a meandering creek. New facilities include the 26,000-square-foot natural sciences building and the Vera M. Long Building, which houses the social sciences.

The general education curriculum is designed to graduate students who are able to write clearly, think across disciplines, work productively with others, analyze, and reason clearly; who are technology savvy, artistically sensitive, adept in scientific and historical thinking; and aware of multiculturalism, the influence of social institutions, and the issues facing women in society. Course requirements cover three broad areas: (1) writing, quantitative, and technological skills; (2) interdisciplinary, gender, and multicultural perspectives; and (3) knowledge of arts, history, natural science, and human behavior.

"The quality of teaching is incredible," says a public policy major. An art history major adds that classes are not only excellent, but often "taught by the professor who designed the course." Most courses are taught by tenured professors, and classes range from small to smaller. "Classes are usually five to seven students, or 15 to 20; my biggest class is 35," says a junior. The professors are highly regarded, friendly, and accessible, and since most of them are women, there's no shortage of strong female role models. Students praise music, art, dance, English, public policy, biology, Spanish, psychology, art, and anthropology. In addition to a premed program, Mills now

> **"Classes are usually five to seven students, or 15 to 20; my biggest class is 35."**

offers a pre-nursing program. Popular among prelaw students is the interdisciplinary program in political, legal, and economic analysis. The fine arts department is Mills' traditional stronghold, and electronic and computer music specializations within the music program are worthy of note.

Mills students are encouraged to explore beyond the Oakland campus, and many take advantage of programs abroad and exchanges with East Coast colleges. Mills has concurrent crossregistration agreements with UC–Berkeley and many other Bay Area colleges, as well as a five-year engineering program in conjunction with USC. Opportunities for internships abound.

More than three-fourths of Mills women are from California, and 80 percent attended public school. African American and Hispanics each make up 12 percent of the student body, and another 11 percent are Asian Americans. An influential subgroup of students consists of "resumers"—women who are returning to college after a break of several years. Nearly three-quarters of students receive merit scholarships. As an NCAA Division III school, Mills does not award athletic scholarships.

Many of the dorm rooms are singles, and the freshwoman dorms get good reviews. The beautiful, Mediterranean-style 1920s and 1930s-era dorms are homey, but "very old and falling apart," says a junior, while another complains, "It takes days to get something fixed." Fifty-five percent of students live on campus. "A large number of students live off campus because they are older and have families," explains a student. The appeal of campus dining seems to have waned in recent

(Continued)
SAT Ranges: V 500–640
  M 470–590
ACT Range: 21–27
Financial Aid: 80%
Expense: Pr $ $
Phi Beta Kappa: Yes
Applicants: 1,122
Accepted: 74%
Enrolled: 28%
Grad in 6 Years: 69%
Returning Freshmen: 75%
Academics: ✍ ✍ ✍
Social: ☎ ☎ ☎
Q of L: ★★★
Admissions: (800) 87-MILLS
Email Address:
  admission@mills.edu

Strongest Programs:
English
Studio Art
Psychology
Chemistry
Economics
Biology
Anthropology and Sociology

One of the most popular traditions is Latina Heritage Month, when the student group Mujeres Unidas plans an "incredible" collection of events that "honor Latinos and all human beings," as one student describes it.

years. Finding too little variety in the main dining hall, many students prefer to eat in the higher-priced, more vegan-friendly tea shop.

Fears about a stunted social life on this tiny campus quietly linger throughout. "Mills throws lousy parties, so if you want to have fun, you need to go off campus," a women's studies major says. Building an active social life requires an "open and adventurous spirit." Students are also concerned about campus security. Although Mills is a gated campus with one entrance, some cite a need for more effective patrolling. What the campus may lack in social options, however, can be found in Berkeley or San Francisco, both accessible via public transportation, beginning with a bus stop outside the front gate. Those with the means can take ski trips to Lake Tahoe or to the sunny Santa Cruz beaches. Back on campus, one of the most popular traditions is Latina Heritage Month, when the student group Mujeres Unidas plans an "incredible" collection of events that "honor Latinos and all human beings," as one student describes it.

> "Mills throws lousy parties, so if you want to have fun, you need to go off campus."

Mills offers no intramural sports, but "many women participate in Division III sports because the teams get to travel to places like Hawaii and New York," says an enthusiastic participant. Tennis, soccer, volleyball, cross-country, crew, and swimming are popular and competitive. A number of students also stay busy volunteering in the local community. The campus happens to be situated between a wealthy neighborhood and a poor neighborhood, so "Mills is in the middle, in a 'bubble,'" a student explains. "Many students want to change the Mills bubble," so they volunteer in the community.

Despite some complaints about campus life, Mills students enjoy small classes with excellent teaching in one of the country's most desirable locations close to San Francisco. One student concludes, "Mills is a place where you learn many life lessons." With an educational program that emphasizes skills critical for understanding today's society and succeeding in the real world, the college's women are sure to graduate as independent thinkers capable of making it on their own.

Overlaps

UC–Berkeley, UC–Santa Cruz, UC–Davis, UC–Santa Barbara, UC–San Diego, San Francisco State University

If You Apply To ➢

Mills: Early action: Nov. 15. Regular admissions: Mar. 1. Meets demonstrated need of 30%. Campus interviews: recommended, informational. Alumnae interviews: optional, informational. SATs or ACTs: required. No SAT IIs. Accepts the Common Application and electronic applications. Writing sample: Submit a graded analytic paper or essay.

Millsaps College

1701 North State Street, Jackson, MS 39210-0001

Millsaps is the strongest liberal arts college in the deep, Deep South. Its largely preprofessional student body typically has sights set on business, law, or medicine. Typically compared to Hendrix, Rhodes, and Sewanee, though less selective than the last two.

Website: www.millsaps.edu
Location: City center
Total Enrollment: 1,154
Undergraduates: 1,085

Millsaps College has long been recognized as a finishing school for well-bred Southern belles and gentlemen. Less well-known outside the Deep South is that this is also one of the region's top liberal arts institutions. Millsaps' motto, *ad excellentiam*, means "promoting excellence"—and that's what the college still does. What differentiates the school is its focus on scholarly inquiry, spiritual growth, and community service, along with its Heritage Program, an interdisciplinary approach to world culture. Like students at Rhodes and Southwestern, those at Millsaps are driven and goal-oriented. "We're laid-back overachievers—nerds by day, partying hard at night," quips a senior.

Millsaps' 100-acre campus sits in the center of Jackson, on the highest point in the city. A mix of modern and traditional buildings is arranged around the Bowl, a sequestered glen surrounded by old-growth trees and shrubs. New additions include the Nicholson Garden,

> "We're laid-back overachievers—nerds by day, partying hard at night."

where students can gather to relax or hold discussion-based classes in nice weather, surrounded by roses, a gurgling fountain, and other flowering plants. Also completed recently: an artificial-turf field for the football and men's and women's soccer teams, and baseball facilities comparable to those used by Major League Baseball's Double-A affiliates.

Millsaps requires students to complete 128 semester hours to earn a degree, all but eight of which must be taken for a letter grade. All students must also complete 10 multidisciplinary courses designed to develop skills in reasoning, communication, quantitative thinking, valuing, and decision-making—including four in the humanities and four in the sciences and math. All freshmen also take a one-hour Perspectives class, led by an academic advisor, to help adjust to college life. New students also take the Introduction to Liberal Studies seminar, which focuses on critical thinking and writing.

Among the best programs at Millsaps are English and education, where students get involved with the Jackson Public Schools from introductory-level classes. Each year, a few select upperclassmen join the Ford Teaching Fellows Program, letting them work closely with a faculty member to learn about teaching—and paying them for their time in the classroom. Also well regarded are history, philosophy, and religious studies. Premed courses, including those in biology and chemistry, are strong; the Wiener Premedical Summer Research Fellowships are available to students seeking careers in medicine, while cooperative agreements allow students to opt for nursing degrees in partnership with the University of Mississippi Medical

> "The campus is very activist and very political—well-split between conservatives and liberals."

Center and Vanderbilt University. Millsaps also offers engineering opportunities in cooperation with Auburn, Columbia, Vanderbilt, and Washington universities; students combine the advantages of a liberal education at Millsaps with the specialized programs of a major university.

No course at Millsaps has more than 50 students, and 83 percent have 25 or fewer. "No graduate students teach classes at all," says a business major. "Everyone gets individual attention and care." Students may do research for credit at the Blue Ridge Center for Environmental Stewardship in Virginia and at Yellowstone National Park, or they may intern for credit with local businesses or in state government offices in Jackson. The college also maintains a "Southern campus" in the Yucatan Peninsula, which hosts courses exploring Mayan culture and the Maya coral reef. Students eager to see how government works may participate in the Washington Semester,* while those seeking passport stamps may spend summers with Millsaps faculty in one of 14 exotic locales, ranging from London, Paris, and Munich to China, Tanzania, and Japan. Cooperative programs are also available through the Associated Colleges of the South* consortium, of which Millsaps is a founding member.

Millsaps has broadened its recruiting efforts, and 49 percent of students now come from out of state, though the campus is still far from diverse. Although Millsaps was the first college in Mississippi to voluntarily open its doors to minority students, and African Americans now make up 12 percent of the student body, Asian Americans add only 4 percent, and Hispanics just 1 percent. "The campus is very activist and

---

(Continued)

Male/Female: 51/49
SAT Ranges: V 540–680
  M 540–650
ACT Range: 23–30
Financial Aid: 55%
Expense: Pr $
Phi Beta Kappa: Yes
Applicants: 1,008
Accepted: 82%
Enrolled: 26%
Grad in 6 Years: 71%
Returning Freshmen: 83%
Academics: 🐾 🐾 🐾
Social: ☎ ☎ ☎
Q of L: ★ ★ ★
Admissions: (601) 974-1050
Email Address:
  admissions@millsaps.edu

Strongest Programs:
Accounting
Business Administration
Biology/Chemistry/Pre-med
English
History
Sociology/Anthropology
Religious Studies
Education

The social scene at Millsaps revolves around the fraternity houses, which are usually open and rocking from Wednesday through Saturday nights.

very political—well-split between conservatives and liberals," says a political science major. "All views are respected; poverty and human rights are hot topics." A whopping 80 percent of Millsaps students get some non-need-based academic scholarships, with the average award totaling $12,410. However, there are no athletic scholarships.

Most Mississippians view Millsaps as a hotbed of liberalism, and the school's co-ed dorms confirm their worst fears, though freshman must still live in single-sex halls. Seventy-three percent of students stay in campus housing—mostly, grouses a sophomore, because those who move off-campus lose 30 percent of their financial aid. The Greek system claims 54 percent of men and 56 percent of women; junior and senior men may live in one of four fraternity houses, but there is no sorority housing. Students say campus chow is decent, and they may choose from three options: the Caf', Kava House (a café and deli), and a new coffeehouse.

"The faculty knows you as a person, not a number."

The social scene at Millsaps revolves around the fraternity houses, which are usually open and rocking from Wednesday through Saturday nights. Greek rush is now held after fall mid-terms instead of during the first hectic week of school, but that hasn't dampened the party spirit. Underage students caught drinking are fined, but students say enforcement of the policy is lax. Major Madness is a favorite annual event, offering a week of open mic nights, hypnotists, and comedians, and culminating in a weekend-long festival in the Bowl, with a crawfish boil, carnival games, and live music. The city of Jackson also offers a wealth of options, including professional symphony, opera, and ballet, and the city is a nexus for Mississippi's legendary blues and "roots rock" musical traditions. Easy road trips include New Orleans, Memphis, and the riverfront casinos in Vicksburg, Mississippi; closer to campus, 10 miles to the north, is a huge reservoir which is popular for weekend water sports.

Millsaps competes in NCAA Division III as a member of the Southern Collegiate Athletic Conference (SCAC), so it isn't nearly as sports-crazy as most Southern campuses. For the men, football, basketball, baseball, and soccer draw the largest crowds; basketball, soccer, softball, and volleyball are the most popular among women's teams. The school's biggest rivalry is with nearby Mississippi College, though as one

"We offer the sort of prestigious education generally only available in New England."

student says: "Lots of MC folks couldn't care less." There are more than 25 intramural sports, plus group exercise classes and sports clubs, so even students with recreational interests and abilities can find a game to play. Everyone benefits from the 65,000-square-foot Hall Activities Center, which has facilities for weight training, aerobics, basketball, racquetball, squash, and volleyball, along with an outdoor pool.

In a state renowned for blues, booze, and barbecue and the tradition of old magnolia trees and grand plantations, progressive Millsaps College is an anomaly. "We offer the sort of prestigious education generally only available in New England," a junior explains. "Millsaps is a magnet for accomplished students from strong backgrounds, and the kind of college not usually found in the South." Small classes ensure plenty of time to get to know fellow students and faculty members. "It's a challenging atmosphere where everyone actually cares about school," agrees a business major. "The faculty knows you as a person, not a number." And that's one tradition that never gets old.

Overlaps

Rhodes, University of Mississippi, Tulane, Louisiana State, Vanderbilt, Trinity (TX)

If You Apply To ➤

Millsaps: Rolling admissions. Early action: Jan. 8. Meets demonstrated need of 26%. Campus interviews: recommended, informational. Alumni interviews: optional, informational. SATs or ACTs: required. SAT IIs: optional. Accepts the Common Application and prefers electronic applications. Essay question: personal statement from the Common Application.

University of Minnesota–Morris

600 East 4th Street, Morris, MN 56267-2199

The plains of western Minnesota may seem an unlikely place to find a liberal arts college—and a public one at that. Morris is cut from the same cloth as UNC–Asheville, St. Mary's of Maryland, and Mary Washington. The draw: private college education at a public university price.

The University of Minnesota–Morris is far more comprehensive than the small size of its student body might suggest. Founded by a Roman Catholic nun to educate Native Americans, Morris has grown into a solid public liberal-arts college. "Students here are very driven and involved," says a sociology and anthropology major. "We immerse ourselves in the campus and activities, making us well-rounded individuals."

The 130-acre Morris campus includes 26 traditional brick-and-mortar buildings, loosely arranged around a central mall. In 2005, the school installed a high-powered wind turbine, which generates nearly half of the power UMM requires each day. In the winter, an on-campus lake is flooded to provide space for ice-skating.

General education requirements at UMM span 60 credits. Everyone starts with the First-Year seminar, an introduction to the liberal arts. Students then move on to as many as five courses under the umbrella of "Skills for the Liberal Arts," in writing, foreign languages, math and symbolic reasoning, and artistic performance. Finally, students take eight courses in "Expanding Perspectives," one each in history, fine arts, social sciences, and humanities, and two each in natural sciences and in "the global village," which encompasses human diversity, international perspectives, and related disciplines. The most popular majors are biology, education, psychology, English, and management; students also give high marks to the music, art, and economics programs.

Students describe the academic climate as challenging, but that's not necessarily a bad thing. "I enjoy competitiveness in the classroom and am proud of the grades I receive because I know I worked hard for them," says a junior. Seventy-nine percent of classes at Morris have 25 students or fewer, and virtually none have more than 100. "Each professor at UMM brings their own style of teaching to the classroom," a student says, "but all work towards a common goal: getting the student to learn." Every freshman is assigned an academic advisor who must approve his or her schedule. And when Morris starts to feel claustrophobic, there's always study abroad—nearly half of UMM students participate, and education majors can even do their student teaching overseas. Morris also offers service-learning projects as part of the classroom experiences. The UMM Honors Program provides high-achievers with various honors courses, a senior honors project, and a core course entitled "Traditions in Human Thought."

> "There is always some event, program, speaker, entertainer, or athletic event going on."

Eighty-six percent of Morris students are Minnesota natives; Native Americans are the largest minority group on campus, comprising 9 percent. Asian Americans add 3 percent and African American and Hispanic students each account for 2 percent. Students are vocal about issues such as the environment and gay rights, says a senior. "The campus is liberal," says a sophomore. "The major issue is the Bush administration, and pretty much every bad policy they have." Still, says a junior, "although we may not all agree with each other, there is a great sense of respect and understanding." Merit scholarships are available (the average award is $1,610), but there are no athletic awards.

Website: www.morris.umn.edu
Location: Small town
Total Enrollment: 1,678
Undergraduates: 1,527
Male/Female: 40/60
SAT Ranges: V 570–680 M 565–680
ACT Range: 22–27
Financial Aid: 83%
Expense: Pub $ $ $ $
Phi Beta Kappa: No
Applicants: 1,097
Accepted: 82%
Enrolled: 43%
Grad in 6 Years: 59%
Returning Freshmen: 86%
Academics: ✑ ✑ ✑ ½
Social: ☎ ☎
Q of L: ★ ★ ★
Admissions: (888) 866-3382
Email Address: admissions @morris.umn.edu

Strongest Programs:
Biology
Elementary and Secondary Education
Psychology
English
Management

The Honors Program provides high-achievers with a core course entitled "Traditions in Human Thought."

Forty-five percent of Morris students live on campus, in one of the five residence halls, or in an apartment complex reserved for upperclassmen. It's easy to get a room, though many opt for less-expensive housing off campus. There is no Greek system, but students say there's always something to do. "There is always some event, program, speaker, entertainer, or athletic event going on," says a political science major. "On campus is where all the action is," another student agrees. "Road-trips are usually taken by small groups of friends, who get together on the weekends to go to Perkins in Alexandria, or to the mall in St. Cloud." Winnipeg, about five hours away, is also a popular destination.

"UMM has a community feel to it. Everyone is friendly around here."

Other campus traditions include the spring hog roast, and pancakes served by professors during final exams. There's also the annual tug-of-war competition between two dorms, Clayton Gay Hall and Indy Hall. "Hundreds of students show up!" says one. In all, there are more than 100 student organizations, including the Morris Campus Student Association, which does everything from lobbying for lower tuition to presenting weekend movie nights. The drinking age is 21-plus, and it's strictly enforced on campus and in local bars. Still, minors find ways to imbibe if they choose.

Morris competes in the Division III Upper Midwest Athletic Conference. Strong programs include women's soccer, men's and women's basketball, men's tennis, volleyball, and football. In addition, 80 percent of Morris students participate in intramural sports—including bowling, dodgeball, broomball, and sand volleyball—or use the Regional Fitness Center or hiking and biking trails.

One of the smaller campuses in the University of Minnesota system, UMM may just epitomize the idea of "Minnesota nice." Tucked away from the state's big cities, some students might find the campus isolated. But the school's location means fewer distractions—and more time for its happy students to focus on independent reading and research, or just getting to know their peers. "UMM has a community feel to it," says a sophomore. "Everyone is friendly around here."

Overlaps
University of Minnesota–Twin Cities, Gustavus Adolphus, St. Olaf, University of Wisconsin–Madison, University of Minnesota–Duluth, Luther

If You Apply To ➤ **Morris:** Early action: Dec. 1. Regular admissions: Mar. 15. Financial aid: Mar. 1. Housing: Aug. 1. Guarantees to meet full demonstrated need. Campus interviews: recommended, evaluative. No alumni interviews. SATs or ACTs: required. SAT IIs: optional. Accepts the Common Application and electronic applications. Essay questions: something about you that isn't elsewhere in the application; what you would contribute to the UMM community.

University of Minnesota–Twin Cities

240 Williamson, 231 Pillsbury Drive SE, Minneapolis, MN 55455

Not quite as highly rated as U of Wisconsin or U of Michigan, but not quite as expensive, either. In a university the size of U of Minnesota, the best bet is to find a niche, such as the honors program in the liberal arts college. Strong programs include engineering, management, and health fields.

Website: www.umn.edu
Location: Urban
Total Enrollment: 35,160
Undergraduates: 26,189

The University of Minnesota, like the nearby Mall of America, can be overwhelming, with its seemingly limitless variety of offerings and gargantuan size. With 147 majors, the U of M offers an abundance of academic choices. Be warned, though—winters can be frigid and it can take a cool customer to navigate the seemingly endless choices here.

The vast Twin Cities campus actually consists of two campuses with three main sections, and within each the architecture is highly diverse. The St. Paul campus encompasses the colleges of agriculture, food, and environmental sciences, natural resources, human ecology, veterinary medicine, and biological sciences. The Minneapolis campus is divided by the Mississippi River into an East Bank and a West Bank that are home to the other colleges and most of the dormitories, as well as most of the fraternities and sororities. Both campuses offer a blend of traditional and modern architecture, with columned buildings seated next to sleek geometric structures. The two campuses are five miles apart and linked by a free bus service. Academic facilities are excellent, beginning with the five-million-volume library system, which is the 14th largest in North America. Every one of the colleges has its own library, many of which are good places to study. A 695-acre arboretum is used for research and teaching. The West Bank Arts Quarter makes the U of M the only public university in the nation with all its art disciplines in the same district.

Minnesota offers 147 undergraduate majors in seven separate schools. The Institute of Technology is notable for the options it offers for tutorials and internships; the electrical and mechanical engineering programs are particularly strong and well subscribed. Psychology, English, and biology are among the most popular majors, as are journalism and mechanical engineering. Undergraduates also have access to more esoteric fields, from aging studies and biometry to therapeutic recreation and mortuary science.

While efforts to limit class size have been stepped up and the university is focusing more on undergraduates, classes can reach 300-plus, with introductory classes typically the largest. "The size can be overwhelming, but oftentimes students figure out classes and find it quite fitting," says a senior. Helpful teaching assistants are abundant, and the excellent honors program in the liberal arts college allows close contact with faculty members as well as leeway to enroll in certain graduate courses and seminars. Students say the academic climate varies by school. "Professors expect the best from students," says a junior, "but I don't feel pressure from classmates."

While undergraduates have had a difficult time enrolling in courses, the use of computer registration has made life a lot easier. One junior reveals, "If a class is closed and somebody really needs it, they can usually get a magic number from the department to be able to register for it." The administration attributes the school's low six-year graduation rate to the fact that students are likely to center their lives in

> **"The size can be overwhelming, but oftentimes students figure out classes and find it quite fitting."**

spheres outside the university—in work and off-campus homes. However, the four-year plan guarantees graduation in four years provided students follow program requirements, including frequent academic counseling and specific coursework.

Professors receive high marks from most students as being approachable and knowledgeable, and freshmen seminars "are a great way to get to know a professor," says a student. "They are small classes and are taught by award-winning faculty on topics that are interesting and fun." Students find plenty of internship opportunities at the many corporations and government agencies in the Twin Cities area. The university is on a semester system, and almost all classes have a pass/fail option (limited to no more than a quarter of a student's courses).

A journalism student says U of M students are a healthy mix of "country and cosmopolitan." Seventy-four percent of students at the university come from the top quarter of their high school class, and 62 percent are from Minnesota. Minorities constitute 15 percent of the student body, with 4 percent African American, 2 percent Hispanic, 1 percent Native American, and 9 percent Asian American. Tuition hikes are a main gripe of students, but there is need-based financial aid, merit scholarships, and athletic awards in all major sports.

(Continued)

Male/Female: 47/53
SAT Ranges: V 540–660
  M 570–690
ACT Range: 23–28
Financial Aid: 77%
Expense: Pub $ $ $ $
Phi Beta Kappa: Yes
Applicants: 20,641
Accepted: 71%
Enrolled: 36%
Grad in 6 Years: 49%
Returning Freshmen: 84%
Academics: ✍ ✍ ✍ ✍
Social: ☎ ☎ ☎
Q of L: ★ ★ ★
Admissions: (612) 625-2008
Email Address: N/A

Strongest Programs:
Chemical Engineering
Management Information
  Systems
Mechanical Engineering

The Institute of Technology is notable for the options it offers for tutorials and internships; the electrical and mechanical engineering programs are particularly strong and well subscribed.

Dorm life at Minnesota follows the big school, wait-in-line theme. Twenty-two percent of all undergraduates live in residence halls; there are eight traditional halls and three university-run apartment facilities. Dorm rooms are hard to obtain, and parking spaces for all those commuters are almost as scarce. Students who have rooms get the chance to keep them for the next year. "The dorms are a great idea for freshman students to meet others," says a junior. "They're comfortable, well maintained, and a mini home-away-from-home for your first year." Once you're there, you're required to join a meal plan. Opinions vary on the quality and variety of food. But "fresh fruit and veggies are always available," notes a junior. Lest anyone fear the dietitians are excessively health-obsessed, she adds, "They have the best chocolate chip cookies." Campus security is adequate. "We have a dedicated police force as well as a free escort program," notes a junior.

> "The dorms are a great idea for freshman students to meet others."

Many U of M students live in apartments and have a thriving social life away from campus. Underage drinking is banned, and students say the policy usually works. The downtown areas of the Twin Cities are easy to get to by bus, and there are scores of good bars, restaurants, nightspots, and movie theaters. This is an athletically inclined bunch of students, as both intramural and varsity sports are popular. Wrestling, baseball, and golf have brought home championship trophies recently, and the hockey team was top in the nation. Students always hope the current season will be one in which the gridiron Gophers take home the roses in a bowl victory, but short of that, a win over Michigan for custody of the Little Brown Jug is cause for celebration. Minnesota's rivalry with the University of Wisconsin is considerable, especially since the U of M has a large Wisconsin population. Intramural competition can go on well past midnight.

Here, being "under the weather" can be a good thing, as campus designers found a way to get around—or under—wet or wintry conditions by linking many of the campus buildings with tunnels. For those who love it, there is the aforementioned Snow Week, and happy skiers and skaters become colorful spots all over the state's white backdrop. In the spring and summer, Minnesota's famed 10,000 lakes offer swimming, boating, and fishing. "Activities are all over campus, all over the Twin Cities, and students have a variety to choose from any day of the week," says a senior. The union's bowling alley, pool tables, movie theater, and live music dance club are good places to meet people. And there are more than 500 student groups on campus. The Carnival Weekend put on by the Greeks each April to raise funds for charity is a huge event. Spring Jam is described by one student as "homecoming in spring—but better," and Campus Kick-off Days in the beginning of the fall quarter is much anticipated.

> "It's big, it's intelligent, it has a ton of school spirit, and it's really fun."

Anonymity is almost a given at a university of this size, but then size does have its virtues in the countless array of campus resources. A junior says, "It's big, it's intelligent, it has a ton of school spirit, and it's really fun." The Univeristy of Minnesota is ideal for those who appreciate an urban setting and a good, old-fashioned, button-up-your-overcoat winter.

Overlaps

University of Wisconsin–Madison, University of Minnesota–Duluth, University of Wisconsin–Eau Claire, University of St. Thomas, Marquette

If You Apply To ➤

Twin Cities: Rolling admissions. Financial aid: March 1. Housing: May 1. Campus and alumni interviews: optional, informational. SATs or ACTs: required. SAT IIs: optional. Prefers electronic applications. No essay question.

University of Missouri at Columbia

130 Jesse Hall, Columbia, MO 65211

Mizzou is renowned for one of the top journalism schools in the nation, but education, agriculture and the health sciences are also standouts. Enrolls only about half the number of out-of-state students as archrival University of Kansas. Columbia is a quintessential college town.

In 1839, the residents of Boone County, Missouri, raised enough money to create the state university in Columbia. Today, Missouri's flagship university has evolved into a top research institution, yet continues to uphold the belief of its founders in the great value of higher education that is accessible to all. Currently in the thick of a $1 billion campaign, the university continues to expand programs and facilities in ways that benefit students.

The oldest public university west of the Mississippi, Mizzou's spacious, tree-filled campus is flanked by mansion-like fraternity and sorority houses. The Francis Quadrangle Historical District, with 19 National Historic Landmark buildings, is the core of the Red Campus (so named for the predominant color of brick). Central to this area are the 60-foot granite columns of the original Academic Hall—the building was destroyed by fire in 1892. To the east of the columns is the original tombstone of Thomas Jefferson. The White Campus consists of vine-covered limestone buildings, symbolized by the Memorial Union Tower. The newest facilities include the $58 million Life Sciences Center, which includes three teaching labs and labs for life scientists, and the $50 million renovation and expansion of the Student Rec Center.

With more than 250 degree programs and 20 schools and colleges, Mizzou offers a comprehensive set of choices for basic and advanced study. Aspiring journalists can get hands-on experience working on the *Columbia Missourian*, the 7,000-circulation local daily paper edited by J-school faculty members and students, or at KOMU-TV, the nation's only university-owned commercial television station. KBIA, MU's National Public Radio station, is popular among journalism students and listeners alike. The J-school has created a convergence sequence to introduce students to new digital technologies, and all journalism students are required to have their own laptops. Agriculture is also nationally ranked, especially in the areas of agricultural economics and applied research for farm communities. The College of Engineering maintains several notable undergraduate segments, including biological and civil engineering. The College of Business is highly competitive and features a five-year bachelor's/master's accounting program. New programs include an information technology major.

Committed preprofessionals will be glad to know that MU offers highly able and directed freshmen guaranteed admission to its graduate-level programs in medicine, law, veterinary medicine, nursing, and health professions. Mizzou is also one of the leading public research institutions in the country for the number and range of lab and scholarly opportunities it offers under-

> **"Classes can be quite difficult, but they are manageable."**

graduates, a task made easier through the new undergraduate research office. More than 350 undergrads do research each year, and about 2,500 enroll in service learning courses. MU sends more students abroad each year—more than 800—than any other higher education institution in the state. Students may choose from nearly 400 programs in 55 countries.

MU undergraduates must meet an array of general education requirements. They include courses in exposition and argumentation, algebra, math reasoning

---

Website: www.missouri.edu
Location: Small city
Total Enrollment: 27,985
Undergraduates: 21,375
Male/Female: 48/52
SAT Ranges: V 540-660
M 540-650
ACT Range: 23-28
Financial Aid: 42%
Expense: Pub $ $ $
Phi Beta Kappa: Yes
Applicants: 12,424
Accepted: 83%
Enrolled: 46%
Grad in 6 Years: 66%
Returning Freshmen: 84%
Academics: ✍ ✍ ✍
Social: ☎ ☎ ☎ ☎
Q of L: ★ ★ ★
Admissions: (573) 882-7786
Email Address:
MU4U@missouri.edu

Strongest Programs:
Journalism
Biology
Psychology
English
Education
Physical Therapy
Food Science and Nutrition
Agriculture

proficiency, and American history or political science. Students also must complete 27 hours in three content areas: social and behavioral sciences, physical and biological sciences and mathematics, and humanistic studies and fine arts. All students must take a course in computer literacy, although the content of those courses varies by degree program. Two writing intensive courses and a senior capstone experience are also required. Full professors teach the lecture courses at Mizzou, supplemented by a weekly discussion session led by a teaching assistant to go over material presented in class. "The professors generally care about their students," says one sophomore. "They are easy to contact and eager to help." Mizzou is one of only six public universities to have law, medicine, and veterinary medicine on one campus.

Students say the courses at Mizzou are challenging but not impossible if you are willing to work hard. "Classes can be quite difficult, but they are manageable," one student says. Owing to MU's size, classes can fill up quickly, but professors do give overrides for students who must take certain credits at specific times. Missouri guarantees the availability of coursework to complete a degree in four years. MU's library system holds more than three million books, almost seven million microfilms, and more than 16,000 periodicals. The Ellis Library includes more than 70 standup computers, 81 workstations with soft seats, and adaptive stations for disabled students.

The Mizzou campus is home mostly to Missourians (83 percent), though every state in the union and more than 100 foreign countries are represented. African Americans account for 6 percent of the student body, while Asian Americans and Hispanics combine for 5 percent. A senior says Mizzou students are "more academically-inclined and have more real-life experiences" than those at rival institutions. To boost its minority population, Mizzou has established several scholarship programs designed especially for minorities. It's also opened a Black Culture Center and an Asian Affairs Center. The Diversity Week program features workshops, speakers, and other events. Merit scholarships averaging $4,170 are available, and student athletes may compete for more than 250 awards in 20 Division I sports.

> "Dorms are extremely nice and new ones are being built."

Thirty-eight percent of MU students live on campus, and freshman under age 20 are required to do so. "Dorms are extremely nice and new ones are being built," a communication major explains. Residence halls have double rooms and are often crowded and noisy—and thus are fun places to be, though single-sex halls, a few single rooms, and round-the-clock quiet floors are also available. Mizzou is in the midst of a plan to upgrade or replace all 19 residence halls; the first phase was the construction of the 721-bed Virginia Avenue Housing and Dining Facility. Dorm choice is first come, first served, and half of the halls offer co-ed living by floor or wing. Students can also choose to live in one of 25 Living/Learning Communities, where residents share a common interest, such as engineering, arts, or nursing. About 75 percent of students choose a Freshman Interest Group—there are more than 100 to choose from—where 15 to 20 students with shared academic interests live in the same residence hall and enroll in three core classes together. Dorm dwellers are required to purchase meal plans, but credits can be used at all-you-can-eat dining halls, coffee bars, and take-out stands, among many options. "The food options are actually very good," says a sophomore. The fraternity and sorority houses are livable (the frat houses less so); 21 percent of Mizzou men and 24 percent of women go Greek.

Students at Mizzou, a champion of tough alcohol policies, have adopted the school's stance and agreed to ban alcohol from all fraternities and sororities, making it one of the largest Greek systems in the nation to go dry. The rule is lifted when alumni come home to visit. Students say MU's social life is packed with options, including movies, shopping, eating out, the usual Greek parties, and great parks and

Mizzou's Tigers compete in the Big 12, and basketball and football games draw big crowds.

With more than 250 degree programs and 20 schools and colleges, Mizzou offers a comprehensive set of choices for basic and advanced study.

The College of Business is highly competitive and features a five-year bachelor's/master's accounting program.

hiking areas on the outskirts of town. "Columbia has a million things to do," says one student. "Of course, there are also frat parties and small house parties here, too." Students support the town by engaging in community service and the community caters to them in return; their concern even goes beyond the borders of campus to the plight of the wild tiger and the preservation of its habitat. Road trips to St. Louis, Kansas City, and Lake of the Ozarks offer a change of scenery.

> **"Football games are never missed by students, no matter how bad the season."**

Mizzou's Tigers compete in the Big 12, and basketball and football games draw big crowds. In fact, the entire town turns out in black and gold for any football game. "Football games are never missed by students, no matter how bad the season," a senior chemical engineering major says. Kansas is their biggest rival. Wrestling, baseball, and volleyball are also competitive. An indoor practice facility for football, baseball, softball, and soccer and a track/soccer complex offer seating for 2,000. MU's popular intramural program has nearly two dozen sports and two skill divisions, attracting more than a quarter of the student body.

Mizzou is a school on the rise. It continues to grow academically and physically, as evidenced by the ambitious capital campaign, while sticking with its longtime traditions. What's more, students are constantly challenged at every turn by quality teaching and ample research opportunities.

Overlaps

University of Kansas, Kansas State, University of Illinois, Indiana University, University of Iowa, Iowa State

If You Apply To ➤   **Mizzou:** Rolling admissions: May 1. No campus or alumni interviews. SATs or ACTs required; ACTs preferred. No SAT IIs. Accepts electronic applications. No essay question.

Montana Tech of the University of Montana

1300 West Park Street, Butte, MT 59701-8997

If you go to Montana for college, you're probably interested in either rocks or trees. Montana Tech covers the former, with strong programs related to mining and geological engineering. Montana Tech is a third bigger than New Mexico Tech and about the same size as Colorado School of Mines.

Students at Montana Tech like to dig into their work, literally and figuratively, and aren't afraid to get their hands dirty. Tech has rightfully earned a reputation as one of the finest science, engineering, and technical colleges in the world. Students get a hands-on education geared toward "things metallic" (as the school's motto loosely translates). In fact, the school's mascot is Charlie Oredigger, and students are affectionately dubbed "diggers." But Tech isn't just for those who like to play in the dirt; the school is strengthening its healthcare programs and has a strong computer science program to boot.

Situated on a shoulder of "the richest hill on Earth" (some of the greatest copper, molybdenum, zinc, and manganese deposits in the world) and virtually on top of the great Continental Divide, Montana Tech's 50-acre campus is composed of 16 buildings of classic college brick architecture. This is exemplified by Main Hall, constructed in 1900, and the Engineering, Laboratory, and Classroom building (the ELC), built in 1987. Other unique features on campus include the Museum Building, which houses

Website: www.mtech.edu

Location: City outskirts

Total Enrollment: 2,188

Undergraduates: 2,089

Male/Female: 53/47

SAT Ranges: V 460–580
 M 510–620

ACT Range: 21–27

Financial Aid: 80%

Expense: Pub $ $

Phi Beta Kappa: No

Applicants: 413

Accepted: 99%

(Continued)

Enrolled: 68%

Grad in 6 Years: 40%

Returning Freshmen: 66%

Academics: ✑ ✑ ✑

Social: ☎ ☎

Q of L: ★ ★ ★

Admissions: (406) 496-4178

Email Address:
admissions@mtech.edu

Strongest Programs:
Engineering
Computer Science

one of the country's largest mineral collections; an Earthquake Studies Office, which records tremors throughout southwestern Montana; and the Montana Bureau of Mines and Geology, a research arm of the college that produces geological and mineralogical maps and publications. The Healthcare Informatics program recently built a fully interactive laboratory that incorporates access-grid-node technology in the classroom (and if you know what that means, start filling out your admissions application).

> **"It is a very competitive environment and the classes are difficult, but the teachers are helpful."**

Montana Tech's degree programs emphasize the study of minerals, energy, and the environment, but students graduate with a well-rounded education. Strong degree programs include general engineering (the most popular major), business information and technology, and several specific engineering fields, including software, mining, geological, and environmental. Faculty is constantly upgrading classes technologically to prepare better job and graduate school candidates; Tech has strengthened its programs in the healthcare fields in a nod to the changes in society. The school also works to place upperclassmen in summer jobs in their fields. Everyone faces general education requirements, including communications, humanities, social sciences, mathematical sciences, writing, and life sciences, although students say the nonscience offerings are weak. For those who want more than just a straight science experience, Tech has a major in science and technology that attempts to relate liberal arts to today's increasingly technological society.

Academic work at Tech is rigorous, with gym the only subject that can be taken pass/fail. "It is a very competitive environment and the classes are difficult," says a senior, "but the teachers are helpful." Faculty members have a genuine interest in teaching and work hard to accommodate students. In addition, the typical professor holds a terminal degree. "The teachers are helpful and are involved," reports one student. Freshmen undergo an extended orientation that continues through their first semester, and can take a two-credit course to help them succeed in college. Students must sit down with their academic advisors each term to discuss their schedules.

"Students tend to be laid-back and friendly," says a junior. Eighty-six percent of Tech's students are from Montana, and the majority were the "brains or the nerds from high school," according to a sophomore. Two-thirds were in the top half of

> **"Students tend to be laid-back and friendly."**

their high school class. Foreign students account for 2 percent of the population. Minorities make up 5 percent, with the largest group coming from Native Americans, at 2 percent. Campus politics lean to the right but students report that morale is high, despite any political disagreements. Students vie for hundreds of merit scholarships averaging $1,000. There are 100 athletic scholarships distributed among football, basketball, volleyball, and golf. Registration and incidental fees are waived for some Montana state residents, including war orphans and those of at least one-fourth Native American blood.

Prospector Hall is "comfortable and spacious," but those who can't fit are often forced to live in married student housing. Prospector includes modern baths, carpeting, exercise rooms, and kitchens, and it is "in the middle of everything." Each room is wired with a microcomputer connected to the campus mainframe. The school likes freshmen to live on campus, but the vast majority of upperclassmen live either in the many nearby apartments or in houses with reasonable rents. Campus grub "could use some work and it's pricey," grumbles a junior.

Butte (population: about 40,000) gets a fair enough rating as a college town, although it is suffering from a collapse of the mining industry. "It's Butte," says one civil engineering major, "It grows on you." One native, a mechanical engineering major, says the community takes a lot of pride in the school. Social activities—many

Butte (population: about 40,000) gets a fair enough rating as a college town, although it is suffering from a collapse of the mining industry.

of which are sponsored by clubs—take place both on and off campus. Underage drinkers have a difficult time getting served in town, but, like any school, can find alcohol if they want it, students say. Butte's police department, however, is known to crack down on raucous parties. The ratio of men to women is evening out: 53 percent male to 47 percent female. There's no Greek system. On weekends, students who don't go home attend music or comedy shows on campus, see movies, go to a game, or frequent the bars in town. Butte's setting—on the slopes of the Continental Divide—is magnif- **"It's Butte. It grows on you."** icent for skiing, fishing, hiking, and camping. Yellowstone Park and Las Vegas are favored road trips. Football games, especially with rivals like Carroll College, draw large crowds.

The student union features a dining area, game room, bookstore, student-owned FM radio station, and a television where students are known to tune in to cartoons in the afternoon. St. Patrick's Day is widely celebrated on and off campus, and on M-Day, part of a three-day festival before spring finals, students whitewash the large stone "M" on a hill above campus and host the largest bonfire in the state.

Athletics at Tech, which competes in the NAIA Division I, are up-and-coming, and jocks are generally considered "cool." The most popular varsity sports are football and volleyball. Tech students take advantage of the excellent intramural program and the facilities of the modern physical education complex. Tech's biggest rival is Western Montana College, and freshman football players from WMC face off with those from Tech in an annual boxing match known as the "Smoker." "Athletics is big here," says a liberal studies major. "Everyone really supports the teams, especially if they are playing a rival of ours."

Montana Tech's students get hands-on, experiential learning and a solid grounding in earth-related engineering disciplines at a reasonable cost. The internship and career center has helped graduates maintain an impressive 97 percent placement rate for graduates over the last 10 years. Students often get job offers—complete with signing bonuses—before they even begin their senior year. Tech doesn't offer your typical college experience, but for would-be miners and engineers, its programs provide mountains of opportunity.

Montana Tech is situated on a shoulder of "the richest hill on Earth" (some of the greatest copper, molybdenum, zinc, and manganese deposits in the world).

Overlaps
University of Montana, Montana State, Colorado School of Mines, Michigan Tech, University of Washington

If You Apply To >

Montana Tech: Financial aid: March 1. Does not guarantee to meet demonstrated need. Campus, alumni interviews: optional, informational. SATs or ACTs: required. SAT IIs optional. Accepts electronic applications. No essay question.

Morehouse College: See page 36.

Mount Holyoke College

50 College Street, South Hadley, MA 01075-1488

One of two women's colleges, with Smith, that are members of the Five College Consortium in western Massachusetts. Less nonconformist than Smith and Bryn Mawr.

MHC is strongest in the natural and social sciences, and one of few colleges to have a program devoted to leadership.

Website: www.mtholyoke.edu
Location: Small town
Total Enrollment: 2,153
Undergraduates: 2,149
Male/Female: 0/100
SAT Ranges: V 620–710
    M 590–680
ACT Range: 27–30
Financial Aid: 62%
Expense: Pr $ $ $ $
Phi Beta Kappa: Yes
Applicants: 3,065
Accepted: 53%
Enrolled: 35%
Graduate in 6 Years: 82%
Returning Freshmen: 92%
Academics: 🖉 🖉 🖉 🖉
Social: ☎ ☎ ☎
Q of L: ★ ★ ★ ★
Admissions: (413) 538-2023
Email Address:
    admission@mtholyoke.edu

Strongest Programs:
English
Biology
Psychology
Economics
International Relations

The women of Mount Holyoke are quick to tell you that the nation's first all-female college is not a girls' school without men, but a women's college without boys. The women who choose MHC value tradition, leadership, and achievement, and eagerly support one another as each strives to meet her goals. Students rave about the quality of teaching and the small classes, and while they complain about the heavy workload, most bring that challenge upon themselves as they seek intellectual fulfillment. "I'm encouraged to explore and be adventurous, to learn about myself and the world," says a junior. "Also, no one fits into one specific category. Athletes are also involved in student government and other organizations—it's a total mix."

Mount Holyoke is located in the heart of New England, on 800 acres of rolling hills dotted with lakes and waterfalls. Modern glass-and-stone buildings stand alongside more traditional ivy-covered sandstone structures. Highlights include the Japanese Meditation Garden and Teahouse, an art building with studios and a bronze-casting foundry, an 18-hole championship golf course, and an equestrian center. A renovation and expansion of the music hall includes a two-story addition that has a 40-seat classroom, three studio offices, and a student lounge. Also updated were the art building and the Blanchard Campus Center, which now holds a performance space, coffee bar, art gallery, and game room. The Unified Science Complex continues to advance the college's reputation as a leader in scientific education.

Despite changes to the campus, curriculum at this 169-year-old institution remains decidedly traditional. Students are required to complete 128 total credits to graduate—32 in their major and 16 in their minor. Required courses include three humanities courses, two courses from science and mathematics disciplines (with at least one lab), two social science courses, and one course in multicultural perspectives. The college offers more than 30 first-year seminars each fall, and more than a dozen in the spring,

"I'm encouraged to explore and be adventurous, to learn about myself and the world."

covering topics and disciplines from biology to women's studies. The focus of these courses is developing skills in analysis and critical inquiry through speaking and writing. Some also include field trips to museums or events in Boston and New York.

The academic climate is rigorous. "The courses are often compelling, and the professors require a high level of engagement with the subject matter," says a senior. Mount Holyoke produces more female PhDs in chemistry and biology than any other liberal arts college. The top-of-the-line chemistry labs, along with a solar greenhouse, a scanning electron microscope, several nuclear magnetic resonance spectrometers, and a linear accelerator, provide the students with state-of-the-art equipment necessary to be the best. Five-year dual-degree programs enable students to combine degrees from MHC with BS degrees in engineering from Dartmouth or Caltech. Newer programs include majors in architectural studies, film, Buddhist studies, and gender studies—many of which are made available at the other four consortium schools.

Although some of Mount Holyoke's intro courses have 50 or more students, most have 25 or fewer. Since the required curriculum is so diverse, there is little trouble getting into the smaller classes and finishing in four years. "Mount Holyoke students are passionate about their academics," says a student. "This creates an environment where you can have a theoretical discussion over milk and cookies, attend lectures about new topics, and take a class in a different department just to expand your horizons." The school's honor code makes possible self-scheduled, self-proctored final exams. After those tests is the optional January winter term, where many students opt for a two-credit, nontraditional course, or an off-campus internship in New York or Washington, D.C. About 40 percent of MHC students seeking a complete change of scenery spend all or part of junior year in another country. The Twelve College Exchange Program* offers opportunities in more than 25 locales, while Mount Holyoke also has its own study abroad programs. Those interested in the sea may be interested in the semester at the Marine Biological Laboratory at Woods Hole, Mass.

Mount Holyoke attracts students from all over the nation and the world; however, one-quarter are Massachusetts natives. "The one thing that unites MHC students is their passion and dedication to their ideals," says one senior. African Americans make up 4 percent of the student body, Asian Americans 12 percent, and Hispanics 5 percent. "People are fairly politically minded and it is very important to be politically correct," says one student. The Student Coalition for Action is a very large and popular campus group dedicated to social change. Merit scholarships are available, ranging from $10,000 to $30,000.

> "The courses are often compelling, and the professors require a high level of engagement with the subject matter."

Nine out of ten Mount Holyoke students live in the 19 residence halls, most of which have their own dining facility. "I feel that dorm life is a large part of the 'Mount Holyoke experience'," says a sophomore. "The dorms themselves are beautiful and well-maintained." Most dorms are also very homey, with living rooms, TV lounges, and baby grand pianos; all serve milk and cookies (as well as healthier fare like hummus and vegetables) most nights at 9:30 p.m. Students from all four classes live together and housing is guaranteed for all four years. Some residence halls also offer apartment-style living.

Students find the Five College Consortium* one of Mount Holyoke's greatest assets. A free bus service runs every 20 minutes between MHC and UMass, Amherst, Smith, and Hampshire, multiplying a Mount Holyoke woman's access to academic, social, and cultural opportunities. "There are five colleges in the area, so there is always something to do," says one sophomore. The majority of social opportunities are on campus, such as parties, plays, concerts, speakers, and cultural events. If on-campus activities aren't appealing, road trips to Boston, Vermont, and New York City are also popular. Closer to campus, the South Hadley Center has eateries, a pub, shops, and apartments, though students say that Amherst and Northampton provide more shopping options. If you are 21, you are allowed to buy and have alcohol on campus; however, if you are underage and caught drinking "a warning or a counseling session may be required," says a sophomore.

Perhaps more than their counterparts at Smith and Wellesley, Mount Holyoke women have made a virtue out of the school's most visible "vice": the lack of men. Women fill all leadership positions, thanks to a strong and supportive community spirit, and boys are just down the road at Amherst or UMass. Like most happy families, Mount Holyoke students take pride in tradition. Upon arrival, each first-year student

> "People are fairly politically minded and it is very important to be politically correct."

is assigned a secret elf (a sophomore), a big sister (a junior), and a disorientation leader (a senior). Each class also has a color and a mascot, and class spirit is huge, especially for the annual Junior Show. Every fall on Mountain Day, students wake up to ringing bells, classes are canceled (even the library is closed), and everyone treks up Mount Holyoke to picnic and see the foliage.

For those breaks in studying, the athletics at Mount Holyoke, such as crew, riding, field hockey, and lacrosse are popular. The successful equestrian team has brought home several national championships recently. The college encourages athletic participation at all levels with a demanding 18-hole golf course, jogging trails, and two lakes. The 20-acre equestrian center includes a 57-stall barn, two riding areas, a training and show area, and seating for 300. Crew regattas and rugby take the place

> ## Overlaps
> Smith, Wellesley, Bryn Mawr, Brown, Tufts, Vassar

of football games, and students come out in force when the opponent is another of the Seven Sisters.

The traditions of academic excellence, modern upgrades, and easy access to New York and Boston provide a small college atmosphere with nearby cultural education. Says a psychology major. "It was the best choice I ever made."

If You Apply To ➤

Mount Holyoke: Early decision: Nov. 15. Regular admissions and financial aid: Jan. 15. Guarantees to meet demonstrated need. Campus and alumnae interviews: recommended, evaluative. ACTs or SATs: optional. SAT IIs: optional. Accepts the Common Application and electronic applications. Essay question. Also requires a two- to five-page paper written in the 11th or 12th grade, with a teacher's comments.

Muhlenberg College

2400 Chew Street, Allentown, PA 18104-5586

There is a definite Muhlenberg type: serious, ambitious, and buttoned-down. It is the only eastern Pennsylvania/New Jersey liberal arts college with a serious religious affiliation—Lutheran. Strong in premed, prelaw, preanything.

Website:
www.muhlenberg.edu
Location: City outskirts
Total Enrollment: 2,457
Undergraduates: 2,396
Male/Female: 42/58
SAT Ranges: V 560–660
M 570–670
ACT Range: 25-31
Financial Aid: 53%
Expense: Pr $ $
Phi Beta Kappa: Yes
Applicants: 4,347
Accepted: 44%
Enrolled: 32%
Grad in 6 Years: 86%
Returning Freshmen: 93%
Academics: ✍ ✍ ✍
Social: ☎ ☎ ☎
Q of L: ★ ★ ★ ★
Admissions: (484) 664-3200
Email Address: admissions
@muhlenberg.edu

Strongest Programs:
Premed/Biology
Prelaw
English/Writing
Theatre Arts and Dance

When a popular school pseudonym is "The Caring College" rather than some line referring to sun, booze, or babes, you know you're in for a different experience. That's the case with Muhlenberg College, a small liberal arts school that nurtures its students. Founded on solid Lutheran roots, the school continues to encourage religious diversity among students and to attract the best and brightest to its top premed school. "Academic excellence is, by far, the top priority," says a junior, "but the people are warm, friendly, compassionate, and truly care about making Muhlenberg the best possible place for everyone."

Set on 91 park-like acres, the Berg campus is a combination of older Gothic stone structures and newer buildings in a variety of architectural styles. Prominent facilities include a lovely chapel, the high-tech Trexler Library, a 40-acre biological field station and wildlife sanctuary, and a 48-acre arboretum with more than 300 species of wildflowers, broadleaf evergreens, and conifer trees. The campus also boasts a football stadium and all-weather track, and the 50,000-square-foot Trexler Pavilion for the Performing Arts that has a dramatic 45-foot glass outer shell and houses a variety of performing spaces. The college is in the midst of a $100-million fundraising campaign that has resulted in a major addition to the sports center, with a new science facility addition yet to come.

Muhlenberg's regional reputation rests on its premedical program, which continues to attract large numbers of students. An agreement with Philadelphia's Drexel University College of Medicine guarantees seats for up to six Muhlenberg students each year. The college's theater arts program is also a national draw, and a few alumni have even gone on to star on Broadway. Science lab equipment at Muhlenberg is cutting-edge, and a comprehensive natural science major allows for a sampling of it all. The Living Writers course is offered every other year and has brought a number of noted authors to campus, including Robert Pinsky, Jay Wright, and Alice Fulton. Muhlenberg sends study groups to Washington, D.C., and students may spend semesters abroad in countries from England, France, Spain, and Germany to Argentina, the Czech Republic, Japan, Australia, and Scotland. Programs sponsored

"Academic excellence is, by far, the top priority."

by the International Student Exchange are also available, and Muhlenberg is member of the Lehigh Valley Association of Independent Colleges.*

(Continued)
Business
Psychology

The college offers three honors programs, the Muhlenberg Scholars Program, the Dana Associates Program, and the R. J. Fellows Program, which focuses on the ramifications of change. Each is limited to 15 students per entering class. They carry an annual $3,000 stipend and culminate in an in-depth mentored senior research project. Psychology is Muhlenberg's most popular major, followed by business, communication, biology, and theatre arts, and these are high profile programs. Still, "I would say that whatever your interests, Muhlenberg offers them as competitive and well-taught courses," says a junior. A major in film studies has been added, and there are joint programs in physical therapy and occupational therapy available with Thomas Jefferson University.

General education requirements are organized into two major groups: Skills (writing, oral expression, reasoning, and foreign language) and Perspectives (literature and the arts, meaning and values, human behavior and social institutions, historical studies, physical and life sciences, and other cultures). Each freshman is assigned a First-Year Advising Team, usually consisting of four students (a student mentor, a student advisor, and two academic advisors) and a faculty member. One Spanish/premed double major says her science advisor helped pick classes to prepare her for medical school, while her Spanish advisor explained how language skills would serve her well as a doctor.

Muhlenberg has produced Fulbright Fellows and Udall Scholars, but it also ensures that every student excels at his or her own pace.

The fun-filled, three-day freshman orientation program carries one requirement: learning the alma mater and then hightailing it to the president's house to serenade him. All freshmen also take a writing-intensive, discussion-intensive First-Year Seminar, with enrollment capped at 15. About two-thirds of freshman courses have fewer than 25 students, and since there are no graduate students, there are no teaching assistants. "The student/faculty relationship is truly unique on this campus," a student says. "Teachers often conduct classes outside on nice days, have classes over to their houses, and can be seen at college-sponsored events."

"There is a niche for everyone at the 'berg," says a senior, "whether your interests are in knitting or theatre, comedy or football, politics or hip-hop dancing." Muhlenberg draws 26 percent of its students from Pennsylvania, and many from adjacent New Jersey. The campus is 92 percent white, but "we have a diverse pool of interests, which creates a lively campus in terms of politics and social awareness," says a biology major. Cultural appreciation is emphasized as

"We have a diverse pool of interests, which creates a lively campus in terms of politics and social awareness."

Muhlenberg strives for a more ethnically and religiously varied campus. Students stay involved in the community by volunteering as tutors and with groups such as Habitat for Humanity and Planned Parenthood. A political science professor has started a student-run polling institute, which is raising campus awareness and activism. Merit scholarships worth an average of $8,000 are available but there are no athletic scholarships.

Muhlenberg encourages students to live on campus and guarantees housing to all undergraduates except transfers, so 88 percent of students live in campus residences. All dorms have computer labs, study lounges, and vending machines. Prosser's co-ed wing makes a good choice for freshmen, while upperclassmen praise the Muhlenberg Independent Living Experience, or MILE, townhouses. Two dorms, Robertson and South, house 140 students in single, air-conditioned rooms overlooking Lake Muhlenberg. Other popular choices include Taylor and Benfer, where students live in eight-person suites that have their own bathrooms. "Dorms are comfortable and clean," says one senior. Freshmen choose from a seven- or five-day meal plan, and students agree that dining options continue to improve. "They try to

The college offers three honors programs, the Muhlenberg Scholars Program, the Dana Associates Program, and the R. J. Fellows Program, which focuses on the ramifications of change.

make provisions for special tastes and the school does a good job to make different foods on different days of the week."

Most social life at Muhlenberg takes place on campus. "The most fun I've had at school has been in my dorm hallway," says a student. "Just hanging out with people I meet is fun." The Muhlenberg Activities Council (MAC) provides comedians every Thursday evening—recent visitors have included Jimmy Fallon—current movies in the Red Door Cafe, live band concerts, and new movies on the lawn. "For those who enjoy different entertainment, it is easy to get together a group of friends, go somewhere, or make your own fun," says one student. City buses stop five minutes from campus for trips to Allentown proper and area malls. There also are daily bus runs to New York City (for clubbing and theater), Philadelphia (for nightlife and cheese steaks), and Baltimore and Washington, D.C. Outdoorsy students can pick up the Appalachian Trail for a little hiking.

"The warm and friendly feeling on campus makes me feel at home."

Eighteen percent of the men and 20 percent of the women pledge their undergraduate years to fraternities and sororities, respectively, but Greek life does not dominate the social scene—and it's becoming less important now that rush doesn't occur until sophomore year. Alcohol is forbidden if you're underage, per Pennsylvania law, and the school takes its policies seriously. Big social events include East Fest, homecoming, Deck Party, the Scotty Wood basketball tournament, the Mr. Muhlenberg awards—which parody the Miss America pageant—and the Henry Awards, the college's version of the Oscars. There's also a candlelight ceremony where freshmen write down their college goals, reexamining them the day before graduation.

For the athletically inclined, intramural sports arouse a great deal of passion. Football, women's basketball, and men's and women's soccer are strong, winning Centennial Conference championships and NCAA tournament bids. Muhlenberg's Life Sports Center offers a pool, basketball court, and other all-purpose courts, and a jogging track. Also popular is Frisbee golf; there's an 18-hole course on campus, where play goes on during all seasons and all hours of the day and night. Students say any contest against Johns Hopkins draws crowds.

Muhlenberg has produced Fulbright Fellows and Udall Scholars, but it also ensures that every student excels at his or her own pace. "The students and staff at Muhlenberg are what continues to make me reluctant to leave and excited to return," a student says. "The warm and friendly feeling on campus makes me feel at home." Guess that slogan does fit after all.

Overlaps

Lafayette, Bucknell, Franklin and Marshall, Dickinson, Gettysburg, Lehigh

If You Apply To ➤

Muhlenberg: Early decision: Feb. 1. Regular admissions and financial aid: Feb. 15. Housing: May 1. Meets demonstrated need of 94%. Campus interviews: recommended, evaluative (required, along with a graded paper, if students choose not to submit SAT scores). Alumni interviews: optional, informational. SATs or ACTs: optional. SAT IIs: optional. Accepts the Common Application and electronic applications. Essay question: significant experience or achievement; issue of personal, local, national, or international concern; influential person; influential fictional character, historical figure, or creative work; topic of your choice.

University of Nebraska at Lincoln

12 Administration Building, Lincoln, NE 68588-0415

Everybody knows Nebraska football, but in other areas UNL has a lower profile. Fewer out-of-staters attend Nebraska than, say, University of Kansas or Iowa.

Agriculture is still the biggest drawing card, and the music program is also strong. Because of the state's demographic makeup, diversity is limited.

Nearly 22,000 students call the University of Nebraska at Lincoln home, and their school pride is contagious. On crisp fall weekends, when spirits are high and the Big Red football arcs through the air, Huskers cheer and paint the town of Lincoln red and white in a show of appreciation for their alma mater. In fact, on home game Saturdays, the stadium is the third largest "city" in the state, holding 5 percent of the population. Away from the stadium, in the classrooms, UNL has reason to cheer with top programs ranging from music to agriculture to journalism.

UNL spreads across two campuses. The East Campus is home to the colleges of agricultural sciences and natural resources, human resources and family sciences, law, and dentistry. Most entering students end up on the larger City Campus, home of six of the eight undergraduate colleges: architecture, arts and sciences, journalism and mass communications, business administration, fine and performing arts, engineering and technology, and the teachers college. On City Campus, the architectural style ranges from the modern Sheldon Art Gallery designed by Philip Johnson to the architecture building, which is on the National Register of Historic Places. There are also several malls, an arboretum, and a sculpture garden.

Nebraska's College of Agricultural Sciences and Natural Resources, known for its outstanding programs in food science and technology, agribusiness, and animal science, is housed in a $19 million complex. The school of music's opera program has received national attention, and the performing arts programs benefit from the $18 million Lied Center for the Performing Arts, which seats 2,300. Business administration, psychology, and marketing are some of the most popular majors. New programs include degrees in landscape architecture and classical languages.

Nebraska's Comprehensive Education Program provides students with a common set of educational experiences across the majors and colleges. It has four components: Information Discovery and Retrieval (one course), Essential Studies (nine courses), Integrative Studies (10 courses), and Co-Curricular Experience. To help freshmen get oriented, a one-semester University Foundations class covers the inner and outer workings of the campus, organized around academic subjects. Big Red Welcome combines entertainment and food in a carnival setting to welcome new students, and the SIPS program (Summer Institute for Promising Scholars) is a six-week preorientation session for incoming minority students. The J.D. Edwards Honors Program gives computer science and management students internships to complement their coursework. The Undergraduate Creative Activity and Research Experience Program provides a stipend for students after freshman year who want to participate in research with a professor. Students also can study abroad in places such as Costa Rica, Germany, Mexico, and Japan.

> "Each teacher excels in his or her own area and is excited to share their knowledge."

Getting into courses in the most popular areas, especially education, business, and engineering, can be a problem, students say; preregistration is a must. UNL can be academically challenging, if students want it to be. "I think the size of UNL is too large to nurture a campus competitiveness," says one business major. "Most of the academic climate is drawn from your peers." Graduate students teach a quarter of the courses, but top professors can be found inside the classroom. "The quality of teaching I have received has been extraordinary," says a senior. "Each teacher excels in his or her own area and is excited to share their knowledge."

The UNL student body is mostly conservative and from the Cornhusker State. Asian Americans make up 3 percent of the student body, with African Americans and Hispanics making up an additional 5 percent combined. Says one junior,

Website: www.unl.edu
Location: Center city
Total Enrollment: 21,675
Undergraduates: 17,037
Male/Female: 53/47
SAT Ranges: V 530–660
  M 540–670
ACT Range: 22–28
Financial Aid: 44%
Expense: Pub $ $
Phi Beta Kappa: Yes
Applicants: 7,474
Accepted: 75%
Enrolled: 63%
Grad in 6 Years: 63%
Returning Freshmen: 84%
Academics: 🖉 🖉 🖉
Social: 🐃 🐃 🐃 🐃
Q of L: ★ ★ ★
Admissions: (800) 742-8800
Email Address:
  http://admissions.unl.edu

Strongest Programs:
Agribusiness and Agronomy
Animal Science
Architecture
Audiology and Speech
  Pathology
Music
Textiles, Clothing, and Design
Bioengineering
Journalism

On home game Saturdays, the stadium is the third largest "city" in the state, holding 5 percent of the population.

"Nebraska is a very conservative state. Thus, on campus, there is a constant struggle between conservative Christian groups and the free thought activist culture that colleges are known for." Merit scholarships are available, and 210 athletic scholarships are awarded in a wide variety of sports.

Thirty-seven percent of students live in the university's single-sex or co-ed dorms, and there's usually no trouble getting a room. Dorm lotteries favor those wanting to stay in the same room or on the same floor. "Dorms are phenomenal," enthuses one senior, who notes that rooms are "huge" and offer a "great community setting." Each room also is wired for the Internet. Freshmen, who must live on campus, are welcomed to the residence halls through the FINK program, which is friendlier than it sounds (the acronym stands for Freshman Indoctrination of New Kids). "Freshmen have been choosing to stay in campus housing instead of leaving because of the quality. Lately, this has caused trouble obtaining rooms."

"Lincoln is a great college town; especially during football season!"

To help freshmen get oriented, a one-semester University Foundations class covers the inner and outer workings of the campus, organized around academic subjects.

UNL is big and there's a group or activity for everyone; fraternity and house parties, roller skating, the movies, eating out, visiting coffee shops and bars (for those of age), and road trips to Omaha or Kansas City are just some of the activities that keep students busy. For many, the fall semester revolves around football weekends and postseason bowl games. "The entire state shuts down on football Saturdays," one junior says. Fraternities draw 14 percent of UNL men and sororities attract 18 percent of the women. They offer both social events and a chance to get involved in the Lincoln community. Homecoming, Greek Week, and Ivy Day are among the most-anticipated campus events, as is The End, new alcohol-free programming at the end of each semester during "dead week" and finals week. For those who want to indulge, plenty of bars are within walking distance of UNL, providing relief to students dissatisfied with the dry campus. So committed is the administration to keeping a distance between students and alcohol that it has obtained a grant to combat high-risk drinking among students. "UNL is a dry campus. [Drinking] is considerably regulated, but does occur and campus police are very passionate about no alcohol or underage drinkers," one advertising major warns.

"Lincoln is a great college town; especially during football season!" exclaims one business administration major. Another student adds, "Thirty bars within a two minute walk of campus." A senior chemical engineering major likes the town for another reason. "Almost everyone I have met on or off campus has been a student at UNL. This gives us a connection—that we belong to the same alma mater and have pride in our school." That said, Omaha is only 45 minutes away. Pachyderm enthusiasts will be delighted by the Nebraska Museum of Natural History's outstanding collection of prehistoric elephant skeletons. Beyond the sidewalks are miles of flat road and plains, ideal for biking, cross-country skiing, and snowmobiling.

"UNL is a good makeup of a variety of attitudes and personalities."

Traditionally strong on the gridiron, UNL has a reputation as a powerhouse in a number of other sports. Men's track and baseball and women's volleyball, gymnastics, track, and softball have all brought home Division I titles. The biggest football rivalries are with Colorado and Oklahoma. Husker fans proclaim that if forced to choose between going to Oklahoma and going to hell after death—well, it would be a tough choice.

At Nebraska, future agriculture experts mingle with techno-whizzes, while teachers-in-training brush elbows with architecture mavens. "UNL is a good makeup of a variety of attitudes and personalities. Overall, most people are "down to earth and very 'Nebraskan'—lots of small-town people," says one senior. Whether studying overseas, immersing themselves in an internship, or going wild on Saturday afternoon, students here know how to make the most of their time as Cornhuskers.

Overlaps

Iowa State, University of Kansas, University of Nebraska–Omaha, University of Nebraska–Kearney, Washington University

New College of Florida

(Formerly New College of the University of South Florida)
5700 North Tamiami Trail, Sarasota, FL 34243-2197

New College is the South's most liberal institution of higher learning—apologies to Guilford. With an enrollment of 761, New College is about one-third the size of a typical liberal arts college. The kicker: It's a public institution and one of the nation's best buys.

The mere existence of New College of Florida is proof that it's possible to find success through individualism. This school has done away with grades and GPAs, so students compete with themselves rather than their classmates. The laid-back student body and rigorous academic program "proves to students that learning can be a self-directed, fun, and productive experience," a sophomore says. Strong programs in psychology, biology, and anthropology complement the liberal arts offerings and boost the reputation of this rising star.

New College began in 1960 as a private college for academically talented students, but when inflation threatened its existence in the mid-1970s, it offered its campus to the University of South Florida. Today, NCF serves as Florida's honors college, but is an academically independent entity. New College's campus is adjacent to Sarasota Bay and consists of historic mansions from the former estate of circus magnate Charles Ringling, abutting modern dorms designed by I.M. Pei. The central quad is filled with palm trees, and sunsets over the bay are spectacular. New College shares its campus with the Sarasota branch of USF, which offers upper-level courses in business, education, and engineering. Newer additions to campus include the $1.2 million Keating Center, which houses the college's alumni association and foundation. In the coming years, a new social sciences building will be constructed, along with a 300-bed dorm.

The administration once had no required core curriculum, but now has added requirements to provide "each student with the depth and breadth of knowledge characteristic of a good liberal arts education." All undergraduates must complete at least eight courses in the liberal arts curriculum, including courses in humanities, social sciences, and natural sciences, and show literacy in math and computers. The school calendar, however, is still unique: the two 14-week semesters are separated by a

> "Students here take academics very seriously, but are only competing with themselves."

month-long January Interterm, during which students devise and carry out their own research or conduct group projects. Students work out a "contract" with their advisor each semester and receive written evaluations instead of grades. The seven semester-long contracts and three independent study projects lead to an area of concentration, capped by a senior project thesis and an oral baccalaureate examination.

Due to the highly individualized nature of the curriculum, getting into some classes can be a challenge, especially for science majors looking to fulfill requirements for grad school admission. Courses are rigorous and challenging, but the

Website: www.ncf.edu
Location: Suburban
Total Enrollment: 761
Undergraduates: 761
Male/Female: 40/60
SAT Ranges: V 630–720
  M 580–670
ACT Range: 25–29
Financial Aid: 36%
Expense: Pub $
Phi Beta Kappa: No
Applicants: 684
Accepted: 60%
Enrolled: 53%
Grad in 6 Years: 69%
Returning Freshmen: 84%
Academics: ✏️✏️✏️✏️
Social: ☎ ☎ ☎
Q of L: ★ ★ ★
Admissions: (941) 359-4269
Email Address:
  admissions@ncf.edu

Strongest Programs:
Psychology
Biology
Literature
Anthropology
Math
Marine biology

environment is not competitive, students say. "Students here take academics very seriously," says a chemistry major, "but are only competing with themselves." The shape of any student's program depends heavily on the outlook of his or her faculty sponsor, and students say advising—both academic and career—is readily available.

New College doesn't offer the specialized courses of a large university, but there's still plenty to choose from across the academic spectrum. A 1,100-gallon sea water system is available for lab experiments in animal behavior and physiology. Anthropology wins raves, and many students gravitate to biology, psychology, political science, and sociology. Academic programs are constantly changing to best suit students' needs; an example is the beefing up of offerings in environmental studies, religion, and Caribbean and Latin American studies. The Jane Bancroft Cook Library makes up for its small size—less than 300,000 volumes—with a language lab, videotape viewing area, an interlibrary loan program with the entire state university system of Florida, and a classroom equipped for teleconferences.

Students praise the personalized attention they receive from professors; graduate students and teaching assistants don't lead classes here and 73 percent of classes have 25 or fewer students. All disciplines provide the opportunity for original research and students also may conduct field research around the globe. "Not only are the teachers brilliant, but they are friendly and always accessible," says a sophomore.

In keeping with the revolution theme, students on this relatively cosmopolitan campus tend to be creative liberal types with '60s nuances. "Students here are thinkers," says a philosophy major. "They are very self-aware and openly, unabashedly analyt-

"Students here are thinkers."

ical." Eighty-three percent hail from Florida, perhaps because New College has yet to make a national name for itself. Minorities account for 15 percent of the student body; Hispanics account for 10 percent, 3 percent are Asian Americans, and African Americans make up 2 percent. Students are active and highly aware of social and political issues, tending to lean to the left. Issues run the gamut from women's rights and the war in Iraq to poverty and commercialism. Merit scholarships, averaging $3,580, are available to qualified students although there are no athletic awards.

Sixty-eight percent of students live in campus housing. "Like most things at New College, the dorms are quirky," says a junior. The Dart dorms, two apartment-style halls, accommodate 140 students in two-bedroom, two-bath suites with a kitchen, living area, and—of course—air-conditioning. Rooms are chosen by lottery and there has been a room shortage as of late, forcing some to find off-campus housing. As for security, the campus is considered exceptionally safe. A sophomore quips, "I only have mosquitoes and raccoons to worry about."

On campus, the "astounding" social life is T-shirts-and-shorts relaxed. "Walls," free-form parties every Friday and Saturday night, can last until 4 or 5 a.m. the following morning. Loudspeakers line Palm Court and students sign up to reserve a night and play whatever music they want. The PCPs (Palm Court Parties) are "blown-out-of-proportion Walls" that occur during Halloween, Valentine's Day, and graduation. While it's not hard for underage students to drink, "this is not a culture of binge drinking," says a junior. While some students call Sarasota "old-timers-ville," it does offer plenty of cultural enrichment (and beautiful beaches). The Ringling Museum of

"We are all highly motivated, but we also love to have fun."

Art and the Asolo State Theater adjoin the campus, and many New College instrumentalists perform with the Florida West Coast Symphony, Sarasota's professionally led symphony orchestra. The open road to Tampa, Gainesville, Key West, Orlando, New Orleans, Atlanta, and even Washington, D.C., ("to protest stuff"), beckons when Sarasota becomes too quiet.

New College is definitely not a haven for jocks; it fields no varsity teams, somewhat of an oddity in football-crazy Florida. But many students take advantage of intramural sports, which range from basketball, soccer, and softball to swimming and sailing. The yearly faculty/student kickball game is popular, and anyone can play. Students also look forward to the Crucial Barbecue in January with music and mud wrestling, the Male Chauvinist Pig Roast, the SemiNormal, a semiformal event on the bay, and the Bowling Ball, a formal-dress occasion at a bowling alley. While the school has a 25-meter swimming pool, students complain that it closes at 10 p.m. The nearby ocean (which is open 24/7 barring hurricanes or red tide) is a bigger draw. "The beaches are gorgeous white sand and blue water," says a student.

Without a Greek scene, grades, or crazy football games, NCF is definitely not your typical southern institution. But the eccentricity doesn't impede students' academic motivation or their love of learning—whether it's belly dancing, biology, origami, or psychology. And students say NCF has much more to offer. "We are all highly motivated, but we also love to have fun," a sophomore says. "There is a sense of community here that's hard to find anywhere else."

Overlaps

University of Florida, Hampshire, Florida State, University of Miami, University of Central Florida, Brown

If You Apply To ➤

New College: Rolling admissions: May 1. Financial aid: May 1. Campus interviews: optional, evaluative. Alumni interviews: optional, informational. SATs or ACTs: required. SAT IIs: optional. Accepts the Common Application and electronic applications; prefers electronic applications to paper ones. Essay question: views on an important issue.

University of New Hampshire

Grant House, 4 Garrison Avenue, Durham, NH 03824-3510

UNH is a public university that looks and feels like a private college, and its tuition hits the pocketbook with similar force. Expensive though it may be, UNH draws nearly than half of its students from out of state. Strong in the life sciences, especially marine biology, and in business and engineering.

Students at the University of New Hampshire know how to get their hands dirty, and this solid public institution provides them with countless opportunities to do so. UNH is one of just eight universities in the nation to receive the land-grant, sea-grant, and space-grant designations. Its research mission has grown dramatically in the last decade. Yet the university remains a moderate sized institution that emphasizes undergraduate instruction. Unlike many large research universities, the UNH faculty teach all students, including freshmen, and value teaching as much as they do their research. A love of the outdoors is a must, as well as the ability to withstand long, cold winters.

The university's wide-open grassy campus hosts a blend of modern facilities and ivy-covered brick buildings. The sprawling lawns are surrounded by nearly 3,000 acres of farms, fields, and woods. During the past few years, UNH has invested in large-scale construction and renovation projects, including Holloway Commons, a dining and conference facility. Kingsbury Hall, home to the College of Engineering and Physical Science, is being renovated to the tune of $52 million, and two new buildings are being added to the Gables, an on-campus apartment building.

Interdisciplinary programs enhance UNH's emphasis on traditional academic programs and the many research opportunities offered by its seven undergraduate schools. Business and engineering are the most respected programs, and the English department has a fine creative writing program. The Whittemore School of Business

Website: www.unh.edu
Location: Small town
Total Enrollment: 12,045
Undergraduates: 10,808
Male/Female: 43/57
SAT Ranges: V 510–610
  M 520–620
Financial Aid: 58%
Expense: Pub $ $ $
Phi Beta Kappa: Yes
Applicants: 12,310
Accepted: 72%
Enrolled: 13%
Grad in 6 Years: 73%
Returning Freshmen: 86%
Academics: ✍ ✍ ✍
Social: 🐨 🐨 🐨 🐨 🐨
Q of L: ★ ★ ★ ★
Admissions: (603) 862-1360

(Continued)
Email Address:
admissions@unh.edu

Strongest Programs:
History
Psychology
Sociology
Performing Arts
English
Biological Sciences
Marine and Agricultural
   Sciences
Engineering

and Economics includes options in entrepreneurial venture creations, information systems, international business, and economics, management, marketing, and accounting; students can also design their own track. Marine biology is also considered stellar, due to UNH's proximity to the ocean and a freshwater bay. Environmental studies, chemistry, kinesiology, and nursing are strong as well. UNH has implemented the Discovery Program, which places a curricular emphasis on the first year experience, a focus on the interdisciplinary learning experience, and integration of the UNH Discovery Program with the academic major and research. An environmental sciences major replaces majors in hydrology, soil science, and water resources management; other programs include majors in justice studies and environmental conservation studies. Qualifying students can begin an honors program featuring small classes in freshman year. Weak spots are largely due to the lack of professional schools such as law, medicine, or pharmacy.

The university's general education requirements apply across the board and mandate completion of 10 courses from eight categories: writing skills; quantitative reasoning; biological, physical, and technological sciences; historical perspectives; foreign cultures; fine arts; social science; and works of philosophy, literature, and ideas. Freshman composition is mandatory as part of a four-course writing intensive requirement. Classes are relatively small, almost always 50 students or fewer, and TAs only facilitate discussion sections or labs. "The academic climate tends to be to work hard Monday through Thursday," says a junior. "The level of difficulty and how rigorous a class is depends on the department, class, and professor."

The school's nickname is the University of No Holidays, since an exceptionally generous winter break limits the number of days off during other seasons.

> "I've had no trouble getting a room and have loved every dorm."

UNH prides itself on producing undergraduates with research experience. The Undergraduate Research Opportunities Program provides about 100 research awards each year for undergraduates to work closely with faculty on original projects. A recent undergraduate research conference drew 200 participants. Budding scientists and sociologists have opportunities to work at research centers for space science and family violence; other students can take advantage of the Institute for Policy and Social Science Research; the Center for Humanities; the Institute for the Study of Earth, Oceans, and Space; and the Center to Advance Molecular Interaction Sciences. The Interoperability Lab enables students to work with professors and businesses on cutting-edge problems of computing equipment compatibility. The Isle of Shoals Marine Laboratory, which operates several research projects with Cornell University, is just seven miles off the coast. And then there's UNH's Technology, Society, and Values Program, designed to address the ethical implications of the computer age. UNH's study abroad program offers exchange programs with more than 170 U.S. colleges via the Center for International Education. Students can even earn a dual major by combining foreign study and classes in international affairs with those of any other program.

While UNH is New Hampshire's major public institution, it has long been popular with out-of-staters, who make up 42 percent of its students. The school is working on becoming more diverse; only 5 percent of the student body are minorities, which students cite as an area of concern. A special task force at the university is working on keeping the minority students in school until graduation and in helping them network. Social issues at UNH include alcohol awareness and—as might be expected in a place with such lush natural beauty—the environment. The school offers hundreds of merit scholarships averaging $5,827; 304 awards are also available for athletic prowess.

While UNH is New Hampshire's major public institution, it has long been popular with out-of-staters, who make up 42 percent of its students.

> "The academic climate tends to be to work hard Monday through Thursday."

Fifty-seven percent of UNHers live in the school's 31 single-sex and co-ed dorms. "I've had no trouble getting a room and have loved every dorm," says a senior. "They are very comfortable and well-maintained throughout the year, thanks to our excellent cleaning staff." The dorms offer special-interest groupings, lounges, fireplaces, TV and study rooms, and kitchenettes. Freshmen and sophomores are guaranteed dorm rooms; most upperclassmen live off campus or in Gables and Woodside, on-campus apartment complexes. Students also gripe that parking is difficult on campus.

Less than a five-minute walk from campus is the beautiful little town of Durham, which caters to the student clientele. Along Main Street, Durham has many restaurants and coffeehouses, a grocery store, an ice cream parlor, and a few bars, which have been divided into separate sections (for legal consumers of alcohol and everyone else). Durham is UNH, students say. "It's the epitome of a college town, with coffee shops, pizza joints, sub places, all with the option of outdoor socialization when the weather finally permits come April," says a senior. Greek groups claim 4 percent of UNH men and 5 percent of women. The Greeks also throw parties, which are subject to the university's no-tolerance alcohol policy that evicts from on-campus housing underage students caught with alcohol more than once. For nondrinkers, the university offers weekend social events including concerts, dances, movies, and coffeehouses. Popular road trips include Boston and the White Mountains, or apple picking at a nearby farm. Late nights at L.L. Bean have also become commonplace, and homecoming, Greek Week, Winter Carnival, Casino Night, and Spring Fling draw crowds every year. And every four years, New Hampshire takes the spotlight when the state holds the nation's earliest presidential primaries.

UNH teams that regularly enjoy national rankings and generate strong spectator interest include men's ice hockey, which recently won the NCAA Championship, and students celebrate the first UNH goal of each game by inexplicably throwing a large fish onto the ice. When the hockey team plays its rival, the University of Maine–Orono, students wear white to "white out" the stadium. Women's hockey, cross-country, swimming, diving, and volleyball are also impressive. The university has a strong intramural sports program involving thousands of students. Broomball—played with brooms, balls, and sneakers on the ice—is very popular.

When there's no game to watch or post-game revelry to indulge in, nature provides UNH students with more than enough to do—if they can find time off. (The school's nickname is the University of No Holidays, since an exceptionally generous winter break limits the number of days off during other seasons.) Skiing, camping, fishing, and hiking in nearby forests are favorite seasonal pastimes, and the Outing Club is among the most popular student activities.

New Hampshire's only major public university offers a huge variety of programs. That's one reason it attracts so many students from out of state. The laid-back atmosphere and multitude of both academic and social opportunities makes the UNH experience worth every dime.

Interdisciplinary programs enhance UNH's emphasis on traditional academic programs and the many research opportunities offered by its seven undergraduate schools.

Overlaps

University of Vermont, University of Massachusetts–Amherst, University of Connecticut, Northeastern, Boston University, University of Rhode Island

If You Apply To >

New Hampshire: Early action: Dec. 1. Regular admissions: Feb. 1. Financial aid: Mar. 1. Housing: Feb 1. No campus interviews. Alumni interviews: optional, informational. SATs or ACTs: required (SATs preferred). SAT IIs: optional. Does not guarantee to meet demonstrated need. Accepts the Common Application and electronic applications. Apply to particular school or program. Essay question: meaningful photograph; how you would contribute to the community; or topic of your choice; recent activities if previous applicant or have been away from school.

The College of New Jersey

(formerly Trenton State College)
P.O. Box 7718, Ewing, NJ 08628-0718

A public liberal arts institution in the mold of William and Mary or UNC–Asheville. Also offers business and education. More than nine-tenths of the students are homegrown Garden Staters. A smaller, more personal alternative to Rutgers.

Website: www.tcnj.edu
Location: Suburban
Total Enrollment: 5,827
Undergraduates: 5,698
Male/Female: 42/58
SAT Ranges: V 570–670
   M 600–700
Financial Aid: 45%
Expense: Pub $ $ $ $
Phi Beta Kappa: No
Applicants: 7,300
Accepted: 45%
Enrolled: 38%
Grad in 6 Years: 81%
Returning Freshmen: 95%
Academics: ✑ ✑ ✑ ✑
Social: 🏠 🏠 🏠
Q of L: ★ ★ ★
Admissions: (609) 771-2131
Email Address:
   admiss@vm.tcnj.edu

Strongest Programs:
Biology
Chemistry
History
Elementary Education
Music
Psychology
Business
Computer science

The College of New Jersey is an up-and-coming public institution with special focus on undergraduates, an emphasis more commonly found at a private school. TCNJ offers professors focused on teaching and a campus physically similar to one found down the road at Princeton University—without the Ivy League price tag. Formerly a teachers' college, TCNJ strives to provide students opportunities in a host of other fields. The small size makes for closeness among students and faculty.

TCNJ is set on 289 wooded and landscaped acres in suburban Ewing Township, six miles from Trenton. The picturesque Georgian Colonial architecture centers on Quimby's Prairie, surrounded by the original academic buildings of the 1930s. A flock of Canada geese makes its home in one of the two campus lakes. Newer facilities include a science complex, spiritual center, and three parking garages.

To graduate, students must earn 120 credits for all BS programs in the School of Business, BA programs except for teacher preparation, and the bachelor of science in nursing. In addition, the list of majors and study abroad opportunities continues to grow. The First Year Experience program is a required two-semester sequence consisting of two courses—From Athens to New York, and Society, Ethics, and Technology. It is designed to ease students into the demanding reality of college life with an approach that integrates academics, individual development, and social understanding. Ten hours of community service is part of the requirement. Also required of incoming freshmen: Expectations, a one-day program to help students and parents understand what they can expect from the college and what the college expects of them; the 10-week College Seminar, to smooth the transition to college; Welcome Week, which gives freshmen a chance to meet their classmates and become acquainted with the campus; and Summer Readings, which exposes students to the kind of scholarship and dialogue they can expect at The College of New Jersey. Other requirements include two semesters each of rhetoric and mathematics, 26 credits of Perspectives on the World, and three semesters of foreign language (arts and sciences students only).

> "The academic climate is somewhat intense."

Consistent with the school's origins as a teachers' college, elementary education is popular, and the business school is strong, as are the natural sciences. Sociology, health, and physical education are considered weak. Academically, TCNJ is competitive and getting more so. "The academic climate is somewhat intense," confides a sophomore. "Some students and professors try to downplay the competitive nature, but overall it's pretty driven." The college offers a combined four-and-one-half-year BS/MA in law and justice, taught jointly by TCNJ and Rutgers; a seven-year BS/MD degree program with the University of Medicine and Dentistry of New Jersey; and a seven-year BS/OD degree with SUNY College of Optometry. TCNJ also offers foreign study in 11 countries and is a member of the International Student Exchange Program, giving students access to 131 colleges and universities across the U.S., including Alaska, the Virgin Islands, Puerto Rico, and Guam. The college has no teaching assistants, and faculty members get high marks, though quality can vary

> "TCNJ is really big on the 'community' feel."

by department. "My professors are very passionate about their work and field of study," says a psychology major. Another student adds, "Some have had a great influence on my life, and some would be better off choosing another profession."

The typical TCNJ student is "overly book smart" with "low social skills," according to one junior. A graduate student says, "TCNJ is really big on the 'community' feel." The school has no cap on out-of-state admissions, but only 5 percent of TCNJ's students are non-Jerseyans; 68 percent of the freshmen graduated in the top tenth of their high school class, and the majority attended public high school. The college has aggressively pursued minority students, and today, African Americans account for 6 percent of the student body, Hispanics 7 percent, and Asian Americans 5 percent. "If you don't leave this school very well-educated in political correctness, then you were obviously unconscious," says a marketing major, who praises the school for its diversity. The school has begun ongoing symposia on the Middle East and Afghanistan in the wake of September 11. Merit scholarships average $4,444. "TCNJ brings in many of the best New Jersey students who are accepted to Ivy League schools but cannot afford them," says a senior.

Dorm housing is only guaranteed for freshmen and sophomores, though 65 percent of all students live on campus. "Most of the dorms are really nice," says a student, "but a few are older and outdated." Freshmen hang their hats in either Travers-Wolfe, a two-building, 10-story hall, or Lakeside, a four-building complex. After that, students can enter the lottery for about 2,100 upperclass spaces in the apartment-style townhouses, Community Commons, and the recently built residence hall, or try one of several local apartment complexes.

Although suburban Ewing doesn't really cater to students, funky New Hope, Pennsylvania, and preppy Princeton, New Jersey, are just up the road; restaurants, bars, movie theaters—and this being New Jersey, many malls—are within a short drive. State alcohol policies are strictly enforced, and the underage shouldn't hope to imbibe at the campus bar, the Rathskeller. Twelve percent of men and 14 percent of women belong to fraternities and sororities, which provide many of the off-campus parties. Campus programming includes dances, concerts, and movies. Road trips to Philadelphia and New York, each about an hour away and accessible by train, are also highly recommended.

The College of New Jersey's 21 varsity teams are big fish in the small pond of NCAA Division III; they've won dozens of Division III crowns and runner-up titles. Students rally around the football and basketball squads, especially when archrival Rowan comes to town, and the women's field hockey, lacrosse, and soccer teams have a faithful following. TCNJers also look forward to several annual events, including homecoming, a Family Fest Day, and—the springtime favorite—Senior Week.

The College of New Jersey is one of the nation's "budget Ivies," with reasonable tuition and a location that offers media types, artists, and budding scientists a relaxed suburban haven within shouting distance of the editors, producers, directors, curators, and pharmaceutical companies of New Jersey, Pennsylvania, and New York.

Although suburban Ewing doesn't really cater to students, funky New Hope, Pennsylvania, and preppy Princeton, New Jersey, are just up the road; restaurants, bars, movie theaters—and this being New Jersey, many malls—are within a short drive.

The First Year Experience program is a required two-semester sequence consisting of two courses—From Athens to New York, and Society, Ethics, and Technology. It is designed to ease students into the demanding reality of college life.

Overlaps

Rutgers, University of Delaware, Villanova, NYU, Lehigh, Rowan

If You Apply To ➤

TCNJ: Early decision: Nov. 15. Regular admissions: Feb. 15. Does not guarantee to meet demonstrated need. Campus interviews: recommended, informational. No alumni interviews. SATs or ACTs: required (SATs preferred). SAT IIs: required (writing). Accepts the Common Applications and electronic applications. Essay question.

New Jersey Institute of Technology

University Heights, Newark, NJ 07102

NJIT is one of the few public technical institutes in the Northeast. It occupies a middle ground between the behemoth Rutgers and smallish Stevens Institute. Offers engineering, architecture, and management. At nearly four to one, NJIT's gender ratio is particularly skewed.

Website: www.njit.edu
Location: Urban
Total Enrollment: 5,408
Undergraduates: 4,082
Male/Female: 80/20
SAT Ranges: V 470–590
   M 540–650
Financial Aid: 58%
Expense: Pub $ $ $ $
Phi Beta Kappa: No
Applicants: 2,562
Accepted: 71%
Enrolled: 42%
Grad in 6 Years: 56%
Returning Freshmen: 85%
Academics: ✍ ✍ ✍
Social: ☎
Q of L: ★ ★
Admissions: (973) 596-3300
Email Address:
   admissions@njit.edu

Strongest Programs:
Architecture
Computer Science
Applied Mathematics
Engineering
Environmental Science

The New Jersey Institute of Technology provides a no-frills technological education that prepares students for a future in an ever-changing global workplace. NJIT's challenging programs emphasize education, research, service, and (not surprisingly) economic development. It's an enticing combination for students seeking a high-tech, low-cost education.

NJIT's urban 45-acre campus is dotted with 24 buildings of diverse architectural styles, ranging from Elizabethan Gothic to contemporary design. Some of New Jersey's greatest cultural institutions are just blocks away, including the Newark Museum, Symphony Hall, and the New Jersey Center for the Performing Arts. Soccer fans can take advantage of the new soccer field with state-of-the-art turf.

NJIT is composed of the Newark College of Engineering, the School of Architecture (the only state-supported one in New Jersey), the School of Management, the College of Science and Liberal Arts, and the Albert Dorman Honors College (which enrolls almost 500 students). Top applicants are offered a spot in Dorman as NJIT freshmen, and they can stay as long as they keep their grades up. Perks of Dorman membership include guaranteed dorm rooms, research opportunities, and acceptance into the BS/MS program after completion of five courses for the undergraduate major.

"It requires hours of study just to keep up."

Engineering, architecture, and computer science garner the most student praise, while mechanical and electrical engineering are especially challenging. The administration acknowledges that business and communication offerings are among the weaker programs.

To graduate, students must fulfill general education requirements in areas ranging from English to management. All freshmen take Calculus I and II, English composition, computer science, physical education, and Freshman Seminar, a course that introduces students to university life. NJIT has worked with Rutgers to create a number of joint-degree programs from biology to history.

Most NJIT courses have 50 students or fewer. While some say the atmosphere can be low-pressure in certain fields, an electrical engineering student relates his experience this way: "The academic climate is extremely competitive. It requires hours of study just to keep up." The administration assists students if they are having a difficult time, arranging for leaves of absence or extra semesters with a lighter courseload. Some have problems getting into classes with enrollment caps that are offered only once a year. But one student confides that if a freshman has all of the prerequisite courses, he or she has a good chance of graduating on time.

Students give teaching quality average to high marks. Since most profs have worked in industry, they can offer job information along with academic assistance. Academic advising isn't as helpful as it could be, say some students. "My advisors just look at the courses I choose and make sure that I am supposed to be taking them. I wish they knew more," says one junior. Career counseling, though, is helpful in preparing students for the job hunt. NJIT's most-favored academic option is the co-op program, which enables juniors to get paid for two six-month periods of work at technical companies.

As New Jersey's comprehensive technological university, NJIT attracts a wide range of students with different interests. But, one sophomore laments, "We need women." African Americans comprise 10 percent of the student body, while Hispanics represent 13 percent. Asian Americans account for 22 percent, and 7 percent hail from other nations. World Week—with cultural performances and ethnic foods—is a popular spring event. Tolerance is not a problem here as it is on some other campuses. One student praises the Educational Opportunity Program, saying, "If it weren't for them, I would not be here. They make it easy to be a minority." But students say it's a shame that ethnic groups tend to stick together and not mingle; as one puts it, "There's not a feeling of oneness." Political issues do come up, such as peaceful political protests against the war in Iraq. NJIT

"They make it easy to be a minority."

has a chapter of Tau Beta Pi, the national engineering honor society. The university does not guarantee to meet the financial aid of all admits, but offers merit scholarships worth an average of $4,210 to qualified students. There are also 139 athletic scholarships in 7 sports.

NJIT's four residence halls can accommodate one-third of the students. Twenty-seven percent of students live on campus. "I love housing," says a junior. "We get free cable, fast Internet connections. We have very big rooms." He does advise, however, that you should get your application in on time in order to get one of those roomy rooms. Freshmen and students living furthest away get first crack at the rooms, and those who get in are guaranteed space the next year. Consensus has it that the best freshman dorms are Redwood Hall and Cypress. Upperclassmen move into fraternity houses or nearby off-campus apartments. The dorms and frats both have kitchen facilities, which many students welcome because the food service fare at times draws a few grumbles from students left with grumbling stomachs after dinner. "Because of its urban location, safety is always a consideration at NJIT. However, students praise the security efforts the school has undertaken. "Public safety officers are always around," says an electrical engineering major.

The four-to-one male/female student ratio definitely puts a crimp in the social life. There are some outlets, though. Seven percent of the men and 5 percent of the women join the Greek system. One of the best annual campus events is Spring Week, which includes bands, novelties, and a semiformal. Diwali, the Indian festival of lights, and Chinese New Year also give undergrads pause to party. "We have a lot of barbecues that bring people together," says an architecture major. Another option is the beach, an hour away, with windsurfing and sailing equipment courtesy of NJIT. Most students agree that the administration's strict alcohol policies are effective.

NJIT students take pride in their athletic prowess. The varsity teams are Division II, except for soccer, which recently moved to Division I. Men's soccer and baseball and women's swimming are the most popular sports on campus, followed by basketball, men's swimming, and tennis. The outstanding athletic facilities are open to all, and include an indoor running track, fitness center, racquetball and squash courts, a six-lane pool, and areas for weight training, archery, or aerobics. Outdoor facilities include lighted tennis

"We get free cable, fast Internet connections. We have very big rooms."

courts, a sand volleyball court, and a multiuse soccer stadium seating 1,000. A proud (and sweaty) tradition is the Hi-Tech Soccer Classic, which pits NJIT athletes against rivals from MIT, RPI, and Stevens Institute of Technology.

"Students here are hard-working, smart, and aggressive," says an engineering management student. NJIT people have chosen their school because they want a topnotch technical education without the topflight price tag. Academics are the priority here, and if the social life is less than electrifying, students deal with it. After all, they know highly skilled jobs will beckon after graduation. Getting through is a challenge,

Overlaps

Drexel, Stevens Institute of Technology, College of New Jersey, Rutgers, Rowan, Penn State

but there's ample compensation available for NJIT alums in the technologically dependent workplaces of today—and tomorrow.

If You Apply To ➤

NJIT: Rolling admissions: Apr. 1. Financial aid: Jan. 5. Does not guarantee to meet full demonstrated need. Campus interviews: optional, informational. No alumni interviews. SATs: required (SATs preferred). SAT IIs: optional. Accepts the Common Application and prefers electronic applications. Architecture applicants must submit portfolio of work.

University of New Mexico

P.O. Box 4895, Albuquerque, NM 87196-4895

UNM is shaped by the encounter between Hispanic, Native American, and white culture. Studies related to Hispanic and Native cultures are strong, and in a land of picture-perfect sunsets, photography is a major deal. Technical programs are fueled by government labs in Albuquerque and Los Alamos.

Website: www.unm.edu
Location: Urban
Total Enrollment: 17,525
Undergraduates: 14,269
Male/Female: 42/58
SAT Ranges: V 480–600
  M 470–600
ACT Range: 19–24
Financial Aid: 70%
Expense: Pub $
Phi Beta Kappa: Yes
Applicants: 7,134
Accepted: 74%
Enrolled: 59%
Grad in 6 Years: 41%
Returning Freshmen: 76%
Academics: 🐜 🐜 🐜
Social: 🐘 🐘 🐘
Q of L: ★ ★ ★
Admissions: (505) 277-2446
Email Address:
  apply@unm.edu

Strongest Programs:
Southwest Hispanic Studies
Photography
Lithography
Geology
Environmental Studies
Laser Optics
Latin American Affairs

The University of New Mexico's heritage goes back to 1889 when New Mexico wasn't even a state, and the university's strengths are still rooted in the rich history of the American Southwest. New Mexico excels in areas such as Latin American affairs and Southwest Hispanic studies. Lest you think it is a typical state school, consider that many students are commuters or of nontraditional age. UNM also boasts New Mexico's only law, medical, and architecture and urban planning schools, as well as its only doctor of pharmacy program.

Seated at the foot of the gorgeous Sandia Mountains in the lap of Albuquerque, the beautifully landscaped campus sports both Spanish and Pueblo Indian architectural influences, with lots of patios and balconies. The duck pond is a favorite spot for sunbathing, and the mountains, which rise majestically to the east, are visible from virtually any point on campus. A new $17.3 million imaging and diagnostic center, Domenici Hall, recently opened, as did the Cornell Parking Garage and Visitors Center. New construction includes the Lobo Center Business Facility.

UNM offers more than 4,000 courses in 11 colleges and two independent divisions, running the gamut from arts and sciences, education, and engineering to management, fine arts, and the allied health fields. Academic and general education requirements vary, but the core curriculum mandates three English courses focused on writing and speaking, two courses each in the humanities, social and behavioral sciences, and physical and natural sciences, and one course in each of the fine arts, a second language, and math. Those reluctant to specialize can spend a few semesters in the broad University College, which also offers the most popular degree, a bachelor of university studies. Freshmen are encouraged to participate in the Freshman Forum and Core Legacy Courses. Engineering and fine arts freshmen can join interest groups who share suites in a dorm. The Tamarind Institute, a nationally recognized center housed at UNM's School of Fine Arts, offers training, study, and research in fine-art lithography. Anthropologists

"We hang out and stuff, but for the most part we are focused on school."

may root around one of New Mexico's many archeological sites, and engineers may join in major solar-energy projects. Other popular majors include biology, education, psychology, and nursing and business management. New programs include majors in health medicine and human values and Native American studies.

The academic climate is "very laid-back and depends on what field of study you are going into," according to a senior. Students are quick to help one another study, and competition for grades is the exception rather than the rule. As for professors, "I would have to give them a B," says a student. "There have been some very good ones and some that were very knowledgeable but didn't know how to teach."

By virtue of its location, UNM enjoys a diverse mix of cultures, even though 89 percent of students are state residents.

By virtue of its location, UNM enjoys a diverse mix of cultures, even though 89 percent of students are state residents. A large minority student enrollment—34 percent Hispanic, 3 percent African American, and 3 percent Asian American—reflects this cultural diversity. A cultural awareness task force and student diversity council work to keep race relations from becoming rancorous, while the student orientation program includes a cultural awareness component, and a full-time human awareness coordinator develops diversity-related programs for the residence halls. UNM also hosts the Arts of the Americas, a broad crosscultural program that involves U.S. and Latin American artists in festivals, classes, and exhibits. "Students here are pretty chill," says a journalism major. "We hang out and stuff, but for the most part we are focused on school." Many classes, and several complete degree programs, are offered in late afternoon and evening sessions, and about half of the student body takes advantage of these after-hours options.

Many UNM students commute, and they say finding parking spots continues to be difficult as a result. Only 11 percent of students live on campus. "This is a commuter campus," one senior says. "The dorms are nice but not too many students live there," preferring instead to look for apartments. "Plus the parking for dorm students is consistently oversold and very expensive," complains a senior. An escort service, emergency phones, good lighting, and police who patrol around the clock help students feel safe. Students are happy with the variety of food available to them, and note that the newly remodeled student union building includes chain restaurants and a restaurant that serves food from different cultures every week.

"This is a commuter campus. The dorms are nice but not too many students live there."

Albuquerque—sometimes referred to as ABQ—is New Mexico's largest city, and it offers a variety of cultural attractions, including the nation's largest hot air balloon fiesta, a growing artists' colony, and concert tours to charm the ears. Santa Fe is an hour away. Those with cars or pickup trucks take advantage of the state's natural attractions: superb skiing in Taos, the Carlsbad Caverns, the Sandias, as well as excellent hiking and camping opportunities. For the historically inclined, numerous Spanish and Indian ruins are within an easy drive. And for those who yearn for more exotic locales, study abroad programs beckon from Mexico, Brazil, Venezuela, Costa Rica, and Scotland.

Those reluctant to specialize can spend a few semesters in the broad University College, which also offers the most popular degree, a bachelor of university studies.

Alcohol, though banned on the UNM campus, is readily available, according to most students, especially at Greek parties. Speaking of the Greeks—a mere 3 percent of men and 4 percent of women join up. Other students find their fun off campus in Albuquerque's clubs and restaurants. "Social life takes place both on and off campus," a junior says. For the more socially conscious, the college sponsors Spring Storm, an outing of roughly a thousand students who volunteer around the city on a Saturday. Annual social events include Welcome Back Days in the fall and Nizhoni Days, a celebration of Native American culture. Each spring, the whole campus turns out for a four-day fiesta with food and live music.

The men's basketball and football squads and the women's softball, soccer, and volleyball teams usually draw crowds. Scholarships are available for male and female athletes in sports ranging from skiing and wrestling to swimming, golf, and track and field.

UNM's campus and educational emphases keep in mind the Indian pueblos that surround the school. For those not concerned about having a "complete" college

Overlaps

Arizona State, New Mexico State, University of Colorado, University of Texas–El Paso, Eastern New Mexico, Highlands

experience, and for students balancing college with a part-time job, UNM offers a sun-drenched location that satisfies—precisely because its academic climate is as relaxed as the rolling desert dunes. "People here are serious and accepting," says a senior, "which makes UNM a comfortable environment."

<table>
<tr><td>If You
Apply
To ➤</td><td>UNM: Regular admissions: Jun. 15. Financial aid: Mar. 1. Guarantees to meet full demonstrated need. Campus interviews: optional, informational. No alumni interviews. ACTs or SATs: required (ACTs preferred). SAT IIs: optional, required for homeschooled students or those at nonaccredited high schools. Prefers electronic applications. Essay question: your educational and career goals and any other information the admissions committee should know.</td></tr>
</table>

New Mexico Institute of Mining and Technology

Campus Station, Socorro, NM 87801

Boutique technical education with a Southwestern flair. Smaller than Rose-Hulman and somewhat larger than Cal Tech and Harvey Mudd, Tech offers intense programs in an ever-widening number of technical fields. Don't count on getting through in only four years.

Website: www.nmt.edu
Location: Rural
Total Enrollment: 1,363
Undergraduates: 1,123
Male/Female: 73/27
SAT Ranges: V 560–670
   M 570–680
ACT Range: 24–29
Financial Aid: 25%
Expense: Pub $
Phi Beta Kappa: No
Applicants: 428
Accepted: 81%
Enrolled: 81%
Grad in 6 Years: 43%
Returning Freshmen: 68%
Academics: ✍ ✍ ✍
Social: ☎ ☎
Q of L: ★ ★
Admissions: (800) 428-TECH
Email Address:
   admission@nmt.edu

Strongest Programs:
Earth Science
Electrical Engineering
Physics

New Mexico Institute of Mining and Technology has evolved so much since its founding that it has outgrown its name. Founded as the New Mexico School of Mines, the college now emphasizes computer science, chemical and electrical engineering, and information technology. New Mexico Tech continues to expand and change with the times, as evidenced by its growing reputation in antiterrorism training and research.

Tech's tree-lined campus, 76 miles south of Albuquerque, is dotted with picturesque, white adobe, red-tiled buildings, and plenty of grassy open spaces that "capture the spirit of the Southwest." The Jones Hall Annex houses classrooms, labs, and offices. NMT owns 20,000 acres adjacent to the town of Socorro (population 9,000), including Socorro Peak, which provides a mother lode of research and testing facilities. A thunderstorm lab sits on another mountaintop 20 miles away. A new student services center provides a central location for all student-related administrative offices, including the registrar and academic counseling. Not surprisingly, mountain bikers, runners, astronomers, hikers, campers, rock climbers, geologists, rock hounds, and scenery enthusiasts feel right at home here.

NMT offers many excellent programs in three main areas: science, engineering, and natural resources. The departments of earth and environmental science, petroleum, and environmental engineering are among Tech's best, as is the program in hydrology, but administrators admit that the mineral engineering department could be bolstered. Freshmen can participate in the First Year Experience Program, in which they are grouped by major under a peer facilitator. New programs have been

"This is a small Western town with a deep Hispanic and Indian culture—it's very relaxed."

added in information technology and mechanical engineering, as well as a master's in electrical engineering and a doctorate in applied mathematics. The arts are not what NMT is about, so don't look for an excess of stellar offerings. The closest thing to a well-regarded program in the soft sciences is technical communication.

The student body is said to contain its share of nerds—75 percent, calculates one senior. "However, they tend to be extremely smart," he adds. As might be expected, then, students are drawn to Tech's spacious library, which holds over 200,000 books. Computer facilities are, naturally, quite good. And teaching gets high marks, though because of Tech's relatively small size, most courses in the technical fields are offered sequentially, and students who don't take a cluster all the way through may wait several semesters before the necessary course is offered again. To graduate, students must take courses in calculus, physics, chemistry, English, technical writing, humanities, social sciences, and foreign language. Students say it's tough to finish all requirements in four years. A chemical engineering major says students get frustrated because "they never have enough time to finish projects and homework."

Tech's student/faculty ratio is low for a technical school, and though professors are research-oriented, they do take teaching seriously. Class sizes vary, though 86 percent have 50 or fewer students. Jobs with mineral industries, research laboratories, and government agencies are available through the five-year cooperative work-study program. Undergraduates can also work part-time at research divisions on campus, including the New Mexico Bureau of Mines and Mineral Resources, the Petroleum Research and Recovery Center, the Energetic Materials Research and Testing Center, and the National Radio Astronomy Observatory's VLA and VLBA facilities. The terrorism training and research comes through NM Tech's association with the Energetic Materials Research and Testing Center. As 99 percent of the faculty does research and most hire undergraduates, opportunities for scientific investigation and independent study are plentiful. Quality advising, on the other hand, is not: "Some advisors really care about their students and try to help, while others sign your forms and can't wait to get back to their research," gripes a biology major.

Only 12 percent of Tech's undergraduates are from out of state, and 2 percent are foreign nationals. Hispanics account for 21 percent of the student body, African Americans 1 percent, and Asian Americans 3 percent. Twenty-seven percent of students are female—a high number for a technical school. "The issues mostly are about how someone got a better grade than so and so," says a chemical engineering student. Tech's housing facilities have improved and expanded since the days when women resided in the school's trailer park. Students can now live in suites with private bedrooms, a kitchen, and a living room. Fifty-nine percent of students live on campus, which one resident describes as "comfortable but a little crowded." A bit of legwork can turn up decent and "incredibly cheap" housing off campus. Plus, dorm-dwellers are required to buy the meal plan, and the food is said to be less than appealing. Luckily, and not surprisingly, the Mexican food available makes the town taco heaven.

Otherwise, the town of Socorro is far from being a student paradise. It is a mining-turned-farming area in one of the most sparsely populated areas in the Southwest that can only be described as tiny. Boredom may be a problem here, especially if you are under 21, says a senior, though he admits that in its own way, Socorro "grows on a person." The good news is the spectacular weather, where something called rain is in danger of becoming a distant memory, and the nearby desert and spectacular mountains provide a wealth of outdoor opportunities. According to one Techie, "This is a small Western town with a deep Hispanic and Indian culture—it's very relaxed." Still, even those who enjoy the scenery and their classmates' company see a direct correlation between sanity and access to a car, which can take them to Albuquerque and El Paso or the Taos ski slopes.

With no Greek system and little excitement in Socorro, it's no wonder students at Tech have always had to work to make their own fun. The alcohol policy—"in your room only, over 21 only"—works in residence halls with active resident advisors. But a senior says, "It is very easy for minors to find alcohol." There are no varsity sports at Tech, but the men's and women's rugby and soccer teams do travel to challenge

The departments of earth and environmental science, petroleum, and environmental engineering are among Tech's best, as is the program in hydrology.

The student body is said to contain its share of nerds—75 percent, calculates one senior.

To graduate, students must take courses in calculus, physics, chemistry, English, technical writing, humanities, social sciences, and foreign language. Students say it's tough to finish all requirements in four years.

Overlaps

University of New
Mexico, New Mexico
State, Colorado School
of Mines, Texas Tech,
MIT

other schools. Many students also enjoy an extensive intramural program and the school's 18-hole golf course. And in the absence of teams to cheer for, Tech's most popular annual events are 49ers Weekend, a homecoming tribute to the miners of yore with gunfighters and a bordello/casino, and Spring Fling, a mini-homecoming. Fall Fest, a new event, welcomes new and old students back to campus.

NMT boasts one of the most intimate and up-to-date technical educations—and certainly some of the best weather—in the nation. NMT is an island of intensity in the otherwise calm New Mexico desert, but those who make it through four years (or five or six) leave with a solid technical education at a rock-bottom price.

If You Apply To ➢ **New Mexico Tech:** Rolling admissions: Aug. 1. Financial aid: Mar. 1. Housing: Jun. 1. Guarantees to meet full demonstrated need. Campus interviews: optional, informational. No alumni interviews. SATs or ACTs: required (ACT preferred). SAT IIs: optional. Accepts electronic applications. No essay question.

New York University

22 Washington Square, New York, NY 10012

Don't count on getting into NYU just because Big Sis did. From safety school to the hottest place in higher education, NYU's rise has been breathtaking. The siren song of Greenwich Village has lured applicants by the thousands. Major draws include the arts, media, and business.

Website: www.nyu.edu
Location: Urban
Total Enrollment: 40,004
Undergraduates: 20,566
Male/Female: 41/59
SAT Ranges: V 620–710
  M 620–710
ACT Range: 27–31
Financial Aid: 49%
Expense: Pr $ $ $
Phi Beta Kappa: Yes
Applicants: 34,509
Accepted: 37%
Enrolled: 37%
Grad in 6 Years: 83%
Returning Freshmen: 92%
Academics: 🖉 🖉 🖉 🖉 ½
Social: ☎ ☎ ☎
Q of L: ★ ★ ★
Admissions: (212) 998-4500
Email Address:
  http://admissions.nyu.edu

With the world at its doorstep, New York University invites its student body to jump right in. Firmly planted in the heart of Greenwich Village, arguably one of the most eclectic and energizing neighborhoods in New York City, NYU is blossoming. Its growing student body, burgeoning new facilities, and multiple opportunities for high-level internships and research projects have made it a top option for a rising number of students. One senior observes, "The prestige of the university has certainly increased." Another boasts, "NYU officials are now very eager to listen to students' ideas for change. Our president holds town hall meetings in residence halls and holds small group dinners in his penthouse. It's hard to complain about NYU's large size when it's normal to have dinner with 20 students and President Sexton."

It doesn't get more real world than New York City. NYU has campuses and centers throughout the city, but is centered on Washington Square. Trendy shops, galleries, clubs, bars, and eateries crowd neighboring blocks; SoHo, Little Italy, and Chinatown are just blocks away. Academic NYU buildings—both modern and historic—blend with 19th-century brick townhouses surrounding Washington Square Park (the closest thing NYU has to a quad), where parades of rappers, punks, Deadheads, and hipsters surround a replica of the Arc de Triomphe in Paris. Furman Hall recently became the School of Law's first new academic building in more than 50 years. Kimmel Center for University Life houses meeting space for NYU's more than 350 student clubs, plus areas for the frequent recruitment fairs and lectures featuring national and international leaders. It houses The Skirball Center for the Performing Arts' 1,000-seat theater, which is the largest performing arts facility south of 42nd Street.

The city scene is a core element of the NYU experience. So, too, is the wide range of academic programs. The Tisch School of the Arts trained such famed directors as Martin Scorcese, Spike Lee, and Oliver Stone, and current undergrads continue to win many national student filmmaker awards. Tisch also boasts excellent drama, dance, photography, and television departments, and it's not uncommon to see students who haven't yet finished BFA degrees performing in Broadway shows.

Wall Street's future bulls and bears make their home at the Stern School of Business, where they benefit from a center for Japanese and American business and economic studies. Another favorite department among students (and New York corporations who recruit them after graduation) is accounting, known for its high job-placement rate. The arts and sciences are strong, with English, journalism, history, political science, and applied math winning highest marks. There's an increased emphasis on foreign exchange and study abroad, with study abroad sites in Paris, Shanghai, London, Florence, Madrid, Prague, Czech Republic, and Accra, as well as exchange agreements with universities in many locations throughout the world. The Gallatin School of Individualized Study provides flexible schedules and freedom from requirements for those wishing to engage in independent study or develop their own programs. An annual under-graduate research conference at the College of Arts and Sciences gives students the chance to pre-

> "It's hard to complain about NYU's large size when it's normal to have dinner with 20 students and President Sexton."

sent findings from their research. The Steinhardt School, the School of Social Work, the College of Nursing, and the Tisch Center for Hospitality and Sports Management offer a plethora of career-based programs, including art, education, nutrition, and sports and leisure studies.

Finding a cheap New York apartment may be easier than sailing through NYU's academics. "The students here work very hard, but it's definitely not any sort of cut-throat competition," says a sophomore. Everyone is very focused on career preparation—it's never enough to just concentrate on your classes. Premed, prelaw, and prebusiness students may encounter packed schedules and competitive classes, while Gallatin and Tisch students may have lots of spare time, students say. "It's hard to avoid the pressure," says a student majoring in drama and politics, with nine hours of acting class a day and writing-intensive academic courses, too. At least the NYU library is accommodating—it's one of the largest open-stack facilities in the country, with more than three million volumes.

Under the Morse Academic Plan, freshmen and sophomores take courses including foreign language, expository writing, foundations of contemporary culture, and foundations of scientific inquiry. The language offerings, though, go beyond the typical Spanish-French-German—among the choices are Cantonese, Hindi/Urdu, Modern Irish, Swahili, and Tagalog—and students are required to show medium proficiency. Despite the university's mammoth size, 78 percent of classes taken by freshmen have 25 or fewer students. Graduate students might lead foreign language sections, writing workshops, and the recitations that accompany lectures, but students still say teaching is top-

> "Students here are fiercely independent, motivated people."

notch. "All courses are taught by faculty members who are committed to getting to know us," raves one drama major. Those qualifying for freshmen honors seminars study in small classes under top faculty and eminent visiting professors.

The variety of degree options here may tempt students to hang around the Village for more than four years. There's a seven-year dental program and a five-year joint engineering program with New Jersey's Stevens Institute of Technology. Freshmen selected as University Scholars travel abroad each year. Point to a spot on a world map and you'll likely hit on a country hosting NYU students. Locally, internships range from jobs on Wall Street to assignments with film industry giants. The career center is "amazingly personal and well-run," says an econ major, and has thousands of listings for on-campus jobs, full-time jobs, and internships.

(Continued)
Strongest Programs:
Drama/Theater Arts
Dance
Business
Art and Design
Film and Television
Music

The Steinhardt School, the School of Social Work, the College of Nursing, and the Tisch Center for Hospitality and Sports Management offer a plethora of career-based programs, including art, education, nutrition, and sports and leisure studies.

The variety of degree options here may tempt students to hang around the Village for more than four years. There's a seven-year dental program and a five-year joint engineering program with New Jersey's Stevens Institute of Technology.

"Students here are fiercely independent, motivated people," says a student. "Do not come here if you need a college that will hold your hand." Thanks in part to the university's investment in new housing, a majority of students (58 percent) now come from outside New York State. The NYU students who are from New York State come primarily from the city and nearby suburbs. African Americans make up 5 percent of the student body, Asian Americans 14 percent, and Hispanics 7 percent. On this generally liberal campus, gender issues, social justice, the Israeli–Palestinian conflict, and rights of all kinds—gay, lesbian, transgender, animal, human, and workers'—are important now, students say.

Because NYU is large and fairly decentralized, the Student Resource Center helps students navigate university resources and services. The university's Wellness Exchange provides students with a hotline that connects them with professionals who can help them address daily challenges or crises they may encounter. New students are urged to attend an all-campus freshman orientation program, a program specifically designed for their school, or both. Students also meet with academic advisors—usually professors in their major department—at least once a semester. Advisors review course selections and give students permission to register, while also helping them stay on track toward graduation.

For concerned parents and students, the Office of Student Life, Protection, and Residence Halls hosts a series of workshops on keeping safe at NYU, and programs like the NYU Trolley and Escort Van Service provide door-to-door service for students until 3:00 a.m. "Campus security is extremely tight," says a student, with "guards in every dorm, safe vans for transportation, and sign-in policies."

While NYU students once had to fend for themselves in New York's outrageous housing market, the university now guarantees four years of housing to all freshmen (and most transfers) who seek it. Twenty-one dorms, ranging from old hotels to a converted monastery, provide a wide range of accommodations. Most rooms have private baths and are larger, cleaner, newer, and better equipped than many city apartments, enticing 54 percent of students to stay on campus. Freshmen are housed largely in freshman residence halls, many of which have theme floors, and rooms are assigned by lottery each spring. Amenities include central air-conditioning, computer centers, musical practice rooms, kitchens, and even, in some buildings, small theaters. The university provides free shuttle buses to dorms that are further uptown or downtown. The dining halls offer extensive choices—from wraps to sushi to Burger King. "The dining facilities are superb! I'm a senior who still eats in the dining hall every day for at least one meal!" says one student. Of course, downtown's array of ethnic restaurants also offers a variety of food at cheap prices.

"Campus security is extremely tight."

While NYU students once had to fend for themselves in New York's outrageous housing market, the university now guarantees four years of housing to all freshmen (and most transfers) who seek it.

Students can't say enough good things about NYU's social life. "It's New York, come on," says a student majoring in playwriting. Another adds: "Greenwich Village is full of students, professors, artists, families…It's the most exciting, alive, cultural part of New York City." On campus, there are concerts, movies, fraternity and sorority events (just 5 percent of the men and 3 percent of the women go Greek), and more than 300 clubs. The springtime Strawberry Festival includes free berries, cotton candy, outdoor concerts, and carnival amusements. Many students march in the city's Halloween Parade, which literally takes over Greenwich Village, while most spring and fall weekends find a city-sponsored street fair somewhere nearby. The Violet Ball, a dinner/dance held each fall in the atrium of Bobst Library, is an excuse to get dressed up. As for alcohol, underage students caught with it in public areas of dorms may lose their housing. The rest take their chances with the notoriously strict bouncers at bars and clubs around Manhattan. "They card like crazy," says one junior.

While sports have not exactly been NYU's strength, women's volleyball and basketball, men's golf and swimming and diving, and men's and women's fencing all won competitions in 2005–06. Other Division III powers include women's swimming and diving and men's volleyball. Thirty-two percent of undergrads participate in intramural sports, which include touch football, bowling, and quickball. The Palladium Athletic Facility boasts a big swimming pool and a 30-foot-high indoor climbing wall. Road trips to Philadelphia, Boston, or Washington are few and far between.

Though it might seem hard to concentrate on schoolwork as the heartbeat of New York City thumps day and night, NYU students thrive on all that energy, and know how to spread it among their studies and social lives. "You can never be bored in Greenwich Village," a junior says.

Overlaps

Columbia, Boston University, University of Pennsylvania, Yale, Harvard, Brown

---

If You Apply To ➤

NYU: Early decision: Nov. 1. Regular admissions: Jan. 15. Financial aid: Feb. 15. Housing: May 1. Does not guarantee to meet demonstrated need. No campus or alumni interviews; campus visit strongly recommended. SATs or ACTs: required. SAT IIs: required for students entering in fall 2007, except applicants to the Tisch School of the Arts or the Steinhardt arts and music programs. Accepts the Common Application (with NYU supplement) and electronic applications. Apply to particular schools or programs. Essay question: important person, place, or event in your life; how a creative work influenced you; hometown experience.

University of North Carolina at Asheville

I University Heights, Asheville, NC 28804-8503

The "other" UNC happens to be one of the best educational bargains in the country. At just over 3,000 students, UNCA is about half the size of fellow public liberal arts college William and Mary and 1,000 students smaller than Mary Washington. Picturesque mountain location in a resort city.

Whether it's the lush environment or the money you're saving, the University of North Carolina at Asheville will have you seeing green. This public liberal arts university offers all of the perks that are generally associated with pricier private institutions: rigorous academics, small classes, and a beautiful setting. And it does it for a fraction of the cost. The university continues to integrate experiential learning into its traditional curriculum, emphasizing internships and service-learning experiences. Any way you look at it, UNCA is a bargain that may have your friends turning green with envy.

Located in the heart of North Carolina's gorgeous Blue Ridge Mountains, the 265-acre campus lies in the middle of one million acres of federal and state forest near the tallest mountain in the East and the most heavily visited national park in the country. The campus was built in the 1960s, and much of the brick architecture reflects the style of that decade, although half of the buildings were added within the past few years. The Botanical Gardens at Asheville, adjacent to the main campus, features thousands of labeled plants and trees, and serves as a wildlife refuge and study center for botany students. Highsmith University Union features modern space for gatherings, as well as a convenience store, bookstore, game room, gallery space, and food court. New Hall is a 32,000-square-foot facility that houses offices, classrooms, and computer labs. The environmentally friendly design features a "green roof" and the use of geothermal pumps for heating and cooling.

> **"Each class on this campus is challenging and the professors do not slack off."**

The university is dedicated to providing a liberal arts education that teaches students to become their own best and lifelong teachers. The newly implemented general education curriculum is known as Integrative Liberal Studies. The program

Website: www.unca.edu
Location: Small city
Total Enrollment: 3,513
Undergraduates: 2,762
Male/Female: 42/58
SAT Ranges: V 540–660
  M 540–640
ACT Range: 22–27
Financial Aid: 39%
Expense: Pub $
Phi Beta Kappa: No
Applicants: 2,362
Accepted: 63%
Enrolled: 32%
Grad in 6 Years: 52%
Returning Freshmen: 76%
Academics: 🖊 🖊 🖊 🖊
Social: 🍷 🍷 🍷
Q of L: ★★★★
Admissions: (828) 251-6481
Email Address:
  admissions@unca.edu

is characterized by first-year and senior capstone liberal arts colloquia; a humanities core that addresses development and beliefs of Western and non-Western cultures; topical cluster courses in natural and social sciences; and courses in written communication, critical thinking, diversity, and information literacy. There are also requirements for foreign language study as well as health promotion and wellness.

The academic climate is demanding and students admit that it can be cutthroat at times. "The school is highly competitive," says a sophomore. "Each class on this campus is challenging and the professors do not slack off." Political science, humanities, and literature receive near-unanimous praise, and one student says the math department "is undoubtedly the strongest on campus." The most popular majors are psychology, management, literature, environmental studies, and art. A joint BS degree in engineering (with a concentration in mechatronics) has been established with North Carolina State University and is the only such program in the state. New majors include a BA in religious studies and an interdisciplinary studies major, which allows students to develop an individual degree program that transcends the scope of a single academic major.

The majority of the students commute from nearby communities, while 34 percent reside on campus.

"UNCA has the best dorms in the UNC system."

Asheville also offers 2–2 programs with NC State in engineering, forestry, and textile chemistry; study abroad is an option in Europe, Asia, Africa, and South America. The UNCA honors program offers special courses—as well as cultural and social opportunities—to motivated students who can make the grade. There are also ample opportunities for undergraduate research; in fact, nearly half of all students will have had an undergraduate research experience by graduation. Professors are given high marks and noted for their passion and experience. "I don't think I can voice how awesome the teaching has been during my four years here," says a senior. "They make class interesting and are always helping students outside of class."

The head count at Asheville has risen dramatically over the past decade, but only 13 percent of the student body comes from out of state. (The school limits its out-of-state admits to 18 percent.) Environmental causes, gay and lesbian issues, campus issues such as parking, and multiculturalism are a few of the buzzwords on campus. "A liberal arts school stimulates a liberal state of mind, but all social and political views are represented," says a sophomore. Currently, the student body is 90 percent white, 2 percent African American, 2 percent Hispanic, and 2 percent Asian American, but Asheville is making special efforts to bring more students who are "underrepresented" to the campus. Asheville offers 90 athletic scholarships in a variety of sports, as well as merit scholarships averaging $3,389.

"Of all the small liberal arts universities, our school is unique in our liberal, free-thinker mindset."

A joint BS degree in engineering (with a concentration in mechatronics) has been established with North Carolina State University and is the only such program in the state.

The majority of the students commute from nearby communities, while 34 percent reside on campus. Students can choose from air-conditioned suites in Mills Hall, double occupancy in the Founders Residence Hall, or singles in the wooded Governors Village complex. There is no lottery, and freshmen are mixed in with upperclassmen. "UNCA has the best dorms in the UNC system," raves one student. For meals, students may eat dining-hall fare or grab a bite at Cafe Ramsey. Vegetarian entrees are available at most meals, in addition to a salad and sandwich bar. Crime is nearly nonexistent on campus, thanks to the school's rural location. Still, emergency phones are available throughout the campus for an added measure of safety.

After class, there are plenty of opportunities for fun, especially for the many Asheville students with a hankering for the great outdoors. The college is surrounded by the Blue Ridge Mountains and the Smokies, where students can hike and rock climb; water buffs can go rafting on the nearby French Broad River. For students with cars, the Blue Ridge Parkway is a short drive away, while Spartanburg and Charlotte are one and two hours away, respectively. Real big-city action takes extra effort, though, since Atlanta is a four-hour trek. Asheville offers a tame but inviting nightlife, with popular hangouts like Boston Pizza, Urban Burrito, Tupelo Honey, and Rosetta's Kitchen. Most parties take place off campus, especially since RAs stalk underage drinkers in the dorms. "There is no tolerance for unsafe, underage, or unwise drinking," says a student. Three percent of the men and 2 percent of the women belong to fraternities and sororities, but their presence is not influential. There are more than 70 campus organizations, including a student newspaper, *The Blue Banner*.

Several campuswide events bring the school together each year, including Founders Day in October, homecoming, a spring fling, and a mock casino night with an auction. Greenfest, a semester-based environment and beautification project, is also very popular. "There are lots of annual events, but the one I feel makes our campus unique is the annual Greenfest," says a student. "All groups on campus—faculty, staff, and students—come together for two or three days to help make a designated section of campus more beautiful."

Involvement is no problem for the athletic teams. The Bulldogs boast Big South conference championship teams in volleyball, women's soccer, and baseball. Other competitive teams include men's soccer and basketball, and women's cross-country. Intramurals are at least as popular as the varsity sports. The Justice Center Sports Complex houses a pool, weight room, racquetball courts, and dance studio.

All the ingredients for a superior college experience lie in wait at Asheville: strong academics, dedicated professors, and an administration that continues to push for excellence. "Of all the small liberal arts universities, our school is unique in our liberal, free-thinker mindset," says a sophomore. "Our school is dedicated to creating well-rounded, intelligent individuals." It's a place to get the kind of liberal arts education usually associated with private colleges—but for a lot fewer greenbacks!

If You Apply To ➢ **UNC–Asheville:** Regular admissions: Feb. 16. Early action: Nov. 10. Financial aid and housing: Mar. 1. Does not guarantee to meet demonstrated need. Campus interviews: optional, evaluative. No alumni interviews. SATs or ACTs: required. SAT IIs: optional. Accepts electronic applications. No essay question.

University of North Carolina at Chapel Hill

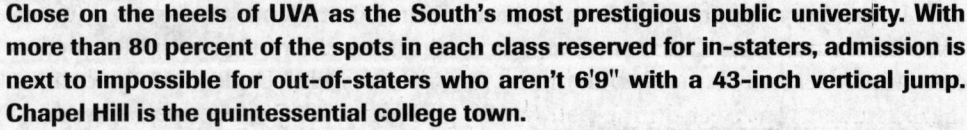

CB 2200, Jackson Hall, Chapel Hill, NC 27599-2200

Close on the heels of UVA as the South's most prestigious public university. With more than 80 percent of the spots in each class reserved for in-staters, admission is next to impossible for out-of-staters who aren't 6'9" with a 43-inch vertical jump. Chapel Hill is the quintessential college town.

Welcome to "the Southern part of heaven," a place where the sky is Carolina Blue and the academics are red-hot. As the flagship campus of the state university system and the first public university in the United States to open its doors, UNC–Chapel Hill has earned its place among the South's most prestigious universities. The atmosphere here is uniquely Southern, a rowdy mixture of hard work, sports fanaticism, and tradition that seems to attract bright, fun-loving students from everywhere.

UNC's gorgeous and comfortable campus occupies 730 acres lush with trees and lawns laced by 30 miles of brick-paved walkways. The architecture ranges from Pal-

Website: www.unc.edu
Location: Suburban
Total Enrollment: 22,395
Undergraduates: 15,898
Male/Female: 42/58
SAT Ranges: V 600–690
  M 610–700

(Continued)

ACT Range: 25–31

Financial Aid: 33%

Expense: Pub $ $

Phi Beta Kappa: Yes

Applicants: 18,414

Accepted: 37%

Enrolled: 56%

Grad in 6 Years: 84%

Returning Freshmen: 97%

Academics: 🏛 🏛 🏛 🏛

Social: 🍺 🍺 🍺 🍺

Q of L: ★ ★ ★ ★

Admissions: (919) 966-3621

Email Address:
uadm@email.unc.edu

Strongest Programs:
Biology
Chemistry
Classics
English
History
International and Area Studies
Philosophy
Political Science
Psychology
Sociology

ladian, Federal, and Georgian to postmodern, and redbrick is the prevailing motif. The original administration building is a replica of the central section of Princeton's gorgeous Nassau Hall, but sleek efficiency defines the latest architectural additions to campus, which include buildings for medical biomolecular research and for bioinformatics. The newly renovated Memorial Hall will serve as an anchor for the new Arts Common, which will encompass an area extending southward from Franklin Street to the Playmakers Theatre. The Rams Head Center, near Kenan Stadium, was designed to be environmentally sustainable, and the roof of its 700-space parking deck is a grassy plaza with brick-lined walkways.

Chapel Hill offers 71 undergraduate degree programs. Some of the strongest are sociology, philosophy, chemistry, business, political science, journalism and mass communications, classics, and biology. One of the most popular on-campus offerings is the small honors seminars open to all undergraduates. UNC's honors program is nationally recognized as among the best in the country. The Carolina Entrepreneurial Initiative (CEI), created in 2003 with a multimillion dollar grant from the Ewing Marion Kauffman Foundation, provides practical experiences that help students become more entrepreneurial; efforts include an entrepreneurship minor and the Carolina Challenge, a student-run competition that awards up to $50,000 in prizes each year for the best business plan.

> **"'College town' in the dictionary should show a picture of Chapel Hill."**

A newly approved general education curriculum requires all students to complete 42 credit hours of coursework in three broad themes: Foundations, Approaches, and Connections. Coursework includes physical and life sciences, social and behavioral sciences, humanities and fine arts, composition and rhetoric, foreign language, quantitative reasoning, and lifetime fitness. The academic climate is challenging but not overwhelming. "I definitely work hard to do well," says one student. Academic and social life are governed by a student-run honor system. The new office of Experiential Education has been established to create and promote the experiential learning available to undergraduates. In addition, "Maymester" was recently established to provide undergraduates with additional opportunities for work and study off campus. The three-week breaks occur during winter break and at the end of spring semester.

Registration is by Web, and seniority determines first access. If you get closed out of a class, "Be persistent," advises a freshman. "Email the professor. You can get in!" For those tired of the classroom rush, Research Triangle Park, a nearby research and corporate community and home of the National Humanities Center, employs many students as research assistants. UNC offers more than 230 study abroad programs in 70 countries. The Carolina faculty is, for the most part, top-notch. Professors keep regular office hours and welcome those students who seek them out. Academic counseling, once less than stellar, has been revamped.

"I am always impressed with how well-rounded students are," says a junior. "Academics are important, of course, but students consider other activities such as community service, art, and sports to be priorities as well." Under state guidelines,

As one of the best college buys in the country, the University of North Carolina at Chapel Hill gives students everything they want, both academically and socially.

> **"I am always impressed with how well-rounded students are."**

82 percent of UNC's freshman class must be state residents, and the admissions office has no problem filling this quota with the cream of the North Carolinian crop. Thus, unless you're an athlete, out-of-state admission is extremely tough. Big social and political issues on campus include multiculturalism, gender roles, local and national elections, and religious issues. A "diversity assessment process" was recently launched to focus on such issues as race and cultural differences. African Americans account for 11 percent of the student body, Asian Americans 6 percent, Native Americans 1 percent, and Hispanics 3 percent. This sports-minded school awards athletic scholarships in all of the major

sports. Students also can vie for thousands of merit scholarships worth an average of $5,308. The innovative Carolina Covenant program enables low-income students to graduate debt-free if they work on campus for 10 to 12 hours weekly in a federal work-study job instead of borrowing. Eligible students must come from families whose income is at or below 200 percent of the federal poverty level.

Forty-two percent of undergraduates live in university housing. "Our housing system is pretty darn good!" says a senior. Freshmen and returning students are guaranteed university housing, and returning students may reserve their rooms for the upcoming academic year. Housing on the north side of campus offers old and newly renovated dorms; the south side offers several brand-new housing options, which are a good hike from classroom buildings (not to worry—there's a free campus shuttle). Students may opt to be part of a living/learning community; house themes include foreign languages, substance-free, wellness, and women's issues. "The pickiest of the picky could be happy with Carolina dining services," says a freshman. "If you don't see it, you can request it. Great vegetarian options. Lots of variety." Campus security is praised for its constant presence on campus. "The university takes precautionary measures," says an education major. These include free bus and shuttle rides, emergency call boxes, and a fully accredited campus police department.

Chapel Hill offers 71 undergraduate degree programs. Some of the strongest are sociology, philosophy, chemistry, business, political science, journalism and mass communications, classics, and biology.

"'College town' in the dictionary should show a picture of Chapel Hill," boasts one senior. Franklin Street, the main drag in town that runs across the northern boundary of campus, offers Mexican and Chinese restaurants, ice cream parlors, coffee houses, vegetarian eateries, bakeries, a disco, and a generous supply of bars. Fraternities and sororities may account for only 11 percent of men and 12 percent of women, but they exert an influence far beyond their numbers. They don't, however, have a monopoly on fun, or on drinking. "Students over 21 can have alcohol in the dorms. If underage students are caught, they can be fined or punished. Some people get caught, lots don't. It's a game of chance. Getting served at the bars when underage is not nearly as easy," reports a journalism major. "Fallfest" kicks off the school year with an emphasis on the idea that you don't have to drink to have fun. Students look forward to several annual festivals: Apple Chill, Festifall, and the Carolina Jazz Festival. The North Carolina Literary Festival is held biannually. Students are involved in the community, many through a unique service-learning program for which they receive academic credit.

The varsity sports teams are extremely popular, especially football. The word "popular" doesn't do justice to the basketball games. A contest between the top NCAA rated Tar Heel Basketball Team with NC State makes any Carolina fan's heart beat faster, but Duke takes the prize as the most hated rival of all. The slam-dunking Tar Heels play in the 21,750-seat Smith Center, named for retired coach Dean Smith, who just happens to be the winningest college basketball coach of all time. In 2005, they sent the entire student body onto Franklin Street to celebrate winning the national Division I championship. Men's and women's basketball, lacrosse, field hockey and men's soccer have won national championships, but the team with one of the best records in college sports history

"Our housing system is pretty darn good!"

is women's soccer, which has won 18 national championships since 1981 and is virtually the only farm team for the U.S. women's team. Women also brought home national championships in field hockey and basketball, and conference championships in volleyball and track. In 2006, the men's varsity baseball team competed in the College World Series. A strong intramural program draws heavy participation. Those less competitive can enjoy the $4.9 million student recreation center, which includes a weight-training facility, an area for aerobic dance, and the student wellness center. Those who crave fresh air can take advantage of the Outdoor Education Center, which offers mountain bike trails, an 18-hole Frisbee golf course, rope courses, and the longest zipline in the U.S.

As one of the best college buys in the country, the University of North Carolina at Chapel Hill gives students everything they want, both academically and socially. The 200-year history of this school creates an atmosphere of extreme pride, a love of tradition, and monumental school spirit. One freshman, full of that school spirit, says, "Southern hospitality blended with a high level of thinking, an overwhelming dose of friendliness and pep, and a spectacularly gorgeous campus make Chapel Hill my favorite place in the world."

If You Apply To ➤

UNC–Chapel Hill: Early action: Nov. 1. Regular admissions: Jan. 15. Financial aid: Mar. 1. Housing: May 1. Guarantees to meet full demonstrated need. No campus or alumni interviews. SATs or ACTs: required (with writing). SAT IIs: optional. Encourages electronic applications. Essay questions.

University of North Carolina at Greensboro

1000 Spring Garden Street, Greensboro, NC 27412

UNCG is a medium-sized alternative in the UNC system—half the size of Chapel Hill and three times bigger than Asheville. UNCG began its life as a women's college and remains about two-thirds female. Residential College program offers a first-rate living/learning option.

Website: www.uncg.edu
Location: Center city
Total Enrollment: 11,783
Undergraduates: 10,569
Male/Female: 32/68
SAT Ranges: V 470–580
  M 470–580
Financial Aid: 40%
Expense: Pub $
Phi Beta Kappa: Yes
Applicants: 8,987
Accepted: 60%
Enrolled: 30%
Grad in 6 Years: 51%
Returning Freshmen: 78%
Academics: ✍ ✍
Social: ☎ ☎
Q of L: ★ ★ ★
Admissions: (336) 334-5243
Email Address:
  admissions@uncg.edu

Strongest Programs:
Art
Dance
Music

At the University of North Carolina at Greensboro, an aggressive campaign is underway to make the school more "student centered." Living Learning Communities have been developed to help meet the needs of freshmen and make the matriculation experience less daunting. In addition, UNCG continues to expand the research opportunities available to undergraduates. "Being a student at UNCG is so much more than going to class, studying, and getting a degree," says a senior. It may be the only school in the country where the acceptance letters come with a gold star on the envelope.

Set on 200 acres sprinkled with magnolia and dogwood trees, the well-landscaped campus features a mix of Colonial, Georgian, brick, and modern architecture. Still standing is the original university building, the Victorian-style Julius Foust Building; built in 1892 and now on the National Register of Historic Places, it is located on a knoll in the center of the campus' original 10 acres. The new, $40 million Science Building offers state-of-the-art facilities for teaching and research. The Hall for Humanities, Research, and Administration houses classrooms and research space. Under construction is a studio arts center, which will house the art and interior architecture departments.

Within the College of Arts and Sciences, psychology, fine arts, and literature are strong. The university's program in human environmental sciences is also highly regarded, and business, nursing, elementary education, and biology are the most popular majors. The School of Music has three ensembles, a symphony orchestra, three choral groups, and has won the National Opera Association's production competition three of the past five years. There are some unusual interdisciplinary programs, such as therapy training, which combines classes from the dance, education, fine arts, and theater departments. Greensboro's program in human environmental sciences is

> **"Being a student at UNCG is so much more than going to class, studying, and getting a degree."**

North Carolina's largest. Core requirements include 37 hours of courses in six categories (humanities and fine arts, historical perspectives, social and behavioral sciences, natural sciences, math, and reasoning and discourse.) All majors also require writing-intensive and speaking-intensive courses.

(Continued)
Creative Writing
Theater
Education
Nutrition
Psychology

The academic climate is rigorous and "very competitive," says a sophomore. Students take their education seriously, and so do the professors. "The teachers push you as well as support you," says a junior. Enrollments in introductory courses sometimes swell to more than 100, but preregistration is done through an online computer system. Those who make it into the residential college program enjoy the atmosphere of an intimate "academic community" with class sizes usually ranging from 15 to 20 students. The school's residential college is among the nation's oldest living/learning programs. The honors college allows tal-

"Here a student is independent, political, and friendly."

ented students the opportunity to tackle a broad interdisciplinary program through small seminars, while undergraduate research assistantships allow 65 students to work with faculty in all fields. To emphasize the importance of writing both as an essential skill and as a tool for learning, all students must take writing-intensive courses.

A summer orientation program allows freshmen to get their feet wet before classes begin. Each academic year starts off with the Fall Kickoff, when campus organizations line College Avenue with the trappings of their activities, creating a festival atmosphere. UNCG has more than 170 student organizations, including club sports, religious groups, service organizations, media groups, national societies, and professional organizations.

Men's basketball and tennis and women's soccer have all brought home conference championships and continue to field solid squads.

Eight percent of undergrads come from outside North Carolina, mostly from the South. African Americans account for 20 percent of the student body, and the Neo-Black Society is very active on campus. Hispanics, Native Americans and Asian Americans combine to make up 6 percent of the student body. According to a junior, "The students at UNCG are an eclectic group. Here a student is independent, political, and friendly. There are so many different types, unlike private or well-known universities." Outstanding students can vie for thousands of scholarships worth an average of $3,909; athletic scholarships are also offered.

Thirty-two percent of the undergraduates live in Greensboro's 23 residence halls, but students note that getting a room can be a challenge. "The past few years have been cramped and we have been experiencing a housing shortage," says a student. "But the rooms are OK." The Tower Village Apartments provide suite-style living on campus for 300 lucky students who get private bedrooms. Rooms in the older buildings are spacious, those in the quad are more attractive, and the most modern, high-rise dorms offer cramped—but air-conditioned—quarters. Four of the 23 dorms are single-sex, and at the beginning of the year, each residence hall votes on guidelines establishing the visitation policy for members of the opposite

"Greensboro is a wonderful old Southern town."

sex. For those who still want to avoid institutionalized living, off-campus housing is plentiful and cheap. Students who live on campus choose from a variety of meal-plan options in the university dining hall or in specialty shops. Administrators boast that the newly renovated student center is one of the nicest centers in the state.

Eight fraternities and 10 sororities attract less than 10 percent of the student population. For others, dances, coffee houses, concerts, movies, and other social activities pick up some of the social slack. "Greensboro is a wonderful old Southern town," reports one student. Women outnumber men nearly two to one, primarily because nearby NC State gets all the engineering students. Bars, restaurants, and stores are within walking distance, and the 23,000-seat Greensboro Coliseum, a scant two miles away, regularly plays host to rock bands and athletic events. Weekend trips are to the beach (three hours) or the mountains (two hours). "UNCG is a

Those who make it into the residential college program enjoy the atmosphere of an intimate "academic community" with class sizes usually ranging from 15 to 20 students.

good college for fun. It is not a party school but there is always something going on," a communication studies major says.

UNCG teams compete in Division I, and the university offers a comprehensive athletic program as part of the Southern Conference. Men's basketball and tennis and women's soccer have all brought home conference championships and continue to field solid squads. Intramurals are also popular.

Some students say that Greensboro is becoming somewhat of a commuter school, though it may be too early to tell. Others say the only problem is that the school, living in the shadow of its big sister at Chapel Hill, lacks the reputation it deserves for providing a first-rate education in such diverse fields as liberal arts, nursing, and education.

If You Apply To ➤ **UNC–Greensboro:** Rolling admissions. Regular admissions: Mar. 1. Does not guarantee to meet demonstrated need. Campus and alumni interviews: optional, informational. SATs or ACTs: required (SATs preferred). SAT IIs: optional. Accepts Common Application and prefers electronic applications. No essay question.

North Carolina State University

Box 7103, Raleigh, NC 27695-7103

It is hard for NC State not to have an inferiority complex next to high falutin' neighbors like Duke and UNC. But having them in the neighborhood is also a blessing—just ask the thousands of graduates who have gotten jobs in the Research Triangle. Engineering and business are the most popular programs.

North Carolina State is one of the bright leaves of the Tobacco Belt. Whether you're looking for a stellar education or a top-rated basketball program, NCSU offers students the benefits of a large school—highly regarded professors, a diverse student body, and plenty to do on weekends—while making sure that no one feels left out. Says one student, "No matter how weird or crazy you are, there is someone just like you on campus."

The 107-year-old, 1,900-acre campus consists of redbrick buildings, brick-lined walks, and cozy courtyards dotted with pine trees. There is no dominant style, but more of an architectural stream-of-consciousness that reveals a campus that grew and changed with time. New campus additions include residence halls, student apartments, and an engineering building.

NCSU excels in the professional areas of engineering, pulp and paper science, statistics, design, agriculture, and forestry, which are the largest and the most demanding divisions. Not surprisingly, given its location in the heart of textile country, the school also boasts a first-rate textile school, the largest and one of the best such programs in the country. Business management tops the list of most popular majors, followed by biological science and engineering. Even the most technical of majors requires students to take a broad range of liberal arts courses, although the humanities are far from the strongest programs on campus. Programs in engineering mechatronics (with UNC Asheville), international studies, sport management, Africana studies, and turfgrass science have been added to the curriculum. General education requirements include 20 hours of math and natural sciences, 21 hours of humanities and social sciences, seven hours of writing, speaking, and information literacy, three hours

Website: www.ncsu.edu
Location: City suburbs
Total Enrollment: 22,875
Undergraduates: 19,226
Male/Female: 57/43
SAT Ranges: V 530–620
   M 560–660
ACT Range: 23–27
Financial Aid: 38%
Expense: Pub $
Phi Beta Kappa: Yes
Applicants: 13,610
Accepted: 66%
Enrolled: 47%
Grad in 6 Years: 71%
Returning Freshmen: 89%
Academics: 🐛🐛🐛
Social: 🐷🐷🐷
Q of L: ★★★
Admissions: (919) 515-2434
Email Address: undergrad
   _admissions@ncsu.edu

of science, technology, and society, two hours of physical education, and computer literacy courses. Students also must demonstrate proficiency in a foreign language.

"The academic atmosphere is competitive," says a senior. "The students are constantly pushing themselves to excel in the lecture hall as well as with undergraduate research." An important feature of NC State's approach to education is the cooperative education program, through which students in all schools can alternate semesters of on-site work with traditional classroom time. There are also domestic and international exchanges with more than 97 countries, and a Residential Scholars program in

"No matter how weird or crazy you are, there is someone just like you on campus."

which academic standouts live together and participate in weekly activities such as guest lectures. A First Year College program provides guidance and counseling for incoming students to introduce them to all possible majors. Many classes at NC State are large, but the faculty gets high grades for being accessible, interested in teaching, and friendly. "Teachers really do care about meeting the students, getting to know them, and offering help in any way that they can."

The university benefits greatly from its relationships with Duke, the University of North Carolina at Chapel Hill, and private industry through the state's high-tech Research Triangle Park. Its star continues to rise as it becomes more selective. The students at NC State are largely hardworking, bright North Carolinians. Some 92 percent are in-state students. "Everyone is accepted and should easily find a group of students to fit in with," says a student. Seventy-eight percent graduated in the top quarter of their high school class, and 91 percent attended public high school. Ten percent of the student body is African American, while Hispanics and Asian Americans make up another 7 percent. Native Americans and foreign students comprise 1 percent each. Amid the public school diversity, conservatism abounds, and the largest political organization is the College Republicans. Jocks and sports fans are visible, and the university offers more than 400 scholarships for men and women. Those with outstanding academic qualifications can compete for one of thousands of merit scholarships averaging $7,344. To be considered for merit awards, students must file a separate

"We have an ongoing rivalry with the University of North Carolina."

application in the early fall of their senior year in high school. The Pack Promise guarantees that the university's neediest students will have 100 percent of their financial need met through a combination of scholarships, grants, work-study employment, and need-based loans.

As for housing, 34 percent choose to stay on campus. All students are guaranteed rooms for all four years. Sullivan and Lee are recommended for freshmen because they provide a mixture of academic and social activities. Rooms range in size from spacious to cramped. Students report that dorm dwelling is actually more expensive than several off-campus units. Off-campus housing and social activities are plentiful. A small percentage of students are housed in fraternities and sororities, and the international house is also an option. The dining halls feed all freshmen and anyone else who cares to join the meal plan. "Dining hall staples include a hamburger grill, salad bar, and pasta and sandwich stations," says a student.

As with many other schools, alcohol policies are not strictly enforced. The 21 fraternities and five sororities attract 9 percent of the men and 10 percent of the women, respectively. The Greek scene provides much of the entertainment, but dorm and suite parties are also popular. Public transportation affords easy access to downtown, with its shops, restaurants, theaters, and nightspots. The university is well integrated into Raleigh, and its proximity to three all-women's colleges helps alleviate the imbalance of the three-to-two male/female ratio. Annual events include Wolfstock (a band party) and an All-Nighter in the student center. Many

(Continued)
Strongest Programs:
Design
Statistics
Engineering
Pulp and Paper Science
Agriculture
Forestry
Textiles

The annual State versus Carolina football game usually packs the stadium, and the never-ending fight to "Beat Carolina!" permeates the campus year-round.

An important feature of NC State's approach to education is the cooperative education program, through which students in all schools can alternate semesters of on-site work with traditional classroom time.

Even the most technical of majors requires students to take a broad range of liberal arts courses, although the humanities are far from the strongest programs on campus.

students also like to head to the beach, which is less than two hours away, or to the mountains for skiing, which is about a three-and-a-half-hour trip.

With home close by for so many students, the campus does tend to thin out on weekends. Those who stay can cheer on the home teams, which do well in men's tennis, wrestling, football, and men's and women's cross-country and track and field. But needless to say, basketball reigns supreme. The Wolfpack plays in the high-powered Atlantic Coast Conference. "We have an ongoing rivalry with the University of North Carolina," says a junior. Some crazy NC State fans have stormed nearby Hillsborough Street following game-day victories. The annual State versus Carolina football game usually packs the stadium, and the never-ending fight to "Beat Carolina!" permeates the campus year-round. Intramurals also thrive, and a particularly popular event is Big Four Day, when NC State's intramural teams compete against their neighbors at Duke, Wake Forest, and UNC–Chapel Hill.

> **"The students are constantly pushing themselves to excel."**

North Carolina State seems to have overcome many of the obstacles associated with large land-grant universities. It has attracted a dedicated and friendly student body independent enough to deal with the inevitable anonymity of a state school, but spirited enough to cheer the Wolfpack to victory. NC State works well for both those who can shoot hoops and those who can calculate the trajectory of the same three-point shot.

Overlaps

UNC–Chapel Hill, East Carolina, UNC–Wilmington, Duke, Wake Forest, Virginia Tech

If You Apply To ➤   **NC State:** Early action: Nov. 1. Regular admissions: Feb. 1. Financial aid: Mar. 1. Does not guarantee to meet full demonstrated need. Campus interviews: optional, informational. No alumni interviews. SATs or ACTs: required. SAT IIs: recommended. Prefers electronic applications. Essay question: personal statement.

Northeastern University

360 Huntington Avenue, 150 Richards Hall, Boston, MA 02115

Northeastern is synonymous with preprofessional education and hands-on experience. By interspersing a co-op job with academic study, students can rake in thousands while getting a leg up on the job market. With Boston beckoning, campus life is minimal.

Website:
www.northeastern.edu
Location: Urban
Total Enrollment: 19,541
Undergraduates: 14,730
Male/Female: 49/51
SAT Ranges: V 560–650
M 580–670
ACT Range: 24–28
Financial Aid: 60%
Expense: Pr $ $
Phi Beta Kappa: No

Northeastern University is revamping its image. Long known for its co-op program and hands-on learning experiences, this private university has set its sights on becoming one of the region's top-tier institutions. NU is more selective than ever, and there is a real push to add new facilities and big-name professors, and to foster an atmosphere that encourages students to stick around despite the sometimes impersonal feel on campus. With a strong emphasis on combining liberal arts requirements with up to 18 months of challenging work placements, students come out of this school not only well rounded, but also ready to take on real world responsibilities.

Northeastern's 67-acre campus is an urban oasis located in the heart of Boston, just minutes away from Fenway Park, shopping centers, nightclubs, cafes, Symphony Hall, and the Museum of Fine Arts. The campus's green spaces are interspersed with brick walkways, outdoor art, and a sculpture garden. Older buildings sport utilitarian gray-brick architecture while newer structures are of modern glass and brick design. During inclement weather, students can be found traversing the underground tunnel system that connects many campus buildings. The newest additions to campus include an updated home for the health and counseling services, an alumni center, and a new building containing housing, classrooms, and the university's African-American Institute.

Northeastern's academic calendar consists of 15-week semesters in the fall and spring and two shorter sessions in the summer. Full-time undergraduates will typically take classes for eight semesters and complete a total of three six-month co-ops during their five years at Northeastern. For students in the College of Arts and Sciences, there are many alternatives to the co-op, such as internships, study abroad, and undergraduate research opportunities. Each of the six colleges presents its own core curriculum, but all freshmen must complete English and diversity requirements. Freshmen are also required to attend the summer orientation program. An honors program is open to top students. "The courses are rigorous and students take academics seriously," explains a sophomore, "but the competition varies from major to major."

Woven into the Northeastern experience is its co-op education program, which enables students to sample professional careers prior to graduation and see how their classroom studies connect with the real world. Co-op assignments are paid, full-time positions related to the student's major and personal interest. **"Our college is a place for go-getters."** Ninety-six percent of students participate in co-op before they graduate and, each year, approximately 5,400 students work at a co-op position. Students warn, however, against becoming overly reliant on co-op advisors for good jobs or good advice. Plus, some students say that when the economy is on a downswing, co-op jobs are tougher to come by. Undergrads say that most of their teachers are concerned with their needs, and most of Northeastern's classes have 49 or fewer students. A student says, "Most professors are receptive, challenging, and fair" Scheduling can be difficult, as students sometimes find that courses they want are offered only when they're scheduled to be away on a job.

"Our college is a place for go-getters," observes a physics major, "students who are ready to take on a big city and the responsibility of 'real world' jobs." Northeastern, which was founded as a YMCA educational program, has traditionally served local students from diverse socioeconomic backgrounds. But that has changed—now, only 34 percent of the students are from Massachusetts. African Americans account for 7 percent of the student population, while Hispanics make up 6 percent, Asian Americans 8 percent, and foreign students 5 percent. Because so many participate in the co-op program, many "don't have a good connection to the university," says one student. Besides the money they can earn in co-op programs, outstanding students can compete for more than 3,900 merit scholarships averaging $13,717, and there are over 250 athletic scholarships—at least one for each sport.

Although 95 percent of freshmen live on campus, upperclassmen have a much harder time finding a place to sleep on campus. Some of the dorms have been showing their age, but NU recently completed several new residence halls, including the West Village complex, a 13-story apartment megaplex that houses up to 600 students. "Freshman dorms are pretty standard compared to any other school," admits one junior. "The upperclassman dorms are almost all new or renovated, and they're fantastic!" **"The upperclassman dorms are almost all new or renovated, and they're fantastic!"** Off-campus options include privately owned apartments or suites located adjacent to the residence halls. The dining halls offer "many, many choices," according to students, who also speak highly of NU Public Safety.

The co-op program puts a strain on campus social life. There are many clubs and activities, but the continuous flow of students on and off the campus tends to be

(Continued)
Applicants: 25,467
Accepted: 47%
Enrolled: 24%
Grad in 6 Years: 61%
Returning Freshmen: 90%
Academics: ✍ ✍
Social: ☎ ☎
Q of L: ★ ★
Admissions: (617) 373-2200
Email Address:
   admissions@neu.edu

Strongest Programs:
Art
Architecture
Psychology
Engineering
Computer Science
Business
Health Sciences
Criminology

Northeastern, which was founded as a YMCA educational program, has traditionally served local students from diverse socioeconomic backgrounds.

Long known for its co-op program and hands-on learning experiences, this private university has set its sights on becoming one of the region's top-tier institutions.

disruptive. "I may see a friend one quarter in class and then not again for six months. It's hard to stay connected," a student explains. Fraternities and sororities attract 4 percent of NU men and women, respectively. Those who are not in the Greek system find Boston with all its attractions to be a fully acceptable substitute. Many of the students call Boston "THE college town."

In the winter, students head to the ski slopes of Vermont, and in balmier weather they're off to the beaches of Cape Cod and the North Shore. Not surprisingly, in the city the Celtics made famous, the basketball team attracts adoring fans. But the biggest sports series of the year is the Beanpot Hockey Tournament, which pits Northeastern against rival teams from Harvard, Boston College, and Boston

> **"We are employees at co-op, students in class, friends and roommates in our free time."**

University. Northeastern's female pucksters have frequently prevailed as champs. And the fleet-footed men's and women's track and cross-country teams, who work out in the Bernard

Solomon Indoor Track Facility, regularly leave their opponents blinking in the dust. One T-shirt reads, "No—we don't want to B.U.", epitomizing the competitive nature of the sports teams, especially toward those in the Boston area.

Perhaps now, more than ever, a school that truly prepares students to enter the working world is needed. "Northeastern students are a different breed," says a senior. "The typical Northeastern student wears many hats. We are employees at co-op, students in class, friends and roommates in our free time. We balance work and play while still meeting deadlines." explains a student. By the time Northeastern students graduate, they have built up a broad reservoir of experiences that they know will serve them well once they start scouring those job listings.

If You Apply To ➤ **Northeastern:** Early action: Nov. 15. Regular admissions: Jan. 15. Financial aid: Feb. 15. Housing: May 1. Does not guarantee to meet full demonstrated need. Campus interviews and alumni interviews: optional, informational. SATs or ACTs: required (SATs preferred). Accepts the Common Application and prefers electronic applications. Essay question: issue of concern and its impact on you; fiction or nonfiction character with admirable values; experience, achievement, or event that has affected your life.

Northwestern University

1801 Hinman Avenue, P.O. Box 3060, Evanston, IL 60204-3060

The Big Ten is not the Ivy League, and NU has more school spirit than its Eastern counterparts. Much more preprofessional than its nearby rival University of Chicago or any of the Ivies except Penn. World renowned in journalism.

Website:
  www.northwestern.edu
Location: Suburban
Total Enrollment: 16,032
Undergraduates: 7,840
Male/Female: 48/52
SAT Ranges: V 640–730
  M 660–750
ACT Range: 28–33
Financial Aid: 60%

On Sunday nights before finals begin at Northwestern University, students are encouraged to let off steam with a campuswide "primal scream." The ear-shattering event illustrates two big themes at NU: students work really hard, but they also know how to have some fun. Regarded as the most elite school in the Midwest, this top-tier university, the only private in the Big Ten, boasts some of the most well respected preprofessional programs in the country. Plus, Northwestern is ideally located just outside of Chicago. "I love being at a place where I can learn and have a great social life," says one student.

Northwestern is situated on 231 acres about a dozen miles north of the Chicago Loop. An eclectic mix of stone buildings with abundant ivy, the leafy campus is set off from the town of Evanston and runs for a mile along the shore of Lake Michigan.

Students migrate between the North Campus (techy) and the South Campus (arty). The newer buildings are located adjacent to a 14-acre lagoon, part of an 85-acre lake-fill addition built in the '60s. This area provides students with a prime location for picnicking, fishing, running, cycling, rollerblading, or just daydreaming. Among the newest additions to the campus are the McCormick Tribune Center, a state-of-the-art broadcast and multimedia center; the 84,000-square-foot Center for Nanofabrication and Molecular Self-Assembly; and the Ford Motor Company Engineering Design Center.

Half of Northwestern's undergraduates are enrolled in arts and sciences, while the other half are spread out among five professional schools, all with national reputations. The Medill School of Journalism, the only such program at a top private university, teaches students how to tell stories across all media platforms and features internships at dozens of top newspapers, magazines, and television stations across the nation. There's also a four-year accelerated BSJ/MSJ program. A dazzling electronic studio centralizes Medill's state-of-the-art broadcast newsroom and the communication school's radio/TV/film

"I love being at a place where I can learn and have a great social life."

department. The McCormick School of Engineering and Applied Science is strong in all aspects of engineering and pairs students with clients with practical problems. Five-year co-op options are available. The School of Music wants students who can combine conservatory-level musicianship with high-level academics. It requires auditions and offers a five-year program from which students emerge with two BA degrees. The School of Communications takes an academic approach to subjects such as communications disorders and has the only theater school at an elite university—liberal arts through the vehicle of theater. It competes head to head with the likes of the University of Southern California and New York University. The School of Education and Social Policy is the only school of its kinds in the country and competes with Vanderbilt for education majors. Students and faculty members alike are encouraged to range across traditional disciplinary barriers—a policy that has lead to the creation of some entirely new fields such as materials science—and even to switch schools once they are enrolled. Consistent with this approach, students say the university's best programs include the Integrated Science Program, the Honors Program in Medical Education, and Mathematical Methods in Social Sciences, a selective program that gives students the technical skills to move into various areas of the social sciences. An undergraduate business certificate is in the works. Each of the undergraduate schools determines its own general education requirements, but the distribution requirements are similar. Each school requires a graduate to have coursework in "the major domains of knowledge"—science, mathematics and technology, individual and social behavior, historical studies, values, the humanities, and the fine arts.

Unlike most schools on a 10-week quarter system, Northwesterners take four (not three) courses each quarter, except in engineering, where five are permitted. "There is certainly a lot of work to do, so budgeting your time is important," a freshman advises. A sophomore adds, "Because of the quarter schedule, more material gets packed into

"Because of the quarter schedule, more material gets packed into one class."

one class." Students can take a break from the campus through any of the numerous field-study programs and programs abroad. Perhaps one reason students study so hard is the motivation provided by their professors. "These teachers are very passionate about what they do and, let me tell you, it's infectious," says a student. In addition, "profs tend to genuinely care about the students," a religion major says. Virtually all undergraduate courses, including required freshman seminars (of 10 to 15 students) in arts and sciences, are taught by regular faculty members. Introductory

(Continued)

Expense: Pr $ $ $
Phi Beta Kappa: Yes
Applicants: 16,221
Accepted: 29%
Enrolled: 41%
Grad in 5 Years: 93%
Returning Freshmen: 97%
Academics: 🖉 🖉 🖉 🖉 🖉
Social: ☎ ☎ ☎
Q of L: ★ ★ ★
Admissions: (847) 491-7271
Email Address: ug-admission
  @northwestern.edu

Strongest Programs:
Chemistry
Engineering
Economics
Journalism
Communications Studies
History
Political science
Theater
Music

Northwestern occupies a unique niche in U.S. higher education. It has the academics of the Ivies, the spirited atmosphere of the Big Ten publics, and combines success in Division I sports with quality instruction.

Fiske Guide to Colleges 2008   **NORTHWESTERN UNIVERSITY 473** courses are larger than most, but the average 100-level class has about 30 students. Strong arts and sciences departments include chemistry, nano sciences, economics, history, and political science. The humanities as a group are less strong.

"Students here work very hard and tend to be incredibly involved on campus," explains a junior. "Most belong to some extracurricular group." More than 40 percent of the student body comes from the Midwest, but California and Florida are heavily represented. Minorities represent 30 percent of the student body, with Asian Americans accounting for 17 percent, African Americans 6 percent, and Hispanics 7 percent. Students tend to be very well-rounded. NU is a liberal-leaning campus, but the abundance of preprofessionals has added to its image as "young corporate America." In recent years, the acceptance rate for medical school applicants has been over 60 percent, while that of graduates at schools of business and law has hovered around 90 percent. There are no academic merit scholarships, but NU does guarantee to meet the full demonstrated need of every admit, and as a Division I school, it offers a number of scholarships for its athletes. In addition to the usual loan sources, the university offers its own loan program.

There are a variety of housing options, the maajority of which are badly in need of refurbishing. "Dorms range from incredibly nice to dilapidated," a religion major confirms. Most rooms are doubles, but there are also singles, triples, and quadruple rooms as well as suites. Several residential colleges bring students and faculty members together during faculty "firesides" or simply over meals. The newer Slivka Hall houses the engineering residential colleges; there also are special dorms for students in communications, international studies, humanities, commerce and industry, performing arts, and

"Students here work very hard and tend to be incredibly involved on campus."

public affairs. Fraternities and sororities also have their own houses. Students can choose to eat at the coffeehouse or at any one of six dining halls on campus. A variety of meal plans are available, including one that provides Sunday brunch and one offering Kosher food. The 30 percent of students who have fled to off-campus apartments are mostly junior and seniors. Students generally feel safe on campus, but crime has been a concern in Evanston, especially after dark.

Nonetheless, "Evanston is enough of a college town," says a freshman. Evanston itself is "the restaurant haven of Chicago's north shore," says a junior, "very quaint with tons of flowers that bloom every spring." For a night out, of course there is that "toddlin' town," Chicago, right across the border. A short stroll off campus brings you to the town's myriad restaurant options, trendy bars, and coffee shops with space to plug in a laptop and study.

Much of the social life on NU's campus is centered on the Greek system, and 30 percent of the men and 40 percent of the women go Greek. For non-Greeks, on-campus entertainment opportunities are numerous, including theater productions, concerts, and movies. The school's alcohol policy is tough, but not always effective, "like the vast majority of campuses nationwide," says a student. The student government and Activities and Organizations Board sponsors an array of campuswide events, such as the very popular 30-hour Dance Marathon and Dillo (Armadillo) Day, an end-of-the-year "festival of music, debauchery, and Greek life," in the words of a journalism major. Another tradition is upheld when representatives of student organizations slip out in the dead of night to paint their colors and slogans on a centrally located rock. Northwestern has the winningest debate team in the country. As a bonus to the social atmosphere as well as to educational hands-on experience, the campus has its own radio station, television studio, and award-winning newspaper, and, says one administrator, "Certainly any student who wishes to act, produce, direct, conduct, build scenery, or play in a musical ensemble has ample opportunity to do so."

Football and tailgate parties are a traditional way of bringing alumni back and rousing the students to support the smallest and only private school in the Big Ten. Although the football team's trip to the 1996 Rose Bowl remains the stuff of legend, NU tends to be strongest in country club sports. Championship teams in recent years have included men's golf and women's tennis. Other competitive women's teams include volleyball, softball, field hockey, swimming, and NU's newest varsity sport: women's lacrosse. As far as facilities, NU is on par with many schools its size and larger, with the beautiful Norris Aquatics Center/Henry Crown Sports Pavilion and the Nicolet Football and Conference Center, used for conditioning of varsity athletes. The student-sponsored intramural program provides vigorous competition among teams from dorms and rival fraternities.

Northwestern occupies a unique niche in U.S. higher education. It has the academics of the Ivies, the spirited atmosphere of the Big Ten publics and, along with Duke, Stanford and perhaps Vanderbilt, combines success in Division I sports with quality instruction. With a strong work ethic, and an equally strong desire for play, Northwestern students bask in their school's balance of challenging academics, preprofessional bent, and myriad opportunities to get off campus to learn and let loose. "I'm able to have fun in a learning environment," says a biomedical engineering major. "The friends I've made here also make this a place that will be hard to leave."

If You Apply To >

Northwestern: Early decision: Nov. 1. Regular admissions: Jan. 1. Financial aid: Feb. 1. Housing: May 27. Guarantees to meet demonstrated need. Campus and alumni interviews: optional, informational. SATs or ACTs: required. SAT IIs: recommended. Accepts electronic applications. Essay question: varies every year, pick one of four.

University of Notre Dame

220 Main Bldg., Notre Dame, IN 46556

The Holy Grail of higher education for many Roman Catholics. ND's heartland location and 82-percent-Catholic enrollment make it a bastion of traditional values. Offers business and engineering in addition to the liberal arts. ND's personality is much closer to Boston College than Georgetown.

Founded 163 years ago by the French priest Edward Sorin, the University of Notre Dame has come a long way from its fledgling days in a rustic log cabin. While described as "a Catholic academic community of higher learning," its students need not be affiliated with the Roman Catholic Church. According to the administration, "What the university asks of all its scholars is not a particular creedal affiliation, but a respect for the objectives of Notre Dame and a willingness to enter into the conversation that gives it life and character." A soft spot for football doesn't hurt either.

With 1,250 acres of rolling hills, twin lakes, and woods, the university offers a peaceful setting for studying. The lofty Golden Dome that rises above the ivy-covered Gothic and modern buildings and the old brick stadium, where in the 1920s Knute Rockne made the Fighting Irish almost synonymous with college football, are national symbols. Newer parts of campus are the Marie P. DeBartolo Center for the Performing Arts, Hammes Mowbray Hall, and the Jordan Hall of Science.

Website: www.nd.edu
Location: City outskirts
Total Enrollment: 11,185
Undergraduates: 8,253
Male/Female: 53/47
SAT Ranges: V 630–730
  M 650–740
ACT Range: 30–33
Financial Aid: 48%
Expense: Pr $ $ $
Phi Beta Kappa: Yes
Applicants: 11,317
Accepted: 32%

From Knute Rockne and the Gipper right on down to modern-day greats such as Joe Montana, the spirit of Notre Dame football reigns supreme.

In the College of Arts and Letters, highly regarded departments include English, theology, and philosophy, while physics and chemistry are tops in the College of Science.

Liberal education is more than just a catchphrase at Notre Dame. No matter what their major, students must take the First Year of Studies, one of the most extensive academic and counseling programs of any university in the nation. The core of the program is a one-semester writing-intensive university seminar limited to 20 students per section. The remainder of each freshman's schedule is reserved for the first of a comprehensive list of general education requirements: one semester each in writing and mathematics and two semesters in natural science, as well as one semester chosen from theology, philosophy, history, social science, and fine arts. It also includes a strong counseling component in which peer advisors are assigned to each student, as are academic advisors and tutors if necessary. Administrators are quick to point out that, due in part to the success of the first-year support program, a whopping 97 percent of the freshmen make it through the year and return for sophomore year. Ninety-six percent of freshmen graduate within six years.

In the College of Arts and Letters, highly regarded departments include English, theology, and philosophy, while physics and chemistry are tops in the College of Science. Within the engineering school, chemical engineering rules. The College of Business Administration's accountancy program is ranked among the nation's best, and the chemistry labs in the Nieuwland Science Hall have first-rate equipment. While weak departments are hard to come by at Notre Dame, some students complain about the creative arts programs. The academic climate at Notre Dame is said to be fairly rigorous. "The workload is very demanding," says a senior. "It requires the student to have very good time-management skills." And while the atmosphere is competitive, students agree that it is not cutthroat by any measure. Faculty members are praised for being dynamic, personable, knowledgeable, and accessible. "The professors here care a great deal about their students and it shows," says a biology major. Students report it can be hard to get all the classes they want during a particular semester, but say it's not difficult to graduate in four years.

"The workload is very demanding."

Notre Dame offers a variety of special academic programs and options. One of the most popular is the Program of Liberal Studies (PLS), in which students study art, philosophy, literature, and the history of Western thought within their Great Books seminars. The Kaneb Center for Teaching and Learning, the university's most recent commitment to teaching, is based in DeBartolo Hall, an 84-classroom complex with state-of-the-art computer and audiovisual equipment. The Arts and Letters Program for Administrators combines a second business major with liberal learning, and the College of Science also gives students the option of pursuing majors in two departments. In addition, Notre Dame offers programs in military and naval science, aerospace studies, and an international study program that allows students to travel to numerous countries.

With a predominantly lay board of trustees and faculty, Notre Dame remains committed to "the preservation of a distinctly Catholic community." The president and several other top administrators are priests of the Congregation of the Holy Cross, and each dorm has its own chapel with daily Masses, though attendance is not required. Nearly 82 percent of the students are Catholic, and students are required to take two theology courses. The main social issues discussed on campus include abortion, gender and racial issues, homosexuality, and faith. Diversity is also a concern and some students feel it is a big problem. But administrators are addressing it, and minority enrollment is growing. African Americans and Hispanics together make up 12 percent of the student body and Asian Americans constitute another 5 percent. Despite its relative cultural homogeneity, Notre Dame recruits from all over the country; 87 percent of the students are from outside Indiana and 4 percent hail from other countries. The university offers competitive academic schol-

"Notre Dame dorm life is extraordinary."

arships to students with outstanding high school records and financial need, and more than 300 athletic scholarships are available.

Dorm life at Notre Dame appeals to three-quarters of the students. "Notre Dame dorm life is extraordinary," says a junior. "The dorm rooms are all very well-kept and very comfortable." Students are assigned to a dorm for their freshman year, and are encouraged to stay in the same one until graduation. Fraternities are banned, and freshmen are spread out among all campus dorms. The single-sex dorms really become surrogate fraternities and sororities that breed a similar spirit of community and family. Parietal rules (midnight on weekdays, 2 a.m. on weekends) are strictly enforced. Boarders eat in either the North Quad or South Quad cafeterias, and must buy a 19-meal plan. For those who tire of institutional cuisine, the Huddle offers plenty of fast-food options as well as a pay-as-you-go snack bar. Students can also reserve the kitchen to cook their own meals.

Notre Dame has been open to female applicants since 1972, and with a nearly perfect gender split in the undergraduate student body, the ratio is now comparable to many other formerly all-male schools. ND's social life isn't as rambunctious as it once was, thanks to the policy that forbids alcohol at campus social events. The rules relating to alcohol in the dorms are a bit more relaxed. For those who choose not to indulge, there are several groups dedicated to good times without alcohol. Most activities take place on campus and include parties, concerts, and movies. Each dorm holds theme dances about twice a month, and there's always the annual Screw Your Roommate weekend, where students are paired with the blind dates selected by their roomies. Another popular event is the An Tostal Festival, which comes the week before spring finals and guarantees to temporarily relieve academic anxiety with its "childish" games such as pie-eating contests and Jell-O wrestling. The annual Sophomore Literary Festival is entirely student-run and draws prominent writers and poets

> "The professors here care a great deal about their students and it shows."

from across the country. Students are involved in the community through volunteer work—more than 10 percent of grads enter community service positions. The best outlet for culture is nearby Chicago, about 90 minutes away.

With its proud gridiron heritage, there's nothing like the Fighting Irish spirit. From Knute Rockne and the Gipper right on down to modern-day greats such as Joe Montana, the spirit of Notre Dame football reigns supreme. It wasn't intentional— at least that's what they say—but the giant mosaic of Jesus Christ on the library lifts his hands toward the heavens as if to signal yet another Irish touchdown. Tailgate parties are also celebrated events, occurring before and after the game. Aside from football, Division I Notre Dame offers one of the strongest all-around athletic programs in the country, with several teams—both men's and women's—bringing home Big East titles in recent years. Diehard jocks who can't make the varsity teams will find plenty of company in ND's very competitive intramural leagues, which attract more than three-quarters of students. The Bookstore Basketball Tournament, the largest five-on-five, single-elimination hoops tournament in the world with more than 700 teams competing, lasts for a month.

Although temperatures here can drop below freezing, few dispute that Notre Dame is red-hot. Everyone at the university, from administrators to students, is considered part of the "Notre Dame family." Traditions are held in high esteem. For those looking for high-quality academics, a friendly, caring environment with a Catholic bent, and an excellent athletic scene, ND could be the answer to their prayers.

The single-sex dorms really become surrogate fraternities and sororities that breed a similar spirit of community and family.

Overlaps

Boston College, Duke, Georgetown, Northwestern, Washington University (MO), Harvard

Oberlin College

101 North Professor Street, Carnegie Building, Oberlin, OH 44074-1075

The college that invented nonconformity. From the Underground Railroad to the modern peace movement, Obies have been front and center. As at Reed and Grinnell, Oberlin's curriculum is less radical than its students. Oberlin is especially strong in the sciences, and its music conservatory is among the nation's best.

Website: www.oberlin.edu
Location: Small town
Total Enrollment: 2,841
Undergraduates: 2,829
Male/Female: 45/55
SAT Ranges: V 650–750
    M 620–710
ACT Range: 27–31
Financial Aid: 60%
Expense: Pr $ $ $ $
Phi Beta Kappa: Yes
Applicants: 6,686
Accepted: 34%
Enrolled: 32%
Grad in 6 Years: 85%
Returning Freshmen: 92%
Academics: ✍ ✍ ✍ ✍ ½
Social: ☎ ☎ ☎ ☎
Q of L: ★ ★ ★ ★
Admissions: (440) 775-8411
Email Address: college
    .admissions@oberlin.edu

Strongest Programs:
Neuroscience
Environmental Studies
Creative Writing
East Asian Studies
Biology
Chemistry
Politics
Music

New and contrasting ideas are a way of life at Oberlin College, a liberal arts school where nonconformity is a long tradition. Tucked away in a small Ohio town, Oberlin was the first American college to accept women and minorities, and it was a stop on the Underground Railroad. That pioneering spirit has not faded. With diverse academic challenges ranging from cinema studies to neuroscience, Obies thrive on higher thinking and exploring their myriad talents. Says a senior: "Come here if you have a desire to change the world."

Oberlin's attractive campus features a mix of Italian Renaissance buildings (four designed by Cass Gilbert), late-19th- and early-20th-century organic stone structures, and some less interesting 1950s barracks-type dorms. The buildings rise over flatlands typical of the Midwest, which do little to stop brutal winter winds. The Allen Memorial Art Museum, sometimes mentioned in the same breath as Harvard's and Yale's, is one of the loveliest buildings on campus, with a brick-paved, flower-laden courtyard and a fountain. The Oberlin College Science Center offers state-of-the-art classrooms, wireless Internet areas, a science library, and laboratory space.

Oberlin has been a leader among liberal arts colleges seeking to promote their science offerings; biology and chemistry are two of the college's strongest departments, and undergraduates may major in interdisciplinary programs like neuroscience and biopsychology. Students also flock to the English, politics, and history departments. Other popular majors include East Asian studies and environmental studies; computer science and anthropology attract relatively few majors. Oberlin students rave about their faculty. "It is clear to students that professors come to Oberlin out of a dedication to their students and subject matter,"

> **"Come here if you have a desire to change the world."**

says a Latin American studies major. Oberlin's conservatory of music holds a well-deserved spot among the nation's most prominent performance schools; the voice, violin, and TIMARA (Technology in Music and Related Arts) programs are especially praised. Interdisciplinary and self-created majors, such as black, Latin American, Russian, Third World, and women's studies, are popular.—not surprisingly at such a liberal school. About 40 percent of students study abroad.

Oberlin's students are as serious about their schoolwork as they are about politics, justice, and other social causes. Courses are rigorous; heavy workloads and the occasional Saturday morning class are the norm. "Obies tend to work together and form study groups for classes," says a senior. "The climate is productive but not competitive." A credit/no-entry policy allows students to take an unlimited number of grade-free courses (if they can get in). Recognizing that students at Oberlin are gifted and want to challenge themselves, most departments offer group and individual independent study opportunities and invite selected students to pursue demanding honors programs, especially during their senior year.

There are no requirements for freshmen at Oberlin, but general education requirements include proficiency in writing and math and nine credit hours in each of the three divisions—arts and humanities, math/natural sciences, and social sciences—plus another nine credit hours in cultural diversity courses, including a foreign language. Students are also required to take one-quarter of the semester hours needed to graduate outside their major's division; those who shy away from math and science can satisfy distribution requirements by taking interdisciplinary courses such as "Chemistry and Crime." Students must also participate in three January terms, during which they pursue month-long projects, traditional or unique, on or off campus. About 60 different first year seminar classes are available, with enrollment limited to 14 students each, and "it's a great way to make friends," says a student. "It also introduces you to the Oberlin academic experience."

One of Oberlin's more unusual offerings is EXCO, an experimental college that offers students and interested townsfolk the chance to learn together. Most classes are taught by students, and topics can range from Fairy Tales to Knitting, Salsa Dancing, X-Files Lovers, and much more. Hot spots on campus include the Mudd Library, with more than 2.1 million volumes—a superb facility for research and studying or socializing; the famous A-level is the place to be on

Oberlin appears to be reversing its losing ways and building a loyal fan base: The men's field hockey team recently brought home the Division III North Coast Athletic Conference championship.

"Oberlin is bursting at the seams with creativity."

weeknights. Even more special are the music conservatory's 150 practice rooms, substantial music library, and Steinway pianos—one of the world's largest collections. Each year, 25 to 35 students enter the dual-degree program, which allows them to earn both a BM and a BA in as little as five years. Dual-degree students must be admitted to both the college and the music conservatory.

"Oberlin is bursting at the seams with creativity," gushes a senior. "The thing that connects everyone is the passion for learning and the genuine desire to make a difference in the world." Ninety percent of students are from out of state, primarily from the Mid-Atlantic states. African Americans account for 5 percent of the student body, Asian Americans 8 percent, Hispanics 5 percent, and foreign students 7 percent. Initiatives to increase diversity at Oberlin include advisors from various ethnic and racial backgrounds and a multicultural resource center with a full-time director. "Though the student body is homogeneously liberal," says a junior, "there are more protests, awareness and advocacy groups, and campaigns here than I can keep track of." A popular annual event is the Drag Ball, in which half the student body comes in full drag. "It's very Oberlin, because it's all about challenging social norms," says a student. Ten percent of undergrads receive merit scholarships worth an average of $11,029.

Oberlin has been a leader among liberal arts colleges seeking to promote their science offerings.

Eighty-three percent of Oberlin's students live on campus in the dorms, including co-ops, several of which focus on foreign languages. Students are guaranteed housing, but choices are widest for upper-classmen. Only one dorm is single-sex; all dorms are four-class except Barrows, which is reserved for freshmen. The best dorms are said to be the program houses, including French House, African Heritage House, Russian House, and Third World House. Seniors and lucky juniors can land the preferred singles (thanks to their standing or good

"Obies tend to work together and form study groups for classes."

lottery numbers), but many move into cheaper off-campus apartments. "The dorms are decent," says a music composition major, "but off campus is way sweeter and also costs half the price." Students may eat in any of seven dining rooms. Appetizing alternatives to institutional fare can be found at the co-ops that comprise the Oberlin Student Cooperative Association (OSCA), a more than $2-million-a-year corporation run entirely by students. Co-op members plan and prepare their own meals, and though only 6 percent of the student body actually live in these houses, nearly 23 percent take their meals there, enjoying everything from homemade bread to whatever's left in the pantry before the next food shipment arrives. "The cop-ops work exceptionally well and the food is excellent," says a student.

Social life, like so much of the Oberlin experience, is what you make of it, students say. Almost all social events are on campus, and students find more than enough to do. Even in the middle of the day, "It's hard to walk across campus without becoming distracted by a pick-up game of Frisbee, the construction of a snow fort, or some students dancing contact improvisation on the lawn," says a senior. House parties, plays, movies, and conservatory performances are planned every other night. And since there's no Greek system, nothing is exclusive. As for drinking, underage students can finagle booze, and of-age students are allowed to imbibe in their rooms.

When the need to wander strikes, Cleveland is only 30 miles away, but students enjoy their small town. "Even though Oberlin is in the middle of cornfields," says one student, "I have never been bored." Another student adds, "Pretty much all of the restaurants downtown are used to working until at least 2 a.m., delivering pizza to hungry late-nighters or having special offers for college students." And Obies are very enthusiastic about giving back to the community through volunteer activities at local schools, hospitals, and nursing homes. "Our motto is 'Think one person can make a difference? So do we!'" says a student.

Oberlin appears to be reversing its losing ways and building a loyal fan base: The men's field hockey team recently brought home the Division III North Coast Athletic Conference championship. As for intramural sports, "As soon as the rain stops in the spring, there are students on every green area between the dorms playing ultimate Frisbee," says a sophomore. Fencing, Aikido, soccer, and women's rugby also have loyal followings.

Oberlin may be small in size, but its emphasis on global learning, undergraduate research, and a vibrant liberal arts education helps it bust those statistical seams. Students are more likely to discuss local poverty than the quality of cereal choices in the dining halls, and can be found playing a Steinway or plugging away at astronomy. One Obie sums it up this way: "Oberlin is all about thinking critically and creatively in order to make a difference. Unconventionality is prized, as is challenging commonly held beliefs and norms."

Overlaps

Wesleyan, Brown, Vassar, Carleton, Macalester, Yale

If You Apply To ➤

Oberlin: Early decision: Nov. 15, Jan. 2. Regular admissions: Jan. 15. Financial aid: Feb. 15. Guarantees to meet full demonstrated need. Campus interviews: recommended, evaluative. Alumni interviews: optional, evaluative. SATs or ACTs: required. SAT IIs: recommended. Accepts the Common Application and prefers electronic applications. Essay question: Common Application; why Oberlin?

Occidental College

1600 Campus Road, Los Angeles, CA 90041

Oxy is a diverse, urban, streetwise cousin to the more upscale and suburban Claremont Colleges. Plentiful internships and study abroad give Oxy students real-world perspectives. Oxy's innovative diplomacy and world affairs program features internships in Washington and at the UN.

Occidental College is one of a handful of small colleges located in a big city, in this case LalaLand. But unlike the sprawling and impersonal City of Angels, Oxy emphasizes a strong sense of community and a decidedly diverse student population.

Set against the backdrop of the San Gabriel Mountains, Oxy's self-contained Mediterranean-style campus is a secluded enclave of flowers and trees between Pasadena and Glendale, minutes from downtown Los Angeles. Construction is underway on the college's largest-ever residence hall—a 273-bed building that will open in late 2007.

Inside this urban oasis resides a thriving community of high achievers who don't for a moment believe that the liberal arts are dead, or even wounded. Required first-year cultural studies seminars include topics in human history and culture, with an emphasis on writing skills. Each seminar has 16 students who also share the same residence hall; it's called the Learning Communities Program. In addition, all Oxy students must show proficiency in a foreign language and complete 12 units of world cultures courses, a fine arts course, preindustrial-era coursework, and 12 units of science and math. Many of Occidental's academic departments are excellent, with English, music, chemistry, and an innovative diplomacy and world affairs program among the strongest, and psychology the most popular. "The best programs are diplomacy and world affairs, English and comparative literary studies, theater, and biology," asserts a sophomore, "because of strong faculty and opportunities available in these fields."

"You can find almost anything except snow."

Oxy encourages diverse learning experiences through internships, independent study, and study abroad—including the only undergraduate program to offer internships with the United Nations Secretariat and the U.S. Mission. Over the past decade, more than 1,000 students have engaged in summer research projects or academic year independent projects, and 14 Occidental students were recently accepted for the National Conference on Undergraduate Research. The student-managed Charles R. Blyth Fund allows students to invest about $150,000 of the college's endowment.

Faculty members are readily available in and out of the classroom, and students say the teaching, in general, is excellent. "The quality of teaching here is good overall, but it is important to talk to current and more experienced students to find out who is worth having," says a junior. The great majority of classes have 25 or fewer students, and as academic advisors are responsible for about four students per class (16 total), personal relationships develop quickly. "I had in-home lunch, dinner, museum outings, and hour-long office discussions with my professors," explains one senior. Students also broaden their horizons in study abroad programs in Western Europe, Japan, China, Mexico, Nepal, Hungary, Costa Rica, and Russia. For politicos, there's Oxy-in-Washington and Oxy-at-the-UN. There are also 3–2 engineering programs with Caltech and Columbia University, exchange programs with Spelman and Morehouse colleges in Atlanta, and crossregistration privileges with Caltech and Pasadena's Art Center College of Design. Students may also take advantage of a 4–2 biotechnology program with Keck Graduate Institute (of the Claremont Colleges).

Fifty-one percent of the students at Occidental are from California, and 4 percent are foreign. African Americans make up 7 percent of the student population, Hispanics 16 percent, and Asian Americans 14 percent. The college's financial aid staff visits local high schools to help families fill out applications, offering sessions in English, Spanish, and Vietnamese. Perhaps not surprisingly, students tend to be liberal. Politics are always a matter of debate, especially "anything and everything the Republicans are doing wrong," says a sophomore. Nearly 27 percent of students receive merit scholarships each year, averaging $16,180. There are no athletic scholarships.

Upperclassmen on the "O-team" plan freshman orientation the week before school starts. Freshmen are required to live on campus and eat in the dining hall

Website: www.oxy.edu
Location: Urban
Total Enrollment: 1,839
Undergraduates: 1,819
Male/Female: 43/57
SAT Ranges: V 600–690
  M 610–690
ACT Range: 27-31
Financial Aid: 55%
Expense: Pr $ $ $
Phi Beta Kappa: Yes
Applicants: 5,114
Accepted: 41%
Enrolled: 21%
Grad in 6 Years: 84%
Returning Freshmen: 92%
Academics: 🖉 🖉 🖉 🖉
Social: ☎ ☎ ☎
Q of L: ★ ★ ★ ★
Admissions: (800) 825-5262
Email Address:
  admission@oxy.edu

Strongest Programs:
Economics
English
Music
Psychology
Chemistry
Biology
Art
Diplomacy and World Affairs

Campus dining fare is "always edible and sometimes tasty," says one student.

(tip from a junior: don't try anything that has an unusual name, like "Mayan tofu.") Campus dining fare is "always edible and sometimes tasty," says one student. The 11 residence halls are small—fewer than 150 students each—and co-ed by floor or room. Students say they are quite comfortable and well maintained. Seventy percent of students live on campus, but what you get depends on your luck in the housing lottery. Students from all four classes live together, many in special-interest houses like the Multicultural Hall, the Women's Center, or the Substance-Free Quad.

While the bright lights of L.A. often beckon on weekends, "social life is largely contained on campus," says a junior. "There are always one or two things happening on the weekends," Fraternities and sororities, though declining on the Oxy social ladder, attract 6 percent of men and 13 percent of the women, but they are neither selective nor exclusive; students choose which to join, rather than being chosen, and the frats must invite everyone to their functions. As for alcohol, "like most other colleges, there is underage drinking even though this is illegal," says another junior. The Senior Smack offers graduates-to-be the chance to smooch whomever they've wanted to during the past four years. Other big events include parties such as Love on the Beach and Da Getaway—a Roaring Twenties bash where students gamble with fake money and Charleston 'til they drop. "Winter Formal is probably my favorite," says a theatre major, Here's a tip: Keep your birthday a secret, or on that unhappy day a roaring pack of your more sadistic classmates will carry you out to the middle of campus and mercilessly toss you in the Gilman Fountain. It's a tradition, after all.

> **"The best programs are diplomacy and world affairs, English and comparative literary studies, theater and biology."**

A student characterizes the surrounding neighborhood of Eagle Rock as "most definitely a family-oriented town that does not cater to the college student, but provides a wonderful array of diverse foods and activities." When students become weary of the incestuous social life in the 57 percent female "Oxy fishbowl," they head for the bars, restaurants, museums, and theaters of downtown Los Angeles and Old Pasadena, where, one student notes, "You can find almost anything except snow." But the ski slopes of the San Gabriel Mountains are not far away, and neither is Hollywood nor the beautiful beaches of Southern California. When they tire of California, students try their luck in Las Vegas—or trek south of the border, into Tijuana. A car (your own or someone else's) is practically a necessity, though the college runs a weekend shuttle service to Old Town Pasadena. The weather is warm and sunny, but the air is often thick with that infamous L.A. smog.

> **"Like most other colleges, there is underage drinking even though this is illegal."**

Oxy's sports teams compete in Division III and draw a modest following. Football is the most popular (Oxy's team won the Southern California Intercollegiate Athletic Conference championship in 2004 and 2005), followed by men's basketball and soccer. Women's water polo is top-notch, too. And don't forget that L.A. is home to the NBA's Lakers, the NHL's Kings, and baseball's Dodgers. Oxy has a small intramural program; popular club sports include rugby, lacrosse, and ultimate Frisbee.

Occidental's creative, motivated, and diverse students are not here for the bright lights and beautiful people of Los Angeles; those are just fringe benefits. Instead, students are drawn to this intimate oasis of learning by professors who hate to see anyone waste one whit of intellectual potential. And students here are only too happy to live up to these lofty expectations.

Overlaps

USC, UCLA, UC–Berkeley, Pomona, Claremont–McKenna, Stanford

Oglethorpe University

4484 Peachtree Road NE, Atlanta, GA 30319

Small wonder that brochures for Oglethorpe trumpet Atlanta as the college's biggest asset. In a region where most liberal arts colleges are in sleepy towns, Oglethorpe has the South's most exciting city at its fingertips. With only 945 undergraduates, Oglethorpe puts heavy emphasis on community.

Each Christmas, students at Oglethorpe University take part in a unique tradition. The Boar's Head Ceremony celebrates a medieval scholar who halted a stampeding wild boar by ramming his copy of Aristotle down the animal's throat. Though some may find it boorish, students at this small Southern school explain that it fosters a sense of family.

Founded in 1835, the school is named for the idealistic founder of the state of Georgia, James Edward Oglethorpe. Its 118-acre campus is located near suburban Buckhead, a ritzy area about 10 miles north of downtown Atlanta. The heavily wooded, slightly rolling terrain is perfect territory for walks or long runs, and the beautiful campus has served as the backdrop for several movies and TV shows. Oglethorpe's academic buildings and some residence halls are in the English Gothic style; the campus is also home of the Georgia Shakespeare Festival.

Oglethorpe's strengths are business administration, English, biology, accounting, psychology, and communications. Weaker bets are the fine arts and foreign language departments, though the latter does offer courses in Japanese, German, French, and Spanish. And whatever isn't offered at Oglethorpe can usually be taken through crossregistration at other schools in the Atlanta area. One student offers this assessment: "The art department is the largest and most fun. The business program is very big and growing—these teachers are the best. The biology department is what students avoid because it's very, very tough."

> "The art department is the largest and most fun. The business program is very big and growing."

Aspiring engineers may take advantage of 3–2 dual-degree programs with Georgia Tech, the University of Southern California, Auburn, and the University of Florida. The school also offers courses and additional resources as a member of the Atlanta Regional Consortium for Higher Education.* Oglethorpe also offers a wide variety of study abroad programs, including a semester at Seigakuin University in Japan and sister-school exchanges in Argentina, the Netherlands, Germany, France, Russia, and Monaco. According to administrators, Oglethorpe "emphasizes the preparation of the humane generalist" and "rejects rigid specialization." That doesn't mean the curriculum is a cakewalk, though. "Academics are pretty competitive and the majority of classes are quite rigorous," explains one sophomore, "but the plus side is that class sizes are small."

The university's guiding principle is the Oglethorpe Idea, which says students should develop academically and as citizens. This philosophy is based on the conviction

Website: www.oglethorpe.edu
Location: City outskirts
Total Enrollment: 1,029
Undergraduates: 945
Male/Female: 35/65
SAT Ranges: V 540–660
  M 530–640
ACT Range: 24–29
Financial Aid: 55%
Expense: Pr $
Phi Beta Kappa: No
Applicants: 746
Accepted: 66%
Enrolled: 23%
Grad in 6 Years: 61%
Returning Freshmen: 87%
Academics: ✍ ✍ ✍
Social: ☎ ☎ ☎
Q of L: ★ ★ ★ ★
Admissions: (404) 364-8307
  or (800) 428-4484
Email Address:
  admission@oglethorpe.edu

Strongest Programs:
Biology
Accounting
Business Administration
English
Psychology

that education should help students make both a life and a living. All students take the sequenced, interdisciplinary Core Curriculum program at the same point in their college careers, providing them with a model for integrating information and gaining knowledge. In addition to the ability to reason, read, and speak effectively, the core asks students to reflect on and discuss matters fundamental to understanding who they are and what they ought to be. The core requires Narratives of the Self (freshmen), Human Nature and the Social Order (sophomores), Historical Perspectives on the Social Order (juniors), and Science and Human Nature (seniors), plus a fine arts core course in music and culture or art and culture, and coursework in modern mathematics or advanced foreign language.

Oglethorpe's faculty may be demanding, but they're also friendly and helpful. "Students are taught to think and learn on their own," says a student. "Most teachers really put themselves out there to help." Classes are generally small, and most students notice few problems at registration. Advising services are said to be helpful. The library's holdings are minuscule, though—just over 131,000 volumes.

What's an Oglethorpian like? The vast majority are smart, semiconservative offspring of middle- and upper-middle-class Southern families. Sixty-six percent ranked in the top quarter of their high school class; most come from public schools and more than half are native Georgians. "Our cam-

"The dorms are big and have nice furniture."

pus is pretty small, however, cliques do form. The students are usually extremely personable and cordial." Oglethorpe prides itself on being one of the first Georgia colleges to admit African American students, and today 30 percent of the students are members of minority groups: roughly 20 percent are African American, 3 percent are Asian American, 3 percent are Hispanic, and 3 percent hail from abroad. There's a level of comfort with racial differences, students report. Some students complain that their peers can be rather cliquish, but say that all in all, everyone gets along well.

Fifty-four percent of Oglethorpe's students choose to live on campus—and love it. "The dorms are big and have nice furniture," says an accounting major. Most rooms are suites with private bathrooms, and some singles are available. Some students commute to campus; a quarter live in Atlanta—not a college town, but where the wild life is. "Some weekends, everyone stays around and life is great, and then there are others when campus is deserted," one student explains. Fraternities and sororities, which claim 33 percent of the men and 25 percent of the women, throw parties that draw big numbers. Officially, the campus is dry, but underage students can find alcohol if they try, students agree. It's rumored that Oglethorpe barflies do more hopping than Georgia bullfrogs, and bars, clubs, and cafes abound within 10 minutes of campus.

Those who tire of the Oglethorpe scene can find excitement on the campuses of the dozen or so other colleges in the area or in downtown Atlanta, which one student describes as "the heart of Atlanta and a few minutes from Buckhead, a party district, and two great malls." Atlanta proper offers everything you can imagine— arts, professional sports (including basketball's Hawks, football's Falcons, and baseball's Braves), and entertainment (ride the Great American Scream Machine at Six

"Students are taught to think and learn on their own."

Flags). Facilities built for the 1996 Olympics also provide a diversion. Oglethorpe always has a big contingent going to Savannah for St. Patrick's Day and to New Orleans for Mardi Gras. The campus celebrates its origins once a year during Oglethorpe Day. Students looking for warmer weather, though, head down to sunny Florida.

Intramurals are important at Oglethorpe, sometimes more so than varsity sports. Perhaps Atlanta's diversions or the relatively small number of students on campus cause varsity sports to be a weak draw. Still, the Stormy Petrels men's golf team has brought home a Southern Collegiate Athletic Conference title, and basketball games against crosscity rival Emory are popular. The Georgia landscape makes possible a plethora of outdoor activities, including hiking at nearby Stone Mountain and boating or swimming in Lake Lanier (named for Georgia poet Sidney Lanier—Oglethorpe Class of 1860).

Though Oglethorpe may lack widespread name recognition, its students get all the attention they need from a caring faculty on a close-knit campus. And being in a large city like Atlanta provides anything else that might be lacking, ranging from great nightlife to internships and postgraduate employment with big-name corporations. In a sea of large Southern state schools, Oglethorpe stands out as a place where students come first.

If You Apply To ➤

Oglethorpe: Rolling admissions. Early action: Dec. 5. Financial aid: Mar. 1. Does not guarantee to meet demonstrated need. Campus interviews: recommended, evaluative. Alumni interviews: optional, informational. SATs or ACTs: required. SAT IIs: optional. Accepts the Common Application. Essay question: more about you.

Ohio State University

3rd Floor, Lincoln Tower, 1800 Cannon Drive, Columbus, OH 43210

Ohio State may be the biggest university in the Big Ten, but it is far from the best. OSU has never achieved the reputation of a Michigan or a Wisconsin—partly because it has three major in-state rivals (Miami, Cincinnati, and Ohio U) that siphon off many top students.

Envision a school with almost 50,000 students and too many opportunities to count. What might come to mind is Ohio State University, located in the heart of the state's capital, offering 19 colleges and more than 10,400 courses in 175 undergraduate majors. If those numbers aren't staggering enough, consider the fact that OSU has 34 varsity teams, 44 intramural sports, and 51 sports clubs. While students cite the school's size as both a blessing and a curse, all seem to agree that at OSU, the sky is the limit for those with a desire to sample its academic and other resources. The term "big" might be an understatement.

This megauniversity stands on 3,200 wooded acres rubbing the edge of downtown Columbus on one side. On the other side, across the Olentangy River, is farmland associated with the College of Agriculture. OSU's architectural style is anything but consistent, yet it's all tied together in one huge redbrick package. "One part of the campus maintains a nostalgic air while another is relatively modern," observes a student. The grounds are nicely landscaped, and a centrally located lake provides a peaceful setting for contemplation. Newer facilities include RPAC—the nation's largest facility dedicated to student fitness, wellness, and recreation—and the Knowlton School of Architecture.

Business, education, geography, industrial design, and engineering are among the school's most celebrated departments. OSU bills itself as the place to go for computer graphics and has a supercomputer center to back up its claim. It also boasts the largest and most comprehensive African American studies program anywhere and turns out more African American PhDs than any other university in the nation. Furthermore, the university has the nation's only programs in welding engineering and

Website: www.osu.edu
Location: City center
Total Enrollment: 50,504
Undergraduates: 37,411
Male/Female: 53/47
SAT Ranges: V 530–640
  M 550–660
ACT Range: 24–28
Financial Aid: 43%
Expense: Pub $ $ $ $
Phi Beta Kappa: Yes
Applicants: 17,566
Accepted: 74%
Enrolled: 46%
Grad in 6 Years: 68%
Returning Freshmen: 86%
Academics: ✑ ✑ ✑
Social: ☎ ☎ ☎ ☎
Q of L: ★ ★ ★
Admissions: (614) 292-OHIO
Email Address:
  askabuckeye@osu.edu

Business, education, geography, industrial design, and engineering are among the school's most celebrated departments.

geodetic science, and the state's only program in medical communications. Although immensely popular, students report that the English program needs improvement.

The university's fundamental commitment to liberal arts learning means all undergrads must satisfy rigorous general education requirements that include at least one course in math, two each in writing and a foreign language, three in social science, four in natural science, and five in arts and humanities. To top it all off, students must complete a capstone requirement that includes a course on Issues of the Contemporary World. A quarterly selective admissions program has replaced OSU's old open-door policy, but a conditional/unconditional admissions policy allows some poorly prepared students to play catch-up in designated areas. Some 10,500 students receive merit-based scholarships while 560 athletes receive scholarships in 17 sports.

> "One part of the campus maintains a nostalgic air while another is relatively modern."

"The classes can be very challenging and are definitely designed to teach you how to think," says a senior. Freshmen, who are grouped together in the University College before entering one of the degree-granting programs, find most of the introductory lectures huge. Teaching assistants, not professors, hold smaller recitation sections and deal on a personal level with students. Students find that class sizes are whittled down as they continue in their fields of study. OSU's honors program allows 2,500 students to take classes that are taught by top professors and limited to 25 students each. "The whole system of honors classes, priority scheduling, honors housing, and co-curricular activities really adds to the overall experience at OSU," a biology major says. Internships are required in some programs and optional in others, and possibilities for study abroad include Japan and the People's Republic of China. A personalized study program enables students to create their own majors. New majors include film studies, Korean, and professional golf management.

Inside OSU's ivy-covered halls and modern additions are some of the best up-to-date equipment and facilities, including a "phenomenal" library system with two dozen branches and nearly four million volumes—all coordinated by computer. Professors "are well-informed about their subjects," according to students. "The majority of my professors have been spectacular," raves one student. Complaints about long registration lines have been answered by BRUTUS, Ohio State's Touch-Tone telephone registration system, which saves on time but does little to ease class overcrowding.

Eighty-eight percent of Ohio State's students come from Ohio, and the remainder come largely from adjacent states. Every type of background is represented, most in huge numbers. The student body is 8 percent African American; Hispanics and Asian Americans make up another 8 percent. Several programs are aimed specifically at "enhancing" efforts to attract and retain minority students, including a statewide Young Scholars Program that yearly guarantees admission and financial aid to seventh graders once they complete high school.

> "The classes can be very challenging and are definitely designed to teach you how to think."

Freshmen, who are grouped together in the University College before entering one of the degree-granting programs, find most introductory lectures huge.

The residence halls that house 24 percent of the Ohio State masses are located in three areas: North, South, and Olentangy (that is, those closest to the Olentangy River). Freshmen—required to live either at home or in the dorms—are scattered among each of OSU's 27 residence halls. "The dorms are very comfortable," says a student. "Plus, you get a lot of living options." Upperclassmen, when they don't head for off-campus life in Columbus, find the South campus section among the most desirable (it's more sociable, louder, and full of single rooms). The Towers in the Olentangy section have gained more popularity since their conversion to eight-person suites. All in all, students have a choice of single-sex, co-ed (by floor or by room), or married-couples apartments if they want to live in campus housing. Computer labs are located in each residence area. A system of variable room rates based on frills (air-conditioning, private bath, number of roommates), as well as a choice of four meal-plan options, give students flexibility in determining their housing costs. Dormitory students have a choice of five dining halls, but others cook for themselves or eat in fraternity houses. "The food at OSU is fantastic," chirps one student.

Such a large student market has, of course, produced a strip of bars, fast-food joints, convenience stores, bookstores, vegetarian restaurants, and you-name-it along the edge of the campus on High Street, and downtown Columbus is just a few minutes away. The fine public transportation system carries students not only throughout this capital city but also around the sprawling campus. In addition to the usual shopping centers, restaurants, golf courses, and movie theaters, Columbus boasts a symphony orchestra and ballet, and its central location in the state makes it easily accessible to Cleveland and Cincinnati. Outdoor enthusiasts can ski in nearby Mansfield, canoe and sail on the Olentangy and Scioto rivers, hike around adjacent quarries, or camp in the nearby woods.

OSU is a bustling place on weekends. "Student involvement is overwhelming," says one student. Various social events are planned by on-campus housing groups—floors, dorms, or sections of the campus. The Michigan–Ohio State football game inspires the best partying of the year, and other annual events include a Renaissance Festival and River Rat Day. Two student unions run eateries as well as movies on Friday and Saturday nights, and High Street's zillion bars, saloons, restaurants, and discos come to life. Campus policies prohibit underage drinking in dorms, but one partier discloses, "I can get served in almost any bar on campus." Just 6 percent of OSU men and women go Greek. By one account, these students make the Greek system "a way of life and isolate themselves from the rest of the student population."

"Get football tickets every year, join a few clubs, use the rec facilities, and learn the bus routes early!"

Ohio State operates the mother of all college sports programs—an $85 million operation once led by the only athletic director in the country with his own Bobble-head doll for sale. The Buckeyes field teams in 36 sports, from women's rifle to men's football. Nonrecruited students should not expect to make any varsity team as walk-ons. Rivalries abound and one student asserts (without a hint of irony), "We party harder, we play harder, and we're more humble [than our rivals]." Many take advantage of an ambitious intramural program that boasts a dozen basketball courts and 26 courts for handball, squash, and racquetball. "It rained one day and 200 softball games were rained out," one student reports. For diehard basketball fans, the first official day of practice, Midnight Basketball, is a favored ritual.

OSU's sheer size is sometimes overwhelming to be sure, but students seem to thrive on the challenge and excitement of a big university. For those who really want to be Buckeyes, jump in with both feet and heed advice of one seasoned student: "Get football tickets every year, join a few clubs, use the rec facilities, and learn the bus routes early!"

Overlaps

Ohio University, Miami (OH), Bowling Green State, Kent State, University of Cincinnati, University of Toledo

If You Apply To ➤ **OSU:** Rolling admissions. Does not guarantee to meet full demonstrated need. No campus or alumni interviews. SATs or ACTs: required. SAT IIs: optional. Accepts the Common Application and electronic applications. Essay question.

Ohio University

Chubb Hall 120, Athens, OH 45701-2979

OU is half the size of Ohio State and plays up its homey feel compared to the cast of thousands in Columbus. The Honors Tutorial College is a sure bet for top students who want close contact with faculty. Communications and journalism top the list of prominent programs.

Website: www.ohiou.edu
Location: Rural
Total Enrollment: 20,461
Undergraduates: 17,207
Male/Female: 47/53
SAT Ranges: V 490–600
  M 490–600
ACT Range: 21–25
Financial Aid: 48%
Expense: Pub $ $ $ $
Phi Beta Kappa: Yes
Applicants: 12,367
Accepted: 89%
Enrolled: 38%
Grad in 6 Years: 71%
Returning Freshmen: 81%
Academics: ✏ ✏ ✏
Social: ☎ ☎ ☎ ☎
Q of L: ★ ★ ★
Admissions: (740) 593-4100
Email Address:
  admissions@ohiou.edu

Strongest Programs:
Journalism
Art
Art History
Home Economics
Business
Engineering
Communications
Film

With top-notch programs in journalism and business, Ohio University has become a competitive public institution without shedding its small-town roots. It has become known as an important research institution, with studies including dinosaur anatomy and rural diabetes rates. Students here love to hit the town for fun but are quick to hit the books, too. "No matter who you are or what you're into, you'll find a group of friends to fit in with," says a freshman.

Established in 1804 as the first institution of higher learning in the old Northwest Territory, Ohio University is located in Athens, about 75 miles from Columbus, the state capital. Encircled by winding hills, the campus features neo-Georgian architecture, tree-lined redbrick walkways, and white-columned buildings all clustered on "greens," which are like small neighborhoods. Long walks are especially nice during the fall foliage season. The university's newest state-of-the-art classroom facility, Margaret M. Walter Hall, is equipped with the latest in educational technology. Other improvements include a $20 million renovation and expansion of Bentley Hall, the social sciences facility, and the Innovation Center, which provides incubation services to biotechnology and IT companies. A new student center is opening this year as well, providing students with state-of-the-art facilities.

> **"No matter who you are or what you're into, you'll find a group of friends to fit in with."**

One of the focal points of an Ohio University education and something that sets the school apart from run-of-the-mill state institutions is the Honors Tutorial College. Founded in 1972, it's the nation's first multi-disciplinary, degree-granting honors program modeled on the tutorial method used in British universities, notably Oxford and Cambridge. It is ranked as one of the best programs on campus, and the most selective: Students must have at least a 1300 on their SAT or a 30 on their ACT, and only about 60 freshmen get in every year. Students take an individualized curriculum in a major field and spend most of their time in one-on-one weekly tutorials with profs. Other top areas are the College of Communication and its three offspring: the schools of telecommunications, visual communication, and journalism, which feature the latest graphics and computer equipment. The Global Learning Community Certificate is an innovative program that prepares students for leadership opportunities in a rapidly changing world. First-year GLC students work in bi-national teams with students from universities in Hungary, Ecuador, Thailand, and elsewhere. Students can take an online journalism course, which teaches skills like computer-assisted reporting and Web design. The bachelor's degree in art education is being phased out.

The academic climate is "somewhere between rigorous and laid-back," offers a senior. "Students work hard and understand the importance of academic success." General education requirements involve a minimum of one course in math or quantitative skills, two courses in English composition, one senior-level interdisciplinary course, plus 30 quarter hours in applied sciences and technology, social sciences, natural sciences, humanities, and cross-cultural perspectives. To lighten the load, you can take electives such as

> **"Local bands and visiting artists are always a big part of the social life."**

The Language of Rock Music (so you can communicate more effectively with fellow moshers?). Study abroad offers worldwide destinations for anywhere from two weeks to one year. Co-op programs are available for engineering students, and nearly anyone can earn credit for an internship.

Students say most of their professors are top-notch. "I have had many wonderful experiences with teachers compared to only a few bad instructors," says a student. Freshmen often are taught by full professors with TAs handling study sessions. Classes of 100-plus students do exist, but the average class size for freshmen is about 25. Budget cutbacks haven't gone unnoticed, but a senior says, "These cutbacks have affected administrative positions more than classes."

You'll find many classmates from the Buckeye State; 92 percent are Ohioans. "There is a large amount of middle- to upper-class students, but diversity is prominent," says a senior. Almost everyone at OU attended public high school and 16 percent graduated in the top tenth of their class. The student body is overwhelmingly white, although administrators are trying to attract more minority students and have established an Office of Multicultural Programs. Four percent are African American; Hispanics and Asian Americans combine for 2 percent. The Gateway Award Program provides financial aid for outstanding students who show academic excellence, financial need, and/or a combination of both.

Campus housing is plentiful (41 dorms) and well liked. "The dorms are efficient, but I wouldn't classify them as cozy," a journalism major says. Almost everyone lives on campus for two years then moves into neighboring dwellings. Campus housing comes with a variety of options: co-ed, single-sex, quiet study, academic interest, and even an international dorm. At the "mods," six men and six women occupy separate wings but share a living room and study room. Upperclassmen usually move to fraternity or sorority houses, nearby apartments, or rental houses. Five different meal plans are available at four cafeterias and include fast-food counters next to regular dorm-food fare.

The social scene is largely off campus. Uptown Athens is dotted with bars and clubs, campus and community activities such as plays, and guest speakers and performers. Some students also choose to participate in Greek life. Eleven percent of men and 13 percent of women join their ranks. "OU offers a very cohesive and communal social life," says a sophomore. "Local bands and visiting artists are always a big part of the social life." The administration and some students have tried **"Athens is its own place. You feel at home."** to downplay OU's party-school image by strictly enforcing the alcohol policy. Freshmen are required to pass an online alcohol education course; if they don't, they might not be able to register for OU classes. But despite their efforts, students say that drinking continues. Athens's fabled Halloween celebration is a huge block party with people from all over the Midwest and wouldn't be missed by many students. Students also look forward to homecoming, International Festival, and block parties like Palmerfest and Millfest. "Athens is its own place," says a senior. "You feel at home. It's beautiful and friendly and like no other." Volunteer opportunities, such as Habitat for Humanity and a local homeless shelter, are available through the Center for Community Service. Students also love to hike and camp at the nearby state parks or trek to Columbus for shopping.

Sports are a big draw at Ohio. The men's basketball team was Mid-American Conference champion in 2004–05, and the men's ice hockey team defeated Penn State to win the 2004 national championship. Women's volleyball is strong and the winner of three consecutive conference championships. Intramural and club sports are popular and include teams for broomball and wallyball (think volleyball using walls).

Students say OU has a lot to offer, from a vibrant social life to quality professors and challenging academics. "I love this school and the city's community," says one student. "I think most students would not regret their decision to come here."

Campus housing is plentiful (41 dorms) and well liked. Almost everyone lives on campus for two years then moves into neighboring dwellings.

One of the focal points that sets OU apart from run-of-the-mill state institutions is the Honors Tutorial College.

The Global Learning Community Certificate is an innovative program that prepares students for leadership opportunities in a rapidly changing world.

Overlaps
Ohio State, Miami University (OH), Bowling Green, Kent State, University of Cincinnati

Ohio Wesleyan University

South Sandusky Street, Delaware, OH 43015

OWU serves up the liberal arts with a popular side dish of business-related programs. In a region of beautiful campuses, Ohio Wesleyan's is nondescript. Like Denison, OWU is working hard to make its fraternities behave. Attracts middle-of-the-road to conservative students with preprofessional aspirations.

Website: www.owu.edu
Location: Small town
Total Enrollment: 1,941
Undergraduates: 1,941
Male/Female: 48/52
SAT Ranges: V 550–660
  M 570–660
ACT Range: 23–28
Financial Aid: 56%
Expense: Pr $ $
Phi Beta Kappa: Yes
Applicants: 2,432
Accepted: 74%
Enrolled: 28%
Grad in 6 Years: 61%
Returning Freshmen: 81%
Academics: ✏ ✏ ✏ ½
Social: 🐾 🐾 🐾 🐾 🐾
Q of L: ★ ★ ★
Admissions: (740) 368-3020
Email Address:
  owuadmit@owu.edu

Strongest Programs:
Psychology
Zoology
Economics and Management
Sociology/Anthropology
English
History
Politics and Government
Biology

Ohio Wesleyan University is a small school with a big commitment to providing its students with a well-rounded education. OWU hallmarks include strong preparation for graduate and professional school, a solid grounding in the liberal arts, and an emphasis on having fun outside the classroom. Once known for its raucous students, this small university has overcome its hard-partying past and now offers its students a rewarding college experience.

Situated smack in the center of the state, OWU's spacious 200-acre campus is peaceful and quaint, albeit with a highway that runs down the middle. Several buildings are on the National Register of Historic Places. The architecture ranges from Greek Revival to Colonial to modern, with ivy-covered brick academic buildings on one side of a busy thoroughfare and dormitories and fraternities on the other side of the highway. Stately Stuyvesant Hall, with its majestic bell tower, is the main campus landmark. The 150,000 square foot Conrades-Wetherell science center with state-of-the-art laboratories, classrooms, and science library, was recently completed, as was a new wireless Internet cafe.

> "A lot of students come here for the zoology program and the chemistry program."

Preprofessional education has always been OWU's forte. Between 85 and 90 percent of students in the premed program are accepted to medical schools. The curriculum includes majors in neuroscience and East Asian studies, and the highly popular zoology and microbiology departments are interesting alternatives to the traditional premed route. The Woltemade Center for Economics, Business, and Entrepreneurship caters to budding entrepreneurs, and the music and fine arts programs offer both professional and liberal arts degrees. "A lot of students come here for the zoology program and the chemistry program," a senior says. "I also think the English department is especially good."

A member of the Great Lakes College Association* consortium, Ohio Wesleyan offers numerous innovative curricular programs. The most prominent is the National Colloquium, a yearlong series of lectures on a timely issue. Speakers have included David Wetherell, president and CEO of CMGI, and novelist Gloria Naylor. The honors program offers qualified students one-on-one tutorials and a chance to conduct research with faculty members in areas of mutual interest. The Special Languages program offers the opportunity for self-directed study and tutoring by native speakers in languages such as Arabic, Chinese, Japanese, and Modern Greek. Students can travel to Mexico for a community service experience during spring break, while fine arts, theater, and music majors can spend a semester in New York City to study with professionals there.

"The courses are challenging," says a sophomore. "Students need to work hard to get As, but it's possible." Students also point out that peers tend to support one another rather than compete for grades. To graduate, OWU students must take a year of foreign language; three courses in each of the social sciences, natural sciences, and humanities; one course in the arts; and one course in cultural diversity. Students must also pass three mandatory writing classes to sharpen their written communication skills, but these aren't burdensome. Students universally laud OWU's faculty for ability and accessibility. "We have incredible profs who are fully devoted to their subjects," a senior says. "They are accessible if you have problems or just want to discuss something that grabbed your attention in class that day."

Forty-nine percent of students come from Ohio; the student body represents 43 states and 40 nations. Students agree that diversity is valued on campus, but African Americans make up only 5 percent of the student body, Hispanics 1 percent, and Asian Americans 2 percent. "Students are friendly and pretty laid-back," says one student. "They love the school and have a great spirit." Liberals and conservatives are well-represented on campus and hot topics include racial, gender, and sexual equality. Merit scholarships worth an average of $11,439 are available to qualified students; there are no athletic scholarships.

"Students need to work hard to get As, but it's possible."

Eighty-three percent of OWU students live in university-sponsored housing. All but one of the dorms are co-ed, and rooms are mostly apartment-style, four-person suites or doubles. Fraternities, unlike sororities, offer a residential option. Special-interest houses, such as Creative Arts House, House of Black Culture, House of Spirituality, and Women's House, are available, as is Welch Hall, which is for students with GPAs over 3.2. "Most students live on campus and there are generally not many room selection issues," says one senior. Seniors are now permitted to live off campus, a policy change that has received praise. Each meal eaten on the college plan subtracts a certain number of points (far too many in the opinion of most students) from students' accounts, but there are numerous culinary choices, from all-you-can-eat in the three dining halls to pizza and snacks from the college grocery store. "Food service is surprisingly good," cheers one Spanish major. What's more, "students feel very safe on our campus," thanks to a friendly and visible Public Safety department.

Does the buttoned-down seriousness of recent years mean that OWU has forsaken its heritage of raucous partying? Administrators certainly hope so. Trying to stamp out drunken binges, OWU slaps fines of up to $150 on all underage students caught drinking and puts them on probation after the fourth offense. One student says, "The campus works to educate students on alcohol and to guide students' decision-making." A sophomore adds, "We have substance-free dorms and halls and clubs that provide an alternative to entertainment by booze. They are pretty popular." Part of OWU's commitment to mend its partying ways includes dry rush for all fraternities and an armband policy at parties. Greek membership, however, still attracts 39 percent of men and 26 percent of women. Among OWU's best-loved traditions are Fallfest and Monnett Weekend in the spring—campuswide bashes for students, parents, and alumni that include bonfires, a Fun Run, and open houses for the Greeks. Romantics will enjoy the President's Ball the weekend before finals in the winter, and the famous Little Brown Jug harness race provides offbeat fun.

"Over 80 percent of OWU students are involved with community service efforts in the community."

Delaware, a town of 25,000, is "a great little town," reports one senior. "There are so many friendly and accommodating people and the restaurants are very good." Another student notes, "Over 80 percent of OWU students are involved with community service efforts in the community." Approximately 85 percent of the students volunteer in the community for Habitat for Humanity and other charitable organizations. Ohio's capital and largest city, Columbus, is only 30 minutes away by car and offers many job and internship opportunities. Lakes, farms, and even ski slopes are within a few hours' drive.

OWU sports rank in the top 10 in the nation, according to the Sears Director's Cup Standings, which rates top all-around programs. Recent NCAC champions include women's soccer and men's soccer and golf. Sports fever carries over into single-sex and co-ed intramurals (nearly 80 percent of students take part), and a massive annual game of Capture the Flag begins at 11:00 one night and lasts until the wee hours.

Ohio Wesleyan University offers a solid liberal arts education devoid of bells and whistles. "Ohio Wesleyan is a place where the education goes well beyond the classroom," explains one student. "The family atmosphere and the opportunities that the college provides you in a diverse environment enrich the entire college experience."

If You Apply To ➤ **Ohio Wesleyan:** Early decision: Dec. 1. Early action: Dec. 15. Financial aid: Mar. 1. Housing: May 1. Does not guarantee to meet full demonstrated need. Campus interviews: recommended, evaluative. Alumni interviews: optional, informational. SATs or ACTs: required. SAT IIs: recommended. Accepts the Common Application and electronic applications. Essay question.

University of Oklahoma

1000 Asp Avenue, Room 127, Norman, OK 73019

Football aside, OU has historically been outclassed by neighbors like the University of Texas at Austin and the University of Kansas. But the signs of improvement are there. Check out the Honors College, which boasts a living/learning option. OU is strong in engineering and geology-related fields.

Website: www.ou.edu
Location: Suburban
Total Enrollment: 23,921
Undergraduates: 18,288
Male/Female: 49/51
ACT Range: 23–28
Financial Aid: 17%
Expense: Pub $ $
Phi Beta Kappa: Yes
Applicants: 7,829
Accepted: 81%
Enrolled: 51%
Grad in 6 Years: 56%
Returning Freshmen: 85%
Academics: ✍ ✍ ✍
Social: ☎ ☎ ☎
Q of L: ★ ★ ★
Admissions: (405) 325-2252

The University of Oklahoma has more to brag about than football. Indeed, President David Boren—a former U.S. Senator and Oklahoma governor—has made it a priority to improve both academics and the physical appearance of the state's flagship campus. "There are new buildings everywhere, because President Boren has found the resources to upgrade our university in every way," says a satisfied senior. Couple that with a rigorous honors program and a genuine friendliness among the student body, and it's easy to understand this favorite saying: "Sooner born and Sooner bred, when I die, I'll be Sooner dead!"

Located about 18 miles south of Oklahoma City, OU's 2,000-acre Norman campus features tree-lined streets and predominantly redbrick buildings. Many are historic in nature, and built in the Cherokee or Prairie Gothic style. The Norman campus houses 13 colleges; seven medical and health-related colleges are located on the OU Health Sciences Center campus in Oklahoma City, and programs from colleges on both campuses are also offered at OU's Schusterman Center in Tulsa. Since 1994, nearly $1 billion in construction projects have been completed, including the new National Weather Research Center, which happens to be the largest weather research center of its kind in the nation.

All Oklahoma freshmen start out in University College before choosing among OU's degree-granting institutions, including colleges of Architecture, Education, and Fine Arts. The College of Engineering offers aerospace, civil, mechanical and environmental engineering. In January 2006, the new College of Earth and Energy was chartered, consolidating resources related to petroleum and geological engineering, meteorology, geology and geophysics, and geography. OU's petroleum program ranks among the best in the nation, and in the College of Arts and Sciences, the natural sciences, notably chemistry, are strong. The Michael F. Price College of Business, named for the superstar investment manager, class of '73, offers a major in entrepreneurship and venture management.

Other well-recognized programs at OU include majors in Native American studies and energy management; the OU Native American studies program teaches more Native American languages for college credit than any other institution. The College of Education's rigorous five-year teacher-certification program, Teacher Education Plus (TE-PLUS),

> "There are new buildings everywhere, because President Boren has found the resources to upgrade our university in every way."

incorporates field experience, mentoring, and instruction from 30 full-time professors. Newer majors include architectural engineering, Chinese, East Asian studies, and Latin American studies; there are also new minors in aviation, enterprise studies, Hebrew, naval science, and non-profit organizational studies.

OU's general education requirements consist of three to five courses in symbolic and oral communication, including English composition; two courses in natural science; two courses in social science; four humanities courses; an upper-division general education course outside the major; and a Senior Capstone Experience. The Honors College offers small classes with outstanding faculty members, and independent study, along with its own dorm. Top students may also apply for the Scholarship-Leadership Enrichment Program, under which well-known lecturers given seminars at the university for academic credit.

Although OU is one of the smaller Big 12 schools, it can still overwhelm a harried freshman. Classes are large, and 36 percent of those taken by undergraduates are taught by graduate students, rather than full-fledged faculty. However, OU has made efforts to address the issue and is now one of the nation's few public universities to cap the class size of first-year English comp courses at no more than 19. A junior says, "Class sizes range from 300 in the most basic general education classes to about 20 students in upper division and honor courses." In the past decade, increased private support has helped OU nearly quadruple its endowed faculty positions, helping the school to attract and retain talented professors. "I believe that the university has some of the best professors in the country," says a student. "Each one has an apparent love for their subject and has acquired a great deal of experience through hands-on work."

OU's student body is primarily homegrown; 75 percent hail from within the Sooner state. African Americans and Asian Americans each make up 5 percent of the student body, while Hispanic students comprise 4 percent, and Native Americans add 7 percent. "Students who attend the University of Oklahoma are driven and care about their future," says an advertising major. "They value traditions." Seventeen percent of OU students receive scholarships

> "Every football game day, the town swells to over 500,000 people, and I consider game days an all-day festival."

based on academic merit, with awards averaging $1,038. There are also 362 athletic scholarships in a variety of sports, and the Presidential Travel and Study Abroad Scholarships, which offer $250,000 for students and faculty to study and conduct research around the globe.

Twenty-eight percent of students live in the university's residence halls, most of which are co-ed by floor. The three tower dorms, the most popular choice for freshmen,

(Continued)

Email Address:
admrec@ou.edu

Strongest Programs:
Meteorology
Finance and Accounting
History of Science
Chemistry and Biochemistry
Petroleum and Geological
  Engineering
Music
English
History

The College of Education's rigorous five-year teacher-certification program, Teacher Education Plus (TE-PLUS), incorporates field experience, mentoring, and instruction from 30 full-time professors.

Although OU is one of the smaller Big 12 schools, it can still overwhelm a harried freshman. Classes are large, and 36 percent of those taken by undergraduates are taught by graduate students, rather than full-fledged faculty.

are currently undergoing renovations, and many students already benefit from new carpet and tile in the bathrooms, and modular furniture that can be arranged in a variety of ways. Upperclassmen may also choose the OU Traditions Square apartment community. Furnished units in the complex include in-unit washers and dryers and full kitchens, plus high-speed wired and wireless Internet access and access to a fitness area, a pool, sand volleyball courts, and a putting green. There are a number of adequate dining options and, according to one senior, "The desserts are catered by an outside bakery and are to die for!" Students say they feel safe on campus. "The campus police department is well-staffed," observes a senior, and the landscape is dotted with call boxes.

Fourteen percent of men and 23 percent of women go Greek, but only 9 percent of students live in Greek housing. Although the dorms and Greek houses are dry, fraternity parties are the highlight of weekends at OU. There are plenty of other options, too, including mixers and movies in the student union, and trips across the street to Campus Corner for shopping, dining, and entertainment. The town of Norman, population 100,000, is Oklahoma's third-largest city, and Oklahoma City is just 20 minutes away. Other popular diversions include the annual road trip to Dallas for the OU–Texas game, the first-of-the-football-season Big Red Rally, the Medieval Festival, and the University Sing, a talent show. The Big Event also brings thousands of students into the community for a day of service each year.

> **"Students who attend the University of Oklahoma are driven and care about their future."**

OU sports powerhouse athletic teams. Sooner football remains strong under coach Bob Stoops. "OU–Texas is insane," says a letters major. "I can't accurately describe it in words. Every football game day, the town swells to over 500,000 people, and I consider game days an all-day festival." The women's basketball squad went undefeated during the 2005–06 regular season and won the Big 12 Tournament. Other recent winners include the men's golf team and the men's and women's gymnastics teams.

Students at Oklahoma have a lot to brag about. In fact, says a senior, "I've spent most of my life living in cities where people are obsessed with looking down on people from everywhere else. The first few days in Norman opened my eyes to how much better it would be if the rest of the U.S. was filled with a few more people from the University of Oklahoma." If you're searching for a school with plenty of spirit and a feeling of family, OU may be worth a look—Sooner, rather than later.

Overlaps

Oklahoma State, Baylor, Texas A&M, Texas Tech, University of Texas–Austin

If You Apply To ➤ **OU:** Rolling admissions: Apr. 1. Financial aid: Mar. 1. Meets demonstrated need of 45%. Campus interviews: optional, informational. No alumni interviews. SATs or ACTs: required. No SAT IIs. Accepts electronic applications. No essay question.

Franklin W. Olin College of Engineering

Olin Way, Needham, MA 02492

The new kid on the block among engineering schools, Olin opened its doors in 2002 with an innovative curriculum, a personalized approach to instruction, and the best students that money can buy. Every accepted student gets a full-ride merit scholarship. Clearly the best-kept secret in U.S. higher education—but not for long.

Founded in 1997, the Franklin W. Olin College of Engineering admitted its inaugural freshman class in 2002 and held its first graduation ceremony in May 2006. Sure, the college lacks the rich tradition and reputation of more established institutions. But that doesn't seem to bother the more than 300 students who call this campus home. Instead, they've embraced the college's innovative engineering curriculum and are working hard to create history of their own. In the past five years, Olin "has built its campus, opened its doors to its first students, and developed into the most dynamic engineering school in the nation," says a junior. What's more, the college aims to lure super-bright students away from MIT and Caltech with full-ride scholarships. Says one happy senior: "Olin has quite literally gone from a hole in the ground to a home."

Olin's 70-acre campus is located adjacent to Babson College in a pleasant suburb less than 20 miles west of Boston. The campus design is an innovative blend of the traditional and futuristic. Five buildings curve around a central green space, creating a sense of community and echoing the design of the traditional New England college. The entire campus is wired for high-tech communications, and designed for easy updating to stay on the cutting edge.

> "Olin has quite literally gone from a hole in the ground to a home."

The classrooms make use of state-of-the-art instructional media, while the residence halls offer unparalleled desktop connectivity. There is also plenty of meeting and public space to encourage the kind of collaboration called for in modern-day engineering. East Hall opened in 2005. It is a residence hall that features single- and double-occupancy rooms, as well as music practice rooms, an exercise room, study rooms, and lounge areas.

Olin's innovative curriculum is based on the "Olin Triangle," which emphasizes science and engineering, business and entrepreneurship, and the liberal arts. Students choose from three majors—general engineering, electrical and computer engineering, and mechanical engineering—and a number of concentrations including bioengineering, computing, materials science, and systems. In addition, students must complete 30 credits of math and science (10 of which must be in math) and 28 credits of arts, humanities, social science (AHS), and entrepreneurship, 12 of which must be in AHS. Every student also completes a major, yearlong senior capstone project known as SCOPE (Senior Consulting Program for Engineering).

Because the academic menu is so narrowly focused, weak programs are virtually nonexistent. While the courses are demanding, students say that the intellectual environment fosters cooperation rather than competition. "The academic climate is intense but non-competitive," says a junior. "There is not a lot of pressure to keep a certain GPA and no ranking is ever published." A senior warns, "Long nights and weekend work is a given." Eighty percent of classes have 50 or fewer

> "Long nights and weekend work is a given."

students and are led by professors. "The faculty is absolutely incredible," says one student. "They are all engaging and have interesting ideas for projects and new ways to study the material."

First-year students have an opportunity to participate in an interactive, weeklong orientation program that includes team-building exercises, meetings, and meals with faculty and advisors, as well as a trip into Boston. The college also encourages students to engage in "Passionate Pursuits," by enabling them to pursue artistic or humanistic interests via non-degree credit projects. Recent student projects include trapeze lessons, exploring Kung Fu, video game design, and conversational Arabic.

Olin students are "self-motivated, friendly, and generally happy," says a mechanical engineering major. "They are not typical dorks," adds a junior. "They are sociable, fashionable, communicative, compassionate, sophisticated, and fun-filled." Only 9 percent hail from Massachusetts, and an overwhelming majority graduated in the top tenth of their high school class. African Americans comprise 2 percent of the

Website: www.olin.edu
Location: Urban
Total Enrollment: 304
Undergraduates: 304
Male/Female: 58/42
SAT Ranges: V 690–780
  M 720–800
ACT Range: N/A
Financial Aid: 4%
Expense: Pr $
Phi Beta Kappa: No
Applicants: 784
Accepted: 18%
Enrolled: 63%
Grad in 6 Years: N/A
Returning Freshmen: 97%
Academics: ✍ ✍ ✍
Social: ☎ ☎ ☎
Q of L: ★ ★ ★
Admissions: (781) 292-2203
Email Address:
  Allison.bahme@olin.edu

Strongest Programs:
General Engineering
Electrical and Computer
  Engineering
Mechanical Engineering

All Olin students live on campus and "housing is given on a seniority basis," says a senior.

student body, Asian Americans 9 percent, and Hispanics 4 percent.," says a senior. Hot-button issues include environmental awareness, homosexuality, abortion, and the war in Iraq. "Students are generally liberal," notes a senior. Every accepted student receives a merit-based scholarship—valued at more than $130,000—that covers tuition for all four years.

All Olin students live on campus and "housing is given on a seniority basis," says a senior. "The rooms are comfortable, large, and well-furnished," reports one student. "The halls include lounges, kitchens, laundry and exercise rooms, as well as space dedicated for homework and team study." The dining hall provides students with a tasty—if repetitive—meal selection that includes "a fresh salad and cold cuts bar, pizza and pasta, and usually two or three entrees for every meal." Campus security is described as "unbelievable" and "officers are amazingly quick to respond to any call," according to a junior. "One look at the campus crime statistics will show you that there's not much to worry about."

When students aren't laboring over the latest engineering project or Structural Biomaterials assignment, they tend to congregate on campus for fun. "There are parties in the residence halls and the Student Activities Committee hosts some sort of schoolwide event every weekend," says a junior. There are no Greek organizations, and students say the social scene does not revolve around alcohol. "The campus alcohol policies are quite reasonable," says a student, "relying largely on student responsibility." When the campus scene grows tiresome, students often travel to nearby Babson or Wellesley to mingle. There are also frequent roadtrips to Boston ("the best college town in the nation"), Vermont, or the beaches of Maine.

"Students are generally liberal."

One student describes the surrounding town of Needham as "a quaint bedroom community that tries to be a college town." An engineering major says, "There are good restaurants nearby, but stores in town close early and there isn't much nightlife." Still, students take advantage of volunteer opportunities and "we also try to have some sort of larger event with the Needham community every month or so," says a student. Such events have included a Halloween labyrinth for local school children and "an event where Olin students auctioned off their skills or services to community members for charity."

Although Olin does not offer varsity sports, "a number of pick-up leagues have evolved for soccer, Frisbee, football, and basketball," says a junior. "Some students also play on teams at Babson or Wellesley for sports such as lacrosse, rugby, and cross-country." Other popular intramural sports include ultimate Frisbee, soccer, volleyball, and dodgeball.

For those who have what it takes, Olin College offers a top-notch engineering degree at a bargain price. Olin students have watched their school grow up before their eyes. "Five years ago we had no students, no buildings, no classes, and no curriculum," says a senior. "Olin now has 300 students, over 30 faculty, two residence halls, an innovative curriculum, and a supportive, enthusiastic community." Says a student, "Students who are passionate about things and excited about engineering would do well here."

Overlaps

MIT, Harvey Mudd, Carnegie Mellon, Caltech, Stanford, Cornell

If You Apply To ➤

Olin: Regular admissions: Jan. 1. Guarantees to meet full demonstrated need. No campus or alumni interviews. SATs or ACTs: required. SAT IIs: required (math, science, and one other). Prefers electronic applications. Essay question: discuss a core value; what are your passions?

University of Oregon

Box 1226, Eugene, OR 97403-1226

UO may be the best deal in public higher education on the West Coast. Less expensive than the UC system and less selective than the University of Washington, UO is a university of manageable size in a great location. The liberal arts are more than just a slogan, and programs in business and communication are strong.

Blend two vegetarians, one track star, one frat brother, two tree huggers, three hikers, and one conservative. What have you got? Ten OU students. Sure, the joke's hokey, but its offbeat humor is typical of the laid-back, slightly eccentric attitude that prevails here in Eugene, where bicycles are the main form of transportation, recycling is a requirement, and littering is *déclassé*.

UO's buildings date from as early as 1876 and are surrounded by the university's lush 295-acre arboretum-like campus, which boasts 2,000 varieties of dew-kissed trees. Most academic buildings were built before World War II and represent a blend of classical styles, including Georgian, Second Empire, Jacobin, and Lombardic. Residential facilities range from 19th-century Colonials to modern high-

> "Journalism, natural sciences, and business are competitive, while social sciences are more laid-back."

rises. Current construction includes the Living Learning Center, a new residential facility that integrates spaces for academic classes, study groups, faculty advising, informal musical and theatrical performances, dining, and living.

While a liberal arts emphasis underlies Oregon's entire curriculum, general education requirements are not highly structured. The calendar is composed of quarters, and students must take two terms of English composition, two years of foreign language (for a BA), one year of math (for a BS), and two courses exploring American or international culture, identity, pluralism, and tolerance, plus six courses in each of three areas: arts and letters, social sciences, and natural sciences. Freshman seminars introduce students to top professors in small-group settings, and profs have to apply to teach them, a process students applaud. Freshman Interest Groups help new students develop close working and advising relationships with faculty members.

OU's professional schools—journalism, architecture and allied arts, education, law, business, and music—are highly regarded, with journalism, education, and architecture drawing much student praise. The School of Architecture and Allied Arts is the home of Oregon's only accredited degrees in architecture, landscape architecture, and interior architecture. Of the more than 35 departments in the College of Arts and Sciences, students give high marks (and high enrollments) to psychology and biology, and the science departments within this particular college offer many opportunities for research. A new undergraduate major in marine biology leads to either a BA or a BS, and a major in mathematics and computer science has been added.

Highly motivated undergraduates may join the Honors College, a small liberal arts college with its own courses. The student-run community internship programs provide credit for community volunteer work, while on-campus internships allow students to earn credit for work with university organizations and academic departments. In keeping with Oregon's eco-friendly reputation, the Green Chemistry Laboratory and Instrumentation Center was the first in the nation to use nontoxic materials in experiments. Pine Mountain Observatory, a field-study resource for astronomy and physics students located high in the Cascade Mountains, and the

Website: www.uoregon.edu
Location: Small city
Total Enrollment: 20,033
Undergraduates: 15,985
Male/Female: 47/53
SAT Ranges: V 498–617
  M 503–615
Financial Aid: 40%
Expense: Pub $ $
Phi Beta Kappa: Yes
Applicants: 10,012
Accepted: 90%
Enrolled: 31%
Grad in 6 Years: 64%
Returning Freshmen: 84%
Academics: ✏ ✏ ✏ ½
Social: ☎ ☎ ☎
Q of L: ★ ★ ★ ★
Admissions: (800) BE A DUCK
Email Address:
  uoadmit@uoregon.edu

Strongest Programs:
Architecture
Music
Creative Writing
Business
Chemistry
Journalism and
  Communication

The student-run community internship programs provide credit for community volunteer work, while on-campus internships allow students to earn credit for work with university organizations and academic departments.

Oregon Institute of Marine Biology give students a chance for intimate studies in their major.

"Certain majors are more competitive than others," warns a junior. "Journalism, natural sciences, and business are competitive, while social sciences are more laid-back." Professors are lauded for their skills behind the lectern, although it's not uncommon to find TAs handling some of the teaching duties.

Students are "different and unique," says a senior, "from the most hard-line Republican to the freest hippie spirit." In general, every part of the white Anglo-Saxon spectrum is well represented, with an especially heavy dose of the athletically inclined. There is a noticeable contingent of international students, who account for 6 percent of the student body. Asian Americans account for another 6 percent, African Americans 2 percent, and Hispanics 3 percent. The 24 percent of students who come from out of state are mostly from California. Numerous merit scholarships worth an average of $1,870 and 304 athletic scholarships are awarded to qualified students.

> "Our residence halls have a reputation for being old and cramped, but they are the best way to meet people as a UO freshman."

Twenty-one percent of UO students choose to live in university housing. There are a number of thematic living arrangements—a crosscultural dorm, an academic-pursuit residence hall, and a music dorm. Students recommend Walton, Carson, and Barnhart Halls. Rooms in the residence halls tend to be small but clean and comfortable, and they have Internet connections. "Our residence halls have a reputation for being old and cramped, but they are the best way to meet people as a UO freshman," explains a junior math major. Another plus is "the convenience of living on campus," adds one student. "I could wake up 15 minutes before class and still be on time while living on campus." Any student can sign up for the meal plan in the two main dining halls; restaurants in the student union and off-campus fast food round out the menu. "The dining halls are so good. There are many different choices as well as options for vegan and vegetarian."

Eugene is "the best college town ever! Everything about Eugene is based around the Ducks!" one student raves. Popular hangouts include Old Taylor's and Rennies, and community and public service projects also draw crowds. The one drawback to all this fun is Oregon's weather: it rains and rains. "Eugene gets some sunny days in early fall, late spring, and summer," reports a veteran. Still, the moist climate rarely dampens enthusiasm for the many expeditions available through the university's well-coordinated outdoor program, from rock climbing to skiing. An hour to the west, the rain turns to mist on the Pacific Coast; an hour to the east, it turns to snow in the Cascade Mountains. Students can also escape the weather year-round in the recreation center, complete with rock climbing wall and juice bar.

> "Our campus is totally dry by policy."

Oregon's professional schools—journalism, architecture and allied arts, education, law, business, and music—are highly regarded, with journalism, education, and architecture drawing much student praise.

Eight percent of UO men and women join Greek organizations, which provide living space and interesting social diversions. Oregon's 21-year-old drinking age means that alcohol is banned from college-owned dorms, but students claim this rule can be broken. One student explains that, "Our campus is totally dry by policy. The only students that are allowed to drink must be of age and do so with their doors closed. Of course, these rules are broken and the Department of Safety does punish students who get caught." Another student concurs and adds, "I feel that this works very well because every staff member goes the extra mile to provide other fun, non-alcoholic events." Major events include the Eugene Celebration, the Oregon Country Fair, the Martin Luther King, Jr., Festival, and weekly street fairs attended by local vendors. University Day, which happens each spring, offers students, faculty, and staff an opportunity to clean up their campus. The Willamette Valley Wine Festival is a fun road trip.

Second only to parties, athletic activities head the list of favorite free-time activities. The Ducks' biggest athletic rival is the Oregon State Beavers, and each year the Civil War game in football is huge. When students aren't on the tracks and fields themselves, they're trooping down to the stadium to join the Quacker Backers in cheering on the successful basketball and football teams. Track is also prominent; Nike is headquartered in nearby Beaverton.

A recent University of Oregon Orientation Week T-shirt sported a picture of a duck and a simple exhortation: "Let your future take flight." UO offers ample opportunities for those with lofty ambitions to succeed. Indeed, UO's caring faculty, excellent academics, and abundance of social activities reveal that UO is all it's quacked up to be.

Overlaps
University of Washington, University of Colorado, Oregon State, UC–Santa Barbara, UC–Davis

If You Apply To ➤

UO: Rolling admissions. Financial aid: Feb 1. Housing: Mar. 31 (guaranteed space). Does not guarantee to meet demonstrated need. Campus interviews: optional, informational. No alumni interviews. SATs or ACTs: required. SAT IIs: required. Prefers electronic applications. Essay question.

Oregon State University

Corvallis, OR 97331-2106

The biggest dilemma facing the typical 18-year-old Oregonian is whether to be a Beaver or a Duck. Choose Duck and hang with the ex-hippies in cosmopolitan Eugene. Choose Beaver and get small-town life with professional programs in business, engineering, and life sciences in Corvallis.

With a wide range of academic programs, Oregon State University could very well be the setting of its own movie, titled *Planes, Trains, and Submarines*. You see, Oregon State is one of just a handful of universities in the country with land, sea, and space grant designations. Though the school might be happy to forget its many years of being called Moo U, that doesn't mean its agriculture department should go unnoticed. In fact, many of the contributions made by Oregon State researchers center on the field of agriculture. Still, there's more to OSU than fruits and vegetables. The school is strong in many departments, including biotechnology, forestry, and engineering. Says one satisfied freshman, "If you are looking for a college that is close-knit and has a friendly 'college town,' OSU is the place to go!"

Located in the pristine but rainy Willamette Valley, OSU's campus is a mix of older, ivy-covered buildings and more modern structures. In addition to the 500-acre main campus, OSU owns 13,000 acres of forestland near campus and numerous agricultural tracts throughout Oregon. Thousands of azalea and rhododendron bushes welcome springtime on campus with their colorful blooms, and summers are unfailingly sunny. (All that rain during the rest of the year has to be good for something, right?) The most recent additions to the campus include a forest resource lab, a residence hall, and a baseball field.

OSU's College of Liberal Arts ranks with Business and Engineering as the largest on campus, but there are many more preprofessionals than poets. With the exceptions of history and English, the liberal arts—including such standard fare as sociology, psychology, economics, and philosophy—play second fiddle to more practical, technical fields. The departments of engineering (with its up-to-date electrical and

Website:
  www.oregonstate.edu
Location: Small city
Total Enrollment: 19,236
Undergraduates: 15,106
Male/Female: 50/50
SAT Ranges: V 470–590
  M 490–610
ACT Range: 20–26
Financial Aid: 51%
Expense: Pub $ $
Phi Beta Kappa: No
Applicants: 6,552
Accepted: 93%
Enrolled: 48%
Grad in 6 Years: 62%
Returning Freshmen: 80%
Academics: ✍ ✍
Social: ☎ ☎ ☎
Q of L: ★ ★ ★
Admissions: (541) 737-4411

(Continued)

Email Address:
osuadmit@oregonstate.edu

Strongest Programs:
Agriculture
Biotechnology
Forestry
Engineering
Business

computer engineering building) and forestry are major drawing cards, and even though agriculture doesn't lure as many students as it used to, those who do come find excellent programs. Business administration is the most popular major, followed by exercise and sport science, liberal studies, general science, and psychology. The business school offers some of the finest business-related programs in the state, while the health and human performance program—a euphemism for home economics—has expanded its offerings.

"If you are looking for a college that is close-knit and has a friendly 'college town,' OSU is the place to go!"

OSU's extensive Baccalaureate Core requires courses in a variety of areas, including skills; perspectives; and difference, power, and discrimination. One writing-intensive course is required as well. Perhaps the core's most innovative facet is its "synthesis" requirement, in which upperclassmen take two interdisciplinary courses on global issues in the modern world. The campus's global awareness is also evident in the new international degree, which can be coupled with any other course of study. Thus, students can earn a BS in forestry and a BA in international studies in forestry simultaneously. The level of academic pressure varies by major, but even those in the various honors programs say they don't feel overworked. "The courses are very rigorous, but if you apply yourself you will surely succeed," says a communications major. Most professors are "so captivated by their area of study that they are eager to enlighten others with their knowledge and interests," says a sophomore. Freshman have full professors for most courses, excluding recitation.

Of particular note is OSU's Experimental College, where undergraduates spice up their semesters with noncredit courses in a range of imaginative subjects—everything from wine tasting to the art of bashing (a medieval war technique). Those who can afford a semester abroad may study at universities in England, France, Australia, Mexico, New Zealand, China, and Japan—or participate in individual exchange programs in still more countries. (Oregon State returns the favor, playing host to about 1,500 foreign students from more than 90 nations each year.) The university's small-town location (population 45,000) makes it difficult to find much career-oriented, part-time employment, and term-time internships are hard to come by. (OSU operates on a quarter system.) Students in almost all majors, however, can participate in the cooperative education program, which allows them to alternate terms of study with several months of work in a relevant job.

"The students are a good mixture of city and rural students."

Statistically, Oregon State certainly doesn't offer the most diverse student body. Eighty percent of the students are from Oregon, and the minority population is 1 percent African American, 4 percent Hispanic, and 8 percent Asian American. "The students are a good mixture of city and rural students," says a senior. Big campus issues include abortion and gay/lesbian rights. Most Oregon Staters are conservative and "very all-American—not cowboys and not city slickers, but very middle-of-the-road in all respects," a business major observes. Says an engineering major, "The college has its intelligentsia, its social butterflies, its determined athletes, and any combination of those."

The departments of engineering (with its up-to-date electrical and computer engineering building) and forestry are major drawing cards.

Freshmen are expected to live in college housing, though fraternity pledges have the option of living in their houses. Co-ed and single-sex options are available in the comfortable and well-maintained dorms, which house 22 percent of the students. "The residence halls are comfortable and well-maintained, for the most part," says a forestry major. "They also offer many opportunities to get involved and meet new people." In addition to standard rooming situations, a new "wellness" hall offers an exercise room and low-calorie meals. Students can also choose life in one of eight cooperatives or the plentiful off-campus apartments. Foraging for food on your own generally beats so-so dorm grub; the better chow at frat and sorority houses is one motivation for students to go Greek; 9 percent of the students do so. Alcohol still flows freely at Greek affairs, but crackdowns by the administration and local police have begun to curb the most wanton debauchery.

Cheering for the Beavers' nationally ranked wrestling team, which has had several top 10 national finishes in the past few years, demands a lot of students' time and energy here, as does participation in the well-rounded intramural program. Benny Beaver, the school's former (and somewhat benign) mascot has been replaced by a more aggressive beaver that students have dubbed the "angry beaver." Other talented varsity squads include women's gymnastics, basketball, and volleyball, and men's football and baseball; the basketball team has the eighth-winningest program of all time among NCAA Division I schools. As for rivalries, one student says, "Civil War games between OSU and U of Oregon are a big part of every season."

Another popular student activity is complaining about the Willamette Valley weather: "People in the valley don't tan, they rust," warns one native. One reward for this sogginess, however, is the abundance of flowers that bloom in every color and shape each

"The residence halls are comfortable and well-maintained, for the most part."

May. Many students consider Corvallis a good-size town; a number of bars and cheap theaters cater to their entertainment needs. Beautifully rugged beaches are less than an hour away, and some of the best skiing in the country can be found in the Cascade Mountains, two hours east. Hiking and rafting are nearby, too, and trips to powwows in the area and camping on the coast provide other good times. Popular campus traditions include the Greeklife Sing (featuring musical numbers staged by fraternity and sorority members) and the annual Fall Festival.

OSU continues to build on its solid reputation as an agricultural institution, marching forward as a faithful part of the state's university system. OSU doesn't scream for attention. Instead, it's content to be a "nice" college, in "a safe and pleasant little town," where professors are "helpful" and, even if everyone doesn't know your name, they'll lend you an umbrella whenever the skies open up.

Popular campus traditions include the Greeklife Sing (featuring musical numbers staged by fraternity and sorority members) and the annual Fall Festival.

Overlaps
University of Oregon, Portland State, Western Oregon, University of Washington, Washington State

If You Apply To ➤ **Oregon State:** Early action: Nov. 1. Regular admissions: Feb. 1 (priority). Does not guarantee to meet full demonstrated need. Campus interviews: optional, informational. No alumni interviews. SATs or ACTs: optional. No SAT IIs. Accepts electronic applications. No essay question.

University of the Pacific

3601 Pacific Avenue, Stockton, CA 95211

The university's name dates from a time when there were no other universities near the Pacific. Pacific is the only small, independent university in California north of LA and offers an eye-popping array of programs for an institution its size, including business, engineering, pharmacy, and education. The student body is equally diverse.

University of the Pacific looks like more than 100 acres of New England plunked down in California wine country. With its stately combination of redbrick and ivy, it could be mistaken for an East Coast liberal arts college. But instead of a blanket of snow, Pacific is surrounded by the lush greenery of the San Joaquin Valley. On

Website: www.pacific.edu
Location: Suburban
Total Enrollment: 6,196

campus, this increasingly competitive bastion of learning offers its more than 3,400 undergrads a solid and diverse academic program and scores of things to do when not hitting the books.

With majestic evergreens and flowering trees, Pacific is home to six undergraduate schools and the College of the Pacific, the university's liberal arts and sciences division. There is also a school of law in Sacramento and a superlative school of dentistry in San Francisco. Recent campus construction includes an expansion and upgrade to the library, renovations of residence halls, and a new baseball field.

Strong departments abound in the schools of engineering, pharmacy, and business (with special programs in the arts/entertainment management and entrepreneurship) as well as the sciences, English, communication, and international studies. Business is the most popular major, followed by engineering, pharmacy, biology, and education.

The university-wide general education program has three components: the Pacific seminars, the breadth program, and fundamental skills. All entering students must complete Pacific Seminar I ("What is a Good Society?") and Pacific Seminar II ("Topical Seminars on a Good Society") in sequence during their first year and Pacific Seminar III ("The Ethics of Family, Work, and Citizenship") in their senior year. In addition to the seminars, students must complete six or nine courses in the breadth program, and must demonstrate competence in writing, math, and reading.

"Because the class size is small, the students receive a personalized academic experience."

A number of internship and co-op programs are available, and students are guaranteed to have the opportunity for some type of experiential learning. Students may also design their own majors with faculty approval. An extensive study abroad program offers 200 choices in dozens of countries, and international studies majors are required to complete one.

The academic climate varies by department, with students citing the science fields as the most rigorous. Studying accounts for anywhere from 10 to 40 hours a week. The university guarantees graduation in four years (assuming the student follows all university guidelines), or it will pay for the extra schooling. Professors at Pacific get raves for accessibility and personal attention. "The teaching is tremendous, with the experience and care they have for their students," says one senior. "Because the class size is small, the students receive a personalized academic experience." Classes average 20 students, and TAs teach labs only.

"Students at Pacific come from a variety of settings and backgrounds. Some can be very studious, but they know how to balance coursework and a personal life very well," says a senior. Eighty-three percent of Pacific students are Californians. Hawaii and Colorado are also strongly represented, while foreign students make up 2 percent of the student body. As for ethnic diversity, Asian Americans account for 29 percent, while African Americans and Hispanics together make up 13 percent. Students come from a mix of economic backgrounds; 67 percent receive financial aid. The school is middle-of-the-road to conservative, though politics in general play a small role on campus. "We're very open to all political, religious, sexual orientations, etc.," says a senior. "We have a little bit of everything." Though not unusually

"We're very open to all political, religious, sexual orientations, etc."

expensive by national standards, the university price tag can seem steep when compared to the University of California system. So Pacific has stepped up efforts to compete using merit scholarships as well as scholarships to athletes in a long list of sports. The financial aid packages generally get good reviews.

Freshmen and sophomores are required to live on campus, and the few complaints mostly center on aging facilities, some of which have undergone much-needed renovations. Grace Covell Hall, with 350 people, is the largest residence, and two additional

(Continued)
Undergraduates: 3,457
Male/Female: 42/58
SAT Ranges: V 530–630
  M 550–670
ACT Range: 22–27
Financial Aid: 87%
Expense: Pr $ $
Phi Beta Kappa: No
Applicants: 5,869
Accepted: 56%
Enrolled: 24%
Grad in 6 Years: 65%
Returning Freshmen: 85%
Academics: ✑ ✑ ✑
Social: ☎ ☎ ☎
Q of L: ★ ★ ★ ★
Admissions: (209) 946-2211
Email Address:
  admissions@pacific.edu

Strongest Programs:
Business
Pharmacy
Engineering

The university guarantees graduation in four years (assuming the student follows all university guidelines), or it will pay for the extra schooling.

suite-style residence halls recently opened. Fifty-eight percent of undergrads make their home on campus. In terms of quality of life, however, residential life is praised for plentiful social programming. The school occasionally hosts big-name concerts and other campuswide events. Students complain, however, that much social life centers on the Greek scene; 19 percent of men and 18 percent of women go Greek. With three meal plans, two dining halls, and one fast-food-type facility, residents are well fed.

For weekend excitement, Pacific students love to hit the road: Within about two hours, they can be skiing, shopping in San Francisco, or surfing in Monterey. Stockton itself (population 280,000) offers shopping and plenty of fast-food joints, as well as numerous volunteer opportunities. Social opportunities on campus are offered by the Residence Hall Association, intramural and club sports, conservatory and drama/dance programs, Division I athletics, campus movies, sororities and fraternities, and more than 100 student clubs. The school enforces state law, prohibiting all students under 21 from drinking. Students caught violating that policy must take an online course in alcohol education. The majority of Greek houses are designated substance-free, following a national trend enforced by their national organizations. "There are opportunities for those who are interested, and our officers are here more to protect the students rather than bust students," notes a senior. Annual campus festivities include Diversity Week, International Spring Festival, and the popular Fall Festival and Greek Week. In sports, basketball, baseball, softball, volleyball, swimming, golf, and tennis were all recent Big West champions.

"The school has a New England feel to it on a West Coast campus."

Pitted against the state's immense public university system, Pacific stands out for offering major university opportunities in a small-college setting. And the administration is striving to place more focus on its student body, which is becoming more top-notch and diverse. A senior says, "The school has a New England feel to it on a West Coast campus. Even though it's a small institution, it has a lot to offer in terms of academics and extracurricular activities."

> ## Overlaps
> UC–Davis, UC–Berkeley, Santa Clara, UCLA, St. Mary's, Loyola Marymount

If You Apply To ➤ **Pacific:** Early action: Nov. 15. Regular admissions: Jan. 15. Does not guarantee to meet demonstrated need. Campus interviews: optional, informational. No alumni interviews. SATs or ACTs: required. SAT IIs: recommended. Accepts the Common Application and prefers electronic applications. Essay question: significant experience, what will make your college experience a success, special personal or academic situation.

University of Pennsylvania

1 College Hall, Philadelphia, PA 19104-6376

An Ivy League institution in name, Penn has more in common with places like Georgetown and Northwestern—where the liberal arts share center stage with preprofessional programs. At Penn, that means business, engineering, and nursing. Penn has something else other Ivies don't: school spirit.

Benjamin Franklin would be proud of the way his university has surged in recent years. Once relegated to the bottom of the Ivy League (and confused with Penn State), the University of Pennsylvania is now the first choice for top students who see no conflict between high-level academics and having a life. The undergraduate

> **Website:** www.upenn.edu
> **Location:** Urban
> **Total Enrollment:** 18,913

Undergraduates may work hard during the week, but in contrast to most Ivy League achievers, they leave it behind them on weekends.

School of Arts and Sciences—once on the university's back burner—is now central not only to its undergraduates, but also to the remaining three undergraduate schools that tap into its programs and course offerings. Penn established the nation's first medical school, the first business school, the first journalism curriculum, and the first psychology clinic. In her inaugural address, a former president paid tribute to Franklin as "the ultimate visionary and pragmatist. Franklin thought education should be for the body as well as for the soul—that it should enable a graduate to be a breadwinner as well as a thinker, that it should produce socially conscious citizens as well as conscientious bankers and traders," she said.

Penn is situated in a tree-shaded, totally self-contained, 260-acre nest called University City, which is adjacent to downtown Philadelphia. Its 116 buildings range from Victorian Gothic to postmodern. There are very old structures, such as College Hall, and newer ones, such as Wharton's Huntsman Hall and Skirkanich Hall, home to Penn's bioengineering programs. While many students thrive on Philadelphia's cultural abundance, the school is located in the western part of town, once considered to be dangerous. Even so, a sophomore explains, "The whole area surrounding the campus has gotten a lot nicer. There are better restaurants, shops, theaters, and bowling alleys." The campus continues to evolve as well, with the recent addition of Fisher-Bennett Hall, home of Penn's English department and the McNeil Center for Early American Studies.

Penn's reputation is primarily wrapped up with its 12 graduate schools, especially the prestigious Wharton School of Business, the Annenberg School of Communication, and the well-known law, medical, and veterinary schools. Three of four undergraduate schools—engineering, nursing, and the undergraduate division of Wharton—are also professionally oriented and offer an education that's hard to beat anywhere. The undergraduate College of Arts and Sciences (a.k.a. "The College") has come into its own in the past decade or so, and provides students with high-quality instruction as well as the chance to run into a Nobel laureate here and there.

Finance is the most popular undergrad major, followed by nursing and economics. Penn's anthropology department ranks with Chicago's as perhaps the best in the country, while programs in management and technology are also outstanding. Penn has earned applause in the field of cognitive and computer sciences because of its special program linking psychology, linguistics, and computers with philosophy. Another popular *crème de la crème* interdisciplinary major, Biological Basis of Behavior, combines psychology, biology, and anthropology. Students are allowed to design their own individualized majors, and they can hop from school to school—undergraduate or graduate—in doing so. Penn is also welcoming its first freshman class of 25 students into the Vagelos Program in Life Science and Management. Students will pursue studies in both The College of Arts and Sciences and The Wharton School, exposing them to research and development, biotech start-ups, managed care, and other related issues.

"The whole area surrounding the campus has gotten a lot nicer. There are better restaurants, shops, theaters, and bowling alleys."

At the Wharton School, officials have introduced the Joseph Wharton Scholars program, which emphasizes breadth in the arts and sciences. All undergraduates must attain proficiency in one of the 45 foreign languages taught at Penn. Another added plus that comes with a Penn undergraduate education is the opportunity for early entry (submatriculation) into the university's graduate programs. Juniors may apply to any master's program (continuing into the Wharton MBA program is especially popular) and begin completing graduate requirements during their senior year. Penn offers no co-op programs and discourages full-time internships for credit, remaining true to the Ivy League belief that learning should be based in the classroom. Those who want to explore more exotic classrooms may study abroad at Penn's programs in Italy, Scotland, Japan, France, England, China, Nigeria, Spain, Germany, and Russia, among others. Freshmen are encouraged (but not required) to participate in a seminar program that explores various areas of academic interest, and also in the Penn Reading Project, which involves student and faculty discussion of a common text.

While professors at Penn take their research responsibilities seriously, they are surprisingly accessible to freshman. "My freshman year I was taught economics by the adjuster to Ariel Sharon's administration," says a senior. "I've also been taught history by a Pulitzer Prize winner." Another student adds, "Many of the professors are involved with cutting-edge research and are the leaders in their fields of study." The academic program at Penn is well supplemented by its huge and busy library, which houses more than three million volumes.

Despite all the preprofessional programs, Penn never lets its undergraduates stray too far from the liberal arts. The general education requirements mandate that students take at least one course in each of seven "sectors": society, history and tradition, arts and letters, living world, physical world, humanities and social sciences, and natural sciences and mathematics. Students must also complete one course in each of five "skills and methods" areas, including the study of a foreign culture as well as a mandatory writing requirement. Strict academic policies and demanding professors exacerbate the academic pressure, but while students find the school competitive, not all are complaining. "Peers are allies," explains a senior, "people you can count on to help you succeed." Each year students evaluate every class themselves and publish their findings in a guide.

Thousands of faculty and students give expression to Benjamin Franklin's adage that service to humanity is "the great aim and end of all learning." Penn is a national leader in service learning and service research. Students work with local public school students as part of academic coursework in disciplines as diverse as history, anthropology, and mathematics. There are tons of opportunities to volunteer—from tutoring to Big Brothers Big Sisters to the Ronald McDonald House. "Penn students have historically been extremely involved with the local community and have taken the experiences they've had in the neighborhood with them to the real world," an economics and history double major says.

Penn is a diverse campus; 18 percent of the population is Asian American, 7 percent African American, and 6 percent Hispanic. Ninety-four percent of the students ranked in the top tenth of their high school class and 52 percent come from public high school. "The students are very 'preprofessional,'" says a student. "Many students come to Penn with a clear idea of what they want to major in and where they want to be in ten years." Penn's Diversity Awareness Program for freshmen helps students from various backgrounds blend more harmoniously. Freshman orientation includes skits that depict students in various situations of possible conflict and include discussions about each situation. "Penn students are notoriously down-to-earth," says a student. "I have friends who are world-famous concert pianists, have started their own companies, or are nationally ranked sports players. Everyone is so gifted that no one acts pretentious or thinks they are better than anyone else here."

Penn admits students regardless of need but does not offer any merit, athletic, or academic scholarships. Through its innovative Penn Plan, the university strives to ensure that virtually every family with a college-bound son or daughter, no matter what its income, can benefit from financial assistance. For some this means subsidized loans and other traditional forms of aid. Other possibilities under the Penn Plan, however, include prepayment and borrowing options designed with the rate of inflation in mind and aimed at families in the higher-income brackets. In all, Penn students have six payment-plan options.

There are living/ learning programs in most College Houses for those who are interested in the arts, Asian studies, etc., and want to be surrounded by others with the same interests.

Penn is a national leader in service learning and service research.

"My freshman year I was taught economics by the adjuster to Ariel Sharon's administration."

Nearly all freshmen and 59 percent of all undergraduates live on campus and enjoy a wide range of living options in Penn's 11 "College Houses." Dorms are co-ed. The Quad, home to three of the 11 houses, seems to be the hot spot, described as "beautiful, self-enclosed, and social" and "the greatest place in the world for freshmen." Upperclassmen reluctantly move to the high-rises across campus that look like "prefabricated, 24-story monsters" but do offer more space, as well as kitchens. There are living/learning programs in most College Houses for those who are interested in the arts, Asian studies, etc., and want to be surrounded by others with the same interests. Rather than compete in the lottery for rooms, many juniors and seniors simply head off campus—"for the freedom, plus it's a lot cheaper," a junior says. Some end up in nearby renovated three-story houses in the neighborhood. "Off-campus living is just an extension of student neighborhoods, as we tend to stay in large groups," explains a senior. Like housing, the meal plans are optional (though strongly recommended in the freshman year as an important source of social life), and the food isn't all that bad for institutional fare.

Undergraduates may work hard during the week, but in contrast to most Ivy League achievers, they leave it behind them on weekends. "Penn is known as the 'social Ivy,'" says a senior. "There is so much to do on campus and off campus." The university prohibits underage drinking, but "parties freely serve alcohol to underage drinkers," says one student. More than two dozen fraternities attract 24 percent of the men and provide "your basic meat market scene." Similarly, sororities claim 17 percent of the women. The frats' exclusive claim to the houses along Locust Walk, the main artery on campus, has been undone: After some controversy, it was determined that non-Greeks, too, must be able to live at the social nexus of the campus.

Two big annual events at Penn are Spring Fling, a three-day weekend "nothing short of absolutely incredible fun," and Hey Day, when juniors, donning Styrofoam hats and bearing the president of the university on their shoulders, march down Locust Walk to officially become seniors, taking chomps out of each other's hats as they go. A less formal tradition is the Quad Streak, when uninhibited undergraduates run through the crowd naked.

> "Penn students have historically been extremely involved with the local community."

Road trips include New York City, Washington, D.C., Atlantic City, and even Maine and Florida. Downtown Philadelphia, only a few minutes away by foot, cab, or public transportation, offers enough social and cultural activities to make up for the less-attractive aspects of city living. "Philadelphia is actually a great college city," a senior says. Students frequent sporting events, malls, South Street ("a miniature Greenwich Village"), and, of course, myriad bars and dancing joints.

Penn is more sports-minded than most Ivy schools, and football is the biggie. The team has grown accustomed to sitting on the top of the Ivy League and has sparked a widespread revival of school spirit. Tickets are free for those with a student ID. The Penn–Princeton rivalry is always a crowd pleaser. At the end of the third quarter of each home game, everyone in the stands begins belting out the lyrics of the Penn fight song, and when they get to "Here's a toast to dear old Penn," the students shower the field with burnt toast, "a moment that makes all Penn students proud," gushes a senior. Aside from football, recent Ivy championship teams include men's basketball and wrestling, and women's field hockey, basketball, and fencing. A bevy of intramural sports bring thousands of less-seasoned athletes out to play each year, and all types of athletes benefit from the swanky track and weight-lifting facilities. Each spring, Penn hosts the prestigious Penn Relays, a track-and-field extravaganza that attracts the nation's best track athletes.

While its students work hard, Penn lacks the intellectual intensity of some of the other top Ivies, and you can detect preprofessional undercurrents. But most

Overlaps

Harvard, Columbia, Cornell, Yale, Brown, Stanford

accept it for what it is: a first-rate university where you can live a relatively normal life. Penn is one Ivy League university where no one apologizes for having fun!

If You Apply To ➤

Penn: Early decision: Nov. 1. Regular admissions: Jan. 1. Financial aid: Feb. 15. Guarantees to meet full demonstrated need. No campus interviews. Alumni interviews: optional, evaluative. SATs or ACTs: required. SAT IIs: two required. Essay question: how you and Penn are a good match.

Pennsylvania State University

201 Old Main, University Park, PA 16802

Aside from UVA and UNC, there are precious few public universities more selective than Penn State. With a student body the size of a small city, the university is strong in everything from meteorology to film and television. The Schreyer Honors College is one of the nation's elite honors programs.

Living it up with nearly 40,000 schoolmates in Happy Valley is the perfect college experience for some, though it might not fit everyone's idea of a good time. Those who can muster the energy to take advantage of Penn State's legendary school spirit and immense academic offerings will find plenty of top-notch programs, including stellar choices in the sciences.

Sporting an eclectic architectural mix, including white-columned brick, stone, and some modern apartments, this land grant university continues to experience growth as major renovation and expansion projects continue. Newer facilities at the University include the School of Forest Resources Building and an expansion to the Rec Hall Wrestling/Fitness Center.

Penn State maintains strong programs in the scientific and technical fields such as earth sciences, engineering, agricultural sciences, and life sciences, as well as nutrition and family studies. The meteorology program boasts alumni worldwide, including the founder of AccuWeather, an internationally renowned private forecasting firm that is headquartered here. As a matter of fact, one in every four meteorologists in the United States is a Penn State graduate. The School of Information Sciences and Technology is designed to prepare students for the digital age. The College of Agricultural Science has extensive facilities that include huge livestock barns. Its Food Sciences program is one of the best in the nation. Dairy products from the school's cows are sold at an on-campus store, and courses are offered in the production of its famous ice cream. Students can choose from 261 baccalaureate programs, and 162 graduate fields spread over 24 locations statewide, including the College of Medicine and the Dickinson School of Law, both located near Harrisburg. At a school of Penn State's size, there are bound to be some weaknesses, but the curriculum is in a seemingly constant state of change, with programs added and dropped on a regular basis.

Penn State's general education requirements consist of 45 credits that include several communications and quantification courses as well as humanities, arts, natural sciences, social and behavioral sciences, and health and physical education courses. The incorporation of critical thinking skills has been a priority in redesigning

> "A majority of the students get involved in community service to maintain and constantly improve town relations."

Website: www.psu.edu
Location: Small city
Total Enrollment: 38,262
Undergraduates: 33,208
Male/Female: 55/45
SAT Ranges: V 530–630
  M 570–670
Financial Aid: 80%
Expense: Pub $ $ $ $
Phi Beta Kappa: Yes
Applicants: 29,904
Accepted: 62%
Enrolled: 35%
Grad in 6 Years: 84%
Returning Freshmen: 93%
Academics: ✍ ✍ ✍ ✍ ½
Social: 🏮 🏮 🏮 🏮 🏮
Q of L: ★ ★ ★
Admissions: (814) 865-5471
Email Address:
  admissions@psu.edu

Strongest Programs:
Earth Sciences
Engineering
Agricultural Sciences
Life Sciences
Nutrition
Family Studies

the general curriculum. About 1,800 of the university's best and brightest are invited to participate in the Schreyer Honors College, which offers opportunities for independent study and graduate work, as well as honors options in regular courses. According to one student, "The courses are rigorous and challenging but allow you to be competitive." Students who are part of the honors college take both honors classes and regular university classes. In addition, undergrads must enroll in "diversity-focused" courses that encourage awareness of minority concerns. One helpful program offered to freshmen is LEAP (Learning Edge Academic Program), which gives new students the benefit of a big university while making it seem small. Students in LEAP take a team approach by taking classes together and living together.

Some of the intro-level lecture courses draw up to 400 students at University Park, yet most of the students seem to agree—classes are excellent and require your full attention. For cramming outside of class, the Penn State library contains approximately 4.6 million volumes. A large number of students study abroad, and combined undergraduate/graduate degree options are available, as are co-op programs in engineering, distance learning, and student-designed majors.

Most undergraduates are residents of Pennsylvania, with 23 percent hailing from out of state and 2 percent hailing from outside the nation. Seventy-eight percent ranked in the top quarter of their high school class. More than half of Penn State's undergraduates who finish at University Park began their education at one of the university's 19 campuses across the state. Many students note that race and diversity issues can be pronounced on a campus that is still pretty homogenous for a public university: Asian Americans

"It's very difficult for underage students to be served at the bars."

make up 6 percent of the undergrad population, and African Americans and Hispanics combine for another 7 percent. A whopping 487 athletic scholarships are available, covering all NCAA-approved sports, as are a number of merit awards worth an average of $4,182.

Freshmen must live in the dorms, which students say are affordable and extremely comfortable. More than 60 percent of students live off campus, often in downtown apartments. The meal plan operates on a point system where you pay for what you eat.

University Park students take advantage of the picturesque and peaceful locale by engaging in outdoorsy activities, including skiing and snowboarding at a nearby slope, and sailing, canoeing, hiking, and renting cabins in Stone Valley. State College offers cultural events such as symphonies, theatrical shows, and ballets, while the Bryce Jordan Center hosts top-notch performers, including Sheryl Crow, Dave Matthews, Alan Jackson, and Britney Spears. The town may be small, but according to a biochem major, "a majority of the students get involved in community service to maintain and constantly improve town relations."

Partying at Penn State is almost as legendary as the football team. Students mostly go to bars and parties off campus in State College, where "it's very difficult for underage students to be served at the bars," according to a student. "Students with fake IDs have almost a zero percent chance of gaining admittance." While the

"Imagine a family of 40,000."

administration strives to keep booze out of underage hands on campus, "there are places students can get alcohol, mostly at other people's apartments," says a junior. The HUB (the campus union building) offers nonalcoholic entertainment. The Greeks—which include 12 percent of Penn State men and 10 percent of women—have parties that may or may not include alcohol.

When thousands of alumni converge to cheer on their Nittany Lions in blue and white, the festivities include tailgate parties replete with marshmallow throwing, pregame parties, and postgame revelry. As a member of the Big Ten, Penn State's foes include Michigan and Ohio State, both of which make great road trips, in addition to

Philadelphia, Pittsburgh, and New York. Other popular events include the mid-July arts festival, the Dance Marathon, and, of course, homecoming. Penn State teams have won nearly 60 national titles in a wide variety of sports, including men's and women's volleyball, women's soccer, field hockey, and fencing. There are three large gyms, a competitive-size pool, an indoor ice rink, and an extensive intramural program for the recreational athlete.

Penn State's nationally recognized academics, sense of pride, community spirit, and sports are a tough combination to beat. "Imagine a family of 40,000—the excitement, pride, compassion, and sense of belonging, which is unparalleled at any institution," says one proud Nittany Lion.

If You Apply To ➤

PSU: Rolling admissions. Does not guarantee to meet full demonstrated need. Campus interviews: optional, evaluative (required for some programs). No alumni interviews. SATs or ACTs: optional. SAT IIs: optional. Prefers electronic applications. Personal statement. Apply to particular school or program.

Pepperdine University

24255 Pacific Coast Highway, Malibu, CA 90263-4392

With apologies to Wellesley and Furman, Pepperdine has the most beautiful campus in America. Small wonder that its acceptance rate hovers at about one in three. Students must come to Pepperdine ready to embrace an evangelical Christian emphasis much stronger than, say, the Roman Catholicism of U of San Francisco.

With picturesque surroundings, it's easy to confuse Pepperdine University with its nicknames—Pepperdine Resort and Club Med. Surrounded by the beautiful Southern California seashore, Pepperdine University might seem paradise found for students seeking sunshine rather than studies at this conservative, Christian-affiliated university, though students take their work and their worship seriously. "The philosophy of the school is that God and the academic experience must be married," says a senior telecommunications major. "This creates an intimate learning environment that prides itself on moral integrity and a high academic standard." Business and communications are the blessed programs, though other departments deserve recognition, too. Undergrads praise their educational opportunities, the strength of their school's spiritual community, the vast sandy beaches beckoning below their hilltop campus (and God's good grace in creating cute little bikinis).

There's no denying that Pepperdine's location—high in the Santa Monica Mountains, about 25 miles northwest of Los Angeles—is a strong selling point. The 830-acre Malibu campus, to which the school moved in 1972, overlooks the Pacific Ocean and features fountains, hillside gardens, mountain trails, and a 20-minute walk to the beach. Cream-colored, Mediterranean-style buildings topped with red terra cotta roofs dot the landscape. A 125-foot-tall white stucco cross stands near the center of campus, reminding students and faculty of the school's affiliation with the Churches of Christ.

Pepperdine was founded in 1937 by George Pepperdine, a devout Christian who had amassed a fortune through his mail-order auto parts supply company. The church's continued influence on the school pervades many aspects of campus life, from the prohibition of overnight dorm-room visits by members of the opposite sex

Website:
  www.pepperdine.edu
Location: Suburban
Total Enrollment: 7,476
Undergraduates: 3,068
Male/Female: 42/58
SAT Ranges: V 570–670
  M 580–680
ACT Range: 25–29
Financial Aid: 54%
Expense: Pr $ $ $
Phi Beta Kappa: No
Applicants: 5,393
Accepted: 36%
Enrolled: 35%
Grad in 6 Years: 68%
Returning Freshmen: 89%
Academics: ✍ ✍ ✍
Social: ☎ ☎
Q of L: ★ ★ ★ ★
Admissions: (310) 506-4392
Email Address: admission-seaver@pepperdine.edu

to the requirement that students attend convocation—similar to chapel—14 times each semester. Students at Seaver College, Pepp's undergraduate school, must also take three religion courses. While drinking is officially prohibited on campus, the administration has lifted the ban on dancing, and now allows students to choose their own seats at convocation. Though restrictions like this would drive the average kid up a wall, most at Pepperdine like the "highly moral" atmosphere. Says one student: "In comparison to other schools, Pepperdine students generally have a more religious foundation and thus have high standards of moral integrity."

Seaver's academic programs aim to provide students "with a liberal arts education in a Christian environment and relate it to the dynamic qualities of life in the 20th century." Individual classes are demanding, as is the required General Studies program, which includes a freshman seminar, a physical education course, three courses in Western heritage, two courses each in American heritage and English composition, and one class each in a foreign language and non-Western culture. However, faculty members are said to be accessible and responsive—not surprising when the average class has 17 students. "The quality of teaching is very personal and exceptional," says an art major. Another student adds: "Because it is a small school, professors don't accept excuses or laziness. They demand a lot from their students and expect a high standard and quality of work."

> **"The philosophy of the school is that God and the academic experience must be married."**

> *Faculty members are said to be accessible and responsive—not surprising when the average class has 17 students.*

The Business Administration department is unequivocally the strongest and most popular department at Pepperdine, and it tends to set the tone on campus. The Communications department, with majors including advertising, public relations, and journalism, is also highly touted, especially now that it boasts radio and television broadcasting studios. Biology and computer science are said to be strong, and sports medicine, rare at the undergraduate level, is both popular and well-respected. Although music studios and a $10 million humanities and visual arts center have been built to enhance the Fine Arts Division, students still say art and music are relatively weak. Recent additions to the academic menu include minors in African American studies, women's studies, and multimedia design. Juniors interested in European culture may spend a year at Pepperdine's own facilities near Heidelberg Castle, or in London or Florence. Other study abroad programs are available in Japan and Australia; locations for summer study include France, Spain, Israel, Asia, and Russia.

Back on campus, students trying to complete term papers can use one of the 292 public computer terminals and the collections of Pepp's eight-facility library system, which boast more than 515,000 volumes and 375,000 titles on microfilm. The well-organized campus Career Center allows students to sign up for job fairs, interviews, and individual and group career-counseling sessions.

One might expect students at this religiously oriented school to be politically conservative and they most definitely are. Many come from well-to-do Republican, California families; there is also a relatively high percentage of wealthy international students. Students joke that there's never a shortage of Porsches and BMWs on campus,

> **"Because it is a small school, professors don't accept excuses or laziness."**

but there is a shortage of places to park them. Hispanics account for 10 percent of the students, Asian Americans 7 percent, and African Americans 7 percent. The Republican influence is felt far and wide. Pepperdine has received millions of dollars from conservative Pittsburgh financier Richard Mellon Scaife. One student sums up the political climate gently: "Pepperdine tends to shy away from political activism."

Some say flashy student vehicles fit into the small, very wealthy community of Malibu better than the students themselves; the city sees the university as a catalyst for development, and that hurts town/school relations. Because the social scene in Malibu is pretty slack, with a 10 p.m. noise curfew and high price tags for everything, students typically head to L.A., Hollywood, Westwood, and Santa Monica for fun. "For a large proportion of students, academics and their social lives take priority over religious matters," says a public relations major. "Parties on weekends are well-attended, and probably draw a larger portion of students than church on Sunday." Twenty-four percent of men and 29 percent of women join one of six national fraternities or eight national sororities, which are playing a larger role in social life. Along with student government, they sponsor dances, movies, and other typical college activities, including the occasional illicit drink. "Pepperdine enforces a 'dry' campus, but 'damp' would be a better way of describing the residential community," says one student. "I think most of us would like to see Pepperdine get out of the dark ages in these matters."

Except for commuters, students are required to live on campus if they are single and under 21. That's a good thing, says one senior, who declares that Pepperdine's dorms are "comfortable, convenient, and really quite nice." The single-sex dorms and apartments are connected to the campus computer network. Rooms are assigned on a first-come, first-served basis, and the housing stock consists of 22 dorms with 26 rooms each, a 135-room tower, and a 75-unit apartment complex for juniors and seniors. Freshmen are typically assigned to suites with bath-

"Parties on weekends are well-attended."

rooms, living rooms, and four double bedrooms. Some consider these arrangements crowded, but a junior says they "connect freshmen instantly to seven suitemates and friends." Despite the above-average cost of living in the Malibu area, many upperclassmen choose to live off campus. The relatively new student union serves as the main campus social center, and annual events including Songfest, Family Weekend, and Midnight Madness draw crowds.

Sports receive a lot of attention at Pepperdine, with athletic scholarships offered in multiple sports, and a tennis pavilion and recreation center drawing varsity jocks and weekend warriors alike. The men's golf team and men's water polo team have recently won NCAA championships. Men's volleyball is always a contender for the national championship, and a Pepperdine men's tennis singles player has won the national title. Eleven club and intramural sports, including lacrosse, rugby, cycling, surfing, and soccer keep students busy, as does the physical education department, with classes in everything from surfing to horseback riding.

Students love to tease their well-manicured university with T-shirts proclaiming, "Pepperdine. 8-month party. 20K cover charge." But most seem to think the solid, values-oriented education they receive is worth the stiff price tag. Pepperdine faces an unusual challenge in trying to marry the Christian focus of a Bible college with the academic rigor of a secular university—all in a location not known for the strength of its moral fiber.

Sports receive a lot of attention at Pepperdine, with athletic scholarships offered in multiple sports.

Overlaps

University of Southern California, UCLA, UC–San Diego, UC–Santa Barbara, Stanford

If You Apply To ➤

Pepperdine: Early action: Nov. 15. Regular admissions: Jan. 15. Financial aid: Feb. 15. Housing: May 1. Does not guarantee to meet demonstrated need. Campus interviews: recommended, evaluative. No alumni interviews. SATs or ACTs: required. SAT IIs: optional. Accepts electronic applications. Essay question: current topics, ethical dilemma, or most embarrassing moment.

University of Pittsburgh

Alumni Hall, 4227 Fifth Avenue, Pittsburgh, PA 15260

As its home city has risen in stature, Pitt has become a hot commodity. A state-related university in the mold of University of Cincinnati—not the state flagship, but strong in a host of mainly preprofessional programs. Curiously, Pitt is among the nation's best in philosophy.

Website: www.pitt.edu
Location: City
Total Enrollment: 25,559
Undergraduates: 17,024
Male/Female: 48/52
SAT Ranges: V 560–660
  M 570–670
ACT Range: 24–29
Financial Aid: 52%
Expense: Pub $ $ $ $
Phi Beta Kappa: Yes
Applicants: 18,153
Accepted: 53%
Enrolled: 34%
Grad in 6 Years: 70%
Returning Freshmen: 90%
Academics: ✍ ✍ ✍
Social: ☎ ☎
Q of L: ★ ★
Admissions: (412) 624-PITT
Email Address: oafa@pitt.edu

Strongest Programs:
Philosophy
Chemistry
Economics
English
History
Physics
Political Science
Psychology

Pittsburgh's idled steel-factory stigma is a thing of the past. This city has joined the ranks of the most livable cities in the United States. The University of Pittsburgh has matured, too, becoming a formidable public research institution. The school offers numerous opportunities for students pursuing technical, medical, and engineering careers, but leaves a great deal of room for exploration in the liberal arts. Students are encouraged to be individuals and carve out their own academic niche, either with multiple majors or certificate programs. Combine this with a satisfying social life, and Pitt has discovered one formula for a rewarding college experience. "Pitt students are able to find the perfect balance between social and academic pursuits," a freshman says.

Pitt began as a tiny, private educational academy in the Allegheny Mountains in 1787. Oh, how times have changed. The university, which became state-related in 1966, is now part of the landscape of shops, parks, museums, galleries, and apartment complexes that make up Oakland, the heart of Pittsburgh's cultural center. Spacious,

> **"Pitt students are able to find the perfect balance between social and academic pursuits."**

light-filled, contemporary buildings and generic modern office buildings make up the Pitt campus, but the architectural delight is a 42-story, neo-Gothic academic building, appropriately called the Cathedral of Learning, a national historic landmark. The stately and towering cathedral, with its unique Nationality Rooms, attracts 100,000 visitors annually. And contrary to images you may hold of inner-city Pittsburgh, the campus borders a 456-acre city park.

With 10 undergraduate schools and more than 100 baccalaureate programs, Pitt rightfully claims to accommodate students with diverse needs. The academically motivated can take advantage of the excellent University Honors College, which one sophomore gushes is "the best thing at Pitt. It can do so much for students taking a serious interest in their education." It offers "small, intensive classes so students can work with professors and do independent research." Honors students publish the *Pittsburgh Undergraduate Review*, which receives submissions from students nationwide. Pitt's extensive research programs are its finest asset. The University of Pittsburgh is one of the top 10 institutions in the nation in terms of the annual research support awarded by the National Institutes of Health, and for good reason. Pitt astronomers discovered what appear to be two new planets orbiting a nearby star, and its physicians were the first to utilize gene therapy on a person with rheumatoid arthritis. The schools of engineering and nursing are excellent and attract high-caliber students. Premed students can even watch transplants at the famed University of Pittsburgh Medical Center, one of the world's leading organ-transplant centers. On the arts side, faculty and students in the music department consistently rake in more awards and fellowships than their counterparts at other U.S. universities.

Pitt offers guaranteed admission into graduate programs in communication disorders, dietetics, education, nursing, occupational therapy, social work, public health, public and international affairs, dental medicine, law, medical school, and physical therapy for outstanding freshman applicants. In two national studies, Pitt had four programs—philosophy, information science, history and philosophy of science, and nursing—ranked in the top six of their fields, and several others rated very high on the list. The School of Arts and Sciences (SAS) has academic requirements that include skill requirements in writing, quantitative and formal reasoning, foreign languages, and distribution requirements in the humanities, social and natural sciences, and foreign cultures. Psychology ranks among the most popular majors, and a heavily subscribed option permits students to combine a business major with an SAS degree.

"All of the courses certainly keep you on your toes."

For undergraduates who want to travel, the university offers numerous study abroad options. Closer to home, Pitt is a partner in the Pittsburgh Supercomputer Center, one of five such centers nationwide established by the National Science Foundation. The center houses the fastest computer in the world. There is also the Engineering Co-op, a program in which students alternate terms of study with work experience.

Pitt students take advantage of the school's flexible scheduling, which includes a strong evening program and summer sessions, and take on double—and even triple—majors, ensuring themselves plenty of education and a degree that's well worth the money. "All of the courses certainly keep you on your toes," one student says. First-year students undergo an extensive orientation that includes three days in the summer focused on academics, another five days before the term starts, and a one-credit freshman studies seminar. Out of more than 2,000 courses, 71 percent have fewer than 29 students enrolled. Most students agree that professors are very approachable and knowledgeable, despite the occasional less-than-stellar faculty member. "Some of the professors I've had were extremely influential on my future career path," says a senior.

Pitt students are described as well rounded, friendly, active, and hardworking. "We're urban and interested in city activities, music, sports, and being a good citizen," says a freshman. Eighty-four percent of all undergraduates are from Pennsylvania, including a substantial number from the Pittsburgh area. African American students account for 9 percent of the undergraduate student body, Asian Americans 4 percent, and Hispanics 1 percent. Incoming freshmen discuss diversity during orientation; a cultural diversity fair is held at the beginning of the school year and every student must pledge to promote civility on campus. "The student community is very accepting and open-minded," says a freshman majoring in anthropology and Arabic. Pitt also offers merit awards averaging $9,699 and athletic scholarships are available in 19 varsity sports.

"The student community is very accepting and open-minded."

While student housing may have been scarce in the past, Pitt is working to increase the amount of on-campus living space. "We have a three-year housing guarantee and most students stay on campus that long," says a senior. "Housing is by lottery, so it is luck that determines where you get to live." Fifteen percent of undergraduates live in on-campus university housing, which features 11 co-ed and single-sex dorms with liberal visitation hours and all kinds of rooming situations, from singles to seven-person suites. The Bouquet Gardens apartments have been expanded, and Pennsylvania Hall, a dorm opened in 2004, houses more than 400 students in four-person suites with private baths. Sutherland Hall boasts a view of Oakland and the rest of the city and offers its own computing lab. Suites in McCormick and Brackenridge halls have two-story windows in the living rooms. Lothrop Hall tends to attract the quieter students and features attractive singles. Six dining halls offer "pretty good" fare, including vegetarian options. Students feel safe on campus, considering the extensive network of campus lighting, emergency phones, the shuttle-bus route, strong police presence, and an on-demand van system.

Honors students publish the Pittsburgh Undergraduate Review, which receives submissions from students nationwide.

Psychology ranks among the most popular majors, and a heavily subscribed option permits students to combine a business major with an SAS degree.

Pitt's urban location provides a wide variety of social activities. Within minutes of campus are shops, parks, museums, and sporting events. "Pittsburgh is simply amazing," boasts one student. "There are tons of cultural events, including museums, parks, and sports." Though only 9 percent of the men and women belong to the Greek system, the fraternities and sororities play a vital role on campus. Many students say alcohol policies on campus are effective. "Students somehow find a way to bring alcohol to campus," one student confides. "However, the penalties are severe when students get caught." A senior notes that underage students caught drinking are referred to the Pittsburgh police department. Pitt students can ride the Port Authority of Pittsburgh bus system for free to nearby neighborhoods to shop and go to coffeehouses, bookstores, or movie theaters. Adjacent Schenley Park offers ice skating, golfing, a pool, jogging trails, and tennis courts. Ski slopes and mountain trails are not far away, and road trips to Penn State and Philadelphia, Boston, and New York City are popular.

In keeping with the university's resurgence, Pitt Panther football has dramatically improved and has earned a trip to postseason bowl games several times in recent years. Men's basketball is also popular, and competition is heated in Big East hoops—and to get Pitt basketball tickets. "Big East athletics are highly competitive," says a sophomore, "so any event involving those is great."

Pittsburgh boasts innumerable resources and opportunities for its students in the sciences and the arts, and is improving not only in its academics, but in the caliber of its student body. Pitt is raising its sights and, in many cases, breaking records. That potential for excellence, in the midst of a happening college town, makes for a very happy student body.

Overlaps

Boston University, Duquesne, Penn State, University of Delaware, University of Maryland

If You Apply To ➤ **Pitt:** Rolling admissions. Financial aid: Mar. 1. Housing: May. 1. Does not guarantee to meet full demonstrated need. Campus interviews: recommended, informational. Alumni interviews: optional, informational. SATs or ACTs: required (SATs preferred). SAT IIs: optional. Accepts the Common Application and electronic applications. Essay questions.

Pitzer College: See page 141.

Pomona College: See page 142.

Presbyterian College

503 Broad Street, Clinton, SC 29325

A South Carolina liberal arts college that competes head-to-head with Wofford for students who want their education served up with plenty of personal attention. Programs in business and engineering complement those in the liberal arts. Lacks the urban allure of Oglethorpe or Furman.

Presbyterian College students live up to the motto "While we live, we serve." More than 85 percent of students participate in some kind of community service while at PC. And while they're living in the "PC Bubble," they're also learning. Current PC students seem a far cry from the orphans for whom William Plumer Jacobs founded the school way back in 1880. And although today's Presbyterians largely come from stable, economically secure families, they continue to pursue personal, spiritual, and academic growth.

The Presbyterian campus sits on 234 acres in the South Carolina piedmont. The redbrick buildings are largely Georgian in style, with tall, white columns and lots of shade trees; many structures are listed on the National Register of Historic Places. The campus resembles Thomas Jefferson's University of Virginia, with buildings grouped around three plazas just perfect for reading, studying, or throwing a Frisbee.

Presbyterian's curriculum emphasizes the traditional liberal arts, combined with requiring freshmen experiences for all students, an electronic portfolio, cross-cultural education, and experiential learning. Course requirements include work on English composition and literature, fine arts, history, math, science,

> "The Sunday buffet is awesome—after church, people from the community pay to eat in our cafeteria."

physical education, religion, foreign language and a social science. Students must also participate in a senior capstone course. The most popular majors are business administration, history, biology, and political science, and psychology. The college has added majors in medical physics and philosophy.

Academically, PC is rigorous, and students strive to do well, says an education major. "Students work hard, but there is no competitive atmosphere," adds a senior. Presbyterian doesn't use teaching assistants, and students say faculty members are among the school's strongest asset. "The professors have been the best part of my college experience," a student says. "I appreciate the personal attention they give to students."

For those who tire of the bubble, options for off-campus study include Oxford's Corpus Christi College, in England, as well as programs in France, Austria, Spain, Mexico, China, Japan, Australia, and New Zealand. Those who can't bear to be away for an entire semester can take part in research and study trips to the Galapagos and Hawaiian islands during the May "fleximester." The Hansard Society for Parliamentary Government in London offers scholarships to students wishing to study and intern in the U.K., while the Russell Program enables those who remain on campus to focus on the media and society, with guest lecturers such as former White House Press Secretary Dee Dee Myers. The honors program is open to sophomores, juniors, and seniors with GPAs of 3.2 or above. It incorporates meetings with deans and dignitaries; a series of seminars on great issues, thinkers, and literary works; and independent research. "All of the [honors] programs are valuable because they allow the top students to share ideas, participate in debates, and learn from other scholars," a biology major says.

PC students are "well rounded, kind, and usually from Southern states," observes one student. "Most are very involved in different activities and fun to get to know." Sixty-eight percent of PC students hail from South Carolina, and minorities are a small but increasing presence on campus: African Americans make up 5 percent of the population, and Hispanics and

> "Students work hard, but there is no competitive atmosphere."

Asian Americans add 1 percent each. Foreign students account for another 1 percent. To address the relative lack of cultural diversity, administrators have created the Southeastern Intercultural Studies Center to promote global awareness; the most recent theme was Islam. And while you don't have to be Presbyterian to attend PC, it does help, since roughly one third of the students define themselves as such.

Website: www.presby.edu
Location: Small town
Total Enrollment: 1,133
Undergraduates: 1,133
Male/Female: 48/52
SAT Ranges: V 510–620
  M 520–620
ACT Range: 21–26
Financial Aid: 48%
Expense: Pr $
Phi Beta Kappa: No
Applicants: 1,110
Accepted: 76%
Enrolled: 37%
Grad in 6 Years: 72%
Returning Freshmen: 83%
Academics: ✐ ✐ ✐
Social: ☎ ☎ ☎
Q of L: ★ ★ ★ ★
Admissions: (800) 960-7583
Email Address:
  admissions@presby.edu

Strongest Programs:
Biology
Business Administration
History

PC's church affiliation keeps them focused on service, and on bettering the broader world, giving their classroom experiences added dimension.

Campus politics tend to be conservative, and some debate centers around religious affiliation, one student says. Other issues under discussion include the war in Iraq and gay marriage.

Ninety-two percent of Presbyterian students live on campus, where all dorms are air-conditioned, and accommodations range from traditional rooms with hall baths to suites and apartments. Where you live is determined by a random lottery, though everyone is guaranteed a bed. "The rooms are comfortable and the maintenance team is prompt in solving any issues," a freshman says. There are two dining halls on campus, both overseen by Sodexho. "Around Thanksgiving and Easter, they serve wonderful meals," says an English major. "The Sunday buffet is awesome—after church, people from the community pay to eat in our cafeteria." Security "works hard to make the campus a safe place to live," says one student. "The campus police are a constant presence."

Forty-four percent of PC's men and 36 percent of the women join Greek groups, and "since there aren't bars in Clinton, people are typically at the frat houses or off campus," says one student. "The weekend consists of going out Thursday, Friday, and Saturday nights around 11 p.m." Still, the school has taken a firm stand against underage drinking, and three violations lead to a two-semester suspension. For the nondrinkers, there's a once-a-month campus film series. Students with cars also enjoy heading to Greenville, Columbia, or Spartanburg for a bite to eat or some shopping—or, if there's more time, to Charleston or Atlanta. PC is equidistant from South Carolina's mountains and beaches, providing many opportunities to enjoy the outdoors. "Any road trip is a good road trip," says a junior. And many students join campus clubs, which typically have at least one off-campus retreat each semester.

Owing to PC's aforementioned motto, Student Volunteer Services is the largest organization on campus. The group routinely sends students to local orphanages, nursing homes, schools, and other facilities where their time and talents can be helpful. Favorite annual traditions include Spring Fling, a weekend carnival featuring four or five bands, and the Student Government Association's midnight breakfast, served during exam weeks. Students also look forward to the candlelight Christmas service, and to the outdoor graduation ceremony under the oaks, with bagpipes and trumpets heralding students and faculty in full academic regalia.

PC's 10 varsity sports teams compete in Division II, and everyone gets excited for the annual football game against archrival Newberry, since the winner gets the sacred Bronze Derby until the teams meet again. The college's mascot is the Blue Hose, a reference to the stockings of their Scottish ancestors. (While some students wear kilts during athletic events, most are more conservative, says a junior.) Men's and women's tennis, volleyball, and men's golf have all brought home tournament titles;

"Since there aren't bars in Clinton, people are typically at the frat houses or off campus."

the football and basketball teams are strong as well. All students may take advantage of PC's 31-acre recreational facility, with lighted softball, football, and soccer fields, volleyball and horseshoe pits, a driving range, a basketball court, a track, and an amphitheater. Recreational sports are divided into three divisions, depending on how competitive you are.

Presbyterian College students take pride in the school's history and traditions, including its very own tartan. PC's church affiliation keeps them focused on service, and on bettering the broader world, giving their classroom experiences added dimension. Says one student, "Students at PC have class. They genuinely care about one another and develop a close sense of family."

Overlaps

Wofford, Clemson, University of South Carolina, College of Charleston, Furman, University of Georgia

Prescott College

220 Grove Avenue, Prescott, AZ 86301

Not a place where students fresh out of high school typically go. Those who succeed here love the outdoors and are looking for an alternative college experience. College of the Atlantic is the only remotely comparable college in the *Fiske Guide*. If you loved Outward Bound, consider Prescott.

Future *Survivor* contestants take note: This tiny outpost in the wilderness of central Arizona is a perfect spot for the nature lover who seeks adventure, wants to learn survival skills, and likes studying outdoors. Where else but Prescott College could you major in adventure education or take courses like mountain search and rescue, ecopsychology, and wilderness rites of passage? Before any Prescott student sets foot in a classroom, the college sends him or her to the outback for three weeks of hiking and camping. Wilderness Orientation is an introduction to everything Prescott stands for: hands-on experience, personal responsibility, cooperative living, and stewardship of the environment.

Founded in 1966, Prescott retains the air of a 1960s commune. Surrounded by national forest, the college's "campus" consists of a two-block-long handful of buildings in the small town of Prescott. The architectural style of the campus ranges from the historic to the modern. The largest of the college's buildings is the Crossroads **"Classes don't exceed 12 people."** Center, an all-green building, which houses the library, computer labs, classrooms, conference centers and the Crossroad's Cafe. The administrative building was once a convent; its chapel is now used for meetings, art shows, and performances.

Prescott bills itself as a college "for the liberal arts and the environment," and most students envision themselves becoming teachers, researchers, park rangers, or wilderness guides. Adventure education, a major including everything from alpine mountaineering to sea kayaking, is a specialty. Also popular is environmental studies, which provides offerings of impressive breadth and depth for such a small school. The major in social and human development—a hodgepodge of sociology, psychology, and New Age mysticism—includes such unorthodox courses as Dreamwork Intensive. Integrative studies has been separated into Cultural and Regional Studies and Human Development, and is "home" for core humanities and liberal arts areas such as religion, philosophy, and social sciences, as well as an "incubator" for programs such as peace studies. Among the college's few concessions to practicality is the Teacher Education Program, which offers students teaching credentials in elementary, secondary, special, and bilingual education, and English as a second language. Prescott does not offer a comprehensive program in advanced math, chemistry, physics, or foreign languages other than Spanish. New programs include geography and women's studies.

Prescott's requirements for graduation are characteristically unorthodox. Instead of grades, faculty members give narrative evaluations, although students

Website: www.prescott.edu
Location: Small city center
Total Enrollment: 1,036
Undergraduates: 805
Male/Female: 40/60
SAT Ranges: V 540–680
  M 490–610
ACT Range: 20–28
Financial Aid: 68%
Expense: Pr $
Phi Beta Kappa: No
Applicants: 147
Accepted: 88%
Enrolled: 42%
Grad in 6 Years: 51%
Returning Freshmen: 68%
Academics: ✍ ✍ ✍
Social: ☎ ☎
Q of L: ★ ★ ★ ★
Admissions: (877) 350-2100
Email Address:
  admissions@prescott.edu

Strongest Programs:
Adventure Education
Environmental Studies
Human Development
Cultural and Regional Studies

may elect to receive grades. And rather than accruing credits, students design individualized "degree plans" that outline the competence (major) and breadth (minor) areas they will pursue, and the Senior Project (thesis) they will complete to demonstrate competence (graduate). Students must also obtain two levels of writing certification (college level and thesis level), and math certification, showing knowledge of college-level algebra.

"Students are very involved in keeping PC a green campus."

Prescott's calendar is divided into three periods, each with one 10-week quarter and one four-week block. During the quarters, students follow a traditional schedule, studying liberal arts and spending time doing fieldwork and student teaching. During the blocks, students pursue intense immersion in one course, most likely in the field, perhaps the backcountry of Baja, California, the alpine meadows of Wyoming, or even a local service clinic. Students can even take a one-month rafting trip down the Colorado River for credit. Summers may be spent studying field methods in agroecology at Prescott's 30-acre Wolfberry Farm. Though Prescott does not offer a traditional study abroad program, it encourages students to take courses at the Kino Bay Center for Cultural and Ecological Studies in Mexico.

While Prescott carefully studies the global problems of the environment, it does so in an intimate, localized setting. "Classes don't exceed 12 people; therefore you have more one-to-one interactions," says a freshman. "You're not a number." There's no tenure track at Prescott, so publishing and research take a backseat to teaching. According to a student, "Each instructor prides themselves on their unique and well-informed approach to their subject," says a senior. The one complaint is that the excellent but small classes mean that it is difficult to graduate in four years.

Environmental issues predominate and one student declares, "Students are very involved in keeping PC a green campus and encouraging that same stewardship in our community as well." PC students "love the earth" and are "social activists" according to a senior. Prescott's unconventional approach entices many well beyond Arizona. Indeed, 35 percent of students come from the Northeast; only 6 percent are in-staters. The minority population is small, with African Americans, Hispanics, and Asian Americans making up only 6 percent of the total. Part of the problem is the college's meager supply of financial aid, although Prescott now offers academic and transfer grants.

Because the college has no housing, students fend for themselves in the town of Prescott, a rapidly growing community of approximately 35,000 where almost everything is accessible by bicycle. The college assists with the apartment hunt by providing lists of available properties and by cosigning leases when necessary. As for the townsfolk, students

"PC students are a unique and self-motivated group of individuals."

describe them as "a mix of artists, activists, students, locals, retirees, and ranchers. Prescott recently set up a meal plan and students seem to love it. "The best food I've ever eaten—no joke!" says a student. A senior adds, "Dining is phenomenal!"

Prescott social life is informal and spontaneous. Aside from environmental activities, Prescott offers a nationally recognized literary magazine, *Alligator Juniper*, and a chapter of Amnesty International. Those looking for nightlife can hit Whiskey Row, the town bar scene, or drive to Flagstaff (90 minutes) or Phoenix (two hours). Though the college offers no athletics, students often participate in city sports leagues. Prescott's personal touch extends to graduation, a unique experience where a faculty member speaks about each student personally and then the student speaks on his or her own behalf.

Prescott may not have a huge campus or financial resources that are typically associated with larger schools; however, the small classes and interesting programs appeal to a student that would not be interested in your "typical" college. A human development senior explains, "PC students are a unique and self-motivated group of individuals. Our passion and dedication to education springs from a deep and inner desire to affect positive change in the world."

If You Apply To ➤	**Prescott:** Regular admissions: Mar. 1. Meets demonstrated need of 35%. Campus interviews: recommended, informational. Alumni interviews: optional, informational. SATs or ACTs: required. SAT IIs: optional. Accepts the Common Application and electronic applications. Apply to particular school or program. Essay question: autobiography, past academic experiences, and reasons for attending Prescott.

Princeton University

110 West College, Princeton, NJ 08540

More conservative than Yale and a third the size of Harvard, Princeton is the smallest of the Ivy League's Big Three. That means more attention from faculty and plenty of opportunity for rigorous independent work. Offers engineering but no business major. The affluent suburban location contrasts with New Haven and Cambridge.

Princeton occupies a distinctive niche among America's super-elite universities. It is a major research university with a world-class corps of professors who, in the absence of lots of graduate and professional students, lavish their attention on a relatively modest number of undergraduates. Princeton has an engineering school, but the academic atmosphere across the campus is dominated by commitment to the liberal arts—with a carefully structured set of core requirements and a heavy emphasis on independent study, including a mandatory senior thesis. For better or worse, Princeton has been known as a bastion of exclusivity. Its undergraduates are just as racially and ethnically diverse as any other Ivy League school, but somehow its image as a denizen of modern-day F. Scott Fitzgeralds refuses to go away. One reason, of course, is that the social scene continues to be dominated by the admittedly exclusive eating clubs.

But Princeton is changing. New four-year residential colleges are being established, aimed at broadening the range of social options available to students, and the administration is making major investments in the creative and performing arts and the sciences. Sensitive to faculty complaints that Princeton draws too many bright students whose main claim to fame is that they have learned to work the system, the admissions office is on the lookout for more students with demonstrated intellectual curiosity—including more high ability/low-income students and creative types. The university has also taken the lead in the national war on grade inflation—earning an "A" is harder than ever. With such changes in mind, Princeton is gradually increasing the size of its undergraduate body by 500 students; and, since Princeton has replaced all loans in its financial packages with grants, just about any

> **"I can open the newspaper and read my professor's article or turn on the TV and see him giving a speech."**

qualified student should be able to afford the place. Princeton's leaders are looking to make the university's particular brand of high-powered undergraduate liberal arts education available to an increasingly diverse group of students.

Cloistered in the secluded but upscale New Jersey town, Princeton's architectural trademark is Gothic, from the cavernous and ornate university chapel to the four-pronged Cleveland Tower rising majestically above the treetops. Interspersed among

Website: www.princeton.edu
Location: Small town
Total Enrollment: 6,935
Undergraduates: 4,790
Male/Female: 54/46
SAT Ranges: V 680–770
  M 690–790
Financial Aid: 55%
Expense: Pr $ $ $
Phi Beta Kappa: Yes
Applicants: 17,564
Accepted: 10%
Enrolled: 69%
Grad in 6 Years: 97%
Returning Freshmen: 98%
Academics: ✍ ✍ ✍ ✍ ✍
Social: ☎ ☎ ☎
Q of L: ★ ★ ★
Admissions: (609) 258-3060
Email Address:
  uaoffice@princeton.edu

Strongest Programs:
Physics
Molecular Biology
Public Policy
Economics
Philosophy
Romance Languages

Since Princeton has replaced all loans in its financial packages with grants, just about any qualified student should be able to afford the place.

the Gothic are examples of Colonial architecture, most notably historic Nassau Hall, which served as the temporary home of the Continental Congress in 1783 and has defined elegance in academic architecture ever since. A host of modern structures, some by leading American architects Robert Venturi and IM Pei, add variety and distinction to the campus, but the ambiance is still quintessential Ivy League at its best. Princeton's campus is self-contained, but those who venture outside its walls will find the surroundings quite pleasing. One side of the campus abuts quaint Nassau Street, which is dominated by chic (and pricey) boutiques and restaurants, most out of the range of student budgets, although coffee shops and affordable restaurants are becoming more prevalent. The other side of campus ends with a huge, man-made lake that was financed by Andrew Carnegie so that Princetonians would not have to forgo crew. A new, ultra-sleek science library by Frank Gehry is currently under construction.

Princeton is distinctive in its scale (among the Ivies, only Dartmouth has a lower total enrollment) and its emphasis on undergraduates. For a major research institution, the university offers its students unparalleled faculty contact. "We have some of the most brilliant professors in the world here," says a junior. An economics major adds, "I can open the newspaper and read my professor's article or turn on the TV and see him giving a speech, then go to a lecture to hear him speak, then go to his office to speak with him one-on-one." With fewer graduate students to siphon off resources or consume faculty time than at large research universities, undergraduates get the lion's share of both; at last count, 70 percent of Princeton's department heads taught introductory undergraduate courses. Special opportunities to work closely with senior faculty members come with the freshman seminar program, taken by two-thirds of new students, that offers the opportunity to go deep into 65 topics ranging from the physics of music to "The Search for Life in the Universe." Lovers of literature can study with Joyce Carol Oates, Paul Muldoon, Chang-rae Lee, or Toni Morrison, and nearly every other department has a few stars of its own. Senior professors lead at least one or two of the small discussion groups that accompany each lecture course. Every liberal arts student must fulfill distribution requirements in epistemology and cognition, ethical thought and moral values, historical analysis, literature and the arts, quantitative reasoning, social analysis, and science and technology. Students must also take writing courses. During their junior year, liberal arts students work closely with a faculty member of his or her choice in completing two junior papers—about 30 pages of independent work each semester in addition to the normal courseload. Princeton is also one of the few colleges in the country to require every graduate to complete a senior thesis—an enterprise that serves as a culmination of their work in their field of concentration. As a result, "seniors develop close personal relationships with their thesis advisor," says one student. Alumni often note the thesis as one of their best experiences at Princeton.

"We've got our share of legacies and athletes."

As one might expect, Princeton's small size means the number of courses offered is smaller than at other Ivies. But lack of quantity does not beget lack of quality. Princeton's math and philosophy departments are among the best in the nation, and English, physics, economics, molecular biology, public policy, and romance languages are right on their heels. Princeton is one of the few top liberal arts universities with equally strong engineering programs, most notably chemical, mechanical, electrical, and aerospace engineering, and computer sciences (which has its own facilities). In fact, the department of civil engineering and operations research has split into two departments: civil and environmental engineering and operations research and financial engineering, with the latter becoming one of the most popular majors. One of Princeton's best-known programs is the prestigious Woodrow Wilson School of Public and International Affairs ("Woody Woo" to the students), which admits undergraduates on a selective basis. The university has undertaken a major effort to become a national center in the field of molecular biology, with a laboratory for teaching and research staffed by 28 faculty members, including one who shared the Nobel Prize in medicine in 1995.

Princeton's semester system gives students a two-week reading period before exams in which to catch up, with first-term exams postponed until after New Year's, much to the dismay of many ski buffs and tropical sun worshipers. The university honor code, unique among the Ivies, allows for unproctored exams. The outstanding library facilities embrace five million volumes and provide 500 private study carrels for seniors working on their theses; there are another 700 enclosed carrels in other parts of the campus.

About 10 percent of the student body take advantage of the opportunity to study abroad. Except for students with sufficient advanced standing to complete their degree requirements in three and a half years, leaves of absence must be taken by the year, not the semester, an impediment to "stopping out." The university offers an intriguing five-year program that includes intense language study in an Asian country. There is also a five-year program that leads to a BSE and ME in mechanical and aerospace engineering. A limited number of courses can be taken on the pass/fail option and the University Scholars program provides especially qualified students with what the administration calls "maximum freedom in planning programs of study to fulfill individual needs and interests." Although the faculty gets high ratings for its academic advising, students are rather cool on the university's nonacademic counseling programs.

Sensitive to faculty complaints that Princeton draws too many bright students whose main claim to fame is that they have learned to work the system, the admissions office is on the lookout for more students with demonstrated intellectual curiosity—including more high ability/low-income students and creative types.

"We've got our share of legacies and athletes," says a senior, and most students are "driven, passionate, and hard-working." African Americans account for 9 percent of the student body, Hispanics 8 percent, and Asians 14 percent. While diversity is present, mixing sometimes isn't. "As an African American, I can say that even the African Americans are subdivided based on economics, place of origin, and whether you went to public or private school," explains one senior. And the campus remains socially conservative, with tweed and penny loafers adorning many students. Students note that political causes generate little concern. Instead, there is a "general apathy in matters concerning active political involvement," according to a junior.

Princeton undergraduates are admitted to the university without regard to their financial need, and those who qualify for aid get an appropriate package of benefits. There are no merit or athletic scholarships, but Princeton's financial aid package is generous to middle-class families. Each student's Princeton experience begins with a week of orientation; 500 each year participate in Outdoor Action, a few days of wilderness activities immediately preceding orientation.

In an attempt to improve the quality of life for freshmen and sophomores, Princeton has grouped many of its dorms into residential colleges, each with its own dining hall, faculty residents, and an active social calendar. Under this system, nearly all the freshmen and sophomores live and dine with their residential college unit, alleviating the formerly fragmented social situation. However, by providing a separate social sphere for these students, the system "creates a gulf between underclassmen and upperclassmen." And all too often the upper-level eating clubs steal the thunder from the college's social events. As a result, "the underclassmen spend too much time pining for the day when they, too, can join the closest thing Princeton has to cliques,"

As one might expect, Princeton's small size means the number of courses offered is smaller than at other Ivies. But lack of quantity does not beget lack of quality.

"Seniors develop close personal relationships with their thesis advisor."

says one student. Under Princeton's new plan, upperclassmen will keep an affiliation to the residential colleges, taking many meals there and even continuing residence there if they prefer.

The university's turn-of-the-century Gothic dorms may look like crosses between cathedrals and castles, and some halls have amenities like living rooms and bay windows. But conditions on the inside are sometimes less glamorous. All dorms are renovated on a rotation schedule, and new dorms have helped ease the housing crunch. Less than 3 percent of the students live off campus. The modern and roomy Spelman dorms, which come complete with kitchens, are the best on campus and fill up quickly every year with seniors who do not belong to eating clubs.

Ah, yes, the eating clubs. Princeton's most firmly entrenched bastions of tradition. Run by students and unaffiliated with the school, they line Prospect Avenue, and have, for more than a century, assumed the dual role of weekend fraternity and weekday dining hall. Of the 10, six admit members through an open lottery, but the others still use a controversial selective admissions process called bicker (because of the wrangling over whom to admit), to the embarrassment of the administration and most of the students. While many of the clubs opened their doors to women back when Princeton went co-ed, two of the oldest and most exclusive—the Ivy Club and the Tiger Inn—remained all-male until 1991, when a court decision compelled them to admit women. Now, all the clubs are co-ed.

Catering exclusively to upperclassmen, the eating clubs provide a secure sense of community for their members. More than half of all sophomores join one of the clubs at the end of the year, becoming full-fledged members by the fall of their junior year. Annual dues vary; the most expensive is the Ivy Club, which charges its members almost $5,000 a year. Unfortunately, the social options for those who choose not to join may feel limited. Some opt for life in independent dormitories or join the handful of Greek fraternities and sororities (not sanctioned by the administration) that have sprung up on campus over the past few years and have become feeders to particular eating clubs. The new residential college system will create themes to improve social opportunities.

Students rarely venture much farther than New York or Philadelphia, each one hour away (in opposite directions) on the train. Few students complain about boredom, and many praise the affluent town of Princeton for the parks, woods, bike trails, and, most important, the quiet and safety it offers students. McCarter Theatre, adjacent to campus, is the nation's seventh-busiest performing arts center and houses Princeton's Triangle Club, which counted Jimmy Stewart and Brooke Shields as members. The roundup of annual campus events includes Communiversity Day, an international festival, and the P Party in the spring, which features a big-name band. Each year about 3,000 students engage in volunteer activities such as tutoring, working in soup kitchens, or helping the elderly.

Princeton has the oldest licensed college radio station in the nation, plenty of journalistic opportunities, a prestigious debating and politics society (Whig-Clio) whose ranks included James Madison and Aaron Burr, and a plethora of arts offerings. Athletics are a big deal at Princeton, both varsity and intramural. The men's football team became Ivy League champs in 2006 and the men's basketball team, known for its cerebral style of play, usually makes it to the early rounds of the NCAA tournament. The women's soccer team made it to the Final Four in 2004. Other strong sports programs include men's lacrosse, men's squash, men's and women's crew, and women's softball. The women's rugby club and women's hockey team are also outstanding. Teams from the eating clubs and residential colleges take part in well-attended intramural events ranging from aikado and ballroom dancing to triathalon. Every fall the freshman and sophomore classes square off in Cane Spree, an intramural Olympics that has been a tradition since 1869.

Princeton's unofficial motto is "Princeton in the nation's service and the service of all nations," and the oft-repeated notion that with privilege comes responsibility lives on as part of its culture. It's easy to be humbled at Princeton. Even the most jaded students must be awed and inspired when they think of those who've traversed the campus paths before them, including former U.S. presidents James Madison and Woodrow Wilson. While some may find the ambiance too insular, not many turn down membership in this very rewarding club.

If You Apply To ➤ **Princeton:** Regular admissions: Jan. 2. Financial aid: Feb. 1. Guarantees to meet demonstrated need. Campus tours: optional, informational. Alumni interviews: recommended, informational. SATs: recommended. SAT IIs: recommended (engineering applicants required to take either physics or chemistry and math I or II). Accepts the Common Application. Essay question: changes every year.

Principia College

Elsah, IL 62028

Prin is a tiny college in a tiny town about an hour from St. Louis. All students have ties to Christian Science. Prin is mainly liberal arts, though its most popular program is business administration. More than two-thirds of the students travel abroad.

Students come to Principia College with a common bond—ties to Christian Science. They shun smoking, drinking, drugs, and sex in favor of their love of God and learning. Prin graduates are culturally, spiritually, and intellectually well-rounded, the product of a liberal arts education that promotes critical thinking and a broad world view. As the only college for Christian Scientists, Prin attracts a lot of international students. The historic campus is reminiscent of Harry Potter's Hogwarts. But the fictional school of wizardry never had a woolly mammoth to unearth between the dormitories, as Prin does.

Principia's 2,600-acre campus, on limestone bluffs above the mighty Mississippi River, was designated a National Historic Landmark in 1993. The dominant architectural influences are colonial American, Tudor, and medieval, and many buildings—including most dormitories—were designed by California architect Bernard Maybeck. A contemporary of Frank Lloyd Wright, Maybeck urged Principia trustees to bring the college to its current spot when they relocated from St. Louis in 1935. The College Chapel, whose bells ring out hymns every evening, is the symbolic center of campus. A 68,000-square-foot indoor athletic facility, complete with a six-lane track and eight-lane pool, was also recently completed.

Students say Principia's strongest programs include biology, music, political science, religion, and business. Art students benefit from a 1,000-square-foot studio in the Voney Building for Studio Arts, where they can immerse themselves in painting and drawing. Mass communication majors can hone their craft at the school's working television studio and FM radio station. The science center provides an aviary, greenhouse, 13 lab rooms, and a computer weather center. In addition to 10 to 12 courses in their major, students must complete one course each in foreign language and the arts, and two courses each in literature, history, religion and philosophy, social science, lab science, and math, computer science, or natural science. Students must take four physical education courses and pass a swimming test. Prin freshman come to campus two weeks early for the First-Year Experience writing seminar, "during which they explore different forms of writing and practice formal thesis writing," explains one senior.

Principia operates on the quarter system, which means that students get only 10 weeks to master the material in each of their courses. Academics are challenging,

> "We might not be in a big city, but Principia has a lot of engaged world citizens."

Website:
    www.prin.edu/college
Location: Rural
Total Enrollment: 543
Undergraduates: 543
Male/Female: 47/53
SAT Ranges: V 530–660
    M 520–650
ACT Range: 21–29
Financial Aid: 58%
Expense: Pr $
Phi Beta Kappa: No
Applicants: 242
Accepted: 89%
Enrolled: 64%
Grad in 6 Years: 73%
Returning Freshmen: 79%
Academics: ✏️ ✏️ ✏️
Social: 🐻 🐻 🐻
Q of L: ★★★★
Admissions: (618) 374-5181
Email Address:
    collegeadmissions
    @principia.edu

Strongest Programs:
Studio Art
History
Biology/Environmental
    Science
Business Administration
Elementary Education

but students can count on each other and their professors for help. "Some of the courses are very rigorous and competitive," a junior says. Most of Prin's faculty members receive high marks. "The quality of instruction varies," says a biology major, "though a large percentage of the faculty express a love of their subject, a deep interest in each student, and incredible knowledge and experience." Over a given four years, more than two-thirds of Principia students participate in the five or six study abroad programs the school organizes each year. Each program enrolls 18 to 22 students, either during a quarter or the six-week inter-term, from mid-November to January. Sites are determined by academic subject and focus, and have included France, Germany, Nepal, South Africa, China, and Italy. A geology field-study course enables students to help excavate the remains of a woolly mammoth found on campus. Others participate in a prairie restoration program, gather data for the study of the Mississippi River's aquatic life or build solar cars to be entered in races around the world.

"Principia is full of Renaissance students."

"Principia is full of Renaissance students," says a senior. "They play sports, do music, participate in clubs, take student government positions, work, and go abroad." While African Americans, Asian Americans, and Hispanics together account for only 4 percent of the population, 12 percent of students arrive from abroad, the third largest percentage of international students on a college campus in the country. The Black Student Union is one of the most active groups on campus, and a Latin Student Union has been formed. "We might not be in a big city," muses one student, "but Principia has a lot of engaged world citizens." Hot topics include homosexuality, censorship, and the war on terror. "Most students address these issues with an open mind and earnest desire to discover the truth," a freshman says. Merit scholarships averaging $8,422 are available, but no athletic scholarships are offered.

Principia's Panthers compete in Division III, and men's and women's soccer are particularly popular.

All students live in Prin's dorms, except for the few who are married or live locally with their parents. The rooms are large and comfortable, and those that aren't air-conditioned have ceiling fans. All have new or relatively new furniture. "With only 60–70 students in each dorm, there is a great sense of community," says a junior. Students are expected not to be in the wings of dorms where the opposite sex lives during "house hours" every night. Freshmen live in two modernized dorms with upperclass resident advisors trained to help new students adjust to college life. "They are without a doubt the best dorms on campus," says one student. In the dining halls, students can submit recipes to the dining service, and in addition to the grill station and a sit-down pub and restaurant, there's always salad, fruit, cereals, and other choices. "There are some remarkably good days and other mediocre ones," a student says, "but, all in all, there's good variety."

Settled in the 1800s, Elsah, Illinois, is a small village. "Elsah is not a college town. I think it was established in the 1800s and not much has changed," a junior says. Stores, restaurants, and movie theaters are about 30 minutes away, and St. Louis is about an hour's drive. As Christian Scientists, students eschew alcohol, tobacco, and drugs, and

"With only 60–70 students in each dorm, there is a great sense of community."

Principia operates on the quarter system, which means that students get only 10 weeks to master the material in each of their courses.

Prin also asks them to sign a pledge of abstention from premarital and extramarital sexual relationships. "Those who sign the contract are committed to those morals for religious reasons," a political science major says. "It's a wonderful thing not to have to deal with alcohol on campus," says one senior. Instead of partying, students keep busy at school-sponsored concerts, movies, dances, or intramural sporting events that pit one dorm against another. Each dorm organizes its own annual celebration, and international students show off their native cuisines at the Whole World Festival. The Public Affairs Conference is the oldest student-run event of its type, bringing in big-name speakers such as Colin Powell and Jimmy Carter.

Principia's Panthers compete in Division III, and men's and women's soccer are particularly popular. About 60 percent of the student body participates in intramural sports like soccer, basketball, and softball. On sunny afternoons, students can be found playing ultimate Frisbee. Campus athletic facilities include a four-court indoor tennis center, a field house with gym and pool, and outdoor courts and running trails.

Prin students don't mind the conservative environment at their Christian Scientist school. In fact, they seem to revel in it. "Most students are committed to leading lives governed by morality and integrity," a freshman says. Gone are the pressures that take hold of most college students. Prin graduates leave their collegiate bubble ready to take on the world. "There is a commitment to the development of the students academically, socially, morally, and spiritually that is unique," a freshman says.

If You Apply To ➤

Principia: Rolling admissions. Early action: Nov. 15. Guarantees to meet full demonstrated need. Campus interviews: recommended, evaluative. No alumni interviews. SATs or ACTs: required (SATs preferred). SAT IIs: recommended (foreign language test). Accepts electronic applications only. Essay question: book, movie, painting, poem, or other creative endeavor that's affected you, and why; person who has had a significant influence on you; describe a situation where your values or beliefs were challenged and how you handled the situation. Only college in the world that admits only Christian Scientists.

University of Puget Sound

1500 North Warner, Tacoma, WA 98416

Ask anyone in Tacoma about UPS—the university, not the parcel service—and they'll tell you that Puget Sound delivers solid liberal arts programs with a touch of business. Within easy reach of the Sound, the university specializes in all things Asia. Compare to Willamette and Whitman.

An ambitious building program and revised core curriculum have been helping raise the profile of the University of Puget Sound, transforming it from a regional liberal arts college in comfortable Tacoma into an undergraduate institution with growing national recognition. As part of the new core, students will be required to demonstrate foreign language proficiency and pass two freshman seminars and a capstone seminar. What won't change are the school's close-knit community and its emphasis on Asia. Every three years, a nine-month Pacific Rim–Asian studies education tour takes 35 lucky students to Far Eastern nations.

Founded in 1888, UPS is cradled by the Cascade Range and the rugged Olympics, with easy access to the urban energy of Seattle and the natural beauty of Mount Rainier. The 97-acre campus boasts carefully maintained lawns, native fir trees, and plenty of other greenery, thanks to the moist climate. Most buildings, with distinctive arches and porticos, were built in the 1950s and 1960s. The 57,000 square foot Trimble Hall dorm was completed in 2002, with suite-style apartments for students. More recent additions include an all-weather track and a 3,850-square-foot Sculpture House, with facilities for welding, woodwork, and painting. Harned Hall, a $25 million science building, was completed in 2006.

UPS students must complete an eight-course core curriculum, which includes a freshman seminar in writing and rhetoric, and another in scholarly and creative inquiry. In their first three years at Puget Sound, students also study five "Approaches to Knowing"—fine arts, humanities, math, natural sciences, and social sciences. An upper-level integrative course, Connections, challenges traditional

Website: www.ups.edu
Location: Suburban
Total Enrollment: 2,749
Undergraduates: 2,565
Male/Female: 42/58
SAT Ranges: V 580–690
  M 560–660
ACT Range: 24–29
Financial Aid: 88%
Expense: Pr $ $ $ $
Phi Beta Kappa: Yes
Applicants: 5,230
Accepted: 71%
Enrolled: 20%
Grad in 6 Years: 74%
Returning Freshmen: 87%
Academics: ✏️ ✏️ ✏️ ½
Social: ☎ ☎ ☎
Q of L: ★ ★ ★ ★
Admissions: (253) 879-3211

(Continued)

Email Address:
admission@ups.edu

Strongest Programs:
Business Administration
Biology
Psychology
English
Music
Politics and Government
International Political Economy

In their first three years at Puget Sound, students also study five "Approaches to Knowing"—fine arts, humanities, math, natural sciences, and social sciences.

Popular school-sponsored activities include the Log Jam BBQ, which kicks off the school year, Midnight Breakfasts, and Foolish Pleasures, a festival of short student-produced films.

disciplinary boundaries and examines the benefits and limits of an interdisciplinary approach to learning. There's also a senior capstone course. Special offerings include a classics-based honors program, the Business Leadership Program, residence-based humanities programs, and the Social Justice Residence Program. "The orientation is definitely not pushed as much as it should be," says one student. "It is a tremendous program."

After navigating Puget Sound's requirements, students may pursue a BA, BS, or bachelor of music degree. The most popular majors are business, English, psychology, and biology, followed by international political economy. The university has also developed a reputation as a jumping-off point to Asia—both literally and figuratively.

> **"Freshman advising sessions will often be over pizza and soda, with the help of both a professor and an upperclassman."**

Its curriculum stresses two of the fastest-growing fields in the region: Asian studies and Pacific Rim economics. Nearly one-third of Puget Sounders take at least one Asian studies course, and once every three years, there's a nine-month school-sponsored trip through Japan, Thailand, Korea, India, China, and Nepal, where participants study native art, architecture, politics, population, and philosophy. UPS has added majors in biochemistry and molecular and cell biology, as well as minors in environmental studies and African American studies. The minor in psychology has been dropped, and the exercise science program can lead to the graduate program in physical therapy.

Regardless of the department or program, students say the academic climate here is very challenging. "Courses tend to be rigorous at all levels, from intro courses to senior seminars," says one student. Eighty percent of the classes taken by freshmen have 25 students or fewer, and you won't find grad students leading classes here. "Classes are always taught by full professors," says a junior. "They encourage students to work hard in classes, but also to develop a passion for the topics and ideas in their classes." Several first-year programs are residence-based, and "freshman advising sessions will often be over pizza and soda, with the help of both a professor and an upperclassman," a junior says. "The Academic and Career Advising center helped me blend my interests in technology and business to develop a four-year plan double majoring in business and computer science."

While most UPS students come from western states, only 30 percent hail from Washington. "There is a wide variety of students on campus—from Greeks to 'hippies' to athletes," says a senior, but most are "middle-upper class whites." African Americans and Hispanics make up 6 percent of the student body. Asian Americans constitute 9 percent, and an active Hawaiian student organization sponsors a number of events, including a luau each spring, with "great food and lots of traditional dances." "We have a range of political attitudes on campus," says a junior. "Right now we are focused on fair trade and sustainability." There are no athletic scholarships.

Sixty percent of Puget Sound students live on campus, and housing is guaranteed for freshmen, who are sprinkled among the 10 dorms. "The dorms are by far the largest and nicest I saw at any college campus when I was visiting," says a senior. The

> **"The dorms are by far the largest and nicest I saw at any college campus when I was visiting."**

halls and bathrooms are cleaned daily. After the first year, students may go Greek and live in chapter housing, pursue a single room in the dorms, or apply for one of 60 university-owned theme houses, which focus on substance-free lifestyles, social justice, or outdoor adventures. Other options include four foreign language houses. One student calls the highly ranked food "amazing." Aside from the main campus dining area—which is open from 7 a.m. to 10 p.m. daily and offers Mexican, Chinese, and Italian stations, plus salad and deli bars—students, faculty, and staff can chow down at the Pizza and at the student-run Diversions Cafe.

Twenty-two percent of the men and 23 percent of Puget Sound women go Greek, though fraternities and sororities don't dominate the social scene. "There are the traditional house parties," says a student, "but I really take for granted the number of events that are offered on campus so regularly." Popular school-sponsored activities include the Log Jam BBQ, which kicks off the school year, Midnight Breakfasts, and Foolish Pleasures, a festival of short student-produced films. Students under 21 may not drink, but determined students usually manage to find booze anyway. Tacoma's "art scene is terrific, and the numerous restaurants are a definite perk," says an economics major, but "it is a city I would love to live in and raise a family," not a college town. Still, there's no concern about boredom. With the mountains and beaches so close—Seattle is 30 minutes away by car, Portland is two hours south, and Vancouver, British Columbia, is three hours north—road trips are *de rigueur*. That's especially true during ski season, and the school rents out all the necessary equipment.

"Individuality is prized here."

Students are fond of saying that Puget Sound's Division III varsity teams, the Loggers, "Kick Axe." Among the teams that have brought home conference championships in recent years are women's basketball, crew, soccer, and swimming, and men's basketball and crew. The school's archrival is Pacific Lutheran University; "football and basketball games against PLU are a big deal and always packed," says a student-fan.

Don't let the students' slacker-chic casual clothes fool you. UPS means business and serious study for students seeking immersion in the liberal arts and the natural beauty of the outdoors. "The location, student autonomy, quality of education, and size of the student body make this college a wonderful place," says a senior. Those considering Puget Sound may do well to heed one junior's advice: "Be yourself. Let yourself show in your application. Individuality is prized here."

Overlaps

University of Washington, Lewis and Clark, Occidental, Whitman, Willamette, Colorado College

If You Apply To ➤ **Puget Sound:** Early decision: Nov. 15, Dec. 15. Regular admissions and financial aid: Feb. 1. Housing: May 1. Meets demonstrated need of 23%. Campus interviews: recommended, evaluative. Alumni interviews: optional, informational. SATs or ACTs: required. SAT IIs: optional. Accepts the Common Application and electronic applications. Essay question.

Purdue University

1080 Schleman Hall, West Lafayette, IN 47907-1080

Purdue is Indiana's state university for science and technology—with side helpings of business, health professions, and liberal arts. Compare to Kansas State and Big Ten rival Michigan State.

Successful Indiana colleges have three things in common: a strong agricultural program, a powerhouse basketball team, and a conservative student body. Purdue University has all of these—plus one of the nation's strongest engineering programs, and the distinction of having awarded more bachelor's degrees in the field than any other institution. Purdue is also home to the nation's first computer science department, and its programs in pharmacy, nursing, and management are likewise strong. Budding classicists, dramatists, and vocalists probably should look elsewhere, as liberal arts are not Purdue's forte. But those seeking small-school friendliness with big-school spirit may be very happy here. "It is a very friendly and laid-back atmosphere," explains a pharmacy major.

Website: www.purdue.edu
Location: Small city
Total Enrollment: 34,926
Undergraduates: 29,169
Male/Female: 60/40
SAT Ranges: V 500–610
  M 530–650
ACT Range: 23–28
Financial Aid: 41%

(Continued)

Expense: Pub $ $ $

Phi Beta Kappa: Yes

Applicants: 24,052

Accepted: 85%

Enrolled: 35%

Grad in 6 Years: 66%

Returning Freshmen: 85%

Academics: ✑ ✑ ✑ ½

Social: ☎ ☎ ☎

Q of L: ★ ★ ★

Admissions: (765) 494-1776

Email Address:

admissions@purdue.edu

Strongest Programs:

Engineering

Technology

Science

Agriculture

Purdue is the main attraction in the small industrial town of West Lafayette, where the population triples when students return each fall. The campus features redbrick and limestone buildings arranged around lush shaded courtyards. A number of facilities have been recently added to campus, including buildings to house the biomedical engineering, computer sciences, and visual and performing arts programs.

Students apply to and enroll in one of Purdue's 10 schools, and academic requirements vary with the school and the major. Typically, they include English, math, a lab science, and perhaps speech or foreign language proficiency. Engineering is the most popular program, followed by management, psychology, and elementary education. Students flock to the five-year engineering co-op program, one of the most competitive on campus, because it marries classroom study with real-world work. Purdue also offers a strong undergraduate program in flight technology, which includes hands-on training at the university's own airport.

> **"It is a very friendly and laid-back atmosphere."**

Purdue has graduated 21 astronauts, including Neil Armstrong and Gus Grissom. Also strong are programs in veterinary medicine and hospitality and tourism.

Purdue students are focused on life after graduation; a senior says, "The engineering programs teach tools for problem-solving that are directly applicable to skills needed on the job. Plus, there are lots of opportunities for paid research, internships, or cooperative education." And despite the university's size, about half of freshman classes are seminar-style, taught by graduate students and academic advisors who help answer students' questions and provide career advice. "Some professors are outstanding," says a senior, "but too many seem more interested in their research."

Purdue's student body is fairly homogeneous, with two-thirds from Indiana and only 12 percent minority: 4 percent African American, 5 percent Asian American, and 3 percent Hispanic. Although some complain that the campus looks like the world's biggest Abercrombie & Fitch catalog, Boilermaker pride does stretch across boundaries of race, gender, and socioeconomic background. Thousands of merit scholarships are awarded annually to qualified students; athletes vie for 369 scholarships.

Thirty-six percent of students live in Purdue's dorms; the numbers may be so low because of rules governing male and female visitation hours. (The notion of a "co-ed dorm" here means that both sexes share a dining hall and a lobby.) Almost all freshmen live on campus, though they aren't required to, and Harrison Hall is said to be a good pick for newbies. "The residence halls are very nice," says a student, thanks to "decent room sizes, new furniture, and a great social environment." Most upperclassmen find inexpensive housing just off campus, where walking and riding escorts, blue-light phones, and more than 40 campus police officers help them feel safe.

Officially, Purdue is as dry as Death Valley, and "people have been kicked out of the residence halls for being caught with alcohol," says a sophomore. Still, underage students get served at the frats, or—as at most schools—when friends over 21 are buying. Those of age may also frequent Harry's Chocolate Shop—a longtime bar, not a candy store. Overall, Greek life draws 16 percent of men and 15 percent of women and offers many social opportunities. But there are other options, too, says an animal science major, including football, basketball, soccer, and baseball games. "Outside of class,

> **"People have been kicked out of the residence halls for being caught with alcohol."**

you can do anything from skydiving, paintball, choir, hiking rock climbing, salsa dancing...anything; it's up to you," encourages a senior mechanical engineering major. Purdue's more than 600 organizations range from the BBQ society to professional development clubs.

What the happy students here have already discovered is that learning is fun when academics are mixed with a healthy dose of school spirit and general carousing.

As far as college towns go, West Lafayette "would not exist if it weren't for Purdue," one student says. Another adds, "This is a college town that has everything you need within walking distance." As far as community service goes, Purdue students tend to find the "various options very rewarding." "But they should do more," complains one sophomore. Chicago and Indianapolis are favored weekend destinations for students with cars, and each spring, a week of fun and parties leads up to the Grand Prix go-cart races. Students also look forward to the Slayter Slammer, a free Labor Day weekend concert on the school's main hill, and the Bug Bowl, an annual celebration sponsored by Purdue's entomology department, including cricket-spitting and cockroach races.

Purdue's all-purpose athletic facility offers opportunities for weekend warriors and varsity athletes alike. Boilermaker pride manifests itself at Division I games of all types, especially when the opposing team is Indiana University, known derisively as "that school down south," in the annual struggle for the Old Oaken Bucket. Every year, the winner adds a link to a chain on the bucket, in the shape of either an "I" or "P." Women's volleyball and softball and men's and women's basketball are among the most popular sports on campus. Also notable is Purdue's rivalry with Notre Dame.

A strategic plan developed several years ago has Purdue focused on reaching "the next level of preeminence," through discovery, learning, and engagement—adding more professors and scholarship funding and decreasing the number of teaching assistants. What the happy students here have already discovered is that learning is fun when academics are mixed with a healthy dose of school spirit and general carousing. "The students here are all highly motivated and hard workers," says one senior. "We're also friendly and supportive to each other because there's the feeling that we are all in it together."

> "This is a college town that has everything you need within walking distance."

Overlaps

Indiana University, Ball State, Indiana State, University of Illinois, University of Michigan, Ohio State

If You Apply To ➤ **Purdue:** Rolling admissions. Financial aid and housing: Mar. 1 (priority consideration). Does not guarantee to meet demonstrated need. Campus interviews: optional, informational. No alumni interviews. Apply to particular schools or programs. SATs or ACTs: required. SAT IIs: optional (English composition, math, and lab science for home-schooled students). Accepts electronic applications. No essay question.

Queen's University: See page 335.

Randolph College

(formerly Randolph-Macon Woman's College)
2500 Rivermont Avenue, Lynchburg, VA 24503

Newly coed Randolph College has a new name but a traditional mission. Once one of the premier women's colleges in the nation, it fields strong programs in the liberal and fine arts. Suburban location on the James River is rich in history.

Like many of its former sister institutions, Randolph-Macon Woman's College has made the decision to go coed despite protests by students and alumnae. Long known for providing women with a forward-looking education in the liberal arts

Website: www.rmwc.edu
Location: Residential

The students of Randolph aren't shy about their academic goals, career drive, or sense of campus unity.

and sciences, the newly christened Randolph College now faces the challenge of integrating men into the mix. With rich traditions, homey dorms, and challenging, seminar-based classes, the school will seek to preserve the best elements of its past, while evolving into an institution that remains relevant today. In addition to going coed, the college will place emphasis on a global honors curriculum.

The college's 100-acre campus sits in a historic neighborhood of Lynchburg (population 80,000), on the banks of the James River. Graceful old buildings are covered with purple wisteria and linked by glass corridors called trolleys; the surrounding trees burst into riotous bloom each spring. Main Hall, dating from 1893, houses dorm rooms, classrooms, and faculty and administrative offices. The Maier Museum of Art has one of the best college collections of American art in the country, including works by Andrew Wyeth, Mary Cassatt, Georgia O'Keefe, and James McNeil Whistler. The school's Riding Center offers indoor and outdoor arenas in the nearby Blue Ridge foothills. Current construction includes a new student center that will house athletic programs, the dance department, and recreation activities.

Randolph's general education requirements are reflected in a matrix of study areas, with artistic expression, cultural inquiry, global issues, gender issues, and quantitative literacy and analysis on one axis, and arts and literature, humanities, natural sciences and math, wellness, and interdisciplinary courses on the other. All freshmen take the First-Year Seminar, which examines how to maximize academic success. Programs in museum studies and American culture combine classroom study with visits to historic sites; the latter has focused on the Deep South, New England, and the Lewis and Clark expedition, depending on the year. "The courses are very rigorous," says one student. Another adds, "Although the minimum requirement is 12 credits, they expect you to take 15 to 18 every semester." The student-run Honor System has been in effect for over a hundred years.

> "Although the minimum requirement is 12 credits, they expect you to take 15 to 18 every semester."

Biology, English, psychology, and global studies are some of the most popular majors at Randolph—and some of the school's best departments. Since there are no TAs, it's easy for students to form friendships with their professors. "The quality of teaching is excellent," says a junior, "with great communication between the students and professors." Students interested in research may compete to assist professors with ongoing projects during an eight-week summer session.

Though the Randolph community is small, students need not worry about claustrophobia; worldliness is a part of life here. The Visiting International Professor program has brought scholars from China, Nigeria, Croatia, India, and Malaysia to campus. In addition, the school offers $1 million in scholarship money for study abroad, prompting 40 percent of students to study in another country before graduation. Some programs are thematic in nature, examining peace studies in Nagasaki and Hiroshima, or theater in London and Stratford; others offer Randolph students the option of studying at the University of Reading in England, the Universidad de las Americas in Mexico, or the University of Economics in Prague. The college also participates in the Seven-College Exchange,* the Tri-College Exchange, and in American University's Washington Semester program. Two-thirds of Randolph students secure off-campus internships, with organizations from the Chicago Lyric Opera and London's Imperial College to businesses such as clothing-maker Tommy Hilfiger.

Thirty-nine percent of Randolph students hail from Virginia. Minorities have a sizable presence on campus, with Hispanics and Asian Americans each contributing 4 percent of the student body, and African Americans making up 10 percent. The Black Women's Alliance encourages awareness of and respect for differences, and Randolph is also a member of the International 50, a group of colleges committed to multiculturalism. "The most polarized rivalry is between Odd and Even classes, based on the year of a student's graduation," says a biology major. "It takes precedence over political, religious, and social groups, all of which are visible and welcoming, instead of exclusionist." Randolph awards merit scholarships each year, worth an average of $14,855; there are no athletic awards.

Students interested in research may compete to assist professors with ongoing projects during an eight-week summer session.

Eighty-eight percent of Randolph students live in the dorms, which have carpeting and plenty of storage space. Everyone gets a room, eventually, though one student cautions that room draw in the spring can take a long time to sort out. Main Hall, once nicknamed "the Hilton," is the largest dorm, and its central location makes it the most convenient. All rooms are linked to the campus computer network and wired for cable TV. "The dining hall provides great food, buffet-style, and aims to please everyone from vegans to Atkins dieters," says a senior. "I believe the vegans have it best here," muses one student. Security officers patrol continuously between 4 p.m. and 8 a.m. and take pride in knowing students by name.

When the weekend comes, the Macon Activities Council makes sure no one is bored. "On-campus social events include visiting comedians, bands, and other entertainers, as well as self-created fun like our 'No Talent' talent show and lip-sync contest, with cash prizes!" says a senior. "Other activities include shopping and visiting the dollar theater, or traveling to nearby colleges like Hampden-Sydney and Washington and Lee. Class trips sometimes venture to Washington, D.C., and the Wintergreen ski slopes"—or even as far north as New York City. Charlottesville, Virginia, an hour away, is also home to the University of Virginia, another popular destination for those seeking frat parties and football. Underage drinking is prohibited at Randolph, in accordance with state and federal law, and one student reports, "It is looked down upon."

The town of Lynchburg, which hosts two other colleges, has a Wal-Mart, but is otherwise "less than exhilarating," says a biology major. A classics major explains. "We get along well enough, but the town is not very receptive." Indeed, most clubs in the area are 21 and over, and the restaurants, stores, and movie theaters that do operate are closed by 10 p.m. Thankfully, the college's upscale bookstore stays open late to satisfy students' caffeine cravings.

Randolph's Wildcats compete in Division III, and the school's top rival is nearby Sweet Briar. The swim team won over 80 percent of their meets last year and placed third in the conference, while the riding program offers instruction for students of all abilities. Riders may also compete against athletes from 12 area colleges, as well as in regional and national competitions. But more than athletic contests, students look forward to Randolph traditions, such as Ring Week (in which a freshman anonymously decorates the door of a junior and leaves her small gifts all

"The quality of teaching is excellent."

week, culminating with a scavenger hunt for her class ring), and the Pumpkin Parade (during which sophomores and seniors, not to be left out, scour campus in search of carved pumpkins). The Never-Ending Weekend each fall includes both a formal and the annual Tacky Party, for which tasteless attire is *de rigueur*.

The students of Randolph aren't shy about their academic goals, career drive, or sense of campus unity. The challenge before them now is learning how to accept men on campus while retaining the traditions and sense of community that are such a vital part of the Randolph scene.

Overlaps

Hollins, Sweet Briar, University of Mary Washington, Virginia Tech, Smith, UNC–Chapel Hill

If You Apply To ➤

Randolph: Early decision: Nov. 15. Meets demonstrated need of 35%. Campus and alumni interviews: recommended, informational. SATs or ACTs: required. SAT IIs: optional. Accepts the Common Application and electronic applications. Essay question: topic of your choice, or a photocopy of a graded essay written in the 11th or 12th grade.

University of Redlands

1200 East Colton, P.O. Box 3080, Redlands, CA 92373-0999

If you like the thought of palm trees against a backdrop of snow-covered peaks, Redlands may be your place. Though a "university," Redlands is about the same size as Occidental and a hair bigger than Whittier. The alternative Johnston Center for Integrative Studies makes an odd contrast to the buttoned-down conservatism of the rest of Redlands.

Website: www.redlands.edu
Location: Small town
Total Enrollment: 2,407
Undergraduates: 2,313
Male/Female: 45/55
SAT Ranges: V 540–630
  M 540–630
ACT Range: 22–26
Financial Aid: N/A
Expense: Pr $ $
Phi Beta Kappa: Yes
Applicants: 3,792
Accepted: 63%
Enrolled: 30%
Grad in 6 Years: 63%
Returning Freshmen: 92%
Academics: 🖋 🖋 🖋
Social: 🎉 🎉 🎉
Q of L: ★ ★ ★
Admissions: (800) 455-5064
Email Address:
  admissions@redlands.edu

Strongest Programs:
Government
English
Creative Writing
Biology
Chemistry
Music
Environmental Studies

Amid the dozens of gigantic and well-known universities in the state of California stands the University of Redlands. With its innovative living/learning college and strong preprofessional emphasis, this versatile school is one of higher education's better-kept secrets, and a place where students receive all the personal attention they could want. One student describes it as a "small liberal arts college in sunny Southern California with great financial aid packages."

The University of Redlands' 160-acre campus, covered in majestic oak trees, is designed around "The Quad," a group of dorms that face one another. The two main landmarks are the Memorial Chapel and the administration building. Redlands' facilities are a mixture of older, more historical columned buildings and more modern, renovated ones. The view from the college can only be described as breathtaking. Mountain ranges form the backdrop, and neighboring Big Bear Lake and Arrowhead ski resorts give endless getaway opportunities. Also nearby are the San Gorgonio Wilderness and Joshua Tree National Park. For those looking for big city adventures, Los Angeles is only an hour away.

One of Redlands' most distinctive attributes is its experimental living/learning college, where students create their own course of study and are judged by self- and professor evaluations rather than grades. The Johnston Center for Integrative Studies was established in 1969 to function as an "alternative" college within a traditional setting; now a program rather than a separate college, about 10 percent of the students here take advantage of this opportunity. The program offers unusual academic freedom; there are no departments, majors, or distribution requirements. Instead, students "contract" with professors for their entire plan of study. At the beginning of each course, students make up the syllabus by consensus and then set their own research

"The dorms are comfortable and complete with Ping-Pong and pool tables."

and writing goals. Each student develops four-year goals—which are reviewed by a student/faculty board for direction and breadth—within one or more broad areas: the social sciences, behavioral sciences, humanities, and fine and performing arts.

Compared with other Redlands students, Johnston undergrads have higher test scores, with average SATs that are 50 to 75 points above the Redlands median. One student explains, "Johnston center students tend to be independent thinkers, self-motivated, and [don't] take classes just because they have to." Johnston's enrollment declined in the late '70s and '80s as students became very career-oriented and found that alternative education no longer fit their needs. Fortunately, student interests and needs have once again changed, and Johnston is seeing its highest enrollment in nearly two decades.

Aside from Johnston, Redlands is unusual among liberal arts institutions mainly in that it also offers professional programs. The schools of education and music provide strong career training, as does the excellent program in communicative disorders. Business is the most popular major, followed by integrative studies, psychology, government and international relations, and English. The environmental studies program consists of courses in natural science, humanities, and social science, focusing on values-based environmental problem solving. Students can receive degrees in environmental studies, environmental science, and environmental management. Redlands has also emerged as a national leader in science curriculum reform.

The liberal arts foundation gives students the fundamental skills essential to effective learning and scholarship by challenging them to examine their own values and the values of society. With its 4–4–1 calendar, Redlands has students take one intensive course each May. Students may also choose from among 100 study abroad options in Europe, Asia, Africa, and Latin America. The highly acclaimed freshman seminar program places small groups of first-year students with some of the school's best professors, while the selective honors program enables outstanding students to work individually with professors, who are very accessible outside of class and occasionally even come by the dorms for "fireside chats." "Freshmen are always taught by full professors," says one sophomore, and a writing major declares that at Redlands, "the professors are the school's biggest asset." Most students agree that Redlands' laid-back academic atmosphere is much appreciated.

"The professors are the school's biggest asset."

Fifty-seven percent of the student body comes from within the state, creating a mellow, Southern California atmosphere on campus. The climate is a definite plus, with temperatures rarely below 50 degrees. The typical Redlands student tends to be fairly conservative, although one student reports that the minority of liberals are quite vocal. The racial makeup of the school is fairly diverse, with Hispanics representing a solid 11 percent, Asian Americans 7 percent, and African Americans 4 percent. About a third of the graduating class each year moves on to graduate schools, while 60 percent head into the workforce. Redlands annually awards a variety of merit scholarships, ranging from $2,500 to about half-tuition. There are talent awards in art, writing, music, and debate, but there are no athletic scholarships.

Students have nothing but rave reviews for the dormitories, in which 70 percent of the students live all four years. "Each dorm has a personality of its own," exclaims one student. "The dorms are comfortable and complete with Ping-Pong and pool tables," raves a sophomore. Most of the dorms are co-ed, though there is an all-woman dorm. Students agree that freshmen should check out Merriam Hall first. Other housing options include on-campus apartments and student-run co-ops. Students enjoy their food at the Irvine Commons or Plaza Cafe, part of the Hunsaker University Center. Students can mingle in the "town square" atmosphere of the center's bookstore, cafe, and student life offices.

The Southern California heat and smog can become unpleasant, and students with wheels often flee Redlands on weekends for healthier pleasure spots along the California coast or in the mountains. But overall, social life is centered on campus activities. Local fraternities and sororities claim 15 percent of the student body, and their parties are open to all, but even these parties are rarely raucous. Road trips to Hollywood, Palm Springs, San Francisco, the beach, and even Mexico are common. Despite the school's enforcement efforts, alcohol is accessible. "It is very easy. Students operate bars out of their rooms," reports one student.

The NCAA Division III sports program injects a measure of excitement into the social scene. Students compete in 21 intercollegiate varsity sports in the Southern California Intercollegiate Athletic Conference (SCIAC). Men and women compete in basketball, cross-country, golf, soccer, swimming, tennis, track, and water polo. Men also compete in football, while women's teams include lacrosse, softball, and volleyball. In addition, more than 40 percent participate in at least one intramural sport. Redlands has won three recent national championships in women's water polo. Also competitive are football, men's soccer and basketball and women's soccer and softball.

One of Redlands' most distinctive attributes is its experimental living/learning college, where students create their own course of study and are judged by self- and professor evaluations rather than grades.

The environmental studies program consists of courses in natural science, humanities, and social science focusing on values-based environmental problem solving.

Overlaps

Chapman, Loyola Marymount, UC–San Diego, UC–Riverside, UC–Santa Barbara, University of Southern California

The University of Redlands is a lot of different things to a lot of different people. With 170 faculty members, it manages to be a preprofessional institute, a liberal arts college, and an alternative school all in one. The Johnston Center is clearly a path to travel for the innovative individualist, but even those who don't join Johnston will likely find what they want and need at Redlands.

Reed College

3203 S.E. Woodstock Boulevard, Portland, OR 97202-8199

Reed is a West Coast version of Oberlin or Grinnell, mixing nonconformist students with a traditional and rigorous curriculum. Only 75 percent of first-year students graduate in six years, due to Reed's demands and the fact that its students often live close to the edge.

Website: www.reed.edu
Location: City outskirts
Total Enrollment: 1,273
Undergraduates: 1,272
Male/Female: 45/55
SAT Ranges: V 660–760
  M 620–710
ACT Range: 29–32
Financial Aid: 52%
Expense: Pr $ $ $ $
Phi Beta Kappa: Yes
Applicants: 2,646
Accepted: 45%
Enrolled: 29%
Grad in 6 Years: 70%
Returning Freshmen: 85%
Academics: 🐜 🐜 🐜 🐜 ½
Social: ☎ ☎ ☎
Q of L: ★ ★ ★ ★
Admissions: (503) 777-7511
Email Address:
  admission@reed.edu

Strongest Programs:
Biology
Chemistry
Psychology
English

Reed College is one of the most intellectual colleges in the country. In fact, this hotbed of liberalism has produced 31 Rhodes scholars, 63 Fulbright scholars, 53 Guggenheim Fellowships, 108 National Science Foundation Fellowships, and two winners each of the Pulitzer Prize and the MacArthur "genius grant." Reed is a place where students complain that the library, which closes its doors at midnight on Fridays and Saturdays, shuts down too early. "Reedies are highly motivated, intellectually curious individuals who want to challenge themselves," says a student. Letter grades are deemphasized as a form of evaluation. Instead, students receive lengthy and detailed commentaries from professors, which fosters continued dialogue.

Located just five miles from downtown Portland, Reed's 115-acre campus boasts rolling lawns, winding lanes, a canyon creek, and protected wetlands. A fish ladder was recently installed to help salmon reach their spawning grounds, and nonnative plants are being removed from the area to protect the natural habitat. In addition to the canyon, the campus hosts 125 different species of trees. Two thousand majestic arbors shade a mix of original campus buildings, constructed of brick, slate, and limestone in the Tudor Gothic style, as well as lodges in the homey Northwest Timber style and some more modern facilities. The 23,000-square-foot Educational Technology Center and a $5.5 million expansion of the Eric V. Hauser library were recently completed.

Although Reed emphasizes personal freedom and responsibility, especially through its Honor Principle, the curriculum and requirements are quite traditional. All freshmen must complete Humanities 110, a yearlong interdisciplinary course focused on society and culture in archaic and classical Greece and imperial Rome. The course, which has been taught for more than a half-century, draws on the expertise of 25 professors, including some of Reed's most senior and distinguished faculty. Students must also take courses in four "breadth" areas: literature, philosophy, religion, and the arts; history, the social sciences, and psychology; the natural sciences; and mathematics, logic, linguistics, or foreign languages. Juniors must pass a qualifying examination in their major departments and all seniors must submit a thesis to graduate. On the due date, just after spring classes have ended, seniors march from the library steps to the registrar's office in the Thesis Parade. This marks the beginning of Renn Fayre, a weekend-long celebration that involves "a bug-eating contest, Glow Opera, and the ball drop," says a history major.

Despite Reed's small size—89 percent of the courses taken by freshman have 25 or fewer students, and it just gets better from there—the school offers unsurpassed research opportunities in the liberal arts and sciences. Budding physicists and environmental scientists can work with college staff at the 250 kilowatt Triga nuclear reactor, after passing an eight-hour Atomic Energy Commission examination. Study abroad programs attract about 30 percent of each graduating class, taking students to 15 countries, from Germany and China to Ecuador and Russia. Reed also offers domestic exchange programs with Howard, Sarah Lawrence, and the Woods Hole Oceanographic Institute. Dual-degree (3–2) programs include engineering, computer science, and forestry/environmental science.

Reed students take full advantage of all the academic options, which often keep them tied to their computers and study carrels. "Reed is incredibly rigorous," one sophomore says. "I've pulled all-nighters and a near all-weeker. I've seen the sun rise over the library." You'll never find a TA at the lectern here or leading a group discussion, so students rarely attend class unprepared for the lively **"Be prepared to study like your life depends on it."** intellectual banter that typically ensues between inquiring and active minds. "Access to full professors is one of Reed's strongest draws," a student says. "Professors here are extremely knowledgeable," adds a junior. Over the years, a quarter of Reed's grads have gone on for PhDs, the highest percentage of any liberal arts college in the country.

While Reed is located in Oregon, its quirky brand of intellectualism means only 15 percent of the students are in-staters. Twenty-two percent are minorities—2 percent Native American, 9 percent Asian American, 6 percent Hispanic, and 3 percent African American. Reedies can't be stereotyped, says a student. "They come in all shapes, sizes, flavors, and colors. There are those who love to shop at J. Crew and others who have forgotten the meaning of wearing shoes." As for politics, "Reedies are mostly left-leaning," says a sophomore.

Sixty-five percent of Reed students live on campus in coed housing or in one of the many "theme" dorms, which usually house fewer than 30 people each. Some rooms feature such homey touches as fireplaces or balconies. Freshmen are guaranteed housing, and usually get divided doubles; upperclassmen get singles, though "recent resurgence in the desire to live on campus makes rooms hard to come by," says a history major. Old Dorm Block and Anna Mann are said to be the most popular choices for first-years, and all rooms are connected to the campus computer network. A wireless network extends throughout campus. For those who lose out in the housing lottery, or upperclassmen seeking a taste of post-college independence, off-campus houses are cheap and plentiful. On-campus students must buy the meal plan, and students say outside caterer **"Reedies are highly motivated,** Bon Appetit does a good job, with salad **intellectually curious individuals."** and sandwich bars and grill and entree stations available each day at lunch and dinner. Vegetarian and vegan choices are offered, too. "Reed has some of the best college food around," a junior says.

Students say Portland is the perfect college town. "Portland rocks!" exclaims one senior. "It's filled with lots of young people. There is good food, arts, music, and cultural events." The vibe is similar to Seattle, with plenty of technology, coffee—and rain. "Portland is a vibrant, bike-friendly, eco-conscious city of pine trees, coffee shops, bookstores, and live music," an English major says. Aside from low-key parties and gatherings with friends on and near campus, Reed students love to road trip—whether to the mammoth Powell's bookstore downtown (about a 15-minute

(Continued)
Philosophy
History
Physics

Reed is definitely not for everyone—emblematic items sold in the bookstore bear the slogan "Atheism, Communism, Free Love."

drive) or to Oregon's coastal beaches, mountains, or high desert, all about two hours away. The school also owns a ski cabin on Mt. Hood that sleeps 15.

Where there are college students, there is also alcohol, though drinking is not a high priority here. Reed follows federal and state laws, meaning that those under 21 may not imbibe. However, use of alcohol or drugs is governed primarily by the Honor Principle, which "makes for an atmosphere in which those who use substances are safely in view," one student says. Although the Honor Principle has no formal definition, according to popular characterization it bids students, staff, and faculty to "not cause unnecessary embarrassment, discomfort, or injury to other individuals or to the community as a whole." Students look forward to Paideia (which means "education" in Ancient Greek), a week-long program of wacky alternative classes before spring semester begins. Paideia's noncredit workshops from range from Skittle appreciation and a South Park marathon to how to get into law school or make Shrinky Dinks. Each April, students celebrate Nitrogen Day, honoring our atmosphere's most plentiful—and underappreciated—element.

"Portland rocks! There is good food, arts, music, and cultural events."

The closest thing Reed has to a school mascot is the Doyle Owl, a 300-pound concrete sculpture that dorms regularly plot to steal from one another. While Reed doesn't have competitive athletics, clubs such as rugby and ultimate Frisbee do compete with other clubs in the area. Reedies must also fulfill a three semester-long physical education requirement. Ballroom dancing, telemark skiing, and juggling are just a few of the many courses offered.

Reed is definitely not for everyone—emblematic items sold in the bookstore bear the slogan "Atheism, Communism, Free Love." But if you're an intellectual who leans left politically and prefers to spend Saturday nights with your nose in a book, this Portland school is worth a look. "Anyone who is thinking of applying to Reed should do so if they're looking for a solid education, taught by talented professors with high expectations for every student," a senior says. But another student warns: "Be prepared to study like your life depends on it."

Overlaps

University of Chicago, UC–Berkeley, Oberlin, Carleton, Stanford, UC–Santa Cruz

If You Apply To ➤ **Reed:** Early decision: Nov. 15, Jan. 2. Regular admissions and financial aid: Jan. 15. Housing: Jun. 15. Meets demonstrated need of 98%. Campus interviews: recommended, evaluative. Alumni interviews: optional, evaluative. SATs or ACTs: required. SAT IIs: recommended. Accepts the Common Application and electronic applications. Essay question: personal statement of student's choice or from the Common Application; why Reed; and a graded writing sample.

Rensselaer Polytechnic Institute

110 Eighth St., Troy, NY 12180-3590

If you can spell Rensselaer, you've already got a leg up on many applicants. RPI is one of the nation's great technical universities, along with Caltech, MIT, Worcester Polytech, and Harvey Mudd. The beauty of RPI is the chance for hands-on learning and synergy between technology and management.

Website: www.rpi.edu
Location: City outskirts
Total Enrollment: 6,055

It would be an exaggeration to say that technology is God at RPI, though the school's conversion of a Gothic chapel into a computer lab does hint in that direction. Even if it's not deified, technology remains omnipresent at this school, which pioneered the teaching of calculus via computer in the early '90s and has been named one of the most-wired campuses. Students attend class in fully-wired studio classrooms where they work on team projects and collaborate to solve real-world problems. Since 1998, the institute has more than doubled its research funding, and it's one of six original National Science Foundation Nanotechnology Centers in the country. Students who may have been known as geeks in high school come to Rensselaer to find a home among their own.

Set high on a bluff overlooking Troy, New York, Rensselaer's 260-acre campus mixes modern research facilities and classical, ivy-covered brick buildings dating to the turn of the century. The Experimental Media and Performing Arts Center will be completed in 2008. Almost the entire campus is wired (or wireless), allowing students to study and collaborate with each other from anywhere on campus. A new East Campus Athletic Village is in the works and will include a new football field, gym, 50-meter pool, indoor track and field area, as well as indoor and outdoor tennis courts.

Not surprisingly, all students are required to have their own laptop computer. Studio courses foster small-group interaction between students and professors, after which students return to their workstations for collaborative problem solving. RPI, which made its reputation as one of the nation's premier engineering schools, continues to excel in traditional favorites such as chemical and electrical engineering, as well as newer specialties like environmental and computer systems engineering. Computer science is the most popular major, followed by mechanical engineering, management, electrical engineering, and the notoriously difficult five-year architecture program. The nuclear engineering department has its own linear accelerator, while graduate and undergraduate students participate in research at the Center for Industrial Innovation. RPI is a national leader in the study and application of electronic media and offers a BS program in electronic arts. RPI offers more than 25 doctoral programs including cognitive science, biochemistry/biophysics, architectural sciences, and game and simulation arts and sciences.

RPI's Lally School of Management and Technology combines elements of a business school with the latest technical applications, and coursework has been redesigned into yearlong classes to capture the complexity of the business world. Entrepreneurship is one of its specialties; budding entrepreneurs may participate in the Rensselaer Business Incubator, a support system for start-up companies run by

"Hockey at RPI equals insanity."

Rensselaer students and alumni. RPI students are required to take courses in entrepreneurship or have an "entrepreneurial experience" before graduation. The BS program in information technology continues to attract top students who often combine it with coursework in e-commerce or the arts. Majors in the humanities and social sciences are limited, and their quality is directly related to their applicability to technical fields. Still, all students must complete at least 24 credits in these areas, as well as at least 24 credits in physical, life, and engineering sciences, a minimum of 30 credits in their majors, and a writing or writing-intensive course. Additionally, students must complete an international requirement by participating in research or an internship in another country. "The courses are pretty rigorous," admits a senior, "but because of the high level of group assignments, competition is eliminated and you learn a lot from your peers."

About two-thirds of Rensselaer students are undergraduates, a high percentage for a top engineering school; because of this, RPI has worked hard to ensure that classes are smaller and more attention is paid to individual needs. Professors are extremely knowledgeable and the quality of teaching is excellent. "The professors I've had were excellent lecturers and very helpful outside class," a sophomore says. Juniors and seniors enjoy self-paced courses and occasionally paid positions helping with faculty research. Career counseling is helpful; academic advising has added an "early warning" system to target students who are having trouble. All of the effort

(Continued)
Undergraduates: 4,911
Male/Female: 76/24
SAT Ranges: V 580–690
  M 640–730
ACT Range: 24–28
Financial Aid: 70%
Expense: Pr $ $ $
Phi Beta Kappa: No
Applicants: 5,574
Accepted: 78%
Enrolled: 29%
Grad in 6 Years: 81%
Returning Freshmen: 92%
Academics: ✍ ✍ ✍ ✍
Social: ☎ ☎ ☎
Q of L: HHH
Admissions: (518) 276-6216
Email Address:
  admissions@rpi.edu

Strongest Programs:
Engineering
Multimedia/Visual
  Communications
Applied Mathematics
Technological
  Entrepreneurship

Social life mostly takes place on campus, but the male/female ratio is a major hassle, forcing lovelorn men to haunt Russell Sage (next door) or Skidmore (40 minutes away) in hopes of finding a mate.

pays off: 64 percent of a recent RPI senior class went on to jobs after graduation, and 32 percent entered graduate or professional school.

For students who can't wait to start working, popular co-op programs in more than a dozen fields help them earn both money and credit. Those who already know what field they'll pursue may enter a seven-year dual degree program in medicine, a six-year program in law, and four- and five-year master's programs in biology, geology, and mathematical science. Although most engineering schools discourage studying abroad, Rensselaer offers exchange programs around the world.

Forty-five percent of RPI students are New Yorkers, and 95 percent ranked in the top quarter of their high school class. RPI is fairly diverse, with Asian Americans comprising 11 percent of the student body, African Americans 4 percent, and Hispanics 5 percent. Students from 70 countries and nearly every state attend RPI; males continue to outnumber females three to one. Students tend to be bright, studious, and technically savvy. "At a technical school we have our fair share of computer geeks, video game crazies, sci-fi fanatics," says a management major. "But mostly they are all intelligent, hard-working students. We have many normal, fun-loving college kids, too." RPI is far from a center of political activism; students say they're just too busy. "Politics? Not so much on this campus," says a student. The biggest campus issue may be choosing the Grand Marshal, who oversees a boisterous weeklong carnival celebrating campus elections, during which professors are forbidden from giving tests.

> "The professors I've had were excellent lecturers and very helpful outside class."

Fifty-six percent of students live in university residence halls; freshmen are required to live on campus and, for the fall term, must buy the meal plan. "The dorms are copious and for the most part, well-maintained," says a senior. All bedrooms have fast Internet connections and most residence halls are wireless. In Barton Hall, even the laundry room is wired for fast access to the Internet and RPI's campus network. Upperclassmen may keep their current room, enter the lottery to get something better, or live in college-owned apartments off campus, widely considered the nicest option. Thirty-nine percent of men and 18 percent of women go Greek.

Social life mostly takes place on campus, but the male/female ratio is a major hassle, forcing lovelorn men to haunt Russell Sage (next door) or Skidmore (40 minutes away) in hopes of finding a mate. Other than Greek parties, weekend options at RPI include sporting events, live entertainment, concerts, movies, and a half-dozen local pubs—some of which students find easily accept fake IDs. Those under 21 can't have alcohol in the dorms, and fraternities aren't allowed to have "containers of mass distribution" (i.e., kegs) at parties; students say alcohol isn't an issue on campus. Extracurricular clubs, organized around such interests as chess, dance, judo, and skiing, are chartered and funded by a student-managed body that doles out more than $8 million annually.

Free shuttle buses run regularly from campus to downtown Troy, a former industrial revolution town, but there aren't many good reasons to make the trip. "Troy is not a college town," says a senior. "It's in the middle of urban renewal right now, so it's really neat, but nothing for college students." Students and Greek groups do get involved with community service projects, though, and the town offers opportunities for internships. Rensselaer has

> "Students at RPI are all a little bit nerdy, and proud of it."

helped generate economic growth in Troy by investing in the downtown area and providing grants to homebuyers. A six-screen movie theater is within easy reach, and for a taste of bigger-city nightlife, Albany is a half-hour drive. For scenic excursions, the Berkshires, Catskills, Adirondacks, Lake George, Lake Placid, the Saranac Lakes, Montreal, and Boston are popular destinations.

RPI's Lally School of Management and Technology combines elements of a business school with the latest technical applications, and coursework has been redesigned into yearlong classes to capture the complexity of the business world.

The athletic scene at Rensselaer revolves around hockey, hockey, hockey—the school's only team playing in Division I. One of the biggest weekends of the year is Big Red Freakout, when all festivities center around cheering on the beloved Big Red. "Hockey at RPI equals insanity," one student says. "If you go to one hockey game all season, go to the men's hockey season opener. The place is packed with rowdy RPI students who scream and chant in unison." There are many intramural sports to choose from, but the most popular may be the D-level hockey team (meaning "I really don't know how to play this," says a junior). Other varsity teams play in Division III, and the football, baseball, men's and women's basketball, and women's field hockey teams are most popular. Women's ice hockey now competes in Division I. Each year, Rensselaer's football players vie with rival Union College for the coveted Dutchman's Shoes, while a fitness center helps even nonvarsity players sculpt six-pack abs.

Computer geeks and video game junkies aren't the only ones who will find a home at Rensselaer. Students who thrive on teamwork and collaboration will also find RPI to their liking. RPI provides cutting-edge technology to students constantly wondering how things work. RPI students work hard—sometimes to the detriment of a social life. "Students at RPI are all a little bit nerdy, and proud of it," a computer engineering major says. "The types of things that might be looked at strangely at other colleges are accepted here."

If You Apply To ➤

Rensselaer: Early decision: Nov. 1. Regular admissions: Jan. 15. Financial aid: Feb. 15. Does not guarantee to meet demonstrated need. Campus interviews: optional, informational. No alumni interviews. SATs or ACTs: required. SAT IIs: optional (required for all accelerated program applicants). Accepts the Common Application and prefers electronic applications. Apply to particular programs. Essay question: depends on program.

University of Rhode Island

Kingston, RI 02881

URI is a smallish alternative to UMass and UConn. With Boston, Providence, and vacation hot-spot Newport within easy reach, there is plenty to do. Strong programs include engineering, marine science, and pharmacy. Almost half of URI's students are out-of-staters.

Once known as an unabashed party school, the University of Rhode Island is building a new reputation for challenging academics and improved town/gown relationships. In the years since President Robert Carothers cracked down on the wild drinking scene, the school has become more focused on Monday morning and less on Saturday night. Carothers also challenged faculty to create a new culture for learning; the result is an environment in which students engage in service learning, do research with top faculty, and find a much heavier emphasis on alternative styles of learning.

URI's 1,200-acre campus is located in the small town of Kingston. Surrounded by farmland and only six miles from the coast, it is also within easy driving distance of cities such as Providence—the Renaissance City and home to Brown, the Rhode Island School of Design, and several other colleges and universities—Boston, and New York. The main academic buildings at URI, a mixture of modern and "old New England granite," surround a central quad on Kingston Hill. At the foot of Kingston Hill lie the athletic buildings and agricultural fields. Current construction includes a $22 million dining hall and three new residence halls.

Website: www.uri.edu
Location: Small town
Total Enrollment: 15,095
Undergraduates: 11,546
Male/Female: 43/57
SAT Ranges: V 510–600
  M 520–610
ACT Range: N/A
Financial Aid: 52%
Expense: Pub $ $ $
Phi Beta Kappa: Yes
Applicants: 13,388
Accepted: 77%
Enrolled: 24%

(Continued)

Grad in 6 Years: 56%

Returning Freshmen: 80%

Academics: ✍ ✍

Social: 🐨 🐨 🐨 🐨

Q of L: ★ ★ ★

Admissions: (401) 874-7100

Email Address:
uriadmit@etal.uri.edu

Strongest Programs:
Pharmacy
International Engineering
Textiles, Fashion
    Merchandising and Design
Oceanography
Marine and Environmental
    Sciences

After a year or two in University College, students choose more specialized colleges, such as the well-regarded College of Pharmacy.

The university offers exchange programs with universities in Australia, England, France, Germany, Japan, Korea, Mexico, Spain, Venezuela, Quebec, and Nova Scotia.

New students are first enrolled in the University College, which offers academic and career guidance as well as advice on selecting from among required general education courses in communication skills, fine arts and literature, natural and social sciences, letters, mathematics, foreign language, and culture requirements. Undecided or exploratory students are assisted in following their intellectual curiosity to find the perfect fit for a major. All new students take URI 101, a one-credit course intended to acquaint students with support services, co-curricular activities, and academic majors and career options. After a year or two in University College, students choose more specialized colleges, such as the well-regarded College of Pharmacy. The university also offers a marine and environment program, landscape architecture, and African and African-American studies. Some majors, such as a professional degree like the doctor of pharmacy, require students to stay for five or six years, but most students graduate in four years. Newer programs include bachelor's degrees in film media and computer science, a doctorate in audiology, and programs through the Center for Personal Financial Education and the URI Feinstein Center for Hunger-Free America.

> **"The teaching I have received has been of the highest quality."**

Freshmen often have full professors. "The teaching I have received has been of the highest quality," says a human development major. The university offers exchange programs with universities in Australia, England, France, Germany, Japan, Korea, Mexico, Spain, Venezuela, Quebec, and Nova Scotia. Students can also participate in the international engineering program, spend a year of the five-year program abroad, and major in both engineering and Spanish, German, or French. The Academic Enhancement Center, along with its Writing Center, provides free tutorial assistance to anyone who wants assistance with a class or feedback on any kind of college or extracurricular writing.

Students mostly hail from the Northeast, and URI gives preference to in-state students who meet requirements. More than half the entering freshmen are from outside Rhode Island. Minority enrollment stands at 4 percent African American, 5 percent Hispanic, and 3 percent Asian. The academic profiles of the incoming class have risen steadily over the past 10 years, in large measure due to the Centennial Scholarship Program for outstanding freshmen.

Though most freshmen live on campus, 61 percent of all undergraduates do not. Because of ongoing renovations, on-campus housing is tight. "Students are tripled up and once you move off campus you can't come back," says a senior. "There is no room." The good news is they live in a sizable off-campus community near the beach. Students complain about a chronic shortage of parking but say that even freshmen are allowed to have cars. "Commuters park near freshmen dorms and freshmen park near all the classes—explain that one!" says a senior.

> **"Commuters park near freshmen dorms and freshmen park near all the classes—explain that one!"**

There is ample public transit and a convenient on-campus shuttle system. The university now has a strict alcohol policy, and students note that underage drinkers have a hard time finding an adult beverage on campus.

While the hearty New England winters are invigorating for some, many students wind up catching colds, earning URI the unflattering nickname Upper Respiratory Infection. As for Kingston, it's a sleepy New England college town. "It's just a regular town," says a senior. Says another student: "It's a quiet place with lots of trees. I wouldn't say it's a college town, but it keeps you from being distracted from city life." Students get involved in the town through clubs or URI 101, which requires volunteer work. URI is also within striking distance of both Rhode Island's famous beaches and the major New England ski slopes. Newport, with its heady social scene, is just 20 minutes away, and those feeling lucky can get to the Foxwoods Resort and Casino in 45 minutes. Other fun road trips include Providence (30 minutes), Boston (90 minutes), and New York City (four hours). Travel is facilitated by the fact that there is an Amtrak stop on campus. Many natives find going home on weekends a pleasant diversion, but on campus there are always movies and guest speakers, plus the usual Greek parties. The campus coffeehouse hosts open mic nights, and movie theaters, clubs, and malls beckon just off campus. The student newspaper that covers it all has one of the most original names anywhere: *The Good 5¢ Cigar* (as in, "what this country needs is").

Sports are big at Rhode Island, and basketball games are especially exciting. Midnight Madness (the team's first sanctioned practice of the year) is always well attended, and URI fans love it when the team defeats archrival Providence College. Another student favorite is Oozeball, an April volleyball tournament played in about two feet of mud. As befits the school's locale, sailing draws much interest, and the team regularly produces All Americans. Students can find unique opportunities running businesses like a flower shop, a sound and lighting group, and a coffeehouse in the Memorial Union under full-time supervision but with a lot of independence.

URI offers students a large-school feel in a small state. Centrally located in dense New England, students benefit from having the ocean at their back door. But instead of spending all their time at the beach or on nearby ski slopes, students work hard to achieve good grades and lay the foundation of lifelong learning.

If You Apply To ➤

URI: Rolling admissions. Early action: Dec. 15. Regular admissions: Nov. 15. Financial aid: March 1. Meets demonstrated need of 63%. Campus interviews: recommended, evaluative. Alumni interviews: optional, evaluative. SATs or ACTs: required. Essay question: personal statement (optional).

Rhode Island School of Design

2 College Street, Providence, RI 02903

The nation's best-known arts specialty school, RISD sits on a hillside adjacent to Brown. The campus offers easy access to downtown Providence, but it can't match the location of rival Parsons in New York's Greenwich Village. Offers an artsy architecture major in addition to programs in the visual arts.

Founded in the late 19th century to address the country's need for more artisans and craftsmen, the Rhode Island School of Design has grown into a premier arts incubator. It's a place where today's artists and designers gather to share ideas and create tomorrow's masterpieces and architectural icons. RISD grants degrees in virtually every design-related topic, and like the varied curriculum, the students and their creations are as diverse as the colors on an artist's palette. "It is competitive but not mean-spirited," says a senior architecture major. "The environment really fosters production, creativity, and a high level of learning," says another student.

Though you might expect an art school like RISD to occupy funky, futuristic buildings, the predominant look here is Colonial New England. Set on the upgrade of College Hill, RISD sits at the edge of Providence's beautifully preserved historic district across the street from Brown University. Many campus buildings date from the 1700s and early 1800s; the mostly redbrick-and-white-trim group includes converted homes, a bank, and even an old church with the campus pub in what used to be its attic. The six-story Industrial Design building, designed by faculty member

Website: www.risd.edu
Location: City center
Total Enrollment: 2,282
Undergraduates: 1,883
Male/Female: 34/66
SAT Ranges: V 530–660
  M 550–670
Financial Aid: 66%
Expense: Pr $ $
Phi Beta Kappa: No
Applicants: 2,511
Accepted: 32%
Enrolled: 48%
Grad in 6 Years: 90%

(Continued)

Returning Freshmen: 92%

Academics: ✑ ✑ ✑ ✑

Social: ☎ ☎

Q of L: ★ ★ ★ ★

Admissions: (401) 454-6300

Email Address:
admissions@risd.edu

Strongest Programs:
Architecture
Illustration
Graphic Design
Industrial Design

Jim Barnes, occupies the old Roitman furniture company warehouse. Its 50,000 square feet are designed for wood- and metal-working and prototype making. Recently, the school purchased buildings to house graduate student studio spaces and plans are currently underway for a new RISD center complex.

While RISD looks traditionally New England on the outside, behind its historic walls lies something else entirely. Students give the architecture, graphic design, and industrial design programs top marks, while liberal arts are considered more of a "joke." One apparel major remarks, "I think most people try to avoid the art history programs." That said, bachelor's and master's liberal arts concentrations in art history, English and literature, history, philosophy, and social sciences are available, complementing offerings in furniture design and architecture and interior architecture (for grad students). Landscape architecture is no longer available to undergraduates. RISD also offers crossregistration at adjacent Brown University for students seeking more diverse courses. The institute's highly specialized library contains 662,000 nonbook items (prints, etc.) and 106,000 volumes, and students are likely to be found at their personally assigned studio carrels. Perhaps RISD's most prized facility is its art museum, which boasts more than 85,000 pieces, a superlative collection that includes everything from Roman and Egyptian art to works by Monet, Matisse, and Picasso.

> **"The department faculty is made up of practicing artists in the field, which really makes a difference."**

To graduate, students must be in residence for at least two years and must complete a final-year project. They must also finish 126 credit hours—54 in their major, 18 in the Foundation Studies program (an integrated year of "functional and conceptual experiences" that leads to "an understanding of visual language"), 42 in the liberal arts (art and architectural history, English, history/philosophy/social sciences, some electives), and 12 in nonmajor electives. Hands-on studio courses abound, and most classes have fewer than 20 students. Still, a sophomore says, "Sometimes classes fill very quickly because they are really interesting, but there are always ample alternatives to choose from." During RISD's winter session, six weeks between the first and second semesters, students are encouraged to take courses outside their major. And each year about 30 juniors and seniors venture to Rome for the European Honors Program, which offers independent study, projects with critics, and immersion in Italian culture. There is also an International Exchange program where students study abroad at approved art institutions.

While students at RISD don't "hit the books" in the traditional sense, the in-studio workload is tremendous. "Our Freshman Foundation program is nicknamed the RISD boot camp," says a senior industrial design student. Students praise faculty members' knowledge and accessibility. "The quality of teaching is good," a junior says. "The department faculty is made up of practicing artists in the field, which really makes a difference." Career and academic counseling both get high marks.

> **"We're artists. We're a bunch of weirdos!"**

Students give the architecture, graphic design, and industrial design programs top marks.

RISD students (referred to as "RISDoids" or "Rizdees") come to Providence to form a largely urban mix of styles and personalities. In a word, the school is diverse, and that can create tension. "We're artists," a sophomore says. "We don't believe in being politically correct. We're a bunch of weirdos!" Indeed, only 8 percent of students are native Rhode Islanders, not surprising given the state's small size. Many of the rest are from the vicinity of other East Coast cities, notably New York and Boston. Though highly selective, RISD will often take a chance on students who did not perform well in high school by the usual academic criteria, but make up for that with special artistic talent. The racial makeup of the campus is fairly mixed, with Asian Americans making up 14 percent of the student body, African Americans 2 percent, and Hispanics 5 percent. Issues of ethnicity and sexuality top the campus agenda, though a sophomore says, "I see Brown as more political, but RISD gets involved as well." Tuition and fees here are steep; though RISD offers some merit scholarships, athletic scholarships are nonexistent, and the school does not guarantee to meet financial need.

All noncommuting freshmen must live in co-ed dorms that are comfortable and feature common studio areas and connections to the campus computer network. The vast majority of upperclassmen move off campus to nearby apartments, many of which occupy floors of restored homes; RISD also owns an apartment building and some renovated Colonial and Victorian houses. All boarders buy the meal plan, which students say has improved as of late. "The food is actually pretty good," admits a senior.

Despite a student body that looks like it could have been plucked from the streets of New York's Greenwich Village, RISD is not the place to come for wild and funky nightlife. "The social life is mostly centered around friends you work in studio with," explains one sophomore. With three eight-hour studios each week, plus two other classes (and that's just freshman year), "typical social activity is running out for a cup of coffee. If you're desperate, there's always Brown University." Providence also provides some social outlets. The Taproom, shown on the campus' map, went dry years ago, and each year, those 21 and over vote on whether to allow drinking in their residences. But underage students can swill. "It is generally easy for underage students to be served," an architecture major says. Though RISD isn't much for traditions, one big annual event is the Artist's Ball, a November formal where dress is "formal or festive, which has been interpreted as everything from chain mail to buck naked," says a senior. When claustrophobia sets in, students can flee to the RISD farm, a 33-acre recreation area on the shores of nearby Narragansett Bay. Boston and New York are one and four hours away by train, respectively.

Though jocks are an endangered species at RISD, recreation opportunities abound. There is no intercollegiate sports program in the ordinary sense, though there is a hockey team, called the Nads (which, of course, leads to RISDoids hollering "Go Nads!"). Students do get involved in intramural sports, ranging from football and baseball to sailing and cycling. There's also a weight room for those who thrive on pumping iron.

Students come to RISD committed to their crafts, and most march to the beat of their own drummer as they rush from studio courses to gallery openings to exhibitions. But this professional preoccupation is not a problem. Students are confident the endless studio hours are starting them on the path to success. Says a student, "Everyone here is really dedicated to what they are doing, lending a kind of general excitement that I have not seen anywhere else."

> *Tuition and fees here are steep; though RISD offers some merit scholarships, athletic scholarships are nonexistent, and the school does not guarantee to meet financial need.*

> **"The social life is mostly centered around friends you work in studio with."**

> *Though jocks are an endangered species at RISD, recreation opportunities abound.*

Overlaps

Pratt Institute, Parsons School of Design, Maryland Institute College of Art, School of Visual Arts, Cooper Union

---

If You Apply To ➢

RISD: Early action: Dec. 15. Regular admissions and financial aid: Feb. 15. Does not guarantee to meet demonstrated need. Campus interviews: optional, informational. No alumni interviews. SATs or ACTs: required. SAT IIs: optional. Accepts electronic applications. Essay question: statement of purpose.

Rhodes College

2000 North Parkway, Memphis, TN 38112-1690

Goes head-to-head with Sewanee for the top spot in the pecking order of mid-South liberal arts colleges. While Sewanee has a gorgeous rural campus, Rhodes has Memphis. Economics and international studies head the list of strong programs.

Website: www.rhodes.edu
Location: City residential
Total Enrollment: 1,633
Undergraduates: 1,618
Male/Female: 42/58
SAT Ranges: V 580–680
  M 580–670
ACT Range: 25–30
Financial Aid: 36%
Expense: Pr $ $
Phi Beta Kappa: Yes
Applicants: 3,695
Accepted: 49%
Enrolled: 24%
Grad in 6 Years: 81%
Returning Freshmen: 88%
Academics: ✍ ✍ ✍ ½
Social: ☎ ☎ ☎
Q of L: ★ ★ ★ ★
Admissions: (800) 844-5969
Email Address:
  adminfo@rhodes.edu

Strongest Programs:
Biology
Chemistry
Physics
International Studies
English
Economics
Political Science
Business Administration

Since 1848, Rhodes College has been instilling the timeless values of truth and trust in Southern sons and daughters, and today increasing numbers of students from the rest of the country are discovering its charms. The school's honor code means exams are not proctored and backpacks are left unattended in the cafeteria. Its small size gives everyone an opportunity to take on leadership roles in campus clubs and organizations, and people are generally "friendly and helpful," says one student. Throw in the college's proximity to Memphis' world-famous Beale Street, barbecue, and the blues, and it's clear that Rhodes offers a winning combination. "Students are intelligent, determined, and poised to succeed," says a sophomore. "We spend time volunteering and in internships, and often form study groups to facilitate learning."

Rhodes was founded as a Presbyterian school in Clarksville, Tennessee. It moved to a 100-acre campus in Memphis in 1925. Located in the residential midtown section of the city, Rhodes is across from a 175-acre park housing the city's largest art museum, a golf course, and the Memphis Zoo, which now has two giant pandas. Whether new or old, all campus buildings are Gothic in style, constructed of Arkansas fieldstone with leaded-glass windows and slate roofs. Thirteen of the original buildings are on the National Register of Historic Places. The $40 million Paul Barret, Jr., Library, the largest construction project in Rhodes history, was completed in 2005. It offers wireless Internet access and a coffee shop, along with group study rooms.

Rhodes will implement a new curriculum in the fall of 2007, though it will retain its highly regarded three-course sequences known as Search for Values in the Light of Western History and Religion, and Life: Then and Now. The Search sequence has been part of the Rhodes curriculum for 60 years. To receive a Rhodes degree, students must demonstrate proficiency in 12 areas that form the foundation of the liberal arts. These include being able to critically examine questions of meaning and value, developing excellence in written communication, and understanding how historical forces have shaped human cultures.

> **"Students are intelligent, determined, and poised to succeed."**

Academically, Rhodes is especially strong in the natural and social sciences, thanks to labs equipped with state-of-the-art equipment. A partnership with St. Jude Children's Research Hospital lets students conduct research there in the summers and continue their projects during the next school year. Their mentors are "some of the world's best scientists," says a junior. Biology is the most popular major, followed by English, business administration, history, and political science, which has won the National Intercollegiate Mock Trial Tournament four times.

Aside from traditional lecture-style classes, the college offers seminars, honors programs, one-on-one Directed Inquiry tutorials, and interdisciplinary majors. Students who can't find what they want on campus may tap into the Greek and Roman studies program, which offers scholarships for a 24-day travel and study trip to Greece. There's also study abroad in a variety of locations around the world. The Buckman International Fellows program offers summer internships in Madrid, Hong Kong, and Johannesburg. Rhodes is a member of the Associated Colleges of the South,* and it participates in a dual-degree program for engineers with Washington

University (MO). Rhodes also offers master's degrees in secondary education and in nursing, both in cooperation with Vanderbilt.

The academic climate at Rhodes is "very competitive and the courses are rigorous," a senior says. "The professors make it a point to challenge the students' ways of thinking." Ninety-six percent of all classes taken by freshmen have fewer than 25 students, which means professors are more than talking heads. "Freshmen are frequently taught by our most experienced faculty," says an international studies and political science major, and at the end of each semester, students provide anonymous feedback to their profs. Registration is completed online, and though students can get the courses they need to graduate on time, "we often cannot get classes with certain professors until senior year," a junior says. Another student gripes that because classes have been shortened to 50 minutes from an hour, more homework is now required.

> **"The professors make it a point to challenge the students' ways of thinking."**

Like Davidson and Hendrix, Rhodes tends to attract white, Southern, middle- and upper-middle-class students, although minority enrollment is on the rise. Twenty-six percent are Tennessee natives; African Americans comprise 6 percent of the student body, Asian Americans 4 percent, and Hispanics add 2 percent. "Representation for minority students is always a hot topic," a student says. Forty-two percent of students receive scholarships based on academic merit, and the average award is almost $10,367 a year. There are no athletic scholarships, however.

Seventy-seven percent of Rhodes students live on campus, where all dorms are air-conditioned and clean. "Some halls are on the National Register of Historic Places, and can be quite drafty," one student says, and most are single-sex. Freshmen live in Glassell or Williford, and upperclassmen vie for rooms in the East Village apartments during the yearly lottery. The complex houses 200 students in four-person units of 1,000 square feet each. Each apartment has a full kitchen, and there are laundry and cookout facilities on the premises as well. "There is a housing crunch because of the large entering classes in the past few years," warns one senior. Students eat in the Refectory or the Lynx Lair; the former has hot food lines for meat-eaters and vegetarians, and the latter offers fare such as wraps, sandwiches, and burgers, along with a well-stocked salad bar.

Lively and energetic Memphis gets so-so marks as a college town, with three other four-year institutions in the area and a number of community colleges as well. There are plenty of clubs and bars, along with live music and arts organizations, volunteer opportunities, and internships. "Service is an integral part of the Rhodes experience," a student says. Fraternities draw 48 percent of the men and sororities sign up 53 percent of the women. Chartered buses provide rides to off-campus parties, many of which are sponsored by the Greeks, though independents are welcome to attend. While it's illegal for students under 21 to drink, students say those who are determined can usually find booze—if not at parties, then in other students' rooms. In April, everyone looks forward to the three-day Rites of Spring concert, and to the Rites of Play carnival that precedes it, which brings underprivileged kids to campus for a day of food, fun, and games.

> **"Service is an integral part of the Rhodes experience."**

Rhodes fields 11 women's and 10 men's varsity teams, all of which compete in Division III. The men's and women's tennis and cross-country teams and the women's golf team are among the college's strongest. Half of all students participate in intramurals, where the most popular games are flag football and five-on-five basketball. There are also a number of club sports, including cheerleading, crew, fencing, equestrian, and men's and women's lacrosse. Everyone can use the $22.5 million Bryan Campus Life Center, which boasts squash and racquetball courts and a suspended indoor track.

Academically, Rhodes is especially strong in the natural and social sciences, thanks to labs equipped with state-of-the-art equipment.

Like Davidson and Hendrix, Rhodes tends to attract middle-class and upper-middle-class white Southerners, although minority enrollment is on the rise.

Overlaps

Vanderbilt, University of the South, Furman, Tulane, Emory, Washington University

Rhodes College students adore the school's solid academics and rich Southern tradition. "We have beautiful, consistent Gothic architecture, a great location in a large metropolitan area, and a small, intimate community," says a sophomore. The school's reputation is rising, within and outside the southeast, leading to what one student calls "explosive growth in enrollment." What hasn't changed is the friendly vibe on campus, and the eagerness of students, faculty, and staff to make sure that—just like on Cheers—everybody knows your name.

If You Apply To ➤

Rhodes: Early decision: Nov. 1, Jan. 1. Regular admissions: Feb. 15. Financial aid: Mar. 15. Meets full demonstrated need of 56%. Campus interviews: recommended, evaluative. No alumni interviews. SATs or ACTs: required. SAT IIs: optional, required for home-schooled students. Accepts the Common Application and electronic applications. Essay question: significant person or experience who has had a profound effect on your life; risk you have taken, and how it's helped you; how you dealt with an ethical dilemma; or how a quote from a favorite movie is personally meaningful.

Rice University

6100 Main Street MS-17, Houston, TX 77005-1892

One of the few elite private colleges that makes the list of best buys. Rice is outstanding in engineering, architecture, and music. With fewer than 3,000 undergraduates, Rice is smaller than many applicants realize. In lieu of frats, Rice has a residential college system like Yale and the University of Miami.

Website: www.rice.edu
Location: Urban
Total Enrollment: 4,971
Undergraduates: 2,988
Male/Female: 52/48
SAT Ranges: V 660–760
 M 670–780
ACT Range: 30–34
Financial Aid: 35%
Expense: Pr $
Phi Beta Kappa: Yes
Applicants: 7,890
Accepted: 25%
Enrolled: 37%
Grad in 6 Years: 76%
Returning Freshmen: 96%
Academics: ✍ ✍ ✍ ✍ ✍
Social: ☎ ☎ ☎
Q of L: ★ ★ ★ ★
Admissions: (713) 348-7423
Email Address:
 admi@rice.edu

Strongest Programs:
Architecture

Founded nearly a century ago by Texas cotton mogul William Marsh Rice, Rice University has stayed true to its mission of providing unsurpassed programs in science, engineering, the arts, and humanities—all with a price tag families can afford. With it's top-notch programs in the liberal arts and sciences and huge endowment (used to keep tuition low), Rice University is one of the best buys around. It is the dominant university in the Southwest and second only to Duke in the entire South. Add to that a strong football team, a spirited student body, and an impressive success rate for graduates, and you've got yourself an incredible deal.

Rice was modeled after such disparate institutions as progressive, tuition-free Cooper Union and the more traditional Princeton University. Despite its resemblance

"The social life is centered around campus parties and pubs."

to other institutions, Rice maintains distinctive characteristics of its own. The predominant architectural theme of the campus, situated three miles from downtown Houston, is Spanish Mediterranean, and it's surrounded by a row of hedges—the singular buffer between the quiet campus and the sounds of the city.

The students tend to put a lot of pressure on themselves. "Everyone here is very focused and goal-oriented," says a senior. "The courses are very rigorous and require much more than simply going to class." Science and engineering are the strongest programs: Competition in the engineering and premed programs is especially intense, and each year a good number of students who start in these fields retreat to the humanities, which in general are less demanding. In fact, Rice has a long tradition of encouraging double, and even triple, majors in such seemingly opposite fields as electrical engineering and art history (nearly 40 percent of students earn a double major). Biosciences is the most popular major, followed by economics, political science, English, and psychology.

The university has traditionally excelled in the sciences and engineering, and SEs (as these students are called) still dominate the student body. Architecture is one of the finest undergraduate programs in the nation, and the space physics program works closely with NASA. Under the Mellon Fellow program, selected humanities and social sciences majors may work with a faculty mentor on an academic project that offers a summer research stipend.

(Continued)

Under the area-major program, students can draw up proposals for independent interdisciplinary majors. An additional option is the "coherent minor" program, which can replace distribution requirements. Class size rarely presents a problem; 78 percent have 25 or fewer students. Faculty members for the most part are friendly and accessible, sometimes providing upperclassmen with research opportunities. "The professors make themselves available to students and take the time to learn your name," a sophomore says. Everyone operates under the honor system and most exams go unsupervised. Those who want to take their education on the road can visit Swarthmore, and there are internships for engineering and architecture students. The library is stocked with most needed materials and, thanks to a renovation, has become more inviting. Nearly 40 percent of students participate in study abroad programs, which are available in 34 countries.

Rice was founded to serve "residents of Houston and the state of Texas," but Texans no longer dominate the student body. Today, 47 percent of Rice students come from out of state, with high percentages transplanted from California, Florida, the Northeast, and other Southern states. Another 3 percent are from foreign countries. Sixteen percent of the student body is Asian American, 12 percent is Hispanic, and 7 percent is African American. Generally, students are too busy and too caught up with life inside Rice's hedges to worry about the outside world. "Students are fairly apathetic about social issues," says a student. Another adds, "Students here tend to get very caught up in their studies and campus life, and therefore tend to forget what's going on outside of Rice." Because much of the university's $3.2 billion endowment is dedicated to keeping tuition low, Rice costs thousands of dollars less than most other selective, private universities. Rice guarantees to meet the full demonstrated need of every admit and there are a variety of merit and athletic scholarships available each year.

Students don't have time to plan anything more formal on weekends than the traditional TGIF lawn parties on Friday afternoons, but campuswide parties sponsored by one of the residential colleges spring up from time to time.

Fraternities and sororities are forbidden on campus—Rice's founder did not approve of elitist organizations—but their functions are largely assumed by the nine residential colleges, Rice's version of dorms. Each college has a "faculty master" and houses about 225 students who remain affiliated with it for all four years. Freshmen are randomly assigned to one of the co-ed dorms. Air-conditioning is a standard weapon against Houston's muggy climate and you may want to keep a can of bug spray handy to ward off mosquitoes during the first few months of the school year. Everyone is guaranteed a room for at least three of the four years, though students sit through a lottery system. Thirty-one percent

"Don't worry about how you will fit in at Rice—you just will."

of the students go packing, some because they were "bumped" from the colleges to make room for more students, but many seeking quieter surroundings and cheaper rents. Students can eat at any of the college dining halls, where there are plenty of good options—and plenty of food (it's "all you can eat").

Houston has plenty of nightlife, but to enjoy it bring a car; even though the new light rail system makes it easier to get to the city, it's still tough to get around. Parking on campus became a bit more complicated with the installation of parking gates, which are highly unpopular among students. Galveston's beaches on the Gulf of Mexico are only 45 minutes away, and heading for New Orleans, especially in February, can make a great weekend trip.

"While Rice students can write algorithms and social commentary, they still know how to throw a good party," says one junior. Students don't have time to plan anything

Under the area-major program, students can draw up proposals for independent interdisciplinary majors.

more formal on weekends than the traditional TGIF lawn parties on Friday afternoons, but campuswide parties sponsored by one of the residential colleges spring up from time to time. Night of Decadence is Rice's Halloween party, where it is reported that the students appear in "lingerie or less." "The social life is centered around campus parties and pubs," says a student. A classmate adds, "If the themed public parties aren't your thing, you can swing by Willy's Pub for a beer or Lovett Undergrounds for some live music and a cup of coffee." If nothing else, there's always the campus movie.

Ardent football fans abound at Rice; tearing down the goalposts after home victories remains a happy tradition. Other strong programs include baseball (2006 conference champs), women's cross-country, men's indoor track, and both men's and women's tennis. Rice students go really wild for intramurals—the most popular pits the colleges against each other in a Beer-Bike Race in which co-ed teams of 20 chug cans of beer and speed around a bicycle track, which gives them a chance to let off academic steam.

> **"Everyone here is very focused and goal-oriented."**

Because Rice is so small, social cliques are nonexistent; everyone is accepted, no matter what their quirks. Says one professor, "Worry about the heat, humidity, the next exam, or getting enough time to relax. But don't worry about how you will fit in at Rice—you just will." And after they venture outside the hedges for the last time, students' Rice diplomas open doors to the corporate world. After all, they've had a terrific academic experience and a decent social life for four years, and their wallets haven't been emptied thanks to a pint-sized tuition.

Overlaps

Harvard, Stanford, Washington (Mo.), Duke, University of Texas–Austin

If You Apply To >

Rice: Early decision: Nov. 1. Interim decision: Dec. 1. Regular admissions: Jan. 10. Housing: May 1. Guarantees to meet full demonstrated need. Campus or alumni interviews: recommended, evaluative. SATs or ACTs required. SAT IIs: required (writing plus two additional in subjects related to proposed area of study). Essay question.

University of Richmond

28 Westhampton Way, Richmond, VA 23173

Richmond provides students with a unique gender-based coordinate college system. Offers students a pre-professional climate with touches of the Old South. Benefits from the coordinate college system—Richmond College for men and Westhampton College for women.

Website: www.richmond.edu
Location: Suburban
Total Enrollment: 3,685
Undergraduates: 2,830
Male/Female: 49/51
SAT Ranges: V 610–690
  M 630–700
ACT Range: 26–30
Financial Aid: 34%
Expense: Pr $ $ $ $
Phi Beta Kappa: Yes

Students at the University of Richmond enjoy a healthy mix of Southern ambience and intellectual rigor that includes small classes, close friendships, and lots of teamwork. Thanks to UR's unique coordinate system, men and women take advantage of separate student governments and traditions, as well as a 10:1 student/faculty ratio, extensive undergraduate research opportunities, and strong study abroad options.

UR's 350-acre campus is nestled in a group of rolling hills about 15 minutes from the center of Richmond, which is also the state capital. Stately pines and a 10-acre lake surround the Gothic buildings. Weinstein Hall, dedicated in 2003, offers 53,000 square feet for the departments of journalism, political science, sociology, and rhetoric and communication studies. The Gottwald Science Center has been expanded and renovated at a cost of $37 million to include the latest technology and research equipment.

All Richmond undergraduates complete general education requirements in communications (including writing and a foreign language) and in five other fields of study: history, literary studies, natural sciences, social analysis, symbolic reasoning, and visual and performing arts. There's also a team-taught first-year seminar known as CORE, which includes intensive discussion and is linked to other campus activities, such as musical performances. Internships and study abroad are encouraged; nearly half of Richmond students take a term away through one of 70 exchange programs in 30 countries. The university has added about two dozen programs to the curriculum in recent years, including degrees in anthropology and cognitive science and a minor in creative writing. The professional accounting degree and the political science minor have been discontinued.

UR is a highly preprofessional campus; business is the most popular major, and 66 percent of recent graduates went right into the working world, while 25 percent enrolled in graduate or professional school. Biology, political science, and psychology are also popular and well regarded. One of the school's more unique offerings is the Jepson School of Leadership Studies, dedicated to helping students understand **"If you come, plan to work hard."** how they can best serve society. Every two years, students are invited to embark on the "Richmond Quest" and submit a challenging question and rationale for in-depth exploration by the entire campus. The student with the winning question receives $25,000.

Aside from praising the classes, Richmond students rave about the faculty. "The quality of teaching I have received is greater than what I could have imagined," gushes a Spanish major. "The professors are here because they desire to teach, work closely with, and mentor undergraduates." Although the climate is cooperative, students are quick to note that classes can be quite challenging. "'Rigorous' is a good word," says a freshman. Male students may participate in one of two first-year leadership programs, Spinning Your Web and RC Extreme. Female students are eligible to join WILL (Women Involved in Living and Learning), which helps build relationships with administrators. WILL taps into UR's renamed program in women, gender, and sexuality studies as well.

Only 15 percent of Richmond students are native Virginians, though many of the rest hail from elsewhere in the Southeast. Foreign students account for 6 percent of the student population. Diversity is growing, albeit slowly—4 percent of students are African American, 3 percent are Asian American, and 2 percent are Hispanic. Programs such as Dialogues in Black, White and Beyond, and the Common Ground Commission, which includes students, faculty, and staff members, help emphasize the importance of crosscultural understanding. Hot political issues include abortion rights and **"The campus is split pretty much right down the middle between conservatives and liberals."** diversity. "The campus is split pretty much right down the middle between conservatives and liberals," says an art history major. The Richmond Scholars program awards 50 full tuition merit scholarships for every entering class and student athletes vie for 170 athletic awards.

Ninety-two percent of Richmond students live on campus, where the dorms are "generally comfortable" and all have air-conditioning. Most buildings are either newly built or recently remodeled, though upperclassmen get the best digs: on-campus townhouses and apartments. The dining hall, or "D-hall," offers hot entrees as well as made-to-order subs, pizza, pasta, salad and wrap bars, cereal, and of course burgers and fries. "Great desserts!" says one happy student.

Owing to Richmond's Southern heritage, traditions play a major role in campus life. First-year students have Proclamation Night (for women) and Investiture (for men), during which they sign the honor code and write a letter to themselves,

(Continued)
Applicants: 5,778
Accepted: 47%
Enrolled: 28%
Grad in 6 Years: 84%
Returning Freshmen: 93%
Academics: ✍ ✍ ✍ ½
Social: ☎ ☎ ☎
Q of L: ★ ★ ★
Admissions: (800) 700-1662
Email Address:
    admissions@richmond.edu

Strongest Programs:
Business
Biology
Political Science
English
International Studies
Leadership Studies

UR is a highly preprofessional campus; business is the most popular major, and 66 percent of recent graduates went right into the working world, while 25 percent enrolled in graduate or professional school.

which is not opened until senior year. During junior year, there is the Ring Dance, a debutante-style ball at the elegant Jefferson Hotel, where junior women don white gowns and receive their class rings from their fathers. Alas, the annual Pig Roast has been discontinued in favor of Festivus, a weekend of carnival games that includes socializing with alumni and brunch on the intramural fields.

Forty-five percent of men join fraternities, and 45 percent of women pledge sororities. For independent students, or just those tired of the frat-party scene, the Campus Activities Board sponsors movie nights, karaoke, and concerts. Although Richmond is more of a small city than a college town, it does have great restaurants, beautiful historic neighborhoods, and plenty of internship opportunities at local corporations and government agencies. Two-thirds of students volunteer, donating over 100,000 hours a year. UR is one of the leading college partners of Habitat for Humanity and the school recently held a Build It, where students, faculty, alumni and the community built a house and renovated three public schools in a week. For those wishing to get away, Williamsburg, Virginia Beach, and Washington, D.C., are not far, and nature buffs also like the river and the nearby backpacking, thanks to the proximity of the Blue Ridge Mountains.

Sports are just one more way to get involved at Richmond, and the school's varsity teams compete in the Division I Atlantic 10 Conference. Men's and women's basketball participated in recent national championship tournaments and women's field hockey has been conference champs for three years straight. Women's lacrosse, men's and women's tennis, and men's soccer also have strong teams. Any game against William and Mary draws a crowd, as the schools have one of the oldest rivalries in Virginia. Fall Saturdays also find throngs of students throwing tailgate parties, with the men in ties and other preppy attire, and the sundress-wearing women looking like they've stepped out of a J. Crew catalog. Club sports teams are more like fraternities, as members live, eat, and compete together, and also host parties. "They are the way to go!" cheers a senior.

There's so much to do on campus, with activities and organizations, that students may be inclined to neglect life beyond the campus walls. Richmond students seem to embody the maxim, "We work better when we have fun." A biochemistry major offers this advice: "If you come, plan to work hard."

If You Apply To ➤ **Richmond:** Early decision: Nov. 15, Jan. 15. Regular admissions: Jan. 15. Guarantees to meet demonstrated need. Campus and alumni interviews: optional, informational. SATs or ACTs: required. SAT IIs: optional. Accepts the Common Application and electronic applications. Essay question.

Ripon College

300 Seward Street, P.O. Box 248, Ripon, WI 54971

Ripon is more conservative than Beloit and Lawrence and more similar in atmosphere to places like DePauw and Knox. With only 1,000 students, Ripon is the smallest of the five. Strengths are science, education and history.

Everything about Ripon College is small, aside from perhaps its academic ambitions. The school is in a tiny east-central Wisconsin town, and there are fewer than 1,000 students, meaning "if you don't go to class, your professor will know," a freshman says. The weather can be a downer, with bitter cold and lots of snow come winter, but the warmth of personal relationships with peers and professors helps to compensate. "The best part of going to a small college like Ripon is that you feel at home," a student says. "It's like having a huge family."

Ripon's 250-acre campus sits in a town of 7,000, just off Highway 41 between Fond du Lac and Oshkosh. "The old-fashioned Main Street offers cute shops and restaurants, including a bakery and pizza parlor, and the town's trails and parks are good for recreation," says one student. The college itself features tree-lined walks, wetlands, prairie, and woods, and a mixture of 19th- and 20th-century architecture lends a "majestic" feel.

Ripon's distribution requirements are arranged around themes: Explore, Select, and Connect. Freshmen take an interdisciplinary seminar and a writing course, as well as one course each in the fine arts, humanities, natural sciences, and social and behavioral sciences. A physical education course is also required, and students must meet a global and cultural studies requirement. Students lay out

"If you don't go to class, your professor will know."

the rest of their Ripon careers in an individualized learning plan, which may include internships, research, or a senior-year capstone experience. Academics take priority. "I find the courses here to be challenging," a sophomore says, "but not to the point where you are pulling your hair out." There are no teaching assistants, and student-initiated study groups are common. Faculty members "are dedicated to their students and specialties," a student says.

Ripon's strengths include the sciences, education, and history; students also give high marks to the business curriculum, the arts, communications, and psychology. "One of the reasons I decided to come to Ripon was because of the incredible psychology department," says a (surprise!) psychology major. Less popular, though not necessarily weaker, are philosophy, music, and languages. Motivated students with AP credits—or just the stamina to take an extra class each term—may finish in three years, thanks to Ripon's accelerated degree program. The forensics team slugs it out in Division I and consistently ranks in the nation's top 20.

Ripon students delight in their small classes; 77 percent of those taken by freshmen have 25 or fewer students. Students may take terms away from campus through Ripon programs in Chicago and overseas, or programs sponsored by the Associated Colleges of the Midwest.* An exchange program with Fisk University, a Southern school with mostly African American students, has helped to improve diversity and race relations.

Ripon students are "genuine, caring, and friendly," says a sophomore. Three-fourths come from Wisconsin, and most of the rest hail from elsewhere in the Midwest, though a freshman says there are some from as far away as California, Texas, and Hawaii. Most come from middle-class homes, and many are "classic over-achievers," says a senior. African Americans comprise 3 percent of the student body; Asian Americans make

"Alcoholic, nonalcoholic, dance, board game, or movie parties can all be found on a typical night."

up 2 percent and Hispanics add 2 percent. Hot-button political and social issues include abortion and the war in Iraq. Ripon offers merit scholarships averaging $14,496 but no athletic awards.

Eighty-one percent of Ripon students live on campus, since you must petition to live off campus. Freshmen are housed together; students may choose co-ed or single-sex halls, with doubles, singles, or suites. "When people hear the words 'college dorms' they normally think of them as being dirty, old, or just plain gross," says a student. "However, the dorms at Ripon are just the opposite." Ripon has three dining halls—the cafe-style Terrace, the Pub, with a grill and *a la carte* options, and the Commons, with a traditional hot-food line. Special tastes are accommodated and the variety of food gets high marks.

(Continued)
Total Enrollment: 934
Undergraduates: 934
Male/Female: 50/50
SAT Ranges: V 480–650
  M 500–620
ACT Range: 21–27
Financial Aid: 64%
Expense: Pr $
Phi Beta Kappa: Yes
Applicants: 976
Accepted: 81%
Enrolled: 33%
Grad in 6 Years: 70%
Returning Freshmen: 84%
Academics: ✍ ✍ ✍
Social: ☎ ☎ ☎
Q of L: HHH
Admissions: (920) 748-8114
Email Address:
  adminfo@ripon.edu

Strongest Programs:
Sciences
Education
History

Ripon's strengths include the sciences, education, and history; students also give high marks to the business curriculum, the arts, communications, and psychology.

Ripon's claim to fame is its status as the birthplace of the Republican Party, founded on campus on February 28, 1854, to be exact. Other than that, there's not much to see. Twenty-seven percent of the men join fraternities and 16 percent of the women pledge sororities, but the Greeks aren't a major social force. Instead, "alcoholic, nonalcoholic, dance, board game, or movie parties can all be found on a typical night," says a religious studies and history major. "Entertainment on campus has included comedians, musical groups, improv troupes, speakers, and even hypnotists," says a freshman. Ripon students have a "very hectic social life," says a student. Per Wisconsin law, students under 21 aren't permitted to drink, and those caught doing so face fines. The best road trips include nearby Oshkosh and Appleton, or even Milwaukee and Madison. Chicago is a three-hour drive. Favorite annual traditions include the Springfest concert and carnival and homecoming. At the start of each year, all the churches in Ripon come together to host a homemade dinner for students. Though Ripon's winters can be bitterly cold, the college's location in the rugged North Woods means frozen lakes and a blanket of snow are a natural part of the winter landscape. When they tire of their books, students may let off steam with cross-country and downhill skiing, tobogganing, and ice skating—or some fervent cheering for dogsled and iceboat races. And when it's not winter ("for one month during the year," warns one student), nearby Green Lake offers boating, fishing, and other water sports. Ripon's varsity teams compete in Division III and

"It is a very friendly, welcoming place to be."

matches against Lawrence University usually draw excited crowds, as the Lawrence–Ripon rivalry is one of the oldest in Wisconsin. The volleyball, men's tennis and men's and women's basketball have been successful in recent years. The baseball team went undefeated in the conference last season. Three-quarters of students participate in intramural sports, which range from basketball and softball to football and innertube water polo. "The variety provided caters to athletes, former athletes and people who just want to have fun," a junior says.

Ripon College offers a strong grounding in the liberal arts, along with a peaceful, quaint, historical, and friendly community where you'll be much more than a number. "I am on a first-name basis with our dean and president, and it's not because I'm in trouble," a junior says. Though being at such a small place can be stifling, Ripon students aren't complaining. "We take pride in our school and really care about the institution and the people," one student says. "It is a very friendly, welcoming place to be."

Overlaps

St. Norbert, University of Wisconsin–Stevens Point, University of Wisconsin–Madison, University of Wisconsin–Oshkosh, University of Wisconsin–Eau Claire, Carroll

If You Apply To ➤

Ripon: Rolling admissions. Meets demonstrated need of 93%. Campus interviews: recommended, informational. No alumni interviews. SATs or ACTs: required. SAT IIs: optional. Accepts the Common Application and electronic applications. Essay question: your reasons for pursuing a college education at Ripon.

University of Rochester

Rochester, NY 14627

The name may conjure up a nondescript public university, but Rochester is a top-notch private university in the orbit of Carnegie Mellon, Case Western Reserve, Johns Hopkins, and Washington U (MO). The university has a scientific bent and is known as a haven for premeds.

The University of Rochester is not afraid of change. In 1996, this distinguished private university implemented its unique Rochester Renaissance Plan, and it has never looked back. The ten-year plan included reducing class size; new investments in the library, classrooms, and computer networking facilities; and launching a new curriculum that eliminates entry-level general education courses to allow students to design their own paths.

Founded in 1850, the University of Rochester occupies a snug little 90-acre campus, which nestles up to a bend in the Genesee River. One student acknowledges that the university has "perpetually gray [read: winter] skies," but finds comfort that "it's great for winter sports or studying or even sleeping late on a snowy Saturday." Although a few buildings are modern—the Wilson Commons student center designed by IM Pei, for example—most of the older structures come in Greek Revival and Georgian Colonial styles. There is an aesthetically pleasing contrast between old and new, and the Eastman Quadrangle, with the library and original academic buildings, adds to Rochester's stately look. Recent construction projects include a 100,000-square-foot home for the Institute of Optics and the Department of Biomedical Engineering—a prime research space just across the street from Rochester's medical center.

> "Rochester students are a hard-working, fun-loving, positive group."

Degree requirements for all majors lack general education courses, but all are designed to ensure that students are exposed to the full range of liberal arts. The curriculum—appropriately but unimaginatively known as the Rochester Curriculum—focuses on three classic divisions of learning: humanities and arts; social science; and natural science, mathematics, and engineering. Students choose a major from one of these areas and also complete a cluster of three courses in each of the remaining two divisions. These clusters give students the opportunity for integrated study in diverse fields and the chance to participate in three very different types of learning. Freshmen have the option of taking seminar-style Quest courses, which teach them how to learn and how to make learning a lifetime habit. Quest courses can involve extensive work with original materials, existing and experimental data, and primary texts. Orientation Rochester-style includes a weeklong fall festival called Yellow Jacket Days designed to help new students "become fully integrated in the university community."

The university's 175 degree programs span the standard fields of study, but Rochester takes special pride in its famed Eastman School of Music. It also excels in the engineering and scientific fields—competition is keen to "beat the mean" among science majors. The Brain and Cognitive Science program is one of the most popular, as is the certificate in management sciences. Another new major in financial economics attracts Wall Street-minded students.

> "The Greek scene plays a large factor in social life here."

The Institute of Optics, the nation's first center devoted exclusively to optics, is a leader in basic optical research and theory. The most popular majors include psychology, political science, economics, and biology.

The academic climate at Rochester is challenging, owing much of that to its energetic professors, who are described as "extremely knowledgeable" and "enthusiastic about teaching undergraduates." No matter what their field of interest, students who are sufficiently advanced may combine undergraduate with graduate study. The Rochester Early Medical Scholars program offers highly qualified first-year students guaranteed admission to the medical school after four years. In addition, Rochester offers a tuition-free fifth year that allows students to explore interests outside their major. The Center for Work and Career Development receives praise for its vigorous preparation of seniors for the job market, and the university sponsors foreign study programs that include what one student calls a "chance of a lifetime"

Website: www.rochester.edu
Location: Small city
Total Enrollment: 8,231
Undergraduates: 4,502
Male/Female: 52/48
SAT Ranges: V 600–700
   M 630–730
ACT Range: 27–30
Financial Aid: 65%
Expense: Pr $ $ $
Phi Beta Kappa: Yes
Applicants: 11,361
Accepted: 45%
Enrolled: 22%
Grad in 6 Years: 80%
Returning Freshmen: 94%
Academics: ✐ ✐ ✐ ✐
Social: ☎ ☎ ☎
Q of L: ★ ★ ★
Admissions: (800) 822-2256
   or (585) 275-3221
Email Address: admit
   @admissions.rochester.edu

Strongest Programs:
Premedicine
Engineering
Music
Optics
Biology
Psychology
Economics
Political Science

As far as housing is concerned, there's virtually nothing but praise from the 84 percent of students who live on campus.

British Parliament internship program. A work-study program called Reach for Rochester provides students with on- or off-campus jobs, summer employment options, and individually tailored "experienceships."

Fifty-five percent of the students hail from New York State. Many also come from New England, and there's been a large jump in the numbers from Florida, the Midwest, California, and overseas. Asian Americans make up 11 percent of the student body, while African Americans account for 6 percent, and Hispanics another 5 percent. "I can say that Rochester students are a hard-working, fun-loving, positive group that puts a high value on academics," says a student. Campus issues include political correctness and antisweatshop campaigns, though one student notes political involvement is "lower than might be expected" due to a "lack of free time."

As far as housing is concerned, there's virtually nothing but praise from the 84 percent of students who live on campus. "Dorms are comfortable, modern, high-tech, very generously sized, and well maintained," a senior says. Some of the housing units offer such benefits as computer terminals, telephones with voicemail features, oak floors, and marble trim. All housing offers Internet access. New students are assigned to rooms—usually doubles—and upperclass students can usually get singles or suites through the lottery. Susan B. Anthony comes highly recommended. Dormitories are available single-sex, co-ed by floor, and co-ed by room. Though few students choose to live off campus, a new shuttle bus runs to and from the major off-campus living areas. Dormitory students may eat meals in the cafeteria, where a credit system ensures they pay per

"This school's policy in no way curtails underage drinking."

meal instead of in one lump sum. The fare served in the dining halls receives high ratings from students, especially the *a la carte* options such as tacos and burritos and the deli bar. Other meal options available include a kosher deli, the Common Ground Coffeehouse, and a submarine sandwich shop.

Twenty percent of the men and 14 percent of the women go Greek. "The Greek scene plays a large factor in social life here, but it is by no means the only social outlet," says one student. Yet, fraternities still contribute heavily to the social life of Greeks and independents alike by sponsoring parties and concerts. UR does have its own set of movie theaters that charge $3 or less per ticket, and campus concerts always draw a crowd. Despite the university's best efforts to enforce a stricter alcohol policy, underage drinking still occurs. "This school's policy in no way curtails underage drinking unless students take it upon themselves to obey," one student says. Many students take the free campus shuttle into "Rochchacha," where they may entertain themselves on the beaches of Lake Ontario, in the International Photography Museum at the George Eastman House, or at the Rochester Philharmonic Orchestra. Favored out-of-town ventures are Niagara Falls, about 70 miles westward, and, for the more venturesome, Toronto, 125 miles farther westward.

The Rochester Early Medical Scholars program offers highly qualified first-year students guaranteed admission to the medical school after four years.

Other popular activities include cappuccino at the student union, frequent ski trips, Yellow Jacket Days, and a spring fling known as Dandelion Day. The Viennese Ball and the Boar's Head Dinner are also popular events. An unofficial Rochester tradition calls for each student to eat a "garbage plate" at the infamous dive called Nick's before graduating. Students are involved in the community through projects on and off campus, and Rochester was recently recognized nationally for its high percentage of student volunteers.

The varsity sports teams are coming of age at Rochester, which competes in the University Athletic Association. For those who want something to cheer about, the golf team, the basketball teams, men's and women's soccer, tennis, and cross-country are all quite successful. Intramurals are a popular outlet for "ex-jocks from high school who miss their glory days gone by." Even if intramurals aren't your bag, Rochester has an $8 million sports complex, complete with basketball, tennis, squash, and volleyball courts; lighted rooftop tennis courts; a Nautilus fitness center; the Speegle-Wilbraham Aquatic Center with an eight-lane pool; and an indoor track.

In the past, students bemoaned the fact that the school didn't have a wider academic reputation, but that's changing due, in part, to the Rochester Renaissance Plan. Improvements have been made in the curriculum, the facilities, and just about anywhere you look on campus. "Prospective students who are unsure where their academic careers will take them will have no problem finding what excites them intellectually," says a junior. "But more importantly, they will build relationships with others that will last their entire lives." Rochester seems to be winning its battle for a spot among the nation's leading private universities. Now if they could only do something about all that snow.

<table>
<tr><td>**Overlaps**
Cornell University,
Brown, Northwestern,
Washington University,
Tufts</td></tr>
</table>

If You Apply To ➤ **Rochester:** Early decision: Nov. 1. Regular admissions: Jan. 15. Financial aid: Feb. 1. Housing: June 1. Guarantees to meet demonstrated need. Campus and alumni interviews: optional, evaluative (required of scholarship candidates). SATs or ACTs: required. SAT IIs: recommended. Musicians apply directly to the Eastman School of Music. Essay question. Looks closely at recommendations, activities, and "indications of intellectual curiosity and a zest for college life."

Rochester Institute of Technology

60 Lomb Memorial Drive, Rochester, NY 14623-5604

RIT is the largest of New York's three major technological universities—about double the size of Rensselaer. The school is strong in anything related to computing, art and design, and engineering, and in the city built by Eastman Kodak, photography is among the tops in the nation.

Unlike many liberal arts colleges that prefer that students test the academic waters before deciding on a major or future job plans, RIT focuses on career-oriented and technology-based academics. And unlike many big universities where the academic luminaries shine from research-oriented graduate schools, RIT's spotlight is very definitely on undergraduates. Students seeking up-to-date technological preparation will be at home at RIT. Those who are geared up and ready to "go professional" will be happy to know that the school places 3,300 juniors and seniors in full-time paid positions through its co-op program.

While the town of Rochester may sometimes seem like a reluctant host to weekend fun-seekers, it can hardly deny that it is in fact a college town; RIT shares the city with six nearby colleges. RIT's main campus, located on 1,300 suburban acres six miles from downtown Rochester, has its own distinctive style—redbrick buildings with sharp, contemporary lines. The Gordon Field House and Activities Center opened in 2004, and a new 35,000-square-foot biosciences facility will open later this year.

RIT specializes in carving out niches for itself with unusual programs, and majors are offered in more than 200 fields, from basic electrical and mechanical engineering to packaging science and bioinformatics. Fortunately, applicants narrow the range of choices to a manageable size by applying to one of eight undergraduate colleges: applied science and technology, business, computing and information sciences, engineering, imaging arts and sciences, liberal arts, science, or the National Technical Institute for the Deaf (NTID). As home to NTID, RIT is a leader in providing access services for deaf and hard-of-hearing students.

Website: www.rit.edu
Location: Suburban
Total Enrollment: 14,627
Undergraduates: 12,210
Male/Female: 68/32
SAT Ranges: V 520–630
  M 560–670
ACT Range: 23–28
Financial Aid: 75%
Expense: Pr $
Phi Beta Kappa: No
Applicants: 10,219
Accepted: 65%
Enrolled: 36%
Grad in 6 Years: 65%
Returning Freshmen: 90%
Academics: ✍ ✍ ✍
Social: ☎ ☎ ☎
Q of L: ★ ★ ★
Admissions: (585) 475-6631
Email Address:
  admissions@rit.edu

Predictably, engineering is the most popular major, but one might be surprised to know that the third-most popular major is photography (Rochester, New York, is home to Eastman Kodak). Information technology, business, and art and design also head the list. Academic programs include aerospace engineering, environmental management, hotel and tourism management, engineering technology, and a physician assistant program. The RIT School for American Crafts offers excellent programs in ceramics, woodworking, glass, metalcraft, and jewelry making, and students have the run of Bevier Gallery, where visiting artists provide firsthand instruction. The College of Liberal Arts has added new programs in public policy, advertising/public relations, and international studies, and is gaining recognition from prospective students.

RIT's general education program has been revised to provide students with more flexibility in meeting requirements. The upshot of this is that the number of required liberal arts credits has been reduced and students may now schedule academic minors—more than 50 have been added within the past two years. Unlike many universities, RIT allows freshmen to schedule significant coursework in their majors early on, and spreads out liberal arts requirements over a more extended period. "The courses are very competitive and rigorous," says a junior. Undergraduates being the school's top priority, the faculty develops new academic programs to fit career needs. "Professors here are extremely friendly and helpful and will go out of their way for students," an imaging science major says. Applied research initiatives with extensive ties to industrial partners are a new thrust of the administration. Academic counseling is very good, as is the co-op's office support for career counseling, says a senior.

"The courses are very competitive and rigorous."

RIT specializes in carving out niches for itself with unusual programs, and majors are offered in more than 200 fields, from basic electrical and mechanical engineering to packaging science and bioinformatics.

Although RIT has its fair share of "typical computer geeks," there are also plenty of outgoing students, especially in the campus clubs and organizations, according to a junior. Half of the students are from New York State, the remainder coming largely from New Jersey, Pennsylvania, and Connecticut. Five percent of the student body is African American, 4 percent Hispanic, and 7 percent Asian American. Pre-professionalism is a common bond, but beyond that interests vary. The unique mix of art, engineering, business, and science students, along with the large number of deaf students, creates a diverse atmosphere on campus. If the men here could change anything, it would likely be the lopsided male/female ratio. That accomplished, they might wish to raise the nightlife wattage in Rochester, along with the temperature in winter. RIT admits without regard to student financial need, and it meets the demonstrated need of 90 percent of the students for as long as the funds allow. RIT offers merit scholarships, averaging $6,500 per year.

"Campus policies on alcohol are very strict."

Sixty-five percent of RIT students live in college dorms and apartments, and students report that getting a room is not that difficult. Freshmen are required to live in the dorms, while upperclassmen can vie for campus apartments on a first come, first served basis. But that shouldn't be too tough because, according to the administration, RIT has the one of the largest number of on-campus apartments in the country. Dorms are well maintained and offer a variety of living styles: single-sex, co-ed by room, or co-ed by floor. Special-interest floors range from nonsmoking to "mainstream" (with hearing-impaired students). Vegetarians, vegans, and carnivores will find on-campus meal options to be reasonably diverse. Those who choose to live off campus take advantage of areas serviced by the school shuttle bus. Five percent of men and 5 percent of women choose to go Greek and live and eat in RIT's 27 fraternity and seven sorority houses. Campus security is "courteous and responsive," reports one student.

Unlike many universities, RIT allows freshmen to schedule significant coursework in their majors early on, and spreads out liberal arts requirements over a more extended period.

"Campus policies on alcohol are very strict," warns a senior, adding, "The policies work because there isn't a lot of partying on campus." The only facilities within walking distance of this sedate suburban campus are a variety of shopping plazas, including one of the largest between New York and Cleveland. Students take road trips to Buffalo, Syracuse, Rochester, and Canada. For those without transportation, there's always something to do on campus. Drama and other creative arts are less common than parties and movies, but a fine jazz ensemble and a chorus perform regularly. RIT is home to the first ESPN Entertainment Zone area on a U.S. campus as part of student union renovations. Brick City Bash, a favorite way to celebrate the end of a long winter, is held each spring.

RIT offers 22 NCAA Division III athletic teams; the men's hockey program recently moved to Division I. Men's and women's hockey are always at the top of the rankings and the overwhelming sports favorite, a real crowd pleaser that draws even the campus commuters and local residents to the rink. Men's soccer and lacrosse have been recent conference champions. Other sports, such as men's basketball and lacrosse and women's volleyball, have all won ECAC championships in the last two years. More than half of RIT undergrads participate in 13 intramural sports.

Organized and focused, RIT students have their eye on the future. "Students are dedicated and career-oriented, and many go for high degrees," says an imaging science major. And best of all, with all the co-op education opportunities, "They graduate with lots of lab/field/hands-on experience."

> "They graduate with lots of lab/field/hands-on experience."

Overlaps

University at Buffalo, Rensselaer Polytechnic, Clarkson, Syracuse, Cornell University, Binghamton

If You Apply To ➤

RIT: Early decision: Dec. 1. Regular admissions: Feb. 1. Financial aid: Mar. 1. Meets demonstrated need of 90%. Campus interviews: optional, informational. No alumni interviews. SATs or ACTs: required. SAT IIs: optional. Accepts the Common Application. Prefers electronic applications. Essay question: applicant's choice.

Rollins College

1000 Holt Avenue, Box 2720, Winter Park, FL 32789-4499

Rollins is the marriage of a liberal arts college and a business school. A haven for Easterners who want their ticket punched to Florida, Rollins attracts conservative and affluent students and world-class water-skiers.

Although Central Florida is best known as the home of Mickey Mouse, Shamu, and the Space Shuttle, these aren't Central Florida's only attractions. For students looking to hit the books under the ever-present Florida sunshine, there's also Rollins College. Located in Winter Park, a quite suburb of Greater Orlando, Rollins College offers students plenty of places to have fun when making the grade gets to be too much.

Rollins' lakefront campus is nationally recognized for its beauty and signature Spanish Mediterranean architecture, featuring stucco buildings, tile roofs, carved woodwork, and decorative balconies. Capitalizing on its location on beautiful Lake Virginia, campus planners have succeeded in combining the natural beauty of the lakeside with consistent architecture. Recently, Keene Hall (which houses the music department) was renovated and expanded, as was the Cornell Fine Arts Museum.

The general education requirements expose Rollins students to various perspectives and areas of knowledge. Students may take as many classes as they want in foreign

Website: www.rollins.edu
Location: Suburban
Total Enrollment: 2,493
Undergraduates: 1,719
Male/Female: 40/60
SAT Ranges: V 540–650
  M 540–640
ACT Range: 22–27
Financial Aid: 45%
Expense: Pr $ $
Phi Beta Kappa: No
Applicants: 2,958
Accepted: 53%

(Continued)

Enrolled: 29%

Grad in 6 Years: 65%

Returning Freshmen: 84%

Academics: ✐ ✐ ✐

Social: %%%%

Q of L: ★ ★ ★

Admissions: (407) 646-2161

Email Address:
admission@rollins.edu

Strongest Programs:
Organizational Communication
Psychology
Economics
International Affairs
English

languages, quantitative reasoning, communication across the curriculum, decision making and valuation, and writing and writing reinforcement. Other requirements include art, Western and non-Western cultures, contemporary American society, and science. For freshmen acclimating to college, the fall-semester Rollins Conference Course eases the transition by placing them into groups of 15 to discuss themes such

> **"Almost every professor I have had has been enthusiastic about the class they are teaching."**

as imaginary voyages, contemporary ethical issues, and the environment. Each group has a professor-advisor and two upper-class peer mentors. Rollins has begun developing living/learning communities where students can live near each other and enroll together in multiple courses. Fifty percent of freshmen are enrolled in the recent initiative. Students also have the opportunity to pursue independent research, travel abroad, and take service-learning classes where they earn credit by volunteering in the community.

Students give the psychology department their highest marks, and also praise English and international business. The chemistry department turned out a Nobel Prize winner. Some students say they avoid the sciences, but only because instructors are so tough and science majors are so competitive. The Annie Russell Theatre hosts productions staged by the active theater department, which takes pride in having set the stage for such stellar actors as alumni Buddy Ebsen and Tony Perkins. An Accelerated Management Program allows qualified freshmen to gain guaranteed admission to the Roy E. Crummer Graduate School of Business when they enter Rollins, leading to BA and MBA degrees in five rather than six years.

While the workload at Rollins varies by major, academics are important. "Rollins College is a challenging school and it is clear that professors want all of their students to do well and succeed," says a sophomore. Students find there's always help from the professors, with whom they have close relationships. "Almost every professor I have had has been enthusiastic about the class they are teaching and their main goal is for the students to learn," says a student. There are no TAs here; teaching is the responsibility of professors. Many students take advantage of Rollins' study abroad program, which offers pro-

> **"Dorm living is a positive experience."**

grams in Australia, Germany, and Spain, as well as internships in England, for regular tuition costs. A minor in Latin American and Caribbean studies was recently added and the European Studies major has been discontinued.

For freshmen acclimating to college, the fall-semester Rollins Conference Course eases the transition by placing them into groups of 15 to discuss themes such as imaginary voyages, contemporary ethical issues, and the environment.

Rollins students are "relaxed, laid-back, and focused on class," according to a senior. Forty-six percent of Rollins students are from outside Florida. The student body is 73 percent Caucasian, 5 percent African American, 8 percent Hispanic, and 4 percent Asian American. While many of the students come from affluent families, adequate financial aid is available, with merit scholarships for qualified students and 58 athletic scholarships given to male and female standouts.

Sixty-five percent of the college's students live on campus in spacious co-ed dorms. "Dorm living is a positive experience," says an English major. "The rooms are big enough for two people and the hallways and bathrooms are cleaned everyday by a cleaning service." Dining facilities are located in the campus center and food is charged on a credit card system, so students eat when they want and pay only when they eat. All the students seem to agree the food is good. And what about safety? "Campus security is top of the line and super friendly," a senior says. Security officers patrol the campus around the clock and call boxes are available throughout the campus.

The high-powered Greek scene claims 37 percent of the women and 35 percent of the men, respectively, so there's always a party somewhere. And with Cocoa Beach, Orlando, Miami, and Tampa nearby, there's always something to do off campus. On campus, there are movies on Mills Lawn or "dive-in" movies at the pool, lip sync contests, and live bands in the campus center. The administration has clamped down on the social scene, with party monitors checking IDs and a student activity director attending each on-campus party. Underage students caught drinking alcohol will be referred to Rollins' judicial affairs. A senior says, "If you want booze, you can find it, but when you get caught, you'll deal with the consequences."

Fox Day is "a sacred tradition"—the president cancels classes for the day by placing a fox statue on the front lawn. Students look forward to the tradition every spring and, though they never know exactly which day the president will choose, almost everyone heads for the beach once the day arrives. Many students volunteer with programs such as Habitat for Humanity and tutoring at local schools. Orlando's offerings include entertainment complexes like Downtown Disney and Universal Citywalk, and amusement parks such as Walt Disney World, Epcot Center, and Universal Studios.

> "Campus security is top of the line and super friendly."

Sports are an integral part of the Rollins scene. Its teams have claimed 20 national championships, the most recent being women's golf in 2005. The National Collegiate Waterski team is also well known, winning championships as recently as 2003. Intramural sports are popular, too, with more than 20 leagues and events during the school year. Teams have been formed for table tennis, bowling, and other sports.

Rollins' students enjoy sand and sun, as well as a tough academic climate and a strong sense of history. As the oldest recognized college in the state of Florida, Rollins enjoys the warmth of a blazing sun, a rich legacy and smooth-as-silk southern character. Says one senior, "Students can expect to find a safe haven, a real college experience, and a good escape from the hustle and bustle" of the real world.

Many students take advantage of Rollins' study abroad program, which offers programs in Australia, Germany, and Spain, as well as internships in England, for regular tuition costs.

Overlaps

University of Central Florida, University of Florida, University of Miami, Stetson, University of Richmond, University of Tampa

If You Apply To ➤

Rollins: Early decision: Nov. 15. Regular admissions: Feb. 15. Financial aid: Mar. 1. Meets demonstrated need of 29%. Campus interviews: recommended, informational. No alumni interviews. SATs or ACTs: optional. SAT IIs: optional. Accepts the Common Application and prefers electronic applications. Essay question: Evaluate a significant experience; indicate a person of influence.

Rose-Hulman Institute of Technology

5500 Wabash Avenue, Terre Haute, IN 47803

Co-ed since 1995, Rose-Hulman offers the rare combination of technical education and personal attention. Only Caltech, Clarkson, and Harvey Mudd offer comparable intimacy and a technical academic environment. Nearby Indiana State and St. Mary's of the Woods help mitigate the skewed gender ratio.

The Rose-Hulman Institute of Technology may not be as well known as Caltech, MIT, or even Carnegie-Mellon, but it was the first private college to offer an undergraduate degree in chemical engineering, and it continues to innovate today. If you can stomach the lopsided male/female ratio (a legacy of only just over a decade of coeducation) and the limited list of just 15 majors (all in engineering and the sciences), Rose offers an outstanding technical background and bright prospects for future employment. Students are smart, motivated, and highly competitive, and love using their computers for work and play. "We are all dorks," says a senior. "Some of us just hide it better than others."

Established in 1874, Rose-Hulman is the oldest private engineering school west of the Alleghenies. Its benefactors were Chauncey Rose, an entrepreneur who

Website: www.rose-hulman .edu
Location: City outskirts
Total Enrollment: 1,804
Undergraduates: 1,766
Male/Female: 82/18
SAT Ranges: V 600–670
  M 640–720
ACT Range: 28–32

(Continued)

Financial Aid: 69%

Expense: Pr $ $

Phi Beta Kappa: No

Applicants: 3,294

Accepted: 70%

Enrolled: 20%

Grad in 6 Years: 82%

Returning Freshmen: 92%

Academics: ✍ ✍ ✍

Social: ☎

Q of L: ★ ★

Admissions: (800) 248-7448

Email Address: admissions
@rose-hulman.edu

Strongest Programs:
Civil Engineering
Chemical Engineering
Mechanical Engineering
Electrical Engineering
Computer Science

Students committed to careers in engineering or the sciences will find a top-flight education at this Midwestern technical school.

brought the railroad to Indiana, and the Hulman family, owners of the Indianapolis Speedway, who gave their fortune to the institution in 1970. The 200-acre campus includes numerous trees, two small lakes, and a student apartment complex.

General education requirements at Rose include math, physics, chemistry, and humanities and social sciences. Humanities professors "are very eager to educate and expose science and engineering-oriented people to a different way of thinking," an applied biology major says. In the first quarter, freshmen must take a life skills course that covers such topics as time management and study skills. Students in every program except math must work in a team and complete a project for an outside company. Rose-Hulman Ventures is a business incubator program that allows students to gain experience in product development, while Fast-Track Calculus compresses three quarters of calculus into five intense weeks before the start of freshman year. Newer majors at R-H include biomedical engineering and software engineering.

> **"We are all dorks. Some of us just hide it better than others."**

Regardless of which discipline you choose, odds are you'll find faculty members eager to help, plus plenty of free tutoring if you need it. "Our professors are enthusiastic and want the students to learn as much as possible," says a sophomore. A junior adds, "The professors are here to teach—not do research—and it really shows." Ninety-five percent of classes have 25 or fewer students, and you won't find TAs leading classes. "There is no such thing as a teaching assistant at Rose," a software engineering major says.

The typical Rose-Hulman student is "intelligent, hardworking, and inventive," says a senior. "We have everything from multi-sport athletes to introverted gamers," adds a junior, "but since we're all engineers it kind of keeps everyone together." African Americans comprise 3 percent of the student body and Asian Americans make up 4 percent, while Hispanics add 1 percent. Hot issues tend to center on campus issues rather than those of global concern. "This isn't a very political campus," admits a junior. Merit scholarships are available, averaging $5,673, but there are no athletic awards.

Sixty percent of students live on campus. Freshmen are guaranteed rooms in the dorms, as are any sophomores who want them. The residence halls boast weekly maid service, and "we're allowed to do almost anything to the rooms, like add lofts or decks to gain space," says a civil engineering major. (As this an engineering school, would you expect anything less?) All but two dorms have air-conditioning. Most upperclassmen move

> **"The professors are here to teach—not do research—and it really shows."**

into Greek houses or find other off-campus digs, students say. There's a traditional cafeteria, as well as a restaurant-style dining facility, and students say you won't go hungry, as there are typically at least five entree choices at each meal plus fresh fruit, and steak and seafood on Friday nights. "They certainly make provisions for vegans, vegetarians, and students who are allergic to certain types of food, and will even make sack lunches if you are unable to attend lunch during the day," a senior says.

The town of Terre Haute has some restaurants, a mall, a Starbucks, and two movie theaters, but generally, it's "sleepy and lacking in nightlife, so we create our own," says a physics major. Various groups, including the Greek organizations and Habitat for Humanity, help the town out with various service projects. There are eight fraternities and two sororities, which draw 37 percent of the men and 43 percent of the women. "The girl/guy ratio here definitely hurts social life," confides a student, "but overall it's not bad." When it comes to alcohol, the Rose-Hulman campus is officially dry, but the unofficial policy is "out of sight, out of mind," says a senior. "As long as you aren't doing anything stupid or being disruptive, you will be left alone." Students say the best weekend activities are usually road trips to

Chicago, Cincinnati, Indianapolis, or St. Louis, all within a few hours' drive. Everyone looks forward to the homecoming bonfire, and to basketball games against DePauw and Purdue. "Homecoming is very big," says a sophomore. "Freshmen build up and guard the bonfire the week preceding homecoming while upperclassmen try to sabotage it," explains a junior.

If you're envisioning Rose-Hulman students as pasty-faced lab dwellers, you're sorely misinformed. Eighty-five percent of students participate in intramurals, with softball, basketball, and volleyball the most popular, and the Indianapolis Colts even use Rose's facilities for month-long summer camps. Varsity teams play in Division III, and the baseball, basketball, softball, soccer, women's volleyball, and men's soccer teams are the most competitive.

"Where else do you know the Dean of Students by his first name?"

Students committed to careers in engineering or the sciences will find a top-flight education at this Midwestern technical school. While Rose "doesn't have that big-school pride" so common in this part of the country, students appreciate the feel created by the small classes and school's small size. "Our community atmosphere makes us different than anywhere else," a junior says. "Where else do you know the Dean of Students by his first name? Or talk to your professors while you are working out?"

Regardless of which discipline you choose, odds are you'll find faculty members eager to help, plus plenty of free tutoring if you need it.

Overlaps

Purdue, Case Western Reserve, MIT, University of Illinois, Carnegie Mellon

If You Apply To ➤ | **Rose-Hulman:** Regular admissions: Mar. 1. Housing: Jun. 1. Campus interviews: recommended, informational. Alumni interviews: optional, informational. SATs or ACTs: required. SAT IIs: optional. Accepts the Common Application and prefers electronic applications. No essay question.

Rutgers–The State University of New Jersey

65 Davidson Road, Piscataway, NJ 08854-8097

Rutgers is a huge institution spread over three regional campuses and 29 colleges or schools. Rutgers College on the New Brunswick campus is the most prominent. Literally everything is available: engineering, business, pharmacy, the arts, and the nation's largest women's college (Douglass College in New Brunswick).

Life at Rutgers University is all about choice. Choices between the more than 100 undergraduate majors and 4,000 courses offered among its campuses in New Brunswick, Newark, and Camden. Choices about which of the more than 400 student organizations to join. Even choices about which library to visit, as there are 18 branches with holdings of more than three million volumes universitywide. "Rutgers' best quality is its wide variety of majors, classes, and social activities," says one junior.

Rutgers University has three regional campuses (Camden, Newark, and New Brunswick/Piscataway). Rutgers–New Brunswick/Piscataway, which has the largest concentration of students, is composed of five smaller campuses located along the Raritan River. The campuses are connected by a free university bus system and students travel among campuses to take classes. Rutgers–Newark is in a downtown section of Newark, giving the campus neighborhood a collegiate feel. The smallest campus in the Rutgers system is in Camden, located one stop away from the shopping and cultural offerings of downtown Philadelphia. For those seeking a broad-based education, the university has eight liberal arts schools spread out among its campuses. Seven colleges

Website: www.rutgers.edu
Location: Small city
Total Enrollment: 50,016
Undergraduates: 37,072
Male/Female: 46/54
SAT Ranges: V 540–640
  M 570–680
Financial Aid: 50%
Expense: Pub $ $ $ $
Phi Beta Kappa: Yes
Applicants: 41,542
Accepted: 56%
Enrolled: 27%
Grad in 6 Years: 69%

(Continued)

Returning Freshmen: 88%

Academics: 🐔 🐔 🐔 🐔

Social: ☎ ☎ ☎

Q of L: ★ ★ ★

Admissions: (732) 932-4636
(New Brunswick);
(856) 225-6104 (Camden);
(973) 353-5205 (Newark)

Email Address: N/A

Strongest Programs:
Accounting
History
Pharmacy
Biological Sciences
Political Science
Psychology
Engineering
Chemistry

cater to the needs of students wanting a preprofessional school (business, nursing, life and environmental studies, fine and performing arts, engineering, and pharmacy).

Among the nearly 100 majors, the three most popular are psychology, biological sciences, and accounting. Especially strong academic programs include accounting, history, political science, and chemistry. The workload is steady for most students; science majors and pharmacy students can expect the heaviest load. "The courses here require a great deal of thought and outside preparation if you want to be successful," says a political science major. Recently added majors include cell biology and neuroscience, genetics and microbiology, biomedical engineering, evolutionary anthropology, and allied health technology. In addition, Douglass, the women's undergraduate college, no longer exists as a separate degree-granting institution. As at any big state university, registration can sometimes be a headache, but the advent of online registration has helped.

> **"Rutgers' best quality is its wide variety of majors, classes, and social activities."**

In an effort to reverse the traditional exodus of New Jersey high school superstars from the state, Rutgers offers a variety of honors programs, including special seminars, internships, independent projects, and research opportunities with the faculty. Rutgers also provides its undergraduates with a chance to study abroad in Britain, Costa Rica, France, Germany, India, Ireland, Israel, Italy, Mexico, Switzerland, and Spain. Biology students have the run of the 370-acre Rutgers Ecological Preserve and Natural Teaching Area. In addition, Rutgers is also home to more than 100 specialized research centers and institutes dedicated to the study of topics ranging from ancient Roman art to mountain gorillas. Professors generally get high marks. "I completely revere most of my professors," says a junior. "They are intelligent, well-respected in their fields, and present dynamic lectures."

Although the administration has been attempting to increase the number of out-of-staters, 92 percent of Rutgers students hail from New Jersey. Nevertheless, the student population is as diverse as that of the state, with a good proportion of students from cities, suburbs, farms, and seaside communities. Minorities account for nearly half the students: 12 percent are African American, 10 percent are Hispanic, and 21 percent are Asian American. "I feel I've grown so much here and learned so much about being a member of a rich and diverse community," explains one senior. The school's administration takes pride in its Committee to Advance Our Common Purpose, for students who want to reduce prejudice and promote diversity on campus. In the past, the committee developed a Web page for multicultural resources and submitted a proposal for the creation of an Intercultural Relations Study Group. The school does not guarantee to meet the full demonstrated need of every admit. About 400 students receive athletic scholarships in a wide range of sports, and more than 6,500 receive merit awards.

> **"The courses here require a great deal of thought and outside preparation."**

Varsity, intramural, and club sports fill whatever gap is left by the social scene. In fact, the university fields more than 1,000 athletes—the highest number of any university in the nation.

On-campus housing in New Brunswick accommodates 39 percent of full-time students. "There has been a big push recently to renovate the dorms, so most of them are really nice," says a history major. Another student says, "With the exception of a few mediocre dorms for freshmen, most are extremely large and have air-conditioning; some have free cable TV; and a good number are directly hard-wired with fiber-optic cables into the Internet." On-campus housing options include conventional dorms, special-interest areas, and apartment complexes with kitchens and living rooms. The university also offers a special dormitory for students who are trying to overcome addictions to drugs and alcohol.

The city of New Brunswick is an attractive place to go for a drink or dinner on the town. Just don't stray too far from the campus. True, Rutgers has its own police department that possesses the same training and powers as the New Jersey state police. "I personally don't feel safe in New Brunswick, so I restrict my outings to on-campus locations," one student admits. For those who want to hit the road for fun, New York City and Philadelphia are each only about an hour drive or train ride, and students flood the Jersey shore in springtime. The Rutgers College Program Council offers trips ranging from white-water rafting to mountain climbing to skiing. "The variety of activities at Rutgers provides you with the opportunity to have fun any way you desire," says one student. "There are lots of on-campus social activities," a senior explains, including "movies, coffeehouses, local and bigger bands, lectures, parties. Off-campus activity includes frat parties and bars." Over a recent five-year period, students enrolled in the Citizenship and Service Education Program at Rutgers contributed more than 90,000 hours of service to communities across New Jersey. "Students definitely get involved in the surrounding community and do a lot of volunteer work," says a senior.

> *Rutgers is also home to more than 100 specialized research centers and institutes dedicated to the study of topics ranging from ancient Roman art to mountain gorillas.*

The Greek system, which attracts only a small percentage of men and women, is entirely off campus. While the school neither owns nor administers any of the Greek organizations, it does have a university office for Greek affairs, which oversees the welfare of those belonging to fraternities and sororities. Students say a lot of the nightlife for the New Brunswick campus takes place at the Greek houses. Reportedly, it is "difficult to drink in dorms," but underage students drink if they really want to. Each college has its own student center with pinball machines, pool tables, bowling alleys, and a snack bar. Major social events include Reggae Day at Livingston, Agricultural Field Day at Cook, and Oktoberfest for the campus as a whole. Pioneer Pride Night is Camden's big party. During homecoming, tailgate parties are held in the stadium parking lot, featuring tons of food (including roast pigs and whole sides of beef), continuous music, and thousands of revelers.

> **"I completely revere most of my professors."**

Varsity, intramural, and club sports fill whatever gap is left by the social scene. In fact, the university fields more than 1,000 athletes—the highest number of any university in the nation. The Rutgers baseball team is competitive, as are many other sports, including football, basketball, tennis, lacrosse, soccer, cross-country, and track. Women's basketball, fencing, soccer, softball, field hockey, tennis, and track teams are also strong. A member of the Big East in football, Rutgers faces a tough schedule that includes Cincinnati, Connecticut, Louisville, Pittsburgh, Syracuse, the University of South Florida, and West Virginia. Big East Conference competition makes up for not getting to play Princeton, which in 1980 bowed out of what was then the oldest football rivalry in the nation.

Rutgers has a plethora of people and programs characteristic of a large state university. It also has a lot more, including loyal support from the state's legislature and private sector and tuition that is relatively affordable. Says one satisfied student: "From the diversity of its majors and courses to the hundreds of student organizations on campus, Rutgers gives me a chance to explore a world of options."

Overlaps

College of New Jersey, Montclair State, NYU, Penn State, Rowan University

If You Apply To ➢

Rutgers: Rolling admissions. Does not guarantee to meet full demonstrated need. No campus or alumni interviews. SATs or ACTs: required. No SAT IIs. No essay. Apply to particular school.

College of St. Benedict and St. John's University

P.O. Box 7155, Collegeville, MN 56321-7155

The College of St. Benedict (CSB) and St. John's University (SJU) are throwbacks to the way colleges were 50 years ago: men and women on separate campuses and copious amounts of school spirit. Roman Catholics comprise about 65 percent of the students, and monastic communities are very active on both campuses.

Website: www.csbsju.edu
Total Enrollment: 3,862
Undergraduates: 3,825
Male/Female: 48/52
SAT Ranges: V 530–650
　M 530–650
ACT Range: 23–28
Financial Aid: 62%
Expense: Pr $
Phi Beta Kappa: No
Applicants: 2,639
Accepted: 87%
Enrolled: 45%
Grad in 6 Years: 82%
Returning Freshmen: 88%
Academics: 🐂🐂🐂
Social: 🐂🐂🐂
Q of L: ★★★
Admissions: (800) 544-1489
Email Address:
　admissions@csbsju.edu

Strongest Programs:
Biology
Chemistry
Economics
Music
Psychology
Education
Nursing
Management

Remember when women's colleges had nearby brother schools, when dorms were single-sex, and when visitors of the opposite gender were only welcome at certain times? No? Well, you might ask your parents. Or you could visit the College of St. Benedict and St. John's University. These two single-sex campuses are five miles apart, but they share a common heritage and mission: that students and faculty work together to better understand and use the liberal arts, guided by principles of their Benedictine founders. The schools' small sizes and strong sense of tradition give rise to a tight-knit community. "Everyone smiles or says hello to one another and opens doors for each other, even if they don't know them," says a junior economics major. "There is a great sense of community, and everyone is so warm and welcoming."

Owned and operated by the largest men's Benedictine monastery in the world, St. John's occupies 2,400 pristine acres in rural Minnesota, an area filled with forests, lakes, and the wide-open spaces perfect for outdoorsy types. Alongside an ancient quadrangle erected by monks is a strikingly modern church designed by Marcel Breuer. St. Benedict is a cohesive campus comprising redbrick buildings and cobblestone walks. Together, the colleges have invested more than $50 million in facilities in recent years, including new and renovated facilities for art, music, theater, and athletics, two new science centers, and two new student centers. There is also a new dining hall. The two colleges are connected by a free and frequent shuttle bus.

St. Benedict and St. John's share a joint academic program through which students take classes together on both campuses. The core curriculum requires a first-year seminar and a senior capstone and students must fulfill requirements in a number of areas, including the humanities, natural sciences, social sciences, fine arts, theology, ethics, and foreign language. An honors program serves exceptional freshmen, and upperclass students may also apply. The most popular major is management. Also popular are biology, psychology, communication, and nursing. Students give high marks to science programs such as premed, chemistry, and the environmental studies major, as well as to education. The environmental studies program is one of the few in the country that is interdisciplinary, and it benefits from the access to area's natural resources. Undergraduate research is becoming more prevalent, and there is an endowed summer research program in the health and medical areas. The theology program benefits from many resources, including the Hill Museum & Manuscript Library, one of the foremost microfilm collections of centuries-old handwritten manuscripts. A mathematics/computer science major has been renamed; it's now numerical computation, and the management, accounting, and finance offerings have been revised.

> "There is a great sense of community, and everyone is so warm and welcoming."

"Many of my courses have been rigorous and it is very competitive," says a senior. No class has more than 50 students and some fill up fast. The small classes encourage strong student/faculty ties. "The faculty are spectacular," says one student, "with a passion for subject matter and a willingness to go the extra mile for students." Members of the monastic communities make up 10 percent of the CSB/SJU faculty.

For those who tire of the Minnesota winters, which can start in October and run until April, faculty-led international study programs are offered in several countries, including one in Chile. Each program is limited to about 30 students, and more than half of CSB/SJU students take part. Numerous shorter trips are offered during semester break and summer break. CSB/SJU is among the top three liberal arts colleges in the nation for the number of students who study overseas.

"Our sense of community is an extremely important and core part of the CSB/SJU experience," a sophomore says. "Students truly identify with what it means to be a 'Bennie' or a 'Johnnie.'" Though 65 percent of students are Roman Catholic, few resemble priests-in-training. Eighty-two percent are from Minnesota, and most are white. African Americans and Hispanics each constitute 1 percent of the student body, and Asian Americans add 2 percent. International students account for 4 percent of the student population. Politically, "we encompass all viewpoints," a junior says, "from ultraconservative to ultraliberal," and hot topics include environmental issues, homosexuality, and abortion. Merit scholarships averaging $7,927 are available although there are no athletic scholarships.

> **"We encompass all viewpoints, from ultraconservative to ultraliberal."**

Ninety-three percent of students live in campus housing; freshman and sophomores are required to stay in the dorms. They are staffed partly by members of the monastic communities, but students aren't made to feel like a nun is watching their every move. "Housing is great!" gushes a junior. "The buildings are kept up quite well, much better than others I've seen." Each room has a sink, plus an Internet connection for each student. One thing they don't have is overnight visits from members of the opposite sex. Other options include on-campus apartments, such as an earth-sheltered complex on the shore of a lake. Upperclassmen may live off campus, but usually stay nearby and commute by foot, car, or college bus. Students can choose from four dining halls. "All menus are posted online so menus can be checked before going there," says a junior who notes there is "lots of variety."

Since there are no fraternities or sororities here, students have learned to make their own fun. SJU's Stephen B. Humphrey Fine Arts Theater and CSB's Benedicta Arts Center host a wide variety of cultural events. For those seeking typical college fun, "the Joint Events Council plans dances, game shows, concerts, and a number of other activities," says a sophomore. "Off campus, there are many things to do in St. Cloud and also in St. Joseph, such as parties, coffee shops, restaurants, shopping, and movies." St. Cloud (metro population more than 100,000) is less than 10 minutes away by car, and the Twin Cities are about an hour away. Each year, students look forward to the Festival of Cultures, Fruit-at-the-Finish Triathlon, the Senior Farewell, and spring break trips involving community service. Also popular is the annual Pinestock music festival, which welcomes the new season with a day of concerts featuring popular national acts.

> **"The campus is beautiful, but the people make it what it is."**

The football team is a perennial Division III powerhouse, and its rivalry with St. Thomas is as strong as ever. Curiously enough, a team of guys known as The Rat Pack gets students psyched up for games. The St. Benedict basketball team is a regular conference champion. Men's hockey and golf and women's soccer also brought home conference championships recently. Nonvarsity students can participate in a variety of club and intramural sports. CSB/SJU students can enjoy such activities as kayaking and indoor rock climbing through the Outdoor Leadership Center, and St. Benedict women can go on outdoor recreation trips with the Women's Expedition program.

Students at these two Catholic schools "are genuinely friendly, relaxed, motivated, and outgoing," a junior says. Those who attend CSB/SJU revel in the schools' small-town setting, their traditions, and the grounding that comes from their shared

The football team is a perennial Division III powerhouse, and its rivalry with St. Thomas is as strong as ever.

The theology program benefits from many resources, including the Hill Museum & Manuscript Library, one of the foremost microfilm collections of centuries-old handwritten manuscripts.

Overlaps
University of St. Thomas, University of Minnesota–Twin Cities, University of Minnesota–Duluth, Gustavus Adolphus, St. Cloud State, St. Olaf

Benedictine values. "Our school is welcoming," a sophomore says. "The campus is beautiful, but the people make it what it is."

If You Apply To ➤

St. Benedict and St. John's: Rolling admissions: Dec. 1 (priority). Financial aid: Mar. 15 (priority). Meets full demonstrated need of 41%. Campus interviews: recommended, informational. No alumni interviews. SATs or ACTs: required. Accepts the Common Application and electronic applications. Essay question: previously written paper; issue of personal, local, national, or international concern; how your culture and ethnic identity or your leadership and service roles have fostered your appreciation of difference; a significant experience, achievement, risk, or ethical dilemma; or a topic of interest to you.

St. John's College

Annapolis campus: P.O. Box 2800, Annapolis, MD 21404-2800
Santa Fe campus: 1160 Camino Cruz Blanca, Santa Fe, NM 87505-4599

Books, books, and more books is what you'll get at St. John's—from Thucydides to Tolstoy, Euclid to Einstein. St. John's attracts smart, intellectual, and nonconformist students who like to talk (and argue) about books. Easy to get in, not so easy to graduate.

Annapolis
Website:
 www.stjohnscollege.edu
Location: City center
Total Enrollment: 608
Undergraduates: 515
Male/Female: 55/45
SAT Ranges: V 650–760
 M 580–680
Financial Aid: 63%
Expense: Pr $ $ $ $
Phi Beta Kappa: No
Applicants: 426
Accepted: 75%
Enrolled: 44%
Grad in 6 Years: 68%
Returning Freshmen: 82%
Academics: 🐾 🐾 🐾 🐾 ½
Social: ☎ ☎ ☎
Q of L: ★ ★ ★ ★
Admissions: (800) 727-9238
Email Address:
 admissions@sjca.edu

Santa Fe
Website:
 www.stjohnscollege.edu
Location: City outskirts
Total Enrollment: 533
Undergraduates: 435

With no traditional professors, departments, or majors, few lectures, and less than 1,000 students on its two campuses, St. John's College is about as far from the typical post-secondary experience as you can get. Or maybe it's much closer to what college used to be—after all, the Annapolis campus traces its roots to King William's School, founded in 1696. More than two centuries later, in 1964, St. John's opened a second campus in Santa Fe, New Mexico, to facilitate a doubling of enrollment and offer its super-serious students a change of scenery. While the campuses may be a thousand miles apart, the Johnnies who populate them share an all-consuming quest for knowledge in the classical tradition. Their true teachers are the Great Books, about 150 of the most influential works of Western civilization. "Perhaps the defining characteristic of the student body is an intense studiousness," says a sophomore.

Physically, the two St. John's campuses are more than just three time zones from one another. The Colonial brick structures of the small urban campus in Annapolis, where the central classroom building dates from 1742, are squeezed into the city's historic district. With the Maryland state capitol and the U.S. Naval Academy in the neighborhood, this campus exudes old-world charm, and its location at the confluence of the Severn River and the Chesapeake Bay allows students to participate in sailing, crew, and individual sculling. The Santa Fe campus sits on 250 acres on the outskirts of the sun-drenched capital of New Mexico. The adobe-style buildings reflect the Spanish and Native American traditions, and from their perch in the Sangre de Cristo Mountains offer beautiful views of the city below. Students at St. John's in Santa Fe can get back to nature in several nearby national parks, which offer hiking, mountain biking, snowboarding and skiing. Students may attend both campuses during their academic careers, and about a quarter do so.

The St. John's curriculum, known as "the program," has every student read the Great Books in roughly chronological order. All students major in liberal arts, discussing the books in seminars, writing papers about them, and debating the riddles of human existence they raise. Classes are led by tutors, who would be tenured professors anywhere else, but here are just the most advanced students. Each tutor is required to teach any subject within the curriculum; as a group, the tutors help students divine wisdom from each other and from great philosophers and thinkers, from

Thucydides and Tolstoy to Euclid and Einstein. Both campuses follow a curriculum that would have delighted poet and educator Matthew Arnold, who argued that the goal of education is "to know the best which has been thought and said in the world."

There are no registration or scheduling hassles at St. John's; the daily course of study is mapped out before students set foot on campus. The curriculum includes four years of mathematics, two years of ancient Greek and French, three years of laboratory science, a year of music, and, of course, four years of Great Books seminars. Freshmen study the Greeks, sophomores advance through the Romans and the Renaissance, juniors cover the 17th and 18th centuries, and seniors do the 19th and 20th

"Perhaps the defining characteristic of the student body is an intense studiousness."

centuries. Readings are from primary sources only: Math from Euclid and Ptolemy, physics from Einstein, psychology from Freud, and so on. For about seven weeks in the junior and senior years, seminars are suspended and students study a book or topic one-on-one with a tutor. The assumption is that the Great Books can stand on their own, representing the highest achievements of human intellect. "While I wouldn't describe it as competitive," says a sophomore, "the climate is fairly serious and the courses are extremely rigorous."

Speaking of difficult, students say the junior year, with its advanced curriculum in math and the natural sciences, is the most challenging. Still, the climate is anything but cutthroat. "Small-group, discussion-based classes and the lack of letter grades don't allow for academic competition," says a junior. While there are no multiple-choice tests and no formal exams, since everyone's doing the same thing, there's a lot of peer pressure not to slack off. "Bonding over readings in Plato or Euclid is inevitable," says one student. Many St. John's students find they need a year off between the sophomore and junior years to decompress; some switch from Annapolis to Santa Fe or vice versa, and more than 30 percent of students take more than six years to graduate.

Admission of qualified students to St. John's is first come, first served, and when the campuses are full—total enrollment is capped at 500 students in each place—admissions begin for the following semester. A fifth of the students are transfers from more conventional colleges—a true act of devotion, since St. John's requires everyone to begin as freshmen. Minorities represent 11 percent of the student body in Santa Fe and 6 percent in Annapolis. Though the reasons students choose St. John's are never simple, the common thread is a fierce love of

"Bonding over readings in Plato or Euclid is inevitable."

learning. "We are here because of our passion for learning and for group discussion and explanation, rather than lecture," says a freshman. "We do not accept what we are told without questioning and exploring it for ourselves." Most students are bright, opinionated, and have no use for the status quo. "It is often loud here, but only with the debates of the students, and often serious, but only when appropriate, and often comic, but only when the issue is too serious to be grave," adds another student. Politically, the student body is apathetic: "We are preoccupied trying to decipher the meaning of life," quips one.

Seventy-five percent of students in Annapolis and Santa Fe live on campus in the co-ed dorms; freshmen are guaranteed a room. In Annapolis, the six "historic" residence halls are arranged around a central quad, while the two modern halls, which opened in January 2006, face College Creek. (Students warn that "historic" is code for "old," and complain about "schizophrenic heating and cooling" and a lack of hot water for morning showers.) In Santa Fe, the dorms are small modern units, clustered around courtyards. Most students get singles or divided double rooms. Upperclassmen typically live off campus in apartments and group houses; those who stay in the dorms usually get single rooms. There are no fraternities or sororities.

(Continued)
Male/Female: 53/47
SAT Ranges: V 650–750
 M 580–670
ACT Range: 28–31
Financial Aid: 65%
Expense: Pr $ $ $ $
Phi Beta Kappa: No
Applicants: 318
Accepted: 83%
Enrolled: 55%
Grad in 6 Years: 67%
Returning Freshmen: 85%
Academics: ✍ ✍ ✍ ✍ ½
Social: ☎ ☎ ☎
Q of L: ★ ★ ★ ★
Admissions: (505) 984-6060
Email Address:
 admissions@sjcsf.edu

Strongest Program:
The Great Books Program

Popular annual events on both campuses include Lola's, a casino night sponsored by the junior class to raise money for Reality, a three-day festival of food, games, and general debauchery thrown for the seniors the weekend before commencement.

Drinking is a favored release for Johnnies, who have, of course, read Plato's Symposium and are familiar with the likes of Rabelais. Still, hard liquor is not allowed on campus, and parties and kegs must be registered. And although no one under 21 may be legally served at college-sponsored events, which are patrolled to prevent underage drinking, youngsters tip their share of brew at smaller gatherings and in their rooms. Only students who are extremely rowdy or disruptive are reported to the dean's office to face penalties. "Campus security really only makes it an issue if you do first," one student explains. Students also keep busy with movie nights, lectures, extra study groups and "just hanging out on the quad, if you aren't reading or preparing for class," says a junior.

Santa Fe undergraduates plunge into the outdoorsy activities made possible by their mountaintop location, while Annapolis students limit their adventures to intra-mural teams with names like the Druids and the Furies. The annual croquet tournament against the U.S. Naval Academy attracts 5,000 spectators, and the Johnnies usually win. Road trips to Washington, D.C., Baltimore, New York, and Assateague State Park are options for students with cars; the annual spring break trip to the Santa Fe campus is popular as well, and known as Wagons West. In Santa Fe, some students venture south of the border on weekends to Juarez, Mexico, and others head to the hot springs of White Sands, New Mexico. Nearby blues and jazz clubs are also popular, though one student cautions that town shuts down around 9 p.m. Santa Fe also tends to be pricey, especially for students on a budget.

Popular annual events on both campuses include Lola's, a casino night sponsored by the junior class to raise money for Reality, a three-day festival of food, games, and general debauchery thrown for the seniors the weekend before commencement.

"Apart from new dorms and the yoga craze at the gym, it's the same St. John's as always."

There's Melee, the club where "combatants" fight epic battles with foam weapons of their own creation, and Tuesday Nite Fites, where students vote for line-ups like Funk versus Dinosaurs and then debate the winner. There's also Fasching, a 1930s-style formal dance; the Ark party, held to celebrate the sophomores' completion of the Old Testament; and Oktoberfest. Senior Prank is a daylong surprise party for the whole college community. Intramural sports in Annapolis include flag football (with rules like you'll see nowhere else), basketball, soccer, team handball, and softball. Participation is voluntary, but anyone who wants to play is included. There are also intercollegiate club teams in crew, fencing, croquet, and women's soccer. In Santa Fe, the nearby Rio Grande and Chama rivers offer excellent whitewater canoeing, kayaking and rafting; the Outdoor Programs Office organizes trips and makes athletic equipment available for use.

Students at St. John's are as passionate about learning as their peers at other schools are about basketball rivalries or blowout parties. And while those larger colleges and universities try desperately to grow and change, St. John's cherishes its tradition—including the mandate that seniors wear formal academic dress to their oral examinations, which are open to the public. "I don't know if you could have another college where life was as determined by what takes place in class," muses a junior. "The liberal arts/classics thing is pretty well-preserved, so apart from new dorms and the yoga craze at the gym, it's the same St. John's as always, and most likely will be five years from now, too."

St. John's University and College of St. Benedict: See page 564.

St. Lawrence University

Canton, NY 13617

St. Lawrence is perched far back in the north country, closer to Ottawa and Montreal than to Syracuse. Isolation breeds camaraderie, and SLU students have a special bond similar to that at places like Dartmouth and Whitman. Environmental studies is the crown jewel.

St. Lawrence University seeks snow lovers who place equal value on their experiences inside and outside the classroom. Its upstate New York location offers quick access to both pristine ski slopes and rugged hiking trails—and to the bright lights of Ottawa and Montreal. Classes are small and there are no TAs, meaning it's as easy to form friendships with faculty members as it is with fellow students. A flood of new facilities has helped to make the campus almost as breathtaking as the natural beauty that surrounds it. "St. Lawrence students are very well-rounded—involved in many different activities," says a psychology major. "People come here to get the best out of their college experience, even if it means overloading their schedules."

Hiking trails, a river, and a golf course surround SLU's 30 buildings, which sit on a 1,000-acre tract; facilities are clustered, so even the most distant buildings are only 10 minutes from one another. Many buildings date from the late 19th century, and though their exteriors have been preserved, their interiors are fully modernized. In all, the school has invested $100 million to beautify and better its campus over the past five years. "The facilities improvements are truly staggering," says a sophomore. "We have the new indoor track, weight room, climbing wall, training room, bookstore, townhouses, turf field, football stadium, and outdoor track—and that doesn't even count the renovations." The new Torrey Center for Health and Counseling opened recently and the Johnson Hall of Science will open in late 2007.

> **"You can't be shy. You must be ready to go out there and do."**

St. Lawrence offers a classical liberal arts education, placing a premium on small classes and team teaching. General education requirements include one course each in arts and expression, humanities, social science and math, or a foreign language, as well as two courses each in diversity and natural science, including one with a lab. Everyone also participates in the two-semester First-Year Program, which emphasizes critical thinking, communications, and interdisciplinary content. "Students

Website: www.stlawu.edu
Location: Village
Total Enrollment: 2,134
Undergraduates: 2,101
Male/Female: 48/52
SAT Ranges: V 520–620
　M 530–630
ACT Range: N/A
Financial Aid: 68%
Expense: Pr $ $ $
Phi Beta Kappa: Yes
Applicants: 2,989
Accepted: 59%
Enrolled: 30%
Grad in 6 Years: 75%
Returning Freshmen: 90%
Academics: ✍ ✍ ✍
Social: ☎ ☎ ☎
Q of L: ★ ★ ★
Admissions: (800) 285-1856
Email Address:
　admissions@stlawu.edu

Strongest Programs:
Biology
Economics

are separated into first-year 'colleges' for housing, and each college is enrolled together in a unique course," an economics major explains. There are 12 "colleges," each with 30 to 45 students, and FYP professors also serve as academic advisors. "Some of my best friends on campus are from my FYP," explains a student, "and normally these tight-knit communities tend to stay close all four years at SLU."

Psychology and economics are the most popular majors at SLU, followed by English, government, and sociology. Newer offerings include programs in neuroscience and biochemistry. Administrators say the music department is small but growing, especially with the addition of an instrumental music professor. St. Lawrence is also well known for its environmental studies department. In an effort "to make the world our classroom," St. Lawrence

"The facilities improvements are truly staggering."

encourages students to spend time away from campus. More than half do so, and while some participate in one of the school's 13 international programs, others choose the nearby "Adirondack semester" at Saranac Lake, or programs in Canada and Washington, D.C. The St. Lawrence University Fellows Program offers stipends for summer research.

The academic climate at SLU is a "healthy competitive environment," says a senior. "Classes are fun and are presented in a very laid-back manner." Since there are no teaching assistants, full professors teach even the introductory courses; nearly three quarters of the courses taken by freshmen have 25 students or fewer. "As a senior, I must say the professors were the best part of SLU," a student says. "I have formed lasting friendships and learned life skills from them."

"Students who do well at SLU are those that are outgoing," confides one senior, adding that "someone who isn't afraid to try new things and wants to be involved" is likely to enjoy the SLU experience. Forty-five percent of the students at St. Lawrence are New Yorkers; 71 percent graduated from public high school and the vast majority are white. In fact, just 2 percent of the student body is African American, 2 percent is Hispanic, and 2 percent is Asian American. "Diversity is an important issue as the university works diligently to attract more and more students, faculty, and staff from a

The new Torrey Center for Health and Counseling opened recently and the Johnson Hall of Science will open in late 2007.

variety of backgrounds," says a speech and theater major. The number of merit scholarships varies each year, as SLU

"Classes are fun and are presented in a very laid-back manner."

sets aside a chunk of money, then allocates it among deserving students. The school also hands out 35 athletic scholarships for Division I men's and women's ice hockey. (Other SLU teams compete in Division III.)

Ninety-nine percent of SLU students live in the dorms, and seniors definitely have it best, with access to "new and spacious townhouses that sit along the golf course," says a sophomore. Everyone else makes do in the "adequate" residence halls, which have mostly double rooms featuring Internet links, cable TV, computer and study lounges, vending machines, laundry rooms, and kitchens. In the dining halls, there are themed dinners once a month, as well as ethnic foods, vegetarian options, and even organic items. Other options include the pub and convenience store, both in the student center, and a cafe in the physical education building. In addition, 21 percent of the women and 8 percent of the men go Greek; fraternity and sorority members may live and eat in their chapter houses.

In addition, 21 percent of the women and 8 percent of the men go Greek; fraternity and sorority members may live and eat in their chapter houses.

When it comes to social life, "because we are in the middle of nowhere, everything must take place on campus," says a biology major. "You can't be shy. You must be ready to go out there and do." University-sponsored activities include a campus nightclub, four different first-run movies each week, and the campus coffeehouse—a great place to hear a live band, acoustic guitarist, or comedian. Still, students say, the most popular pastimes include skiing, hiking, and road trips to Ottawa and Montreal, where there's better shopping and dining and the drinking age is lower, too.

On campus, students under 21 aren't permitted to drink, per New York State law. That said, "People still do it, because they think they can get away with it," says a senior. "Not all do," and those who are caught are required to complete counseling and educational programs.

The "charming" town of Canton is "a speck on the state map," says one student. "We are surrounded by farmland, then by forest and mountains. We do not live in a college town at all." Still, Canton does have everything from bagels to handmade jewelry to bars and restaurants, and Potsdam, 10 minutes away, offers more. Favorite annual traditions include Winterfest, a two-week celebration of the season, and Peak Weekend, during which the SLU Outing Club tries to "put St. Lawrence students on every peak in the Adirondacks."

Most St. Lawrence students enjoy sports; the school's fine athletic facilities include two field houses, one with 10 squash courts and a second with three indoor tennis courts, a pool, a three-story climbing wall, and a ropes course. The school's golf course doubles as a running route in warmer weather and a cross-country ski trail during the winter. Hiking and rock climbing are also popular, as is canoeing down the St. Lawrence River (at least when it's not frozen over). In varsity sports, the Division I Skating Saints are the top draw, especially when the opponent is archrival Clarkson. Other competitive teams include men's and women's soccer, basketball, golf, and equestrian. Over 90 percent of all students participate in intramurals at some point; available sports range from basketball and broomball to ruckus and wiffle ball.

St. Lawrence makes up for frigid winters with the warmth of a close-knit, caring community. As the frenzied pace of construction winds down and academic standards are ratcheted up, SLU is a school on the rise, especially for those wanting to get back to nature.

If You Apply To ➤ | **St. Lawrence:** Early decision: Nov. 15, Jan. 15. Regular admissions, financial aid and housing: Feb. 15. Does not guarantee to meet demonstrated need. Campus interviews: recommended, evaluative. Alumni interviews: optional, evaluative. SATs or ACTs: optional. SAT IIs: optional. Accepts the Common Application and electronic applications. Essay question: significant experience, achievement, risk or ethical dilemma, and its impact; issue of personal, local, national, or international concern; influential person, fictional character, historical figure, or creative work; importance of diversity, or how you would bring diversity to the college community; or a topic of your choice.

St. Mary's College of Maryland

St. Mary's City, MD 20686

A public liberal arts institution of the same breed as Mary Washington, UNC–Asheville, and William and Mary. St. Mary's historic but sleepy environs are 90 minutes from D.C. and Baltimore. With the Chesapeake Bay close at hand, St. Mary's is a haven for sailors.

Twenty-five years ago, St. Mary's College of Maryland was just another public college, albeit one with a gorgeous waterfront campus in the oldest continuously inhabited English settlement in the New World. Then, the state of Maryland decided to make St. Mary's its public "honors college"—and the rest, as they say, is history. With a student/faculty ratio of 12.2 to 1, students here can easily design their own majors, undertake independent research projects, or work closely with professors to investigate whatever interests them.

Website: www.smcm.edu
Location: Rural
Total Enrollment: 2,013
Undergraduates: 2,004
Male/Female: 43/57

(Continued)

SAT Ranges: V 570–690
   M 560–650

Financial Aid: 52%

Expense: Pub $ $ $ $

Phi Beta Kappa: Yes

Applicants: 2,200

Accepted: 68%

Enrolled: 33%

Grad in 6 Years: 82%

Returning Freshmen: 89%

Academics: ✍ ✍ ✍ ✍

Social: ☎ ☎ ☎

Q of L: ★ ★ ★ ★

Admissions: (800) 492-7181

Email Address:
   admissions@smcm.edu

Strongest Programs:
Biology
English
Economics
Political Science
Psychology

The ultimate Frisbee club team has a heated rivalry with Navy and even attracts considerable alumni interest.

St. Mary's has never been owned by any church denomination and takes its name from its location. It began in 1840 as a boarding school for women, intended as a monument to the colonial birthplace of the state. The campus sits on a peninsula in southern Maryland where the Potomac River meets the Chesapeake Bay. Not surprisingly, it has an excellent center for estuary research, as well as a strong working relationship with the Chesapeake Biological Laboratory; the school even has its own marina right on the St. Mary's River, with a shoreline that gets beautiful sunset views. Architectural styles range from Colonial to modern buildings, though the land on which the campus is built belongs to a 1,100-acre national historic landmark, commemorating Maryland's first Colonial settlement. For that reason, students may step over archeological digs as they stroll to class—an annoyance, perhaps, but also an unparalleled opportunity for primary research. The athletics and recreation center was expanded in 2005, providing students with an Olympic-size swimming pool and expanded facilities.

> **"This school feels like a little town—everyone is kind and polite."**

General requirements at St. Mary's include courses in English composition, history, art, literature, math, natural sciences, behavioral science, economics or political science, and values inquiry. Students must also demonstrate foreign language proficiency, take part in first-year seminars, and complete a senior capstone experience. Thirty-four percent of students are from the top tenth of their high school class. "Last year on my hall, there were eight valedictorians," says a sophomore. "Everyone works hard, but everyone finds a way to balance that with fun." Biology is the third-most popular major, and also one of the more difficult programs; students can spend time on the college's research boat when they tire of the lab. Students also sign up in droves for psychology, English, economics, and political science. The strong music department includes prize-winning pianist Brian Ganz, and aside from the 21 established majors and five cross-disciplinary study areas, more freethinking types may design their own programs. Newer programs include biochemistry and neuroscience; all science majors benefits from a state-of-the-art facility, with 55,000 square feet of classrooms, labs, and research space.

Students with exceptional academic potential may participate in the Nitze Scholars Program, which offers special seminars and opportunities for the study of foreign culture. St. Mary's also offers study programs at Oxford's Center for Medieval and Renaissance Studies, the University of Heidelberg in Germany, China's Fudan University, and the Institute for the Study of Politics in Paris. There are also study tours to Gambia, Belize, Greece, and Costa Rica, while a partnership with the National Student Exchange further expands the number of potential destinations. Back on campus, 80 percent of classes have fewer than 26 students, but "getting into classes can be competitive," says a philosophy major.

> **"If you love the water, this is the place to be."**

Eighty-two percent of St. Mary's students come from Maryland, which helps give the campus a homegrown feel. "This school feels like a little town—everyone is kind and polite," says a biology major. "Everyone knows each other, making it a comfortable place to be." More minority students are also finding their way to the peninsula; 10 percent of the student body is now African American, 5 percent is Hispanic, and 5 percent is Asian American. There are no big social or political disagreements, one student says: "Attitudes are wonderful here, very open—not what I experience at home." Adds a classmate: "Every student acts very friendly and nice." Nearly a quarter of students receive merit scholarships, worth an average of $3,500. There are no athletic awards, since teams compete in Division III.

Eighty-six percent of full-time students live on campus, and most residence halls are co-ed. Apartments and townhouses with kitchens are reserved for upperclassmen, and students select rooms based on the number of credits they have.

Older students get preference if they decide to retain a room or want a different one nearby; there are plenty of choices off campus, too, including old farmhouses and riverside cottages for rent. Food in the dining hall, known as the Great Room, is "decent," but one student says there aren't as many options as there would be at a larger school. Those who tire of the chow line may join the vegetarian co-op, or maybe just go catch their own fish and crabs.

St. Mary's doesn't have fraternities or sororities, and its secluded location—about two hours from Washington, D.C., Annapolis, and Baltimore—means there's little nightlife off campus. ("A lot of great places to eat down here, but not many to shop," says a sophomore.) Students say they don't mind the isolation, though: "Point Lookout is only 10 or 15 minutes away," one explains. "If you love the water, this is the place to be," with sailing and other pursuits close at hand. "It's virtually impossible to graduate without knowing how to sail," says one student. The waterfront also becomes the focus of campuswide activities, including the cardboard boat race held each fall, Earth Day in April, and the end-of-year World Carnival. For the culture-starved, there's also theaters, an art gallery, lectures, and films on campus.

> **"Everyone works hard, but everyone finds a way to balance that with fun."**

Although St. Mary's isn't known for its athletic prowess, the men's soccer team is frequently regionally ranked. Men's and women's lacrosse are also very competitive, and the women's team has placed in the top 10 nationally for the past five seasons. The varsity sailing team won the Inter-Collegiate Sailing Association/Layline North American Team Race Championship in 2004. The ultimate Frisbee club team has a heated rivalry with Navy and even attracts considerable alumni interest. Overall, about a third of the students at St. Mary's take part in intramurals; coed soccer, indoor soccer, and men's dodgeball are especially popular.

St. Mary's has been working hard to establish itself as one of the nation's premier public liberal arts colleges. Though its name sounds religious and its small size can be stifling, students leave with a solid grounding in the liberal arts—and the close bonds that they forge with friends during peaceful days on the bay.

The strong music department includes prize-winning pianist Brian Ganz, and aside from the 21 established majors and five cross-disciplinary study areas, more freethinking types may design their own programs.

Overlaps

University of Maryland–College Park, Washington College (MD), University of Maryland–Baltimore County, College of William and Mary, University of Mary Washington, Towson State University

If You Apply To ➤ | **St. Mary's:** Early decision: Dec. 1. Regular admissions: Jan. 15. Financial aid: Feb. 15. Housing: May 1. Does not guarantee to meet demonstrated need. Campus interviews: recommended, informational. Alumni interviews: optional, informational. SATs or ACTs: required (SAT preferred). Prefers electronic applications. Essay question (choose one): interesting characteristics you would bring to St. Mary's; what aliens would learn about Earth by observing you for one week; where the U.S. will find itself in 50 years; who you would change places with for a year; why you would run for president and your three most important agenda items.

St. Olaf College

1520 St. Olaf Avenue, Northfield, MN 55057-1098

Lutheran to the core, the well-scrubbed undergraduates at St. Olaf are a stark contrast to the grunge of crosstown rival Carleton. The music program is world famous, and 69 percent of students study abroad. Daily chapel is not mandatory, but many students go.

Northfield, Minnesota, which bills itself the city of "Cows, Colleges, and Contentment," is home to St. Olaf College—and the blondest student body this side of Oslo. This small Midwest school, founded by Norwegian Lutheran immigrants, was named for

Website: www.stolaf.edu
Location: Small town

Total Enrollment: 3,058

Undergraduates: 3,007

Male/Female: 42/58

SAT Ranges: V 590–700
  M 580–690

ACT Range: 25–30

Financial Aid: 62%

Expense: Pr $ $

Phi Beta Kappa: Yes

Applicants: 2,991

Accepted: 73%

Enrolled: 35%

Grad in 6 Years: 84%

Returning Freshmen: 92%

Academics: ✏️✏️✏️✏️

Social: ☎ ☎ ☎

Q of L: ★ ★ ★ ★

Admissions: (507) 800-3025

Email Address:
  admissions@stolaf.edu

Strongest Programs:
Biology
Psychology
Economics
English
Mathematics
Music

Most students are high achievers from Midwestern public schools, drawn in part by hundreds of merit scholarships.

the country's patron saint and aims to be "an excellent liberal arts college of the church." One student describes her peers at St. Olaf as "Minnesota nice," adding, "You can't go anywhere without saying 'hi' to people. Even if you don't know a person, there is still a feeling of belonging to the same community. We are friendly, outgoing, hardworking, and we love to have fun."

St. Olaf's meticulously landscaped 350-acre campus, featured in several architectural journals, is located on Manitou Heights, overlooking the Cannon River valley and Northfield. More than 10,000 trees, native prairie, and a wetlands wildlife area surround the 34 native limestone buildings that form the campus. The student union, the Buntrock Commons, offers dining and food services, a bookstore, a post office, conference and banquet facilities, a movie theater, and a game room.

All students at St. Olaf complete a general education requirement that covers three areas: foundation studies, core studies, and integrative study. A first-year seminar emphasizing writing, a foreign language, math, oral communication, and physical education fulfill the first area. Two courses in each of six disciplines—Western culture, multicultural studies, artistic and literary studies, biblical studies and theological studies, natural sciences, and human behavior and society—complete the second requirement. A course in ethical issues and perspectives fulfills the third area. Typically, 14 to 16 courses satisfy the general education requirements; some courses may fulfill requirements in more than one area.

Chemistry, biology, and psychology are hailed as the best programs. Economics, English and religion are also popular. The music department draws high praise; it offers many performance opportunities with five school choirs, a band, and an orchestra. The choirs are often featured at church services and other religious events, and can be heard singing with the Minnesota Orchestra. In the last quarter-century, the college has cultivated an international agenda for its students and faculty, and has created the largest international studies program in the country among liberal arts colleges. Programs are available in more than 40 countries, including China, Japan, Scotland, and India, and some are offered through membership in the Associated Colleges of the Midwest* consortium. There are also several opportunities to spend a semester studying elsewhere in the country. The Finstad Office for Entrepreneurial Studies conveys knowledge about the challenges, risks, rewards, opportunities, and responsibilities of being an entrepreneur. Research is available in the sciences and psychology—36 percent of students take advantage of it—and the Center for Integrative Studies allows students to form their own majors. All students get faculty advisors, and the vast majority of students graduate in four years. Students can participate in one of three "conversation" programs, intensive sequences of courses in the classic works, East Asia, or liberal arts.

> **"We are friendly, outgoing, hardworking, and we love to have fun."**

Faculty members are highly praised by students. "Every professor is an expert in his or her area, is extremely available, and is genuinely interested in the well-being of students," says a senior. Instructors participate in and out of the classroom, reportedly having as many as 10 hours of open office time a week. Still, just because professors want to see them succeed doesn't mean students don't have their work cut out for them. "St. Olaf is academically rigorous," says one student. "The atmosphere is not competitive," adds a freshman. "Students study together, go over each other's papers, and generally encourage each other to do well." There's also an annual study break where professors serve their stressed-out pupils ice cream.

Fifty-seven percent of students hail from Minnesota. Asian Americans make up 5 percent, Hispanics 2 percent and African Americans 1 percent. Most students are high achievers from Midwestern public schools, drawn in part by hundreds of merit scholarships averaging $7,210. St. Olaf students "are liberal, white, upper-middle-class, and intelligent," says a nursing major. "Students study together, go over each other's papers, and generally encourage each other to do well." Students are also politically aware and "the general idea is that this is a very liberal campus," says an English major. "I have met a good number of conservatives, though." Hot-button issues include sexual orientation, the environment, and diversity on campus.

Ninety-six percent of undergrads live in on-campus housing, with college-owned houses available off campus. "Dorms are decent and get better as you increase in seniority," advises one student. All freshmen live on campus, and are assigned double rooms in 20-student "corridors," each with two junior counselors. Dorms are co-ed by floor, and each has its own personality. "All the dorms are beautiful on the inside and outside," a sophomore says. "Nice rooms and huge lounges provide a real sense of community." Rooms are selected by lottery and it can be difficult to get a room at times. Students eat in a large modern cafeteria, where the food is considered above average for college fare. "The cooks can get pretty creative sometimes," says a freshman. Every December, the dining hall serves a special meal of traditional Norwegian cuisine, including lutefisk.

> "Students study together, go over each other's papers, and generally encourage each other to do well."

St. Olaf's social life takes place mostly on campus, and weekend activities include a nightclub called Lion's Pause and a coffeehouse, the Alley. The fine arts department provides many music, theater, and dance performances. "On any given weekend there is usually a play, a concert, and a handful of sporting events," notes a senior. The campus is officially dry, and there are no fraternities or sororities at St. Olaf, but that doesn't mean students don't imbibe. Alcohol is available to those determined enough to seek it out, students report, and can also be found at local bars. The Student Activities Committee sponsors frequent dances, speakers, and cultural events, covered by student fees and at-the-door ticket sales. Daily chapel services, though not mandatory, are heavily attended.

Northfield is "small and cute," but there is little of social interest for St. Olaf students aside from Carleton. The most talked-about annual event, which has been running for more than 75 years, is the four-day Christmas Festival during which choirs, orchestras, and bands combine in televised concerts celebrating the birth of Jesus. A fall concert has recently featured such performers as George Clinton and Ben Folds. Many students volunteer in Northfield and report a friendly rapport with the community. "Students are actively involved in community mentor, volunteer, and outreach programs," one student says. For those with wanderlust, buses leave regularly for the twin cities of Minneapolis and St. Paul, less than an hour drive, where one can experience a shopper's paradise at the huge Mall of America.

> "All the dorms are beautiful on the inside and outside."

St. Olaf has outstanding Division III athletic programs. Women's cross-country, swimming, and diving and men's soccer and cross-country secured recent conference titles. There is also an extensive intramural program, and broomball—ice hockey played with brooms instead of sticks, and shoes rather than skates—is the sport of choice in the winter. The St. Olaf football team competes against Carleton for the honor of having the statue in the town's square face the winning campus. The game, called the Cereal Bowl, is sponsored by Malt-O-Meal, which has a factory in town. The chorus of St. Olaf's fight song is "Um Ya Ya," which has become a popular chant on campus.

For those yearning for a school where spirituality and scholarship exist on the same exalted plane, St. Olaf could be the right place to spend four years. It's a school where students work hard, are encouraged by good teachers, toughened by Minnesota winters, and nourished by strong moral values—in addition to hearty Scandinavian food. Says one satisfied senior: "It is an incredible experience that will change your life and challenge you in ways you never thought possible."

In the last quarter-century, the college has cultivated an international agenda for its students and faculty, and has created the largest international studies program in the country among liberal arts colleges.

Overlaps

Gustavus Adolphus, Carleton, Luther, University of Wisconsin–Madison, University of Minnesota–Twin Cities

Saint Louis University

221 North Grand Boulevard, St. Louis, MO 63103

This is not your father's SLU. The campus and surrounding neighborhood have been spiffed up in the past two decades, and SLU's campus is a pleasant oasis in the bustle of midtown St. Louis. In addition to strengths in premed and communication, SLU has an unusual specialty in aviation science.

Website: www.slu.edu
Location: Urban
Total Enrollment: 9,464
Undergraduates: 6,541
Male/Female: 43/57
SAT Ranges: V 550–650
   M 550–670
ACT Range: 24–29
Financial Aid: 67%
Expense: Pr $
Phi Beta Kappa: Yes
Applicants: 8,105
Accepted: 78%
Enrolled: 24%
Grad in 6 Years: 75%
Returning Freshmen: 86%
Academics: 🖎 🖎 🖎
Social: ☎ ☎
Q of L: ★ ★ ★
Admissions: (314) 977-2500
Email Address:
   admitme@slu.edu

Strongest Programs:
Biology
Chemistry
Nursing
Criminal Justice/Forensic
   Science
Philosophy
Psychology
Theology

Within sight of Saint Louis's famed Gateway Arch, the historical gateway to the American West, sits Saint Louis University, the first university established west of the Mississippi River. SLU offers students many nationally recognized programs, from premed to aviation and, of course, theology. The school's academic atmosphere is shaped by its Jesuit tradition; administrators ensure that each student receives personal care and attention and expect graduates to contribute to society and lead efforts for social change. In return, students find an atmosphere where their faith is encouraged.

The SLU campus has undergone an $870 million renovation and features pedestrian walkways, lush greenery, fountains, and sculptures, as well as the signature Saint Louis University gates at all entrances. Cupples House, a beautiful old mansion in the middle of campus, houses 19th-century furniture and an art gallery—and is just a short walk from a new, modern focal point on campus, the expanded and renovated Busch Student Center. The center is home to a bookstore, copy center, eateries, lounges, and conference facilities. Numerous other renovations and additions have been completed over the last few years, including the John and Lucy Cook Hall, which doubled the size of the business school facilities; the Saint Louis University Museum of Art (SLUMA); and the Salus Center, which houses the School of Public Health. Plans are underway for a $70 million on-campus arena, which will be used by the university's basketball and volleyball teams, as well as for concerts and events. An $80 million health science research facility is also in the works.

In keeping with its strong Jesuit commitment to education in the broader sense, all SLU undergrads must complete distribution requirements in cultural diversity, world history, fine arts, literature, science, social science, mathematics, languages, and foundations of discourse. Additionally, students must take philosophy and theology courses, such as SLUVision, which integrates community service with the philosophy and theology component. The most popular majors are psychology, nursing, marketing, finance, and communications. Premed, aviation science, philosophy, and theology are also outstanding programs. SLU attracts scholars from around the globe with one of the world's most complete microfilm collections of Vatican documents. Parks College, America's first certified college of aviation, offers degree programs in aviation science where students can become professional pilots. The College of Arts and Science's meteorology program provides students with an opportunity to study with specialists in satellite, radar, and mesoscale meteorology.

"SLU students are very down to earth."

The College of Public Service offers majors in communication disorders, educational studies, and urban affairs, which encourage students to put research into action.

Two-thirds of SLU freshmen come from the top quarter of their high school class. "There is a sense of camaraderie throughout the campus, though competition within programs is evident and significant," says a theology major. Full professors teach most classes and "they are very knowledgeable, professional, and engaging," says a senior. More than 300 SLU students currently study outside of the United States. In Madrid, Spain, SLU has one of the largest and most charming of any American campus in Europe. The Micah House Program is a living-learning program integrated around themes of peace and justice—it takes its name from the Biblical prophet Micah, who spoke out against social injustice in ancient Israel.

"SLU students are very down to earth," says a junior, who adds, "They are also very focused on service." "SLU students are extremely involved in community service projects and typically have genuine concern for one another," adds a sopho-more. Many students come from private, religiously affiliated high schools. Forty-four percent of SLU students hail from the Show-Me State; the balance represents all

> **"The entire student body is much more active in service than I ever expected college students to be."**

50 states and more than 70 foreign countries. African Americans comprise 8 percent of the student body, Asian Americans 5 percent, and Hispanics 3 percent. Race relations tend to be nonconfrontational, and political correctness does not seem to be an issue for most students. That doesn't mean students aren't interested in politics, and because SLU is a Jesuit institution, human rights and abortion are prominent debates. "SLU has a big social justice emphasis and clubs such as Amnesty International are large," says one student. "However, the students are diverse in their views and opinions." The Residential Life department trains diversity advocates who serve as programmers and facilitators for the dorms; they get in-depth training about diversity, racism, and oppression. Billiken World Festival is a weeklong celebration of diversity that includes a citywide festival of African and Caribbean culture and music. The school offers 92 athletic scholarships, along with merit scholarships worth an average of $11,579.

Fifty-seven percent of students live on campus. Upperclassmen can move into spacious courtyard-style apartments, but many opt for less expensive apartments off campus. "The majority of students live on campus, which creates a fun environment in the residential halls," says a senior. "The biggest hall is for freshmen, which only makes the transition easier and creates a unique first-year experience on campus." The dining options are "constantly improving," says a junior, with an emphasis on "new restaurants and better cafeteria food." SLU has all the advantages and problems usually associated with being in

> **"There is a sense of camaraderie throughout the campus."**

the middle of a city, but most students feel safe on campus. "Campus security officers are as prevalent as white-tailed deer in a Montana forest," says a senior.

Social life at SLU includes campus events, such as movies in the Quad, dances, and Greek parties, and the plethora of restaurants and coffee shops in St. Louis, as well as movie theatres, museums, bars, sporting events, and night life. A student says, "Because [St. Louis] is an urban environment, it has a vibrant nightlife and many things college students enjoy." Road trips to Kansas City, Chicago, and nearby schools like the University of Illinois and Indiana University are also popular. Greek life at SLU—highly unusual for a Jesuit institution—claims 19 percent of the men and 15 percent of the women. Students say most parties takes place in off-campus apartments or at an off-campus fraternity house, and despite the rules limiting alcohol on campus, it's common in the apartments. Spring Fever and Fall Homecoming, both annual events, feature bands, club-sponsored booths, and vendors. True to tradition,

Social life at SLU includes campus events, such as movies in the Quad, dances, and Greek parties, and the plethora of restaurants and coffee shops in St. Louis, as well as movie theatres, museums, bars, sporting events, and night life.

SLU attracts scholars from around the globe with one of the world's most complete microfilm collections of Vatican documents.

Sunday evening Mass is usually packed with students of all beliefs, and "practically everyone" participates in community service and outreach projects.

SLU has no varsity football team, but other Billiken squads more than compensate for this deficit. A billiken was a common good-luck charm in the early 1900s. A popular sportswriter of the time said the charm resembled the then-football coach, and the name stuck. Men's and women's basketball have been highly successful, along with women's volleyball and soccer. The SLU Division I men's soccer team has won seven straight Conference USA championships and been in the NCAA quarterfinals two of the past three seasons. The most popular intramural sports include flag football, dodgeball, basketball, softball, and soccer. For weekend warriors, the Simon Recreation Center boasts a 40-meter pool, six racquetball courts, and loads of equipment.

Saint Louis University is winning students' devotion and increasing its national visibility by offering a slew of strong programs. The Jesuit education prepares students to work for a more just and humane world. One sophomore says, "The entire student body is much more active in service than I ever expected college students to be. SLU is a school that focuses greatly on its relationship with the community."

Overlaps

University of Missouri, University of Illinois, Truman State, Marquette, Loyola (IL)

If You Apply To ➤

SLU: Rolling admissions. Financial aid: Mar. 1. Housing: May 1. Meets full demonstrated need of 11%. Campus interviews: recommended, informational. No alumni interviews. SATs or ACTs: required. SAT IIs: optional. Apply to particular programs. Accepts the Common Application and prefers electronic applications. Essay question: significant experience, achievement, risk, or ethical dilemma and its impact on you; a person's significant influence on you; why Saint Louis University?

University of San Francisco

2130 Fulton Street, San Francisco, CA 94117-1080

Talk about prime real estate: USF is next door to the legendary Haight–Ashbury district, catty-corner to Golden Gate Park, and within five miles of the Pacific Ocean. Though USF is a Jesuit institution, only about half of its students are Roman Catholic. Pacific Rim studies is a stand-out.

Website: www.usfca.edu
Location: Urban
Total Enrollment: 8,568
Undergraduates: 4,796
Male/Female: 34/66
SAT Ranges: V 510–610
  M 510–610
ACT Range: 21–26
Financial Aid: 49%
Expense: Pr $ $
Phi Beta Kappa: No
Applicants: 7,105
Accepted: 72%
Enrolled: 21%
Grad in 6 Years: 67%
Returning Freshmen: 86%
Academics: ✍ ✍ ✍

In the heart of one of the nation's most liberal cities is a thriving Jesuit university that has become an integral part of its community. Instead of shunning the city's reputation, the University of San Francisco embraces it. With an incredibly diverse student body and an emphasis on programs such as nursing and business, students encounter a broad set of cultures, academic challenges in a liberal arts setting, and the chance to put all that experience to good use.

USF's 55 well-kept acres, spotted with beautiful basilica-type buildings and modern facilities, are, as one student puts it, "wedged into the heart of San Francisco." The campus stands atop one of San Francisco's seven hills, adjacent to Golden Gate Park, overlooking San Francisco Bay and the city skyline. The new enrollment center combines registrar, financial aid, and payment services in one convenient location.

Although liberal arts programs draw the most majors, there is also a strong emphasis on preprofessional programs, especially nursing, health studies, communications, and business. The Center for the Pacific Rim allows students to do interdisciplinary majors with an Asian focus, as does the Asian studies program. The 44-unit core curriculum requires students to take courses in six major categories: foundation of communication; math and sciences; humanities; philosophy, theology, and ethics;

social sciences; and visual and performing arts. The St. Ignatius Institute program offers an integrated four-year curriculum based on the great books of Western civilization presented in an unusual seminar/lecture combination. The St. Ignatius program is not restricted to top students, and participants can spend their junior year studying in Oxford, Rome, or Budapest. Some of the preprofessional majors are demanding, and students say that competition is common, but not cutthroat. "Classes are rigorous, and the university has a competitive feel, especially within the sciences and nursing programs," says a biology major. Extensive and mandatory academic advising ensures that students' courseloads are manageable. "Overall, the professors are very knowledgeable and passionate about their fields," says a sophomore. Classes average 25 students and are always taught by full professors.

> "The university has a competitive feel, especially within the sciences and nursing programs."

(Continued)
Social: ☎ ☎ ☎
Q of L: ★ ★ ★ ★
Admissions: (415) 422-6563
Email Address:
admissions@usfca.edu

Strongest Programs:
Business Administration
Hospitality Management
Psychology
Communication
Nursing
Biology
Pacific Rim Studies

The university operates on the basis of fall and spring semesters, with optional three-week courses during the five-week winter break. New majors include architecture and community design, art history and management, Internet engineering, and comparative literature and culture. There's also a joint BA/BS–JD program. The visual arts program provides courses in art education, graphic and fine art, drawing, painting, art history, and museum studies. Five computer labs are available to students, and every residence hall is equipped with high-speed Internet connections. "We attract open-minded students with diverse backgrounds," says a junior. Another undergrad says students are also "very aware of the world outside the campus." Community-minded students take advantage of volunteer programs, mostly through University Ministries. USF is also the official host of the Human Rights Watch Festival, which is integrated into the curriculum.

Community-minded students take advantage of volunteer programs, mostly through University Ministries. USF is also the official host of the Human Rights Watch Festival, which is integrated into the curriculum.

Sixty-nine percent of the students are from California, and 45 percent attended private or parochial high schools. Asian Americans account for 25 percent of the population, African Americans 5 percent, and Hispanics 14 percent. One of the university's missions is to "prepare men and women to shape a multicultural world with creativity, generosity, and compassion." A student remarks, "As a black student, I feel very comfortable at the University of San Francisco. Students from different backgrounds mix." Admissions are need-blind, and USF offers 295 merit scholarships averaging $12,679 as well as athletic scholarships in all Division I sports.

> "We attract open-minded students with diverse backgrounds."

On-campus housing ("extremely expensive but conveniently located," says a senior) is guaranteed for the first two years. After that, at least half the students brave San Francisco's rental market. On-campus students say the dining facilities offer a range of vegetarian and other choices.

USF's greatest asset is undoubtedly its location. San Francisco is a cosmopolitan city where students can take advantage of reliable public transportation, including the famous cable cars, to get to a variety of cultural attractions, ranging from Chinatown to the symphony. Nightlife is great for those who want to dance at the clubs or meet in the bars. "Social life usually takes place off campus," a sophomore says. "After all, it is San Francisco." Campus activities include the Hawaiian Club's annual luau and the Barrio Festival held by the Filipino American Club. The College Players is the oldest continuously performing college theater group in the West. Fraternities and sororities attract 1 percent of men and women, although they don't have houses. Though underage drinking is officially prohibited, the policy "doesn't really work," a senior says.

> "Social life usually takes place off campus. After all, it is San Francisco."

The university operates on the basis of fall and spring semesters, with optional three-week courses during the five-week winter break.

Varsity athletics provide a popular diversion, and USF touts a conference-winning national powerhouse in soccer (2005 conference champs). Basketball and baseball are among the most popular sports for male athletes, while the women have formed strong volleyball and basketball teams. Athletes are pleased with the health and recreation center, which touts an Olympic-size swimming pool, exercise rooms, and courts.

The core mission of USF is to use the Jesuit tradition, which views "faith and reason as complementary resources in the search for truth and authentic human development," as the backdrop for a solid liberal arts and preprofessional education. With a diverse student body that encourages social interaction and a home city that virtually demands it, students at USF make sure they have fun while they learn and grow.

If You Apply To ➤

USF: Rolling admissions. Regular admissions and financial aid: Feb. 1. Does not guarantee to meet demonstrated need. Campus and alumni interviews: optional, informational. SATs or ACTs: required (SAT preferred). ACT optional writing test required for placement only. Accepts the Common Application and prefers electronic applications. Essay question: personal statement that includes how students can contribute to the university mission.

Santa Clara University

500 El Camino Real, Santa Clara, CA 95053-1500

Popular because of its Bay Area location, Santa Clara is one of the few middle-sized universities in California that is not impossible to get into. A well-developed core curriculum keeps students focused on the basics, and Santa Clara offers engineering and business in addition to the liberal arts.

Website: www.scu.edu
Location: Suburban
Total Enrollment: 6,196
Undergraduates: 4,743
Male/Female: 44/56
SAT Ranges: V 550–650
  M 570–670
ACT Range: 24–28
Financial Aid: 42%
Expense: Pr $ $
Phi Beta Kappa: Yes
Applicants: 8,904
Accepted: 61%
Enrolled: 22%
Grad in 6 Years: 84%
Returning Freshmen: 94%
Academics: ✑ ✑ ✑
Social: ☎ ☎ ☎ ☎
Q of L: ★ ★ ★ ★
Admissions: (408) 554-4700
Email Address:
  ugadmissions@scu.edu

Steeped in history and tradition, Santa Clara University was founded with a Jesuit mission that emphasizes a commitment to academics and the community. Classes stay small and intimate, while the revised curriculum focuses on an expanding global society. The class schedule is based on quarters (10 weeks), and classes are challenging, but not too competitive. According to one senior, the environment and school "promotes group work and learning as a community."

SCU's Old World charm includes 106 acres complete with lush green lawns, palm trees, and luscious rose gardens, accented by authentic Spanish architecture. The Mission Gardens, with many olive trees, are a beautiful escape from the pressures of school. The famous classic mission church was rebuilt in 1926 in the design of the six previous churches that were destroyed by disasters ranging from fires to floods. Stephen Schott Stadium opened in 2005, while a residence hall and the Commons on Kennedy Mall—which features classrooms, meeting rooms, and lounges—opened in 2006. Current projects include a new library building.

The core curriculum, whose theme is Community and Leadership for a Global Society, is meant to give the students broad knowledge in three main themes: community, global societies, and leadership. Courses include composition, Western culture, world culture, the United States, ethics, religious studies, mathematics and the natural sciences, technology, social sciences, and foreign language. There is also an emphasis on professional programs like engineering and business. Students can opt for the 3–2 engineering program, which allows them to get a bachelor's and master's degree in five years, and the Leavey School of Business is renowned along the West Coast, with accounting, agribusiness, and retail management being particularly strong. In the College of

"Missing a couple of classes can be devastating due to the fast-paced quarters."

Arts and Sciences, psychology remains popular, as do biology, communications, and English. The Combined Sciences program allows students who desire a broad curriculum to include courses from both the natural and social sciences. The newest program is a general engineering concentration in bioengineering.

For those students looking for more of a challenge, the honors program places 45 to 50 selected freshmen in special classes, and an endowed scholarship sponsors one student's junior year at Mansfield College, Oxford University. Also, an extensive study abroad program, in which approximately one-third of the students participate, allows travel and study in Europe, Central and South America, the Caribbean, Canada, Africa, Asia, New Zealand, and Australia.

As at any small school, resources and facilities are limited. Students used to report that the library is not always adequate for research needs, forcing many to go to nearby Stanford to get the materials they need; however, a portion of a $350 million fund-raising campaign will be donated to a new library. Students complain that finding a seat in some of the classes is a challenge. "Communications is the biggest major and therefore

"There are always activities which bring a wide array of students together."

they are the hardest classes to get into," says one student. But a senior explains that if your first choice isn't available "there is always a required course available to take instead." Bottom line: Stay flexible and you'll likely graduate on time.

Small classes taught by full professors mean "the quality of teaching is good and professors are really there to help the students rather than do research," says a senior. "They truly have a passion for students and want to see us succeed," adds a sophomore. The academic climate is described as challenging. "Students truly have to work for their grades and there is no faking it," warns a freshman. "Missing a couple of classes can be devastating due to the fast-paced quarters."

More than half the students are Roman Catholic, and religion, while not intrusive, is a factor in many aspects of campus life. Campus ministry provides counseling and opportunities for spiritual development, and many students are active in local volunteer organizations. "Most SCU students are white, upper-class prep school kids from California," says a junior. Fifty-seven percent of the student body is comprised of undergraduates from California, and the rest are from the West Coast or at least the West—Oregon, Hawaii, Washington, and Arizona. Split fairly evenly between parochial and public schools, 72 percent of the students are from the top quarter of their graduating class. The diversity on campus is impressive; 18 percent of the students are Asian American, 3 percent are African American, and 13 percent are Hispanic. The Santa Clara Community Action Program, a student volunteer outreach program, provides community service opportunities for the entire campus. "There are clubs on campus concerned with social justice, like SCAP and the GREEN club, which send volunteers to help out in the community," says one freshman. A variety of academic and athletic scholarships are available to those who qualify.

Since housing is guaranteed to nearly all who apply, "housing can be a pain when trying to find where you'll live next year," says a senior. Almost all freshmen and sophomores live on campus, but soon they are packing up and heading for apartments. Most residence halls are co-ed by floor. "Housing here is good because it caters to the partiers, the studiers, and everyone in between," says one student. The Freshman Residential Learning Community program gives incoming students the opportunity to room with students who share similar academic or social interests. All students, including commuters and those living off campus, belong to RLCs and take courses with other members of the group. In all, there are nine RLC programs with emphases such as Italian arts and culture, environmental sustainability, diversity, and social justice. Dining is described as "adequate" and security is rarely an issue. "We live in a protective bubble," says one freshman.

(Continued)
Strongest Programs:
Communications
English
Biology
Psychology
Finance
Marketing
Mechanical Engineering

Santa Clara sports compete in and with the Division I teams and players that are among the best in the nation.

More than half the students are Roman Catholic and religion, while not intrusive, is a factor in many aspects of campus life.

Greek organizations no longer exist at Santa Clara, but don't let this fool you: Social activity is alive and well. With such beautiful weather and great California locations and events nearby, how could anyone resist partying and road trips? "There are always activities which bring a wide array of students together," says one student. Only 20 miles away is Santa Cruz, for those who want to bask in the sun. San Francisco is 45 minutes away and other short road trips to popular hot spots include Napa Valley, Monterey, and Palo Alto. Those who chose to party on or near campus can take advantage of the school's attempt at enforcing "no drinking and driving" by providing the students with transportation called the Bronco Bus. "Underage drinking is not tolerated," warns an accounting major. Alternate forms of leisure activity include a fitness center that is open for long hours and The Bronco, a sports and recreation bar where students over 21 can drink beer or wine.

"Underage drinking is not tolerated."

Santa Clara sports compete in Division I. For two consecutive years, the women's basketball team won the West Coast Conference championship. The women's soccer team has also brought home conference and national championship trophies. The women's volleyball team made the NCAA Final Four in 2005 as an originally unseeded team.

Santa Clara University is a warm place—in every sense of the word—and a comfortable and beautiful setting where morality and ethics are infused into the curriculum of strong academics. It's this blend of traditional values and progressive academics that turns out people who want to make a difference in the world.

Overlaps

UC–Davis, UCLA, UC–San Diego, UC–Berkeley, UC–Santa Barbara, University of Southern California

If You Apply To ➤

SCU: Rolling admissions. Regular admissions: Jan. 15. Financial aid: Feb 1. Housing: Guaranteed for freshmen. Does not guarantee to meet full demonstrated need. Campus interviews: recommended, informational. Alumni interviews: optional, informational. SATs or ACTs: required (SATs preferred). SAT IIs: optional. Accepts the Common Application and electronic applications. Essay question.

Sarah Lawrence College

1 Mead Way, Bronxville, NY 10708-5999

The free-spirited sister of East Coast alternative institutions. Though SLC is co-ed, women outnumber men nearly three to one. Strong in the humanities and fine arts with a specialty in creative writing. Full of quirky, head-strong intellectuals who hop the train to New York City every chance they get.

Website:
  www.sarahlawrence.edu
Location: Suburban
Total Enrollment: 1,502
Undergraduates: 1,266
Male/Female: 26/74
SAT Ranges: V 620–710
  M 550–640
ACT Range: 25–30
Financial Aid: 50%
Expense: Pr $ $ $ $

Sarah Lawrence College attracts creative, highly motivated individuals who are both critical thinkers and devotees of independent learning. They love literature and the fine arts and take pride in their academic prowess. Indeed, freedom and exploration are valued more highly than any tradition here, save perhaps May Fair. And although some students lament SLC's move toward the mainstream, they also appreciate the things that haven't changed on campus, such as the emphasis on close, personal attention from faculty members. "All classes are capped at 15 students," a writing major explains. "Every student has equal access to amazing professors, who often become surrogate parents and mentors for years to come."

Founded in 1926, Sarah Lawrence sits on a quaint, 40-acre tract in Bronxville, a wealthy Westchester County community, where even the public library boasts Oriental rugs and fireplaces. "Bronxville consists mostly of little old ladies with dogs in designer raincoats," quips a junior. On campus, the prevailing architectural theme is English Tudor, including mansions from converted estates, but more modern structures are present as well. The landscape is hilly and green, with more than a hundred types of trees and abundant rock outcroppings. Because the school's founders believed that there should be as little physical separation as possible between life and work, classrooms, dormitory suites, and faculty offices are all housed in the same ivy-covered buildings. A 60,000-square-foot Visual Arts Center opened in 2004 and renovations to Bates Hall bring all aspects of student life together in one place, including an art gallery, student lounge, dining hall, and Office of Student Affairs.

General education requirements at Sarah Lawrence include 120 credits in at least three of four academic areas, leaving lots of room for students to dabble in whatever strikes their fancy. "There is no math or science requirement, so I'll never have to take them again!" crows one. In fact, it's tough to find two Sarah Lawrence students studying the same thing, because every student designs his or her own program of study, and almost no subject is out of bounds. And, since

"All classes are capped at 15 students."

there are no mandatory or required courses, competition is virtually nonexistent, says a senior: "Competition in the classroom is rare. However, the amount of work we have per class is extensive so there is stress." Though there are formal grades, more important is the student's portfolio of work, accompanied by in-depth, written evaluations from professors, filed twice a year. Perhaps not surprisingly, SLC no longer requires students to submit SAT scores as part of the admissions process.

Regardless of what they focus on, all students become intimately acquainted with the written word; writing begins in the first year at SLC, and continues relentlessly "across the curriculum" for the next three. Courses at Sarah Lawrence "are very rigorous, but the climate is pretty laid-back," says a history major. That's because everyone takes only three courses per semester. Still, professors meet with their students weekly or biweekly, in a system modeled after Oxford University's tutorials, so there's no time to slack off—or fall behind. To ease the transition to college, all first-years take a First-Year Studies seminar. Typically, more than 30 subjects are available, and "you pick the topic before school starts," says one student. "The teacher of the class will be your don, or advisor, for the next four years—so choose wisely!" Courses can last just one semester or for the full year.

Perhaps because of SLC's emphasis on personal relationships with professors, even the registration process requires deep thought: Students interview teachers to ensure that courses fit into their academic plans, and to confirm the instructor is someone they respect and want to study with. And although getting into popular classes can

"There is no math or science requirement, so I'll never have to take them again!"

be a problem, administrators guarantee students at least two of their first three choices each semester. "I cannot imagine better teachers than the ones I have had here—they are immensely talented, involved in their fields, usually brilliant, and often have amazing connections for internships and jobs," says a writing major. "It seems about a third of the campus does theater, which is also amazing—I would go into class, only to find my prof had been on *Law and Order* the night before." Study abroad programs are offered in Florence, Paris, Oxford, London, and Havana.

The most popular concentrations at Sarah Lawrence include literature, history, psychology, writing, and visual arts. Aspiring psychologists—also a significant group on campus—may participate in fieldwork at the college's Early Childhood Center. Historians benefit both from the college's proximity to New York City and from expanded offerings in filmmaking and film history. "Students here are, by nature, very eclectic and our academic program reflects that." Weaker departments include Russian, Japanese, and German, administrators say, since each has only one professor.

(Continued)
Phi Beta Kappa: No
Applicants: 2,634
Accepted: 45%
Enrolled: 32%
Grad in 6 Years: 74%
Returning Freshmen: 91%
Academics: ✍ ✍ ✍ ✍
Social: ☎ ☎
Q of L: ★ ★ ★
Admissions: (914) 395-2510
 or (800) 888-2858
Email Address: slcadmit
 @sarahlawrence.edu

Strongest Programs:
History
Literature
Psychology
Writing
Visual and Performing Arts

SLC competes in the Hudson Valley Conference, and the women's swim team has brought home conference titles in recent years.

The premed program, more structured than other offerings, places nearly all eligible graduates into medical school. Though science majors are few and far between, those who focus on biology, chemistry, and physics have access to a state-of-the-art science center, with 22,500 square feet of classrooms, lab benches, and computer technology. Though its holdings are small—just 224,000 volumes—the charming library takes the sting out of studying, with an area for eating and a pillow room for occasional naps.

Sarah Lawrence students are not your run-of-the-mill adolescents. "We're very liberal, but open-minded," says one. "Bush-hating is near-universal, and the school provides vans to antiwar protests." Twenty-three percent are natives of New York State—the bulk from nearby New York City—and minorities account for 14 percent of the student body; 5 percent African American, 4 percent Asian American, 4 percent Hispanic, and 1 percent Native American. As for hot topics on campus, "the gay and lesbian issue has quieted down considerably," says a junior. "Though this is still a very supportive environment, it's just not discussed as much, or as visible, as it used to be." There are no merit or athletic scholarships.

Eighty-four percent of Sarah Lawrence students live on campus, where "dorms are excellent for everyone who is not a freshman," says a sophomore. "If you ask for nonsmoking and quiet housing, you're pretty much guaranteed an amazing room," agrees a junior. "If you don't, you may end up in the one bad dorm, with three people per room and constant noise. It's gross." Upperclassmen may get to live in one of four townhouses, each with seven

"I cannot imagine better teachers than the ones I have had here."

single rooms. Students on a budget often commute from Westchester County's lower-rent districts, or from their parents' homes in New York City. Dining services caters to the quarter of SLC students who are vegetarians, but the meal plan is "ridiculously overpriced," students say. The student center's greasy spoon is also an option, and most dorms have their own kitchens as well.

When the weekend comes, SLC has a modicum of social life—free dances and movies, plays, poetry readings, guest lectures, and tea and coffeehouses—but most students head south to New York City, just a half-hour away by train. Theater fans and aspiring actors flock to discounted Broadway shows, and clubs, bars, museums, and concert halls also beckon. Drinking isn't such a big deal here; campus policies require hosts that serve alcohol to register. "For the most part, campus policies work," says a student. Favorite traditions include May Fair, the annual Coming Out Dance ("everyone drinks copious amounts of alcohol, then stumbles down the hall for near-naked dancing," says a junior), the "actually interesting" sex education program called Sleaze Week, and midnight breakfast, served during the last week of each semester.

Sarah Lawrence competes in the Hudson Valley Conference, and the women's swim team has brought home conference titles in recent years. The intramural program revolves around invitational events—road races, basketball and squash tournaments, fitness challenges—rather than league play. The men's basketball and women's equestrian teams attract the most fans. And if you can't find the team or activity you want? "The school is so generous with money," says one student. "Start a club, propose it to Student Senate, and you'll get $2,000 for whatever project you choose."

Sarah Lawrence offers a close-knit community for writers and artists, in a lush setting just outside the hustle and bustle of Manhattan. "If you want an intensely creative, small liberal arts college with amazing professors and opportunities, and you're okay without a large party scene, then this is very likely the best place for you," explains a junior. One caveat, though—all that personal attention doesn't come cheap, and admissions at SLC are need-aware. "Assume your financial aid will decrease, and the tuition will increase, by at least a few thousand dollars each year," a writing major says.

Scripps College: See page 145.

Skidmore College

815 North Broadway, Saratoga Springs, NY 12866

Founded in 1903 as the Young Women's Industrial Club of Saratoga, Skidmore College still excels in the fine and performing arts that were then deemed proper for ladies. Little else remains the same. Compare to Vassar, Connecticut College, and Wheaton (MA).

Skidmore College serves up solid academics with a decidedly nontraditional flair. Although these politically liberal and free-spirited students complain about the cold weather and frosty relations with the conservative locals, Skidmore students are a happy lot. Classes are small, faculty members are available, and "the college has a charm, sort of like a summer camp," says a junior.

In 1961, as enrollment surpassed 1,300 and many of Skidmore's turn-of-the-century Victorian buildings grew obsolete, Skidmore traded its campus in the heart of Saratoga Springs for 750 acres on the northwest edge of town. Since then, the campus has grown to more than 50 buildings on 890 acres, and the student body has doubled in size (and welcomed men). While contemporary in style, the buildings on Skidmore's Jonsson campus—named after the donor who made it possible—are human in scale, and their aesthetic features reflect the Victorian heritage of the school's original Scribner campus. Covered walkways connect the residential, academic, and social centers, and the prevailing views are of surrounding mountains, woods, and fields.

In 2005, Skidmore replaced its 20-year-old Liberal Studies program with the new First Year Experience. FYE includes a class-wide summer reading project—this year's was Gregory Howard Williams' *Life on the Color Line*—and the selection of one of 50 Scribner seminars. Each seminar is capped at 15 students, and taught by a professor who also becomes the mentor and advisor for that group. Seminar topics are broad and varied, ranging from Molecular Frontier to Post-Wall German Cinema to Hard Times in the Big Easy. Students in each seminar live near one another, and themes raised in the summer reading crop up again during the year in campuswide programming.

> "The college has a charm, sort of like a summer camp."

The most popular majors at Skidmore are English, management and business, psychology, art, and government—not coincidentally, students say these are some of the college's best programs as well. Biology majors may conduct fieldwork in the marsh at the northern end of campus. Through the Hudson-Mohawk Association of Colleges and Universities, students may also take courses at most other colleges

Website: www.skidmore.edu
Location: Small city
Total Enrollment: 2,691
Undergraduates: 2,637
Male/Female: 42/58
SAT Ranges: V 580–670
  M 580–670
ACT Range: 26–29
Financial Aid: 43%
Expense: Pr $ $ $ $
Phi Beta Kappa: Yes
Applicants: 6,032
Accepted: 48%
Enrolled: 24%
Grad in 6 Years: 85%
Returning Freshmen: 94%
Academics: ✑ ✑ ✑ ✑
Social: ☎ ☎ ☎
Q of L: ★ ★ ★
Admissions: (800) 867-6007
Email Address:
  admissions@skidmore.edu

Strongest Programs:
Drama
Studio Art
English
Biology
Government

nearby. There's also a cooperative program in engineering with Dartmouth, a Washington semester with American, a semester at the Marine Biological Laboratory in Woods Hole, Massachusetts, a master of arts in teaching with Union, an MBA program with Clarkson, and an accelerated JD program with Cardozo Law School at New York's Yeshiva University.

Skidmore augments liberal arts and sciences offerings with pre-professional training. Studio art and art history majors frequently enroll in an introductory business class, where they work with business majors and make presentations to operating executives. Another much-praised option is junior year abroad, especially through Skidmore-run programs in France, England, Spain, India, or China. Some 50 percent of students spend at least one semester abroad. Students who just can't get enough time on campus may compete for summer grants to fund research with the "very accessible" professors, a history major says. "Classes are usually discussions rather than lectures." And speaking of those classes, "they aren't easy, though they're not impossible, either." "There's enough work to keep us busy, but kids aren't particularly stressed out; we have enough time to relax," says an economics major. "Everyone supports one another," a theater major adds.

"Classes are usually discussions rather than lectures."

Given Skidmore's hefty price tag—only a bit less than Harvard's—students are typically well-off; they hail primarily from New York, New Jersey, and New England. Asian Americans constitute 7 percent of the student body, Hispanics 4 percent, African Americans 3 percent and Native Americans 1 percent. "People are friendly as long as you initiate conversation," says a business major. "Students are either preppy jocks or hippie Frisbee players." Hot topics include the rights of women, minorities, and gays and lesbians, though a junior says, "It's kind of useless since all the kids are quite liberal." Athletic scholarships are not offered, though there are $10,000 awards for academic merit—four annually in music and five annually in math and science.

Eighty-three percent of Skidmore students live in the dorms, where rooms are guaranteed, and most students get singles after freshman year. Dorms are integrated by class and co-ed by floor or suite, with kitchenettes and lounges on every floor. "There are some dorms that are fabulous, and others that are just OK," says a psychology and women's studies major. Most buildings have carpet, air-conditioning, and cozy window seats. Juniors and seniors typically get single rooms in the centrally located residence halls, or move to apartments—whether on campus in Scribner Village, off campus in Saratoga Springs, or in Skidmore's new Northwoods Village apartments. The Murray-Aikins dining center, which opened in 2006, provides students with fresh food choices in a state-of-the-art facility.

"Students are either preppy jocks or hippie Frisbee players."

With no fraternities or sororities, Skidmore students flock to concerts, lectures, dances—and dorm parties, especially if there's live music. "Students are very easygoing, and they drink whenever they get a chance," says a freshman. The school's three-strikes policy means fines don't kick in until the third time an underage student is busted with alcohol, causing some students to call the policy "a joke." The best road trips include Albany, New York City, Boston—and especially Montreal, where you don't need a fake ID to drink at 19. Every year there are also two or three formal dances, often in the elegant and formal Hall of Springs, plus "Fun Day" in the spring, with games and an inflatable obstacle course on the college green.

The nearby Adirondacks make Skidmore a haven for backpackers, skiers, and members of the popular Outdoors Club, while the old resort town of Saratoga Springs, with its healing waters and antique shops, offers plenty of culture, including the Saratoga Performing Arts Center and the country's oldest thoroughbred racetrack. Saratoga is also the summer home of the New York City Ballet, the New York

With no fraternities or sororities, Skidmore students flock to concerts, lectures, dances—and dorm parties, especially if there's live music.

Given Skidmore's hefty price tag—only a bit less than Harvard's—students are typically well-off.

City Opera, and the Philadelphia Orchestra. Students reach out to the community through BenefAction, a volunteer group connected to several local agencies and schools. Skidmore's more traditional activities, which have continued even after a quarter-century of coeducation, include Junior Ring Week, when juniors receive their class rings and a dance is held in honor of their initiation.

Skidmore's 19 men's and women's varsity teams compete in NCAA Division III; men's and women's tennis and the equestrian team have claimed several national championships in recent years. Intramurals are also popular, especially basketball and racquetball. The program has more than 20 teams and a student commissioner, and teams can duke it out on 250 acres of newly renovated playing fields. Varsity athletes and weekend warriors alike enjoy the 400-meter, all-weather track and the athletic center, which includes a fitness center and locker rooms.

Skidmore continues to win the hearts of motivated students with gorgeous scenery, caring faculty, and its flexibility, openness, and receptivity to change and growth. Students here are also a bit quirky, says a sophomore, "wearing shorts in the winter, for example." They're more likely to cheer on the fall of a foreign dictator than a goal by the lacrosse team. The point is, there's room for—and encouragement of—all types. Says a women's studies major, "The campus is beautiful, and the professors and students actually create bonds that last a lifetime."

> Students reach out to the community through BenefAction, a volunteer group connected to several local agencies and schools.

Overlaps

Vassar, Connecticut College, Hamilton, Wesleyan, Colgate, Tufts

If You Apply To ➤

Skidmore: Early decision: Nov. 15, Jan. 15. Regular admissions: Jan. 15. Meets demonstrated need of 91%. Campus interviews: recommended, evaluative. No alumni interviews. SATs or ACTs: required. SAT IIs: recommended. Accepts the Common Application and prefers electronic applications. Essay question: significant experience; important issue of personal, local, or national concern; important person; or significant fictional or historical character.

Smith College

College Lane, Northampton, MA 01063

The furthest left-leaning of the nation's leading women's colleges. Liberal Northampton provides big-city social life, and the Five College Consortium adds depth and breadth all around. With a total enrollment of more than 3,000, Smith is the biggest of the leading women's colleges, and the only one to offer engineering.

Heaven only knows what Sophia Smith would think of the women's college she founded in 1871 with the hope it would be "pervaded by the Spirit of Evangelical Christian Religion." There are still evangelicals at Smith, but today they join the rest of their schoolmates in crusading against societal "isms" such as racism, classism, sexism, and heterosexism. Though the all-female school remains strongly committed to its liberal arts mission, it is also focused on placing women at the forefront of science and technology. Students here have the opportunity to become leaders in the male-dominated field of engineering, or pursue interdisciplinary fields such as landscape studies.

Smith is in the small city of Northampton, an artsy oasis in the foothills of the Berkshire Mountains. The 125-acre campus resembles a medieval fortress from the front gate, but inside it sparkles with many gardens, Paradise Pond, and a plant house. Buildings cover a range of styles from late-18th-century to modern, and the college has successfully retained its historic atmosphere while keeping facilities up to date. In 2003, the college celebrated the openings of the $35 million Brown Fine Arts Center and Smith's first campus center, a 60,000-square-foot, $23 million facility that has

Website: www.smith.edu
Location: Small city
Total Enrollment: 3,093
Undergraduates: 2,642
Male/Female: 0/100
SAT Ranges: V 580–710 M 570–670
ACT Range: 25–31
Financial Aid: 60%
Expense: Pr $ $ $
Phi Beta Kappa: Yes
Applicants: 3,408
Accepted: 48%
Enrolled: 37%

(Continued)

Grad in 6 Years: 86%

Returning Freshmen: 89%

Academics: 🐿🐿🐿🐿½

Social: ☎ ☎ ☎

Q of L: ★ ★ ★ ★

Admissions: (413) 585-2500
or (800) 383-3232

Email Address:
admission@smith.edu

Strongest Programs:
Government
Art
Psychology
Biological Sciences
Economics

become the focal point for campus life. The $4 million Olin Fitness Center opened in 2004, and ground will be broken in 2007 on a $65 million molecular science and engineering building.

Be ready to hit the books with your newfound sisters at Smith. "You learn a ton in a short period of time, but it is usually interesting, so academics tend to be more fun than grueling," a government major says. "It's competitive and courses are difficult, but the professors are very helpful," an art history major adds. Smith's student-run honor system, which covers everything from exams to library checkout, is widely praised and enforced. Students generally refrain from discussing grades, choosing instead to focus on helping one another.

Government is the most popular major on campus, followed by art, psychology, economics, and English. One in four Smith women majors in science and thereby enjoys numerous opportunities to assist professors with their research. The STRIDE program allows freshmen and sophomores to become paid research assistants to professors. Science students also benefit from a spacious, state-of-the-art science center. Two electron microscopes are available for student use, and, through the Five College Consortium,* students have access to one of the best radio astronomy facilities in the world. Smith's art history department is among the best in the nation and enjoys access to the college's superb museum. In addition, newer majors are available in Italian studies and East Asian studies, and there are minors in digital art and digital music within the computer science program. Students cite sociology as a program that many avoid.

> "Academics tend to be more fun than grueling."

With the exception of at least one writing course, Smith women have unusual freedom to plan a course of study. They must take half of their credits outside of their major, and first-year students can take small seminars on topics such as "Biography in African History." Qualified students may enter the Smith Scholars program and embark on one or two years of independent or extra college research for full credit. About 230 older students are enrolled in the Ada Comstock Scholars program for women going back to college. The Picker Program in Engineering and Technology—which graduated its first class in 2004—is the first of its kind at a women's college. The program offers students the opportunity to pursue an ambitious engineering program taught within the full depth and breadth of the liberal arts. Administrators hope the curriculum will lead to greater gender parity in engineering. Members of the first graduating class headed to prestigious graduate programs at Harvard, MIT, Cornell, Princeton, and other colleges, and two received highly competitive National Science Foundation fellowships. Others were quickly commandeered by employers.

> "There is never a lack of nightlife."

All courses are taught by full professors, and students seem to be pleased with the quality of teaching and their access to the professors. "Professors know a lot about the areas in which they teach and really push the students to get into the material," a sophomore says. Smith's four libraries have almost two million holdings among them, making it one of the largest collections of any liberal arts college in the country. Students can study for a semester or two at one of 12 well-known New England colleges through the Twelve College Exchange Program,* or take advantage of the innovative Maritime Studies Program.* Students are also enthusiastic about the opportunity to take part in Smith's well-known study abroad program, which usually sends half the junior class abroad for at least a semester in a number of countries. The Praxis program lets each student participate in at least one summer internship funded by the college.

One in four Smith women majors in science and thereby enjoys numerous opportunities to assist professors with their research.

Sixty-one percent of Smith first-years ranked in the top quarter of their high school class. Seventy-nine percent of Smithies are from out of state or abroad.

African Americans account for 6 percent of the student body, Asian Americans 11 percent, Hispanics 6 percent, and Native Americans 1 percent. With an endowment of more than a billion dollars, Smith has deeper pockets than many of its competitors. And though it's got a hefty price tag, the school manages to recruit a diverse group of women who aren't afraid to say what they think. "Students are passionate, involved, and committed to social change," a government major says. Nobody disputes that Smith is a liberal place, with the social issues of the day dominating conversations, though some students are surprised to find themselves in such a freewheeling atmosphere—one where

"Students are passionate, involved, and committed to social change."

potentially divisive topics such as lesbianism are nonissues. Several students say budget cuts have forced a few changes to be made, including consolidated dining halls, a reduced kitchen staff, and fewer courses.

Housing at Smith, which consists of houses, not dorms, is unabashedly adored, and home to 97 percent of students. "The houses are unique and not institutional like dorms," one student says. "The rooms are large with lots of natural light." Students note that they often eat breakfast in their pajamas and that the bathrooms are probably much cleaner than if there were men around. Each of the 35 houses, accommodating from 13 to 100 students, is a self-governing unit, responsible for everything from visiting hours to weekend parties and concerts. Accouterments in each house include a living room, a TV room, and a study room, many with a fireplace and a grand piano. The head resident, selected by the administration, is the leader of the house, but the elected house council runs day-to-day affairs.

Professors are often invited to Thursday dinners, which are served family-style, by candlelight, to add a touch of graciousness midweek.

The atmosphere is less that of a sorority than of an extended family. Except for one senior house, classes are mixed in each house, and first-year students easily mingle with seniors. Incoming students can indicate a preference for size and location of their first house, and changes are possible by entering a lottery. All undergraduates except for Ada Comstock Scholars must live on campus—a bone of contention among some juniors and seniors. Two alternatives offered are a vegetarian cooperative and an apartment complex. A house system is also used for the dining halls, and the food is highly praised. "There is a house that always has pasta, an Asian-themed house, Mediterranean, comfort food and even a house that does breakfast for dinner," a student says. Professors are often invited to Thursday dinners, which are served family-style, by candlelight, to add a touch of graciousness midweek.

You will not be greeted with a rocking social scene at Smith, but there are plenty of parties to be had and great places to visit. "There is always something to do here," a senior says. "It depends on what you want to make of it." Meeting men (or non-Smith women) is made easier by the five-college system. In addition, each house throws an average of two parties a semester. For special weekends, a whole fraternity may be invited from Dartmouth or another nearby college, an arrangement that is only slightly more civilized than the typical college bar scene. Students must be 21 to drink alcohol at campus parties; IDs are checked and

"The houses are unique and not institutional like dorms."

hands are stamped. Students say the alcohol policies are getting stricter. There is a free bus service to the other four members of the consortium, which among them offer a broad range of social and cultural opportunities.

Northampton, known as NoHo after New York City's SoHo neighborhood, is a college town of about 30,000 that is known for funky bohemianism. The town is home to multiple subcultures, and is generally tolerant of everyone. "Northampton is one of my favorite places," says a senior. "It's small and artsy, has two independent movie theaters, multiple venues for music and dance, a dance club, bowling alley, and a lot of great restaurants. There is never a lack of nightlife." Service Organizations of Smith (SOS) arranges for students to volunteer in about 600 placements in

Northampton, the surrounding communities, and on campus. Smith also offers time-honored traditions like Mountain Day, when the president cancels class for a day of hiking and female bonding, complete with brown-bag lunches. The New England countryside has numerous special charms, including ski slopes only an hour away. The best road trips are to Boston (two hours) or New York City (three hours).

Smith has a long tradition of success in athletics; the college was the first women's college to join the NCAA, and still places a premium on recruiting scholar-athletes. Top teams include crew, softball, soccer, equestrian, and skiing. The rowing and volleyball teams have brought home recent championships. Smith's multimillion-dollar sports complex includes indoor tennis and track facilities, a six-lane swimming pool, and a riding ring. Interhouse competitions include everything from kickball to inner tube water polo to rugby.

The strict evangelism is gone, and today's Smith women are far from Sophia Smith wannabes. But her namesake and spirit lives on at this eclectic, open-minded institution where women don lab coats, power suits, combat boots, and even white dresses at graduation. This "community of close, intelligent, interesting and compassionate women" readies them to be and do just about anything.

Overlaps

Mount Holyoke, Wellesley, Bryn Mawr, Brown, Harvard, Wesleyan

If You Apply To ➤

Smith: Early decision: Nov. 15 and Jan. 2. Regular admissions, financial aid and housing: Jan. 15. Guarantees to meet full demonstrated need. Campus and alumnae interviews: recommended, evaluative. SATs or ACTs: required. SAT IIs: optional. Accepts the Common Application and electronic applications. Essay question: Common Application questions.

University of the South (Sewanee)

735 University Avenue, Sewanee, TN 37383-1000

Sewanee is like a little bit of Britain's Oxford plunked down in the highlands of Tennessee. More conservative than Rhodes and Davidson, Sewanee is a guardian of the tried and true. Affiliated with the Episcopal Church, Sewanee still draws heavily from old-line Southern families.

Website: www.sewanee.edu
Location: Village
Total Enrollment: 1,496
Undergraduates: 1,410
Male/Female: 47/53
SAT Ranges: V 588–670
   M 570–660
ACT Range: 25–29
Financial Aid: 46%
Expense: Pr $ $
Phi Beta Kappa: Yes
Applicants: 2,027
Accepted: 66%
Enrolled: 31%
Grad in 6 Years: 82%
Returning Freshmen: 88%

Tradition is revered at University of the South, known simply as Sewanee after the school-owned village where it's located. Leonidas Polk, an Episcopal bishop and later a Confederate general, founded the school in 1857, envisioning it as a distinguished center of learning in the region. When Sewanee's cornerstone was destroyed during the Civil War, Anglican parishes in England gave money to restart the school, and Oxford and Cambridge donated the library's first volumes. Sewanee opened again in 1868, with nine students and four professors. Though some traditions remain alive and well, students say the school is modernizing and some traditions are disappearing as the school emphasizes a broader national appeal. "Previously totally removed from the rest of the world, Sewanee has had to upgrade, crack down, and modernize," a sophomore says.

Sewanee is located atop Tennessee's Cumberland Plateau, between Chattanooga and Nashville. "The atmosphere on the mountain is unparalleled," a sophomore says. Stately English Gothic buildings are carved from beige-and-pink sandstone native to the region, and each has plenty of space, as the school spreads out over a 10,000-acre plot fondly known as "the Domain." Particularly noteworthy structures are St. Luke's and All Saints Chapel, and Convocation Hall, built in 1886. The campus has strong ties to the Episcopal Church, even calling its semesters Advent and Easter, and the student body is overwhelmingly Christian. Recently completed construction includes a new studio art building. Gailor Hall, once a dorm and dining facility, was recently converted into a center for languages and literature, with offices for the *Sewanee Review*—the oldest continuously published literary quarterly in the U.S.— and the Sewanee Writers' Conference.

To graduate, all Sewanee students must take at least 32 courses, achieve a GPA of at least 2.00, and spend at least four semesters in residence, including both semesters of the final year. They are also required to take at least 21 courses outside their major, including English and one other course in literature or English and a writing-intensive course; foreign language at the third-year level or higher; math; lab science; history; religion or philosophy; anthropology, economics, or political science; fine arts; and physical education. The First Year Program attracts two-thirds of the freshman class with small seminars and academic advising that help ease the transition to college. In keeping with European tradition, Sewanee seniors must also pass comprehensive exams in their majors to earn their diplomas. While students are tested, friends decorate their cars to celebrate.

> "Previously totally removed from the rest of the world, Sewanee has had to upgrade, crack down, and modernize."

Sewanee's English department is nationally recognized, thanks in part to a bequest from playwright Tennessee Williams. The sciences are also strong, especially variations on environmental studies. The school's unique natural resources department focuses on geology and forestry, and offers a cooperative master's program with Duke and Yale. Premed and preprofessional programs are also highly regarded; about nine of every 10 students applying to medical, dental, and veterinary schools in the last decade have been admitted. A new major—international and global studies—replaces several former programs, and weaker departments include Russian and Japanese, as each has only two professors. One side note about pets on campus: The dogs and cats you may see running around are said to be the reincarnated souls of deceased professors.

Speaking of those professors, most wear black academic gowns when they teach, as do members of Sewanee's signature honor society, the Order of the Gownsmen. Administrators say students and professors voluntarily observe these traditions to demonstrate their commitment to teaching and learning. And that's not all. Typically, women wear dresses or skirts to class, and men wear jackets and ties. Sewanee also takes its honor code very seriously. Viola-

> "Plain and simple, academics at Sewanee are tough."

tions—such as lying, cheating, or stealing—usually result in expulsion. Sewanee is "steeped in tradition that almost all students abide by, no matter how archaic," a sophomore says. Not everything here is so serious, though. "Plain and simple, academics at Sewanee are tough," says a senior. "However, students do not compete with others, it's about proving to yourself that you are a dedicated student."

Sewanee kids are "preppy, laid-back, smart, fun, and not prissy at all," says an art history major. Less than a quarter of Sewanee's students are Tennessee natives, though many of the rest come from the Southeast. Southern culture is strong here and the atmosphere can be quite familial—more than a quarter of entering freshmen are legacies. Minorities have a small but growing presence on campus, with African Americans making up 4 percent of the student body, Asian Americans 2 percent, and Hispanics 2 percent. Sewanee is attempting to boost diversity with financial aid packages and merit awards, including the Tutu Scholars program for students from South Africa. Each year, the school hands out academic scholarships averaging $11,581 each, but no athletic awards. Students who maintain a 3.0 cumulative GPA may have the loan portion of their financial aid award replaced with grant money.

(Continued)
Academics: 🕮 🕮 🕮 🕮
Social: ☎ ☎ ☎
Q of L: ★ ★ ★ ★
Admissions: (800) 522-2234
Email Address:
admiss@sewanee.edu

Strongest Programs:
English
History
Psychology
Mathematics
Forestry and Geology

Typically, women wear dresses or skirts to class, and men wear jackets and ties.

The First Year Program attracts two-thirds of the freshman class with small seminars and academic advising that help ease the transition to college.

Ninety-four percent of Sewanee students live in the dorms, and the most sought-after bunk is Humphreys Hall, which houses 119 students from all classes in singles, doubles, and suites. And it's air-conditioned, too. "Although the dorms are old, they are in great shape," says a student. Language houses are also available. McClurg Dining Hall serves a wide variety of food and accommodates students' requests, students say. Still, some students complain about being forced to buy a meal plan as long as they live on campus.

Greek life is a big deal here, with 70 percent of the men and 68 percent of the women signing up. "Sewanee students know how to have a good time," says one student. "The majority go out on Thursday, Friday, and Saturday nights," often to fraternity parties, which are "widespread and wild." Indeed, drinking is a fact of life, even though it's against the law for anyone under 21. "Alcohol policies are enforced in the dorms and glass bottles are a huge no-no," says a senior. Annual Fall and Spring Party Weekends draw alumni and friends back to campus, and students also enjoy the Shakespeare Festival and blues fest. Popular road trips include Atlanta, Nashville, and Chattanooga, so it helps to have a car. Nearby lakes, waterfalls, and caverns also offer rafting, hiking, camping, and other active day trips.

Sports are popular at Sewanee, where virtually no one gets cut from varsity squads because they compete in Division III. The most popular sport on campus is probably football—not so much because the team is any good, but because games are important social events, where everyone shows up in coats, ties, and dresses (presumably, not all at once). And then there's the cheer: "Sewanee, Sewanee, leave 'em in the lurch. Down with the heathens and up with the Church.

> **"Sewanee students know how to have a good time."**

Yea, Sewanee's right." The equestrian team has reached nationals during eight of the past ten years, and the swimming, diving, and tennis teams are also competitive. About 60 percent of students participate in intramural sports, ranging from tennis and touch football to handball, Ping-Pong, and pool.

Sewanee's small size means it offers students plenty of opportunity to really make a difference. And though academics are a focus, they're not the only game in town. Is the campus climate formal? You bet. Anachronistic? Perhaps. But, a freshman says, "the resulting intimate community and peculiar micro-culture is treasured by many."

If You Apply To ➤ **Sewanee:** Early decision: Nov. 15., Jan. 2. Regular admissions: Feb. 1. Financial aid: Mar. 1. Campus and alumni interviews: recommended, informational. SATs or ACTs: required; SAT preferred. SAT IIs: required. Accepts the Common Application and electronic applications. Essay question: most enlightening experience; educational influences; in what ways do you want to have a positive impact?

University of South Carolina

Columbia, SC 29208

In the state that started the Civil War, USC is still trying to fight off the image of being one step behind UNC–Chapel Hill. The university has paid big money to attract star professors and boasts one of the top international business programs in the nation. Criminal justice is also a specialty.

Whether it's football or international business, students at the University of South Carolina are game—after all, they're the Gamecocks and, like their mascot, they've got plenty of fighting spirit. Students love to cheer on the school's football and basketball teams, especially if the opponent is longtime rival Clemson. And while 80 percent of the student body comes from within the state, South Carolina is working hard to give its campus a more global feel, through programs such as SEED, which stands for Students Educating and Empowering for Diversity. "Diversity and loyalty are two words to describe Gamecock students," says a chemical engineering major.

South Carolina's mostly modern campus is located in the heart of Columbia (population 450,000), which also happens to be the state capital. Government buildings and downtown businesses are within an easy walk, allowing students to secure internships or even part-time jobs during the school year. The old section of the campus, which dates to the school's 1801 founding, includes the glorious oak-lined Horseshoe; 10 of its 19th-century buildings are now listed in the National Register of Historic Places. Mild winters are typical here, and snow—even just a few flurries—is a traffic-stopping event.

South Carolina offers 70 undergraduate degree programs; biology is the most popular major, followed by experimental psychology, nursing, criminal justice, and political science. Students in the journalism and mass communications program benefit from an excellent film library right on campus, while budding marine scientists may study and do research at a 17,000-acre facility about three hours away. Because South Carolina's coastal economy depends on foreign trade, the university has also developed a top-notch international business program. Art students, neglected at many universities, here have access to the latest cameras, editing stations, and computers, as well as pottery kilns and other necessary equipment. Musicians benefit from a four-level building with classrooms, a music and performance library, rehearsal rooms, recording studios, and a 250-seat lecture hall. Students also give high marks to USC's engineering programs and its major in sports and exercise science. An unusual minor in medical humanities gives doctors-to-be an introduction to the ethical, cultural, legal, economic, and political factors that affect medical practice today.

Regardless of the program in which they enroll, students must take six credits each in English and in numerical and analytical reasoning, 12 credits in the liberal arts, and seven credits in the natural sciences. Foreign language proficiency is required for graduation, as is the three-hour University 101 seminar, designed to help freshmen adjust to college. To

"Diversity and loyalty are two words to describe Gamecock students."

build community, there's the Freshman Reading Experience, in which entering students read the same book before coming to campus, then discuss it in small groups upon arrival. Nearly two-thirds of the classes taken by freshmen have 25 or fewer students, which helps students establish personal relationships with faculty members. And the English program benefits from sizable collections of research material on F. Scott Fitzgerald and Ernest Hemingway, both held in the on-campus library.

Though the vast majority of Gamecock students are South Carolinians, they're not all alike. USC draws students from all 50 states and from more than 100 countries and also has the highest number of minority students of any public university in the state; 5 percent are African American, 3 percent are Asian American, and 2 percent are Hispanic. Hot-button issues include gay rights and smoking, says a junior, but since there are about 300 religious, social, athletic, and professional clubs on campus, you'll probably be able to find a niche, no matter where you fall on the political spectrum. The university awards more than 100 merit scholarships, as well as a variable number of athletic awards in 14 sports, including cheerleading and diving.

Fourteen percent of South Carolina's men and 15 percent of the women go Greek, and their chapters provide much of the weekend social life on campus.

Website: www.sc.edu
Location: City center
Total Enrollment: 25,596
Undergraduates: 17,689
Male/Female: 46/54
SAT Ranges: V 520–620
  M 530–630
ACT Range: 22–27
Financial Aid: 40%
Expense: Pub $ $ $
Phi Beta Kappa: Yes
Applicants: 12,379
Accepted: 67%
Enrolled: 42%
Grad in 6 Years: 64%
Returning Freshmen: 85%
Academics: ✍ ✍ ✍
Social: %%%
Q of L: ♥♥♥
Admissions: (803) 777-7700
Email Address: admissions-ugrad@sc.edu

Strongest Programs:
Biology
English
International Business
Psychology
Criminal Justice
Nursing
Business

Musicians benefit from a
four-level building with
classrooms, a music and
performance library,
rehearsal rooms, recording
studios, and a 250-seat
lecture hall.

Just 40 percent of USC students live on campus, because housing can be expensive and difficult to get, students say. The best rooms in the stately old Horseshoe section of campus, for example, cost considerably more than traditional double rooms with hall baths located elsewhere. The best advice? Apply early, since the system is first come, first served, and freshmen compete with upperclassmen for space. And the surest way to beat the housing system? Get into the Honors College, which entitles you to some of the best rooms. Dining options range from fast-food stands (Chik-fil-A, Einstein Bros. Bagels, Burger King) to all-you-can-eat lines, with plenty of vegetarian and healthy choices—and of course, some junk food, too.

Nearly two-thirds of the
classes taken by freshmen
have 25 or fewer students,
which helps students
establish personal
relationships with faculty
members.

Fourteen percent of South Carolina's men and 15 percent of the women go Greek, and their chapters provide much of the weekend social life on campus. While students under 21 may not legally drink, some "sneak it in, and are not bothered if they behave themselves," says a speech major. Still, administrators have become more concerned about binge and underage drinking, and funding for alternative activities during high-risk times for alcohol has recently tripled. Those activities include films, dance performances, theatrical productions, concerts, and comedy shows. Downtown Columbia offers more theaters, a comedy club, a performing arts center, and Five Points, a strip boasting six different bars. Outdoorsy types will appreciate Myrtle Beach, just three hours away, and the mountain ranges four hours north, for hiking, skiing, and camping.

Fall football weekends are always a big deal at South Carolina, and recent successes with coach Steve Spurrier have given them lots of reasons to cheer, although an NCAA probation due to violations has muted those cheers somewhat. The enduring USC–Clemson rivalry is one of the oldest and most colorful in college sports, with festivities beginning weeks in advance; aside from the Tigerburn parade and bonfire and the all-night tailgating parties, the schools compete in a blood drive. Winter weekends bring another of USC's strong sports, basketball, now played in the 342,000-square-foot Colonial Center. All students may dip into the indoor and outdoor pools at the Strom Thurmond Fitness and Wellness Center, which also features an indoor track, volleyball and basketball courts, a climbing wall, and racquetball courts.

The pace of change is picking up at South Carolina's flagship university. SC is "more global, more diverse, and more open-minded" than its closest rival, though it retains the lively, sports-oriented culture typical of Southern schools, says a junior. With a campus beautification initiative underway, along with scholarships and small, seminar-style courses working to draw more-capable students, it seems that no place could be finer, indeed.

South Carolina: Regular admissions: Dec. 1. Financial aid: Apr. 1. Meets demonstrated need of 40%. Campus interviews: optional, informational. No alumni interviews. SATs or ACTs: required. SAT II: optional. Accepts electronic applications. No essay question.

University of Southern California

University Park, Los Angeles, CA 90089

USC's old handle: "The University of Spoiled Children." USC's new handle: highly selective West Coast university with preeminent programs in arts and media. The

difference: a deluge in applications of historic proportions as students flock to L.A. and the region's only major private university.

Once dismissed as little more than an academic bastion of privilege, the University of Southern California has come into its own as a West Coast destination for students seeking entry into the elite Southern California business community. The school's lush campus and prime Los Angeles location has led to a flood of applicants, making it continually tougher to win admission. Students cheer on national championship teams, solid engineering and communications programs, and give high marks to the Trojan alumni network as well.

The USC campus offers a mix of traditional ivy-covered and modern structures, arranged around fountains and reflecting pools, well-shaded from the Southern California sun. Sitting on 235 park-like acres, just minutes from downtown Los Angeles, the school is a veritable urban oasis. And though nearby areas can be gritty, USC has its own police department, and most students say they've never felt unsafe. "When you are at USC, you feel like you're at the beach: hot girls tanning in the quad, people playing Frisbee," says a sophomore. New facilities include the 143,000-square-foot Molecular & Computational Biology Building and Ronald Tutor Hall, which serves as a gathering place for students and the center of cutting-edge research.

USC's Core Curriculum requires nine courses: six general education, two intensive writing, and one diversity. Together, administrators say, they "provide a coherent approach to fundamental areas of learning, and give students the tools to think critically, communicate clearly, and locate themselves in history." Students with high GPAs and test scores may choose the Thematic Option—a.k.a. the "Traumatic Option"—in place of regular general education courses. The 200 or so who do get smaller classes with some of the university's best teachers, and a hand-picked group of writing instructors. Freshmen may also join one of the school's Learning Communities, groups of 20 students with common academic interests, such as business, medicine, technology, or languages. Each community takes four common courses during the first year and meets with a dedicated faculty mentor and staff advisor three to six times a semester.

The academic climate is challenging and new students "often have to adjust because either they are not the best in their class anymore or they have to work harder to be the best," says a senior. Aside from a strong alumni network, USC offers undergraduates the chance to pursue degrees not only in the College of Letters, Arts, and Sciences, but also at any of its 17 professional schools and schools of the arts. This means business majors may minor in bioethics or Russian,

> "When you are at USC, you feel like you're at the beach: hot girls tanning in the quad, people playing Frisbee."

that international relations majors may double major in urban planning or international urban development, and art history majors may study cinema and television, the music industry, or business, too. The Renaissance Scholars program recognizes those who excel in two or more disparate areas of study by finishing their majors and minors in no more than five years, and achieving a GPA of 3.5 or higher in those disciplines. New programs include majors in animation and digital arts, architectural studies, and interdisciplinary archeology. The quality of teaching varies, especially in some introductory courses for freshmen, which can be huge. "I have come across professors who seemed to care more about research than the students," says an account major, but, overall, "the professors are truly amazing."

USC students are a healthy mix of "artistic brilliance and ambitious drive," says one senior. Thirty-one percent of USC freshmen come from out of state, and 61 percent went to public high schools, though the university also boasts more foreign students than any other institution in the country—nearly 6,000. Aside from all those

Website: www.usc.edu
Location: City center
Total Enrollment: 28,823
Undergraduates: 16,072
Male/Female: 49/51
SAT Ranges: V 620–710
  M 650–730
ACT Range: 28–32
Financial Aid: 45%
Expense: Pr $ $ $
Phi Beta Kappa: Yes
Applicants: 31,634
Accepted: 27%
Enrolled: 33%
Grad in 6 Years: 83%
Returning Freshmen: 95%
Academics: ✑ ✑ ✑ ½
Social: ☎ ☎ ☎
Q of L: ★ ★ ★
Admissions: (213) 740-1111
Email Address:
  admitusc@usc.edu

Strongest Programs:
Business
Cinema/Television
Engineering
Communication
Art

Aside from a strong alumni network, USC offers undergraduates the chance to pursue degrees not only in the College of Letters, Arts, and Sciences, but also at any of its 17 professional schools and schools of the arts.

foreigners, this campus is one of the more diverse in the U.S., with African Americans making up 6 percent of the student body, Hispanics adding 13 percent, and Asian Americans contributing 21 percent. Regardless of their nationality, most students here are "outgoing and friendly," says a music industry major—and most pride themselves on their ability to multitask, maintaining decent grades along with an active social life. Hundreds of merit scholarships are awarded each year (averaging $12,874), as are athletic awards in nine men's and nine women's sports.

Freshmen are guaranteed university housing, and students say dorm rooms, which come with microwaves and refrigerators, are comfortable. Halls are co-ed, and, "New 1 North is definitely the social dorm that everyone in surrounding residence halls wishes they lived in," says a sophomore. Since swimming pools, tennis courts, carpeting, and air-conditioning are just some of the luxuries to be found in USC dorms, it's no wonder more upperclassmen would like to stay on campus. But because there isn't enough space for everyone, sophomores, juniors, and seniors typically move to fraternity and sorority houses or apartments, which are just a short walk away. (Sixteen percent of the men and 20 percent of the women go Greek.) Dining halls offer plenty of options, including an international buffet in the new Parkside

"We are drawing an academically competitive and involved student body."

complex—and just off campus, everything from burgers at Carl's Jr. to hot original glazed doughnuts at Krispy Kreme. USC's urban location means personal safety can be an issue. That said, blue-light phones, tram and taxi services, and escorts from the campus police mean most students don't worry too much about crime.

Though Los Angeles is hardly a "college town" in the traditional sense, "it does allow you to experience a wide variety of cultures," says an accounting major. Whether you're looking for an internship at a law firm or a movie studio, you want to learn to surf, or you're eager to check out a new band before they get signed to a major label, L.A. delivers. Famous Venice Beach is just a few miles from USC's campus, and in the winter months, students can reach the San Gabriel Mountains (and its ski resorts) in less than an hour (by car, not by skis). USC students are also active in the community, tutoring in five local schools through the Joint Educational Project.

On campus, sports are pretty much the biggest thing going. Football mania has reached a new level since the team brought home two consecutive national titles and played for a third, and the men's and women's water polo and women's volleyball teams are also championship-caliber. Indeed, two of USC's biggest schoolwide traditions revolve around the ol' pigskin. The first is Troy Week—the week leading up to the UCLA game—which culminates with a pep rally and concert in the middle of campus. Then there's the Weekender, when USC students take off en masse for northern California to see their beloved Trojans face off against Stanford or Berkeley. Throngs of USC undergrads, alumni, and fans gather in San Francisco's Union Square for a huge pep rally, featuring the band, cheerleaders, and university personalities.

USC is a university on the move. "We are drawing an academically competitive and involved student body," says a geography and communications major. "We have received major donations. And USC athletics are back on par." Pack your sunscreen, flip-flops, and some assertiveness, and you'll fit right in. Shrinking violets, on the other hand, should probably look elsewhere.

P.O. Box 750181, Dallas, TX 75275-0181

SMU is all but the official alma mater of the Dallas business and professional elite. The university is best known for business, performing arts, and upscale conservatism. Though tuition is moderate by national standards, SMU is pricey compared to rivals Texas Christian and Rice.

Southern Methodist University is a training ground for the business elite of Dallas, and for those who may want to lead the state of Texas someday. With admissions standards on the rise, both for entering freshmen and for current students who want to transfer into the Cox School of Business and SMU's other top programs, "the stereotypical fraternity or sorority member can't just breeze by," says a senior. "I have seen a shift from social to studious since arriving."

SMU's lush, well-landscaped campus is located in the affluent suburb of University Park, "five minutes from downtown Dallas and within 30 minutes of everything else," according to one student. Flowerbeds, fountains, and neatly trimmed lawns surround stately brick buildings, most in the collegiate Georgian style. Dallas Hall, with its four-story rotunda, is the centerpiece. The Dedman Lifetime Sports Center opened in 2006, and a state-of-the-art motion capture studio was completed in 2004.

SMU's 41-credit-hour general education curriculum includes courses in rhetoric (writing), math, information technology, wellness, cultural formations (interdisciplinary humanities, social sciences, and natural science options, emphasizing writing), perspectives (one course each in five of six categories: arts, literature, religious and philosophical thought, history, politics and economics, and behavioral sciences), science and technology (understanding their social, legal, and ethical implications), and human diversity. In all, the school offers more than 100 majors, and many students pursue more than one.

> "I have seen a shift from social to studious since arriving."

The most popular field of study at SMU is finance, followed by advertising, psychology, political science, and business. Engineers have access to an extensive co-op program, thanks to the proximity of more than 800 high-tech companies, including Nokia and Texas Instruments, which have facilities in the Dallas suburbs. The John Goodwin Tower Center for Political Studies, named for the former senator, focuses on international relations and comparative politics, while the on-campus Tate Forums provide informal question-and-answer sessions with national and international figures. The Meadows School of the Arts, also home to the Temerlin Advertising Institute, shines just as brightly. Its facilities include the Bob Hope and Greer Garson theaters, funded by their namesake performers.

SMU's humanities programs are notable, too, with English and history particularly strong. The annual literary festival brings to campus poets and playwrights such as Adrienne Rich and Edward Albee for readings and seminars with aspiring authors. SMU also publishes *Southwest Review*, one of the four oldest continuously published literary quarterlies in the nation. Communications programs have been strengthened with new curricula, faculty, and state-of-the-art studio facilities. "My professors know my name and care about me," gushes one junior. "I have never had a TA teach a class."

SMU prides itself on small classes; 90 percent of the courses taken by freshmen have 50 students or fewer. Teaching assistants are available for extra help, students say, but they never teach classes. "Courses really work to take theory and apply it to interactive and real-world situations," a sophomore says. The Honors Program enables about 850 students to take seminars on topics not offered broadly, with

Website: www.smu.edu
Location: Suburban
Total Enrollment: 11,152
Undergraduates: 6,489
Male/Female: 45/55
SAT Ranges: V 560–660
  M 570–670
ACT Range: 24–28
Financial Aid: 37%
Expense: Pr $ $
Phi Beta Kappa: Yes
Applicants: 6,981
Accepted: 58%
Enrolled: 20%
Grad in 6 Years: 71%
Returning Freshmen: 87%
Academics: 🖉 🖉 🖉
Social: 🍷 🍷 🍷 🍷
Q of L: ★ ★ ★ ★
Admissions: (214) 768-2000
Email Address:
  enrol_serv@smu.edu

Strongest Programs:
History
Anthropology
Political Science
Business
Natural Sciences/Pre-med
Performing Arts

enrollment in each course capped at 15 to 20 students. Twenty-two study abroad programs take students from Austria to Australia; about 450 people participate each year. SMU also operates a second campus near Taos, New Mexico, on the grounds of historic Fort Burgwin, a mid-19th-century army outpost, and an excavated 13th-century pueblo. Each year, about 100 exceptional students are named President's Scholars; they get full-tuition scholarships, plus opportunities to study abroad and attend a retreat in Taos, being matched with corporate mentors, and meeting with the world leaders who visit campus for the Tate Distinguished Lecture Series.

Although it was founded by what is now the United Methodist Church, SMU is nondenominational and welcomes students of all faiths. In fact, while about 25 percent of students are Methodist, 21 percent are Roman Catholic, and the remainder are affiliated various Protestant denominations—or with Judaism, Buddhism, or Islam. Almost two-thirds of the total are native Texans, and politically, they lean

"The new dorms are nicer than the house I own off campus."

right, with a great deal of excitement around the possibility of hosting George W. Bush's presidential library. "There's not tons of ethnic diversity," admits one junior. "Students are mostly white and from middle- to upper-class families." Hispanics account for 8 percent of the student body, while Asian Americans comprise 6 percent and African Americans another 6 percent. SMU offers a loan program for middle-income families, as well as 265 athletic scholarships in six men's and eight women's sports. In addition, a quarter of the students receive awards based on academic merit, averaging $5,311 a year.

Forty percent of SMU undergrads live on campus, and freshmen are required to do so. "The new dorms are nicer than the house I own off campus," sighs one student. Sophomores typically live in Greek houses—40 percent of the women join sororities and 29 percent of the men pledge fraternities—and then move to off-campus apartments for their junior and senior years. All residence halls are co-ed by floor, and theme floors are also available, focused on honors, wellness, community service, multiculturalism, or the arts. Options include single and double rooms, some with their own bathrooms, and all with high-speed wired and wireless Internet access. The Inter-Community Experience house allows students to live and provide community service in East Dallas, a struggling section of the city. SMU's two dining halls offer hot entrees, salad and sandwich bars, and plenty of desserts. The traditional, all-you-can-eat meal plans also include dining dollars that can be used at the on-campus Subway and Chick-fil-A outposts, as well as at the Java City Cyber Cafe.

When the weekend comes, nearly 200 student groups sponsor speakers and other diversions. Still, with Dallas so close and the campus officially dry—except on football game days—most social life takes place off campus. (Regardless of their age, students say it's easy to find beer at the fraternity houses. Those under 21 and caught drinking are referred to the Judicial Council, but "if you act like a responsible adult, you will be treated like one," says a senior.) Highlights of the campus calendar include the Mane Event, honoring the Mustang mascot, and the Celebration

"Students are mostly white and from middle- to upper-class families."

of Lights, when students gather to enjoy holiday lights and carols at Dallas Hall. Formals and other Greek parties are by invitation only, and students are often bussed to them. SMU's neighborhood "has the feel of a small town, but the convenience of a big city," says one student, noting that museums, amusement parks, big-league sports, entertainment, and shopping are all within a quick drive. Cars are necessary for road trips to New Orleans (for Mardi Gras), Shreveport (for gambling), and Austin (where the Sixth Street bars and music halls stay open until the wee hours).

Football games are a big deal here—after all, this is Texas—and SMU students get riled up for the annual battle against Texas Christian University. When the Mustangs play at home, there's tailgating on the Boulevard, with tents, family activities, music, and food on the main quad. SMU competes in the Division I-A of Conference USA, and has recently added women's rowing and equestrian programs. Men's golf and track and field and women's cross-country, soccer, swimming, and diving are among the varsity teams that have brought home conference titles in recent years. Intramural options attract 40 percent of students and include water polo, billiards, dodgeball, and even tug of war.

Southern Methodist is known for beautiful people and a beautiful campus, but there's more to the story than that. The school offers solid pre-professional training along with an active social life, and offers students the opportunity to give back to the city of Dallas—with more than just their bar tabs. Close-knit friendships are easily formed with other students and the supportive faculty and staff. All in all, the school is "a place I'll want to return in years to come," says one student.

If You Apply To ➤

SMU: Early action: Nov. 1. Rolling admissions: March 15. Financial aid: Feb. 15. Meets demonstrated need of 37%. Campus interviews: optional, informational. No alumni interviews. SATs or ACTs: required. SAT IIs: optional. Accepts the Common Application and electronic applications. Essay question: the title of your autobiography, and a page or two from it; the origin of your e-mail address and how it reflects your personality; your definition of diversity and its role in your learning and development; or some of the things you are currently thinking, laughing, or talking about.

Southwestern University

1001 E. University Avenue, Georgetown, TX 78626

Camouflaged beneath a nondescript name, Southwestern is the top liberal arts college in Texas—a private-college alternative to Austin and Trinity (TX). Southwestern is about half the size of the latter and prides itself on individual attention and down-to-earth friendliness. Strong programs include communications and international studies.

In a state known for political conservatism, and for the prevailing view that bigger is better, Southwestern University stands out. The school attracts "academically oriented leaders with a desire to make a difference and be socially involved," says a psychology and English major. "Gay and lesbian rights and social inequalities" are hot topics on this "very liberal" campus, where "we are really empowered to be active—our faculty encourage it," adds a sociology major. And while SU is the oldest college in Texas, it fields no varsity football team, so homecoming festivities revolve around Sing!, a night of songs and skits poking fun at the school.

The Southwestern campus sits on 700 acres at the edge of the rolling Texas Hill Country. The limestone buildings, built in the Romanesque style, date from the early 20th century, and there are plenty of open spaces, including a nine-hole golf course and outdoor tennis courts. The Alma Thomas Theater, part of the Sarofim School of Fine Arts, is currently being renovated, with completion expected in 2007; the project will improve classrooms in the building, enhance the audience experience with seating and other upgrades, and eliminate environmental interference to enhance sound quality.

To graduate, Southwestern students must complete eight hours of Foundation courses (First-Year Seminar, Writing and Critical Thinking, and Math), and 30 hours of Perspectives on Knowledge courses, chosen from a range of subject areas, such as American and Western Cultural Heritage or The Natural World. Students must also

Website:
www.southwestern.edu
Location: Suburban
Total Enrollment: 1,310
Undergraduates: 1,310
Male/Female: 40/60
SAT Ranges: V 570–670
M 580–690
ACT Range: 24–29
Financial Aid: 45%
Expense: Pr $
Phi Beta Kappa: Yes
Applicants: 1,956
Accepted: 65%
Enrolled: 26%
Grad in 6 Years: 78%
Returning Freshmen: 89%
Academics: ✑ ✑ ✑ ½
Social: ☎ ☎ ☎

(Continued)

Q of L: ★★★

Admissions: (800) 252-3166

Email Address:

admission@southwestern
.edu

Strongest Programs:

English

Biology

Chemistry

Music

Psychology

History

Spanish

demonstrate computer proficiency, and must complete an integrative or capstone experience, such as a research project or thesis. Finally, there are requirements related to writing, fitness, and recreation. SU encourages undergraduate research, and each year holds a symposium to showcase students' scholarly endeavors. The King Creativity Fund, established by an alumnus with a $500,000 donation, sup-

> **"Students here are all high achievers, which can make one feel overwhelmed at first."**

ports up to 20 "innovative and visionary projects" each academic year, with grants of up to $1,500 each. The Paideia Program allows students to compare, contrast, and integrate knowledge and skills gained in various areas of study, through leadership, collaborative-learning, service and intercultural experiences in the sophomore, junior and senior years. Southwestern pays $1,000 of program-related expenses.

"Students here are all high achievers, which can make one feel overwhelmed at first," says a freshman. Still, while the workloads are significant, the atmosphere isn't cutthroat. Ninety-nine percent of classes have 25 students or fewer, and "if you send a message via e-mail, the prof will usually respond within five minutes to 24 hours," says a sophomore. The most popular majors are English, business, psychology, history, and political science. While engineering attracts fewer students, those who do enroll benefit from the option of 3–2 master's programs with schools including Arizona State, Texas A&M, and Washington University (MO).

Over half of Southwestern's students choose to study abroad, and each year, SU faculty lead programs in London, Mexico, Hungary, Germany, and Honduras, plus a service-learning program in Jamaica. Internships are available, too, offering the chance to study politics, foreign policy, journalism, or architecture in Washington, D.C., or to apprentice with professional artists, designers, actors, and filmmakers in New York. Southwestern is also affiliated with the United Methodist Church, and a member of the Associated Colleges of the South.*

"Most SU students are very intelligent," says a sophomore. "They are well-rounded, have a sense of humor, and are very welcoming." Hispanics are the largest minority group at SU, comprising 14 percent of the student body; African Americans add 3 percent and Asian Americans 5 percent. "There's a broad spectrum of views, but only a few perspectives are represented when it comes to activism and speaking

> **"You do not have to be Greek to have a fabulous social life."**

out," says a political science and Spanish major. SU's tuition is markedly less than at institutions of similar quality elsewhere in the U.S. In addition, 55 percent of students receive scholarships based on academic performance, averaging $9,500; talent awards are also available for fine arts majors, though there are no athletic scholarships.

Eighty-five percent of SU students live in the residence halls, where options improve as you get older—juniors and seniors usually get apartment-style facilities with their own bedrooms, bathrooms, and kitchens. "Some of the dorms are old and need repairs," says one student. Some halls are co-ed and others are single-sex; a sophomore says that housing for first-year women is far superior than that for first-year men. Most rooms are suite-style, with housekeepers cleaning up several times a week. Dining halls are all you can eat, and also serve Sunday brunch. "Food is very diverse for all diets and ethnic choices," says a senior. "Vegans always have options prepared instead of having to make do with a simple salad."

Southwestern has a strong Greek system, drawing 28 percent of the men and 30 percent of the women, "but you do not have to be Greek to have a fabulous social life," says a sociology major. SU is officially a dry campus—no kegs are allowed, and anyone under 21 caught imbibing is fined "something like $75," says a sophomore. Other options include live music and movies every week, sponsored by the Union Program Council and Student Activities Office.

The King Creativity Fund, established by an alumnus with a $500,000 donation, supports up to 20 "innovative and visionary projects" each academic year, with grants of up to $1,500 each.

Georgetown itself reminds one student of Andy Griffith's Mayberry—"beautiful, small, and inviting," but catering primarily to families and retirees, rather than college students. Things are changing, though—a new mall just opened, and the town will have two new movie theaters and several restaurants within the year. The bars and clubs of Austin's Sixth Street are just a half-hour away, and San Antonio, College Station and Houston aren't that much further. Mall Balls begin and end the year with food, frivolity, music, and games on SU's Academic Mall. The Blue Hole, popular for swimming and cliff jumping, is within walking distance of campus as well. And the Pirate Bike Program has made 30 new yellow bicycles available, for free, to students, faculty and staff—riders simply pick up a bike when they need one, then leave it for the next rider when they reach their destination.

Southwestern offers 14 varsity sports, and its Pirates compete in NCAA Division III; the school also belongs to the Southern Collegiate Athletic Conference. Men's basketball and soccer are the most popular sports for spectators, especially games against archrival Trinity University in San Antonio, and the men's basketball team boasted the SCAC player

"There are so many differing views here, yet everyone is accepting and so nice."

and coach of the year last year. A professional sports specialist and eight undergraduate assistants oversee 25 intramural leagues and tournaments each year. Teams and individuals compete in everything from tennis and bowling to dodgeball, soccer, and inner tube water polo, but some students grumble about the meager facilities. "There is only one treadmill for the entire campus!" exclaims a freshman.

"SU students are laid-back and easy to get along with," asserts a biology and communications studies major. "Many were in the top tier in their high school classes, but don't act like they are smarter or better than anyone else." A sociology major agrees: "There are so many differing views here, yet everyone is accepting and so nice." Couple that feeling of friendliness with "phenomenal" professors and a campus full of natural beauty, and the attraction of this Hill Country college becomes clear. In a state where things tend to be huge and overwhelming, Southwestern University proves that good things can—and do—come in small packages.

Over half of Southwestern's students choose to study abroad, and each year, SU faculty lead programs in London, Mexico, Hungary, Germany, and Honduras, plus a service-learning program in Jamaica.

Overlaps

Trinity (TX), University of Texas–Austin, Texas A&M, Austin College, Rhodes, Hendrix

If You Apply To > **Southwestern:** Early decision: Nov. 15. Rolling admissions: Feb. 15. Financial aid: Mar. 1. Meets demonstrated need of 75%. Campus interviews: recommended, evaluative. Alumni interviews: optional, informational. SATs or ACTs: required. SAT IIs: optional. Accepts the Common Application and electronic applications. Essay question: influential personal experience or achievement; most important challenge facing your generation; biggest misconception people have of you, and why.

Spelman College: See page 37.

Stanford University

Stanford, CA 94305-3005

If you're looking for an Eastern version of Stanford, think Duke with a touch of MIT mixed in. Stanford's big-time athletics, preprofessional feel, and laid-back atmosphere differentiate it from Ivy League competitors. In contrast to the hurly-burly of Bay Area rival Berkeley, Stanford's aura is upscale suburban.

Website: www.stanford.edu

Location: Suburban

Total Enrollment: 14,881

Undergraduates: 6,705

Male/Female: 52/48

SAT Ranges: V 670–770
   M 690–780

ACT Range: 29–33

Financial Aid: 46%

Expense: Pr $ $ $

Phi Beta Kappa: Yes

Applicants: 20,195

Accepted: 12%

Enrolled: 67%

Grad in 6 Years: 94%

Returning Freshmen: 98%

Academics: ✍ ✍ ✍ ✍ ✍

Social: ☎ ☎ ☎ ☎

Q of L: ★ ★ ★ ★ ★

Admissions: (650) 723-2091

Email Address:
   admission@stanford.edu

Strongest Programs:
Biology
Computer Science
International Relations
Engineering
Political Science
Economics

Think the only difference between Stanford and the Ivy League is a couple hundred extra sunny days each year? Think again. From the red-tiled roofs to the lush greenery and California vibe, Stanford is a world away from the Gothic intellectual culture of the Ivies. Virtually all the great Eastern universities began as places to ponder human existence and the meaning of life, using European institutions as their models. Stanford, by contrast, built its academic reputation around science and engineering, fields characterized by American ingenuity, and only later cultivated excellence in the humanities and social sciences. Stanford is, without a doubt, the nation's first great "American" university.

The differences between Stanford and other institutions it competes against for the country's top high school seniors are evident everywhere, from the architecture to the curriculum. The school's mission-style buildings look outward to the world at large, rather than inward to ivy-covered courtyards. And unlike Yale and Princeton, Stanford—founded in 1885 by Leland and Jane Stanford—has been co-ed from the beginning. During its centennial, the school became the first U.S. university to successfully launch a billion-dollar capital campaign; today Stanford's endowment is $12 billion. Some architectural critics say the campus looks like the world's biggest Mexican restaurant, even though Frederick Law Olmsted, designer of New York City's Central Park, planned many of the buildings. The campus stretches from the foothills of the Santa Cruz Mountains to the edge of Palo Alto in the heart of Silicon Valley, smack in the middle of earthquake country. Newer facilities include Ziff Center for Jewish Life and the new Stanford Stadium.

Biology and human biology, the quintessential premed preparation, are the most popular programs on campus, followed by economics and computer science. Stanford has also developed a particularly interesting set of interdisciplinary programs. For students who are able to study abroad, programs are offered in Australia, Japan, Chile, England, China, Germany, Italy, Russia, and France. In fact, more than one-quarter of each graduating class takes advantage of these programs. Closer to home, the Stanford-in-Washington program allows 60 students to live, study, and intern in the nation's capital each quarter. The Haas Center for Public Service offers service-learning courses in a wide range of disciplines, while the communications department sponsors the Rebele internship, which offers paid positions at various California newspapers. The Stanford Hopkins Marine Station is located on a mile of coastland in Pacific Grove, next to the Monterey Bay Aquarium, and offers courses in marine and biological sciences.

The general education requirements are extensive. All students take an Introduction to the Humanities course each quarter of the freshman year; they are designed to hone skills through close reading and critical investigation of a limited number of works. Students must also take five courses in Disciplinary Breadth (including engineering and applied sciences, humanities, math, natural sciences, and social sciences), and courses in ethical reasoning, global community, and cultures. In addition, there are writing and foreign language requirements. The Stanford Introductory Studies program includes Freshman Seminars, courses limited to 16 students to enable close discourse with professors. The Freshman-Sophomore College pairs 180 freshmen and sophomores for special programs with faculty and advising staff. The new Schwab Learning Center—named after alum Charles Schwab—offers services for students with learning disabilities and Attention Deficit Hyperactivity Disorder. There's also Summer Research College, designed to create community among undergraduates engaged in full-time summer research on campus, and three honors programs. The new Potter College provides a residential facility for young researchers.

Don't let Stanford's California location fool you into thinking studying is optional—it's more like a full-time job. "Stanford offers an array of courses," says a

> **"Stanford students are taught by the best of the best."**

senior, "from hair-pulling, 'I'm about to drop out' rigor to 'I can't believe this is being taught here' ease." Stanford's faculty ranks among the best in the nation, with most departments boasting a nationally known name or two. Professors are considered outstanding scholars with outstanding credentials, and 95 percent of classes are taught by faculty, as opposed to graduate students. "The teaching and professors have been phenomenal," says a graduate student. "Stanford students are taught by the best of the best." Seventy-nine percent of classes have 30 or fewer students.

Despite its upscale image, Stanford tends to have the same demographic profile as its state-supported neighbor in Berkeley. Sixty percent attended public high school; 87 percent of a recent freshman class graduated in the top tenth of their class. Forty-four percent of students are from California, while foreign students account for 6 percent of the student population. Minority enrollment is far above average, with Asian Americans accounting for 24 percent of the student body, Hispanics 11 percent, African Americans 10 percent and Native Americans 2 percent. "Stanford is the sunny, palm-tree-laced, Spanish-inspired Ivy of the West," boasts a communications major. "Just being in California provides a lax attitude."

About 20 percent of Stanford's students take advantage of its liberal stop-out policy, which lets students take some time off along the way, rather than staying in school for four straight years. Admissions are need-blind, and the university guarantees to meet the full demonstrated financial need of every domestic admit. While there are no merit scholarships, 76 percent of students receive some sort of internal or external financial aid; the university also awards 400 athletic scholarships annually in 34 sports. A recent fund-raising effort added more than $300 million to the scholarship pool, and Stanford has eliminated any required financial contribution for families making less than $45,000 annually.

Freshmen must live on campus, and Stanford guarantees housing for four years; 95 percent of students stay on campus, with most of the others attending Stanford-in-Washington or other off-campus study programs. One percent of students commute from home, in part because of the lack of affordable off-campus options in extraordinarily expensive Silicon Valley. "Much of life at Stanford is about the dorms," a senior history major says. "They are a source of community and are well-run." As students gain seniority, a lottery system decides where they'll live. "Junior year I lived in an old faculty mansion for 30 students that had a Thai chef," one student says. The multimillion-dollar Governor's Corner complex includes all-oak fixtures, homey rooms with views of the foothills, microwave ovens in the kitchenettes, and Italian leather sofas in the lounges. Dorm dwellers must sign up for a meal plan. Campus security is good, students say, and includes an escort service. Bike theft is the biggest complaint.

"Stanford is the sunny, palm-tree-laced, Spanish-inspired Ivy of the West."

When academic pressures become too great, students seek refuge in the outdoors. Nearby hills are perfect for jogging and biking, and a small lake is great for sailing and windsurfing (the most popular spring physical education class). Palo Alto "has fun bars, but isn't a true college town because of its prices and older population," a senior says. Trips to the Sierra Nevada mountains (four hours away) or to the Pacific coast (45 minutes) are popular, as are jaunts to San Francisco (much more of a college town), Los Angeles, or the Napa Valley.

Like most things at Stanford, activities and social life vary a great deal, although most take place on campus. As tradition goes, freshmen aren't "true Stanford students" until they've been kissed at midnight in the quad by a senior. Full Moon on the Quad occurs at the first full moon, and features a bevy of first-year students eager to receive their initiation (courtesy of a well-timed entrance by upper-class students). Greek organizations claim 13 percent of students, and provide their share of happy hours and weekend beer bashes, which are open to all. Underage drinking

Some architectural critics say the campus looks like the world's biggest Mexican restaurant, even though Frederick Law Olmsted, designer of New York City's Central Park, planned many of the buildings.

The Stanford Hopkins Marine Station is located on a mile of coastland in Pacific Grove, next to the Monterey Bay Aquarium, and offers courses in marine and biological sciences.

happens, but is kept under control, students say. "The RAs are not required to report instances of drinking, so it's safe and open," a senior says. Another Stanford tradition is the Viennese Ball, a February event that may make you wish you'd taken ballroom dancing lessons. Halloween finds students partying at the Mausoleum, the Stanfords' final resting place.

Stanford has a proud athletic tradition that includes 73 NCAA championships since 1990. Cardinal teams recently won their 12th Director's Cup, which recognizes the best overall collegiate athletic program in the country. The baseball team has been to the College World Series. In 2004–05, Stanford had three national team championships and 11 teams ranked in the top five nationally. The football team has pulled off its share of upsets, and the annual contest against Cal (Berkeley) is the "Big Game." The Leland Stanford Junior University Marching Band proudly revels in its raucous irreverence, to the delight of students and the dismay of conservative types. For those not inclined to varsity play, Stanford offers a full slate of intramurals, and its vast sports complex includes 26 tennis courts, two gymnasiums, a stadium, an 18-hole golf course, and four swimming pools. The equestrian team is housed at the newly renovated Red Barn, where horseback riding lessons are also offered.

> "Much of life at Stanford is about the dorms."

Stanford's sunny demeanor and infectious West Coast optimism offer an appealing alternative to the gloom and gray weather that seem to hang over some of its East Coast counterparts, with the same high-caliber academics and deep athletic traditions that have made them great. "The unique combination of top-tier academics, beautiful weather, championship athletics, fascinating people and self-deprecating humor—you won't find that anywhere else," one student says.

State University of New York

As the largest university system in the world, the State University of New York provides more than 400,000 students with a vast landscape of educational opportunities—both figuratively and literally. Encompassing 66 campuses and 21,000 acres of property, SUNY's staggering physical presence is exceeded only by the scope of its academic offerings.

The statistics of SUNY (pronounced "SOOney") are awesome. The university has an annual operating budget of billions, greater than the gross national product of many countries and larger than the budget of more than a dozen American states. It has more than 400,000 students, 4,000 academic programs, and 24,500 faculty members, and maintains more than 2,200 buildings. Every year it awards approximately 65,000 degrees, from associate to PhD, in thousands of different academic fields. And—you're not going to believe this one—it has more than one million living graduates.

Such figures are all the more remarkable considering that until 1948, New York had no state university at all. That year, the legislature created the State University around a cluster of 32 existing public institutions, the best of which focused on the training of teachers, to handle the flow of returning World War II veterans. But a "gentleman's agreement" not to compete with the state's private colleges (which for generations had enjoyed a monopoly on higher education in New York) hindered SUNY's movement into the liberal arts. Not until Nelson A. Rockefeller became governor in 1960 and made the expansion of the university his major priority did SUNY begin its dramatic growth.

SUNY has now ripened into a network of four research-oriented "university centers," 13 arts and sciences colleges, six agricultural and technical colleges, five "statutory" colleges, four specialized colleges, 30 locally sponsored community colleges, and four health science centers. Annual costs at institutions like SUNY–Albany have traditionally been well below those at such hoary and prestigious publicly supported flagship campuses as California at Berkeley and the University of Michigan, but they are now slightly above the national average. The SUNY system faces continuing budget cuts and, under the heavy political hand of Governor George Pataki, is rigorously replacing taxpayer funds with student contributions. Tuition and fees have risen by more than 75 percent since 1995, and the state's share of financing for core academics has declined from 69 to 50 percent.

Prospective students apply directly to the SUNY unit they seek to attend. Forty-six of the colleges, though, use a "common form" application that enables a prospective student to apply to as many as four SUNY campuses at the same time. The central administration runs a SUNY Admissions Assistance Service that helps rejected students find places at other campuses. Students who earn associate degrees at community or other two-year colleges are guaranteed the chance to continue their education at a four-year institution, though not necessarily at their first choice. The level of selectivity varies widely. Most community colleges guarantee admission to any local high school student, but the university centers, as well as some specialized colleges, are among the most competitive public institutions in the nation. As part of a recent "standards revolution," SUNY trustees voted to adopt a new budgeting model designed to financially reward campuses that increase enrollment. Undergraduates at all liberal arts colleges and university centers pay the same tuition, but the rates at community colleges vary (and are lower). Out-of-state students, who make up only 4 percent of SUNY students, pay about double the amount of in-state tuition.

Mainly for political reasons, the State University of New York chose not to follow the model of other states and build a single flagship campus the likes of an Ann Arbor, Madison, or Chapel Hill. Instead, it created the four university centers with undergraduate, graduate, and professional schools and research facilities in each corner of the state. When they were created in the 1960s, each one hoped to become fully comprehensive, but there has been a certain degree of specialization from the beginning. They also decided not to establish a Division I football program, something that has lowered their visibility to out-of-state students, as has the lack of prestigious PhD programs.

Albany is strongest in education and public policy, Binghamton is best known for undergraduate arts and sciences, and Stony Brook is noted for its hard sciences. Buffalo, formerly a private university, maintains a strong reputation in the life sciences and geography and comes the closest of any of the four to being a fully comprehensive university. Critics of the system say that the decision to forgo a flagship campus guarantees a lack of national prominence, and the lack of big-time football or other sports programs has affected SUNY's reputation as well. Still, many insist that somewhere in the labs and libraries of these four university centers are lurking the Nobel Prize winners of this century. To these supporters, it's only a matter of time before SUNY achieves excellence in depth as well as breadth.

The 13 colleges of arts and sciences likewise vary widely in size and character. They range from the 26,000-student College at Buffalo, whose 125-acre campus reflects the urban flavor of the state's second-largest city, to the rural and highly selective College at Geneseo, where half the students nearly outnumber the year-round residents of the small local village. Still others are suburban campuses, such as Purchase, which specializes in the performing arts, and Old Westbury, which was started as an experimental institution to serve minority students, older women, and others who have been "bypassed" by more traditional institutions.

With the exception of Purchase and Old Westbury, which were started from scratch, the four-year colleges are all former teachers' colleges that have, for the most part, successfully made the transition into liberal arts colleges on the small, private New England model. Now they face a new problem: the growing desire of students to study business, computer science, and other more technically oriented subjects. Some have adjusted to these demands well; others are trying to resist the trend.

SUNY's technical and specialized colleges, while not enjoying the prominence of the colleges of arts and sciences, serve the demand for vocational training in a variety of two- and four-year programs. Five of the six agricultural and technical colleges—Alfred, Canton, Cobleskill, Delhi, and Morrisville—are concerned primarily with agriculture, but also have programs in engineering, nursing, medical technology, data processing, and business administration. The sixth, Farmingdale, offers the widest range of programs, from ornamental horticulture to aerospace technology. A new upper-division technical campus at Utica-Rome now provides graduates of these two-year institutions with an opportunity to finish their education in SUNY instead of having to head for Penn State University, Ohio State, the University of Massachusetts, or destinations in other directions.

Four of the five statutory schools are at Cornell University—agriculture and life sciences, human ecology, industrial and labor relations, and veterinary medicine—while the internationally known College of Ceramics is housed at Alfred University, another private university. In addition to Utica-Rome, the specialized colleges consist of the College of Environmental Sciences and Forestry at Syracuse, the Maritime College at Fort Schuyler in the Bronx, the College of Optometry in New York City, and the Fashion Institute of Technology, whose graduates are gobbled up as fast as they emerge by employers in the Manhattan Garment District.

The 29 community colleges have traditionally been the stepchildren of the system, but the combination of rampant vocationalism and the rising cost of education elsewhere is rapidly turning them into the most robust members of the family. Students once looked to the community colleges for terminal degrees that could be readily applied in the marketplace. Now, with the cost of college soaring, a growing number of students who otherwise would have been packed off to a four-year college are saving money by staying home for the first two years and then transferring to a four-year college—or even a university center—to get their bachelor's degree.

Following are full-length descriptions of SUNY–Purchase, which is the liberal arts institution best known beyond New York's borders, SUNY–Geneseo, and the four university centers.

SUNY–University at Albany

1400 Washington Avenue, Albany, NY 12222

SUNY–Albany won't win any awards for campus beauty, but it does have some attractive programs. Albany is strong in anything related to politics, public policy, and criminal justice. Study abroad programs in Europe and Asia are also strengths. About 10 percent of undergrads are international or out-of-state students.

Website: www.albany.edu
Location: Suburban
Total Enrollment: 13,315
Undergraduates: 11,125
Male/Female: 50/50
SAT Ranges: V 500–590
    M 520–610
Financial Aid: 54%
Expense: Pub $ $
Phi Beta Kappa: Yes
Applicants: 16,725
Accepted: 63%
Enrolled: 25%
Grad in 6 Years: 62%
Returning Freshmen: 85%
Academics: ✑ ✑ ✑ ✑
Social: ☎ ☎ ☎
Q of L: ★ ★ ★
Admissions: (518) 442-5435
Email Address:
    ugadmissions@albany.edu

Strongest Programs:
Criminal Justice
Atmospheric Science
Physics
Accounting
Business

Founded in 1844 to train teachers, SUNY–Albany offers outstanding programs in arts and sciences, public policy, and human services, along with cutting edge research in nanoscience and biotechnology. Preprofessional programs are strong, and students who long to see the world can hit the books in exotic locales around the globe.

Designed by Edward Durrell Stone, who also designed the Kennedy Center and Lincoln Center, SUNY–Albany's campus is modern and suburban. Almost all the academic buildings are clustered in the center of the campus, while students are housed in symmetrically situated quads so similar in appearance that it usually takes a semester to figure out which one is yours. (Hint: the quads are named for periods in New York history—Indian, Dutch, Colonial, State, and Freedom—and progress clockwise around the campus.) Two new all-weather fields provide much-needed space for varsity teams and serve as a multipurpose recreational field as well.

Most of the preprofessional programs are among the best of any SUNY branch. Students in the public administration and social welfare programs may take advantage of their proximity to the state government to participate in internships. Biology, physics, sociology, and psychology are other notable majors, and undergrads are clamoring for admittance to the university's business administration program, which is especially strong in accounting. The New York State Writers' Institute is the least traditional of Albany's offerings, and with William Kennedy as head of the institute, the university's dream of becoming distinguished for its creative writing is well on its way. The recently established College of Nanoscale Science and Engineering is the first college in the world devoted exclusively to the study of nanoscale science.

All undergraduates must fulfill Albany's 30-credit general education program, which includes courses in disciplinary prospectives, national and international perspectives, mathematics and statistics, pluralism and diversity, communication and reasoning competencies, and foreign language. If this liberal arts exposure whets your appetite for interdisciplinary study, try your hand at human biology, information science, or urban studies. The more career-minded can sign up for one of 40 BA/MA programs or opt for a law degree, with the bachelor's in only six years. Many students take advantage of SUNY–Albany's superior offerings in foreign study. The university was one of the first in the nation to develop exchange programs with Russia (and China, for that matter).

(Continued)
Political Science
Social Welfare
Computer Science

Undergraduates may also study in several European countries, as well as in Brazil, Costa Rica, Israel, Japan, and Singapore. Project Renaissance brings together groups of 300 freshmen with a team of instructors in a shared academic and living community. Participants engage in a yearlong, unified course of study covering 12 hours of the university's general education requirements, and have access to special perks including

"Courses are rigorous and to be successful requires much hard work."

housing and faculty mentors. Qualified students can take part in the new Honors College, which allows freshmen and sophomores to enroll in up to six introductory courses that have been designed by distinguished faculty. The courses emphasize research, service learning, and a creative component. Senior honors students design and complete a year-long research or creative project.

The academic climate is challenging and courses tend to be demanding. "The climate can be very, very competitive," says a junior. "Courses are rigorous and to be successful requires much hard work." Students form study groups to help one another through the coursework, and professors are always available to offer support. "I've seen that most professors are willing to bend over backward for students if they just make an effort to see them," a business major says. Registration doesn't receive the same accolades. "When it is time to pick classes, it is very difficult," says a student, "because most of the seats are taken."

Qualified students can take part in the new Honors College, which allows freshmen and sophomores to enroll in up to six introductory courses that have been designed by distinguished faculty.

One undergraduate describes his peers as "intelligent, assertive, hardworking, urban—generally pretty fast company." Albany students tend to spend a lot of time thinking about their future. But, lest you get the wrong impression, "They're also highly motivated to party at every available moment," another adds. The student body comprises "bits and pieces of every Long Island high school and a dash of upstate, topped off with a Big Apple or two." All but 8 percent of the students are native New Yorkers. African American and Hispanic enrollment now stands at 15 percent combined, while Asian Americans make up another 6 percent. Fifty-five percent of the students are from the top quar-

"You can find an outlet here for even the most obscure interest."

ter of their high school class. SUNY–Albany makes available merit scholarships ranging from $1,000 to $6,500. More than 250 athletic scholarships are offered in a dozen different sports such as soccer, tennis, and golf, to name a few.

Sixty-one percent of students live in university housing; freshmen and sophomores are required to live in dorms. "Housing isn't that great," admits a junior. "Some of the rooms are very tiny and if you get stuck having a common bathroom, it's not very enjoyable." Others report that the co-ed quads are exceptionally friendly, surprisingly quiet, and comfortable. Each floor is divided into four- to six-person suites. But the word from most students is that the best dorms are in the Alumni Quad on the downtown campus. "Alumni has much more attractive rooms and ambiance overall," says a business major. Many students move off campus because "the transportation system to and from campus is convenient and the cost of apartments is as cheap (or cheaper) than living on campus," says a junior. Students on the main campus take their meals at any of the four dorms or at the campus center that includes a food court and bookstore, while downtowners haunt the cheap local eateries as well as their own cafeterias.

The more career-minded can sign up for one of 40 BA/MA programs or opt for a law degree, with the bachelor's in only six years.

While most people are serious about their work, a SUNY–Albany weekend starts on Thursday night for many. Students go to parties or go bar-hopping about town. Students warn that alcohol policies forbidding underage drinking are strict and well-enforced. "It is easy for underage residents to drink on campus, but if you're caught there are severe penalties," says a student. Fraternities and sororities attract 2 percent of the men and 5 percent of the women, and have become the main party-throwers on campus. Albany students tend to be traditional, but rites-of-spring festivals, mandatory after enduring the miserable upstate winters, have produced Guinness records for the largest games of Simon Says, Twister, and Musical Chairs, as well as the all-school pillow fight. Fountain Day brings thousands of students together for the spring turn-on of the infamous podium fountain. Mayfest is a huge all-school concert party that brings in well-known as well as up-and-coming bands.

> **"The transportation system to and from campus is convenient."**

The natural resources of the upstate region keep students busy skiing and hiking. Treks to Montreal and Saratoga are popular. Plus, the student association owns and operates Dippikill, a private camp in the Adirondacks.

Men's basketball, lacrosse, and football, and women's softball, volleyball, and track and field are strong, and the school is a member of the Division I America East Conference. Meanwhile, intramurals engender a great deal of student enthusiasm, and participation numbers in the thousands.

SUNY–Albany is not the concrete, sterile diploma mill it may appear to be. It's a place of opportunity for those willing to put in the hours and hard work. As one veteran warns, "You can find an outlet here for even the most obscure interest, but this is not a school that will educate you when you're not looking."

Overlaps

SUNY–Binghamton, NYU, Boston University, Syracuse, Cornell

If You Apply To ➤

SUNY–Albany: Early action: Nov. 15. Regular admissions: Mar. 1. Financial aid: Apr. 15. Does not guarantee to meet demonstrated need. Campus interviews: recommended, informational. No alumni interviews. SATs: required. SAT IIs: optional. Accepts the Common Application and prefers electronic applications. Essay question (optional): personal statement.

SUNY–Binghamton University

P.O. Box 6001, Binghamton, NY 13902-6000

If 100,000 screaming fans on a Saturday afternoon tickles your fancy, head 200 miles southwest to Penn State. Binghamton has become the premier public university in the Northeast because of its outstanding academic programs, such the Binghamton Scholars and Discovery Initiative, and its commitment to undergraduates.

Website:
   www.binghamton.edu
Location: Suburban
Total Enrollment: 14,018
Undergraduates: 11,174
Male/Female: 52/48
SAT Ranges: V 560–660
   M 600–690
ACT Range: 25–29
Financial Aid: 46%
Expense: Pub $ $
Phi Beta Kappa: Yes

Binghamton University offers a private-school feeling at a public-school price, even for out-of-staters. With more than 130 clubs, nearly 350 study abroad opportunities, and an emphasis on small classes—81 percent of those taken by freshmen have fewer than 50 students—it's no wonder that students who apply here are also considering schools such as Cornell and NYU. Binghamton offers an environment that's intellectually challenging, but manageable—and fun. Students are "well-balanced, and very active on campus, in addition to being very studious," says a nursing major.

Binghamton's campus sits on almost 800 acres of open grassy space, and includes a nature preserve, trails, fountains, and a pond. The oldest buildings date from 1958, so the prevailing architectural style is modern and "functional." Some students say that from the air, the circular campus bears a striking resemblance to the human brain, but administrators say that's merely a coincidence.

Binghamton students apply to one of the university's five schools: the Decker School of Nursing, the Harpur College of Arts and Sciences, the School of Education and Human Development, the School of Management, or the Watson School of Engineering and Applied Science, named for IBM founder Thomas J. Watson, Sr. Regardless of the school they choose, students face the same the general education requirements, which span five thematic areas: language and communication, creating a global vision, sciences and mathematics, aesthetics and humanities, and physical activity and wellness. Management is a popular major, and qualified students may sub-matriculate into Binghamton's MBA program to earn undergraduate and graduate degrees in five years. Other popular choices include psychology, English, biology, computer science, and economics. A bioengineering major has been introduced, and Asian and Asian American studies have been added to the academic menu as well.

> "Students are often encouraged to work together rather than compete on projects and assignments."

Binghamton's academic reputation is enhanced by a tough grading policy, which includes plusses and minuses as well as straight letter grades, and Fs on the transcripts of failing students rather than no credit. "The courses and academic atmosphere are both challenging," says a student. "However, students are often encouraged to work together rather than compete on projects and assignments." Professors receive high marks for classroom presentations and accessibility. "Freshmen are almost always taught by full professors," a senior says. "The teachers really want students to learn and really make an effort to get to know the students." Faculty-supervised independent research, often culminating in a senior honors thesis, is common in Harpur College. Aside from study abroad, students may take a term away from Binghamton at one of more than 170 institutions in the U.S. and Canada through the National Student Exchange program.

The Binghamton Scholars Program offers financial aid, special seminars, and leadership training to exceptional students, along with opportunities for experiential learning and junior- and senior-year capstone projects. The Evolutionary Studies and Global Studies integrated curricula allow students to supplement their major coursework with interdisciplinary exploration. And programs in

> "I would prefer eating here rather than at home."

global and international affairs, information systems, and management bring more than 300 students a year to Binghamton from four of Turkey's most prestigious universities. Binghamton administrators hope the programs will also encourage more of their students to travel to Turkey and the Middle East.

Students at Binghamton are "go-getters," with a liberal bent, says a senior. Although Binghamton offers a top-notch liberal arts and sciences education, word of its excellence has been slow to cross state lines: Only 6 percent of students come from outside of the Empire State. By other measures, though, Binghamton's student body is rather diverse: African Americans make up 6 percent of the total, while Hispanics add 7 percent and Asian Americans 18 percent. Hot political issues range from the global (the war in Iraq) to the local (state funding for campus programs). Binghamton offers merit scholarships worth an average of $3,895, as well as 225 full or partial athletic scholarships in 20 sports.

Fifty-eight percent of Binghamton's students live in the dorms, where options range from traditional double rooms with bathrooms down the hall to suites and apartments. "The rooms are a decent size and well-kept," says a junior. Although securing a room on campus is generally not a problem, a senior warns, "The main housing problem is that there is a very large discrepancy between the best and worst housing." To help make the university seem a little smaller, residence halls are grouped into five areas, each with its own personality and reputation. Dickinson,

(Continued)
Applicants: 21,658
Accepted: 43%
Enrolled: 23%
Grad in 6 Years: 79%
Returning Freshmen: 90%
Academics: ✍ ✍ ✍ ✍ ½
Social: ☎ ☎ ☎
Q of L: ★ ★
Admissions: (607) 777-2171
Email Address:
admit@binghamton.edu

Strongest Programs:
Anthropology
Asian and Asian American
Studies
Chemistry
Comparative Literature
History
Human Development
Nursing
Political Science

Binghamton's academic reputation is enhanced by a tough grading policy, which includes plusses and minuses as well as straight letter grades, and Fs on the transcripts of failing students rather than no credit.

Binghamton teams compete in Division I, but the school doesn't field a football squad. As a result, some of the most significant rivalries are with Cornell, in men's lacrosse and co-rec football.

for example, is the oldest and most stately. Dining halls have plenty of options, including sushi. "I would prefer eating here rather than at home," one student says.

When the weekend comes, Binghamton students know how to let off steam. Aside from the usual concerts and plays, a program called Late Night Binghamton brings free movies, games, a coffee bar, and other nonalcoholic fun to campus, from 10 p.m. until 2 a.m. every Friday and Saturday. There's also a bowling alley, video arcade, and billiard room on the school grounds. "Off campus," says a psychology and sociology major, "there's a zoo, ice-skating, skiing, and a museum." Frat parties also occur off campus; 8 percent of the men and 9 percent of the women go Greek. Only those 21 and over may drink, per New York State law. While some underage students do manage to find alcohol, any caught violating the policy "will be taken care of accordingly," says a senior.

Binghamton itself is "very much a college town," says a sophomore. "We're the only people keeping this place in business." The downtown area offers some restaurants and bars, and "many students volunteer with local groups, such as food drives and mentoring children," says a nursing major. Students also get involved in Special Olympics and various fund-raising walks, and "there is even a co-ed service fraternity." Popular road trips include Ithaca, for parties at Ithaca College and Cornell, and Syracuse, for the

"I've made lifelong friends, gotten a top education, and grown as a person."

Carousel Mall, as well as Cortland and Oneonta, about an hour away by car. The toughest part about going away may be finding a parking space when you return, as permits currently outnumber spaces by about three to one. Annual campus traditions include the Passing of the Vegetables to welcome winter, and Jumping on the Coat during the Spring Fling carnival, to celebrate the arrival of warm weather. Picnic in the Park is the annual senior barbecue.

Binghamton teams compete in Division I, but the school doesn't field a football squad. As a result, some of the most significant rivalries are with Cornell in men's lacrosse and co-rec football, where teams have three men and three women and always a female quarterback. Men's tennis, men's golf, men's and women's soccer, and swimming are also strong. Given the weather in this part of New York, a popular winter pastime is "traying," or downhill sledding on cafeteria trays.

With a four-year graduation rate that is among the highest of any public university, Binghamton has a reputation for an excellent education at a reasonable price that continues to draw smart New Yorkers to its vibrant and growing campus. Despite the hubbub of city life, the university maintains a cozy feel. Says a human development major, "I've made lifelong friends, gotten a top education, and grown as a person."

Overlaps

NYU, Cornell University, Syracuse, Boston University, SUNY–Stony Brook

If You Apply To ➢

Binghamton University: Early action: Nov. 15. Financial aid: Feb. 1. Guarantees to meet demonstrated need of in-state students. Campus and alumni interviews: optional, informational. SATs or ACTs: required. SAT IIs: optional. Accepts the Common Application and prefers electronic applications. Essay question: personal statement.

SUNY–University at Buffalo

15 Capen Hall, Buffalo, NY 14260

Glamorous it is not, but the University at Buffalo offers solid programs in everything from business and engineering to geography and English. The majority of students

come from western New York and a high percentage commute from home. The largest of the SUNY campuses.

Although part of the mammoth State University of New York system, the University at Buffalo ensures it doesn't get overlooked. Very few universities share its strength in medicine, engineering, and computer science, and UB is one of the world's leading supercomputer sites. Its resources are large enough to warrant two campuses, North and South. Considered one of the nation's "most wired" campuses, UB makes sure its students are at the cutting edge of technology. In addition to the sciences, the former private university is strong in the arts, humanities, and professional schools. Students interested in pharmacy and architecture find Buffalo has the only schools in the SUNY system.

The North campus of the University at Buffalo, less than 25 years old and home to most undergraduate programs, stretches across 1,100 acres in the suburbs just outside the city line and boasts buildings designed by world-renowned architects such as IM Pei. Meanwhile, the South campus, along Main Street, favors collegiate ivy-covered buildings and the schools of architecture and health sciences, including the highly rated programs in medicine and dentistry. The university provides connecting bus service—known as the UB Stampede—between the North and South Campuses. A major addition that quadrupled the size of the Multidisciplinary Center for Earthquake Engineering is complete, as is construction of Creekside Village, a dormitory for grad and professional students on the North campus. The newest facilities include the Center of Excellence in Bioinformatics and the Alfiero Center, an addition to the Jacobs Management Center.

On the academic front, the engineering and business management schools are nationally prominent, and architecture is solid. Occupational and physical therapy programs are also quite good, while the English department is notable for its emphasis on poetry. Well-known poets visit the campus frequently, and students not only compose and read poetry, but study the art of performing it as

> "Students here often combine disciplines, work in a variety of fields, and have more developed skills."

well. French, physiology, geography, and music are highly regarded, but other humanities vary in quality. The academic climate varies greatly by major but students report that competition is a given. "There tends to be pressure to do better in class in comparison to your classmates," says a senior.

UB has a multitude of special programs, joint degrees (such as a five-year BS/MBA), and interdisciplinary majors, as well as opportunities for self-designed majors and study abroad. Students accepted into the honors program enjoy smaller classes, priority in class registration, individual faculty mentors, and special scholarships regardless of financial need. Freshmen are encouraged to take University Experience 101, which orients students to UB's academic life, general social experience, and resources. About two-thirds of students do so. General education requirements are standardized and include courses such as writing skills, math sciences, natural sciences, foreign language, world civilizations, and American pluralism. The university's newest programs include majors in biochemical pharmacology, Asian studies, bioinformation and computational biology, film studies, geography, and occupational science. Also available are programs for adult health, geriatric, and acute care nursing. More than $2.2 million in merit scholarships are available through the Honors and University Scholars programs.

Class size can be a problem, especially for freshmen. Scheduling conflicts are not unusual, and required courses are often the most difficult to get into. But registration is available online or by phone. Students seem to accept that some degree of faculty unavailability is the necessary trade-off for having professors who are

Website: www.buffalo.edu
Location: Suburban
Total Enrollment: 22,494
Undergraduates: 16,807
Male/Female: 54/46
SAT Ranges: V 510–600
  M 540–640
ACT Range: 23–28
Financial Aid: 53%
Expense: Pub $ $ $
Phi Beta Kappa: Yes
Applicants: 18,391
Accepted: 57%
Enrolled: 31%
Grad in 6 Years: 59%
Returning Freshmen: 88%
Academics: ✍ ✍ ✍ ✍
Social: ☎ ☎
Q of L: ★ ★
Admissions: (716) 645-6900
Email Address:
  ub-admissions@buffalo.edu

Strongest Programs:
Business Administration
Engineering
Psychology
Architecture
Communications
English
Computer and Information
  Sciences

Students without cars can get trapped when the intercampus bus stops running after 2:00 a.m. on weekends.

experts in their fields at a school where graduate education and research get lots of the attention.

"In more cases than not, the professors are generally interested in their students and are more than willing to work with them and help them out on their own time," a student says. The academically oriented student body spends plenty of time in UB's six main libraries or one of the several branches, which are for the most part comfortable and well-stocked at three million volumes.

Nearly 60 percent of UB students ranked in the top quarter of their high school class and 90 percent hail from New York. African Americans and Hispanics combined account for 11 percent of the student body, and Asian Americans represent

"You can meet so many people and have the time of your life."

another 8 percent. "Students here often combine disciplines, work in a variety of fields, and have more developed skills"

than those at rival institutions, says a theater major. UB's considerable efforts in increasing awareness of diversity include a Committee on Campus Tolerance, Office of Student Multi-Cultural Affairs, and Multi-Cultural Leadership Council. Merit scholarships worth an average of $2,791 are available for qualified students; student athletes vie for 114 athletic awards.

Students seem to accept that some degree of faculty unavailability is the necessary trade-off for having professors who are experts in their fields at a school where graduate education and research get lots of the attention.

Thirty-eight percent of students live on campus; the rest commute from home or find apartments near the Main Street campus. Students warn that potential renters should shuffle off to Buffalo a couple of months early to secure a place. UB has added several apartment-style complexes over the last few years for upperclassmen who wish to live on campus. Those complexes feature cable, high-speed Internet connections, and central air-conditioning. Most of the on-campus dwellers are housed on the Amherst campus. Governors is known as the nicest, smallest, and quietest dorm on campus, while social butterflies prefer Ellicott. The Main Street campus dorms are smaller, older, and of a more traditional collegiate design, which upperclassmen tend to prefer. Three all-freshmen dorms house extremely sociable freshmen. "Sometimes three students are squeezed into two-person rooms," says a freshman. Security on campus is adequate, with a full-time police station comprised of trained officers and emergency call lights all around. Some of the cafeteria food gets mixed reviews, but there are a variety of good choices.

Despite the large number of commuters and the split campus, "UB is infamous for its social life," says a junior. "There is so much going on, both on and off campus. You can meet so many people and have the time of your life." Fraternities attract 3 percent of the men, and 4 percent of UB women join sororities. Drinking is banned in the "dry" dorms, but students over 21 are allowed to drink in the "wet" dorms. Activities during the week are mostly on campus, but students tend to gravitate to downtown Buffalo on the weekends. Friday-night happy hour centers on beer and the chicken wings that spread the fame of Buffalo cuisine. Also popular are the Albright-Knox Art Gallery, with its world-renowned collection of modern art,

"The best road trip is 10 minutes to Canada."

and the Triple-A baseball Bisons, who play downtown. The two major pro teams, the Buffalo Bills in football and the Sabres in hockey, are both top draws.

Although most students are content to stay in Buffalo, those who want a change of scene can drive to Niagara Falls, Toronto, Rochester, or Cleveland. "The best road trip is 10 minutes to Canada," says an anthropology and geology double major.

Having a car might be a good idea, though parking can be a problem on campus. Students without cars can get trapped when the intercampus bus stops running after 2:00 a.m. on weekends. The winters are cold in Buffalo, but students can take refuge inside a series of tunnels that connect the buildings. The flip side is that the outlying areas of the city offer great skiing, skating, and snowmobiling—and the ski club even offers free rides to the slopes. UB supports more than 500 other student organizations ranging from jugglers to math enthusiasts. Students can preview their honeymoons by darting over to Niagara Falls, just a few minutes away, or flee the country altogether by driving to Canada, where the drinking age is lower.

"The on-campus events revolve around football and basketball," says a senior. UB is the only major SUNY unit to field a Division I-A football team, which competes in the Mid-American Conference. The men's basketball team won a school-record 23 games in 2004–05 and finished second at the MAC tournament. The men's and women's cross-country teams are among Buffalo's championship-caliber squads, as are women's basketball and men's and women's swimming. School spirit is sometimes generated at the student union and UB's impressive sports complex, which boasts the fourth-largest pool in the world, a 10,000-seat arena, squash and racquetball courts, and other amenities. Intramural sports are popular, and earthy types appreciate the annual Oozefest—a mud-volleyball tournament.

UB students love the size of their school, with its huge range of academic programs, social events, and people to meet. Yes, students are exposed to the long Buffalo winter, but they also get exposed to some top-notch professors. And they get to meet a diverse mix of native New Yorkers, other East Coast residents, and international students drawn to the university's outstanding science programs.

Overlaps

SUNY–Albany, SUNY–Binghamton, SUNY–Stony Brook, Cornell University, NYU, Syracuse

If You Apply To ➢

University at Buffalo: Rolling admissions. Early decision: Nov. 1. Regular admissions: Nov. 1 (priority). Financial aid: Mar. 1 (priority). Housing: May 1. No campus or alumni interviews. SATs or ACTs: required. Prefers electronic applications. No essay question.

SUNY–College at Geneseo

1 College Circle, Geneseo, NY 14454

Geneseo is a preferred option for New Yorkers who want the feel of a private college at a public-university price. It is similar in scale to William and Mary and Mary Washington in Virginia, smaller than Miami of Ohio. Offers business and education in addition to the liberal arts.

The SUNY College at Geneseo offers a seriously academic environment at a state-school price. Students here "tend to be friendly, liberal, hardworking, and most like to have fun on the weekends," says a junior. Indeed, this public institution attracts high achievers from around the nation. Responsive, attentive professors help compensate for the long winters and somewhat isolated location. And excellent preprofessional programs in disciplines such as education and business have been making it harder to win admission to this most bucolic campus of the New York State university system.

Geneseo sits in the scenic Genesee Valley of western New York. Campus architecture ranges from Gothic to modern, and the surrounding community has been designated a National Historic Landmark Community by the U.S. Department of the Interior. An elementary education major calls the town "small and inviting" and says the historic buildings and nearby forests and mountains make for "lots of beautiful scenery." Education is the most popular major at Geneseo, followed by biology, business administration, psychology, and English. Regardless of their course of study, all students must demonstrate foreign language proficiency. General education requirements also include two courses each in the natural sciences, social sciences,

Website: www.geneseo.edu
Location: Small town
Total Enrollment: 5,242
Undergraduates: 5,174
Male/Female: 41/59
SAT Ranges: V 600–670
  M 600–670
ACT Range: 26–29
Financial Aid: 47%
Expense: Pub $ $
Phi Beta Kappa: No
Applicants: 10,448
Accepted: 41%
Enrolled: 24%
Grad in 6 Years: 81%

Students may enroll in a one-credit First Year Seminar focused on a topic such as genealogy, Chinese medicine, or Frederick Douglass.

If Greek parties don't appeal, students can hit the campus dance club (the Knight Spot) or partake in school-sponsored late-night activities at the College Union.

and fine arts; a two-semester sequence in Western Humanities; and one course each in non-Western traditions, U.S. history, and numeric and symbolic reasoning. During the freshman year, all students take a seminar in critical writing and reading in a small class focusing on a unique topic related to the instructor's discipline. Students may also enroll in a one-credit First Year Seminar focused on a topic such as genealogy, Chinese medicine, or Frederick Douglass. Cooperative programs with other SUNY campuses in dentistry, optometry, business, physical therapy, and other disciplines allow students to finish their graduate degrees a year ahead of schedule. New additions include dual degree programs with Hacettepe University in Turkey (economics) and Universidad de la Americas in Mexico (international relations).

As for the academic climate, "Some classes are easier than others, but for the most part, they are very rigorous and demand a lot of work," says a business and computer science major. "The quality of teaching has been amazing," says a junior majoring in education. Another education major adds, "Professors are always willing to help and tutoring is easily accessible. If you put in the work, you will get the grade you deserve." One benefit of Geneseo's size: "You will never have a TA, except

"You will never have a TA, except in labs."

in labs," says a freshman. And even then, "labs are overseen by a professor." Twenty students from each class are invited to join the prestigious Honors Program; membership includes a $1,400 annual scholarship and five courses designed specifically for the program, plus a thesis during the senior year.

Ninety-six percent of Geneseo students are from New York State, and 80 percent attended public high school. "Students are tight-knit. They know most other students and are very friendly. They're always willing to help others," says a business and computer science major. They're also "smart and hard-working," says an elementary education student. Asian Americans make up 8 percent of the student body, while Hispanic students comprise 4 percent and African Americans make up 2 percent. "There are a lot of concerns and issues dealing with diversity," says a junior. Admissions are not need-blind; Geneseo offers merit scholarships averaging $1,900 each, but no athletic awards.

Fifty-six percent of Geneseo students live in the dorms, and rooms are guaranteed for four years. "The dorms are comfortable, well-maintained, and clean," says a junior; however, "Students tend to live off campus past their sophomore year." Students say they feel safe at Geneseo, and while campus chow can't compare to mom's home cooking, "It has improved over the years," says a senior. "Now it's actually pretty good." Provisions are made for vegans and vegetarians, and special meals are served on holidays.

Fraternities draw 10 percent of the men and sororities claim 12 percent of the women; if Greek parties don't appeal, students can hit the campus dance club (the Knight Spot) or partake in school-sponsored late-night activities at the College Union. Some dorms are dry; even in those that aren't, only students over 21 are permitted to have or consume alcohol, and they may not do either in the presence of underage students. Still, this is college, says a sophomore: "Rules are made to be bro-

"Once you are here, you must work hard."

ken." Circle K and Alpha Phi Omega, the co-ed service fraternity, allow students to make a difference in the community. In addition, there are nearly 200 other student-run organizations, including a newspaper and radio and television stations.

The "gorgeous small town" of Geneseo "really caters to college students," says an education major. There's a supermarket and a Wal-Mart, and restaurants and shops on Main Street. Outdoorsy types will appreciate the nearby mountain ranges, which offer plenty of opportunities to hike and ski; boaters will enjoy beautiful Conesus Lake, only a 10-minute drive from campus. Popular road trips include

Rochester, 30 miles north, and Buffalo, 60 miles west; don't forget your hat, mittens, and parka!

Half of Geneseo students participate in intramural sports, and "broomball is most popular," says a sophomore. Basketball and volleyball are also very popular. Geneseo's varsity teams compete in NCAA Division III, and the men's and women's cross-country, indoor track, and swimming teams have brought home SUNY Athletic Conference championships in recent years. But

> **"The dorms are comfortable, well-maintained, and clean."**

hockey stirs up the most school spirit. "Hockey is huge here," says a senior. Students also look forward to Spring Fest and MidKnight Madness, the pep rally that precedes the basketball season.

The SUNY College at Geneseo gives students the best of two worlds. Given its size, professors can provide the kind of personal attention normally seen only at private liberal arts colleges; because of its public status, that attention comes at bargain-basement cost. Those factors have made it more and more difficult to get in—and, it turns out, getting in is only half the battle. "Students are very serious about their education," says a sophomore. "Once you are here, you must work hard."

Overlaps

SUNY–Binghamton, Rochester, Hamilton, Cornell, Boston College, Colgate

If You Apply To ➤ **SUNY–Geneseo:** Rolling admissions. Early decision: Nov. 15. Financial aid: Feb. 15. Housing: May 1. Does not guarantee to meet demonstrated need. Campus interviews: recommended, informational. No alumni interviews. SATs or ACTs: required. Accepts electronic applications. Essay question (choose one): type of injustice that makes you mad; scary childhood experience that you now find humorous; Do you believe media are the primary shapers of public opinion?; unpublished writing sample on a topic of your choice, but not a research paper or book report.

SUNY–Purchase College

735 Anderson Hill Road, Purchase, NY 10577-1400

One of the few public institutions that is also an arts specialty school. The visual and performing arts are signature programs, though Purchase has developed some liberal arts specialties in areas like environmental science.

SUNY–Purchase College is a dream come true for aspiring artists of all kinds—an academic environment that provides a strong sense of community and support, yet celebrates individuals for their unique talents and contributions; it's OK to be an individual here. A senior says the best thing about Purchase is "the freedom to focus on whatever you want without feeling pressured to join anything to fit in."

Set on a 500-acre wooded estate in an area of Westchester's most scenic suburbia, Purchase has a campus described by one student as "sleek, modern, ominous, and brick." The college has earned a national reputation for its instruction in music, dance, visual arts, theater, and film. Almost all the faculty members in the School of the Arts are professionals who perform or exhibit regularly in the New York metropolitan area, and the spacious, dazzling facilities rank among the best in the world. Purchase College boasts of the Neuberger Museum, the sixth-largest public college museum. The four-theater Performing Arts Center is huge, and dance students, whose building contains a dozen studios, whirlpool rooms, and a "body-correction" facility, may never again work in such splendid and well-equipped surroundings. Recent campus improvements include townhouse-style student apartments that allow more than 400 additional students to hang their hats on campus and a new student services building.

Website: www.purchase.edu
Location: Suburban
Total Enrollment: 3,291
Undergraduates: 3,163
Male/Female: 45/55
SAT Ranges: V 510–620
  M 480–580
ACT Range: 20-25
Financial Aid: 54%
Expense: Pub $ $
Phi Beta Kappa: No
Applicants: 6,946
Accepted: 31%
Enrolled: 33%
Grad in 6 Years: 47%
Returning Freshmen: 78%

(Continued)

Academics: ✍ ✍ ✍ ½
Social: ☎ ☎ ☎
Q of L: ★ ★ ★
Admissions: (914) 251-6300
Email Address:
admissn@purchase.edu

Strongest Programs:
Acting
Art History
Dance
Environmental Science
Film
Liberal Studies
Music
Women's Studies

The college has earned a national reputation for its instruction in music, dance, visual arts, theater, and film. Almost all the faculty members in the School of the Arts are professionals who perform or exhibit regularly in the New York metropolitan area.

Mingling with highly motivated and talented performers and artists can make some students in the liberal arts and sciences feel a little drab and out of place. "Dancers, actors, visual artists, and music students pull the most weight as far as campus life is concerned," says a student. Still, Purchase is a fine place to study humanities and the natural sciences, particularly literature, psychology, art history, environmental studies, and biology. Most of the shaky liberal arts and sciences programs are confined to some majors in the social sciences and language and culture, where offerings are limited.

Students in the liberal arts and sciences spend one-third of their time at Purchase fulfilling the general education requirements, which include 10 knowledge areas. Students now take mathematics, natural sciences (fulfilled by the Science in the Modern World courses), social science, American history, Western civilization, other world civilizations, the arts, humanities, foreign language, and basic communications. In addition to these core requirements, they also must take critical thinking and information management as skill areas, and everyone must complete a senior project. Students in the arts divisions usually have many more required courses, culminating in a senior recital or show.

"We believe strongly in animal rights, civil rights, women's rights, etc."

There are two separate sets of degree requirements, one for the liberal arts and sciences and one for the performing and visual arts. BFA students in the performing and visual arts are required to sample the liberal arts; BA and BS students in the liberal arts and sciences are required to sample fine arts courses. The college also offers several certificate programs including computer science, arts management, and early child development. New majors include arts management and environmental studies.

The atmosphere at Purchase varies between programs. "Our classes are rigorous and competitive, but on a level that demands a lot from all students," reports a senior. Students tend to be very serious about their own personal achievements. Professors tend to be accessible and friendly. "Everyone comes from actual industry experience and knowledge," says a senior.

Purchase is comprised of "all the students from high school who weren't the cheerleaders and football players," says one student, describing classmates as "artsy, creative, hippies, gay, vegans, and open-minded, liberal activists." Eighty-one percent of the students are from New York State, most from New York City and Westchester County. Others are from Long Island, New Jersey, and Connecticut, but all are different and very political (that means very liberal). Minorities make up 22 percent of the student body. "Our student body is full of scholars and artists,

"A car is a definite must at Purchase."

most of whom are heavily driven by our politically active campus," says a women's studies/sociology major. Indeed, the largest student organization on this politically active campus is the gay and lesbian union. "We are diverse, but politically, we are a very liberal college," says a senior. "We believe strongly in animal rights, civil rights, women's rights, etc." In addition to need-based aid, merit scholarships are awarded each year on the basis of academic achievements, auditions, and portfolios.

The living facilities have undergone renovations, but continue to receive mixed reviews. "We have a huge new dorm being built that I'm sure will solve housing space issues," says one student. "Most freshmen will live in converted triples," says a senior, "meaning a room meant for two." The two eating facilities offer decent fare, and for those who tire of institutional cuisine, there is a student-run co-op that specializes in health food. Forty-two percent of the student body commutes from nearby communities, though housing in the surrounding suburbs is expensive and hard to find.

The campus is a neighbor to the world headquarters of IBM, Texaco, AMF, General Foods, and Pepsico, but while sharing the "billion-dollar mile" with a few Fortune

500s might excite students at some other SUNY schools, "it doesn't do much for us except provide convenient antiapartheid demonstration locations," admits one Purchase activist. No college town exists *per se* at Purchase. "Because so many students are from New York, the weekends are pretty dead," complains one student. Still, the Big Apple provides a regular weekend distraction that inhibits the formation of a tight campus community. Since the campus shuttle bus runs only once on the weekend, students started their own van service, which goes into Manhattan three times a day. Still, "a car is a definite must at Purchase," counsels one student. The Performing Arts Center is host to at least two student or faculty performances every weekend, and there is a constant flow of New York artists and celebrities. Notes a literature major, "Most of the social life on campus takes the form of parties thrown by apartment residents," and the over-21 crowd often frequents the Pub. Fraternities and sororities are definitely out. "The closest we come to Greek are the two guys from Athens who go here," quips a staunch independent. Besides, as one artist explains, "individuality is far more important to the artist than being part of a group." Despite the unconventional aura of the

> **"The closest we come to Greek are the two guys from Athens who go here."**

place, Purchase is not without its traditions. There's an annual autumn dance, a Spring Semiformal, and an April Showers campus festival, and on Halloween there are ghost stories told at a historic graveyard on campus.

Purchase is a member of the NCAA (Division III), and the few competitive teams include men's basketball, women's volleyball, and women's soccer. Intramural programs draw 30 percent of students, but informal Frisbee-tossing remains more popular than organized sports. Says a student: "Our 'teams' are our dancers, our vocalists, our musicians, and our theater companies."

Despite a conspicuous lack of a college-town atmosphere, Purchase is a perfect place to study the arts and still be able to indulge in academics of all kinds, or vice versa. "It's a great place to come and devote yourself to your craft," says a senior. Those willing to put up with what a student calls "an isolated campus full of ugly architecture," may find that Purchase offers the opportunity for a personalized, diverse education unique within the SUNY system.

Overlaps
SUNY–New Paltz, NYU, SUNY–Binghamton, SUNY–Stony Brook, SUNY–Albany, Ithaca

If You Apply To ➤ **SUNY–Purchase College:** Rolling admissions. Does not guarantee to meet full demonstrated need. Campus interviews: optional, evaluative (required of theatre design/technology and film program applicants). No alumni interviews. SATs or ACTs: required. SAT I preferred. SAT IIs: optional. Accepts the Common Application and prefers electronic applications. Essay question: personal or historical event of impact; personal influences; describe a time you took a leadership role. Apply to particular school or program. Auditions held for acting, dance, and music.

SUNY–Stony Brook University

118 Administration Building, Stony Brook, NY 11794-1901

Strategically located 90 minutes from New York City, Stony Brook has risen a few notches in the SUNY pecking order. The natural sciences, engineering, and health fields are the major drawing cards. Situated in the lap of Long Island luxury, Stony Brook offers easy access to beachfront play lands.

Stony Brook, one of the academic leaders in the SUNY system, aims to be the model of a student-centered research university. The six undergraduate colleges provide a small

Website: www.stonybrook.edu
Location: City outskirts
Total Enrollment: 16,401
Undergraduates: 13,117
Male/Female: 50/50
SAT Ranges: V 520–620
  M 560–660
Financial Aid: 59%
Expense: Pub $ $
Phi Beta Kappa: Yes
Applicants: 18,206
Accepted: 51%
Enrolled: 27%
Grad in 6 Years: 59%
Returning Freshmen: 87%
Academics: ✍ ✍ ✍ ✍
Social: ☎ ☎ ☎
Q of L: ★ ★
Admissions: (631) 632-6868
Email Address:
  enroll@stonybrook.edu

Strongest Programs:
Anthropology
Biology
Computer Science
Engineering Science
English
Nursing
Marine Sciences
Music

The longtime ban on fraternities and sororities has been lifted, so a fledgling Greek system is another option.

college community experience with all the assets of a leading research university. The public university has made a name for itself with its top-notch programs in the hard sciences. It has also become known for its highly competitive learning environment and the high quality of its professors.

The school's location on Long Island's plush North Shore (Gatsby's stomping grounds) is a wonderful drawing point. Sitting on about 1,000 wooded acres just outside of the small, picturesque village of Stony Brook, and only 90 minutes from New York City and half an hour from the beaches of the South Shore, the campus is a conglomeration of redbrick Federal-style buildings interspersed with several modern brick and concrete designs. Campus beautification is a priority, and grass and trees have replaced much of the uninspiring campus concrete. A new humanities building is the most recent addition to campus, and the main entrance has been reconfigured to be more functional and attractive.

Coming of age in the high-tech era, Stony Brook quickly became widely known and respected for its science departments. Facilities are extensive, and the science faculty includes a number of internationally known researchers. The comprehensive university hospital and research center make health sciences strong, especially physical therapy. The hospital, which has been ranked among the nation's best for teaching, attracts grants to the campus and offers a lot of opportunities for various research programs for undergrads as well as graduate students. Engineering is also strong, along with business administration and psychology. The school's arts program has been strengthened with a fine arts center, complete with studios and a reference library. The center complements Stony Brook's beautiful five-theater Staller Center for the Arts. The music department faculty boasts the American pianist Gilbert Kalish. Students spend a lot of time studying. "The classes are challenging but not impossible," says a junior.

Stony Brook's Diversified Education Curriculum requires students to demonstrate competency in math, communication, and critical thinking. Students must also show an understanding of natural and social sciences, and knowledge of American history, Western civilization, other world civilizations, the humanities, and the arts. Students in the College of Arts and Sciences, the College of Business, and some within the School of Health Technology and Management are required to demonstrate basic foreign language proficiency and must know how to use computers. Freshmen participate in theme-based academic and cocurricular programs, which include two small seminar courses. The first course is an introduction to the university; the second is up to the faculty member teaching it to decide. Its purpose is to introduce students to what the faculty is studying and researching. New degree programs include majors in Asian and Asian American studies, marine sciences, marine vertebrate biology, European studies, and technological systems management.

> "The classes are challenging but not impossible."

Although professors spend much time with their own research projects, they make time for their students. "Most of the professors are excellent," says a student. "They are very clear and helpful." TAs are rarely used, according to a freshman, and "all lecture classes are taught by professors." Freshmen are the last to register, so they sometimes have to wait a semester or two to get into the most popular electives.

An Undergraduate Research and Creative Activities program (URECA) offers undergraduates the opportunity to work on research projects with faculty members from the time they are freshmen until they graduate. Women in Science and Engineering (WISE) is a multifaceted program for women who show promise in math, science, or engineering. All freshmen—residents and commuters alike—now enter the university as members of one of six undergraduate colleges. Each college has its own faculty director, as well as both academic and residential advisors. Some students take advantage of one of Stony Brook's wonderful travel programs (France, England, Italy, Japan, and Tanzania are some possibilities), while others choose established internships in the fields of policy analysis, political science, psychology, foreign language, and social welfare. Combined BA/MA or BS/MS programs are available in engineering, the teaching of math, and management and policy.

Ninety-one percent of Stony Brook students hail from in-state, and about half commute from Long Island homes. Sixty-nine percent graduated in the top quarter of their high school class, and more than half of Stony Brook's graduates go on to graduate and professional schools. The student body is 9 percent African American and 9 percent Hispanic, and Stony Brook enrolls 22 percent Asian Americans. The university has a Campus Relations Team, composed of university police officers who educate the community on topics ranging from personal safety to rape prevention to drug and alcohol awareness. Students are also required to take classes focusing on different cultures. The university does not meet the demonstrated financial need of all accepted applicants. Approximately 59 percent of those who apply, however, do receive financial aid. Merit scholarships averaging $2,972 are given out each year, in addition to 248 athletic scholarships.

"Social life on campus is very organized. There are major events held every few weeks."

Stony Brook, which has one of the largest residential facilities in the SUNY system, has in the past several years completed a major rehabilitation, giving students access to state-of-the-art fitness centers, computing facilities, Internet access, and widescreen TVs. Fifty-four percent of undergrads live in university housing. "Some dorms are nice, some are gross," says a psychology major, adding that the "suites are better than corridor style." While residential freshmen must take a meal plan, upperclassmen who live on campus either opt for a flexible food-service plan or pay a nominal fee to cook for themselves. The suites come equipped with dishwashers and ranges, and each hall has a lounge and kitchen area, all of which, students say, could be kept a lot cleaner. Kosher and vegetarian food co-ops keep interested students well supplied with cheap eats.

An Undergraduate Research and Creative Activities program (URECA) offers undergraduates the opportunity to work on research projects with faculty members from the time they are freshmen until they graduate.

Most activities take place on campus, since there isn't too much to do in Stony Brook. "Social life on campus is very organized," says one student. "There are major events held every few weeks." The university has fairly strict policies on alcohol consumption, and "the policies are as effective as possible with young students in college," a senior reports. The longtime ban on fraternities and sororities has been lifted, so a fledgling Greek system is another option. Current and classic movies are screened during the week, and other entertainment is available in the form of frequent concerts, plays, and other performances. Annual festivals in the fall and spring and the football game with Hofstra are among the biggest social events of the year. Because many students go home on the weekends, Thursday is the big party night.

"All lecture classes are taught by professors."

The students who remain on the weekends often go beachcombing on the nearby North Shore or the Atlantic Ocean shore of Long Island, or head into New York City. "You absolutely need a car if you want to get around the town at all," says a junior. Still, many students make do with trains, and a station is conveniently located at the edge of campus. "Stony Brook is a wealthy residential town that cannot be categorized as a 'college town,'" one student says. "It doesn't appreciate the large campus located within its limits." Nearby Port Jefferson offers small shops and interesting restaurants. Sports facilities have been upgraded, and all 20 varsity teams compete in Division I. Intramurals provide one of the school's greatest rallying points, and competition in oozeball (a mud-caked variant of volleyball) is especially fierce.

Though Stony Brook is not old enough to have ivy-covered walls, it does offer some of the best academic opportunities in the SUNY system. Students have to

Overlaps

SUNY-Albany, SUNY-Binghamton, NYU, SUNY-Buffalo, Penn State, Cornell

maneuver around lots of rough spots, including increasing class sizes and decreasing course offerings. Yet despite these budget-crisis-induced problems, students share in the promise of Stony Brook's future. In the meantime, they boast of their school's diversity and creativity as well as the feeling of hospitality that pervades campus life.

If You Apply To ≫

SUNY–Stony Brook: Rolling admissions. Early Action: Nov. 15. Financial aid: Mar. 1. Does not guarantee to meet demonstrated need. Campus interviews: recommended, informational. No alumni interviews. SATs or ACTs: required. SAT IIs: recommended (math and one additional). Prefers electronic applications. No essay question.

Stetson University

421 N. Woodland Boulevard, DeLand, FL 32723

Stetson keeps company with the likes of Baylor and Furman among prominent Deep South institutions with historic ties to the Baptist Church. The common thread is conservatism, and business is easily the most popular program. Stetson is also strong in music and has a specialty in sport and integrative health sciences.

Website: www.stetson.edu
Location: Suburban
Total Enrollment: 3,097
Undergraduates: 2,131
Male/Female: 42/58
SAT Ranges: V 520–620
  M 520–615
ACT Range: 22–27
Financial Aid: 54%
Expense: Pr $ $
Phi Beta Kappa: Yes
Applicants: 2,782
Accepted: 69%
Enrolled: 29%
Grad in 6 Years: 65%
Returning Freshmen: 77%
Academics: ✍ ✍ ✍
Social: ☎ ☎ ☎
Q of L: ★ ★ ★
Admissions: (386) 822-7100
Email Address:
  admissions@stetson.edu

Strongest Programs:
Biology
Business
English
Music
Political Science

Stetson University, named for the maker of the famed 10-gallon hat, draws students from around the Southeast with its small size and emphasis on the liberal arts. Long a bastion of conservatism, students say the school has become more liberal since cutting ties with the Southern Baptists. With top-notch business courses and surprising strengths in music and Russian studies, this Florida university continues to attract students who aren't afraid to wear a variety of hats during their stay.

Located halfway between Walt Disney World and Daytona Beach, Stetson's 170-acre campus features mainly brick structures in styles from Gothic to Moorish to Southern Colonial. While some modern wood buildings are scattered about, the theme is decidedly old-fashioned, complete with royal palms and oak trees. A $12.6 million renovation brought high-resolution cameras, new computers, and state-of-the-art sound and projection systems to classrooms in the Lynn Business Center, along with a 144-seat auditorium for multimedia presentations. New student housing opened last year and accommodates 335 students among six buildings.

Stetson has three undergraduate colleges—music, business administration, and arts and sciences—and its general education requirements apply to all of them. Those requirements include two freshman English courses and one course each in religious heritage, oral communications, and math. Students must also demonstrate proficiency with computers and in a foreign language, and take courses in the natural and social sciences and in civilization and contemporary culture. Stetson has also added a First Year Studies program, including living-learning communities and seminars specifically for first-year students, to help with the transition from high school to college.

Business is Stetson's most popular program, and would-be money managers benefit from the Roland George Investments program, where they oversee a $2.8 million cash portfolio. Students who hope to work for themselves can tap into the Prince Entrepreneurial Program, a mentoring initiative that connects them with successful business owners. In addition, there's a major in family business, plus the Family Business Center, which offers tips on how to balance work and family obligations, especially when they're one and the same. Education is the second most popular major, and that department has partnered with Walt Disney and the Osceola County School Board to develop a state-of-the-art school and teaching academy in the Disney-created city of Celebration, Florida. Stetson's music department is notable (no pun intended) for its programs in brass instruments, organ, and voice.

Classes are "extremely competitive with very rigorous coursework," says a senior. Students in the college of arts and sciences benefit from the Sullivan Writing Program and the Lawson Program in Philosophy. They also complete a research project before graduation; those who choose to do so in the summer may get funding from the Stetson Undergraduate Research Experience (SURE) program. Professors are always willing to help; a management major says many have worked in the field they are teaching before stepping in front of the lectern. Sixty-nine percent of classes have 25 or fewer students and there are no graduate students behind the lectern. "The quality of teaching is really high," a junior says. Students with wanderlust may study abroad in England, Mexico, Russia, Germany, Spain, France, and Hong Kong; Stetson's honors program incorporates foreign study, community service and a senior colloquium, and also allows students to create their own majors. In addition, there are internship opportunities in Germany, England, and Latin America, and each year, professors lead study trips to places such as Turkey, Greece, and the Czech Republic.

Seventy-six percent of Stetson "Hatters" are native Floridians; they tend to be white, wealthy, and friendly, a sociology major says. African Americans constitute 4 percent of the student body, Hispanics make up 7 percent, and Asian Americans add 2 percent. Students report little interest in political or social issues such as campus diversity. Seventy-six percent of Stetson students receive financial aid and scholarships based on academic merit, rather than need, and the average award is $9,975. Stetson also hands out athletic scholarships in baseball, basketball, golf, soccer, tennis, softball, cross-country, and volleyball.

"The quality of teaching is really high."

Sixty-three percent of Stetson students live in the school's dorms, all but two of which are co-ed. "The dorms are nice but the rooms and beds are way too small," grumbles one student. "You feel like a sardine sometimes." Students say Emily Hall, Chaudoin, Stetson, and Conrad are in the best shape, but they add that more people would probably move off campus if the school didn't cut financial aid awards for doing so. Stetson's traditional cafeteria is known as the Commons, and features a made-to-order deli sandwich line, a pizza bar, and a grill serving burgers and fries. There's also the Hat Rack food court, with a bagel stand, a smoothie shop, and another fast-food grill. "The food is edible if you like bricks and diverse if you like pasta," quips a junior.

Social life at Stetson occurs mainly off campus, where students usually drift to local bars or seek entertainment further abroad. Fraternities attract 23 percent of the men and sororities draw 22 percent of the women, but after the administration cracked down on Greek parties, they happen "maybe two times a semester, at the most," says a senior. The Council for Student Activities brings in big-name acts, such as former Saturday Night Live comedian Jimmy Fallon and the band Less Than Jake. Music majors also stage concerts. Students look forward to annual events such as Greenfeather, aimed at promoting community service, and Greek Week, when sorority and fraternity chapters compete in a lip-synch contest and other events to raise money for charity. When your birthday rolls around, don't forget to wear your bathing suit—it's a tradition for fellow students to toss you into the mid-campus Holler Fountain.

As for the "adorable, small Southern town" of DeLand, it boasts "shops, galleries, and cafes," but only three bars, so students often head to Orlando (40 minutes from campus) or Daytona Beach (20 minutes) to eat out, shop, or dance the night away. In addition to the omnipresent beaches, Blue Springs and DeLeon Springs offer canoeing

(Continued)
Psychology
Russian Studies
Teacher Education

Stetson has added a First Year Studies program, including living-learning communities and seminars specifically for first-year students, to help in the transition from high school to college.

There are internship opportunities in Germany, England, and Latin America, and each year, professors lead study trips to places such as Turkey, Greece, and the Czech Republic.

In 2005, the women's basketball team won the A-Sun Championship and competed in the NCAA tournament.

and nature-watching. Popular road trips include Miami for clubbing, the Keys for camping, and Gainesville, to see the University of Florida Gators play, especially since Stetson lacks a varsity football team.

Though the gridiron contests here are limited to intramurals, Stetson's other teams compete in NCAA Division I. The school is a perennial powerhouse in baseball, and the team has been to the NCAA Regional Tournament for the past five years. The soccer team has also been in the Conference tournament for six years running. In 2005, the women's basketball team won the A-Sun Championship, and competed in the NCAA tournament; the softball team benefits from a fast-pitch field completed in 2003. For those not up to intercollegiate competition, the Hollis Wellness Center includes a field house, outdoor pool, game room, dance studio, and exercise room. In addition, about 40 percent of students participate in one of 15 intramural sports, ranging from bowling and flag football to Ping-Pong, inner tube water polo, and sand volleyball.

Stetson students savor the one-on-one attention freely given at this small Sunshine State university. After four years spent enjoying great weather and forming close friendships with peers and professors, they emerge with solid academic foundations for future work or study.

If You Apply To ➤

Stetson: Early decision: Nov. 1. Regular admissions: Mar. 1. Financial aid: Mar. 15 (priority). Housing: Jun. 1. Meets demonstrated need of 39%. Campus interviews: recommended, evaluative. No alumni interviews. SATs or ACTs: required (SATs preferred). SAT IIs: optional. Accepts the Common Application and electronic applications. Essay question: significant experience or achievement with special meaning; issue of personal, local, or national concern and its importance; influential person; or why a commitment to a spiritual life, environmental responsibility, diversity, community service, gender equity, or ethical decision making is important.

Stevens Institute of Technology

Castle Point on the Hudson, Hoboken, NJ 07030

Stevens ranks with Clarkson and Worcester Polytech among East Coast technical institutes that offer intimacy and personalized education. Youth-oriented Hoboken is a major plus and a quicker commute to Manhattan than most places in Brooklyn.

Website: www.stevens.edu
Location: Urban
Total Enrollment: 4,689
Undergraduates: 1,789
Male/Female: 75/25
SAT Ranges: V 560–660
    M 620–710
Financial Aid: 57%
Expense: Pr $ $ $
Phi Beta Kappa: No
Applicants: 2,418
Accepted: 47%
Enrolled: 43%
Grad in 6 Years: 72%
Returning Freshmen: 88%

Stevens Institute of Technology doesn't offer a "normal" college experience, says a mechanical engineering major. "If you're interested in parties, look into going elsewhere." On the other hand, says a classmate, "If you are looking for a strong technical education in a great environment, Stevens is where you should be." Just across the Hudson River from Manhattan, students here get a top-notch background in engineering and the sciences, with the cultural, athletic, and gastronomic resources of the Big Apple at their fingertips. The problem is finding time to take advantage of everything New York City offers. (This is the city that never sleeps, nor it seems, do students at Stevens.) "The workload is a love-hate relationship," sighs a freshman. "We hate it while we do it, but it benefits us in the end."

"If you're interested in parties, look into going elsewhere."

There's an eclectic mix of architectural styles on Stevens' 55-acre campus. Many of the residence halls and administrative buildings are redbrick; classroom and lab facilities range from traditional, ivy-covered brownstones to modern glass-and-steel structures. The Lawrence T. Babbio Jr. Center for Technology Management opened in fall 2005 and undergrads now have access to new suite-style housing.

Stevens is organized into three schools—the Charles V. Schaefer Jr. School of Engineering, the Wesley J. Howe School of Technology Management, and the Arthur E. Imperatore School of Sciences and Arts. Engineering has long been king of the hill at Stevens; programs in biomedical, chemical, civil, electrical, environmental, and computer engineering are all highly regarded, as is the major in mechanical engineering, not surprising since Stevens was the first school to award a degree in the field. The climate tends to be competitive and demanding. "The courses are definitely challenging and require a solid investment of your time," says a sophomore. Professors are "incredibly knowledgeable and experienced," according to a math major, "and the only real complaint is that sometimes research is emphasized more than teaching."

Each major at Stevens has its own requirements, but most programs require calculus, chemistry, physics, humanities courses, and physical education. In the engineering school, the core curriculum is followed by technical electives that culminate in a senior design project. Two five-year programs are also available; one allows students to incorporate internships into their studies, and the other enables stu-

> "The courses are definitely challenging and require a solid investment of your time."

dents to take fewer courses per term, without extra tuition charges, to make the workload a little more manageable. Among several new programs are cybersecurity, music and technology, information systems, and naval engineering. Study abroad opportunities include exchange programs with universities in Australia, Scotland, and Turkey, a chemical ecology program in the Dominican Republic, and programs run by the International Student Exchange. Stevens also has an active on-campus recruiting program with major corporations, start-up firms, and the government.

Stevens students "represent a pretty wide spectrum," says a student. "There are students who are into computers or video games, students who are sports-obsessed, and students who practically live in the theater." Sixty-one percent of the students at Stevens are from New Jersey; another third come from the greater New York City portion of New York State. Nine percent of the student body is Asian American, 11 percent is Hispanic, and 5 percent is African American. Women comprise only 25 percent of the student body. Politics doesn't receive a lot of attention on campus, although the Stevens Political Awareness Committee can be quite active. Admissions officers look more closely at high school grades, especially in math and science, than at test scores; evaluative, on-campus inter-

> "There is no excuse for having nothing to do on a Friday night."

views are also required of applicants. In addition to need-based financial aid, 15 percent of students at Stevens receive scholarships based on academic merit, averaging $10,106 per year.

Campus housing is guaranteed for a student's entire four or five-year stay at Stevens, which is fortunate because the Hoboken housing market is almost as tight—and expensive—as the one in Manhattan. Seventy-five percent of Stevens students live on campus. "Campus housing is great for a city school," says a senior. In addition to standard dorms, there are fraternities, old brownstones off campus, and "very nice" university-owned apartments. Students say that the dining hall has improved greatly in recent years. "I enjoy the campus food, and there are many options off campus that are part of the meal plan," says a junior.

Thirty-four percent of Stevens men join fraternities, and 31 percent of the women pledge sororities, so Greek groups control much of the social life on campus. That is changing, though, with the formation of an Entertainment Committee that plans weekly events, from comedy nights to hypnotists and musical guests. "Social life has increased and has been much better, with something to do almost every night. If not, there is always New York City," says a junior. Greenwich Village, Times

(Continued)
Academics: 🖊 🖊 🖊
Social: ☎ ☎
Q of L: ★ ★ ★ ★
Admissions: (201) 216-5194
Email Address:
    admissions@stevens.edu

Strongest Programs:
Mechanical Engineering
Chemical Biology
Computer Science
Business

Stevens boasts a powerhouse men's lacrosse team, which has brought home the Knickerbocker championship multiple times; the women's team isn't bad, either, having captured the same title for three of the past four years.

Each major at Stevens has its own requirements, but most programs require calculus, chemistry, physics, humanities courses, and physical education.

Square, and the bright lights of Broadway are just 15 minutes away on the PATH train. There is also Hoboken, which offers popular pubs and clubs right next to campus. Road trips, often taken by train, include Yankee Stadium in the Bronx and Six Flags Great Adventure, a New Jersey amusement park. Beaches and ski slopes are both within a 90-minute drive. "There is no excuse for having nothing to do on a Friday night," says a mechanical engineering major—except, perhaps, having too much work.

Stevens boasts a powerhouse men's lacrosse team, which has brought home the Knickerbocker championship multiple times; the women's team isn't bad, either, having captured the same title for three of the past four years. Both men's and women's soccer have won Skyline championships. Stevens also has a winning women's equestrian team and championship teams in men's tennis and volleyball. For those without the time or talent to play at the varsity level, there are intramurals in everything from basketball, soccer, and football to dodgeball and even bowling.

With its lopsided male/female ratio and its emphasis on technical disciplines, Stevens Tech isn't for everyone. But that doesn't mean it's not worth a look. "You'll work hard at Stevens," admits one sophomore, "but there are plenty of ways to enjoy yourself." Stevens' urban location and relatively small size help make for fun times once the studying is done.

If You Apply To ➤ **Stevens:** Early decision: Nov. 15. Regular admissions and financial aid: Feb. 15. Housing: Jun. 15. Meets demonstrated need of 27%. Campus interviews: required for local applicants, evaluative. No alumni interviews. SATs or ACTs: required (SATs preferred). SAT IIs: optional. Accepts the Common Application and prefers electronic applications. Essay question: personal statement.

Susquehanna University

Selinsgrove, PA 17870

Susquehanna offers welcome relief from the plodding, unimaginative education at many universities. The university's innovative core curriculum includes personal development and transition skills (e.g., computer proficiency) in addition to more conventional topics. Best known for its business program.

"Susquewho?" That's the question many students ask when they're first introduced to this undergraduate institution. While it may not be a household name, Susquehanna University is earning a reputation as an innovator. Friendly faculty, personal attention, and an increasing emphasis on community make SU a good place to expand your mind and indulge your senses. "SU is gorgeous year-round," says a junior. "It's a place that makes me feel at home and puts me at ease when I'm stressed."

Susquehanna's campus is beautiful, set on more than 200 lush acres on the Susquehanna River. Most of the 50 buildings on campus are brick, with Georgian the predominant architectural style. Selinsgrove Hall, built in 1858, and Seibert Hall, built in 1901, are on the National Register of Historic Places. The campus is compact and serene. The newest campus addition is Trax, a social space for students that hosts live music, dancing, and comedy nights, and offers food and beverages. In addition, renovation of the Degenstein Campus Center in 2005 has significantly improved the dining facilities.

SU's best academic programs are in business and the sciences. The Sigmund Weis School of Business is not only one of the most striking buildings on campus, but also a prestigious accredited business program that attracts the most majors. The Weis School also sponsors a semester in London exclusively for its junior business majors. The business school, along with the other majors, encourages Susquehanna students to take summer internships as a crucial part of their education and future job search. Susquehanna is becoming increasingly recognized for its science programs, especially biology (another popular major), biochemistry, and environmental science. Weaker departments at Susquehanna include classical languages, which are hampered by their small size and lack of funds. Minors in advertising and Asian studies have been added and the Latin minor has been dropped. Graphic design—added only a few years ago—has attracted a large number of students and is already producing award-winning designers.

"SU is gorgeous year-round."

Susquehanna's unusual core curriculum consists of four components: personal development (wellness/fitness, career planning, and orientation to university life); intellectual skills (including computer proficiency, logical reasoning, writing, and foreign languages); perspectives on the world (social sciences, humanities, sciences, and fine arts); and a senior capstone experience. All freshmen must take a writing seminar (or its honors equivalent) that involves small-group readings and discussion of a particular author. Reading centers on a common contemporary work, with the author often visiting campus to partake in the seminar. An "unbelievably helpful" seven-week orientation experience is offered to first-year students; topics include study skills, stress management, and interpersonal communication. Freshmen also have access to Leaders Inc., a series of workshops and mentoring followed by workshops to help students develop their leadership potential.

For about a tenth of each class, the academic experience is defined by the Susquehanna Honors Program. Unlike other programs in schools of similar size, SU's program does not separate its students from the rest of the campus. Instead, it allows students to take most of their classes in other classes with the general student body, thus creating a balance between freedom of choice and a challenging education. "The honors program significantly widens the range of classes open to students and allows them to fulfill their core requirements in creative ways," says participant. Honors student or not, most agree that SU offers a moderately rigorous academic climate. "There is certainly a lot of work, but students are not very cutthroat and do not let academics dominate their lives," says a history major.

Student/faculty interaction is one of Susquehanna's strong points, and students have high praise for their professors. "As a political science major, I enjoy a department where I know all of the professors, have been to most of their houses for dinner and class, and know all of my classmates," says a junior. SU

"Dorm rooms are big with great closets."

students may also take classes at nearby Bucknell University and study abroad on almost every continent, under any of 100 programs offered. The school also offers several study programs in Washington, D.C., and a cooperative dentistry program with Temple University. An assistantship program for outstanding first-year students combines a $16,000 scholarship with hands-on work with a professor or staff member (10 hours per week). Past positions have included academic research, university publications, the Writers' Institute, and marketing research.

Sixty percent of SU students are from Pennsylvania, and 85 percent attended public high school. At a school in which 92 percent of the students are white, and which is "dominated by middle- and upper-class conservative Republicans," most agree that ethnic diversity is lacking. A communications major explains, "Diversity and multiculturalism are big issues. The university wants the student body to be more diverse." To wit, the Center for Teaching and Learning focuses on educating students on issues of diversity, including "invisible differences" such as socio-economic

(Continued)
Returning Freshmen: 85%
Academics: 🖾 🖾 🖾
Social: ☎ ☎ ☎
Q of L: ★ ★ ★ ★
Admissions: (570) 372-4260
Email Address:
  suadmiss@susqu.edu

Strongest Programs:
Accounting
Biology/Chemistry
Business Administration
Music
Psychology
English/Writing
Communication
Environmental Science

Susquehanna is becoming increasingly recognized for its science programs, especially biology, biochemistry, and environmental science.

The business school, along with the other majors, encourages SU students to take summer internships as a crucial part of their education and future job search.

status, religion, and sexual orientation. Merit scholarships averaging $9,655 are available for resident brainiacs, but there are no athletic scholarships.

Residence halls are described as "comfortable." A senior says, "Dorm rooms are big with great closets. All housing is pretty equal." Students must get permission to live off campus, and the 20 percent of students who have that privilege—mostly seniors—are selected by lottery. Construction of a new dorm will help to alleviate recent crowding of three-to-a-room for some freshmen. And how about the food? "The good news is that the dining facilities are newly renovated and beautiful," says a student. "The bad news is that the diversity and quality of the food is awful." Campus security is good, according to most, and includes well-lit walkways and emergency call boxes.

Twenty percent of the men and women belong to fraternities or sororities, respectively, but when Greek parties began getting out of hand on the Susquehanna campus, SU became a dry campus and Greek life was moved off campus. Fall Weekend, homecoming, and Spring Weekend are the big annual events. Favorite campus traditions include a candlelight Christmas service and a Thanksgiving dinner at which faculty members serve students "the best meal of the year." Sports are popular among Susquehanna students, especially when the football team plays Lycoming College. Men's golf won the last 10 Middle Atlantic Conference championships, and men's and women's track, women's soccer, and volleyball have made their mark in recent years. Intramural sports are very popular, with more than two dozen programs offered.

> "It's not uncommon to see an Amish family go by in their horse and buggy."

Outside the university, Selinsgrove is "a small, rural, quaint town" with several restaurants and stores. In the surrounding countryside, "it's not uncommon to see an Amish family go by in their horse and buggy or for them to come to your off-campus house selling home-baked goods," says a student. For those with cars, it's a short drive to additional shopping and entertainment options. Penn State is an hour's drive. SU began by preparing students for the ministry, and the university's commitment to the community has remained strong. Each year, two-thirds of the student population volunteer on major community service projects.

At Susquehanna, "the professors care about their students," a biochemistry major says. A classmate adds that the "personal atmosphere, great faculty/student relationships, and beautiful campus" make Susquehanna worthwhile—and a name worth learning.

Overlaps

Elizabethtown, Muhlenberg, Gettysburg, Bucknell, Ithaca, Penn State

If You Apply To ➤

Susquehanna: Rolling admissions. Early decision: Nov. 15, Jan. 1. Regular admissions: Mar. 1. Financial aid: May 1. Meets demonstrated need of 24%. Campus interviews: recommended, evaluative. Alumni interviews: optional, informational. SATs or ACTs: optional; students may submit two graded writing samples instead. SAT II: optional. Accepts the Common Application and electronic applications. Essay question: significant experience; issue of concern; accomplishments since high school (if applicable).

Swarthmore College

500 College Avenue, Swarthmore, PA 19081-1397

Don't mistake Swarthmore for a miniature version of an Ivy League school. Swat is more intellectual (and liberal) than its counterparts in New Haven and Cambridge. The college's honors program gives hardy souls a taste of graduate school, where many Swatties invariably end up.

Swarthmore College's leafy green campus may be just 11 miles from Philadelphia, but students often don't have the time or the inclination to make the jaunt. That's because they have opted for one of the country's most self-consciously intellectual undergraduate environments. Swatties are bright, hardworking, and eclectic in their interests, and campus life is fabled for its intensity. But the intensity doesn't come from huge amounts of coursework (*a la* Yale) as much as the self-imposed drive of talented students who want to do lots of things simultaneously—from academics to social protest to rugby—and do them well. "Swat is a truly intellectual place where people love ideas with all of their hearts," a senior philosophy major says. "But that doesn't prevent them from having an eye for activism and a knack for partying hard."

Swarthmore's 357-acre campus is a nationally registered arboretum, distinguished by rolling wooded hills. Multistory buildings with natural stone exteriors from local quarries, shaped roofs, and cornices are the norm, fostering a quiet, collegiate atmosphere. A $77 million science center provides students and faculty with 80,000 square feet of lab space, state-of-the-art lecture halls, and flexible computer-friendly workstations. New on campus are a 75-bed dormitory and renovations to Parrish Hall, including a new student lounge.

Swarthmore's student/faculty ratio is quite low—89 percent of classes have 25 or fewer students—so personal attention is the norm. "The professors are dedicated, engaging, accessible, and interested in getting to know each student," a junior says. Students are required to take three courses in each of its three divisions—humanities, natural sciences and engineering, and social sciences—and at least two of the three must be in different departments. Students must also complete 20 courses outside their majors, demonstrate foreign language competency, and fulfill a physical education requirement, which includes a swimming test. A new writing requirement requires students to take three writing courses from at least two divisions. Freshman seminars emphasize close interaction with faculty members in a seminar format; about 70 percent of students participate. The acclaimed honors program features small

> "Swat is a truly intellectual place where people love ideas with all of their hearts."

seminars or independent study, collegial relationships between student and professor, and written and oral examinations by external reviewers at the end of the senior year accompanied by festive banquets. About a third of Swat's juniors and seniors take the honors option, after demonstrating—through their academic records—that they can handle the work.

The college has boosted the number of departments in which students may pursue honors to include studio and performing arts, as well as study abroad. Forty-five percent of Swarthmore students study abroad in countries such as France, Japan, Poland, and Spain, and the Office of Foreign Study helps arrange programs in other countries. Crossregistration is also offered with nearby Haverford, Bryn Mawr, and Penn, and a semester exchange program includes Harvey Mudd, Middlebury, Mills, Pomona, Rice, and Tufts. Through the Venture Program,* students who tire of staring at the chalkboard may take time off for short-term jobs in areas of academic or professional interest.

While the academic climate at Swarthmore is intense, it is not competitive. There is no class rank or dean's list, and there is a big emphasis on group projects. "The courses here are really rigorous, but the academic climate is collaborative," says a student, "and that's encouraged by the professors and administrators." Indeed, the administration has encouraged this spirit of collegiality by sprinkling small lounges and cappuccino bars around the dorms and academic spaces, and it even publishes the code needed to get access to the faculty lounge. Aside from teaching, Swarthmore professors also serve as advisors, each helping a small group of students choose their classes each semester. Students are likewise assigned to Student

Website:
  www.swarthmore.edu
Location: Suburban
Total Enrollment: 1,479
Undergraduates: 1,479
Male/Female: 48/52
SAT Ranges: V 680–770
  M 670–760
Financial Aid: 50%
Expense: Pr $ $ $
Phi Beta Kappa: Yes
Applicants: 4,085
Accepted: 22%
Enrolled: 42%
Grad in 6 Years: 92%
Returning Freshmen: 96%
Academics: 🖾 🖾 🖾 🖾 🖾
Social: ☎ ☎ ☎
Q of L: ★ ★ ★ ★
Admissions: (610) 328-8300
Email Address: admissions
  @swarthmore.edu

Strongest Programs:
Biology
Economics
English Literature
Sociology/Anthropology
History
Political Science
Physics

Students' biggest complaints include lack of sleep and too much work.

Academic Mentors, who shepherd them through the transition to college and the first year on campus.

Consistent with their school's Quaker roots, the student body at Swarthmore pays huge attention to social and political issues, and the Eugene Lang Center for Civic and Social Responsibility has made Swarthmore a national force in the area of service learning. "Swarthmore is characterized by a genuine will to do good in the world," a senior engineering major says. The school encourages students to be as educated as possible on issues of cultural, racial, and socioeconomic pluralism, and the entire community is brought into decisions on issues such as the socially responsible investments and pay scale of campus workers. Liberals outnumber conservatives, students say, but political correctness doesn't get out of hand. "Students here are globally conscious and committed to social justice, thereby creating a great environment for political discussions and activism," says a political science major. Those discussions include the genocide in Darfur, the war in Iraq, and workers' rights. Swarthmore is home to a diverse student body; 11 percent are native Pennsylvanians. The student body is 7 percent African American, 17 percent Asian American, and 12 percent Hispanic. A "Diversity Workshop" exists for all first-year students in the week after orientation. Swarthmore's Quaker roots come through in the concern and respect students have for others.

"Swarthmore is characterized by a genuine will to do good in the world."

Most social life at Swarthmore takes place on campus, and it often begins late, since students hit the books until 10 or 11 p.m. and then head out for fun. Annual activities include Primal Scream, a tradition where everyone screams at midnight the night before exams; the Mile Run, where everyone decorates the McCabe Library with toilet paper around shelves (the founder of Scott Tissue is an alumnus and contributor); and last, but certainly not least, Screw Your Roommate, where roommates pair each other with other roommates and are forced to meet each other in "really crazy" manners. More regular activity options range from parties, dances, and movies to performances and concerts by student troupes. There's also a student-run cafe and Pub Night every Thursday.

When it comes to alcohol, Swarthmore follows Pennsylvania law, which states that you must be 21 to drink. But, says a senior, "the college is more concerned with student health than with catching every last underage drinker. The college treats students as responsible adults, and the students respond by acting as such." Swarthmore's two fraternities attract 6 percent of the men; women are out of luck, since there are no sororities. The Greeks and other campus groups volunteer in both Philadelphia and the nearby smaller town of Chester.

Students' biggest complaints include lack of sleep and too much work. If they're not studying, Swatties are volunteering or out pursuing a personal whim. "Swatties are quirky," says a junior honors history major. "We're passionate about what we do—whether that's mini-golf, studying endangered languages in Alaska, or dissent from the 17th century Anglican Church in Virginia. Swatties don't just do their academic work for the grades or to be first in the class, but because they want to." The village of Swarthmore, known as the "'Ville," has some stores, a pizza parlor and a Chinese restaurant.

"The courses here are really rigorous, but the academic climate is collaborative."

The environment fosters a feeling of safety and security, but students say there's not much in the way of off-campus social activity. For that, students hop the commuter rail into Philadelphia from the on-campus station, where many temptations await, including concerts, dance clubs, museums, and four professional sports teams. The King of Prussia mall, with a movie theater and department stores, isn't far, either.

With Swarthmore's focus on academics, athletics aren't a high priority. The school recently scrapped its football program because the need to recruit enough males to remain competitive in the increasingly intense Division III environment was undermining efforts to recruit students with other interests and talents. However, the University desires to improve the quality of athletics and the facilities devoted to athletics. Women's sports are the true powerhouses, with swimming, softball, lacrosse, and tennis bringing home several conference and national championships. Badminton and men's tennis have also produced champions. Any victory over archrival Haverford will have Swatties swelling with pride. Intramurals are also popular, with the rugby team's Dash for Cash fund-raiser a favored annual event. Players streak through the halls of the main administration building, where spectators—including faculty and administrators—hold out money for them to grab. In the Crum Regatta, student-made boats float in nearby Crum Creek—Swarthmore's answer to the America's Cup.

Swarthmore is an institution where the administration supports the student body completely and students are given a voice in a variety of issues ranging from faculty hiring decisions to making campuswide policies. Students who want to take an active role in their education beyond the classroom door may find the right fit here. "Swarthmore presents you with a unique opportunity to design and shape your path with so many of its resources," says a student. "Utilize them to the fullest."

> ### Overlaps
> **Yale, Brown, Harvard, Princeton, Stanford, Amherst**

If You Apply To ➤ **Swarthmore:** Early decision: Nov. 15, Jan. 2. Guarantees to meet full demonstrated need. Campus or alumni interviews: recommended, evaluative. Required: SAT with writing section and two SAT IIs, ACT with writing section, or SAT or ACT without writing section and three SAT IIs, including the writing test. Accepts the Common Application and electronic applications. Essay questions: why Swarthmore, activities or personal interests which have the most meaning to you, personal statement.

Sweet Briar College

Box B, Sweet Briar, VA 24595

Sweet Briar offers the pure women's college experience—served up with plenty of tradition and gift-wrapped in one of the nation's most beautiful campuses. SBC is the country girl next to in-state rivals Randolph and Hollins. Academic standouts include English and the life sciences, and it now offers engineering.

Indiana Fletcher Williams, who founded Sweet Briar College in 1901, envisioned a school that would educate young women "to be useful members of society." These days, the college—in the heart of beautiful, rural Virginia—produces more career women than homemakers, and a few men even make an appearance as nondegree or exchange students. But this remains a place where, in the words of a popular bumper sticker, "Women are leaders and men are guests."

Set on 3,300 acres of rolling green hills, dotted with small lakes and surrounded by the Blue Ridge Mountains, Sweet Briar's campus of early 20th-century redbrick charmers is a picture of pastoral beauty. Sweet Briar House, now the president's residence, was once the 18th-century home of the college's founder and is listed on the National Register of Historic Places. The Florence Elston Inn and Conference Center includes high-tech meeting rooms, and the school's old dairy farms have been converted into facilities for the studio arts programs.

Sweet Briar's general education program has four components: an English course called Thought and Expression, Skills Requirements (oral and written communication, and quantitative reasoning), Experience Requirements (self-assessment, physical

Website: www.sbc.edu
Location: Rural
Total Enrollment: 555
Undergraduates: 538
Male/Female: 5/95
SAT Ranges: V 530–640
  M 500–590
ACT Range: 22–27
Financial Aid: 65%
Expense: Pr $
Phi Beta Kappa: Yes
Applicants: 623
Accepted: 80%
Enrolled: 37%
Grad in 6 Years: 67%

(Continued)

Returning Freshmen: 75%

Academics: ✍ ✍ ✍

Social: ☎ ☎

Q of L: ★ ★ ★ ★

Admissions: (800) 381-6142

Email Address:
admissions@sbc.edu

Strongest Programs:
Psychology
Biology
Chemistry
Government
English/Creative Writing
History
International Affairs
Business

When the weekend rolls around, the Sweet Briar Social Committee's annual fee is put to good use, covering mixers with other schools, theatrical performances, formal dances, and yearbooks.

The college recently became only the second women's college in the nation (after Smith) to offer engineering as an undergraduate degree. Students may choose either a BS in engineering science or a BA in Integrated Engineering and Management.

activity, and a major), and Knowledge Area Requirements (various courses, including Western and non-Western culture, foreign language, the arts, economics, politics, and law). Seniors must pass a culminating exercise in their majors, which may include comprehensive exams. Students give high marks to Sweet Briar's science programs, which benefit from state-of-the-art equipment such as a digital scanning electron microscope, modular laser lab, gas chromatograph/mass spectrograph, and a nuclear magnetic resonance spectrometer. Students may take classes in these disciplines at nearby Lynchburg College and Randolph College. The college recently became only the second women's college in the nation (after Smith) to offer engineering as an undergraduate degree; students may choose either a BS in engineering science or a BA in Integrated Engineering and Management. The Spanish business major has been discontinued.

> **"Professors make an effort to get to know each student individually."**

Sweet Briar's academic climate is rigorous and competitive. "SBC is an academically rigorous environment full of challenging, exciting, and interesting courses," says a student. Most professors have terminal degrees in their fields, and the quality of teaching is exceptional, students report. "Professors make an effort to get to know each student individually and they truly want all to succeed," claims one theater major. And with 40 percent of the faculty living on campus, "You get to know the faculty's spouses, kids, and dogs," says one woman. The Sweet Briar Honor Pledge, which states that "Sweet Briar women do not lie, cheat, steal, or violate the rights of others," makes possible self-scheduled exams and take-home tests. An honors program and self-designed majors allow some students to further challenge themselves.

Sweet Briar's Junior Year in France is the oldest and best known of its study abroad programs. And the younger Junior Year in Spain program is gaining in popularity, as are exchanges with Germany's Heidelberg University and Oxford University in the U.K. It's also common for faculty to offer short courses abroad during semester breaks, such as a theater course in London or an antiquities course in Italy. Students can also spend time on other campuses through the Seven-College Exchange, the Tri-College Exchange, or 3–2 liberal arts and engineering programs. Classes end in early May, providing ample opportunity for internships. Sweet Briar's unusually strong alumnae network is helpful in arranging positions and housing in cities across the country.

> **"Sweet Briar girls as a whole are smart, funny, girls-next-door kind of gals."**

"Sweet Briar girls as a whole are smart, funny, girls-next-door kind of gals," says an English major. Thirty-eight percent of Sweet Briar's student body hail from Virginia, and 64 percent graduated in the top quarter of their high school class. Three percent are African American, 4 percent Hispanic, and 2 percent Asian American. An increasing number of older "turning point" students also contribute a valued perspective. Campus issues often revolve around gay and lesbian rights, women's issues, and the usual Republican versus Democrat rancor. Various merit scholarships are available for up to $11,213 each.

Ninety percent of Sweet Briar's students live in the college's vintage dorms, which are more like stately antebellum homes with sweeping wooden staircases, fireplaces, and furnished parlors. Thanks to lots of attention from the college, they have aged gracefully, with "hardwood floors and functioning ceiling fans," says a senior. With a few exceptions, students are required to live on campus all four years. Freshmen are advised to choose Reid, Grammer, or Randolph, though every hall has a fair share of students from all four classes. Student leaders are given the first shot at singles, and upperclassmen choose rooms in a lottery. Several dorms have 24-hour male visitation, and administrators are considering the addition of newer, more independent housing for upperclassmen. All residents may eat in the common dining hall or may choose to eat at the short-order type restaurant called LeBistro.

When the weekend rolls around, the Sweet Briar Social Committee's annual fee is put to good use, covering mixers with other schools, theatrical performances, formal dances, and yearbooks. The committee's events attract men "like flies," one student says—and if students don't like the ones who show up, frat parties beckon at Washington and Lee, Hampden-Sydney, and the University of Virginia (an hour away). Other popular road trips include Washington, D.C., and Virginia's beaches (three hours). Campus alcohol policies are "very strict," a sophomore says, and heavy drinking is rare. SBC students volunteer at local schools and with Habitat for Humanity. And the Sweet Briar Outdoor Program (SWEBOP) has introduced hundreds of girls to the joys of backpacking, canoeing, and white-water rafting

"This place will give you an education with a capital 'E.'"

with weekly expeditions. Students also lovingly nurture traditions such as Founder's Day, lantern bearing, step singing, and "tapping" for clubs. Students receive their class rings at the annual junior banquet; they are worn on the left pinkie.

SBC's Vixens compete in Division III. Hockey, lacrosse, soccer, softball, swimming, tennis, and volleyball offer ample opportunity for the athlete to excel. And backed by the largest private indoor ring in the country, the equestrienne squad has snagged eight national championships. That's led to another Sweet Briar slogan: "Where women are athletes, and men are spectators."

Women come here for a well-balanced mix of academics, friendliness, and career preparation. "This school is not for the immature students who think college is about drinking beer and going out," asserts an English major. "This place will give you an education with a capital 'E'." While some complain about the remoteness of the school's rural location, most say the beauty of the campus and the sense of family that prevails are more than satisfactory compensation.

Overlaps

Randolph College, Hollins, Mount Holyoke, University of Virginia, University of Washington, Lynchburg College

If You Apply To ➤	**Sweet Briar:** Early decision: Dec. 1. Regular admissions: Feb. 1. Financial aid: Mar. 1. Does not guarantee to meet demonstrated need. Campus and alumni interviews: recommended, informational. SATs or ACTs: required. SAT IIs: optional. Accepts the Common Application and electronic applications. Essay question.

Syracuse University

201 Tolley Administration Bldg., Syracuse, NY 13244-1140

Syracuse has recast itself to make undergraduate education a top priority. Offerings such as the Gateway program provide small classes for first-year students. World-famous in communications, Syracuse is also strong in engineering and public affairs. Big East basketball provides solace during long winter nights.

Anyone who has watched college sports on TV is familiar with the bright orange color associated with Syracuse University. They have seen the screaming fans and the stadiums overflowing with cheering hordes. Beyond all the athletic fanfare is passion of another sort: Syracuse has set out to become a student-centered research university. By fostering close relationships between students and faculty, expanding course offerings, and pouring loads of money into facility upgrades, Syracuse's reputation as an academic assembly line with killer sports teams is rapidly changing.

The Syracuse campus is located on a hill overlooking the town of Syracuse in central New York State. The Carrier Dome sits on the hillside like an oversized alien

Website: www.syr.edu
Location: City center
Total Enrollment: 17,266
Undergraduates: 11,000
Male/Female: 44/56
SAT Ranges: V 570–650
  M 570–670
ACT Range: N/A

(Continued)

Financial Aid: 60%

Expense: Pr $ $

Phi Beta Kappa: Yes

Applicants: 16,260

Accepted: 64%

Enrolled: 31%

Grad in 6 Years: 79%

Returning Freshmen: 92%

Academics: ⚖ ⚖ ⚖

Social: ☎ ☎ ☎

Q of L: ★ ★ ★

Admissions: (315) 443-3611

Email Address:
orange@syr.edu

Strongest Programs:
Aerospace Engineering
Architecture
Entrepreneurship
Information Management and
    Technology
Inclusive Education
Political Science
Policy Studies

The Newhouse School, which has produced such media celebrities as NBC sportscaster Bob Costas and Steve Kroft of 60 Minutes, is undoubtedly Syracuse's flagship.

spacecraft. The character and mixture of architectural styles depict a continuously changing campus, which is grassy, full of trees, and bordered by residential neighborhoods. Fifteen of SU's 140 buildings are listed in the National Register of Historic Places. Many schools and colleges have restructured facilities to accommodate more faculty/student research, as well as social interaction between the two groups. All the dorms and campus-owned apartments are wired with high-speed data ports. The university is embarking on a multimillion-dollar campuswide upgrade. A new School of Management building was completed in 2005, and the School of Information Studies has moved into a new building. Residence halls are being upgraded, and the Newhouse School of Public Communications has broken ground on a third building, which will include facilities for web-based reporting.

> **"The social life at Syracuse is a dynamic one."**

Syracuse has a diverse set of academic offerings. The Newhouse School, which has produced such media celebrities as NBC sportscaster Bob Costas and Steve Kroft of *60 Minutes*, is undoubtedly Syracuse's flagship. It offers a dual program where students can major in any one of the eight Newhouse majors and have a dual major in information management and technology. Also well known is the Maxwell School of Citizenship and Public Affairs, whose faculty members teach sought-after undergraduate economics, history, geography, political science, and social sciences. Teaming with NASA, the school now has a $3 million virtual aerospace engineering facility—one of three in the nation—where students have helped design a new reusable space launch vehicle. SU students have also participated in NASA's reduced-gravity student flight programs. The College of Arts and Sciences is the largest college at Syracuse and offers recognized programs in creative writing, philosophy, geography, and chemistry. Three-quarters of students do some type of undergraduate research and 40 percent study abroad. The most popular majors are political science, psychology, architecture, marketing, and television/radio/film.

General education requirements vary, but all Syracuse students are expected to take writing courses. Several schools and colleges subscribe to the Arts and Sciences core requirements, which include coursework in the sciences, math, social sciences, humanities, and contemporary issues. Entering freshmen must complete a writing seminar, and each school and college offers a small-group experience course, known as the Freshman Forum, to share common first-year experiences and stimulate discussion of academic and personal issues.

> **"I have received a high quality education from very intelligent, experienced professors."**

The Gateway program allows freshmen to take introductory classes with senior faculty members in a small classroom setting. For upperclassmen, Syracuse offers a strong honors program based on seminars and independent research. Newer additions include a sport management program and a health and wellness major.

Students at Syracuse can expect challenging courses and competitive peers. "A high number of students come from college prep high schools," says a junior, who notes "the climate is competitive but not to the point where it impedes learning." Despite the school's large size, students say professors are friendly and accessible. "I have received a high quality education from very intelligent, experienced professors," a senior says. Classes are usually small (fewer than 25 students) and registration can be easy. "Students do not have difficulty getting into the course they would like to or are required to take," assures one junior.

Admissions standards differ among the various schools and are most rigorous in the professional schools, especially architecture, communications, and engineering. Forty-four percent of Syracuse's undergraduates come from the top tenth of their high school class, and 75 percent attended public high school. The number of students of color has been steadily increasing; African Americans and Asian Americans each account for 6 percent of the student body, and Hispanics make up another 5 percent. "I would characterize SU students as intelligent individuals who love their university and like to have fun," says one senior. Big debates include diversity issues and the war in Iraq. Forty-one percent of the students are from New York State, and most of those hail from New York City and Long Island. There are nearly 300 athletic scholarships in sports ranging from football and basketball to rowing and lacrosse. Merit scholarships are also available, averaging $8,800 each.

Housing on campus is clean and comfortable, and is provided all four years in modern and well-maintained halls. "The dorms are clean and safe," says one student. Seventy-five percent of the undergraduates live in university housing. Syracuse is continually upgrading dorm facilities, and students appreciate the efforts. Freshmen and sophomores are required to live in the dorms, and should check out Brewster-Boland, Day, and Flint halls. Living and dining in fraternity or sorority houses is another option because 18 percent of the men and 20 percent of the women go Greek. As for campus safety, SU has emergency alarms throughout the campus, a card-key access system in all dorms, and bus service for students studying late on campus. "Public safety officers are constantly on patrol," notes a student.

The social life tends to stay on campus for freshmen and sophomores and move off campus for upperclassmen. "The social life at Syracuse is a dynamic one," says a senior. There are always activities available such as movies, bowling, skating, and dancing. Students over 21 spend many an **"The dorms are clean and safe."** evening barhopping on Marshall Street, a lively strip near campus. For underage students, there is a campus club where student bands play and nonalcoholic drinks and snacks are free. Drama productions are frequent on the weekends, and popular road trips include Ithaca, Niagara Falls, Montreal, and Rochester.

Students generally enjoy the town of Syracuse, which offers a variety of off-campus retreats. "Everything you need is within walking distance," a junior says. Many students are involved in the community through internships in the corporations and SU education students give more than 40,000 hours in community service in Syracuse. Downtown is within easy reach on foot or by convenient public transportation. Once there, opportunities include an excellent art museum, a resident opera company, a symphony, and a string of movie theaters and restaurants. If you tire of the city life, several quaint country towns, complete with orchards, lakes, and waterfalls, are nearby, as are several ski resorts. The Turning Stone casino is a big draw. The six-story Carousel Mall, about 10 minutes away, has an 18-theater cinema. Erie Boulevard is home to big chain stores, and the city's Armory Square is flanked with coffee shops, a great music store, clubs, and eateries.

The spacious Carrier Dome rocks every time the Orange (formerly the Orangemen) take the field—or the court. Football games against Miami and a basketball rivalry with Georgetown make for great fun and much enthusiasm during the year. The men's basketball team brought home a national Division I championship several years ago, as did the men's lacrosse team. The "painters" are famous at SU—they are the students who paint each letter of the school's name on their bare chests and run through rain, sleet, or snow to each home game in the Carrier Dome. Though the Dome seats 33,000 for basketball—enough to shatter NCAA attendance records—tickets must still be parceled out by a lottery, to the disdain of some. About 1,200 students participate in 20 intramural sports.

From special partnerships with NASA to opportunities to study abroad or help out right at home, students at Syracuse know they've got something unique. The place itself can be enough to inspire school spirit. The wintry climate is less than popular, but a junior says, "SU students are the ones that tough out the cold to gain from everything available here."

Admissions standards differ among the various schools and are most rigorous in the professional schools, especially architecture, communications, and engineering.

Football games against Miami and a basketball rivalry with Georgetown make for great fun and much enthusiasm during the year.

Overlaps

NYU, Boston University, Penn State, Boston College, University of Maryland, University of Michigan

If You Apply To ➤

Syracuse: Early decision: Nov. 15. Regular admissions: Jan. 1. Financial aid: Feb. 1 (regular decision). Housing: May 1. Campus interviews: recommended, evaluative. Alumni interviews: optional, evaluative. SATs or ACTs: required. Sat I preferred. SAT IIs: optional. Apply to particular program; can apply to single, dual, or combined programs. Accepts the Common Application and electronic applications. Essay questions: who or what influenced you to apply to Syracuse; academic and career aspirations; most meaningful activity outside the classroom and why.

University of Tennessee at Knoxville

Knoxville, TN 37996-0230

UT is in the middle of the pack among its southeastern rivals—behind UNC, U of Georgia, and Florida; ahead of Arkansas, Alabama, and Ole' Miss. As the only major public university in Tennessee, UT comes close to being all things to all students. Strong in business, engineering, and communications.

Website: www.tennessee.edu
Location: City center
Total Enrollment: 25,457
Undergraduates: 20,232
Male/Female: 48/52
SAT Ranges: V 520–630
   M 530–640
ACT Range: 23–28
Financial Aid: 38%
Expense: Pub $ $
Phi Beta Kappa: Yes
Applicants: 12,251
Accepted: 74%
Enrolled: 43%
Grad in 6 Years: 57%
Returning Freshmen: 80%
Academics: ✍ ✍ ✍
Social: ☎ ☎ ☎ ☎
Q of L: ★ ★ ★
Admissions: (865) 974-2184
Email Address:
   admissions@tennessee.edu

Strongest Programs:
Business
Engineering
Communications
Education
Nursing
Architecture
Social Work
Veterinary Medicine

Students at the University of Tennessee put a premium on school spirit, athletics, and academics—usually in that order. In the fall, more than 100,000 boisterous fans pack into one of the nation's largest on-campus football stadiums to watch the Volunteers play against national powerhouses like Florida, Alabama, and Arkansas. Also competitive are the SEC-dominating women's basketball and soccer teams and men's baseball teams. Amid this excitement, it's easy to forget that UT also prides itself on having a strong academic program.

Set in the foothills of the Great Smoky Mountains, UT is in the heart of east Tennessee's urban hub and only a few miles away from Oak Ridge, Tennessee, home to the prominent Oak Ridge National Laboratory. The 511-acre campus has an array of architectural styles ranging from Gothic to Georgian to modern. Particularly noteworthy is the John C. Hodges Library—the largest one in the state—built in the shape of a ziggurat. Current projects include a new College of Business Administration building, a public policy building, a new women's soccer stadium, and improvements to the biology building.

Many strong academic programs are in preprofessional fields, most notably business, architecture, accounting, and engineering. On the liberal arts side, psychology and communications are popular majors. Several majors in French, German, and Spanish incorporate a concentration in international business. The Ready for the World initiative seeks to improve students' cultural competency by revamping general education courses and creating an endowment for study abroad scholarships. The five-year goal calls for devoting $1.5 million to faculty recruiting, campus programming, and support of campus diversity efforts. UT is the managing partner of Oak Ridge National Laboratory—the federal government's largest nonweapons lab—which bolsters science and technology offerings, and involves more than 400 students and faculty in majors as diverse as English and physics. The honors program at UT is a campuswide program that offers qualified students scholarships, honors courses, seminars, and a chance to complete an original research project in collaboration with a faculty member. Students who participate in the Whittle Scholars Program are encouraged to pursue their leadership skills through campus and community organizations, and are given the opportunity to travel to a variety of different countries including Argentina, Australia, France, Mexico, and the Netherlands.

Academic competition varies, depending on the class, as does course difficulty. "UT has both laid-back and rigorous courses, with the difficulty increasing each year," says a junior. UT faculty gets a mixed rating. "Most of my freshman classes were taught by teaching assistants and graduate students," says a chemistry major. Students report occasional problems with registration because preference is given to seniors, but none that would extend a four-year stay. "During the past two years, courses were difficult to get into because there were too many students attending UT," says a senior. Advising also gets mixed reviews, depending on the field of study, and students are expected to meet with their advisors each semester before registering. UT's general education requirements are fairly extensive and include two courses each in English composition, math, humanities, history, social sciences, and natural sciences, plus intermediate proficiency in a foreign language or multicultural studies. Business majors are immersed in a broad liberal arts program, including a foreign language requirement, during their first two years of study.

Eighty-one percent of the student body is made up of homegrown Tennesseans, and 34 percent of the freshman class graduated in the top tenth of their high school class. Minority enrollment is increasing; African Americans account for 9 percent of the students, while Asian Americans and Hispanics combine for 7 percent. Students also warn that the campus's size can lead to a phenomenon called the "Big Orange Screw,"

> "UT has both laid-back and rigorous courses, with the difficulty increasing each year."

in which the impersonal bureaucratic system makes students' lives miserable. Financial aid opportunities are generous: More than 3,000 merit scholarships are available, from $500 to $8,000. An additional 360 athletic scholarships are awarded in 20 sports.

Dorm rooms are average, and 67 percent of UT students live off campus. "They could be bigger, but couldn't they all?" says one psychology major. "They're better compared to most." The dorms are for the most part comfortable and well-maintained, and about the only hitch is that some don't have air-conditioning (which can be brutal in August). Each of the dorms has a residence hall association, which for a token fee provides check-out of sports equipment, games, cooking utensils, and other useful items. The university goes out of its way to ensure the security of the campus and the students. To this end, UT has installed remote alarm units that allow students to report a crime from anywhere on campus. "There are cops everywhere," a junior says. "As long as you're not stupid, you're fine."

Students say that the social life is "very active" both on and off campus. The social calendar is dotted with numerous major events, including River Fest on the nearby Tennessee River, Saturday Night on the Town, and the Dogwood Arts Festival. Greek life is growing more popular—16 percent of men and 22 percent of women go Greek. Alcohol flows freely, even for underage students. But nothing compares to the sea of orange that engulfs the campus on Saturday afternoons in the fall. More than 100,000 people jam the football stadium to see their

> "During the past two years, courses were difficult to get into because there were too many students attending UT."

Vols take on Southeast Conference rivals ("'Alabama' is a four-letter word" in these parts). Denizens liken football to religion in Knoxville. Coach Pat Summit's Lady Vols basketball team has won a record-breaking number of NCAA championships—including the 2007 title—and enjoys tremendous crowds (drawing more fans than many NBA teams). In 2005-06, the Lady Vols captured the SEC title in cross-country and soccer and all 11 varsity women's sports advanced to postseason play. The men's program sent six sports to postseason competition and five finished the year ranked in the top 20.

The University of Tennessee is well known for its athletics, and administrators and students are hoping that it can develop the same reputation for academics. Though the oft-sluggish bureaucracy may turn some off, many will find the myriad opportunities here at the "Big Orange" to be well worth the squeezing.

The honors program at UT is a campuswide program that offers qualified students scholarships, honors courses, seminars, and a chance to complete an original research project in collaboration with a faculty member.

The Lady Vols basketball team draws more fams than some NBA teams.

Overlaps

Middle Tennessee State, Tennessee Tech, Auburn, Georgia, University of Memphis

University of Texas at Austin

John Hargis Hall, Austin, TX 78712-1157

UT is on anybody's list of the top 10 public universities in the nation. The Plan II liberal arts honors program is one of the nation's most renowned. Though it is also the capital of Texas, Austin ranks among the nation's best college towns.

Website: www.utexas.edu
Location: Urban
Total Enrollment: 49,696
Undergraduates: 33,063
Male/Female: 48/52
SAT Ranges: V 540–670
   M 570–690
ACT Range: 23–29
Financial Aid: 47%
Expense: Pub $ $ $
Phi Beta Kappa: Yes
Applicants: 23,925
Accepted: 51%
Enrolled: 56%
Grad in 6 Years: 75%
Returning Freshmen: 93%
Academics: ✍ ✍ ✍ ✍ ½
Social: 🕯 🕯 🕯 🕯
Q of L: ★ ★ ★ ★
Admissions: (512) 475-7399
Email Address: N/A

Strongest Programs:
Engineering
Business
Law
Education
Pharmacy

The University of Texas at Austin has come a long way from where it began as a small school with only one building, eight teachers, two departments, and 221 students. Today, the UT campus is Texas-sized—home to more than 45,000 students. From its extensive academic programs to its powerful athletic teams to its location in one of the nation's ultimate college towns, the University of Texas has it all.

A 400-acre oasis near downtown Austin, replete with rolling hills, trees, creeks, and fountains, the campus features buildings ranging from "old, distinguished" limestone structures to "contemporary" Southwest architecture. Statues of famous Texans line the mall, and the fabled UT Tower is adorned with a large clock and chimes (a lifesaver for the disorganized). From the steps of the tower, one can see the verdant Austin hills and the state capitol. The outstanding library system at the University of Texas has more than seven million volumes located in 19 different libraries across campus, and is the sixth-largest academic library system in the United States. New additions to campus include the Nano Science and Technology Building and a new residence hall.

Many UT classes are extremely large, and smaller sections fill up quickly. UT is a research-oriented institution, so the professors are often busy in the laboratories or the library. They do, however, have office hours. "All of my freshman courses were taught by full professors," says a senior. "They are very passionate about their individual fields."

The list of academic strengths at University of Texas is impressive for such a large school. Undergraduate offerings in accounting, architecture, botany, biology, business, foreign languages, and history are first-rate. The engineering and computer science departments are excellent and continue to expand. The English department is huge (95 tenure-track professors) and students give it high marks, but say the art/photography department needs improvement. A molecular biology building and a telescope have been completed for the prestigious Institute for Fusion Studies, which already boasts one of the world's largest telescopes. A universitywide committee is reviewing its curriculum, so expect changes in the next few years.

The Plan II liberal arts honors program, a national model, is one of the oldest honors programs in the country and one of the best academic deals anywhere. It offers qualified students a flexible curriculum, top-notch professors, small seminar courses, and individualized counseling, and provides them with all of the advantages of a large university in a small-college atmosphere. Business and natural sciences honors programs are also available. Engineering majors can alternate work and study in the co-op program, while education and health majors hold term-time internships. Being in the capital city should have its advantages, and it does. Almost 200 UT undergrads work for lawmakers in the Texas State House, only a 20-minute walk from campus. A strong Reading and Study Skills Lab services students in need of remedial help. UT recently opened a new college: the John A. and Katherine G. Jackson School of Geosciences.

Students say the academic climate is competitive and demanding. "Most of the courses are fairly rigorous," says a student. Freshmen can take University 101, which covers everything from major requirements to healthy lifestyle choices and cultural diversity. In addition, the colleges within the university have established basic requirements for all majors: four English courses, with two writing-intensive; five courses in social sciences; five courses in natural sciences and math; and one course in the fine arts or humanities. Urban studies, biomedical engineering, and women's gender studies are new programs.

UT students are "intelligent, involved, and proactive in their education," says a senior. Ninety-two percent of UT students are Texans. Students say there is no dominant political pattern on campus—despite the fact that historically UT has been integral in the careers of big-time (conservative) Texas politicians. The liberals are anything but hiding out on this huge campus. Political issues can get students pretty riled up on one side or the other here, such as the war in Iraq and tuition deregulation. Hispanics account for 16 percent of students, Asian Americans 17 percent, and African Americans 4 percent. The university offers special "welcome programs" for African American and Hispanic students, with social and educational events and peer mentoring. The university also provides thousands of merit scholarships based on academic performance, worth an average of $3,120, as well as 402 athletic scholarships in a range of sports.

"Most of the dorms are old, but they have nice facilities."

University housing can accommodate only 6,500 students. It can range from functional to plush, and dormies have a variety of living options based on common social and educational interests. "Most of the dorms are old," says a student, "but they have nice facilities." Most students live off campus; apartments and condos close to campus are lovely—and very expensive. More reasonably priced digs can be found in other parts of town, a free shuttle ride away. But be forewarned: UT life requires lots of walking, especially for commuters, with 110 buildings and bus stops and parking lots scattered about. As for food, bigness is an asset here as well; the word "variety" hardly does justice to the array. "There are so many options for food," says an advertising major. "Two all-you-can-eat cafeterias, a dining hall food court with different choices such as tacos, Chinese, chicken, make-your-own sandwich, stores with sushi, smoothie bars, bakeries, candy stores, pizza places, Greek food, and so much more. The student union has a food court with over nine places to eat. The meal plan extends all around campus and off (surrounding places and delivery)."

As the state capital, Austin is not a typical college town, but it is one of the best ones. "I love it," exclaims a junior. "It was really the deciding factor on going to UT. It has a great live music scene and is beautiful." Nightlife centers on nearby Sixth Street, full of pubs and restaurants of all types, and the well-known music scene that features everything from jazz to rock to blues to folk. Numerous microbreweries have opened their doors in the past few years. Halloween draws an estimated 80,000 costumed revelers to Sixth Street (and sometimes up its lampposts). Annual festivals include 40 Acres, a sprawling carnival of all

"All of my freshman courses were taught by full professors."

the campus organizations, and Eeyore's birthday party, where students pay homage to the A. A. Milne character with food and live music. Two pep rallies get students psyched before the Longhorns play Texas A&M or Oklahoma, their biggest rivals. And Texas Independence Day provides an occasion for celebration in March.

On campus, the Texas Union sponsors movies and social events and boasts the world's only collection of orange-top pool tables. For those more interested in octaves than eight-balls, the Performing Arts Center has two concert halls that attract nationally known performers. Students hang out at the union's coffee shop or cafe, and the on-campus pub draws top local talent to the stage (but you must be 21 to drink). When the weather gets too muggy (quite often in spring and summer), students head for off-campus campgrounds, lakes, and parks. The most popular road trips are to San Antonio or Dallas. For Spring Break, the students travel to Padre Island, if not New Orleans. Although only 12 percent of the men and 16 percent of the women go Greek, members of fraternities and sororities are "probably highest on the social totem pole," says one student. The chapters tend to be choosy, have high visibility, and have increased in size dramatically over the past year.

Athletics are as important as oxygen to most Texans. In fact, the UT Tower is lit in Longhorn orange whenever any school team wins. The students look forward to the annual Texas–Oklahoma football game played in the Cotton Bowl in Dallas, and the Texas A&M–UT game is an incredibly noisy experience you have to see to believe. "Football games pull the student body together and give us a chance to show our school spirit," says one student. Basketball is also popular, and the men's and women's teams regularly reach their respective NCAA tournaments. The baseball program has many alumni in the major leagues, and the annual spring game between UT's baseball alumni and the current college squad is quite a contest. The men's golf, swimming, baseball, and outdoor track and field teams recently brought home conference championships, as did women's golf, swimming and diving, softball, basketball, and indoor and outdoor track and field teams.

"Most of the courses are fairly rigorous."

The chess team is a perennial powerhouse, too. UT's intramural program draws 86% of the students and is the largest in the nation. It offers weekend athletes access to the same great facilities that the big-time jocks use.

The University of Texas may seem overwhelming because of its imposing size, but students say the school spirit and sense of community found here make it feel smaller. UT prides itself in having one of the most reasonably priced tuitions in the country. It offers one of best all-around educational experiences a student could ask for.

Texas A&M University

College Station, TX 77843-0100

Coming to A&M is like joining a fraternity with 45,000 members. In addition to fanatical school spirit, Texas A&M offers leading programs in the natural sciences, business, and engineering. To succeed in this mass of humanity, students must find their academic niche.

Since its inception as a military academy, Texas A&M University has become known for its top-notch engineering program and its unsurpassed school spirit. This school of over 39,000 students boasts a massive endowment and more traditions than Vatican City. When they're not studying for rigorous technical classes, Aggies are likely to be found at "yell practice" before each home football game or yelling—the administration says sternly, "Aggies don't cheer, they yell!"—for their teams at other high-energy athletic events.

Texas A&M is now the largest university campus in the country, in terms of acreage—something made obvious to students every time they walk to class. The A&M campus combines historic brick buildings from the turn of the century with newer structures in more modern styles, and it is pulled together by its heavy cover of live oak trees.

Texas A&M is best known for its agriculture and engineering colleges, and for veterinary medicine, although the university is cultivating a strong liberal arts program and an even stronger business school. Aggies also stand by science programs, especially chemistry and physics. Technical programs of virtually all kinds are heartily supported at A&M, especially nuclear, space, and biotechnical research. A&M has become a sea grant college due to its outstanding research in oceanography, and is also a space grant college. Add that to the college's land grant status, and the whole universe seems covered by A&M. Coursework sometimes takes students far from Aggieland. Participants in the Nautical Archaeology Program conduct research all over the world. The Academy for Future International Leaders trains 15 students per year in international business and cultural issues, followed by a summer international internship. The school is trying to build up its weaker programs in fine arts. Newer majors include music and telecommunication media studies.

> **"Freshmen must request a dorm as soon as they are accepted in order to get a dorm of choice."**

Incoming Aggies can expect some heavy coursework in general education requirements, which consist of communications, math, natural sciences, humanities, social/behavioral sciences, kinesiology, visual/performing arts, and U.S. history/political science. They are also expected to have at least two years of a foreign language, complete a cultural diversity requirement, and demonstrate computer literacy. Students generally agree that academics are taken seriously at A&M. "It's highly competitive," says a senior biomedical science major. Professors receive rave reviews, although TAs and grad students are often behind the lectern. "Every teacher I have encountered has gone above and beyond the call of duty that you would expect from a large university such as Texas A&M," a classmate reports. Because of the school's size, it's sometimes hard to enroll in a required class. "You just have to go to the professor and get forced in," one student advises. The administration hopes to alleviate the situation by adding nearly 450 faculty members over the next few years as a part of the "faculty reinvestment program." Still, many classes remain large.

Ninety-five percent of students are from Texas and the political climate is decidedly conservative. The school's departments of Multicultural Services and Student Life offer numerous programs to enhance minority student recruitment and retention. African Americans comprise 3 percent of the student body, Asian Americans 4 percent, and Hispanics 11 percent. For many students the unifying characteristic of the A&M experience is the university spirit. "Aggies look out for other Aggies. Everyone is an Aggie," a senior history major says. Athletes compete for hundreds of scholarships parceled out each year, while scholars vie for thousands of merit awards averaging $1,745.

A&M's 38 single-sex and co-ed dorms range from the cheap and not-so-comfortable to the expensive and cushy (with air-conditioning and private bathrooms). "Freshmen must request a dorm as soon as they are accepted in order to get a dorm of choice—or any dorm, for that matter," a senior says. Because of the ever-growing student population, the school's dorms provide only enough space for a quarter of

Website: www.tamu.edu
Location: Small city
Total Enrollment: 39,607
Undergraduates: 32,984
Male/Female: 51/49
SAT Ranges: V 530–640
  M 560–670
ACT Range: 23–28
Financial Aid: 29%
Expense: Pub $ $ $
Phi Beta Kappa: Yes
Applicants: 17,871
Accepted: 70%
Enrolled: 57%
Grad in 6 Years: 76%
Returning Freshmen: 92%
Academics: ✍ ✍ ✍ ✍
Social: ☎ ☎ ☎
Q of L: ★ ★ ★
Admissions: (979) 845-3741
Email Address:
  admissions@tamu.edu

Strongest Programs:
Engineering
Business
Veterinary medicine
Agriculture
Architecture
Biomedical sciences

Texas A&M is best known for its agriculture and engineering colleges, and for veterinary medicine, although the university is cultivating a strong liberal arts program and an even stronger business school.

those enrolled. "The majority of students live on campus for only a year and then they usually move off campus," a junior says. Most upperclassmen live in the numerous apartments and houses in College Station or its twin city, Bryan. But they needn't fear being cut off from campus life, though, as the entire town is filled with Aggies, and the university offers a shuttle bus to and from scores of apartment complexes throughout the community. Several meal plans are offered in the dining halls, and snack shops are all over campus. Campus security is lauded, but the school is not incident-free.

Although College Station may appear uninspiring at first glance, most students fall in love with it. "College Station is a model college town," a senior explains. "Every restaurant has an 'Aggie special,' the radio stations play our 'war hymn' at

"Aggies look out for other Aggies."

times throughout the day." And when the school empties out for a holiday, the town does too. Students are actively involved in the community, including the largest single-day service project in the nation annually, the Big Event. Although the Texas Alcoholic Beverage Commission is stationed in town, "those who choose to drink have no problem getting alcohol." Those with more sophisticated tastes can drive an hour and a half to either Houston or Austin or three hours to Dallas. Greeks draw 6 percent of the men and 12 percent of the women, and students say Greek life is less important than at other schools. More than 700 organizations are available to meet everyone's interests.

Athletics, whether on the varsity level or for recreation, are tops on anyone's list here. Football fans rock Kyle Field with cries of "Gig 'em, Aggies," or "Hump it, Ags." After touchdowns are scored, Aggie fans kiss their dates, and the annual game against the University of Texas stirs up the Aggies and their fans all season long. Men's baseball routinely fields outstanding teams, and the women's golf and soccer teams have brought home a few championships of their own. The well-organized and extensive intramural program includes hundreds of softball teams. Tennis has become a popular sport as well, along with flag football, basketball, and soccer. Aggie jokes abound, much to the chagrin of A&M students, who don't take too kindly to being the object of ridicule. Example: "How do you get a one-armed Aggie out of a tree?" "Wave."

Favorite traditions include "Twelfth Man," in which all students stand for the entirety of every football game as a symbol of their loyalty and readiness to take part, and the Aggie Muster, a memorial service for A&M alumni around the world who died within the year. There's also the 300-plus member Fightin' Texas Aggie Band and the senior "boot line" at the end of the halftime show. The treasured Corps of Cadets,

"College Station is a model college town."

one of the largest military training programs in the country, is structured like a military unit; students lead other cadets. While less than 10 percent of the school belong to the Corps, it remains the single most important conservator of the spirit and tradition in Aggieland. Although the administration has banned the traditional on-campus bonfire before the football game against the University of Texas because of a tragic incident in 1999, student groups have organized their own bonfires—albeit smaller—off campus.

Texas A&M, while extremely large, is uniquely familial. Being a student here is being a part of something seemingly so much bigger, which is what the Aggie Spirit embodies. Students get the best of two intense worlds at A&M—a large school with tons of people surrounded by a small community. A&M's no longer just a military school, it's a potpourri of varied educational opportunities worth cheering. Boasts one senior, "We have the same traditions as were created years ago and, young or old, Aggies know them all."

Texas Christian University

TCU Box 297013, Fort Worth, TX 76129

The personalized alternative to the behemoth state universities of Texas. Tuition is about $2,500 less than that at archrival SMU. Though affiliated with the Disciples of Christ, the atmosphere at TCU goes lighter on religion than, say, Baylor. Strengths include the fine arts, business, and communications.

You know a school has spirit when its students paint themselves purple to cheer raucously for a horny frog. Texans know these folks are TCU fans cheering for the home team (known officially as the Texas Christian University Horned Frogs) at a Saturday afternoon football game. There's a true sense of solidarity and school spirit here. "TCU is welcoming," says a senior. "That's why I came as a freshman and stayed until graduation."

The spacious 268-acre campus is kept in almost perfect condition and features tree-lined walkways and grassy areas. Nearby is a lovely residential neighborhood not too far from the shops and restaurants of downtown Fort Worth. The campus features an eclectic mix of architecture, ranging from neo-Georgian to contemporary. Newer facilities include the Campus Recreation Center and the Walsh Center for Performing Arts, a 56,000-square-foot performance hall and theater complex. The Campus Commons project will add four new residence halls and a university union to the center of campus.

Students can choose their majors from 100 disciplines, with the core curriculum embodying the base of the liberal arts education. The core emphasizes critical thinking and is divided into three areas: essential competencies; human experience and endeavors; and heritage, mission, vision, and values. There are freshman seminar courses, along with a new student orientation and Frog Camp (an optional summer camp that emphasizes team building and school spirit).

TCU's stand-out programs are business, nursing, communications, engineering, and fine arts. In the Neeley School of Business, some students manage a $1.5 million investment portfolio that is one of the largest student-run investment funds in the nation. The university also offers an innovative dance program with a ballet major and a strong theater internship program. The most recent addition to the curriculum is a major in ranch management. The communications program offers hands-on experience in various areas including broadcast news, reporting, and advertising/public relations. The campus also features a geological center for remote sensing, a nuclear magnetic resonance facility, and an art gallery, along with a state-of-the-art electrical engineering lab.

> "We all really love our school and what it stands for."

The academic climate at TCU is challenging, but not overwhelming. "TCU students are definitely concerned about academics and grades, but there is also a feeling of support and open cooperation between students," says an English major. Professors are well-liked and respected. "The quality of teaching is unbeatable," enthuses one

Website: www.tcu.edu
Location: Suburban
Total Enrollment: 8,749
Undergraduates: 7,171
Male/Female: 41/59
SAT Ranges: V 520–630
  M 540–640
ACT Range: 23–28
Financial Aid: 70%
Expense: Pr $
Phi Beta Kappa: Yes
Applicants: 8,155
Accepted: 67%
Enrolled: 29%
Grad in 6 Years: 67%
Returning Freshmen: 84%
Academics: ✍ ✍ ✍
Social: ☎ ☎ ☎
Q of L: ★ ★ ★
Admissions: (817) 257-7490
Email Address:
  frogmail@tcu.edu

Strongest Programs:
Business
Nursing
Communications
Fine Arts
Engineering
Premed

student. Academic advising received high marks, too. "My advisor is so eager to help," says a student. "I think he really enjoys giving me advice."

TCU's student body is fairly homogeneous; 76 percent are from Texas, many from affluent, conservative families. TCU is affiliated with the Christian Church (Disciples of Christ), but the atmosphere is not overtly religious. This is hardly an activist campus, but it is definitely correct to be politically correct. In fact, "TCU should be called PCU," says a junior. The undergrad student body is 14 percent minority. TCU offers scholarships and financial assistance to 70 percent of students.

"The quality of teaching is unbeatable."

Forty-six percent of the student body lives on campus, and dorm life is a good experience. Cable and free Internet access are available in all rooms. Students describe them as comfortable and clean. An evening transportation service, Froggy Five-O, takes you wherever you want to go on campus. "There are also plenty of lights and emergency phones," one freshman says, "so students feel physically safe." Many juniors and seniors move off campus, and fraternity and sorority members may live in their Greek houses after freshman year. Dorm residents must take the meal plan, which does not receive high marks. A variety of other dining options are offered at various locations around campus.

Greek life is important at TCU; 43 percent of the men and 38 percent of the women join Greek organizations. They party in the *esprit de corps* tradition, but there's plenty of fun left in Fort Worth and on campus to keep the non-Greek frogs hopping. The alcohol rules on campus are fairly strict for minors: Resident advisors even perform occasional "fridge checks" looking for alcohol in students' rooms. "Alcohol violations are a big deal," one student says.

"Fort Worth is cultured and has plenty of things to do," says a senior. "Downtown has tons of bars, clubs, theaters, comedy shows and is safe to walk around in. The stockyards let you get in touch with the inner country in you, and no one should

"Fort Worth is cultured and has plenty of things to do."

miss a visit to Billy Bob's, the world's largest honky tonk." Dallas is only 45 minutes to the east. Parents' Weekend, Siblings' Weekend, homecoming, and the traditional lighting of the Christmas tree are all special events. Road trips include Austin, San Antonio, the Gulf Coast, and Shreveport, Louisiana.

As for athletics, the school competes in Conference USA. The football team's resurgence and a 2005 Mountain West championship title have given hungry fans a reason to cheer, and big rivalries include SMU and Rice. Championship teams in the past few years include football, baseball, and men's and women's tennis. On-campus sports facilities feature two indoor pools, weight rooms, a track, and tennis, basketball, sand volleyball, and racquetball courts.

Those seeking a personalized college experience that's heavy on the academics and light on the religious influence may want to consider TCU. "We all really love our school and what it stands for," says one student. Like the beloved Horned Frog, TCU graduates have taken a giant leap toward their futures.

If You Apply To ➤ TCU: Regular admissions: Feb. 15. Financial aid and housing: May 1. Campus interviews: recommended, evaluative. No alumni interviews. SATs or ACTs: required. SAT IIs: optional. Accepts the Common Application and electronic applications. No essay question.

Texas Tech University

Lubbock, TX 79409

A child of the remote West Texas plains, Texas Tech is finally emerging from the shadow of Texas A&M. Though engineering is its specialty, Tech is a full-service university. It takes big-time sports to be on the map in Texas, and the Red Raiders are making a name in both football and basketball.

Texas Tech University has come a long way from its humble beginnings. First proposed in 1923 as a branch of Texas A&M in the West Texas city of Lubbock, Tech opened its doors two years later as an independent institution, with fewer than 1,000 students, and courses in the liberal arts, agriculture, engineering, and home economics. Today, Tech hosts more than 28,000 students, about 400 student organizations, hundreds of undergraduate programs, and schools of medicine and law.

Tech's 1,839-acre campus features expansive lawns, impressive landscaping and Spanish Renaissance-style red-tile-roofed buildings. The school has completed over $1 billion in construction projects in recent years, including the 15,000-square-foot United Spirit Arena, a new building for the English, education, and philosophy departments, and an expansion and renovation of the Student Center. Tech also has a slew of other facilities around Texas, such as a 16,000-acre agricultural facility and research farm.

Even if new money is going toward building projects, the smorgasbord of academic options at Tech hasn't gotten short shrift. The university's 10 undergraduate colleges and schools boast more than 150 degree programs, and the College of Agricultural Sciences and Natural Resources "has the strongest financial foothold on campus, with the largest endowment," says an economics major. "Within the College of Human Sciences lies the department of personal and family financial planning, the best of its type in the country." The BS in integrated pest management is now a BS degree in plant biotechnology.

Tech's general education requirements span all of the colleges and schools, and include courses in written and oral communication, math, natural science, technology and applied science, humanities, visual and performing arts, social and behavioral sciences and multiculturalism. Despite Tech's massive size, three-quarters of the classes taken by freshmen have 50 or fewer students. "Tech's academic climate is outstanding," says one freshman. "The courses, while difficult, are all structured with the student being the first priority." Professors "are personable and approachable," says a junior. "I have never had a bad teacher during my three years here," adds a student. Graduate assistants may lead discussion sections or labs, but they aren't the main force at the lectern. A one-credit freshman seminar helps with the transition from high school to college, and 43 faculty positions were recently added to enhance teaching and research.

> **"I have never had a bad teacher during my three years here."**

Outstanding students may enroll in Tech's Honors College, where they sit on committees, help with recruiting, make decisions about course content, and evaluate faculty. They can also work on research projects, either independently (with a professor's guidance), or as part of a student/faculty team. Tech invested $2 million in undergraduate research in one recent year, and about 2,500 students participated, investigating topics such as pain management, wind engineering, and sick-building syndrome. Those yearning to leave the hardscrabble plains of Texas may study abroad in over 40 countries; Tech also has its own campus in Seville, Spain, where courses are taught by Tech professors (in English and Spanish), and students live with local families.

Website: www.ttu.edu
Location: Suburban
Total Enrollment: 24,220
Undergraduates: 20,803
Male/Female: 55/45
SAT Ranges: V 510–600
  M 530–620
ACT Range: 22–26
Financial Aid: 41%
Expense: Pub $ $ $
Phi Beta Kappa: Yes
Applicants: 12,583
Accepted: 71%
Enrolled: 42%
Grad in 6 Years: 55%
Returning Freshmen: 84%
Academics: ✍ ✍ ✍
Social: ☎ ☎ ☎
Q of L: ★ ★ ★
Admissions: (806) 742-1480
Email Address:
  admissions@ttu.edu

Strongest Programs:
Agriculture Education and
  Communications
Business Administration
Chemistry
Computer Science
Engineering
Human Development and
  Family Studies
Technical Writing

The Tech student body is overwhelmingly white and homegrown; only 5 percent of students hail from outside the Lone Star State. Asian Americans account for 2 percent of the total population, African Americans 3 percent, Hispanics 12 percent, and Native Americans 1 percent. Students are "very laid-back and friendly," and more humble than their counterparts at other big-name Texas schools, students say. Tech offers merit scholarships worth an average of $1,638, as well as 216 athletic awards. Elite chess players can vie for a handful of scholarships as well.

Only 24 percent of the students at Tech live in the dorms and freshmen are required to do so. Co-ed, single-sex, and quiet study dorms are available; one men's dorm lacks air-conditioning, which might be considered cruel and unusual punishment given the West Texas heat. Most students do move off campus by their sophomore year. Blue-light emergency phones and a safe-ride shuttle service help students feel safe, and dining options are plentiful. "Whether you're a vegetarian or on a protein diet, you'll eat well here," says a junior. About 50 restaurants in the area also take the university's Tech Express debit card, and many have student specials one night a week.

> **"Any alcohol found on your person results in police charges, so don't bring it on campus!"**

Tech is officially dry—"Any alcohol found on your person results in police charges, so don't bring it on campus!" warns a junior—and since so many students live off campus, that's where the weekend action is. Eleven percent of the men pledge fraternities, and 17 percent of the women join sororities, so a sizable contingent heads to the parties at Greek Circle, where students say it's fairly easy for the underage to be served. The Depot District is also a popular destination, as most bars and clubs admit anyone 18 and over. The city of Lubbock (population 200,000) offers many opportunities to get involved with the community, through work with the Boys and Girls Club, Habitat for Humanity, animal shelters, or Bible study at local churches. Annual traditions include homecoming, complete with a chili cook-off; the Carol of Lights during the first weekend in December; and Arbor Day, when hundreds of students fan out across campus to plant flowers and trees for the spring. Popular road trips include any of the four nearby lakes (for picnicking, boating, or camping), skiing in New Mexico (four hours away), and anywhere the Red Raiders are playing, especially if it's against the Texas A&M Aggies, or the University of Texas Longhorns.

The Division I Red Raiders compete in the Big 12 Conference, and the football, baseball, and men's track and field teams are among the school's best. (At football games, when players take the field, the Masked Rider, replete with red and black cape and cowboy hat, motivates the crowd by galloping up and down the sidelines.) The Lady Raiders basketball team has won the conference championship a number of times in the past few years,

> **"We have a good mix of 'small town' students and 'big town' students."**

and the men's program has gotten a lot of attention since Tech hired explosive former Indiana coach Bobby Knight. Intramural sports, which about half of the undergraduates take part in, include everything from the typical soccer and flag football to mud volleyball, table tennis, dodge ball, and inner-tube water polo.

Texas Tech may have begun as a stepchild of Texas A&M, but over the past 80 years the school has done much to carve out its own niche. "We have a good mix of 'small town' students and 'big town' students," says a communications major. If you can stand the heat and relative isolation of the West Texas Plains, and the effort it takes to be more than a number at a school of this size, Texas Tech may be worth a look.

Overlaps

University of Texas–Austin, University of North Texas, University of Houston, Texas A&M, Texas State, Baylor

Trinity College

300 Summit Street, Hartford, CT 06106

While many small colleges have been treading water, Trinity has had a notable increase in applications and selectivity in recent years. It is taking advantage of its urban setting through a $175 million community revitalization initiative. Trinity joins Lafayette, Swarthmore, and Smith among small colleges offering engineering.

For students at Trinity College, the learning experience doesn't stop at the campus borders. At first glance, the small liberal arts college and the large, gritty city of Hartford, Connecticut, seem like an uneasy match. But instead of insulating itself from outside problems, Trinity takes advantage of its surroundings by using Hartford as its classroom. The centerpiece of the college's community revitalization effort is the Learning Corridor, an eclectic mix of 16 schools in the neighborhood surrounding the campus. On campus, academic standards continue to rise, and students graduate with a strong liberal arts background.

Splendid Gothic-style stone buildings behind wrought-iron fences decorate Trinity's 100-acre campus. The large, grassy quadrangle is home to pick-up games of hackeysack and lazy relaxation on warm spring and fall afternoons. Along with revitalizing the neighborhood that surrounds it, Trinity's campus is undergoing its own revitalization. The $35 million Library and Information Technology Center is one of the leading small-college library facilities in the nation with an array of multimedia stations.

Trinity's general education requirements include one course each in the arts, humanities, natural sciences, numerical and symbolic reasoning, and social sciences. Students must also demonstrate proficiency in writing and mathematics. The First-Year Program includes a seminar emphasizing writing, speaking, and critical thinking; the seminar instructor serves as students' academic advisor. Freshmen may also choose one of three guided-studies programs in the humanities; natural sciences; or the history, culture, and future of cities, which one student calls "phenomenal—very challenging and rewarding."

Students give rave reviews to Trinity's English, economics, and computer science departments, and say the school's small but accredited engineering program is likewise strong. That department sponsors the Fire-Fighting Home Robot Contest, the largest public robotics competition in the U.S., open to entrants of any age, ability, and experience. Newer majors at Trinity include public policy and law; women, gender, and sexuality; and environmental science, which benefits from the 256-acre Field Station at Church Farm in Ashford, Connecticut. Through the BEACON program, biomedical engineering students can take courses at UConn, the UConn Health Center, and the University of Hartford while conducting research at three area health centers. Trinity's close ties to the community also are apparent in the curriculum; students can take courses on urban development and history of the city of Hartford. Thirty Community Learning courses are offered each year that provide an out-of-classroom learning experience.

Website: www.trincoll.edu
Location: Urban
Total Enrollment: 2,184
Undergraduates: 2,165
Male/Female: 50/50
SAT Ranges: V 610–700
   M 610–700
ACT Range: 25–29
Financial Aid: 37%
Expense: Pr $ $ $ $
Phi Beta Kappa: Yes
Applicants: 5,744
Accepted: 39%
Enrolled: 25%
Grad in 6 Years: 85%
Returning Freshmen: 92%
Academics: ✎ ✎ ✎ ✎
Social: ☎ ☎ ☎ ☎
Q of L: ★ ★ ★
Admissions: (860) 297-2180
Email Address: admissions .office@trincoll.edu

Strongest Programs:
Economics
Political Science
History
Biology
Chemistry
Engineering
Modern Languages

Faculty/student collaboration is a tradition at Trinity. Nearly 30 percent of a recent graduating class worked with professors on research and scholarly papers, and many students join their mentors to present findings at symposia.

The academic climate can be challenging. "I think Trinity is a very rigorous school," says a senior. "Even its study abroad programs struck me as more demanding than those of schools like Middlebury." Professors get high marks, as they do the lion's share of teaching—graduate assistants only lead review sessions before exams. "My professors have all been wonderful," a senior English major says. "They are always available to talk whether it's about schoolwork or any kind of problem you might have."

Nearly two-thirds of the students seek internships with businesses and government agencies in Hartford (the insurance capital of the world), and some also take terms at other schools through the Twelve College Exchange.* The Global Sites program enables students to study with Trinity professors in seven exotic locations, from South Africa and Trinidad to Chile, Nepal, and China. Other enticing choices include the

> **"It's gotten much more diverse and is larger now than it was when I matriculated."**

Mystic Seaport term* for marine biology enthusiasts, studying Italian language and art history at Trinity's campus in Rome, or learning Spanish at the University of Cordoba. Back on campus, the Department of Theater and Dance offers an unusual integrated major, and sponsors a study-away program with LaMaMa in New York City and Europe. Administrators caution that the educational studies department offers students a richer understanding of the field, but that to get teacher certification, students must tap into the Hartford higher education consortium.

Eighteen percent of Trinity students are Connecticut natives; many of the rest hail from Massachusetts and other nearby states. One student describes the student body as "highly aggressive in all ways: socially, academically, athletically." African Americans and Hispanics each constitute 5 percent of the student body, and Asian Americans add 6 percent. "It's gotten much more diverse and is larger now than it was when I matriculated," a senior says. "Trinity is traditionally a conservative school, however the college Democrats have more members," a sophomore notes. The college does not offer merit or athletic scholarships, but the majority of admissions decisions are made without regard to financial considerations, and the college guarantees it will meet the full demonstrated need for four years.

Ninety-one percent of Trinity's students live in the co-ed dorms. The best bets for freshmen are said to be Jones or Jarvis because of their central location on the quad. "There aren't any dorms that are horrible," says one student. "As a freshman you might get some not-so-great housing. But after that, all rooms are pretty nice." After freshman year, rooms are assigned by lottery; seniors pick first, then juniors and sophomores, though everyone is guaranteed a bed. Options include singles, doubles, quads, units with kitchens, and theme houses for those interested in

> **"Trinity's theme parties are famous."**

music, community service, wellness, art, and quiet. Freshmen must eat in Trinity's dining hall; the Bistro, an upscale but reasonable cafe, is another choice for students on the meal plan. Others hibernate in the Cave, which offers sandwiches and grilled fare. "The dining facilities change their menus daily," an Italian major says, "and are very good in terms of service and quality."

When it comes to Trinity's social scene, the options are plenty. "Trinity is a small college that is quiet during the week, but has a social scene to rival large colleges on the weekends," a sophomore says. Some students praise the Trinity College Activities Council, which brings in comedians and musical performers and organizes parties, study breaks, and community service days. The Underground Coffeehouse and the Bistro's weekly comedy nights are also popular. A college-sponsored "culture van" takes students to downtown Hartford, to catch a show at the Busnell or visit the Wadsworth Atheneum, the nation's oldest public art museum. But the action on Thursday, Friday, and Saturday nights is mostly on campus and mostly at the co-ed Greek houses (20 percent of men and 16 percent of women join up). "The social life is definitely on campus," according to one student. "We have several frats and sororities, plus a very active activities council. Trinity's theme parties are famous." Alcohol is not hard to come by, but "the campus policies on alcohol are fairly severe on underage drinkers and abusers," says a student.

The First-Year Program includes a seminar emphasizing writing, speaking, and critical thinking; the seminar instructor serves as students' academic advisor.

Hartford may not be the typical college town, but Trinity students are working to change that. "During the past few years Trinity has dedicated a lot of efforts to giving back to the community surrounding it," says one sophomore. It "offers unique opportunities for internships, mentoring, and community service," another sophomore says. The $175 million community revitalization campaign includes a $5.1 million grant from the W.K. Kellogg Foundation and draws on existing community resources. The initiative will generate $130 million in new construction. The Office of Community Service and Civic Engagement offers

"Our average student is self-motivated and active."

opportunities for students to work and learn in the city. Students have created and run organizations that provide housing, tutoring, meals, and other services to youth, families and senior citizens.

On campus, drinking is officially taboo for students under 21, and at official college functions, including the frat parties, students must present two forms of ID to be served. "Very few students venture off campus to bars," one student says, "but there are room parties on campus." Spring Weekend brings bands to campus for a three-day party outdoors. Popular road trips include Montreal, Boston, New York City, and the beaches and mountains of Maine.

Trinity's Bantams compete in Division III, and both men's and women's squash are powerhouses. The men's squad has won the national championship for eight years running, and the women's team has brought home several titles. Homecoming typically brings Wesleyan or Amherst to campus for a football game, which gets underway after Trinity students burn the opposing school's letter on the quad before the game. Three-quarters of students take part in the intramural program, which offers a variety of sports; about 800 of them play softball.

Trinity is ahead of the curve in liberal arts education, and, a sophomore says, "you can appreciate it much better if you are interested in more than one subject." Even more importantly, Trinity students have taken their civic responsibility to heart. "Our average student is self-motivated and active, whether it be in sports or in clubs or community service," a senior says. "It's a place where everyone can have their voice heard."

Overlaps
Tufts, Boston College, Brown, Georgetown, Middlebury

If You Apply To ➤

Trinity College: Early decision: Nov. 15, Jan. 1. Regular admissions: Jan. 1. Financial aid: Feb. 1. Guarantees to meet full demonstrated need. Campus and alumni interviews: recommended, evaluative. SATs, ACT or three SAT IIs: required. SAT IIs: optional. Accepts the Common Application and electronic applications. Essay question: topic of personal significance.

Trinity University

One Trinity Place, San Antonio, TX 78212-7200

One of the few quality Southwestern liberal arts colleges found in a major city. Trinity is twice as big as nearby rivals Austin and Southwestern, and offers a diverse curriculum that includes business, education, and engineering in addition to the liberal arts. Upscale and conservative.

Website: www.trinity.edu
Location: Urban
Total Enrollment: 2,756
Undergraduates: 2,524
Male/Female: 46/54
SAT Ranges: V 600–690
  M 600–690
ACT Range: 27–31
Financial Aid: 77%
Expense: Pr $
Phi Beta Kappa: Yes
Applicants: 3,684
Accepted: 63%
Enrolled: 27%
Grad in 6 Years: 74%
Returning Freshmen: 89%
Academics: ✍ ✍ ✍
Social: ☎ ☎ ☎
Q of L: ★ ★ ★
Admissions: (800) 874-6489
Email Address:
  admissions@trinity.edu

Strongest Programs:
Business Administration
Modern Languages and
  Literature
English
Political Science
Communication
Economics
History

Trinity University is a small school with big bucks. Thanks to the oil boom of the 1970s, Trinity has one of the nation's largest and fastest-growing educational endowments at a school its size. The wealth is used unashamedly to lure capable students with bargain tuition rates and to entice talented professors with Texas-sized salaries. The result? A student body comprised of smart, ambitious men and women, and a knowledgeable and caring faculty. Students here enjoy challenges, but still manage a laid-back Texas attitude. "It's friendly, warm, personal, engaged, and academically stimulating," a senior says.

Trinity was founded in a small central Texas town just after the end of the Civil War. In 1952, the school moved to its current location, a residential area about three miles from downtown San Antonio, one of the most beautiful cities in the Southwest. The 117-acre campus, filled with the Southern architecture of O'Neill Ford, is located on what was once a rock quarry. Everything fits the school's somewhat well-to-do image, from the uniform redbrick buildings to the stately pathways that wind along gorgeous green lawns and through immaculate gardens spotted with Henry Moore sculptures. Trinity's most dominant landmark is Murchison Tower, which rises in the center of campus and is visible from numerous vantage points throughout San Antonio. Recent

"It's friendly, warm, personal, engaged, and academically stimulating."

construction includes the new Northrup Hall, which houses the Modern Languages and Literature and English departments as well as various administrative offices. The 70,000-square-foot Ruth Taylor Art and Music Complex is currently undergoing a $20 million renovation.

The school makes a strong effort to maintain its admissions standards, keeping a small enrollment, tightening the grading system, and recruiting high achievers with greater energy than is possible from larger universities. The university has set its sights on becoming the premier small liberal arts school in the Southwest. Students report that the academics are very rigorous, but the climate is not competitive.

The professors at Trinity are there because "they are committed to teaching and like to engage with the students," a senior religion major says. "The quality is excellent." Another student adds, "Professors teach all the classes and labs and are always available during office hours to answer questions." Trinity has a highly praised education department, with a five-year MAT program and a good premed program. Other strong departments include economics, business, and engineering and chemistry. The communications department offers students hands-on training with television equipment or the chance to produce a newscast. Accounting majors are offered a chance to serve an internship with the Big Four accounting firms in San Antonio, Houston, Dallas, or Austin while earning a salary and receiving college credit.

Trinity's approach to general education requirements is the Common Curriculum, under which students must take courses from five areas: cultural heritage, art and literature, human social interaction, quantitative reasoning, and science and technology. Students must also have proficiency in a foreign language and computer skills, although those requirements can be fulfilled in high school. First-year students must take a topical seminar and a writing workshop, or an intensive, six-credit, one-semester Readings from Western Culture. As a member of the Associated Colleges of the South*, Trinity approves a number of study abroad programs and encourages premed and prelaw students, as well as history and English majors, to take advantage of them. Another feature of Trinity's curriculum is an opportunity for students to participate in research projects with faculty mentors. New degree programs include a major in neuroscience and minors in computing and film

"The campus as a whole leans to the left."

studies. Trinity has the fastest-growing Chinese language program in the country, and the Languages Across the Curriculum program features classes such as religion and anthropology taught in languages including German, Spanish, and Russian.

The classes at Trinity are small; almost all have 50 or fewer students, and freshmen are assigned to mentor groups of 10 to 15 students for academic and guidance counseling, as well as peer tutoring from upperclassmen. "The advisors do a good job of helping students because each advisor does not have more than 10 advisees most of the time," says a sociology major. However, students say the career counseling area could use a little work.

"The students at Trinity are a very vivacious bunch who welcome everyone with warm smiles and honest advice," says one student. Sixty-nine percent of the students are Texans. The school is fairly diverse, with approximately 21 percent of the student body comprised of minority groups; Hispanics alone account for 11 percent, Asian Americans another 6 percent, African Americans 3 percent, and Native Americans 1 percent. As with most universities, there is some attention paid to social and political issues, including the war in Iraq. "The campus as a whole leans to the left but there is definitely a conservative presence as well," a senior says. Merit scholarships averaging more than $7,000 are available to academically gifted students, but student athletes must fend for themselves.

Trinity requires students to live on campus until their junior year. In fact, 78 percent of the students live in the residence halls, which are described as "awesome." "Every dorm room/suite has its own balcony, bathroom, microwave oven, and is large in size," says one student. "When they vacuum and clean the room, they leave mints." Dorms are co-ed, with one single-sex residence hall that is off-limits to freshmen; all dorms are now wired for cable and the Internet. Most seniors move off campus so they can have single rooms.

"There are no rivalries in sports since we win at everything."

San Antonio receives a "thumbs up" from students. "San Antonio has an eclectic downtown beat," says one student. With a number of colleges nearby, young people abound and frequent the many outdoor shops and cafes at the Riverwalk, as well as touristy hangouts such as Sea World. There are also many cultural and musical attractions. Students get involved in city life through the Trinity University Volunteer Action Center.

Trinity's typical social life is off-campus bars or parties or on-campus meeting places and events such as lectures, coffeehouses, or bands. "The social life on campus is constantly buzzing and we tend to have as much fun as possible outside of studying," a junior says. Most of the parties are within walking distance of the school and are primarily hosted by the fraternities and sororities—but open to all students. Twenty-four percent of the men and 29 percent of the women in the student body join the local fraternity and sorority organizations. Country line dancing is also a popular activity, and the university sponsors an excellent lecture series that brings notable politicians and public figures to campus, including Colin Powell, Michael Moore, and Margaret Thatcher. Alcohol may be consumed on Trinity's campus by those of legal age in any of the upperclass dorms, except for designated substance-free floors.

As a member of the Associated Colleges of the South, Trinity approves a number of study abroad programs and encourages premed and prelaw students, as well as history and English majors, to take advantage of them.

The school makes a strong effort to maintain its admissions standards, keeping a small enrollment, tightening the grading system, and recruiting high achievers with greater energy than is possible from larger universities.

The beautiful, warm weather of San Antonio provides plenty of activities for the students year round, but there are also plenty of fun road trips that the students enjoy. The funky state capital of Austin is 90 miles north, and students can also road trip internationally to nearby Mexico. An annual event most students look forward to is Fiesta, a weeklong celebration of San Antonio's mixed culture that features bands, dancing, food, and drink. They also anticipate the Tigerfest dance and parade on homecoming weekend and the Chili Cook-Off that pits Greek and other clubs against one another. The school year kicks off with a party at the school's bell tower, which students can climb to get a knockout view of San Antonio.

Sports are another popular activity at Trinity, where they compete in the Division III and SCAC Championships. Since 2000, Trinity has won several national championships and claimed six consecutive SCAC Presidents' Trophies, which are awarded to the school with the best overall athletic program. In fact, one student brags, "There are no rivalries in sports since we win at everything." Successful sports include football, soccer, basketball, tennis, and women's volleyball. Sixty-five percent of students take advantage of the intramural program, and flag football is the most popular sport.

A big state and big money give students at this small university many of the advantages of a larger school. If you are looking for a small community feel with quality professors, look no further than Trinity. "It is a very nice campus, students hold a lot of school pride, and the teachers are very welcoming," one satisfied student says.

Overlaps

University of Texas, Texas A&M, Rice, Southwestern, Texas Christian, Baylor

If You Apply To ➤

Trinity University: Early decision: Nov. 1. Early action: Nov. 1., Dec. 15. Regular admissions: Feb. 1. Financial aid: Feb. 15 preferred. Meets demonstrated need of 56%. Campus interviews: optional, evaluative. No alumni interviews. SATs or ACTs: required. SAT IIs: optional. Accepts the Common Application and electronic applications; prefers electronic applications. Essay question: Common application personal statement.

Truman State University

McClain Hall, Room 205, Kirksville, MO 63501

Formerly Northeast Missouri State University, Truman changed its name to emphasize that it has more in common with private institutions than with nondescript regional publics. Truman is looking for a public ivy niche like Miami of Ohio and William and Mary. A new residential college program will increase cocurricular learning.

Website: www.truman.edu
Location: Small town
Total Enrollment: 5,674
Undergraduates: 5,486
Male/Female: 41/59
SAT Ranges: V 570–670
  M 570–670
ACT Range: 25–30
Financial Aid: 39%
Expense: Pub $ $ $
Phi Beta Kappa: Yes
Applicants: 4,567
Accepted: 84%

Truman State University, Missouri's only public liberal arts college, attracts overachievers from across the Show-Me State. Indeed, since 1996—when the school changed its name from Northeast Missouri State University—Truman has worked to become a "public ivy" on the order of Miami University (OH) or the College of William and Mary. True, the tiny town of Kirksville, Missouri, is no Williamsburg, Virginia—or even Oxford, Ohio. But the school's relative isolation makes it easier to focus on academics.

Truman is located in the northeastern corner of Missouri, about 200 miles from both Kansas City and St. Louis. The flower-laden campus includes 39 buildings on 140 acres, many of which are Georgian in style—in fact, the oldest portion of the campus, dating to 1873, is based on Thomas Jefferson's University of Virginia. A $26 million expansion of Magruder Hall, the science building, wrapped up in 2005 and a new residence hall was completed in 2006.

The most popular major at Truman is business administration; biology, English, psychology, and communications round out the list of programs with the highest enrollment. An interdisciplinary major allows students to combine coursework from two or more disciplines to create a specialized major. There are also five-year programs for students interested in education or accounting, which culminate in the awarding of bachelor's and master's degrees. "Despite the rigorous nature of courses, students here tend to work together to help each other succeed," says a junior, "by creating their own study groups and by working on the large number of group projects assigned by professors."

Truman's general education requirements revolve around the liberal arts. All students must take courses in fine arts, math and science, religion, humanities, foreign language, and social science. In the junior year, students complete an interdisciplinary, writing-enhanced seminar. The required Truman Experience program includes a week of freshman orientation, just before classes start, and a one-semester course designed to help first-year students transition to college life. The vast majority of courses taken by freshmen have 25 students or fewer—about a third have 25 to 50 students. "With a few exceptions, the quality of teaching is out-standing," says a student. "Truman has very few teaching assistants and the vast majority of classes are taught by professors with the highest degree in their field." About 9 percent of students go abroad annually, visiting approximately 50 countries through the College Consortium for International Studies and the Council on International Educational Exchange. Twenty percent of students conduct independent research, or collaborate with faculty members on research projects.

> "Students here tend to be more interested in learning and studying than at other public schools."

"With high ACT scores, students here tend to be more interested in learning and studying than at other public schools," says a junior. Seventy-six percent of Truman students hail from Missouri and 80 percent ranked in the top quarter of their high school class. Hispanics and Asian Americans each make up 2 percent of the Truman student body, and African Americans add 4 percent. Students tend to be liberal, and the biggest social and political issues on campus include the war in Iraq, gay marriage, and the death penalty, according to one student. Sixty-eight percent of Truman students receive scholarships based on academic merit, and the average award is $4,002. The school also hands out 295 athletic scholarships each year.

Fifty-one percent of Truman students live in the dorms, and the recent housing crunch has been alleviated somewhat by the addition of a new residence hall. The Residential College program creates an integrated living and learning environment, with each dorm run by a senior faculty member known as a College Rector. Dining-hall menus are on a five-week rotation, and "the food is average, with a great dessert selection countering the lack of edible meat,"

> "Be ready to meet people who will impact the rest of your life."

says a finance and economics major. "Grilled and baked food is hard to find, as everything is deep-fried." Campus security is good and "the campus is well-lit, emergency phones are located all over campus, and the Department of Public Safety patrols regularly," says a political science major.

The town of Kirksville (population 17,000) leaves much to be desired but grows on you, say students. "Most students dread coming to school in Kirksville," says a student, "but soon love the laid-back and friendly nature of the town." Rock-climbing enthusiasts can drive 90 minutes to Columbia, Missouri—both to indulge their passion and to visit the much larger, more-happening main campus of the University of Missouri. Truman's SERVE Center keeps a list of projects seeking volunteers, and the Big Event is a campus-wide day of service that sends vast numbers of Truman students out to better the community.

Because of Kirksville's size, the Greek system dominates Truman's social life; 31 percent of men and 22 percent of women sign up, though the campus is officially

(Continued)
Enrolled: 39%
Grad in 6 Years: 67%
Returning Freshmen: 86%
Academics: ✍ ✍ ✍
Social: ☎ ☎ ☎
Q of L: ★ ★ ★
Admissions: (660) 785-4114
Email Address:
  admissions@truman.edu

Strongest Programs:
Accounting/Business
  Administration
Biology
Education
Chemistry
Foreign Languages
Political Science

The required Truman Experience program includes a week of freshman orientation, just before classes start, and a one-semester course designed to help first-year students transition to college life.

dry. Other diversions are provided by the Student Activities Board, which brings in comedians, bands, movies, and plays. "Social life is strictly divided between the alcohol-swilling Greek community, and the nerds who stay in the dorms and have fun there," one student explains. Aside from Columbia, popular road trips include St. Louis, various destinations in Iowa, and Quincy, Illinois. Everyone looks forward to homecoming in the fall and the Dog Days carnival in the spring, with wacky games such as the rubber-band race and sumo wrestling (in inflatable suits). Other enjoyable traditions include snowball fights in the winter, and "late-night runs to Pancake City after playing Capture the Flag."

When not competing academically, Truman's Bulldogs are succeeding on the playing field and in the pool. The women's swim team recently won its sixth straight NCAA Division II title, while the volleyball team finished in fourth place in the national championship tournament. The women's soccer team also captured a seventh straight conference championship. Intramural sports here began in the 1920s, and include everything from bowling to basketball, golf, and running. Truman also plays Northwest Missouri State every year for the Hickory Stick, one of the oldest college rivalry trophies west of the Mississippi.

"With a few exceptions, the quality of teaching is outstanding."

Truman State offers a winning combination of challenging academics and a close-knit community. Though Truman's rural Missouri location isn't for everyone, its bargain-basement price is certainly worth considering. "Be ready to work," advises a junior, "and be ready to meet people who will impact the rest of your life."

Overlaps

University of Missouri–Columbia, University of Illinois, Southwest Missouri State, St. Louis University, Washington University (MO), Drake

If You Apply To ➤

Truman State University: Early action: Nov. 15. Rolling admissions: Mar. 1. Financial aid: Apr. 1. Housing: May 1. Meets full demonstrated need of 62%. Campus interviews: recommended, informational. No alumni interviews. SATs or ACTs: required. SAT IIs: optional. Accepts the Common Application and electronic applications. Essay question: one- to three-page writing sample on a topic important to you.

Tufts University

Bendetson Hall, Medford, MA 02155

Tufts will always be a second banana to Harvard in the Boston area, but given the Hub's runaway popularity among college students, second is not so bad. Best known for international relations, Tufts is also strong in engineering and health-related fields. In the Experimental College, students can take off-the-wall courses for credit.

Website: www.tufts.edu
Location: Suburban
Total Enrollment: 8,863
Undergraduates: 4,971
Male/Female: 48/52
SAT Ranges: V 660–740
     M 670–740
ACT Range: 28–32
Financial Aid: 38%
Expense: Pr $ $ $ $

Some academic superstars used to consider Tufts University a safety school, a respectable place to go if you didn't get into Penn or Cornell. But Tufts isn't so safe anymore, at least not when it comes to admissions. Applications are up dramatically, propelling Tufts into the ranks of the most selective schools in the country. With its strong academics, high-achieving student body, and attractive setting, some might say that not much more separates Tufts University from its illustrious neighbors, Harvard and MIT, than a few stops on the T.

Tufts's 150-acre, tree-lined hilltop campus overlooks the heart of nearby Boston, and is a striking scene. The main campus, with its brick and stone buildings, sits on the Medford/Somerville boundary. Medford, the fifth-oldest city in the country, was a powerful shipbuilding center during the 19th century. Somerville, the historic

Revolutionary powder house, lies adjacent to the Tufts campus, and in 1776, the first American flag was raised on its Prospect Hill. For years, Tufts has devoted resources to traditional areas of graduate strength—medicine, dentistry, law, and diplomacy—as well as new ventures, such as a Nutrition Research Center. Such additions had only a peripheral impact on the liberal arts and engineering colleges, but Tufts has made a noticeable commitment to facilities that primarily benefit undergraduates. Recent additions include the high-tech project development laboratory for student design projects, a $21 million addition to the Tisch library, which doubled its size, and the renovation of the chemistry research building, which allowed more lab space for undergraduate research.

Despite the recent flurry of expansion, undergraduate teaching is what attracts students. They get highly personalized attention from faculty, and they enjoy wide freedom to design their own majors, pursue independent study, and do research and internships for credit. "The classes here are challenging," says a student, "but I find that students come together to collectively help one another." Strong departments include international relations, political science, biology, engineering, drama, and languages, and there is an excellent child-study program. The most popular major is international relations, followed by economics, political science, psychology, and child development. While upper-level courses are reasonably sized (with an average of about 25 students), intro lectures can be quite large.

> "The classes here are challenging, but I find that students come together to collectively help one another."

Tufts has two popular programs in which students who need a break from being students can develop and teach courses: the 31-year-old Experimental College, which annually offers more than 100 nontraditional, full-credit courses taught by students, faculty, and outside lecturers; and the Freshman Explorations seminars, each taught by two upperclassmen and a faculty member to between 10 and 15 students. With topics ranging from media and politics to juggling, Exploration courses are a way for freshmen to get to know each other and ease into the college experience, since the teachers double as advisors. Newer programs include an undergraduate major in biomedical engineering.

Tufts students also get a healthy diet of traditional academic fare. Distribution requirements include a new World Civilization course in addition to art, English and foreign languages, social sciences, humanities, natural sciences, and math. Engineers only have an English requirement in addition to the standard math, science, and technically oriented curriculum, but they must complete 38 credits compared to the liberal arts students' 34. Full professors, who are praised for their knowledge, teach most courses. "They make themselves readily available to students and are extremely approachable," says a senior. Tufts offers the Washington Semester,* the Mystic Seaport program,* an exchange with Swarthmore, and cross-registration at a number of Boston schools. Tufts is among the top 10 research universities for the percentage of undergrads who study abroad and frequently ranks as one of the top Peace Corps suppliers.

The biggest homeland of the student body is Massachusetts (25 percent). New Jersey, New York, and California are also well represented, but students hail from all 50 states and 61 countries. The university's reputation in international relations attracts a substantial number of foreign students (9 percent) and Americans living abroad. Asian Americans make up 13 percent of the population, Hispanics 7 percent, and African Americans 7 percent. Political liberals outnumber conservatives. "I think students who attend Tufts are looking for the benefits of a liberal arts education and the benefits of a top tier research university," one student says. No merit or athletic scholarships are available, but several prepayment and loan options are, and in the past the school has met the full demonstrated need of all admits.

(Continued)
Phi Beta Kappa: Yes
Applicants: 15,532
Accepted: 28%
Enrolled: 31%
Grad in 6 Years: 90%
Returning Freshmen: 95%
Academics: 🐔 🐔 🐔 🐔 ½
Social: ☎ ☎ ☎
Q of L: ★ ★ ★ ★
Admissions: (617) 627-3170
Email Address: admissions .inquiry@ase.tufts.edu

Strongest Programs:
Engineering Technology Center
Center for Environmental Management
Center for Materials and Interfaces
Electro-Optics Technology Center
Experimental College
International Relations

With topics ranging from media and politics to juggling, Exploration courses are a way for freshmen to get to know each other and ease into the college experience, since the teachers double as advisors.

Tufts is in the midst of a modern-day renaissance, or what many universities know as a capital campaign. Money raised already is allowing Tufts to improve campus facilities and financial aid for students.

Accommodations in the Uphill and Downhill (the two quads joined by a great expanse of grass and trees) campus dorms vary from long hallways of double rooms to apartment-like suites, old houses, and co-ops. A good-natured rivalry exists between the two areas; Uphill is closer to the humanities and social sciences classrooms and supposedly a little more social, while Downhill is nearer the science facilities. Freshmen and sophomores must live on campus in the dorms, while upperclassmen compete in a lottery. "Getting a room can be a hassle," complains one sophomore. Students and administration agree that the addition of South Hall makes housing available to just about anyone who wants a room. Apartments are plentiful and, according to at least one student, affordable. Still, 25 percent of the students live off campus. All but one of the dorms is co-ed by floor, suite, or alternating rooms. Food plans for five, 10, 14, or 20 meals per week are offered to everyone but freshmen, who must choose one of the latter two options. Kosher and vegetarian meals are available, and occasional special meals (e.g., Italian night and Mexican night) spice up standard college cuisine.

While suburban Medford is not very exciting for those of the college class, the T metro system extends to the Tufts campus, so it's easy to make a quick jaunt to "student city" (aka Boston) for work or play. Harvard Square is even nearer and provides plenty of restaurants, nightlife, and music stores. For those with valid IDs, the campus pub has become an "in" place to hang out, especially Monday through Thursday nights. Tufts, incidentally, has earned a national reputation for its programs to promote the "responsible" use of alcohol.

A small band of 13 fraternities and three sororities provides many of the on-campus weekend parties, though only 15 percent of the men and 4 percent of the women join the Greek system, and there is talk of getting rid of it altogether. University-sponsored activities include concerts, plays (there are 15 to 20 productions each year at Aidekman Arts Center), and parties, and there are two-dollar movies on Wednesday and weekend nights. Major campus events in the fall include homecoming and Halloween on the Hill, the latter of which is a carnival for children in the community. At the end of finals week in December, the "students turn out by the hundreds and watch and participate in the Naked Quad Run!" confesses one student. Another adds, "Who wouldn't want to see a bunch of cold, naked kids running around?" In the spring, there is Tuftsfest, a month-long affair with festivals, interdorm Olympics, a semiformal dance, and Spring Fling, an end-of-year hurrah. Of all student activities,

> **"Who wouldn't want to see a bunch of cold, naked kids running around?"**

the largest by far, with more than 500 students, is the Leonard Carmichael Society, the umbrella group for all volunteer activities. The students are involved in programs of adult literacy, blood drives, elderly outreach, teaching English as a second language, hunger projects, tutoring, low-income housing construction, and active work with the homeless and with battered women.

Tufts students take athletics seriously. For the 2004–05 year, nine Tufts teams were represented at NCAA championship and tournament events, including men's and women's indoor track and field, women's tennis, men's and women's swimming and diving, and women's rowing. The intramural program serves nearly half the student body via 10 intramural athletic leagues. "Participation is pretty popular among the student body," a student says.

Tufts is in the midst of a modern-day renaissance, or what many universities know as a capital campaign. Money raised already is allowing Tufts to improve campus facilities and financial aid for students. This, along with a swelling applicant pool, makes Tufts a much hotter school than it was just a few years ago. And its proximity to Boston, an intellectual and educational mecca, makes it even more attractive than if it were it in, say, Detroit. Tufts gives every indication that it's going

Overlaps

Brown, University of Pennsylvania, Cornell, Georgetown, Yale, Duke

to keep scaling the university ranks until it reaches the summit—and that's not too far from Prospect Hill.

Tulane University

6823 St. Charles Avenue, New Orleans, LA 70118

The map may say that Tulane is in the South, but Tulane has the temperament of an East Coast institution. The university is trying to shoehorn its way into the front rank of Southeastern universities, though it still trails Emory and Vanderbilt. High achievers should shoot for the Tulane Honors program.

In an effort to cope with the devastating effects of Hurricane Katrina, Tulane University has opted for an "emergency surgery" of sorts: The school has laid off more than 200 faculty members, dropped several undergraduate majors, and suspended nearly a dozen athletic programs. By taking such painful measures, administrators hope to bring new life to this recovering university.

The school's 110-acre campus is located in an attractive residential area of uptown New Orleans, about 15 minutes from the French Quarter and the business district. Tulane's administration building, Gibson Hall, faces St. Charles Avenue, where one of the nation's last streetcar lines still clatters past mansions. Across the street is Audubon Park, a 385-acre spread where students jog, walk, study, or feed the ducks in the lagoon. The buildings of gray stone and pillared brick are modeled after the neocollegiate/Creole mixture indigenous to Louisiana institutional-type structures. One particular point of pride is the university's 13 Tiffany windows, one of the largest collections in existence. Severe flooding damaged two-thirds of the campus, including residence halls and the Health Sciences Center, and caused $200 million of damage. Fortunately, the university has rebuilt and restored much of the campus and renovations are expected to continue for some time to come.

Hurricane Katrina not only ravaged the university campus, but also took a severe toll on program offerings. Five undergraduate science and engineering majors have been eliminated, including civil and electrical engineering. In addition, the graduate school has closed. Tulane's strength lies in the natural sciences, environmental sciences, and the humanities; international studies in general and Latin American studies in particular are especially

> "There is a large Northeastern constituency here."

strong. The Stone Center for Latin American studies includes the 200,000-volume Latin American Library and offers more than 150 courses taught by 80 faculty members. An interdisciplinary program in political economy (economics, political science, and philosophy) stands out among the social sciences and is very popular with prelaw students. Environmental studies majors benefit from the Tulane/Xavier Center for Bioenvironmental Research, where faculty members and students work together on research projects that include hazardous-waste remediation and the ecological effects of environmental contaminants.

Website: www.tulane.edu
Location: Urban
Total Enrollment: 13,214
Undergraduates: 7,976
Male/Female: 47/53
SAT Ranges: V 610–730
  M 630–690
ACT Range: 28–32
Financial Aid: 73%
Expense: Pr $ $ $ $
Phi Beta Kappa: Yes
Applicants: 14,107
Accepted: 55%
Enrolled: 22%
Grad in 6 Years: 74%
Returning Freshmen: 86%
Academics: ⚖ ⚖ ⚖ ½
Social: ☎ ☎ ☎ ☎
Q of L: ★ ★ ★
Admissions: (504) 865-5260
Email Address: undergrad.admission@tulane.edu

Strongest Programs:
Premed
Prelaw
Political Economy
Biomedical Engineering
Engineering
Business
Anthropology

Tulane offers several study abroad options, including one-semester programs to locations such as Japan to study sociology and culture, Mexico City to delve into the language, and London to study liberal arts. In addition, the Tulane/Newcomb Junior Year Abroad program is one of the country's oldest and most prestigious programs, in which the student is fully immersed in the language and culture of the particular country. For students looking to go to medical or law school, approximately 66 percent of graduates are accepted. Helping freshman make the transition are several programs. One is TIDES, where students can join groups on such topics as Understanding Your Classmates, World Religions, and Cultures. Another offering for freshmen is the First Year Experience, one-credit courses on such subjects as Metacognition (Thinking about Thinking), Campus Life, and Women and Leadership.

Sixty-one percent of the classes at Tulane have fewer than 25 students, while an additional one-fourth have fewer than 50, making it difficult for students to get into the classes of their choice. About 60 percent of those classes are taught by full professors. Graduate instructors are most likely to teach the beginning-level classes in English, foreign languages, and math, and are rarely found in the schools of business, architecture, or engineering. Overall, students praise Tulane's faculty, and the academic atmosphere can be very intense, depending on the class.

All Tulane liberal arts majors must complete a rigorous set of general education requirements. Besides demonstrating competency in English, math, and a foreign language, these requirements mandate that students take distribution requirements in the humanities and fine arts, the social sciences, and mathe-

"New Orleans itself never stops partying!"

matics and the sciences. In the process of satisfying the requirements, students must take at least one course in Western and non-Western civilization, as well as a writing-intensive class, although freshmen with high SATs can place out of some classes. Each year the university's highly acclaimed honors program invites about 700 outstanding students, known as Tulane Scholars, to partake in accelerated courses taught by top professors. These select scholars also have the opportunity to design their own major and spend their junior year abroad.

While Tulane has a somewhat Southern feel, it is a sophisticated and cosmopolitan institution. Says one student, "There is a large Northeastern constituency here who have brought their Type-A personalities and racial tolerance down to a Southern city. If you're a Northerner, it's impossible to escape the Southern influence of the city, and if you're a Southerner, it's impossible to escape the Northern influence that exists on campus." Sixteen percent of the students are minorities, half of whom are African American. Tulane awards hundreds of merit scholarships, ranging from $16,000 up to full tuition, and athletic scholarships for student athletes. Although student enrollment took a severe hit in the months following the storm, nearly 88 percent of students had returned by Spring 2006, and that figure is expected to rise.

In the aftermath of Katrina, securing adequate housing has proved difficult for many students. Although most residence halls have been rebuilt and refurnished, the housing crunch has forced the university to find innovative solutions. To wit, the school has leased an 1,100-passenger cruise ship to house 200 professors, students, and staff members. Back on land, non-local freshmen must live on campus and leave their cars at home. After freshman year, housing is by lottery, and choices include Stadium Place, a student apartment complex. Many students opt to move off campus, claiming that it's much cheaper than university housing, but others are concerned about the safety factor of living in New Orleans. Some men live in their fraternity houses, but sororities only have social halls due to an old New Orleans law that makes it illegal to have more than four unrelated women living in one house. Freshmen have to stomach the cost of Tulane's meal plan, but alternatives exist at the University Center food court.

Hurricane Katrina not only ravaged the university campus, but also took a severe toll on program offerings. Five undergraduate science and engineering majors have been eliminated, including civil and electrical engineering.

Although most residence halls have been rebuilt and refurnished, the housing crunch has forced the university to find innovative solutions. To wit, the school has leased an 1,100-passenger cruise ship to house 200 professors, students, and staff members.

Social life at Tulane goes almost without saying. "New Orleans itself never stops partying!" boasts a junior. Fraternities and sororities are a presence—33 percent of the men and 37 percent of the women join—but do not dominate the social life. Though you're supposed to be 21 to buy alcohol or enjoy the bar scene in the cafes and clubs that dot the French Quarter, a sophomore explains that "alcohol is accessible." Mardi Gras is such a celebration that classes are suspended for two days and students from all over the country pour in to celebrate. An annual Jazzfest in the spring also draws wide participation. Road-trip destinations include the Gulf Coast, Mississippi, Houston, Atlanta, and Memphis.

While schoolwork is taken seriously at Tulane, so are sports. The campuswide acclaim for men's basketball borders on hysteria. Unfortunately, that spirit has been dampened somewhat by the temporary suspension of eight varsity sports due to the university's post-Katrina reorganization. Those cut include men's and women's golf, women's soccer, women's swimming and diving, men's and women's tennis, men's track, and men's cross-country. Still, club sports are big, and students can also opt for weight work, squash, or swimming among other options at the Reily Recreational Center.

Among those hardest hit by Hurricane Katrina, Tulane has been forced to make tough decisions about the future of the university. By cutting faculty and staff, as well as some academic athletic programs, the university hopes to emerge with a renewed vigor and focus on undergraduates. Through it all, Tulane remains a forward-looking school where the possibilities seem endless.

If You Apply To ⟩ **Tulane:** Early decision: Nov. 1. Regular admissions: Jan. 15. Does not guarantee to meet demonstrated need. No campus interviews. Alumni interviews: optional, informational. SATs or ACTs: required. SAT IIs: recommended (home-schooled applicants only). Accepts electronic applications. Essay question: personal statement

University of Tulsa

600 South College Avenue, Tulsa, OK 74104

Tulsa is a notch smaller than Texas Christian and Washington U, but bigger than most liberal arts colleges. The university has a technical orientation rooted in Oklahoma oil, but it has a much more diverse curriculum than Colorado School of Mines. Tulsa has an innovative program allowing undergraduates to do research beginning in their first semester.

The University of Tulsa is a small, private liberal arts school with a growing international reputation. Known for its engineering programs, including petroleum and geosciences, it attracts students from around the world. With an emphasis on undergraduate research, hands-on work experience, and a diverse array of course offerings, TU has found its niche. "TU students are self-motivated and academically centered, with a healthy focus on fun," a senior says.

TU's 210-acre campus is just three miles from downtown Tulsa, and there's a striking view of the city's skyline from the steps of the neo-Gothic McFarlin Library. The university's more than 50 buildings run the architectural gamut from 1930s-vintage neo-Gothic to contemporary, all variations on a theme of yellow Tennessee limestone dubbed "TU stone." The new Collins Hall houses student services, financial

Website: www.utulsa.edu
Location: Urban
Total Enrollment: 3,511
Undergraduates: 2,635
Male/Female: 51/49
SAT Ranges: V 540–700
  M 550–710
ACT Range: 23–30
Financial Aid: 50%
Expense: Pr $

(Continued)

Phi Beta Kappa: Yes

Applicants: 2,687

Accepted: 75%

Enrolled: 31%

Grad in 6 Years: 61%

Returning Freshmen: 84%

Academics: 🖉 🖉 🖉

Social: ☎ ☎ ☎

Q of L: ★ ★

Admissions: (918) 631-2307

Email Address:
admission@utulsa.edu

Strongest Programs:
Engineering
Computer Science
Communication Disorders
Psychology
Anthropology
Biological Sciences
History
Art

All of the dorms are equipped with free cable television and connections to the campus mainframe and the Internet.

aid, and undergraduate admissions in one central location, while the Case Athletic Complex provides a variety of football-related facilities. New student apartments are also in the works.

In addition to its well-established and internationally recognized petroleum and geosciences engineering programs, TU offers solid majors in computer science, the natural sciences, and many social sciences. The rapidly growing English department has some impressive resources at its disposal in McFarlin Library's special collections. The collections boast original works by 19th- and 20th-century American and British authors, including books, letters, manuscripts, a stained necktie that once belonged to James Joyce, and more than 50,000 items representing Nobel Laureate V.S. Naipaul's life and work from the 1950s to the present. Newer courses have been added to the English department, and the first classes have completed the Internet-based MBA program. The exercise science program has received national accreditation and a newer program allows advanced students in history, chemical engineering, and applied mathematics to earn a bachelor's degree and master's degree in five years.

"No one minds sharing notes and studying in groups."

In accordance with the Tulsa Curriculum, the cornerstone of the school's emphasis on liberal arts, all undergraduates take two writing courses, at least one mathematics course, and one or two years of foreign language, depending on the degree. In addition, each student completes at least 25 credit hours of general curriculum classes in aesthetic inquiry and creative experience, historical and social interpretation, and scientific investigation. Academics are a top priority at TU, but students say the atmosphere is cooperative. "The academic climate is challenging, yet rewarding," says one student. "No one minds sharing notes and studying in groups." Professors generally receive high marks from students. "The low professor-to-student ratio creates a unique learning environment that fosters classroom discussion and greater academic achievement," says a senior.

Honors students take exclusive seminars, complete a thesis or advanced project, and can live together in a computer-equipped house. The Tulsa Undergraduate Research Challenge offers outstanding opportunities for cutting-edge scientific research, and has produced 38 Goldwater Scholarship winners, 24 National Science Foundation Graduate Fellowships, and four Fulbright Grants. Thirty-five to 50 percent of engineering and natural science students do research with faculty. "The College of Engineering and Natural Sciences has the best programs and departments," offers a student. "All of the students there receive excellent internship opportunities and most have jobs lined up before graduating." Twelve percent travel abroad for programs including language immersion in Spain, studio art in Italy, business integration in Germany, and environmental study in Costa Rica. Tulsa is also one of 20 schools in the nation that train America's "Cyber Corps," the first line of defense against computer hackers and terrorists.

"Get involved! It'll make or break the experience."

Fifty-nine percent of Tulsa's students are from Oklahoma; most others are from the Midwest and Southwest, with many hailing from Dallas and St. Louis. Eight percent are foreign, coming from the Middle East, East Asia, and Scandinavia. "TU has outstanding student leaders and scholars with an array of interests," says one student. The student body is 7 percent African American, 2 percent Asian American and 4 percent Hispanic. Political issues do not seem to be prevalent and political correctness is not often an issue. "A lot of opinions are represented at TU," says a student. "Overall, students are open to hearing other points of view." Athletes compete for 346 scholarships in six men's sports and eight women's sports. TU also offers merit scholarships, averaging $9,800.

At Tulsa, freshmen and sophomores are required to live on campus. Sixty-four percent of the students use campus housing, and that number has been on the rise, making it harder to get a room. Students have plenty of "extremely livable" options, including three mixed-sex dorms (two co-ed by wing and the other co-ed by suite), one women's dorm, one men's dorm, fraternity and sorority houses, and campus apartments. All of the dorms are equipped with free cable television and connections to the campus mainframe and the Internet. The single-sex dorms are quieter and more attractive to upperclassmen. "Rooms are spacious and the residence hall associations take great effort to make sure students feel comfortable and at home," says a psychology major. Dining options receive a passing grade, although "the food can get tiresome," says a junior, "but so would my favorite restaurant if I ate there three times a day, seven days a week."

The social life at TU is surprisingly robust thanks to a "healthy Greek scene and well-funded Residence Hall Association," according to one student. There are scheduled events and students enjoy simply hanging with friends at small parties, too. The Bricktown section of Oklahoma City and nearby casinos (for those 21 and older), along with more distant Dallas, St. Louis, and Kansas City, are popular road trips. The Greek organizations claim 21 percent of TU men and 23 percent of the women, and the frats host campuswide house parties. Student-initiated policies govern drinking on campus. Administrators say this self-policing has led to responsible imbibing.

Campus traditions include the ringing of the college bell in the Alumni Center cupola by each senior after his or her last class and during Springfest. Other big events include Reggaefest, homecoming, and Greek events such as the Kappa Sigma Olympics, the Sigma Chi Derby Days, and the Delta Gamma Anchor Splash. Nearby parks, lakes, and a huge recreational water park please outdoor enthusiasts. Downtown Tulsa offers symphony, ballet, opera, and an annual Oktoberfest. Students are very active in community service. "Students can

> **"Overall, students are open to hearing other points of view."**

get involved in a variety of community organizations such as churches, Habitat for Humanity, Big Brothers Big Sisters, and Tulsa's Day Center for the Homeless," says one student.

Here in Oklahoma, sports are important. Students get riled up when the football team is pitted against rivals Oklahoma and Oklahoma State, and when the basketball team suits up against Arkansas and OSU. Tulsa competes in Conference USA and recent conference champs include the women's softball and basketball teams and the men's tennis squad. At least 70 percent of undergrads participate in at least one of the 30 intramural sports; the most popular by far is flag football.

TU is trying to do some things differently: be a small liberal arts school in a part of the country most known for sprawling public universities, and incorporate professional preparation with an emphasis on broad intellectual challenges. Those considering the University of Tulsa might do well to heed the advice of this junior: "Get involved! It'll make or break the experience."

TU is trying to do some things differently: be a small liberal arts school in a part of the country most known for sprawling public universities.

Overlaps

Texas Christian, Southern Methodist, Washington University (MO), University of Oklahoma, Oklahoma State

---

If You Apply To ➤

Tulsa: Rolling admissions. Financial aid: Feb. 15 (priority). Housing: May 1 (priority). Meets demonstrated need of 26%. Campus interviews: recommended, evaluative. No alumni interviews. SATs or ACTs: required. SAT IIs: optional. Accepts the Common Application and electronic applications. Essay question: challenge you have faced; how you would change the world; significant life event.

Union College

807 Union Street, Schenectady, NY 12308

Union is split down the middle between liberal arts and engineering. That means its center of gravity is more toward the technical side than places like Trinity, Lafayette, and Tufts, but less so than Clarkson and Rensselaer. Schenectady is less than exciting, but there are outdoor getaways in all directions.

Website: www.union.edu
Location: City outskirts
Total Enrollment: 2,180
Undergraduates: 2,180
Male/Female: 55/45
SAT Ranges: V 570–660
   M 590–690
ACT Range: 25-29
Financial Aid: 48%
Expense: Pr $ $ $ $
Phi Beta Kappa: Yes
Applicants: 4,230
Accepted: 47%
Enrolled: 29%
Grad in 6 Years: 84%
Returning Freshmen: 91%
Academics: ✑ ✑ ✑ ✑
Social: ☎ ☎ ☎
Q of L: ★ ★ ★
Admissions: (518) 388-6112
Email Address:
   admissions@union.edu

Strongest Programs:
Political Science
Mathematics
Mechanical Engineering
Chemistry
Philosophy
Psychology
Modern Languages

Founded in 1795, Union College is so named thanks to its emphasis on the union of the sciences and humanities. More than 200 years later, this independent liberal arts college is known for its interdisciplinary studies and its study abroad programs. At Union, engineering and the arts go hand-in-hand. Undergraduate research has deep roots at Union, starting in the 20th century when a chemistry professor began involving students in his colloid chemistry investigations. Union College has shed its partying reputation in favor of hard studies and community involvement.

Union's 100-acre campus, designed in 1813 by French architect and landscaper Joseph Jacques Ramée, sits on a hill overlooking Schenectady. Ramée's vision took shape in brownstone and redbrick, with plenty of white arches, pilasters, and lacy green trees; the campus plan also includes eight acres of formal gardens and woodlands. The 16-sided Nott Memorial, a National Historic Landmark, is a meeting, study, and exhibition center—and the site of a naked run each year. Parcels of land adjacent to College Park Hall, a hotel-turned-dorm, have become soccer fields for college and community use. Five of the seven Minerva House System buildings have been renovated and a new 12,000-square-foot fitness center features copious space for training, dance, and locker rooms.

Union's general education requirements "provide the sort of education that will allow students to flourish in a rapidly changing world by analyzing and integrating knowledge from a wide variety of areas," according to administrators. Students must take two core courses in their first and second years that promote reading, writing, and analytical skills. They also must take three interdisciplinary courses in an approved cluster and eight courses spread among social science, humanities, linguistic and cultural competency, quantitative and mathematical reasoning, and natural and applied science or engineering.

Among Union's most popular majors are political science and psychology. Students also flock to economics, biology, and history; the latter department is home to Union's most esteemed lecturer, Stephen Berk, whose course on the Holocaust and Twentieth-Century Europe is a hot ticket. Chemistry and engineering also get high marks from students. Each year, about 50 incoming freshmen are named Union Scholars. The designation extends the First-Year Preceptorial to two terms from one, allows students to work on independent study projects, and gives them access to departmental honors programs and expanded study abroad options. Each spring, Union cancels classes one afternoon for the Charles Steinmetz Symposium, so that students can present scholarly projects to their peers and professors. More than 300 students participated in 2005.

> **"Due to the trimesters, we have to cover a lot of information in a small period of time."**

Interdisciplinary study is the norm at Union, with established programs in Russian and Eastern European studies, industrial economics, and law and public policy. Newer majors include astronomy; neuroscience; and science, medicine and technology in culture. New minors include astrophysics, dance, and nanotechnology. The Educational Studies program allows aspiring teachers to complete courses and fieldwork required for secondary school certification in 14 subjects, along with a strong liberal arts grounding. The Leadership in Medicine program, a joint program with the Graduate College of Union University and Albany Medical College, gives students the opportunity to earn a bachelor's degree, a master of science in health management of business administration in health systems administration, and a medical degree in eight years.

Union operates on a trimester system, which means thrice-a-year exams and a late start to summer jobs—but also the opportunity to concentrate on just three courses a term. More terms also means more opportunities for independent study and internships, either in the state capital of Albany, 20 minutes away, or in Washington, D.C. By the time graduation rolls around, over half of each class has studied abroad. "The coursework is rigorous," a freshman confides. "Due to the trimesters, we have to cover a lot of information in a small period of time."

Back on campus, students give the faculty high marks. "Profs are always willing to accommodate you," says a student. "The quality of teaching has been, for the most part, above average." Seventy-three percent of all freshmen classes have 25 or fewer students and students can expect to see full professors at the lecterns rather than TAs.

"The majority of students come from middle-class families from the Northeast," says a freshman. "We tend to have strong family bonds, good work ethics, and a drive to live life to the fullest." Forty-three percent of Union students are New Yorkers, and 70 percent went to public high schools. Three percent of the student body is African American, another 4 percent is Hispanic, and 6 percent is Asian American. The school

> **"We tend to have strong family bonds, good work ethics, and a drive to live life to the fullest."**

has been working to boost these numbers, through its membership in the Consortium for High Achievement and Success, and through the establishment of the President's Commission on Diversity. Union offers a number of merit scholarships but no athletic scholarships. The Chester Arthur Undergraduate Support for Excellence (CAUSE) award, which takes its name from the former U.S. president (and distinguished Union graduate), offers loans to students interested in public service. They're forgiven at 20 percent per year if the student pursues a service-oriented career.

Eighty-seven percent of Union students live in the dorms. The Minerva house system (named after the Roman goddess of wisdom) is aimed at getting students and faculty members to contribute to Union's social, residential, and intellectual life—and, students say, at decreasing the influence of the Greek system, which draws 24 percent of the men and 30 percent of the women. (Greek chapters are not permitted to offer housing.) Union's dorms "have singles, doubles, triples, and suites (quads)," says a human geography major. "Freshmen will get quads, but you can get singles as a sophomore." Students recommend West, which is co-ed by room, and thus very social, as well as Fox and Davidson, where freshmen and sophomores live in suites: two double bedrooms and a "huge" common room. Dining options consist of four main eating areas and "the food is certainly edible and quite diverse," according to one student.

Social life at Union stays on campus; the Greeks dominate, and "we do have a strong drinking culture," says a biology major. That said, there are a variety of campus events, including comedians, concerts, and speakers. The administration established a committee of students, staff, and faculty members to oversee programming about alcohol abuse. Favorite

> **"You need a car."**

annual traditions include the lobster bake (each student gets his or her own crustacean), Party in the Garden, "Painting the Idol" (a really ugly campus statue) and Spring Fest, a huge concert.

Off campus, Schenectady is an old-line industrial city that's "a slum on one half, and a green paradise on the other," says sophomore. "There are essentially no college-town amenities—like bookstores—in walking distance. You need a car." What Schenectady lacks can be found in Saratoga Springs, which boasts restaurants, jazz clubs, horse racing, and Skidmore College, or in the nearby Adirondacks and Catskills. Popular road trips include Boston, Montreal, and New York, and the ski slopes of nearby Vermont. And students are trying to help Schenectady rebound, through tutoring programs in local schools and work with Big Brothers Big Sisters. There's also a service project during freshman orientation. "My group had to paint one of the bridges in Schenectady," says one student. "It was actually a good time."

Union's athletic teams compete in Division III, aside from men's and women's ice hockey, both of which are Division I. Men's basketball and women's lacrosse and softball have recently brought home league championships. There's a full-time director of intramural sports, and students may field teams in everything from tennis and volleyball to broomball and lacrosse. Rugby and ultimate Frisbee are among the sports that enjoy club status, and about 60 percent of students participate in intramural or club athletics.

Union's mission is constantly evolving, as the college struggles to meet the needs and interests of students and faculty. It retains its commitment to a strong, core liberal arts curriculum while acknowledging the increasing effect of globalization and technology. Union College has plenty to offer—a small, friendly place full of eager intellectual exchange. You just have to seek it out.

If You Apply To ➤

Union: Early decision: Nov. 15, Jan. 15. Regular admission and housing: Jan. 15. Financial aid: Feb. 1. Guarantees to meet full demonstrated need. Campus, alumni interviews: recommended, evaluative. SAT I or ACT: required. SAT IIs: required if neither SAT I nor ACT is submitted (writing and two others). Accepts the Common Application and prefers electronic applications. Essay question (choose one): significant experience, achievement, risk, or ethical dilemma; issue of personal, local, national, or international concern; influential person, fictional character, historical figure, or creative work; or a topic of your choice.

Ursinus College

Box 1000, Collegeville, PA 19426

Ursinus is the smallest of the cohort of eastern Pennsylvania liberal arts colleges that includes Franklin and Marshall, Muhlenberg, and Lafayette. The plus side is more attention from faculty and more emphasis on independent learning. Although Philly is within arm's reach, the setting is quiet.

Zacharias Ursinus established his namesake college with the directive that students "examine all things, and retain what is good." So, for many years, the school focused on training in practical fields ranging from business administration to sports science. Recently, though, Ursinus has returned to its liberal arts roots—even expanding its offerings, with new majors in theater and dance, and restructuring its core curriculum to emphasize the philosophical underpinnings of modern thought. What hasn't changed is the close-knit feel of the school. "Ursinus is distinctive because of the small, mostly accepting environment," says a biology and sociology major. "Everyone knows everyone else and we help each other through difficult times, much like a family."

Ursinus is located in Collegeville, about 40 minutes west of Philadelphia, and 10 miles from the green, rolling hills of Valley Forge National Park. Buildings on the 170-acre campus are mostly constructed of Pennsylvania fieldstone; many have had their interiors upgraded and their exteriors preserved and restored. Actors and dancers benefit from rehearsal and exhibition space in the Kaleidoscope performing-arts center. Recent additions include a 180-bed dorm and a complete renovation of historic Bomberger Hall.

General education requirements at Ursinus fall under the college's Plan for Liberal Studies, and they are grounded in the assumption that while individuals have intrinsic value, they also live in a community. Therefore, the core includes two semesters of the Common Intellectual Experience—a course that explores topics from Plato to Buddhist scripture to Nietzsche—as well as requirements in science, culture, and history. Students also choose one of 28 majors, and engage in at least one Independent Learning Experience—a research project, an internship, study abroad, or student teaching—before graduation. For honors students, the requirements are more stringent; their independent projects are evaluated by outside examiners. And up to 30 percent of rising seniors get fellowships from the school to fund full-time summer research projects with a faculty member. The library has 400,000 volumes, and the school is connected to OCLC, a consortium of more than 18,000 libraries, and most books are accessible within a few days. Ursinus students may also use the Penn libraries.

> "Ursinus is distinctive because of the small, mostly accepting environment."

The most popular majors here are biology, economics, psychology, exercise and sport science, and English. The college also offers a strong program in East Asian studies, and was the first Pennsylvania institution to win state approval to certify secondary teachers of Japanese. "Courses are very rigorous and challenging," says a freshman, "but I do not feel overwhelmed." Professors draw praise for their skills in the classroom. "I have learned how to be a much better student," says a communications major. Classes are small—88 percent have 25 students or less—and students say they are pleased with the quality of teaching, especially since there are no TAs. Study abroad is available around the globe, thanks to the college's membership in the Bradley University Consortium. Ursinus also offers programs with its own faculty in Japan, Mexico, Spain, Italy, England, Costa Rica, and Germany. Finally, students can take a term at other U.S. universities, such as Howard and American in Washington, D.C., while prospective engineers may choose 3–2 programs at Columbia University and elsewhere.

Ursinus students are "extremely open-minded, ambitious, diverse, and diligent," according to one freshman. Sixty-one percent are from Pennsylvania, with most others hailing from New York, New Jersey, and other mid-Atlantic and New England states. African Americans make up 7 percent of the student body, Asian Americans 4 percent, and Hispanics 3 percent, and students report little tension among the groups. Race relations, sexual harassment,

> "Students are much more liberal and open than at other schools."

and physical safety are covered in peer education programs included in freshman orientation. The gay and lesbian organization on campus is very active, but otherwise, students aren't politically active. There are merit scholarships available each year, averaging $9,000 each, but no athletic scholarships are offered.

Ninety-five percent of students at Ursinus live in the dorms, which adds to the feeling of community. Upperclassmen quickly grab the Main Street houses, a string of Victorian-era homes across the street from campus, while freshmen are clustered in The Quad and BWC, short for Brodbeck-Wilkinson-Curtis, which have generously sized rooms. Richter-North Hall, a 143-bed dorm, opened in 2002 and more housing will become available in 2007. What does all that mean? You'll get a room, but "it's often hard to get the exact room you want," says a biology and sociology major. There are four meal plans—19, 14, or 10 meals per week, or 220 meals per semester—and all of them give students access to the main dining room and to Zack's, a snack

(Continued)
Accepted: 74%
Enrolled: 32%
Grad in 6 Years: 78%
Returning Freshmen: 93%
Academics: ✍ ✍ ✍
Social: ☎ ☎ ☎
Q of L: ★ ★ ★
Admissions: (610) 309-3200
Email Address:
   admissions@ursinus.edu

Strongest Programs:
Biology
Chemistry
Political Science
Economics
History
Premed
Prelaw
Writing and Modern
   Languages

The college offers a strong program in East Asian studies, and was the first Pennsylvania institution to win state approval to certify secondary teachers of Japanese.

bar. "There are so many different choices," cheers one student, "I never run out of things to eat."

Security is not much of a concern here, as the town of Collegeville is just 10 blocks long—and Ursinus takes up five of those. Still, the town is growing, and is adding the amenities that students want and need, such as a sushi restaurant, a sports bar, an ice cream and coffee shop, four pizza parlors—and, most importantly, late-night pizza delivery. On campus, Greek life draws 17 percent of the men and 28 percent of the women, and fraternity parties are a popular weekend diversion. Registered on-campus parties are monitored by student "social hosts" who check IDs and make sure things don't get out of hand. "We have a zero-tolerance policy on underage drinking, period, and any drinking in the freshman centers," explains one student. "These rules are not heavily enforced." Other options include free movies, lectures, and dances—or the shops and restaurants of central Philadelphia, less than an hour away. Many students with cars escape to the Jersey shore during warmer months, and each spring, students look forward to Airband, "a big charity lip-synching and performing event," says an English major. Finally, there are two mega-malls within a 20-minute drive of the campus, including the King of Prussia complex, the second-largest mall in the country.

Ursinus students love sports, and 60 percent play on a Division III varsity or an intramural team. And for women seeking post-collegiate careers in coaching, Ursinus is a well-known stepping stone to those posts; more than 50 colleges have hired Ursinus alumnae, including the University of Virginia and Old Dominion. The school's lacrosse team is a perennial contender for the national championship, and the women's field hockey team is also strong. Sadly, the longest-running football rivalry in the Philadelphia area—against rival Swarthmore—is no more, since Swarthmore dropped its football program. Among other contributions in the world of sports, Ursinus is mentioned in Trivial Pursuit for having a tree in the end zone of its football field.

> "Courses are very rigorous and challenging, but I do not feel overwhelmed."

Ursinus is on the rise. It may not be in the center of a big metropolitan area, and it doesn't offer big-time sports, but the college compensates for its lack of size with the feeling that students, faculty, and staff are one big family. New majors and facilities for aspiring artists are helping to diversify the student population, while development in the town of Collegeville is adding excitement to life outside the classroom. "It's a very small and intimate campus," says a senior, comparing the Ursinus experience to boarding school. "Students are much more liberal and open than at other schools."

If You Apply To ➤ **Ursinus:** Early decision: Jan. 15. Regular admissions, financial aid, and housing: Feb. 15. Meets full demonstrated need of 60%. Campus interviews: recommended, informational. Alumni interviews: optional, informational. SATs or ACTs: required (optional for those in top 10%). SAT IIs: recommended. Accepts the Common Application and electronic applications. Essay question: how your unique experiences during the past four years have prepared you to contribute to the Ursinus community; also requires submission of a graded research paper.

University of Utah

250 SSB, Salt Lake City, UT 84112

While the true-blue Mormons generally head for BYU, University of Utah attracts a diverse crowd that is drawn to the region's only major city. The majority of Utah

students hail from the Salt Lake City region, and many commute from home. Science and professional programs such as business and engineering are the most popular.

In addition to being the flagship institution of the public Utah System of Higher Education, the University of Utah is a major national scientific research center. Founded in 1850, the university is unusual in its ability to offer students the advantages of living in a city while at the same time maintaining a connection with nature.

Set in the foothills of the Wasatch Mountains near the shores of the Great Salt Lake, the university enjoys a picturesque location a half-hour drive from "the greatest snow on earth." Occupying 1,500 well-landscaped acres with nearly as many different kinds of trees as undergraduates, the campus is the state's arboretum. The architectural style of the university's structures ranges from 19th-century, ivy-covered buildings to state-of-the-art athletic facilities. The lake, snow, and beautiful mountain location are among the reasons why this city was chosen to host the 2002 Winter Olympics; the relatively few students who live on campus reside in beautiful dorms, which were used originally as Olympic Village.

> "There are lots of great social activities on campus."

While the professional degrees are quite popular, the U does not skimp on general education requirements. Students must fulfill classes in writing, American institutions, math, statistics, and intellectual explorations, which includes two courses in the humanities, sciences, social sciences, or fine arts, as well as an international course requirement. There are also bachelor's degree requirements, which include upper-division writing, a course in diversity, and either foreign language or quantitative courses. Renowned for its research in biomedical engineering, Utah hosted the first mechanical heart transplant. The Undergraduate Research Opportunities Program offers semester grants to students who join faculty members in scholarly pursuits. Two thousand students are enrolled in the honors program, where class size averages 17 students. Newer minors include Latin American studies, international studies, and an interdisciplinary minor in animation studies.

The academic climate is challenging. "The courses are pretty rigorous," says a senior. "It can be competitive within departments." Introductory courses often enroll hundreds of students, and freshmen often find themselves in very large lectures with smaller discussion sections led by graduate student teaching assistants. Utah's professors generally receive high marks from the students. "I have had great teachers that have passion for their subject," says a political science major.

Utah's students are a mostly middle-class, fairly homogeneous lot; of the 81 percent who are from Utah, all but 3 percent attended public schools. "'U' students are mostly commuter students who come from the Salt Lake Valley or out-of-staters who come to ski or snowboard," a junior says. African American and Native American students together make up 2 percent of the student population. Asian Americans account for 5 percent and Hispanics 4 percent. There is an active student government and most students agree that students here are less conservative than those found at rival BYU. Utah offers a handful of scholarships for academic achievement and 49 athletic scholarships.

> "The dorms are great."

Only 8 percent of students live on campus. "The dorms are great," cheers one student. "They are big with a private bathroom." Students are also generally satisfied with the food, though edibility varies based on which campus eatery you choose.

Social life is low-key, due in large part to the high number of commuters. Still, "There are lots of great social activities on campus," says a senior, including "over 200 student groups, lots of service opportunities, concerts, movies, festivals, speakers, and performing artists." Favorite road trips take students to Las Vegas, Lake Powell, and nearby ski resorts (with slopeside bus service available from the school). Only 1 percent of the men and women go Greek.

Website: www.utah.edu
Location: Urban
Total Enrollment: 29,012
Undergraduates: 15,413
Male/Female: 55/45
SAT Ranges: V 495–630
  M 500–630
ACT Range: 21–26
Financial Aid: 25%
Expense: Pub $
Phi Beta Kappa: Yes
Applicants: 6,687
Accepted: 85%
Enrolled: 50%
Grad in 6 Years: 55%
Returning Freshmen: 83%
Academics: ✿ ✿ ✿
Social: ☎ ☎ ☎
Q of L: ★ ★ ★
Admissions: (801) 581-7281
Email Address:
  admissions@sa.utah.edu

Strongest Programs:
Business
Psychology
Communications
Ballet/Dance
Chemistry
Engineering

Introductory courses often enroll hundreds of students, and freshmen often find themselves in very large lectures with smaller discussion sections led by graduate student teaching assistants.

Salt Lake City isn't exactly a college town, but a junior says, "The night life in SLC downtown is great if you are over 21." Some students get very involved in community service or political activities. Adjacent to campus, the Latter Day Saints Institute of Religion sponsors dances and other social activities with a decidedly conservative bent. Cultural activities include the respected Utah Symphony, several dance companies, opera, and, of course, the Mormon Tabernacle Choir. For basketball fans, there's the Utah Jazz.

Football and basketball bring students together in the MUSS—Mighty Utah Student Section—where the cheering is loudest in games against rival BYU. Utah's women's gymnastics team has won 10 NCAA titles; the men's basketball team also won the 2004–05 conference championship. In addition to the university's dozens of intramurals and club sports, the Outdoor Recreation Program offers backpacking, river running, canoeing, mountain biking, and skiing trips.

> **"I have had great teachers that have passion for their subject."**

Students say that diversity and academic quality are both on the rise at Utah. "The best thing about students at the U is that they are friendly to all, supportive of each other, and you can do whatever you want," says a computer animation major. It's also one of the few places where you can find nationally recognized professional programs within easy reach of nationally recognized skiing.

If You Apply To ➤ **University of Utah:** Regular admissions: Apr. 1. Does not guarantee to meet demonstrated need. Campus and alumni interviews: optional; informational. SATs or ACTs: required (ACT preferred). SAT II: optional. Prefers electronic applications. No essay question.

Vanderbilt University

2305 West End Avenue, Nashville, TN 37240

More "Southern" than Emory or Duke, Vandy has traditionally been a preferred choice in the Deep South suburbs of Atlanta and Birmingham. Along with Tulane, it is the only leading Southern private institution with both business and engineering (and one of the South's leading music programs to boot).

Once a quiet, conservative school in the heart of the South, Vanderbilt University is working hard to diversify its student body. In contrast to institutions such as Duke, Emory, and Tulane, which are merely located in the South, Vanderbilt remains a bastion of Southern culture. Indeed, despite the arrival of the 21st century, football games here continue to require coats, ties, pearls, and dates, though presumably not at all once. The flip side of that formality is a relaxed attitude and a friendly culture that make the rigorous academic environment easier to handle.

Founded in 1873 by Cornelius Vanderbilt, the university's 330-acre tract in Nashville was named a national arboretum in the late 1980s and includes Peabody College, the central section of which is listed on the National Register of Historic Places. On the main campus, art and sculptures dot the landscape, and architectural styles range from Gothic to modern glass and brick. In the past five years, the school has added the Ben Schulman Center for Jewish Life, a 48,000 square foot studio arts center, and the Bishop Joseph Johnson Black Cultural Center.

Undergraduates choose one of four schools—arts and science, engineering, music, or education and human development—but everyone takes their core liberal arts courses in arts and science, where the writing program is a standout. "I've taken a wide variety of liberal arts courses, and come away with the impression that all of Vanderbilt's academic offerings are well above average," a senior reports. Freshman seminars allow students to explore various topics in small groups with close faculty interaction; recent seminars have included everything from New York, New York: Film and Literature, to Americans in Paris and Ethics of Life and Death. Popular majors include social sciences, engineering, psychology, foreign languages and literature, and mathematics. Education majors, who enroll at Vanderbilt's Peabody College, are required to double major, usually in a liberal arts field. They also have the opportunity to spend a summer at England's Cambridge University while teaching at a British school. Though Vanderbilt has no undergraduate major in business administration, many students interested in financial careers declare an economics major, and pursue a business minor; there's also a 3–2 program with Vandy's Owen Graduate School of Business, which lets talented undergraduates save a year on the path to

"Probably the most infamous students are the affluent, socially oriented, designer-clad types."

their MBAs. Newer programs include a major and an honors program in women's studies, a minor in Italian studies, a revised history major, and a concentration in collaborative arts for pianists. The ENGAGE program offers select incoming undergraduates the opportunity to receive conditional admission to the Vanderbilt graduate or professional school of their choice at the same time they are admitted as freshmen.

Vanderbilt's study abroad program offers students the chance to spend a year or semester in one of 19 countries, including England, Ireland, Scotland, China, Chile, Mexico, and Japan. The optional May session allows students to spend four weeks on a single project, helpful for double majors, or those who'd like to spend some time in a foreign country but can't commit to being away for an entire term. During the "May Mester," drama classes may tour London theaters while archaeology students work on digs. The state-of-the-art library also has one of the country's best archives of network evening newscasts, a boon for historians and political scientists.

In the classroom, Vanderbilt students are governed by the school's honor system, which dates from 1875. The system governs all aspects of academic conduct, and makes it possible for Vandy to give unproctored exams. Students rave about the faculty, who "love interacting with students," says a biomedical engineering major. "I recently passed a professor in the dining room on campus and he interrupted his morning newspaper to talk to me about his fascination with Mark Twain."

Vanderbilt students are split more or less down the middle politically between conservatives and liberals. "Probably the most infamous students are the affluent, socially oriented, designer-clad types," says one student. "But another group—the probably-affluent-but-not-flaunting-it, socially conscious, intellectually minded—is probably a better representation of the student body." The number of minorities on campus is also increasing, with Asian Americans now making up almost 6 percent of the student body, African Americans 7 percent, and Hispanics 4 percent. In addition to need-based financial aid, Vanderbilt offers merit scholarships and more than 200 athletic awards.

Eighty-four percent of Vanderbilt undergraduates live in the dorms. Beginning in 2008, all first-year students will come together in 10 houses on one of the most historic areas of campus. Older students may choose 10-person townhouses, six-room suites, theme dorms, and school-owned apartments. Seniors may move off campus but must obtain a special waiver. Vanderbilt has a whopping 13 dining facilities, and freshmen are required to buy a dinner plan.

Fifty percent of the women and 34 percent of the men join the Greek system; while many Greek parties are open to the entire campus, the effort to encourage

(Continued)
Accepted: 34%
Enrolled: 39%
Grad in 6 Years: 89%
Returning Freshmen: 96%
Academics: 🖉 🖉 🖉 🖉
Social: 🐨 🐨 🐨 🐨
Q of L: ★ ★ ★ ★
Admissions: (615) 322-2561
   or (800) 288-0432
Email Address:
   admissions@vanderbilt.edu

Strongest Programs:
Music and Arts
Education
English
Social Sciences
Engineering

The ENGAGE program offers select incoming undergraduates the opportunity to receive conditional admission to the Vanderbilt graduate or professional school of their choice at the same time they are admitted as freshmen.

mixing between the groups is not always successful. Dating is big; the most coveted invitation is the Accolade formal that precedes homecoming, tickets to which are both expensive and limited. (Funds from the Accolade benefit scholarships for minority students.) Another favorite Vanderbilt tradition is the Rites of Spring festival, a music festival that takes place on the main lawn. Though Vanderbilt students seem to drink just as much as those at any other school, they have to work to get their hands on booze, since open containers are banned in public and kegs are also taboo. As at many colleges, though, "underage students can find loopholes," says one student.

Vanderbilt's proximity to Music City USA provides "something for pretty much everyone"—a rich supply of bluegrass, country, and rock music, and an abundance of restaurants, theaters, and brewpubs, all within walking distance of campus. Country music fans shouldn't miss the Hall of Fame. Beyond Nashville's borders are the Great Smoky Mountains, and state parks with picnic facilities, beautiful lakes, and skiing in the winter. The best road trips are to Memphis (home of Elvis!), New Orleans (for Mardi Gras), Louisville (for the Kentucky Derby), and Atlanta.

Vanderbilt is the smallest—and the only private—institution in the competitive and football-crazy Southeastern Conference (Division IA). The Commodores have developed several competitive programs and the football team—traditionally an SEC underdog—has recorded big wins recently over schools such as Georgia, Tennessee, and Arkansas. The men's and women's basketball squads are perennial

> **"I've taken a wide variety of liberal arts courses, and come away with the impression that all of Vanderbilt's academic offerings are well above average."**

contenders and the baseball team has drawn national attention. Other solid programs include men's and women's tennis and golf, women's lacrosse, and the new women's bowling team.

Vanderbilt is no longer merely a regional powerhouse. The school is capitalizing on its unique blend of Southern charm and scholarly achievement to attract students from around the country. Four years here do carry a steep price tag—witness a tongue-in-cheek campus slogan, "Vanderbilt: It Even Sounds Expensive." But for many, investing in a Vanderbilt education is money well spent.

Overlaps

Duke, Washington University (MO), University of Virginia, Emory, Yale, Harvard

If You Apply To ➤   **Vanderbilt:** Early decision: Nov. 1, Jan. 3. Regular admissions: Jan. 3. No campus interviews. Alumni interviews: optional, informational. SATs or ACTs: required. SAT IIs: optional. Apply to particular school or program. Accepts the Common Application and electronic applications. Essay question: applicant's choice; also required are an academic interest statement, discussing why the applicant has applied to his or her chosen school at Vanderbilt, and a brief essay on why Vanderbilt is a good fit.

Vassar College

Poughkeepsie, NY 12604

It is hard to imagine that Vassar once considered picking up and moving to Yale in the 1960s rather than become a co-ed institution. Thirty-five years after admitting men, still on its ancient and picturesque campus, Vassar is a thriving, highly selective, avant-garde institution with an accent on the fine arts and humanities.

Website: www.vassar.edu
Location: Small town

Are you a scientist who composes music in your spare time? Or perhaps an actor who also enjoys dissecting Plato and Aristotle? You may feel at a home at Vassar, a small liberal arts college just 70 miles north of New York City. Once known as the most liberal of the Seven Sisters, and still a bastion of the left, Vassar prides itself on curricular flexibility, tolerance, and diversity. "We range from jocks to hippies, musicians to activists," says a senior. A junior adds, "What ties us together is our intellectual curiosity and general respect for the world."

The college's 1,000-acre campus, just outside Poughkeepsie, New York, includes two lakes and plenty of trees. Daffodils bloom in the spring, and foliage is omnipresent in the fall. Encircled by a fieldstone wall, the campus also boasts an astronomical observatory with one of the largest telescopes in the Northeast, a state-of-the-art chemistry building, a farm with an ecological field station, and an art center with 13,500 works, from Ancient Egyptian to modern times.

> **"Vassar is a vibrant community where students, faculty, administration, and staff all take an interest in each other."**

The architecture is predominantly neo-Gothic, with buildings also designed by notables such as Marcel Breuer, Eero Saarinen, and James Renwick. Kenyon Hall has been renovated to provide new classrooms, plus practice and performance space for the dance program, and the campus recently went wireless.

Vassar has no core curriculum, and no general education or distribution requirements. Indeed, academic flexibility is paramount. That said, all students must take a Freshman Course, a small seminar emphasizing oral and written expression, as well as one course that requires significant quantitative analysis. Finally, students must demonstrate intermediate-level proficiency in a foreign language, by studying one of the 18 languages taught at Vassar, or through a satisfactory score on an AP or SAT II test. Since Vassar has no graduate students or research-only faculty, all classes are taught by professors, who are "fantastic, engaging, and fun to work with," says a science, technology, and society major. Small classes and tutorials are the norm, and exams are given under an honor system.

The most popular majors include English, political science, and psychology; the newest programs are full majors in Chinese and Japanese. Regardless of their course of study, students find the academic climate cooperative rather than competitive. "Self-motivation keeps the students here on track," says one student, "and despite the lack of competition or grade awareness, we all work hard." The biology building houses two electron microscopes, while music students are spoiled by a grand collection of Steinway pianos. Each year, 35 to 40 percent of students study abroad, on Vassar programs in Germany, France, Italy,

> **"We range from jocks to hippies, musicians to activists."**

Spain, Morocco, England, or Ireland, or through programs run by other institutions. Vassar is also one of a rapidly dwindling number of colleges that allow students to use their financial aid packages to support study away from campus.

Back in the U.S., Vassar students interested in urban education benefit from the school's agreement with New York's Bank Street College. In addition, there are programs with the other 11 members of the Twelve College Exchange* and at four historically black colleges, as well as a drama program at the Eugene O'Neil Theater and a maritime studies program at the Mystic Seaport*, both in Connecticut. Also highly regarded is the college's Undergraduate Research Summer Institute (URSI), which offers stipends for students to work one-on-one with faculty members on scientific projects, either at Vassar or off campus. Finally, the Ford Scholars program offers opportunities for student-faculty collaboration in the humanities and social sciences.

Sixty-seven percent of Vassar students were in the top tenth of their high school classes; 24 percent are native New Yorkers. Minorities account for a substantial subset of the student population, with Asian Americans making up 9 percent, African Americans 5 percent, and Hispanics 6 percent. The school's ALANA Center supports and recognizes students of color and other ethnic and cultural groups. Hot-button political issues include diversity, same-sex marriage, and immigration, one student says.

(Continued)
Total Enrollment: 2,330
Undergraduates: 2,330
Male/Female: 40/60
SAT Ranges: V 660–740
  M 650–710
ACT Range: 28–32
Financial Aid: 60%
Expense: Pr $ $ $ $
Phi Beta Kappa: Yes
Applicants: 6,314
Accepted: 29%
Enrolled: 36%
Grad in 6 Years: 90%
Returning Freshmen: 96%
Academics: ✍ ✍ ✍ ✍ ½
Social: ☎ ☎ ☎
Q of L: ★ ★ ★ ★
Admissions: (845) 437-7300
Email Address:
  admissions@vassar.edu

Strongest Programs:
English
Political Science
Psychology
Biology
History
Economics
Film and Theater
Art History

While Vassar continues to offer a menu of high-quality liberal arts courses, emphasizing interdisciplinary connections, the college has also embraced technology and diversity.

Housing is guaranteed for four years and 98 percent of students live on campus, where there's an eclectic mix of nine dorms. All but one of is co-ed. "Some dorms are really modern and sparkling," says a junior. "Others have a vintage college feel, with traditional wood paneling, trim, and floors." The word is that Lathrop is the best dorm for freshmen, but no halls are reserved strictly for first-year students. Juniors and seniors favor the college-owned townhouses (five-person suites), and the four-person Terrace Apartments, both with kitchens and living rooms. "Dorm spirit is pretty big here," confides a junior. When those hunger pangs strike, students take advantage of two dining facilities that provide a wide assortment of vittles.

Vassar doesn't have a Greek system, so social life revolves around films, lectures, parties, concerts, and the like. Students who are 21 or older can party at the on-campus dance club Matthew's Mug, named for school founder Matthew Vassar. Those not old enough to legally imbibe can drop by the college's non-alcoholic coffeehouse. And while alcohol is permitted on campus, "underage drinking and open containers are not," says a political science major. Poughkeepsie itself is an old industrial town—not too exciting after dark, but with plenty of opportunities for fieldwork, internships for academic credit with local politicians, businesses, and community organizations. Restaurants and shops are within walking distance of campus, and malls and movie theaters aren't much farther away. In warmer weather, Mohonk State Park offers hiking and other outdoor diversions. Also close by are Franklin Roosevelt's Hyde Park (for history) and the Culinary Institute of America (for cheap gourmet meals prepared by fellow students). Popular road trips include New York City and Boston, both easily reached by train.

Owing to its heritage as one of the Seven Sisters, traditions are big at Vassar. "The year starts with Serenading—a time when the freshmen pay tribute to the seniors by singing them songs," one student explains. "The day ends with fireworks over the lake, and a trophy is presented to the dorm with the best song." On Founder's Day, in May, the entire community celebrates Matthew Vassar's birthday with music, carnival rides, food, and a day out on the grass. Fireworks and a movie cap off the festivities. "It's basically a big carnival for everyone at Vassar," says a junior. Students can still unwind after a hard day of classes with afternoon tea in the Rose Parlor of historic Main Building. And the number of a cappella singing groups has now reached double digits, one student says.

"Self-motivation keeps the students here on track."

Vassar's varsity squads compete in the Liberty League, and the women's tennis and volleyball teams brought home conference championships last year. Also in 2005–06, the women's rugby team was ranked 13th in the U.S., while the women's squash team was ranked 18th. Men's and women's soccer and crew have also done well. Intramural sports are offered at two levels, competitive and recreational, and roughly half the student body participates. Teams face off in everything from basketball and softball to billiards, golf, Frisbee, Ping-Pong, and water polo.

Vassar is no longer an elite women's college, and the gender of its students isn't the only thing that's changed since the school's founding. While Vassar continues to offer a menu of high-quality liberal arts courses, emphasizing interdisciplinary connections, the college has also embraced technology and diversity. "Vassar is a vibrant community where students, faculty, administration, and staff all take an interest in each other," says a senior. "People are happy and optimistic," a junior adds. "We take the time to enjoy college for what it is—a serious, but not too serious, time of life for learning and development."

Overlaps
Brown, Wesleyan, Yale, Tufts, Dartmouth, Cornell

University of Vermont

194 South Prospect Street, Burlington, VT 05401

For an out-of-stater sizing up public universities, there could hardly be a more appealing place than UVM. The size is manageable, Burlington is a fabulous college town, and Lake Champlain and the Green Mountains are on your doorstep. UVM feels like a private university, but alas, it is also priced like one.

With its beautiful setting, wide academic offerings, and abundance of clubs and cocurricular pursuits, the University of Vermont draws students from around the country. And, says a math major, they're not all granola types with a penchant for soy milk and snowboarding: "From hippies to preppies, UVM has it all, and they all coexist together." (The acronym is short for *universitas viridis montis*, Latin for "University of the Green Mountains.") While it's a public school, UVM's academics, research opportunities, and price tag are more akin to those of a private institution. Generous financial aid packages, and the growth plans of President Daniel Mark Fogel, are helping to ensure that Vermont remains both affordable and relevant, amid increasing competition with schools of both types. "It's turning into the school I always wanted it to be," says a business major.

Chartered in 1791, UVM was established as the fifth college in New England. UVM's picturesque campus sits on the shores of Lake Champlain in Burlington, virtually a stone's throw from the Canadian border. Architectural styles range from Colonial to high Victorian Gothic and functional modern; the oldest structures, in the center of the campus, are recognized on the National Registry of Historic Places. New construction is a way of life these days. The 164,000-square-foot Dudley H. Davis student center—scheduled to open this year—will provide a bistro and space for entertainment, multiple dining options, and meeting space. A new residence hall houses 800 students as well as a learning community focused on environmental issues.

> "It's turning into the school I always wanted it to be."

The product of a merger of a private college and public university, UVM attained quasi-public status in 1862 with the passage of the Morrill Land Grant College Act. Today, the university blends the traditions of both a private and public university. UVM's seven undergraduate colleges and schools set their own curricula and general education requirements, but administrators say most students need to take at least 30 credits—or 10 courses—in the arts, humanities, social sciences, languages, literature, math, and physical sciences. Also mandatory are a credit-bearing course in Race and Culture or a course that explores race relations and ethnic diversity in the U.S., and two credits of physical education. The most popular majors are business, psychology, English, biology, and political science; students also give high marks to environmental studies and engineering programs. While administrators say the fine arts suffer from a relative lack of breadth and depth, students say the campus and surrounding area offer an "immense art and music scene."

Website: www.uvm.edu
Location: Small city
Total Enrollment: 10,490
Undergraduates: 8,784
Male/Female: 45/55
SAT Ranges: V 520–630
  M 520–630
ACT Range: 22–26
Financial Aid: 55%
Expense: Pub $ $ $
Phi Beta Kappa: Yes
Applicants: 17,723
Accepted: 65%
Enrolled: 21%
Grad in 6 Years: 65%
Returning Freshmen: 88%
Academics: ✑ ✑ ✑ ½
Social: 🐾 🐾 🐾 🐾 🐾
Q of L: ★ ★ ★ ★ ★
Admissions: (802) 656-3370
Email Address:
  admissions@uvm.edu

Strongest Programs:
Biology
Health Sciences
Environmental Studies
Business

Premed, nursing, and prevet students benefit from the research and teaching capabilities of UVM's fine medical school, as well as from a seven-year program with the vet school at Tufts. There's also a 3–3 program that guarantees admission to UVM's graduate program in physical therapy, and a similar option with Vermont Law School. The newest majors include athletic training, early childhood special education, and public communication. A life skills program for athletes emphasizes health and wellness, academic skills, and moral and ethical reasoning. How tough is the academic environment? "Laid-back," says a business major. "There is definitely no pressure from anyone to do well." Still, classes can be challenging and professors expect much from their students. "The professors have a lot of knowledge to offer,"

"Everyone is open-minded and extremely accepting of each other."

says a freshman. "All of my professors have been enthusiastic and passionate about what they teach." Special programs for freshmen include the five-day TREK programs, in which students go hiking or biking, do community service, or take a leadership-skills development course before classes start. It's led by upperclass student mentors. Students are especially enthused about the Teacher Advisory Program, in which groups of 10 to 15 freshmen participate in a writing-intensive, discussion-oriented seminar taught by a professor who is also each student's advisor. Each year, about 400 students go abroad; UVM has programs in South America, Western Europe, and Asia and the Pacific Rim.

"There are a number of preppy students, hippies, and outdoorsy students," says a sophomore. "I think the students are all very accepting." Thirty-six percent of UVM students are native Vermonters and 91 percent graduated in the top half of their high school class. Hispanic and Asian American students each make up 2 percent of the student body, while African Americans represent 1 percent. Despite the low minority representation, "everyone is open-minded and extremely accepting of each other," says a senior. Liberals outnumber conservatives, if only slightly, and "the school is big on social justice," explains a senior. UVM offers merit scholarships averaging $12,157 each, and 229 athletic awards.

Freshmen and sophomores must live on campus, where a student ID is required to enter the dorms—a security precaution that students appreciate, but say is probably unnecessary. "The dorms are clean, well-sized, and ideally located," says a senior. Most juniors and seniors move off campus, "which creates a great UVM community in downtown Burlington," says a student, but the new dorms for upperclassmen encourage more of them to stay on campus. The administration also recently declared all residence halls alcohol-free. UVM has more than a dozen dining halls, and the quality and variety of the food varies. Many restaurants in downtown Burlington also take the school's CatScratch debit card.

Just 7 percent of UVM men and 5 percent of the women join Greek groups, so when the weekend comes, college-sponsored movies, dances, bars, and coffeehouses are big draws. "The social life is great and very active," says one student. Much of the fun also happens on nearby ski slopes and in Burlington itself, where the music scene

"Burlington is a quaint, fun town with a significant number of things to do."

rivals that of Athens, Georgia, birthplace of R.E.M. and the B-52s. A popular campus T-shirt warns: "If you want to party, come to UVM. If you want to stay, study!" Per Vermont law, students under 21 may not drink, so the most popular road trip is Montreal—90 minutes away—with even bigger concerts and a drinking age of 18.

"Burlington is a quaint, fun town with a significant number of things to do," says a student, "such as shopping, sightseeing, and concerts." The energetic downtown boasts symphonies, art galleries, chic shopping, and lively bars and restaurants, plus there's Lake Champlain only five minutes away, not to mention the mountains and ski slopes. As much as they love their little city, students also look forward to getting out of town. The Outing Club is one of UVM's most popular student organizations, as the nearby Green Mountains, White Mountains, and Adirondacks offer prime hiking, backpacking, and rock climbing.

Vermont's ice hockey team is the school's pride and joy, having ranked as high as second nationally. Students get access to tickets before the general public, a nice perk since Catamount games are always sold out. Soccer draws crowds in the fall, since there is no football team. Men's basketball brought home conference titles in 2003, 2004, and 2005. The ski team is a perennial powerhouse and recently finished fifth at the NCAA championships; the baseball team and men's and women's soccer are also very competitive. Most students participate in at least one intramural sport, with broomball being most popular. Couch potatoes can sign up for intramural video game tournaments.

Students at "Groovy UV" may be laid-back, but they're also curious, caring, open-minded, active, and willing to work hard. They enjoy giving their all, both in the classroom and outside—viewing extracurricular involvement as critical to the undergraduate experience. "There is something about this place that makes you want to do more, whether it be in the school, the community, the environment, or your activities," says a senior.

Overlaps

University of New Hampshire, Northeastern, Boston University, University of Connecticut, University of Massachusetts, Boston College

If You Apply To ➤ **UVM:** Early action: Nov. 1. Regular admissions: Jan. 15. Financial aid: Feb. 10 (priority). Meets full demonstrated need of 78%. Campus and alumni interviews: optional, informational. SATs or ACTs: required (including optional ACT Writing Test). SAT IIs: optional. Apply to individual schools or programs. Accepts the Common Application and prefers electronic applications. Essay questions: significant experience, contribution to diverse student body, topic of your choice.

Villanova University

800 Lancaster Avenue, Villanova, PA 19085-1672

Set in an upscale suburb, Villanova is Philadelphia's counterpart to Boston College. As at BC, about 80 percent of the students are Roman Catholic (compared to about half at Georgetown). The troika of business, engineering, and premed are popular at 'Nova, as is nursing.

Villanova University takes pride in its Augustinian roots, even basing its admissions essay on one of St. Augustine's teachings about transforming "hearts and minds." The school has all the trappings of a typical Roman Catholic university, from strong academics to deeply rooted traditions and rivalries, and students firmly dedicated to their faith. The emphasis on community creates an "inviting and friendly atmosphere," says a sophomore.

Villanova's lush campus of more than 250 acres once served as the estate of a Revolutionary War officer and is situated along Philadelphia's suburban "Main Line." The campus continues to keep some of its historical roots with ivy-covered buildings, well-kept lawns, and secluded, tree-lined walkways. Villanova continues to grow, and a new law school, nursing school, and athletic training center are on the horizon.

Nursing, biology, and finance are especially popular at Villanova, as are the communication and engineering programs. Undergraduates may enroll in the College of Commerce and Finance, the College of Engineering, the College of Liberal Arts and Sciences, or the College of Nursing. In fact, the engineering school enjoys strong

Website: www.villanova.edu
Location: Suburban
Total Enrollment: 7,850
Undergraduates: 6,494
Male/Female: 49/51
SAT Ranges: V 580–660
  M 600–690
ACT Range: 27–30
Financial Aid: 40%
Expense: Pr $ $ $
Phi Beta Kappa: Yes
Applicants: 10,394
Accepted: 51%
Enrolled: 30%
Grad in 6 Years: 85%

New programs include degrees in biochemistry, physics, and business honors.

Weekend social life centers around campus events and parties, some sponsored by Greek groups, which claim 11 percent of the men and 31 percent of the women.

regional recognition. General education requirements vary by school, but all students take the two-semester Core Humanities Seminar, which includes theology, English, math, social science, philosophy, and, for most students, science and a foreign language. These courses stress discussion, intensive writing, readings from primary texts, and close student/faculty relationships. They also let students study a topic or theme from several perspectives, with an emphasis on Christian thought and values.

"The academic climate varies, but a good deal of the classes are competitive," says a senior. "A lot of courses require group work, which creates teamwork, but also competition among the groups." Another student says, "I have had incredible teachers who are both intelligent and personable, making the daily classroom experience enjoyable." An honors program is available to about 200 students, by invitation only. But even outside the honors program, a biology major says that in recent years, "the academics have become more demanding, and the admissions process has become more competitive." New programs include degrees in biochemistry, physics, and business honors.

Villanova students are "typically white, upper-middle-class, and Catholic," says a communication major. Many are from the East Coast, and 82 percent are white. African Americans account for 4 percent of the student body; Hispanics comprise 6 percent, and Asian Americans contribute another 6 percent. A philosophy major says,

"The academics have become more demanding, and the admissions process has become more competitive."

"One of the primary social concerns on campus revolves around diversity. Although the student body is not noticeably diverse, it is accepting of other faiths, races, and ethnicities." Thirteen percent of undergraduates receive merit scholarships averaging $10,971, and there are athletic scholarships in a number of sports.

Housing on campus is guaranteed for three years and 67 percent of the student population calls the dorms home. Freshmen live primarily on the South Campus Circle, while upperclassmen take their chances in the lottery system. Most seniors move to houses and apartments in the surrounding towns. The meal plan offers more than a dozen campus eateries, which serve everything from pizza to Chinese food, wraps, and vegetarian and vegan menus.

Weekend social life centers around campus events and parties, some sponsored by Greek groups, which claim 11 percent of the men and 31 percent of the women. If you are under 21 and caught drinking on campus, "you see the Dean and a fine is given," says a junior. On Fridays and Saturdays there is something called "Late Night at Villanova" where comedians, bands, open mic, and dance parties provide an opportunity for students to get together with friends at the Student Center. The annual Novafest draws "great bands—Starting Line, Ataris, Lifehouse," says a student. Despite the tough courses, students take pride in finding the time to socialize and stay involved in the community. "Students don't have to have any prior experience to do Habitat for Humanity," says a junior, "nor do they have to be Catholic to go on a retreat. We send about 400 people away for a week in the fall and another 400 in the spring for Habitat or Mission trips." Each fall, Villanova hosts the largest student-run Special Olympics, which draws people from the local community as well as the campus.

"Everyone works hard, plays hard, and still finds time to give back to the community."

"Academics is only one facet of a Villanova student," concludes a human services major, who says students are "involved with many activities, sports, schoolwork, campus ministry, and friends. Time management is a skill learned by all here. We are multi-faceted and are 'people-persons.'"

Juniors and seniors tend to spend evenings at bars along the local Main Line or in Philadelphia, just 12 minutes away by train. The city's entertainment and cultural opportunities include museums and pro sports, as well as events at numerous other colleges and universities, from La Salle and Temple to St. Joseph's, Drexel, and Penn.

When they're not out socializing, Villanova students are cheering for their Wildcats. Students frequently "unleash the Wildcat within" against their rivals. The men's and women's track teams sometimes see their members going for Olympic gold against the best runners in the world, and the football team plays to sold-out crowds in Division IAA. Other top teams include women's basketball and soccer, and men's and women's cross-country. The school's fitness complex has a swimming center and an indoor track, and 75 percent of students participate in intramural sports.

> **"Academics is only one facet of a Villanova student."**

Despite the changes in the world around them, Villanova continues to be a Catholic university devoted to its students, community, and strong traditions, both spiritually and academically. "Everyone works hard, plays hard, and still finds time to give back to the community," says a senior. While the school takes pride in tradition, it recognizes the technology improvements, continuing upgrades to facilities, and changing and improving educational programs will help its students remain competitive in the workplace and the world beyond.

If You Apply To ➤ **Villanova:** Early action: Nov. 1. Regular admissions: Jan. 7. Does not guarantee to meet demonstrated need. No campus or alumni interviews. SATs or ACTs: required. SAT IIs: optional. Apply to a particular school or program. Accepts the Common Application and prefers electronic applications. Essay question.

University of Virginia

P.O. Box 400160, Charlottesville, VA 22904

Is it Thomas Jefferson? The Romanesque architecture? The Charlottesville air? Whatever it is, students nationwide go ga-ga for UVA, where competition for out-of-state admission has hit the Ivy League level. UVA combines old-line conservatism with a touch of rowdy frat boy.

The University of Virginia may be located south of the Mason-Dixon Line, but it's far from a sleepy southern backwater. Easily one of the most prestigious public schools in the nation, UVA is known to all in Charlottesville as Mr. Jefferson's University. Not just any Mr. Jefferson, mind you, but *the* Mr. Jefferson, author of the Declaration of Independence. Though he passed away more than 150 years ago, he is referred to here as if he ran down to the apothecary shop for a bit of snuff and will be back in a moment. Of all his accomplishments, Jefferson was arguably proudest of UVA—he even asked that his epitaph speak to his role in creating the university rather than his presidency of the United States.

Located just east of the Blue Ridge Mountains in central Virginia, UVA's campus is dotted with historic buildings designed by Jefferson himself, and still in use today. At the core is Jefferson's "academical village," with majestic white pillars, serpentine walls, and extensive brickwork. The village rises around a rectangular terraced green, known as the Lawn, which is flanked by two rows of identical one-story rooms reserved for undergraduate student leaders. Five pavilions, each in a different style, are arranged on either side of the Lawn; all of them open onto a colonnaded

Website: www.virginia.edu
Location: Small city
Total Enrollment: 18,996
Undergraduates: 13,151
Male/Female: 46/54
SAT Ranges: V 610–710
  M 620–720
ACT Range: 25–31
Financial Aid: 25%
Expense: Pub $ $ $
Phi Beta Kappa: Yes
Applicants: 15,657
Accepted: 38%
Enrolled: 53%
Grad in 6 Years: 92%

Forty-five percent of students at Virginia live on campus, including 1,100 who bunk in the three residential colleges: Brown College at Monroe Hill, Hereford College, and International Residential College.

walkway. Behind the buildings are public gardens, while the Rotunda, a half-scale model of the Roman Pantheon, overlooks the Lawn.

UVA isn't just an elite public school; this university holds its own against the best private schools as well. The most popular majors are psychology, commerce, economics, history, and interdisciplinary studies, and students say the latter are some of the best departments at Virginia. Foreign languages and politics also get high marks, but that's a double-edged sword: "Spanish, economics, and politics have university-provided wait-lists because their class sizes never accommodate the demand of students wanting to enroll," says a senior. Sociology and the life sciences draw praise as well, but students caution that math and other quantitative disciplines can be tough—and not just because of the subject matter. "Half of the professors in statistics don't speak English," gripes a junior. Most students matriculate into the College of Arts and Sciences, but undergraduates may also enroll in the schools of Engineering, Nursing, or Architecture. After their second year, about 320 external transfer students and students from other UVA schools transfer into the McIntire School of Commerce, UVA's undergraduate business school. Not surprisingly, competition for these spots is tough. "Some individual courses are easy, but most require more than six hours of study per week to do great," says a foreign affairs major.

> **"Professors also pay a lot of attention to individual students, even in 400-person lectures."**

Virginia requires students in the College of Arts and Sciences and the McIntire School to master a foreign language before graduation. Arts and Sciences students must also take courses in English composition, humanities and fine arts, social science, natural sciences and mathematics, non-Western studies, and composition. Special programs for freshmen include University Seminars, designed to develop critical-thinking skills in an environment that encourages interactive learning and intensive discussion, and limited to 20 students each. "These classes are based on interesting, current topics that provide somewhat of a break from your rigorous, everyday class," says a sociology major. They're also taught by some of the university's best faculty. A five-year program for aspiring teachers yields a BA from the College of Arts and Sciences and a Master of Teaching degree from UVA's Curry School of Education.

Highly capable students may win admission to the Echols Scholars program, which allows about 200 top entering freshmen the chance to pursue academic exploration without the constraints of distribution or major-field requirements. Echols students also live together for their first year. The Rodman Scholars program in the School of Engineering and Applied Science selects its members based on financial need, leadership, and scholarship. Students who qualify for the Distinguished Majors program may pursue independent study during their third and fourth years. "TAs teach some lower-level mathematics and economics courses, but the quality of teaching is consistent—two of my favorite instructors have been TAs," says a junior. "Professors also pay a lot of attention to individual students, even in 400-person lectures."

> **"We play just as hard as our rivals, but we work a lot harder, too."**

Students instituted Virginia's notable honor system in 1842, after no one owned up to shooting a professor on the Lawn. The residence halls, student council, and Judiciary Committee remain student-run to this day—and they really put the brakes on lying, cheating, or stealing. "This gives us a real sense of ownership of our school, and also makes UVA students ultra-active in all sorts of extracurricular activities," one student explains. "We are extremely self-reliant." And don't take the policies lightly—breaching the codes means a swift dismissal from campus. After a number of controversial cases recently, discussions continue about the appropriateness of the single-sanction system. But rest assured, some form of the honor code will remain a way of life here, as many students say the policies keep them feeling comfortable leaving backpacks and calculators unattended.

Because UVA is a state school, 70 percent of students are Virginians; admission for out-of-staters is more competitive, and many of those students come from New York, New Jersey, Pennsylvania, and Maryland. "We play just as hard as our rivals, but we work a lot harder, too," says a junior. "We are more grounded, outgoing, and well rounded," another student adds. Nine percent of UVA students are African American, 11 percent are Asian American, and 4 percent are Hispanic. UVA boasts the highest African American graduation rate of any public university. The school hands out 419 athletic scholarships each year, along with merit awards, including 25 to 30 highly prized Jefferson Scholarships, given annually by the alumni association and good for full tuition, room, and board. UVA has stopped requiring low-income students to take out loans as part of their financial aid packages; instead, these students now receive grants, which do not have to be repaid.

The Echols Scholars program allows about 200 top entering freshmen the chance to pursue academic exploration without the constraints of distribution or major-field requirements.

Forty-five percent of students at Virginia live on campus, including 1,100 who bunk in the three residential colleges: Brown College at Monroe Hill, Hereford College, and International Residential College. Hereford's contemporary architecture has been described by the *New York Times* as "proudly, almost defiantly modern," in contrast to most of the other campus buildings. There are foreign language houses for students who want to work on their Russian, Spanish, French, or German—and one house where students can find groups of peers speaking Arabic, Chinese, Hindi-Urdu, Italian, Japanese, or Persian. "First-year dorms are OK, but most have no air-conditioning," says one student. "For upperclassmen, getting a decent on-campus apartment is slightly more difficult," prompting many students to move off campus. Meal plans are required for first-years. Upperclassmen either cook for themselves or take meals at the Greek houses; fraternities draw 30 percent of Virginia's men and sororities sign up an equal portion of the women.

"As long as students understand that there is no pressure to drink, and that there are consequences to their actions, then they will be safer."

Mr. Jefferson founded UVA as a place where students could come together to "drink from the cup of knowledge," and now that fraternity rush is dry and parties must have guest lists, there's a lot less quaffing of other brews going on. Still, determined Virginians haven't stopped metamorphosing into Rowdy Wahoos when the sun goes down—the nickname comes from a school cheer about a fish that can drink twice its weight. Alcohol policies "are geared toward education, rather than punishment," says one student. "That seems to work. As long as students understand that there is no pressure to drink, and that there are consequences to their actions, then they will be safer." For non-drinkers, and those under 21, the student-run University Union and more than 500 clubs and other organizations offer movies, concerts, social hours, and other booze-free options.

UVA isn't just an elite public school; this university holds its own against the best private schools as well.

As for Charlottesville, it's "one of the best towns in the country," gushes a foreign affairs major. "Malls and historical places are located close by. The community is rather affluent, safe, and extremely friendly." There are amazing restaurants, gorgeous vineyards and wineries, and plenty of bars, shops, theaters, and other cultural attractions. Students also tend to immerse themselves in community service in the area; UVA's nationally recognized Madison House coordinates the activities of a host of volunteer groups. Outdoorsy folks can hike, bike, ski, and sightsee in the nearby Blue Ridge Mountains, or simply daydream while strolling Skyline Drive. Popular road trips include Washington, D.C., Richmond, and anywhere the Cavaliers are playing football, basketball, or soccer.

While big-time Atlantic Coast Conference basketball has long been an integral part of UVA life, the men's and women's swimming, women's rowing, and men's and women's lacrosse teams are the perennial powerhouses here, with a slew of NCAA and ACC titles between them. There are also 16 intramural sports leagues or tournaments, in everything from flag football to inner-tube water polo, and 63 club sports organizations. Favorite traditions include Foxfield, each April, in which students dress up and host catered parties prior to attending a steeplechase horse race; jackets and ties also come out for football games, a relic of when UVA was all-male, and gridiron contests were an opportunity to meet women. May brings Beach Week, and we could tell you more about the various secret societies, but we won't—they're secret, after all!

UVA offers a top-notch education at a bargain-basement price, at least for in-state students. The social life is rowdy, the academics are rigorous, and the friendships that are formed here last far beyond the college years—much as Mr. Jefferson's legacy continues to be felt on campus, years after his death.

Overlaps

Virginia Tech, College of William and Mary, James Madison, Duke, Princeton, UNC–Chapel Hill

If You Apply To ➤

UVA: Regular admissions: Jan. 2. Financial aid: Mar. 1. Guarantees to meet full demonstrated need. Campus interviews: optional, informational. No alumni interviews. SATs or ACTs: required (SAT preferred). Students taking the ACT must take the optional Writing Test. SAT IIs: required (two tests of the student's choice). Prefers electronic applications. Essay question: one question depends on the college to which you are applying (Arts and Sciences, Architecture, Engineering or Nursing). All applicants must also submit a half-page essay on their favorite word, the kind of diversity they will bring to UVA, their thoughts on a Nobel Prize winner's assertion about the laws of the universe, how the world they come from has shaped who they are, why pop culture is making us smarter, or their favorite natural phenomenon; and a full-page essay on a topic of their choice.

Virginia Polytechnic Institute

Blacksburg, VA 24061

Offers a unique blend of high tech and Southern hospitality. Engineering has always been its calling card, but business and architecture are popular. Blacksburg is a nice college town, but is far from the population centers near the coast. Hokie Nation loves its football team.

Website: www.vt.edu
Location: Small town
Total Enrollment: 27,979
Undergraduates: 21,627
Male/Female: 59/41
SAT Ranges: V 540–630
  M 570–660
ACT Range: 22–27
Financial Aid: 33%
Expense: Pub $ $ $
Phi Beta Kappa: Yes
Applicants: 17,681
Accepted: 72%
Enrolled: 40%
Grad in 6 Years: 76%
Returning Freshmen: 88%
Academics: ✍ ✍ ✍

On April 16, 2007, we all became Hokies. That was the day Virginia Tech experienced one of the worst shootings in U.S. history. Millions of Americans donned orange and maroon clothing, lit candles, and otherwise expressed support for a university whose spirit was tested but unbowed. An email posted on the university website said it all: "A tragic event occurred, yes, but what has not changed is that Virginia Tech is a family, and the Hokie Nation is proud and strong."

Virginia Tech's full name is the Virginia Polytechnic Institute and State University. Its campus, set on a plateau in the scenic Blue Ridge Mountains, occupies 3,000 acres and comes complete with a duck pond, hiking trails, and a 200-year-old plantation that is a local landmark. Students enjoy unlimited outdoor recreation thanks to the proximity of the Jefferson National Forest, the Appalachian Trail, the scenic Blue Ridge Parkway, and the majestic old New River. The campus buildings are an attractive mix of gray limestone structures, Colonial-style brick, and modern cement buildings. Newer facilities include the 150,000-square-foot Advanced Communications and Information Technology Center (ACITC).

Virginia Tech is best known for its first-rate technical and professional training. For undergrads with an appetite for engineering, Tech has programs for every taste, including aerospace, ocean, biological systems, civil, chemical, computer, electrical, industrial and systems, materials, mechanical (the most popular), and mining. The Pamplin College of Business is also prominent, and the five-year architecture program is considered one of the nation's best. Though no longer Tech's centerpiece, the College of Agriculture and Life Sciences remains strong, especially in animal science. Students in the College of Natural Resources can choose from such concentrations as environmental conservation, fisheries science, forestry, and wildlife management. The College of Education merged with the College of Human Resources, where Hospitality and Tourism is the best-known program.

Students in the sciences reap the benefits of their high-tech environment; other disciplines do not fare so well. The humanities have been hard-hit by budget cuts. In particular, "religious studies, foreign languages, and philosophy are dwindling away," says a

> "On campus you can do anything from handing in homework to checking grades to checking out what movies are playing, all by computer."

senior. One bright spot is internationally known poet Nikki Giovanni, who teaches creative writing and advanced poetry. The university also has a tradition of excellence in the performing arts, and the school's theater group has received more awards from the American College Theater Arts Festival than any other college in the Southeast.

Introductory class size tends to be large—sometimes well into the hundreds—and the budget ax has only made matters worse. Most of the big lecture classes are taught by full-time faculty, though discussions and grading are generally handled by TAs. Nevertheless, a communications major says her professors "keep you on the edge of your seat." An accounting major adds, "Freshmen are taught by professors and grad students." All students are required to take courses in English, math, humanities, and social and natural science. There is also a foreign language requirement, though high school coursework may cover this. The 1,500 or so students who participate in the university honors program are guaranteed access to top faculty and research opportunities. Tech is among the nation's leaders in the integration of computers into all facets of life. All freshmen are required to own a computer. Says an engineering major, "On campus you can do anything from handing in homework to checking grades to checking out what movies are playing, all by computer."

Each year, more than 1,000 students take advantage of Tech's co-op opportunities available in almost all majors. The nationally acclaimed Small Business Institute program enables faculty-led groups of business majors to work with local merchants, analyze their problems, and make suggestions on how to increase profits. The Corps of Cadets, a tradition once on the verge of extinction, has made a comeback. Cadets earn a minor in leadership and can choose from three tracks: military/ROTC, civic professions, or a combination of the two.

Students looking at pricey Northeastern technical schools will find Tech a real bargain. Not surprisingly, the admissions office is inundated with out-of-state applicants, which means stiff competition for the 25 percent of the slots available to non-Virginians. Tech's relative isolation from major cities is a drag on minority recruitment: African Americans and His-

> "Rooms are on the small side, but they provide all you need to live."

panics account for only 7 percent of the student body. Asian Americans account for another 7 percent. Seventy-nine percent graduated in the top quarter of their high school class. Students with financial need who apply for aid before the deadline receive priority consideration; those who apply later are likely to be out of luck. Tech hands out a few hundred athletic scholarships, and there are merit scholarships available to qualified students. The "Funds for the Future" program aims to provide additional aid to low-income undergraduates.

Tech housing is nothing to write home about, but adequate to meet the needs of most students. "Rooms are on the small side, but they provide all you need to live,"

(Continued)

Social: ☎ ☎ ☎
Q of L: ★ ★ ★ ★
Admissions: (540) 231-6267
Email Address:
vtadmiss@vt.edu

Strongest Programs:
Engineering
Architecture and Urban
  Studies
Business
Sciences
Human Resources and
  Education
Mathematics
Forestry and Wildlife
  Resources

It will take time for the wounds of April 16, 2007, to heal, but Virginia Tech remains committed to pursuing its particular blend of high-tech learning and Southern hospitality. As the school website proclaims, "We are the Hokies. We will prevail."

says a freshman. Some 25 undergraduate dorms serve 8,400 students; 41 percent of the student body lives on campus, though only freshmen and the Corps of Cadets are required to. Most upperclassmen live off campus in nearby apartment complexes. Dietrick's Depot, the largest dining hall on campus, was recently renovated. Three specialty lines supplement the standard dining-hall fare to create an intimate, cafe-style atmosphere. Students who are committed to a healthy lifestyle can opt to reside in the WELL (Wellness Environment for Living and Learning), which will provide them with a substance-free atmosphere, and includes special healthy-living courses and is overseen by a specially trained wellness staff.

Leisure-time favorites include school-sponsored plays, jazz concerts, arts and crafts fairs, and dances. The nearby Cascades National Park is an especially popular retreat for lovers and camping jocks alike, and tubing down the New River is a ritual for summer students. Thirteen percent of the men and 15 percent of the women join fraternities and sororities, which set the tone of the social life. If going Greek isn't

"I think Tech is pretty personal considering the number of students it serves."

for you, don't worry, as one student says, "Most students go downtown to shoot pool, dance, or go to a bar." The most important annual event is the Ring Dance (when the juniors receive their school rings), the German Club's Midwinter's Dance, and the Corps of Cadets military ball. Blacksburg offers the usual city fare and one student says the town "revolves around Tech." For real big-city action, Washington, D.C., and Richmond are four and three hours away by car, respectively.

A longtime member of the Big East, Tech now competes in the Atlantic Coast Conference. Tech's varsity athletics program struggled for years to make the big time—and never succeeded—but the football team's recent appearances in post-season bowl games have cheered alumni and hiked applications by several thousand. The annual "big game" pits the backwoodsy Hokies against the aristocratic (snobby?) Cavaliers of the University of Virginia. Tech has one of the nation's most extensive intramural programs, with everything from football to horseshoes and underwater hockey—a recent rage—and more than 400 softball teams each spring, many of them co-ed. Weekend athletes benefit from the addition of a fitness center.

It will take time for the wounds of April 16, 2007, to heal, but Virginia Tech remains committed to pursuing its particular blend of high-tech learning and Southern hospitality. As the school website proclaims, "We are the Hokies. We will prevail."

Overlaps

University of Virginia, James Madison, George Mason, University of Maryland, Penn State

If You Apply To ➤

Tech: Early decision: Nov. 1. Regular admissions: Jan. 15. Financial aid: Mar. 11. Does not guarantee to meet demonstrated need. No campus or alumni interviews. SATs: required. SAT IIs: optional (required for applicants from non-accredited schools). Prefers electronic applications. No essay question.

Wabash College

301 West Wabash, Crawfordsville, IN 47933

Wabash and Hampden-Sydney in Virginia are the last of the all-male breed. With steady enrollment and plenty of money in the bank, Wabash shows no signs of changing. Intense bonding is an important part of the Wabash experience, and few co-ed schools can match the loyalty of Wabash alumni.

Wabash is a traditional, conservative school and its roots rest in providing young men with quality education. At all-male Wabash College, students "are intelligent and hard-working," one junior says. There is a great emphasis on academics but the majority of students have strong social ties to the school as well, with an overwhelming number who join campus fraternities and participate in intramural sports. Wabash was founded in 1832 as a "classical and English high school, rising into a college as soon as the wants of the country demand" by transplanted Ivy Leaguers who most certainly held a positive view of a man's future.

The Wabash campus is characterized by redbrick, white-pillared Federal-style buildings (three are originals from the 1830s). Located in the heart of Crawfordsville, a small town of about 15,000, Wabash is surrounded by grass and tall trees that are part of the gorgeous Fuller Arboretum. More recent additions to the campus include a renovated physics, mathematics, and computer science building and five new or renovated fraternity houses.

The Wabash educational program has certainly proved itself over the years. This small college has amassed an impressive list of alumni: executives of major corporations, doctors, lawyers, and a large number of PhDs. Most Wabash alumni are faithful to their school in the form of generous donations. On a per capita basis, the school's $300 million endowment makes it one of the wealthiest in the nation. This financial security enables Wabash to refuse any federal aid, with the exception of Pell Grants, which go directly to students.

History and English draw the most majors at Wabash, and the highest accolades go to the biology (premed) and chemistry departments, which are among the most challenging and produce many successful grads. Wabash has an electron microscope and a laser spectrometer, a 180-acre biological field station, and a cell culture lab. Political science and the psychology department are also popular, but students say the "speech [recently changed to rhetoric] department is not as academically stringent as others." A 42,600-square-foot fine arts center provides more studio space and practice rooms and is a pleasant addition to the music and art departments. Wabash emphasizes its mission as a liberal arts college and the traditional programs have been augmented with crosscultural immersion learning courses with travel components—at no cost—during spring break or at the end of the semesters to countries such as Ecuador, Turkey, and Greece.

> "There is one rule on campus: 'Act as a gentleman at all times.'"

The studying at Wabash is intense. "The courses are very competitive and the professors always challenge students and push them to expand their thinking," explains a freshman. General education requirements include courses from a wide variety of fields—natural and behavioral sciences, literature and fine arts, mathematics, language studies, and a course on cultures and traditions. In addition, a freshman tutorial is designed to focus students on reading, writing, and class participation with no more than 15 students in a class. A special writing center is available for all Wabash students who demonstrate a weakness in written communication skills. Juniors are encouraged to study on continents throughout the world or in various domestic programs through the Great Lakes College Association.* Those who can't satisfy their high-tech interests at Wabash can opt for a 3–2 program in engineering with Columbia University or Washington University in St. Louis. Wabash also offers a tuition-free semester after graduation to train students to become teachers. Students use the words "tremendous" and "excellent" to describe their professors. "Our professors have distinguished backgrounds and reputations and always push us to learn both in and out

> "The courses are very competitive and the professors always challenge students and push them to expand their thinking."

Website: www.wabash.edu
Location: Small town
Total Enrollment: 853
Undergraduates: 853
Male/Female: 100/0
SAT Ranges: V 530–650
  M 550–660
ACT Range: 23–28
Financial Aid: 67%
Expense: Pr $
Phi Beta Kappa: Yes
Applicants: 1,376
Accepted: 49%
Enrolled: 37%
Grad in 6 Years: 68%
Returning Freshmen: 82%
Academics: ✍ ✍ ✍ ½
Social: ☎ ☎
Q of L: ★ ★ ★
Admissions: (800) 345-5385
Email Address:
  admissions@wabash.edu

Strongest Programs:
Prebusiness
Economics
Biology
Chemistry
Premed

Wallies are tough in athletics. The football, baseball, outdoor track, and wrestling teams are very competitive.

of the classroom," says a student. Advising is also considered excellent, and students say they have no problems getting into required courses.

Most of Wabash's students come from public high schools in Indiana, and 68 percent were in the top quarter of their high school class. Many were active in athletics and student government and continue that tradition in college. The campus is mostly conservative, though both Republican and Democratic student organizations are strong. The administration is working with a grant to improve diversity on campus. It has expanded the freshman orientation program to include diversity and community issues, focusing on making choices and accepting the consequences. African Americans comprise 7 percent of the student body, Hispanics 4 percent, Asian Americans 3 percent, and foreign students 4 percent. Merit scholarships are available to qualified students, although there are no athletic scholarships.

Residential life for the temporary denizens of small-town Crawfordsville revolves around the 10 fraternities, each with its own house. Sixty-eight percent of the students join up, and many end up living with their brothers. As an alternative to Greek life, there are three modern dorms, two of which have all single rooms. Dorm

"Road trips to big state schools (with women) are key, but we also have big parties."

residents must eat in the dining hall. Those living in the dorms (and the 8 percent who live off campus) may feel excluded from what there is of campus social life because the fraternities "ship in" sorority members from Purdue, Indiana, DePauw, and Butler for parties. Campus security is rated well by students, who feel safe.

The surrounding town of Crawfordsville leaves much to be desired, and students say most (if not all) social life takes place on campus. As for drinking on campus, students agree that policies are loose. "There is one rule on campus: 'Act as a gentleman at all times,' and this applies to alcohol consumption as well," says a student. Significant campus events include the Monon-Keg rugby game and homecoming. Another popular, though less sweaty, event is Chapel Sing, where all the freshmen sing the lengthy school song in unison. "Road trips to big state schools (with women) are key, but we also have big parties," a senior says.

Wallies are tough in athletics. The football, baseball, outdoor track, and wrestling teams are very competitive. When they're not studying or partying, students are likely to be found working out in the gym or running. Eighty percent of students participate in intramurals, which encompass 20 sports, including flag football and softball. School spirit is abundant, especially when the opponent is long-standing rival DePauw.

Traditions have not changed much since the school's founding in the 1830s, and still play an important part in the lives of the men at Wabash. Some students complain about the lack of women and culture in the surrounding area, but many are very happy with the education they have received. It's a boys club and the students know they will be able to focus on their classes without female distractions during the school week. The college is focused on academics and prides itself on intensive and rigorous programs.

Overlaps

Indiana, Butler, Purdue, Hanover, DePauw

If You Apply To ➤ **Wabash:** Rolling admissions: Jan. 31. Early decision: Nov. 15. Early action: Dec. 15. Regular admissions and financial aid: Mar. 1. Housing: Jun. 1. Guarantees to meet full demonstrated need. Campus interviews: recommended, informational. Alumni interviews: optional, informational. SATs or ACTs: required. Accepts the Common Application and prefers electronic applications. Essay question: discuss an event, job, or person with significant impact; career goals; high school paper.

Wake's Baptist heritage and Winston-Salem location give it a more down-home flavor than Duke or Emory. With only 4,115 undergraduates, Wake is small compared to its ACC rivals but bigger than most liberal arts colleges. The strong Greek system dominates the social scene.

Already one of the top private schools in the Southeast, Wake Forest is working hard to transform its regional recognition into a national reputation. Long known for basketball—at least half the student body attends every home game, one junior says—Wake's solid academics are worthy of a look as well. Students work hard, hence the nickname "Work Forest," but the university's size and strong Greek system means it's also easy to establish close friendships. "We have all the benefits of a small private school—I am someone here, not just a number," says a psychology and German major. "But Division I athletics and the medical and graduate schools give us a lot of resources that other schools our size don't have."

Located in the Piedmont region of North Carolina, Wake's 340-acre campus features flowers, wooded trails, and stately magnolias. There are more than 40 Georgian-style buildings constructed of old Virginia brick with granite trim. That new mortar and stone complements the school's lush surroundings, including the 148-acre Reynolda Gardens annex, with a formal garden and greenhouses. The garden features one of the first collec-

> "Division I athletics and the medical and graduate schools give us a lot of resources that other schools our size don't have."

tions of Japanese cherry trees in the U.S. In 2004, Wake completed an expansion and renovation of Calloway Hall, home to the Calloway School of Business and Accountancy, as well as the departments of math and computer science. The main campus cafeteria has been renovated and reopened in 2006.

To graduate from Wake Forest, students must complete two courses in health and exercise science, one foreign language course, a writing seminar, and a first-year seminar. In addition, students must take three courses in history, religion, and philosophy; two in literature; one in fine arts; three in the social and behavioral sciences; and three in the natural sciences, math, and computer science. Students must also satisfy quantitative reasoning and cultural diversity requirements. There's a new major in East Asian languages and cultures, as well as new minors in film studies, entrepreneurship and social enterprise, and Middle East and South Asian studies. The most popular programs by enrollment are business, political science, communication, psychology, and biology. "The workload at Wake is extremely rigorous," says a junior. "There is definitely no grade inflation here!" The Pro Humanitate Center allows students to put their skills and knowledge to work helping the community; the center takes its name from the school's motto, which means "In Service to Humanity." Exceptionally able students may qualify for the "Honors in Arts and Sciences" distinction by taking three or more honors seminars during their first three years. Faculty members get high marks; graduate assistants may teach labs, but professors are at the lectern during larger classes and at the head of the table in many seminars. "Professors are highly approachable and readily accessible," a student says. Richter Fellowships fund collaborative research between students and faculty members. And for those who get claustrophobic in Winston-Salem, Wake's own residential study centers on the Grand Canal in Venice and in London and Vienna beckon; more than half of the student body studies abroad.

Website: www.wfu.edu
Location: City outskirts
Total Enrollment: 6,462
Undergraduates: 4,115
Male/Female: 49/51
SAT Ranges: V 620–700
  M 640–710
Financial Aid: 36%
Expense: Pr $ $ $
Phi Beta Kappa: Yes
Applicants: 7,484
Accepted: 39%
Enrolled: 39%
Grad in 6 Years: 86%
Returning Freshmen: 93%
Academics: ✍ ✍ ✍ ✍
Social: ☎ ☎ ☎
Q of L: ★ ★ ★
Admissions: (336) 758-5201
Email Address:
  admissions@wfu.edu

Strongest Programs:
Health and Exercise Science
Chemistry
English
Psychology
Mathematics
Economics
Political science
Romance languages

While Wake students travel the globe sampling foreign cultures, life on campus is fairly homogeneous. "The typical Wake student is a preppy conservative who parties as hard as he works and is involved in Greek life," offers one history major. Twenty-eight percent of the student body hails from North Carolina; many others come from nearby Southern states. African Americans make up 7 percent of Wake's student population, Asian Americans add 4 percent, and Hispanics contribute 2 percent, though Wake's efforts to boost diversity continue. Fourteen percent of students receive scholarships based on academic merit, and the average award is $10,586. There are also 210 athletic scholarships.

Seventy-one percent of Wake students live on campus; freshmen are required to do so, and everyone else is guaranteed housing, unless they move off campus and later decide to return. "Living in Wake dorms has been the best experience I've had

"There is definitely no grade inflation here!" so far," a student says. "The rooms are small but very comfortable with many amenities." Dining options have improved, with higher quality and more diverse food offered in the renovated cafeteria, one student says. There's also a convenience store, a food court with chains such as Subway, and the classier Magnolia Room.

Thirty-three percent of men and 53 percent of women go Greek; fraternities have open parties on and off campus, and sororities have date functions and other invite-only events. "The social scene is centered around the Greek system," says a junior. Popular roadtrips are to the beach or the mountains; Chapel Hill, Durham, and Raleigh are each 100 miles away, and Atlanta and the Washington/Baltimore areas are about five hours' drive. The Blue Ridge Mountains, for hiking and camping, are just two hours from campus.

Back at Wake, underage students say it's relatively easy to get alcohol at frat parties, even though all students must sign in before entering. The school's honor code helps to keep rowdy behavior in check; as one student says, "If you are caught with alcohol, and you are under 21, there are strict consequences." Everyone enjoys the annual homecoming festivities, and after the Demon Deacons score a victory on the basketball court, students roll the quad in toilet paper to celebrate. Other favorite events include a midnight concert by the school orchestra every Halloween, with members in full costume, and the Lilting Banshees comedy troupe, which helps students laugh off their stressful workloads.

The city of Winston-Salem is rich in culture, with a symphony, a Christmastime "Moravian love feast," and the well-known Carolina School of the Arts. It's also home to the corporate headquarters of Krispy Kreme Doughnuts. In addition to Wake's on-

"The social scene is centered around the Greek system." campus Museum of Anthropology, a number of art museums—Reynolds House, the Museum of American Art, and Southeastern Center for Contemporary Art—are within three miles of campus. The town also has a strong music scene, with live bands playing at Zing's and The Garage. Popular volunteer activities include Project Pumpkin, a trick-or-treat night on campus for underprivileged children.

When it comes to sports, basketball is the undisputed king at Wake. The men's team competes in the incredibly tough Atlantic Coast Conference and has qualified for the NCAA tournament or NIT for 16 consecutive years—the longest active post-season streak in the conference. The Division I football team draws big crowds, too, especially for games against Appalachian State, and the field hockey team won the NCAA championship in 2002, 2003, and 2004. Of course, virtually any contest against in-state rival UNC–Chapel Hill is guaranteed to get students excited. Intramural and club sports are also offered; the most popular intramurals include soccer, basketball, and flag football, though volleyball, water polo, and bowling are available, too. "Fun supersedes talent," a senior says. "Students are rarely left out of a particular game if they express interest in participating."

That spirit of involvement and dedication to the community pervades the Wake Forest experience. Whether students are serving the less fortunate or chipping away at their heavy workloads, they benefit from motivated peers, dedicated faculty, and gorgeous surroundings. "Learning to work under pressure and time constraints is pivotal to your future success," one student explains. "Wake students receive not only the advantage of small classes, but also the benefits of a big-name school for the next step in their lives."

If You Apply To ≫ **Wake Forest:** Early decision: Nov. 15. Regular admissions: Jan. 15. Guarantees to meet full demonstrated need. Campus and alumni interviews: optional, informational. SATs: required. SAT IIs: optional. Accepts the Common Application and electronic applications. Essay questions: favorite literary character, this generation's greatest challenge.

Warren Wilson College

P.O. Box 9000, Asheville, NC 28815-9000

The best of schools where students combine academics, community service, and on-campus work. Roots in the culture of Appalachia combine with a strong international orientation to give Warren Wilson its unique flavor. Setting in the mountains of western North Carolina is tough to beat.

Warren Wilson, a small liberal arts college, is flush with engaging little quirks. It promotes global perspectives, puts students to work on the campus farm, and makes service-learning a central part of the educational experience. And at what other college is whitewater paddling considered the most important intercollegiate sport? "The unique and eclectic bunch of characters on campus make for an interesting experience," a freshman says.

Founded by the Presbyterian Church in 1894 as the Asheville Farm School, Warren Wilson College initially provided formal schooling for "mountain boys." By 1966, it had evolved into a four-year, co-ed liberal arts college that, while still maintaining its Presbyterian ties, welcomes students of all backgrounds. WWC is located 15 minutes from downtown Asheville in the lush Swannanoa Valley of the Blue Ridge Mountains. Its 1,132-acre campus features formal gardens, fruit and vegetable gardens, a 300-acre farm, and a myriad of hiking trails. Consistent with campus culture, the wood and stone buildings are small in scale and done in an architectural style that emphasizes natural earth tones accented by extensive stonework by traditional Appalachian stone masons. The campus is also home to one of the most important Cherokee archeological sites in the Southern Appalachian Mountains, dating from as early as 5,000 B.C. Two new 40-bed residence halls are more recent campus additions. WWC is also home to North Carolina Outward Bound, an organization with which it has close ties, and the Mountain Area Child and Family Center, which serves as a laboratory school for students of education, psychology, and social work.

The heart of the WWC curriculum is its unique Triad Education Program, which combines liberal arts coursework, community service, and work. Students may choose from 46 majors and concentrations. The most popular programs are environmental studies, outdoor leadership, biology, art, political science, and history. To

> "We are very laid-back, but that doesn't mean we don't have tough classes."

Website: www.warren-wilson.edu
Location: Small town
Total Enrollment: 896
Undergraduates: 827
Male/Female: 40/60
SAT Ranges: V 550–670 M 510–620
ACT Range: 24–29
Financial Aid: 55%
Expense: Pr $
Phi Beta Kappa: No
Applicants: 857
Accepted: 77%
Enrolled: 37%
Grad in 6 Years: 48%
Returning Freshmen: 70%
Academics: ✐ ✐ ✐ ½
Social: ☎ ☎ ☎ ☎
Q of L: ★ ★ ★ ★ ★
Admissions: (800) 934-3536
Email Address: admit@warren-wilson.edu

Strongest Programs:
Art

graduate, students have to perform at least a hundred hours of service-learning through organizations such as Habitat for Humanity or environmental organizations. Warren Wilson is also one of only a half dozen four-year colleges in the nation that requires all residential students to work on campus. To fulfill their weekly 15-hour work requirement, students do electrical work, plumbing, and landscaping; clean the dorms, tend the farm animals, and maintain the campus gardens.

To meet general education requirements, WWC students take at least one class within each of the school's eight liberal arts areas: language and global issues, literature, history and political science, natural science, mathematics, social science, philosophy and religion, and artistic expression. They must also take two college composition courses. The First-Year Experience Program lays a foundation for academic success by introducing students to the study–serve–work trinity through small, interactive group activities. All first-year students enroll in the First-Year Seminar, which includes field experience and/or a service-learning component that takes students and faculty off campus for a day or a weekend.

Almost every student spends some time getting the taste of a foreign culture. Juniors take a semester-long course and then spend several weeks on an international "field experience"—with the cost built into their regular tuition—and qualified students may also study for a semester or two in countries such as Guatemala, Costa Rica, Italy, Scotland, Greece, and Vietnam. The college offers honors programs in biology, chemistry, English, environmental studies, math, and computer science. Internship opportunities are available in most undergraduate programs and there are dual-degree programs in pre-environmental management, pre-forestry, engineering, and applied science. Newer programs include majors in Spanish, creative writing, women's studies, philosophy and religious studies, and an interdisciplinary major in global studies. The Appalachian Studies program serves as a catalyst for local cultural activities, including numerous musical groups. And where else does the music department offer you the choice of "finger-picking" or "flat-picking" guitar?

> "There are no big sports teams or other rivalries."

Students say hard work is the norm, but the atmosphere is relaxed. "We are very laid-back," says a junior, "but that doesn't mean we don't have tough classes. It just means we don't have to compete for a particular program." Creative writing and the natural sciences receive high marks from students. Business and math aren't as well-received. Music and physics could also use improvement, administrators say. Nearly all classes have 25 or fewer students, and students say the quality of teaching is a mixed bag. "Some teachers are more slacker than the students," quips a sociology major, "but we have some really phenomenal teachers that are very interested."

The typical WWC student is "very liberal and outspoken," according to a senior. A junior adds, "There's the hippies, the Trust-afarians (rich students who believe they are Rastafarian), the regular kids, the punk kids, and about two Republicans." Upon graduation, most go into service professions such as teaching or working for environmental or other nongovernmental organizations. The first step for many is into the Peace Corps. Eighty-one percent of the student body hail from out of state and about 20 percent attended public school. Students of color account for only 4 percent of the student population: African Americans and Asian Americans each make up 1 percent, and Hispanics account for 1 percent. Another 3 percent are foreign. But a student asserts, "While we are not racially diverse, we are very geographically diverse and I think students come here with a variety of life experiences." The campus is a "hive of activism," a freshman says. Hot campus issues include diversity, sexual assault, and global concerns such as the war in Iraq. Merit scholarships averaging $2,050 are available, but there are no athletic scholarships.

Eighty-seven percent of students live in dorms, which are described as "very nice, intimate places." The 36-bed EcoDorm incorporates solar heating and natural

Almost every student spends some time getting the taste of a foreign culture.

ventilation, and is made of hardwoods milled on campus. Overall, housing is plentiful and well-maintained (unless the student workforce slacks off). Dining is a treat, and there are plenty of edible options, including vegetarian and vegan meals. "The menu in the cafeteria is like clockwork—very predictable," says a student. "We do use local food that we've grown on our own farm, though."

Social life "can be very wild," a freshman says. "There are often bonfires and parties in a certain pasture." Despite the absence of Greek organizations, students find plenty of ways to have fun and blow off steam, including open mic nights, concerts, and plays—in addition to the parties. The outing club is the largest on campus and sponsors weekly hiking, skiing, or other excursions. Students of legal drinking age are allowed to imbibe indoors, and students say it's very easy for underage students to obtain alcohol. Popular events include Mayhem (a parade in May), The Bubba (a huge party and bonfire), and homecoming, which features live bluegrass music, a barbeque, hay rides, and dancing.

Where else does the music department offer you the choice of "finger-picking" or "flat-picking" guitar?

Asheville is "definitely not a 'traditional' college town, but it has tons to offer," says an elementary education major. Museums, cafes, theaters, music clubs, and the symphony are only 15 minutes away. "The downtown scene is amazing for the relative small size of the town," says a student. Thanks to the college's service requirement, students take an active role in the community through volunteer work. WWC sponsors short-term service projects during vacation breaks, and students camp out overnight in order to qualify for the annual trip to Cumberland Island off the coast of Georgia. Popular road trips include Atlanta and the beaches of South Carolina, but the best excursions "are to protests and political events or camping and backpacking trips," according to a political science major.

In a state famed for its rabid sports fans, Warren Wilson students are decidedly laid-back about athletics. "There are no big sports teams or other rivalries," says a freshman. "Students come here largely because they don't like those rivalries. There is a great sense of school spirit, nonetheless." The college is a member of the United States Collegiate Athletic Association and competes against small universities and colleges in the South. Com-

"The downtown scene is amazing for the relative small size of the town."

petitive sports include mountain biking, cross-country, soccer, and swimming. Thirty-five percent of WWC students participate in intramural or recreational sports, which range from rock-climbing to tennis to basketball. Students have access to myriad facilities, including soccer fields, basketball courts, an indoor pool, and a fitness center.

Success at Warren Wilson is measured not only by grades, but by community service and a sense of stewardship. "Students come here for all different reasons and are allowed to shine in all different areas," a junior says. Those who aren't afraid to get their hands dirty will see this small liberal arts college as a valuable place that combines the notion of thinking globally with acting locally.

Overlaps

Earlham, Hampshire, UNC–Asheville, Guilford, Appalachian State, Northland

If You Apply To ➤ **Warren Wilson:** Early decision: Nov. 15. Regular admissions: Feb. 28. Financial aid: Mar. 15. Housing: May 1. Campus interviews: recommended, evaluative. Alumni interviews: optional, informational. SATs or ACTs: required. SAT IIs: optional. Accepts the Common Applications and prefers electronic applications. Essay question: personal statement; why Warren Wilson?

University of Washington

1410 N.E. Campus Parkway, Seattle, WA 98195

UDub wows visitors with its sprawling park-like campus in hugely popular Seattle. Washington is tougher than University of Oregon for out-of-state admission but not as hard as UC heavyweights Berkeley or UCLA. Location near the coast and mountains makes for strong marine and environmental studies programs.

Website:
www.washington.edu

Location: Urban

Total Enrollment: 39,251

Undergraduates: 27,488

Male/Female: 48/52

SAT Ranges: V 530–650
M 570–670

ACT Range: 23–28

Financial Aid: 48%

Expense: Pub $ $

Phi Beta Kappa: Yes

Applicants: 15,923

Accepted: 67%

Enrolled: 46%

Grad in 6 Years: 74%

Returning Freshmen: 93%

Academics: ✑ ✑ ✑ ✑ ½

Social: ☎ ☎ ☎

Q of L: ★ ★ ★

Admissions: (206) 543-9686

Email Address:
www.freshman.washington
.edu/apply

Strongest Programs:
Business
English
Psychology
Drama
Engineering
Computer Science
Architecture
Biology

Washington has cemented its reputation as a solid research institution. In fact, UW has been the number one public university in federal research funding since 1974. Students here understand that anonymity and size are the prices that must be paid for the wealth of opportunities that await them. Those looking for an extra, personal touch might want to investigate the school's two branch campuses in Tacoma and Bothell, where class sizes average 25 students. But if the Seattle campus is your focus, one senior hints, just "learn to work the system."

Washington's Seattle campus features a number of distinctive landmarks. Red Square sits atop the Central Plaza parking garage and features the Broken Obelisk, a 26-foot-high steel sculpture gifted to the university by the Virginia Wright Fund.

Many of Washington's diverse undergraduate strengths correspond with its excellent graduate programs. The competitive business major, for example, benefits from the university's highly regarded business school and is the most popular undergraduate major, followed by biology, art, and English. Similarly, students majoring in public health, community medicine, pharmacy, and nursing profit from access to facilities and faculty at the medical school, an international leader in cancer and heart research, cell biology, and organ transplants. Also recommended for undergraduates are biological and life sciences, and most engineering programs, especially computer science, computer engineering, and bioengineering. Reflecting the focus on natural resources in Washington's economy, the program in fisheries is excellent, as are earth and atmospheric sciences, including oceanography.

Undergraduates in both professional and liberal arts programs must take five credits in English composition, seven credits in writing beyond composition, and one course in quantitative and symbolic reasoning. Students must also fulfill 40 credits in general education requirements, including the arts, individuals and societies, and the natural world. Schools and colleges also have their own requirements that must be met. Many Washington professors are tops in their field, but students may have to be patient about seeing professors after class. As for academic advising, a student advises, "The key is to take the bull by the horns and find out which advisors are better than others." While students once complained that classes were difficult to get into, that problem seems to have been reduced. Eighty-three percent of the classes have fewer than 50 students. "It is very common to be taught by teaching assistants, but overall the quality is good in both senior-faculty and graduate-student teachers," a business major says.

For those interested in skirting the masses, UW sports an honors program that offers small classes on interesting subjects taught by fine professors. The academic environment at UW is "very much centered on learning. Part of that comes from the fact that this is a research institution." And if students get the itch to see some different scenery, there are 60 different study abroad programs offered in 20 countries, including China, Denmark, and Russia. A program in experiential learning encourages students to find internships and participate in community service. This fits in with a variety of

"Take the bull by the horns and find out which advisors are better than others."

classes that give students the opportunity to volunteer as part of their coursework. A senior says, "There's simply no better way to learn than by combining challenging courses with real-world experience. When you're learning in the classroom, you can't always apply it. Experience adds to your learning, and it helps you remember it." Freshmen are given special attention via the Freshman Interest Group (FIG) program, which offers freshmen a chance to meet, discuss, and study with other freshmen who have similar interests. Each FIG consists of 20 to 24 students who share a cluster of classes (which meet graduation requirements), and includes a weekly seminar led by a junior or senior peer advisor. Also of interest to freshmen are General Studies 101, an optional two-credit course designed to help students meet the demands and expectations of college life, and Freshman Seminars, one-credit courses with 10 or 12 other freshmen.

Part of the reason for UW's national anonymity is the fact that it turns away large numbers of out-of-state applicants, preferring to keep its focus on the home folks. Eighty-six percent of undergraduates are state residents, and an unusually large proportion are over the age of 25. The student body is 26 percent Asian American, with Hispanics, Native Americans, and African Americans combining for another 8 percent. Students say the school strives for diversity by offering Valuing Diversity workshops to foster increased awareness of and sensitivity to individual differences. The campus is very active politically, as one junior reports: "Political correctness is very big here and students find very creative ways to make their points." Students say budget cuts have decreased the number of course offerings and programs like Society and Justice. Merit scholarships are awarded to Washington residents with good high school records and test scores. Athletic scholarships are awarded to men and women in a wide variety of sports, including swimming, women's gymnastics, golf, tennis, and track and field.

> **"Political correctness is very big here and students find very creative ways to make their points."**

More than anything else, the great outdoors define the University of Washington. The campus offers breathtaking views of Lake Washington and the Olympic Mountains.

Fifty-five percent of freshmen live in the school's eight co-ed dorms. Hagget provides a comfortable setting for freshmen, and McMahon is recommended for those inclined to party. Housing is also available for married students, and the fraternity and sorority organizations are home to another 12 percent of the men and 11 percent of the women. The rest live off campus in Seattle or other parts of King County. Each dorm has its own cafeteria and fast-food line based on a debit card system. The Husky Union Building also offers a dining hall, espresso bar (don't forget, this is Seattle!), writing center, sun deck, and lounges. "Great" is how one student describes campus security, while others also report that they feel "100 percent safe" on campus.

For those interested in skirting the masses, UW sports an honors program that offers small classes on interesting subjects taught by fine professors.

Given the large number of commuters, it's no surprise that most of Washington's social life takes place away from campus, except for the Greeks (members of a combined total of 48 fraternities and sororities). "The dorms and fraternities seem to contain the most social activity, and what they lack is made up for by the proximity to downtown Seattle," says one student. There are also free movies and drama productions for those who must find entertainment on campus. Neither dormies nor Greeks are supposed to drink if they're under 21. Sooner or later most students hit "The Ave," University Way, where shops and restaurants await them.

> **"The dorms and fraternities seem to contain the most social activity, and what they lack is made up for by the proximity to downtown Seattle."**

And that's true of Seattle, too. A 10-minute bus ride connects students to a full array of urban offerings. The Seattle Center and other venues host outstanding operas, symphonies, and touring shows, while Qwest Field houses the NFL's Seahawks, Safeco Field the MLB Mariners, and Key Arena the NBA Sonics. But who needs pro sports with

Washington's Huskies around? Husky Fever breaks out on every football weekend, and the stands are always packed for UW's team. The team won the Rose Bowl in 2001 and coach Tyrone Willingham promises a return to glory. While students get fired up for the trip to Pasadena, they're equally excited when Washington State comes to town to vie for the coveted Apple Cup. Other strong UW teams include women's basketball, crew, cross-country, and tennis, and men's crew, basketball, baseball, soccer, and tennis.

More than anything else, the great outdoors define the University of Washington. The campus offers breathtaking views of Lake Washington and the Olympic Mountains. Outdoor pastimes for students include boating, hiking, camping, and skiing, all found nearby, and Canada is close enough for road trips to Vancouver. The weather is consistently temperate, and natives insist that the city's reputation for rain is undeserved. Then again, the sports stadium has an overhang to protect spectators from showers.

While some students will not appreciate the no-nonsense and occasionally impersonal academic programs and the lack of a centralized social life, many students can overlook these obstacles for the big picture of the up-and-coming University of Washington—one that takes in more than just the beautiful scenery.

If You Apply To ➤

Washington: Regular admissions. Does not guarantee to meet full demonstrated need. Campus interviews: optional, informational. No alumni interviews. SATs or ACTs: required. SAT IIs: optional. Essay question: personal statement. Primarily committed to state residents.

Washington and Jefferson College

60 South Lincoln Street, Washington, PA 15301

Premed Central would be as good a name as any for W&J, which has one of the nation's highest proportion of students who go on to medical school. Law school and business school are also popular destinations. The tenor of life is conservative and the Greek system dominates the social life.

Website: www.washjeff.edu
Location: Small town
Total Enrollment: 1,394
Undergraduates: 1,394
Male/Female: 52/48
SAT Ranges: V 520–610
M 530–620
ACT Range: 23–26
Financial Aid: 73%
Expense: Pr $ $
Phi Beta Kappa: Yes
Applicants: 4,477
Accepted: 39%
Enrolled: 22%
Grad in 6 Years: 68%
Returning Freshmen: 86%

Wannabe doctors and lawyers would be well-advised to give Washington and Jefferson College a look. This small Pennsylvania college is renowned for its preprofessional programs and graduates are almost guaranteed acceptance into medical or health-related graduate programs. Classes remain small here and, despite the somewhat rural location, students enjoy an active social life thanks to a hearty Greek scene and the nearby city of Pittsburgh.

The campus, like the student body, is tight-knit: just over 30 buildings sit on 51 acres in a small town about 30 miles outside of Pittsburgh. W&J is the 11th-oldest college in the country, and houses the eighth-oldest college building, which was built in 1793. Famous songwriter Stephen Foster was a student here until he was kicked out. The prevailing architectural style is traditional Colonial/Georgian, though modern structures have been added at a rapid pace during the past two decades. The four-story Howard J. Burnett Center houses the economics, business and accounting, modern languages, education, and entrepreneurial studies departments, and the Office of Life-Long Learning. The Technology Center includes computer labs and a videoconferencing center. Two new dorms—with 96 and 100 beds—as well as 10 theme houses opened recently.

Graduation requirements call for students to complete 34 courses and demonstrate proficiency in writing, speaking, reading, quantitative reasoning, foreign language, and use of information technology. Students must take credits in culture and intellectual tradition, fine arts, language and literature, science and mathematics, and social sciences. Students must satisfy language, cultural diversity, and physical education and wellness requirements. A thematic major allows students to design their own course of study, while double majors produce such types as a biologist well versed in literature. Rare among liberal arts colleges are the 3–4 programs with the Pennsylvania Colleges of Optometry and Podiatry. More technically minded students can take advantage of the 3–2 engineering programs with Case Western Reserve and Washington University in St. Louis. Business administration, accounting, psychology, history, and political science are the most popular majors. There are a slew of newer academic programs: environmental studies; neuroscience; biochemistry; biophysics; information technology leadership; international business; gender and women's studies; and mind, brain, and behavior. More than 90 percent of W&J graduates who apply to medical or health-related programs are accepted for admission, and the acceptance rate of W&J graduates to law schools is 90 percent.

> **"There are no easy courses at W&J."**

W&J's formula for success starts with individual attention in small classes; 91 percent of freshman classes have 25 students or fewer. "With small classes, it becomes a rigorous competition for good grades," a sophomore says. Sixty-five percent of students ranked in the top quarter of their high school class, and most agree the academic climate at W&J is tough, especially for those on the premed and prelaw tracks. An accounting major warns, "There are no easy courses at W&J." Full professors teach most classes. "The professors are excellent and go out of their way to engage and teach students not only on the lessons, but lifelong values," says a junior.

During the January intersession, students find brief apprenticeships in prospective career areas, take a school tour abroad (marine biology trips to the Bahamas or Australia, English theater trips to London, history trips to Russia or China, and biology trips to Africa), or engage in nontraditional coursework. There is also a semester- or year-abroad option in Germany, Moscow, Russia, or Bogotá, Colombia. Other study abroad opportunities are also available in Australia, Europe, Asia, and Latin America.

Diversity is not W&J's strong suit: 76 percent of the students hail from Pennsylvania and many are from neighboring states in the Northeast. Two percent of students are African American, 1 percent Hispanic, and 2 percent Asian American. Students get a bargain at this somewhat pricey school if they win one of the several hundred academic scholarships that average $9,297. There are no athletic scholarships, and the school doesn't guarantee to meet students' full demonstrated need, but

> **"The college makes every effort to financially help students get here and remain here for four years."**

the same package is promised for a student's four years at the college. One English and Russian major says, "The college makes every effort to financially help students get here and remain here for four years."

Students can live in either co-ed or single-sex dorms, and 81 percent of students live on campus. Housing is guaranteed for four years and getting a room is no trouble "unless you are trying to get into the new dorms," a student says. Students also say the freshman dorms are fair, but the choices get better with academic rank, and on-campus apartments are available based on GPA, activities, and need. Campus grub ranges from "edible" to "great" and students say security is visible and reliable. "I always feel safe on campus," says a junior.

Without a doubt, the social life at W&J centers on the Greeks, which attract 43 percent of the men and 37 percent of the women. The school has seven national

(Continued)
Academics: 📖 📖 📖
Social: ☎ ☎ ☎
Q of L: ★ ★
Admissions: (888) 926-3529
Email Address:
admission@washjeff.edu

Strongest Programs:
Premed
Prelaw
Business Administration
Biology
English

Students can also head home on the weekends or explore dating opportunities at nearby colleges, most notably Penn State and the University of Pittsburgh.

fraternities and four national sororities. A crackdown on alcohol and noise violations has somewhat quieted the school's tradition of enormous parties, but the Student Activity Board has begun filling the gap with more on-campus events, such as free movies, and a student-run coffeehouse provides another nonfrat option. "Campus life has died down since the 'Quads' were moved," confirms a sophomore. "There are not enough parties." A nonalcoholic pub called George & Tom's has become quite a popular diversion with comedy, musical, and novelty/variety acts. During the course of the year, the Spring Street Fair and Spring Concert are the most popular events. "Social life is nice because the college is small and you have a chance to build close friendships," a freshman says.

Students can also head home on the weekends or explore dating opportunities at nearby colleges, most notably Penn State and the University of Pittsburgh. One of the most popular excursions is a 30-minute commute to Pittsburgh. Still, not all students share the administration's appreciation for "the unique characteristics of the

"I always feel safe on campus."

western Pennsylvania milieu." Many complain that there is nothing to do in this former steel/mining town, now hit by hard times. "There is little interaction between the city and the school," says a student. "Most people just take the drive to Pittsburgh." Townies tend to be a bit resentful of dressy W&J undergrads, but students try to assuage this attitude by actively volunteering in the community.

Just about anyone has a shot at varsity sports. Football, baseball, water polo, soccer, men's golf, and women's softball, soccer, basketball, and water polo have all won recent championships. Hundreds of students participate in intramural sports, with flag football, basketball, and bowling attracting the most interest. "The IM sports are great, really fun and competitive and there are many of them," a freshman says. There are also numerous clubs to join, from the equestrian club to the student theater company.

With a revised curriculum, start-of-the art facilities, and expanding academic options, the leadership at W&J is opening more and more doors for students. Students praise the education they receive, the professors who give it to them, and the climate in which they receive it. "W&J has momentum and will continue to grow in the coming years," a junior history and English major says.

Washington and Lee University

Lexington, VA 24450

Coeducation came to W&L, and the pillars of the Colonade did not come tumbling down. Nearly 20 years after the coming of women, W&L is the most selective small college in the South, rivaled only by Davidson. W&L supplements the liberal arts with strong programs in business and journalism.

Washington and Lee University, which shares the town of Lexington, Virginia, with the Virginia Military Institute, is as genteel as a Southern school can be. The Fancy

Dress Ball is a highlight of each year, and a "speaking tradition," tracing its roots to Robert E. Lee, results in at least casual communication between students and professors when they pass one another on the well-manicured grounds. And behind the frills and fun lies a student-run Honor System that students cite as one of their school's defining features.

W&L's wooded campus sits atop a hill of lush green lawns, sweeping from one national landmark to another. Redbrick buildings feature white Doric columns and the prevailing architectural style is Greek Revival, although the physical face is changing. The school boasts the John W. Elrod University Commons, which contains a dining hall, movie theater, and bookstore. Reid Hall, home of the journalism and mass communication department, underwent a $6 million renovation recently and a new "green" art and music building has opened.

Though Washington and Lee is a school steeped in tradition, it is also responsive to change. Women's numbers are growing—they now make up half of the student population—as is their influence on the campus academic and social life. Although a standard liberal arts program remains the foundation of the school's curriculum, it offers excellent preprofessional programs, particularly business, economics, and accounting through the Williams School of Commerce, Economics, and Politics. Journalism and mass communications is popular, as are biology and English. W&L also has bachelor's degree programs in fields as diverse as Russian Area studies and East Asian languages and literature. Nearly half of all students spend time abroad, and the Center for International Education is increasing international study opportunities.

"The academic climate is challenging, but invigorating and supportive."

General education requirements account for one-third of a student's coursework. Distribution requirements at W&L include English composition; courses in literature, fine arts, history, philosophy, and religion; three courses in science and math; social science courses; four terms of physical education; proficiency in swimming; and two years of a foreign language.

"The academic climate is challenging," a junior says, "but invigorating and supportive." Classes tend to be small, with an average size of 16, and freshmen can count on getting professors; there are no teaching assistants. "The quality of teaching is exceptional," a student says. "Profs are engaging, knowledgeable, and interesting." The famous Honor System ("The foundation of all areas of the school," according to one student) lends a relaxed feeling to the otherwise rigorous academic climate. Tests and final exams are taken without faculty supervision; doors remain unlocked, laptops stay on desks, and library stacks are open 24 hours a day. The modern library, like most W&L facilities, is superb and offers 800 individual study areas as well as private rooms for honors students. Well-qualified students can apply for the Robert E. Lee Undergraduate Research Program, which offers students paid fellowships for assisting professors in research or doing their own.

This is a friendly campus. "The speaking tradition is such that W&L students say 'hi' to people they pass on campus, which definitely contributes to the collegial atmosphere and sense of community here," says a premed student. Prospective students should be aware that there is little urgency on this campus to change the world. "You would still be hard-pressed to find a more conservative school," a student says. Although students are generally "rich, white, conservative Republican Southerners," according to a junior, "this has changed a lot in recent years." African Americans account for a mere 4 percent of the student body, despite the school's argument that it is strongly committed to recruiting African American students; Hispanics make up 2 percent and Asian Americans 4 percent. Fourteen percent of students are native Virginians, but the university's geographic diversity has not prompted much other differentiation. The number of applications continues to rise

Website: www.wlu.edu
Location: Small town
Total Enrollment: 2,134
Undergraduates: 1,745
Male/Female: 50/50
SAT Ranges: V 660–730
   M 660–720
ACT Range: 28–31
Financial Aid: 33%
Expense: Pr $ $
Phi Beta Kappa: Yes
Applicants: 3,905
Accepted: 29%
Enrolled: 41%
Grad in 6 Years: 87%
Returning Freshmen: 95%
Academics: 🏛🏛🏛🏛½
Social: 🎭🎭🎭
Q of L: ★ ★ ★ ★
Admissions: (540) 458-8710
Email Address:
   admissions@wlu.edu

Strongest Programs:
Business
History
Politics
English
Journalism
Economics

Women's numbers are growing—they now make up half of the student population—as is their influence on the campus academic and social life.

dramatically, as do the median SAT scores of each freshman class, and the acceptance rate is below 30 percent. The school offers a number of merit scholarships averaging $12,600 but no athletic awards.

Sixty-one percent of students reside on campus. Students spend their first year at W&L in co-ed dorms which are "spacious but not air-conditioned," according to one

"You would still be hard-pressed to find a more conservative school."

student. Freshmen are required to live in freshman dorms and buy the meal plan, and the food is said to be good. Upperclass dorms and apartments are available, and many students move into country houses when they are juniors or seniors. The Honor System and campus security contribute to the students' feelings of safety on campus.

Seventy-nine percent of the men join fraternities; sororities claim 78 percent of the women. "Greeks are huge and run the social scene," a junior says. Underage drinking is banned in the dorms but students insist "underage drinking is rampant." Greek bashes often feature live bands, although the Fancy Dress Ball, or "$100,000 prom," also draws raves. A lot of creative energy goes into fraternity parties—one recently featured a real mechanical bull, another was decorated like the Playboy Mansion. Another student sighs that fraternities pretty much have a monopoly on revelry here, but that some outgrow them. "They are cool at the beginning, but they get repetitive." W&L's mock political convention for the party out of power, held every four years, has predicted past presidential nominees with uncanny accuracy. The Foxfield races near Charlottesville and the Kentucky Derby are popular road trip destinations. Washington, D.C., Richmond, and Roanoke are easily reached by car for weekend trips, though because there is seldom a dull moment, most students like to stay on campus during weekends.

With its scenic location in the midst of the Appalachian Mountains, the university provides an abundance of activities for nature lovers, including hunting, fishing, camping, skiing, and tubing on the rivers. "I adore Lexington," says one student.

"Greeks are huge and run the social scene."

Lexington, a "quiet, friendly town that has much history to offer," also offers a few bars, two movie theaters, and several restaurants. "Dominated by the beauty of surrounding Shenandoah Valley and the Blue Ridge Mountains, Lex is small, but has everything one might need," a senior says.

Football sparks some interest in the fall, with its attendant tailgate parties, but W&L students live for the spring and the Lee-Jackson Lacrosse Classic against VMI. The tennis, lacrosse, women's soccer, and volleyball teams are most competitive, and W&L's teams regularly bring home conference championships and participate in post-season tournaments.

Many graduates are proud to look back on their college careers and point to the progress their alma mater has made in the intervening years, but not those from Washington and Lee. The culture here doesn't embrace change, and students here grin and bear the teasing of those who say W&L really stands for "White and Loaded." But strong academics and respect for tradition have their appeal. "Washington and Lee students are proud to be W&L students and this is evident in their happiness and commitment to pursue their academics," concludes a sophomore.

Overlaps

University of Virginia, College of William and Mary, Davidson, Vanderbilt, University of Richmond, Wake Forest

If You Apply To ➤ **W&L:** Early decision: Nov. 15, Jan. 2. Regular admissions: Jan. 15. Does not guarantee to meet demonstrated need. Campus interviews: recommended, informational. Alumni interviews: optional, informational. ACTs or SATs: required. SAT IIs: required (writing and two nonrelated subjects). Accepts the Common Application and prefers electronic applications. Essay question: significant experience or achievement with special meaning; best advice received; issue of personal, national, or local concern and its importance.

Washington University in St. Louis

Campus Box 1089, One Brookings Drive, St. Louis, MO 63130-4899

In the space of little more than a decade, Wash U has gone from Midwestern backup school to elite private university close on the heels of Northwestern. Wash U is strong in everything from art to engineering. The halo effect of the university's medical school attracts a slew of aspiring doctors.

Washington University has made a name for itself as one of higher education's rising stars. Though it's always been well recognized regionally, Wash U has now firmly established itself as a truly national institution—with a relaxed Midwestern feel that differentiates it from the high-strung eastern Ivies. Applications have skyrocketed, and with a hefty endowment, strong preprofessional programs, and an emphasis on research, it's not hard to see why. "Washington U is a very academically centered community," says a freshman. But applicants beware: For some reason, the admissions office has recently taken to relegating huge numbers of wannabes to waitlist purgatory.

The school's 169-acre campus adjoins Forest Park, one of the nation's three largest urban parks. Buildings are constructed in the Collegiate Gothic style, mostly in red Missouri granite and white limestone, with plenty of climbing ivy, gargoyles, and arches. The Sam Fox Arts Center encompasses two new structures and three existing art facilities, and an addition to the psychology building adds 15,000 square feet of new space.

Undergraduates enroll in one or more of Washington U's five schools—arts and sciences, architecture, art, business, or engineering. Double and interdisciplinary majors, such as environmental studies, are encouraged and easily arranged—so easily arranged, in fact, that about 60 percent of students earn either a major and minor, more than one major, and sometimes more than one degree. General education requirements vary by school and program. For liberal arts students, they include courses in quantitative reasoning, physical and life sciences, social or behavioral sciences, minority or gender studies, language or the arts, and English composition.

> **"Washington U is a very academically centered community."**

Washington U's offerings in the natural sciences, especially biology and chemistry, have long been notable, especially among premeds. The outstanding medical school runs a faculty exchange program with the undergraduate biology department, which affords bio majors significant opportunities to conduct advanced laboratory research. The University Scholars Program allows students to apply for undergraduate and graduate admission before entering college. If accepted, students can begin exploring their chosen career path earlier—though they aren't obligated to follow through if their interests change.

The freshman FOCUS program helps balance the preprofessional bent of some of Washington U's best programs with the school's desire to provide a broad and deep educational experience and a smaller class size. In FOCUS, students use a weekly seminar to explore topics of contemporary significance, such as law and society. The program lets first-year students work closely with professors—including at least one Nobel Prize winner—and sample offerings from various departments. Other notable options include one-, two-, and four-year interdisciplinary programs such as Text and Tradition, International Leadership Program, and Medicine and Society.

As Washington U's applicant pool has gotten bigger, the admissions committee has become more selective—and classes have gotten tougher, students report. "Everyone is very intelligent and gives academics their top priority," says a senior.

Website: www.wustl.edu
Location: Suburban
Total Enrollment: 10,669
Undergraduates: 6,080
Male/Female: 49/51
SAT Ranges: V 670–750
   M 690–770
ACT Range: 30–33
Financial Aid: 43%
Expense: Pr $ $ $
Phi Beta Kappa: Yes
Applicants: 21,515
Accepted: 19%
Enrolled: 34%
Grad in 6 Years: 91%
Returning Freshmen: 96%
Academics: 🖉 🖉 🖉 🖉 ½
Social: ☎ ☎ ☎ ☎
Q of L: ★ ★ ★ ★
Admissions: (314) 935-6000
   or (800) 638-0700
Email Address:
   admissions@wustl.edu

Strongest Programs:
Biology and Natural Sciences
Premedicine
Psychology
Foreign Languages
English
Accounting and Business
Political Science
Earth and Planetary Sciences

Washington U's offerings in the natural sciences, especially biology and chemistry, have long been notable, especially among premeds.

"However, WU is a very nurturing community and students consistently help one another out and encourage each other's success." Those who are struggling will find plenty of help from teaching assistants who conduct help sessions or study groups of their peers. What really sets WU's quality of teaching apart is that undergrads have access to one-on-one mentoring relationships with the top faculty. "The education at WU has been outstanding," says a senior. "I have gained so much from this university."

The school's population is relatively heterogeneous, with 64 percent Caucasian and African Americans comprising 9 percent. Hispanics represent 3 percent, Asian Americans 10 percent, foreign students 5 percent, and Native Americans less than 1 percent. "Our students are extremely liberal and laid-back," says one student. "They are all open-minded, respect diversity, and respect others' opinions." There's also a large contingent from Eastern states like New York and New Jersey, many of whom are Jewish, leading to much debate about the Arab-Israeli conflict. Organizations such as STAR (Students Together Against Racism) and ADHOC (Against Discrimination and Hatred on Campus) also help enlighten students.

Seventy-five percent of Wash U students live in the school's dormitories, known as residential colleges. "The dorms are extremely plush," says a senior. Most are clustered in an area called South 40, a 40-acre plot next to campus. All dorms are co-ed and air-conditioned, and some have suites for six to eight students of the same gender. Freshmen are guaranteed rooms, and students who stay in the dorms after that are promised rooms for the following year, says a junior. Upperclassmen may live in university-owned apartments, and there's always an argument as to which dorm is better. Some juniors and seniors do choose true off-campus digs in the nearby neighborhoods of University City and Clayton, where apartments are reasonably priced. Dorm-dwellers and others who buy the meal plan may use their credits in any of 14 dining centers. "Dining halls and cafes around every corner and something for everyone's taste. There's even a 24-hour cybercafe in our main library."

> **"Our students are extremely liberal and laid-back."**

Washington U students pride themselves on being able to balance work and play, and on weekends, movies, fraternity parties, and concerts tear them away from their books. Every spring, the whole campus turns out for the century-old Thurtene Carnival, the oldest student-run philanthropic festival in the country. Diwali, a Hindu festival, draws huge crowds. Student groups—especially fraternities and sororities, which attract approximately a quarter of the men and 15 percent of the women, respectively—build booths, sell food, and put on plays, and profits are donated to a children's charity. Another big event is Walk-In-Lay-Down (WILD) Theater, held the first and last Fridays of the academic year. Everyone brings a blanket to the main quad, assumes a horizontal position, and listens to live bands until the wee hours. Alcohol policies emphasize responsible drinking, though students younger than 21 aren't supposed to drink at all, per Missouri law. "Students appreciate the administration's trust and are very respectful of each other. If students find themselves in trouble, they're not afraid to ask for help."

Every spring, the whole campus turns out for the century-old Thurtene Carnival, the oldest student-run philanthropic festival in the country.

Aside from an active campus social life, Wash U offers incredible recreational options because of its location abutting Forest Park: A golf course, an ice-skating rink, a zoo, a lake with boat rentals, art and history museums, an outdoor theater, and a science center are all within a short walk. The St. Louis Rams, Blues, and Cardinals attract pro football, hockey, and baseball fans, and the city is also home to the addictive Ted Drewes frozen custard. The school runs a free shuttle service to parts of St. Louis not within walking distance, and community service programs such as "Each One Teach One," in partnership

> **"I love St. Louis! It is a wonderful place to be a student."**

with the city's schools, attract a sizable number of students. "I love St. Louis!" says a senior. "It is a wonderful place to be a student. It features the benefits of a major metropolitan area but retains a strong sense of neighborhood. It is very accessible to students—most major cultural institutions are either free or very cheap." The best road trips include Chicago; Nashville and Memphis, Tennessee; Lake of the Ozarks; and Columbia, Missouri—home of the University of Missouri.

Washington U competes in Division III, and is a women's basketball powerhouse, bringing home NCAA championships four times and UAA titles 16 times. The women's softball team won its fourth University Athletic Association title. Other strong programs include women's volleyball, with eight NCAA national championships and 17 UAA championships in its history, and football, racking up their eighth UAA title. Intramural sports range from badminton, arm wrestling, and floor hockey to pocket billiards and ultimate Frisbee.

With demanding courses and excellent programs in the sciences and art, Wash U is no longer a sleepy little Midwest school but has become a competitive educational institution. The curriculum is becoming increasingly focused on global and international perspectives. But while academics are challenging, students find there's always a peer or a professor willing to help. "Everybody is so nice! I feel that most everyone has a genuine care for those around them." Just try to avoid that long waitlist.

> **Overlaps**
> Yale, Duke, Harvard, Northwestern, Stanford, University of Pennsylvania

If You Apply To ➤

Washington U: Early decision: Nov. 15. Regular admissions: Jan. 15. Financial aid: Nov. 15 (early decision applicants), Feb. 15. Housing: May 1. Meets demonstrated need of 99%. Campus and alumni interviews: optional, evaluative. SATs or ACTs: required. SAT IIs: optional. Apply to one of five undergraduate schools. Accepts the Common Application and electronic applications. Essay question: personal statement.

Wellesley College

Wellesley, MA 02481

There is no better recipe for popularity than a postcard-perfect campus on the outskirts of Boston. That formula keeps Wellesley at the top of the women's college pecking order—along with superb programs in economics and the natural sciences. Nearly a quarter of the students are Asian American, the highest proportion in the East.

Wellesley is not just the best women's college in the nation—it's one of the best colleges in the nation, period. With an alumnae roster that includes Senator Hillary Rodham Clinton, Madame Chiang Kai-shek, Madeleine Albright, and Diane Sawyer, Wellesley College should be at the top of the list for those who are seeking an all-women's education. Wellesley women excel in whatever field they choose, including traditional male bastions like economics and the sciences. "We have a very strong community made up of very strong personalities," says a junior.

Nestled in a Boston suburb, the Wellesley campus, one of the most beautiful anywhere, occupies 500 rolling acres of cultivated and natural areas, including Lake Waban. Campus buildings range in architectural style from Gothic (with stone towers and brick quadrangles) to state-of-the-art science, arts, and sports facilities. A 22-acre arboretum and botanical garden features a wide variety of trees and plants.

With its hefty endowment and lavish facilities, Wellesley offers a top-of-the-line educational experience. The most popular majors are psychology, English, political science, and international relations, though economics is known as the biggest

Website: www.wellesley.edu
Location: Suburban
Total Enrollment: 2,300
Undergraduates: 2,300
Male/Female: 0/100
SAT Ranges: V 600–750
 M 650–720
ACT Range: 29–31
Financial Aid: 58%
Expense: Pr $ $ $
Phi Beta Kappa: Yes
Applicants: 4,098
Accepted: 35%
Enrolled: 41%

(Continued)

Grad in 6 Years: 90%

Returning Freshmen: 95%

Academics: ✍ ✍ ✍ ✍ ✍

Social: ☎ ☎ ☎

Q of L: ★ ★ ★ ★

Admissions: (781) 283-2270

Email Address:

admission@wellesley.edu

Strongest Programs:

Economics

Political Science

Psychology

English

International Relations

Neuroscience

Anything Wellesley women find lacking in their facilities or curriculum can probably be found at MIT, where they have full crossregistration privileges.

Under the honor system, students may take their finals, unsupervised, at any time during exam week.

powerhouse. In fact, Wellesley has produced virtually all of the country's high-ranking female economists. Students in molecular biology work with faculty on DNA research, and a high-tech science center houses two electron microscopes, two NMR spectrometers, ultracentrifuges, two lasers, and other such equipment. A comprehensive renovation of the social science building added videoconferencing, computer labs, and a research facility.

The Ruth Nagel Jones Theater provides performance space for experimental theater and mainstage productions. The Davis Museum and Cultural Center houses 11 galleries, a cinema, and a cafe. The students at Wellesley will find an academic art museum to their benefit, along with more than a million volumes in the campus libraries. The five libraries sport a computerized catalog system that is accessible from on or off campus. Anything Wellesley women find lacking in their facilities or curriculum can probably be found at MIT, where they have full crossregistration privileges. Wellesley students can also take courses at Brandeis University, Babson College, and nearby Olin College of Engineering, or participate in exchange programs with Spelman College in Atlanta or Mills College in Oakland, California.

> **"We have a very strong community made up of very strong personalities."**

Wellesley has distribution requirements that include three units drawn from language and literature and visual arts, music, theater, film, and video; one unit from social and behavioral analysis; a unit each from two of the following: epistemology and cognition, religion, ethics, and moral philosophy, and historical studies; and three units from natural and physical science and mathematical modeling and problem solving. In addition, students must take a first-year writing class, a foreign language, and a course on multiculturalism. Academics are taken very seriously at Wellesley. "The courses are extremely rigorous," says a senior. "I've never worked harder in my life." Professors are highly respected and make themselves available through email, voicemail, office hours, and by appointment. "Full professors teach at all levels and are easily accessible," a student says. A chemistry major adds, "The quality of teaching is fantastic."

Grants from private foundations have allowed Wellesley to add other innovative programs, including independent research tutorials for advanced science students and fellowship funding for joint student/faculty projects. Students can participate in the Twelve College Exchange*, including the National Theater Institute, and the Maritime Studies Program, or they can travel and study abroad through one of Wellesley's recently expanded international programs, including the summer internship program in Washington, D.C.

> **"I've never worked harder in my life."**

Under the honor system, students may take their finals, unsupervised, at any time during exam week. Class sizes are almost always small (they average 18 to 23 students per class). First-years (as they are exclusively called here) and upperclasswomen alike have faculty advisors. First-years also have a dean of first-year students to offer additional advice on courses and other academic matters. "There are good counseling resources available, but students must seek them out," explains a sophomore.

"It's not uncommon to sit down at dinner with a group of students and discuss Plato," explains a senior. "Sometimes I wish people would lighten up a little and talk about the latest reality TV show!" Nearly 50 percent of Wellesley's students are racial or ethnic minorities. Although the Northeast is the best-represented geographical area (though only 11 percent are from Massachusetts), students also come from every state and more than 75 countries. Eighty-five percent ranked in the top tenth of their high school class. Whatever their background, most have a fair amount of social aplomb. Issues on campus run the gamut from multiculturalism and racism to gender questions and national politics. "You have to be politically correct," says one student.

Residence life at Wellesley is a step ahead of most institutions, to say the least. Virtually every student lives on campus, in rooms that are described as "immaculate." Residence halls feature high-ceilinged living rooms, hardwood floors, fireplaces, computers and laser printers, television annexes, walk-in closets, kitchenettes with microwaves, and even grand pianos. The halls are renovated every five years or so and all are well-maintained, students say. "The dorms are absolutely gorgeous and no two rooms look alike," gushes one student. All residence halls are smoke-free. There are no halls specifically for first-years—all classes live on all floors. Peer tutors also live in each hall. These students, called APT advisors, are trained to tutor in specific subjects and in study skills and time management. Juniors and seniors are generally granted single rooms. Two co-ops, one with a feminist bent, present an educational housing option. Meal cards are valid in every dining hall, and at the campus snack bar, which is stocked with everything from milk and flour to Twinkies.

When it comes to weekend fun, Wellesley is in a prime location. Not even half an hour away, Boston is the place where Wellesley women can mingle with lots of other students—specifically male—from Harvard and MIT. Cambridge—with Harvard Square, MIT frat parties, and lots of jazz clubs—is accessible by an hourly school shuttle that runs on weekdays and weekends. There is also a commuter rail station

"It's not uncommon to sit down at dinner with a group of students and discuss Plato."

located a short walk from school. Cape Cod, Providence, and the Vermont and New Hampshire ski slopes are close by car. "Wellesley is not a party school," says a senior. "That said, there is tons of stuff to do on weekends."

The town of Wellesley is an upper-crust Boston suburb without many amenities for students. "Be forewarned," cautions a student, "Wellesley is a snobby town of rich people." Wellesley is a dry town, but there is a student-run pub in the new campus center. When alcohol is served, campus police check IDs. The new Lulu Chow Wang Campus Center, referred to affectionately as "the Lulu," is a hub of activity day and night. Students enjoy going to the student-run Cafe Hoop, the campus coffeehouse, to sip tea or share a fro-yo or a chocolate croissant. The closest thing Wellesley has to sororities are societies for arts and music, literature, Shakespeare, and general lectures. These societies also sometimes hold parties.

Wellesley is chock-full of traditions, but the most endearing ones include Flower Sunday, step-singing (an all-campus sing-along on the chapel steps), the sophomore class planting a tree, a junior class variety show, Spring Weekend (with a big-name band and comedian), and a hoop-rolling contest by seniors in their graduation robes. The winner of this contest will supposedly be the first in her class to achieve her goals, whatever they may be, and she gets off to a flying start when her classmates toss her in the lake. Speaking of the lake, students mention their unofficial campus event, Lake Day, where students take breaks between (or from) classes to enjoy a festival held on the lawn near the lake.

Many students balance their academic schedule with athletics and club sports. Lacrosse and tennis are among the top varsity sports and cross-country made the NEWMAC Conference championships for four consecutive years. Field hockey, soccer, and volleyball have also claimed championships. The sports center, named the Nannerl Keohane Sports Center in honor of Wellesley's 11th president (who went on to run Duke), offers an Olympic-size pool; squash, racquetball, and tennis courts; dance studios; a weight room; and an indoor track. Harvard's Head of the Charles crew race and the

"Wellesley is not a party school. That said, there is tons of stuff to do on weekends."

Boston Marathon—Wellesley's "Scream Tunnel" is legendary among runners worldwide—share honors as the most popular spectator sport of the year. The big athletic rival is Smith College, another of the Seven Sisters group of great women's colleges.

Overlaps
Stanford, UC–Berkeley, Washington University, Northwestern, Duke, Rice

When it comes to academics, Wellesley women are serious. Their school is competitive with all but the top three Ivies. Many of them enjoy the traditions of the school and appreciate the idyllic atmosphere for contemplation, but know they are poised to dominate whatever field they enter. As one contented senior says, "It's a wonderful place to grow as individuals, as students, and as women."

If You Apply To > **Wellesley:** Early decision: Nov. 1. Regular admissions and financial aid: Jan. 15. Guarantees to meet demonstrated financial need. Campus and alumni interviews: recommended, evaluative. SATs and SAT IIs (writing and two others) or ACTs: required. Accepts the Common Application and electronic applications. Essay question: significant experience or achievement, issue of personal concern, influential person. Students participate on admissions board.

Wells College

Aurora, NY 13026

Co-ed as of fall 2005. A family atmosphere is the hallmark of Wells, right down to the dinner bell that calls everyone to the evening meal. Wells is big on interdisciplinary study and internships during the January term.

Website: www.wells.edu
Location: Small town
Total Enrollment: 405
Undergraduates: 402
Male/Female: 9/91
SAT Ranges: V 520–630
  M 480–580
ACT Range: 20–26
Financial Aid: 74%
Expense: Pr $
Phi Beta Kappa: Yes
Applicants: 1,036
Accepted: 64%
Enrolled: 20%
Grad in 6 Years: 67%
Returning Freshmen: 71%
Academics: ✍ ✍ ✍
Social: ☎ ☎
Q of L: ★★★★
Admissions: (800) 952-9355
Email Address:
  admissions@wells.edu

Strongest Programs:
Psychology
Biological and Chemical
  Sciences
English
Performing Arts
Education

Due to a lingering cash crunch and declining appeal of single-sex education, Wells College has opened its doors to men for the first time since its founding in 1868. More students means more money to help fix old buildings, boost financial aid packages, and improve faculty salaries, or at least replace professors who retire. Still, the school hasn't abandoned its storied history. Whether it's riding to graduation in an old Wells Fargo stagecoach or showing off in the annual Odd-Even basketball game between freshmen and sophomores, the tradition of Wells is apparent at every turn. With just more than 400 students, anonymity is nonexistent, and close relationships with professors and peers come with the territory.

Wells's 365-acre campus sits on the shores of Cayuga Lake—and on the National Register of Historic Places. Most buildings are old, massive, and covered with ivy—the way college should look, you might say. The lakeside location affords beautiful sunsets as well as boating and fishing opportunities—a welcome relief from the sometimes arduous studying that also goes on here. The Schwartz Athletic Center has received extensive renovations and the school recently broke ground on a 45,000-square-foot science facility that is slated to open for the 2007–08 academic year.

Wells aims to give students a solid grounding in the liberal arts. To that end, general education requirements include two courses in a foreign language, one course in formal reasoning, three courses in the arts and humanities (with at least one in each area), three courses in the natural or social sciences (with at least one lab), and four courses in physical education, including one semester of swimming and one focused on wellness. Students are also required to compete a senior thesis. Wells 101, a core course required for all first-years, covers the basics of college writing, speaking, and analytical thinking.

The most popular majors include psychology, biological and chemical sciences, English, performing arts, and sociology and anthropology; students also give high marks to education and women's studies, though they note that computer science suffers from outdated equipment. There's an unusual minor in bookbinding, while a minor in indigenous studies and first nations has been added. Regardless of major, "the classes and expectations are very rigorous," says a junior.

Students with broad interests may choose one of Wells' integrated majors, such as American studies, which require coursework across traditional disciplinary boundaries. Individualized majors are available for students whose needs are not met by established programs. There are also dual-degree programs in community health and business administration with the University of Rochester; in veterinary medicine with Cornell University; and in engineering with Case Western Reserve, Cornell, Columbia, and Clarkson. For a change of pace, students may take one nonmajor course each semester on a pass/fail basis. They may also crossregister for up to four courses each at Cornell and at Ithaca College.

Wells uses a semester calendar with elements of the 4–1–4 plan, with internships, research and study abroad taking place in January. Though the school doesn't focus on business and other professional fields, there is a corporate-affiliate program aimed at preparing women to work in the financial world, through special courses and lectures, and portfolio-management experience. Foreign study is available, too. Almost all courses taken by freshmen have 25 students or fewer, and most are discussion-based, with the professor present to moderate and focus the conversation. The student-run collegiate association enforces the honor system, and take-home and self-scheduled tests are the rule rather than the exception.

Sixty-nine percent of Wells students are state residents, and 61 percent are white; Asian Americans make up 3 percent of the student body, African Americans 7 percent, and Hispanics 4 percent. Courses, workshops, and a support network for new students help to educate the campus on the importance of multiculturalism. The campus is very liberal, and women's rights, feminism, and homosexuality are key topics of discussion. Despite the fact that fewer than one in 10 students is male, coeducation continues to be a hot topic. There are numerous merit scholarships, averaging nearly $5,000 each, but no athletic awards.

"My room is beautiful."

All students are guaranteed college housing, and 80 percent do live in the dorms, since only seniors are permitted to move off campus. There are plenty of single rooms, though first-years are typically assigned to doubles. Aside from complaints that the heat is too high during the winter, students rave about their residences, some of which have bay windows and winding staircases, and all of which offer lake views. "My room is beautiful," crows a psychology major. "I have a lot of space and all of the necessities: closet, wardrobe, mirror, bed, desk, Internet connection, and chairs." Lounges are professionally decorated, and one dorm is a converted mansion dating from the 19th century. The food gets decidedly less stellar reviews—"horrible" to "disgusting" and the kindest description, "not horrible"— but at least it's served in a magnificent Tudor-style dining hall, with two working fireplaces. Security is "adequate," says one senior.

Though there are no Greek organizations at Wells, the school doesn't need them, given its bevy of other traditions. For example, bells are rung every evening to announce dinner, and also to celebrate the first snowfall of the season. Additionally, tea and coffee are served every weekday afternoon. Though the long dresses and china cups have long since disappeared, tea is still a great time to hang out with friends, faculty members, and staff, as well as a welcome break from long afternoon seminars. On the last day of classes, there's a celebration around the sycamore tree, where sophomores present roses to their senior-class big sisters. Then, the president of the college and her staff serve breakfast to the graduating class.

As for social life on campus—well, there isn't much. The Student Activities Club sponsors comedians, dances, movie and spa nights, guest speakers, and poetry slams, but if you want to party, head to Cornell or Ithaca College in the Wells van. Underage students can get alcohol, and no one seems to bother them if they're drinking, as long as they're not bothering anyone else—and frankly, drinking isn't

All students are guaranteed college housing, and 80 percent do live in the dorms, since only seniors are permitted to move off campus.

Though there are no Greek organizations at Wells, the school doesn't need them, given its bevy of other traditions.

For a change of pace, students may take one nonmajor course each semester on a pass/fail basis. They may also crossregister for up to four courses each at Cornell and at Ithaca College.

much of a focus here, anyway. The town of Aurora has a pizza parlor, ice cream shop, bar and hotel—and that's about it. "It's a small village, not a college town at all," says a sociology major. As a result, students enjoy getting out and about, whether they head to Auburn and Syracuse for shopping, or to New York City and Montreal for shows and other big-city perks. Given the beautiful, hilly terrain, Wells students also enjoy camping in the warmer months, and cross-country or downhill skiing in the winter, especially with the slopes of Greek Peak less than an hour away.

Varsity teams at Wells compete in Division III, and the soccer, tennis, and field hockey squads are the strongest. Also popular are swimming, lacrosse, and softball—and since the college is so small, "everyone that I know of has been able to play," says a psychology major. Wells recently added men's intercollegiate cross-country, swimming, and soccer, and a men's lacrosse team is in the works. Anyone may use the golf course and the college's tennis and paddle-tennis courts, while the field house offers a pool and other exercise equipment.

Wells students aren't ones to shy away from a challenge, so odds are they'll weather the school's coed status. As for the men, well, they'll be hard-pressed to change the "liberal, progressive, feminist, and independent" spirit of this place.

Overlaps

Hobart and William Smith, Lemoyne, Elmira, Mount Holyoke, Smith, SUNY–Cortland

If You Apply To ➤ **Wells:** Early action and early decision: Dec. 15. Regular admissions: Mar. 1. Financial aid: Feb. 15. Meets demonstrated need of 25%. Campus interviews: recommended, informational. Alumni interviews: optional, informational. SATs or ACTs: required. SAT IIs: optional. Accepts the Common Application and electronic applications. Essay question: significant experience or achievement; issue of personal, local, or national concern; or influential person.

Wesleyan University

North College, Middletown, CT 06457

Usually compared to Amherst or Williams, Wesleyan is really more like Swarthmore. The key difference: Wesleyan is twice as big. Wes students are progressive, politically minded, and fiercely independent. Exotic specialties like ethnomusicology and East Asian studies add spice to the scene.

Website: www.wesleyan.edu
Location: Small town
Total Enrollment: 3,217
Undergraduates: 2,777
Male/Female: 47/53
SAT Ranges: V 660–750
    M 650–740
ACT Range: 28–32
Financial Aid: 48%
Expense: Pr $ $ $ $
Phi Beta Kappa: Yes
Applicants: 7,241
Accepted: 28%
Enrolled: 36%
Grad in 6 Years: 91%
Returning Freshmen: 96%

Whether they're engrossed in academics, debating and demonstrating over various issues, or engaged in community service, Wesleyan students seem to do things with a passion and intensity that helps set this school apart from tamer institutions. "Wes students take an in-your-face approach to life," says a government major. "There's an energy on this campus; for me, it's a spirit of creativity and political energy," a sophomore explains. In recent years, a significant number of Wesleyan alumni have gone on to make their mark in the entertainment industry and the high-tech world on the West Coast.

This New England college offers more academic and extracurricular options than almost any school its size, and the Wesleyan experience means liberal learning in a climate of individual freedom. "The courses are very rigorous but the academic climate is rather laid-back," an American studies major says. The freedom at Wesleyan requires motivated students who stay on task despite the laid-back atmosphere. There are abundant opportunities open to students willing to take advantage of them, which is precisely what these doers do.

It begins with the Wesleyan campus architecture, which is as diverse as the student body. The nucleus of this stately university is a century-old row of lovely ivy-covered brownstones that look out over the football field. The rest of the buildings can be described as "eclectic" and range from mod-looking dorms of the '50s and '60s to the early-19th-century architecture of many academic buildings to the beautiful and modern Center for the Arts. The Wesleyan-owned student residences look freshly plucked from Main Street, U.S.A. The most recent additions to campus include classroom renovations and the initiation of a 10-year program to renovate facilities and classrooms across the campus, including an addition to the athletic center, the Center for Film Studies, and a new university center.

Wesleyan has used its considerable wealth to attract highly rated faculty members who are expected to be scholar-teachers: academic superstars who juggle ground-breaking research, enthusiastic lectures, and personal student attention at the same time—and they seem to pull it off. "The teaching here is incredibly engaging," says a senior. Among Wesleyan's strongest departments are music, astronomy, economics, molecular biology and biochemistry, American studies, earth and environmental sciences, and English, which for years has been the most popular degree. But even the smaller departments attract attention. Ethnomusicology, including African drumming and dance, is a particular specialty; students can be found reclining on the wide, carpeted bleachers at the World Music Hall or watching a dozen musicians play the Indonesian gamelan. The film department is first-rate. The East Asian Studies Center has both a strong program and an authentic Japanese tea room. The math department emphasizes problem solving in small groups rather than interminable lectures dedicated to theory. Undergraduates in the sciences work alongside faculty in their research laboratories and frequently earn the opportunity to publish in scientific journals.

> "There's an energy on this campus; for me, it's a spirit of creativity and political energy."

Wesleyan's curriculum renewal program ensures the relevance of liberal arts education in the 21st century by adding focused inquiry courses for first- and second-year students, clustering courses to help students reach their academic goals, and requiring an electronic portfolio from each student that allows students to compile their work and set goals with their advisers. Students are expected in their first two years to take a minimum of two courses in each of three areas—humanities and the arts, social and behavioral sciences, and natural sciences and mathematics. During their second two years, students must take one course in each of the three areas. At the end of their freshman year, Wesleyan students can apply to major in one of three competitive, interdisci-plinary seminar colleges: the College of Letters (literature, politics, and history), the College of Social Studies (politics, economics, history), and the Science in Society program (concerned with the humane use of scientific knowledge, *a la* Buckminster Fuller). New students can take First-Year Initiative courses designed just for them. The university implements a Web-based course selection and registration system and a pre-registration system allocating course slots according to student preferences, class year, and major requirements.

> "The teaching here is incredibly engaging."

Wesleyan students are marked by an unusual commitment to debate, from political to cultural to intellectual. "One thing about Wesleyan, people aren't afraid to speak their mind or challenge someone else's idea," a student reports. "This creates an environment that is constantly debating and discussing things." It is hands-down an activist campus. One sophomore lists rallies and protests alongside parties and concerts as part of the campus social life. The politics generally take a leftist bent. "We are liberal. Period," says a music major. "Conservatives and intolerant individuals will not do well here." But students level their smarts against topics close to home, mostly the university administration.

Wesleyan strives to keep its classes small and only 2 percent of the courses have 50 or more students. Some students claim they sometimes have trouble getting into

(Continued)

Academics: ✎ ✎ ✎ ✎
Social: ☎ ☎ ☎
Q of L: ★ ★ ★
Admissions: (860) 685-3000
Email Address:
admissions@wesleyan.edu

Strongest Programs:
Astronomy
Classical Studies
East Asian Studies
Earth and Environmental
Sciences
Economics
English/Creative Writing
Molecular Biology and
Biochemistry
Music
Film

The ultimate Frisbee club (the "Nietzsch Factor," named after a former star player's dog, not the philosopher) almost always whips challengers, and intramurals are extremely popular.

the "hot" courses. "With popular classes, you have to be persistent, but you can get in," a history major says. If beseeching is not your style, studying abroad may be a temporary tonic to registration headaches. Programs are available in Israel, Germany, Africa, Japan, Latin America, France, Spain, and China. Students can also participate in the Venture Program,* study at Mystic Seaport,* or take a semester at another Twelve College Exchange* school. Internships are popular, and students can also take advantage of 3–2 engineering programs with Columbia and Caltech.

Wesleyan likes to describe itself as "a small college with university resources." The libraries have more than a million volumes, practically unheard of at a school this size. Students claim that whenever you happen to walk past the brightly lit, glass-walled study room of Sci-Li (the science library), you're apt to see numerous students huddled over their books. Wesleyan's excellent reputation and strong recruiting network attract students from all over, ensuring the clash of viewpoints that makes it such a vital place. "Here you'll find everything from your eccentric personalities to your artists, your athletes and your fashion conscious, and a good number of common, everyday people," says a senior. The student body is 7 percent African American, 7 percent Hispanic, and 10 percent Asian American. Students report that diversity is cherished

"With popular classes, you have to be persistent, but you can get in."

at Wes. "Wes students are tolerant because there's an open dialogue both in and outside of the classroom that creates a high level of awareness and acceptance," says a student. Sixty-six percent graduated in the top tenth of their high school class. No academic or athletic merit scholarships are offered, but Wesleyan does guarantee to meet the financial need of all admits. Freshman orientation, which gets rave reviews, consists of a week of standard preregistration fare, plus comedy nights, movies, and square dancing.

For housing, most freshmen are consigned to singles or doubles in the campus dorms. Popular opinion indicates that the Butterfield complex is the choice for quiet study, while Clark Hall is where the party people go. Ninety-four percent of undergraduates live in university housing, and housing is guaranteed for four years. Juniors and seniors enjoy numerous housing options: townhouses for four or five students, fraternities, college-owned houses and apartments, or special-interest houses organized around concerns such as ecology, feminism, or minority student unity. Upperclass students who want to live off campus must apply for permission. Those who move off have the option of eating at home, in the school grill, or at the fraternity eating clubs. Everyone else takes meals in the new university center.

Middletown is a small city within easy driving distance of Hartford and New Haven, but it is off the beaten track of steady public transportation. It has undergone a renaissance in recent years, and students love the myriad of ethnic restaurants available. Wes students contribute a great deal of time to community service, and help maintain a peaceful, beneficial relationship with the town. And Wesleyan's rural surroundings afford the much-appreciated opportunity to jog through the countryside, swim at nearby Wadsworth Falls, or pick apples in the local orchards. Good road trips include New York and Boston, each two hours away, and decent ski areas and beaches just under an hour away. Wesleyan sports tend to be for scholar-athletes rather than spectators. In any sport, annual encounters with "Little Three" rivals Williams and Amherst get even the most bookwormish student out of the library and into the heat of the action. The ultimate Frisbee club (the "Nietzsch Factor," named after a former star player's dog, not the philosopher) almost always whips challengers, and intramurals are extremely popular. Athletics are enhanced by a complex that comes complete with a 200-meter indoor track, a fitness center, and a 50-meter pool.

Although two former fraternities have turned into co-ed literary societies, Greek life at the remaining three is a jock preserve. Only 5 percent of the men and 1 percent of the women go Greek. Wesleyan's enforcement of the 21-year-old drinking age is moderate compared with most schools. Consistent with the university's encouragement of independence, students bear a large part of the responsibility for policing themselves. "To the university's credit, they are really stressing alcohol awareness, so to speak," says a neuroscience and behavioral studies major. Students who throw a party for 75 or more people must attend a workshop that stresses safe drinking. Still, most students concur that it is quite easy for the underaged to imbibe. "You can get alcohol if you want to," notes a sophomore. It seems, however, that drinking alcohol is a minor event when compared to the multitude of other things happening on campus. Activities abound from comedy performances to a cappella groups, films, plays, bands, lectures, parties, and events planned by the more than 200 student groups. "All social life takes place on campus," says a senior, "with shows and concerts and parties all in on-campus venues." Major events on the social calendar include Spring Fling and Fall Ball—two outdoor festivals—and Uncle Duke Day and Zonker Harris Day, two similar events with a more psychedelic, 1960s flavor, in which students pay tribute to the infamous Doonesbury characters.

With so much to do and learn, students at Wesleyan take it all in stride. The key to Wesleyan's success seems to be the fostering of an intellectual milieu where independent thinking and an appreciation of differences are omnipresent. So different, in fact, that a sophomore describes the typical Wesleyan student as "sarcastic, dramatic, poetic, athletic, creative, proactive, open-minded, caring, loud, and polite." And if you're still not convinced, take it from a veteran: "We would make any liberal arts college proud, but you can only find us here."

> ### Overlaps
>
> **Brown, Yale, Columbia, Amherst, Harvard**

If You Apply To ➤ **Wesleyan:** Early decision: Nov. 15, Jan. 1. Regular admissions: Jan. 1. Financial aid: Feb. 15. Guarantees to meet demonstrated need. Campus and alumni interviews: recommended, evaluative. SATs and SAT IIs (writing and two others) or ACTs: required. Accepts the Common Application and electronic applications. Essay question: influential person and why, intellectual experience, and community influence.

West Virginia University

P.O. Box 6009, Morgantown, WV 26506-6009

Many are drawn to WVU for its one-of-a-kind forensics program. The honors program is a must for top students, and the university has solid programs in professional fields ranging from journalism to engineering.

West Virginia University earned the right to be the state's flagship land grant college by being the only one in the state to offer research and doctoral-degree programs. With 178 degree programs, approximately 270 student organizations, and 16 intercollegiate varsity athletic programs, it's no wonder students flock to this large university. With strong academic programs and student organizations, the school has become a solid choice for scholars, researchers, and athletes, as well as party animals.

WVU is situated in the picturesque mountains of north-central West Virginia, a few miles from the Pennsylvania border and overlooking the Monongahela River. A driverless rail system bridges the school's two campuses—the older Morgantown and the more modern Evansdale, which are a mile and a half apart. Ten of the ivy-covered Morgantown buildings, dating mainly from the 19th century, are listed on the National Register of Historic Places; many of their interiors have been restored

Website: www.wvu.edu
Location: Small city
Total Enrollment: 26,051
Undergraduates: 19,510
Male/Female: 54/46
SAT Ranges: V 470–560
    M 480–580
ACT Range: 20–26
Financial Aid: 49%
Expense: Pub $
Phi Beta Kappa: Yes

(Continued)

Applicants: 10,957
Accepted: 92%
Enrolled: 45%
Grad in 6 Years: 54%
Returning Freshmen: 81%
Academics: ✏️✏️
Social: 🐻🐻🐻🐻
Q of L: ★★★
Admissions: (304) 293-2121
   or (800) 322-WVU1
Email Address:
   go2wvu@mail.wvu.edu

Strongest Programs:
Forensic and Investigative
   Science
Biometric Systems
Engineering
Political Science
Pharmacy
Psychology
Allied Health

Operation Jump-Start helps students adjust to college with New Student Convocation, dorm-based Freshman Interest Groups, and Resident Faculty Leaders, who live next door to the dorms and serve as mentors and friends.

There are hundreds of intramural teams, the most popular being basketball, flag football, dodgeball, and indoor soccer.

or renovated. Campus construction is booming, with new facilities—including two residence halls and a 38,000-square-foot agricultural sciences building—and renovations dotting the landscape.

West Virginia's degree programs span 13 schools, the best of which are engineering (particularly energy-related) and the allied health sciences (medical technology, physical therapy, nursing, and occupational therapy). The most popular majors are business, engineering, journalism, biology, and the health fields. Regardless of major, students must complete the General Education Curriculum (GEC), designed to ensure all students have a foundation of skills and knowledge necessary to reason clearly, communicate effectively, and contribute to society. Graduates are expected to possess knowledge and experience in nine objective areas—communication, math and science, issues of contemporary society, history, artistic expression, the individual in society, American culture, non-Western culture, and Western culture. New programs include undergraduate majors in management information systems, business for foreign languages, art history, music, and theater.

> **"The overall academic climate is generally laid-back, but certain classes and a segment of the student population create a competitive climate."**

While WVU is a big school, administrators say that 76 percent of classes taken by freshmen have 50 or fewer students. Students say the difficulty of WVU academics depends largely on the classes they take. "The overall academic climate is generally laid-back," says a senior, "but certain classes and a segment of the student population create a competitive climate." Operation Jump-Start helps students adjust to college with New Student Convocation, dorm-based Freshman Interest Groups, and Resident Faculty Leaders, who live next door to the dorms and serve as mentors and friends. A one-credit course, University 101–First Year Experience, also helps students understand the academic, social, and emotional expectations of the college experience, covering study skills, university and community support services, goal setting, and career planning. On the other end of the spectrum, an honors program offers small classes, special housing, and early registration to the top 5 percent of WVU students.

Though WVU attracts students from all U.S. states and nearly 90 countries, it is primarily regional. Fifty-seven percent of undergrads are in-staters, and a sizable contingent arrives from western Pennsylvania and southern New Jersey—so many that the university has been dubbed "New Jersey University: West Virginia campus." Says one student: "WVU has a unique conglomerate of students, ranging from Appalachian natives to a lot of inner-city NYC kids and New Jersey natives." African Americans comprise 4 percent of the student body, and Hispanics and Asian Americans account for nearly 2 percent each. The school has produced 25 Rhodes Scholars, 29 Goldwater Scholars, and 17 Truman Scholars. The university offers merit scholarships averaging $3,036, along with 246 athletic scholarships.

Twenty-seven percent of WVU's undergraduates live on campus, where dorms are mediocre but fill up fast because of the increase in students flocking to the university. Most are co-ed; the older ones are known for their character, while the newer residential complexes on the Evansdale campus have larger rooms and luxuries like air-conditioning. Many upperclassman opt for nearby apartments. Nine percent of the men

> **"This town revolves around the university and provides so much for the students."**

and 10 percent of the women go Greek and may live in their respective chapter houses. Each dorm has its own cafeteria, and students may buy meal plans regardless of where they live. Fraternities and sororities have their own cooks. The biggest complaint is the lack of parking—the campus is so large students say they need cars to get around, but there isn't any place to park. Students can ride the Morgantown

Mountain Line buses for free from 10 p.m. to 3 a.m. Thursday through Saturday. It stops at all university housing.

Morgantown is a small city with a college-town feel and plenty of community service opportunities, students say. "This town revolves around the university and provides so much for the students," says a senior. The school has worked hard to curtail underage drinking, banning alcohol in the dorms and limiting each frat to three social events per semester with 300 members and guests. Students report that it's now near impossible for those under 21 to be served on campus, and say off-campus bars have also cracked down. That said, social life is still centered on campus, often focused on the free food, movies, bands, and comedians offered Thursday through Saturday by the school-sponsored Up All Night program. Spring Fest and Fall Fest provide stress relief each semester, and Mountaineer Week showcases the customs of Appalachia. For those with cars, road trips to Columbus, Washington, D.C., or Pittsburgh are quick and easy. Football rivalries with Syracuse, Maryland, and Pittsburgh (the "Backyard Brawl") take students farther afield. "Every football game is a festival in some way," says a senior.

Aside from Mountaineer football, which has achieved national prominence, and a Big East championship, West Virginia fields very competitive women's soccer, riflery, gymnastics, and basketball teams, and men's soccer, basketball, rowing, and wrestling. There are hundreds of intramural teams, the most popular being basketball, flag football, dodgeball, and indoor soccer.

As WVU grows, the university continues to be dedicated to research that will improve the lives of citizens not only in West Virginia, but across the globe. At the same time, WVU's mission is changing to be more student-centered. And the campus is forever growing and changing. "WVU is a school," says a senior, "that won't look the same for very long."

New programs include undergraduate majors in management information systems, business for foreign languages, art history, music, and theater.

Overlaps

Marshall, Penn State, University of Pittsburgh, University of Maryland, Virginia Tech, Ohio State

---

If You Apply To ➤

West Virginia: Rolling admissions. Financial aid: Mar. 1. Does not guarantee to meet full demonstrated need. Campus and alumni interviews: optional, informational. SATs or ACTs: required. SAT IIs: optional. Accepts the Common Application and prefers electronic applications. No essay question.

Wheaton College (IL)

501 College Avenue, Wheaton, IL 60187

Wheaton is at the top of the heap in evangelical education, rivaled only by Pepperdine (with its Malibu digs) and traditional competitors such as Calvin and Hope. Students pledge to live a Christian life. Wheaton's low tuition makes it a relative bargain.

Wheaton College combines academic rigor and evangelical orthodoxy with a firm commitment to the liberal arts, preparing students "to help build the church and improve society worldwide...For Christ and His Kingdom." The Community Covenant prohibits the use of alcohol, tobacco, or drugs; professors are allowed to drink and smoke, but are discouraged from doing so, especially in front of students. Though most adolescents would chafe under such restrictions, Wheaties take it all in stride. "Wheaton students are required to have a personal relationship with Jesus Christ, so everything centers around that fact," says a freshman. "We like to have good, clean fun."

Website: www.wheaton.edu
Location: Suburban
Total Enrollment: 2,697
Undergraduates: 2,365
Male/Female: 50/50
SAT Ranges: V 620–720
　M 610–700
ACT Range: 27–31

(Continued)

Financial Aid: 53%

Expense: Pr $

Phi Beta Kappa: No

Applicants: 2,115

Accepted: 55%

Enrolled: 49%

Grad in 6 Years: 84%

Returning Freshmen: 94%

Academics: ✑ ✑ ✑

Social: ☎ ☎ ☎

Q of L: ★ ★ ★ ★

Admissions: (630) 752-5005
or (800) 222-2419

Email Address:
admissions@wheaton.edu

Strongest Programs:
Biology
Business/Economics
English
Music
Bible
International Relations
Political Science
Psychology

Favorite traditions include Missions in Focus week, which brings missionary organizations and Christian speakers to campus, and the individual dorm floors own traditions, one of which includes an annual root beer kegger.

Wheaton is non-denominational and its verdant, 80-acre campus is an oasis of sorts, in the midst of one of Chicago's oldest and most established suburbs. The castle-like Blanchard Hall, built in the last century, keeps watch over the community from atop the front campus hill; when couples get engaged, they climb to the top of the tower to share their news by ringing the bell. Nearby is the $13.5 million Billy Graham Center, a museum and library that has become a hub for research on American evangelicalism. (Graham is a Wheaton alumnus, as is former U.S. House of Representatives Speaker Dennis Hastert, R-IL.) The $21 million Todd M. Beamer Student Center, named for an alumnus who died aboard United Flight 93 on Sept. 11, 2001, was finished in 2004.

General education requirements at Wheaton include intermediate-level foreign language study; proficiency in quantitative skills, writing, and oral communication; and varying amounts of credit hours in nine thematic areas: kinesiology, philosophy, Biblical studies, world history, social science, lab science, natural science, literature, and art and music. In the classroom, there's an emphasis on teamwork, and "while Wheaton students are passionate about serving Jesus with their lives, they are also bright and driven to succeed." Another student says Christian perspective is integrated across the curriculum: "The professors seem to love the subjects they teach, and want to pass this love on to their students. They also take their relationship with God seriously." And according to a senior, "Challenge is the goal, and growth the result. These professors are experts in their fields, and oftentimes literally wrote the book pertaining to the coursework."

Business and economics is Wheaton's most popular major, followed by music, communication, and psychology. The arts and science facilities will be updated as part of a giving campaign, and students have the opportunity to perform research at Argonne National Laboratory, just down the road. Wheaton has added a gender studies certificate; for those seeking a truly global experience, there's study abroad in East Asia, England, France, Germany, Spain, Russia, Latin America, and Israel.

Those concerned with social justice, a definite focus at Wheaton, may be interested in the "life-changing" Human Needs and Global Resources (HNGR) program, which sends students to Third World countries for six months of work on development projects such as roads and schools. Wheaties may also spend a semester at one of the 12 other evangelical schools that belong to the Council for Christian Colleges and Universities.* Motivated students may opt for 3–2 programs that allow them to combine an undergraduate degree in the liberal arts with a master's in nursing or engineering, saving a year in the process. Summer study in botany and zoology is available at the Black Hills Science Station, while leadership training takes place at Honey Rock, Wheaton's campus in the Wisconsin North Woods. For incoming freshmen, the 18-day Wheaton Passage wilderness trek provides "an experience in team-building, self-discovery, and physical challenge that should not be missed," says a sophomore.

> **"While Wheaton students are passionate about serving Jesus with their lives, they are also bright and driven to succeed."**

Twenty-nine percent of Wheaton students hail from the Land of Lincoln, and 82 percent graduated in the top quarter of their high school class, a fact that adds to both the intellectual energy and the academic pressure on campus. Caucasian students comprise 87 percent of the student body and African Americans a little more than 2 percent. Hispanics make up 3 percent, Asian Americans more than 7 percent, Native Americans less than 1 percent and foreign students 1 percent. Important social and political issues include the AIDS crisis, abortion, sustainable development, civil rights, and the war in Iraq. "There is a good dialogue ongoing, and multiculturalism is continuously discussed," says one student. Many students lead church youth groups, and others get involved in the community through tutoring or volunteer work at homeless shelters, prisons, or hospitals.

Eighty-eight percent of Wheaton students live in the single-sex dorms, and accommodations range from traditional double rooms with bathrooms down the hall to college-owned houses and apartments. Opposite-sex visitation is limited to certain hours on certain days, though "each semester, dorms are allowed two 'raids' to their opposite floor," says an education major. While everyone is guaranteed a bed, "housing is determined by lottery, so there can be some surprise in where you end up," says a sophomore. That said, "Most people get their first or second choice." Everyone eats in Anderson Commons, where an outside vendor called Bon Appetit does the cooking. "They offer such diversity—Cajun, Asian, Italian, pizza, sandwiches," says a sophomore. "There is always a salad bar, pasta station, meat station, vegetarian station—and of course, ice cream and dessert!" a freshman adds.

Since Wheaton lacks fraternities and sororities, and because students agree to abstain from alcohol, drugs, and tobacco, the social life revolves around pursuits such as movies, on-campus dances, live music, and coffeehouse evenings. A commuter train near campus whisks students to downtown Chicago in 45 minutes, where restaurants, blues clubs, theaters, museums, shopping, and professional sports are in abundance. Favorite traditions include Missions in Focus week, which brings missionary organizations and Christian speakers to campus, and the individual dorm floors own traditions, one of which includes an annual root beer kegger.

Football, basketball, and soccer games "are always exciting and well-attended," says one student, especially if the opponent is Augustana, which is Lutheran. And while it's not an athletic competition *per se*, juniors and seniors do get excited about decorating "the Bench," a reinforced concrete slab that is the subject of an ongoing and often rough-and-tumble game of keep-away.

> "There is a good dialogue ongoing, and multiculturalism is continuously discussed."

Wheaton's teams, now known as the Thunder, compete in Division III, and the school fields its best squads in men's and women's soccer, basketball, and swimming. The name change occurred after critics claimed the old moniker—the Crusaders—glorified medieval Christians who killed thousands of people in the name of Jesus. The new mascot was chosen from 1,300 suggestions because it sounds strong and is one of the natural phenomena associated with God. For recreational players, the sports center boasts new basketball courts, a climbing wall, and a weight-lifting facility.

Even after 147 years, Wheaton remains "committed to the principle that truth is revealed by God through Christ, in whom is hidden all the treasures of wisdom and knowledge." While this kind of education isn't for everyone, students believe their school's dedication to Christianity only strengthens the bonds they develop with one another—and their understanding of the broader world.

Wheaton has added a gender studies certificate; for those seeking a truly global experience, there's study abroad in East Asia, England, France, Germany, Spain, Russia, Latin America, and Israel.

Overlaps

Taylor, Calvin, Northwestern, Grove City

If You Apply To ➤

Wheaton: Early action: Nov. 1. Regular admissions: Jan. 10. Financial aid: Feb. 15. Campus interviews: optional, evaluative. No alumni interviews. SATs or ACTs: required. SAT IIs: optional. Accepts electronic applications. Essay question: your personal experience of coming to faith in Christ, and how your faith has developed and grown; how culture or family history has shaped who you are; what draws you to your desired area of study; issue of local, national, or international concern; experience in which your values were tested.

Wheaton is the most recent convert to coeducation among prominent East Coast institutions, and women still outnumber men by nearly two to one. Although it's in Massachusetts, Wheaton is actually closer to Providence than Boston. One of the few moderately selective liberal arts institutions in the area.

Website:
www.wheatoncollege.edu
Location: Suburban
Total Enrollment: 1,555
Undergraduates: 1,555
Male/Female: 38/62
SAT Ranges: V 600-690
M 580-660
ACT Range: 26-30
Financial Aid: 49%
Expense: Pr $ $ $ $
Phi Beta Kappa: Yes
Applicants: 3,697
Accepted: 44%
Enrolled: 28%
Grad in 6 Years: 72%
Returning Freshmen: 86%
Academics: ✐ ✐ ✐
Social: ☎ ☎ ☎
Q of L: ★ ★ ★ ★
Admissions: (508) 286-8251
Email Address: admission
@wheatoncollege.edu

Strongest Programs:
Social Sciences
Fine Arts
Humanities
Chemistry
Italian
French
Mathematics
Computer Science

Wheaton College first admitted men in 1988 and continues to attract students with solid programs in the social sciences, fine arts, and humanities. Selection has gotten more competitive over the years, with each incoming class even stronger than the one that came before. When students aren't hitting the books, they can be found socializing or enjoying one of the college's many unique traditions. For example, "only seniors can go through the front door of the chapel or sit on the front steps of the library," says a sophomore. "The students at Wheaton are motivated and involved," says a junior. "Everyone has something they are passionate about."

Wheaton's rural location offers few distractions from intellectual pursuits. Its 385-acre campus blends Georgian brick buildings and modern structures set among beautiful lawns and shade trees. The two halves of the campus are separated by Peacock Pond, which probably qualifies as the only heated duck pond on any American campus. Wheaton completed a $20 million arts project in recent years, culminating in new and renovated space for studio art, art history, music, and theater arts. An additional $8 million has been spent on Beard Hall, a 100-bed dormitory, which boasts wireless Internet connections throughout. Improvements have also been made on landscaping and existing dormitories.

> **"Everyone has something they are passionate about."**

Students say Wheaton's best programs include psychology, political science, English, international relations, economics, and history. Programs in the arts are well recognized, impressive given the school's small size, and the chemistry department is also strong, producing Wheaton's Rhodes Scholarship winner and two Fulbright scholars. The major in Hispanic studies benefits from its affiliation with a study abroad program in Cordoba, and a Mellon Foundation grant supports native speakers in Spanish and other languages. Those interested in interdisciplinary or crossdisciplinary work may design an independent major or add classes taken at other schools in the Twelve College Exchange Program.* Students may also take classes not offered at Wheaton at nearby Brown University. For those tired of studying on land, Wheaton offers the Maritime Studies Program.* Dual-degree programs offer motivated students the chance to earn a bachelor's degree in engineering or graduate degrees in business, communication, religion, optometry, or studio art in conjunction with schools ranging from Dartmouth to the School of Museum of Fine Arts in Boston.

Wheaton's curriculum shows students how to make linkages across disciplines. Rather than simply checking off required courses, all Wheaton students study across the major disciplines, developing a fully dimensional view of the world. In practical terms, this means that every student must take a series of courses on a single topic from various departments. The curriculum links experiential learning to each department and requires a capstone senior project. Core classes include English, quantitative skills, foreign language, natural science, and non-Western history. Students also choose a first-year seminar from among 25 sections, each focused on "controversies" that have generated debate or heralded changes in how they experience or understand the world.

Wheaton's small classes encourage close ties between students and faculty. "The professors are understanding, intellectually stimulating, and push students to perform their best," a student says. Aside from a faculty advisor, students get a staff mentor and two peer advisors, known as preceptors. "The courses are challenging and usually stimulating," one student explains. There is little competition, although "junior and senior year can get hectic because of scholarship competitions," a women's studies major says. The Filene Center for Work and Learning gets high marks from students seeking internships—and jobs after graduation.

About a third of Wheaton's students come from Massachusetts, and 63 percent went to public high schools. The student body is largely white; African Americans account for 4 percent, Hispanics 5 percent, and Asian Americans 4 percent. Students are friendly and open to differing views. "There are preps, jocks, and hippies," explains a psychology major. "There really isn't one type of student who attends Wheaton." Politically, liberals and conservatives are both well represented on campus. The administration has undertaken a massive capital campaign, using some of the proceeds to hire more minority scholars—and recruit more minority students. Thirty-one percent of all undergraduates receive merit awards averaging $10,760. There are no athletic awards.

Social life at Wheaton includes dances, concerts, lectures, and parties on campus, or road trips to Boston (35 miles away) and Providence (15 miles).

As might be expected on this small, secluded campus, virtually everyone (96 percent) lives in one of Wheaton's dorms or houses. Gebbie Hall regularly hosts panels, presentations, and colloquia on gender issues. All students are guaranteed housing for four years; freshmen live in doubles, triples, or quads, and upperclassmen try their luck in the lottery system.

"There really isn't one type of student who attends Wheaton."

"The older dorms have a lot of character and are quite pretty," says a senior, although "housing 'crunches' are common." The bright and spacious dining halls offer unlimited chow, and the biggest winners of all are the ducks, which thrive on the leftover bread students toss into Peacock Pond. Students say they feel safe on campus and that security is visible and active. "They patrol 24 hours a day to ensure things are okay," says a sophomore.

Social life at Wheaton includes dances, concerts, lectures, and parties on campus, or road trips to Boston (35 miles away) and Providence (15 miles). The town of Norton, just outside campus, draws some students with a Big Brother Big Sister program, hospital visits, and opportunities to tutor and mentor children, but there's little to do otherwise, students say. "The town seems to resent Wheaton," says a senior. "Luckily, inside the Wheaton bubble there is always something going on, and Providence and Boston are not far," adds a student. Students head to the college's student center, which offers a cafe, dance studio, and sun deck for afternoon study breaks. There are no sororities and fraternities here, which helps cut down on underage drinking, though students say that as on most campuses, those who want to find alcohol will do so.

As might be expected on this small, secluded campus, virtually everyone (96 percent) lives in one of Wheaton's dorms or houses.

When they're not immersed in classwork, Wheaton students love to put on their dancing shoes—whether for the Boston Bash party on a boat in Boston Harbor, for the Valentine's Dance, or for any number of events at Rosecliff, a mansion in Newport, Rhode Island. Spring Weekend features the Head of the Peacock boat race, where students build ships and race them across the pond, as well

"Inside the Wheaton bubble there is always something going on, and Providence and Boston are not far."

as live bands and outdoor barbecues. When it rains, students get dirty as they slide around the craters on Wheaton's lawns, an activity known as "dimple diving."

Wheaton competes in Division III of the NCAA, and its strongest teams include men's and women's soccer—both of which have won numerous championships. Men's and women's indoor track, baseball, softball, and men's and women's lacrosse

Overlaps

Skidmore, Connecticut College, Clark, Boston University, Bates, Brandeis

are also competitive. The athletic facility boasts an eight-lane swimming pool, a field house, and an 850-seat arena for basketball or volleyball. Approximately 45 percent of the student body participates in intramural sports.

Students at Wheaton understand that there is an education that needs to be attained in college in addition to an academic one. That explains why so many students are involved in campus planning and college operations, as well as in their community. Students here take pride in their achievements inside and outside the classroom, while striving to preserve the school's friendly, small-town feel.

If You Apply To ➢

Wheaton: Early decision: Nov. 15, Jan. 15. Regular admissions: Jan. 15. Financial aid: Feb 1. Campus and alumni interviews: recommended, evaluative. SATs or ACTs: optional. SAT IIs: optional. Accepts the Common Application and electronic applications. Essay question: significant experience or achievement; issue of personal, local, national, or international concern; influential person; influential fictional character, historical figure, or creative work; or a topic of your choice.

Whitman College

345 Boyer Avenue, Walla Walla, WA 99362-2083

Whitman has quietly established itself as one of the West's leading liberal arts colleges. Don't sweat the umbrella: Walla Walla is in arid eastern Washington. Whitman's isolation breeds community spirit and alumni loyalty. True to its liberal arts heritage, Whitman has no business program.

Website: www.whitman.edu
Location: Small city
Total Enrollment: 1,455
Undergraduates: 1,455
Male/Female: 46/54
SAT Ranges: V 620–720
  M 620–690
ACT Range: 28–32
Financial Aid: 45%
Expense: Pr $ $
Phi Beta Kappa: Yes
Applicants: 2,821
Accepted: 47%
Enrolled: 30%
Grad in 6 Years: 86%
Returning Freshmen: 94%
Academics: ✐ ✐ ✐ ✐
Social: ☎ ☎ ☎
Q of L: ★ ★ ★ ★
Admissions: (509) 527-5176
Email Address:
  admission@whitman.edu

Strongest Programs:
Politics

You don't have to own a Frisbee to succeed at Whitman, but if you've got one, bring it along—you'll find a campus full of students eager to toss it back to you. Though it isn't well known outside the Pacific Northwest, Whitman offers a top-notch liberal arts education, along with plenty of fun for outdoorsy types. Students are down-to-earth and friendly and feel a deep loyalty to one another—and to their school. "You never know what to expect from students here," says a sociology and music theory major. "All of them are intelligent in their own way, but individual activities differ. I have debate friends, lazy friends, and artistic friends, but we all unite in the classroom."

Whitman was founded in 1882 and named in honor of Marcus and Narcissa Whitman, who served the Cayuse Indians and immigrants on the Oregon Trail. Even today, everything important is within walking distance of campus, including the main drag of Walla Walla ("Walla Squared"), which recently won a national Best Main Street award. The 117-acre campus, where Colonial buildings and modern facilities sport New England ivy, sits at the foot of the Blue Mountains, surrounded by golden wheat fields and vineyards. Beyond Walla Walla (which means "many waters" in the Cayuse Indian language) are gorgeous mountains, rivers, and forests. New campus additions include the state-of-the-art Baker Ferguson Fitness Center and Harvey Indoor Pool and Welty Health and Wellness Center, which is open 24/7. A new visual arts center will open in 2008.

> **"I have debate friends, lazy friends, and artistic friends, but we all unite in the classroom."**

All Whitman students complete the General Studies Program, which includes both a first-year core and distribution requirements in various disciplines. The first-year program revolves around a seminar in "Antiquity and Modernity" that emphasizes analytical reading of texts—such as Plato's *Symposium*, Homer's *Odyssey*, the

Bible, and Mary Shelley's *Frankenstein*—as well as effective writing. Students then take at least six credits in social sciences, humanities, fine arts, and science, as well as three or more credits in quantitative analysis, and two courses that focus on alternative voices. Seniors must pass comprehensive written and oral exams in their major; Whitman is the first U.S. college or university to require seniors to do so.

Academics are rigorous, but students say the competition is internal. "We help each other by forming study groups," says an environmental studies and geology major. "Last semester, I wrote 26 papers for a single English course." Biology, English, psychology, and politics are some of the best (and most popular) departments at Whitman; the school also boasts an astronomy program, unusual among small colleges. Though Whitman lacks a business program, its alumni include John Stanton, a cell phone pioneer who made millions of dollars building and selling VoiceStream and then Western Wireless. The newest majors include gender studies; religion, race, and ethnic studies; and rhetoric and film studies; there is also a new minor in Latin American Studies. Whitman also has an extensive Asian art collection, and additional coursework in Chinese language and Asian studies is offered through a summer program in China. One unusual offering is the Parents Core, which allows parents to do the same readings as their children in core classes, join in online discussions, and get together twice a year for face-to-face meetings.

(Continued)
English
Psychology
History
Biology

Whitman boasts an astronomy program, unusual among small colleges.

Whitman has 3–2 or 3–3 programs in engineering and computer science (with the University of Washington, Caltech, Columbia, Washington University, and Duke), international studies and international business (Monterey Institute of International Studies), computer science and oceanography (University of Washington), forestry or environmental management (Duke), and law (Columbia). Students may also pursue teacher certification through partnerships with the University of Puget Sound.

"The stereotypical student here is a white, Frisbee-playing, indie-rock-loving hippie."

Foreign study is available in countries ranging from Japan and Sri Lanka to England, Italy, Ireland, and Egypt, and students may take terms within the U.S. at the Chicago Urban Studies Program, the Philadelphia Center, the Washington Semester, and Biosphere 2. "Semester in the West is the biggest up-and-coming program," adds a psychology major. "Students travel for one semester along the West Coast, learning from different professors in variable environments." Speaking of those professors, "If I ever had any questions, they would make time in their day to meet with me and help," one student says. "I have been to three of my professors' homes, and have had lunch with my core professor."

"The stereotypical student here is a white, Frisbee-playing, indie-rock-loving hippie," says a student, "but of course we are not all like that." Forty-one percent of Whitties are from Washington State, and many of the rest hail from the suburbs of Western cities, notably San Francisco and Portland. Nine percent of the student body is Asian American, 4 percent is Hispanic, and 3 percent is African American. "Politics lean left, but most students find themselves more towards the middle," says a junior. There are numerous merit scholarships, averaging more than $8,000 each, but no athletic awards.

One unusual offering is the Parents Core, which allows parents to do the same readings as their children in core classes, join in online discussions, and get together twice a year for face-to-face meetings.

Seventy-five percent of students live in campus housing; freshmen and sophomores are required to do so, and many others stay because "frat houses and our 11 theme houses count, too," says a sociology and music theory major. "Halls are recently renovated, comfortable, and not trashy." Prentiss Hall and Lyman House, both built in 1926, have received multimillion-dollar face-lifts; theme houses are available for students interested in foreign languages, fine arts, writing, community service, Asian studies, environmental studies, multiculturalism, and the outdoors. Dining halls are run by Bon Appetit, and students say the company is responsive to special requests. "Food here is really good," says a freshman.

The social life at Whitman revolves around fraternity parties—34 percent of the men and 32 percent of the women go Greek—and other on-campus events, such as theatrical productions (the drama department stages about a dozen a year) and the spring Renaissance Faire. There are also free movies and a weekly student coffeehouse with live music. The town of Walla Walla supports a symphony, community playhouse, art galleries, two rodeos, and a hot-air balloon festival. "Downtown is two blocks away—there is a tiny Bon Marche, and mostly family-owned shops," says one student. "Traffic signals turn off at night because there are so few drivers." Annual traditions include Duckfest, in which students create ducks, place them all over campus, and then design a map indicating where they are, so that peers, professors, and community members can take a walking tour—and college president George Bridges can rank the best efforts. Dragfest is a popular dance party, and the end of the year brings "the Beer Mile, where students celebrate on Ankeny Field, and the administration is thankful we're safe," says a junior. That said, "social life does not revolve around drinking, which is awesome," says an environmental studies major.

Outdoor pursuits are important in this part of the country, where autumn is gorgeous, winter sporadically snowy, and spring delightfully warm. Walla Walla (population 30,000) is located in the center of agricultural southeastern Washington, in an arid valley. Hiking, biking, and backpacking are minutes away, and white-water rafting and rock climbing

> "Social life does not revolve around drinking, which is awesome."

are popular on weekends. Two ski centers and other recreational areas are within an hour's drive, and Seattle (260 miles) and Portland (235 miles) offer a welcome change of scenery.

Whitman scrapped varsity football in the late 1970s, but the "Missionaries" maintain an active interest in physical activity. At the varsity level, teams compete in Division III; men's tennis and women's basketball won Northwest Conference titles in 2003–2004. Club lacrosse and rugby tournaments draw crowds, as does "Onionfest," the ultimate Frisbee competition, and "Anchor Smash," the spring intramural football tournament. Rock climbers can challenge themselves on two walls.

Students seeking a solid liberal arts education with a healthy dose of outdoor fun would do well to consider Whitman. Professors here know their stuff and care about teaching and their students. Combine the college's beautiful campus with the close friendships nurtured by its small size, and it's easy to see why Whitties remain loyal to each other—and their school—long after they've left campus.

Overlaps

University of Washington, Pomona, Colorado College, Carleton, University of Puget Sound, Willamette

If You Apply To ➤

Whitman: Early decision: Nov. 15, Jan. 1. Regular admissions: Jan. 15. Financial aid: Feb. 1. Does not guarantee to meet full demonstrated need. Campus interviews: recommended, informational. No alumni interviews. SATs or ACT's: required. SAT IIs: optional. Accepts the Common Application and electronic applications. Essay question: critical analysis of a book you've read in the last two years, or how would you contribute to diversity at Whitman?

Whittier College

13406 East Philadelphia, P.O. Box 634, Whittier, CA 90608-4413

Whittier's Quaker heritage brings a touch of the east to this suburban campus on the outskirts of LA. Less selective than Occidental and the Claremont Colleges, Whittier lures top students with an arsenal of academic scholarships. Beware the sneaky October 15 deadline for the best of them.

Founded in 1887 by members of the Society of Friends, Whittier College is fast becoming a global training ground. Whittier students can be found all around the world, studying in 30 foreign countries. And when they return to the Whittier campus, they have access to caring faculty and a close-knit environment.

Located just 18 miles away from downtown Los Angeles, the college is perched on a hill overlooking the town of Whittier, California, with the San Gabriel Mountains rising up from the horizon. The 73-acre campus is a pleasant mixture of modern buildings tucked between the red-roofed, white-walled Spanish traditionals. Its landmark building, Deihl Hall, has been updated to include a digital audio/video computer lab for languages. Renovation and expansion of the Bonnie Bell Wardman Library is complete. A four-foot-high granite monument stands on the north campus lawn, honoring Whittier's most famous alum, former president Richard Nixon.

Whittier officially ended its affiliation with the Quakers in the 1940s, but the prevailing spirit of community hearkens back to their traditions. Faculty wins high marks for their concern and accessibility. One junior says, "The climate is competitive in most of the fields but it depends on your major. The courses (mostly) are excellent due to the proficient faculty." Freshman are taught by full professors, "which raises the standards and challenges your thinking," a business major says. "For the most part, the quality of the teaching has been very intense, but very rewarding." While classes are somewhat competitive, students still work together. "Whittier students tend to be open and friendly," says a student majoring in comparative cultures. "You aren't afraid to sit with someone you haven't met before."

Whittier offers its students two major programs: the liberal education program and the Whittier Scholars Program. About 80 percent of the students take the revised liberal education track, in which they fulfill distribution requirements in writing skills, mathematics, natural sciences, global perspectives, comparative knowledge, and creative and kinesthetic performance. The emphasis of the liberal education program is on an interdisciplinary focus, globalism, and critical and quantitative thinking. These liberally educated Whittierans next choose a major from among 26 departments, the strongest and most popular of which include English, biology, psychology, political science, and business administration, and programs that focus on teaching certification.

"You aren't afraid to sit with someone you haven't met before."

Whittier's strongest reputation lies in the Whittier Scholars Program, a path taken by 20 percent of the undergraduates who choose to bypass the traditional liberal education program. They are relieved of most general requirements and start from square one with an "educational design" process. With the help of an academic advisor, the scholars carve their majors out of standard offerings by taking a bit of this and a bit of that. Majors have included such names as symbol systems, visual studies and business, and dynamics of politics and urban life. The program is highly regarded because of the more active role it allows students to play and the freedom it affords them in pursuing their interests. All students, no matter which curriculum they choose, must fulfill a yearlong freshman writing requirement. In an attempt to help freshmen develop both their critical-thinking skills and their ability to communicate clearly in writing, Whittier lets students choose their preferences from a variety of seminars. They are also encouraged to take an additional writing course, mathematics, and lab science during their freshman year. First-years also must attend a series of speakers who discuss topics relevant to student coursework and must take part in the Exploring Los Angeles series, which includes trips to museums and cultural events.

Study abroad options include programs in Denmark, India, Mexico, and Asia, and undergraduates may also take foreign study tours during the January interim. Forty percent of the students come from California, and the rest are from all over

Website: www.whittier.edu
Location: Suburban
Total Enrollment: 1,427
Undergraduates: 1,317
Male/Female: 46/54
SAT Ranges: V 490–590
  M 490–600
ACT Range: 21–26
Financial Aid: 82%
Expense: Pr $ $
Phi Beta Kappa: No
Applicants: 3,120
Accepted: 58%
Enrolled: 19%
Grad in 6 Years: 62%
Returning Freshmen: 72%
Academics: ✍ ✍ ✍
Social: ☎ ☎ ☎
Q of L: ★ ★ ★ ★
Admissions: (562) 907-4238
Email Address:
  admission@whittier.edu

Strongest Programs:
Child Development
Education
Social Work
Psychology
Business
Biology
Political Science
English

Whittier offers its students two major programs: the liberal education program and the Whittier Scholars Program.

the United States and the world. Diversity plays a major role on this campus. While African Americans make up only 5 percent of the students, Hispanic enrollment is

"You can get severely disciplined or expelled for being intolerant."

an impressive 26 percent, and other minorities constitute 16 percent. The Black Student Union and Hispanic Student Association are vocal on campus. An on-campus cultural center focuses on diversity programming and resources. "Tolerance is a big watchword on campus," a junior says. "You can get severely disciplined or expelled for being intolerant." In addition to need-based aid, the college grants some students merit scholarships ranging from $3,000 to $20,000.

Thirty-eight percent of Whittier students seek off-campus shelter, but the Turner Residence Hall entices many students to stay on campus and vie for a chance to get a room with a panoramic view of Los Angeles and access to the campus computer network in every room. Most freshmen are assigned rooms, though Whittier Scholars, athletes, and members of Whittier's social societies tend to cluster in selected dorms and houses. All dorms are equally suited for freshmen, says one student, because each is its own little community. All campus residents must take at least 10 meals at the Campus Inn dining hall, where the food is said to be typical college fare. "The dining facilities are of average quality, but the menu is diverse," a junior says. The Spot (Whittier's popular campus coffeehouse) includes a state-of-the-art nightclub called—what else?—the Club.

Nine social societies (they're not called fraternities or sororities here) attract 12 percent of the men and 17 percent of the women but hardly dominate the social scene. However, their dances, which frequently feature live entertainment, are welcomed by all. For many, entertainment takes the form of road trips, everything from Disneyland to the California beaches. Other common destinations include Las Vegas, Mexico, Joshua Tree, Hollywood, San Diego, and northern California. Whittier itself is a spot for community-minded students to get involved. "Whittier is a friendly town," one junior says. "The college is involved a lot with the community."

Whittier has a fairly strict alcohol policy, and underage drinking is not permitted. Students say it's difficult for underage drinkers to get served at campus events. "Underage students are not allowed to even be in the same proximity as alcohol," a senior says. Popular annual events include a Spring Sing talent show, the football game against archrival Occidental College—dubbed the Battle of the Shoes—and Sportsfest, which is a campuswide competition in which dorms compete in a variety of athletic, intellectual, and wacky games and events. A favorite among students is Mona Kai, a Hawaiian party put on by the Lancer Society, when tons of sand are shipped in for the event. Also favored is the Midnight Breakfast served by professors during second-semester finals. The most important campus landmark is the Rock,

"It's a small, friendly place where everybody knows everybody."

which sits near the front of campus and is given a fresh coat of paint by countless aspiring artists. The beach is a frequent destination, and for nightlife, Los Angeles looms large. The local community, known as Uptown Whittier, offers quaint shops, restaurants, and cobblestone sidewalks, but little in the way of entertainment.

Men's lacrosse is the most successful sports team on campus. Men's football and soccer, women's soccer, softball, and track are the most popular. Facilities have gotten an upgrade, including a new football field and a fully equipped fitness center.

The students at Whittier have created a supportive, intimate environment where people work together and celebrate their diversity. "It's a small, friendly place where everybody knows everybody," a junior math major says. With the opportunity to design their own majors, students are active in their own education.

Whittier: Rolling admissions. Early action: Dec. 1. Regular admissions: Feb. 1 (priority). Financial aid: Feb. 15. Does not guarantee to meet demonstrated need. Campus interviews: recommended, evaluative. No alumni interviews. SATs or ACTs: required. SAT IIs: optional. Accepts the Common Application and electronic applications. Essay question: achievement in 25 years, agenda for conversation with president of United States, one thing to change about high school experience, issue of concern, person you emulate.

Willamette University

900 State Street, Salem, OR 97301

Willamette is strategically located next door to the Oregon state capitol and 40 minutes from Portland. Bigger than Whitman, smaller than U of Puget Sound, and more conservative than Lewis and Clark, Willamette offers extensive study abroad enhanced by ties to Asia.

Willamette University, founded in 1842, was the first university in the Pacific Northwest. Students can take advantage of their proximity to the state's legislative offices and a nearby hospital for internships, jobs, or off-campus learning experiences. Students at WU find a more personal atmosphere than larger universities nearby. The campus has become increasingly diverse.

The 61-acre campus lies across the street from the Oregon state capitol building and the state Supreme Court, providing a perfect avenue for student internships and political involvement. Willamette is home to full trees (thanks to Oregon's omnipresent rain), small wildlife, and occasionally steelhead salmon, which splash around in the Mill Stream that runs between WU's redbrick academic buildings. The university has embarked on a multiyear plan to transform dorms into residential colleges. The goal is to make the campus more of a series of neighborhoods, to further build and nurture community.

Willamette (pronounced "Will-AM-it") offers bachelor of arts and bachelor of music degrees, and the most popular majors are English, psychology, and politics—especially strong because of the school's location in Oregon's capital. Students say other good bets include biology and chemistry. All students complete the freshman World Views seminar, four writing-centered courses, two courses in quantitative and analytical reasoning, study in a language other than English, and course work in six modes of inquiry—the natural world; the arts; arguments, reasons, and values; thinking historically; interpreting texts; and understanding society. Students also take capstone senior seminars, often culminating in research or thesis projects. More than half of the student body participates in a robust study abroad program, and Willamette also benefits from its proximity to the United States campus of Tokyo International University.

Classes are small; 96 percent of those taken by freshmen have 29 or fewer on the roster. Students work hard but don't compete for grades. "Willamette's academic climate is competitive," says a freshman, "but most students and faculty help struggling students reach their fullest potential." A peer agrees, saying, "Many professors promote group projects in addition to individual papers." Outside the classroom, the computerized card catalog and spacious study lounges in the Mark O. Hatfield Library (named for the former U.S. senator) make it easier to shoulder the workload. If students aren't reading or writing papers, numerous

> **"Most professors I've had prefer to be called by their first name, and they all still remember mine."**

Website: www.willamette.edu
Location: Center city
Total Enrollment: 2,663
Undergraduates: 1,977
Male/Female: 45/55
SAT Ranges: V 580–680
  M 580–660
ACT Range: 25–30
Financial Aid: 43%
Expense: Pr $ $
Phi Beta Kappa: Yes
Applicants: 2,489
Accepted: 76%
Enrolled: 27%
Grad in 6 Years: 72%
Returning Freshmen: 91%
Academics: ✍ ✍ ✍ ½
Social: ☎ ☎ ☎
Q of L: ★ ★ ★ ★
Admissions: (503) 370-6300
Email Address:
  libarts@willamette.edu

Strongest Programs:
Politics
Biology
Economics
English

If students aren't reading or writing papers, numerous undergraduate research opportunities beckon.

undergraduate research opportunities beckon. Support has more than doubled with the creation of the Carson Undergraduate Research Awards, the Science Collaborate Research Program, and a humanities center.

Students give Willamette's professors high marks. "Most professors I've had prefer to be called by their first name, and they all still remember mine," a sophomore says. "They are excited about what they know and what they teach and they want to share that information with you." Without a doubt, the low student/faculty ratio helps foster this personal atmosphere.

"It is so exciting to go to a school where students aren't only intelligent, but also creative and motivated!" exclaims one student. Thirty-nine percent of WU students are native Oregonians, and much of the remainder come from Western states, notably California and Washington. African Americans make up 2 percent of the student body, Hispanics 4 percent, and Asian Americans 6 percent. Women's issues, race relations, and the war in Iraq spark discussion on campus. "Students are respectful of others and socially aware," a sophomore says. "We have thoughtful, provocative social and political dialogue." Willamette offers talent and academic merit scholarships each year, ranging from $2,500 to full tuition; there are no athletic awards.

> **"It is so exciting to go to a school where students aren't only intelligent, but also creative and motivated!"**

Seventy-one percent of students live in campus housing, which is social and convenient to classes and parties; doing so is required for freshmen and sophomores. "Students cannot live off campus until junior year; most students seem to stay on campus anyway," a sophomore reports. "I think that is a great testament to the quality of on-campus housing." All housing is co-ed, and theme wings or floors are available, focused on community service, the outdoors, wellness, or substance-free living. "Willamette's wellness floors are very popular and build very strong relationships," says a politics major. The student-owned and -operated Bistro offers a coffeehouse atmosphere and is a popular alternative to cafeteria fare. "The dining facilities? They are great. Otherwise, the president of the university, faculty, staff, or even the governor of the state wouldn't eat here," a senior says. "The food is good and is focused around what the students want." When it comes to security, students feel very safe and take advantage of the college's escort service when they need it.

Willamette may be the best little school you've never heard of, especially if you're from outside the California–Oregon–Washington corridor.

Most of WU's social life takes place on campus, whether it's free movies and lectures, open-mic nights at the Bistro, dance parties (salsa or swing), or performances by the music and theater departments. Thirty-two percent of men and women go Greek. The Ram Brewery draws big crowds on Thursdays. Annual social highlights include the spring Wulapalooza, celebrating art and music, and the Hawaiian Club Luau, where students chow down on spit-roasted pig. Each fall, students from Tokyo International organize the Harvest Festival. Other wacky traditions include being Mill-Streamed—dumped into the campus brook on your birthday. When it comes to drinking, Willamette abides by state law, which says no one under 21 can imbibe—but students say anyone who wants booze can find and consume it behind closed doors.

> **"We have thoughtful, provocative social and political dialogue."**

Downtown Salem is a short walk from campus, and while students say it isn't a college town, it does have movies, shopping, restaurants, and coffeehouses. Also nearby are the Cascade Mountains and rugged beaches of Lincoln City and Coos Bay (an hour's drive), skiing and snowboarding on Mount Hood or in the high desert town of Bend (three hours), and the cosmopolitan cities of Portland (40 minutes) and Seattle (about four hours north). San Francisco is an eight- to nine-hour drive. Willamette students remain true to the school motto, "Not unto ourselves alone are we born," when they go "Into the Streets" for a day of service each fall.

Willamette competes in Division III, and football, women's rowing, and track (women's and men's) are strong. Women's and men's cross-country and track and field, and men's golf have claim to recent championships. The university made history in October 1997, when junior soccer star Liz Heaston kicked her way into the record books as the first woman to play intercollegiate football. The annual football game against Pacific Lutheran usually has conference championship implications, and games against Linfield are also well-attended.

Willamette may be the best little school you've never heard of, especially if you're from outside the California–Oregon–Washington corridor. "There are few other schools where I would have the support, attention, and opportunities to enjoy my two majors, study abroad for a semester, intern for a legislator, work on campus, volunteer in the community, participate in the Greek system, and graduate in four years with 300 equally involved and committed students," marvels a sophomore. The school's close-knit community is strengthened by its emphasis on service, and by warm, supportive faculty members who push students to achieve.

If You Apply To ➤

Willamette: Early action: Nov. 1, Dec. 1. Regular admissions and financial aid: Feb. 1. Does not guarantee to meet demonstrated need. Campus interviews: recommended, evaluative. Alumni interviews: optional, evaluative. SATs or ACTs: required. SAT IIs: optional. Accepts the Common Application and electronic applications. Essay question: how did you learn about Willamette; why Willamette is a good match for you; and choice of one: how you strive to serve others; source of racial tension; how does music reflect who you are; define success; responding to failure; personal experience.

College of William and Mary

P.O. Box 8795, Williamsburg, VA 23187

Founded in 1693, William and Mary is the original public Ivy. History, government, and international studies are among the strongest departments. With more than 5,600 undergraduates, W&M is larger than Richmond and Mary Washington, and smaller but more intellectual than the University of Virginia.

Though the physical campus might seem stuck in a time warp, students say everything about William and Mary—from the amazing faculty to the picturesque grounds—is up to date. Traditions abound, yet this historic public university—the second oldest in the nation—continues to evolve in its pursuit of academic excellence. It has graduated three former United States presidents—Jefferson, Monroe, and Tyler. Rival UVA prides itself on being "Mr. Jefferson's," but W&M has the distinction of having educated Mr. Jefferson in the first place.

A profusion of azaleas and crape myrtle adds splashes of color to William and Mary's finely manicured campus, located about 150 miles southeast of Washington, D.C. The campus is divided into three sections and includes Lake Matoaka, the largest man-made lake in Virginia, and a wooded wildlife preserve, which is filled with trails and widely used by the science departments. The Ancient Campus is a grouping of three Colonial structures, the oldest being Wren Hall, which has been in continuous use since 1695 and is one of the most visually pleasing buildings in American higher education. The Old Campus, where the buildings date from the '20s and '30s, is a little farther out, and next to it is New Campus, where ground was first broken in the '60s. The W&M campus boasts one of the most romantic spots of any in the nation: Crim Dell, a wooded area with a small pond spanned by an old-

Website: www.wm.edu
Location: Small city
Total Enrollment: 6,985
Undergraduates: 5,501
Male/Female: 45/55
SAT Ranges: V 630–730
  M 630–710
ACT Range: 28–31
Financial Aid: 27%
Expense: Pub $ $ $ $
Phi Beta Kappa: Yes
Applicants: 10,610
Accepted: 31%
Enrolled: 40%
Grad in 6 Years: 91%
Returning Freshmen: 95%
Academics: ✦ ✦ ✦ ✦ ✦

(Continued)

Social: ☎ ☎ ☎
Q of L: ★ ★ ★
Admissions: (757) 221-4223
Email Address:
admiss@wm.edu

Strongest Programs:
Business Administration
Government
History
English
Biology
Psychology

style wooden bridge. The 95,000-square-foot university center includes a store, auditorium, game room, post office, conference rooms, and student lounge. The most recent addition is a 390-room residence hall.

William and Mary created Phi Beta Kappa in December of 1776, and the honor code, established by Thomas Jefferson in 1779, demands much from the college's students. There are no "easy A" classes at the college, and the academic climate is intense. "There's no lie that William and Mary is a tough school, but if you got accepted, you can handle the classes," says a senior. W&M is one place where grade inflation is not an issue.

Fittingly, the history department, a joint sponsor with Colonial Williamsburg of the Institute for Early American History and Culture, is among William and Mary's best departments. Business, psychology, government, English, and history are the most popular majors. The accounting program ranks in the top 20 nationwide; employers seek out graduates from the program. State-mandated restructuring eliminated "master's only" programs in English, government, mathematics, and sociology, but undergraduate programs haven't yet felt the pinch. There are summer and yearlong study abroad programs around the globe, from Europe to China, the Philippines, Australia, and Mexico, and summer field schools in archeology, including one in St. Eustatius in the Caribbean. The College's International Relations center is internationally acclaimed. The top 5 percent of freshmen are designated Monroe Scholars and receive a $3,000 summer research stipend, which is typically used after their sophomore or junior year. One student used his to distribute his band's CD; another traveled to Paris to sketch and study.

Relations between students and professors are warm. "I continue to be impressed by the fact that not only do the professors love the subjects that they are teaching, they also truly want you to learn," says a freshman. "Professors are more than happy to meet with you at almost any scheduled time—which sometimes might be for dinner at their house." More than half of all classes have 25 or fewer students, although a few introductory lectures may have a couple hundred. Virtually every class is taught by a full professor, and TAs are used for grading or lab purposes only. The college established freshman seminars, limited to 15 students each, which provide even closer faculty interaction. A computerized registration system has taken the headaches out of the once-hellish scheduling process.

W&M graduation requirements are thorough and include proficiency in a foreign language, writing, and computing (concentration-specific). More specific general education requirements include a course in mathematics and quantitative reasoning, two courses in the natural sciences, two in the social sciences, one each in literature and history of the arts and creative and performing arts,

> **"There's no lie that William and Mary is a tough school, but if you got accepted, you can handle the classes."**

and one course in philosophical, religious, and social thought. The Center for Honors and Interdisciplinary Studies allows outstanding students four semesters of intensive liberal arts seminars, with lectures by top scholars from around the country, and also facilitates interdisciplinary majors like American studies, environmental science, and women's studies.

Its appeal to tourists notwithstanding, Williamsburg leaves much to be desired as a college town. Nightlife is a hit-or-miss affair (mostly miss), although volunteer opportunities abound and many students participate.

Because William and Mary is a state-supported university, two-thirds of its students are Virginians. Competition for the nonresident spots—mostly taken by students from the mid-Atlantic and farther north—is stiff. Ninety-seven percent of freshmen ranked in the top quarter of their high school class. The college has made a major effort to recruit and retain more minorities; Asian Americans now account for 7 percent of the students, Hispanics make up 5 percent, and African Americans contribute 6 percent. An ongoing series of programs in the residence halls addresses safety issues as well as diversity and gender communication. W&M has its share of eagerly recruited athletes; nearly 300 athletic scholarships are offered annually.

Seventy-six percent of the undergraduates live on campus in mostly co-ed dorms that range from stately old halls with high ceilings to modern buildings equipped with air-conditioning. All freshmen are guaranteed a room on campus (with both cable and Internet connections), but after that students try their luck with the infamous lottery ("stressful but efficient"). Some students, usually sophomore men, draw the Dillard Complex—two dorms located a couple miles off campus. But Dillard soon will be obsolete, due to on-campus construction of new housing facilities. "Dorms are nice and well-maintained," says one student. "Most students choose to live on campus, and those that move 'off campus' are really not that far off at all." Special-interest housing is available—there are seven language houses and an International Studies House—and life in a fraternity or sorority house is also an option. Students give the three campus cafeterias mixed reviews, but all freshmen must purchase a 19-meal plan. Others have a variety of options, including cooking in the dorms and dinner plans open to Greeks and non-Greeks alike in sorority and fraternity houses. "Speaking as a child who loves food, W&M is by no means disappointing," raves one student.

> The accounting program ranks in the top 20 nationwide; employers seek out graduates from the program.

W&M isn't known as a social school, but there's always something to do on campus. On any given weekend, students can enjoy the soothing voices of one of the many a cappella groups on campus, dance the night away at fraternity parties, grab a midnight milkshake at the University Center, or watch the **"Dorms are nice and well-maintained."** latest dance or theater performance at Phi Beta Kappa Theater. "There is always something to do," says a senior. Twenty-four percent of the men and 27 percent of the women join Greek organizations, which host most of the on-campus parties. The few local bars pick up the rest. The college has strict policies against underage drinking, but students say as long as they are safe, they stay out of trouble. The Student Association sponsors mixers, band and tailgate parties, and a film series. Campus security is regarded as tight, although crime is not a big issue. A student-run escort service will pick you up in a golf cart and take you to your desired location, and three police departments watch over students.

Anyone who gets restless can always step across the street to Colonial Williamsburg to picnic in the restored area, walk or jog down Duke of Gloucester Street (called "Dog Street"), or study in one of the beautiful gardens. Its appeal to tourists notwithstanding, Williamsburg leaves much to be desired as a college town. Nightlife is a hit-or-miss affair (mostly miss), although volunteer opportunities abound and many students participate. Richmond and Norfolk, each an hour's drive, are top road trips; the University of Virginia, although an archrival, is also popular; and Virginia Beach, a favorite springtime mecca, is a little farther away.

Traditions are the stuff of which William and Mary is made, and perhaps the most cherished is the annual Yule Log Ceremony in Wren Hall, where students sing carols and hear the president, dressed in a Santa Claus outfit, read the Dr. Seuss story, *How the Grinch Stole Christmas.* Grand Illumination is a great Christmas fireworks display, and on Char- **"The concept of community has great weight for us at William and Mary."** ter Day, bells chime and students celebrate the distinguished history of their 300-year-old institution. Each year, freshmen walk to the Wren Building for Opening Convocation, where they're greeted by cheering upperclassmen and faculty. Four years later, as they graduate, they walk through Wren in the other direction. Romantics will be happy to learn that any couple who kisses at the top of Crim Dell Bridge will be married by the end of the year. The 13 Club, a secret society of students dedicated to the college, provides students and professors with a helping hand (like missed class notes) or a pat on the back when they're feeling low—all

> Virtually every class is taught by a full professor, and TAs are used for grading or lab purposes only.

delivered anonymously. One activity, though illegal, is always popular: jumping the wall at the Governor's Mansion at the end of Dog Street.

W&M isn't a football powerhouse like most Southern state schools—a fact of life that makes it tough to lure athletic-minded males from UVA—but the athletic program is strong nonetheless. Women's soccer, volleyball, tennis, and track and field have won championships, and the men's and women's cross-country squads each brought home a conference championship in 2005. Men's gymnastics has dominated the state, winning 27 championships. The football team is a nationally ranked program in Division I-AA and always stirs enthusiasm, especially on homecoming weekend, while basketball and soccer are other popular men's sports. Intramurals and clubs such as skydiving and ultimate Frisbee—the No. 1 team in North America in 2004—attract two-thirds of the student body. A $6 million recreational athletic complex is being enhanced to better serve the health and fitness needs of students and faculty. The old gym is now the home of the graduate school of business and undergraduate student services, including admissions and career planning.

From Thomas Jefferson to Jon Stewart, William and Mary has educated some of the nation's most famed and infamous. William and Mary's traditions stretch back to the dawn of this nation, and its grand old campus and stirring history make it a distinguished and cherished part of many students' lives. "The word 'community' has become somewhat of a buzzword," says a junior. "But that is the word that best sums up what makes my school special. The concept of community has great weight for us at William and Mary and we work hard to ensure that both our academic and social lives are supported by a strong campus network."

Overlaps

University of Virginia, Duke, Georgetown, University of North Carolina–Chapel Hill, Virginia Tech

If You Apply To ➤

William and Mary: Early decision: Nov. 1. Regular admissions: Jan. 1. Financial aid: Feb. 15. Housing: Feb. 17. Meets demonstrated need of 89%. Campus interviews: optional, informational. No alumni interviews. ACTs or SATs: required. SAT IIs: recommended. Accepts the Common Application and electronic applications. Essay question: personal statement.

Williams College

Williamstown, MA 01267

Running neck and neck with Amherst on the selectivity chart, Williams occupies a campus of surpassing beauty in the foothills of the Berkshires. Williams has shaken the preppy image, but still attracts plenty of well-toned, all-around jock-intellectuals. The splendid isolation of Williamstown is either a blessing or a curse.

Website: www.williams.edu
Location: Small town
Total Enrollment: 2,070
Undergraduates: 2,017
Male/Female: 50/50
SAT Ranges: V 670–770
 M 670–760
ACT Range: N/A
Financial Aid: 43%
Expense: Pr $ $ $

Williams College vies with rival Amherst for possession of both the color purple—they each use it on team uniforms and in their logos—and the title of most selective liberal arts college in the U.S. While both schools have large numbers of "preppy, white, rich kids," students at Williams tend to be more "athletic, well-rounded, driven, friendly, and liberal," says a junior. At this isolated hamlet in the Berkshires, school spirit abounds, and the stunning natural backdrop helps keeps everyone in a good mood. When not gazing at the purple mountains' majesty, students at Williams are digging into their studies with fervor. And they do more than just hit the books, says a Chinese and political science major: "Williams students are characteristically passionate about one or several things, from national politics, to hockey, to writing poetry."

The college's buildings constitute a virtual arboretum of architectural styles, from the elegantly simple Federal design of the original West College to contemporary structures by Charles Moore and William Rawn. The brick and gray-stone buildings are arranged in loosely organized quads, which are both enclosed and open to nature. A theater and dance center has recently opened, and students may also take advantage of WCMA (the Williams College Museum of Art), the Clark Art Institute, and MASS MoCA, a nearby center for contemporary visual, performing, and media arts. The school is in the midst of a $400 million fundraising campaign.

The Williams curriculum emphasizes interdisciplinary studies and personalized teaching, made easier by the fact that the majority of courses taken by freshmen enroll 25 students or fewer. Distribution requirements include at least three courses in each of the school's three divisions, languages and arts, social studies, and sciences and mathematics; two must be completed by the end of the sophomore year. Students must also fulfill requirements in writing and in quantitative and formal reasoning, pick a major from the available 33, pass four quarters of phys ed, and spend at least six semesters in residence. The optional First-Year Residential Seminars allow one group of students taking a common course to live in the same hall, helping to integrate the social and intellectual aspects of college life. Williams has recently increased the number of its courses taught in the Oxford tutorial format: Two students and a faculty member meet each week with the students alternating who has to write a paper and who gets to critique it. Students also must complete four winter study projects or courses during the January Winter Study. Students with wanderlust may accompany faculty members on relatively inexpensive study tours, to exotic locales such as India, the former Soviet Union, and West Africa during Winter Study.

In addition, in January, the Free University, designed by students, is in operation and students may sign up for enrichment courses in which they teach each other everything from how to make wontons to how to do the jitterbug. If the walls of campus threaten to close in, especially during the bitter and blustery winter, take a term away, through the college's study away office, the Twelve College Exchange*, or the Williams at Mystic Seaport* program. There's also an innovative program organized with Oxford's Exeter College in England and two new programs, Williams in New York and Williams in Africa.

"Williams students are characteristically passionate about one or several things, from national politics, to hockey, to writing poetry."

One of Williams's greatest strengths is art history, which benefits from one of the finest college art museums in America, along with economics, psychology, political science, math, and history, students say. Environmental studies majors can perform fieldwork in 2,000-acre, college-owned Hopkins Forest. Future engineers may opt for 3–2 master's programs with Columbia University in New York, and Washington University in St. Louis. Students agree that the Romance languages are weaker, and say artsy disciplines such as music are definitely less popular.

"The academic workload at Williams definitely falls on the intense end of the spectrum," says a senior. "Luckily, though, the atmosphere is such that students are not forced to compete against one another for grades." There are only two small graduate programs at Williams—in art history and in development economics—so graduate students are few and far between and you'll never find them at the lectern. "Teaching is one of the key things that makes Williams great," says a junior. "I have encountered some absolutely wonderful professors who are smart, engaging, and interesting." Faculty members are quick to return email, says a sophomore, and may invite students over for informal get-togethers and home-cooked meals. The college also provides a stipend for advisors to take their charges out for a bite to eat.

(Continued)
Phi Beta Kappa: Yes
Applicants: 5,822
Accepted: 19%
Enrolled: 49%
Grad in 6 Years: 96%
Returning Freshmen: 98%
Academics: ✍ ✍ ✍ ✍ ✍
Social: ☎ ☎ ☎
Q of L: ★ ★ ★ ★
Admissions: (413) 597-2211
Email Address:
admission@williams.edu

Strongest Programs:
Economics
Psychology
History
Political Science
Biology
Art History
English

The Williams curriculum emphasizes interdisciplinary studies and personalized teaching, made easier by the fact that the majority of courses taken by freshmen enroll 25 students or fewer.

The typical Williams student is bright, enthusiastic, energetic, and well informed about current events. "The majority come from pretty affluent backgrounds on the East Coast, so there is a touch of elitist culture," says one student. A junior adds, "There is a reputation of Williams as a jock school, which is true to some extent." African Americans make up 10 percent of the student body, while Asian Americans add nearly 9 percent, and Hispanics comprise 8 percent. Politically, Williams is liberal, says an American studies and Spanish major: "We have learned to state our opinions and listen to what others say." Still, the most important issues seem to be not the war in Iraq or gay marriage, but the lack of dating on campus, and the administration's review of campus alcohol policies. And while there are no athletic or merit scholarships at this expensive school, Williams does guarantee to meet the demonstrated financial need of every admit, and recently announced it will lower the amount of loans it expects students from low-income families to take out, in some cases to zero.

Most students remain in the dorms for all four years because they're guaranteed a bed—and because only seniors are eligible to move out. Some freshmen luck into large single rooms in one of the school's dorms, one of which was converted from a charming inn. "Dorms are excellent," says a chemistry major, who explains that dorms are cleaned daily and are well-maintained. First-years live in groups of about 20 students each (known as "entries") along with junior advisors, who serve as big siblings, mentors, and sounding boards. After the first year, students have an affiliation with one of four upperclass residential neighborhoods and enter their housing draw. Small co-ops are available for seniors who want to cook and play house.

"The academic workload at Williams definitely falls on the intense end of the spectrum."

Fraternities and sororities were abolished in the '60s, but that hasn't stopped Williams students from partying hard. "Social life is very much on campus," says a student, "since there isn't anywhere else to go." Drinking is a popular pastime on weekends, but most students report no pressure to imbibe. The college is taking steps to make sure that whatever drinking does go on happens safely, by offering a first-year education program at the beginning of the school year, outlawing drinking games, requiring registration of parties over a certain size, mandating availability of food and nonalcoholic beverages whenever alcohol is present, and regulating tailgating. Further measures are under review by a faculty/student committee. Films and concerts abound on most weekends, often organized by the student-run All Campus Entertainment, and distract keg-seeking students with alcohol-free alternatives. Favorite traditions include Winter Carnival, Spring Fling, the Harvest Dinner ("steak and lobster everywhere," says one salivating student) and Mountain Day, one of the first three Fridays in October. Which day it will be is a well-kept secret, broken only when the college president sends out an email canceling classes, and church bells begin tolling at 8 a.m. Students hike to the top of Mount Greylock, where hot cider and donuts are waiting on the summit.

The small village of Williamstown is "small and extremely friendly and engaged," says a student. When students aren't holed up in the library, nearby slopes and trails beckon, offering skiing, cycling, and backpacking. The main street

"Complaints are few and far between."

in town boasts "several strange clothing stores that I've never been in, a bunch of overpriced antique shops and pharmacies, and three ethnic restaurants," says a political science major. The Clark Art Institute, within walking distance of campus, possesses one of the finest collections of Renoir and Degas in the nation, as well as a great library. The modern college music center attracts top classical musicians, and the college theater is home to a renowned summer festival that often features Broadway stars. There's a Wal-Mart 15 minutes away and a Stop and

Shop grocery store even closer, and civilization—in the form of Albany, New York—is just an hour's drive. Other popular destinations include New York City (five hours by car, bus, or train).

Sports are more like a religion here than an extracurricular activity, and Williams is a perennial winner of the Division III Sears Cup, awarded annually to the school with the strongest overall athletic program. Any contest with archrival Amherst ensures a big crowd. After all, Amherst was founded in 1821 by a breakaway group of Williams students, along with the school's then-president. The men's and women's swim teams are nationally ranked, and the men's tennis, basketball, and track and field teams are among the many that have brought home conference titles. Softball and field hockey are likewise strong, and the ski teams are so good they compete in Division I.

It takes a special kind of student to be happy at Williams. Those who delight in the life of the mind, who can take or leave the creature comforts found at more urban schools, will no doubt bleed purple by the time that they leave. "You feel safe here—isolated, yet shielded from the outside world," says a Chinese major. Certainly, what's warm and fuzzy for some is claustrophobic for others. Generally, though, among students who stick it out, "complaints are few and far between," reports a recent graduate. "We love it!"

Overlaps
Amherst, Dartmouth, Yale, Middlebury, Harvard, Brown

If You Apply To ➤   **Williams:** Early decision: Nov. 10. Regular admissions: Jan. 1. Financial aid: Feb. 1. Guarantees to meet full demonstrated need. Campus and alumni interviews: optional, informational. SATs or ACTs: required. SAT IIs: required (any two). Accepts the Common Application and the electronic application. Essay question: significant experience, achievement, or risk; issue of importance; an influential person; a character in fiction or history who has influenced you; what you would bring to Williams; topic of your choice.

University of Wisconsin–Madison

Armory and Gymnasium, 716 Langdon Street, Madison, WI 53706-1481

Madison draws nearly 30 percent of its students from out of state, the highest proportion among leading Midwestern public universities. Why brave the cold? Reasons include top programs in an array of professional fields and several innovative living/learning programs.

At the University of Wisconsin–Madison, 27,000 undergraduates take advantage of world-class academics and a surprisingly rich array of resources. Professional programs are strong and all that's required is a desire to learn—and a very warm coat.

Described by one Madison student as "architecturally olden with a modern touch," the mainly brick campus is distinctive. It spreads out over 903 hilly, tree-covered acres and across an isthmus between two glacial lakes, Mendota and Monona, named by prehistoric Native Americans who once lived along their shores. From atop Bascom Hill, the center of campus, you look east past the statue of Lincoln and the liberal arts buildings, down to a library mall that was the scene of many a political demonstration during the '60s. Farther east you see rows of State Street pubs and restaurants and the bleached dome of the Wisconsin state capitol. On the other side of the hill, another campus, dedicated to the sciences, twists along Lake Mendota. But students from both sides of the hill congregate in the old student union, the Rathskeller, where political arguments and backgammon games can rage all night. Outside on the union's veranda, students can look out at the sailboats

Website: www.wisc.edu
Location: City center
Total Enrollment: 36,493
Undergraduates: 27,134
Male/Female: 46/54
SAT Ranges: V 560–670
  M 600–700
ACT Range: 26–30
Financial Aid: 39%
Expense: Pub $ $ $
Phi Beta Kappa: Yes
Applicants: 21,682
Accepted: 68%
Enrolled: 42%

(Continued)

Grad in 6 Years: 78%

Returning Freshmen: 94%

Academics: ⚲ ⚲ ⚲ ⚲ ½

Social: 🐿 🐿 🐿 🐿

Q of L: ★ ★ ★ ★

Admissions: (608) 262-3961

Email Address: onwisconsin
@admissions.wisc.edu

Strongest Programs:
Political Science
Psychology
English
Economics
History

Professors at Madison are certainly among the nation's best, with Nobel laureates, National Academy of Science members, and Guggenheim fellows scattered liberally among the departments.

First-Year Interest Groups consist of 20 first-year students who live in the same residence hall or "residential neighborhood" and who also enroll in a cluster of three classes together.

in summer or iceboats in winter. A new microbial sciences building is in the works, and a new residence hall is in the final phase of construction.

Madison's academic climate is demanding. Predictably, grading is tough and inflexible and often figured on a strict curve. "There are a lot of smart people studying here," notes one student with a firm grasp of the obvious. A list of first-rate academic programs at Madison would constitute a college catalog elsewhere. There are 70 programs considered in the top 10 nationally. Some highlights include education, agriculture, communications, biological sciences, and social studies. The most popular majors are political science, psychology, English, economics, and history. Due to overcrowding, some popular fields, such as engineering and business, have had to restrict entry to their majors by requiring high GPAs.

> **"There are a lot of smart people studying here."**

Distribution requirements vary among the different schools and academic departments, but they are uniformly rigorous, with science and math courses required for BA students, and a foreign language for virtually everyone. All students must fulfill a two-part graduation requirement in both quantitative reasoning and communication. For students who prefer the academic road less traveled, options include the Institute for Environmental Studies and the Integrated Liberal Studies (ILS), which consists of related courses introducing the achievements of Western culture. A variety of internships are available, as are study abroad programs all over the world, including Europe, Brazil, India, Israel, and Thailand.

Professors at Madison are certainly among the nation's best, with Nobel laureates, National Academy of Science members, and Guggenheim fellows scattered liberally among the departments. While the university's size can be daunting, harried freshmen aren't left to fend for themselves. The university offers a number of first-year programs designed to ease the transition into college life. A first-year seminar encourages students to examine learning strategies, connect with faculty, staff, and peers, and become familiar with campus resources. First-Year Interest Groups consist of 20 first-year students who live in the same residence hall or "residential neighborhood" and who also enroll in a cluster of three classes together. Each FIG cluster of courses has a central theme; the central or "synthesizing" course integrates content from the other two classes.

If there is one common characteristic among Madison undergraduates, it is aggressiveness. "It's easy to get lost in the crowd here, so you have to be fairly strong and confident," declares one student. "No one holds your hand." The flip side is that "anyone can fit in, you just have to find your own niche." Just more than two-thirds of the students are from Wisconsin. The school is a heartland of progressive politics, and Madison's reputation as a haven for liberals remains intact. "Students here are called liberal because they are eager and willing to change and are continually looking for newer and better ideas," explains an activist. African Americans and Hispanics currently make up 6 percent of the student body, while Asian Americans constitute another 5 percent. Academic merit scholarships are awarded each year, and most of the sports on campus offer full scholarships. The two-day orientation program, known as SOAR, welcomes incoming freshmen in groups staggered throughout the summer.

> **"It's easy to get lost in the crowd here, so you have to be fairly strong and confident."**

Thirty-nine percent of students reside in university housing. Dorms are either co-ed or single-sex and come equipped with laundry facilities, game rooms, and lounges. Most also have a cafeteria. The student union also offers two meal plans, and there are plenty of restaurants and fast-food places nearby. Campus safety is always an issue, but the school offers a variety of services for those on campus. There are escort services for those walking and those needing a ride, and a free shuttle system that operates seven days a week. Madison (a.k.a. Madtown) has been the stomping ground for many fine rock 'n' roll or blues bands on the road to fame.

There are more film clubs than anyone can follow, and everyone has a favorite bar. Nine percent of the men and women go Greek. "Frat parties are a very popular break from the bar scene," reports one expert on both options. One old standby that is still as popular as ever is the student union, which hosts bands, shows, and so forth and provides a great atmosphere in which to hang out. Volunteering is another popular option; for a decade the university has provided the Peace Corps with the most entrants of any college or university in the nation. Nature enthusiasts can lose themselves in the university's 12,000-acre nature preserve or hit nearby ski slopes.

The students at this Big Ten school show "tons of interest" in sports, especially hockey and football, and especially when the Badgers try to rout the University of Minnesota's Gophers. Bucky Badger apparel, emblazoned with slogans ranging from the urbane to the decidedly uncouth, is ubiqui-

"You feel you're accepted for who you are no matter what."

tous. However, the much-acclaimed marching band known as Fifth Quarter may outdo all the teams in popularity. The Badgers are recent Big Ten champions in a number of sports, notably men's cross-country, men's indoor and outdoor track, and women's hockey. Intramurals are also popular.

One of the best and most well rounded state schools anywhere, Madison is a school that students sum up as "diverse, intellectual, fashionable, and moderately hedonistic." And these are the qualities that attract bright and energetic students from everywhere. "You feel you're accepted for who you are no matter what," says one student. "It's so nice to just be yourself."

If You Apply To ➤

Wisconsin: Rolling admissions. Does not guarantee to meet full demonstrated need. Campus and alumni interviews: optional, informational. SATs or ACTs: required. SAT IIs: optional. Essay question: personal statement. Special consideration given to students from disadvantaged backgrounds. Apply to particular school.

Wittenberg University

P.O. Box 720, Springfield, OH 45501

Wittenberg is an outpost of cozy Midwestern friendliness. Less national than Denison or Wooster, Witt has plenty of old-fashioned school spirit and powerhouse Division III athletic teams. Top students should aim for the honors and fellows programs, the latter providing a chance for undergraduate research.

Founded in 1845 by German Lutherans, Wittenberg University remains true to its faith by emphasizing strong student/faculty relationships—and making sure that students don't become too comfortable in the campus bubble. In fact, before granting their diplomas, Wittenberg requires students to complete 30 hours of community service in the surrounding town of Springfield (population 65,000). Students at Wittenberg work hard at their studies, but they also know how to play hard.

The Wittenberg campus is classic Midwestern collegiate, with a mixture of 1800s and Gothic-inspired buildings on 70 rolling acres in southwestern Ohio. The redbrick Myers residence hall, with picturesque white pillars and an open-air dome dating from the 19th century, stands at the center. Wittenberg formally opened Post 95, a cutting-edge cafe-style dining area in the Benham-Pence Student Center, in 2005. There's also a new fitness center. A new residence hall opened in 2006.

Website: www.wittenberg.edu
Location: City outskirts
Total Enrollment: 2,177
Undergraduates: 1,992
Male/Female: 43/57
SAT Ranges: V 510–630
  M 510–620
ACT Range: 21–27
Financial Aid: 73%
Expense: Pr $ $
Phi Beta Kappa: Yes

When the weekend rolls around, social life centers on parties in houses, dorm rooms, and apartments on and near campus.

Wittenberg's general education requirements emphasize a solid liberal arts background. The school's Wittenberg Plan includes 17 learning goals, ranging from experience with writing and research to exposure to the natural sciences and foreign languages. Students select courses from a variety of disciplines to meet the goals, and also must fulfill requirements in religion or philosophy, non-Western cultures, and physical education. Wittenberg has a special leadership development program for first-year students, and new students are matched with returning students in a Student Impact mentoring program. There is also a University Scholars Program for outstanding freshmen, and the Wittenberg Fellows Program provides opportunities to work on research with faculty members. A minor program in marine science is available as well.

Wittenberg students give high marks to the school's education department, which recently added a master's degree program. Other strong programs include biology, English, and business—so perhaps it's not surprising that the most popular majors on campus are in those fields. "Students tend to major in whatever their interest is and personalize their majors through internships, study abroad, and research," a psychology major says. "Courses are fairly rigorous and challenging," says a junior, "but professors are always willing to help." A psychology major adds, "Our professors truly show concern for students as individuals."

Despite Wittenberg's small size, students say they have no trouble registering for needed courses and graduating in four years. So long as students declare their major on time and complete the proper coursework in the correct order, the college guarantees a degree in four years—and will pay for any additional necessary courses. Wittenberg also encourages students to take a semester or a year away from campus, either in the U.S. or abroad. Options include the International Student Exchange Program, field study in the Bahamas or Costa Rica, work with the National Institutes of Health in Washington, D.C., or a term at the United Nations in New York. Wittenberg also offers 3–2 engineering programs with Columbia, Case Western Reserve, and Washington University in St. Louis.

> **"Courses are fairly rigorous and challenging, but professors are always willing to help."**

"Wittenberg students are friendly and outgoing," says a junior. "We are a very social campus with students constantly involved in clubs, sports, and academic teams." Seventy-four percent of Wittenberg students are native Ohioans and 5 percent hail from other countries. Many others are from nearby states like Indiana, Michigan, and Pennsylvania. African Americans comprise 6 percent of the student body, and Hispanics and Asian Americans together account for 3 percent. The school's multicultural affairs director is working to boost those numbers through changes in minority recruiting and advising. Students say the campus is fairly evenly split between conservatives and liberals and both groups are vocal. However, "it's not a tense or mean-spirited division," says a student. Wittenberg offers more than 500 merit scholarships but no athletic scholarships.

Wittenberg students are required to live on campus their first two years. After that, most chose nearby houses and apartments owned by the school. "The dorms are a great place to live," a junior says. Sophomore students have first pick of rooms, and then the freshmen are assigned to the remaining rooms, so almost everyone gets a double when he or she first comes to campus. Greek groups draw 16 percent of the men and 29 percent of the women; members may live in chapter houses. Sodexho provides the chow in Witt's dining hall, which has been

> **"We are a very social campus with students constantly involved in clubs, sports, and academic teams."**

rated the company's best college food service. Choices range from burgers to made-to-order breakfasts to monthly theme dinners. All students on a meal plan eat at

CDR—which stands for Central Dining Room—or Post 95, which offers extended hours and pizza, Mexican food, sandwiches, and other specialties. "Food used to be poor, but now it's really good," reports one economics major. Students say they feel safe on campus thanks to an active and visible security presence.

When the weekend rolls around, social life centers on parties in houses, dorm rooms, and apartments on and near campus. "Every weekend, something is going on," says a junior. Greek groups, the Union Board, and the Residence Hall Association also bring in guest speakers, movies, comedians, and concerts. Favorite annual events include Greek Week, homecoming ("the alumni involvement is incredible"), and Wittfest in May, a music festival with games, food, prizes, and socializing before finals. "It is open to the community, but all the students go," a senior says. "It resembles a carnival, and at night there's a big concert on the lawn." There's also "W" Day, a day during the week preceding Wittfest in which many students skip classes to party outside. Springfield has movie theaters, a mall, restaurants, and a $15 million performing arts center. "It doesn't have the characteristics of a small college town, but it does have the benefits of a city," a senior says. "That's what makes Witt unique—a small college in a big town." Says another student, "It has the stuff you need, but a car is helpful." Popular road trips include Dayton (30 minutes), Columbus (45 minutes), and Cincinnati (90 minutes), and for those with more time, Washington, D.C., New York City, and Chicago. Nearby state parks also offer swimming, camping, biking trails, and picnics in the warmer months, and skiing in the winter.

While not as well known as many of its bigger Midwestern brethren, Wittenberg's athletic teams are competitive in Division III. Recent championship teams include men's football, basketball, and track, and women's field hockey, volleyball, soccer, and tennis. Rivalries with the College of Wooster, Wabash, and Allegheny really get students riled up. Intramurals are a huge draw, too, with sports ranging from ice hockey to water polo. Weekend warriors

> "That's what makes Witt unique—a small college in a big town."

may take advantage of the Bill Edwards Athletic and Recreational Complex, which boasts a stadium and eight-lane track, football and soccer fields, 12 lighted tennis courts, a brand-new, state-of-the-art fitness center, and a weight room, plus a pool and racquetball courts.

Wittenberg stands out among small liberal arts colleges for several reasons, one of which is that Wittenberg students defy stereotyping. Says a student, "For a small, liberal arts college, Wittenberg is very diverse socioeconomically, ethnically, and religiously. I do not feel the students can be categorized into one group."

Wittenberg also encourages students to take a semester or a year away from campus, either in the U.S. or abroad.

Overlaps
College of Wooster, Denison, Ohio Northern, Ohio Wesleyan

If You Apply To ➤ **Wittenberg:** Rolling admissions. Early decision: Nov. 15, Dec. 1. Early action: Jan. 15. Does not guarantee to meet demonstrated need. Campus interviews: recommended, evaluative. Alumni interviews: optional, informational. SATs or ACTs: required. SAT IIs: optional. Accepts the Common Application and electronic applications. Essay question: significant experience or relationship; issue of concern.

Wofford College

429 North Church Street, Spartanburg, SC 29303-3663

Wofford is about one-third as large as Furman and roughly the same size as Presbyterian. With more than a few gentleman jocks, Wofford is one of the smallest

institutions to compete in NCAA Division I football. Fraternities and sororities dominate the traditional social scene.

Website: www.wofford.edu
Location: Small city
Total Enrollment: 1,158
Undergraduates: 1,153
Male/Female: 52/48
SAT Ranges: V 570–660
   M 580–680
ACT Range: 22–27
Financial Aid: 54%
Expense: Pr $
Phi Beta Kappa: Yes
Applicants: 1,871
Accepted: 66%
Enrolled: 26%
Grad in 6 Years: 78%
Returning Freshmen: 89%
Academics: ✍ ✍ ✍
Social: ☎ ☎
Q of L: ★ ★ ★
Admissions: (864) 597-4130
Email Address:
   admissions@wofford.edu

Strongest Programs:
Life Sciences (Biology and
   Psychology)
Prelaw
Business
English
Computer Science
Economics

Wofford students take pride in the Wofford Way, combining a well-rounded curriculum with career-related internships and study abroad. Students study hard under the "eyes of Old Main" and often visit their professors' homes for dinner, forming lasting friendships with peers and faculty members. "Wofford combines an open and friendly atmosphere with high academic standards and expectations that position students for success," says a senior.

Wofford is near the heart of Spartanburg, a midsize city in the northwest corner of South Carolina. Founded in 1854, it's one of fewer than 200 existing American colleges that opened before the Civil War—and it still operates on its original campus. Azaleas, magnolias, and dogwoods surround the distinctive, twin-towered Main Building and four original faculty homes on the 150-acre campus, which has been designated an arboretum. Nearly 4,500 trees have been planted since 1992. The Main Building has been completely renovated and construction has begun on an apartment-style housing development for seniors.

Traditionally, Wofford's strongest and most attractive programs have been in the life sciences, which account for more than one-third of its graduates and a large share of its Phi Beta Kappas. Business programs, especially when combined with a

"If you are fortunate enough to get an A, you earned it."

second major in foreign languages, and English with its emphasis on creative writing are solid and getting better. Students rave about Wofford's programs in biology and economics. About two dozen of the school's 260 graduates go on to graduate medical or dental programs within two years of graduation; another two dozen go on to prestigious law schools. For aspiring entrepreneurs, Wofford's alumni network has clout: More than 1,200 of its 15,000 living graduates serve as presidents or owners of corporations or organizations. Thirteen hundred practice medicine, dentistry, or other health care professions, and three of South Carolina's five Supreme Court Justices are Wofford grads. Prospective engineers may apply for 3–2 programs with Clemson or New York's Columbia University. Religion, English, foreign language, and study abroad programs also get high marks, especially for the one lucky junior chosen as the Presidential International Scholar. This student is sent around the world, all expenses paid, to study an issue of global importance for a year. Students in the Creative Writing Sequence end by writing a novella, the best of which is given the Benjamin Wofford Prize and is published in paperback.

"Wofford is extremely challenging academically," says a junior. "If you are fortunate enough to get an A, you earned it." At the end of his first year in office in 2001, college President Benjamin Dunlap asked the faculty what they would do if they had the time and resources to teach the courses they've always dreamed of. Since then, more than 50 new courses and interdisciplinary course sequences have been added to the curriculum. Among the newest are programs in African and African American studies and a program in Chinese language and culture. Courses are required in English, fine arts, foreign languages, humanities, science, history, philosophy, cultural

Every year before finals, student musicians and readers perform a Festival of Nine Lessons and Carols, perhaps praying as well for luck on their exams.

"You don't come to Wofford, you join it."

perspectives, math, and physical education. Wofford also boasts a unique program called the Novel Experience, part of the first-year orientation. The program consists of a common reading, which is then discussed during a meeting of 22 humanities classes at different restaurants throughout Spartanburg.

Professors have high standards, and students feel pressure to meet them because they have personal relationships with the faculty, students say. "I would not trade any of my professors," says a senior. "They are scholarly, engaging, understanding, and grounded." Wofford's professors put teaching ahead of research, says another student.

The Wofford College Success Initiative is a scholarship-based learning community that complements the curriculum. While not for academic credit, it reinforces traditional liberal arts concepts such as critical thinking and communications through intellectual explorations and experiences. Students take one exploration each fall semester and propose an independent of small-group exploration that could form the basis of team projects. Recent programs include learning American Sign Language with faculty and students at the South Carolina School for the Deaf and Blind.

Three-quarters of the student body graduated from public high school—more than half in the top tenth of their class. "Most students tend to be rather affluent," says a student. "The Wofford 'uniform' includes polos, khakis, seersucker, a North Face jacket, and a monogrammed tote bag," a freshman says. African Americans comprise 6 percent of the student population, Asian Americans 2 percent, and Hispanics 1 percent. Politically, the campus is fairly conservative but all viewpoints are represented, students say. "Students do sometimes discuss issues of diversity and religion, but generally they are too caught up in schoolwork and campus events," says a psychology major. Merit and athletic scholarships are available to qualified students.

Eighty-nine percent of Wofford's students live in the dorms, where first-year women get doubles in Greene Hall, and their male counterparts have similar digs in Marsh Hall. "Living on campus is important to staying in the know," says a senior. Another says "dorms are comfortable and more than adequate. Our maintenance staff is friendly, and I usually carry on conversations with the ones in my hall." Students give pretty good marks to the college cafeteria and praise the Canteen, a lunchtime option that serves Southern cooking. Campus safety is good; there are call boxes and constant patrols.

The Greek system is a huge force in Wofford's social life, with fraternities attracting 51 percent of the men and 59 percent of the women. Each fraternity has a house and most host parties every Friday and Saturday—with some kicking off the weekend on Thursday. "Social life revolves around weekends at the Row," a junior says. "If the Row is closed for a weekend, folks tend to venture off campus." The campus is dry, except for certain events, like

"Living on campus is important to staying in the know."

Spring Weekend, where alcohol for those of age is allowed on Fraternity Row, a student says. The Student Affairs Committee offer campus-wide events like comedians and music for those uninterested in the Greek system.

Off campus, Spartanburg is home to Converse College and a few other schools. Students say it's not a great college town, but that is changing with the addition of coffee shops and other fun hangouts. Students can head to Greenville, Atlanta, and Charlotte. And almost every Wofford student participates in some type of volunteer work. Terrier Play Day brings kids from the community to campus for a fair with booths and games. Bid Day, when the fraternities tap their new members, is another annual tradition, involving lots of mud, and then a bath in the college fountain. Every year before finals, student musicians and readers perform a Festival of Nine Lessons and Carols, perhaps praying as well for luck on their exams. The Greeks clean up their acts in time for Spring Weekend, a sort of campus Olympics.

Wofford is the smallest member of the NCAA Division I that fields a football program. Wofford has beaten former rival the Citadel in football eight straight years, and have found a new rival in Furman. The baseball teams have been strengthened by new facilities on campus, and the men's and women's tennis teams are doing well in their conference. Men and women compete side-by-side on Wofford's riflery team.

Wofford's former chaplain was fond of saying, "You don't come to Wofford, you join it." And students say that's true, citing the close-knit community and intimate student/faculty relationships fostered by the school's small size. "I've often said that there is a close community at Wofford, but it is really more of a family," says a

For aspiring entrepreneurs, Wofford's alumni network has clout: More than 1,200 of its 15,000 living graduates serve as presidents or owners of corporations or organizations.

Overlaps

Furman, Clemson, University of South Carolina, Wake Forest, University of the South, Presbyterian

senior. "Administrators are not merely the authority but are real friends to students. Wofford is a family of seekers with learning as its focus."

If You Apply To >

Wofford: Early decision: Nov. 15. Regular admissions: Feb. 1. Financial aid: Mar. 15. Housing: May 1. Does not guarantee to meet full demonstrated need. Campus interviews: recommended, informational. Alumni interviews: optional, informational. SATs or ACTs: required. SAT IIs: optional. Accepts the Common Application and prefers electronic applications. Essay question: event, interest, experience, goal, or person who reveals something about you.

The College of Wooster

Wooster, OH 44691

Though not well-known to the general public, Wooster is renowned in academic circles and a number of foreign countries. Getting admitted is not difficult, but graduating takes work. All students complete an independent study project in their last two years. More intellectually serious than competitors such as Denison.

Website: www.wooster.edu
Location: Small town
Total Enrollment: 1,819
Undergraduates: 1,819
Male/Female: 49/51
SAT Ranges: V 540–670
   M 560–650
ACT Range: 23–29
Financial Aid: 57%
Expense: Pr $ $
Phi Beta Kappa: Yes
Applicants: 2,504
Accepted: 80%
Enrolled: 25%
Grad in 6 Years: 73%
Returning Freshmen: 87%
Academics: ✑ ✑ ✑½
Social: ☎ ☎ ☎
Q of L: ★ ★ ★
Admissions: (330) 263-2322
Email Address:
   admissions@wooster.edu

Strongest Programs:
Biology
Economics
English
Communication
Political Science
Psychology
History
Sociology & Anthropology

Instead of teaching students what to think, the College of Wooster focuses on teaching students how to think. From the first courses of the freshman year seminar to the final day when seniors hand in their hard-won theses, the college paves each student's path to independence. The emphasis here is on global perspectives and the heritage that stems from its origin as a college founded by Presbyterians. The one-to-one attention from faculty makes Wooster an intellectual refuge in the rural countryside of Ohio.

Located in the city of Wooster, Ohio, COW's hilltop campus is spread over 240 acres, many designed in the English–Collegiate Gothic style and constructed of cream-colored brick. More recent buildings are trimmed in Indiana limestone or Ohio sandstone. The central arch and two towers of Kauke Hall, the central building in Quinby Quadrangle (the square around which the college grew), make it stand out. The Gault Library for Independent Study offers a private carrel for each senior in the humanities and social sciences. The new Bornhuetter residence hall has two wings of double rooms and a state-of-the-art multipurpose room with kitchen facilities, audio-visual equipment, and a stunning view of the oak trees and lawn on the northeast side of the building.

"Professors get the chance to study abroad and this brings more experience to the classrooms."

What goes on behind the facades of Wooster's attractive buildings is more impressive than the structures themselves. The required first-year seminar in critical inquiry, limited to 15 students per section, invites students to engage in issues, questions, or ideas drawn from classical readings. The seminar is linked to the Wooster Forum, a series of first-semester lectures and events focused on a single, broad issue. Students enrolled in the First-Year Living and Learning Project not only take the seminar but live together in specially designated housing. "The classes are rigorous courses, however I did not consider the environment to be extremely competitive," says a senior. "People are always willing to assist other students."

Wooster's curriculum is built around the required Independent Study, which lets students explore subjects they're passionate about with faculty guidance. Independent Study has become such a part of COW that each year seniors celebrate IS Monday—the day they turn in their projects—with a campuswide parade. "We are led around campus by the bagpipers," says a senior. "The whole campus shows up." Completion of the IS earns you a Tootsie Roll, to eat or keep for posterity next to your diploma. "It's a day all Wooster graduates will always remember!" a senior says. The college even awards $60,000 each year for student research, travel, materials purchases, or conference registration fees.

In addition to the critical inquiry seminar, three semesters of Independent Study, and six cross-discipline courses, Wooster mandates courses in writing, global and cultural perspectives, religious perspectives, and quantitative reasoning; foreign language proficiency; and seven to nine courses in the students' major. Students praise faculty members for their devotion to teaching and mentoring; only a few introductory courses have teaching assistants, who run review sessions and offer extra help. "The level of teaching here is exceptional," says a senior. "Professors get the chance to study abroad and this brings more experience to the classrooms." It's not uncommon for science majors to coauthor faculty papers, while students in the Jenny Investment Club manage a portion of the college's assets, with professors serving as advisors. Wooster's small size hasn't placed it at a technological disadvantage; WoosterNet links every academic and administrative building and residence hall, and labs and kiosks offer 24-hour access to more than 150 microcomputers and terminals.

The most popular majors are psychology, history, political science, communication, and sociology and anthropology. The foreign language departments are small but provide individual attention to each student. Students in the new biochemistry/molecular biology major program take a foundation of core science courses and can choose between upper-level chemistry and/or biology classes along with their Independent Study. For those who tire of

> **"The campus encourages liberal views and acceptance for everyone."**

rural Ohio, a leadership and liberal learning program includes a seminar class and a weeklong acquaintanceship, where participants shadow a prominent politician, executive, or other professional. Wooster also sponsors overseas programs on five continents, by itself and through the Great Lakes Colleges Association.*

Each year, Wooster's admissions office strives to assemble a diverse group of scholars, and as the college's reputation spreads, it's becoming more selective, with acceptance rates falling and freshman-retention rates improving. Still, diversity mainly extends to academic and extracurricular interest; African Americans constitute 4 percent of the student body, and Asian Americans and Hispanics are just 2 percent each. Half of the students are from Ohio. That said, COW students aren't politically apathetic. Gay rights, women's issues, the Middle East conflict, U.S.–Afghanistan relations, and even free-trade coffee pepper campus table talk. "The campus encourages liberal views and acceptance for everyone," says a senior.

Ninety-nine percent of students live on Wooster's smoke-free campus in nine co-ed and two single-sex dorms, where rooms are small but maid service is regular. Housing is okay; one political science major says the rooms are "comfortable and well maintained. There always seems to be shortage in the spring, though everyone ends up having a room in the fall." Students stay connected with the community through the Wooster Volunteer Network. Students seriously committed to service may apply to live in one of the college's 26 residential program houses, each of which is affiliated with a

> **"The Scot Bagpipers playing 'Amazing Grace' is beyond compare."**

community group. Food in the two campus dining halls is getting better, more parking has been added, and all dorms are now wired for cable television. Given Wooster's location "in the middle of corn fields," security isn't an issue, students say, although emergency phones are strategically located just in case.

Wooster social life is campus-based, thanks in large part to the school's isolated location. Students do travel to colleges such as Denison, Oberlin, Kenyon, and Ohio

Wooster social life is campus-based, thanks in large part to the school's isolated location. Students do travel to colleges such as Denison, Oberlin, Kenyon, and Ohio State to combat cabin fever.

The school's Scottish heritage can be seen in its band, which performs in bagpipes and kilts, and in its Scottish dancers, who trot on stage during the fall's Scot Spirit Day.

State to combat cabin fever. Other popular road trips are Cleveland's Flats or warehouse district, and just across the border to Windsor, Canada, where there's legal gambling and a lower drinking age. That's important to some because "it is getting harder and harder for underage students to be served" on campus, says a senior. The college has no national Greek organizations, but local "sections" draw 13 percent of men and "clubs" attract 17 percent of women. One major weekend hangout is the Underground, a bar and dance club that hosts well-known bands, as well as the campus's bowling alley, pool hall, and game room. The school's Scottish heritage can be seen in its bagpipe band, which performs in kilts, and in its Scottish dancers, who trot on stage during the fall's Scot Spirit Day. "The Scot Bagpipers playing 'Amazing Grace' is beyond compare," says a junior. Other annual traditions include the outdoor Party on the Green, a fall concert, and the formal Winter Gala, where students, faculty, and staff dance the night away to the sounds of a swing band. When it snows—which it does quite often in Wooster—the student body descends upon the central arch and fills it with snow.

Wooster fields a number of competitive Division III teams. Men's basketball is a spectator favorite—it and the men's baseball team have both come home with the North Coast Athletic Conference Championship title. Any match versus rival Wittenberg usually brings out the fan in Wooster students (you have to get the free tickets well ahead of time), and men's and women's soccer, swimming, and women's lacrosse have been strong in recent years.

The College of Wooster is nationally recognized for its commitment to independent study and global perspectives. Wooster students are proud to be "fighting Scots" and independent-thinkers. They make do during the long, cold winters by immersing themselves in study and finding ways to have a good time.

If You Apply To ➤ **Wooster:** Early decision: Dec. 1. Regular admissions and financial aid: Feb. 15. Housing: May 1. Meets demonstrated need of 100%. Campus, alumni interviews: recommended, informational. SATs or ACTs: optional. SAT IIs: optional. Requires optional writing test for ACT. Accepts the Common Application and electronic applications. Essay question: personal statement.

Worcester Polytechnic Institute

100 Institute Road, Worcester, MA 01609-2280

Small, innovative, and undergraduate-oriented, WPI is anything but a stodgy technical institute. The WPI Plan is hands-on and project-based and takes a humanistic view of engineering. Teamwork is emphasized instead of competition. WPI is half the size of Rensselaer and a third as big as MIT.

Website: www.wpi.edu
Location: City outskirts
Total Enrollment: 3,817
Undergraduates: 2,764
Male/Female: 77/23
SAT Ranges: V 570–670
  M 630–720
Financial Aid: 74%

As a pioneer in engineering education, Worcester Polytechnic Institute has built a solid reputation on the sciences. But with its ever-expanding academic curriculum, surprising devotion to music and theater, and dedication to hands-on undergraduate experiences, WPI is not content to rest on its laurels. Students must complete several extensive projects, endure seven-week semesters, and engage in real-world experiences. But it's WPI's humanistic approach to engineering that really sets it apart.

WPI is the third-oldest independent science and engineering school in the nation. Its compact 80-acre campus is set atop one of Worcester's "seven hills" on the residential outskirts of town. Worcester is an industrial city (third largest in New

England) and the home of 14 colleges, most notably WPI, Clark University, and Holy Cross. These schools are brought together by the 13-member Colleges of Worcester Consortium.* WPI is also home to the Social Web, which is a one-stop academic and social gathering place on the World Wide Web where students can discover what is happening at local college and nonprofit organizations. Newer additions to the campus include the Fuller Chemistry Complex, the Bartlett Center for administrative offices, and a life sciences research center.

The city is also a base for the rapidly growing Northeastern biotechnology and biomedicine industry. The area is one of the nation's most successful high-technology regions, which supports WPI's project and research programs. WPI's campus borders two parks and the historic Highland Street District, where local merchants and students come together to form the neighborhood community. Old English stone buildings, complete with creeping ivy, are focal points of the architecture, but modern facilities have moved in to claim their own space on the immaculately kept grounds.

There are four terms per academic year at WPI, each lasting seven weeks, which students say changes the climate. "WPI has rigorous academics," says a sophomore, "but the students aren't competitive with each other." A senior adds that "classes are very hands-on with an emphasis on learning by doing." The

> **"WPI is pretty committed to having a strong relationship between faculty and students in every class."**

Interactive Qualifying Project has students apply technical knowledge to one of society's problems and the Major Qualifying Project represents a student's first chance to work on a truly professional-level problem. Courses provide the information students need to complete their projects, reemphasizing WPI's curriculum as one driven by knowledge and not credit. When students are not completing projects, they take three courses per term. Some say it's difficult to graduate in four years in majors such as chemical engineering.

The intent of WPI's unique grading system and educational philosophy is to polish social skills, build self-confidence, produce well-rounded students, and nurture young people interested in using their knowledge to improve the world. WPI focuses especially on developing teamwork abilities. The curriculum remains remarkably flexible for a high-powered technological university. Standard course distribution requirements vary by major but include courses in engineering, math, and science. Every WPI student must complete a humanities and arts sufficiency project. To promote cooperation and cohesiveness, the only recorded grades are A, B, C, or No Record. Failing grades do not appear on transcripts, and the school does not compute GPAs or class ranks.

Technical writing has grown. The most popular departments are, not surprisingly, mechanical engineering, electrical and computer engineering, computer science, chemical engineering, and biology and biotechnology. "Departments like chemical engineering, mechanical engineering, electrical engineering, and computer science tend to be more rigorous and challenging," a senior reports. There are also a number of interdisciplinary programs such as prelaw and international studies. A major in Interactive Media and Game Development requires coursework in computer science and the humanities and arts. Many biomedical engineering majors do their projects at UMass Medical and Tufts Veterinary, as well as at local hospitals. And WPI has a joint PhD Program with UMass medical school. WPI also offers a rare fire protection engineering program and a system dynamics major and minor. Math and science types can pick up middle or high school teaching credentials on the way. The theater technology major requires projects in set, lighting, or audio design, which means school shows often highlight cutting-edge production techniques. Well over 300 students participate in 13 musical ensembles, one of the largest music programs among technological universities. Professors are praised for their availability and willingness to

(Continued)
Expense: Pr $ $ $
Phi Beta Kappa: No
Applicants: 3,708
Accepted: 75%
Enrolled: 28%
Grad in 6 Years: 80%
Returning Freshmen: 92%
Academics: ✍ ✍ ✍ ½
Social: ☎ ☎ ☎
Q of L: ★ ★ ★ ★
Admissions: (508) 831-5826
Email Address: admissions@wpi.edu

Strongest Programs:
Computer Science
Mechanical, Electrical, and
    Computer Engineering
Biology/Biotechnology
Biomedical Engineering

There are four terms per academic year at WPI, each lasting seven weeks, which students say changes the climate.

establish relationships. "WPI is pretty committed to having a strong relationship between faculty and students in every class," says a student. "Professors are totally accessible by email or by appointment and ready to help you learn."

In light of an increasingly interdependent global economy, WPI offers a unique Global Perspectives Program that spans five continents. Nearly 70 percent of WPI students visit locations including France, Ireland, Spain, Australia, Costa Rica, Hong Kong, and Namibia. The school's residential project centers in New York, Silicon Valley, Washington, D.C., England, Denmark, Italy, Puerto Rico, or Thailand provide students with the opportunity to tackle current problems for a sponsor and spend the term working independently on a specific sociotechnical assignment under the direction of one or more faculty members. The co-op program offers upperclassmen two eight-month work experiences and adds an extra half or full year to the degree program.

"The biggest issues are ethics in engineering and the environment."

Eighty-six percent of students ranked in the top quarter of their high school class. African Americans and Hispanics together account for only 5 percent of the students, and Asian Americans represent an additional 6 percent. Foreign students comprise 5 percent of the student body. "Students are driven to do well," says a mechanical engineering major, "but are also very involved in sports and Greek life." Politics don't factor into campus debates too often because "students are too busy and are, for the most part, apathetic," reports a senior. "The biggest issues are ethics in engineering and the environment." Typical complaints include the lack of parking and the need for better recreational facilities. Twenty percent of undergrads receive merit scholarships with an average award of more than $18,000, but there are no athletic scholarships.

Only first-year students are guaranteed spots in the university residence halls; however, the room-assignment process usually allows upperclassmen their first, second, or third choice. There are 12 co-ed halls, including a spacious 230-bed residence hall that houses students in suites of four or six, on-campus apartments, and smaller houses for more homelike living. "Most people live off campus in surrounding apartments," says a student. Still, 48 percent of the students live on campus, and they can choose from the multiple meal plans of the dining halls. "The dining hall serves a variety of hot meals everyday and the food is typically good," a sophomore says.

Twenty-nine percent of the men join fraternities, and 28 percent of the women enter sororities. Of-age students may have alcohol, provided it is kept in their rooms, but underage students do find a way to imbibe. In addition to Greek parties, there are student-organized coffeehouses, concerts, poetry readings, movies, and pub shows. "There's always something to do on campus," says a biotechnology major. Nearby

"Students are driven to do well, but are also very involved in sports and Greek life."

colleges such as Holy Cross, Assumption, and Clark University are linked to WPI through shuttle buses, which provide even more social and academic opportunities. Boston and Hartford are both an hour's drive, as are ski resorts and beaches.

One of WPI's more notable campus traditions is the Goat's Head Rivalry, a grudge match between the freshman and sophomore classes that includes the Pennant Rush, a rope pull next to Institute Pond, and a WPI trivia competition. The prize? A 100-year-old bronze goat's head trophy (the winning class's year is engraved on it). There is also an annual festival of international culture and QuadFest, complete with carnival rides. Students say the new campus center encourages cohesiveness across the campus and offers a viable alternative to the Greek domination of the social scene. "There is constantly something going on," says a senior.

While not particularly scenic, Worcester does offer a large number of clubs and restaurants and an art museum. A large multipurpose arena, the recently renamed

DCU Center, is host to frequent concerts (like Phish and P. Diddy) and occasional visits from Boston's Bruins and Celtics and a minor league hockey team. Men's crew and basketball and women's soccer, cross-country, and track had great seasons. The women's tennis and men's wrestling are titleholders among the 18 varsity sports. Crew is also popular with both sexes, the women's team having finished the season ranked second in the nation in Division III.

One of WPI's chants is fittingly mathematic: "E to the x, d-y,d-x, e to the x, d-x, cosine, secant, tangent, sine, 3.14159, e-i, radical, Pi, fight 'em fight 'em WPI!" If you know what any of that stuff means, you'll fit right in.

If You Apply To ➤

WPI: Early action: Nov. 15, Jan.1. Regular admissions: Feb. 1. Financial aid: Mar. 1. Housing: Jun. 1. Does not guarantee to meet demonstrated need. Campus and alumni interviews: optional, informational. ACTs or SATs: optional. SAT IIs: optional. Accepts the Common Application and electronic applications. Essay questions: personal choice.

Xavier University of Louisiana

1 Drexel Drive, New Orleans, LA 70125

Xavier's strategic location in New Orleans is its biggest drawing card. XU is bigger than a small college, but smaller than most universities. Just more than one-fourth of the students are Roman Catholic. Competes with other leading historically black institutions such as Spelman, Morehouse, and Howard.

As the nation's only historically African American Roman Catholic college, Xavier University is no stranger to adversity. In the aftermath of Hurricane Katrina, this small New Orleans university is struggling to regain its footing. Despite the setback, the school continues to prepare students for great careers while providing a strong foundation in the liberal arts. With its stellar reputation for graduating a wealth of scientists, the school's primary focus is much loftier: "The promotion of a more just and humane society."

Founded in 1915, Xavier is located near the heart of New Orleans in a quiet neighborhood dotted with bungalows. The focal point of the campus is the Library Resource, which, with its green roof and stately neo-Gothic architectural style, has become a landmark for those traveling by car from the New Orleans airport to the French Quarter. A closed campus green mutes the urban feel of the encroaching city, and yellow brick buildings have been erected among the older limestone structures. The ground floor of nearly every building on campus has been restored and all storm debris has been cleared. Many facilities, including the University Center, the library, and the residence halls have been completely refurbished.

Although many predicted major enrollment decreases in the aftermath of Katrina, nearly three out of four Xavier students returned to campus for the Spring 2006 semester. Fortunately, they returned to find all academic programs intact. However, student and faculty research capabilities were dealt a severe blow. Lab equipment was severely damaged and the Center for Undergraduate Research was totally destroyed. As the university works to restore its research programs, it maintains its reputation as one of the most effective teaching institutions anywhere; the National Science Foundation has designated it as one of

> **"If you want to go into medicine, I would recommend Xavier."**

Website: www.xula.edu
Location: City center
Total Enrollment: 3,012
Undergraduates: 2,272
Male/Female: 25/75
SAT Ranges: V 410–540
   M 380–520
ACT Range: 18–23
Financial Aid: 0%
Expense: Pr $
Phi Beta Kappa: No
Applicants: 1,692
Accepted: 62%
Enrolled: 42%
Grad in 6 Years: 48%
Returning Freshmen: 81%
Academics: ✑ ✑ ✑
Social: ☎ ☎ ☎
Q of L: ★ ★ ★
Admissions: (504) 520-7388
Email Address:
   apply@xula.edu

(Continued)

Strongest Programs:
Biology
Chemistry
English
Psychology
Art

In addition to the many internships available, Xavier offers cooperative education programs in all fields and study abroad programs throughout Europe, Africa, Japan, and North, Central, and South America.

Computer engineering is the newest concentration, and the school is trying to strengthen its small history and philosophy departments.

only a few Model Institutions for Excellence. Political science is a small but good department, and the psychology and education departments have been traditional strengths. "If you want to go into medicine, I would recommend Xavier," explains a student, "They have one of, if not the best, premed programs for minority students." Nearly two-thirds of the student body majors in a science-related field. Xavier has frequently led the nation in the number of undergraduate physical science degrees awarded to African Americans, as well as the number of African Americans placed into medical school. Xavier is also credited with educating 25 percent of all African American pharmacists nationally. Forty-four percent of its alums go on to professional or grad schools. Computer engineering is the newest concentration, and the school is trying to strengthen its small history and philosophy departments. Economics and German are no longer offered as majors, and departments consistently revise their curriculums to accommodate the evolving demands of the students. In addition to the many internships available, Xavier offers cooperative education programs in all fields and study abroad programs throughout Europe; Africa; Japan; and North, Central, and South America. The Center for the Advancement of Teaching works to improve pedagogy across the curriculum and encourages African American students to become teachers and researchers.

Xavier's undergraduate curriculum is centered on the liberal arts. All students are required to take a core of prescribed courses in theology and philosophy, the arts and humanities, communications, history and the social sciences, mathematics, and the natural sciences. Freshmen also take a mandatory seminar titled The Student and the University to help them adjust to college life. The academic climate is competitive and challenging. "The academic climate is very competitive," says a senior. Priests and nuns teach and help run the school, though the faculty and staff are composed mainly of multiracial laypeople. "Teaching has been excellent due to small class sizes," says a sophomore. "Students are able to interact closely with teachers and each other." Academic and career advising are well received, and registration doesn't present any major concerns.

"Students are able to interact closely with teachers and each other."

There is a strong emphasis on community service and it is required with some on-campus organizations. Students are also concerned about social issues outside of the university. "Many information sessions or forums are developed so that Xavier student can be kept abreast of the current events of the world that affect us," a senior says.

For a historically African American college, Xavier's student body is quite diverse. Sixty-three percent of the students come from public schools and 22 percent are not African American. Students come mostly from the Deep South; 55 percent are from Louisiana, and a high percentage are second- or third-generation Xavierites. Xavier has achieved a national reputation for its programs to reach out to local high schools to identify and nurture talented minority group students. More than half graduated in the top quarter of their high school class.

Sixty percent of Xavier students flock to the residence halls. "Housing is great," says a student. Three of the four contemporary-looking residence halls are single-sex. Though the housing situation is far from perfect, students admit that dorm rooms are comfortable and well kept. New Orleans has one of the highest crime rates in the nation, so campus security is always an issue. Although campus security is highly visible, it "definitely needs to be tightened up," says a chemistry major.

When students tire of microscopes and mass, they can trek into the Crescent City. Back on campus, fraternities and sororities play a leading role in extracurricular and social life, though only 1 percent of the men and 5 percent of the women go Greek. Popular events include Bayou Classic, Spring Fest, and Oktoberfest. But don't get caught drinking—Xavier is a dry campus. "They do room inspections and confiscate prohibited items," explains a sophomore. Underage drinkers return to the city where nearly anything goes. When it comes to road trips, students head to Baton Rouge, Shreveport, Tallahassee, Houston, and Miami.

Varsity sports include men's and women's basketball, tennis, and cross-country, and the teams are enthusiastically supported, especially when the opponent is rival Dillard. All Xavier athletic programs were suspended in the wake of Hurricane Katrina, but they have been restored. Intramural sports are also offered with high participation from men in basketball, tennis, cross-country, and flag football.

Students at Xavier focus on their studies, community service, and school spirit for basketball. With a mind for the future, Xavier stays true to its beginnings as a historically black and Catholic University and to a mission of preparing students for a future that will "promote a more just and humane society."

If You Apply To ➤

Xavier: Early action: Jan. 15. Regular admissions: Jul. 1. Does not guarantee to meet demonstrated need. Campus interviews: optional, informational. No alumni interviews. Recommendations are very important. SATs or ACTs: required. SAT IIs: optional. Accepts the Common Application. No essay.

Yale University

38 Hillhouse Avenue, New Haven, CT 06511

Yale is the middle-sized member of the Ivy League's big three: bigger than Princeton, smaller than Harvard. Its widely imitated residential-college system helps Yale strike a balance between research university and undergraduate college. Gritty New Haven pales next to Cambridge or Morningside Heights.

Tradition is more than just a buzzword at Yale. Founded in 1701 by Connecticut Congregationalists concerned about "backsliding" among their counterparts at a certain school in Cambridge, Massachusetts, Yale has long been recognized as one of the nation's—and the world's—finest private universities, and one of the few Ivy League schools focused on undergraduates. In recent years, the school has undertaken the Herculean task of renovating its aging campus. The university is in the midst of the largest construction and renovation project ever undertaken at an American university. Despite the changes, students here remain as focused on their studies and as humble of their achievements as ever. And thanks to Yale's residential college system, this huge research university feels like more of a home for its students.

Yale's campus looks like the traditional archetype—magnificent courtyards, imposing quadrangles, Gothic buildings designed by James Gamble Rogers, and Harkness Tower, a 201-foot spire once washed with acid to create its aged, stately look. A massive, $2.6 billion construction and renovation project, to be completed in 2010, is slowly transforming the university's historic buildings into state-of-the-art facilities. All the residential colleges, some of which date back to the 1930s, will be or have been renovated, thanks to a $300 million effort. Yale also has committed $500 million to promote its basic science, engineering, and biomedical research programs. The class of 1954 Environmental Science Center opened two years ago, and the Daniel C. Malone Engineering Center has been completed.

Website: www.yale.edu
Location: Urban
Total Enrollment: 11,276
Undergraduates: 5,350
Male/Female: 51/49
SAT Ranges: V 700–780
  M 700–780
ACT Range: 31–34
Financial Aid: 42%
Expense: Pr $ $ $
Phi Beta Kappa: Yes
Applicants: 19,451
Accepted: 10%
Enrolled: 71%
Grad in 6 Years: 97%
Returning Freshmen: 98%
Academics: ✍ ✍ ✍ ✍ ✍
Social: ☎ ☎ ☎
Q of L: ★ ★ ★
Admissions: (203) 432-9300

(Continued)

Email Address:
    student.questions@yale.edu

Strongest Programs:
Art and Architecture
History
English
Biology
Economics
Political Science
Psychology
Music
Drama

The prominence of the arts makes for an interesting juxtaposition: while tradition is ever-present on campus, today's Yale attracts one of the most liberal and forward-thinking student bodies in the Ivy League.

Inside Yale's wrought-iron gates, academic programs are superb across the board. Arts and humanities programs are especially outstanding. The prominence of the arts makes for an interesting juxtaposition: While tradition is ever-present on campus, today's Yale attracts one of the most liberal and forward-thinking student bodies in the Ivy League. Still, the ancient Puritan work ethic remains. Graduating from Yale requires 36 credits, or nine courses a year, rather than the 32 courses required at most other colleges. Students say there's virtually no competition between students; rather, competition is with oneself. "Yale students pride themselves on building an atmosphere that is academically rigorous but not directly competitive," says a student.

Although Yale has 10 professional schools and a Graduate School of Arts and Sciences, Yale College—the undergraduate arts and sciences division—remains the university's heart. Virtually all professors teach undergraduates, and the professional schools' resources—especially architecture, fine arts, drama, and music—are available to them as well. Yale's superb history department offers the most popular undergraduate major, followed by political science, economics, psychology, and English. History offers one of the most demanding programs, including a mandatory 30- to 50-page senior essay. The English department is routinely at the vanguard of literary theory, while an outstanding interdisciplinary humanities major includes the study of the medieval, Renaissance, and modern periods. While some science majors grumble about the walk up Science Hill, where most labs and science classrooms are, they agree it's worth the trip. The biological science department is excellent, where student interests range from biomedical engineering research to preparation for medical school. "The quality of Yale's undergraduate science education is largely underestimated," says a senior. "I have found research to be an integral part of education, which is only possible at an institution with Yale's strength in research and development." Architecture and modern languages, especially French and Chinese, are top-notch, and the school's Center for the Study of Globalization is renowned as well. Elite students with a particularly strong appetite for the humanities can enroll in Directed Studies, which examines the literature, philosophy, history, and politics of Western tradition. Prospective DSers should be prepared for some serious bonding with their books—they don't call it "Directed Suicide" for nothing. Freshmen with strong science backgrounds can enroll in Perspectives on Science, their version of DS.

> **"Yale students pride themselves on building an atmosphere that is academically rigorous but not directly competitive."**

Despite its reverence for tradition, Yale doesn't require any specific courses for graduation, and it doesn't have a core curriculum. Instead, students must take two classes in humanities and arts, social sciences, and natural sciences and math, along with two courses that emphasize writing and another two that emphasize quantitative reasoning. Instead of preregistering, students spend two weeks "shopping" at the beginning of each term, sampling morsels of the various offerings before finalizing their schedules. Yale also mandates intermediate-level mastery of a foreign language, and students must study a language for one to three semesters, depending on their proficiency. Yale offers study abroad options, too: Each year, about a hundred juniors participate in the Junior Term/Year Abroad program, while others study abroad during the summer, on a leave of absence, or after graduation. Yale also sponsors the Yale World Fellowships, in which burgeoning world leaders spend a semester in a new global leadership program.

Introductory-level classes at Yale are usually large lectures, accompanied by small recitation and discussion sections, typically led by graduate teaching assistants. Some of the most popular courses, such as John Gaddis' Cold War history class, seem more like performances, students say. Upper-level seminars are small and plentiful, and underclassmen can usually get into the ones that interest them. Of the 1,000 classes offered each semester, 75 percent have 20 or fewer students, and 29 percent of classes have 10 students or fewer. "Professors are stellar," gushes a junior. A sophomore adds, "Professors are readily available to help undergraduates." Newly-expanded freshman seminars allow freshmen to interact with professors and peers in small groups.

Yale's libraries hold more than 10,000,000 volumes, second in size only to Harvard (library envy, anyone?). The 26,000-square-foot Irving S. Gilmore Music Library holds the university's extensive music collection and features electronic access to music information, a historic sound recording room, and a place to listen to records. The flashcube-shaped Beinecke Rare Book Library houses many extraordinary manuscripts, including a Gutenberg Bible and some music manuscripts penned by Bach, while the most frequently used books are found underground in the Cross Campus Library. Also located in CCL are rows of study carrels, tiny beige boxes that look like phone booths with desks in them. To Yalies, these are "weenie bins," where many a tired student has dozed on an open book.

"Yalies are happy, enthusiastic, and passionate about what they do," says a student. Ninety-two percent of the student body is from out of state; a large majority of the student body hails from the Northeast. Yale is also consistently more popular with women than many of its rivals, most notably Princeton; the student body is evenly split along gender lines. Most traditions unique to Yale—all-male singing groups like the Whiffenpoofs, drinking at "the tables down at Mory's"—have female counterparts, like Whim 'n Rhythm, if they haven't gone co-ed. Yale strives to increase the diversity of its student body; African Americans make up 8 percent of students, Hispanics 7 percent, Asian Americans 14 percent, and Native Americans 1 percent. Foreign students account for 8 percent of the student body. Its best-known alumnus notwithstanding, Yalies are less conservative than their counterparts at Harvard and Princeton, and they aren't shy about expressing their opinions.

The residential colleges that serve as Yale's dorms—and as the focal points for undergraduate social life—are "one of the greatest attractions in a Yale education," says one student. Endowed by Yale graduate Edward S. Harkness (who also began the house system at Harvard) and modeled on those at Oxford and Cambridge, Yale's colleges provide intimate living/learning communities, creating the atmosphere of a small liberal arts college within a large research university. Each college has a library, dining hall, and special facilities such as photography darkrooms or tree swings—one is even said to have an endowment used solely for whipped cream. Dining halls serve great food, students say, and one has a completely organic menu. All colleges also have their own dean and affiliated faculty members, a few of whom live in the college, who can help undergraduates struggling to adapt to the rigors of college life. Residential college masters organize social and cultural events, including master's teas, where prominent public

"Yalies are happy, enthusiastic, and passionate about what they do."

figures meet with groups of students. College-sponsored seminars, along with plays, concerts, lectures, and other events, add to the cultural life of the university as a whole. Yale strives to make an education accessible to everyone; families making less than $45,000 a year don't pay any portion of the cost of their child's education.

Much of each residential college's distinctive identity comes from its architecture. Some are fashioned in a craggy, fortress-like Gothic style, while others are done in the more open Colonial style, with redbrick and green shutters as the prevailing motif. All colleges have their own special nooks and crannies with cryptic inscriptions paying tribute to illustrious Yalies of generations past. Most freshmen live together in the Old Campus, the historic 19th-century quadrangle, before moving into their colleges as sophomores, who, along with juniors, generally live in suites of single and double rooms. Many seniors get singles. Some upperclassmen move into New Haven,

Elite students with a particularly strong appetite for the humanities can enroll in Directed Studies, which examines the literature, philosophy, history, and politics of Western tradition.

although 88 percent choose to stay on campus all four years. "Almost everyone lives on campus," reports a student, "and how could you not when you have gothic suites with common rooms? Our dorms are great for study sessions, but just as great for gathering friends together."

In addition to identifying with their colleges, many Yale students identify strongly with extracurricular groups, clubs, and organizations, spending most of their waking hours outside class at the newspaper, radio station, or computer center. Particularly clubby are the a cappella singing groups, whose members do everything from drinking together on certain weeknights to touring together during spring break. Many of Yale's mysterious secret societies, such as Skull and Bones (which counts President George W. Bush and former presidential candidate John Kerry as members), have their own mausoleum-like clubhouses and issue invitations to those with the right qualifications. There are also the Yale Anti-Gravity Society and improv comedy groups.

Though studying takes the lion's share of their time, students here also find ways to unwind. The university doesn't have a strict policy on alcohol, students say. "It's nice that the administration trusts students and I think for the most part students respond really well to that trust," says a junior. Still, the drinking age of 21 is enforced at larger, university-sponsored bashes, pushing most socializing to private parties in the colleges or off-campus apartments. A handful of Greek organizations have yet to make their mark. For the artistically inclined, local film societies offer numerous weekend screenings and York Square Cinemas show indie films, foreign films, and the occasional blockbuster. The Palace Theater and the Shubert Performing Arts Center host touring Broadway musicals, dance companies, musicians, and popular singers. The Tony Award–winning Yale Repertory Theater is an excellent, innovative professional company that depends heavily on graduate school talent but always brings in a few top stage stars each season. "Yale is self-contained and, on the weekend (that is, Thursday through Saturday), very active," says a senior. Natural history and art museums on and near campus, especially the British Art Center, are excellent. For those who want more excitement, the typical Yalie refrain on New Haven—"It's halfway between New York and Boston"—tells it all. Metro North trains run almost hourly to New York, and visiting Boston is nearly as easy. Storrs, Connecticut, home of UConn, is also a popular road trip.

> "Our dorms are great for study sessions, but just as great for gathering friends together."

New Haven isn't a typical college town, students say. It's an urban area with all the problems a city has, including crime and poverty. But students say they feel relatively safe there, and plenty of good comes from being in a city—arts, culture, and great restaurants. "New Haven and Yale are inextricably linked; one could not survive without the other," a sophomore says. A summer jazz festival brings thousands to the historic town green, and there are outdoor ethnic food fairs and theatrical performances. The city's long-standing theatrical tradition—it was once the place to try out plays headed for Broadway—has been revived with the reopening of two grand old theater and concert halls a block from campus. Locals will swear that Pepe's on Wooster Street was the first (and best!) pizza parlor in the country, while Louis's Lunch was the first true hamburger joint. On Saturdays, "all roads lead to Toads," a popular dance club. Relations between students and locals are improving, and more than two-thirds of Yale undergrads do volunteer work in town through Dwight Hall, the largest college community service organization in the country.

Yale fields a full complement of athletic teams, which play in the NCAA Division I. More than half the student body takes part in intramural competition among the residential colleges; the winning college gets the coveted Tying Cup. The annual Harvard–Yale football game is the hottest event on campus.

Overlaps

Harvard, Princeton, Stanford, Columbia, University of Pennsylvania

Yale is one of America's oldest institutions of higher learning, and students and graduates here take seriously the intonation, "For God, for country, and for Yale." For proof, just remember that among its alumni, Yale counts the presidents or former presidents of about 70 other colleges and universities, as well as the last three presidents of the United States. As the university passes its three-hundredth anniversary, its past and former students continue to make their marks on the world.

| If You Apply To ➤ | **Yale:** Non-binding early action: Nov. 1. Regular admissions: Dec. 31. Financial aid: Mar. 1. Guarantees to meet full demonstrated need. Campus and alumni interviews: optional, evaluative. SATs or ACTs: required. SAT IIs: required if ACT is submitted (any three). Accepts the Common Application and prefers electronic applications. Essay question: personal essay; meaningful interest or activity. |

Consortia

Students who feel that attending a small college might limit their college experiences should realize that many of these schools have banded together to offer unusual programs that they could not support on their own. Offerings range from exchange programs—trading places with a student on another campus—to a semester or two anywhere in the world on one of the seven continents or somewhere out at sea.

The following is a list of some of the largest and oldest of these programs, some sponsored by groups of colleges and others by independent agencies. An asterisk (*) after the name of a college indicates that the institution is the subject of a write-up in the *Fiske Guide*. An asterisk following the name of a program in the college write-ups means that it is described below.

The **Associated Colleges of the Midwest** (www.acm.edu) comprises 14 institutions in five states: Beloit,* Lawrence,* and Ripon* in Wisconsin; Carleton,* Macalester,* and St. Olaf* in Minnesota; the University of Chicago,* Knox,* Lake Forest,* and Monmouth in Illinois; Coe, Cornell,* and Grinnell* in Iowa; and Colorado College.*

The consortium offers its students semester-long programs to study art in London and Florence; culture and society in Florence, the Czech Republic, and Zimbabwe; language and culture in Costa Rica and Russia; and tropical field research in Costa Rica. Year-long programs include Chinese studies in Hong Kong, India studies, and study in Japan. The Arts of London and the Florence program are the most popular with students. Language study is a component of all the ACM overseas programs. Prior language study is required for the programs in Costa Rica, Japan, and Russia. Domestic off-campus programs include Humanities at the Newberry Library (an in-depth research project) or a semester in Chicago in the arts, urban education, or urban studies. Scientists can study at the Oak Ridge National Laboratory in Tennessee or at a wilderness field station in northern Minnesota.

Living arrangements vary with the program and region. Students in programs in Chicago live in apartments and residential hotels; Minnesota's wilderness enthusiasts must rough it in cabins, and Oak Ridge scientists are on their own. There are no comprehensive costs for any of the ACM programs, domestic or foreign, and tuition is based on the home school's standard fees. The programs are open to sophomores, juniors, and seniors majoring in all fields. The only programs that tend to be especially strict with admissions are the Oak Ridge, Newberry, and Russian arrangements. For information, contact Associated Colleges of the Midwest, 18 South Michigan Ave., Chicago, IL 60603, (312) 263-5000.

The **Associated Colleges of the South** (www.colleges.org), incorporated in 1991, is composed of 12 Southern schools (Birmingham-Southern,* Centenary, Centre,* Millsaps,* Rhodes,* University of the South,* Furman University, Hendrix College,* Morehouse College,* Southwestern University, Trinity University,* and the University of Richmond*). Established to strengthen liberal education in the South, the consortium focuses on academic program development (with attention to international programs), and faculty, staff, and student development. Overseas courses are offered year-round. Affiliated and ACS-managed programs are offered at Oxford, in Central Europe, and in Brazil. For information, contact the Associated Colleges of the South, 17 Executive Park Dr., Suite 420, Atlanta, GA 30329, (404) 636-9533.

The **Atlanta Regional Consortium for Higher Education** (www.atlantahighered.org) comprises 20 public and private colleges and universities in the Atlanta area, as well as several specialized institutions of higher education. Members are Agnes Scott College,* Atlanta College of Art, Clark Atlanta University,* Clayton College and State University, Columbia Theological Seminary, Emory University,* Georgia Institute of Technology,* Georgia State University, State University of West Georgia, Institute of Paper Science and Technology, Interdenominational Theological Center, Kennesaw State University, Mercer University, Morehouse College,* Morehouse School of Medicine, Morris Brown College,* Oglethorpe University,* Southern Polytechnic State University, Spelman College,* and the University of Georgia.*

Students from member colleges and universities may register for approved courses at any of the other institutions, including those with highly specialized courses. The consortium's interlibrary lending program uses a daily truck delivery service to put more than 10,000,000 books and other resources at students' disposal.

The **Christian College Consortium** (www.ccconsortium.org) comprises 13 of the nation's top evangelical liberal arts schools: Asbury, Bethel (MN), George Fox, Gordon,* Greenville, Houghton,* Malone, Messiah, Seattle Pacific, Taylor, Trinity (IL), Westmont, and Wheaton (IL).*

The consortium offers a "student visitors program" whereby students can spend a semester—with little paper pushing—at any of the member schools. More than 100 students (not including freshmen) participate each year, and the cost is strictly the home school's regular fees. Other than a reasonably good grade average, there are no special requirements. Consortium schools share a wide array of international programs on a space-available basis, and the consortium has cooperative arrangements with Daystar University College in Nairobi, Kenya, and Han Nam University in Taejon, Korea.

The **Five College Consortium** (www.fivecolleges.edu) is a nonprofit organization that comprises Amherst,* Hampshire,* Mount Holyoke,* Smith,* and the University of Massachusetts at Amherst,* and is designed to enhance the social and cultural life of the 30,000 students attending these Connecticut Valley colleges. Legally known as Five Colleges Inc., this cooperative arrangement allows any undergraduate at the four private liberal arts colleges and UMass to take courses for credit and use the library facilities of any of the other four schools. A free bus service shuttles among the schools.

The consortium sponsors joint departments in dance and astronomy, as well as a number of interdisciplinary programs, including black studies, East Asian languages, coastal and marine sciences, Near Eastern studies, peace and world security studies, Canadian studies, and Irish studies. Certificate programs are available in African studies and Latin American studies. There are five college centers for East Asian studies, women's studies research, and foreign language resources. Students from the four smaller colleges benefit from the large number of course choices available at the university. The undergrads from UMass, in turn, take advantage of the small-college atmosphere as well as particularly strong departments such as art at Smith, sculpture at Mount Holyoke, and film and photography at Hampshire. There is also a Five College Orchestra and an open theater auditions policy that allows students to audition for parts in productions at any of the colleges. The social and cultural aspects of the Five College Consortium are more informal than the academic structure. The consortium puts out a calendar listing art shows, lectures, concerts, and films at the five schools, as well as the bus schedules. In addition, student-sponsored parties are advertised on all campuses, and there is a good deal of informal meeting of students from the various schools.

For those students taking courses on other campuses, one's home-school meal ticket is valid on any of the five member campuses for lunch. Dinners are available with special permission. Taking classes at other schools is encouraged, but not usually for first-semester freshmen. The consortium is a big drawing card for all schools involved.

The **Great Lakes Colleges Association** (www.glca.org) comprises 12 independent liberal arts institutions in three states: Antioch,* Denison,* Kenyon,* Oberlin,* Ohio Wesleyan,* and the College of Wooster* in Ohio; DePauw,* Earlham,* and Wabash* in Indiana; and Albion,* Hope,* and Kalamazoo* in Michigan. Like ACM, the Great Lakes group offers students off-campus opportunities both in the U.S. and overseas.

For adventures abroad, there are African studies programs in Sierra Leone, Senegal, and Kenya. Students can spend a year studying in Scotland or Japan, or a semester comparing socioeconomic changes in Poland, the United Kingdom, and Germany (European Academic Term). GLCA cosponsors five programs mentioned in the ACM write-up above, but these sometimes cost more: a fall or a year at a People's Republic of China university; study in Hong Kong, Russia, or the Czech Republic; and programs at the Newberry Library in Chicago and Oak Ridge National Laboratory in Tennessee. Other domestic programs include a one-semester arts internship in New York City and a liberal arts urban-study semester in Philadelphia.

Primarily juniors participate, but the programs are open to sophomores and seniors. New York and Philadelphia are the most popular domestic plans, and Scotland is the largest of those abroad. There are language requirements to meet in several of the programs, such as a year of Mandarin for China, a year of Japanese for Japan, and two years of Russian for Russia. Sometimes, however, an intensive summer language program can be substituted. Contact GLCA, 2929 Plymouth Rd., Suite 207, Ann Arbor, MI 48105-3206, (313) 761-4833.

The **Lehigh Valley Association of Independent Colleges** (www.lvaic.org) is a 23-year-old cooperative effort among six colleges in the same area of Pennsylvania: Allentown College of St. Francis de Sales, Cedar Crest College, Lafayette College,* Lehigh University,* Moravian College, and Muhlenberg College.*

Approximately 400 students each year crossregister at member campuses, although the bulk of the activity occurs between schools that are closest to each other. A Jewish studies program, headquartered at Lehigh, draws on the faculties of Lehigh, Lafayette, and Muhlenberg. Faculty members travel from college to college in order to offer students a variety of courses in this field. The association's Consortium Professors program puts faculty on two other member campuses each year to teach unusual or special-interest courses. Several members exchange courses by video conference. Special seminars are arranged at central locations for selected students, with transportation provided. The association offers a cooperative cultural program sponsoring nationally known visiting dance companies. Students at each college are eligible for reduced-rate tickets to plays and other events on campuses of association schools. But the most frequently used service of the association is its interlibrary loan program, which permits students at one institution to use the research facilities of the others. Summer study abroad programs take students to Germany, Spain, Mexico, or Israel.

The **Maritime Studies Program** of Williams College and Mystic Seaport Museum is an interdisciplinary semester designed for 22 undergraduates (primarily juniors, but some second-semester sophomores and seniors) who are eager to augment liberal arts education with an in-depth study of the sea. Participants take four Williams College courses (maritime history, literature of the sea, marine policy, and oceanography or marine ecology). Classes are taught with an emphasis on independent research in the setting of the Mystic Seaport Museum. Classroom lectures are enhanced by hands-on experience in celestial navigation, boat building, sailing, blacksmithing, and other historic crafts. Students spend two weeks offshore in deep-sea oceanographic research aboard a traditionally rigged schooner highlighting the purpose of the program: to understand our relationship with the sea—past, present, and future.

Most students are drawn from 20 affiliate colleges: Amherst,* Bates,* Bowdoin,* Colby,* Colgate,* Connecticut,* Dartmouth,* Hamilton,* Middlebury,* Mount Holyoke,* Oberlin,* Smith,* Trinity,* Tufts,* Union,* Vassar,* Wellesley,* Wesleyan,* Wheaton (MA),* and Williams.* Credit is granted through Williams College, and financial aid is transferable. Students from all four-year liberal arts colleges are encouraged to apply. Write to the Maritime Studies Program, Box 6000 Mystic Seaport Museum, Mystic, CT 06355-0990, (302) 572-5359.

Sea Semester (www.sea.edu, not to be confused with Semester at Sea) is a similar venture for water lovers, but it is designed for students geared more toward the theoretical and practical applications of the subject. Five 12-week sessions are offered each year, and there are 48 students in each session. One prerequisite for the program is a course in college-level lab science or the equivalent. All majors are considered as long as they're in good academic standing, submit transcripts and recommendations, and have an interview with an alumnus in their area.

Students spend the first half of the term living on Sea's campus in the Woods Hole area and immersing themselves in oceanography and maritime and nautical studies. Independent-study projects begun ashore are completed during the sea component aboard either a schooner or a brigantine, which cruises along the eastern seaboard and out into the Atlantic, North Atlantic, or Caribbean, depending on the season. Six weeks on the ocean is when theory becomes reality, and the usual mission consists of enough navigation, oceanographic data collection, and recordkeeping to keep even Columbus on the right course.

Students from affiliated colleges (Boston University,* College of Charleston,* Colgate,* Cornell,* Drexel,* Eckerd,* Franklin and Marshall,* the University of Pennsylvania,* and Rice University*) receive a semester's worth of credit directly through their school. Students from other schools must receive credit through Boston University. Write to the Sea Education Association, P.O. Box 6, Woods Hole, MA 02543.

Semester at Sea (www.semesteratsea.com) takes qualified students from any college and whisks them around the globe on a study/cruise odyssey. Based at the University of Pittsburgh, this nonprofit group takes 450 students each term (from second-semester freshmen to grads) and puts them on a ship bound for almost everywhere. The vessel itself is a college campus in its own right. Sixty courses are taught by two dozen professors in subjects ranging from anthropology to marketing, and usually stressing the international scene as well as the sea itself. What's more, art, theater, music, and other extras can be found on board. When students aren't at sea, they're in port in any of 12 foreign countries throughout India, the Middle East, the Commonwealth of Independent States (Russia), the Far East, Africa, South America, and the Mediterranean, and it's not uncommon for leaders and diplomats to meet them along the way.

Students must be in good standing at their home colleges to be considered, which often means a GPA of 2.5 or better. Some financial aid is available in the form of the usual federal grants and loans, and 30 eligible students can use a work/study plan to pay for half the trip. Most colleges do recognize the Semester at Sea program and will provide participating students with a full term's worth of credits. Information may be obtained by writing to Semester at Sea, University of Pittsburgh, 811 William Pitt Union, Pittsburgh, PA 15260, (800) 854-0195 or (412) 648-7490.

The **Seven-College Exchange** consists of four women's colleges (Hollins,* Mary Baldwin, Randolph-Macon Woman's,* and Sweet Briar*), one men's school (Hampden-Sydney*), and two coed schools (Randolph-Macon College and Washington and Lee*). The exchange program was more popular when it began almost two decades ago and was utilized mainly for social reasons. Today the exchange program is used mainly for academic reasons and enables students to take advantage of courses offered on the other campuses. Eligibility for participation is determined by the home institution, and except for special fees, rates are those of the home institution. Designed primarily for juniors, the program also considers sophomores and seniors as applicants. Several participating members sponsor study-abroad programs.

The **Twelve-College Exchange Program** comprises a dozen selective schools in the Northeast: Amherst,* Bowdoin,* Connecticut College,* Dartmouth,* Mount Holyoke,* Smith,* Trinity (CT),* Vassar,* Wellesley,* Wesleyan,* Wheaton (MA),* and Williams.*

The federation means that students enrolled in any of these schools can visit for a semester or two (usually the latter) with a minimum of red tape. Approximately 300 students utilize the opportunity each year; most of them are juniors. Placement is determined mainly by available space, but students must also display good academic standing. While the home college arranges the exchange, students must meet the fees and standards of the host school. Financial-aid holders can usually carry their packages with them. Also available through this exchange is participation in the Williams College–Mystic Seaport Program in American maritime studies or study at the Eugene O'Neill National Theater Institute.

The **Venture Program** is based at Brown University, but has at various times counted many of the most prestigious East Coast and Midwestern colleges and universities in its membership. The eight current member institutions include Bates,* Brown,* Connecticut College,* Hobart and William Smith,* College of the Holy Cross,* Swarthmore College,* Vassar,* and Wesleyan.*

Venture, established in 1973, places students who want to take time off from college in short-term, full-time jobs in many fields of interest and geographic locations. Venture provides students with an opportunity to test academic, career, and personal interests on the job. There is no cost to students or employers for participation. Venture is supported by member institutions. About 200 students apply to the program, and about half of them are eventually placed in positions. All students attending a member college are eligible to participate. Venture also works with students who want to take time off between high school and college. In 1987, the consortium initiated the Venture II program, which encourages graduating seniors from member schools to explore work opportunities in the not-for-profit sector. The consortium also operates the Urban Education Semester in collaboration with the Bank Street College of Education and Community School District Number 4 in New York City, introducing liberal arts undergraduates to issues and practice in urban education. All of Venture's programs aim to foster social awareness and responsibility among students and build connections between higher education and the community.

The **Washington Semester of American University** takes about 750 students each year from hundreds of colleges across the country (who meet minimum academic qualifications of a 2.75 GPA) and gives them unbeatable academic and political opportunities in the nation's capital. The program is the oldest of its kind in Washington.

Students take part in a semester of seminars with policymakers and lobbyists, an internship, and a choice between an elective course at the university or a self-designed, in-depth research project. Students live in dorms on the campus, and are guided by a staff of 20 American University professors.

Ninety percent of the students are drawn from 192 affiliated schools. Although admissions competition depends on the home school and how many it chooses to nominate, the average GPA hovers around a 3.3. Most who participate are juniors, but second-semester sophomores and seniors get equal consideration. The cost is either American University's tuition, room, board, and fees or that of the home school. Just over a third of the affiliated colleges are profiled in the *Fiske Guide*.

The **Worcester Consortium** is made up of 10 institutions nestled in and about Worcester, Massachusetts: Anna Maria, Assumption, Becker Junior, Clark University,* Holy Cross,* Quinsigamond Community, Tufts University of Veterinary Medicine, the University of Massachusetts Medical Center, Worcester Polytechnic Institute,* and Worcester State. Member schools coordinate activities ranging from purchasing light bulbs to sharing libraries, and a bus transports scholars to the various campuses as well as public libraries. Academic cross-registration is offered, as are two special programs: a health studies option and a certificate in gerontology. The consortium calendar lists upcoming events on each campus and encourages community service with a special emphasis on college/school collaboration. The consortium also provides free academic and financial-aid counseling to low-income, first-generation students thinking about college. Write the Educational Opportunity Center, 26 Franklin St., Worcester, MA 01608.

Index

Acknowledgments

Fiske Guide to Colleges Staff

Editor: Edward B. Fiske
Managing Editor: Robert S. Logue
Contributing Editor: Bruce G. Hammond
Production Coordinator: Julia Fiske Hogan

Fiske Guide to Colleges reflects the talents, energy, and ideas of many people. Chief among them are Robert Logue, the managing editor, and Julia Fiske Hogan, the production coordinator. I am also grateful for the continuing valuable contributions of Bruce G. Hammond, my coauthor on the *Fiske Guide to Getting into the Right College* and other resources in the field of college admissions. We are all grateful for the dedicated, formidable editorial assistance of Dominique Raccah, Todd Stocke, Peter Lynch, Erin Nevius and their talented colleagues at Sourcebooks.

In the final analysis, the *Fiske Guide* is dependent on the contributions of the thousands of students and college administrators who took the time to answer detailed and demanding questionnaires. Their candor and cooperation are deeply appreciated; and while I, of course, accept full responsibility for the final product, the quality and usefulness of the book is a testimony to their thoughtful reflections on their colleges and universities.

Edward B. Fiske
Durham, NC
May 2006

Editorial Advisory Group

Nancy Beane, Atlanta, GA
Eileen Blattner, Shaker Heights, OH
Kevin Callaghan, Montreal, Quebec
Susan Case, Wellesley, MA
Angela Connor, Raleigh, NC
Anne Ferguson, Shaker Heights, OH
Carol Gill, Dobbs Ferry, NY
Marsha Irwin, San Francisco, CA
Margaret Johnson, San Antonio, TX

Gerimae Kleinman, Shaker Heights, OH
William Mason, Southborough, MC
Jane McClure, San Francisco, CA
Judy Muir, Houston, TX
Susan Moriarty Paton, New Haven, CT
Alice Purington, Andover, MA
Jan Russell-Cebull, Danville, CA
Rod Skinner, Milton, MA
Phyllis Stein, Westport, CT

College Counselors Advisory Group

Marilyn Albarelli, Moravian Academy (PA)
Scott Anderson, St. George's Independent School (TN)
Christine Asmussen, St. Andrew's-Sewanee School (TN)
Bruce Bailey, Lakeside School (WA)
Amy E. Belstra, Cherry Creek H.S. (CO)
Greg Birk, Kinkaid School (TX)
Susan T. Bisson, Advocates for Human Potential (MA)
Robin Boren, Education Consultant (CO)
Clarice Boring, Cody H.S. (WY)
John B. Boshoven, Community High School &
Jewish Academy of Metro Detroit (MI)
Mimi Bradley, St. Andrew's Episcopal School (MS)
Nancy Bryan, Pace Academy (GA)
Claire Cafaro, Clear Directions (NJ)
Nancy Caine, St. Augustine H.S. (CA)
Mary Calhoun, St. Cecilia Academy (TN)
Jane M. Catanzaro, College Advising Services (CT)
Mary Chapman, St. Catherine's School (VA)
Anthony L. Clay, Durham Academy (NC)
Kathy Cleaver, Durham Academy (NC)
Jimmie Lee Cogburn, Independent Counselor (GA)
Teresa A. Corrigan, Chapel Hill-Chauncy Hall School
(MA)
Alison Cotten, Cypress Falls H.S. (TX)
Alice Cotti, Polytechnic School (CA)
Rod Cox, St. Johns Country Day School (FL)
Kim Crockard, Crockard College Counceling (AL)
Carroll K. Davis, North Central H.S. (IN)
Mary Jo Dawson, Academy of the Sacred Heart (MI)
Christy Dillon, Crystal Springs Uplands School (CA)
Tara A. Dowling, Saint Stephen's Episcopal School (FL)
Dan Feldhaus, Iolani School (HI)
Ralph S. Figueroa, Albuquerque Academy (NM)
Emily E. FitzHugh, The Gunnery (CT)
Larry Fletcher, Salesianum School (DE)
Nancy Fomby, Episcopal School of Dallas (TX)
Daniel Franklin, Eaglecrest High School (CO)
Laura Johnson Frey, Vermont Academy (VT)
Phyllis Gill, Providence Day School (NC)
H. Scotte Gordon, Moses Brown School (RI)
Freida Gottsegen, Pace Academy (GA)
Molly Gotwals, Suffield Academy (CT)
Kathleen Barnes Grant, The Catlin Gabel School (OR)
Madelyn Gray, John Burroughs School (MO)
Amy Grieger, Northfield Mount Hermon School (MA)
Mimi Grossman, St. Mary's Episcopal School (TN)

Elizabeth Hall, Education Consulting Services (TX)
Andrea L. Hays, Education Consultant (GA)
Rob Herald, Cairo American College (Egypt)
Darnell Heywood, Columbus School for Girls (OH)
Bruce Hunter, Rowland Hall-St. Mark's School (UT)
Deanna L. Hunter, Shawnee Mission East H.S. (KS)
John Keyes, The Catlin Gabel School (OR)
Linda King, College Connections (NY)
Sharon Koenings, Brookfield Academy (WI)
Joan Jacobson, Shawnee Mission South H.S. (KS)
Diane Johnson, Lawrence Public Schools (NY)
Gerimae Kleinman, Shaker Heights H.S. (OH)
Laurie Leftwich, Brother Martin High School (LA)
MaryJane London, Los Angeles Center for Enriched
Studies (CA)
Martha Lyman, Deerfield Academy (MA)
Brad MacGowan, Newton North H.S. (MA)
Robert S. MacLellan, Jr., The Pingry School (NJ)
Margaret M. Man, La Pietra-Hawaii School for Girls (HI)
Susan Marrs, The Seven Hills School (OH)
Karen A. Mason, Wyoming Seminary (PA)
Lisa Micele, University of Illinois Laboratory H.S. (IL)
Corky Miller-Strong, The Culver Academies (IN)
Janet Miranda, Prestonwood Christian Academy (TX)
Richard Morey, Dwight-Englewood School (NJ)
Joyce Vining Morgan, White Mountain School (NH)
Daniel Murphy, The Urban School of San Francisco
(CA)
Judith Nash, Highland High School (ID)
Stuart Oremus, Wellington School (OH)
Deborah Robinson, Mandarin H.S. (FL)
Julie Rollins, Episcopal H.S. (TX)
William C. Rowe, Thomas Jefferson School (MO)
Bruce Scher, Chicagoland Jewish H.S. (IL)
David Schindel, Sandia Preparatory School (NM)
Kathy Z. Schmidt, St. Mary's Hall (TX)
Joe Stehno, Bishop Brady H.S. (NH)
Bruce Stempien, Weston H.S. (CT)
Paul M. Stoneham, The Key School (MD)
Audrey Threlkeld, Forest Ridge School of the Sacred
Heart (WA)
Ted de Villafranca, Peddie School (NJ)
Scott White, Montclair H.S. (NJ)
Linda Zimring, Los Angeles Unified School District
(CA)

About the Authors

In 1980, when he was education editor of the New York Times, **Edward B. Fiske** sensed that college-bound students and their families needed better information on which to base their educational choices. Thus was born the *Fiske Guide to Colleges*. A graduate of Wesleyan University, Fiske did graduate work at Columbia University and assorted other bastions of higher learning. He left the Times in 1991 to pursue a variety of educational and journalistic interests, including a book on school reform, *Smart Schools, Smart Kids*. When not visiting colleges, he can be found playing tennis, sailing, or doing research on the educational problems of South Africa and other Third World countries for UNESCO and other international organizations. Fiske lives in Durham, North Carolina, near the campus of Duke University, where his wife, Helen Ladd, is a member of the faculty. They are coauthors of *When Schools Compete: A Cautionary Tale* and *Elusive Equity: Education Reform in Post-Apartheid South Africa*.

Robert Logue has served on the *Fiske Guide to Colleges* staff for more a decade, first as a staff writer and subsequently in his current position as managing editor. As a freelance author and editor, Logue has written for dozens of corporate and educational clients, including USNews.com. In addition, he serves as webmaster for FiskeGuide.com and develops websites for corporate and educational clients.

Off to college with the complete line of Fiske books!

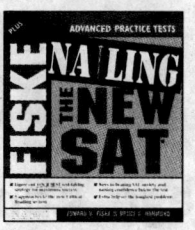

FISKE'S
College Admission Pledge
for Students

I have accepted the fact that my parents are clueless. I am serene. I will betray not a tremor when they offer opinions or advice, no matter how laughable. My soul will be light as a feather when my mother elbows her way to the front of my college tour and talks the guide's ear off. I am serene.

Going to college is a stressful time for my parents, even though they are not the ones going. I recognize that neurosis is beyond anyone's control. Each week, I will calmly reassure them that I am working on my essays, have registered for my tests, am finishing my applications, have scheduled my interviews, am aware of all deadlines, and will have everything done in plenty of time. I will smile good-naturedly as my parent asks four follow-up questions at College Night.

I will try not to say "no" simply because my parents say "yes," and remain open to the possibility, however improbable, that they may have a point. I may not be fully conscious of my anxieties about the college search—the fear of being judged and the fear of leaving home are both strong. I don't really want to get out of here as much as I say I do, and it is easier to put off thinking about the college search than to get it done. My parents are right about the importance of being proactive, even if they do get carried away.

Though the college search belongs to me, I will listen to my parents. They know me better than anyone else, and they are the ones who will pay most of the bills. Their ideas about what will be best for me are based on years of experience in the real world. I will seriously consider what they say as I form my own opinions.

I must take charge of the college search. If I do, the nagging will stop, and everyone's anxiety will go down. My parents have given me a remarkable gift—the ability to think and do for myself. I know I can do it with a little help from Mom and Dad.

FISKE'S

College Admission Pledge
for Parents

I am resigned to the fact that my child's college search will end in disaster. I am serene. Deadlines will be missed and scholarships will be lost as my child lounges under pulsating headphones or stares transfixed at a Game Cube. I am a parent and I know nothing. I am serene.

Confronted with endless procrastination, my impulse is to take control—to register for tests, plan visits, schedule interviews, and get applications. It was I who asked those four follow-up questions at College Night—I couldn't help myself. And yet I know that everything will be fine if I can summon the fortitude to relax. My child is smart, capable, and perhaps a little too accustomed to me jumping in and fixing things. I will hold back. I will drop hints and encourage, then back off. I will facilitate rather than dominate. The college search won't happen on my schedule, but it will happen.

I will not get too high or low about any facet of the college search. By doing so, I give it more importance than it really has. My child's self-worth may already be too wrapped up in getting an acceptance letter. I will attempt to lessen the fear rather than heighten it.

I will try not to say "no" simply because my son or daughter says "yes," and remain open to the possibility, however improbable, that my child has the most important things under control. I understand that my anxiety comes partly from a sense of impending loss. I can feel my child slipping away. Sometimes I hold on too tightly or let social acceptability cloud the issue of what is best.

I realize that my child is almost ready to go and that a little rebellion at this time of life can be a good thing. I will respect and encourage independence, even if some of it is expressed as resentment toward me. I will make suggestions with care and try to avoid unnecessary confrontation.

Paying for college is my responsibility. I will take a major role in the search for financial aid and scholarships and speak honestly to my child about the financial realities we face.

I must help my son or daughter take charge of the college search. I will try to support without smothering, encourage without annoying, and consult without controlling. The college search is too big to be handled alone—I will be there every step of the way.

Notes

Notes

Notes

Notes

Notes